Creatine Supplementation for Health and Clinical Diseases

Creatine Supplementation for Health and Clinical Diseases

Editors

Richard B. Kreider

Jeffrey R. Stout

MDPI • Basel • Beijing • Wuhan • Barcelona • Belgrade • Manchester • Tokyo • Cluj • Tianjin

Editors
Richard B. Kreider
Texas A&M University
USA

Jeffrey R. Stout
University of Central Florida
USA

Editorial Office
MDPI
St. Alban-Anlage 66
4052 Basel, Switzerland

This is a reprint of articles from the Special Issue published online in the open access journal *Nutrients* (ISSN 2072-6643) (available at: https://www.mdpi.com/journal/nutrients/special_issues/creatine_supplementation).

For citation purposes, cite each article independently as indicated on the article page online and as indicated below:

LastName, A.A.; LastName, B.B.; LastName, C.C. Article Title. *Journal Name* **Year**, *Volume Number*, Page Range.

ISBN 978-3-0365-2155-8 (Hbk)
ISBN 978-3-0365-2156-5 (PDF)

Cover image courtesy of Alzchem Group AG

Contents

About the Editors

Richard B. Kreider serves as a Professor, Executive Director of the Human Clinical Research Facility (HCRF), and Director of the Exercise & Sport Nutrition Lab in the Department of Health & Kinesiology at Texas A&M University. He has conducted numerous studies on nutrition and exercise, including many on creatine supplementation, and has published 8 books, 250 peer-reviewed publications, and delivered over 650 research presentations. He currently serves as the Chair of the Scientific Advisory Board on Creatine for Health sponsored by Alzchem.

Jeffrey R. Stout is a Pegasus Professor and Founding Director for the School of Kinesiology and Physical Therapy in the College of Health Professions and Sciences. Throughout his career, Dr. Stout has co-authored more than 300 peer-reviewed publications, 8 books, and 12 book chapters, and delivered over 300 national and international presentations, focusing on nutrition, exercise performance, and body composition in youth and older populations.

Preface to "Creatine Supplementation for Health and Clinical Diseases"

Creatine supplementation is one of the most studied and effective ergogenic aids for athletes. Creatine monohydrate supplementation increases high-intensity exercise capacity, leading to increases in performance and muscle mass during training. Creatine may also serve as an adjunctive nutritional strategy in the treatment and/or management of health-related conditions such as diabetes, sarcopenia, osteoporosis, cancer, rehabilitation, cognition, and cardiovascular health. There has been increased interest in creatine use as a nutritional strategy to help maintain functional and mental capacity, reduce risk to chronic disease as we age, and/or serve as an adjunctive intervention to help manage disease and/or promote recovery. This book organizes invited papers from leading creatine scholars who contributed to the Nutrients Special Issue on Creatine for Health and Clinical Diseases. It includes a general overview of the state of the science on the role of creatine on health and disease. This overview is followed by more detailed chapters on the metabolic basis of creatine in health and disease, as well as the potential role of creatine in reproductive health, pregnancy, and newborn health; children and adolescents; exercise and performance; medical rehabilitation; women's health; aging, sarcopenia, and osteoporosis; brain health and neuroprotection; glucose management and diabetes; immunity, cancer protection and therapy; cardiovascular health; inflammatory bowel disease; chronic dialysis patients; and chronic and post-viral fatigue. We hope that this book will help readers and medical practitioners to better understand the safety and efficacy of creatine supplementation in a variety of populations and provide recommendations about future research needs.

Acknowledgments

Publication costs for the invited papers published in the Nutrients Special Issue on creatine supplementation for health and clinical diseases and published in this book were provided by Alzchem Group AG, Trostberg GmbH.

We would like to thank all of the contributors to this Special Issue, as well as all of the participants and researchers who have contributed to the publications cited in this book.

Richard B. Kreider, Jeffrey R. Stout

Editors

Review

Creatine in Health and Disease

Richard B. Kreider [1,*] and Jeffery R. Stout [2]

[1]   Human Clinical Research Facility, Exercise & Sport Nutrition Lab, Department of Health & Kinesiology, Texas A&M University, College Station, TX 77843, USA
[2]   Physiology of Work and Exercise Response (POWER) Laboratory, Institute of Exercise Physiology and Rehabilitation Science, School of Kinesiology and Physical Therapy, University of Central Florida, 12494 University Blvd., Orlando, FL 32816, USA; Jeffrey.Stout@ucf.edu
*   Correspondence: rbkreider@tamu.edu

Abstract: Although creatine has been mostly studied as an ergogenic aid for exercise, training, and sport, several health and potential therapeutic benefits have been reported. This is because creatine plays a critical role in cellular metabolism, particularly during metabolically stressed states, and limitations in the ability to transport and/or store creatine can impair metabolism. Moreover, increasing availability of creatine in tissue may enhance cellular metabolism and thereby lessen the severity of injury and/or disease conditions, particularly when oxygen availability is compromised. This systematic review assesses the peer-reviewed scientific and medical evidence related to creatine's role in promoting general health as we age and how creatine supplementation has been used as a nutritional strategy to help individuals recover from injury and/or manage chronic disease. Additionally, it provides reasonable conclusions about the role of creatine on health and disease based on current scientific evidence. Based on this analysis, it can be concluded that creatine supplementation has several health and therapeutic benefits throughout the lifespan.

Keywords: ergogenic aids; cellular metabolism; phosphagens; sarcopenia; cognition; diabetes; creatine synthesis deficiencies; concussion; traumatic brain injury; spinal cord injury; muscle atrophy; rehabilitation; pregnancy; immunity; anti-inflammatory; antioxidant; anticancer

Citation: Kreider, R.B.; Stout, J.R. Creatine in Health and Disease. *Nutrients* **2021**, *13*, 447. https://doi.org/10.3390/nu13020447

Academic Editors: Roberto Iacone
Received: 8 December 2020
Accepted: 27 January 2021
Published: 29 January 2021

1. Introduction

Creatine supplementation is one of the most studied and effective ergogenic aids for athletes [1]. The multifaceted mechanisms by which creatine exerts its beneficial effect include increasing anaerobic energy capacity, decreasing protein breakdown, leading to increased muscle mass and physical performance [1]. While these well-recognized creatine effects benefit the athlete, creatine may also serve as a potential clinical and therapeutic supplementary treatment to conventional medical interventions [2–10]. In this regard, over recent years, researchers have been investigating the potential therapeutic role of creatine supplementation on health-related conditions such as diabetes [11], sarcopenia [4,6,12,13], osteoporosis [2,14], cancer [10,15–18], rehabilitation [4,19–26], cognition [3,27–29], and cardiovascular health [5,6,8,30–32], among others. This work has increased interest in creatine use as a nutritional strategy to help maintain functional and mental capacity and, as we age, reduce risk to chronic disease, and/or serve as an adjunctive intervention to help manage disease and/or promote recovery. This special issue aims to provide comprehensive reviews of the role of creatine in health and clinical disease. To do so, we have invited a number of top creatine scholars to contribute comprehensive reviews as well as encouraged colleagues to submit meta-analyses and original research to this special issue.

As an introduction about creatine's potential role in health and disease, the following provides a general overview of creatine's metabolic role, purported benefits throughout the lifespan, and potential therapeutic applications. Additionally, we provide reasonable conclusions about the state of the science on creatine supplementation. This overview will be accompanied by separate, more comprehensive, literature reviews on the metabolic basis of creatine in health and disease as well as the potential role of creatine in pregnancy; children and adolescents; exercise and performance; physical therapy and rehabilitation; women's health; aging, sarcopenia, and osteoporosis; brain neuroprotection and function; immunity, cancer protection and management; heart and muscle health; and, chronic and post-viral fatigue. We hope that this review and special issue will help readers and medical practitioners better understand the safety and efficacy of creatine supplementation in a variety of populations and provide recommendations about future research needs.

2. Methods

A systematic review of the scientific and medical literature was conducted to assess the state of the science related to creatine supplementation on metabolism, performance, health, and disease management. This was accomplished by doing keyword searches related to creatine supplementation on each topic summarized using the National Institutes for Health National Library of Medicine PubMed.gov search engine. A total of 1322 articles were reviewed with relevant research highlighted in this systematic review.

3. Metabolic Role

Creatine (*N*-aminoiminomethyl-*N*-methyl glycine) is a naturally occurring and nitrogen-containing compound comprised from amino acids that is classified within the family of guanidine phosphagens [1,33]. Creatine is synthesized endogenously from arginine and glycine by arginine glycine amidinotransferase (AGAT) to guanidinoacetate (GAA). The GAA is then methylated by the enzyme guanidinoacetate N-methyltransferase (GAMT) with S-adenosyl methionine (SAMe) to form creatine [34]. The kidney, pancreas, liver, and some regions in the brain contain AGAT with most GAA formed in the kidney and converted by GMAT to creatine in the liver [35–37]. Endogenous creatine synthesis provides about half of the daily need for creatine [35]. The remaining amount of creatine needed to maintain normal tissue levels of creatine is obtained in the diet primarily from red meat and fish [38–41] or dietary supplements [1,42,43]. About 95% of creatine is stored in muscle with the remaining amount found in other tissues, like the heart, brain, and testes [44,45]. Of this, about 2/3 of creatine is bound with inorganic phosphate (Pi) and stored as phosphocreatine (PCr) with the remainder stored as free creatine (Cr). The total creatine pool (Cr + PCr) is about 120 mmol/kg of dry muscle mass for a 70 kg individual who maintains a diet that includes red meat and fish. Vegetarians have been reported to have muscle creatine and PCr stores about 20–30% lower than non-vegetarians [46,47]. The body breaks down about 1–2% of creatine in the muscle per day into creatinine which is excreted in the urine [46,48,49]. Degradation of creatine to creatinine is greater in individuals with larger muscle mass and individuals with higher physical activity levels. Therefore, a normal-sized individual may need to consume 2–3 g/day of creatine to maintain normal creatine stores depending on diet, muscle mass, and physical activity levels. In fact, Wallimann and colleagues [50] noted that since creatine stores are not fully saturated on vegan or normal omnivore diets that generally provide 0 or 0.75–1.5 g/day of creatine, daily dietary creatine needs may be in the order of 2–4 g/person/day to promote general health [1,50]. The most effective and rapid way to increase muscle creatine stores is to ingest 5 g of creatine monohydrate four times daily for 5–7 days (i.e., 0.3 g/kg/day) [46,49]. However, some studies have shown that consuming 2–3 g/day of creatine for 30 days can also effectively increase muscle creatine stores [46,49]. Dietary supplementation of 20–30 g/day of creatine monohydrate for up to 5 years has also been studied in some clinical populations who need higher levels to increase brain concentrations of creatine, offset creatine synthesis deficiencies, or influence disease states [51–53].

Creatine and phosphagens play a critical role in providing energy through the creatine kinase (CK) and PCr system [50,54,55]. In this regard, the free energy yielded from the enzymatic degradation of adenosine triphosphate (ATP) into adenosine diphosphate (ADP) and Pi by CK serves as a primary fuel to replenish ATP for cellular metabolism. Breaking down PCr into Pi and Cr with the enzyme CK yields about 10.3 kcals of free energy that can be used to resynthesize ADP -+ Pi into ATP [38,39,56,57]. The ability to replenish depleted ATP levels during high-energy demand states like intense exercise or in conditions where energy production is either impaired (e.g., ischemia, hypoxia) or insufficient due to increased demand (e.g., mental fatigue, some disease states) is important in maintaining ATP availability.

Creatine enters the cytosol through creatine transporters (CRTR) [58–61]. In the cytosol, creatine and associated cytosolic and glycolytic CK isoforms help maintain glycolytic ATP levels, the cytosolic ATP/ADP ratio, and cytosolic ATP-consumption [50]. Additionally, creatine diffuses into the mitochondria and couples with ATP produced from oxidative phosphorylation and the adenine nucleotide translocator (ANT) via mitochondrial CK. PCr then diffuse back into the cytosol and help meet energy needs. This coupling reduces the formation of reactive oxygen species (ROS) and therefore creatine acts as a direct and/or indirect antioxidant [18,21,62,63]. The creatine phosphate shuttle is important in translocating ATP produced from oxidative phosphorylation in the mitochondrial to the cytosol and areas within the cell needing ATP for energy metabolism [50,56,57]. The creatine phosphate shuttle thereby serves as an important regulator of cellular metabolism. The role of creatine in energy metabolism and impact that creatine has on maintaining energy availability in diseases that depend on the CK/PCr system provides the metabolic basis on how creatine can affect health, disease, and provide therapeutic benefit [6,9,21,41,50,64–71]. The role of creatine in energy metabolism will be discussed in greater detail in another paper in this special issue.

4. General Health Benefits

Most creatine research initially focused on creatine's role in exercise performance, training adaptations, and safety in untrained and trained healthy individuals [1]. Creatine supplementation has been reported to increase muscle creatine and PCr levels, enhance acute exercise capacity, and improve training adaptations [44,66,69,72–96]. The improvement in performance has generally been 10–20% on various high-intensity exercise tasks [97] that include lifetime fitness activities like fitness/weight training [77,84,91,98–108], golf [109], volleyball [110], soccer [82,111,112], softball [113], ice hockey [114], running [115–119], and swimming [73,74,120–123], among others. Ergogenic benefits have been reported in men and women from children to elderly populations, although the majority of studies have been conducted on men [74,111,113,124–128]. After comprehensively reviewing the literature, the International Society of Sports Nutrition (ISSN) concluded that creatine is "the most effective ergogenic nutritional supplement currently available to athletes in terms of increasing high-intensity exercise capacity and lean body mass during training" [1,42,44,89]. The American Dietetic Association, Dietitians of Canada, and the American College of Sports Medicine have come to similar conclusions in their position stands [129,130]. Thus, there is a strong scientific consensus that creatine supplementation is an effective ergogenic nutrient for athletes as well as individuals starting a health and fitness program.

As performance-related studies assessed health and safety markers, evidence began to accumulate that creatine supplementation may also offer some health and/or therapeutic benefits as we age [4,12,14,67,69–71,131]. In this regard, creatine supplementation has been reported to help lower cholesterol, triglycerides and/or manage blood lipid levels [77,132,133]; reduce the accumulation of fat on the liver [133,134]; decrease homocysteine thereby reducing risk of heart disease [30,135]; serve as an antioxidant [30,136–139]; enhance glycemic control [1,11,140–143]; reduce the progress of some forms of cancer [8,17,18,135,144–147]; increase strength and muscle mass [2,9,13,67,70,71,93,99,101,148–154]; minimize bone loss in some studies [2,4,14,16,99,150,155–160]; improve functional capacity in osteoarthritic and fibromyalgia patients [22,161,162]; enhance cognitive function particularly in older populations [3,27,28,69,94,127,131,159,163–168]; and, in some instances, improve the efficacy of some anti-depressant medications [5,29,169–172]. These findings support contentions that it is prudent for individuals to consume at least 3 g/day of creatine to support general health as one ages [1,50]. Therefore, although more research is needed, it can be reasonably concluded based on current evidence that creatine supplementation can increase cellular energy availability and support general health, fitness, and well-being throughout the lifespan.

5. Role of Creatine in Aging Populations

Several studies have evaluated the effects of creatine supplementation in older populations in an attempt to prevent sarcopenia, maintain strength, and/or reduce the risk of chronic disease. The following discusses some of these potential applications.

5.1. Muscle Mass, Strength, Bone and Body Composition

Sarcopenia is an age-related muscle condition characterized by a reduction in muscle quantity, muscle strength, and functional capacity. Although multifactorial, sarcopenia may be caused by changes in muscle protein kinetics (synthesis and breakdown), neuromuscular function, inflammation, physical activity, and nutrition [12,14]. We also generally lose strength, muscle mass, bone mass, balance while increasing body fat as we age, whether clinically diagnosed with sarcopenia or not [3,69,131]. A number of nutritional and exercise interventions have been suggested to counteract sarcopenia in older individuals, including creatine supplementation during resistance training [12,14]. For example, Brose and colleagues [173] were among the first to report that creatine supplementation (5 g/day for 14 weeks) during heavy resistance training promoted greater gains in muscle mass and isometric muscle strength in older adults (>65 years). Chrusch and coworkers [106] reported that older participants (60–84 years) who supplemented their diet with creatine (0.3 g/kg/day for 5 days and 0.07 g/kg/day for 79 days) during supervised resistance training (3 days/week for 12 weeks) experienced greater gains in lean tissue mass, lower-body maximal strength, and endurance, and isokinetic knee flexion/extension power compared to controls. Candow and colleagues [99] reported that creatine (0.1 g/kg/day) and protein (0.3 g/kg/day) supplementation increased muscle mass and strength while decreasing protein degradation and bone resorption markers in older men. Chilibeck and associates [150] found that creatine supplementation (0.1 g/kg/day) during 12 months of resistance training increased strength and bone density in postmenopausal women. Gualano and coworkers [98] reported that creatine supplementation (20 g/day for 5 days; 5 g/day for 161 days) during resistance training improved appendicular lean mass and muscle function in older vulnerable women and that creatine supplementation alone resulted in similar gains in muscle mass compared to those engaged in resistance training alone. Aguiar and coworkers [96] also found that creatine supplementation (5 g/day for 12 weeks) combined with resistance training improved muscle endurance, ability to perform functional tasks, maximal strength, and muscle mass in older women.

Additionally, McMorris et al. [174] reported that creatine supplementation (20 g/day for 7 days) after sleep deprivation improved balance measures. Bernat and colleagues [175] reported that creatine supplementation (0.1 g/kg/day) during 8 weeks of high-velocity resistance training in untrained healthy aging men promoted significantly greater gains in leg press and total lower-body strength, muscle thickness, and some measures of peak torque and physical performance. Moreover, a meta-analysis revealed that older individuals participating in resistance training experienced greater gains in muscle mass, strength, and functional capacity when supplementing their diet with creatine [91]. A similar meta-analysis conducted by Candow and colleagues [9] found that older individuals who took creatine during resistance training experienced significantly greater gains in muscle mass and upper body. While not all studies report statistically significant effects, the preponderance of available research supports contentions that creatine supplementation, when combined with resistance exercise, can help maintain or increase muscle mass, strength, and balance in older individuals and therefore serve as an effective countermeasure to attenuate sarcopenia. The role of creatine supplementation during resistance training in sarcopenic populations will be discussed in more detail in this paper series on aging, sarcopenia, and bone health.

In addition, people often experience adult-onset obesity as they age, prompting them to diet to promote weight loss. Unfortunately, this often leads to loss of muscle mass and strength, which would be counterproductive in older individuals. Creatine supplementation while following an energy-restricted diet may be an effective strategy to maintain muscle mass, promote fat loss, and help manage adult-onset obesity. In support of this contention, Forbes and colleagues [176] recently conducted a meta-analysis on the effects of creatine on body composition and found that creatine supplementation may not only help maintain muscle mass but also promote fat mass loss. This strategy could be helpful in preventing or managing adult-onset obesity. Thus, although more research is needed, it can be reasonably concluded based on available literature that creatine supplementation, particularly when combined with resistance training, can promote gains in strength and help maintain or increase muscle mass and bone density in older individuals. Further, creatine supplementation during energy-restriction-induced weight loss interventions may be an effective way to preserve muscle mass, promote fat loss, and thereby help manage adult-onset obesity.

5.2. Cognitive Function

Creatine supplementation has been reported to increase brain PCr content by 5–15% and thereby enhance brain bioenergetics [21,53,69,131,171]. Consequently, research has examined whether creatine supplementation affects cognition, memory, and/or executive function in older individuals as well as patients with mild cognitive impairment [94,168,174,177,178]. Several studies have found that creatine supplementation attenuates mental fatigue [27,28,127] and/or can improve cognition, executive function, and/or memory [28,94,127,168,177,179]. For example, Watanabe and associates [180] found that creatine supplementation (8 g/day for 5 days) increased oxygen utilization in the brain and reduced mental fatigue in participants performing repetitive mathematical calculations. Rae et al. [177] found that working memory and processing speed increased with creatine supplementation (5 g/day for 6 weeks). McMorris and colleagues [174] reported that sleep-deprived participants better maintained random movement generation, time to react to choices, mood state, and balance when supplemented with creatine (20 g/day for 7 days). These researchers also reported that random number generation, forward spatial recall, and long-term memory tasks were significantly improved in elderly participants when supplemented with creatine. Ling et al. [178] also reported that cognition on some tasks was improved with creatine ethyl ester supplementation (5 g/day for 15 days). More recently, VAN Cutsem and coworkers [27] reported that creatine supplementation (20 g/day for 7 days) prior to performing a simulated soccer match improved muscular endurance and prolonged cognitive performance. While more research is needed and not all studies show benefit [127,167], it can be reasonably concluded based on current scientific evidence that creatine supplementation may increase brain creatine content and/or support cognitive function, particularly as one ages.

5.3. Glucose Management and Diabetes

Creatine uptake into tissue is influenced by glucose and insulin [142,181,182]. Creatine supplementation has also been reported to prevent declines in the GLUT-4 transporter during immobilization while increasing GLUT-4 by 40% during rehabilitation after atrophy [140]. Moreover, co-ingestion of creatine with carbohydrate [47,183] or creatine with carbohydrate and protein [184] has been reported to increase creatine uptake and/or muscle glycogen levels [47,184,185]. Consequently, research has evaluated whether creatine supplementation may influence glucose management [10,11,140–143]. For example, Gualano et al. [141] evaluated the effects of creatine supplementation (5 g/day for 12 weeks) during training in participants with type 2 diabetes. The researchers found that creatine supplementation improved glucose tolerance to ingesting a standard meal, increased GLUT-4 translocation, and promoted a significant reduction in HbA1c levels.

Moreover, the AMPK-alpha protein content tended to be higher after Cr supplementation and was significantly related to the changes in GLUT-4 translocation and Hb1Ac levels, suggesting that AMPK signaling may be implicated in the effects of supplementation on glucose uptake in type 2 diabetes [143]. Thus, there is evidence to suggest that creatine supplementation enhances glucose uptake and insulin sensitivity and, therefore, can help individuals manage glucose and HbA1c levels, particularly when initiating an exercise program [10,11,186]. Based on this literature, it can be reasonably concluded that creatine supplementation may support healthy glucose management.

5.4. Heart Disease

Coronary artery disease limits blood supply to the heart, thereby increasing susceptibility to ischemic events, arrhythmias, and/or heart failure. Creatine and PCr play an important role in maintaining myocardial bioenergetics during ischemic events [21]. For this reason, there has been interest in assessing the role of creatine or PCr administration in reducing arrhythmias, ischemia-related damage, and/or heart function in individuals with chronic heart failure [187–197]. For example, Anyukhovsky et al. [195] reported that intravenous administration of PCr and phosphocreatinine (300 mg/kg) in canines prevented the accumulation of lysophosphoglycerides in the ischemic zone of the heart, which is associated with an increased prevalence of arrhythmias. The researchers concluded that this might explain the antiarrhythmic action of PCr and phosphocreatinine in acute myocardial ischemia. Sharov and coworkers [194] reported that exogenous PCr administration protected against ischemia in the heart. Likewise, Balestrino and coworkers [21] evaluated the effects of adding PCr to cardioplegic solutions on energy availability during myocardial ischemia. The researchers found that PCr administration improved energy availability to the heart, reduced the incidence of arrhythmias, and improved myocardial function. As noted below, there is also evidence that creatine supplementation may maintain energy availability during brain ischemia and reduce stroke-related damage. Moreover, several studies have reported some benefit of oral creatine supplementation in heart failure patients participating in rehabilitation programs [198–201]. While not all studies report benefit from oral creatine supplementation [23,202] and more research is needed, current evidence suggests that phosphocreatine administration and possibly creatine supplementation support heart metabolism and health, particularly during ischemic challenges.

6. Potential Therapeutic Role of Creatine Supplementation

Given the metabolic role of creatine and the PCR/CK system, particularly during ischemia and in some disease states, there has been interest in examining the potential therapeutic role of creatine in a number of clinical populations. The following provides a brief overview of some of this work as an introduction to topics that will be reviewed in greater detail in other papers in this special issue.

6.1. Creatine Synthesis Deficiencies

Some individuals are born with rare deficiencies in creatine-related enzymes or transporters (e.g., AGAT, GAMT, and CRTR) that reduce the ability to transport creatine into the cell or synthesize creatine endogenously [203]. There is also recent evidence that the human genome encodes 19 genes of the solute carrier 6 (SLC6) family and that nonsynonymous changes in the coding sequence give rise to mutated or misfolded transporters that cause diseases in affected individuals [204]. This includes the creatine transporter (CT1, SLC6A8) in which deficiencies have been reported to account for about 2% of intellectual disabilities in boys [205]. Individuals with creatine synthesis deficiencies and creatine transporter mutations typically present with low brain Cr and PCr levels [53,61,204,206–210]. Low brain creatine content has been associated with muscle myopathies (e.g., weakness), voluntary or involuntary movement disorders that can affect muscle function and coordination, speech development, epilepsy, cognitive and motor development delays, and/or autism [53,61,203,204,206–210]. Individuals with these conditions have a greater dependence on dietary creatine. For this reason, high-dose, long-term creatine supplementation (e.g., 0.3–0.8 g/kg/day) throughout the lifespan is a nutritional strategy of increasing brain creatine content in these populations [53,61,203,204,206–214]. This research has generally found that long-term creatine supplementation can improve clinical outcomes, particularly in patients with AGAT and GAMT deficiencies [207].

For example, Bianchi et al. [215] found that creatine supplementation (200–800 mg/kg/day divided into 5 servings per day) significantly increased brain creatine and PCr levels in patients with GAMT-d and AGAT-d creatine synthesis deficiencies. Battini et al. [216] reported that a patient diagnosed at birth with AGAT deficiency who was treated with creatine supplementation beginning at four months of age experienced normal psychomotor development at eighteen months compared to siblings who did not have the deficiency. Stockler-Ipsiroglu and coworkers [217] evaluated the effects of creatine monohydrate supplementation (0.3–0.8 g/kg/day) in 48 children with GMAT deficiency with clinical manifestations of global developmental delay/intellectual disability (DD/ID) with speech/language delay and behavioral problems ($n = 44$), epilepsy ($n = 35$), or movement disorder ($n = 13$). The median age at treatment was 25.5 months, 39 months, and 11 years in patients with mild, moderate, and severe DD/ID, respectively. The researchers found that creatine supplementation increased brain creatine levels and improved or stabilized clinical symptoms. Moreover, four patients treated younger than nine months had normal or almost normal developmental outcomes. Long-term creatine supplementation has also been used to treat patients with ornithine aminotransferase (OAT) deficiency that causes gyrate atrophy of the choroid and retina due to secondary creatine depletion that is characterized by progressive vision loss [218–222]. These findings and others provide promise that high-dose creatine monohydrate supplementation is well tolerated and may be an effective adjunctive therapy for infants, children, and adults, particularly with AGAT deficiency [207,223–226]. Thus, it can be reasonably concluded that long-term, high-dose creatine supplementation in individuals with creatine synthesis can increase brain creatine and PCr levels and reduce the severity of deficits associated with these disorders.

6.2. Neurodegenerative Diseases and Muscular Dystrophy

Several studies have investigated the short- and long-term therapeutic benefit of creatine supplementation in animals, children, and adults with various neuromuscular diseases like Huntington's disease (HD) [51,227–232]; Parkinson's disease (PD) [51,66,100,227,233–235]; mitochondria-related diseases [58,235–239]; amyotrophic lateral sclerosis (ALS) [227,240–246]; spinal and bulbar muscular atrophy [247]; and, muscular dystrophies (MD) [248–253]. Several of these investigations, particularly in animal models, reported improved exercise tolerance and/or clinical outcomes. However, a large multi-site clinical trial conducted by Bender and coworkers [51] on PD, HD, and ALS patients did not find promising results. In this regard, they monitored 1687 participants who supplemented their diet with creatine (9.5 g/day for up to 5 years). The researchers did not observe statistically significant improvement in PD or ALS patient outcomes. However, in patients with HD, there was some evidence that creatine supplementation attenuated brain atrophy, suggesting some potential clinical benefit in this population. The reason animal studies may have yielded more promising results may be due to the fact that people typically do not present with symptoms of neurodegenerative disorders (e.g., ALS, HD, PD, etc.) until they have lost 70% or more of their alpha neurons. On the other hand, results in muscular dystrophy populations have been more promising because the muscle is the primary target. To support this contention, Kley and coworkers [254] conducted a Cochrane systemic review of the literature and found that high-quality evidence from randomized clinical trials (RCTs) demonstrated that short- and medium-term creatine supplementation increases muscle strength in muscular dystrophies and functional performance in muscular dystrophy and idiopathic inflammatory myopathy. However, assessment of high quality RCTs found no significant improvement in muscle strength in metabolic myopathies [254]. Thus, while creatine supplementation has been shown to have neuroprotective properties and improve muscle strength and endurance in patient populations, the efficacy of long-term, high-dose creatine supplementation in individuals with neurodegenerative diseases is currently equivocal, while promising, in patients with muscular dystrophy.

6.3. Brain and Spinal Cord Neuroprotection

It is well known that creatine supplementation increases brain bioenergetics [21,166, 215,235,255,256] and has neuroprotective benefits, particularly in response to injury and/or ischemic conditions [58,64,66,257]. Consequently, there has been interest in determining the effects of creatine supplementation on cerebral ischemia, stroke, traumatic brain injury (TBI), and spinal cord injury (SCI). For example, Adcock and associates [258] prophylactically administered neonatal rats creatine (3 g/kg for 3 days) and assessed brain bioenergetics in response to a cerebral ischemic event. The researchers found that creatine feeding significantly increased the ratio of brain PCr to Pi and promoted a 25% reduction in the volume of brain damage. Prass and coworkers [259] found that creatine administration decreased ischemia-induced brain infarction size by 40%. Zhu and colleagues [260] reported that oral creatine feeding in mice decreased the size of ischemia-induced brain damage and attenuated neuronal cell death, thereby providing neuroprotection. Allah and colleagues [261] found that neonatal mice fed creatine monohydrate for 10 weeks experience less ischemia-induced brain damage, as well as had better learning/memory during recovery. Finally, Turner and coworkers [166] reported that 7 days of creatine supplementation increased brain creatine content by 9.2%, increased corticomotor excitability, and prevented the decline in attention during hypoxia in healthy adults. Collectively, these findings suggest that prophylactic creatine supplementation may reduce the severity of brain ischemia and therefore may have some therapeutic benefits in individuals at risk to stroke [8,21,197].

Several studies have also evaluated the impact of creatine supplementation on mild traumatic brain injury (TBI) and spinal cord injury (SCI) outcomes in animals [3,6,171,262–266]. For example, Sullivan and coworkers [264] found that provision of creatine in the diet for 5 days prior to TBI decreased the amount of cortical brain damage by 36% in rats and 50% in mice. The researchers attributed the reduction in cortical damage to an improved energy availability. Hausmann and associates [265] reported that rats fed creatine (5 g/100 g dry food) prior to and following moderate SCI experienced less scar tissue and improved locomotor function test performance compared to controls. Moreover, Rabchevsky et al. [267] reported that rats fed a diet with 2% creatine for 4–5 weeks prior to and following SCI experienced less loss of gray matter. While these types of studies could not be performed in humans, they support contentions that creatine supplementation may reduce the severity of TBI and/or SCI. In humans, creatine supplementation has also been reported to enhance training adaptations in patients recovering from SCI. For example, Jacobs et al. [268] reported that creatine supplementation (20 g/day for 7 days) enhanced aerobic exercise capacity and ventilatory anaerobic threshold in patients with cervical SCI. Moreover, Amorim et al. [266] reported that individuals with SCI who consumed creatine (3 g/day for 8 weeks) with vitamin D (25,000 IU/day) while participating in a resistance-training program experienced significantly greater improvements in arm muscle area, strength, and functional capacity. While some studies report no benefit of creatine supplementation in patients with SCI [269,270], there is compelling evidence that creatine supplementation may reduce the severity of mild concussions, TBI, and/or SCI in animal models [21,263]. In fact, this evidence was so strong that the International Society of Sports Nutrition recommended that all athletes who are involved in sports with risk to TBI and/or SCI should take creatine to reduce the severity of these types of injury [1]. Based on this literature, it can be reasonably concluded that creatine supplementation can enhance energy availability during ischemic events and provide neuroprotection from TBI and/or SCI.

6.4. Enhanced Rehabilitation Outcomes

Since creatine supplementation has been reported to increase resistance-training adaptations, a number of studies have examined whether creatine supplementation may enhance physical therapy outcomes from musculoskeletal injury [25,159,171,247]. For example, Hespel and associates [26] reported that creatine supplementation (20 g/day and reduced to 5 g/day during immobilization, 15 g/day during the first 3 weeks of rehabilitation, and 5 g/day for the remaining 7 weeks) promoted increases in myogenic regulating factor 4 (MRF4) and myogenic protein expression, which was associated with greater muscle fiber area (+10%) and peak strength (+25%) during rehabilitation. Jacobs et al. [268] reported that creatine supplementation (20 g/d for 7 days) increased peak oxygen uptake and ventilatory anaerobic threshold in patients with cervical-level spinal cord injury (SCI). Moreover, several studies reported that creatine supplementation in chronic heart failure and chronic obstructive pulmonary disease (COPD) patients enhanced rehabilitative outcomes [23,198–200,202,271–273]. For example, Andrews and colleagues [199] found that creatine supplementation (20 g/day for 5 days) in chronic heart failure patients augmented skeletal muscle endurance and attenuated the abnormal skeletal muscle metabolic response to exercise. Fuld et al. [271] reported that creatine supplementation (17.1 g/day for 2 weeks prior to rehabilitation and 5.7 g/day for 16 weeks during rehabilitation) increased fat-free mass, peripheral muscle strength, and endurance, and health status in patients with COPD. Hass and colleagues [100] reported that creatine supplementation (20 g/day for 5 days and 5 g/day for 12 weeks) during resistance training in PD patients promoted greater muscle strength and ability to perform the functional chair sit-to-rise test. Cooke and assistants [274] reported that creatine supplementation prior to (0.3 g/kg/day for 5 days) and following (0.1 g/kg/day for 14 days) performing an eccentric-resistance-only exercise bout designed to promote muscle injury significantly reduced markers of muscle damage and hastened recovery of muscle function. Finally, Neves et al. [22] reported that creatine supplementation (20 g/day for 5 days and 5 g/day for 79 days) improved physical function, lower-limb lean mass, and quality of life in postmenopausal women with knee osteoarthritis undergoing strengthening exercises. Conversely, some studies have found no statistically significant effects of creatine supplementation during recovery from orthopedic injury. For example, Roy et al. [275] reported that creatine supplementation (10 g/day for 10 days before surgery and 5 g/day for 30 days after surgery) did not improve body composition, muscle strength, or enhance recovery in osteoarthritic patients who underwent total knee arthroplasty. Likewise, Tyler et al. [276] reported that creatine supplementation (20 g/day for 1 week and 5 g/day for 11 weeks) after anterior cruciate ligament (ACL) reconstruction had no significant effects on isokinetic strength measures during or following rehabilitation. Although more research is needed, there is evidence that creatine supplementation prior to and following injury may reduce immobilization-related atrophy and/or enhance rehabilitative outcomes in a number of populations.

6.5. Pregnancy

Since creatine supplementation has been shown to improve cellular bioenergetics during ischemic conditions and possess neuroprotective properties, there has been interest in creatine use during pregnancy to promote neural development and reduce complications resulting from birth asphyxia [7,277–285]. The rationale for creatine supplementation during pregnancy is that the fetus relies upon placental transfer of maternal creatine until late in pregnancy, and significant changes in creatine synthesis and excretion occur as pregnancy progresses [7,280]. Consequently, there is an increased demand for and utilization of creatine during pregnancy. Maternal creatine supplementation has been reported to improve neonatal survival and organ function following birth asphyxia in animals [277–279,281–283,285]. In humans, there is evidence that the creatine needs of the mother increase during pregnancy [7,280]. Consequently, there has been interest in determining the role of creatine during pregnancy on fetal growth, development, and health of the mother and child [7,280,286–288]. Available literature suggests that creatine metabolism may play an essential role in the bioenergetics of successful reproduction and that creatine supplementation may improve reproductive and/or perinatal outcomes [7,277–280,283,284,286,288]. However, it should be noted that research on the role of creatine supplementation in pregnant women is limited. While creatine supplementation has been reported to be safe in a number of populations [10,42,171,289,290] and there is no evidence that creatine supplementation poses a risk for women of reproductive age or preterm infants [287,288,291], additional safety and tolerability studies in pregnant women and those trying to conceive are needed. Consequently, although there is emerging evidence that creatine supplementation may help support the mother and child's nutritional needs and health, due to the limited studies in pregnant humans, caution should be exercised when recommending use during human pregnancy.

6.6. Immune Support

One of the more novel potential uses of creatine is its influence on the immune system. A number of in vitro and animal studies indicate that creatine has immunomodulatory effects [6]. In this regard, several studies have reported that creatine supplementation may alter production and/or the expression of molecules involved in recognizing infections like toll-like receptors (TLR) [6]. For example, Leland and colleagues [292] reported that creatine down-regulated expression of TLR-2, TLR-3, TLR-4, and TLR-7 in a mouse macrophage cell line (RAW 254.7). While this could reduce the ability to sense some infections in immunocompromised individuals, TLR-4 downregulation may also alter Parkinson's disease pathology and inhibit neuronal death as the disease progresses [293,294]. There is also evidence that creatine influences cytokines possibly via the NF-κB signaling pathway, thereby affecting cytokines, receptors, and/or growth factors that can positively or negatively influence immune response [6,292]. A creatine-induced reduction of pro-inflammatory cytokines (e.g., IL-6) and other markers of inflammation (e.g., TNFα, PGE2) may help explain some of the neuroprotective benefits observed in patients with central nervous system-related diseases [6]. It may also explain reports that creatine supplementation attenuates inflammatory and/or muscle damage in response to intense exercise [274,295–297]. On the other hand, there have been several studies in mice suggesting that creatine supplementation may impair airway inflammation, thereby exacerbating exercise-induced asthma [298,299]. However, other studies suggest that creatine attenuates the pulmonary and systemic effects of lung ischemia in reperfusion injury in rats [300]; improves rehabilitative outcomes in patients with cystic fibrosis [301] and COPD [271]; or, has no statistically significant effects on pulmonary rehabilitation outcomes [24,273] and youth soccer players with allergies [302]. Additional research is needed to understand creatine's anti-inflammatory and immunomodulating effects, but it is clear that creatine can affect these pathways. Thus, there is evidence to suggest that supplementation may have anti-inflammatory and immunomodulating effects.

6.7. Anticancer Properties

Another emerging area is related to the potential anticarcinogenic effects of creatine supplementation. As noted above, creatine and phosphagens play an important role in maintaining energy availability [38,39,56,57], particularly related to the role of the CK/PRr system and shuttling of ATP, ADP, and Pi in and out of the mitochondria for cellular metabolism [50,54,55]. Prior studies have shown that creatine content and energy availability are low in several types of malignant cells and T cells that mediate the immune responses against cancer [17,18,144,145,147]. Additionally, the creatine transport *SLC6A8* gene expression encodes a surface transporter controlling the uptake of creatine into a cell, markedly increases in tumor-infiltrating immune cells [17]. It has been well established that creatine and its related compound cyclocreatine have anticancer properties [144,303,304]. For example, Patra et al. [144] also noted that the efficacy of the anticancer medication methylglyoxal (MG) is significantly augmented in the presence of creatine and that administration of creatine, methylglyoxal, and ascorbic acid provided greater efficacy and eliminated visible signs of tumor growth. Moreover, creatine and CK, which were very low in sarcoma tissue, were significantly elevated with the concomitant regression of tumor cells. Similarly, Pal and colleagues [147] reported that MG efficacy was improved with co-administration of creatine and ascorbic acid in muscle cells in vitro and in sarcoma animal model in vivo, suggesting that creatine supplementation may serve as an adjunctive anticancer therapeutic intervention with MG. Di Biase and coworkers [17] also reported that creatine uptake deficiency severely impaired CD8 T cell responses to tumor challenge in vivo and to antigen stimulation in vitro, while supplementation of creatine through either direct administration or dietary supplementation significantly suppressed tumor growth in multiple mouse tumor models. Moreover, the energy-shuttling function of creatine goes beyond regulating CD8 T cells, in that reduced energy capacity has also been reported in multiple immune cells in various mouse tumor models in creatine transporter knockout mice [17]. The researchers concluded that creatine is an important metabolic regulator controlling antitumor T cell immunity and that creatine supplementation may improve T cell–based cancer immunotherapies [17]. Collectively, these findings indicate that creatine supplementation may have anticancer properties. Thus, it can be reasonably concluded based on available evidence that creatine is an important energy source for immune cells, can help support a healthy immune system, and may have some anticancer properties.

6.8. Improve Functional Capacity in Patients with Chronic Fatigue?

Chronic fatigue syndrome (CFS), also known as post-viral fatigue syndrome (PFS) or myalgic encephalomyelitis (ME), is characterized by fatigue and associated symptoms (e.g., muscle and joint pains, anxiety, cognitive and sleep disorders, intolerance to physical exertion) persisting more than six months in duration [305]. Although the etiology of these conditions are unknown, there has been some recent interest in whether creatine may help improve functional capacity and thereby help people with CFS conditions better manage this condition. Although controversial, there is some evidence that a lack of creatine availability and/or impaired creatine metabolism may play a role in CFS-related diseases. For example, Malatji et al. [306] reported a significant relationship between urinary creatine levels and symptoms of pain, fatigue, and energy levels in patients with CFS-related chronic pain syndrome, fibromyalgia. Mueller and associates [307] reported that creatine levels in the left parietal cortex was significantly lower in patients with ME/CFS, while higher in the left putamen and not affected in 45 other areas examined. Moreover, when using creatine as the denominator to normalize values, significant differences were observed in the ratio of N-acetylasparte/creatine, choline/creatine, lactate/creatine, and myo-inositol/creatine ratios between CFS and controls. In a similar study, van der Schaaf et al. [308] reported that greater pain levels inversely related to the N-acetylaspartylglutamate/creatine ratio in the dorsolateral prefrontal cortex of a group of 89 women with CF compared to controls. While it is unclear how changes in brain metabolites, including creatine, are involved in the pathology or symptomology of CFS, creatine and GAA supplementation have been reported to increase brain creatine content and might thereby help normalize some of these ratios. Although this is highly speculative and needs additional research, it is interesting to note that alterations in the ratio of brain metabolites to creatine have been implicated in CFS.

With that said, several studies have investigated the role of creatine or creatine-related compounds on patient outcomes in CFS patients. For example, Amital and coworkers [309] reported that creatine supplementation (3 g/day for 7 days and 5 g/day for 21 days) in a patient presenting with post-traumatic stress disorder, depression, and fibromyalgia showed improvement in symptoms of depression, pain measures, and quality of life. The patient continued supplementation for another 4 weeks and retained these benefits. Leader et al. [310] conducted an open-label study to assess the effects of creatine supplementation (3 g/day for 3 weeks and 5 g/day for 5 weeks) as an adjunctive nutritional therapy in 16 patients with Fibromyalgia Syndrome. The researchers found that creatine supplementation significantly improves markers related to the severity of fibromyalgia, disability, pain, sleep quality, and quality of life. The improvements observed returned toward baseline after 4 weeks after stopping creatine therapy. Alves and colleagues [162] reported that creatine supplementation (20 g/day for 5 days; 5 g/day for 107 days) increased intramuscular phosphorylcreatine content and improved lower- and upper-body muscle function, with minor changes in other fibromyalgia features. The authors concluded that creatine supplementation may serve as a useful dietary intervention to improve fibromyalgia patients' muscle function. Finally, Ostojic and colleagues [311] reported that GAA supplementation (2.4 g/day for 3 months) positively affected creatine metabolism and work capacity in women with CFS but did not affect general fatigue symptoms musculoskeletal soreness. While all studies do not report benefits, these findings provide some support that creatine and/or GAA may have some therapeutic benefit for patients with CFS, PFS, ME, and/or fibromyalgia. However, it should be noted that the improvements in functional capacity observed in these studies are similar to those observed in healthy individuals who take creatine and that pain indices were not significantly affected in all of these studies. Nevertheless, although more research is needed, it can be reasonably concluded that creatine and/or GAA may improve functional capacity in patients with chronic fatigue-related syndromes such as post-viral fatigue syndrome (PFS) and myalgic encephalomyelitis (ME).

6.9. Antidepressive Effects

Reports since the early 1980s have suggested that creatine metabolism and/or availability may have antidepressive effects [312–318]. These studies and others have provided the basis for assessing the effects of creatine and/or creatine precursors like S-adenosyl-L-methionine (SAMe) and GAA affect brain phosphagen levels, markers of depression, and/or the therapeutic efficacy of antidepressant medications [8,169,170]. For example, the creatine precursor SAMe has been reported to be an effective treatment for clinical depression. Silveri et al. [316] reported that SAMe supplementation (1600 mg/day) increased brain creatine and PCr levels and lowered transverse relaxation time (T2RT) using magnetic resonance spectroscopy (^{31}P MRS) in nondepressed subjects; this effect was larger in women compared to men. Allen and colleagues [319] reported that rats fed creatine diets (4%) for 5 weeks altered depression-like behavior in response to forced swim training in a sex-dependent manner, with female rats displaying an antidepressant-like response. Ahn and coworkers [320] reported that a single treatment of creatine or exercise has partial effects as an antidepressant in mice with chronic mild stress-induced depression and that combining creatine and exercise promoted greater benefits. Pazini et al. [321] reported that creatine administration (21 days, 10 mg/kg, p.o.) abolished corticosterone-induced depressive-like behaviors in mice. Similarly, Leem and colleagues [322] reported that mice exposed to mild chronic stress for 4 weeks had a greater effect on hippocampal neurogenesis via the Wnt/GSK3beta/beta-catenin pathway activation when creatine and exercise were combined compared with each treatment in chronic mild stress-induced behavioral depression. There is some support in human trials that creatine supplementation may affect depression [171,323]. For example, Bakian et al. [324] recently assessed the dietary patterns from the National Health and Nutrition Examination Survey (NHANES) database and found a significant negative relationship between dietary creatine intake and depression among adults in the United States. Roitman et al. [169] reported in an open-label study that creatine monohydrate supplementation (3–5 g/day for 4 weeks) improved outcomes in a small sample of patients with unipolar depression. Toniolo et al. [29] evaluated the effects of creatine supplementation (6 g/day for 6 weeks) in bipolar patients and reported on Montgomery–Asberg Depression Rating Scale (MADRS) remission rates (i.e., 66.7% remission in the creatine group vs. 18.2% in the placebo group). In a similar study [29], this group reported that adjunctive creatine therapy (6 g/day for 6 weeks) in patients with bipolar depression improved verbal fluency tests. Moreover, in a proof-of-concept study [172], these researchers reported that creatine supplementation (6 g/day for 6 weeks) in patients with bipolar disorder type I or II enhanced remission MADRS scores in participants who completed the study. Although more research is needed, there is some evidence suggesting that creatine may help individuals manage some types of depression and/or anxiety disorders, particularly when combined with choline [325,326]. Thus, there is evidence that creatine supplementation may support mental health.

6.10. Fertility

Since sperm motility is dependent on ATP availability and CK activity has been associated with greater sperm quality and function [50,327–329], there has been some interest in whether creatine supplementation and/or administration might improve fertility. For example, creatine has been added to medium during intrauterine insemination to increase the viability of sperm and the success of fertility treatments [327–332]. Although more research is needed, these findings suggest that creatine may play an important role in fertility and support reproductive health.

6.11. Skin Health

Since creatine availability has been reported to affect energy status in the dermis and is an antioxidant, several studies have evaluated whether creatine's topical application influences skin health and/or may serve as an effective anti-wrinkle intervention [333]. For example, Lenz et al. [333] reported that stress decreases CK activity in cutaneous cells and that topical creatine application improved cellular energy availability and markedly protected against a variety of cellular stress conditions, like oxidative and UV damage, which are involved in premature skin aging and skin damage. Peirano and coworkers [334] found that topically applied creatine rapidly penetrates the dermis, stimulates collagen synthesis, and influences gene expression and protein. Additionally, the topical application of a creatine-containing formulation for 6 weeks significantly reduced the sagging cheek intensity in the jowl area, crow's feet wrinkles, and wrinkles under the eyes. The researchers concluded that creatine represents a beneficial active ingredient for topical use in the prevention and treatment of human skin aging. Thus, there is evidence that creatine supports skin health.

7. Conclusions

The benefits of creatine monohydrate supplementation go well beyond increasing muscle Cr and PCr levels and thereby enhancing high-intensity exercise and training adaptations. Research has clearly shown several health and/or potential therapeutic benefits as we age and in clinical populations that may benefit by enhancing Cr and PCr levels. Although additional research is needed to explore further the health and potential therapeutic benefits of creatine supplementation, many of these topics will be described in more detail in other papers within this special issue. Based on the available evidence, the following can be reasonably concluded based.

1. Creatine supplementation can increase cellular energy availability and support general health, fitness, and well-being throughout the lifespan.
2. Creatine supplementation, particularly with resistance training, can promote gains in strength and help maintain or increase muscle mass in older individuals. Additionally, creatine supplementation during energy-restriction-induced weight loss may be an effective way to preserve muscle while dieting and thereby help manage adult-onset obesity.
3. Creatine supplementation may support cognitive function, particularly as one ages.
4. Creatine supplementation may support healthy glucose management.
5. Phosphocreatine administration and possibly creatine supplementation may support heart metabolism and health, particularly during ischemic challenges.
6. Long-term, high-dose creatine supplementation in individuals with creatine synthesis deficiencies can increase brain creatine and PCr levels and may reduce the severity of deficits associated with these disorders.

7. Although creatine supplementation has been shown to have neuroprotective properties and improve strength and endurance, the efficacy of long-term, high-dose creatine supplementation in individuals with neurodegenerative diseases is equivocal, while promising, in patients with muscular dystrophy.
8. Creatine supplementation may increase brain creatine content, enhance energy availability during ischemic events, and provide neuroprotection from TBI and/or SCI.
9. Creatine supplementation prior to and following injury may reduce immobilization-related atrophy and/or enhance rehabilitative outcomes in a number of populations.
10. Creatine supplementation during pregnancy may help support the mother and child's nutritional needs and health; however, due to the limited studies in pregnant humans, caution should be exercised when recommending use during human pregnancy.
11. Creatine supplementation may have anti-inflammatory and immunomodulating effects.
12. Creatine is an important energy source for immune cells, can help support a healthy immune system, and may have some anticancer properties.
13. Creatine and/or GAA may improve functional capacity in patients with chronic fatigue-related syndromes such as post-viral fatigue syndrome (PFS) and myalgic encephalomyelitis (ME).
14. Creatine may support mental health.
15. Creatine may support reproductive health.
16. Creatine may support skin health.

Author Contributions: Conceptualization, R.B.K. and J.R.S.; writing—original draft preparation, R.B.K.; writing—review and editing, R.B.K. and J.R.S.; funding acquisition, R.B.K. All authors have read and agreed to the published version of the manuscript.

Funding: The APC of selected papers of this special issue are being funded by AlzChem, LLC. (Trostberg, Germany), which manufactures creatine monohydrate. The funders had no role in the writing of the manuscript, interpretation of the literature, or in the decision to publish the results.

Institutional Review Board Statement: Not applicable.

Informed Consent Statement: Not applicable.

Data Availability Statement: Not applicable.

Acknowledgments: The authors would like to thank all of the research participants, scholars, and funding agencies who have contributed to the research cited in this manuscript.

Conflicts of Interest: R.B.K. has conducted industry sponsored research on creatine, received financial support for presenting on creatine at industry sponsored scientific conferences, and has served as an expert witness on cases related to creatine. Additionally, he serves as Chair of the Scientific Advisory Board for AlzChem who sponsored this special issue. J.R.S. has conducted industry-sponsored research on creatine and other nutraceuticals over the past 25 years. Further, J.R.S. has also received financial support for presenting on the science of various nutraceuticals, except creatine, at industry-sponsored scientific conferences.

References

1. Kreider, R.B.; Kalman, D.S.; Antonio, J.; Ziegenfuss, T.N.; Wildman, R.; Collins, R.; Candow, D.G.; Kleiner, S.M.; Almada, A.L.; Lopez, H.L. International Society of Sports Nutrition position stand: Safety and efficacy of creatine supplementation in exercise, sport, and medicine. *J. Int. Soc. Sports Nutr.* **2017**, *14*, 18. [CrossRef] [PubMed]
2. Stares, A.; Bains, M. The Additive Effects of Creatine Supplementation and Exercise Training in an Aging Population: A Systematic Review of Randomized Controlled Trials. *J. Geriatr. Phys. Ther.* **2020**, *43*, 99–112. [CrossRef] [PubMed]
3. Dolan, E.; Gualano, B.; Rawson, E.S. Beyond muscle: The effects of creatine supplementation on brain creatine, cognitive processing, and traumatic brain injury. *Eur. J. Sport Sci.* **2019**, *19*, 1–14. [CrossRef] [PubMed]
4. Dolan, E.; Artioli, G.G.; Pereira, R.M.R.; Gualano, B. Muscular Atrophy and Sarcopenia in the Elderly: Is There a Role for Creatine Supplementation? *Biomolecules* **2019**, *9*, 642. [CrossRef]
5. Wallimann, T.; Riek, U.; Moddel, M. Intradialytic creatine supplementation: A scientific rationale for improving the health and quality of life of dialysis patients. *Med. Hypotheses* **2017**, *99*, 1–14. [CrossRef]

6. Riesberg, L.A.; Weed, S.A.; McDonald, T.L.; Eckerson, J.M.; Drescher, K.M. Beyond muscles: The untapped potential of creatine. *Int. Immunopharmacol.* **2016**, *37*, 31–42. [CrossRef]
7. Ellery, S.J.; Walker, D.W.; Dickinson, H. Creatine for women: A review of the relationship between creatine and the reproductive cycle and female-specific benefits of creatine therapy. *Amino Acids* **2016**, *48*, 1807–1817. [CrossRef]
8. Smith, R.N.; Agharkar, A.S.; Gonzales, E.B. A review of creatine supplementation in age-related diseases: More than a supplement for athletes. *F1000Research* **2014**, *3*, 222. [CrossRef]
9. Candow, D.G.; Chilibeck, P.D.; Forbes, S.C. Creatine supplementation and aging musculoskeletal health. *Endocrine* **2014**, *45*, 354–361. [CrossRef]
10. Gualano, B.; Roschel, H.; Lancha, A.H., Jr.; Brightbill, C.E.; Rawson, E.S. In sickness and in health: The widespread application of creatine supplementation. *Amino Acids* **2012**, *43*, 519–529. [CrossRef]
11. Pinto, C.L.; Botelho, P.B.; Pimentel, G.D.; Campos-Ferraz, P.L.; Mota, J.F. Creatine supplementation and glycemic control: A systematic review. *Amino Acids* **2016**, *48*, 2103–2129. [CrossRef] [PubMed]
12. Candow, D.G.; Forbes, S.C.; Chilibeck, P.D.; Cornish, S.M.; Antonio, J.; Kreider, R.B. Variables Influencing the Effectiveness of Creatine Supplementation as a Therapeutic Intervention for Sarcopenia. *Front. Nutr.* **2019**, *6*, 124. [CrossRef] [PubMed]
13. Chilibeck, P.D.; Kaviani, M.; Candow, D.G.; Zello, G.A. Effect of creatine supplementation during resistance training on lean tissue mass and muscular strength in older adults: A meta-analysis. *Open Access J. Sports Med.* **2017**, *8*, 213–226. [CrossRef] [PubMed]
14. Candow, D.G.; Forbes, S.C.; Chilibeck, P.D.; Cornish, S.M.; Antonio, J.; Kreider, R.B. Effectiveness of Creatine Supplementation on Aging Muscle and Bone: Focus on Falls Prevention and Inflammation. *J. Clin. Med.* **2019**, *8*, 488. [CrossRef]
15. Fairman, C.M.; Kendall, K.L.; Newton, R.U.; Hart, N.H.; Taaffe, D.R.; Chee, R.; Tang, C.I.; Galvao, D.A. Examining the effects of creatine supplementation in augmenting adaptations to resistance training in patients with prostate cancer undergoing androgen deprivation therapy: A randomised, double-blind, placebo-controlled trial. *BMJ Open* **2019**, *9*, e030080. [CrossRef]
16. Fairman, C.M.; Kendall, K.L.; Hart, N.H.; Taaffe, D.R.; Galvao, D.A.; Newton, R.U. The potential therapeutic effects of creatine supplementation on body composition and muscle function in cancer. *Crit. Rev. Oncol Hematol* **2019**, *133*, 46–57. [CrossRef]
17. Di Biase, S.; Ma, X.; Wang, X.; Yu, J.; Wang, Y.C.; Smith, D.J.; Zhou, Y.; Li, Z.; Kim, Y.J.; Clarke, N.; et al. Creatine uptake regulates CD8 T cell antitumor immunity. *J. Exp. Med.* **2019**, *216*, 2869–2882. [CrossRef]
18. Campos-Ferraz, P.L.; Gualano, B.; das Neves, W.; Andrade, I.T.; Hangai, I.; Pereira, R.T.; Bezerra, R.N.; Deminice, R.; Seelaender, M.; Lancha, A.H. Exploratory studies of the potential anti-cancer effects of creatine. *Amino Acids* **2016**, *48*, 1993–2001. [CrossRef]
19. Dover, S.; Stephens, S.; Schneiderman, J.E.; Pullenayegum, E.; Wells, G.D.; Levy, D.M.; Marcuz, J.A.; Whitney, K.; Schulze, A.; Tein, I.; et al. The effect of creatine supplementation on muscle function in childhood myositis: A randomized, double-blind, placebo-controlled feasibility study. *J. Rheumatol.* **2020**. [CrossRef]
20. Balestrino, M.; Adriano, E. Creatine as a Candidate to Prevent Statin Myopathy. *Biomolecules* **2019**, *9*, 496. [CrossRef]
21. Balestrino, M.; Sarocchi, M.; Adriano, E.; Spallarossa, P. Potential of creatine or phosphocreatine supplementation in cerebrovascular disease and in ischemic heart disease. *Amino Acids* **2016**, *48*, 1955–1967. [CrossRef] [PubMed]
22. Neves, M., Jr.; Gualano, B.; Roschel, H.; Fuller, R.; Benatti, F.B.; Pinto, A.L.; Lima, F.R.; Pereira, R.M.; Lancha, A.H., Jr.; Bonfa, E. Beneficial effect of creatine supplementation in knee osteoarthritis. *Med. Sci. Sports Exerc.* **2011**, *43*, 1538–1543. [CrossRef] [PubMed]
23. Cornelissen, V.A.; Defoor, J.G.; Stevens, A.; Schepers, D.; Hespel, P.; Decramer, M.; Mortelmans, L.; Dobbels, F.; Vanhaecke, J.; Fagard, R.H.; et al. Effect of creatine supplementation as a potential adjuvant therapy to exercise training in cardiac patients: A randomized controlled trial. *Clin. Rehabil.* **2010**, *24*, 988–999. [CrossRef] [PubMed]
24. Al-Ghimlas, F.; Todd, D.C. Creatine supplementation for patients with COPD receiving pulmonary rehabilitation: A systematic review and meta-analysis. *Respirology* **2010**, *15*, 785–795. [CrossRef] [PubMed]
25. Hespel, P.; Derave, W. Ergogenic effects of creatine in sports and rehabilitation. *Subcell Biochem.* **2007**, *46*, 245–259. [PubMed]
26. Hespel, P.; Op't Eijnde, B.; Van Leemputte, M.; Urso, B.; Greenhaff, P.L.; Labarque, V.; Dymarkowski, S.; Van Hecke, P.; Richter, E.A. Oral creatine supplementation facilitates the rehabilitation of disuse atrophy and alters the expression of muscle myogenic factors in humans. *J. Physiol.* **2001**, *536*, 625–633. [CrossRef]
27. Van Cutsem, J.; Roelands, B.; Pluym, B.; Tassignon, B.; Verschueren, J.O.; De Pauw, K.; Meeusen, R. Can Creatine Combat the Mental Fatigue-associated Decrease in Visuomotor Skills? *Med. Sci. Sports Exerc.* **2020**, *52*, 120–130. [CrossRef]
28. Avgerinos, K.I.; Spyrou, N.; Bougioukas, K.I.; Kapogiannis, D. Effects of creatine supplementation on cognitive function of healthy individuals: A systematic review of randomized controlled trials. *Exp. Gerontol.* **2018**, *108*, 166–173. [CrossRef]
29. Toniolo, R.A.; Fernandes, F.B.F.; Silva, M.; Dias, R.D.S.; Lafer, B. Cognitive effects of creatine monohydrate adjunctive therapy in patients with bipolar depression: Results from a randomized, double-blind, placebo-controlled trial. *J. Affect. Disord.* **2017**, *224*, 69–75. [CrossRef]
30. Van Bavel, D.; de Moraes, R.; Tibirica, E. Effects of dietary supplementation with creatine on homocysteinemia and systemic microvascular endothelial function in individuals adhering to vegan diets. *Fundam. Clin. Pharmacol.* **2019**, *33*, 428–440. [CrossRef]
31. Zervou, S.; Whittington, H.J.; Russell, A.J.; Lygate, C.A. Augmentation of Creatine in the Heart. *Mini Rev. Med. Chem.* **2016**, *16*, 19–28. [CrossRef] [PubMed]
32. Clarke, H.; Kim, D.H.; Meza, C.A.; Ormsbee, M.J.; Hickner, R.C. The Evolving Applications of Creatine Supplementation: Could Creatine Improve Vascular Health? *Nutrients* **2020**, *12*, 2834. [CrossRef] [PubMed]

33. Jager, R.; Purpura, M.; Shao, A.; Inoue, T.; Kreider, R.B. Analysis of the efficacy, safety, and regulatory status of novel forms of creatine. *Amino Acids* **2011**, *40*, 1369–1383. [CrossRef] [PubMed]
34. Paddon-Jones, D.; Borsheim, E.; Wolfe, R.R. Potential ergogenic effects of arginine and creatine supplementation. *J. Nutr.* **2004**, *134*, 2888S–2894S. [CrossRef] [PubMed]
35. Brosnan, M.E.; Brosnan, J.T. The role of dietary creatine. *Amino Acids* **2016**, *48*, 1785–1791. [CrossRef]
36. da Silva, R.P.; Clow, K.; Brosnan, J.T.; Brosnan, M.E. Synthesis of guanidinoacetate and creatine from amino acids by rat pancreas. *Br. J. Nutr.* **2014**, *111*, 571–577. [CrossRef]
37. da Silva, R.P.; Nissim, I.; Brosnan, M.E.; Brosnan, J.T. Creatine synthesis: Hepatic metabolism of guanidinoacetate and creatine in the rat in vitro and in vivo. *Am. J. Physiol. Endocrinol. Metab.* **2009**, *296*, E256–E261. [CrossRef]
38. Bertin, M.; Pomponi, S.M.; Kokuhuta, C.; Iwasaki, N.; Suzuki, T.; Ellington, W.R. Origin of the genes for the isoforms of creatine kinase. *Gene* **2007**, *392*, 273–282. [CrossRef]
39. Suzuki, T.; Mizuta, C.; Uda, K.; Ishida, K.; Mizuta, K.; Sona, S.; Compaan, D.M.; Ellington, W.R. Evolution and divergence of the genes for cytoplasmic, mitochondrial, and flagellar creatine kinases. *J. Mol. Evol.* **2004**, *59*, 218–226. [CrossRef]
40. Sahlin, K.; Harris, R.C. The creatine kinase reaction: A simple reaction with functional complexity. *Amino Acids* **2011**, *40*, 1363–1367. [CrossRef]
41. Harris, R. Creatine in health, medicine and sport: An introduction to a meeting held at Downing College, University of Cambridge, July 2010. *Amino Acids* **2011**, *40*, 1267–1270. [CrossRef] [PubMed]
42. Kerksick, C.M.; Wilborn, C.D.; Roberts, M.D.; Smith-Ryan, A.; Kleiner, S.M.; Jager, R.; Collins, R.; Cooke, M.; Davis, J.N.; Galvan, E.; et al. ISSN exercise & sports nutrition review update: Research & recommendations. *J. Int. Soc. Sports Nutr.* **2018**, *15*, 38. [CrossRef] [PubMed]
43. Meyers, S. Sports nutrition market growth watch. In *Natural Products Insider*; Informa Exhibitions: Irving, TX, USA, 2016.
44. Buford, T.W.; Kreider, R.B.; Stout, J.R.; Greenwood, M.; Campbell, B.; Spano, M.; Ziegenfuss, T.; Lopez, H.; Landis, J.; Antonio, J. International Society of Sports Nutrition position stand: Creatine supplementation and exercise. *J. Int. Soc. Sports Nutr.* **2007**, *4*, 6. [CrossRef] [PubMed]
45. Kreider, R.B.; Jung, Y.P. Creatine supplementation in exercise, sport, and medicine. *J. Exerc. Nutr. Biochem.* **2011**, *15*, 53–69. [CrossRef]
46. Hultman, E.; Soderlund, K.; Timmons, J.A.; Cederblad, G.; Greenhaff, P.L. Muscle creatine loading in men. *J. Appl. Physiol.* **1996**, *81*, 232–237. [CrossRef]
47. Green, A.L.; Hultman, E.; Macdonald, I.A.; Sewell, D.A.; Greenhaff, P.L. Carbohydrate ingestion augments skeletal muscle creatine accumulation during creatine supplementation in humans. *Am. J. Physiol.* **1996**, *271*, E821–E826. [CrossRef]
48. Balsom, P.D.; Soderlund, K.; Ekblom, B. Creatine in humans with special reference to creatine supplementation. *Sports Med.* **1994**, *18*, 268–280. [CrossRef]
49. Harris, R.C.; Soderlund, K.; Hultman, E. Elevation of creatine in resting and exercised muscle of normal subjects by creatine supplementation. *Clin. Sci.* **1992**, *83*, 367–374. [CrossRef]
50. Wallimann, T.; Tokarska-Schlattner, M.; Schlattner, U. The creatine kinase system and pleiotropic effects of creatine. *Amino Acids* **2011**, *40*, 1271–1296. [CrossRef]
51. Bender, A.; Klopstock, T. Creatine for neuroprotection in neurodegenerative disease: End of story? *Amino Acids* **2016**, *48*, 1929–1940. [CrossRef]
52. Hanna-El-Daher, L.; Braissant, O. Creatine synthesis and exchanges between brain cells: What can be learned from human creatine deficiencies and various experimental models? *Amino Acids* **2016**, *48*, 1877–1895. [CrossRef] [PubMed]
53. Braissant, O.; Henry, H.; Beard, E.; Uldry, J. Creatine deficiency syndromes and the importance of creatine synthesis in the brain. *Amino Acids* **2011**, *40*, 1315–1324. [CrossRef] [PubMed]
54. Wallimann, T.; Schlosser, T.; Eppenberger, H.M. Function of M-line-bound creatine kinase as intramyofibrillar ATP regenerator at the receiving end of the phosphorylcreatine shuttle in muscle. *J. Biol. Chem.* **1984**, *259*, 5238–5246. [CrossRef]
55. Wallimann, T.; Dolder, M.; Schlattner, U.; Eder, M.; Hornemann, T.; O'Gorman, E.; Ruck, A.; Brdiczka, D. Some new aspects of creatine kinase (CK): Compartmentation, structure, function and regulation for cellular and mitochondrial bioenergetics and physiology. *Biofactors* **1998**, *8*, 229–234. [CrossRef] [PubMed]
56. Schlattner, U.; Klaus, A.; Rios, S.R.; Guzun, R.; Kay, L.; Tokarska-Schlattner, M. Cellular compartmentation of energy metabolism: Creatine kinase microcompartments and recruitment of B-type creatine kinase to specific subcellular sites. *Amino Acids* **2016**, *48*, 1751–1774. [CrossRef]
57. Ydfors, M.; Hughes, M.C.; Laham, R.; Schlattner, U.; Norrbom, J.; Perry, C.G. Modelling in vivo creatine/phosphocreatine in vitro reveals divergent adaptations in human muscle mitochondrial respiratory control by ADP after acute and chronic exercise. *J. Physiol.* **2016**, *594*, 3127–3140. [CrossRef] [PubMed]
58. Tarnopolsky, M.A.; Parshad, A.; Walzel, B.; Schlattner, U.; Wallimann, T. Creatine transporter and mitochondrial creatine kinase protein content in myopathies. *Muscle Nerve* **2001**, *24*, 682–688. [CrossRef]
59. Santacruz, L.; Jacobs, D.O. Structural correlates of the creatine transporter function regulation: The undiscovered country. *Amino Acids* **2016**, *48*, 2049–2055. [CrossRef]
60. Braissant, O. Creatine and guanidinoacetate transport at blood-brain and blood-cerebrospinal fluid barriers. *J. Inherit. Metab. Dis.* **2012**, *35*, 655–664. [CrossRef]

61. Beard, E.; Braissant, O. Synthesis and transport of creatine in the CNS: Importance for cerebral functions. *J. Neurochem.* **2010**, *115*, 297–313. [CrossRef]
62. Saraiva, A.L.; Ferreira, A.P.; Silva, L.F.; Hoffmann, M.S.; Dutra, F.D.; Furian, A.F.; Oliveira, M.S.; Fighera, M.R.; Royes, L.F. Creatine reduces oxidative stress markers but does not protect against seizure susceptibility after severe traumatic brain injury. *Brain Res. Bull.* **2012**, *87*, 180–186. [CrossRef] [PubMed]
63. Rahimi, R. Creatine supplementation decreases oxidative DNA damage and lipid peroxidation induced by a single bout of resistance exercise. *J. Strength Cond. Res.* **2011**, *25*, 3448–3455. [CrossRef] [PubMed]
64. Tarnopolsky, M.A. Clinical use of creatine in neuromuscular and neurometabolic disorders. *Subcell Biochem.* **2007**, *46*, 183–204. [PubMed]
65. Kley, R.A.; Tarnopolsky, M.A.; Vorgerd, M. Creatine for treating muscle disorders. *Cochrane Database Syst. Rev.* **2011**. [CrossRef]
66. Tarnopolsky, M.A. Potential benefits of creatine monohydrate supplementation in the elderly. *Curr. Opin. Clin. Nutr. Metab. Care* **2000**, *3*, 497–502. [CrossRef]
67. Candow, D.G.; Vogt, E.; Johannsmeyer, S.; Forbes, S.C.; Farthing, J.P. Strategic creatine supplementation and resistance training in healthy older adults. *Appl. Physiol. Nutr. Metab.* **2015**, *40*, 689–694. [CrossRef]
68. Moon, A.; Heywood, L.; Rutherford, S.; Cobbold, C. Creatine supplementation: Can it improve quality of life in the elderly without associated resistance training? *Curr. Aging Sci.* **2013**, *6*, 251–257. [CrossRef]
69. Rawson, E.S.; Venezia, A.C. Use of creatine in the elderly and evidence for effects on cognitive function in young and old. *Amino Acids* **2011**, *40*, 1349–1362. [CrossRef]
70. Candow, D.G. Sarcopenia: Current theories and the potential beneficial effect of creatine application strategies. *Biogerontology* **2011**, *12*, 273–281. [CrossRef]
71. Candow, D.G.; Chilibeck, P.D. Potential of creatine supplementation for improving aging bone health. *J. Nutr. Health Aging* **2010**, *14*, 149–153. [CrossRef]
72. Cornish, S.M.; Chilibeck, P.D.; Burke, D.G. The effect of creatine monohydrate supplementation on sprint skating in ice-hockey players. *J. Sports Med. Phys. Fit.* **2006**, *46*, 90–98.
73. Dawson, B.; Vladich, T.; Blanksby, B.A. Effects of 4 weeks of creatine supplementation in junior swimmers on freestyle sprint and swim bench performance. *J. Strength Cond. Res.* **2002**, *16*, 485–490. [PubMed]
74. Grindstaff, P.D.; Kreider, R.; Bishop, R.; Wilson, M.; Wood, L.; Alexander, C.; Almada, A. Effects of creatine supplementation on repetitive sprint performance and body composition in competitive swimmers. *Int. J. Sport Nutr.* **1997**, *7*, 330–346. [CrossRef] [PubMed]
75. Juhasz, I.; Gyore, I.; Csende, Z.; Racz, L.; Tihanyi, J. Creatine supplementation improves the anaerobic performance of elite junior fin swimmers. *Acta Physiol. Hung.* **2009**, *96*, 325–336. [CrossRef]
76. Silva, A.J.; Machado Reis, V.; Guidetti, L.; Bessone Alves, F.; Mota, P.; Freitas, J.; Baldari, C. Effect of creatine on swimming velocity, body composition and hydrodynamic variables. *J. Sports Med. Phys. Fit.* **2007**, *47*, 58–64.
77. Kreider, R.B.; Ferreira, M.; Wilson, M.; Grindstaff, P.; Plisk, S.; Reinardy, J.; Cantler, E.; Almada, A.L. Effects of creatine supplementation on body composition, strength, and sprint performance. *Med. Sci. Sports Exerc.* **1998**, *30*, 73–82. [CrossRef]
78. Stone, M.H.; Sanborn, K.; Smith, L.L.; O'Bryant, H.S.; Hoke, T.; Utter, A.C.; Johnson, R.L.; Boros, R.; Hruby, J.; Pierce, K.C.; et al. Effects of in-season (5 weeks) creatine and pyruvate supplementation on anaerobic performance and body composition in American football players. *Int. J. Sport Nutr.* **1999**, *9*, 146–165.
79. Bemben, M.G.; Bemben, D.A.; Loftiss, D.D.; Knehans, A.W. Creatine supplementation during resistance training in college football athletes. *Med. Sci. Sports Exerc.* **2001**, *33*, 1667–1673.
80. Hoffman, J.; Ratamess, N.; Kang, J.; Mangine, G.; Faigenbaum, A.; Stout, J. Effect of creatine and beta-alanine supplementation on performance and endocrine responses in strength/power athletes. *Int. J. Sport Nutr. Exerc. Metab.* **2006**, *16*, 430–446. [CrossRef]
81. Chilibeck, P.D.; Magnus, C.; Anderson, M. Effect of in-season creatine supplementation on body composition and performance in rugby union football players. *Appl. Physiol. Nutr. Metab.* **2007**, *32*, 1052–1057. [CrossRef]
82. Claudino, J.G.; Mezencio, B.; Amaral, S.; Zanetti, V.; Benatti, F.; Roschel, H.; Gualano, B.; Amadio, A.C.; Serrao, J.C. Creatine monohydrate supplementation on lower-limb muscle power in Brazilian elite soccer players. *J. Int. Soc. Sports Nutr.* **2014**, *11*, 32. [CrossRef] [PubMed]
83. Kerksick, C.M.; Rasmussen, C.; Lancaster, S.; Starks, M.; Smith, P.; Melton, C.; Greenwood, M.; Almada, A.; Kreider, R. Impact of differing protein sources and a creatine containing nutritional formula after 12 weeks of resistance training. *Nutrition* **2007**, *23*, 647–656. [CrossRef] [PubMed]
84. Kerksick, C.M.; Wilborn, C.D.; Campbell, W.I.; Harvey, T.M.; Marcello, B.M.; Roberts, M.D.; Parker, A.G.; Byars, A.G.; Greenwood, L.D.; Almada, A.L.; et al. The effects of creatine monohydrate supplementation with and without D-pinitol on resistance training adaptations. *J. Strength Cond. Res.* **2009**, *23*, 2673–2682. [CrossRef] [PubMed]
85. Galvan, E.; Walker, D.K.; Simbo, S.Y.; Dalton, R.; Levers, K.; O'Connor, A.; Goodenough, C.; Barringer, N.D.; Greenwood, M.; Rasmussen, C.; et al. Acute and chronic safety and efficacy of dose dependent creatine nitrate supplementation and exercise performance. *J. Int. Soc. Sports Nutr.* **2016**, *13*, 12. [CrossRef] [PubMed]
86. Volek, J.S.; Kraemer, W.J.; Bush, J.A.; Boetes, M.; Incledon, T.; Clark, K.L.; Lynch, J.M. Creatine supplementation enhances muscular performance during high-intensity resistance exercise. *J. Am. Diet. Assoc.* **1997**, *97*, 765–770. [CrossRef]
87. Volek, J.S.; Mazzetti, S.A.; Farquhar, W.B.; Barnes, B.R.; Gomez, A.L.; Kraemer, W.J. Physiological responses to short-term exercise in the heat after creatine loading. *Med. Sci. Sports Exerc.* **2001**, *33*, 1101–1108. [CrossRef]

88. Volek, J.S.; Ratamess, N.A.; Rubin, M.R.; Gomez, A.L.; French, D.N.; McGuigan, M.M.; Scheett, T.P.; Sharman, M.J.; Hakkinen, K.; Kraemer, W.J. The effects of creatine supplementation on muscular performance and body composition responses to short-term resistance training overreaching. *Eur. J. Appl. Physiol.* **2004**, *91*, 628–637. [CrossRef]

89. Kreider, R.B.; Wilborn, C.D.; Taylor, L.; Campbell, B.; Almada, A.L.; Collins, R.; Cooke, M.; Earnest, C.P.; Greenwood, M.; Kalman, D.S.; et al. ISSN exercise & sport nutrition review: Research & recommendations. *J. Int. Soc. Sports Nutr.* **2010**, *7*, 7. [CrossRef]

90. Branch, J.D. Effect of creatine supplementation on body composition and performance: A meta-analysis. *Int. J. Sport Nutr. Exerc. Metab.* **2003**, *13*, 198–226. [CrossRef]

91. Devries, M.C.; Phillips, S.M. Creatine supplementation during resistance training in older adults-a meta-analysis. *Med. Sci. Sports Exerc.* **2014**, *46*, 1194–1203. [CrossRef]

92. Lanhers, C.; Pereira, B.; Naughton, G.; Trousselard, M.; Lesage, F.X.; Dutheil, F. Creatine Supplementation and Lower Limb Strength Performance: A Systematic Review and Meta-Analyses. *Sports Med.* **2015**, *45*, 1285–1294. [CrossRef] [PubMed]

93. Wiroth, J.B.; Bermon, S.; Andrei, S.; Dalloz, E.; Hebuterne, X.; Dolisi, C. Effects of oral creatine supplementation on maximal pedalling performance in older adults. *Eur. J. Appl. Physiol.* **2001**, *84*, 533–539. [CrossRef] [PubMed]

94. McMorris, T.; Mielcarz, G.; Harris, R.C.; Swain, J.P.; Howard, A. Creatine supplementation and cognitive performance in elderly individuals. *Neuropsychol. Dev. Cogn. B Aging Neuropsychol. Cogn.* **2007**, *14*, 517–528. [CrossRef] [PubMed]

95. Rawson, E.S.; Clarkson, P.M. Acute creatine supplementation in older men. *Int. J. Sports Med.* **2000**, *21*, 71–75. [CrossRef] [PubMed]

96. Aguiar, A.F.; Januario, R.S.; Junior, R.P.; Gerage, A.M.; Pina, F.L.; do Nascimento, M.A.; Padovani, C.R.; Cyrino, E.S. Long-term creatine supplementation improves muscular performance during resistance training in older women. *Eur. J. Appl. Physiol.* **2013**, *113*, 987–996. [CrossRef]

97. Kreider, R.B. Effects of creatine supplementation on performance and training adaptations. *Mol. Cell Biochem.* **2003**, *244*, 89–94.

98. Gualano, B.; Macedo, A.R.; Alves, C.R.; Roschel, H.; Benatti, F.B.; Takayama, L.; de Sa Pinto, A.L.; Lima, F.R.; Pereira, R.M. Creatine supplementation and resistance training in vulnerable older women: A randomized double-blind placebo-controlled clinical trial. *Exp. Gerontol.* **2014**, *53*, 7–15. [CrossRef]

99. Candow, D.G.; Little, J.P.; Chilibeck, P.D.; Abeysekara, S.; Zello, G.A.; Kazachkov, M.; Cornish, S.M.; Yu, P.H. Low-dose creatine combined with protein during resistance training in older men. *Med. Sci. Sports Exerc.* **2008**, *40*, 1645–1652. [CrossRef]

100. Hass, C.J.; Collins, M.A.; Juncos, J.L. Resistance training with creatine monohydrate improves upper-body strength in patients with Parkinson disease: A randomized trial. *Neurorehabilit. Neural Repair* **2007**, *21*, 107–115. [CrossRef]

101. Candow, D.G.; Chilibeck, P.D. Effect of creatine supplementation during resistance training on muscle accretion in the elderly. *J. Nutr. Health Aging* **2007**, *11*, 185–188.

102. Chilibeck, P.D.; Chrusch, M.J.; Chad, K.E.; Shawn Davison, K.S.; Burke, D.G. Creatine monohydrate and resistance training increase bone mineral content and density in older men. *J. Nutr. Health Aging* **2005**, *9*, 352–353. [PubMed]

103. Burke, D.G.; Chilibeck, P.D.; Parise, G.; Candow, D.G.; Mahoney, D.; Tarnopolsky, M. Effect of creatine and weight training on muscle creatine and performance in vegetarians. *Med. Sci. Sports Exerc.* **2003**, *35*, 1946–1955. [CrossRef] [PubMed]

104. Wilder, N.; Gilders, R.; Hagerman, F.; Deivert, R.G. The effects of a 10-week, periodized, off-season resistance-training program and creatine supplementation among collegiate football players. *J. Strength Cond. Res.* **2002**, *16*, 343–352. [PubMed]

105. Izquierdo, M.; Ibanez, J.; Gonzalez-Badillo, J.J.; Gorostiaga, E.M. Effects of creatine supplementation on muscle power, endurance, and sprint performance. *Med. Sci. Sports Exerc.* **2002**, *34*, 332–343. [CrossRef]

106. Chrusch, M.J.; Chilibeck, P.D.; Chad, K.E.; Davison, K.S.; Burke, D.G. Creatine supplementation combined with resistance training in older men. *Med. Sci. Sports Exerc.* **2001**, *33*, 2111–2117. [CrossRef]

107. Becque, M.D.; Lochmann, J.D.; Melrose, D.R. Effects of oral creatine supplementation on muscular strength and body composition. *Med. Sci. Sports Exerc.* **2000**, *32*, 654–658. [CrossRef] [PubMed]

108. Volek, J.S.; Duncan, N.D.; Mazzetti, S.A.; Staron, R.S.; Putukian, M.; Gomez, A.L.; Pearson, D.R.; Fink, W.J.; Kraemer, W.J. Performance and muscle fiber adaptations to creatine supplementation and heavy resistance training. *Med. Sci. Sports Exerc.* **1999**, *31*, 1147–1156. [CrossRef]

109. Ziegenfuss, T.N.; Habowski, S.M.; Lemieux, R.; Sandrock, J.E.; Kedia, A.W.; Kerksick, C.M.; Lopez, H.L. Effects of a dietary supplement on golf drive distance and functional indices of golf performance. *J. Int. Soc. Sports Nutr.* **2015**, *12*, 4. [CrossRef]

110. Lamontagne-Lacasse, M.; Nadon, R.; Goulet, E.D. Effect of creatine supplementation on jumping performance in elite volleyball players. *Int. J. Sports Physiol. Perform.* **2011**, *6*, 525–533. [CrossRef]

111. Ramirez-Campillo, R.; Gonzalez-Jurado, J.A.; Martinez, C.; Nakamura, F.Y.; Penailillo, L.; Meylan, C.M.; Caniuqueo, A.; Canas-Jamet, R.; Moran, J.; Alonso-Martinez, A.M.; et al. Effects of plyometric training and creatine supplementation on maximal-intensity exercise and endurance in female soccer players. *J. Sci. Med. Sport* **2016**, *19*, 682–687. [CrossRef]

112. Yanez-Silva, A.; Buzzachera, C.F.; Picarro, I.D.; Januario, R.S.; Ferreira, L.H.; McAnulty, S.R.; Utter, A.C.; Souza-Junior, T.P. Effect of low dose, short-term creatine supplementation on muscle power output in elite youth soccer players. *J. Int. Soc. Sports Nutr.* **2017**, *14*, 5. [CrossRef] [PubMed]

113. Ayoama, R.; Hiruma, E.; Sasaki, H. Effects of creatine loading on muscular strength and endurance of female softball players. *J. Sports Med. Phys. Fit.* **2003**, *43*, 481–487.

114. Jones, A.M.; Atter, T.; Georg, K.P. Oral creatine supplementation improves multiple sprint performance in elite ice-hockey players. *J. Sports Med. Phys. Fit.* **1999**, *39*, 189–196. [CrossRef]

115. Ahmun, R.P.; Tong, R.J.; Grimshaw, P.N. The effects of acute creatine supplementation on multiple sprint cycling and running performance in rugby players. *J. Strength Cond. Res.* **2005**, *19*, 92–97. [CrossRef]
116. Cox, G.; Mujika, I.; Tumilty, D.; Burke, L. Acute creatine supplementation and performance during a field test simulating match play in elite female soccer players. *Int. J. Sport Nutr. Exerc. Metab.* **2002**, *12*, 33–46. [CrossRef] [PubMed]
117. Preen, D.; Dawson, B.; Goodman, C.; Lawrence, S.; Beilby, J.; Ching, S. Effect of creatine loading on long-term sprint exercise performance and metabolism. *Med. Sci. Sports Exerc.* **2001**, *33*, 814–821. [CrossRef] [PubMed]
118. Aaserud, R.; Gramvik, P.; Olsen, S.R.; Jensen, J. Creatine supplementation delays onset of fatigue during repeated bouts of sprint running. *Scand. J. Med. Sci. Sports* **1998**, *8*, 247–251. [CrossRef]
119. Bosco, C.; Tihanyi, J.; Pucspk, J.; Kovacs, I.; Gabossy, A.; Colli, R.; Pulvirenti, G.; Tranquilli, C.; Foti, C.; Viru, M.; et al. Effect of oral creatine supplementation on jumping and running performance. *Int. J. Sports Med.* **1997**, *18*, 369–372. [CrossRef]
120. Dabidi Roshan, V.; Babaei, H.; Hosseinzadeh, M.; Arendt-Nielsen, L. The effect of creatine supplementation on muscle fatigue and physiological indices following intermittent swimming bouts. *J. Sports Med. Phys. Fit.* **2013**, *53*, 232–239.
121. Selsby, J.T.; Beckett, K.D.; Kern, M.; Devor, S.T. Swim performance following creatine supplementation in Division III athletes. *J. Strength Cond. Res.* **2003**, *17*, 421–424.
122. Leenders, N.M.; Lamb, D.R.; Nelson, T.E. Creatine supplementation and swimming performance. *Int. J. Sport Nutr.* **1999**, *9*, 251–262. [CrossRef] [PubMed]
123. Peyrebrune, M.C.; Nevill, M.E.; Donaldson, F.J.; Cosford, D.J. The effects of oral creatine supplementation on performance in single and repeated sprint swimming. *J. Sports Sci.* **1998**, *16*, 271–279. [CrossRef] [PubMed]
124. Vandenberghe, K.; Goris, M.; Van Hecke, P.; Van Leemputte, M.; Vangerven, L.; Hespel, P. Long-term creatine intake is beneficial to muscle performance during resistance training. *J. Appl. Physiol.* **1997**, *83*, 2055–2063. [CrossRef] [PubMed]
125. Tarnopolsky, M.A.; MacLennan, D.P. Creatine monohydrate supplementation enhances high-intensity exercise performance in males and females. *Int. J. Sport Nutr. Exerc. Metab.* **2000**, *10*, 452–463. [CrossRef] [PubMed]
126. Ziegenfuss, T.N.; Rogers, M.; Lowery, L.; Mullins, N.; Mendel, R.; Antonio, J.; Lemon, P. Effect of creatine loading on anaerobic performance and skeletal muscle volume in NCAA Division I athletes. *Nutrition* **2002**, *18*, 397–402. [CrossRef]
127. Benton, D.; Donohoe, R. The influence of creatine supplementation on the cognitive functioning of vegetarians and omnivores. *Br. J. Nutr.* **2011**, *105*, 1100–1105. [CrossRef]
128. Johannsmeyer, S.; Candow, D.G.; Brahms, C.M.; Michel, D.; Zello, G.A. Effect of creatine supplementation and drop-set resistance training in untrained aging adults. *Exp. Gerontol.* **2016**, *83*, 112–119. [CrossRef]
129. Rodriguez, N.R.; DiMarco, N.M.; Langley, S.; American Dietetic, A.; Dietitians of, C.; American College of Sports Medicine, N.; Athletic, P. Position of the American Dietetic Association, Dietitians of Canada, and the American College of Sports Medicine: Nutrition and athletic performance. *J. Am. Diet. Assoc.* **2009**, *109*, 509–527.
130. Thomas, D.T.; Erdman, K.A.; Burke, L.M. Position of the Academy of Nutrition and Dietetics, Dietitians of Canada, and the American College of Sports Medicine: Nutrition and Athletic Performance. *J. Acad. Nutr. Diet.* **2016**, *116*, 501–528. [CrossRef]
131. Gualano, B.; Rawson, E.S.; Candow, D.G.; Chilibeck, P.D. Creatine supplementation in the aging population: Effects on skeletal muscle, bone and brain. *Amino Acids* **2016**, *48*, 1793–1805. [CrossRef]
132. Earnest, C.P.; Almada, A.L.; Mitchell, T.L. High-performance capillary electrophoresis-pure creatine monohydrate reduces blood lipids in men and women. *Clin. Sci.* **1996**, *91*, 113–118. [CrossRef] [PubMed]
133. da Silva, R.P.; Leonard, K.A.; Jacobs, R.L. Dietary creatine supplementation lowers hepatic triacylglycerol by increasing lipoprotein secretion in rats fed high-fat diet. *J. Nutr. Biochem.* **2017**, *50*, 46–53. [CrossRef] [PubMed]
134. Deminice, R.; de Castro, G.S.; Francisco, L.V.; da Silva, L.E.; Cardoso, J.F.; Frajacomo, F.T.; Teodoro, B.G.; Dos Reis Silveira, L.; Jordao, A.A. Creatine supplementation prevents fatty liver in rats fed choline-deficient diet: A burden of one-carbon and fatty acid metabolism. *J. Nutr. Biochem.* **2015**, *26*, 391–397. [CrossRef] [PubMed]
135. Deminice, R.; Cella, P.S.; Padilha, C.S.; Borges, F.H.; da Silva, L.E.; Campos-Ferraz, P.L.; Jordao, A.A.; Robinson, J.L.; Bertolo, R.F.; Cecchini, R.; et al. Creatine supplementation prevents hyperhomocysteinemia, oxidative stress and cancer-induced cachexia progression in Walker-256 tumor-bearing rats. *Amino Acids* **2016**, *48*, 2015–2024. [CrossRef] [PubMed]
136. Lawler, J.M.; Barnes, W.S.; Wu, G.; Song, W.; Demaree, S. Direct antioxidant properties of creatine. *Biochem. Biophys. Res. Commun.* **2002**, *290*, 47–52. [CrossRef]
137. Rakpongsiri, K.; Sawangkoon, S. Protective effect of creatine supplementation and estrogen replacement on cardiac reserve function and antioxidant reservation against oxidative stress in exercise-trained ovariectomized hamsters. *Int. Heart J.* **2008**, *49*, 343–354. [CrossRef]
138. Rahimi, R.; Mirzaei, B.; Rahmani-Nia, F.; Salehi, Z. Effects of creatine monohydrate supplementation on exercise-induced apoptosis in athletes: A randomized, double-blind, and placebo-controlled study. *J. Res. Med. Sci.* **2015**, *20*, 733–738. [CrossRef]
139. Deminice, R.; Jordao, A.A. Creatine supplementation decreases plasma lipid peroxidation markers and enhances anaerobic performance in rats. *Redox Rep.* **2015**. [CrossRef]
140. Op't Eijnde, B.; Urso, B.; Richter, E.A.; Greenhaff, P.L.; Hespel, P. Effect of oral creatine supplementation on human muscle GLUT4 protein content after immobilization. *Diabetes* **2001**, *50*, 18–23. [CrossRef]
141. Gualano, B.; V, D.E.S.P.; Roschel, H.; Artioli, G.G.; Neves, M., Jr.; De Sa Pinto, A.L.; Da Silva, M.E.; Cunha, M.R.; Otaduy, M.C.; Leite Cda, C.; et al. Creatine in type 2 diabetes: A randomized, double-blind, placebo-controlled trial. *Med. Sci. Sports Exerc.* **2011**, *43*, 770–778. [CrossRef]

142. Op't Eijnde, B.; Jijakli, H.; Hespel, P.; Malaisse, W.J. Creatine supplementation increases soleus muscle creatine content and lowers the insulinogenic index in an animal model of inherited type 2 diabetes. *Int. J. Mol. Med.* **2006**, *17*, 1077–1084. [CrossRef] [PubMed]

143. Alves, C.R.; Ferreira, J.C.; de Siqueira-Filho, M.A.; Carvalho, C.R.; Lancha, A.H., Jr.; Gualano, B. Creatine-induced glucose uptake in type 2 diabetes: A role for AMPK-alpha? *Amino Acids* **2012**, *43*, 1803–1807. [CrossRef] [PubMed]

144. Patra, S.; Ghosh, A.; Roy, S.S.; Bera, S.; Das, M.; Talukdar, D.; Ray, S.; Wallimann, T.; Ray, M. A short review on creatine-creatine kinase system in relation to cancer and some experimental results on creatine as adjuvant in cancer therapy. *Amino Acids* **2012**, *42*, 2319–2330. [CrossRef] [PubMed]

145. Soares, J.D.P.; Howell, S.L.; Teixeira, F.J.; Pimentel, G.D. Dietary Amino Acids and Immunonutrition Supplementation in Cancer-Induced Skeletal Muscle Mass Depletion: A Mini-Review. *Curr. Pharm. Des.* **2020**, *26*, 970–978. [CrossRef]

146. Cella, P.S.; Marinello, P.C.; Borges, F.H.; Ribeiro, D.F.; Chimin, P.; Testa, M.T.J.; Guirro, P.B.; Duarte, J.A.; Cecchini, R.; Guarnier, F.A.; et al. Creatine supplementation in Walker-256 tumor-bearing rats prevents skeletal muscle atrophy by attenuating systemic inflammation and protein degradation signaling. *Eur. J. Nutr.* **2020**, *59*, 661–669. [CrossRef] [PubMed]

147. Pal, A.; Roy, A.; Ray, M. Creatine supplementation with methylglyoxal: A potent therapy for cancer in experimental models. *Amino Acids* **2016**, *48*, 2003–2013. [CrossRef]

148. Canete, S.; San Juan, A.F.; Perez, M.; Gomez-Gallego, F.; Lopez-Mojares, L.M.; Earnest, C.P.; Fleck, S.J.; Lucia, A. Does creatine supplementation improve functional capacity in elderly women? *J. Strength Cond. Res.* **2006**, *20*, 22–28. [CrossRef]

149. Candow, D.G.; Zello, G.A.; Ling, B.; Farthing, J.P.; Chilibeck, P.D.; McLeod, K.; Harris, J.; Johnson, S. Comparison of creatine supplementation before versus after supervised resistance training in healthy older adults. *Res. Sports Med.* **2014**, *22*, 61–74. [CrossRef]

150. Chilibeck, P.D.; Candow, D.G.; Landeryou, T.; Kaviani, M.; Paus-Jenssen, L. Effects of Creatine and Resistance Training on Bone Health in Postmenopausal Women. *Med. Sci. Sports Exerc.* **2015**, *47*, 1587–1595. [CrossRef]

151. O'Bryan, K.R.; Doering, T.M.; Morton, R.W.; Coffey, V.G.; Phillips, S.M.; Cox, G.R. Do multi-ingredient protein supplements augment resistance training-induced gains in skeletal muscle mass and strength? A systematic review and meta-analysis of 35 trials. *Br. J. Sports Med.* **2020**, *54*, 573–581. [CrossRef]

152. Nilsson, M.I.; Mikhail, A.; Lan, L.; Di Carlo, A.; Hamilton, B.; Barnard, K.; Hettinga, B.P.; Hatcher, E.; Tarnopolsky, M.G.; Nederveen, J.P.; et al. A Five-Ingredient Nutritional Supplement and Home-Based Resistance Exercise Improve Lean Mass and Strength in Free-Living Elderly. *Nutrients* **2020**, *12*, 2391. [CrossRef] [PubMed]

153. Gielen, E.; Beckwee, D.; Delaere, A.; De Breucker, S.; Vandewoude, M.; Bautmans, I.; Sarcopenia Guidelines Development Group of the Belgian Society of Geriatrics. Nutritional interventions to improve muscle mass, muscle strength, and physical performance in older people: An umbrella review of systematic reviews and meta-analyses. *Nutr. Rev.* **2020**. [CrossRef] [PubMed]

154. Evans, M.; Guthrie, N.; Pezzullo, J.; Sanli, T.; Fielding, R.A.; Bellamine, A. Efficacy of a novel formulation of L-Carnitine, creatine, and leucine on lean body mass and functional muscle strength in healthy older adults: A randomized, double-blind placebo-controlled study. *Nutr. Metab.* **2017**, *14*, 7. [CrossRef] [PubMed]

155. Sales, L.P.; Pinto, A.J.; Rodrigues, S.F.; Alvarenga, J.C.; Goncalves, N.; Sampaio-Barros, M.M.; Benatti, F.B.; Gualano, B.; Rodrigues Pereira, R.M. Creatine Supplementation (3 g/d) and Bone Health in Older Women: A 2-Year, Randomized, Placebo-Controlled Trial. *J. Gerontol. A Biol. Sci. Med. Sci.* **2020**, *75*, 931–938. [CrossRef] [PubMed]

156. Castoldi, R.C.; Ozaki, G.A.T.; Garcia, T.A.; Giometti, I.C.; Koike, T.E.; Camargo, R.C.T.; Dos Santos Pereira, J.D.A.; Constantino, C.J.L.; Louzada, M.J.Q.; Camargo Filho, J.C.S.; et al. Effects of muscular strength training and growth hormone (GH) supplementation on femoral bone tissue: Analysis by Raman spectroscopy, dual-energy X-ray absorptiometry, and mechanical resistance. *Lasers Med. Sci.* **2020**, *35*, 345–354. [CrossRef] [PubMed]

157. Laskou, F.; Dennison, E. Interaction of Nutrition and Exercise on Bone and Muscle. *Eur. Endocrinol.* **2019**, *15*, 11–12. [CrossRef]

158. Candow, D.G.; Forbes, S.C.; Vogt, E. Effect of pre-exercise and post-exercise creatine supplementation on bone mineral content and density in healthy aging adults. *Exp. Gerontol.* **2019**, *119*, 89–92. [CrossRef]

159. Rawson, E.S.; Miles, M.P.; Larson-Meyer, D.E. Dietary Supplements for Health, Adaptation, and Recovery in Athletes. *Int. J. Sport Nutr. Exerc. Metab.* **2018**, *28*, 188–199. [CrossRef]

160. Forbes, S.C.; Chilibeck, P.D.; Candow, D.G. Creatine Supplementation During Resistance Training Does Not Lead to Greater Bone Mineral Density in Older Humans: A Brief Meta-Analysis. *Front. Nutr.* **2018**, *5*, 27. [CrossRef]

161. Cornish, S.M.; Peeler, J.D. No effect of creatine monohydrate supplementation on inflammatory and cartilage degradation biomarkers in individuals with knee osteoarthritis. *Nutr. Res.* **2018**, *51*, 57–66. [CrossRef]

162. Alves, C.R.; Santiago, B.M.; Lima, F.R.; Otaduy, M.C.; Calich, A.L.; Tritto, A.C.; de Sa Pinto, A.L.; Roschel, H.; Leite, C.C.; Benatti, F.B.; et al. Creatine supplementation in fibromyalgia: A randomized, double-blind, placebo-controlled trial. *Arthritis Care Res.* **2013**, *65*, 1449–1459. [CrossRef] [PubMed]

163. Bell, K.E.; Fang, H.; Snijders, T.; Allison, D.J.; Zulyniak, M.A.; Chabowski, A.; Parise, G.; Phillips, S.M.; Heisz, J.J. A Multi-Ingredient Nutritional Supplement in Combination With Resistance Exercise and High-Intensity Interval Training Improves Cognitive Function and Increases N-3 Index in Healthy Older Men: A Randomized Controlled Trial. *Front. Aging Neurosci.* **2019**, *11*, 107. [CrossRef] [PubMed]

164. Scholey, A. Nutrients for neurocognition in health and disease: Measures, methodologies and mechanisms. *Proc. Nutr. Soc.* **2018**, *77*, 73–83. [CrossRef] [PubMed]

165. Merege-Filho, C.A.; Otaduy, M.C.; de Sa-Pinto, A.L.; de Oliveira, M.O.; de Souza Goncalves, L.; Hayashi, A.P.; Roschel, H.; Pereira, R.M.; Silva, C.A.; Brucki, S.M.; et al. Does brain creatine content rely on exogenous creatine in healthy youth? A proof-of-principle study. *Appl. Physiol. Nutr. Metab.* **2017**, *42*, 128–134. [CrossRef] [PubMed]
166. Turner, C.E.; Byblow, W.D.; Gant, N. Creatine supplementation enhances corticomotor excitability and cognitive performance during oxygen deprivation. *J. Neurosci.* **2015**, *35*, 1773–1780. [CrossRef]
167. Rawson, E.S.; Lieberman, H.R.; Walsh, T.M.; Zuber, S.M.; Harhart, J.M.; Matthews, T.C. Creatine supplementation does not improve cognitive function in young adults. *Physiol. Behav.* **2008**, *95*, 130–134. [CrossRef] [PubMed]
168. McMorris, T.; Harris, R.C.; Howard, A.N.; Langridge, G.; Hall, B.; Corbett, J.; Dicks, M.; Hodgson, C. Creatine supplementation, sleep deprivation, cortisol, melatonin and behavior. *Physiol. Behav.* **2007**, *90*, 21–28. [CrossRef]
169. Roitman, S.; Green, T.; Osher, Y.; Karni, N.; Levine, J. Creatine monohydrate in resistant depression: A preliminary study. *Bipolar Disord.* **2007**, *9*, 754–758. [CrossRef]
170. D'Anci, K.E.; Allen, P.J.; Kanarek, R.B. A potential role for creatine in drug abuse? *Mol. Neurobiol.* **2011**, *44*, 136–141. [CrossRef]
171. Balestrino, M.; Adriano, E. Beyond sports: Efficacy and safety of creatine supplementation in pathological or paraphysiological conditions of brain and muscle. *Med. Res. Rev.* **2019**, *39*, 2427–2459. [CrossRef]
172. Toniolo, R.A.; Silva, M.; Fernandes, F.B.F.; Amaral, J.; Dias, R.D.S.; Lafer, B. A randomized, double-blind, placebo-controlled, proof-of-concept trial of creatine monohydrate as adjunctive treatment for bipolar depression. *J. Neural Transm.* **2018**, *125*, 247–257. [CrossRef] [PubMed]
173. Brose, A.; Parise, G.; Tarnopolsky, M.A. Creatine supplementation enhances isometric strength and body composition improvements following strength exercise training in older adults. *J. Gerontol. A Biol. Sci. Med. Sci.* **2003**, *58*, 11–19. [CrossRef] [PubMed]
174. McMorris, T.; Harris, R.C.; Swain, J.; Corbett, J.; Collard, K.; Dyson, R.J.; Dye, L.; Hodgson, C.; Draper, N. Effect of creatine supplementation and sleep deprivation, with mild exercise, on cognitive and psychomotor performance, mood state, and plasma concentrations of catecholamines and cortisol. *Psychopharmacology* **2006**, *185*, 93–103. [CrossRef] [PubMed]
175. Bernat, P.; Candow, D.G.; Gryzb, K.; Butchart, S.; Schoenfeld, B.J.; Bruno, P. Effects of high-velocity resistance training and creatine supplementation in untrained healthy aging males. *Appl. Physiol. Nutr. Metab.* **2019**, *44*, 1246–1253. [CrossRef]
176. Forbes, S.C.; Candow, D.G.; Krentz, J.R.; oberts, M.D.; Young, K.C. Changes in Fat Mass Following Creatine Supplementation and Resistance Training in Adults ≥50 Years of Age: A Meta-Analysis. *J. Funct. Morphol. Kinesio.* **2019**, *4*, 62. [CrossRef]
177. Rae, C.; Digney, A.L.; McEwan, S.R.; Bates, T.C. Oral creatine monohydrate supplementation improves brain performance: A double-blind, placebo-controlled, cross-over trial. *Proc. Biol. Sci.* **2003**, *270*, 2147–2150. [CrossRef]
178. Ling, J.; Kritikos, M.; Tiplady, B. Cognitive effects of creatine ethyl ester supplementation. *Behav. Pharmacol.* **2009**, *20*, 673–679. [CrossRef]
179. Robinson, J.L.; McBreairty, L.E.; Ryan, R.A.; Randunu, R.; Walsh, C.J.; Martin, G.M.; Brunton, J.A.; Bertolo, R.F. Effects of supplemental creatine and guanidinoacetic acid on spatial memory and the brain of weaned Yucatan miniature pigs. *PLoS ONE* **2020**, *15*, e0226806. [CrossRef]
180. Watanabe, A.; Kato, N.; Kato, T. Effects of creatine on mental fatigue and cerebral hemoglobin oxygenation. *Neurosci. Res.* **2002**, *42*, 279–285.
181. Rooney, K.; Bryson, J.; Phuyal, J.; Denyer, G.; Caterson, I.; Thompson, C. Creatine supplementation alters insulin secretion and glucose homeostasis in vivo. *Metabolism* **2002**, *51*, 518–522. [CrossRef]
182. Newman, J.E.; Hargreaves, M.; Garnham, A.; Snow, R.J. Effect of creatine ingestion on glucose tolerance and insulin sensitivity in men. *Med. Sci. Sports Exerc.* **2003**, *35*, 69–74. [CrossRef] [PubMed]
183. Greenwood, M.; Kreider, R.B.; Earnest, C.P.; Rasmussen, C.; Almada, A. Differences in creatine retention among three nutritional formulations of oral creatine supplements. *J. Exerc. Physiol. Online* **2003**, *6*, 37–43.
184. Steenge, G.R.; Simpson, E.J.; Greenhaff, P.L. Protein- and carbohydrate-induced augmentation of whole body creatine retention in humans. *J. Appl. Physiol.* **2000**, *89*, 1165–1171. [CrossRef] [PubMed]
185. Nelson, A.G.; Arnall, D.A.; Kokkonen, J.; Day, R.; Evans, J. Muscle glycogen supercompensation is enhanced by prior creatine supplementation. *Med. Sci. Sports Exerc.* **2001**, *33*, 1096–1100. [CrossRef] [PubMed]
186. Gualano, B.; Artioli, G.G.; Poortmans, J.R.; Lancha Junior, A.H. Exploring the therapeutic role of creatine supplementation. *Amino Acids* **2010**, *38*, 31–44. [CrossRef]
187. Hultman, J.; Ronquist, G.; Forsberg, J.O.; Hansson, H.E. Myocardial energy restoration of ischemic damage by administration of phosphoenolpyruvate during reperfusion. A study in a paracorporeal rat heart model. *Eur. Surg. Res.* **1983**, *15*, 200–207.
188. Thelin, S.; Hultman, J.; Ronquist, G.; Hansson, H.E. Metabolic and functional effects of creatine phosphate in cardioplegic solution. Studies on rat hearts during and after normothermic ischemia. *Scand. J. Thorac. Cardiovasc. Surg.* **1987**, *21*, 39–45.
189. Osbakken, M.; Ito, K.; Zhang, D.; Ponomarenko, I.; Ivanics, T.; Jahngen, E.G.; Cohn, M. Creatine and cyclocreatine effects on ischemic myocardium: 31P nuclear magnetic resonance evaluation of intact heart. *Cardiology* **1992**, *80*, 184–195.
190. Thorelius, J.; Thelin, S.; Ronquist, G.; Halden, E.; Hansson, H.E. Biochemical and functional effects of creatine phosphate in cardioplegic solution during aortic valve surgery–A clinical study. *Thorac. Cardiovasc. Surg.* **1992**, *40*, 10–13. [CrossRef]
191. Boudina, S.; Laclau, M.N.; Tariosse, L.; Daret, D.; Gouverneur, G.; Bonoron-Adele, S.; Saks, V.A.; Dos Santos, P. Alteration of mitochondrial function in a model of chronic ischemia in vivo in rat heart. *Am. J. Physiol. Heart Circ. Physiol.* **2002**, *282*, H821–H831. [CrossRef]

192. Laclau, M.N.; Boudina, S.; Thambo, J.B.; Tariosse, L.; Gouverneur, G.; Bonoron-Adele, S.; Saks, V.A.; Garlid, K.D.; Dos Santos, P. Cardioprotection by ischemic preconditioning preserves mitochondrial function and functional coupling between adenine nucleotide translocase and creatine kinase. *J. Mol. Cell Cardiol.* **2001**, *33*, 947–956. [CrossRef] [PubMed]

193. Conorev, E.A.; Sharov, V.G.; Saks, V.A. Improvement in contractile recovery of isolated rat heart after cardioplegic ischaemic arrest with endogenous phosphocreatine: Involvement of antiperoxidative effect? *Cardiovasc. Res.* **1991**, *25*, 164–171. [CrossRef] [PubMed]

194. Sharov, V.G.; Saks, V.A.; Kupriyanov, V.V.; Lakomkin, V.L.; Kapelko, V.I.; Steinschneider, A.; Javadov, S.A. Protection of ischemic myocardium by exogenous phosphocreatine. I. Morphologic and phosphorus 31-nuclear magnetic resonance studies. *J. Thorac. Cardiovasc. Surg.* **1987**, *94*, 749–761. [CrossRef]

195. Anyukhovsky, E.P.; Javadov, S.A.; Preobrazhensky, A.N.; Beloshapko, G.G.; Rosenshtraukh, L.V.; Saks, V.A. Effect of phosphocreatine and related compounds on the phospholipid metabolism of ischemic heart. *Biochem. Med. Metab. Biol.* **1986**, *35*, 327–334.

196. Sharov, V.G.; Afonskaya, N.I.; Ruda, M.Y.; Cherpachenko, N.M.; Pozin, E.; Markosyan, R.A.; Shepeleva, I.I.; Samarenko, M.B.; Saks, V.A. Protection of ischemic myocardium by exogenous phosphocreatine (neoton): Pharmacokinetics of phosphocreatine, reduction of infarct size, stabilization of sarcolemma of ischemic cardiomyocytes, and antithrombotic action. *Biochem. Med. Metab. Biol.* **1986**, *35*, 101–114. [CrossRef]

197. Perasso, L.; Spallarossa, P.; Gandolfo, C.; Ruggeri, P.; Balestrino, M. Therapeutic use of creatine in brain or heart ischemia: Available data and future perspectives. *Med. Res. Rev.* **2013**, *33*, 336–363. [CrossRef]

198. Gordon, A.; Hultman, E.; Kaijser, L.; Kristjansson, S.; Rolf, C.J.; Nyquist, O.; Sylven, C. Creatine supplementation in chronic heart failure increases skeletal muscle creatine phosphate and muscle performance. *Cardiovasc. Res.* **1995**, *30*, 413–418. [CrossRef]

199. Andrews, R.; Greenhaff, P.; Curtis, S.; Perry, A.; Cowley, A.J. The effect of dietary creatine supplementation on skeletal muscle metabolism in congestive heart failure. *Eur. Heart J.* **1998**, *19*, 617–622. [CrossRef]

200. Kuethe, F.; Krack, A.; Richartz, B.M.; Figulla, H.R. Creatine supplementation improves muscle strength in patients with congestive heart failure. *Pharmazie* **2006**, *61*, 218–222.

201. Fumagalli, S.; Fattirolli, F.; Guarducci, L.; Cellai, T.; Baldasseroni, S.; Tarantini, F.; Di Bari, M.; Masotti, G.; Marchionni, N. Coenzyme Q10 terclatrate and creatine in chronic heart failure: A randomized, placebo-controlled, double-blind study. *Clin. Cardiol.* **2011**, *34*, 211–217. [CrossRef]

202. Carvalho, A.P.; Rassi, S.; Fontana, K.E.; Kde, S.C.; Feitosa, R.H. Influence of creatine supplementation on the functional capacity of patients with heart failure. *Arq. Bras. Cardiol.* **2012**, *99*, 623–629. [CrossRef] [PubMed]

203. Sykut-Cegielska, J.; Gradowska, W.; Mercimek-Mahmutoglu, S.; Stockler-Ipsiroglu, S. Biochemical and clinical characteristics of creatine deficiency syndromes. *Acta Biochim. Pol.* **2004**, *51*, 875–882. [PubMed]

204. Freissmuth, M.; Stockner, T.; Sucic, S. SLC6 Transporter Folding Diseases and Pharmacochaperoning. *Handb. Exp. Pharmacol.* **2018**, *245*, 249–270. [CrossRef] [PubMed]

205. van de Kamp, J.M.; Mancini, G.M.; Salomons, G.S. X-linked creatine transporter deficiency: Clinical aspects and pathophysiology. *J. Inherit. Metab. Dis.* **2014**, *37*, 715–733. [CrossRef]

206. Mercimek-Mahmutoglu, S.; Salomons, G.S. Creatine Deficiency Syndromes. In *GeneReviews(R)*; Pagon, R.A., Adam, M.P., Ardinger, H.H., Wallace, S.E., Amemiya, A., Bean, L.J.H., Bird, T.D., Ledbetter, N., Mefford, H.C., Smith, R.J.H., Eds.; University of Washington: Seattle, WA, USA, 1993.

207. Stockler-Ipsiroglu, S.; van Karnebeek, C.D. Cerebral creatine deficiencies: A group of treatable intellectual developmental disorders. *Semin. Neurol.* **2014**, *34*, 350–356. [CrossRef]

208. Joncquel-Chevalier Curt, M.; Voicu, P.M.; Fontaine, M.; Dessein, A.F.; Porchet, N.; Mention-Mulliez, K.; Dobbelaere, D.; Soto-Ares, G.; Cheillan, D.; Vamecq, J. Creatine biosynthesis and transport in health and disease. *Biochimie* **2015**, *119*, 146–165. [CrossRef]

209. Cameron, J.M.; Levandovskiy, V.; Roberts, W.; Anagnostou, E.; Scherer, S.; Loh, A.; Schulze, A. Variability of Creatine Metabolism Genes in Children with Autism Spectrum Disorder. *Int. J. Mol. Sci.* **2017**, *18*, 1665. [CrossRef]

210. Salazar, M.D.; Zelt, N.B.; Saldivar, R.; Kuntz, C.P.; Chen, S.; Penn, W.D.; Bonneau, R.; Leman, J.K.; Schlebach, J.P. Classification of the Molecular Defects Associated with Pathogenic Variants of the SLC6A8 Creatine Transporter. *Biochemistry* **2020**, *59*, 1367–1377. [CrossRef]

211. Longo, N.; Ardon, O.; Vanzo, R.; Schwartz, E.; Pasquali, M. Disorders of creatine transport and metabolism. *Am. J. Med. Genet. C Semin Med. Genet.* **2011**, *157C*, 72–78. [CrossRef]

212. Nasrallah, F.; Feki, M.; Kaabachi, N. Creatine and creatine deficiency syndromes: Biochemical and clinical aspects. *Pediatr. Neurol.* **2010**, *42*, 163–171. [CrossRef]

213. Mercimek-Mahmutoglu, S.; Stoeckler-Ipsiroglu, S.; Adami, A.; Appleton, R.; Araujo, H.C.; Duran, M.; Ensenauer, R.; Fernandez-Alvarez, E.; Garcia, P.; Grolik, C.; et al. GAMT deficiency: Features, treatment, and outcome in an inborn error of creatine synthesis. *Neurology* **2006**, *67*, 480–484. [CrossRef] [PubMed]

214. Stromberger, C.; Bodamer, O.A.; Stockler-Ipsiroglu, S. Clinical characteristics and diagnostic clues in inborn errors of creatine metabolism. *J. Inherit. Metab. Dis.* **2003**, *26*, 299–308. [CrossRef] [PubMed]

215. Bianchi, M.C.; Tosetti, M.; Battini, R.; Leuzzi, V.; Alessandri, M.G.; Carducci, C.; Antonozzi, I.; Cioni, G. Treatment monitoring of brain creatine deficiency syndromes: A 1H- and 31P-MR spectroscopy study. *AJNR Am. J. Neuroradiol.* **2007**, *28*, 548–554. [PubMed]

216. Battini, R.; Alessandri, M.G.; Leuzzi, V.; Moro, F.; Tosetti, M.; Bianchi, M.C.; Cioni, G. Arginine:glycine amidinotransferase (AGAT) deficiency in a newborn: Early treatment can prevent phenotypic expression of the disease. *J. Pediatr.* **2006**, *148*, 828–830. [CrossRef] [PubMed]

217. Stockler-Ipsiroglu, S.; van Karnebeek, C.; Longo, N.; Korenke, G.C.; Mercimek-Mahmutoglu, S.; Marquart, I.; Barshop, B.; Grolik, C.; Schlune, A.; Angle, B.; et al. Guanidinoacetate methyltransferase (GAMT) deficiency: Outcomes in 48 individuals and recommendations for diagnosis, treatment and monitoring. *Mol. Genet. Metab.* **2014**, *111*, 16–25. [CrossRef]
218. Valtonen, M.; Nanto-Salonen, K.; Jaaskelainen, S.; Heinanen, K.; Alanen, A.; Heinonen, O.J.; Lundbom, N.; Erkintalo, M.; Simell, O. Central nervous system involvement in gyrate atrophy of the choroid and retina with hyperornithinaemia. *J. Inherit. Metab. Dis.* **1999**, *22*, 855–866. [CrossRef]
219. Nanto-Salonen, K.; Komu, M.; Lundbom, N.; Heinanen, K.; Alanen, A.; Sipila, I.; Simell, O. Reduced brain creatine in gyrate atrophy of the choroid and retina with hyperornithinemia. *Neurology* **1999**, *53*, 303–307. [CrossRef]
220. Heinanen, K.; Nanto-Salonen, K.; Komu, M.; Erkintalo, M.; Alanen, A.; Heinonen, O.J.; Pulkki, K.; Nikoskelainen, E.; Sipila, I.; Simell, O. Creatine corrects muscle 31P spectrum in gyrate atrophy with hyperornithinaemia. *Eur. J. Clin. Investig.* **1999**, *29*, 1060–1065. [CrossRef]
221. Vannas-Sulonen, K.; Sipila, I.; Vannas, A.; Simell, O.; Rapola, J. Gyrate atrophy of the choroid and retina. A five-year follow-up of creatine supplementation. *Ophthalmology* **1985**, *92*, 1719–1727. [CrossRef]
222. Sipila, I.; Rapola, J.; Simell, O.; Vannas, A. Supplementary creatine as a treatment for gyrate atrophy of the choroid and retina. *N. Engl. J. Med.* **1981**, *304*, 867–870. [CrossRef]
223. Evangeliou, A.; Vasilaki, K.; Karagianni, P.; Nikolaidis, N. Clinical applications of creatine supplementation on paediatrics. *Curr. Pharm. Biotechnol.* **2009**, *10*, 683–690. [CrossRef] [PubMed]
224. Verbruggen, K.T.; Knijff, W.A.; Soorani-Lunsing, R.J.; Sijens, P.E.; Verhoeven, N.M.; Salomons, G.S.; Goorhuis-Brouwer, S.M.; van Spronsen, F.J. Global developmental delay in guanidionacetate methyltransferase deficiency: Differences in formal testing and clinical observation. *Eur. J. Pediatr.* **2007**, *166*, 921–925. [CrossRef] [PubMed]
225. Ganesan, V.; Johnson, A.; Connelly, A.; Eckhardt, S.; Surtees, R.A. Guanidinoacetate methyltransferase deficiency: New clinical features. *Pediatr. Neurol.* **1997**, *17*, 155–157. [CrossRef]
226. Ensenauer, R.; Thiel, T.; Schwab, K.O.; Tacke, U.; Stockler-Ipsiroglu, S.; Schulze, A.; Hennig, J.; Lehnert, W. Guanidinoacetate methyltransferase deficiency: Differences of creatine uptake in human brain and muscle. *Mol. Genet. Metab.* **2004**, *82*, 208–213. [CrossRef] [PubMed]
227. Adhihetty, P.J.; Beal, M.F. Creatine and its potential therapeutic value for targeting cellular energy impairment in neurodegenerative diseases. *Neuromol. Med.* **2008**, *10*, 275–290. [CrossRef]
228. Verbessem, P.; Lemiere, J.; Eijnde, B.O.; Swinnen, S.; Vanhees, L.; Van Leemputte, M.; Hespel, P.; Dom, R. Creatine supplementation in Huntington's disease: A placebo-controlled pilot trial. *Neurology* **2003**, *61*, 925–930. [CrossRef]
229. Dedeoglu, A.; Kubilus, J.K.; Yang, L.; Ferrante, K.L.; Hersch, S.M.; Beal, M.F.; Ferrante, R.J. Creatine therapy provides neuroprotection after onset of clinical symptoms in Huntington's disease transgenic mice. *J. Neurochem.* **2003**, *85*, 1359–1367.
230. Andreassen, O.A.; Dedeoglu, A.; Ferrante, R.J.; Jenkins, B.G.; Ferrante, K.L.; Thomas, M.; Friedlich, A.; Browne, S.E.; Schilling, G.; Borchelt, D.R.; et al. Creatine increase survival and delays motor symptoms in a transgenic animal model of Huntington's disease. *Neurobiol. Dis.* **2001**, *8*, 479–491. [CrossRef]
231. Ferrante, R.J.; Andreassen, O.A.; Jenkins, B.G.; Dedeoglu, A.; Kuemmerle, S.; Kubilus, J.K.; Kaddurah-Daouk, R.; Hersch, S.M.; Beal, M.F. Neuroprotective effects of creatine in a transgenic mouse model of Huntington's disease. *J. Neurosci.* **2000**, *20*, 4389–4397. [CrossRef]
232. Matthews, R.T.; Yang, L.; Jenkins, B.G.; Ferrante, R.J.; Rosen, B.R.; Kaddurah-Daouk, R.; Beal, M.F. Neuroprotective effects of creatine and cyclocreatine in animal models of Huntington's disease. *J. Neurosci.* **1998**, *18*, 156–163.
233. Bender, A.; Samtleben, W.; Elstner, M.; Klopstock, T. Long-term creatine supplementation is safe in aged patients with Parkinson disease. *Nutr. Res.* **2008**, *28*, 172–178. [CrossRef] [PubMed]
234. Bender, A.; Koch, W.; Elstner, M.; Schombacher, Y.; Bender, J.; Moeschl, M.; Gekeler, F.; Muller-Myhsok, B.; Gasser, T.; Tatsch, K.; et al. Creatine supplementation in Parkinson disease: A placebo-controlled randomized pilot trial. *Neurology* **2006**, *67*, 1262–1264. [CrossRef] [PubMed]
235. Duarte-Silva, S.; Neves-Carvalho, A.; Soares-Cunha, C.; Silva, J.M.; Teixeira-Castro, A.; Vieira, R.; Silva-Fernandes, A.; Maciel, P. Neuroprotective Effects of Creatine in the CMVMJD135 Mouse Model of Spinocerebellar Ataxia Type 3. *Mov. Disord.* **2018**, *33*, 815–826. [CrossRef] [PubMed]
236. Komura, K.; Hobbiebrunken, E.; Wilichowski, E.K.; Hanefeld, F.A. Effectiveness of creatine monohydrate in mitochondrial encephalomyopathies. *Pediatr. Neurol.* **2003**, *28*, 53–58. [CrossRef]
237. Tarnopolsky, M.A.; Parise, G. Direct measurement of high-energy phosphate compounds in patients with neuromuscular disease. *Muscle Nerve* **1999**, *22*, 1228–1233. [CrossRef]
238. Tarnopolsky, M.A.; Roy, B.D.; MacDonald, J.R. A randomized, controlled trial of creatine monohydrate in patients with mitochondrial cytopathies. *Muscle Nerve* **1997**, *20*, 1502–1509. [CrossRef]
239. Gowayed, M.A.; Mahmoud, S.A.; El-Sayed, Y.; Abu-Samra, N.; Kamel, M.A. Enhanced mitochondrial biogenesis is associated with the ameliorative action of creatine supplementation in rat soleus and cardiac muscles. *Exp. Ther. Med.* **2020**, *19*, 384–392. [CrossRef]
240. Andreassen, O.A.; Jenkins, B.G.; Dedeoglu, A.; Ferrante, K.L.; Bogdanov, M.B.; Kaddurah-Daouk, R.; Beal, M.F. Increases in cortical glutamate concentrations in transgenic amyotrophic lateral sclerosis mice are attenuated by creatine supplementation. *J. Neurochem.* **2001**, *77*, 383–390. [CrossRef]
241. Choi, J.K.; Kustermann, E.; Dedeoglu, A.; Jenkins, B.G. Magnetic resonance spectroscopy of regional brain metabolite markers in FALS mice and the effects of dietary creatine supplementation. *Eur. J. Neurosci.* **2009**, *30*, 2143–2150. [CrossRef]

242. Derave, W.; Van Den Bosch, L.; Lemmens, G.; Eijnde, B.O.; Robberecht, W.; Hespel, P. Skeletal muscle properties in a transgenic mouse model for amyotrophic lateral sclerosis: Effects of creatine treatment. *Neurobiol. Dis.* **2003**, *13*, 264–272. [CrossRef]

243. Drory, V.E.; Gross, D. No effect of creatine on respiratory distress in amyotrophic lateral sclerosis. *Amyotroph. Lateral. Scler Other Motor. Neuron. Disord.* **2002**, *3*, 43–46. [CrossRef] [PubMed]

244. Ellis, A.C.; Rosenfeld, J. The role of creatine in the management of amyotrophic lateral sclerosis and other neurodegenerative disorders. *CNS Drugs* **2004**, *18*, 967–980. [PubMed]

245. Mazzini, L.; Balzarini, C.; Colombo, R.; Mora, G.; Pastore, I.; De Ambrogio, R.; Caligari, M. Effects of creatine supplementation on exercise performance and muscular strength in amyotrophic lateral sclerosis: Preliminary results. *J. Neurol. Sci.* **2001**, *191*, 139–144. [CrossRef]

246. Vielhaber, S.; Kaufmann, J.; Kanowski, M.; Sailer, M.; Feistner, H.; Tempelmann, C.; Elger, C.E.; Heinze, H.J.; Kunz, W.S. Effect of creatine supplementation on metabolite levels in ALS motor cortices. *Exp. Neurol.* **2001**, *172*, 377–382. [CrossRef]

247. Hijikata, Y.; Katsuno, M.; Suzuki, K.; Hashizume, A.; Araki, A.; Yamada, S.; Inagaki, T.; Ito, D.; Hirakawa, A.; Kinoshita, F.; et al. Treatment with Creatine Monohydrate in Spinal and Bulbar Muscular Atrophy: Protocol for a Randomized, Double-Blind, Placebo-Controlled Trial. *JMIR Res. Protoc.* **2018**, *7*, e69. [CrossRef]

248. Ogborn, D.I.; Smith, K.J.; Crane, J.D.; Safdar, A.; Hettinga, B.P.; Tupler, R.; Tarnopolsky, M.A. Effects of creatine and exercise on skeletal muscle of FRG1-transgenic mice. *Can. J. Neurol Sci.* **2012**, *39*, 225–231. [CrossRef]

249. Louis, M.; Lebacq, J.; Poortmans, J.R.; Belpaire-Dethiou, M.C.; Devogelaer, J.P.; Van Hecke, P.; Goubel, F.; Francaux, M. Beneficial effects of creatine supplementation in dystrophic patients. *Muscle Nerve* **2003**, *27*, 604–610. [CrossRef]

250. Banerjee, B.; Sharma, U.; Balasubramanian, K.; Kalaivani, M.; Kalra, V.; Jagannathan, N.R. Effect of creatine monohydrate in improving cellular energetics and muscle strength in ambulatory Duchenne muscular dystrophy patients: A randomized, placebo-controlled 31P MRS study. *Magn. Reson. Imaging* **2010**, *28*, 698–707. [CrossRef]

251. Felber, S.; Skladal, D.; Wyss, M.; Kremser, C.; Koller, A.; Sperl, W. Oral creatine supplementation in Duchenne muscular dystrophy: A clinical and 31P magnetic resonance spectroscopy study. *Neurol. Res.* **2000**, *22*, 145–150. [CrossRef]

252. Radley, H.G.; De Luca, A.; Lynch, G.S.; Grounds, M.D. Duchenne muscular dystrophy: Focus on pharmaceutical and nutritional interventions. *Int. J. Biochem. Cell Biol.* **2007**, *39*, 469–477. [CrossRef]

253. Tarnopolsky, M.A.; Mahoney, D.J.; Vajsar, J.; Rodriguez, C.; Doherty, T.J.; Roy, B.D.; Biggar, D. Creatine monohydrate enhances strength and body composition in Duchenne muscular dystrophy. *Neurology* **2004**, *62*, 1771–1777. [CrossRef] [PubMed]

254. Kley, R.A.; Tarnopolsky, M.A.; Vorgerd, M. Creatine for treating muscle disorders. *Cochrane Database Syst. Rev.* **2013**. [CrossRef] [PubMed]

255. Pan, J.W.; Takahashi, K. Cerebral energetic effects of creatine supplementation in humans. *Am. J. Physiol. Regul. Integr. Comp. Physiol.* **2007**, *292*, R1745–R1750. [CrossRef] [PubMed]

256. Ipsiroglu, O.S.; Stromberger, C.; Ilas, J.; Hoger, H.; Muhl, A.; Stockler-Ipsiroglu, S. Changes of tissue creatine concentrations upon oral supplementation of creatine-monohydrate in various animal species. *Life Sci.* **2001**, *69*, 1805–1815. [CrossRef]

257. Kley, R.A.; Vorgerd, M.; Tarnopolsky, M.A. Creatine for treating muscle disorders. *Cochrane Database Syst. Rev.* **2007**. [CrossRef]

258. Adcock, K.H.; Nedelcu, J.; Loenneker, T.; Martin, E.; Wallimann, T.; Wagner, B.P. Neuroprotection of creatine supplementation in neonatal rats with transient cerebral hypoxia-ischemia. *Dev. Neurosci.* **2002**, *24*, 382–388. [CrossRef]

259. Prass, K.; Royl, G.; Lindauer, U.; Freyer, D.; Megow, D.; Dirnagl, U.; Stockler-Ipsiroglu, G.; Wallimann, T.; Priller, J. Improved reperfusion and neuroprotection by creatine in a mouse model of stroke. *J. Cereb. Blood Flow Metab.* **2007**, *27*, 452–459. [CrossRef]

260. Zhu, S.; Li, M.; Figueroa, B.E.; Liu, A.; Stavrovskaya, I.G.; Pasinelli, P.; Beal, M.F.; Brown, R.H., Jr.; Kristal, B.S.; Ferrante, R.J.; et al. Prophylactic creatine administration mediates neuroprotection in cerebral ischemia in mice. *J. Neurosci.* **2004**, *24*, 5909–5912. [CrossRef]

261. Allah Yar, R.; Akbar, A.; Iqbal, F. Creatine monohydrate supplementation for 10 weeks mediates neuroprotection and improves learning/memory following neonatal hypoxia ischemia encephalopathy in female albino mice. *Brain Res.* **2015**, *1595*, 92–100. [CrossRef]

262. Ainsley Dean, P.J.; Arikan, G.; Opitz, B.; Sterr, A. Potential for use of creatine supplementation following mild traumatic brain injury. *Concussion* **2017**, *2*, CNC34. [CrossRef]

263. Freire Royes, L.F.; Cassol, G. The Effects of Creatine Supplementation and Physical Exercise on Traumatic Brain Injury. *Mini Rev. Med. Chem.* **2016**, *16*, 29–39. [CrossRef] [PubMed]

264. Sullivan, P.G.; Geiger, J.D.; Mattson, M.P.; Scheff, S.W. Dietary supplement creatine protects against traumatic brain injury. *Ann. Neurol.* **2000**, *48*, 723–729. [CrossRef]

265. Hausmann, O.N.; Fouad, K.; Wallimann, T.; Schwab, M.E. Protective effects of oral creatine supplementation on spinal cord injury in rats. *Spinal Cord* **2002**, *40*, 449–456. [CrossRef] [PubMed]

266. Amorim, S.; Teixeira, V.H.; Corredeira, R.; Cunha, M.; Maia, B.; Margalho, P.; Pires, J. Creatine or vitamin D supplementation in individuals with a spinal cord injury undergoing resistance training: A double-blinded, randomized pilot trial. *J. Spinal Cord Med.* **2018**, *41*, 471–478. [CrossRef] [PubMed]

267. Rabchevsky, A.G.; Sullivan, P.G.; Fugaccia, I.; Scheff, S.W. Creatine diet supplement for spinal cord injury: Influences on functional recovery and tissue sparing in rats. *J. Neurotrauma* **2003**, *20*, 659–669. [CrossRef]

268. Jacobs, P.L.; Mahoney, E.T.; Cohn, K.A.; Sheradsky, L.F.; Green, B.A. Oral creatine supplementation enhances upper extremity work capacity in persons with cervical-level spinal cord injury. *Arch. Phys. Med. Rehabil.* **2002**, *83*, 19–23. [CrossRef]

269. Kendall, R.W.; Jacquemin, G.; Frost, R.; Burns, S.P. Creatine supplementation for weak muscles in persons with chronic tetraplegia: A randomized double-blind placebo-controlled crossover trial. *J. Spinal Cord Med.* **2005**, *28*, 208–213.

270. Perret, C.; Mueller, G.; Knecht, H. Influence of creatine supplementation on 800 m wheelchair performance: A pilot study. *Spinal Cord* **2006**, *44*, 275–279. [CrossRef]
271. Fuld, J.P.; Kilduff, L.P.; Neder, J.A.; Pitsiladis, Y.; Lean, M.E.; Ward, S.A.; Cotton, M.M. Creatine supplementation during pulmonary rehabilitation in chronic obstructive pulmonary disease. *Thorax* **2005**, *60*, 531–537. [CrossRef]
272. Griffiths, T.L.; Proud, D. Creatine supplementation as an exercise performance enhancer for patients with COPD? An idea to run with. *Thorax* **2005**, *60*, 525–526. [CrossRef]
273. Faager, G.; Soderlund, K.; Skold, C.M.; Rundgren, S.; Tollback, A.; Jakobsson, P. Creatine supplementation and physical training in patients with COPD: A double blind, placebo-controlled study. *Int. J. Chron. Obs. Pulmon Dis.* **2006**, *1*, 445–453. [CrossRef] [PubMed]
274. Cooke, M.B.; Rybalka, E.; Williams, A.D.; Cribb, P.J.; Hayes, A. Creatine supplementation enhances muscle force recovery after eccentrically-induced muscle damage in healthy individuals. *J. Int. Soc. Sports Nutr.* **2009**, *6*, 13. [CrossRef] [PubMed]
275. Roy, B.D.; de Beer, J.; Harvey, D.; Tarnopolsky, M.A. Creatine monohydrate supplementation does not improve functional recovery after total knee arthroplasty. *Arch. Phys. Med. Rehabil.* **2005**, *86*, 1293–1298. [CrossRef] [PubMed]
276. Tyler, T.F.; Nicholas, S.J.; Hershman, E.B.; Glace, B.W.; Mullaney, M.J.; McHugh, M.P. The effect of creatine supplementation on strength recovery after anterior cruciate ligament (ACL) reconstruction: A randomized, placebo-controlled, double-blind trial. *Am. J. Sports Med.* **2004**, *32*, 383–388. [CrossRef]
277. Ellery, S.J.; LaRosa, D.A.; Cullen-McEwen, L.A.; Brown, R.D.; Snow, R.J.; Walker, D.W.; Kett, M.M.; Dickinson, H. Renal dysfunction in early adulthood following birth asphyxia in male spiny mice, and its amelioration by maternal creatine supplementation during pregnancy. *Pediatr. Res.* **2017**. [CrossRef]
278. LaRosa, D.A.; Ellery, S.J.; Snow, R.J.; Walker, D.W.; Dickinson, H. Maternal creatine supplementation during pregnancy prevents acute and long-term deficits in skeletal muscle after birth asphyxia: A study of structure and function of hind limb muscle in the spiny mouse. *Pediatr. Res.* **2016**, *80*, 852–860. [CrossRef]
279. Ellery, S.J.; LaRosa, D.A.; Kett, M.M.; Della Gatta, P.A.; Snow, R.J.; Walker, D.W.; Dickinson, H. Dietary creatine supplementation during pregnancy: A study on the effects of creatine supplementation on creatine homeostasis and renal excretory function in spiny mice. *Amino Acids* **2016**, *48*, 1819–1830. [CrossRef]
280. Dickinson, H.; Ellery, S.; Ireland, Z.; LaRosa, D.; Snow, R.; Walker, D.W. Creatine supplementation during pregnancy: Summary of experimental studies suggesting a treatment to improve fetal and neonatal morbidity and reduce mortality in high-risk human pregnancy. *BMC Pregnancy Childbirth* **2014**, *14*, 150. [CrossRef]
281. Bortoluzzi, V.T.; de Franceschi, I.D.; Rieger, E.; Wannmacher, C.M. Co-administration of creatine plus pyruvate prevents the effects of phenylalanine administration to female rats during pregnancy and lactation on enzymes activity of energy metabolism in cerebral cortex and hippocampus of the offspring. *Neurochem. Res.* **2014**, *39*, 1594–1602. [CrossRef]
282. Vallet, J.L.; Miles, J.R.; Rempel, L.A. Effect of creatine supplementation during the last week of gestation on birth intervals, stillbirth, and preweaning mortality in pigs. *J. Anim Sci.* **2013**, *91*, 2122–2132. [CrossRef]
283. Ellery, S.J.; Ireland, Z.; Kett, M.M.; Snow, R.; Walker, D.W.; Dickinson, H. Creatine pretreatment prevents birth asphyxia-induced injury of the newborn spiny mouse kidney. *Pediatr. Res.* **2013**, *73*, 201–208. [CrossRef] [PubMed]
284. Dickinson, H.; Ireland, Z.J.; Larosa, D.A.; O'Connell, B.A.; Ellery, S.; Snow, R.; Walker, D.W. Maternal dietary creatine supplementation does not alter the capacity for creatine synthesis in the newborn spiny mouse. *Reprod. Sci.* **2013**, *20*, 1096–1102. [CrossRef]
285. Ireland, Z.; Castillo-Melendez, M.; Dickinson, H.; Snow, R.; Walker, D.W. A maternal diet supplemented with creatine from mid-pregnancy protects the newborn spiny mouse brain from birth hypoxia. *Neuroscience* **2011**, *194*, 372–379. [CrossRef] [PubMed]
286. De Guingand, D.L.; Ellery, S.J.; Davies-Tuck, M.L.; Dickinson, H. Creatine and pregnancy outcomes, a prospective cohort study in low-risk pregnant women: Study protocol. *BMJ Open* **2019**, *9*, e026756. [CrossRef] [PubMed]
287. de Guingand, D.L.; Palmer, K.R.; Bilardi, J.E.; Ellery, S.J. Acceptability of dietary or nutritional supplementation in pregnancy (ADONS)—Exploring the consumer's perspective on introducing creatine monohydrate as a pregnancy supplement. *Midwifery* **2020**, *82*, 102599. [CrossRef]
288. de Guingand, D.L.; Palmer, K.R.; Snow, R.J.; Davies-Tuck, M.L.; Ellery, S.J. Risk of Adverse Outcomes in Females Taking Oral Creatine Monohydrate: A Systematic Review and Meta-Analysis. *Nutrients* **2020**, *12*. [CrossRef]
289. Jagim, A.R.; Stecker, R.A.; Harty, P.S.; Erickson, J.L.; Kerksick, C.M. Safety of Creatine Supplementation in Active Adolescents and Youth: A Brief Review. *Front. Nutr.* **2018**, *5*, 115. [CrossRef]
290. Rawson, E.S. The safety and efficacy of creatine monohydrate supplementation. *Sport Sci. Exch.* **2018**, *29*, 1–6.
291. Bohnhorst, B.; Geuting, T.; Peter, C.S.; Dordelmann, M.; Wilken, B.; Poets, C.F. Randomized, controlled trial of oral creatine supplementation (not effective) for apnea of prematurity. *Pediatrics* **2004**, *113*, e303–e307. [CrossRef]
292. Leland, K.M.; McDonald, T.L.; Drescher, K.M. Effect of creatine, creatinine, and creatine ethyl ester on TLR expression in macrophages. *Int. Immunopharmacol.* **2011**, *11*, 1341–1347. [CrossRef]
293. Beraud, D.; Maguire Zeiss, K.A. Misfolded alpha-synuclein and Toll-like receptors: Therapeutic targets for Parkinson's disease. *Parkinsonism. Relat. Disord.* **2012**, *18* (Suppl. 1), S17 S20. [CrossRef]
294. De Paola, M.; Sestito, S.E.; Mariani, A.; Memo, C.; Fanelli, R.; Freschi, M.; Bendotti, C.; Calabrese, V.; Peri, F. Synthetic and natural small molecule TLR4 antagonists inhibit motoneuron death in cultures from ALS mouse model. *Pharmacol. Res.* **2016**, *103*, 180–187. [CrossRef] [PubMed]

295. Bassit, R.A.; Curi, R.; Costa Rosa, L.F. Creatine supplementation reduces plasma levels of pro-inflammatory cytokines and PGE2 after a half-ironman competition. *Amino Acids* **2008**, *35*, 425–431. [CrossRef] [PubMed]

296. Deminice, R.; Rosa, F.T.; Franco, G.S.; Jordao, A.A.; de Freitas, E.C. Effects of creatine supplementation on oxidative stress and inflammatory markers after repeated-sprint exercise in humans. *Nutrition* **2013**, *29*, 1127–1132. [CrossRef]

297. Santos, R.V.; Bassit, R.A.; Caperuto, E.C.; Costa Rosa, L.F. The effect of creatine supplementation upon inflammatory and muscle soreness markers after a 30km race. *Life Sci.* **2004**, *75*, 1917–1924. [CrossRef]

298. Garcia, M.; Santos-Dias, A.; Bachi, A.L.L.; Oliveira-Junior, M.C.; Andrade-Souza, A.S.; Ferreira, S.C.; Aquino-Junior, J.C.J.; Almeida, F.M.; Rigonato-Oliveira, N.C.; Oliveira, A.P.L.; et al. Creatine supplementation impairs airway inflammation in an experimental model of asthma involving P2 x 7 receptor. *Eur. J. Immunol.* **2019**, *49*, 928–939. [CrossRef]

299. Vieira, R.P.; Duarte, A.C.; Claudino, R.C.; Perini, A.; Santos, A.B.; Moriya, H.T.; Arantes-Costa, F.M.; Martins, M.A.; Carvalho, C.R.; Dolhnikoff, M. Creatine supplementation exacerbates allergic lung inflammation and airway remodeling in mice. *Am. J. Respir. Cell Mol. Biol.* **2007**, *37*, 660–667. [CrossRef]

300. Almeida, F.M.; Oliveira-Junior, M.C.; Souza, R.A.; Petroni, R.C.; Soto, S.F.; Soriano, F.G.; Carvalho, P.T.; Albertini, R.; Damaceno-Rodrigues, N.R.; Lopes, F.D.; et al. Creatine supplementation attenuates pulmonary and systemic effects of lung ischemia and reperfusion injury. *J. Heart Lung Transplant.* **2016**, *35*, 242–250. [CrossRef]

301. Braegger, C.P.; Schlattner, U.; Wallimann, T.; Utiger, A.; Frank, F.; Schaefer, B.; Heizmann, C.W.; Sennhauser, F.H. Effects of creatine supplementation in cystic fibrosis: Results of a pilot study. *J. Cyst. Fibros.* **2003**, *2*, 177–182. [CrossRef]

302. Simpson, A.J.; Horne, S.; Sharp, P.; Sharps, R.; Kippelen, P. Effect of Creatine Supplementation on the Airways of Youth Elite Soccer Players. *Med. Sci. Sports Exerc.* **2019**, *51*, 1582–1590. [CrossRef]

303. Miller, E.E.; Evans, A.E.; Cohn, M. Inhibition of rate of tumor growth by creatine and cyclocreatine. *Proc. Natl. Acad. Sci. USA* **1993**, *90*, 3304–3308. [CrossRef] [PubMed]

304. Wyss, M.; Kaddurah-Daouk, R. Creatine and creatinine metabolism. *Physiol. Rev.* **2000**, *80*, 1107–1213. [CrossRef] [PubMed]

305. Ostojic, S.M. Postviral fatigue syndrome and creatine: A piece of the puzzle? *Nutr. Neurosci.* **2020**. [CrossRef] [PubMed]

306. Malatji, B.G.; Meyer, H.; Mason, S.; Engelke, U.F.H.; Wevers, R.A.; van Reenen, M.; Reinecke, C.J. A diagnostic biomarker profile for fibromyalgia syndrome based on an NMR metabolomics study of selected patients and controls. *BMC Neurol.* **2017**, *17*, 88. [CrossRef] [PubMed]

307. Mueller, C.; Lin, J.C.; Sheriff, S.; Maudsley, A.A.; Younger, J.W. Evidence of widespread metabolite abnormalities in Myalgic encephalomyelitis/chronic fatigue syndrome: Assessment with whole-brain magnetic resonance spectroscopy. *Brain Imaging Behav.* **2020**, *14*, 562–572. [CrossRef] [PubMed]

308. van der Schaaf, M.E.; De Lange, F.P.; Schmits, I.C.; Geurts, D.E.M.; Roelofs, K.; van der Meer, J.W.M.; Toni, I.; Knoop, H. Prefrontal Structure Varies as a Function of Pain Symptoms in Chronic Fatigue Syndrome. *Biol. Psychiatry* **2017**, *81*, 358–365. [CrossRef]

309. Amital, D.; Vishne, T.; Rubinow, A.; Levine, J. Observed effects of creatine monohydrate in a patient with depression and fibromyalgia. *Am. J. Psychiatry* **2006**, *163*, 1840–1841. [CrossRef]

310. Leader, A.; Amital, D.; Rubinow, A.; Amital, H. An open-label study adding creatine monohydrate to ongoing medical regimens in patients with the fibromyalgia syndrome. *Ann. N. Y. Acad. Sci.* **2009**, *1173*, 829–836. [CrossRef]

311. Ostojic, S.M.; Stojanovic, M.; Drid, P.; Hoffman, J.R.; Sekulic, D.; Zenic, N. Supplementation with Guanidinoacetic Acid in Women with Chronic Fatigue Syndrome. *Nutrients* **2016**, *8*, 72. [CrossRef]

312. Agren, H.; Niklasson, F. Creatinine and creatine in CSF: Indices of brain energy metabolism in depression. Short note. *J. Neural Transm.* **1988**, *74*, 55–59. [CrossRef]

313. Niklasson, F.; Agren, H. Brain energy metabolism and blood-brain barrier permeability in depressive patients: Analyses of creatine, creatinine, urate, and albumin in CSF and blood. *Biol. Psychiatry* **1984**, *19*, 1183–1206. [PubMed]

314. Kato, T.; Takahashi, S.; Shioiri, T.; Inubushi, T. Brain phosphorous metabolism in depressive disorders detected by phosphorus-31 magnetic resonance spectroscopy. *J. Affect. Disord.* **1992**, *26*, 223–230. [CrossRef]

315. Kato, T.; Takahashi, S.; Shioiri, T.; Murashita, J.; Hamakawa, H.; Inubushi, T. Reduction of brain phosphocreatine in bipolar II disorder detected by phosphorus-31 magnetic resonance spectroscopy. *J. Affect. Disord.* **1994**, *31*, 125–133. [CrossRef]

316. Silveri, M.M.; Parow, A.M.; Villafuerte, R.A.; Damico, K.E.; Goren, J.; Stoll, A.L.; Cohen, B.M.; Renshaw, P.F. S-adenosyl-L-methionine: Effects on brain bioenergetic status and transverse relaxation time in healthy subjects. *Biol. Psychiatry* **2003**, *54*, 833–839. [CrossRef]

317. Kondo, D.G.; Forrest, L.N.; Shi, X.; Sung, Y.H.; Hellem, T.L.; Huber, R.S.; Renshaw, P.F. Creatine target engagement with brain bioenergetics: A dose-ranging phosphorus-31 magnetic resonance spectroscopy study of adolescent females with SSRI-resistant depression. *Amino Acids* **2016**, *48*, 1941–1954. [CrossRef] [PubMed]

318. Yoon, S.; Kim, J.E.; Hwang, J.; Kim, T.S.; Kang, H.J.; Namgung, E.; Ban, S.; Oh, S.; Yang, J.; Renshaw, P.F.; et al. Effects of Creatine Monohydrate Augmentation on Brain Metabolic and Network Outcome Measures in Women With Major Depressive Disorder. *Biol. Psychiatry* **2016**, *80*, 439–447. [CrossRef]

319. Allen, P.J.; D'Anci, K.E.; Kanarek, R.B.; Renshaw, P.F. Chronic creatine supplementation alters depression-like behavior in rodents in a sex-dependent manner. *Neuropsychopharmacology* **2010**, *35*, 534–546. [CrossRef]

320. Ahn, N.R.; Leem, Y.H.; Kato, M.; Chang, H.K. Effects of creatine monohydrate supplementation and exercise on depression-like behaviors and raphe 5-HT neurons in mice. *J. Exerc. Nutrition Biochem.* **2016**, *20*, 24–31. [CrossRef]

321. Pazini, F.L.; Cunha, M.P.; Azevedo, D.; Rosa, J.M.; Colla, A.; de Oliveira, J.; Ramos-Hryb, A.B.; Brocardo, P.S.; Gil-Mohapel, J.; Rodrigues, A.L.S. Creatine Prevents Corticosterone-Induced Reduction in Hippocampal Proliferation and Differentiation: Possible Implication for Its Antidepressant Effect. *Mol. Neurobiol.* **2017**, *54*, 6245–6260. [CrossRef]
322. Leem, Y.H.; Kato, M.; Chang, H. Regular exercise and creatine supplementation prevent chronic mild stress-induced decrease in hippocampal neurogenesis via Wnt/GSK3beta/beta-catenin pathway. *J. Exerc. Nutrition Biochem.* **2018**, *22*, 1–6. [CrossRef]
323. Kious, B.M.; Kondo, D.G.; Renshaw, P.F. Creatine for the Treatment of Depression. *Biomolecules* **2019**, *9*, 406. [CrossRef] [PubMed]
324. Bakian, A.V.; Huber, R.S.; Scholl, L.; Renshaw, P.F.; Kondo, D. Dietary creatine intake and depression risk among U.S. adults. *Transl. Psychiatry* **2020**, *10*, 52. [CrossRef] [PubMed]
325. Lyoo, I.K.; Demopulos, C.M.; Hirashima, F.; Ahn, K.H.; Renshaw, P.F. Oral choline decreases brain purine levels in lithium-treated subjects with rapid-cycling bipolar disorder: A double-blind trial using proton and lithium magnetic resonance spectroscopy. *Bipolar Disord.* **2003**, *5*, 300–306. [CrossRef] [PubMed]
326. Lyoo, I.K.; Kong, S.W.; Sung, S.M.; Hirashima, F.; Parow, A.; Hennen, J.; Cohen, B.M.; Renshaw, P.F. Multinuclear magnetic resonance spectroscopy of high-energy phosphate metabolites in human brain following oral supplementation of creatine-monohydrate. *Psychiatry Res.* **2003**, *123*, 87–100. [CrossRef]
327. Sbracia, M.; Sayme, N.; Grasso, J.; Vigue, L.; Huszar, G. Sperm function and choice of preparation media: Comparison of Percoll and Accudenz discontinuous density gradients. *J. Androl.* **1996**, *17*, 61–67.
328. Huszar, G.; Vigue, L.; Corrales, M. Sperm creatine kinase activity in fertile and infertile oligospermic men. *J. Androl.* **1990**, *11*, 40–46.
329. Fakih, H.; MacLusky, N.; DeCherney, A.; Wallimann, T.; Huszar, G. Enhancement of human sperm motility and velocity in vitro: Effects of calcium and creatine phosphate. *Fertil. Steril.* **1986**, *46*, 938–944. [CrossRef]
330. Oehninger, S.; Alexander, N.J. Male infertility: The focus shifts to sperm manipulation. *Curr. Opin. Obstet. Gynecol.* **1991**, *3*, 182–190. [CrossRef]
331. Gergely, A.; Szollosi, J.; Falkai, G.; Resch, B.; Kovacs, L.; Huszar, G. Sperm creatine kinase activity in normospermic and oligozospermic Hungarian men. *J. Assist. Reprod. Genet.* **1999**, *16*, 35–40. [CrossRef]
332. Froman, D.P.; Feltmann, A.J. A new approach to sperm preservation based on bioenergetic theory. *J. Anim. Sci.* **2010**, *88*, 1314–1320. [CrossRef]
333. Lenz, H.; Schmidt, M.; Welge, V.; Schlattner, U.; Wallimann, T.; Elsasser, H.P.; Wittern, K.P.; Wenck, H.; Stab, F.; Blatt, T. The creatine kinase system in human skin: Protective effects of creatine against oxidative and UV damage in vitro and in vivo. *J. Investig. Dermatol.* **2005**, *124*, 443–452. [CrossRef] [PubMed]
334. Peirano, R.I.; Achterberg, V.; Dusing, H.J.; Akhiani, M.; Koop, U.; Jaspers, S.; Kruger, A.; Schwengler, H.; Hamann, T.; Wenck, H.; et al. Dermal penetration of creatine from a face-care formulation containing creatine, guarana and glycerol is linked to effective antiwrinkle and antisagging efficacy in male subjects. *J. Cosmet. Dermatol.* **2011**, *10*, 273–281. [CrossRef] [PubMed]

 *nutrients*

Review

Metabolic Basis of Creatine in Health and Disease: A Bioinformatics-Assisted Review

Diego A. Bonilla [1,2,3,4,*], Richard B. Kreider [5], Jeffrey R. Stout [6], Diego A. Forero [7], Chad M. Kerksick [8], Michael D. Roberts [9,10] and Eric S. Rawson [11]

[1] Research Division, Dynamical Business & Science Society–DBSS International SAS, Bogotá 110861, Colombia
[2] Research Group in Biochemistry and Molecular Biology, Universidad Distrital Francisco José de Caldas, Bogotá 110311, Colombia
[3] Research Group in Physical Activity, Sports and Health Sciences (GICAFS), Universidad de Córdoba, Montería 230002, Colombia
[4] kDNA Genomics®, Joxe Mari Korta Research Center, University of the Basque Country UPV/EHU, 20018 Donostia-San Sebastián, Spain
[5] Exercise & Sport Nutrition Laboratory, Human Clinical Research Facility, Texas A&M University, College Station, TX 77843, USA; rbkreider@tamu.edu
[6] Physiology of Work and Exercise Response (POWER) Laboratory, Institute of Exercise Physiology and Rehabilitation Science, University of Central Florida, Orlando, FL 32816, USA; jeffrey.stout@ucf.edu
[7] Professional Program in Sport Training, School of Health and Sport Sciences, Fundación Universitaria del Área Andina, Bogotá 111221, Colombia; dforero41@areandina.edu.co
[8] Exercise and Performance Nutrition Laboratory, School of Health Sciences, Lindenwood University, Saint Charles, MO 63301, USA; ckerksick@lindenwood.edu
[9] School of Kinesiology, Auburn University, Auburn, AL 36849, USA; mdr0024@auburn.edu
[10] Edward via College of Osteopathic Medicine, Auburn, AL 36849, USA
[11] Department of Health, Nutrition and Exercise Science, Messiah University, Mechanicsburg, PA 17055, USA; erawson@messiah.edu
* Correspondence: dabonilla@dbss.pro; Tel.: +57-320-335-2050

Citation: Bonilla, D.A.; Kreider, R.B.; Stout, J.R.; Forero, D.A.; Kerksick, C.M.; Roberts, M.D.; Rawson, E.S. Metabolic Basis of Creatine in Health and Disease: A Bioinformatics-Assisted Review. *Nutrients* **2021**, *13*, 1238. https://doi.org/10.3390/nu13041238

Academic Editor: Susanne Klaus

Received: 23 March 2021
Accepted: 7 April 2021
Published: 9 April 2021

Publisher's Note: MDPI stays neutral with regard to jurisdictional claims in published maps and institutional affiliations.

Abstract: Creatine (Cr) is a ubiquitous molecule that is synthesized mainly in the liver, kidneys, and pancreas. Most of the Cr pool is found in tissues with high-energy demands. Cr enters target cells through a specific symporter called Na^+/Cl^--dependent Cr transporter (CRT). Once within cells, creatine kinase (CK) catalyzes the reversible transphosphorylation reaction between $[Mg^{2+}:ATP^{4-}]^{2-}$ and Cr to produce phosphocreatine (PCr) and $[Mg^{2+}:ADP^{3-}]^-$. We aimed to perform a comprehensive and bioinformatics-assisted review of the most recent research findings regarding Cr metabolism. Specifically, several public databases, repositories, and bioinformatics tools were utilized for this endeavor. Topics of biological complexity ranging from structural biology to cellular dynamics were addressed herein. In this sense, we sought to address certain pre-specified questions including: (i) What happens when creatine is transported into cells? (ii) How is the CK/PCr system involved in cellular bioenergetics? (iii) How is the CK/PCr system compartmentalized throughout the cell? (iv) What is the role of creatine amongst different tissues? and (v) What is the basis of creatine transport? Under the cellular allostasis paradigm, the CK/PCr system is physiologically essential for life (cell survival, growth, proliferation, differentiation, and migration/motility) by providing an evolutionary advantage for rapid, local, and temporal support of energy- and mechanical-dependent processes. Thus, we suggest the CK/PCr system acts as a dynamic biosensor based on chemo-mechanical energy transduction, which might explain why dysregulation in Cr metabolism contributes to a wide range of diseases besides the mitigating effect that Cr supplementation may have in some of these disease states.

Keywords: creatine kinase; energy metabolism; cell survival; bioinformatics; systems biology; cellular allostasis; dynamic biosensor

1. Introduction

Creatine (Cr) is a ubiquitous non-protein amino acid (PubChem CID: 586) that is synthesized mainly in the liver, kidneys, and pancreas [1]. However, other tissues (e.g., brain and testes) are also able to produce Cr [2–4]. Endogenous Cr synthesis begins with the transfer of the amidino group of L-arginine to the N^{α}-amine group of L-glycine following a ping-pong mechanism that is catalyzed by L-Arginine-Glycine amidinotransferase (AGAT-EC 2.1.4.1) [5]. This first reaction yields L-ornithine and guanidinoacetate (GAA), which is then methylated at the original nitrogen of glycine using S-adenosyl-L-methionine as the donor of the methyl group by means of the Guanidinoacetate N-Methyltransferase (GAMT-EC 2.1.1.2). This reaction follows the formation of a strong nucleophile on the deprotonated glycine-derived N of GAA that interacts with the methyl group from the positively charged sulfonium ion of S-adenosyl-L-methionine [6] to produce Cr and S-adenosyl-L-cysteine (Figure 1).

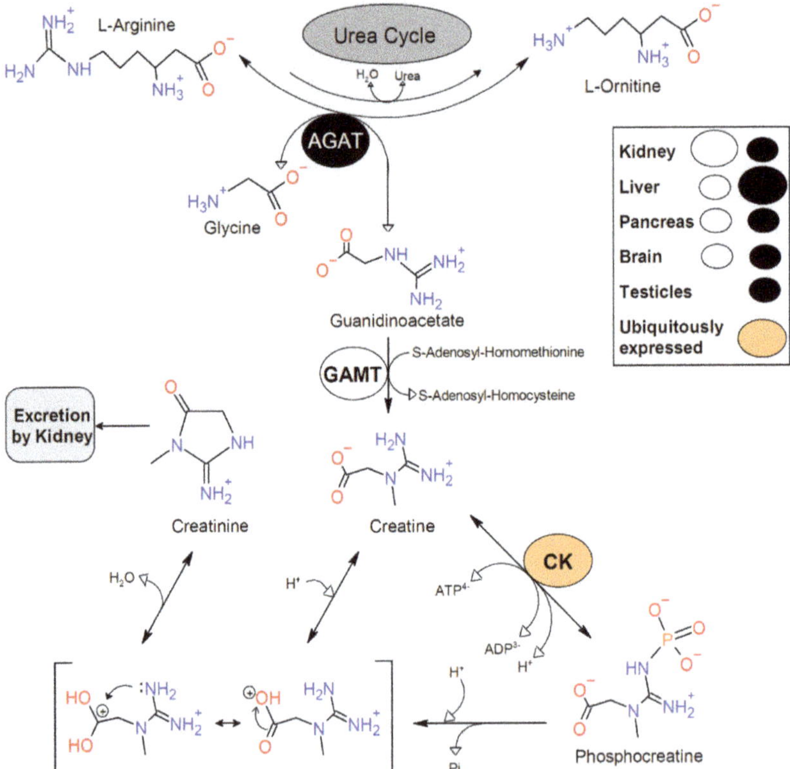

Figure 1. Creatine synthesis/excretion and the creatine kinase reaction. Enzymes are represented by ovals. Once synthesized from L-arginine, glycine, and S-adenosyl-L-methionine, creatine (Cr) is converted to phosphocreatine (PCr) by means of the creatine kinase (CK), which catalyzes the reversible transference of a phosphoryl group (PO_3^{2-}), not a phosphate (PO_4^{3-}), from ATP. The kinetic rate of the non-enzymatic conversion of Cr (or PCr) to creatinine (Crn) depends on the H^+ concentration of the media. It is important to note that neither Crn nor PCr are substrates of the sodium- and chloride-dependent creatine transporter (not shown). Oval size represents the expression level of AGAT (black), GAMT (white), and CK (orange) in some tissues. For more details related to expression in different tissues or conditions (i.e., pathologies) use the following BioGPS ID numbers: AGAT–2628; GAMT–2593. AGAT: L-Arginine-Glycine amidinotransferase; GAMT: Guanidinoacetate N-Methyltransferase; H^+: hydrogen ion; Pi: inorganic phosphate. Modified with permission from Bonilla and Moreno [7] using the Freeware ACD/ChemSketch 2021 (Advanced Chemistry Development, Inc., Toronto, ON, Canada).

High Cr concentrations are found in skeletal muscle and the brain [8]. High Cr levels are also found in other cells with high energy demands such as the cardiomyocytes, hepatocytes, kidney cells, inner ear cells, enterocytes, spermatozoa, and photoreceptor cells [9,10]. However, approximately 95% of the Cr pool in the body is found in skeletal muscle [11–13]. After synthesis, Cr reaches target tissues through the bloodstream, and intracellular transport mediated by a solute carrier protein called sodium- and chloride-dependent creatine transporter (CRT, also known as SLC6A8) [14]. This symporter belongs to a family of neurotransmitter transporters known as solute carrier family 6, which has shown a high affinity to Cr in the plasmalemma (low Km, 15–77 μM) [15–17]. Cr is one of the main osmolytes of the central nervous system, which may play important roles in pathophysiological conditions of the brain [18,19]. Currently, some consider Cr a neurotransmitter that may be released in the synapse, re-uptaken by presynaptic CRT, and might either depress post-synaptic GABAergic neurotransmission or stimulate post-synaptic glutamatergic pathways [20]; nevertheless, more studies are needed to generate consensus, in particular by discovering a so far unknown specific post-synaptic Cr receptor [21]. Although some of the aforementioned tissues might synthesize Cr, CRT is necessary to transport endogenous and exogenous Cr to cells with high and fluctuant energy demands for proper physiological function [22].

Cr exists as a zwitterion, with the positive charge on the resonance structures of the guanidinium moiety and the negative charge on the carboxylate oxygen atoms. Thus, it forms a monoclinic crystal system with one water molecule of crystallization [23,24]. These crystals are well-known as creatine monohydrate (CrM), which dehydrates at 110 °C [25]. In Figure 1, the Cr molecule is shown in the zwitterionic form as found in the crystal structure of CrM (where both H-atoms of the water act as hydrogen bond donors—not shown) [23]. It is important to note that the solubility of CrM in water increases with temperature (e.g., $8.5 \text{ g} \cdot \text{L}^{-1}$ at 4 °C and $14 \text{ g} \cdot \text{L}^{-1}$ at 25 °C) [26]. It is also notable that CrM has been extensively studied as a nutritional supplement. In this regard, CrM supplementation has been deemed as a safe and effective ingredient across various disciplines ranging from sports nutrition to health and disease [27–43]. Although other forms of Cr have been studied, such as Cr nitrate [44–46], there is no evidence that these ingredients are more efficacious relative to CrM [47]. Readers are encouraged to refer to the outstanding invited reviews of this book/special issue on "Creatine Supplementation for Health and Clinical Diseases" to learn more about the effects of CrM supplementation [48].

Cr and its phosphorylated form, phosphocreatine (PCr), have a critical and centralized role in maintaining adenosine triphosphate (ATP) concentrations in tissues with high-energy demands, such as skeletal muscle, heart, and brain [28]. Alterations in Cr concentrations due to CRT, AGAT, or GAMT deficiencies may produce functional changes in these tissues, leading to a wide range of diseases [14,22,49–51] that are grouped into the Cr deficiency syndrome [52]. For example, CRT malfunction results in low levels of intracellular Cr, which, while not lethal, induces an impairment in brain energy metabolism to the same extent as deficiencies in the Cr biosynthesis enzymes [22,53]. A dysregulation in Cr metabolism has also been implicated in various pathological conditions including muscle dysfunction, cardiomyopathy, and cancer, among others [48,54]. Given the aforementioned evidence, a systems biology approach is needed to deepen our comprehension of the molecular, cellular, tissue and systemic effects of Cr and its applications to health and disease. Therefore, the aim of this bioinformatics-assisted review was to highlight the most recent findings and up-to-date literature concerning Cr metabolism.

2. Methods

To summarize the basis and to report the most recent findings of creatine metabolism, we performed a search of articles indexed in PubMed/MEDLINE, ScienceDirect, Cochrane, SciELO, and Google Scholar databases using terms related to 'creatine metabolism'. A bioinformatics-assisted analysis was performed for functional annotations within the literature review. To this end, we accessed public databases and repositories such as

UniProtKB (https://www.uniprot.org/), PDB (https://www.rcsb.org/), Ensembl (https://www.ensembl.org/index.html), The Gene Ontology Resource (http://geneontology.org/), and the BioGPS–Gene Portal System (http://biogps.org/). Additionally, we used the freely available Search Tool for the Retrieval of Interacting Genes (STRING: https://string-db.org/) to report the experimentally validated interacting proteins. The following options were activated in the STRING tool to obtain the protein–protein interactions network: (i) search—by multiple proteins; (ii) network type—full STRING network; (iii) meaning of network edges—evidence; (iv) minimum required interaction score—high confidence (0.700); and, (v) max number of interactors to show—1st shell = 30, and 2nd shell = no more than 20 interactors. To cluster the most similar nodes of the network into an easily distinguishable function-based classification, we used the Markov Cluster Algorithm for graphs, which is based on simulation of stochastic flow in the obtained graph. The inflation factor was set at 1.5 to balance sensitivity and selectivity. Databases/repositories and bioinformatics tools were accessed from 11 November 2020 to 14 February 2021.

The idea of complexity in biological systems was addressed from a reformulated insight that followed development (self-organizing) to cellular dynamics (functional and structural stability through change–allostasis). Therefore, the retrieved references were summarized and discussed in this review's narrative to answer certain pre-specified questions: (i) What happens when creatine is transported into cells? (ii) How is the CK/PCr system involved in cellular bioenergetics? (iii) How is the CK/PCr system compartmentalized throughout the cell? (iv) What is the role of creatine amongst different tissues? and (v) What is the basis of creatine transport?

3. Findings

3.1. What Happens When Creatine Is Transported into Cells?

Once in the intracellular environment, the creatine kinase (CK, ATP:creatine phosphotransferase, EC 2.7.3.2) catalyzes the reversible transphosphorylation reaction between $[Mg^{2+}:ATP^{4-}]^{2-}$ and Cr to produce PCr and $[Mg^{2+}:ADP^{3-}]^{-}$ following a bimolecular nucleophilic substitution reaction [55]. The average concentration of total Cr (free Cr + PCr) in skeletal muscle is around 120 mmoL·kg^{-1} dry mass ($\approx$40 mM) [56] although PCr is found in higher concentration (80–85 mmoL·kg^{-1} dry mass or $\approx$27 mM, $\approx$67%) than free Cr ($\approx$40 mmoL·kg^{-1} dry mass or $\approx$13 mM, $\approx$33%) [8]. Besides the difference in the free energy change ($\Delta G°$) for the hydrolysis of PCr and ATP at pH 7.0 (-44.58 kJ·moL^{-1} versus -31.8 kJ·moL^{-1}, respectively) [57], PCr and Cr are smaller in molecular size, less negatively charged, and more abundant than ATP and adenosine diphosphate (ADP) in cells expressing CK, which represents a thermodynamic and functional improvement to energy metabolism due to a higher intracellular flux of high-energy phosphates [8]. Importantly, in tissues that require large and intermittent amounts of energy, several CK isozymes are ubiquitously expressed in different cellular compartments (e.g., sarcomere, cytosol, mitochondria) connecting places of ATP synthesis with sites of ATP consumption. This is known as the CK/PCr system [11].

Cr is spontaneously degraded to creatinine (Crn) in a monomolecular and non-enzymatic reaction that depends on temperature and pH [58]. Crn might diffuse out of the cells to be excreted by the kidneys into the urine with a mean excretion rate of 23.6 mg·kg^{-1}·day^{-1} (about 1.7% of the total Cr pool per day) [8]. As more than 90% of Cr and PCr molecules are found in skeletal muscle, Crn excretion is $\approx$20% less in women and the peak urinary excretion rate is found between 18 to 29 years old [1]. Hence, the daily requirement of Cr from either diet or endogenous synthesis for a 70-kg male is approximately 2 g·day^{-1} [59]. This has raised concerns in vegan and vegetarian population who have been reported to have lower Cr concentrations in different tissues [60,61] since Cr is naturally found in animal products [62,63]. Figure 1 represents the basis of Cr, PCr, and Crn metabolism.

CrM supplementation increases serum and muscular Cr levels [59,64,65], as well as brain Cr levels [66], although no effect is seen with ATP concentrations [67]. While this increase is very significant in serum and skeletal muscle, Cr is not as permeable through the blood-brain barrier as it is in other tissues, so it typically takes higher doses of Cr over a longer periods of time (e.g., 15–20 g per day for 2–4 weeks) to significantly increase Cr content in the brain in healthy individuals [40]. Patients with AGAT and GAMT deficiencies are more dependent on dietary sources of Cr and may need to consume 20–30 g·day^{-1} of CrM habitually to increase and maintain elevations in brain Cr content [68]. For example, in AGAT-deficient patients, it has been shown that after nine months of CrM supplementation (400 mg·kg^{-1}·day^{-1}), brain Cr levels were increased to 80% of Cr [69]; whereas, GAMT-deficient patients have a slower rise of brain Cr with a nearly complete replenishment after more than two years [70]. Conversely, in response to 20 g CrM in healthy individuals, serum Cr concentration increases by 50-fold (peak value of serum Cr is approximately 2.17 ± 0.66 mM) 2.5 h following ingestion [71]. However, in response to lower doses ($\approx$2 g CrM), Cr increase in blood is less significant [72]. In skeletal muscle, total Cr levels increase by about 25% after CrM supplementation, while increases up to 37% occur if the ingestion is accompanied with exercise training [1]. It has been reported that CrM supplementation increases muscle PCr content by $\approx$20%, generally from 80 to 95 mmoL·kg^{-1} dry mass [64,65]. Brault et al. (2007) demonstrated that CrM does not alter the PCr/total Cr ratio and hence the $\Delta G°$ for the hydrolysis of ATP at rest. The authors reported a linear increase of PCr and total Cr concentrations in the *vastus lateralis* after five days of CrM supplementation (0.43 g·kg body mass^{-1}·day^{-1}) using ^{31}P and ^{1}H magnetic resonance spectroscopy [73]. This increase in muscle PCr concentration and the maintenance of the PCr/total Cr ratio are critical in regulating the skeletal muscle bioenergetics due to the crucial role of the CK/PCr system [74]. It is well-established that PCr concentration and oxygen uptake (VO$_2$) vary with similar kinetic profiles from the start-up of the exercise until a new state of energy production by oxidative metabolism [74,75], which has been explained as a function of the mitochondrial resistance and the metabolic capacitance of the CK reaction [76,77]. The regulation of mitochondrial respiration is intimately linked to the CK/PCr system, where changes in the time constant (τ) for the decrease in muscle PCr concentration become critical, as it has been shown in both the "electrical" [78] and "hydraulic" [79] analog models of oxidative metabolism. In fact, recent findings have reinforced the notion that the decline in mitochondrial function due to the aging process is closely related to the muscular performance (i.e., post-exercise PCr recovery rate) [80]. In accordance with these models, an increase in the muscle metabolic capacitance (determined by the augmentation in total Cr) after five days of CrM supplementation (20 g per day) has resulted in a longer τ (slower PCr kinetics) [81], and a slower VO$_2$ response [82]. Thus, the rise in PCr levels following the CrM supplementation optimizes the cellular thermokinetics of energy transduction by regulating the cellular ATP/ADP ratio [7].

3.2. How Is the CK/PCr System Involved in Cellular Bioenergetics?

Cell growth and survival depend on constant ATP regeneration in order to sustain motor proteins (e.g., muscle contraction, vesicle trafficking), ion pumping, protoplasmic streaming, cytoskeletal rearrangement, among others. ATP is synthesized either through substrate-level phosphorylation or through oxidative phosphorylation [83]; however, to guarantee it is mostly used in contraction machinery, ATPase pumps, and other organelles (i.e., nucleus, endoplasmic reticulum, etc.), the cell relies on a phosphotransfer network that is based on the CK/PCr system [84]. This system encompasses two cytoplasmic and two mitochondrial CK (MtCK) isozymes. MtCK is functionally associated with oxidative phosphorylation by co-localization with the adenine nucleotide translocase (ANT, also known as SLC25A4), and by the formation of a proteolipid complex (physical interaction) with the voltage-dependent anion channel (VDAC) and other biological structures in the mitochondrial inner membrane (e.g., cardiolipin-rich domains and other anionic phospholipids) [85]. This system allows ATP to be generated in mitochondria, and this ATP can be subsequently utilized by MtCK to synthesize PCr. This newly-synthesized PCr can then be transported to the cytosol where isozymes of CK resynthesize ATP from ADP [86]. Bessman and Carpenter (1985) initially called such transfer of high-energy phosphates the Cr-PCr shuttle [87]. Thus, in cells that require constant energy for metabolic reactions, PCr acts as an abundant energy buffering molecule that facilitates phosphotransfer reactions by CK parallel to ATP diffusion. Because of fluctuating energy requirements in skeletal muscle and other tissues, the CK/PCr system not only serves as "spatial" but also "temporal" energy buffer where PCr follows closely the energy-requiring processes (e.g., force generation throughout the contraction cycle) while ATP remains constant [88]. The CK/PCr system also improves the thermodynamic efficiency of ATP hydrolysis by maintaining low intracellular ADP concentration and a high ATP/ADP ratio at those subcellular sites where CK is functionally coupled to ATP-requiring processes [13]. In this sense, the CK/PCr system's ATP generation capacity is quite high and exceeds both ATP utilization and ATP resynthesis by other energy-producing pathways (e.g., oxidative phosphorylation and glycolysis) [89,90]. For example, the maximal rate of ATP synthesis by the CK reaction in rat cardiac muscle is 30 $\mu mol \cdot s^{-1} \cdot g^{-1}$, which is much higher than ATP synthesis by oxidative phosphorylation (2.5 $\mu mol \cdot s^{-1} \cdot g^{-1}$) or by de novo pathways (0.39 $\mu mol \cdot s^{-1} \cdot g^{-1}$) [11]. This small reduction in net energy balance (work done per hydrolyzed ATP) makes CK system become crucial for survival, from an evolutionary point of view; in fact, these phosphagen kinase systems date back to several hundred millions years to early metazoan and bacteria [91,92].

CK Isozymes

As mentioned before, the CK isozymes are the core of the CK/PCr system during the process of energy transduction in tissues with high and intermittent energy demands (i.e., skeletal muscle, brain, heart, etc.). Cytosolic CK may be assembled as a protein hetero- or homodimer after binding the M-CK and B-CK subunits to form the MM-, MB-, and BB-CK isozymes, which have an approximate relative mass of 80,000–86,000 [93]. MM- is the major isoform in muscle and heart, MB- is mainly present in the myocardium, and BB-CK exists in many tissues, especially the brain. In skeletal muscle, besides being specifically located at the sarcomeric M-band, a significant proportion of MM-CK is in close proximity to the sarcoendoplasmic reticulum Ca^{2+}-ATPase (SERCA) and sarcolemma. This guarantees the thermodynamic efficiency of ATP hydrolysis ($\Delta G°$ is kept high) [94]. Interestingly, Ramírez et al. (2014) have reported specific phosphorylation of the B-CK isoform at Ser6 can be facilitated by different AMP-activated protein kinase (AMPK) isoforms [95]. This does not affect enzymatic activity, but causes its localization to specific subcellular compartments (e.g., endoplasmic reticulum) as well as its co-localization with the highly energy-demanding SERCA. Moreover, it has been shown that a decrease in intracellular pH in muscle activates MM-CK to facilitate ATP regeneration [96], which might be expected after heightened muscle activity if we consider the optimum pH of this enzyme is between 6.5 and 6.7 [97].

There are two mitochondrial CK isoenzymes: the striated muscle specific, sarcomeric MtCK or sMtCK and the ubiquitous MtCK or uMtCK [11]. Although there is a high degree of sequence homology between these two, sMtCK and uMtCK differ in many biochemical and biophysical parameters. For example, in comparison to M- and B-CK isoenzymes, which are protein dimers, sMtCK and uMtCK are homooctamers (relative mass of ≈340,000) composed of four dimers as the stable building blocks [98,99]. The MtCK is localized between inner and outer mitochondrial membranes in co-localization with ANT, but is also anchored to the cytoskeleton via VDAC and the mitochondrial interactosome [100]. The different characteristics and expression patterns of the CK isozymes account for the cell-compartmentalized and tissue-specialized functions as might be expected (Table 1).

Table 1. Characteristics of the creatine kinase isozymes.

Enzyme Name and Commission Number	Isozyme	Gene Name	Ensembl ID †	Gene Location	UniprotKB	Subunit Structure and PDB Entry	Cellular Location	Tissue Location *
Creatine kinase EC: 2.7.3.2	M-type	*CKM*	ENSG00000104879	Chromosome 19: 45,306,414–45,322,977 Reverse strand.	P06732	Dimer of identical or non-identical chains (1I0E)	Cytosol	Skeletal muscle & heart
	B-type	*CKB*	ENSG00000166165	Chromosome 14: 103,519,659–103,523,111 Reverse strand.	P12277	Dimer of identical or non-identical chains (3B6R)	Cytosol, dendrite, extracellular exosome, extracellular space, mitochondrion, myelin sheath, neuronal cell body and nucleus	Mainly brain, but also in testes, retina, bone, among several others
	U-Type	*CKMT1A*	ENSG00000223572	Chromosome 15: 43,692,886–43,699,222 Forward strand.	P12532	Octamer of four CKMT dimers (1QK1)	Mitochondrial inner membrane and Extracellular exosome	Brain, heart, brown adipose tissue, among several others
	S-type	*CKMT2*	ENSG00000131730	Chromosome 5: 81,233,285–81,266,397 Forward strand.	P17540	Octamer of four CKMT dimers (4Z9M)	Mitochondrial Inner Membrane	Mainly skeletal muscle

Data extracted from Ensembl, UniProtKB, PDB, and Gene Ontology. The heterodimer MB-CK exists mainly in heart. * For more details related to expression in different tissues or conditions (i.e., pathologies) visit BioGPS (http://biogps.org/), a database of gene expression profiles for human tissues [101], using the following ID numbers: *CKM*-1158; *CKB*-1152; *CKMT1A*-548596; and *CKMT2*-1160. † Use the cross-reference from Ensembl to BioGrid, IntAct, MINT or STRING databases in order to analyze protein–protein interactions. Many other bioinformatic tools are currently available. Databases/repositories were accessed on 11 November 2020.

3.3. How Is the CK/PCr System Compartmentalized throughout the Cell?

3.3.1. Mitochondrial Reticulum

Energy-demanding cells have a high hydrolase activity (e.g., ATPases) throughout the entire protoplasm and membranes. The purpose of this is to release the chemical energy stored in the covalent bonds of phosphagen compounds and thereby cover the requirements for survival and growth. At that point, cellular organelles should not be viewed as isolated compartments but, instead, should be seen as a super-connected network of subsystems that maintain cellular allostasis. In this regard, the mitochondrial reticulum has been proposed to exist as a conductive pathway for energy distribution, based on energy distribution across the cell via a much more rapid direct electrical conduction of the mitochondrial membrane potential [102] and constant metabolite diffusion [103]. As a conductive network for skeletal muscle energy distribution, the mitochondrial reticulum helps to cover more surface area and minimize distances for metabolites to support the rapid energy transduction over large cell regions. This connectivity puts the energy distribution system at risk though, because damaged elements could compromise the entire network. Nevertheless, it has been shown that several intermitochondrial junctions exist, which limits the cellular impact of localized dysfunction. However, the dynamic disconnection of damaged mitochondria allows the remaining mitochondria to resume normal function within seconds [104]. In this context, wherever the mitochondrial reticulum is extending, MtCK and PCr are likely present to support energy transduction between metabolic microcompartments [103].

Octameric MtCK has membrane-binding properties, and it acts as a typical peripheral membrane protein. More specifically, it is anchored to cristae and the peripheral intermembrane space of mitochondria, showing a high affinity for acidic phospholipids, especially cardiolipin (diphosphatidylglycerol) in the inner membrane, and to VDAC in the outer membrane [11]. Hence, because of its size and its binding properties, MtCK can bridge the intermembrane space [105]. As previously mentioned, there is also enough evidence to suggest that MtCK is functionally close to the transmembrane ANT in the inner mitochondrial membrane [85]. This proteolipid complex comprising ANT, ATP synthasome, MtCK, VDAC, membrane phospholipid compounds, and β-tubulin in cytoskeleton contact sites has been named as mitochondrial interactosome and is an important regulator of mitochondrial oxidative metabolism [106]. It has been shown that endogenous ADP is a crucial regulator of oxidative phosphorylation but only in the presence of Cr and MtCK, which is strongly amplified by the co-localization with ANT due to the continuous recycling of adenine nucleotides within the mitochondrial interactosome [107]. The MtCK transfers the phosphoryl group from mitochondrial ATP to Cr producing PCr and recycling ADP in mitochondria. Recycled ADP is returned to F_oF_1-ATP synthase complex due to its functional coupling with MtCK while PCr leaves mitochondria due to the high selective permeability of VDAC for this compound [100]. The remarkably high affinity of MtCK for both Cr and PCr, and the metabolic channeling of ATP and ADP via ANT, show that PCr is the main carrier for energy flux carried out from mitochondria reticulum [108]. To highlight, Karo et al. [109] developed a coarse-grained model to simulate the molecular dynamics of the MtCK system, including MtCK, transmembrane ANT, and a membrane composed of phosphatidylcholine, phosphatidylethanolamine, and cardiolipin (2:1:1). The model was validated against many structural and dynamical experimental properties, which makes it useful for future developments. For a recent and comprehensive review of the molecular characteristics and essentials of the mitochondrial proteolipid complexes of CK please refer to Schlattner et al. [85].

Recent studies have proposed that Cr metabolism might have a potential role in thermogenesis. This heat production process occurs in mitochondria through the uncoupling proteins (UCPs), which serve as H^+ carriers from intermembrane space to matrix and thereby shunt energy from electron transport chain during ATP synthesis [110]. In general, this process releases the oxidation energy as heat and decreases ATP synthesis rates. Initially called thermogenin, UCPs belong to the solute carrier family 25 (SLC25), with UCP1 (also known as SLC25A7) as the isoform only expressed in the brown adipose tissue (BAT) [111]. Notwithstanding, several UCP isoforms have been reported in humans. UCP2 (SLC25A48) is expressed in various tissues, such as skin, muscle, pancreas, adipose tissue [112]. UCP3 (SLC25A9) is mainly found in cardiac and skeletal muscle, and UCP4 (SLC25A27) and UCP5 (SLC25A14, also called brain mitochondrial carrier protein-1) are expressed in the central nervous system [113]. Although these UCP isoforms have high homology and structural similarities (i.e., C- and N-terminal chains are found towards the intermembrane space) [114,115], their biological role and the H^+ transport mechanism seem to be different according to the cell/tissue where they are expressed [116]. After stimulation and in presence of fatty acids, UCPs allow the passive movement of H^+ from intermembrane space to mitochondrial matrix via two putative mechanistic models including: (i) the fatty acid cycling model, which is based on a "flip-flop" mechanism, where the UCPs can also transport anions (e.g., fatty acids derivatives) outside the intermembrane space in order to allow them to protonate and get back to matrix [117,118]; and, (ii) the fatty acid buffering model, in which UCPs are proton carriers with fatty acids working as co-factors that interact with carboxyl groups of negatively charged amino acids to mediate the H^+ transport through a hypothetic channel [119]. An alternative modification of the latter model is called the fatty acid shuttling model, where the fatty acid anions bind inside the UCP cavity resulting in a conformational change that shuttles the H^+ [120]. Taking into account differences in molecular mechanisms among isoforms, UCPs possess negative regulation sites for nucleotides (ADP, GDP, etc.) and Pi, which can bind to the cavity and allosterically displace fatty acids from the peripheral site and consequently prevent H^+ transport [116,121]. Therefore, it is plausible that the metabolism of high-energy phosphates regulates this mitochondrial energy dissipation.

Interestingly, CK activity and genes related to Cr metabolism are coordinately elevated by cold-exposure in beige/brite adipocytes [122]. Additionally, according to Kazak et al. [123] the genetic-induced depletion of Cr in mice significantly blunts β3-adrenergic activation and affects whole-body oxygen consumption. These authors also reported an obese phenotype in mice lacking the capability of the adipose tissue to synthesize Cr, and Cr supplementation rescues aspects of thermogenesis in these animals. Bertholet et al. [124] implemented patch-clamp and bioenergetics analyses to characterize wild-type and *UCP1*-negative beige/brite adipocytes from C57BL/6J mice. These authors found that UCP1 appeared non-essential for the process of browning (because robust mitochondrial biogenesis was still observed in cells lacking UCP1 expression), as well as higher *CKMT2* expression in the *UCP1*-negative model, which supported Cr cycling as a UCP1-independent thermogenic mechanism. Since *UCP1*-negative adipocytes are unable to exhibit a rapid adaptive thermogenic response [123], the ATP-dependent thermogenic pathways may play a key role in diet-induced thermogenesis [125]. Nowadays, it is hypothesized that Cr metabolism may also provide an alternative mechanism of heat production following a futile cycle [126] (also called Cr-driven thermogenesis or Cr-dependent substrate cycling [127]) that coexists with the ATP-dependent Ca^{2+} cycling by SERCA as the main UCP1-independent thermogenic pathways in BAT and beige adipocytes [128]. While the existence of a novel mitochondrial phosphocreatine phosphatase has been hypothesized to explain this highly unusual type of Cr utilization in thermogenic adipocytes [126,128], Wallimann et al. [129] proposed that Cr may operate as part of the classical CK/PCr system by providing ATP to other thermogenic pathways, such as the previously mentioned ATP-dependent Ca^{2+} cycling by SERCA. In spite of these findings, a recent study by Connell et al. [130] showed that CrM supplementation (20 g·day^{-1} for nine consecutive days) did not enhance BAT activation after acute cold exposure in young, healthy, lean, and vegetarian adults. Thus, future clinical research is needed to determine if Cr metabolism plays a role in beige/brite adipose tissue thermogenesis.

3.3.2. Cytosol and Cytoskeleton

In the cytosol, CK is functionally coupled to the enzymatic machinery of glycogenolysis and glycolysis to form an efficient subsystem of energy production and transduction [131]. Several proteins of the glycolytic machinery are located at the I-band and associated with the thin filaments in the sarcomere. Similarly, most of the soluble MM-CK is located at I-band, and, thus, serves to maintain the efficiency of the extramitochondrial ATP production [11]. During periods of high energy demand, the net result of the CK reaction includes the breakdown of PCr to Cr and Pi while ATP and ADP concentrations remain almost constant [132]. This net release of Pi is a seldom-recognized consequence of the CK reaction and is proportional to the amount of PCr hydrolyzed [13]. In this sense, besides buffering ATP concentrations, the CK/PCr system also provides a source of increasing Pi with elevations in work rate [133]. The reaction has a regulatory effect on glycogenolysis and glycolysis since Pi can stimulate glycogen phosphorylase and phosphofructokinase [13]. In fact, anchoring of MM-CK to the I-band via phosphofructokinase has been shown to be strongly pH-dependent and taking place below pH 7.0 [131]. It is important to note that several glycolytic enzymes, glycogen phosphorylase, CK, and adenylate kinase, bound to phosphofructokinase [134], as a key enzymatic complex to regulate glycolysis [135]. Moreover, M-CK has also been shown to bind β-enolase as an anchor for glycolytic complexes on the sarcomere [136].

Overall, while mt-CK activity lowers cytosolic Pi levels, cytosolic CK isozymes have the opposite effect [137]. This not only supports the notion that CK/PCr system acts as an important regulator of mitochondrial ATP synthesis with Pi as a primary controller of oxidative phosphorylation [138] but also demonstrates its interconnectivity with glycolysis. According to the molecular system bioenergetics-part of the systems biology approach [139], in vivo regulation of cellular respiration and energy fluxes (i.e., system level properties) depend on intracellular interactions between mitochondrial reticulum, cytoskeleton, intracellular ATPases, and cytoplasmic glycolytic machinery (i.e., system's components) [140]. For example, hexokinase and β-tubulin (important proteins for glycolysis and cytoskeleton modulation, respectively) have been shown to regulate the mitochondrial outer membrane permeability via interaction with VDAC within the large intermembrane protein supercomplex of the mitochondrial interactosome [141].

Hexokinase binds to VDAC to regulate mitochondrial function while stimulating glycolysis considering that ATP from oxidative phosphorylation will be guided directly to active sites of the glycolytic machinery (like hexokinase-2) [142]. In cancer cells, this functional and structural proximity leads to a common metabolic phenotype where there is a higher glycolysis rate rather than oxidative metabolism for energy production, known as the "Warburg effect" [143]. Besides the direct antioxidant properties [144], the potential anti-tumor progression that has been associated to Cr and cyclocreatine administration [126] might be partially explained by a less glycolytic rate in tumor cells. Based on the Warburg hypothesis, it has also been discussed that high-intensity exercise may inhibit glycolysis and have a stronger anti-tumor growth effect in comparison to moderate-intensity exercise [145]. Since immune-based manipulation of glucose metabolism are a subject of high interest to ameliorate cancer progression [146,147], further research might evaluate the effects and regulation of high-intensity exercise plus CrM supplementation (and derivatives) on tumor growth. Several authors have reported lower lactate accumulation after Cr administration in different conditions both in vivo (human and animal models) and in vitro studies [148–153]. This reduction in lactate concentration, especially during circumstances requiring high amounts of ATP, has been attributed to less reliance on glycolytic ATP production due to higher intracellular PCr levels after Cr administration. Interestingly, PCr not only inhibits phosphofructokinase [154] and pyruvate kinase [155] activity, but this molecule also stimulates fructose-1,6-biphosphatase [156]. The enzymatic regulation and the frequent rest lapses of intermittent exercise (that contribute to the maintenance of ATP, PCr, and malate levels) may consequently inhibit glycolysis. Although the exact mechanism is still unknown, PCr has also been proposed to modulate AMPK by regulating intracellular PCr concentration. Ponticos et al. [157] reported in vitro that an increase in the intramuscular concentrations of PCr inhibits AMPK activity while free Cr antagonizes this inhibition. A decrease in the AMP/ATP ratio also inhibits this metabolic regulator [158].

Because AMPK activation occurs in response to a reduction in energy availability, an increase in the energy availability by optimization of the phosphagen system after Cr supplementation would favor a direct inhibition and/or delay of AMPK activation during periods of high-energy demand. Recently, Zhang et al. [159] showed that dietary addition of CrM (1200 mg·kg^{-1}) inhibited the AMPKα pathway and reduced muscle glycolysis, which improved meat quality in transport-stressed broilers. In spite of the above, Taylor et al. [160] found that PCr neither inhibited phosphorylation of AMPK by LKB1 (AMPKK), nor inhibited recombinant or highly purified rat liver AMPK. Moreover, Eijnde et al. [161] reported that CrM supplementation during two weeks of immobilization (15 g·day^{-1}) and subsequent six-week rehabilitation training (2.5 g·day^{-1}) did not affect the expression of AMPK α1, α2, or β2 subunits or the phosphorylation status of AMPK α1. Thus, while certain evidence suggests that changes in PCr concentrations might regulate AMPK activity, other studies do not support these findings. Therefore, future studies are needed to better comprehend the mechanisms by which CrM supplementation modulates glycolysis at high work rates as well as AMPK activity.

Cr metabolism may also regulate cellular processes by being involved with cytoskeletal dynamics. Aside from serving as a scaffold to maintain cellular integrity by cross-linking microtubules (tubulin), microfilaments (actin) and intermediate filaments (lamin), the cytoskeleton possesses architectural, mechanical, and signaling functions that connect cellular subsystems (e.g., sarcomere) to other organelles (e.g., mitochondrial reticulum, membrane and nucleus) [162]. In this regard, it has been shown that the interaction between cytoskeletal proteins and mitochondria (e.g., β-tubulin-VDAC interaction) modulates cellular energy metabolism by contributing to the switch from oxidative phosphorylation to glycolysis [163]. The proteins of the mitochondrial interactosome, including the MtCK, are responsible for this regulation [164]. Furthermore, in myocytes, the Four-and-a-Half Lim 2 (FHL2) not only binds to titin and serves as an important mechanosensor that triggers hypertrophy in response to strain (via mitogen-activated protein kinases, MAPKs) but also docks key metabolic enzymes involved in the energy transduction process like M-CK, adenylate kinase, and phosphofructokinase [165]. Refer to Henderson et al. [166] for a comprehensive review regarding cytoskeleton architecture and proteins functions. Maintaining a close interaction between mitochondrial reticulum and myofibrils through a highly structured cytoarchitecture seems critical for optimal energetic regulation, especially by compartmentalized phosphotransfer enzymes and glycolytic machinery [167]. Hence, energetic interactions between subcellular organelles in high-energy demanding cells depend largely on phosphotransfer kinases, the most important being CK, and their connections to cytoskeleton proteins [168]. It is not surprising that energy disturbances due to the dysfunction of mitochondria and mitochondria-cytoskeleton connections/interactions can lead to various congenital and age-associated diseases [169–173].

The extensive cytoskeletal reorganization that occurs before and during cell fusion (e.g., myoblast fusion during muscle development) is highly dependent on ATP hydrolysis, and the polymerization and dissociation of actin monomers may require up to 50% of cellular energy expenditure [174]. As an ATP-consuming process, actin cytoskeleton polymerization can be also optimized by higher phosphagen availability. This was demonstrated by O'Connor et al. (2008) by assessing the in vitro and in vivo effects of Cr administration on myoblast fusion. The authors concluded that Cr enhanced both myotube growth and myonuclear addition in a CK- and actin polymerization-dependent manner [175]. Current available evidence also suggest that ATP produced by cytosolic CK isoforms near the ends of myotubes plays a key role in myoblast fusion during myogenesis [176,177].

3.3.3. Nucleus

The role of the cytoskeleton is not limited to maintaining the structural integrity of the cell, but is also closely involved in gene expression. The linker of nucleoskeleton and cytoskeleton (LINC) complex has been described as an important system of proteins that provides structural support to maintain the nuclear morphology and genome integrity by means of the interaction between the nucleoskeleton with the cytoskeleton [178]. Also, the LINC complex regulates dynamic events including DNA replication and gene transcription [179], and miRNA processing [180]. Briefly, the LINC complex contains three proteins: (i) lamins, which are the basic subunit of intermediate filaments as previously mentioned; (ii) SUN domain proteins, which correspond to Sad1 and UNC-84 proteins; and, (iii) nuclear envelope spectrin repeat proteins (nesprins) [181]. Here, various FHL isoforms (mainly FHL1) have been reported to interact with different transcription factors in the nucleus (e.g., NFAT proteins or RBP-J) that are involved in cell proliferation and differentiation, as well as with the pro-apoptotic protein Siva where it is involved in cell survival [182].

Nuclear migration is seemingly critical for muscle development, fertilization, neuronal development, and cellular polarization, with the ATP-binding protein known as torsinA as the main candidate that mediates these processes [183]. It has been identified that the ATPase activation of torsinA involves two stimulatory co-factors, LAP1 and LULL1 [183]. Accordingly, DNA replication, chromatin remodeling, gene transcription and active transport of macromolecules across the nuclear envelope are highly dependent upon constant ATP generation [86]. While principles governing nuclear energetics and energy support for nucleocytoplasmic communication are still poorly understood, it has been demonstrated that mitochondrial ATP production is required to support energy-consuming processes at the nuclear envelope, while glycolysis by itself might be insufficient to perform such a function [184]. In addition, inhibition of nuclear transport by disruption of the adenylate kinase might be rescued through upregulation of alternative phosphotransfer pathways, such as the CK/PCr system, underscoring the plasticity of the cellular energetic network [185]. For instance, nucleoside-diphosphate kinase (NDPK), which is localized in mitochondria, cytosol, and nucleus, is in charge of nucleoside triphosphates synthesis other than ATP [186]. The γ-phosphate of the ATP molecule is transferred to the β-phosphate of NDP via a ping-pong mechanism, using a phosphorylated active-site intermediate [187]. In addition, NDPK possesses several enzymatic activities, acting as serine/threonine-specific protein kinase, geranyl and farnesyl pyrophosphate kinase, histidine protein kinase, and $3'$-$5'$ exonuclease (UniprotKB ID: P15531). Therefore, NDPK facilitates channeling nucleoside triphosphates into protein synthesis/DNA replication complexes, and GTP/GDP exchange on Ran GTPase as an essential factor in nuclear transport through importins and exportins [188]. Particularly, CK is essential for energy distribution in the nucleus because of its buffering ATP concentrations. Thus, the interaction between these systems (adenylate kinase, CK, and NDPK) secure proper nucleotide ratios at and across the nuclear envelope, sustaining the high energy demand of ATP and GTP hydrolysis [86].

3.3.4. Ion Pumps

MM-CK is functionally coupled to SERCA to favor Ca^{2+} handling (optimal uptake rate and sarcoendoplasmic reticulum content) [189]. Despite the presence of high levels of cytosolic ATP, depletion of PCr impairs Ca^{2+} uptake [190]. This clearly shows the importance of MM-CK in rapid rephosphorylation of local ADP produced in the SERCA reaction, independently from the cytoplasmic environment, demonstrating that bound MM-CK acts in a non-equilibrium manner [94]. On the other hand, co-localization and/or functional coupling of CK isoforms with the Na^+/K^+-ATPase [191,192], the ATP-gated K^+-channel [13], the H^+/K^+-ATPase [191] and the Na^+/Ca^{2+} exchanger [193] have been reported in different tissues.

3.3.5. Motor Proteins

Cellular processes involving contractile machinery for cell division and fusion (e.g., satellite cell proliferation and myoblast fusion, respectively), cell motility (e.g., sperm motility), organelle and cytoskeletal rearrangement (e.g., morphology remodeling after virus infection), membrane transport and clathrin-mediated vesicular trafficking (e.g., GLUT4 endo- and exocytosis), and signaling transduction (e.g., the MAPK pathway c-Jun NH_2-terminal kinase [JNK]) rely vastly on motor proteins. These large mechanochemical

ATPases traverse the cytoskeleton by producing a force that propels them and their cargo forward by transforming chemical energy into mechanical movement via ATP hydrolysis [194]. There are three classes of motor proteins: (i) myosin isoforms, dyneins, and kinesins. Approximately 40 isoforms have been reported in humans, and these proteins traverse on actin filaments to translocate their cargo via anterograde transport (i.e., outward movement from the cell body toward the axon or the cell membrane). Various myosin isoforms are involved with muscle movement, cytokinesis, and transporting cargo along microfilaments [195]. Dyneins traverse cargo on microtubules mostly via retrograde transport (i.e., towards the cell center). Sixteen mammalian classes of these motor proteins exist, and can be divided into cytoplasmic dyneins (vesicle trafficking) and axonemal dyneins (movement of cilia or flagella) [196]. Kinesins usually traverse anterogradely on microtubules, and are in charge of transporting cargos such as vesicles, organelles, mRNA, proteins, and chromosomes (14 classes have been described) [197].

Motor proteins act by hydrolyzing ATP, which results in conformational changes that propel them and the cargo towards its destination. Given the high amounts of ATP involved in these processes, it is logical to link the CK/PCr system to these mechanical processes. The roles and importance of M- and B-CK in different tissues have been well-described [11,13,198]. As mentioned previously, MM-CK is bound to M-line and some relevant proportions of this isozyme are in the I-bands of sarcomeres. This position of the MM-CK is crucial for maintaining the efficiency of ATP regeneration in actomyosin ATPases during muscle contraction. Conversely, PCr accelerates the muscle relaxation from rigor tension by decreasing the necessary ATP concentration possibly due to co-localization of M-CK and the very rapid ADP rephosphorylation [199]. On the other hand, various myosin-associated motor mechanisms involved in the formation of the specialized structures at the phagosome may also be B-CK dependent (i.e., B-CK co-localizes transitorily with F-actin at the nascent phagosome), given that actin polymerization and particle adhesion are highly controlled by the ATP/ADP ratio [200]. It is important to note that cytoskeletal regulators of myofibrillogenesis, rearrangement of mitochondrial reticulum, intracellular signaling, and gene expression, such as desmin, can interact with actin, tubulin, plectin (cytolinker protein), and dynein to facilitate these biological processes [169]. In other cells (e.g., astrocytes and fibroblasts), B-CK facilitates actin-driven cell spreading and migration by localizing in peripheral cellular structures [201]. Indeed, animal models deficient in B-CK, M-CK, or Cr have shown a significant decline in brain, muscle, heart, and sensory organs function. These models have been critical to study how disturbances in Cr metabolism affects various tissues and/or involved with certain disease states [2,15,22,198,202].

Hu et al. [203] examined protein–protein interactions using several experimental databases to describe CK-associated networks in *homo sapiens*. In short, these authors reported more than 120 proteins interacted with B-CK, and approximately 90 proteins interacted with M-CK. The identification of NFKB1, FHL2, MYOC, and ASB9 as hub proteins associated with CK further suggest an important interaction with cytoskeletal- and motor-related proteins. NFKB1 is a functionally cytoskeleton-dependent protein while FHL2 was already described as an important scaffold protein involved in mechanosensing and glycolysis. MYOC is a motor protein classified as class-I myosin, and ASB9 is a protein involved in the ubiquitination-mediated proteolysis pathway. To group the most relevant and recent CK-interacting proteins into an easily distinguishable classification based on function, we submitted various CK isoforms (CKMT1B, CKM, CKB, CKMT2, and CKMT1A) to STRING. Subsequently, we performed a clustering analysis using the Markov Cluster Algorithm for graphs. As shown in Figure 2, two main clusters were identified through this bioinformatics analysis. One cluster of proteins is enriched with enzymes involved in extra- and intramitochondrial ATP production. The second cluster contains proteins that are involved in cellular mechanical allostasis such as cytoskeletal and contractile machinery.

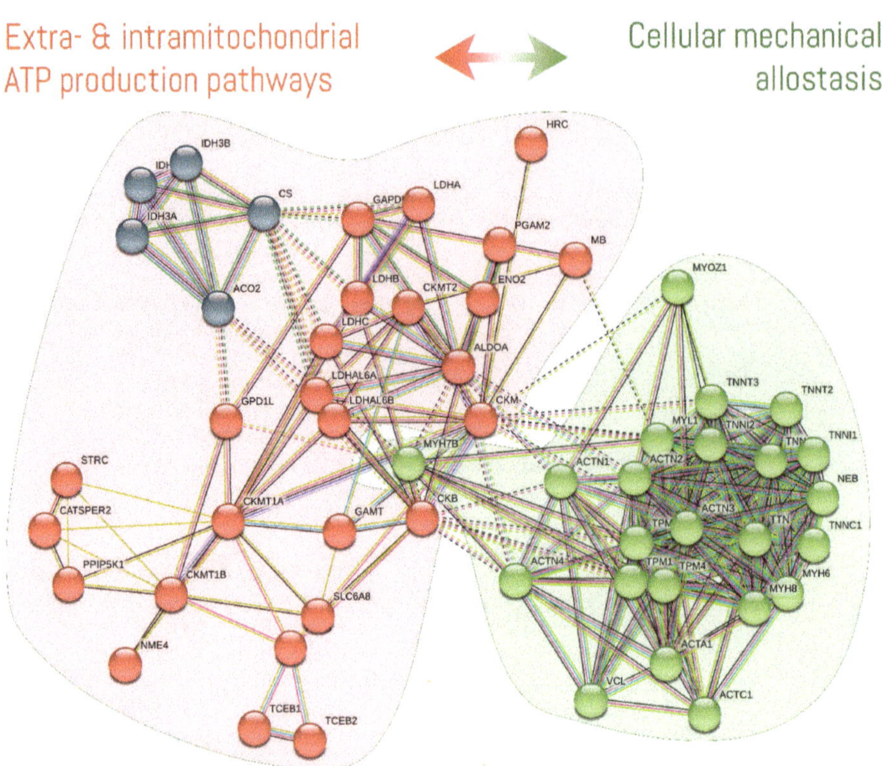

Figure 2. Clustering of CK-interacting proteins using the Markov Cluster Algorithm. Network nodes represent proteins while edges represent protein–protein associations. The red cluster includes a subgroup of enzymes participating in the tricarboxylic acid cycle that are represented in the graph with blue nodes. To visualize our interactive network access to this permanent link: https://version-11-0b.string-db.org/cgi/network?networkId=bu20zAE45PpB (accessed on 14 February 2021).

Intriguingly, the results of our clustering analysis of CK-interacting proteins highly agree with the contention suggesting cellular allostasis is regulated through a complex balance of subcellular energy production and cellular mechanics. This highlights the critical role of force-sensitive cytoskeleton [204]. In this sense, the CK/PCr system could be viewed as a dynamic biosensor of cellular allostasis, and this may explain various positive benefits of CrM supplementation. On this basis, a biosensor is a system composed by a receptor (that interacts with the environment) and a transducer (that converts the biological response into an energy signal) to elicit a physiologically relevant function [205]. The CK/PCr system encompasses a molecular network made of enzymes and metabolites capable of sensing multi-input physiological changes to produce a broad spectrum of specific energy signals (e.g., chemical, electric, mechanical, heat) with biological significance (e.g., muscle contraction, cell motility, human vision, thermogenesis). The CK/PCr system is dynamic in nature but can also operate within adjustable ranges and sensitivities based on the potential alterations in Cr and PCr concentrations (e.g., via CrM supplementation or disease). For example, increases in myoblast fusion (shown in vitro [206] and in vivo [207]) and subsequent myotube growth after CrM administration [47] might involve the cellular mechanical energy properties and the optimization of cytoskeleton dynamics. Cr has a well-documented energy buffering effect [28]. Moreover, it has been shown that Cr enhances actin polymerization [175] and regulates scaffolding and motor proteins that control mechanosensing MAPKs [206,208]. This dynamic biosensor activity of CK/PCr system under the cellular allostasis model also provides a possible mechanistic basis as to why CrM supplementation favorably affects glucose management [126,209]. Specifically, the possible optimization of motor proteins (i.e., cellular mechanics) participating in the transport of GLUT4-containing vesicles to the plasma membrane (i.e., kinesins [KIF3 and KIF] and myosins [MYO5 and MYO1C]) and activation of energy-sensing signaling pathways due to the higher energy availability following CrM supplementation could facilitate improvements in glucose metabolism. This is supported by the fact that even though glucose tolerance is improved, several studies have failed to show a higher muscle content of GLUT-4 protein after CrM administration [209]. Additionally, cytolinker and motor proteins are important components that regulate signaling pathways like MAPKs [208], which in turn might trigger the IGF-I/Akt1/AS160 and/or the mTORC2/Akt1/AS160 pathways to promote GLUT-4 translocation [210–212]. This dynamic biosensor activity will be discussed in further detail according to the results of the convergent functional genomics analysis in an upcoming paper in this special issue.

To summarize, the CK/PCr system can operate in a variety of capacities including: (i) acting as a spatio-temporal energy buffer (this would avoid the inactivation of ATPases and a net loss of adenine nucleotides by preventing the rise in intracellular ADP); (ii) preventing localized acidification through buffering [H^+], which seems especially relevant in the early phase of physical exercise; (iii) becoming a source of increasing Pi at high work rates, which might reduce glycolytic activity; (iv) operating as a low-threshold ADP sensor that increases the thermodynamic efficiency of ATP hydrolysis. Finally, based on the model of predictive regulation [213], Cr metabolism should be seen as a noteworthy mechanism for cell survival and growth if we consider that the CK/PCr system behaves as a hub of chemo-mechanical energy transduction (i.e., dynamic biosensor) during a given allodynamic process. This complex balance of energy and mechanics may provide a manner to better understand the formation onset and progression of certain diseases and aging [204]. Figure 3 depicts a general overview of the CK/PCr system with the muscle cell as a model.

3.4. What Is the Role of Creatine among Tissues?

It has been mentioned that cytosolic and organelle-associated CKs constitute an intricate cellular energy buffering and transport system that connects PCr with sites of energy consumption, especially in tissues with high-energy needs. However, the function of the CK/PCr system as a chemo-mechanical energy transducer are different according to the biological process in non-muscle tissues. Table 2 summarizes the function of different CK isozymes according to the expression location. Additionally, Figure 4 summarizes the importance of CK/PCr system and Cr metabolism in tissues beyond skeletal muscle.

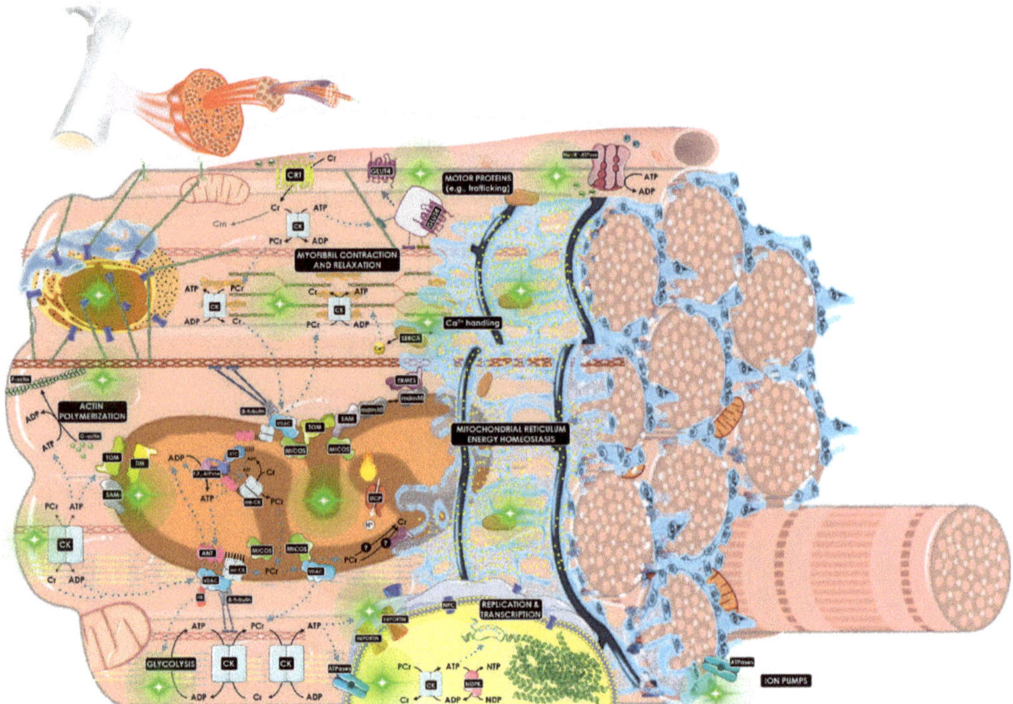

Figure 3. General overview of the CK/PCr system. The diagram represents the super-connected subcellular energy production and cellular mechanics of Cr metabolism. This chemo-mechanical energy transduction network involves structural and functional coupling of the mitochondrial reticulum (mitochondrial interactosome and oxidative metabolism), phosphagen and glycolytic system (extramitochondrial ATP production), the linker of nucleoskeleton and cytoskeleton complex (nesprins interaction with microtubules, actin polymerization, β-tubulins), motor proteins (e.g., myofibrillar ATPase machinery, vesicles transport), and ion pumps (e.g., SERCA, Na$^+$/K$^+$-ATPase). The cardiolipin-rich domain is represented by parallel black lines. Green sparkled circles represent the subcellular processes where the CK/PCr system is important for functionality (see the previous sections for rationale and citations). Several proteins of the endoplasmic reticulum–mitochondria organizing network (ERMIONE), the SERCA complex, the TIM/TOM complex, the MICOS complex, the linker of nucleoskeleton and cytoskeleton complex, and the architecture of sarcomere and cytoskeleton are not depicted for readability. ANT: adenine nucleotide translocase; CK: creatine kinase; Cr: creatine; Crn: creatinine; CRT: Na$^+$/Cl$^-$-dependent creatine transporter; ERMES: endoplasmic reticulum-mitochondria encounter structure; ETC: electron transport chain; GLUT-4: glucose transporter type 4; HK: hexokinase; mdm10: mitochondrial distribution and morphology protein 10; MICOS: mitochondrial contact site and cristae organizing system; NDPK: nucleoside-diphosphate kinase; NPC: nuclear pore complex; PCr: phosphocreatine; SAM: sorting and assembly machinery; SERCA: Sarco/Endoplasmic Reticulum Ca^{2+} ATPase; TIM: translocase of the inner membrane complex; TOM: translocase of the outer membrane complex; UCP: uncoupling protein; VDAC: voltage-dependent anion channel. Source: designed by the authors (D.A.B.) using figure templates developed by Servier Medical Art (Les Laboratoires Servier, Suresnes, France), licensed under a Creative Common Attribution 3.0 Generic License. http://smart.servier.com/ (accessed on 14 January 2021).

Given length restrictions, in-depth discussion of Cr metabolism in each tissue is not provided in-text. However, we aim to give particular attention to Cr metabolism and gut physiology given that this has been vastly understudied. Over 100 trillion microbes reside in the human intestine, and most are located in the colon. A high proportion of gut microbiota are bacteria, but it is notable that protozoans, fungi, archaea, and viruses might be also present. From an evolutionary point of view, these microbes fulfill relevant functions in human metabolism (e.g., vitamin production, fiber digestion, immune system regulation) [214]. Analyses of the collective genomes of these microbiota have led to intense interest regarding how the gut microbiome affects human physiology [215]. Relevant to this review, human Cr and Crn are important markers of microbiota given that they are also eliminated from the host by the action of intestinal microorganisms [8]. Additionally, underexpression of GAMT (rate-limiting step of Cr biosynthesis) can be linked to a colitis phenotype, among other conditions, although CrM administration in homozygous GAMT mutants may ameliorate the symptoms [216]. This illustrates the relevance of Cr in vivo for rapid replenishment of cytoplasmic ATP within colonic epithelial cells in the maintenance of the mucosal barrier after injury. It is also worth noting that Marcobal et al. [217] showed that fecal levels of Cr and Crn were elevated in germ-free versus wild-type mice, which is consistent with previous studies showing an increase of these molecules in biofluids of antibiotic-treated mice. In this way, low Cr concentrations might negatively impact mucosal barrier integrity, which postulates this metabolite as an early functional biomarker of inflammatory bowel disease [218]. Furthermore, Cr and Crn degradation has been shown to be heightened in the gut microbiomes of older mice compared to the middle-aged and younger mice [219]. Although research on the potential of gut microbiota in sports nutrition is in its infancy, it seems that Cr concentrations might be regulated by the microbiome which highlights the potential effects of CrM supplementation in this regard. This might be relevant if we consider the microbial diversity in elite athletes [220] and the effect of gut microbiota on GAA (an intermediary compound of the Cr synthesis) concentrations via guanidinoacetase [221].

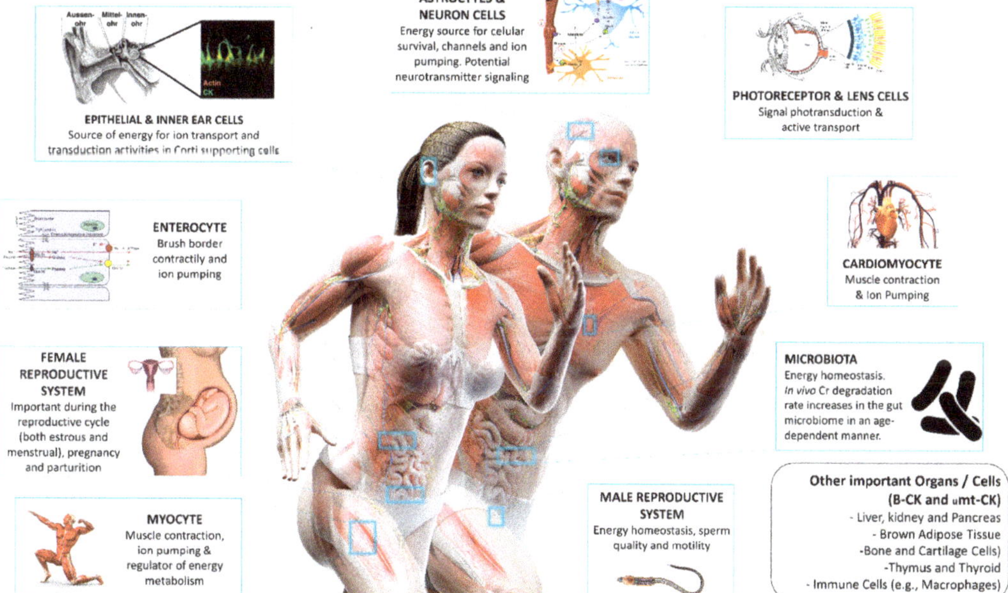

Figure 4. Importance of Cr metabolism in whole-body physiology. The CK/PCr system is essential for the chemo-mechanical energy transduction of cells/tissues with high, fluctuant, and constant energy demands. Source: designed by the authors (D.A.B.) using an anatomy template developed by 3dMediSphere (https://www.turbosquid.com/), licensed 3D standard Vray 3.60. accessed on 14 February 2020.

Table 2. Creatine kinases and creatine among tissues.

Tissue	CK Isozyme	Function
Brain	BB-CK uMtCK	Supports brain cells energy production and buffers ATP and ion pumping during electrical activity in neurons [50]. Oral Cr supplementation has been shown to improve memory in healthy adults, and potential benefits for aging and stressed individuals have been described [222]. Additionally, Cr supplementation seems beneficial in reducing the severity or enhancing recovery from mild traumatic brain injury, but further studies are needed not only as a post-injury therapy but also as a neuroprotective agent in populations at high risk of mild traumatic brain injury [223].
Heart	MB-CK sMtCK	PCr provides about 80% of the energy needed for contraction and ion pumping, and about 20% of energy is transported into the cytoplasm via adenylate kinase and glycolytic phosphotransfer pathways [133,224]. MB-CK is an acute myocardial infarction marker [225].
Testes	BB-CK uMtCK	Energy production and ATP buffer at axoneme, where microtubules and dynein use direct energy for sperm motility [13,226]. Cr concentrations and CK activity are potential indicators of sperm quality [227].
Uterus	BB-CK uMtCK	Special attention should be paid to the increased Cr demand during pregnancy due to the important role of the PCr/CK system in the uterus and placenta for the maintenance and termination of gestation [34,228,229].
Sensory organs	BB-CK MM-CK MB-CK uMtCK sMtCK	Visual system: important role in phototransduction by providing energy for the visual cycle, maintaining high local ATP/ADP ratios and consuming H$^+$ produced by ATPases located in the outer segment and, thereby, preventing acidification [230]. Auditory system: MM-CK is located in the strial marginal cells and dark cells while BB-CK in the inner hair cells. High levels of CK are also found in the cochlea's inner and outer phalangeal cells. This provides a source of energy for ion transport and transduction activities in the inner ear [231]. Olfactory system: Olfactory sensory neurons express BB-CK in the cilia [232]. In large cells within the olfactory neuroepithelium and ventral spinal cord, differential compartmentation of CK isoforms has been evident, with B-CK localized primarily in cell nuclei, whereas uMtCK is present in the cell body (but not within nuclei). In olfactory bulb neuroepithelium, both isoforms are expressed in the middle zone of the germinal layer associated with DNA synthesis [233]. Tactile and skin system: BB-CK co-expresses with low amounts of uMtCK in suprabasal layers of the epidermis (cell of hair follicles, sebaceous glands, and the subcutaneous panniculus carnosus muscle). MM-CK and sMtCK were restricted to panniculus carnosus [234]. Epidermal CK is very important for cellular energy metabolism and might decline under oxidative stress conditions, including skin-aging processes; interestingly, application of Cr to skin cells in vitro and in vivo can refuel these cells energetically, and, thus, protect them against free radical-induced cell damage [235]. Gustatory system: crucial for optimal cell and motor development and function [236]. CK is also involved in the control of maturation and maintenance of myofibers in the distal tongue [237,238].
Intestines	BB-CK uMtCK	Distributed in the brush border web region, specifically at the contractile-ring myosin, to supply energy for contraction [239,240].
Miscellaneous	BB-CK MB-CK uMtCK	CK has been associated with the clotting cascade by means of thrombin receptor signaling [241]. The CK/PCr system has also been implicated in the function of the immune cells [126]. Finally, Cr metabolism has been implicated in UCP-independent thermogenesis in the brown and beige adipose tissue [129,242], and B-CK has been shown to be a key effector of the futile Cr cycle [243].

3.5. What Is the Basis of Creatine Transport?

The CRT (SLC6A8) is the solute carrier responsible for the $2Na^+/Cl^-$-dependent co-transport of Cr into the cells. However, SLC16A12 has also been identified as a transporter of guanidino compounds (including Cr, Crn and GAA) that affect plasma, urinary and renal concentrations although its physiological function is unknown [244–246]. As previously mentioned, CRT has shown a high affinity to Cr in the plasmalemma but neither Crn nor

PCr act as substrates. It has been shown that SLC6A8 also mediates the GAA transport, particularly in brain cells [247]. The main reason for this high substrate specificity is the separation by no more than 2–3 carbon atoms (4.5–5 Å) between the carboxyl group (to possibly interact with G73 and the Na$^+$) and the guanidine group (to establish a hydrogen bond with C144), which suggests the presence of a dipole moment in the binding site that facilitates orientation and accommodation of the ligand molecules [248]. The most efficient competitive inhibitor on Cr transport is the β-guanidinopropionic acid [249]. In humans, the gene encoding CRT is located in chromosome Xq28, and this gene is made up of 3747 base pairs and 13 exons (GenBank Accession Number L31409–official symbol *SLC6A8*, also known as *CRT1*) [17]. Notably, the localization of the *SLC6A8* gene is in close proximity to genes responsible for several neuromuscular disorders [250]. *SLC6A10P* (also known as *CRT2*) is a pseudogene located in the 16p11.2 genomic region [251]. *SLC6A10P* contains ≈97% nucleotide sequence similarity to *SLC6A8*, but has been suggested to have an early stop codon [252]. Although there are reports of mRNA expression for the *SLC6A10P* in testes [253] and the brain [254], there is no evidence in publications or databases about its translation to a protein and additional information is needed about the functional effects of the respective transcribed RNA. Interestingly, microdeletions in 16p11.2 are one of the most common recurrent genomic disorders associated with autism [255]. Please refer to the following BioGPS ID for more details about gene expression patterns in different tissues: *SLC6A8*-6535, and *SLC6A10P*-386757.

Cr is transported into the muscle cells exclusively by CRT1. This protein consists of 635 amino acids (≈70.5 kDa) [256], it has 12 membrane-spanning domains with the N- and C- termini facing the cytoplasm, and it contains a large extracellular loop between the third and fourth transmembrane domains with sites for N-linked glycosylation [257]. The current literature suggests at least four isoforms of the CRT1 are transcribed from the *SLC6A8* gene by alternative splicing, and these include SLC6A8A, SLC6A8B, SLC6A8C and SLC6A8D. The first splice variant of the full-length SLC6A8A, called SLC6A8B, was identified by cloning and sequencing two cDNAs from a human hippocampal library with a rat CRT cDNA-specific probe. Compared to the fully homologous protein, the authors found a novel protein sequence with four different segments [258]. Prior to this report, González and Uhl [259] reported two different sequences of the SLC6A8 mRNA (4.0–4.3 and 2.2–3.0 kb) using Northern Blot analysis. Additionally, Guerrero-Ontiveros and Wallimann [260] found two polypeptides that were ≈70 kDa and ≈50 kDa with identical amino- and carboxy-terminal regions, which were linked to the variant of the full-length transcript due to alternative splicing. More recently, in an attempt to characterize the SLC6A8B mRNA and protein, Martínez-Muñoz et al. [261] identified a new splice variant called SLC6A8C that contained 270 amino acids (≈27.6 kDa) in humans and mice. Ndika and colleagues subsequently identified a new variant that was identical to SLC6A8C with the exception of an in-frame deletion of exon 9 in human and animal cells, and this protein (SLC6A8D) contained 224 amino acids (≈15 kDa) [262]. Interestingly, these authors also demonstrated that these splice variants (SLC6A8C and SLC6A8D), while lacking transport function, increased Cr transport through co-expression with the full-length CRT. Previous research has similarly shown that splice isoforms of the Na$^+$/Cl$^-$-dependent neurotransmitter transporter family may facilitate trafficking of full-length transporters [263].

While increasing Cl$^-$ concentration significantly augments Cr influx in vitro [264], research has focused mainly on the Na$^+$-dependent regulation. For example, a series of hormones that increase the sodium gradient across the muscle cell membrane (via Na$^+$/K$^+$-ATPase) influence the net Cr uptake into skeletal muscle cells in vivo and in vitro. It has been shown that insulin (at supraphysiological concentrations), insulin-like growth factor 1 (IGF-1), 3,3′,5-triiodothyronine, and certain catecholamines (noradrenaline, isoproterenol and clenbuterol) can stimulate Cr transport through membrane receptor activation mechanisms [250,265]. Tyrosine phosphorylation is a conserved mechanism for regulating the transport of neurotransmitters via SLC6 Na$^+$-dependent transporters [266,267], and Cr uptake can also be affected by this mechanism. CRT has amino acid residues in the amino-terminal, carboxy-terminal and intracellular domains that can be phosphorylated by different kinases including the cAMP-dependent protein kinase (PKA) and the Ca^{2+}-dependent protein kinase (PKC) [17]. In addition, CRT is post-translationally modified and has two N-glycosylation sites, located in domains 3–4 and 11–12 in the extracellular space [268]. Phosphorylation and glycosylation might be important in the regulation of CRT activity and localization. Derave et al. [269] demonstrated that electrical stimulation of incubated rodent skeletal muscles stimulates rapid Cr transport possibly by endosomal translocation of CRT from an intracellular pool to the sarcolemma, rather than *de novo* protein synthesis. It is interesting to note that proteins that have been associated with regulation of CRT [270], such as the serine/threonine-protein kinases 1 and 3 (also known as serum and glucocorticoid-regulated kinases, SGK1/3), are activated upon H_2O_2 accumulation [271], which was observed after the electrical stimulation protocol of Derave et al. [269]. Other in vitro and animal studies have found that several kinases regulate CRT activity [14,268,272]. Additionally, Almeida et al. (2006) demonstrated in vitro that Cr is synthesized and taken up by central neurons and released by exocytosis depending on an action potential, which implies certain mechanisms of vesicular translocation are responsible for CRT localization [273]. This is supported by the fact that human and animal studies have shown that Cr saturation (by CrM supplementation) or depletion (by β-GPA administration) result in variations in the maximum rate of transporter activity (V_{max}) rather than changes in the total CRT levels [274,275]. For instance, in cardiomyocytes, these changes in V_{max} correlate with CRT decreases in the cell surface fraction, indicating that changes in the cell surface are associated with the cellular responses to changes in Cr availability [268].

Finally, it is worth noting that congenital CRT deficiency is associated with autism, epilepsy, neurological defects, and intellectual disabilities [276,277]. This neurometabolic disorder is part of the Cr deficiency syndrome [52]. Thus, examining structural determinants of substrate binding in the CRT will provide a deeper understanding of the regulation of Cr uptake as well as novel therapeutic ligands [248,278]. For a more detailed coverage, both on human pathology and on their different in vivo models (KO and KI mice and rats), of the genetic conditions (AGAT, GAMT, and SLC6A8 deficiencies) of the Cr deficiency syndrome please refer to [18,22,279].

4. Limitations/Strengths and Future Directions

This review should be read in the light of various limitations/strengths. First, data from in vitro and in vivo animal models should be interpreted with caution given they might not fully reflect cellular behavior in humans. Second, we did not describe how Cr metabolism affects immunity, cancer, and certain conditions through lifespan (i.e., elderly, pregnancy) since these conditions extend beyond the main scope of this review and will be covered in other invited reviews of this book/special issue on "Creatine Supplementation for Health and Clinical Diseases". This bioinformatics-assisted review should be seen as an up-and-coming method to address the lack of systematization in narrative reviews that aim to describe and analyze potential mechanisms of action. For example, besides cross-referencing the query results from several databases, we performed a clustering of CK-interacting proteins based on the Markov Cluster Algorithm using an open-source bioinformatics tool. This enriched the biological significance behind the Cr metabolism under a systems biology approach with experimentally-validated information that would be cumbersome to manually extract. The Research Division of the Dynamical Business & Science Society—DBSS International SAS is leading an initiative to develop and standardize the reporting guidelines of bioinformatics-assisted reviews.

Future studies about Cr metabolism should examine the implications of the CK/PCr system on thermogenic futile cycles considering the novel findings that have been reported regarding the role of AMPK in regulating the UCP-independent thermogenesis in white adipose tissue. Future research should also address the age-dependent changes that occur in the microbiome that cause higher Cr degradation rates in vivo, and whether this could be counteracted through CrM supplementation. More research is also needed to evaluate the effects of CrM supplementation during low-carbohydrate high fat diets [280] since preclinical evidence has revealed a suppression of the positive effects on muscle performance after CrM administration (by downregulation of the IGF1/Akt/mTOR pathway) during high-fat diet in rats [281]. It is also worth noting that dynamic simulations are important tools that can be used to predict how molecules potentially affect physiology. In this regard, new models could be developed considering the recent methodologies for kinetic analysis of the transphosphorylation reactions of the CK [282]. This allows testing and iteratively improving the prediction models before the experimental verification of systems perturbations might occur.

5. Conclusions

Cr and PCr play an essential role in the optimal functioning of tissues with high and fluctuating energy demands (e.g., muscle, brain, and heart). Moreover, alterations in Cr and PCr concentrations produce marked functional changes that lead to various types of diseases (e.g., cancer or pathologies associated with Cr deficiency syndrome). After performing a comprehensive and bioinformatics-assisted review, and under the cellular allostasis paradigm, the current scientific evidence suggest that the CK/PCr system is physiologically essential for life (i.e., cell survival, growth, proliferation, differentiation, and migration/motility), and provides an evolutionary advantage for rapid and localized support of energy- and mechanical-dependent processes. In this sense, the CK/PCr system could be viewed as a dynamic biosensor of the cellular chemo-mechanical energy transduction, which may explain various positive benefits of CrM supplementation and cellular pathophysiology of the Cr deficiency syndrome. Given this centralized role of Cr metabolism in whole-body physiology, further studies are needed in order to further examine how Cr supplementation may affect other unidentified aspects of health and disease.

Author Contributions: Conceptualization, D.A.B. and R.B.K.; Methodology, formal analysis and visualization, D.A.B.; writing—original draft preparation, D.A.B.; writing—critical review, R.B.K., D.A.F. and E.S.R.; writing—editing, J.R.S., C.M.K. and M.D.R.; project administration, D.A.B. All authors have read and agreed to the published version of the manuscript.

Funding: The APC was funded was funded by Edelife™ and AlzChem Tostberg GmbH, that manufactures Creapure®. However, they had no intervention whatsoever in the collection, analysis and interpretation of the data, or in writing this manuscript.

Institutional Review Board Statement: Not applicable.

Informed Consent Statement: Not applicable.

Data Availability Statement: Not applicable.

Acknowledgments: Authors would like to thank fellows of DBSS International (Jorge Petro and Yurany Moreno) for their feedback during the construction of the draft. Special thanks to Theo Wallimann for providing comments and suggestions on this review article.

Conflicts of Interest: D.A.B. serves as science product manager for MTX Corporation®, a company that produces, distributes, sells and does research on dietary supplements (including creatine) in Europe, has acted as scientific consultant for MET-Rx and Healthy Sports in Colombia, and has received honoraria for speaking about creatine at international conferences. R.B.K. has conducted industry sponsored research on creatine, received financial support for presenting on creatine at industry sponsored scientific conferences, and has served as an expert witness on cases related to creatine. Additionally, he serves as Chair of the "Creatine in Health" Scientific Advisory Board for AlzChem Tostberg GmbH who sponsored this special issue. J.R.S. has conducted industry-sponsored research on creatine and other nutraceuticals over the past 25 years. Further, J.R.S has also received financial support for presenting on the science of various nutraceuticals, except creatine, at industry-sponsored scientific conferences. D.A.F. has been previously supported by grants from MinCiencias but not related to creatine. C.M.K. have consulted with and received external funding from companies who sell certain dietary ingredients, and have received remuneration from companies for delivering scientific presentations at conferences. M.D.R. has received academic and industry funding related to dietary supplements, served as a non-paid consultant for industry and received honoraria for speaking at various conferences. E.S.R. has conducted industry-sponsored research on creatine and received financial support for presenting on creatine at industry-sponsored scientific conferences. R.B.K. acts as chair of the "Creatine in Health" scientific advisory board for AlzChem Tostberg GmbH while all other authors serve as members (except D.A.F.).

References

1. Brosnan, J.T.; Brosnan, M.E. Creatine: Endogenous Metabolite, Dietary, and Therapeutic Supplement. *Annu. Rev. Nutr.* **2007**, *27*, 241–261. [CrossRef]
2. Béard, E.; Braissant, O. Synthesis and transport of creatine in the CNS: Importance for cerebral functions. *J. Neurochem.* **2010**, *115*, 297–313. [CrossRef] [PubMed]
3. Moore, N.P. The distribution, metabolism and function of creatine in the male mammalian reproductive tract: A review. *Int. J. Androl.* **2000**, *23*, 4–12. [CrossRef]
4. Brosnan, J.T.; da Silva, R.P.; Brosnan, M.E. The metabolic burden of creatine synthesis. *Amino Acids* **2011**, *40*, 1325–1331. [CrossRef] [PubMed]
5. Humm, A.; Fritsche, E.; Steinbacher, S. Structure and reaction mechanism of L-arginine:glycine amidinotransferase. *Biol. Chem.* **1997**, *378*, 193–197. [PubMed]
6. Komoto, J.; Yamada, T.; Takata, Y.; Konishi, K.; Ogawa, H.; Gomi, T.; Fujioka, M.; Takusagawa, F. Catalytic Mechanism of Guanidinoacetate Methyltransferase: Crystal Structures of Guanidinoacetate Methyltransferase Ternary Complexes. *Biochemistry* **2004**, *43*, 14385–14394. [CrossRef] [PubMed]
7. Bonilla, D.A.; Moreno, Y. Molecular and metabolic insights of creatine supplementation on resistance training. *Rev. Colomb. Química* **2015**, *44*, 11–18. [CrossRef]
8. Wyss, M.; Kaddurah-Daouk, R. Creatine and Creatinine Metabolism. *Physiol. Rev.* **2000**, *80*, 1107–1213. [CrossRef]
9. Brosnan, M.E.; Edison, E.E.; da Silva, R.; Brosnan, J.T. New insights into creatine function and synthesis. *Adv. Enzyme Regul.* **2007**, *47*, 252–260. [CrossRef]
10. Bonilla, D.A. A Systems Biology Approach to Creatine Metabolism. In *Creatine: Biosynthesis, Health Effects and Clinical Perspectives*; Hogan, L., Ed.; Nova Science Publishers Inc.: New York, NY, USA, 2017.
11. Wallimann, T.; Wyss, M.; Brdiczka, D.; Nicolay, K.; Eppenberger, H.M. Intracellular compartmentation, structure and function of creatine kinase isoenzymes in tissues with high and fluctuating energy demands: The 'phosphocreatine circuit' for cellular energy homeostasis. *Biochem. J.* **1992**, *281*, 21–40. [CrossRef]
12. Hemmer, W.; Wallimann, T. Functional Aspects of Creatine Kinase in Brain. *Dev. Neurosci.* **1993**, *15*, 249–260. [CrossRef] [PubMed]
13. Wallimann, T.; Hemmer, W. Creatine kinase in non-muscle tissues and cells. *Mol. Cell. Biochem.* **1994**, *133–134*, 193–220. [CrossRef] [PubMed]
14. Balestrino, M.; Gandolfo, C.; Perasso, L. Controlling the Flow of Energy: Inhibition and Stimulation of the Creatine Transporter. *Curr. Enzym. Inhib.* **2009**, *5*, 223–233. [CrossRef]
15. Speer, O.; Neukomm, L.J.; Murphy, R.M.; Zanolla, E.; Schlattner, U.; Henry, H.; Snow, R.J.; Wallimann, T. Creatine transporters: A reappraisal. *Mol. Cell. Biochem.* **2004**, *256*, 407–424. [CrossRef]
16. Christie, D.L. Functional Insights into the Creatine Transporter. In *Creatine and Creatine Kinase in Health and Disease*; Salomons, G.S., Wyss, M., Eds.; Springer: Dordrecht, The Netherlands, 2007; pp. 99–118. [CrossRef]
17. Nash, S.R.; Giros, B.; Kingsmore, S.F.; Rochelle, J.M.; Suter, S.T.; Gregor, P.; Seldin, M.F.; Caron, M.G. Cloning, pharmacological characterization, and genomic localization of the human creatine transporter. *Recept Channels* **1994**, *2*, 165–174. [PubMed]
18. Hanna-El-Daher, L.; Braissant, O. Creatine synthesis and exchanges between brain cells: What can be learned from human creatine deficiencies and various experimental models? *Amino Acids* **2016**, *48*, 1877–1895. [CrossRef] [PubMed]
19. Braissant, O.; Rackayová, V.; Pierzchala, K.; Grosse, J.; McLin, V.A.; Cudalbu, C. Longitudinal neurometabolic changes in the hippocampus of a rat model of chronic hepatic encephalopathy. *J. Hepatol.* **2019**, *71*, 505–515. [CrossRef]

20. Joncquel-Chevalier Curt, M.; Voicu, P.-M.; Fontaine, M.; Dessein, A.-F.; Porchet, N.; Mention-Mulliez, K.; Dobbelaere, D.; Soto-Ares, G.; Cheillan, D.; Vamecq, J. Creatine biosynthesis and transport in health and disease. *Biochimie* **2015**, *119*, 146–165. [CrossRef]

21. Marques, E.P.; Wyse, A.T.S. Creatine as a Neuroprotector: An Actor that Can Play Many Parts. *Neurotox. Res.* **2019**, *36*, 411–423. [CrossRef]

22. Wallimann, T.; Harris, R. Creatine: A miserable life without it. *Amino Acids* **2016**, *48*, 1739–1750. [CrossRef]

23. Frampton, C.S.; Wilson, C.C.; Shankland, N.; Florence, A.J. Single-crystal neutron refinement of creatine monohydrate at 20 K and 123 K. *J. Chem. Soc. Faraday Trans.* **1997**, *93*, 1875–1879. [CrossRef]

24. Arlin, J.-B.; Bhardwaj, R.M.; Johnston, A.; Miller, G.J.; Bardin, J.; MacDougall, F.; Fernandes, P.; Shankland, K.; David, W.I.F.; Florence, A.J. Structure and stability of two polymorphs of creatine and its monohydrate. *CrystEngComm* **2014**, *16*. [CrossRef]

25. Dash, A.K.; Mo, Y.; Pyne, A. Solid-state Properties of Creatine Monohydrate. *J. Pharm. Sci.* **2002**, *91*, 708–718. [CrossRef]

26. Pischel, I.; Gastner, T. Creatine—its Chemical Synthesis, Chemistry, and Legal Status. In *Creatine and Creatine Kinase in Health and Disease*; Salomons, G.S., Wyss, M., Eds.; Springer: Dordrecht, The Netherlands, 2007; pp. 291–307. [CrossRef]

27. Kreider, R.B.; Kalman, D.S.; Antonio, J.; Ziegenfuss, T.N.; Wildman, R.; Collins, R.; Candow, D.G.; Kleiner, S.M.; Almada, A.L.; Lopez, H.L. International Society of Sports Nutrition position stand: Safety and efficacy of creatine supplementation in exercise, sport, and medicine. *J. Int. Soc. of Sports Nutr.* **2017**, *14*. [CrossRef] [PubMed]

28. Wallimann, T.; Tokarska-Schlattner, M.; Schlattner, U. The creatine kinase system and pleiotropic effects of creatine. *Amino Acids* **2011**, *40*, 1271–1296. [CrossRef]

29. Bonilla, D.A.; Pérez-Idárraga, A.; Odriozola-Martínez, A.; Kreider, R.B. The 4R's Framework of Nutritional Strategies for Post-Exercise Recovery: A Review with Emphasis on New Generation of Carbohydrates. *Int. J. Environ. Res. Public Health* **2020**, *18*, 103. [CrossRef]

30. Mielgo-Ayuso, J.; Calleja-Gonzalez, J.; Marqués-Jiménez, D.; Caballero-García, A.; Córdova, A.; Fernández-Lázaro, D. Effects of Creatine Supplementation on Athletic Performance in Soccer Players: A Systematic Review and Meta-Analysis. *Nutrients* **2019**, *11*, 757. [CrossRef]

31. Kaviani, M.; Shaw, K.; Chilibeck, P.D. Benefits of Creatine Supplementation for Vegetarians Compared to Omnivorous Athletes: A Systematic Review. *Int. J. Environ. Res. Public Health* **2020**, *17*, 3041. [CrossRef]

32. Bakian, A.V.; Huber, R.S.; Scholl, L.; Renshaw, P.F.; Kondo, D. Dietary creatine intake and depression risk among U.S. adults. *Transl. Psychiatry* **2020**, *10*. [CrossRef]

33. Forbes, S.C.; Candow, D.G.; Smith-Ryan, A.E.; Hirsch, K.R.; Roberts, M.D.; VanDusseldorp, T.A.; Stratton, M.T.; Kaviani, M.; Little, J.P. Supplements and Nutritional Interventions to Augment High-Intensity Interval Training Physiological and Performance Adaptations—A Narrative Review. *Nutrients* **2020**, *12*, 390. [CrossRef] [PubMed]

34. De Guingand, D.L.; Palmer, K.R.; Bilardi, J.E.; Ellery, S.J. Acceptability of dietary or nutritional supplementation in pregnancy (ADONS)—Exploring the consumer's perspective on introducing creatine monohydrate as a pregnancy supplement. *Midwifery* **2020**, *82*. [CrossRef]

35. Candow, D.G.; Forbes, S.C.; Chilibeck, P.D.; Cornish, S.M.; Antonio, J.; Kreider, R.B. Effectiveness of Creatine Supplementation on Aging Muscle and Bone: Focus on Falls Prevention and Inflammation. *J. Clin. Med.* **2019**, *8*, 488. [CrossRef]

36. Stares, A.; Bains, M. The Additive Effects of Creatine Supplementation and Exercise Training in an Aging Population: A Systematic Review of Randomized Controlled Trials. *J. Geriatr. Phys. Ther.* **2020**, *43*, 99–112. [CrossRef] [PubMed]

37. Rawson, E.S.; Miles, M.P.; Larson-Meyer, D.E. Dietary Supplements for Health, Adaptation, and Recovery in Athletes. *Int. J. Sport Nutr. Exerc. Metab.* **2018**, *28*, 188–199. [CrossRef]

38. Clarke, H.; Kim, D.-H.; Meza, C.A.; Ormsbee, M.J.; Hickner, R.C. The Evolving Applications of Creatine Supplementation: Could Creatine Improve Vascular Health? *Nutrients* **2020**, *12*, 2834. [CrossRef] [PubMed]

39. Machek, S.B.; Bagley, J.R. Creatine Monohydrate Supplementation: Considerations for Cognitive Performance in Athletes. *Strength Cond. J.* **2018**, *40*, 82–93. [CrossRef]

40. Dolan, E.; Gualano, B.; Rawson, E.S. Beyond muscle: The effects of creatine supplementation on brain creatine, cognitive processing, and traumatic brain injury. *Eur. J. Sport Sci.* **2018**, *19*, 1–14. [CrossRef]

41. Forbes, S.C.; Candow, D.G.; Ferreira, L.H.B.; Souza-Junior, T.P. Effects of Creatine Supplementation on Properties of Muscle, Bone, and Brain Function in Older Adults: A Narrative Review. *J. Diet. Suppl.* **2021**, 1–18. [CrossRef]

42. De Souza e Silva, A.; Pertille, A.; Reis Barbosa, C.G.; Aparecida de Oliveira Silva, J.; de Jesus, D.V.; Ribeiro, A.G.S.V.; Baganha, R.J.; de Oliveira, J.J. Effects of Creatine Supplementation on Renal Function: A Systematic Review and Meta-Analysis. *J. Ren. Nutr.* **2019**, *29*, 480–489. [CrossRef] [PubMed]

43. Forbes, S.C.; Candow, D.G.; Krentz, J.R.; Roberts, M.D.; Young, K.C. Changes in Fat Mass Following Creatine Supplementation and Resistance Training in Adults ≥50 Years of Age: A Meta-Analysis. *J. Funct. Morphol. Kinesiol.* **2019**, *4*, 62. [CrossRef] [PubMed]

44. Galvan, E.; Walker, D.K.; Simbo, S.Y.; Dalton, R.; Levers, K.; O'Connor, A.; Goodenough, C.; Barringer, N.D.; Greenwood, M.; Rasmussen, C.; et al. Acute and chronic safety and efficacy of dose dependent creatine nitrate supplementation and exercise performance. *J. Int. Soc. Sports Nutr.* **2016**, *13*. [CrossRef] [PubMed]

45. Dalton, R.; Sowinski, R.; Grubic, T.; Collins, P.; Coletta, A.; Reyes, A.; Sanchez, B.; Koozehchian, M.; Jung, Y.; Rasmussen, C.; et al. Hematological and Hemodynamic Responses to Acute and Short-Term Creatine Nitrate Supplementation. *Nutrients* **2017**, *9*, 1359. [CrossRef] [PubMed]

46. Ostojic, S.M.; Stajer, V.; Vranes, M.; Ostojic, J. Searching for a better formulation to enhance muscle bioenergetics: A randomized controlled trial of creatine nitrate plus creatininevs.creatine nitratevs.creatine monohydrate in healthy men. *Food Sci. Nutr.* **2019**, *7*, 3766–3773. [CrossRef]

47. Antonio, J.; Candow, D.G.; Forbes, S.C.; Gualano, B.; Jagim, A.R.; Kreider, R.B.; Rawson, E.S.; Smith-Ryan, A.E.; VanDusseldorp, T.A.; Willoughby, D.S.; et al. Common questions and misconceptions about creatine supplementation: What does the scientific evidence really show? *J. Int. Soc. Sports Nutr.* **2021**, *18*. [CrossRef] [PubMed]

48. Kreider, R.B.; Stout, J.R. Creatine in Health and Disease. *Nutrients* **2021**, *13*, 447. [CrossRef]

49. Patra, S.; Bera, S.; SinhaRoy, S.; Ghoshal, S.; Ray, S.; Basu, A.; Schlattner, U.; Wallimann, T.; Ray, M. Progressive decrease of phosphocreatine, creatine and creatine kinase in skeletal muscle upon transformation to sarcoma. *FEBS J.* **2008**, *275*, 3236–3247. [CrossRef]

50. Bender, A.; Klopstock, T. Creatine for neuroprotection in neurodegenerative disease: End of story? *Amino Acids* **2016**, *48*, 1929–1940. [CrossRef]

51. Cheng, Y.; Chen, Y.; Shang, H. Aberrations of biochemical indicators in amyotrophic lateral sclerosis: A systematic review and meta-analysis. *Transl. Neurodegener.* **2021**, *10*. [CrossRef]

52. Salazar, M.D.; Zelt, N.B.; Saldivar, R.; Kuntz, C.P.; Chen, S.; Penn, W.D.; Bonneau, R.; Koehler Leman, J.; Schlebach, J.P. Classification of the Molecular Defects Associated with Pathogenic Variants of the SLC6A8 Creatine Transporter. *Biochemistry* **2020**, *59*, 1367–1377. [CrossRef]

53. Salomons, G.S.; van Dooren, S.J.; Verhoeven, N.M.; Cecil, K.M.; Ball, W.S.; Degrauw, T.J.; Jakobs, C. X-linked creatine-transporter gene (SLC6A8) defect: A new creatine-deficiency syndrome. *Am. J. Hum. Genet.* **2001**, *68*, 1497–1500. [CrossRef]

54. Shearer, J.; Weljie, A.M. Biomarkers of skeletal muscle regulation, metabolism and dysfunction. In *Metabolomics and Systems Biology in Human Health and Medicine*; Jones, O., Ed.; CABI: Oxfordshire, UK, 2014; pp. 157–170. [CrossRef]

55. McLeish, M.J.; Kenyon, G.L. Relating Structure to Mechanism in Creatine Kinase. *Crit. Rev. Biochem. Mol. Biol.* **2008**, *40*, 1–20. [CrossRef]

56. Stout, J.R.; Antonio, J.; Kalman, D. *Essentials of Creatine in Sports and Health*; Humana Press: New York, USA, 2008. [CrossRef]

57. Ellington, W.R. Phosphocreatine represents a thermodynamic and functional improvement over other muscle phosphagens. *J. Exp. Biol.* **1989**, *143*, 177–194.

58. Uzzan, M.; Nechrebeki, J.; Zhou, P.; Labuza, T.P. Effect of water activity and temperature on the stability of creatine during storage. *Drug Dev. Ind. Pharm.* **2009**, *35*, 1003–1008. [CrossRef] [PubMed]

59. Harris, R.C.; Söderlund, K.; Hultman, E. Elevation of creatine in resting and exercised muscle of normal subjects by creatine supplementation. *Clin. Sci.* **1992**, *83*, 367–374. [CrossRef] [PubMed]

60. Vermeulen, A.; Wieme, R.; Robbrecht, J.; De Buyzere, M.; De Slypere, J.P.; Delanghe, J. Normal reference values for creatine, creatinine, and carnitine are lower in vegetarians. *Clin. Chem.* **1989**, *35*, 1802–1803. [CrossRef]

61. Blancquaert, L.; Baguet, A.; Bex, T.; Volkaert, A.; Everaert, I.; Delanghe, J.; Petrovic, M.; Vervaet, C.; De Henauw, S.; Constantin-Teodosiu, D.; et al. Changing to a vegetarian diet reduces the body creatine pool in omnivorous women, but appears not to affect carnitine and carnosine homeostasis: A randomised trial. *Br. J. Nutr.* **2018**, *119*, 759–770. [CrossRef]

62. Balsom, P.D.; Söderlund, K.; Ekblom, B. Creatine in Humans with Special Reference to Creatine Supplementation. *Sports Med.* **1994**, *18*, 268–280. [CrossRef]

63. Wu, G. Important roles of dietary taurine, creatine, carnosine, anserine and 4-hydroxyproline in human nutrition and health. *Amino Acids* **2020**, *52*, 329–360. [CrossRef]

64. Casey, A.; Constantin-Teodosiu, D.; Howell, S.; Hultman, E.; Greenhaff, P.L. Creatine ingestion favorably affects performance and muscle metabolism during maximal exercise in humans. *Am. J. Physiol. Endocrinol. Metab.* **1996**, *271*, E31–E37. [CrossRef]

65. Greenhaff, P.L.; Bodin, K.; Soderlund, K.; Hultman, E. Effect of oral creatine supplementation on skeletal muscle phosphocreatine resynthesis. *Am. J. Physiol. Endocrinol. Metab.* **1994**, *266*, E725–E730. [CrossRef] [PubMed]

66. Dechent, P.; Pouwels, P.J.W.; Wilken, B.; Hanefeld, F.; Frahm, J. Increase of total creatine in human brain after oral supplementation of creatine-monohydrate. *Am. J. Physiol. Regul. Integr. Comp. Physiol.* **1999**, *277*, R698–R704. [CrossRef] [PubMed]

67. Kreider, R.; Willoughby, D.; Greenwood, M.; Parise, G.; Payne, E.T. Effects of serum creatine supplementation on muscle creatine and phosphagen levels. *J. Exerc. Physiol. Online* **2003**, *6*, 24–33.

68. Schulze, A. Creatine deficiency syndromes. *Mol. Cell. Biochem.* **2003**, *244*, 143–150. [CrossRef] [PubMed]

69. Stockler-Ipsiroglu, S.; Apatean, D.; Battini, R.; DeBrosse, S.; Dessoffy, K.; Edvardson, S.; Eichler, F.; Johnston, K.; Koeller, D.M.; Nouioua, S.; et al. Arginine: Glycine amidinotransferase (AGAT) deficiency: Clinical features and long term outcomes in 16 patients diagnosed worldwide. *Mol. Genet. Metab.* **2015**, *116*, 252–259. [CrossRef] [PubMed]

70. Stöckler-Ipsiroglu, S.; Battini, R.; DeGrauw, T.; Schulze, A. Disorders of Creatine Metabolism. In *Physician's Guide to the Treatment and Follow-Up of Metabolic Diseases*; Blau, N., Leonard, J., Hoffmann, G.F., Clarke, J.T.R., Eds.; Springer: Berlin/Heidelberg, Germany, 2006; pp. 255–265. [CrossRef]

71. Mesa, J.L.M.; Ruiz, J.R.; Gonzalez-Gross, M.M.; Gutierrez Sainz, A.; Castillo Garzon, M.J. Oral Creatine Supplementation and Skeletal Muscle Metabolism in Physical Exercise. *Sports Med.* **2002**, *32*, 903–944. [CrossRef] [PubMed]

72. Harris, R.C.; Almada, A.L.; Harris, D.B.; Dunnett, M.; Hespel, P. The creatine content of Creatine Serum™ and the change in the plasma concentration with ingestion of a single dose. *J. Sports Sci.* **2004**, *22*, 851–857. [CrossRef]
73. Brault, J.J.; Towse, T.F.; Slade, J.M.; Meyer, R.A. Parallel Increases in Phosphocreatine and Total Creatine in Human Vastus Lateralis Muscle during Creatine Supplementation. *Int. J. Sport Nutr. Exerc. Metab.* **2007**, *17*, 624–634. [CrossRef]
74. Broxterman, R.M.; Layec, G.; Hureau, T.J.; Amann, M.; Richardson, R.S. Skeletal muscle bioenergetics during all-out exercise: Mechanistic insight into the oxygen uptake slow component and neuromuscular fatigue. *J. Appl. Physiol.* **2017**, *122*, 1208–1217. [CrossRef]
75. Burnley, M.; Jones, A.M. Oxygen uptake kinetics as a determinant of sports performance. *Eur. J. Sport Sci.* **2007**, *7*, 63–79. [CrossRef]
76. Sweeney, H.L. The importance of the creatine kinase reaction: The concept of metabolic capacitance. *Med. Sci. Sports Exerc.* **1994**, *26*, 30–36. [CrossRef] [PubMed]
77. Francescato, M.P.; Cettolo, V.; di Prampero, P.E. Influence of phosphagen concentration on phosphocreatine breakdown kinetics. Data from human gastrocnemius muscle. *J. Appl. Physiol.* **2008**, *105*, 158–164. [CrossRef] [PubMed]
78. Meyer, R.A. A linear model of muscle respiration explains monoexponential phosphocreatine changes. *Am. J. Physiol. Cell Physiol.* **1988**, *254*, C548–C553. [CrossRef] [PubMed]
79. Willis, W.T.; Jackman, M.R.; Messer, J.I.; Kuzmiak-Glancy, S.; Glancy, B. A Simple Hydraulic Analog Model of Oxidative Phosphorylation. *Med. Sci. Sports Exerc.* **2016**, *48*, 990–1000. [CrossRef]
80. Gonzalez-Freire, M.; Scalzo, P.; D'Agostino, J.; Moore, Z.A.; Diaz-Ruiz, A.; Fabbri, E.; Zane, A.; Chen, B.; Becker, K.G.; Lehrmann, E.; et al. Skeletal muscle ex vivo mitochondrial respiration parallels decline in vivo oxidative capacity, cardiorespiratory fitness, and muscle strength: The Baltimore Longitudinal Study of Aging. *Aging Cell* **2018**, *17*. [CrossRef] [PubMed]
81. Jones, A.M.; Wilkerson, D.P.; Fulford, J. Influence of dietary creatine supplementation on muscle phosphocreatine kinetics during knee-extensor exercise in humans. *Am. J. Physiol. Regul. Integr. Comp. Physiol.* **2009**, *296*, R1078–R1087. [CrossRef] [PubMed]
82. De Andrade Nemezio, K.M.; Bertuzzi, R.; Correia-Oliveira, C.R.; Gualano, B.; Bishop, D.J.; Lima-Silva, A.E. Effect of Creatine Loading on Oxygen Uptake during a 1-km Cycling Time Trial. *Med. Sci. Sports Exerc.* **2015**, *47*, 2660–2668. [CrossRef]
83. Rigoulet, M.; Bouchez, C.L.; Paumard, P.; Ransac, S.; Cuvellier, S.; Duvezin-Caubet, S.; Mazat, J.P.; Devin, A. Cell energy metabolism: An update. *Biochim. Biophys. Acta BBA Bioenerg.* **2020**, *1861*. [CrossRef]
84. Sumien, N.; Shetty, R.A.; Gonzales, E.B. Creatine, Creatine Kinase, and Aging. In *Biochemistry and Cell Biology of Ageing: Part I Biomedical Science*; Harris, J., Korolchuk, V., Eds.; Springer: Singapore, 2018; pp. 145–168. [CrossRef]
85. Schlattner, U.; Kay, L.; Tokarska-Schlattner, M. Mitochondrial Proteolipid Complexes of Creatine Kinase. In *Membrane Protein Complexes: Structure and Function*; Harris, J., Boekema, E., Eds.; Springer: Singapore, 2018; pp. 365–408. [CrossRef]
86. Dzeja, P.P.; Terzic, A. Phosphotransfer networks and cellular energetics. *J. Exp. Biol.* **2003**, *206*, 2039–2047. [CrossRef]
87. Bessman, S.P.; Carpenter, C.L. The Creatine-Creatine Phosphate Energy Shuttle. *Annu. Rev. Biochem.* **1985**, *54*, 831–862. [CrossRef]
88. Kongas, O.; van Beek, J. Creatine kinase in energy metabolic signaling in muscle. *Nat. Preced.* **2007**. [CrossRef]
89. Fiedler, G.B.; Schmid, A.I.; Goluch, S.; Schewzow, K.; Laistler, E.; Niess, F.; Unger, E.; Wolzt, M.; Mirzahosseini, A.; Kemp, G.J.; et al. Skeletal muscle ATP synthesis and cellular H+ handling measured by localized 31P-MRS during exercise and recovery. *Sci. Rep.* **2016**, *6*. [CrossRef]
90. Barclay, C.J. Energy demand and supply in human skeletal muscle. *J. Muscle Res. Cell Motil.* **2017**, *38*, 143–155. [CrossRef] [PubMed]
91. Uda, K.; Ellington, W.R.; Suzuki, T. A diverse array of creatine kinase and arginine kinase isoform genes is present in the starlet sea anemone Nematostella vectensis, a cnidarian model system for studying developmental evolution. *Gene* **2012**, *497*, 214–227. [CrossRef] [PubMed]
92. Bertin, M.; Pomponi, S.M.; Kokuhuta, C.; Iwasaki, N.; Suzuki, T.; Ellington, W.R. Origin of the genes for the isoforms of creatine kinase. *Gene* **2007**, *392*, 273–282. [CrossRef] [PubMed]
93. Eppenberger, H.M.; Dawson, D.M.; Kaplan, N.O. The comparative enzymology of creatine kinases. I. Isolation and characterization from chicken and rabbit tissues. *J. Biol. Chem.* **1967**, *242*, 204–209. [CrossRef]
94. Wallimann, T.; Tokarska-Schlattner, M.; Neumann, D.; Epand, R.M.; Epand, R.F.; Andres, R.H.; Widmer, H.R.; Hornemann, T.; Saks, V.; Agarkova, I.; et al. The Phosphocreatine Circuit: Molecular and Cellular Physiology of Creatine Kinases, Sensitivity to Free Radicals, and Enhancement by Creatine Supplementation. In *Molecular System Bioenergetics*; Saks, V., Ed.; Wiley-VCH: Weinheim, Germany, 2007; pp. 195–264. [CrossRef]
95. Ramírez Ríos, S.; Lamarche, F.; Cottet-Rousselle, C.; Klaus, A.; Tuerk, R.; Thali, R.; Auchli, Y.; Brunisholz, R.; Neumann, D.; Barret, L.; et al. Regulation of brain-type creatine kinase by AMP-activated protein kinase: Interaction, phosphorylation and ER localization. *Biochim. Biophys. Acta BBA Bioenerg.* **2014**, *1837*, 1271–1283. [CrossRef]
96. McFarland, E.W.; Kushmerick, M.J.; Moerland, T.S. Activity of creatine kinase in a contracting mammalian muscle of uniform fiber type. *Biophys. J.* **1994**, *67*, 1912–1924. [CrossRef]
97. Wallimann, T.; Schlösser, T.; Eppenberger, H.M. Function of M-line-bound creatine kinase as intramyofibrillar ATP regenerator at the receiving end of the phosphorylcreatine shuttle in muscle. *J. Biol. Chem.* **1984**, *259*, 5238–5246. [CrossRef]
98. Fritz-Wolf, K.; Schnyder, T.; Wallimann, T.; Kabsch, W. Structure of mitochondrial creatine kinase. *Nature* **1996**, *381*, 341–345. [CrossRef]

99. Eder, M.; Fritz-Wolf, K.; Kabsch, W.; Wallimann, T.; Schlattner, U. Crystal structure of human ubiquitous mitochondrial creatine kinase. *Proteins* **2000**, *39*, 216–225. [CrossRef]

100. Guzun, R.; Gonzalez-Granillo, M.; Karu-Varikmaa, M.; Grichine, A.; Usson, Y.; Kaambre, T.; Guerrero-Roesch, K.; Kuznetsov, A.; Schlattner, U.; Saks, V. Regulation of respiration in muscle cells in vivo by VDAC through interaction with the cytoskeleton and MtCK within Mitochondrial Interactosome. *Biochim. Biophys. Acta BBA Biomembr.* **2012**, *1818*, 1545–1554. [CrossRef]

101. Wu, C.; Orozco, C.; Boyer, J.; Leglise, M.; Goodale, J.; Batalov, S.; Hodge, C.L.; Haase, J.; Janes, J.; Huss, J.W.; et al. BioGPS: An extensible and customizable portal for querying and organizing gene annotation resources. *Genome Biol.* **2009**, *10*. [CrossRef] [PubMed]

102. Glancy, B.; Hartnell, L.M.; Malide, D.; Yu, Z.-X.; Combs, C.A.; Connelly, P.S.; Subramaniam, S.; Balaban, R.S. Mitochondrial reticulum for cellular energy distribution in muscle. *Nature* **2015**, *523*, 617–620. [CrossRef]

103. Wallimann, T. The extended, dynamic mitochondrial reticulum in skeletal muscle and the creatine kinase (CK)/phosphocreatine (PCr) shuttle are working hand in hand for optimal energy provision. *J. Muscle Res. Cell Motil.* **2015**, *36*, 297–300. [CrossRef]

104. Glancy, B.; Hartnell, L.M.; Combs, C.A.; Femnou, A.; Sun, J.; Murphy, E.; Subramaniam, S.; Balaban, R.S. Power Grid Protection of the Muscle Mitochondrial Reticulum. *Cell Rep.* **2017**, *19*, 487–496. [CrossRef]

105. Saks, V.; Schlattner, U.; Tokarska-Schlattner, M.; Wallimann, T.; Bagur, R.; Zorman, S.; Pelosse, M.; Santos, P.D.; Boucher, F.; Kaambre, T.; et al. Systems Level Regulation of Cardiac Energy Fluxes Via Metabolic Cycles: Role of Creatine, Phosphotransfer Pathways, and AMPK Signaling. In *Systems Biology of Metabolic and Signaling Networks*; Aon, M., Saks, V., Schlattner, U., Eds.; Springer: Berlin/Heidelberg, Germany, 2014; pp. 261–320. [CrossRef]

106. Timohhina, N.; Guzun, R.; Tepp, K.; Monge, C.; Varikmaa, M.; Vija, H.; Sikk, P.; Kaambre, T.; Sackett, D.; Saks, V. Direct measurement of energy fluxes from mitochondria into cytoplasm in permeabilized cardiac cells in situ: Some evidence for mitochondrial interactosome. *J. Bioenerg. Biomembr.* **2009**, *41*, 259–275. [CrossRef]

107. Guzun, R.; Saks, V. Application of the Principles of Systems Biology and Wiener's Cybernetics for Analysis of Regulation of Energy Fluxes in Muscle Cells in Vivo. *Int. J. Mol. Sci.* **2010**, *11*, 982–1019. [CrossRef] [PubMed]

108. Saks, V.; Guzun, R.; Timohhina, N.; Tepp, K.; Varikmaa, M.; Monge, C.; Beraud, N.; Kaambre, T.; Kuznetsov, A.; Kadaja, L.; et al. Structure–function relationships in feedback regulation of energy fluxes in vivo in health and disease: Mitochondrial Interactosome. *Biochim. Biophys. Acta BBA Bioenerg.* **2010**, *1797*, 678–697. [CrossRef] [PubMed]

109. Karo, J.; Peterson, P.; Vendelin, M. Molecular Dynamics Simulations of Creatine Kinase and Adenine Nucleotide Translocase in Mitochondrial Membrane Patch. *J. Biol. Chem.* **2012**, *287*, 7467–7476. [CrossRef] [PubMed]

110. Bonilla, D.A.; Marín, E.; Pérez, A.; Carbone, L.; Kammerer, M.; Vargas, S.; Lozano, J.; Barale, A.; Quiroga, L.; Mata, F.; et al. Thermogenesis and Obesity; A Brief Review and rs104894319 Polymorphism in Venezuelan Population. *EC Nutr.* **2018**, *13*, 4–16.

111. Rousset, S.; Alves-Guerra, M.C.; Mozo, J.; Miroux, B.; Cassard-Doulcier, A.M.; Bouillaud, F.; Ricquier, D. The Biology of Mitochondrial Uncoupling Proteins. *Diabetes* **2004**, *53*, S130–S135. [CrossRef]

112. Brand, M.D.; Esteves, T.C. Physiological functions of the mitochondrial uncoupling proteins UCP2 and UCP3. *Cell Metab.* **2005**, *2*, 85–93. [CrossRef]

113. Ramsden, D.B.; Ho, P.W.L.; Ho, J.W.M.; Liu, H.F.; So, D.H.F.; Tse, H.M.; Chan, K.H.; Ho, S.L. Human neuronal uncoupling proteins 4 and 5 (UCP4 and UCP5): Structural properties, regulation, and physiological role in protection against oxidative stress and mitochondrial dysfunction. *Brain Behav.* **2012**, *2*, 468–478. [CrossRef]

114. Krauss, S.; Zhang, C.-Y.; Lowell, B.B. The mitochondrial uncoupling-protein homologues. *Nat. Rev. Mol. Cell Biol.* **2005**, *6*, 248–261. [CrossRef] [PubMed]

115. Busiello, R.A.; Savarese, S.; Lombardi, A. Mitochondrial uncoupling proteins and energy metabolism. *Front. Physiol.* **2015**, *6*. [CrossRef] [PubMed]

116. Pohl, E.E.; Rupprecht, A.; Macher, G.; Hilse, K.E. Important Trends in UCP3 Investigation. *Front. Physiol.* **2019**, *10*. [CrossRef]

117. Skulachev, V.P. Fatty acid circuit as a physiological mechanism of uncoupling of oxidative phosphorylation. *FEBS Lett.* **1991**, *294*, 158–162. [CrossRef]

118. Ježek, P.; Engstová, H.; Žáčková, M.; Vercesi, A.E.; Costa, A.D.T.; Arruda, P.; Garlid, K.D. Fatty acid cycling mechanism and mitochondrial uncoupling proteins. *Biochim. Biophys. Acta BBA Bioenerg.* **1998**, *1365*, 319–327. [CrossRef]

119. Klingenberg, M.; Huang, S.-G. Structure and function of the uncoupling protein from brown adipose tissue. *Biochim. Biophys. Acta BBA Biomembr.* **1999**, *1415*, 271–296. [CrossRef]

120. Fedorenko, A.; Lishko, P.V.; Kirichok, Y. Mechanism of Fatty-Acid-Dependent UCP1 Uncoupling in Brown Fat Mitochondria. *Cell* **2012**, *151*, 400–413. [CrossRef] [PubMed]

121. Macher, G.; Koehler, M.; Rupprecht, A.; Kreiter, J.; Hinterdorfer, P.; Pohl, E.E. Inhibition of mitochondrial UCP1 and UCP3 by purine nucleotides and phosphate. *Biochim. Biophys. Acta BBA Biomembr.* **2018**, *1860*, 664–672. [CrossRef]

122. Kazak, L.; Chouchani, E.T.; Jedrychowski, M.P.; Erickson, B.K.; Shinoda, K.; Cohen, P.; Vetrivelan, R.; Lu, G.Z.; Laznik-Bogoslavski, D.; Hasenfuss, S.C.; et al. A Creatine-Driven Substrate Cycle Enhances Energy Expenditure and Thermogenesis in Beige Fat. *Cell* **2015**, *163*, 643–655. [CrossRef]

123. Kazak, L.; Chouchani, E.T.; Lu, G.Z.; Jedrychowski, M.P.; Bare, C.J.; Mina, A.I.; Kumari, M.; Zhang, S.; Vuckovic, I.; Laznik-Bogoslavski, D.; et al. Genetic Depletion of Adipocyte Creatine Metabolism Inhibits Diet-Induced Thermogenesis and Drives Obesity. *Cell Metab.* **2017**, *26*, 660–671.e663. [CrossRef]

124. Bertholet, A.M.; Kazak, L.; Chouchani, E.T.; Bogaczyńska, M.G.; Paranjpe, I.; Wainwright, G.L.; Bétourné, A.; Kajimura, S.; Spiegelman, B.M.; Kirichok, Y. Mitochondrial Patch Clamp of Beige Adipocytes Reveals UCP1-Positive and UCP1-Negative Cells Both Exhibiting Futile Creatine Cycling. *Cell Metab.* **2017**, *25*, 811–822.e814. [CrossRef] [PubMed]
125. Kazak, L.; Roesler, A. UCP1-independent thermogenesis. *Biochem. J.* **2020**, *477*, 709–725. [CrossRef]
126. Kazak, L.; Cohen, P. Creatine metabolism: Energy homeostasis, immunity and cancer biology. *Nat. Rev. Endocrinol.* **2020**, *16*, 421–436. [CrossRef] [PubMed]
127. Ikeda, K.; Yamada, T. UCP1 Dependent and Independent Thermogenesis in Brown and Beige Adipocytes. *Front. Endocrinol. (Lausanne)* **2020**, *11*. [CrossRef] [PubMed]
128. Chouchani, E.T.; Kajimura, S. Metabolic adaptation and maladaptation in adipose tissue. *Nat. Metab.* **2019**, *1*, 189–200. [CrossRef] [PubMed]
129. Wallimann, T.; Tokarska-Schlattner, M.; Kay, L.; Schlattner, U. Role of creatine and creatine kinase in UCP1-independent adipocyte thermogenesis. *Am. J. Physiol. Endocrinol. Metab.* **2020**, *319*, E944–E946. [CrossRef]
130. Connell, N.J.; Doligkeit, D.; Andriessen, C.; Kornips-Moonen, E.; Bruls, Y.M.H.; Schrauwen-Hinderling, V.B.; van de Weijer, T.; van Marken-Lichtenbelt, W.D.; Havekes, B.; Kazak, L.; et al. No evidence for brown adipose tissue activation after creatine supplementation in adult vegetarians. *Nat. Metab.* **2021**, *3*, 107–117. [CrossRef] [PubMed]
131. Kraft, T.; Hornemann, T.; Stolz, M.; Nier, V.; Wallimann, T. Coupling of creatine kinase to glycolytic enzymes at the sarcomeric I-band of skeletal muscle: A biochemical study in situ. *J. Muscle Res. Cell Motil.* **2000**, *21*, 691–703. [CrossRef]
132. Westerblad, H.; Allen, D.G.; Lännergren, J. Muscle Fatigue: Lactic Acid or Inorganic Phosphate the Major Cause? *Physiology* **2002**, *17*, 17–21. [CrossRef]
133. Wu, F.; Beard, D.A. Roles of the creatine kinase system and myoglobin in maintaining energetic state in the working heart. *BMC Syst. Biol.* **2009**, *3*. [CrossRef]
134. Gerlach, G.; Hofer, H.W. Interaction of immobilized phosphofructokinase with soluble muscle proteins. *Biochim. Biophys. Acta BBA Gen. Subj.* **1986**, *881*, 398–404. [CrossRef]
135. Mor, I.; Cheung, E.C.; Vousden, K.H. Control of Glycolysis through Regulation of PFK1: Old Friends and Recent Additions. *Cold Spring Harb. Symp. Quant. Biol.* **2011**, *76*, 211–216. [CrossRef]
136. Foucault, G.; Vacher, M.; Merkulova, T.; Keller, A.; Arrio-Dupont, M. Presence of enolase in the M-band of skeletal muscle and possible indirect interaction with the cytosolic muscle isoform of creatine kinase. *Biochem. J.* **1999**, *338*, 115–121. [CrossRef] [PubMed]
137. Brown, K.S.; Hettling, H.; van Beek, J.H.G.M. Analyzing the Functional Properties of the Creatine Kinase System with Multiscale 'Sloppy' Modeling. *PLoS Comput. Biol.* **2011**, *7*. [CrossRef]
138. Bose, S.; French, S.; Evans, F.J.; Joubert, F.; Balaban, R.S. Metabolic Network Control of Oxidative Phosphorylation. *J. Biol. Chem.* **2003**, *278*, 39155–39165. [CrossRef] [PubMed]
139. Saks, V.; Monge, C.; Guzun, R. Philosophical Basis and Some Historical Aspects of Systems Biology: From Hegel to Noble—Applications for Bioenergetic Research. *Int. J. Mol. Sci.* **2009**, *10*, 1161–1192. [CrossRef] [PubMed]
140. Guzun, R.; Timohhina, N.; Tepp, K.; Monge, C.; Kaambre, T.; Sikk, P.; Kuznetsov, A.V.; Pison, C.; Saks, V. Regulation of respiration controlled by mitochondrial creatine kinase in permeabilized cardiac cells in situ. *Biochim. Biophys. Acta BBA Bioenerg.* **2009**, *1787*, 1089–1105. [CrossRef]
141. Klepinin, A.; Ounpuu, L.; Mado, K.; Truu, L.; Chekulayev, V.; Puurand, M.; Shevchuk, I.; Tepp, K.; Planken, A.; Kaambre, T. The complexity of mitochondrial outer membrane permeability and VDAC regulation by associated proteins. *J. Bioenerg. Biomembr.* **2018**, *50*, 339–354. [CrossRef]
142. Anflous-Pharayra, K.; Cai, Z.-J.; Craigen, W.J. VDAC1 serves as a mitochondrial binding site for hexokinase in oxidative muscles. *Biochim. Biophys. Acta BBA Bioenerg.* **2007**, *1767*, 136–142. [CrossRef] [PubMed]
143. Pedersen, P.L. Warburg, me and Hexokinase 2: Multiple discoveries of key molecular events underlying one of cancers' most common phenotypes, the "Warburg Effect", i.e., elevated glycolysis in the presence of oxygen. *J. Bioenerg. Biomembr.* **2007**, *39*, 211–222. [CrossRef] [PubMed]
144. Sestili, P.; Martinelli, C.; Colombo, E.; Barbieri, E.; Potenza, L.; Sartini, S.; Fimognari, C. Creatine as an antioxidant. *Amino Acids* **2011**, *40*, 1385–1396. [CrossRef]
145. Hofmann, P. Cancer and Exercise: Warburg Hypothesis, Tumour Metabolism and High-Intensity Anaerobic Exercise. *Sports* **2018**, *6*, 10. [CrossRef]
146. Fadaka, A.; Ajiboye, B.; Ojo, O.; Adewale, O.; Olayide, I.; Emuowhochere, R. Biology of glucose metabolization in cancer cells. *J. Oncol. Sci.* **2017**, *3*, 45–51. [CrossRef]
147. Marchesi, F.; Vignali, D.; Manini, B.; Rigamonti, A.; Monti, P. Manipulation of Glucose Availability to Boost Cancer Immunotherapies. *Cancers Basel* **2020**, *12*, 2940. [CrossRef]
148. Balsom, P.D.; Söderlund, K.; Sjödin, B.; Ekblom, B. Skeletal muscle metabolism during short duration high-intensity exercise: Influence of creatine supplementation. *Acta Physiol. Scand.* **1995**, *154*, 303–310. [CrossRef]
149. Balsom, P.D.; Ekblom, B.; Söerlund, K.; Sjödn, B.; Hultman, E. Creatine supplementation and dynamic high-intensity intermittent exercise. *Scand. J. Med. Sci. Sports* **2007**, *3*, 143–149. [CrossRef]
150. Dos Santos, M.G. *Estudio Del Metabolismo Energético Muscular Y De La Composición Corporal De Atletas Por Métodos No Destructivos*; Universitat Autònoma de Barcelona: Barcelona, Spain, 2001.

151. Ceddia, R.B.; Sweeney, G. Creatine supplementation increases glucose oxidation and AMPK phosphorylation and reduces lactate production in L6 rat skeletal muscle cells. *J. Physiol.* **2004**, *555*, 409–421. [CrossRef] [PubMed]

152. Dobgenski, V.; Santos, M.; Kreider, R. Effects of creatine supplementation in the concentrations of creatine kinase, creatinine, urea and lactate on male swimmers. *J. Nutr. Health* **2016**, *2*, 1–5.

153. Oliver, J.M.; Joubert, D.P.; Martin, S.E.; Crouse, S.F. Oral Creatine Supplementation's Decrease of Blood Lactate During Exhaustive, Incremental Cycling. *Int. J. Sport Nutr. Exerc. Metab.* **2013**, *23*, 252–258. [CrossRef]

154. Storey, K.B.; Hochachka, P.W. Activation of muscle glycolysis: A role for creatine phosphate in phosphofructokinase regulation. *FEBS Lett.* **1974**, *46*, 337–339. [CrossRef]

155. Kemp, R.G. Inhibition of muscle pyruvate kinase by creatine phosphate. *J. Biol. Chem.* **1973**, *248*, 3963–3967. [CrossRef]

156. Fu, J.Y.; Kemp, R.G. Activation of Muscle Fructose 1,6-Diphosphatase by Creatine Phosphate and Citrate. *J. Biol. Chem.* **1973**, *248*, 1124–1125. [CrossRef]

157. Ponticos, M.; Lu, Q.L.; Hardie, D.G.; Partridge, T.A.; Carling, D. Dual regulation of the AMP-activated protein kinase provides a novel mechanism for the control of creatine kinase in skeletal muscle. *EMBO J.* **1998**, *17*, 1688–1699. [CrossRef] [PubMed]

158. Jørgensen, S.B.; Richter, E.A.; Wojtaszewski, J.F.P. Role of AMPK in skeletal muscle metabolic regulation and adaptation in relation to exercise. *J. Physiol.* **2006**, *574*, 17–31. [CrossRef] [PubMed]

159. Zhang, L.; Wang, X.; Li, J.; Zhu, X.; Gao, F.; Zhou, G. Creatine Monohydrate Enhances Energy Status and Reduces Glycolysis via Inhibition of AMPK Pathway in Pectoralis Major Muscle of Transport-Stressed Broilers. *J. Agric. Food Chem.* **2017**, *65*, 6991–6999. [CrossRef] [PubMed]

160. Taylor, E.B.; Ellingson, W.J.; Lamb, J.D.; Chesser, D.G.; Compton, C.L.; Winder, W.W. Evidence against regulation of AMP-activated protein kinase and LKB1/STRAD/MO25 activity by creatine phosphate. *Am. J. Physiol. Endocrinol. Metab.* **2006**, *290*, E661–E669. [CrossRef] [PubMed]

161. Eijnde, B.O.; Derave, W.; Wojtaszewski, J.F.P.; Richter, E.A.; Hespel, P. AMP kinase expression and activity in human skeletal muscle: Effects of immobilization, retraining, and creatine supplementation. *J. Appl. Physiol.* **2005**, *98*, 1228–1233. [CrossRef] [PubMed]

162. Gautel, M.; Djinović-Carugo, K. The sarcomeric cytoskeleton: From molecules to motion. *J. Exp. Biol.* **2016**, *219*, 135–145. [CrossRef]

163. Puurand, M.; Tepp, K.; Timohhina, N.; Aid, J.; Shevchuk, I.; Chekulayev, V.; Kaambre, T. Tubulin βII and βIII Isoforms as the Regulators of VDAC Channel Permeability in Health and Disease. *Cells* **2019**, *8*, 239. [CrossRef]

164. Kuznetsov, A.V.; Javadov, S.; Guzun, R.; Grimm, M.; Saks, V. Cytoskeleton and regulation of mitochondrial function: The role of beta-tubulin II. *Front. Physiol.* **2013**, *4*, 82. [CrossRef]

165. Raskin, A.; Lange, S.; Banares, K.; Lyon, R.C.; Zieseniss, A.; Lee, L.K.; Yamazaki, K.G.; Granzier, H.L.; Gregorio, C.C.; McCulloch, A.D.; et al. A Novel Mechanism Involving Four-and-a-half LIM Domain Protein-1 and Extracellular Signal-regulated Kinase-2 Regulates Titin Phosphorylation and Mechanics. *J. Biol. Chem.* **2012**, *287*, 29273–29284. [CrossRef]

166. Henderson, C.A.; Gomez, C.G.; Novak, S.M.; Mi-Mi, L.; Gregorio, C.C. Overview of the Muscle Cytoskeleton. *Compr. Physiol.* **2017**, *7*, 891–944. [CrossRef]

167. Kaasik, A.; Veksler, V.; Boehm, E.; Novotova, M.; Minajeva, A.; Ventura-Clapier, R.e. Energetic Crosstalk Between Organelles. *Circ. Res.* **2001**, *89*, 153–159. [CrossRef] [PubMed]

168. Piquereau, J.; Veksler, V.; Novotova, M.; Ventura-Clapier, R. Energetic Interactions Between Subcellular Organelles in Striated Muscles. *Front. Cell Dev. Biol.* **2020**, *8*. [CrossRef] [PubMed]

169. Kuznetsov, A.V.; Javadov, S.; Grimm, M.; Margreiter, R.; Ausserlechner, M.J.; Hagenbuchner, J. Crosstalk between Mitochondria and Cytoskeleton in Cardiac Cells. *Cells* **2020**, *9*, 222. [CrossRef] [PubMed]

170. Muñoz-Lasso, D.C.; Romá-Mateo, C.; Pallardó, F.V.; Gonzalez-Cabo, P. Much More Than a Scaffold: Cytoskeletal Proteins in Neurological Disorders. *Cells* **2020**, *9*, 358. [CrossRef] [PubMed]

171. Ross, J.A.; Levy, Y.; Ripolone, M.; Kolb, J.S.; Turmaine, M.; Holt, M.; Lindqvist, J.; Claeys, K.G.; Weis, J.; Monforte, M.; et al. Impairments in contractility and cytoskeletal organisation cause nuclear defects in nemaline myopathy. *Acta Neuropathol.* **2019**, *138*, 477–495. [CrossRef]

172. Dowling, P.; Gargan, S.; Murphy, S.; Zweyer, M.; Sabir, H.; Swandulla, D.; Ohlendieck, K. The Dystrophin Node as Integrator of Cytoskeletal Organization, Lateral Force Transmission, Fiber Stability and Cellular Signaling in Skeletal Muscle. *Proteomes* **2021**, *9*, 9. [CrossRef] [PubMed]

173. Lai, W.-F.; Wong, W.-T. Roles of the actin cytoskeleton in aging and age-associated diseases. *Ageing Res. Rev.* **2020**, *58*. [CrossRef] [PubMed]

174. Perry, R.L.; Rudnick, M.A. Molecular mechanisms regulating myogenic determination and differentiation. *Front. Biosci.* **2000**, *5*, D750–D767. [CrossRef]

175. O'Connor, R.S.; Steeds, C.M.; Wiseman, R.W.; Pavlath, G.K. Phosphocreatine as an energy source for actin cytoskeletal rearrangements during myoblast fusion. *J. Physiol.* **2008**, *586*, 2841–2853. [CrossRef]

176. Simionescu-Bankston, A.; Pichavant, C.; Canner, J.P.; Apponi, L.H.; Wang, Y.; Steeds, C.; Olthoff, J.T.; Belanto, J.J.; Ervasti, J.M.; Pavlath, G.K. Creatine kinase B is necessary to limit myoblast fusion during myogenesis. *Am. J. Physiol. Cell Physiol.* **2015**, *308*, C919–C931. [CrossRef] [PubMed]

177. Lehka, L.; Rędowicz, M.J. Mechanisms regulating myoblast fusion: A multilevel interplay. *Semin. Cell Dev. Biol.* **2020**, *104*, 81–92. [CrossRef]

178. Stroud, M.J.; Banerjee, I.; Veevers, J.; Chen, J. Linker of Nucleoskeleton and Cytoskeleton Complex Proteins in Cardiac Structure, Function, and Disease. *Circ. Res.* **2014**, *114*, 538–548. [CrossRef]

179. Spichal, M.; Fabre, E. The Emerging Role of the Cytoskeleton in Chromosome Dynamics. *Front. Genet.* **2017**, *8*. [CrossRef]

180. Loo, T.H.; Ye, X.; Chai, R.J.; Ito, M.; Bonne, G.; Ferguson-Smith, A.C.; Stewart, C.L. The mammalian LINC complex component SUN1 regulates muscle regeneration by modulating drosha activity. *eLife* **2019**, *8*. [CrossRef]

181. Piccus, R.; Brayson, D. The nuclear envelope: LINCing tissue mechanics to genome regulation in cardiac and skeletal muscle. *Biol. Lett.* **2020**, *16*. [CrossRef] [PubMed]

182. Brull, A.; Morales Rodriguez, B.; Bonne, G.; Muchir, A.; Bertrand, A.T. The Pathogenesis and Therapies of Striated Muscle Laminopathies. *Front. Physiol.* **2018**, *9*. [CrossRef]

183. Starr, D.A.; Rose, L.S. TorsinA regulates the LINC to moving nuclei. *J. Cell Biol.* **2017**, *216*, 543–545. [CrossRef]

184. Dzeja, P.P.; Bortolon, R.; Perez-Terzic, C.; Holmuhamedov, E.L.; Terzic, A. Energetic communication between mitochondria and nucleus directed by catalyzed phosphotransfer. *Proc. Natl. Acad. Sci. USA* **2002**, *99*, 10156–10161. [CrossRef] [PubMed]

185. Dzeja, P.P.; Terzic, A. Adenylate kinase and creatine kinase phosphotransfer in regulation of mitochondrial respiration and cellular energetic efficiency. In *Creatine Kinase*; Vial, C., Ed.; Nova Science Publishers: London, UK, 2006; pp. 195–221.

186. Adam, K.; Ning, J.; Reina, J.; Hunter, T. NME/NM23/NDPK and Histidine Phosphorylation. *Int. J. Mol. Sci.* **2020**, *21*, 5848. [CrossRef]

187. Attwood, P.V.; Muimo, R. The actions of NME1/NDPK-A and NME2/NDPK-B as protein kinases. *Lab. Invest.* **2017**, *98*, 283–290. [CrossRef] [PubMed]

188. Macara, I.G. Transport into and out of the Nucleus. *Microbiol. Mol. Biol. Rev.* **2001**, *65*, 570–594. [CrossRef] [PubMed]

189. de Groof, A.J.C.; Fransen, J.A.M.; Errington, R.J.; Willems, P.H.G.M.; Wieringa, B.; Koopman, W.J.H. The Creatine Kinase System Is Essential for Optimal Refill of the Sarcoplasmic Reticulum Ca2+ Store in Skeletal Muscle. *J. Biol. Chem.* **2002**, *277*, 5275–5284. [CrossRef] [PubMed]

190. Duke, A.M.; Steele, D.S. Effects of creatine phosphate on Ca2+regulation by the sarcoplasmic reticulum in mechanically skinned rat skeletal muscle fibres. *J. Physiol.* **1999**, *517*, 447–458. [CrossRef]

191. Sistermans, E.A.; Klaassen, C.H.W.; Peters, W.; Swarts, H.G.P.; Jap, P.H.K.; De Pont, J.J.H.H.M.; Wieringa, B. Co-localization and functional coupling of creatine kinase B and gastric H+/K+-ATPase on the apical membrane and the tubulovesicular system of parietal cells. *Biochem. J.* **1995**, *311*, 445–451. [CrossRef]

192. Grosse, R.; Spitzer, E.; Kupriyanov, V.V.; Saks, V.A.; Repke, K.R.H. Coordinate interplay between (Na$^+$ + K$^+$)-ATPase and creatine phosphokinase optimizes (Na+/K+)-antiport across the membrane of vesicles formed from the plasma membrane of cardiac muscle cell. *Biochim. Biophys. Acta BBA Biomembr.* **1980**, *603*, 142–156. [CrossRef]

193. Yang, Y.-C.; Fann, M.-J.; Chang, W.-H.; Tai, L.-H.; Jiang, J.-H.; Kao, L.-S. Regulation of Sodium-Calcium Exchanger Activity by Creatine Kinase under Energy-compromised Conditions. *J. Biol. Chem.* **2010**, *285*, 28275–28285. [CrossRef]

194. Kato, Y.; Miyakawa, T.; Tanokura, M. Overview of the mechanism of cytoskeletal motors based on structure. *Biophys. Rev.* **2017**, *10*, 571–581. [CrossRef]

195. Jena, B.P. Myosin: Cellular Molecular Motor. In *Cellular Nanomachines*; Jena, B.P., Ed.; Springer: Cham, Switzerland, 2020; pp. 79–89. [CrossRef]

196. Wickstead, B. The evolutionary biology of dyneins. In *Dyneins*; King, S.M., Ed.; Academic Press: New York, NY, USA, 2018; pp. 100–138. [CrossRef]

197. Ali, I.; Yang, W.-C. The functions of kinesin and kinesin-related proteins in eukaryotes. *Cell Adhes. Migr.* **2020**, *14*, 139–152. [CrossRef] [PubMed]

198. Schlattner, U.; Klaus, A.; Ramirez Rios, S.; Guzun, R.; Kay, L.; Tokarska-Schlattner, M. Cellular compartmentation of energy metabolism: Creatine kinase microcompartments and recruitment of B-type creatine kinase to specific subcellular sites. *Amino Acids* **2016**, *48*, 1751–1774. [CrossRef] [PubMed]

199. Krause, S.M.; Jacobus, W.E. Specific enhancement of the cardiac myofibrillar ATPase by bound creatine kinase. *J. Biol. Chem.* **1992**, *267*, 2480–2486. [CrossRef]

200. Kuiper, J.W.P.; Pluk, H.; Oerlemans, F.; van Leeuwen, F.N.; de Lange, F.; Fransen, J.; Wieringa, B. Creatine Kinase–Mediated ATP Supply Fuels Actin-Based Events in Phagocytosis. *PLoS Biol.* **2008**, *6*. [CrossRef]

201. Aziz, S.A.; Kuiper, J.W.P.; van Horssen, R.; Oerlemans, F.; Peters, W.; van Dommelen, M.M.T.; te Lindert, M.M.; ten Hagen, T.L.M.; Janssen, E.; Fransen, J.A.M.; et al. Local ATP Generation by Brain-Type Creatine Kinase (CK-B) Facilitates Cell Motility. *PLoS ONE* **2009**, *4*. [CrossRef]

202. Duran-Trio, L.; Fernandes-Pires, G.; Simicic, D.; Grosse, J.; Roux-Petronelli, C.; Bruce, S.J.; Binz, P.-A.; Sandi, C.; Cudalbu, C.; Braissant, O. A new rat model of creatine transporter deficiency reveals behavioral disorder and altered brain metabolism. *Sci. Rep.* **2021**, *11*. [CrossRef]

203. Hu, W.-J.; Zhou, S.-M.; Yang, J.S.; Meng, F.-G. Computational Simulations to Predict Creatine Kinase-Associated Factors: Protein-Protein Interaction Studies of Brain and Muscle Types of Creatine Kinases. *Enzym. Res.* **2011**, *2011*, 1–12. [CrossRef]

204. Wang, Q.; Qian, W.; Xu, X.; Bajpai, A.; Guan, K.; Zhang, Z.; Chen, R.; Flamini, V.; Chen, W. Energy-Mediated Machinery Drives Cellular Mechanical Allostasis. *Adv. Mater.* **2019**, *31*. [CrossRef]

205. Lee, J.H.; Jin, H.E.; Desai, M.S.; Ren, S.; Kim, S.; Lee, S.W. Biomimetic sensor design. *Nanoscale* **2015**, *7*, 18379–18391. [CrossRef]
206. Deldicque, L.; Theisen, D.; Bertrand, L.; Hespel, P.; Hue, L.; Francaux, M. Creatine enhances differentiation of myogenic C2C12 cells by activating both p38 and Akt/PKB pathways. *Am. J. Physiol. Cell Physiol.* **2007**, *293*, C1263–C1271. [CrossRef]
207. Sestili, P.; Barbieri, E.; Stocchi, V. Effects of Creatine in Skeletal Muscle Cells and in Myoblasts Differentiating Under Normal or Oxidatively Stressing Conditions. *Mini Rev. Med. Chem.* **2015**, *16*, 4–11. [CrossRef] [PubMed]
208. Gyoeva, F.K. The role of motor proteins in signal propagation. *Biochem. Mosc.* **2014**, *79*, 849–855. [CrossRef] [PubMed]
209. Solis, M.Y.; Artioli, G.G.; Gualano, B. Potential of Creatine in Glucose Management and Diabetes. *Nutrients* **2021**, *13*, 570. [CrossRef] [PubMed]
210. Somwar, R.; Kim, D.Y.; Sweeney, G.; Huang, C.; Niu, W.; Lador, C.; Ramlal, T.; Klip, A. GLUT4 translocation precedes the stimulation of glucose uptake by insulin in muscle cells: Potential activation of GLUT4 via p38 mitogen-activated protein kinase. *Biochem. J.* **2001**, *359*. [CrossRef]
211. Niu, W.; Huang, C.; Nawaz, Z.; Levy, M.; Somwar, R.; Li, D.; Bilan, P.J.; Klip, A. Maturation of the Regulation of GLUT4 Activity by p38 MAPK during L6 Cell Myogenesis. *J. Biol. Chem.* **2003**, *278*, 17953–17962. [CrossRef] [PubMed]
212. Kleinert, M.; Parker, B.L.; Fritzen, A.M.; Knudsen, J.R.; Jensen, T.E.; Kjøbsted, R.; Sylow, L.; Ruegg, M.; James, D.E.; Richter, E.A. Mammalian target of rapamycin complex 2 regulates muscle glucose uptake during exercise in mice. *J. Physiol.* **2017**, *595*, 4845–4855. [CrossRef]
213. Sterling, P. Allostasis: A model of predictive regulation. *Physiol. Behav.* **2012**, *106*, 5–15. [CrossRef] [PubMed]
214. Rankin, A.; O'Donavon, C.; Madigan, S.M.; O'Sullivan, O.; Cotter, P.D. 'Microbes in sport'—The potential role of the gut microbiota in athlete health and performance. *Br. J. Sports Med.* **2017**, *51*, 698–699. [CrossRef]
215. Hiergeist, A.; Gläsner, J.; Reischl, U.; Gessner, A. Analyses of Intestinal Microbiota: Culture versus Sequencing: Figure 1. *ILAR J.* **2015**, *56*, 228–240. [CrossRef] [PubMed]
216. Turer, E.; McAlpine, W.; Wang, K.-w.; Lu, T.; Li, X.; Tang, M.; Zhan, X.; Wang, T.; Zhan, X.; Bu, C.-H.; et al. Creatine maintains intestinal homeostasis and protects against colitis. *Proc. Natl. Acad. Sci. USA* **2017**, *114*, E1273–E1281. [CrossRef] [PubMed]
217. Marcobal, A.; Kashyap, P.C.; Nelson, T.A.; Aronov, P.A.; Donia, M.S.; Spormann, A.; Fischbach, M.A.; Sonnenburg, J.L. A metabolomic view of how the human gut microbiota impacts the host metabolome using humanized and gnotobiotic mice. *ISME J.* **2013**, *7*, 1933–1943. [CrossRef] [PubMed]
218. Savidge, T. Predicting Inflammatory Bowel Disease Symptoms Onset: Nitrous Take on Gut Bacteria Is No Laughing Matter. *Cell. Mol. Gastroenterol. Hepatol.* **2021**, *11*, 661–662. [CrossRef]
219. Langille, M.G.I.; Meehan, C.J.; Koenig, J.E.; Dhanani, A.S.; Rose, R.A.; Howlett, S.E.; Beiko, R.G. Microbial shifts in the aging mouse gut. *Microbiome* **2014**, *2*. [CrossRef]
220. O'Sullivan, O.; Cronin, O.; Clarke, S.F.; Murphy, E.F.; Molloy, M.G.; Shanahan, F.; Cotter, P.D. Exercise and the microbiota. *Gut Microbes* **2015**, *6*, 131–136. [CrossRef]
221. Ostojic, S.M. Human gut microbiota as a source of guanidinoacetic acid. *Med. Hypotheses* **2020**, *142*. [CrossRef]
222. Avgerinos, K.I.; Spyrou, N.; Bougioukas, K.I.; Kapogiannis, D. Effects of creatine supplementation on cognitive function of healthy individuals: A systematic review of randomized controlled trials. *Exp. Gerontol.* **2018**, *108*, 166–173. [CrossRef]
223. Roschel, H.; Gualano, B.; Ostojic, S.M.; Rawson, E.S. Creatine Supplementation and Brain Health. *Nutrients* **2021**, *13*, 586. [CrossRef] [PubMed]
224. Aliev, M.; Guzun, R.; Karu-Varikmaa, M.; Kaambre, T.; Wallimann, T.; Saks, V. Molecular System Bioenergics of the Heart: Experimental Studies of Metabolic Compartmentation and Energy Fluxes versus Computer Modeling. *Int. J. Mol. Sci.* **2011**, *12*, 9296–9331. [CrossRef]
225. Mair, J.; Artner-Dworzak, E.; Dienstl, A.; Lechleitner, P.; Morass, B.; Smidt, J.; Wagner, I.; Wettach, C.; Puschendorf, B. Early detection of acute myocardial infarction by measurement of mass concentration of creatine kinase-MB. *Am. J. Cardiol.* **1991**, *68*, 1545–1550. [CrossRef]
226. Hoag, G.N.; Singh, R.; Franks, C.R.; DeCoteau, W.E. Creatine kinase isoenzymes in testicular tissue of normal subjects and in a case of lymphoblastic lymphosarcoma. *Clin. Chem.* **1980**, *26*, 1360–1361. [CrossRef]
227. Nasrallah, F.; Hammami, M.; Omar, S.; Aribia, H.; Sanhaji, H.; Feki, M. Semen Creatine and Creatine Kinase Activity as an Indicator of Sperm Quality. *Clin. Lab.* **2020**, *66*. [CrossRef] [PubMed]
228. Philip, M.; Snow, R.J.; Gatta, P.A.D.; Bellofiore, N.; Ellery, S.J. Creatine metabolism in the uterus: Potential implications for reproductive biology. *Amino Acids* **2020**, *52*, 1275–1283. [CrossRef] [PubMed]
229. Muccini, A.M.; Tran, N.T.; de Guingand, D.L.; Philip, M.; Della Gatta, P.A.; Galinsky, R.; Sherman, L.S.; Kelleher, M.A.; Palmer, K.R.; Berry, M.J.; et al. Creatine Metabolism in Female Reproduction, Pregnancy and Newborn Health. *Nutrients* **2021**, *13*, 490. [CrossRef]
230. Hemmer, W.; Riesinger, I.; Wallimann, T.; Eppenberger, H.M.; Quest, A.F. Brain-type creatine kinase in photoreceptor cell outer segments: Role of a phosphocreatine circuit in outer segment energy metabolism and phototransduction. *J. Cell Sci.* **1993**, *106*, 671–683. [PubMed]
231. Spicer, S.S.; Schulte, B.A. Creatine kinase in epithelium of the inner ear. *J. Histochem. Cytochem.* **1992**, *40*, 185–192. [CrossRef]
232. Acevedo, C.; Blanchard, K.; Bacigalupo, J.; Vergara, C. Possible ATP trafficking by ATP-shuttles in the olfactory cilia and glucose transfer across the olfactory mucosa. *FEBS Lett.* **2019**, *593*, 601–610. [CrossRef]

233. Chen, L.; Roberts, R.; Friedman, D.L. Expression of brain-type creatine kinase and ubiquitous mitochondrial creatine kinase in the fetal rat brain: Evidence for a nuclear energy shuttle. *J. Comp. Neurol.* **1995**, *363*, 389–401. [CrossRef]
234. Schlattner, U.; Möckli, N.; Speer, O.; Werner, S.; Wallimann, T. Creatine Kinase and Creatine Transporter in Normal, Wounded, and Diseased Skin. *J. Invest. Dermatol.* **2002**, *118*, 416–423. [CrossRef]
235. Lenz, H.; Schmidt, M.; Welge, V.; Schlattner, U.; Wallimann, T.; Elsässer, H.-P.; Wittern, K.-P.; Wenck, H.; Stäb, F.; Blatt, T. The Creatine Kinase System in Human Skin: Protective Effects of Creatine Against Oxidative and UV Damage In Vitro and In Vivo. *J. Invest. Dermatol.* **2005**, *124*, 443–452. [CrossRef]
236. Lyons, G.E.; Muhlebach, S.; Moser, A.; Masood, R.; Paterson, B.M.; Buckingham, M.E.; Perriard, J.C. Developmental regulation of creatine kinase gene expression by myogenic factors in embryonic mouse and chick skeletal muscle. *Development* **1991**, *113*, 1017–1029.
237. Yamane, A.; Mayo, M.; Shuler, C.; Crowe, D.; Ohnuki, Y.; Dalrymple, K.; Saeki, Y. Expression of myogenic regulatory factors during the development of mouse tongue striated muscle. *Arch. Oral Biol.* **2000**, *45*, 71–78. [CrossRef]
238. Nguyen, Q.G.; Buskin, J.N.; Himeda, C.L.; Fabre-Suver, C.; Hauschka, S.D. Transgenic and tissue culture analyses of the muscle creatine kinase enhancer Trex control element in skeletal and cardiac muscle indicate differences in gene expression between muscle types. *Transgenic Res.* **2003**, *12*, 337–349. [CrossRef]
239. Gordon, P.V.; Keller, T.C., 3rd. Functional coupling to brush border creatine kinase imparts a selective energetic advantage to contractile ring myosin in intestinal epithelial cells. *Cell Motil. Cytoskelet.* **1992**, *21*, 38–44. [CrossRef]
240. Sistermans, E.A.; de Kok, Y.J.; Peters, W.; Ginsel, L.A.; Jap, P.H.; Wieringa, B. Tissue- and cell-specific distribution of creatine kinase B: A new and highly specific monoclonal antibody for use in immunohistochemistry. *Cell Tissue Res.* **1995**, *280*, 435–446. [CrossRef]
241. Mahajan, V.B.; Pai, K.S.; Lau, A.; Cunningham, D.D. Creatine kinase, an ATP-generating enzyme, is required for thrombin receptor signaling to the cytoskeleton. *Proc. Natl. Acad. Sci. USA* **2000**, *97*, 12062–12067. [CrossRef]
242. Pollard, A.E.; Martins, L.; Muckett, P.J.; Khadayate, S.; Bornot, A.; Clausen, M.; Admyre, T.; Bjursell, M.; Fiadeiro, R.; Wilson, L.; et al. AMPK activation protects against diet-induced obesity through Ucp1-independent thermogenesis in subcutaneous white adipose tissue. *Nat. Metab.* **2019**, *1*, 340–349. [CrossRef]
243. Rahbani, J.F.; Roesler, A.; Hussain, M.F.; Samborska, B.; Dykstra, C.B.; Tsai, L.; Jedrychowski, M.P.; Vergnes, L.; Reue, K.; Spiegelman, B.M.; et al. Creatine kinase B controls futile creatine cycling in thermogenic fat. *Nature* **2021**, *590*, 480–485. [CrossRef]
244. Takahashi, M.; Kishimoto, H.; Shirasaka, Y.; Inoue, K. Functional characterization of monocarboxylate transporter 12 (SLC16A12/MCT12) as a facilitative creatine transporter. *Drug Metab. Pharmacokinet.* **2020**, *35*, 281–287. [CrossRef]
245. Jomura, R.; Tanno, Y.; Akanuma, S.-i.; Kubo, Y.; Tachikawa, M.; Hosoya, K.-i. Monocarboxylate transporter 12 as a guanidinoacetate efflux transporter in renal proximal tubular epithelial cells. *Biochim. Biophys. Acta BBA Biomembr.* **2020**, *1862*. [CrossRef]
246. Verouti, S.N.; Lambert, D.; Mathis, D.; Pathare, G.; Escher, G.; Vogt, B.; Fuster, D.G. Solute carrier SLC16A12 is critical for creatine and guanidinoacetate handling in the kidney. *Am. J. Physiol. Ren. Physiol.* **2021**, *320*, F351–F358. [CrossRef]
247. Tachikawa, M.; Fujinawa, J.; Takahashi, M.; Kasai, Y.; Fukaya, M.; Sakai, K.; Yamazaki, M.; Tomi, M.; Watanabe, M.; Sakimura, K.; et al. Expression and possible role of creatine transporter in the brain and at the blood-cerebrospinal fluid barrier as a transporting protein of guanidinoacetate, an endogenous convulsant. *J. Neurochem.* **2008**, *107*, 768–778. [CrossRef]
248. Colas, C.; Banci, G.; Martini, R.; Ecker, G.F. Studies of structural determinants of substrate binding in the Creatine Transporter (CreaT, SLC6A8) using molecular models. *Sci. Rep.* **2020**, *10*. [CrossRef] [PubMed]
249. Saks, V.; Oudman, I.; Clark, J.F.; Brewster, L.M. The Effect of the Creatine Analogue Beta-guanidinopropionic Acid on Energy Metabolism: A Systematic Review. *PLoS ONE* **2013**, *8*. [CrossRef]
250. Snow, R.J.; Murphy, R.M. Creatine and the creatine transporter: A review. *Mol. Cell. Biochem.* **2001**, *224*, 169–181. [CrossRef]
251. Xu, W.; Liu, L.; Gorman, P.A.; Sheer, D.; Emson, P.C. Assignment of the human creatine transporter type 2 (SLC6A10) to chromosome band 16p11.2 by in situ hybridization. *Cytogenet. Genome Res.* **1997**, *76*, 19. [CrossRef]
252. Eichler, E. Duplication of a gene-rich cluster between 16p11.1 and Xq28: A novel pericentromeric-directed mechanism for paralogous genome evolution. *Hum. Mol. Genet.* **1996**, *5*, 899–912. [CrossRef]
253. Iyer, G.S.; Krahe, R.; Goodwin, L.A.; Doggett, N.A.; Siciliano, M.J.; Funanage, V.L.; Proujansky, R. Identification of a Testis-Expressed Creatine Transporter Gene at 16p11.2 and Confirmation of the X-Linked Locus to Xq28. *Genomics* **1996**, *34*, 143–146. [CrossRef]
254. Bayou, N.; M'Rad, R.; Belhaj, A.; Daoud, H.; Zemni, R.; Briault, S.; Helayem, M.B.; Ben Jemaa, L.; Chaabouni, H. The Creatine Transporter Gene Paralogous at 16p11.2 Is Expressed in Human Brain. *Comp. Funct. Genom.* **2008**, *2008*, 1–5. [CrossRef]
255. Kumar, R.A.; KaraMohamed, S.; Sudi, J.; Conrad, D.F.; Brune, C.; Badner, J.A.; Gilliam, T.C.; Nowak, N.J.; Cook, E.H., Jr.; Dobyns, W.B.; et al. Recurrent 16p11.2 microdeletions in autism. *Hum. Mol. Genet.* **2008**, *17*, 628–638. [CrossRef] [PubMed]
256. Mayser, W.; Schloss, P.; Betz, H. Primary structure and functional expression of a choline transporter expressed in the rat nervous system. *FEBS Lett.* **1992**, *305*, 31–36. [CrossRef]
257. Dodd, J.R.; Christie, D.L. Substituted Cysteine Accessibility of the Third Transmembrane Domain of the Creatine Transporter. *J. Biol. Chem.* **2005**, *280*, 32649–32654. [CrossRef]
258. Barnwell, L.F.S.; Chaudhuri, G.; Townsel, J.G. Cloning and sequencing of a cDNA encoding a novel member of the human brain GABA/noradrenaline neurotransmitter transporter family. *Gene* **1995**, *159*, 287–288. [CrossRef]

259. Gonzalez, A.M.; Uhl, G.R. 'Choline/orphan V8-2-1/creatine transporter' mRNA is expressed in nervous, renal and gastrointestinal systems. *Mol. Brain Res.* **1994**, *23*, 266–270. [CrossRef]

260. Guerrero-Ontiveros, M.L.; Wallimann, T. Creatine supplementation in health and disease. Effects of chronic creatine ingestion in vivo: Down-regulation of the expression of creatine transporter isoforms in skeletal muscle. *Mol. Cell. Biochem.* **1998**, *184*, 427–437. [CrossRef]

261. Martínez-Muñoz, C.; Rosenberg, E.H.; Jakobs, C.; Salomons, G.S. Identification, characterization and cloning of SLC6A8C, a novel splice variant of the creatine transporter gene. *Gene* **2008**, *418*, 53–59. [CrossRef]

262. Ndika, J.D.T.; Martinez-Munoz, C.; Anand, N.; van Dooren, S.J.M.; Kanhai, W.; Smith, D.E.C.; Jakobs, C.; Salomons, G.S. Post-transcriptional regulation of the creatine transporter gene: Functional relevance of alternative splicing. *Biochim. Biophys. Acta BBA Gen. Subj.* **2014**, *1840*, 2070–2079. [CrossRef]

263. Sitte, H.H.; Farhan, H.; Javitch, J.A. Sodium-dependent neurotransmitter transporters: Oligomerization as a determinant of transporter function and trafficking. *Mol. Interv.* **2004**, *4*, 38–47. [CrossRef] [PubMed]

264. Peral, M.J.; García-Delgado, M.; Calonge, M.L.; Durán, J.M.; Horra, M.C.; Wallimann, T.; Speer, O.; Ilundáin, A.A. Human, rat and chicken small intestinal Na$^+$-Cl$^-$-creatine transporter: Functional, molecular characterization and localization. *J. Physiol.* **2002**, *545*, 133–144. [CrossRef]

265. Odoom, J.; Kemp, G.; Radda, G. The regulation of total creatine content in a myoblast cell line. *Mol. Cell. Biochem.* **1996**, *158*. [CrossRef]

266. Pramod, A.B.; Foster, J.; Carvelli, L.; Henry, L.K. SLC6 transporters: Structure, function, regulation, disease association and therapeutics. *Mol. Aspects Med.* **2013**, *34*, 197–219. [CrossRef]

267. Rudnick, G.; Krämer, R.; Blakely, R.D.; Murphy, D.L.; Verrey, F. The SLC6 transporters: Perspectives on structure, functions, regulation, and models for transporter dysfunction. *Pflügers Archiv Eur. J. Physiol.* **2013**, *466*, 25–42. [CrossRef]

268. Santacruz, L.; Darrabie, M.D.; Mishra, R.; Jacobs, D.O. Removal of Potential Phosphorylation Sites does not Alter Creatine Transporter Response to PKC or Substrate Availability. *Cell. Physiol. Biochem.* **2015**, *37*, 353–360. [CrossRef]

269. Derave, W.; Straumann, N.; Olek, R.A.; Hespel, P. Electrolysis stimulates creatine transport and transporter cell surface expression in incubated mouse skeletal muscle: Potential role of ROS. *Am. J. Physiol. Endocrinol. Metab.* **2006**, *291*, E1250–E1257. [CrossRef] [PubMed]

270. Shojaiefard, M.; Christie, D.L.; Lang, F. Stimulation of the creatine transporter SLC6A8 by the protein kinases SGK1 and SGK3. *Biochem. Biophys. Res. Commun.* **2005**, *334*, 742–746. [CrossRef] [PubMed]

271. Kobayashi, T.; Cohen, P. Activation of serum- and glucocorticoid-regulated protein kinase by agonists that activate phosphatidylinositide 3-kinase is mediated by 3-phosphoinositide-dependent protein kinase-1 (PDK1) and PDK2. *Biochem. J.* **1999**, *339*, 319–328. [CrossRef] [PubMed]

272. Fezai, M.; Warsi, J.; Lang, F. Regulation of the Na+,Cl- Coupled Creatine Transporter CreaT (SLC6A8) by the Janus Kinase JAK3. *Neurosignals* **2015**, *23*, 11–19. [CrossRef] [PubMed]

273. Almeida, L.S.; Salomons, G.S.; Hogenboom, F.; Jakobs, C.; Schoffelmeer, A.N.M. Exocytotic release of creatine in rat brain. *Synapse* **2006**, *60*, 118–123. [CrossRef] [PubMed]

274. Brault, J.J.; Abraham, K.A.; Terjung, R.L. Muscle creatine uptake and creatine transporter expression in response to creatine supplementation and depletion. *J. Appl. Physiol.* **2003**, *94*, 2173–2180. [CrossRef]

275. Tarnopolsky, M.; Parise, G.; Fu, M.H.; Brose, A.; Parshad, A.; Speer, O.; Wallimann, T. Acute and moderate-term creatine monohydrate supplementation does not affect creatine transporter mRNA or protein content in either young or elderly humans. *Mol. Cell. Biochem.* **2003**, *244*, 159–166. [CrossRef] [PubMed]

276. Jangid, N.; Surana, P.; Salmonos, G.; Jain, V. Creatine transporter deficiency, an underdiagnosed cause of male intellectual disability. *BMJ Case Rep.* **2020**, *13*. [CrossRef]

277. Wang, Q.; Yang, J.; Liu, Y.; Li, X.; Luo, F.; Xie, J. A novel SLC6A8 mutation associated with intellectual disabilities in a Chinese family exhibiting creatine transporter deficiency: Case report. *BMC Med. Genet.* **2018**, *19*. [CrossRef]

278. Dunbar, M.; Jaggumantri, S.; Sargent, M.; Stockler-Ipsiroglu, S.; van Karnebeek, C.D. Treatment of X-linked creatine transporter (SLC6A8) deficiency: Systematic review of the literature and three new cases. *Mol. Genet. Metab.* **2014**, *112*, 259–274. [CrossRef] [PubMed]

279. Sharer, J.D.; Bodamer, O.; Longo, N.; Tortorelli, S.; Wamelink, M.M.; Young, S. Laboratory diagnosis of creatine deficiency syndromes: A technical standard and guideline of the American College of Medical Genetics and Genomics. *Genet. Med.* **2017**, *19*, 256–263. [CrossRef]

280. Kaviani, M.; Izadi, A.; Heshmati, J. Would creatine supplementation augment exercise performance during a low carbohydrate high fat diet? *Med. Hypotheses* **2021**, *146*. [CrossRef]

281. López Lluch, G.; Ferretti, R.; Moura, E.G.; dos Santos, V.C.; Caldeira, E.J.; Conte, M.; Matsumura, C.Y.; Pertille, A.; Mosqueira, M. High-fat diet suppresses the positive effect of creatine supplementation on skeletal muscle function by reducing protein expression of IGF-PI3K-AKT-mTOR pathway. *PLoS ONE* **2018**, *13*. [CrossRef]

282. Mine, M.; Mizuguchi, H.; Takayanagi, T. Kinetic analysis of the transphosphorylation with creatine kinase by pressure-assisted capillary electrophoresis/dynamic frontal analysis. *Anal. Bioanal. Chem.* **2021**, *413*, 1453–1460. [CrossRef] [PubMed]

Review

Creatine Metabolism in Female Reproduction, Pregnancy and Newborn Health

Anna Maria Muccini [1,2], Nhi T. Tran [1,3], Deborah L. de Guingand [1,2], Mamatha Philip [4], Paul A. Della Gatta [4], Robert Galinsky [1,2], Larry S. Sherman [5,6], Meredith A. Kelleher [7], Kirsten R. Palmer [2], Mary J. Berry [8], David W. Walker [3], Rod J. Snow [4] and Stacey J. Ellery [1,2,*]

[1] The Ritchie Centre, Hudson Institute of Medical Research, Clayton, VIC 3168, Australia; ammuc2@student.monash.edu (A.M.M.); nhi.tran2@student.rmit.edu (N.T.T.); deborah.deguingand@hudson.org.au (D.L.d.G.); robert.Galinsky@hudson.org.au (R.G.)

[2] Department of Obstetrics and Gynaecology, Monash University, Clayton, VIC 3800, Australia; kirsten.Palmer@monash.edu

[3] School of Health & Biomedical Sciences, RMIT University, Bundoora, VIC 3082, Australia; david.walker@rmit.edu.au

[4] Institute for Physical Activity and Nutrition, School of Exercise and Nutrition Sciences, Deakin University, Geelong, VIC 3220, Australia; mphilipgabriel@deakin.edu.au (M.P.); paul.dellagatta@deakin.edu.au (P.A.D.G.); rod.snow@deakin.edu.au (R.J.S.)

[5] Division of Neuroscience, Oregon National Primate Research Center, Beaverton, OR 97006, USA; shermanl@ohsu.edu

[6] Department of Cell, Developmental and Cancer Biology, Oregon Health & Science University, Portland, OR 97239, USA

[7] Division of Reproductive & Developmental Sciences, Oregon National Primate Research Center, Beaverton, OR 37009, USA; kellehem@ohsu.edu

[8] Capital and Coast District Health Board, Department of Paediatrics, University of Otago Wellington, Wellington 6242, New Zealand; max.berry@otago.ac.nz

* Correspondence: stacey.ellery@hudson.org.au

Citation: Muccini, A.M.; Tran, N.T.; de Guingand, D.L.; Philip, M.; Della Gatta, P.A.; Galinsky, R.; Sherman, L.S.; Kelleher, M.A.; Palmer, K.R.; Berry, M.J.; et al. Creatine Metabolism in Female Reproduction, Pregnancy and Newborn Health. *Nutrients* 2021, 13, 490. https://doi.org/10.3390/nu13020490

Academic Editor: Richard B. Kreider
Received: 15 January 2021
Accepted: 30 January 2021
Published: 2 February 2021

Publisher's Note: MDPI stays neutral with regard to jurisdictional claims in published maps and institutional affiliations.

Abstract: Creatine metabolism is an important component of cellular energy homeostasis. Via the creatine kinase circuit, creatine derived from our diet or synthesized endogenously provides spatial and temporal maintenance of intracellular adenosine triphosphate (ATP) production; this is particularly important for cells with high or fluctuating energy demands. The use of this circuit by tissues within the female reproductive system, as well as the placenta and the developing fetus during pregnancy is apparent throughout the literature, with some studies linking perturbations in creatine metabolism to reduced fertility and poor pregnancy outcomes. Maternal dietary creatine supplementation during pregnancy as a safeguard against hypoxia-induced perinatal injury, particularly that of the brain, has also been widely studied in pre-clinical in vitro and small animal models. However, there is still no consensus on whether creatine is essential for successful reproduction. This review consolidates the available literature on creatine metabolism in female reproduction, pregnancy and the early neonatal period. Creatine metabolism is discussed in relation to cellular bioenergetics and de novo synthesis, as well as the potential to use dietary creatine in a reproductive setting. We highlight the apparent knowledge gaps and the research "road forward" to understand, and then utilize, creatine to improve reproductive health and perinatal outcomes.

Keywords: creatine; nutritional supplements; fertility; pregnancy; newborn; development; brain injury

---

1. Introduction

Female reproductive organs are some of the most regenerative and highly energetic tissues within the body [1]. As such, there is an undeniable link between energy metabolism and reproductive success. By understanding the intricacies of energy metabolism and adenosine triphosphate (ATP) production throughout female reproduction, we are best placed to address irregularities that may contribute to infertility and poor pregnancy outcomes.

To sustain high energy levels, cells, including those of our reproductive tissues, are equipped with high energy phosphagens [2]. While invertebrates use a variety of different phosphagen systems, the creatine kinase circuit is the sole phosphagen system of higher vertebrates [3]. Ultimately, the creatine kinase circuit serves as an immediate temporal energy buffer, maintaining ATP turnover and the intracellular ATP/adenosine diphosphate (ADP) ratio. It also provides spatial energy buffering by providing for the transport of high-energy phosphates from sites of ATP production (i.e., oxidative phosphorylation and glycolysis) to sites of ATP utilization within the cytosol (Figure 1) [4].

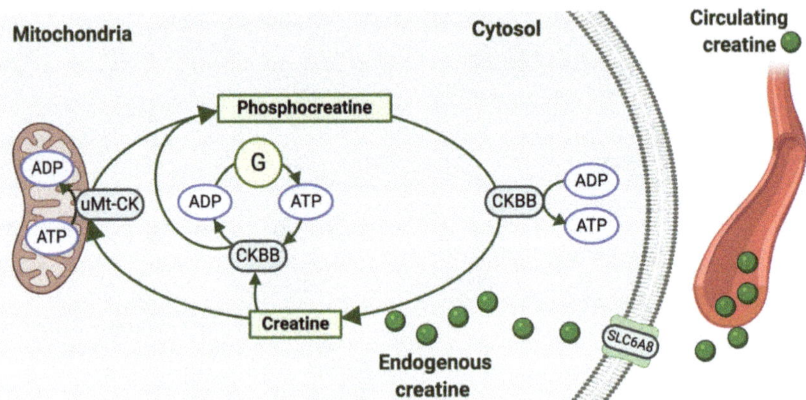

Figure 1. The creatine kinase circuit. Creatine can be produced endogenously by some cells or taken up from the circulation via the creatine transporter (*SLC6A8*). Creatine is then phosphorylated from adenosine triphosphate (ATP) to form phosphocreatine and adenosine diphosphate (ADP) in a reaction catalyzed by ubiquitous mitochondrial creatine kinase (uMt-CK). Isoforms of creatine kinase in the cell cytosol (mainly brain-type creatine kinase CKBB) are also linked to glycolytic enzymes (G) to generate phosphocreatine and ADP from glycolytic ATP. Then, when required for energy-dependent cellular processes, cytosolic isoforms of creatine kinase (CKBB) hydrolyze the bond between creatine and the phosphate group stored as phosphocreatine, thus regenerating ATP and creatine. Note: muscle-type cytosolic creatine kinase (CKMM) expression has also been detected in mouse oocytes.

Creatine is the substrate for the creatine kinase circuit. It can be endogenously synthesized by the body or acquired from a diet containing meat, fish, dairy products or over-the-counter nutritional supplements [5]. De novo creatine synthesis involves a two-step process (Figure 2). First, the enzyme L-Arginine: glycine amidinotransferase (AGAT, translated from the *GATM* gene) catalyzes the production of guanidinoacetate (GAA), the creatine precursor, and ornithine from arginine and glycine. In the second step, guanidinoacetate N-methyltransferase (GAMT) catalyzes the methylation of GAA producing creatine and S-Adenosyl homocysteine.

Circulating creatine is taken up by cells via a sodium-dependent creatine transporter encoded by the *SLC6A8* gene (Figure 1) [6,7]. Reversible phosphorylation of intracellular creatine by the ubiquitous mitochondrial creatine kinase (uMt-CK) or cytosolic creatine kinases linked to glycolytic enzymes produces the high energy compound, phosphocreatine [8]. Promoted by shifts in the intracellular ADP/ATP ratio, cytosolic isoforms of creatine kinase then hydrolyze the bond between creatine and the stored phosphate group, regenerating ATP (Figure 1) [4]. There are two cytosolic isoforms of creatine kinase that have been identified in female reproductive tissues. The most prominent is the brain-type creatine kinase (CKBB), with the muscle-type creatine kinase (CKMM) isoform also being identified in the mouse oocyte [9]. Overall, the creatine kinase circuit produces ATP more rapidly than any other metabolic system, with the interplay between the different components of creatine metabolism being essential to sustain the bioenergetic demands of a cell [3].

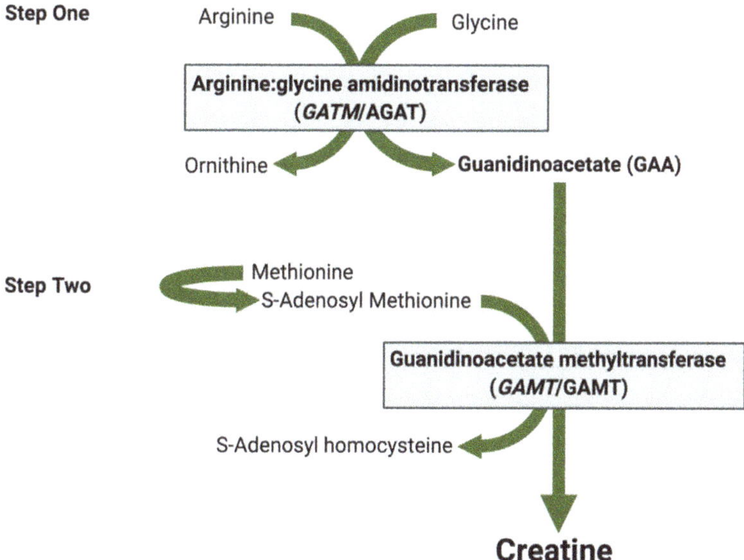

Figure 2. Schematic of de novo creatine synthesis. Step one—Arginine and glycine combine to form guanidinoacetic acid (GAA) and ornithine in a reaction catalyzed by arginine:glycine amidinotransferase (AGAT translated from the *GATM* gene). Step two—GAA is methylated to form creatine and S-Adenosyl homocysteine. S-adenosyl methionine (SAM) is the primary methyl donor for this reaction catalyzed by guanidinoacetate methyltransferase (*GAMT*/GAMT).

Our understanding of creatine metabolism in reproduction and pregnancy is growing. Indeed, review of the literature sees creatine metabolism increasingly appear throughout studies of both male and female reproduction. This association extends from observational studies published in the early 20th century through to more recent non-targeted metabolomic screens [10,11]. There is now strong evidence that alterations to creatine homeostasis occur with the normal progression of a healthy female reproductive cycle and with pregnancy [11,12]. Altered creatine metabolism has also been linked to reduced fertility and specific pregnancy-related pathologies [13,14]; however, there is no clear consensus as to whether creatine metabolism is indeed an essential component of bioenergetics for successful reproduction.

This review aims to consolidate the available literature, old and new, pertaining to creatine metabolism in reproduction, pregnancy, fetal brain development, and the early neonatal period. Creatine metabolism will be discussed in relation to its capacity to maintain cellular bioenergetics, the ability of reproductive tissues to synthesize creatine de novo, and the potential to use dietary creatine as a protective treatment against the effects of in utero fetal hypoxia and perturbations in newborn brain metabolism. We will highlight the apparent knowledge gaps and the research "road forward" to ultimately understand, and then potentially harness, creatine metabolism to improve reproductive health and perinatal outcomes.

2. Creatine Metabolism in the Female Reproductive System

2.1. Oocytes and Surrounding Cells

The oocyte requires large amounts of energy during its development in preparation for fertilization. There are also energy reserves stored within the mature oocyte to facilitate the initial period of embryogenesis [15]. Creatine metabolism occurs within human and mouse oocytes. These cells contain phosphocreatine (~4 to 5 mmol.kg^{-1} dry mass) with an equal level of creatine [16]. They also express creatine kinase (CK) genes and proteins, displaying high levels of in vitro CK activity [16,17]. The available data on CK expression and activity suggest that the use of the creatine kinase circuit to generate ATP by the oocyte may be species-specific. In mice, cytosolic creatine kinase (CKBB) activity increased 5-fold during oocyte maturation. This same study found that CKBB activity increased after fertilization, up to the stage of an eight-cell embryo, before a steep decline as the embryo reached the blastocyst phase [17]. These findings were confirmed by a further study of mouse embryos completed by Forsey et al. (2013) [18]. A more recent study, also in mice, found that mature oocytes have increased cytosolic creatine kinase (*CKMM*) gene expression when exposed to human chorionic gonadotropin (hCG) stimulation; the hormone produced by trophoblast cells after fertilization [9]. However, these same cells lacked expression of the *uMt-CK* gene, bringing into question the functional capacity of the creatine kinase circuit to produce phosphocreatine and then ATP, within the experimental paradigm used [9]. Further studies focused on protein expression and CK enzyme activity are still required. In contrast, both uMt-CK and cytosolic (CKBB) gene and protein expression have been detected in bovine oocytes. Scantland et al. (2014) observed that, compared to mature bovine oocytes, immature oocytes had higher gene expression of the creatine kinases, and that when oocytes were matured in a medium containing specific CK inhibitors, they displayed an elevated intra-oocyte ADP:ATP ratio [19]. These findings indicate that CK activity is present in oocytes and that the creatine kinase circuit is used to help maintain intracellular ATP levels.

The source of creatine for the oocyte (endogenous synthesis or cellular up-take) remains unclear. A study by Fezai et al. (2015) found that oocytes from Xenopus laevis (African clawed frog) do not transport creatine across their plasma membrane, concluding that these cells do not express creatine transporter proteins [20]. However, caution should be taken when extrapolating amphibian to mammalian physiology as high levels of creatine transporter gene (*SLC6A8*) expression have been reported in the ovaries of rats, suggesting that at least some ovarian cells may contain the transporter [21]. It is also important to note that for some metabolites, passive diffusion into the oocyte from attached surrounding support cells (e.g., cumulus cells) has been documented and one cannot rule out the possibility that the same mechanism supplies the oocyte with creatine.

Indeed, creatine metabolism has been reported in the specialized somatic cells that surround the mammalian oocyte. Collectively, these cells support oocyte maturation, facilitate fertilization and subsequent development into a viable embryo. Human ovarian stromal cells, in particular, may contribute to de novo creatine synthesis via GAA production, with these cells having detectable levels of the *GATM* gene and AGAT protein [22]. However, *GAMT* gene but not protein expression was detected in these same cells, indicating that they are unable to methylate GAA to produce creatine [22]. The enzymes of the creatine kinase circuit are present within cumulus cells signifying the presence of creatine metabolism in these specialized granulosa cells which lie directly adjacent to the oocyte. The activity of the CK enzymes appears to be low and remain unchanged with hormone stimulation, so there is no current evidence for creatine metabolism varying across the ovarian cycle in these specific cells [9]. Interestingly, *CKBB* gene expression appears elevated in cumulus cells acquired from women undergoing assisted fertility treatment who were either older than 38 years of age or younger than 28 years of age, with the level of expression positively associated with good-quality embryos in both younger and older women [23]. Finally, research investigating the metabolites secreted from bovine cumulus–oocyte complexes (COCs) during in vitro maturation showed a substantial increase in creatine (~ 450-fold) and a comparatively modest increase (~ 2-fold) in GAA concentration in the maturation medium bathing the cells. Moreover, both creatine and GAA are detectable within bovine cumulus cells [24]. Further experiments demonstrated that the addition of creatine alone to the maturation medium did not affect the developmental competence of the oocyte [24]. The secretion of creatine from the COCs is therefore likely performing another role such as facilitating sperm function. Indeed, there is considerable research on the importance of creatine metabolism for optimal sperm motility, hyperactivation and capacitation in preparation for fertilization [25].

2.2. Follicular Fluid, the Oviduct and Oviductal Fluid

Follicular fluid is derived from plasma and secretions synthesized in the follicle wall [26]. This fluid contains metabolites, which are critical for oocyte growth and development [27]. Creatine is present in human follicular fluid [28,29]. Interestingly, the creatine levels in follicular fluid are significantly lower in women with ovarian endometrioma compared with controls [30]. Umehara et al. (2018) demonstrated that mouse follicular fluid creatine concentrations increased markedly around ovulation [9]. This is in contrast to equine follicular fluid, where creatine levels were unchanged with follicular development or near ovulation [31]. The data from Umehara et al. (2018) also suggest that the increase in follicular fluid creatine levels likely resulted from an increase in creatine synthesizing capacity within the granulosa cells of the ovary because they displayed a significant increase in *GATM* and *GAMT* gene expression with equine chorionic gonadotropin stimulation around the time of ovulation [9]. Further studies are needed to track human follicular fluid creatine levels across the ovarian cycle to better understand the importance of creatine metabolism for oocyte growth and development.

Fertilization, the process of the mature oocyte and sperm fusing to give rise to the embryo, occurs within the oviduct. This structure is lined with an epithelium coated by an oviductal fluid, composed in part by secretions from these cells and in part by blood plasma filtrate. The oviductal fluid composition is species-specific, but overall contains proteins, hormones, growth factors and metabolites that vary depending on the stage of the reproductive cycle and also on the presence of gametes or embryos. To the best of our knowledge, creatine, phosphocreatine, and GAA levels have not been determined in human oviductal fluid. Gene expression of the creatine synthesizing enzymes *GATM* and *GAMT* has been measured in the human and rat oviduct [21,22], but neither protein was detected in a study using immunohistochemical analysis [22]. In partial agreement, GAMT gene and protein were not found to be expressed in the mouse oviduct [32]. There are also no human data on the protein expression levels of the creatine transporter in the oviduct tissue; however, the creatine transporter gene (*SLC6A8*) is expressed in both human and rat oviducts [21,22]. Further analysis is obviously required to adequately characterize creatine metabolism in the human oviduct, and its role in the bioenergetics of fertilization. Studies in other mammalian species provide some further information on creatine metabolism in the oviduct and oviductal fluid. For example, the creatine concentration in equine oviductal fluid is very high (3–4 mM) compared with plasma levels (8–103 µM) and does not change pre- to post-ovulation [33,34]. Creatine levels in mouse oviductal fluid increased with hCG stimulation although this was not associated with an increase in *GATM* or *GAMT* gene expression in oviduct cells [9]. Consequently, the source of the elevated creatine levels (endogenous synthesis or cellular up-take) found in the oviductal fluid is currently unclear. Interestingly, mice sperm cultured in in vitro fertilization (IVF) medium supplemented with creatine displayed elevated ATP levels and increased motility [9]. A similar observation has recently been reported in pig IVF studies [35]. These findings raise the possibility again that increased creatine levels in the female reproductive tract are taken up by sperm, contributing to their hyperactivation and increasing the chance of successful fertilization [25]. Whether the same is true for human sperm and whether the simple addition of creatine to IVF medium can improve outcomes for couples undertaking artificial reproductive therapies warrant further investigation.

2.3. The Endometrium

There is evidence that the creatine kinase circuit is active in uterine tissue [36,37], and that components of this metabolic system change throughout the female reproductive cycle and with pregnancy [38]. The use and regulation of this system in the uterus are not well understood; however, it is likely to be significant given that up-regulation of creatine metabolism in uterine tissues appears to correlate with phases of increased uterine energy demand throughout the female reproductive cycle, pregnancy, and parturition.

Research explicitly examining human endometrial tissue reports that components of creatine metabolism are up-regulated during the secretory phase of the menstrual cycle [39]. During this phase when embryo implantation can occur, endometrial tissue contains creatine and displays increased expression of the creatine transporter gene (*SLC6A8*), as well as an up-regulation of cytosolic creatine kinase (*CKBB*) gene expression and enzyme activity [37,39–41]. This increase in CKBB enzyme activity occurs in both stromal cells and endometrial glands, but is much higher in the latter [37]. Recently, the increased CKBB protein expression has been localized to the apical surface of the human endometrial glandular and luminal epithelial cells [42]. This raises the possibility that CKBB activity, and therefore production of ATP from phosphocreatine stores, may be necessary for regulating energy homeostasis during the receptive phase of the menstrual cycle [11]. Whether endometrial creatine synthesis also changes across the cycle has not been investigated in humans. However, some aspects of creatine synthesis in the endometrium have been linked to reduced fertility in rodents, with AGAT knockout female mice proving to be infertile [43]. It is unclear if infertility is directly linked to endometrial function or another component of the reproductive cycle. It is also unclear if untreated AGAT deficiency leads to infertility or poor pregnancy outcomes, but this warrants further investigation [44].

Components of endometrial creatine metabolism are also altered during pregnancy [36,45–48]. In the pregnant rodent, uMt-CK and CKBB protein are highly expressed in the decidua parietalis and basalis, with these enzymes needed to complete the creatine kinase circuit mainly located within stromal cells close to the multiple sites of placental implantation [36]. Surprisingly, very little is known about creatine kinase gene and protein expression in the human endometrium during pregnancy. Only one study has attempted to investigate this, reporting that creatine kinase activity was present in human decidual explants obtained at term [46]. In regard to creatine synthesizing capacity, one rodent study [48] has reported that AGAT activity was high in the uterine decidua during pregnancy. However, there was little or no GAMT enzyme activity present in the endometrium of these animals. These findings suggest that the decidua in pregnant rodents has a high capacity to produce GAA but does not complete the methylation step to produce creatine. A similar finding of increased uterine GAA production has also been noted in pregnant sheep [49]. It is currently unknown what adaptations in creatine synthesis capacity, if any, occur in the human endometrium with pregnancy. Furthermore, characterization of endometrial creatine, phosphocreatine or GAA levels during human pregnancy remains to be established.

2.4. The Myometrium

The human non-pregnant myometrium displays creatine kinase activity [50] and phosphocreatine at a low level compared to human pregnant myometrium [51]. Cultured human uterine smooth muscle cells are capable of importing extracellular creatine using a mediated process, suggesting that creatine transporter proteins are present in these cells [52]. However, there have been no studies exploring whether the human myometrium in the non-pregnant state can produce GAA or creatine. It is also not known if the myometrial expression of creatine kinase isoforms is altered during the female reproductive cycle.

There is evidence that creatine metabolism is up-regulated in the myometrium during pregnancy [51,53–56]. Phosphocreatine levels are increased in the human pregnant myometrium at term compared to non-pregnant tissue [55]. This likely acts as an increased energy reserve for the uterus during labor [57]. The mechanism(s) leading to the increased phosphocreatine levels in the myometrium during pregnancy remain unclear but are likely due to a concomitant increase in the total creatine content. Currently, no evidence demonstrates the presence of the creatine transporter, or synthesizing enzymes AGAT and GAMT in the human pregnant myometrium. Consequently, it is not known if the myometrium is capable of transporting creatine into cells, or whether myometrial cells can produce GAA and/or creatine during pregnancy. Cytosolic creatine kinase (*CKBB*) gene expression has been measured in human pregnant myometrium and at term is three-fold higher than earlier in gestation [53]. However, the underlying mechanism for this increase is unknown. Additionally, there are no existing data for CKBB protein, nor uMt-CK gene and protein expression in human myometrium during pregnancy, so the functional consequences of these gene expression changes and the overall use of creatine to sustain myometrial ATP production remain unclear. This should be an area of focus for future research, as it is highly plausible that creatine metabolism in the myometrium is important for optimal contractile performance during labor [38]. Studies on creatine metabolism in the female reproductive tract are summarized in Table 1.

Table 1. Summary of creatine metabolism in the female reproductive tract.

Tissue	Species	Creatine and Phosphocreatine Content	Creatine Kinases	Creatine Synthesis and Transport
Oocytes	Mouse	Creatine and phosphocreatine present (~4 to 5 mmol.kg^{-1} dry mass) [16].	CKBB gene, protein and activity reported [16,17]. Activity increased with oocyte maturation and fertilization [18]. *CKMM* detected. Expression levels increased with hCG stimulation [9].	
	Bovine		uMt-CK and CKBB gene and protein expression reported [18]. Use of CK inhibitors elevated intra-oocyte ADP:ATP ratio [19].	
	Human	Creatine and phosphocreatine present (~4 to 5 mmol.kg^{-1} dry mass) [16].		
Ovaries	Rat			High *SLC6A8* gene expression reported [21].
Ovarian stromal cells	Human	/		Detectable levels of the *GATM* gene and AGAT protein, but *GAMT* undetected [22].
Cumulus cells or cumulus–oocyte complexes (COCs)	Human		*CKBB* gene expression detected and elevated in women with good quality embryos undergoing ART [23].	
Follicular fluid	Bovine	Creatine and GAA detected in media bathing cells, with an increase in creatine (~450-fold) and GAA (~2-fold) reported during in vitro maturation [24].		
	Human	Creatine detected and lower in women with endometrioma [30].		
	Mouse	Creatine detected and increases around ovulation [9].		
	Equine	Creatine detected. Remains unchanged with follicular development [31].		
Granulosa cells	Rat			Increase in *GATM* and *GAMT* expression with equine CG stimulation [9].
Oviduct	Human			*GATM* and *GAMT* and *SLC6A8* detected [21,22]
	Rat			*GATM* and *GAMT* and *SLC6A8* detected [21,22]
	Mouse			*GAMT* gene and protein not expressed [32]
Oviductal fluid	Equine	High creatine concentration (3–4 mM) that did not change pre- to post-ovulation [33,34].		

Table 1. *Cont.*

Tissue	Species	Creatine and Phosphocreatine Content	Creatine Kinases	Creatine Synthesis and Transport
Non-pregnant Endometrium	Mouse	Creatine levels detected and increased with hCG stimulation [9].		*GATM* and *GAMT* detected. No change in expression with hCG stimulation [9].
	Human		Up-regulation of *CKBB* expression and enzyme activity in the secretory phase of the menstrual cycle [37,39–41].	Increased *SLC6A8* expression during the secretory phase of the menstrual cycle [37].
Pregnant endometrium	Rat		uMt-CK and CKBB proteins expressed in the decidua parietalis and basalis [36].	AGAT activity high in the decidua. No GAMT enzyme activity present [48].
	Sheep			GAA produced at a higher level than non-pregnant animals [49].
	Human		Creatine kinase activity present in term decidual explants [46].	
Non-pregnant myometrium	Human	Phosphocreatine detected at a low level compared pregnant myometrium [51].	Creatine kinase activity detected [50].	
Pregnant myometrium	Human	Phosphocreatine detected with higher levels at term compared to non-pregnant tissue [55].		*CKBB* gene expression detected. Levels were three-fold higher at term compared with earlier in gestation [53].

Abbreviations—cytosolic brain-type creatine kinase (CKBB), cytosolic muscle-type creatine kinase (CKMM), ubiquitous mitochondrial creatine (uMt-CK), guanidinoacetate (GAA), artificial reproductive therapy (ART), human chorionic gonadotropin (hCG), chorionic gonadotropin (CG).

3. Creatine Metabolism in the Human Placenta

Optimal placental function is required to ensure both the successful maintenance of pregnancy, as well as fetal growth and development [58,59]. As such, the human placenta is a highly metabolic organ, consuming 40–60% of oxygen and glucose transported to the uterine cavity [60]. This energy consumption serves two purposes: [1] growth of the placenta itself (placental tissue turnover is 3–4 g a day, or 1–2% of its total mass); [2] nutrient transfer, waste transport, and peptide and steroid hormone production for fetal growth and development [61]. Consequently, pregnancy encompasses large changes in maternal glucose, carbohydrate, amino acid, lipid, and fatty acid-derived energy metabolism to meet placental and fetal requirements [62,63]. There is growing evidence that creatine metabolism should be added to the list of pathways needed to maintain cellular bioenergetics in both the healthy and metabolically compromised placenta.

The human placenta expresses the mitochondrial (uMt-CK) and cytosolic (CKBB) isoforms of creatine kinase, with expression patterns varying throughout the three trimesters of pregnancy. At a gene level, *uMt-CK* and *CKBB* mRNA expression appear low in the first and second trimester before a large peak closer to term [64]. In this study by Thomure et al. (1996), post-transcriptional regulation of both creatine kinases was apparent with CKBB protein expression remaining consistent throughout gestation and uMt-CK protein rising through to mid-gestation before declining just before term. Overall, this biphasic expression correlates with the metabolic activity of the placenta and suggests that the creatine kinase circuit contributes to placental metabolism during pregnancy [64].

It has also been reported that the human placenta has the enzymatic machinery to synthesize GAA and creatine [65]. The capacity for placental creatine synthesis and transport is likely in place from early gestation, with first-trimester chorionic villous biopsies (10–13 weeks' gestation) expressing *GATM*, *GAMT* and *SLC6A8* mRNA [13]. Assessment of placental tissue collected at term confirmed that the human placenta expresses AGAT, GAMT and SLC6A8 at both the gene and protein level. The AGAT protein is localized to the stromal and endothelial cells of the fetal capillaries, whereas GAMT is predominantly located on the apical side of multinucleated syncytiotrophoblast cells [65]. The creatine transporter is also located on these highly specialized cells at the maternal–fetal interface, which are the site of glucose, amino acid, and fatty acid transfer from maternal blood into the fetal circulation [62,65,66]. This continuous epithelial barrier in the outer layer of all placental villi has a high metabolic demand, particularly during late gestation [67]. Whether the location of the creatine transporter and enzyme responsible for the methylation of GAA to creatine facilitates maternal/fetal transfer of creatine or supports the metabolic activities intrinsic to syncytiotrophoblast function, or both, is yet to be fully determined.

Inadequate placental perfusion and subsequent metabolic compromise are hallmarks of several common obstetric complications, including preeclampsia (PE), gestational diabetes, and fetal growth restriction (FGR). Thus, in addition to the adaptations of increasing energy demands during a healthy pregnancy, the placenta often must respond to homeostatic challenges, such as acute and chronic hypoxic insults, throughout gestation [68]. Investigations of creatine metabolism in metabolically unstable placentae revealed that there might be an increased reliance on the creatine kinase circuit to maintain ATP homeostasis under sub-optimal conditions. Indeed, increased levels of phosphocreatine have been detected in placentae from pregnancies occurring at high altitude, where the oxygen in air is chronically reduced from 21% to ~18% [69]. In a study of FGR placentae, total creatine content was increased by 43%, and creatine transporter (*SLC6A8*) mRNA expression was increased by two-fold in the third trimester compared to gestation-matched healthy controls [13]. These changes occurred despite no differences in creatine synthesizing enzyme (AGAT and GAMT) protein expression, and there were no differences in creatine concentrations in either maternal or venous cord serum at delivery [13]. It should also be noted that expression patterns of the creatine transporter, creatine synthesizing enzymes and creatine kinases did not differ in early gestation between pregnancies that delivered appropriately grown babies in comparison to small for gestational age. As such, the authors postulate that it is the progressive nature of placental insufficiency, and changes in intracellular creatine content or ADP/ATP ratios that likely steer the changes in creatine metabolism observed in the third-trimester placenta [13]. This timing coincides with the peak metabolic rate of the placenta, when placental insufficiency may become most detrimental to the developing fetus and places the fetus at further disadvantage in terms of its ability to tolerate the physiological stress of labor [70].

Similarly, in PE placentae, total creatine content has been reported to increase by 38%, with *GATM*, *GAMT*, *SLC6A8* and *CKBB* mRNA expression also significantly increased compared to gestational age-matched controls, although, again, these differences were not observed at a protein level [14]. There is evidence that, in the case of PE, this additional creatine may be transported to the compromised fetus, with a recent study by Jääskeläinen et al. (2018) reporting an increase in creatine concentration in venous cord plasma from PE pregnancies [71]. Some interesting correlations were also observed in healthy control placentae throughout these collective retrospective studies, with placental *GATM* mRNA expression and GAA tissue content decreasing with advancing gestational age and birth weight. These adaptations associated with placental senescence were not observed in FGR or PE placentae, indicating an ongoing reliance on the creatine kinase circuit for placental bioenergetics in compromised pregnancies [13,14]. It is interesting to note that *GATM*, the gene that expresses AGAT, has been identified as a maternally imprinted gene, and thus is exclusively expressed in placental tissue from the maternal allele [72]. Imprinted genes are often associated with regulation of energy exchange between the mother and developing embryo, and are thought to restrain the over-allocation of maternal resources to the fetus [73]. It is postulated that high levels of AGAT expression in the placenta, and thus potential for creatine synthesis through the production of GAA, may protect the mother from dynamic shifts in the energy required to sustain embryonic or fetal development [73]. Indeed, a study by McMinn et al. (2006) investigating changes to the expression of maternally imprinted genes in the term human placenta identified a down-regulation in *GATM* mRNA in samples from FGR pregnancies (see Table 2) [74].

Table 2. Summary of Creatine Metabolism in the Human Placenta.

Study	Condition	Gestation	Creatine and Phosphocreatine Content	Creatine Kinases	Creatine Synthesis and Transport
Thomure et al. [64]	Healthy	First, second and third trimester		*uMt-CK* and *CKBB* gene expression detected. Expression was low in the first and second trimester before a peak at term. CKBB protein expression consistent throughout gestation. uMt-CK expression rose through to mid-gestation before declining just before term.	
Ellery et al. [13]	Healthy	First trimester (10–13 weeks' gestation)			*GATM, GAMT* and *SLC6A8* detected.
Ellery et al. [13,65]	Healthy	Third trimester			AGAT, GAMT and SLC6A8 gene and protein detected. *GATM* expression and GAA tissue content decreased with advancing gestational age and birth weight.
Tissot et al. [69]	High altitude	Term	Increased phosphocreatine levels detected.		
Ellery et al. [13]	FGR	Third trimester	43% higher total creatine content compared to gestation-matched controls.		2-fold increase in *SLC6A8* expression.
Ellery et al. [14]	PE	Third trimester	38% higher total creatine content compared to gestation-matched controls.	Increased *CKBB* mRNA expression.	Increased *GATM, GAMT, SLC6A8* mRNA expression.
Jääskeläinen et al. [71]	PE	Term	Increase in creatine concentration in venous cord plasma at delivery.		
McMinn et al. [74]	FGR	Term			Down-regulation in *GATM*.

Abbreviations—fetal growth restriction (FGR), preeclampsia (PE), cytosolic brain-type creatine kinase (CKBB), ubiquitous mitochondrial creatine (uMt-CK), the creatine transporter (SLC6A8), guanidinoacetate (GAA), arginine:glycine amidinotransferase (AGAT translated from the *GATM* gene), guanidinoacetate methyltransferase (GAMT).

With consideration of the metabolic demands of the human placenta, it is not hard to rationalize the use of the creatine kinase circuit to support placental bioenergetics. Together, studies to date indicate that the hypoxic placenta may have an increased reliance on creatine and the creatine kinase circuit to buffer spatial fluctuations in ATP homeostasis. This increased capacity to re-phosphorylate ADP via an oxygen-independent pathway may help maintain the high metabolic rate of the third-trimester placenta, and the myriad of ATP-dependent processes required to sustain pregnancy, including synthesis of structural proteins, enzymes, and a wide range of products with important endocrine, hemostatic and immunological functions [75,76]. However, what remains to be understood is the trigger for these adaptations, as well as the consequences they have on both the placenta and the fetus. Defining the mechanisms associated with placental hypoxia in terms of metabolic changes may explain what drives adaptations in placental creatine metabolism. Importantly, placental mitochondrial changes are characteristic of common pregnancy stress conditions such as maternal diabetes, obesity, PE, and hypoxia [77–80]. Hence, the metabolic adaptation to increase intracellular placental creatine stores in the context of mitochondrial function warrants further investigation.

4. Maternal Creatine Metabolism during Pregnancy

As pregnancy progresses and the fetus grows, maternal metabolism shifts with several adaptations required to meet the changing metabolic demand of advancing gestation [64]. As detailed in this review, the role of creatine metabolism in contributing to cellular bioenergetics during reproduction and pregnancy is becoming evident. With provisions of creatine being required to accommodate rapidly expanding tissue beds within the uterus, placenta and fetus, one must also consider the source of the additional creatine required throughout gestation to optimally meet these demands. Data from both human and animal studies have established that pregnancy modifies maternal creatine homeostasis, that maternal characteristics are associated with circulating creatine concentrations during gestation, and that alterations to maternal creatine homeostasis throughout gestation may be linked to the growth and well-being of the offspring.

Adaptations to maternal creatine metabolism throughout pregnancy were first characterized extensively in a study of pregnant spiny mice. This rodent undergoes in utero organ maturation on a similar trajectory to humans [81]. This study mapped changes in maternal circulating creatine, synthesis, excretion, transport, and storage across gestation [12]. They found that maternal plasma creatine concentrations fell progressively from mid to late gestation, with levels in pregnant spiny mice being significantly lower than in non-pregnant controls at all time points analyzed. Urinary excretion of creatine also decreased from mid to late gestation and was significantly lower compared to non-pregnant female spiny mice. Pregnancy was associated with increased *GATM* mRNA and AGAT protein expression in the maternal kidney, considered a primary site of GAA production. This may indicate an up-regulation of creatine synthesis with pregnancy, as renal AGAT activity is also considered a rate-limiting step of creatine production [12]. Increased creatine transporter (*SLC6A8*) mRNA expression was observed in maternal tissues with high-energy demand, such as the heart and skeletal muscle at term. In contrast, creatine transporter expression was decreased in the maternal brain and liver. These changes suggest this may be an adaptive mechanism that ensures creatine is available to maternal tissues where energy expenditure can be high. Indeed, the creatine content of the maternal heart and kidney was increased at term, compared to levels observed in non-pregnant tissues [12]. Overall, these results indicate changes to maternal creatine homeostasis in the spiny mouse may be a fundamental physiological adaptation to pregnancy.

Changes in maternal plasma and urinary creatine have also been reported in human pregnancy. The normative range of plasma creatine is reported to be 35.6 µM ± 15.15 during pregnancy, which is ~35% lower than the normal range of 54.8 µM ± 21.0 for non-pregnant females [82,83]. Conversely, urinary creatine excretion rises from 46 (9–135) µmol/mmol creatinine in a non-pregnant state [84] up to 146.7 (58–273) µmol/mmol creatinine during pregnancy, a 3-fold increase (Ellery et al., unpublished data). Unlike the spiny mouse, multiple human studies have reported that maternal plasma creatine levels, while low, remain stable throughout gestation but consistently between species, the rate of urinary creatine excretion declines with advancing gestation [85,86]. Collectively, these data suggest that there is an increased requirement for maternal creatine due to the rapid growth and increased metabolic requirements of the fetus in the third trimester of pregnancy.

In a retrospective study, associations were identified between key maternal characteristics and circulating and excreted creatine levels during pregnancy [85]. Maternal smoking was positively associated with plasma creatine levels, whereas parity (having previously given birth) had a negative association. The study also found that maternal body mass index (BMI) and asthma were positively associated with urinary creatine, whereas maternal urinary creatine excretion across pregnancy was positively correlated with birth weight centile and birth length, suggesting a relationship between maternal creatine status and fetal growth [85]. This notion is supported by a study dating back to 1913, where increases in newborn body weight were shown to be roughly proportional to the creatine excreted in the urine by the mother at term [10]. Thus, the regulation of creatine acquisition, its loss via urinary excretion, and its delivery to the fetus across the placenta may be important determinants of fetal growth and development. Whether alterations in maternal circulating creatine concentrations are indicative of other poor perinatal outcomes is still to be ascertained. In a retrospective case-controlled study, an 18% reduction in maternal serum creatine concentration during the third trimester of pregnancy was associated with a greater incidence of poor perinatal outcomes, which was defined by a composite measure of small for gestational age, preterm birth and admission to neonatal intensive care [87]. A recent case study of a pregnant female with an AGAT deficiency (inability to synthesize creatine) reported that the patient required adjustment of her dietary creatine treatment during pregnancy when sonographic monitoring at mid gestation indicated a reduction in fetal growth associated with a decline in the patient's plasma and urinary excreted creatine concentrations. After increasing the patient's dietary creatine supplement from 2 g to 3 g daily, she delivered a healthy infant at 35 weeks' gestation on the 25th centile for birth weight, with normal brain creatine levels. The infant achieved typical developmental milestones at one year of age, when the case study ceased [44].

Thus, while clear links are beginning to emerge between maternal creatine homeostasis throughout pregnancy and infant outcomes, further studies are required to better understand adaptations to maternal creatine homeostasis throughout gestation and their association with pregnancy outcomes. In particular, due to the importance of dietary creatine to maintain circulating creatine levels and augmenting endogenous synthetic capability, understanding habitual dietary preferences and basal nutritional status of women is essential to optimize pregnancy wellbeing. A prospective longitudinal cohort study (the Creatine and Pregnancy Outcomes (CPO) study) has been undertaken to address these knowledge gaps [88].

5. Fetal Creatine Metabolism and Use of Supplementary Creatine to Prevent Perinatal Brain Injury

The creatine kinase circuit is thought to play an essential role in energy homeostasis during embryonic development, particularly the development of the central nervous system (CNS) [89,90]. This idea was first explored by Braissant et al. (2005), in a study which described embryonic gene and protein expression of AGAT, GAMT and the creatine transporter (SLC6A8) in numerous tissue types in the rat from early in gestation [91]. From a neurodevelopmental perspective, creatine metabolism has been linked to the growth of dendrites and axons, and migration of neural growth cones [92,93].

The importance of creatine metabolism for cellular bioenergetics in the fetal brain is further supported by a recent in utero fetal magnetic resonance spectroscopy (^{1}H-MRS) study that clearly illustrated cerebral creatine accretion with advancing gestation [94]. This study by Evangelou et al. (2015) examined 204 spectra obtained from fetuses of 129 pregnant women and found fetal cerebral creatine concentrations more than doubled between 18 and 40 weeks' gestation [94]. Indeed, the brain possesses both uMt-CK and CKBB creatine kinases, and a significant amount of cerebral ATP is generated via the creatine kinase circuit [95]. CK isoforms have been found within specific cells of the hippocampus (granular and pyramidal cells), the cerebellum and choroid plexus [96]. This suggests a need for these cell types to use creatine to maintain ATP turnover. Additionally, transfer of creatine between cells is made possible by the expression of the creatine transporter on neurons, oligodendrocytes and astrocytes [92,97]. The entry of circulating creatine into the adult brain appears to be limited to some extent by the blood–brain barrier [98]; however, the creatine synthesizing enzymes, AGAT and GAMT are expressed in varying levels by developing and mature neurons, astrocytes and oligodendrocytes [99], as well as low levels by microglia [100], meaning the brain is able to synthesize creatine endogenously. Indeed, it has been argued that the brain is not reliant on exogenous or systemic endogenous sources of creatine at all, and that cells within the brain can synthesize adequate amounts of creatine to maintain function [101], but whether this holds true for the immature and rapidly developing perinatal brain is yet to be ascertained.

Similar to the axons of neurons in the developing brain, oligodendrocytes, which supply myelin sheaths to axons in the CNS, and oligodendrocytes progenitor cells have high energy demands and are highly susceptible to perinatal energy deprivation (e.g., [102]). Indeed, perinatal brain injury can lead to significant neurological impairments as a result of dysmyelination or delayed myelination linked to the loss of oligodendrocyte progenitor cells that fail to mature into myelinating oligodendrocytes [103]. Although AGAT is expressed by neurons, astrocytes, and oligodendrocytes, GAMT is primarily expressed by oligodendrocytes [92,104], suggesting that oligodendrocytes are a major source of endogenous creatine in the CNS. Indeed, mutations in the creatine synthesizing genes, *GATM* and *GAMT*, demonstrate that impaired myelination [105] and *GAMT*-deficient mice have impaired remyelination and oligodendrocyte apoptosis following a demyelinating insult [104]. These studies highlight the importance of creatine in myelination and remyelination and indicate that oligodendrocytes are likely a major target of supplemental creatine. An initial study completed in pregnant rats found that supplementing the maternal diet with 1% creatine for 10 days prior to delivery improved morphological and electrophysiological development of cornu ammonis (CA1) neurons in the offspring within the first three weeks of life [106].

In addition to creatine metabolism supporting the brains' basal metabolic function, growth and maturation, creatine can maintain ATP turnover, acid-base balance and mitochondrial function. This, together with its antioxidant, vasodilator, and anti-excitotoxic properties, makes it a candidate for the treatment of ischemic–reperfusion brain injuries [107]. In the adult setting, studies have focused on neurodegenerative diseases including Alzheimer's, Parkinson's and Huntington's disease, as well as Amyotrophic lateral sclerosis [108]. Much of this work to date has also focused on fetal and neonatal brain injury [109–112]. For example, in brain slices prepared from neonatal mice and fetal guinea pigs, the ex vivo addition of creatine preserved ATP turnover and reduced neuronal injury [113,114]. Early in vivo studies in rats also demonstrated that a low phosphocreatine/creatine ratio correlated with a higher susceptibility of the immature rat to experience hypoxic seizures early in development [115]. This same study reported that creatine supplementation improved survival and prevented seizure activity [115]. Similar findings were made in a study of rabbit pups [116]. Finally, a study in immature rat pups found that subcutaneous injections of creatine administered in the neonatal period prevented brain edema associated with severe hypoxia–ischemia [117].

Following on from these studies, the neuroprotective role of creatine was investigated in the context of intrapartum (birth) asphyxia using the precocial spiny mouse. In this model, pregnant spiny mice were fed a creatine supplemented diet (5% *w/v*) from mid-gestation (day 20 of gestation), which resulted in creatine loading of the fetal brain, heart, liver and kidney at term (day 39 of gestation) [115,118]. In the offspring of control fed dams, intrauterine hypoxia of 7.5–8 min at term was associated with an increase in pro-apoptotic protein BAX, cytoplasmic cytochrome *c*, and caspase-3 in the fetal brain and high perinatal mortality rates [115]. Conversely, Ireland et al. (2008 and 2011) reported that maternal creatine supplementation mitigated neurological injury in the fetal brain and was associated with increased pup survival and improved postnatal growth [115,118]. In addition to specifically protecting the brain, these studies of intrapartum asphyxia in the spiny mouse model also reported that maternal dietary creatine supplementation during gestation had beneficial effects for other organs involved in the multi-organ pathology of intrapartum asphyxia. For example, maternal creatine supplementation prevented structural and functional damage to the diaphragm [119,120] and skeletal muscle [121], and prevented acute kidney injury in the neonatal period [122], as well as the risk of developing chronic kidney disease later in life (Table 3) [123]. These findings are significant as it has been reported that for 70% of cases of newborn brain injury following intrapartum hypoxia, the primary and direct cause of hypoxic–ischemic encephalopathy (HIE) with neurological injury may not be cerebral oxygen deprivation per se, but rather, injury developing secondary to multi-organ injury [124]. The capacity of maternal dietary creatine supplementation to protect against multiple organ injury following intrapartum hypoxia makes it a unique candidate treatment within the perinatal asphyxia landscape and one worth exploring further. Limitations to the in vivo and rodent studies completed to date include the restricted ability to assess the pathophysiological response to in utero hypoxia with creatine supplementation and the effects of hypoxia and creatine loading in discrete brain regions. Pre-clinical studies in translational large animal models are currently underway to overcome these limitations and fully ascertain the capacity of maternal dietary supplementation during pregnancy as a preventative strategy for hypoxia-induced perinatal brain injury [125].

Table 3. Creatine Treatment in Animal Models of Perinatal Brain Injury.

Species	Developmental Timing	Treatment	Main Outcomes
Guinea pigs and Rats [113]	Fetal guinea pigs (0.9 gestation) Or neonatal rats (P7)	2 h creatine treatment to hippocampal slices in vitro or injection of 3g/kg creatine before and after hypoxic-ischemic insult.	Creatine improved recovery of brain protein synthesis, reduced infarction and neuronal cell injury.
Mice [114]	Neonatal (P0–5) and juvenile (P6–13)	Maternal dietary creatine supplementation (2 g/kg/day) or incubation of brain slices (200 μM) creatine.	Creatine preserved ATP turnover and reduced neuronal injury.
Rat [115]	Neonatal or juvenile (P10–15)	Subcutaneous creatine (3 mg/g of body weight) for 3 days before hypoxic insult.	Low phosphocreatine/creatine ratio led to higher susceptibility of seizures. Creatine improved survival and prevented seizure activity.
Rabbit [116]	5 to 30 day-old pups	Subcutaneous creatine (3 mg/g of body weight) for 3 days before hypoxic insult.	Creatine increased brain PCr/NTP ratio and prevented hypoxic seizures.
Rat [117]	Neonatal (P6)	Subcutaneous creatine (3 g/kg body weight/day) for 3 days before hypoxic insult.	Creatine prevented brain oedema associated with severe hypoxia-ischemia.
Spiny Mouse [118]	Fetal (term) and juvenile (P15)	Maternal dietary creatine supplementation (5% *w/w* from mid-gestation).	Creatine increased pup survival and improved postnatal growth.
Spiny Mouse [118]	Neonatal (P1)	Maternal dietary creatine supplementation (5% *w/w* from mid-gestation).	Creatine reduced perinatal mortality and pro-apoptotic protein BAX, cytoplasmic cytochrome *c*, and caspase-3 in the fetal brain.
Spiny Mouse [119–122]	Neonatal (P1) or juvenile (P35)	Maternal dietary creatine supplementation (5% *w/w* from mid-gestation).	Creatine prevented structural and functional damage to the diaphragm [119,120], skeletal muscle [121], and kidney [122].
Spiny Mouse [123]	Adult (P90)	Maternal dietary creatine supplementation (5% *w/w* from mid-gestation).	Creatine decreased the risk of male offspring developing chronic kidney disease.

Abbreviations—postnatal age (P), phosphocreatine/nucleoside triphosphate ratio (PCr/NTP), weight/weight (*w/w*).

6. Creatine Metabolism in the Neonate, with a Focus on the Potential Consequences of Preterm Birth

Creatine and an effective creatine kinase circuit is critical for brain metabolism [126]. Postnatal studies have identified that an increase in cerebral creatine content occurs during the first three months of life amongst infants born at term [127,128]. The importance of this cerebral creatine accretion for optimal brain development is made evident by those infants diagnosed with inherited creatine deficiency syndromes (CDS). During fetal life, and in the immediate newborn period, these infants are symptom-free as creatine requirements have been met through maternal/placental supply. Following birth, they become progressively creatine deficient [129] with progressive manifestation of neurological symptoms, including impaired psychomotor function and seizures [130]. Importantly, dietary creatine supplementation is proving to be a relatively simple solution to certain forms of CDS, specifically AGAT- and GAMT-deficiency disorders, enabling restoration of cerebral creatine, thereby allowing young children with these conditions to thrive. However, much still needs to be done to improve awareness of CDS in the wider community and promote early screening of at-risk infants [131].

As creatine and phosphocreatine are spontaneously broken down into creatinine at a rate of 1.7%/day, we all have a requirement to replenish our creatine and phosphocreatine stores either through our diet or endogenous synthesis [5]. In adults, the acquisition of creatine via the diet or de novo synthesis is purported to be 50:50 [5]. However, due to low levels of creatine in human breast milk and commercial formulas, a term baby most likely synthesizes 64–93% of their daily creatine requirements [132]. When the human fetus/newborn develops the capacity to synthesize creatine is unknown, but it requires sufficient renal, pancreas and hepatic maturity to express the enzymes necessary for creatine synthesis [132]. Studies conducted in the precocial spiny mouse suggest that preterm infants are unlikely able to synthesize creatine before an age equivalent to ~35 weeks' gestation, due to the developmental immaturity of kidney and liver limiting production of the AGAT and GAMT enzymes [133]. These observations around dietary creatine intake and endogenous synthesis raise questions about whether creatine insufficiency occurs in infants born preterm (before 37 completed weeks of gestation). This is important as the global incidence of preterm birth is reported to be ~11% [134] with rates stable or increasing in many parts of the world. Children and adults born preterm, even those discharged from the neonatal intensive care unit free of any gross cerebral injury have an increased risk of developing neurological disorders that mirror many of those experienced by CDS patients, including impaired executive function, developmental delay, psychiatric and behavioral sequelae [134]. Crucially too, the more preterm the birth, the greater the risk of later neurodevelopmental problems [135]. Finally, maternal mental health is also a powerful modulator of preterm birth outcomes. Psychological distress in otherwise healthy women is known to increase rates of preterm birth, with emergent data also suggesting that the fetuses of these women have altered brain development and reduced cerebral creatine concentrations [136]. Infants born preterm to mothers with mental health difficulties may therefore be at heightened risk for later neurodevelopmental problems.

There have been several small observational studies that have identified perturbations in creatine homeostasis in preterm infants [127,137]. A study by Koob et al. (2016) reported reduced creatine concentrations in the centrum semiovale of preterm infants (born 29.1 ± 2 weeks) when compared to term controls at term-corrected age [137]. When going on to consider systemic creatine levels, a study completed by Lage et al. (2013) found that by the time of hospital discharge, preterm infants had higher urinary GAA and reduced urinary creatine excretion. This was particularly apparent in their very preterm group (28–29 weeks) [138]. The role of systemic creatine homeostasis on cerebral creatine levels has not been adequately evaluated in preterm infants, but this should be explored further, as an inability to methylate GAA to produce creatine could be detrimental to the preterm infant, not merely because of creatine depletion, but also because increased levels of GAA can be neurotoxic [129]. With improvements in perinatal care and lowering gestational age threshold of human viability, the population of children and adults born <28 weeks' gestation will increase. Key periods of brain development, usually occurring during the third trimester of pregnancy, need to be supported in an ex utero environment where there is no longer a custom pipeline of placental/maternally derived nutrients, including creatine. At these gestational ages, nutrition is largely supported by intravenous (parenteral) nutrition, which is creatine deplete until the gut is functionally mature enough to tolerate milk feeds, which are also creatine deficient. Parenteral nutrition is also associated with hepatotoxicity and neonatal cholestasis is a well-recognized complication; whether this further impairs the preterm infant's ability for endogenous creatine synthesis is unknown.

No studies to date have established whether preterm infants develop cerebral creatine deficiency or whether a reduction in cerebral creatine content is associated with neurodevelopmental outcomes. Further to this, no study has monitored systemic creatine levels (both circulating and excreted) in a single preterm population, nor have they assessed nutritional creatine availability in total parenteral nutrition, preterm infant formulas or preterm maternal or donor breastmilk. These are the aims of the Understanding Creatine for Neurological Health (UNICORN) in babies observational cohort study currently underway [139]. The anticipated findings of this study are that low levels of creatine will be detected in preterm breastmilk and infant formula, placing a large burden of de novo creatine synthesis on the preterm infant. It is hypothesized that these babies will not be able to sustain creatine accretion; thus, ^{1}H-MRS examination will show that preterm infants have lower cerebral creatine concentrations compared to those born at term. Whether or not creatine deficiency is associated with neurological deficit will be a secondary outcome of this study. The authors contend that detection and measurement of cerebral creatine perturbations in the preterm infant could provide the basis of early intervention with dietary creatine.

7. Conclusions and Research Road Forward

It is clear that creatine is involved in energy metabolism throughout female reproduction. Specifically, this review highlights the potential importance of creatine metabolism for successful fertilization. While many unanswered questions remain, there is clear evidence that both the endometrium and myometrium can use the creatine kinase circuit for energy homeostasis with indications that adaptations to creatine metabolism occur across the uterine reproductive cycle, and during pregnancy and parturition. What remains to be established is the functional importance of creatine metabolism within the various layers of the human uterus during the menstrual cycle and in the pregnant state. Overall, whether creatine deficiencies can be linked to sub-optimal fertility in females, as well as the capacity to use dietary creatine supplements to improve reproductive outcomes, warrants further investigation. This should be considered for both natural pregnancies and within the realm of artificial reproductive therapies. Indeed, the potential utility of creatine in IVF media should not be overlooked.

A surprising development in recent years has been the degree by which maternal creatine metabolism shifts with pregnancy; in particular, the capacity of the human placenta to synthesize creatine and that these processes are disturbed in pregnancy complications where oxygen, and thus cellular energy, depletion underpin pathology. To advance this research further, there is a need to uncover the mechanisms driving these changes. Specifically, when in gestation energy collapse may be imminent, and which pregnancies may benefit from creatine supplementation to safeguard against subsequent placental dysfunction and fetal compromise. The required increases in maternal creatine concentrations during pregnancy also raise questions about the use of dietary creatine supplements in settings where access to adequate nutrition, particularly animal protein, is limited.

Our understanding of the potential use of dietary creatine supplementation during pregnancy to improve outcomes for the neonate following intrapartum complications is further advanced. The potential use of maternal dietary creatine supplementation during pregnancy as a prophylactic treatment for fetal hypoxia and perinatal brain injury is exciting, as this treatment may prove beneficial in all resource settings globally. Further to this, observational studies underway in preterm infants will soon inform the medical community on whether the simple inclusion of creatine in preterm nutrition may support ex utero brain development and function, ultimately reducing the risk of long-term neurological deficit in these vulnerable babies. Translation of these pre-clinical in vitro, animal and human observational studies are still on the horizon. While it may appear simple to implement the use of an already available nutritional supplement in pregnancy, before initiating clinical trials, we must always consider the consumer perspective. The recently conducted Acceptability of Dietary or Nutritional Supplements in pregnancy (the ADONS Study), explored knowledge of, and acceptance of, introducing creatine as a nutritional supplement in late pregnancy [140]. This study assessed the perspectives of pregnant women, their families and healthcare providers, concluding that creatine would be an acceptable supplement during pregnancy provided they were given evidence-based assurances of efficacy and safety. There is no indication that creatine supplements produced under high-quality manufacturing standards and consumed following manufacturer's directions pose any safety risks or cause adverse side-effects in women of reproductive age or preterm infants [141,142]. However specialized safety and tolerability studies in pregnant women or those trying to conceive are still required. Overall, the available literature supports creatine metabolism being considered an essential component of bioenergetics for successful reproduction, and one may be cautiously optimistic, with further research, about the potential impact of creatine supplementation to improve reproductive and perinatal outcomes.

Author Contributions: S.J.E., A.M.M. and R.J.S. produced the initial draft of this narrative review. M.P. and P.A.D.G. provided specific feedback on Section 2; D.L.d.G. and K.R.P. on Sections 3 and 4; N.T.T., R.G., L.S.S., D.W.W. and M.A.K. on Section 5; M.J.B. on Section 6. All other authors provided written feedback on the final draft on the manuscript. All authors have read and agreed to the published version of the manuscript.

Funding: At the time of writing, S.J.E. (1125539) and R.G. (1090890) were support by Australian National Health and Medical Research Council (NHMRC) Early Career Fellowships. L.S.S. is supported by the Oregon National Primate Research Centre ONPRC (NIH P51 OD011092). The creatine research program being undertaken by the authors of this narrative review is supported by NHMRC project grants to D.W.W., R.J.S. and S.J.E. (1124493) and R.G. (1164954), the Cerebral Palsy Alliance project grants to S.J.E., D.W.W., R.J.S., L.S.S., M.K. and M.J.B. (PG2715 and PG20518), and the Andrea Logan Trust.

Institutional Review Board Statement: Not applicable.

Informed Consent Statement: Not applicable.

Data Availability Statement: Not applicable.

Conflicts of Interest: S.J.E. serves as a member of the Scientific Advisory Broad on creatine in health and medicine (AlzChem LLC.). The company had no oversight of the structure or content presented in this narrative review.

References

1. Wade, G.N.; Schneider, J.E. Metabolic fuels and reproduction in female mammals. *Neurosci. Biobehav. Rev.* **1992**, *16*, 235–272. [CrossRef]
2. Canonaco, F.; Schlattner, U.; Pruett, P.S.; Wallimann, T.; Sauer, U. Functional expression of phosphagen kinase systems confers resistance to transient stresses in Saccharomyces cerevisiae by buffering the ATP pool. *J. Biol. Chem.* **2002**, *277*, 31303–31309. [CrossRef] [PubMed]
3. Wallimann, T.; Wyss, M.; Brdiczka, D.; Nicolay, K.; Eppenberger, H.M. Intracellular compartmentation, structure and function of creatine kinase isoenzymes in tissues with high and fluctuating energy demands: The 'phosphocreatine circuit' for cellular energy homeostasis. *Biochem. J.* **1992**, *281 Pt 1*, 21–40. [CrossRef]
4. Wyss, M.; Kaddurah-Daouk, R. Creatine and creatinine metabolism. *Physiol. Rev.* **2000**, *80*, 1107–1213. [CrossRef] [PubMed]
5. Brosnan, J.; Brosnan, M. Creatine: Endogenous metabolite, dietary, and therapeutic supplement. *Annu. Rev. Nutr.* **2007**, *27*, 241–261. [CrossRef] [PubMed]

6. Boehm, E.; Chan, S.; Monfared, M.; Wallimann, T.; Clarke, K.; Neubauer, S. Creatine transporter activity and content in the rat heart supplemented by and depleted of creatine. *Am. J. Physiol. Endocrinol. Metab.* **2003**, *284*, E399–E406. [CrossRef] [PubMed]

7. Snow, R.J.; Murphy, R.M. Creatine and the creatine transporter: A review. *Mol. Cell. Biochem.* **2001**, *224*, 169–181. [CrossRef]

8. Schlattner, U.; Klaus, A.; Rios, S.R.; Guzun, R.; Kay, L.; Tokarska-Schlattner, M. Cellular compartmentation of energy metabolism: Creatine kinase microcompartments and recruitment of B-type creatine kinase to specific subcellular sites. *Amino Acids* **2016**, *48*, 1751–1774. [CrossRef]

9. Umehara, T.; Kawai, T.; Goto, M.; Richards, J.S.; Shimada, M. Creatine enhances the duration of sperm capacitation: A novel factor for improving in vitro fertilization with small numbers of sperm. *Hum. Reprod.* **2018**, *33*, 1117–1129. [CrossRef]

10. Mellanby, E. The metabolism of lactating women. *Proc. R. Soc. Lond. Ser. B Contain. Pap. A Biol. Character* **1913**, *86*, 88–109.

11. Kao, L.; Tulac, S.; Lobo, S.a.; Imani, B.; Yang, J.; Germeyer, A.; Osteen, K.; Taylor, R.; Lessey, B.; Giudice, L. Global gene profiling in human endometrium during the window of implantation. *Endocrinology* **2002**, *143*, 2119–2138. [CrossRef] [PubMed]

12. Ellery, S.J.; LaRosa, D.A.; Kett, M.M.; Della Gatta, P.A.; Snow, R.J.; Walker, D.W.; Dickinson, H. Maternal creatine homeostasis is altered during gestation in the spiny mouse: Is this a metabolic adaptation to pregnancy? *BMC Pregnancy Childbirth* **2015**, *15*, 92. [CrossRef]

13. Ellery, S.J.; Murthi, P.; Davies-Tuck, M.L.; Gatta, P.D.; May, A.K.; Kowalski, G.M.; Callahan, D.L.; Bruce, C.R.; Alers, N.O.; Miller, S.L. Placental Creatine Metabolism in Cases of Placental Insufficiency and Reduced Fetal Growth. *Mol. Hum. Reprod.* **2019**, *25*, 495–505. [CrossRef] [PubMed]

14. Ellery, S.J.; Murthi, P.; Gatta, P.A.D.; May, A.K.; Davies-Tuck, M.L.; Kowalski, G.M.; Callahan, D.L.; Bruce, C.R.; Wallace, E.M.; Walker, D.W. The Effects of Early-Onset Pre-Eclampsia on Placental Creatine Metabolism in the Third Trimester. *Int. J. Mol. Sci.* **2020**, *21*, 806. [CrossRef] [PubMed]

15. Warzych, E.; Lipinska, P. Energy metabolism of follicular environment during oocyte growth and maturation. *J. Reprod. Dev.* **2020**. [CrossRef] [PubMed]

16. Chi, M.M.-Y.; Manchester, J.K.; Yang, V.C.; Curato, A.D.; Strickler, R.C.; Lowry, O.H. Contrast in levels of metabolic enzymes in human and mouse ova. *Biol. Reprod.* **1988**, *39*, 295–307. [CrossRef]

17. Iyengar, M.R.; Iyengar, C.W.L.; Chen, H.Y.; Brinster, R.L.; Bornslaeger, E.; Schultz, R.M. Expression of creatine kinase isoenzyme during oogenesis and embryogenesis in the mouse. *Dev. Biol.* **1983**, *96*, 263–268. [CrossRef]

18. Forsey, K.E.; Ellis, P.J.; Sargent, C.A.; Sturmey, R.G.; Leese, H.J. Expression and localization of creatine kinase in the preimplantation embryo. *Mol. Reprod. Dev.* **2013**, *80*, 185–192. [CrossRef]

19. Scantland, S.; Tessaro, I.; Macabelli, C.H.; Macaulay, A.D.; Cagnone, G.; Fournier, É.; Luciano, A.M.; Robert, C. The adenosine salvage pathway as an alternative to mitochondrial production of ATP in maturing mammalian oocytes. *Biol. Reprod.* **2014**, *91*, 75. [CrossRef]

20. Fezai, M.; Warsi, J.; Lang, F. Regulation of the Na+, Cl-Coupled Creatine Transporter CreaT (SLC6A8) by the Janus Kinase JAK3. *Neurosignals* **2015**, *23*, 11–19. [CrossRef]

21. Lee, H.; Kim, J.-H.; Chae, Y.-J.; Ogawa, H.; Lee, M.-H.; Gerton, G.L. Creatine synthesis and transport systems in the male rat reproductive tract. *Biol. Reprod.* **1998**, *58*, 1437–1444. [CrossRef] [PubMed]

22. Uhlén, M.; Fagerberg, L.; Hallström, B.M.; Lindskog, C.; Oksvold, P.; Mardinoglu, A.; Sivertsson, Å.; Kampf, C.; Sjöstedt, E.; Asplund, A. Tissue-based map of the human proteome. *Science* **2015**, *347*, 1260419. [CrossRef] [PubMed]

23. Lee, M.-S.; Liu, C.-H.; Lee, T.-H.; Wu, H.-M.; Huang, C.-C.; Huang, L.-S.; Chen, C.-M.; Cheng, E.-H. Association of creatin kinase B and peroxiredoxin 2 expression with age and embryo quality in cumulus cells. *J. Assist. Reprod. Genet.* **2010**, *27*, 629–639. [CrossRef] [PubMed]

24. Uhde, K.; van Tol, H.T.; Stout, T.A.; Roelen, B.A. Metabolomic profiles of bovine cumulus cells and cumulus-oocyte-complex-conditioned medium during maturation in vitro. *Sci. Rep.* **2018**, *8*, 9477. [CrossRef] [PubMed]

25. Fakih, H.; MacLusky, N.; DeCherney, A.; Wallimann, T.; Huszar, G. Enhancement of human sperm motility and velocity in vitro: Effects of calcium and creatine phosphate. *Fertil. Steril.* **1986**, *46*, 938–944. [CrossRef]

26. Karaer, A.; Tuncay, G.; Mumcu, A.; Dogan, B. Metabolomics analysis of follicular fluid in women with ovarian endometriosis undergoing in vitro fertilization. *Syst. Biol. Reprod. Med.* **2019**, *65*, 39–47. [CrossRef]

27. Edwards, R. Follicular fluid. *Reproduction* **1974**, *37*, 189–219. [CrossRef]

28. Huyser, C.; Fourie, F.l.R.; Wolmarans, L. Spectrophotometric absorbance of follicular fluid: A selection criterion. *J. Assist. Reprod. Genet.* **1992**, *9*, 539–544. [CrossRef]

29. Morelli, C.; Iuliano, A.; Schettini, S.C.A.; Petruzzi, D.; Ferri, A.; Colucci, P.; Viggiani, L.; Ostuni, A. Metabolic changes in follicular fluids of patients treated with recombinant versus urinary human chorionic gonadotropin for triggering ovulation in assisted reproductive technologies: A metabolomics pilot study. *Arch. Gynecol. Obstet.* **2020**, *302*, 741–751. [CrossRef]

30. Pocate-Cheriet, K.; Santulli, P.; Kateb, F.; Bourdon, M.; Maignien, C.; Batteux, F.; Chouzenoux, S.; Patrat, C.; Wolf, J.P.; Bertho, G. The follicular fluid metabolome differs according to the endometriosis phenotype. *Reprod. Biomed. Online* **2020**, *41*, 1023–1037. [CrossRef]

31. Gérard, N.; Loiseau, S.; Duchamp, G.; Seguin, F. Analysis of the variations of follicular fluid composition during follicular growth and maturation in the mare using proton nuclear magnetic resonance (1H NMR). *Reprod. Camb.* **2002**, *124*, 241–248. [CrossRef] [PubMed]

32. Lee, H.; Ogawa, H.; Fujioka, M.; Gerton, G.L. Guanidinoacetate methyltransferase in the mouse: Extensive expression in Sertoli cells of testis and in microvilli of caput epididymis. *Biol. Reprod.* **1994**, *50*, 152–162. [CrossRef] [PubMed]
33. Sewell, D.; Harris, R. Effects of creatine supplementation in the Thoroughbred horse. *Equine Vet. J.* **1995**, *27*, 239–242. [CrossRef]
34. González-Fernández, L.; Sánchez-Calabuig, M.J.; Calle-Guisado, V.; García-Marín, L.J.; Bragado, M.J.; Fernández-Hernández, P.; Gutiérrez-Adán, A.; Macías-García, B. Stage-specific metabolomic changes in equine oviductal fluid: New insights into the equine fertilization environment. *Theriogenology* **2020**, *143*, 35–43. [CrossRef]
35. Umehara, T.; Tsujita, N.; Goto, M.; Tonai, S.; Nakanishi, T.; Yamashita, Y.; Shimada, M. Methyl-beta cyclodextrin and creatine work synergistically under hypoxic conditions to improve the fertilization ability of boar ejaculated sperm. *Anim. Sci. J.* **2020**, *91*, e13493. [CrossRef] [PubMed]
36. Payne, R.M.; Friedman, D.L.; Grant, J.W.; Perryman, M.B.; Strauss, A.W. Creatine kinase isoenzymes are highly regulated during pregnancy in rat uterus and placenta. *Am. J. Physiol. Endocrinol. Metab.* **1993**, *265*, E624–E635. [CrossRef] [PubMed]
37. Satyaswaroop, P.; Mortel, R. Creatine kinase activity in human endometrium: Relative distribution in isolated glands and stroma. *Am. J. Obstet. Gynecol.* **1983**, *146*, 159–162. [CrossRef]
38. Philip, M.; Snow, R.J.; Della Gatta, P.A.; Bellofiore, N.; Ellery, S.J. Creatine metabolism in the uterus: Potential implications for reproductive biology. *Amino Acids* **2020**, *52*, 1275–1283. [CrossRef]
39. Subramani, E.; Jothiramajayam, M.; Dutta, M.; Chakravorty, D.; Joshi, M.; Srivastava, S.; Mukherjee, A.; Datta Ray, C.; Chakravarty, B.; Chaudhury, K. NMR-based metabonomics for understanding the influence of dormant female genital tuberculosis on metabolism of the human endometrium. *Hum. Reprod.* **2016**, *31*, 854–865. [CrossRef]
40. Scambia, G.; Kaye, A.; Iacobelli, S. Creatine kinase BB in normal, hyperplastic and neoplastic endometrium. *J. Steroid Biochem.* **1984**, *20*, 797–798. [CrossRef]
41. Borthwick, J.M.; Charnock-Jones, D.S.; Tom, B.D.; Hull, M.L.; Teirney, R.; Phillips, S.C.; Smith, S.K. Determination of the transcript profile of human endometrium. *MHR: Basic Sci. Reprod. Med.* **2003**, *9*, 19–33. [CrossRef] [PubMed]
42. Chen, Q.; Zhang, A.; Yu, F.; Gao, J.; Liu, Y.; Yu, C.; Zhou, H.; Xu, C. Label-free proteomics uncovers energy metabolism and focal adhesion regulations responsive for endometrium receptivity. *J. Proteome Res.* **2015**, *14*, 1831–1842. [CrossRef]
43. Choe, C.-u.; Nabuurs, C.; Stockebrand, M.C.; Neu, A.; Nunes, P.; Morellini, F.; Sauter, K.; Schillemeit, S.; Hermans-Borgmeyer, I.; Marescau, B. L-arginine: Glycine amidinotransferase deficiency protects from metabolic syndrome. *Hum. Mol. Genet.* **2013**, *22*, 110–123. [CrossRef] [PubMed]
44. Alessandrì, M.G.; Strigini, F.; Cioni, G.; Battini, R. Increased creatine demand during pregnancy in Arginine: Glycine Amidino-Transferase deficiency: A case report. *BMC Pregnancy Childbirth* **2020**, *20*, 506. [CrossRef] [PubMed]
45. Zeng, S.; Bick, J.; Ulbrich, S.E.; Bauersachs, S. Cell type-specific analysis of transcriptome changes in the porcine endometrium on Day 12 of pregnancy. *BMC Genom.* **2018**, *19*, 459. [CrossRef] [PubMed]
46. Weisman, Y.; Golander, A.; Binderman, I.; Spirer, Z.; Kaye, A.; Sömjen, D. Stimulation of creatine kinase activity by calcium-regulating hormones in explants of human amnion, decidua, and placenta. *J. Clin. Endocrinol. Metab.* **1986**, *63*, 1052–1056. [CrossRef] [PubMed]
47. Franczak, A.; Wojciechowicz, B.; Kotwica, G. Transcriptomic analysis of the porcine endometrium during early pregnancy and the estrous cycle. *Reprod. Biol.* **2013**, *13*, 229–237. [CrossRef]
48. Walker, J.B.; Gipson, W.T. Occurrence of transamidinase in decidua and its repression by dietary creatine. *Biochim. Biophys. Acta (BBA) Spec. Sect. Enzymol. Subj.* **1963**, *67*, 156–157. [CrossRef]
49. Baharom, S.; De Matteo, R.; Ellery, S.; Della Gatta, P.; Bruce, C.R.; Kowalski, G.M.; Hale, N.; Dickinson, H.; Harding, R.; Walker, D. Does maternal-fetal transfer of creatine occur in pregnant sheep? *Am. J. Physiol. Endocrinol. Metab.* **2017**, *313*, E75–E83. [CrossRef]
50. Emery, A.E.; Pascasio, F.M. The effects of pregnancy on the concentration of creatine kinase in serum, skeletal muscle, and myometrium. *Am. J. Obstet. Gynecol.* **1965**, *91*, 18–22. [CrossRef]
51. Steingrimsdottir, T.; Ericsson, A.; Franck, A.; Waldenström, A.; Ulmsten, U.; Ronquist, G. Human uterine smooth muscle exhibits a very low phosphocreatine/ATP ratio as assessed by in vitro and in vivo measurements. *Eur. J. Clin. Investig.* **1997**, *27*, 743–749. [CrossRef] [PubMed]
52. Daly, M.M.; Seifter, S. Uptake of creatine by cultured cells. *Arch. Biochem. Biophys.* **1980**, *203*, 317–324. [CrossRef]
53. Charpigny, G.; Leroy, M.-J.; Breuiller-Fouché, M.; Tanfin, Z.; Mhaouty-Kodja, S.; Robin, P.; Leiber, D.; Cohen-Tannoudji, J.; Cabrol, D.; Barberis, C. A functional genomic study to identify differential gene expression in the preterm and term human myometrium. *Biol. Reprod.* **2003**, *68*, 2289–2296. [CrossRef] [PubMed]
54. Iyengar, M.; Fluellen, C.; Iyengar, C. Increased creatine kinase in the hormone-stimulated smooth muscle of the bovine uterus. *Biochem. Biophys. Res. Commun.* **1980**, *94*, 948–954. [CrossRef]
55. Noyszewski, E.A.; Raman, J.; Trupin, S.R.; McFarlin, B.L.; Dawson, M.J. Phosphorus 31 nuclear magnetic resonance examination of female reproductive tissues. *Am. J. Obstet. Gynecol.* **1989**, *161*, 282–288. [CrossRef]
56. Clark, J.F.; Khuchua, Z.; Kuznetsov, A.; Saks, V.; Ventura-Clapier, R. Compartmentation of creatine kinase isoenzymes in myometrium of gravid guinea-pig. *J. Physiol.* **1993**, *466*, 553–572.
57. Dawson, M.J.; Wray, S. The effects of pregnancy and parturition on phosphorus metabolites in rat uterus studied by 31P nuclear magnetic resonance. *J. Physiol.* **1985**, *368*, 19–31. [CrossRef]
58. Butte, N.F.; Wong, W.W.; Treuth, M.S.; Ellis, K.J.; O'Brian Smith, E. Energy requirements during pregnancy based on total energy expenditure and energy deposition. *Am. J. Clin. Nutr.* **2004**, *79*, 1078–1087. [CrossRef]

59. Vaughan, O.; Fowden, A. Placental metabolism: Substrate requirements and the response to stress. *Reprod. Domest. Anim.* **2016**, *51*, 25–35. [CrossRef]

60. Maltepe, E.; Fisher, S.J. Placenta: The forgotten organ. *Annu. Rev. Cell Dev. Biol.* **2015**, *31*, 523–552. [CrossRef]

61. Smith, R.; Maiti, K.; Aitken, R. Unexplained antepartum stillbirth: A consequence of placental aging? *Placenta* **2013**, *34*, 310–313. [CrossRef] [PubMed]

62. Herrera, E. Metabolic adaptations in pregnancy and their implications for the availability of substrates to the fetus. *Eur. J. Clin. Nutr.* **2000**, *54*, S47–S51. [CrossRef]

63. Cetin, I.; Alvino, G.; Cardellicchio, M. Long chain fatty acids and dietary fats in fetal nutrition. *J. Physiol.* **2009**, *587*, 3441–3451. [CrossRef] [PubMed]

64. Thomure, M. Regulation of creatine kinase isoenzymes in human placenta during early, mid-, and late gestation. *J. Soc. Gynaecol. Investig.* **1996**, *3*, 322–327. [CrossRef]

65. Ellery, S.J.; Della Gatta, P.A.; Bruce, C.R.; Kowalski, G.M.; Davies-Tuck, M.; Mockler, J.C.; Murthi, P.; Walker, D.W.; Snow, R.J.; Dickinson, H. Creatine biosynthesis and transport by the term human placenta. *Placenta* **2017**, *52*, 86–93. [CrossRef] [PubMed]

66. Knöfler, M.; Haider, S.; Saleh, L.; Pollheimer, J.; Gamage, T.; James, J. Human placenta and trophoblast development: Key molecular mechanisms and model systems. *Cell. Mol. Life Sci.* **2019**, *76*, 3479–3496. [CrossRef] [PubMed]

67. Kingdom, J.C.P.; Kaufmann, P. Oxygen and placental villous development: Origins of fetal hypoxia. *Placenta* **1997**, *18*, 613–621. [CrossRef]

68. Sferruzzi-Perri, A.N.; Camm, E.J. The Programming Power of the Placenta. *Front. Physiol.* **2016**, *7*, 33. [CrossRef]

69. Tissot van Patot, M.C.; Murray, A.J.; Beckey, V.; Cindrova-Davies, T.; Johns, J.; Zwerdlinger, L.; Jauniaux, E.; Burton, G.J.; Serkova, N.J. Human placental metabolic adaptation to chronic hypoxia, high altitude: Hypoxic preconditioning. *Am. J. Physiol. Regul. Integr. Comp. Physiol.* **2010**, *298*, R166–R172. [CrossRef]

70. Krishna, U.; Bhalerao, S. Placental insufficiency and fetal growth restriction. *J. Obstet. Gynecol. India* **2011**, *61*, 505–511. [CrossRef]

71. Jääskeläinen, T.; Kärkkäinen, O.; Jokkala, J.; Litonius, K.; Heinonen, S.; Auriola, S.; Lehtonen, M.; Hanhineva, K.; Laivuori, H.; Kajantie, E.; et al. A Non-Targeted LC-MS Profiling Reveals Elevated Levels of Carnitine Precursors and Trimethylated Compounds in the Cord Plasma of Pre-Eclamptic Infants. *Sci. Rep.* **2018**, *8*, 14616. [CrossRef] [PubMed]

72. Sandell, L.L.; Guan, X.-J.; Ingram, R.; Tilghman, S.M. Gatm, a creatine synthesis enzyme, is imprinted in mouse placenta. *Proc. Natl. Acad. Sci. USA* **2003**, *100*, 4622–4627. [CrossRef] [PubMed]

73. Moore, T.; Haig, D. Genomic imprinting in mammalian development: A parental tug-of-war. *Trends Genet.* **1991**, *7*, 45–49. [CrossRef]

74. McMinn, J.; Wei, M.; Schupf, N.; Cusmai, J.; Johnson, E.; Smith, A.; Weksberg, R.; Thaker, H.; Tycko, B. Unbalanced placental expression of imprinted genes in human intrauterine growth restriction. *Placenta* **2006**, *27*, 540–549. [CrossRef] [PubMed]

75. Chard, T. Placental synthesis. *Clin. Obstet. Gynaecol.* **1986**, *13*, 447.

76. Carter, A. Placental oxygen consumption. Part I: In vivo studies—A review. *Placenta* **2000**, *21*, S31–S37. [CrossRef]

77. Holland, O.; Dekker Nitert, M.; Gallo, L.A.; Vejzovic, M.; Fisher, J.J.; Perkins, A.V. Review: Placental mitochondrial function and structure in gestational disorders. *Placenta* **2017**, *54*, 2–9. [CrossRef]

78. Holland, O.J.; Cuffe, J.S.M.; Dekker Nitert, M.; Callaway, L.; Kwan Cheung, K.A.; Radenkovic, F.; Perkins, A.V. Placental mitochondrial adaptations in preeclampsia associated with progression to term delivery. *Cell Death Dis.* **2018**, *9*, 1150. [CrossRef]

79. Sferruzzi-Perri, A.N.; Higgins, J.S.; Vaughan, O.R.; Murray, A.J.; Fowden, A.L. Placental mitochondria adapt developmentally and in response to hypoxia to support fetal growth. *Proc. Natl. Acad. Sci. USA* **2019**, *116*, 1621–1626. [CrossRef]

80. Myatt, L.; Maloyan, A. Obesity and Placental Function. *Semin. Reprod. Med.* **2016**, *34*, 42–49. [CrossRef]

81. Dickinson, H.; Walker, D. Managing a colony of spiny mice (*Acomys cahirinus*) for perinatal research. *Aust. N. Z. Counc. Care Anim. Res. Train. (ANZCCART) News* **2007**, *20*, 4–11.

82. Bahado-Singh, R.O.; Akolekar, R.; Chelliah, A.; Mandal, R.; Dong, E.; Kruger, M.; Wishart, D.S.; Nicolaides, K. Metabolomic analysis for first-trimester trisomy 18 detection. *Am. J. Obstet. Gynecol.* **2013**, *209*, 65.e1–65.e9. [CrossRef] [PubMed]

83. Marescau, B.; De Deyn, P.P.; Holvoet, J.; Possemiers, I.; Nagels, G.; Saxena, V.; Mahler, C. Guanidino compounds in serum and urine of cirrhotic patients. *Metabolism* **1995**, *44*, 584–588. [CrossRef]

84. Class, S.; Class, S. Human Metabolome Database. *Enzyme* **2006**, *1*, 14.

85. Dickinson, H.; Davies-Tuck, M.; Ellery, S.; Grieger, J.; Wallace, E.; Snow, R.; Walker, D.; Clifton, V. Maternal creatine in pregnancy: A retrospective cohort study. *BJOG Int. J. Obstet. Gynaecol.* **2016**, *123*, 1830–1838. [CrossRef]

86. Pinto, J.; Barros, A.n.S.; Domingues, M.R.r.M.; Goodfellow, B.J.; Galhano, E.l.; Pita, C.; Almeida, M.d.C.; Carreira, I.M.; Gil, A.M. Following healthy pregnancy by NMR metabolomics of plasma and correlation to urine. *J. Proteome Res.* **2015**, *14*, 1263–1274. [CrossRef]

87. Heazell, A.E.; Bernatavicius, G.; Warrander, L.; Brown, M.C.; Dunn, W.B. A metabolomic approach identifies differences in maternal serum in third trimester pregnancies that end in poor perinatal outcome. *Reprod. Sci.* **2012**, *19*, 863–875. [CrossRef]

88. De Guingand, D.L.; Ellery, S.J.; Davies-Tuck, M.L.; Dickinson, H. Creatine and pregnancy outcomes, a prospective cohort study in low-risk pregnant women: Study protocol. *BMJ Open* **2019**, *9*, e026756. [CrossRef]

89. Miller, T.; Hanson, R.; Yancey, P. Developmental changes in organic osmolytes in prenatal and postnatal rat tissues. *Comp. Biochem. Physiol. Part A: Mol. Integr. Physiol.* **2000**, *125*, 45–56. [CrossRef]

90. Kreis, R.; Hofmann, L.; Kuhlmann, B.; Boesch, C.; Bossi, E.; Hüppi, P.S. Brain metabolite composition during early human brain development as measured by quantitative in vivo 1H magnetic resonance spectroscopy. *Magn. Reson. Med.* **2002**, *48*, 949–958. [CrossRef]

91. Braissant, O.; Henry, H.; Villard, A.-M.; Speer, O.; Wallimann, T.; Bachmann, C. Creatine synthesis and transport during rat embryogenesis: Spatiotemporal expression of AGAT, GAMT and CT1. *BMC Dev. Biol.* **2005**, *5*, 9. [CrossRef] [PubMed]

92. Braissant, O.; Henry, H.; Loup, M.; Eilers, B.; Bachmann, C. Endogenous synthesis and transport of creatine in the rat brain: An in situ hybridization study. *Mol. Brain Res.* **2001**, *86*, 193–201. [CrossRef]

93. Braissant, O.; Bachmann, C.; Henry, H. Expression and function of AGAT, GAMT and CT1 in the mammalian brain. In *Creatine and Creatine Kinase in Health and Disease*; Springer: Dordrecht, The Netherlands, 2007; pp. 67–81.

94. Evangelou, I.E.; Du Plessis, A.J.; Vezina, G.; Noeske, R.; Limperopoulos, C. Elucidating metabolic maturation in the healthy fetal brain using 1H-MR spectroscopy. *Am. J. Neuroradiol.* **2015**, *37*, 360–366. [CrossRef] [PubMed]

95. Jost, C.R.; Van Der Zee, C.E.; In 't Zandt, H.J.; Oerlemans, F.; Verheij, M.; Streijger, F.; Fransen, J.; Heerschap, A.; Cools, A.R.; Wieringa, B. Creatine kinase B-driven energy transfer in the brain is important for habituation and spatial learning behaviour, mossy fibre field size and determination of seizure susceptibility. *Eur. J. Neurosci.* **2002**, *15*, 1692–1706. [CrossRef] [PubMed]

96. Hemmer, W.; Zanolla, E.; Furter-Graves, E.M.; Eppenberger, H.M.; Wallimann, T. Creatine Kinase Isoenzymes in Chicken Cerebellum: Specific Localization of Brain-type Creatine Kinase in Bergmann Glial Cells and Muscle-type Creatine Kinase in Purkinje Neurons. *Eur. J. Neurosci.* **1994**, *6*, 538–549. [CrossRef] [PubMed]

97. Carducci, C.; Carducci, C.; Santagata, S.; Adriano, E.; Artiola, C.; Thellung, S.; Gatta, E.; Robello, M.; Florio, T.; Antonozzi, I. In vitro study of uptake and synthesis of creatine and its precursors by cerebellar granule cells and astrocytes suggests some hypotheses on the physiopathology of the inherited disorders of creatine metabolism. *BMC Neurosci.* **2012**, *13*, 41. [CrossRef] [PubMed]

98. Lyoo, I.K.; Kong, S.W.; Sung, S.M.; Hirashima, F.; Parow, A.; Hennen, J.; Cohen, B.M.; Renshaw, P.F. Multinuclear magnetic resonance spectroscopy of high-energy phosphate metabolites in human brain following oral supplementation of creatine-monohydrate. *Psychiatry Res.* **2003**, *123*, 87–100. [CrossRef]

99. Béard, E.; Braissant, O. Synthesis and transport of creatine in the CNS: Importance for cerebral functions. *J. Neurochem.* **2010**, *115*, 297–313. [CrossRef]

100. Tachikawa, M.; Fukaya, M.; Terasaki, T.; Ohtsuki, S.; Watanabe, M. Distinct cellular expressions of creatine synthetic enzyme GAMT and creatine kinases uCK-Mi and CK-B suggest a novel neuron–glial relationship for brain energy homeostasis. *Eur. J. Neurosci.* **2004**, *20*, 144–160. [CrossRef]

101. Dolan, E.; Gualano, B.; Rawson, E.S. Beyond muscle: The effects of creatine supplementation on brain creatine, cognitive processing, and traumatic brain injury. *Eur. J. Sport Sci.* **2019**, *19*, 1–14. [CrossRef]

102. Pantoni, L.; Garcia, J.H.; Gutierrez, J.A. Cerebral white matter is highly vulnerable to ischemia. *Stroke* **1996**, *27*, 1641–1646, discussion 1647. [CrossRef] [PubMed]

103. Back, S.A.; Rosenberg, P.A. Pathophysiology of glia in perinatal white matter injury. *Glia* **2014**, *62*, 1790–1815. [CrossRef] [PubMed]

104. Chamberlain, K.A.; Chapey, K.S.; Nanescu, S.E.; Huang, J.K. Creatine enhances mitochondrial-mediated oligodendrocyte survival after demyelinating injury. *J. Neurosci.* **2017**, *37*, 1479–1492. [CrossRef] [PubMed]

105. Braissant, O.; Béard, E.; Torrent, C.; Henry, H. Dissociation of AGAT, GAMT and SLC6A8 in CNS: Relevance to creatine deficiency syndromes. *Neurobiol. Dis.* **2010**, *37*, 423–433. [CrossRef]

106. Sartini, S.; Lattanzi, D.; Ambrogini, P.; Di Palma, M.; Galati, C.; Savelli, D.; Polidori, E.; Calcabrini, C.; Rocchi, M.; Sestili, P. Maternal creatine supplementation affects the morpho-functional development of hippocampal neurons in rat offspring. *Neuroscience* **2016**, *312*, 120–129. [CrossRef] [PubMed]

107. Ellery, S.J.; Dickinson, H.; McKenzie, M.; Walker, D.W. Dietary interventions designed to protect the perinatal brain from hypoxic-ischemic encephalopathy–creatine prophylaxis and the need for multi-organ protection. *Neurochem. Int.* **2016**, *95*, 15–23. [CrossRef]

108. Beal, M.F. Neuroprotective effects of creatine. *Amino Acids* **2011**, *40*, 1305–1313. [CrossRef]

109. Genius, J.; Geiger, J.; Bender, A.; Möller, H.-J.; Klopstock, T.; Rujescu, D. Creatine protects against excitoxicity in an in vitro model of neurodegeneration. *PLoS ONE* **2012**, *7*, e30554. [CrossRef]

110. Holtzman, D.; Tsuji, M.; Wallimann, T.; Hemmer, W. Functional maturation of creatine kinase in rat brain. *Dev. Neurosci.* **1993**, *15*, 261–270. [CrossRef]

111. Prass, K.; Royl, G.; Lindauer, U.; Freyer, D.; Megow, D.; Dirnagl, U.; Stöckler-Ipsiroglu, G.; Wallimann, T.; Priller, J. Improved reperfusion and neuroprotection by creatine in a mouse model of stroke. *J. Cereb. Blood Flow Metab.* **2006**, *27*, 452–459. [CrossRef]

112. Ireland, Z.; Castillo-Melendez, M.; Dickinson, H.; Snow, R.; Walker, D. A maternal diet supplemented with creatine from mid-pregnancy protects the newborn spiny mouse brain from birth hypoxia. *Neuroscience* **2011**, *194*, 372–379. [CrossRef] [PubMed]

113. Berger, R.; Middelanis, J.; Vaihinger, H.-M.; Mies, G.; Wilken, D.; Jensen, A. Creatine protects the immature brain from hypoxic-ischemic injury. *J. Soc. Gynecol. Investig.* **2004**, *11*, 9–15. [CrossRef] [PubMed]

114. Wilken, B.; Ramirez, J.; Probst, I.; Richter, D.; Hanefeld, F. Creatine protects the central respiratory network of mammals under anoxic conditions. *Pediatr. Res.* **1998**, *43*, 8–14. [CrossRef] [PubMed]

115. Holtzman, D.; Togliatti, A.; Khait, I.; Jensen, F. Creatine Increases Survival and Suppresses Seizures in the Hypoxic Immature Rat. *Pediatr. Res.* **1998**, *44*, 410–414. [CrossRef] [PubMed]

116. Holtzman, D.; Khait, I.; Mulkern, R.; Allred, E.; Rand, T.; Jensen, F.; Kraft, R. In vivo development of brain phosphocreatine in normal and creatine-treated rabbit pups. *J. Neurochem.* **1999**, *73*, 2477–2484. [CrossRef]

117. Adcock, K.H.; Nedelcu, J.; Loenneker, T.; Martin, E.; Wallimann, T.; Wagner, B.P. Neuroprotection of creatine supplementation in neonatal rats with transient cerebral hypoxia-ischemia. *Dev. Neurosci.* **2002**, *24*, 382–388. [CrossRef]

118. Ireland, Z.; Dickinson, H.; Snow, R.; Walker, D.W. Maternal creatine: Does it reach the fetus and improve survival after an acute hypoxic episode in the spiny mouse (*Acomys cahirinus*)? *Am. J. Obstet. Gynecol.* **2008**, *198*, 431–436. [CrossRef]

119. Cannata, D.J.; Ireland, Z.; Dickinson, H.; Snow, R.J.; Russell, A.P.; West, J.M.; Walker, D.W. Maternal Creatine Supplementation From Mid-Pregnancy Protects the Diaphragm of the Newborn Spiny Mouse From Intrapartum Hypoxia-Induced Damage. *Pediatr. Res.* **2010**, *68*, 393–398. [CrossRef]

120. LaRosa, D.A.; Ellery, S.J.; Parkington, H.C.; Snow, R.J.; Walker, D.W.; Dickinson, H. Maternal Creatine Supplementation during Pregnancy Prevents Long-Term Changes in Diaphragm Muscle Structure and Function after Birth Asphyxia. *PLoS ONE* **2016**, *11*, e0149840. [CrossRef]

121. LaRosa, D.A.; Ellery, S.J.; Snow, R.J.; Walker, D.W.; Dickinson, H. Maternal creatine supplementation during pregnancy prevents acute and long-term deficits in skeletal muscle after birth asphyxia: A study of structure and function of hind limb muscle in the spiny mouse. *Pediatr. Res.* **2016**, *80*, 852–860. [CrossRef]

122. Ellery, S.J.; Ireland, Z.; Kett, M.M.; Snow, R.; Walker, D.W.; Dickinson, H. Creatine pretreatment prevents birth asphyxia-induced injury of the newborn spiny mouse kidney. *Pediatr. Res.* **2013**, *73*, 201–208. [CrossRef] [PubMed]

123. Ellery, S.J.; LaRosa, D.A.; Cullen-McEwen, L.A.; Brown, R.D.; Snow, R.J.; Walker, D.W.; Kett, M.M.; Dickinson, H. Renal dysfunction in early adulthood following birth asphyxia in male spiny mice, and its amelioration by maternal creatine supplementation during pregnancy. *Pediatr. Res.* **2017**, *81*, 646–653. [CrossRef] [PubMed]

124. Hankins, G.D.; Koen, S.; Gei, A.F.; Lopez, S.M.; Van Hook, J.W.; Anderson, G.D. Neonatal organ system injury in acute birth asphyxia sufficient to result in neonatal encephalopathy. *Obstet. Gynecol.* **2002**, *99*, 688–691. [CrossRef]

125. Steinbach, R.J.; Ellery, S.J.; Snow, R.J.; Walker, D.W.; Kempton, B.; Brigande, J.V.; Renner, L.; Neuringer, M.; Kroenke, C.D.; Schelonka, R.L. A Non-Human Primate Model of Hypoxic Ischemic Encephalopathy to Evaluate Novel Translational Therapeutics. In *Proceedings of Reproductive Sciences*; Springer: Heidelberg, Germany, 2020; p. 151A.

126. Braissant, O.; Henry, H.; Béard, E.; Uldry, J. Creatine deficiency syndromes and the importance of creatine synthesis in the brain. *Amino Acids* **2011**, *40*, 1315–1324. [CrossRef] [PubMed]

127. Blüml, S.; Wisnowski, J.L.; Nelson Jr, M.D.; Paquette, L.; Panigrahy, A. Metabolic maturation of white matter is altered in preterm infants. *PLoS ONE* **2014**, *9*, e85829. [CrossRef]

128. Blüml, S.; Wisnowski, J.L.; Nelson Jr, M.D.; Paquette, L.; Gilles, F.H.; Kinney, H.C.; Panigrahy, A. Metabolic maturation of the human brain from birth through adolescence: Insights from in vivo magnetic resonance spectroscopy. *Cereb. Cortex* **2012**, *23*, 2944–2955. [CrossRef]

129. Almeida, L.S.; Rosenberg, E.H.; Verhoeven, N.M.; Jakobs, C.; Salomons, G.S. Are cerebral creatine deficiency syndromes on the radar screen? *Future Neurol.* **2006**, *1*, 637–649. [CrossRef]

130. Battini, R.; Alessandri, M.; Leuzzi, V.; Moro, F.; Tosetti, M.; Bianchi, M.; Cioni, G. Arginine:glycine amidinotransferase (AGAT) deficiency in a newborn: Early treatment can prevent phenotypic expression of the disease. *J. Paediatr.* **2006**, *148*, 828–830. [CrossRef]

131. Rostami, P.; Hosseinpour, S.; Ashrafi, M.R.; Alizadeh, H.; Garshasbi, M.; Tavasoli, A.R. Primary creatine deficiency syndrome as a potential missed diagnosis in children with psychomotor delay and seizure: Case presentation with two novel variants and literature review. *Acta Neurol. Belg.* **2020**, *120*, 511–516. [CrossRef]

132. Edison, E.E.; Brosnan, M.E.; Aziz, K.; Brosnan, J.T. Creatine and guanidinoacetate content of human milk and infant formulas: Implications for creatine deficiency syndromes and amino acid metabolism. *Br. J. Nutr.* **2013**, *110*, 1075–1078. [CrossRef]

133. Dickinson, H.; Ireland, Z.J.; LaRosa, D.A.; O'Connell, B.A.; Ellery, S.; Snow, R.; Walker, D.W. Maternal dietary creatine supplementation does not alter the capacity for creatine synthesis in the newborn spiny mouse. *Reprod. Sci.* **2013**, *20*, 1096–1102. [CrossRef] [PubMed]

134. Blencowe, H.; Cousens, S.; Chou, D.; Oestergaard, M.; Say, L.; Moller, A.-B.; Kinney, M.; Lawn, J. Born too soon: The global epidemiology of 15 million preterm births. *Reprod. Health* **2013**, *10*, S2. [CrossRef] [PubMed]

135. Berry, M.J.; Foster, T.; Rowe, K.; Robertson, O.; Robson, B.; Pierse, N. Gestational age, health, and educational outcomes in adolescents. *Pediatrics* **2018**, *142*, e20181016. [CrossRef]

136. Wu, Y.; Lu, Y.-C.; Jacobs, M.; Pradhan, S.; Kapse, K.; Zhao, L.; Niforatos-Andescavage, N.; Vezina, G.; du Plessis, A.J.; Limperopoulos, C. Association of Prenatal Maternal Psychological Distress With Fetal Brain Growth, Metabolism, and Cortical Maturation. *JAMA Netw. Open* **2020**, *3*, e1919940. [CrossRef] [PubMed]

137. Koob, M.; Viola, A.; Le Fur, Y.; Viout, P.; Ratiney, H.; Confort-Gouny, S.; Cozzone, P.J.; Girard, N. Creatine, glutamine plus glutamate, and macromolecules are decreased in the central white matter of premature neonates around term. *PLoS ONE* **2016**, *11*, e0160990. [CrossRef]

138. Lage, S.; Andrade, F.; Prieto, J.A.; Asla, I.; Rodríguez, A.; Ruiz, N.; Echeverría, J.; Luz Couce, M.; Sanjurjo, P.; Aldámiz-Echevarría, L. Arginine-guanidinoacetate-creatine pathway in preterm newborns: Creatine biosynthesis in newborns. *J. Pediatr. Endocrinol. Metab.* **2013**, *26*, 53–60. [CrossRef]

139. Berry, M.J.; Schlegel, M.; Kowalski, G.M.; Bruce, C.R.; Callahan, D.L.; Davies-Tuck, M.L.; Dickinson, H.; Goodson, A.; Slocombe, A.; Snow, R.J. UNICORN Babies: Understanding Circulating and Cerebral Creatine Levels of the Preterm Infant. An Observational Study Protocol. *Front. Physiol.* **2019**, *10*, 142. [CrossRef]

140. De Guingand, D.L.; Palmer, K.R.; Bilardi, J.E.; Ellery, S.J. Acceptability of dietary or nutritional supplementation in pregnancy (ADONS)–Exploring the consumer's perspective on introducing creatine monohydrate as a pregnancy supplement. *Midwifery* **2020**, *82*, 102599. [CrossRef]

141. De Guingand, D.L.; Palmer, K.R.; Snow, R.J.; Davies-Tuck, M.L.; Ellery, S.J. Risk of adverse outcomes in females taking oral creatine monohydrate: A systematic review and meta-analysis. *Nutrients* **2020**, *12*, 1780. [CrossRef]

142. Bohnhorst, B.; Geuting, T.; Peter, C.S.; Dördelmann, M.; Wilken, B.; Poets, C.F. Randomized, controlled trial of oral creatine supplementation (not effective) for apnea of prematurity. *Pediatrics* **2004**, *113*, e303–e307. [CrossRef]

nutrients

Review

Creatine Supplementation in Children and Adolescents

Andrew R. Jagim [1,2,*] and Chad M. Kerksick [1,3]

1   Sports Medicine, Mayo Clinic Health System, La Crosse, WI 54601, USA; ckerksick@lindenwood.edu
2   Exercise & Sport Science Department, University of Wisconsin-La Crosse, La Crosse, WI 54601, USA
3   Exercise & Performance Nutrition Laboratory, Lindenwood University, St. Charles, MO 63301, USA
*   Correspondence: jagim.andrew@mayo.edu; Tel.: +1-608-392-5280

Abstract: Creatine is a popular ergogenic aid among athletic populations with consistent evidence indicating that creatine supplementation also continues to be commonly used among adolescent populations. In addition, the evidence base supporting the therapeutic benefits of creatine supplementation for a plethora of clinical applications in both adults and children continues to grow. Among pediatric populations, a strong rationale exists for creatine to afford therapeutic benefits pertaining to multiple neuromuscular and metabolic disorders, with preliminary evidence for other subsets of clinical populations as well. Despite the strong evidence supporting the efficacy and safety of creatine supplementation among adult populations, less is known as to whether similar physiological benefits extend to children and adolescent populations, and in particular those adolescent populations who are regularly participating in high-intensity exercise training. While limited in scope, studies involving creatine supplementation and exercise performance in adolescent athletes generally report improvements in several ergogenic outcomes with limited evidence of ergolytic properties and consistent reports indicating no adverse events associated with supplementation. The purpose of this article is to summarize the rationale, prevalence of use, performance benefits, clinical applications, and safety of creatine use in children and adolescents.

Keywords: ergogenic aid; dietary supplement; youth; athletes

Citation: Jagim, A.R.; Kerksick, C.M. Creatine Supplementation in Children and Adolescents. *Nutrients* 2021, 13, 664. https://doi.org/10.3390/nu13020664

Academic Editor: Elena Barbieri

Received: 15 January 2021
Accepted: 14 February 2021
Published: 18 February 2021

1. Introduction

Creatine (methyl-guanidine-acetic acid) is a naturally occurring amino acid-like compound that is endogenously produced within the human body and exogenously consumed in food sources such as red meat and seafood [1]. Creatine is primarily stored within skeletal muscle tissue (~95% of total stores) mostly in the form of phosphocreatine, which functions as an energy source through an enzymatic reaction involving creatine kinase, phosphocreatine, and adenosine di-phosphate to yield adenosine tri-phosphate (ATP). The inorganic phosphate and free energy yielded from ATP hydrolysis is then used for cellular work, with increasing demands as the intensity of effort is increased [1]. Research has indicated that internal phosphocreatine stores can be increased by 15–40% through creatine supplementation strategies, which can subsequently have various performance-related benefits in isolation, and when used in conjunction with structured exercise training programs over time [2–6]. The purported ergogenic benefits of creatine are well-supported within the literature for multiple populations [1]. Because of the strong evidence for ergogenic benefits pertaining to high-intensity exercise performance, as well as an increase in strength and skeletal muscle hypertrophy, creatine is a popular dietary supplement of choice among athletic populations. A systematic review by Knapik et al. [7] in 2016 reported creatine use in 50 out of 159 (31%) unique studies examining the prevalence of dietary supplement use by athletes of all ages. Creatine is not only one of the more popular dietary supplements from a performance perspective, but there is also strong evidence to support its use in clinical settings for a variety of patient populations [1,7]. However, the use of creatine among children and adolescent populations still remains somewhat controversial. While the physiological rationale regarding an ergogenic benefit in adolescents is similar to that seen in adults [8], the lack of randomized controlled trials and clinical data supporting the safety of creatine supplementation protocols among adolescent populations has resulted in hesitation regarding its widespread recommendation by some practitioners. Despite these concerns, creatine is still a popular dietary supplement of choice among adolescent populations and has been studied for its ergogenic potential in select athletic populations, albeit mostly in international (non-US) settings. It is clear that more research is warranted to better understand the short and long-term safety of creatine among adolescent athletes, however there is a precedent for its use among certain adolescent populations, both in athletes and special populations. Therefore, the focus of this article is to summarize the prevalence of use, performance benefits, clinical applications, and safety of creatine use in adolescents. For the purposes of the current review, children are defined as individuals between the ages of 0–12 years, while adolescents are defined as individuals between the ages of 13–19 years of age.

2. Effects of Creatine Supplementation on Creatine Content

Although limited in number, select studies have demonstrated that creatine supplementation is an effective nutritional strategy to promote increases in the phosphocreatine content and energy status of the cell among pediatric populations [9,10]; however, some evidence suggests this is potentially so, but to a lesser extent when compared to what is commonly reported in adult populations [1]. Depending on baseline levels of creatine content, which tend to be heavily influenced by exogenous creatine intake [11], increases of 10–40% in creatine or phosphocreatine content within skeletal muscle tissue are routinely reported following periods of creatine supplementation in adult populations [1]. Interestingly, preliminary evidence suggests that an age-dependent effect of creatine supplementation may exist regarding intramuscular creatine uptake, thereby indicating that the development of age-specific supplementation strategies may be warranted [9,10]. Further, research in adult populations [3,12,13] has indicated a moderate degree of variability in tissue uptake response to creatine supplementation protocols. It is currently unknown if similar variabilities (i.e., responder vs. nonresponder effects) are also present among children and adolescents. Due to age and ethical considerations, the majority of research in pediatric populations has relied on magnetic resonance imaging or laboratory markers as indirect measures of creatine and phosphocreatine content rather than through muscle biopsies, which is a common technique used for measuring intramuscular creatine content in adult populations. Additionally, lower doses of creatine are sometimes used among pediatric populations, which is also likely to influence the magnitude of changes observed in creatine content following a supplementation period. Therefore, it is difficult to directly compare the efficacy of creatine supplementation strategies between age groups when different techniques are used to quantify creatine content and different dosing strategies may be employed. However, a study by Solis et al. [10] was able to examine changes in brain and muscle phosphocreatine content across three age groups (children, n = 15, adult omnivores, n = 17, adult vegetarians, n = 14, and elderly adults, n = 18) using ^{31}P-magnetic resonance spectroscopy, following a standard creatine loading protocol (0.3 g/kg/day for 7 days). Results indicated intramuscular phosphocreatine content significantly increased by 13.9% while brain phosphocreatine levels only increased by 2.1% among the group of children [10]. Comparatively, following the same creatine dosing regimen, larger increases in muscle phosphocreatine content were observed in the elderly (22.7%) but not adult omnivores (10.3%) when compared to the children. It is also worth noting that lower baseline levels of muscle phosphocreatine content were observed in the children compared to the adult groups. Using a lower dose (5 g/day), but over an 8-week period, Banerjee et al. [9] reported significantly greater increases in the mean phosphocreatine/inorganic phosphate ratio following creatine supplementation compared to placebo (Creatine: 4.7; 95% CI: 3.9–5.6 vs. Placebo 3.3; 95% CI 2.5–4.2; p = 0.03) in patients with Duchenne Muscular Dystrophy. Alternatively, contradictory reports to these outcomes regarding the efficacy of creatine supplementation have indicated little to no impact on intramuscular phosphocreatine content; however the relative dose used in these studies may have been too low (0.1 g/kg/day for 12 weeks) to elicit any meaningful changes in creatine content [12,13]. Additionally, it is also worth noting these studies were conducted in patients with childhood-onset systemic lupus erythematosus and juvenile dermatomyositis, which may have influenced the efficacy of creatine supplementation in its ability to increase phosphocreatine content. Preliminary evidence among specialized clinical populations with inborn errors of metabolism, such as creatine deficiencies, has also indicated that creatine supplementation can positively influence brain creatine content with a subsequent influence on cognition, natural development, and quality of life [14–16]. However, as highlighted previously by Solis et al. [10], it was reported that a standard creatine loading regimen (0.3 g/kg/day for 7 days) in prepubescent children, omnivore adults, vegetarian adults, and elderly adults was not able to elicit significant changes in brain phosphocreatine content [10]. Similar findings have also been observed in healthy young children between the ages of 10 to 12 years of age [17], which failed to observe significant increases in brain creatine content following creatine supplementation. However, a review by Dolan et al. [18] highlighted multiple studies in adult populations, all of which utilized varying creatine supplementation strategies (2–20 g per day from 5 days to 8 weeks), that were able to demonstrate significant increases in brain creatine and phosphocreatine content following supplementation. It is possible that higher doses of creatine over longer periods of time may be required for meaningful increases in brain creatine content to occur among healthy populations as has been previously suggested [18]. The potential increase in brain creatine content following supplementation may also be age-dependent. Nevertheless, more research is needed to identify the extent to which creatine supplementation can impact brain creatine levels and whether or not modified creatine supplementation strategies for this purpose are warranted.

3. Prevalence of Use among Adolescents

The popularity of dietary supplements among adolescent populations has slowly increased over the past two decades. One of the more popular self-reported dietary supplements used among this population is creatine, with evidence of use for performance-enhancing purposes beginning in the late 1990s [7]. In 2001, when 674 high school athletes within the United States were surveyed on the use and perceptions of oral creatine supplementation, results indicated that 75% of athletes were aware of creatine and 16% reported some level of previous use [19]. Males exhibited much higher (23%) prevalence rates of use when compared to females (2%), with the percentage of creatine use increasing with grade level as 5% of 9th grade survey respondents reported creatine use compared to 22% of those in 12th grade [19]. Interestingly, 97% of athletes indicating creatine use reported a benefit from creatine, while 26% reported side-effects, a value that is far above the typical rate of side-effects from creatine supplementation within the literature [1]. It is possible that underlying confusion or bias regarding the side-effects of creatine supplementation influenced the high rate of side-effects reported in this study. Nevertheless, a more in-depth discussion regarding adverse events and the safety of creatine supplementation can be found later in the review in Section 6. Nearly 70% of creatine users reported ingesting a loading dose (i.e., 5–10 g for a period of 3–5 days) and nearly all creatine users reported the use of a maintenance dose following the loading period. In 2002, 16.7% of 4011 high school student-athletes surveyed from the United States self-reported current or prior creatine use with a higher use rate among males (25.3%) compared to females (3.9%) [20]. Additionally, the authors noted a large range of creatine use across sport-type, with football representing the highest (30.1%) rates of use and female cross-country representing the lowest (1.3%) rates of use. Increased strength was indicated as the most common reason for creatine supplementation. In 2004, among a convenience sample of 333 adolescents in a Midwestern Canadian province, 5.3% of those surveyed self-reported prior creatine use with 6.6% self-reporting they would use creatine in the future [21]. Individuals who self-reported current creatine use spent more time being physically active throughout the week compared with those not reporting creatine use. Additionally, among this population a higher rate of use was observed in males compared to females, which is in accordance with previous findings. Interestingly, 43.1% of athletes did not know if creatine would enhance performance while 14.2% believed that it would not improve performance, which highlights the need for improved education in this area. Shortly thereafter, in 2008, Hoffman et al. [22] surveyed a group of 3248 students in grades 8–12 within the United States and found 7% of survey respondents self-reported creatine use. Boys reported greater use of creatine with progressively increasing rates of creatine use reported at higher grade levels (22% of 12th grade boys). When asked about their primary source of information regarding dietary supplements, teachers (36%) and parents (16%) were the most common responses among all students. In 2012, results from a large subset ($n = 9417$) of the National Health Interview Survey respondents indicated that 34.1% of the children or adolescents (mean age = ~11 years) reporting the use of dietary supplements to enhance sports performance, self-reported using creatine [23]. Most recently, in 2020 among a sample of Australian adolescent boys participating in a variety of sports, 8.4% of survey respondents reported current use of creatine and 25.7% reported the intent to use creatine or other dietary supplements in the near future [24]. Drive for muscularity, participating in weight training, and sport participation were strong predictors of supplement use [24]. It is important to note that there is an underlying concern that creatine use among adolescents may be a predictor of future illegal performance enhancing substance use. However, the survey utilized to examine this relationship [25], inappropriately categorized creatine as an androgenic anabolic agent, rather than an amino-acid compound, thereby introducing bias to the future projection and therefore caution should be used when interpreting these findings.

The prevalence of creatine use tends to be higher among international adolescent athletes competing at the elite level. For example, Petroczi et al. [26] surveyed elite adolescent athletes as part of the United Kingdom Sport 2005 Drug Free Survey and when a subsample of 874 athletes were reanalyzed, 36.1% of athletes reported using creatine. Of note, a strong relationship was present between reasons for creatine use and physiological rationale among creatine users. This relationship indicates that those self-reporting creatine use were able to correctly identify the purported benefits while selecting physiologically appropriate reasons for use, which was not observed for other dietary supplements. Similarly, in 2008 Petroczi and Naughton [27] surveyed a cohort of 403 elite athletes from the United Kingdom and reported 28% of athletes self-reported creatine use. More recently in 2019, Jovanov et al. [28] surveyed 348 male and female adolescent athletes across four different countries who were all competing at an international level for their respective countries and reported that 25.3% of all athletes indicated using creatine. In alignment with previous studies, a significantly higher proportion of male athletes reported using creatine compared to female athletes (72.0% vs. 28.0%). Additionally, a higher proportion of athletes in the 17–18 year old age group reported creatine use, compared to the 15–16 year old group (60.0% vs. 40.0%). It is important to highlight these findings as they are some of the highest creatine use rates reported among a cohort of adolescents, which may be an indication of recent trends for higher creatine use. These outcomes may also be attributed to the continued growth of the dietary supplement market, popularity of competitive youth sports at an elite level, and the increasing reputation of creatine's ergogenic properties. It is also important to note that there has been a significant increase in the prevalence of creatine supplementation among female athletes, in particular, over the past 20 years [19,28]. While creatine appears to be a popular dietary supplement of choice among elite adolescent athletes, creatine is still less frequently used when compared to other dietary supplement categories such as multivitamins, protein powders, and energy products, in which usage rates of these product categories can range anywhere from 60–80% among adolescent athletes [26–32]. More research is needed to better understand the contextual factors underpinning these trends in dietary supplement use among adolescents.

4. Performance Benefits

When compared to adults, a limited number of controlled investigations examining the ability of creatine supplementation to impact measures of exercise performance among adolescent populations exist. All available studies to date, have been completed in only two types of athletes: swimming ($n = 5$) and soccer ($n = 4$). Additionally, studies were completed across all parts of the globe with two studies being completed in Brazil and one study being completed in Hungary, Australia, USA, United Kingdom, Iran, and Yugoslavia. The studies completed on swimmers differed somewhat in the dosing regimen that was employed. Three of the studies utilized a loading phase for the entirety of the supplementation regimen, incorporating doses of 21 g per day for nine days, 20 g per day for five days, and 20 g per day for 4 days [33–35]. The other two studies used a combination of a loading (5 days at 20 g/day or four days at 25 g/day) and a maintenance phase (5 g/day for 22 days or 5 g/day for 2 months) [36,37]. Various parameters of swimming performance were assessed ranging from in-water sprint swimming performance to power outputs completed during a swim bench/ergometer test. All studies that reported outcomes within the initial 4–9 days of supplementation identified an improvement in various performance measures such as swim bench test performance, sprint swimming performance, dynamic strength, and anaerobic exercise performance. The longest study by Theodorou et al. [37], reported improvements in interval swimming exercise performance after the loading phase, but no further improvement after a maintenance dose was administered. Alternatively, Dawson et al. [36] reported improvements in swim bench performance, but not sprint swimming performance after completion of both a loading and maintenance phase. While not directly performance related, Juhasz et al. [38] indicated that creatine supplementation may also be an effective strategy to support the rehabilitation of overuse-associated tendinitis in adolescent swimmers when combined with a targeted physical therapy program. Notably, when indicated by the authors, no adverse events were reported in any of these studies.

The studies that enrolled adolescent soccer athletes as study participants ranged in duration from 7–49 days. Three studies were seven days in duration and employed loading phases that each delivered different loading doses (0.03 g/kg/day, 20 g/day, and 30 g/day) [39–41]. One study [42] was seven weeks in duration and used a seven-day loading phase (20 g/day) followed by a six-week maintenance dose of 5 g/day. All studies were placebo-controlled. Of interest, all three studies that were <7 days in duration, reported statistically significant improvements in various performance outcomes. For example, Mohebbi et al. [39] reported a significant improvement in repeat sprinting and soccer dribbling performance, while Ostojic et al. [40] reported improvements in performance of a soccer dribbling test, countermovement jump, and power production during a sprint. Lastly, Yanez-Silva et al. [41] reported improvements in peak and mean power output as well as total work completed during a Wingate anaerobic capacity test. The remaining study, Claudino et al. [42] failed to report any improvements in lower body power production after 14 male adolescent soccer athletes completed a one-week loading and a six-week maintenance dose phase. Finally, and similar to what was observed with the studies involving swimming, creatine was well tolerated with no adverse events being reported. A summary table of these studies has been included in Table 1.

Table 1. Efficacy of creatine use in adolescents on exercise performance.

Author Year (Country)	Subjects	Design	Duration	Dosing Protocol	Primary Variables	Results	Adverse Events
				Swimming			
Dawson et al. 2002 (Australia) [36]	10 male, 10 female (16.4 ± 1.8 years) swimmers	Matched, placebo-controlled	4 weeks	20 g/day (5 days) 5 g/day (22 days)	Sprint swim performance and swim bench test	↑ swim bench test performance	None reported
Grindstaff et al. 1997 (USA) [33]	18 (11 female, 7 male) adolescent swimmers (15.3 ± 0.6 years)	Randomized, double-blind, placebo controlled	9 days	21 g/day	Sprint swim performance; arm ergometer performance	↑ sprint swimming performance	None reported
Juhasz et al. 2009 (Hungary) [34]	16 male fin swimmers (15.9 ± 1.6 years)	Randomized, placebo-controlled, single-blind trail	5 days	20 g/day	Average power, dynamic strength (swim based tests)	↑ anaerobic performance; ↑ dynamic strength	None reported
Theodorou et al. 1999 (UK) [37]	10 elite female (17.7 ± 2.0 years) and 12 elite male (17.7 ± 2.3 years) swimmers	Randomized, double-blind, placebo-controlled	11 weeks	25 g/day (4 days) 5 g/day (2 months)	Swimming interval performance	↑ interval performance following loading phase; ↔ long-term improvements after maintenance dose	None reported
Theodorou et al. 2005 (United Kingtom) [35]	10 high performance swimmers (males: $n = 6$; females: $n = 4$) (17.8 ± 1.8 years	Randomized, double-blind trial	4 days	20 g/day of CrM or 20 g/day of CrM + 100 g of carbohydrates per serving	High-intensity swim performance during repeated intervals	↑ mean swim velocity for all swimmers; ↔ swim velocity in Cr + Carbohydrate condition	Gastrointestinal discomfort in CrM + Carbohydrate group only
				Soccer			
Claudino et al. 2014 (Brazil) [42]	14 male Brazilian elite soccer players (18.3 ± 0.9 years)	Randomized, double-blind, placebo-controlled	7 weeks	20 g/day (1 week) 5 g/day (6 weeks)	Lower limb muscle power via counter-movement vertical jump	↔ lower body power	None reported
Mohebbi et al. 2012 (Iran) [39]	17 adolescent soccer players (17.2 ± 1.4 years)	Randomized, double-blind, placebo-controlled	7 days	20 g/day	Repeated sprint test, soccer dribbling performance and shooting accuracy	↑ repeat sprint performance; ↑ dribbling performance	None reported
Ostojic et al. 2004 (Yugoslavia) [40]	20 adolescent male soccer players (16.6 ± 1.9 years)	Matched, placebo-controlled	7 days	30 g/day	Soccer specific skills tests	↑ dribble test and endurance times; ↑ sprint power test and countermovement jump	None reported
Yanez-Silva et al. 2017 (Brazil) [41]	Elite youth soccer players (17.0 ± 0.5 years)	Matched, double-blind, placebo-controlled	7 days	0.03 g/kg/day	Muscle power output (Wingate anaerobic power test)	↑ peak and mean power output; ↑ total work	None reported

↔ = Creatine supplementation resulted in no significant ($p > 0.05$) change; ↑ = Creatine supplementation resulted in a significant increase ($p < 0.05$) over control. CrM = creatine monohydrate; g/day = grams per day. Adapted from Jagim et al. 2018 [43].

In summary, limited research is available that has examined the potential of creatine supplementation to impact various aspects of exercise performance. Of the limited work that has been completed, creatine appears to be well-tolerated with no adverse events being reported and consistent improvements in assessments associated with swimming and soccer performance observed in adolescent athletes. Future research is warranted to better evaluate the ability of creatine to influence other types of exercise performance and sport-specific activities as well as the potential for synergistic adaptations to exercise training. Moreover, additional research is urgently needed in adolescent females across all sport types and while research is welcomed in swimming and soccer, future research should examine the potential of creatine in other popular team-based sports where strength and power are key physiological attributes in predicting sporting success.

5. Clinical Applications

Over the past 30 years, the discovery of inborn errors of metabolism and the potential physiological, neurological, and neuroprotective benefits of creatine have led to advancements in the therapeutic use of creatine. As such, several pediatric clinical populations have been shown to benefit from creatine supplementation, which notably includes patients with genetic defects associated with creatine deficiency. Table 2 presents a summary of studies that have examined the therapeutic benefits of creatine supplementation for a variety of clinical disorders. To date, the majority of clinical trials investigating the therapeutic potential of creatine supplementation in pediatric populations have focused on creatine (and/or creatine transporter) deficiencies, inborn errors of metabolism, neuromuscular disorders, and myopathies [9,44–49]. Guanidinoacetate methyltransferase (GAMT) and arginine:glycine amidinotransferase (AGAT) deficiency, are types of inborn errors of creatine metabolism, collectively characterized as cerebral creatine synthesis deficiencies [47,50]. Several studies and case reports have indicated that creatine supplementation can restore tissue creatine content and improve some of the symptoms resulting from creatine deficiencies [51–54]. Since its discovery in 1994, GAMT deficiency has shown to be treatable through creatine supplementation strategies [51,53]. For example, in 1996, Stöckler et al. [54] treated an infant patient with GAMT deficiency using a creatine replacement therapy of 4–8 g/day over a 25-month period and reported substantial clinical improvement, normalization of brain MRI abnormalities, and improvements in electroencephalogram readings post-treatment. Since that time, several additional case reports and reviews have been published, highlighting effective strategies to diagnose and treat cerebral creatine deficiency with consistent improvements in intellectual development reported, especially when early detection and ensuing treatment were employed [51,52]. While similar in nature, AGAT deficiency is extremely rare with only 20 documented cases worldwide [52]. AGAT deficiency is also an autosomal recessive disorder that disrupts the biosynthesis of creatine and is associated with a variety of clinical features such as intellectual development disorder, speech delays, autistics behaviors, and occasional seizures [52]. Thankfully, AGAT also appears to betreatable with creatine supplementation. For example, Ndika et al. [14] treated a 9-year-old female pediatric patient with AGAT deficiency with up to 800 mg/kg/day of creatine over an 8-year period and reported partial recovery of cerebral creatine levels with the patient demonstrating superior nonverbal and academic abilities at age 9, compared to initially presenting with a score of 43% of her chronological age at 16 months when assessed using the Bayley's Infant Development Scale. Although similar, creatine transporter deficiency is another inborn error of metabolism that can result in creatine deficiencies in select tissues, particularly within the brain [51]. Creatine transporters are membrane-bound transport proteins that have been found in a variety of different tissues and are required for tissue uptake of creatine against its concentration gradient. Moreover, creatine transporter 1 (CrT1) is expressed ubiquitously across human tissues and deficiencies of this protein are another type of creatine metabolism disorder that can result in brain atrophy, intellectual disabilities, and developmental delays [51]. However, this defect is not as responsive to exogenous creatine supplementation strategies, as the deficiency is attributable to an inability to transport creatine across the cell membrane, rather than a lack of creatine availability [47]. As such, current research has focused on identifying alternative strategies that may enhance brain creatine content in these populations. For example, recent work has demonstrated early promise with the use of creatine fatty esters and lipid nanocapsules as a nutrition-based therapeutic treatment for creatine transporter deficiency [55,56]. These creatine esters and lipid based nanocapsules are better able to cross the blood−brain barrier, thereby helping to increase brain creatine content and correct the creatine deficiency [56].

Table 2. Efficacy of creatine in clinical settings.

Author Year	Subjects	Design	Duration	Dosing Protocol	Primary Variables	Results	Adverse Events
Sipila et al. 1981 [57]	7 (3 adolescents) patients with gyrate atrophy of retina	Open label treatment intervention	12 months	1.5 g/day	Visual acuity, muscle fiber characteristics, laboratory markers of creatine metabolism	↔ Visual acuity; ↑ Thickness of Type II muscle fibers	No side effects reported
Vannas-Sulonen et al. 1985 [58]	13 patients (9 male, 4 female) between ages of 6–31 years diagnosed with gyrate atrophy of the choroid	Prospective, open-label cohort	36–72 months	0.25–0.5 g dose 3×/day	Morphological and eye function assessments	↔ Cr supplementation did not prevent normal deterioration; ↓ Muscle atrophy, primarily in type II fibers	None reported
Walter et al. 2000 [59]	36 patients with multiple types of muscular dystrophies (overall mean age: 26 ± 16 years) 8 patients with Duchenne dystrophy (mean age: 10 ± 3 years)	Randomized, double-blind, placebo-controlled	8 weeks	10 g/day (adults) 5 g/day (children)	Muscular performance, neuromuscular symptoms score, vital capacity and qualitative assessments	↑ (3%) in muscle strength; ↑ (10%) in neurological symptoms. Children tended to experience greater strength changes.	None reported. Indicated to be well-tolerated.
Braegger et al. 2003 [60]	18 cystic fibrosis patients (7 F, 11 M) ranging in age from 8–18 years	Prospective open-label pilot	Supplemented for 12 weeks; monitored for 24–36 weeks	12 g/day for 1st week; 6 g/day for remaining 11 weeks	Lung function, strength, and clinical parameters	↔ Lung function or sweat electrolytes. ↑ (18%) in peak isometric strength	One patient experienced transient muscle pain; No other side effects
Louis et al. 2003 [61]	15 boys with muscular dystrophy (mean age: 10.8 ± 2.8 years)	Double-blind, placebo-controlled, cross-over study design	3 months, with 2 months washout	3 g/day	Muscle function, densitometry, markers of hepatic and renal function, magnetic resonance spectroscopy	↑ MVC by 15% ↑ TTE (~2×) ↑ TJS ↑ LS and WB BMD in ambulatory patients ↑ NTx/creatinine ratio in ambulatory patients	No changes in liver or kidney markers
Tarnopolsky et al. 2004 [45]	30 boys with Duchenne muscular dystrophy; mean age: 10 ± 3 years; height: 129.2 ± 16.0 cm; weight: 35.3 ± 15.8 kg	Double-blind, randomized, crossover trial	4 months	0.10 g/kg/day	Pulmonary function, strength, body composition, bone health, task function, blood & urinary markers	↑ handgrip strength, fat-free mass, and bone markers ↔ functional tasks or activities of daily living	None

Table 2. *Cont.*

Author Year	Subjects	Design	Duration	Dosing Protocol	Primary Variables	Results	Adverse Events
Escolar et al. 2005 [49]	50 ambulatory steroid naïve boys with Duchenne Muscular Dystrophy (mean age: 6 years)	Double-blind, placebo-controlled, randomized	6 months	5 g/day of creatine powder, 0.3 mg/kg of glutamine (×2 per day), or placebo	Manual muscle performance, quantitative muscle testing, time to rise	↔ primary or secondary outcomes measures	Deemed safe and well-tolerated with no side effects reported.
Sakellaris et al. 2008 [62]	39 children/adolescents following traumatic brain injury	Open-label pilot study	6 months	0.4 g/kg/day	Duration of amnesia, duration of intubation, and intensive care unit stay post traumatic brain injury	↓ Amnesia ↓ Intubation period ↓ Intensive care unit stay	None
Bourgeois et al. 2008 [63]	9 children with lymphoblastic leukemia during chemotherapy (in treatment group); mean age of 7.6 years, 50 healthy children as history controls	Cross sectional, mixed cohort designs	16 weeks	0.1 g/kg/day	Height, weight, BMI, BMD, BMC, FFM, %BF, serum creatinine	↑ %BF and BMI	None reported
Banerjee et al. 2010 [9]	33 ambulatory male patients with Duchenne muscular dystrophy	Randomized, placebo-controlled, single-blind trial	8 weeks	Cr, 5 g/day (*n* = 18)	Cellular energetics, manual muscle test score and functional status	↑ in PCr/Pi ratios	None reported
Van de Kamp et al. 2012 [16]	9 boys with creatine transporter defect	Long-term follow-up investigation	4–6 years	Cr (400 mg/kg/day) and L-arginine (400 mg/kg/day)	Locomotor and personal social IQ subscales	Initial ↑ in locomotor and personal social IQ subscales; No lasting clinical improvement was recorded	No adverse events were reported.
Hyashi et al. 2014 [13]	15 participants with childhood systemic lupus erythematosus	Double-blind, placebo controlled, cross-over design	12 weeks with 8 week washout period	0.1 g/kg/day	Muscle function, body composition, biochemical markers of bone, aerobic conditioning, quality of life	↔ intramuscular PCr, muscle function, and aerobic conditioning parameters, body composition, quality of life	↔ laboratory parameters; No side effects reported
Solis et al. 2016 [12]	Patients with juvenile dermatomyositis (mean age: 13 ± 4 years)	Randomized, double-blind, placebo-controlled, crossover trial	12 weeks	0.1 g/kg/day	Primary: muscle function Secondary: body composition, biochemical markers of bone remodeling, cytokines, laboratory markers of kidney function, aerobic conditioning, and quality of life	↔ Muscle function, intramuscular PCr content, or other secondary outcomes measures	No side efforts reported. ↔ Markers of kidney function
Kalamitsou et al. 2019 [64]	22 children (9 F, 13 M) with refractory epilepsy ranging in age from 10 months to 8 years	Prospective cohort	3–12 months follow-up	0.4 g/kg/day creatine + ketogenic diet	Proportion of responders to ketogenic diet	6/22 (27%) responded to creatine addition to ketogenic diet	None reported, well-tolerated with no exacerbations of underlying pathology

Table 2. *Cont.*

Author Year	Subjects	Design	Duration	Dosing Protocol	Primary Variables	Results	Adverse Events
Dover et al. 2020 [65]	13 (7 F, 6 M) patients ranging in age from 7–14 years with juvenile dermatomyositis; 25.6–64.6 kg; 14.3–22.9 kg/m²	Randomized, double-blind, placebo-controlled	6 months	Up to 40 kg was 150 mg/kg/day >40 kg was 4.69 g/m²/day	Safety and tolerability muscle function, disease activity, aerobic capacity, muscle strength	↔ in muscle function, strength, aerobic capacity, fatigue, physical activity ↓ in muscle pH following exercise	No adverse events reported

↔ = Creatine supplementation resulted in no change in the target outcome; ↑ = Creatine supplementation resulted in an increase in the target outcome; ↓ = Creatine supplementation resulted in a decrease (directional) in the target outcome. BMI = body mass index; FFM = fat-free mass; %BF = body fat percentage; TJS = total joint stiffness; TTE = time to exhaustion; g/d = grams per day; g/kg/d = grams per kilogram of bodyweight per day; mg/kg/d = milligrams per kilogram of bodyweight per day; PCr = phosphocreatine. MVC = maximum voluntary contraction; NTx = N-terminal telopeptide of type I collagen; LS = lumbar spine; WB = whole body; BMD = bone mineral density; BMC = bone mineral content; Pi = inorganic phosphate.

Considerable research has also been dedicated to investigating the therapeutic benefit of creatine in the management of myopathies. As a high percentage of creatine is stored within skeletal muscle tissue, muscle disorder pathologies, such as myopathies, are often associated with reduced intramuscular concentrations of creatine, phosphocreatine, and ATP in addition to subsequent neuromuscular impairments and muscle weakness [66]. As such, a strong underlying physiological rationale exists to support the potential of creatine supplementation as a therapeutic agent in the management of myopathies. For example, Duchenne's muscular dystrophy is one such myopathy that is progressive in nature with no known cure. Patients are often prescribed corticosteroids to slow disease progression, which can have several adverse side effects when used long-term. Because of the catabolic nature of corticosteroid therapy, and musculoskeletal pathology associated with muscular dystrophy, creatine supplementation has been identified as a therapeutic agent to potentially counteract the deleterious effects of both the disease, and comorbidities which arise secondary to the primary corticosteroid treatment. Favorable improvements have been observed for fat-free mass and strength in pediatric patients [45]. A major challenge with clinical trials investigating the therapeutic benefits of creatine supplementation in patients with various types of myopathies is dealing with the heterogeneity of the disease itself. The diversity in how the disease manifests in patients can subsequently influence the time course of disease progression and individualistic nature of active versus remission disease states, all of which can be difficult to account for with an optimal study design.

Gyrate atrophy of the retina and choroid is another creatine deficiency disorder that is characterized as an enzymatic disorder attributable to defects in ornithine aminotransferase, resulting in elevated levels of ornithine, which negatively impacts creatine synthesis [57,67]. As a result, creatine concentrations in serum, urine, erythrocytes, brain, and muscle are reduced in these patient populations [58,67]. These patients present primarily with eyesight problems beginning as early as age five, which progressively deteriorate over time. Results from studies spanning up to 72 months in this patient population indicate that creatine supplementation can slow disease progression while also helping to maintain levels of type II skeletal muscle fiber content [68,69].

Lower amounts of daily exogenous creatine intake have also been associated with depression in young adult populations as Bakian et al. [70] observed a significantly higher prevalence of depression (10.2/100 persons) in the lowest quartile of dietary creatine intake compared to the highest quartile of creatine intake, which had a depression prevalence rate of 6.0/100 persons. This relationship appeared to be strongest in females and those in the 20–39 years of age category and therefore may also extend to adolescents. Additionally, early evidence indicates that creatine may be used as an adjunctive therapy in the actual management of clinical depression [71–73]. Creatine supplementation has also been used as an experimental therapeutic agent for conditions pertaining to hypoxia and energy-related brain pathologies such as traumatic brain injuries or cerebral ischemia in pediatric patients [20,49,74]. There has also been recent interest in examining the potential benefits of creatine supplementation for pregnant women with potential benefits extending to the developing fetus [75,76]. Currently, clinical trials are underway to better understand how creatine may affect both the mother and developing fetus [75]. It is also worth noting that a growing body of evidence exists demonstrating that creatine may also confer a variety of physiological benefits for multiple clinical conditions in adult populations, such as mitochondrial disease, neurological disorders, and autoimmune disorders [47,74], but the extent to which these findings may extend to pediatric populations requires more research due to the limited data currently available. Lastly, and a point that is beyond the scope of this review, all of these findings may hold particular importance for any clinical population (adult or adolescent) who are vegetarians as they may be susceptible to low daily creatine intake through diet alone, as has been reported in adults [3,11,61].

6. Safety

An extensive summary highlighting the safety of creatine supplementation was recently addressed in a 2017 Position Stand by Kreider et al. [1]. However, to date, all published studies involving an a priori research question and utilizing a study designed to examine the safety of creatine supplementation have thus far only been completed in adult populations. Currently, no studies have been published examining safety considerations in healthy young or adolescent athletic populations. Importantly and practically speaking, no indications currently exist as to why a similar safety profile would not be observed in adolescents, and as highlighted throughout the previous sections, multiple creatine supplementation studies have been conducted in adolescent athlete and clinical populations with no adverse events reported. Several of these studies have even closely monitored laboratory markers throughout the supplementation period with no indications of clinically relevant adverse effects observed. While the authors agree, and would like to publicly state that an absence of self-reported adverse events by study subjects is not a confirmation of safety, it does support the hypothesis that creatine is likely safe for this population. Nonetheless, safety studies in youth and adolescent populations using randomized controlled trial designs are desperately needed to help continue building the safety profile for creatine supplementation among these younger age groups. Another way to assess safety-related concerns is to examine adverse event reports that are submitted to the Center for Adverse Event Reports, which is overseen by the Center for Food Safety and Applied Nutrition as part of the United States Food and Drug Administration (FDA). When searching this adverse event reporting system database, which is publicly available, only 22 out of the 15,274 (0.144%) adverse events reports were associated with creatine during the 2018–2020 reporting period, after filtering out food, cosmetics, and adverse events without a known product code (Accessed 5 February 2021; file dates January 2018–March 2020) [77]. In perhaps the strongest testament to the safety of creatine, the United States FDA recently designated creatine as "generally recognized as safe" (GRAS) (https://www.fda.gov/media/143525/download) [78]. Ultimately, this classification indicates that creatine is considered safe under the conditions of its intended use based on the currently available scientific evidence as decided upon by a panel of qualified content experts. Importantly, and a point that is pertinent to the topic of the current article, this designation of safety extends to older children and adolescents.

A unique 2019 study published by Simpson and colleagues [79] may provide some of the first published data involving youth athletes following creatine supplementation with outcomes pertaining to safety implications, and adverse events. In this study 19 elite soccer players ($n = 13$, U18 "Under 18 years" and $n = 6$, U21 "Under 21 years") completed an eight-week supplementation regimen of creatine monohydrate (0.3 g/kg/day for 7 days and 5 g/day for the remaining seven weeks) in a randomized, double-blind, placebo-controlled fashion. Before and after supplementation, study participants had airway inflammation (using exhaled nitric oxide) and airway responsiveness to dry air (hyperpnea) assessed before and after supplementation. Participants with previous pulmonary disease were excluded and all participants were assessed for unknown or undiagnosed allergies prior to the study. A statistically meaningful trend ($p = 0.056$) between the groups with medium to large observed effect sizes ($n = 0.199$) were found regarding the amount of exhaled nitric oxide (an assessment of airway inflammation). There was also as a trend ($p = 0.070$, $n = 0.975$) of forced expired volumes in one second to reduce with creatine supplementation when compared to placebo. The authors concluded that, "we cannot exclude that creatine supplementation has an adverse effect of the airways of elite athletes, particularly in those with allergy sensitization." These isolated and initial findings are some of the first published data to suggest that creatine supplementation may compromise airway health and thus more research is needed to confirm or refute these findings.

7. Practical Recommendations and Future Directions

In conclusion, there appears to be strong evidence of creatine use among adolescents, particularly among male athletes with the highest usage rates evident among international adolescent athletes competing at the elite level. The majority of adolescents who self-report using creatine, appear to get their information from friends, coaches, and parents. However, the need for replication of dietary supplement questionnaire studies continues to be present because of the recent growth in the dietary supplement industry. Additionally, the recent popularity of marketing through social media platforms and online markets for supplement companies has likely altered how adolescents perceive and obtain information surrounding dietary supplements in addition to how they purchase them. Further, there continues to be a need for more education regarding safe and effective dietary supplement strategies among adolescents, rather than strict policies avoiding or advocating against their use. Even young adults pursuing careers in health professions do not appear to have sufficient knowledge of the safety, regulation, and efficacy of various dietary supplements [80], which reinforces the need for more education and dialogue surrounding the topic, particularly when one considers the high prevalence of use across all populations.

A small number of investigations have reported on the ergogenic benefits of creatine supplementation in adolescent athletes (Table 1), with the majority of this data being published in international (non-US) adolescent males competing in swimming or soccer. However, a paucity of observational and no experimental research exists that examines changes in the clinical health markers among healthy adolescents who are supplementing with creatine, especially those who regularly partake in high-intensity exercise training and athletic competition. While data is limited, it is not entirely absent as many studies report on the lack of reported side-effects, nonsignificant changes in laboratory markers of kidney and liver function, and a lack of changes in inflammatory cytokines among clinical populations, therein supporting the hypothesis that creatine supplementation is likely safe for an adolescent population. The lack of adverse event reports in the literature among clinical populations is telling, particularly when considering that several of the patients are on immunosuppressive therapies or have multiple comorbidities that could negatively influence various indictors of health status. Regardless, a dire need exists for prospective randomized, double-blind, placebo-controlled trials examining the safety and efficacy of creatine among children and adolescent populations; both among athletes and the general population. Equally important are well-powered, randomly controlled trials examining the status and changes in body water and cellular hydration status before, throughout, and after standard regimens of creatine supplementation and to what extent these changes impact creatine content, performance and physiological adaptations to regular exercise training. Priority should first be placed on examining the effects of various creatine-dosing strategies on markers of clinical health or any contraindications for use among this population. From there, emphasis should be placed on continuing to explore the potential benefits of creatine among clinical populations where there might be a unique physiological rationale for a therapeutic benefit of creatine. These conditions may include myopathies, muscular dystrophy, muscle wasting conditions, cancer cachexia, clinical depression, traumatic brain injuries, spinal cord injuries, orthopedic injuries, and periods of bed rest or immobilization. Lastly, several of the previously published studies which examined physical performance outcomes in healthy adult populations should be replicated in adolescents to examine if similar ergogenic benefits from creatine supplementation are possible among this population.

It is important for adolescents, coaches, and parents to be aware of evidence-based recommendations regarding the safety and efficacy of creatine supplementation when considering its use. Many misconceptions are present regarding creatine [81], and therefore consumers should seek out expert advice regarding safe and informed use of creatine. Readers are also directed to the most recently published Position Stand on creatine published by the International Society of Sports Nutrition [1] for a complete summary regarding the mechanisms of action, ergogenic benefits, safety, clinical applications and dosing recommendations of creatine. In brief, the statements from the position stand below are most pertinent to the focus of the current review article:

- "Creatine monohydrate is the most effective ergogenic nutritional supplement currently available to athletes with the intent of increasing high intensity exercise capacity and lean body mass during training."
- "Creatine monohydrate supplementation is not only safe, but has been reported to have a number of therapeutic benefits in healthy and diseased populations ranging from infants to the elderly. There is no compelling scientific evidence that the short- or long-term use of creatine monohydrate (up to 30 g/day for 5 years) has any detrimental effects on otherwise healthy individuals or among clinical populations who may benefit from creatine supplementation."
- "If proper precautions and supervision are provided, creatine monohydrate supplementation in children and adolescent athletes is acceptable and may provide a nutritional alternative with a favorable safety profile to potentially dangerous anabolic androgenic drugs. However, we recommend that creatine supplementation only be considered for use by younger athletes who: (a) are involved in serious/competitive supervised training; (b) are consuming a well-balanced and performance-enhancing diet; (c) are knowledgeable about the appropriate use of creatine; and (d) do not exceed recommended dosages."
- "Label advisories on creatine products that caution against usage by those under 18 years old, while perhaps intended to insulate their manufacturers from legal liability, are likely unnecessary given the science supporting creatine's safety, including in children and adolescents. The quickest method of increasing muscle creatine stores may be to consume ~0.3 g/kg/day of creatine monohydrate for 5–7-days followed by 3–5 g/day thereafter to maintain elevated stores. Initially, ingesting smaller amounts of creatine monohydrate (e.g., 3–5 g/day) will increase muscle creatine stores over a 3–4 week period, however, the initial performance effects of this method of supplementation are less supported."

Author Contributions: Conceptualization, A.R.J. and C.M.K.; investigation, A.R.J. and C.M.K.; writing—original draft preparation, A.R.J. and C.M.K.; writing—review and editing, A.R.J. and C.M.K.; visualization, A.R.J. and C.M.K. All authors have read and agreed to the published version of the manuscript.

Funding: The APC was funded by Alzchem Trotsberg, GmbH.

Institutional Review Board Statement: Not applicable.

Informed Consent Statement: Not applicable.

Data Availability Statement: Data sharing not applicable: No new data were created or analyzed in this study.

Conflicts of Interest: Authors of this manuscript received financial remuneration for preparing and reviewing this paper from the sponsor of the special issue, Alzchem. A.R.J. and C.M.K. have consulted with and received external funding from companies who sell certain dietary ingredients, and have received remuneration from companies for delivering scientific presentations at conferences. A.R.J. and C.M.K. also write for online and other media outlets on topics related to exercise and nutrition. None of these entities had any role in the design of the paper, collection, analyses, or interpretation of data; in the writing of the manuscript, or in the decision to publish this paper.

References

1. Kreider, R.B.; Kalman, D.S.; Antonio, J.; Ziegenfuss, T.N.; Wildman, R.; Collins, R.; Candow, D.G.; Kleiner, S.M.; Almada, A.L.; Lopez, H.L. International Society of Sports Nutrition position stand: Safety and efficacy of creatine supplementation in exercise, sport, and medicine. *J. Int. Soc. Sports Nutr.* **2017**, *14*, 18. [CrossRef]
2. Hultman, E.; Soderlund, K.; Timmons, J.A.; Cederblad, G.; Greenhaff, P.L. Muscle creatine loading in men. *J. Appl. Physiol.* **1996**, *81*, 232–237. [CrossRef]
3. Harris, R.C.; Soderlund, K.; Hultman, E. Elevation of creatine in resting and exercised muscle of normal subjects by creatine supplementation. *Clin. Sci.* **1992**, *83*, 367–374. [CrossRef]
4. Jagim, A.R.; Oliver, J.M.; Sanchez, A.; Galvan, E.; Fluckey, J.; Riechman, S.; Greenwood, M.; Kelly, K.; Meininger, C.; Rasmussen, C.; et al. A buffered form of creatine does not promote greater changes in muscle creatine content, body composition, or training adaptations than creatine monohydrate. *J. Int. Soc. Sports Nutr.* **2012**, *9*, 43. [CrossRef]
5. Kreider, R.B. Effects of creatine supplementation on performance and training adaptations. *Mol. Cell. Biochem.* **2003**, *244*, 89–94. [CrossRef]
6. Preen, D.; Dawson, B.; Goodman, C.; Beilby, J.; Ching, S. Creatine supplementation: A comparison of loading and maintenance protocols on creatine uptake by human skeletal muscle. *Int. J. Sport. Nutr. Exerc. Metab.* **2003**, *13*, 97–111. [CrossRef]
7. Knapik, J.J.; Steelman, R.A.; Hoedebecke, S.S.; Austin, K.G.; Farina, E.K.; Lieberman, H.R. Prevalence of Dietary Supplement Use by Athletes: Systematic Review and Meta-Analysis. *Sports Med.* **2016**, *46*, 103–123. [CrossRef]
8. Unnithan, V.B.; Veehof, S.H.; Vella, C.A.; Kern, M. Is there a physiologic basis for creatine use in children and adolescents? *J. Strength Cond. Res.* **2001**, *15*, 524–528.
9. Banerjee, B.; Sharma, U.; Balasubramanian, K.; Kalaivani, M.; Kalra, V.; Jagannathan, N.R. Effect of creatine monohydrate in improving cellular energetics and muscle strength in ambulatory Duchenne muscular dystrophy patients: A randomized, placebo-controlled 31P MRS study. *Magn. Reson. Imaging* **2010**, *28*, 698–707. [CrossRef]
10. Solis, M.Y.; Artioli, G.G.; Otaduy, M.C.G.; Leite, C.D.C.; Arruda, W.; Veiga, R.R.; Gualano, B. Effect of age, diet, and tissue type on PCr response to creatine supplementation. *J. Appl. Physiol.* **2017**, *123*, 407–414. [CrossRef]
11. Burke, D.G.; Chilibeck, P.D.; Parise, G.; Candow, D.G.; Mahoney, D.; Tarnopolsky, M. Effect of creatine and weight training on muscle creatine and performance in vegetarians. *Med. Sci. Sports Exerc.* **2003**, *35*, 1946–1955. [CrossRef]
12. Solis, M.Y.; Hayashi, A.P.; Artioli, G.G.; Roschel, H.; Sapienza, M.T.; Otaduy, M.C.; De Sa Pinto, A.L.; Silva, C.A.; Sallum, A.M.; Pereira, R.M.; et al. Efficacy and safety of creatine supplementation in juvenile dermatomyositis: A randomized, double-blind, placebo-controlled crossover trial. *Muscle Nerve* **2016**, *53*, 58–66. [CrossRef]
13. Hayashi, A.P.; Solis, M.Y.; Sapienza, M.T.; Otaduy, M.C.; de Sa Pinto, A.L.; Silva, C.A.; Sallum, A.M.; Pereira, R.M.; Gualano, B. Efficacy and safety of creatine supplementation in childhood-onset systemic lupus erythematosus: A randomized, double-blind, placebo-controlled, crossover trial. *Lupus* **2014**, *23*, 1500–1511. [CrossRef]
14. Ndika, J.D.; Johnston, K.; Barkovich, J.A.; Wirt, M.D.; O'Neill, P.; Betsalel, O.T.; Jakobs, C.; Salomons, G.S. Developmental progress and creatine restoration upon long-term creatine supplementation of a patient with arginine:glycine amidinotransferase deficiency. *Mol. Genet. Metab.* **2012**, *106*, 48–54. [CrossRef]
15. Clark, J.F.; Cecil, K.M. Diagnostic methods and recommendations for the cerebral creatine deficiency syndromes. *Pediatr. Res.* **2015**, *77*, 398–405. [CrossRef]
16. van de Kamp, J.M.; Pouwels, P.J.; Aarsen, F.K.; ten Hoopen, L.W.; Knol, D.L.; de Klerk, J.B.; de Coo, I.F.; Huijmans, J.G.; Jakobs, C.; van der Knaap, M.S.; et al. Long-term follow-up and treatment in nine boys with X-linked creatine transporter defect. *J. Inherit. Metab. Dis.* **2012**, *35*, 141–149. [CrossRef]
17. Merege-Filho, C.A.; Otaduy, M.C.; de Sa-Pinto, A.L.; de Oliveira, M.O.; de Souza Goncalves, L.; Hayashi, A.P.; Roschel, H.; Pereira, R.M.; Silva, C.A.; Brucki, S.M.; et al. Does brain creatine content rely on exogenous creatine in healthy youth? A proof-of-principle study. *Appl. Physiol. Nutr. Metab.* **2017**, *42*, 128–134. [CrossRef]
18. Dolan, E.; Gualano, B.; Rawson, E.S. Beyond muscle: The effects of creatine supplementation on brain creatine, cognitive processing, and traumatic brain injury. *Eur. J. Sport Sci.* **2019**, *19*, 1–14. [CrossRef]
19. Ray, T.R.; Eck, J.C.; Covington, L.A.; Murphy, R.B.; Williams, R.; Knudtson, J. Use of oral creatine as an ergogenic aid for increased sports performance: Perceptions of adolescent athletes. *South Med. J.* **2001**, *94*, 608–612. [CrossRef]
20. McGuine, T.A.; Sullivan, J.C.; Bernhardt, D.A. Creatine supplementation in Wisconsin high school athletes. *WMJ* **2002**, *101*, 25–30.
21. Bell, A.; Dorsch, K.D.; McCreary, D.R.; Hovey, R. A look at nutritional supplement use in adolescents. *J. Adolesc. Health* **2004**, *34*, 508–516. [CrossRef]
22. Hoffman, J.R.; Faigenbaum, A.D.; Ratamess, N.A.; Ross, R.; Kang, J.; Tenenbaum, G. Nutritional supplementation and anabolic steroid use in adolescents. *Med. Sci. Sports Exerc.* **2008**, *40*, 15–24. [CrossRef]
23. Evans, M.W., Jr.; Ndetan, H.; Perko, M.; Williams, R.; Walker, C. Dietary supplement use by children and adolescents in the United States to enhance sport performance: Results of the National Health Interview Survey. *J. Prim. Prev.* **2012**, *33*, 3–12. [CrossRef]
24. Yager, Z.; McLean, S. Muscle building supplement use in Australian adolescent boys: Relationships with body image, weight lifting, and sports engagement. *BMC Pediatr.* **2020**, *20*, 89. [CrossRef] [PubMed]

25. Nagata, J.M.; Ganson, K.T.; Gorrell, S.; Mitchison, D.; Murray, S.B. Association Between Legal Performance-Enhancing Substances and Use of Anabolic-Androgenic Steroids in Young Adults. *JAMA Pediatr.* **2020**, *174*, 992–993. [CrossRef]
26. Petroczi, A.; Naughton, D.P.; Mazanov, J.; Holloway, A.; Bingham, J. Performance enhancement with supplements: Incongruence between rationale and practice. *J. Int. Soc. Sports Nutr.* **2007**, *4*, 19. [CrossRef] [PubMed]
27. Petroczi, A.; Naughton, D.P. The age-gender-status profile of high performing athletes in the UK taking nutritional supplements: Lessons for the future. *J. Int. Soc. Sports Nutr.* **2008**, *5*, 2. [CrossRef]
28. Jovanov, P.; Dordic, V.; Obradovic, B.; Barak, O.; Pezo, L.; Maric, A.; Sakac, M. Prevalence, knowledge and attitudes towards using sports supplements among young athletes. *J. Int. Soc. Sports Nutr.* **2019**, *16*, 27. [CrossRef]
29. Braun, H.; Koehler, K.; Geyer, H.; Kleiner, J.; Mester, J.; Schanzer, W. Dietary supplement use among elite young German athletes. *Int. J. Sport Nutr. Exerc. Metab.* **2009**, *19*, 97–109. [CrossRef] [PubMed]
30. Calfee, R.; Fadale, P. Popular ergogenic drugs and supplements in young athletes. *Pediatrics* **2006**, *117*, e577–e589. [CrossRef] [PubMed]
31. DesJardins, M. Supplement use in the adolescent athlete. *Curr. Sports Med. Rep.* **2002**, *1*, 369–373. [CrossRef]
32. Diehl, K.; Thiel, A.; Zipfel, S.; Mayer, J.; Schnell, A.; Schneider, S. Elite adolescent athletes' use of dietary supplements: Characteristics, opinions, and sources of supply and information. *Int. J. Sport Nutr. Exerc. Metab.* **2012**, *22*, 165–174. [CrossRef]
33. Grindstaff, P.D.; Kreider, R.; Bishop, R.; Wilson, M.; Wood, L.; Alexander, C.; Almada, A. Effects of creatine supplementation on repetitive sprint performance and body composition in competitive swimmers. *Int. J. Sport Nutr.* **1997**, *7*, 330–346. [CrossRef]
34. Juhasz, I.; Gyore, I.; Csende, Z.; Racz, L.; Tihanyi, J. Creatine supplementation improves the anaerobic performance of elite junior fin swimmers. *Acta Physiol. Hung.* **2009**, *96*, 325–336. [CrossRef]
35. Theodorou, A.S.; Havenetidis, K.; Zanker, C.L.; O'Hara, J.P.; King, R.F.; Hood, C.; Paradisis, G.; Cooke, C.B. Effects of acute creatine loading with or without carbohydrate on repeated bouts of maximal swimming in high-performance swimmers. *J. Strength Cond. Res.* **2005**, *19*, 265–269. [CrossRef]
36. Dawson, B.; Vladich, T.; Blanksby, B.A. Effects of 4 weeks of creatine supplementation in junior swimmers on freestyle sprint and swim bench performance. *J. Strength Cond. Res.* **2002**, *16*, 485–490.
37. Theodorou, A.S.; Cooke, C.B.; King, R.F.; Hood, C.; Denison, T.; Wainwright, B.G.; Havenetidis, K. The effect of longer-term creatine supplementation on elite swimming performance after an acute creatine loading. *J. Sports Sci.* **1999**, *17*, 853–859. [CrossRef]
38. Juhasz, I.; Kopkane, J.P.; Hajdu, P.; Szalay, G.; Kopper, B.; Tihanyi, J. Creatine Supplementation Supports the Rehabilitation of Adolescent Fin Swimmers in Tendon Overuse Injury Cases. *J. Sports Sci. Med.* **2018**, *17*, 279–288.
39. Mohebbi, H.; Rahnama, N.; Moghadassi, M.; Ranjbar, K. Effect of creatine supplementation on sprint and skill performance in young soccer players. *Middle East J. Sci. Res.* **2012**, *12*, 397–401. [CrossRef]
40. Ostojic, S.M. Creatine supplementation in young soccer players. *Int. J. Sport Nutr. Exerc. Metab.* **2004**, *14*, 95–103. [CrossRef]
41. Yanez-Silva, A.; Buzzachera, C.F.; Picarro, I.D.C.; Januario, R.S.B.; Ferreira, L.H.B.; McAnulty, S.R.; Utter, A.C.; Souza-Junior, T.P. Effect of low dose, short-term creatine supplementation on muscle power output in elite youth soccer players. *J. Int. Soc. Sports Nutr.* **2017**, *14*, 5. [CrossRef] [PubMed]
42. Claudino, J.G.; Mezencio, B.; Amaral, S.; Zanetti, V.; Benatti, F.; Roschel, H.; Gualano, B.; Amadio, A.C.; Serrao, J.C. Creatine monohydrate supplementation on lower-limb muscle power in Brazilian elite soccer players. *J. Int. Soc. Sports Nutr.* **2014**, *11*, 32. [CrossRef] [PubMed]
43. Jagim, A.R.; Stecker, R.A.; Harty, P.S.; Erickson, J.L.; Kerksick, C.M. Safety of Creatine Supplementation in Active Adolescents and Youth: A Brief Review. *Front. Nutr.* **2018**, *5*, 115. [CrossRef]
44. Kley, R.A.; Tarnopolsky, M.A.; Vorgerd, M. Creatine for treating muscle disorders. *Cochrane Database Syst. Rev.* **2013**. [CrossRef]
45. Tarnopolsky, M.A.; Mahoney, D.J.; Vajsar, J.; Rodriguez, C.; Doherty, T.J.; Roy, B.D.; Biggar, D. Creatine monohydrate enhances strength and body composition in Duchenne muscular dystrophy. *Neurology* **2004**, *62*, 1771–1777. [CrossRef] [PubMed]
46. Tarnopolsky, M.A. Clinical use of creatine in neuromuscular and neurometabolic disorders. *Subcell. Biochem.* **2007**, *46*, 183–204. [CrossRef]
47. Evangeliou, A.; Vasilaki, K.; Karagianni, P.; Nikolaidis, N. Clinical applications of creatine supplementation on paediatrics. *Curr. Pharm. Biotechnol.* **2009**, *10*, 683–690. [CrossRef]
48. Louis, M.; Poortmans, J.R.; Francaux, M.; Hultman, E.; Berre, J.; Boisseau, N.; Young, V.R.; Smith, K.; Meier-Augenstein, W.; Babraj, J.A.; et al. Creatine supplementation has no effect on human muscle protein turnover at rest in the postabsorptive or fed states. *Am. J. Physiol. Endocrinol. Metab.* **2003**, *284*, E764–E770. [CrossRef]
49. Escolar, D.M.; Buyse, G.; Henricson, E.; Leshner, R.; Florence, J.; Mayhew, J.; Tesi-Rocha, C.; Gorni, K.; Pasquali, L.; Patel, K.M.; et al. CINRG randomized controlled trial of creatine and glutamine in Duchenne muscular dystrophy. *Ann. Neurol.* **2005**, *58*, 151–155. [CrossRef]
50. Yoganathan, S.; Arunachal, G.; Kratz, L.; Varman, M.; Sudhakar, S.V.; Oommen, S.P.; Jain, S.; Thomas, M.; Babuji, M. Guanidinoacetate Methyltransferase (GAMT) Deficiency, A Cerebral Creatine Deficiency Syndrome: A Rare Treatable Metabolic Disorder. *Ann. Indian Acad. Neurol.* **2020**, *23*, 419–421. [CrossRef]
51. Stockler, S.; Schutz, P.W.; Salomons, G.S. Cerebral creatine deficiency syndromes: Clinical aspects, treatment and pathophysiology. *Subcell. Biochem.* **2007**, *46*, 149–166. [CrossRef] [PubMed]

52. Stockler-Ipsiroglu, S.; van Karnebeek, C.D. Cerebral creatine deficiencies: A group of treatable intellectual developmental disorders. *Semin. Neurol.* **2014**, *34*, 350–356. [CrossRef]

53. Stockler, S.; Holzbach, U.; Hanefeld, F.; Marquardt, I.; Helms, G.; Requart, M.; Hanicke, W.; Frahm, J. Creatine deficiency in the brain: A new, treatable inborn error of metabolism. *Pediatr. Res.* **1994**, *36*, 409–413. [CrossRef]

54. Stockler, S.; Hanefeld, F.; Frahm, J. Creatine replacement therapy in guanidinoacetate methyltransferase deficiency, a novel inborn error of metabolism. *Lancet* **1996**, *348*, 789–790. [CrossRef]

55. Trotier-Faurion, A.; Dezard, S.; Taran, F.; Valayannopoulos, V.; de Lonlay, P.; Mabondzo, A. Synthesis and biological evaluation of new creatine fatty esters revealed dodecyl creatine ester as a promising drug candidate for the treatment of the creatine transporter deficiency. *J. Med. Chem.* **2013**, *56*, 5173–5181. [CrossRef]

56. Trotier-Faurion, A.; Passirani, C.; Bejaud, J.; Dezard, S.; Valayannopoulos, V.; Taran, F.; de Lonlay, P.; Benoit, J.P.; Mabondzo, A. Dodecyl creatine ester and lipid nanocapsule: A double strategy for the treatment of creatine transporter deficiency. *Nanomedicine* **2014**, *10*, 185–191. [CrossRef]

57. Sipila, I.; Rapola, J.; Simell, O.; Vannas, A. Supplementary creatine as a treatment for gyrate atrophy of the choroid and retina. *N. Engl. J. Med.* **1981**, *304*, 867–870. [CrossRef] [PubMed]

58. Vannas-Sulonen, K.; Sipila, I.; Vannas, A.; Simell, O.; Rapola, J. Gyrate atrophy of the choroid and retina. A five-year follow-up of creatine supplementation. *Ophthalmology* **1985**, *92*, 1719–1727. [CrossRef]

59. Walter, M.C.; Lochmuller, H.; Reilich, P.; Klopstock, T.; Huber, R.; Hartard, M.; Hennig, M.; Pongratz, D.; Muller-Felber, W. Creatine monohydrate in muscular dystrophies: A double-blind, placebo-controlled clinical study. *Neurology* **2000**, *54*, 1848–1850. [CrossRef]

60. Braegger, C.P.; Schlattner, U.; Wallimann, T.; Utiger, A.; Frank, F.; Schaefer, B.; Heizmann, C.W.; Sennhauser, F.H. Effects of creatine supplementation in cystic fibrosis: Results of a pilot study. *J. Cyst. Fibros.* **2003**, *2*, 177–182. [CrossRef]

61. Louis, M.; Lebacq, J.; Poortmans, J.R.; Belpaire-Dethiou, M.C.; Devogelaer, J.P.; Van Hecke, P.; Goubel, F.; Francaux, M. Beneficial effects of creatine supplementation in dystrophic patients. *Muscle Nerve* **2003**, *27*, 604–610. [CrossRef]

62. Sakellaris, G.; Nasis, G.; Kotsiou, M.; Tamiolaki, M.; Charissis, G.; Evangeliou, A. Prevention of traumatic headache, dizziness and fatigue with creatine administration. A pilot study. *Acta Paediatr* **2008**, *97*, 31–34. [CrossRef] [PubMed]

63. Bourgeois, J.M.; Nagel, K.; Pearce, E.; Wright, M.; Barr, R.D.; Tarnopolsky, M.A. Creatine monohydrate attenuates body fat accumulation in children with acute lymphoblastic leukemia during maintenance chemotherapy. *Pediatr. Blood Cancer* **2008**, *51*, 183–187. [CrossRef] [PubMed]

64. Kalamitsou, S.; Masino, S.; Evangelos, P.; Gogou, M.; Katsanika, I.; Papadopoulou-Legbelou, K.; Aspasia, S.; Spilioti, M.; Evangeliou, A. The effect of creatine supplementation on seizure control in children under ketogenic diet: A pilot study. *Integr. Mol. Med.* **2019**, *6*, 1–6. [CrossRef]

65. Dover, S.; Stephens, S.; Schneiderman, J.E.; Pullenayegum, E.; Wells, G.D.; Levy, D.M.; Marcuz, J.A.; Whitney, K.; Schulze, A.; Tein, I.; et al. The effect of creatine supplementation on muscle function in childhood myositis: A randomized, double-blind, placebo-controlled feasibility study. *J. Rheumatol.* **2020**. [CrossRef]

66. Kemp, G.J.; Taylor, D.J.; Dunn, J.F.; Frostick, S.P.; Radda, G.K. Cellular energetics of dystrophic muscle. *J. Neurol. Sci.* **1993**, *116*, 201–206. [CrossRef]

67. Nanto-Salonen, K.; Komu, M.; Lundbom, N.; Heinanen, K.; Alanen, A.; Sipila, I.; Simell, O. Reduced brain creatine in gyrate atrophy of the choroid and retina with hyperornithinemia. *Neurology* **1999**, *53*, 303–307. [CrossRef] [PubMed]

68. Valtonen, M.; Nanto-Salonen, K.; Jaaskelainen, S.; Heinanen, K.; Alanen, A.; Heinonen, O.J.; Lundbom, N.; Erkintalo, M.; Simell, O. Central nervous system involvement in gyrate atrophy of the choroid and retina with hyperornithinaemia. *J. Inherit. Metab. Dis.* **1999**, *22*, 855–866. [CrossRef]

69. Heinanen, K.; Nanto-Salonen, K.; Komu, M.; Erkintalo, M.; Heinonen, O.J.; Pulkki, K.; Valtonen, M.; Nikoskelainen, E.; Alanen, A.; Simell, O. Muscle creatine phosphate in gyrate atrophy of the choroid and retina with hyperornithinaemia–clues to pathogenesis. *Eur. J. Clin. Invest.* **1999**, *29*, 426–431. [CrossRef]

70. Bakian, A.V.; Huber, R.S.; Scholl, L.; Renshaw, P.F.; Kondo, D. Dietary creatine intake and depression risk among U.S. adults. *Transl. Psychiatry* **2020**, *10*, 52. [CrossRef]

71. Kious, B.M.; Kondo, D.G.; Renshaw, P.F. Creatine for the Treatment of Depression. *Biomolecules* **2019**, *9*, 406. [CrossRef]

72. Cullen, K.R.; Padilla, L.E.; Papke, V.N.; Klimes-Dougan, B. New Somatic Treatments for Child and Adolescent Depression. *Curr. Treat. Options Psychiatry* **2019**, *6*, 380–400. [CrossRef]

73. Toniolo, R.A.; Silva, M.; Fernandes, F.B.F.; Amaral, J.; Dias, R.D.S.; Lafer, B. A randomized, double-blind, placebo-controlled, proof-of-concept trial of creatine monohydrate as adjunctive treatment for bipolar depression. *J. Neural Transm.* **2018**, *125*, 247–257. [CrossRef]

74. Riesberg, L.A.; Weed, S.A.; McDonald, T.L.; Eckerson, J.M.; Drescher, K.M. Beyond muscles: The untapped potential of creatine. *Int. Immunopharmacol.* **2016**, *37*, 31–42. [CrossRef] [PubMed]

75. De Guingand, D.L.; Ellery, S.J.; Davies-Tuck, M.L.; Dickinson, H. Creatine and pregnancy outcomes, a prospective cohort study in low-risk pregnant women: Study protocol. *BMJ Open* **2019**, *9*, e026756. [CrossRef]

76. de Guingand, D.L.; Palmer, K.R.; Bilardi, J.E.; Ellery, S.J. Acceptability of dietary or nutritional supplementation in pregnancy (ADONS)—Exploring the consumer's perspective on introducing creatine monohydrate as a pregnancy supplement. *Midwifery* **2020**, *82*, 102599. [CrossRef]

77. United States Food and Drug Administration. Center for Food Safety & Applied Nutrition Adverse Event Reporting System, 07/29/2020 ed.2021. Available online: https://www.fda.gov/food/compliance-enforcement-food/cfsan-adverse-event-reporting-system-caers#files (accessed on 5 February 2021).

78. United States Food and Drug Administration. Recently Published GRAS Notices and FDA Letters. Available online: https://www.fda.gov/food/gras-notice-inventory/recently-published-gras-notices-and-fda-letters (accessed on 5 February 2021).

79. Simpson, A.J.; Horne, S.; Sharp, P.; Sharps, R.; Kippelen, P. Effect of Creatine Supplementation on the Airways of Youth Elite Soccer Players. *Med. Sci. Sports Exerc.* **2019**, *51*, 1582–1590. [CrossRef] [PubMed]

80. Bukic, J.; Rusic, D.; Bozic, J.; Zekan, L.; Leskur, D.; Seselja Perisin, A.; Modun, D. Differences among health care students' attitudes, knowledge and use of dietary supplements: A cross-sectional study. *Complement. Ther. Med.* **2018**, *41*, 35–40. [CrossRef]

81. Antonio, J.; Candow, D.G.; Forbes, S.C.; Gualano, B.; Jagim, A.R.; Kreider, R.B.; Rawson, E.S.; Smith-Ryan, A.E.; VanDusseldorp, T.A.; Willoughby, D.S.; et al. Common questions and misconceptions about creatine supplementation: What does the scientific evidence really show? *J. Int. Soc. Sports Nutr.* **2021**, *18*, 13. [CrossRef]

Review

Creatine for Exercise and Sports Performance, with Recovery Considerations for Healthy Populations

Benjamin Wax [1,*], Chad M. Kerksick [2,*], Andrew R. Jagim [3], Jerry J. Mayo [4], Brian C. Lyons [5] and Richard B. Kreider [6]

[1] Applied Physiology Laboratory, Department of Kinesiology, Mississippi State University, Mississippi State, MS 39759, USA

[2] Exercise & Performance Nutrition Laboratory, College of Science, Technology, and Health, Lindenwood University, St. Charles, MO 63301, USA

[3] Sports Medicine, Mayo Clinic Health System, La Crosse, WI 54601, USA; Jagim.Andrew@mayo.edu

[4] Department of Nutrition and Family Sciences, University of Central Arkansas, Conway, AR 72035, USA; jmayo@uca.edu

[5] Health, Kinesiology, and Sport Management Department, University of Wisconsin—Parkside, Kenosha, WI 53141, USA; lyons@uwp.edu

[6] Exercise & Sport Nutrition Lab, Human Clinical Research Facility, Department of Health & Kinesiology, Texas A&M University, College Station, TX 77843, USA; rbkreider@tamu.edu

* Correspondence: bw244@msstate.edu (B.W.); ckerksick@lindenwood.edu (C.M.K.)

Citation: Wax, B.; Kerksick, C.M.; Jagim, A.R.; Mayo, J.J.; Lyons, B.C.; Kreider, R.B. Creatine for Exercise and Sports Performance, with Recovery Considerations for Healthy Populations. *Nutrients* **2021**, *13*, 1915. https://doi.org/10.3390/nu13061915

Academic Editor: Elena Barbieri

Received: 19 March 2021
Accepted: 30 May 2021
Published: 2 June 2021

Publisher's Note: MDPI stays neutral with regard to jurisdictional claims in published maps and institutional affiliations.

Abstract: Creatine is one of the most studied and popular ergogenic aids for athletes and recreational weightlifters seeking to improve sport and exercise performance, augment exercise training adaptations, and mitigate recovery time. Studies consistently reveal that creatine supplementation exerts positive ergogenic effects on single and multiple bouts of short-duration, high-intensity exercise activities, in addition to potentiating exercise training adaptations. In this respect, supplementation consistently demonstrates the ability to enlarge the pool of intracellular creatine, leading to an amplification of the cell's ability to resynthesize adenosine triphosphate. This intracellular expansion is associated with several performance outcomes, including increases in maximal strength (low-speed strength), maximal work output, power production (high-speed strength), sprint performance, and fat-free mass. Additionally, creatine supplementation may speed up recovery time between bouts of intense exercise by mitigating muscle damage and promoting the faster recovery of lost force-production potential. Conversely, contradictory findings exist in the literature regarding the potential ergogenic benefits of creatine during intermittent and continuous endurance-type exercise, as well as in those athletic tasks where an increase in body mass may hinder enhanced performance. The purpose of this review was to summarize the existing literature surrounding the efficacy of creatine supplementation on exercise and sports performance, along with recovery factors in healthy populations.

Keywords: supplementation; ergogenic aid; athletic performance; weightlifting; resistance exercise; training; muscular power; recovery; muscular adaptation; muscle damage

1. Introduction

In the area of sports performance and exercise, both athletes and recreational non-athletes are continuously seeking competitive advantages to improve their health and optimize physical performance. Although various activities and considerations interact to achieve this end, many people turn to various exercise and nutritional strategies to augment performance (i.e., enhanced muscular strength, power, and force) [1,2]. One of the most commonly used and scientifically supported ergogenic aids is creatine monohydrate (commonly referred to as creatine) [1,3–5]. Creatine is an amino acid found in relatively high concentrations in skeletal muscle. Since 1992, when the first reports emerged that exogenous creatine monohydrate supplementation increases intramuscular phosphocreatine (PCr) stores [6], and shortly afterwards, when these increases were inextricably linked to augmented exercise performance [7,8], the ability of creatine to function as an ergogenic aid has attracted great interest. Still today, creatine is one of the most popular nutritional ergogenic aids for athletes and recreational performers [1,3,4]. In addition to its popularity in the consumer realm, creatine's ability to enhance or augment some types of exercise performance has arguably been one of the most researched topics in the sport nutrition literature for the past 25 years [1,3,7,9–11]. In this regard, creatine has yielded predominantly positive effects regarding exercise performance measures with no ergolytic effects and minimal to no side effects in populations ranging from adolescents to the elderly [3,9]. The reported ergogenic benefits of creatine monohydrate include enhanced force output, augmented power output, increased strength, increased anaerobic threshold, increased work capacity, enhanced recovery, and enhanced training adaptations [1,3,4,9,12,13].

Although a complete discussion is beyond the scope of this review, several supplementation strategies have been explored to increase intramuscular creatine stores. A loading phase was initially proposed by Harris et al. in 1992 [6] and has subsequently been used in a large number of scientific investigations. This approach requires consuming four separate doses of 5 g/day for five consecutive days and consistently leads to a 20%–40% increase in creatine content [3]. Later, Hultman et al. [14] determined that smaller 'maintenance' doses (2–5 g per dose, 1 ×/day, or 0.03 g/kg/dose) could be used to maintain elevated creatine stores in the muscle. It is now commonly accepted that a loading phase may not be needed, but this approach remains the most rapid means to increase intramuscular PCr levels and, thereby, performance [14,15]. Notably, Law and colleagues [16] compared the efficacy of creatine loading on performance measures using a 2- and 5-day regimen (4 × 5 g/day) in 20 physically active men. They reported significant improvements in maximal leg strength and average anaerobic power following a 5-day creatine loading regimen compared to the placebo group; however, no significance in performance was found following 2 days of loading. Additionally, Sale et al. [17] found that the total ingestion of 20 g of creatine at 1 g per 30 min intervals for 5 days yielded lower urinary excretion of creatine than the typical loading regimen of 4 × 5 g/day over a 5-day period, leading the authors to conclude that this likely resulted in higher intramuscular levels. In this respect, it is without question that increasing intramuscular creatine stores through any number of supplemental approaches can increase intramuscular PCr levels and that these increases are directly linked to various ergogenic outcomes [3,9]. In this respect, Table 1 (adapted from: [3]) outlines the potential ergogenic benefits of creatine supplementation, whereas Table 2 provides examples of sports or sporting events that may be enhanced by creatine supplementation (also adapted from [3]). In addition to these tables, results from previous selected original investigations and review papers surrounding the ergogenic potential of creatine supplementation are summarized throughout this paper in tables. Finally, the interested reader is directed to other reviews that have outlined the impact of creatine supplementation on exercise performance [3,9,10,12,13,18]. The purpose of this review is to summarize the existing literature surrounding the efficacy of creatine supplementation on exercise and sports performance, along with recovery factors in healthy populations.

Table 1. Potential ergogenic benefits of creatine supplementation.

- Increased single and repetitive sprint performance
- Increased work performed during sets of maximal effort muscle contractions
- Increased muscle mass and strength adaptations during training
- Enhanced glycogen synthesis
- Increased anaerobic threshold
- Possible enhancement of aerobic capacity via greater shuttling of ATP from mitochondria
- Increased work capacity
- Enhanced recovery
- Greater training tolerance

Adopted from Kreider et al. 2017 [3].

Table 2. Examples of sports and activities in which performance may be enhanced by creatine supplementation.

Increased PCr
- Track sprints: 60–200 m
- Swim sprints: 50 m
- Pursuit cycling

Increased PCr Resynthesis
- Basketball
- Field hockey
- America Football
- Ice hockey
- Lacrosse
- Volleyball

Reduced Muscle Acidosis
- Downhill skiing
- Water Sports (e.g., Rowing, Canoeing, Kayaking, Stand-Up Paddling)
- Swim events: 100, 200 m
- Track events: 400, 800 m
- Combat Sports (e.g., MMA, Wrestling, Boxing, etc.)

Oxidative Metabolism
- Basketball
- Soccer
- Team handball
- Tennis
- Volleyball
- Interval Training in Endurance Athletes

Increased Body Mass/Muscle Mass
- American Football
- Bodybuilding
- Combat Sports (e.g., MMA, Wrestling, Boxing, etc.)
- Powerlifting
- Rugby
- Track/Field events (Shot put; Javelin; Discus; Hammer Throw)
- Olympic Weightlifting

Adopted from Kreider et al. 2017 [3].

2. Materials and Methods

This review was completed using a narrative, non-systematic approach. A range of databases, including PubMed, Medline, Google Scholar, and EBSCO-host, were searched for this review paper (see Figure 1). A representative but non-exclusive list of keywords for these searches included: creatine, creatine supplementation, exercise, performance, ergogenic aid, exercise performance, athletic performance, athlete, weightlifting, maximal, resistance exercise, resistance training, muscular power, muscular force, skill performance, sports performance, sprinting, jumping, recovery, muscular adaptation, training adaptations, and muscle damage. Articles were chosen for inclusion based on the information they outlined and with a specific focus on exercise, performance, training adaptations, sport-specific skills, or recovery in healthy populations. Further citations were found, evaluated, and incorporated from the bibliographies of the selected literature. Articles with a focus on clinical applications or consisting of study protocols conducted among clinical populations were not included in the current review.

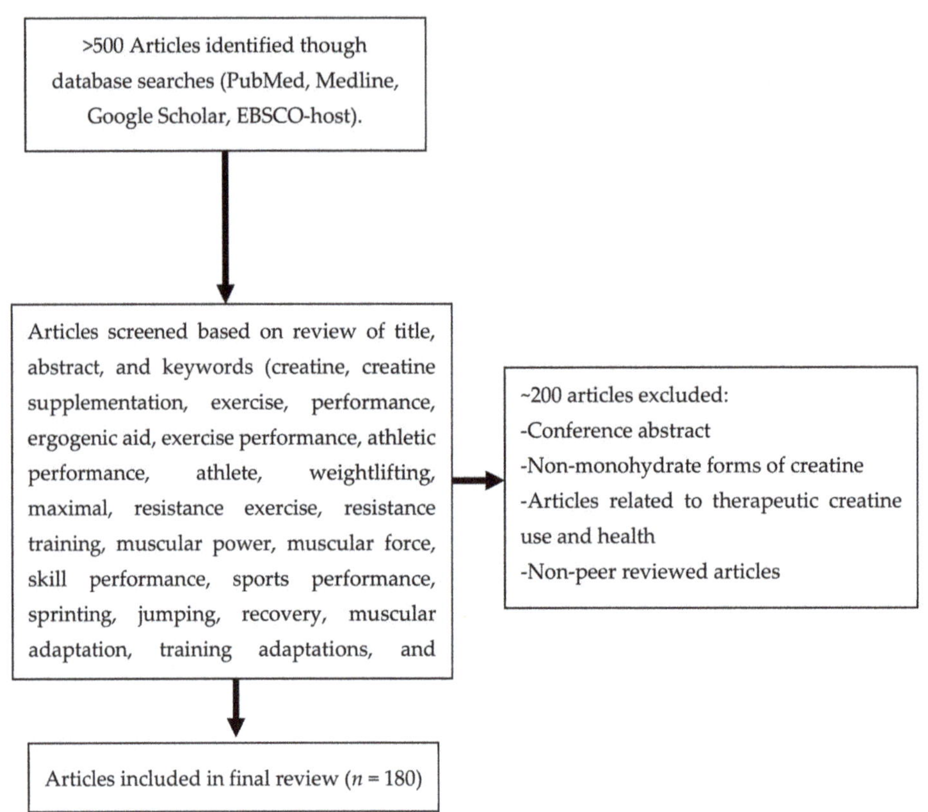

Figure 1. General flow diagram highlighting the selection process for included articles.

3. Exercise and Sports Performance

Creatine's ability to increase various parameters of acute exercise performance is well documented [3,9,10]. A review by Kreider in 2003 summarized the literature and concluded that approximately 70% of these studies had reported an improvement in some aspect of exercise performance [9]. The magnitude of the increase in performance is dependent on a large number of variables, which can include the dosing regimen, training status of the athlete, and any one of a number of acute exercise variables (intensity of exercise, duration of effort, etc.). An overview of this literature reveals that performance increases of 10%–15% are typically observed [9,12]. More specifically, 5%–15% improvements in maximal power and strength, anaerobic capacity, and work performance during repetitive sprint performance are commonly reported, whereas improvements in single-effort sprint performance have been indicated to range from 1%–5% improvements. Alternatively, no consistent reports indicate that supplementation with creatine may have an ergolytic or performance-decreasing response. In this respect, a large number of studies have commonly reported an increase in body mass of 1–2 kg during the first week of loading [3], which may or may not have ergolytic implications, depending on the type of athlete and the phase of training. Finally, research involving various types of endurance activity in conjunction with creatine supplementation has received some attention as well.

3.1. Strength and Power

In most but not all studies, creatine supplementation has been demonstrated to be an effective ergogenic aid for increasing muscular strength and/or power, and these findings have included trained and untrained men and women, including athletes and non-athletes. The primary mechanism behind these ergogenic outcomes for creatine supplementation appears to be attributable, in part, to increases in intramuscular PCr concentrations [3,6,14,19]. Due to its potential not only to enhance strength and power output but also to expedite recovery from intense intermittent exercise, creatine supplementation has been shown to allow for increased volumes of work and increased work output during resistance training, which may then translate into greater strength gains [3,10,20,21].

3.1.1. Short Term Benefits (<2 Weeks)

Improvements in strength and power performance have been observed following creatine supplementation. Generally, loading doses (i.e., 0.3 g/kg/day or ~20 g/day) are often used for short-duration supplementation periods or as part of a loading phase, which is commonly then followed by a more extended period of supplementation at a maintenance dose (i.e., 0.03 g/kg/day or 2–5 g/day). As intramuscular PCr stores increase through supplementation, subsequent improvements of 5%–15% in various performance parameters are commonly reported [10,22], sometimes even after relatively short periods of creatine supplementation (0–14 days). A study by Law et al. [16] reported significant improvements in average anaerobic power during a Wingate test and in back squat strength following a 5-day creatine supplementation protocol combined with a resistance training program. Interestingly, similar benefits were not observed after only 2 days of creatine loading, thereby indicating that perhaps >2 days of creatine loading is required to elicit significant strength and power-related benefits when combined with resistance training. Short-term benefits have also been reported in young adult males who were naïve to resistance training. After 10 days of creatine supplementation, improvements in the bench press and squat strength and power output were observed [23]. It is worth noting that subjects did not participate in a resistance training program during the supplementation period, yet ergogenic benefits were still present [23]. More recently, similar benefits in isometric leg strength were reported in young adult female futsal players following 7 days of creatine supplementation in conjunction with a concurrent resistance training program [24]. In contrast to these outcomes, creatine supplementation has not always yielded statistically significant ergogenic results. For example, following 7 days of creatine supplementation, maximal isometric knee extension strength was not altered in either the creatine or placebo group in 31 resistance-trained individuals who continued resistance training just as they had during the previous week [25]. Similarly, no strength differences were observed for 1 RM bench press or 1 RM leg extension exercises for either group after 7 days of creatine or placebo supplementation with no concurrent training in young adult males [26].

Thus, when attempting to glean meaning from numerous and varied study designs and supplementation protocols, it is crucial to look at how creatine was delivered, the nature of the subjects, whether or not concurrent resistance training occurred, whether the study design allowed for self-progression of training volume, the duration of the study, intended target outcomes, and several other factors. In this respect, a review by Rawson et al. [10] concluded that of the 22 identified studies on resistance training with creatine supplementation, the average increase in relative muscle strength (i.e., 1, 3, or 10 RM) was approximately 8% greater than that from resistance training alone. Furthermore, the average increase in muscular endurance (number of repetitions completed with a fixed load) was 14% greater than that following placebo ingestion and resistance training alone. More recently, a systematic review [18] identified 60 studies that met their criteria of randomized controlled trial study designs that utilized a double-blind, placebo control approach to examine the effects of creatine supplementation on lower limb performance. The outcomes of the review indicated an overall effect size of 0.336 and 0.297 for the strength increases in the back squat and leg press exercises, respectively. Furthermore, the calculated effect size for quadriceps strength was 0.266, with an overall global lower limb strength effect size of 0.266. Interestingly, these outcomes were observed independently of the target population or supplementation regimen, thereby supporting the overall efficacy of creatine supplementation, independently of outside-influencing factors. In a follow-up review, the same group [27] identified 53 studies utilizing similar search criteria but focused on the effects of creatine supplementation for upper limb strength. Results of the meta-analysis indicated an overall effect size of 0.265 and 0.677 for strength changes associated with the bench press and chest press exercises, respectively. For the pectoral exercise and global upper limb strength, effect sizes of 0.289 and 0.317 were found, again with no link with characteristics attributable to the population of supplementation regimen used, supporting the overall efficacy of creatine for upper body exercise performance independently of contextual factors. Table 7 provides a summary of selected studies that have examined the acute (<2 weeks) effects of creatine supplementation on strength and anaerobic performance.

Table 3. Summary of selected studies examining the acute effects of creatine monohydrate on strength performance.

Author and Year	Subjects	Duration	Dosing Protocol	Primary Variables	Results	Adverse Events
Peak Torque/Force Production						
Greenhaff et al. 1993	12 healthy males	5 days	20/d for 5 days	isokinetic performance during repeated intervals	↑ muscle torque ↓ plasma ammonia ↔ BLa	None reported
Casey et al. 1996	9 active males	5 days	20 g/day for 5 days	cycling sprint performance	↑ total work	None reported
Gilliam et al. 2000	23 healthy males	5 days	20 g/day for 5 days	isokinetic performance of quadriceps	↔ muscle torque	None reported
Rossouw et al. 2000	13 trained power-lifters	5 days	9 g/day for 5 days	isokinetic knee extension	↑ peak torque ↑ average power, ↑ total work ↑ work output ↑ deadlift volume	None reported
Kilduff et al. 2002	32 trained males	5 days	20 g/day for 5 days	isometric bench press	↑ peak force ↑ total force	None reported

Table 3. *Cont.*

Author and Year	Subjects	Duration	Dosing Protocol	Primary Variables	Results	Adverse Events
Strength Performance						
Birch et al. 1994	14 healthy males	5 days	20 g/day for 5 days	isokinetic cycling performance during repeated intervals	↑ mean power output ↑ peak power output ↑ total power output ↓ plasma ammonia ↔ BLa	None reported
Barnett et al.1996	17 active males	4 days	280 mg/kg for 4 days	10 s sprints on a cycle ergometer	↔ mean power output ↔ peak power output ↔ blood pH ↔ BLa	None reported
Strength and Performance Outcomes						
Edwards et al. 2000	21 active males	6 days	20 g/d for 6 days	anaerobic intervals on a treadmill	↔ speed test performance ↔ BLa ↓ plasma ammonia	None reported
Rockwell et al. 2001	16 trained males	4 days	20 g/day for 4 days	repeated cycling sprints	↔ total work ↔ maximal power ↔ work capacity	None reported
Volek et al. 2001	10 healthy males	7 days	0.3 g/kg for 7 days	repeated cycling sprints cardiovascular and thermoregulatory responses	↔ cardiovascular or thermoregulatory responses ↑ peak power ↑ mean power	None reported
Finn et al. 2001	16 male triathletes	5 days	20 g/day for 5 days	repeated cycling sprints	↔ peak power ↔ fatigue index ↔ mean power	None reported

Table 3. *Cont.*

Author and Year	Subjects	Duration	Dosing Protocol	Primary Variables	Results	Adverse Events
			Strength and Performance Outcomes			
Ziegenfuss et al. 2002	20 (10 male, 10 female) power athletes	3 days	0.35 g/kg of fat-free mass for 3 days	repeated cycling sprints	↑ peak power ↑ total work	None reported
Yquel et al. 2002	9 healthy males	6 days	20 g/day for 6 days	repeated dynamic plantar flexion muscle pH	↑ mean power ↔ muscle pH	None reported
Delecluse et al. 2003	12 (7 male, 5 female) competitive sprinters	7 days	0.35 g/day for 7 days	single 40-m sprint repeated 40-m sprints	↔ single 40-m sprint ↔ repeated 40-m sprints	None reported
Kocak et al. 2003	20 elite male wrestlers	5 days	20 g/day for 5 days	average and peak power during wingate anaerobic power test	↑ average power ↑ peak power	None reported
Selsby et al. 2004	31 trained males	10 days	2.5 g/day for 10 days	bench press strength bench press endurance	↔ bench press strength ↔ bench press endurance	None reported
Zuniga et al. 2012	22 active males	7 days	20 g/day for 7 days	wingate cycling test lower body strength upper body strength	↑ mean power output ↔ leg extension ↔ bench press	None reported
del Favero et al. 2012	34 untrained males	10 days	20 g/day for 10 days	lower body strength upper body strength	↑ bench press ↑ squat	None reported
Wang et al. 2017	17 high school canoeists	6 days	20 g/day for 6 days	upper body strength overhead medicine ball throw Post-activation potentiation	↑ upper body strength ↔ overhead medicine ball throw ↔ post-activation potentiation	None reported

BLa = blood lactate; CR or no identifier = creatine monohydrate; ↓ indicates decrease; ↑ indicates increase; ↔ indicates no difference.

3.1.2. Longer-Term Training Adaptations (>2 Weeks)

Ergogenic benefits have also been reported following moderate-duration supplementation periods across several different populations. Although short-term benefits from creatine supplementation tend to be more specific to bouts of high-intensity exercise performance in regards to work output and anaerobic capacity, moderate-length supplementation periods appear to confer ergogenic benefits by helping to facilitate increases in training volume, thereby helping to augment training adaptations when combined with a structured training program. This indicates that consistent improvements in training quality and total work completed, underpinned by increasing intramuscular PCr content, are likely driving a lot of the training adaptations and benefits of creatine supplementation. These benefits appear to occur in both those who have been previously participating in resistance training programs and those who have been previously sedentary. For example, greater gains in 3 RM bench press were observed in experienced powerlifters supplementing with creatine compared to those supplementing with a placebo over a 26-day training period. Those in the creatine group also performed more total repetitions than the placebo group at 85% of 3 RM [28]. Similar results were reported in a study by Antonio et al. [29], who observed greater improvements in fat-free mass accrual and bench press 1 RM strength in recreational male body builders following a 5-week creatine supplementation period, in conjunction with a body-building-type resistance training program compared to those in the placebo group. Likewise, following a 6-week training period in experienced resistance-trained males, individuals supplementing with creatine displayed greater increases in bench press 1 RM compared to a placebo [30]. A 12-week concurrent periodized resistance training program paired with creatine supplementation resulted in significant improvements in bench press and back squat in 19 resistance-trained men. Additionally, improved jump squat performance was also observed. Interestingly, representing a protocol that is conducive to the performance-enhancing potential of creatine, subjects were provided with individualized training instruction throughout this study and were encouraged to progressively increase the intensity and volume of training throughout the study. The researchers noted that the creatine group displayed greater increases in exercise intensity and volume as the study progressed, which may partially explain the results [21]. In contrast, when subjects were involved in concurrent resistance training in which the training volume and intensity were controlled (i.e., subjects were prohibited from exceeding pre-determined numbers of repetitions and sets), it was concluded that creatine supplementation with concurrent weight training did not provide a greater ergogenic benefit when compared to the placebo group. This supports the notion that some of the ergogenic benefits are likely underpinned by facilitating increases in training intensity and work volume throughout a training program [31]. Table 4 provides a summary of selected studies that have examined the long-term (>2 weeks) effects of creatine monohydrate on strength and anaerobic performance.

Table 4. Summary of selected studies examining the long-term (>2 weeks) effects on strength performance.

Author and Year	Subjects	Duration	Dosing Protocol	Primary Variables	Results	Adverse Events
			Peak Torque/Force Production			
Stevenson et al. 2001	18 (17 males, 1 female) trained subjects	9 weeks	20 g/day for 7 days 5 g/day 8 weeks	maximal torque on isokinetic dy-namome-ter quadriceps cross-sectional area	$\leftrightarrow$ maximal torque $\uparrow$ cross-sectional area	None reported
Chrusch et al. 2001	30 healthy older males	12 weeks	0.03 g/kg/day for 5 days 0.07 g/kg/day for 11 weeks	lean tissue bench press strength leg press strength knee extension muscle endurance average power	$\uparrow$ lean tissue $\leftrightarrow$ bench press $\uparrow$ leg press $\uparrow$ knee torque $\uparrow$ muscle endurance $\uparrow$ average power	None reported
			Strength			
Vandenberghe et al.1997	19 healthy female subjects	11 weeks	5 g/day for 4 days 2.5 g/day for 10 weeks	arm flexion on isokinetic dy-namome-ter upper and lower body muscle strength body com-position	$\uparrow$ arm torque $\uparrow$ leg press, leg extension, squat $\leftrightarrow$ bench press, leg curl, shoulder press $\uparrow$ lean muscle mass	None reported
Kelly et al. 1998	18 male power-lifters	26 days	20 g/day for 5 days 5 g/day for 21 days	bench press strength bench press endurance total body mass	$\uparrow$ bench press strength $\uparrow$ bench press endurance $\uparrow$ body mass	None reported

Table 4. *Cont.*

				Strength		
Volek et al. 1999	19 trained males	12 weeks	25 g/day for 7 days 5 g/day for 11 weeks	body mass fat-free mass bench press strength squat strength quadriceps cross-sectional area	↔ body mass ↑ fat-free mass ↑ bench press ↑ squat ↑ quadriceps cross-sectional area	None reported
Becque et al. 2000	23 trained males	6 weeks	20 g/day for 5 days 2 g/day for days 6–42	upper body strength body composition	↑ arm flexor strength ↑ body mass ↔ body fat ↑ fat-free mass ↑ upper arm muscle area	None reported
Brenner et al. 2000	20 female college lacrosse players	5 weeks	20 g/day for 7 days 2 g/day for days 8–32	body composition bench press strength knee extension strength knee extension endurance BLa	↔ body composition ↑ bench press strength ↔ knee extension strength ↔ knee extension endurance ↔ BLa	None reported
Larson-Meyer et al. 2000	14 female college soccer players	13 weeks	15 g/day for 5 days 5 g/day 5 days/week for 12 weeks	bench press strength squat strength vertical jump body composition	↑ bench press strength ↑ squat strength ↔ vertical jump ↑ body mass ↑ lean mass ↔ body fat	1 subject reported nausea

Table 4. *Cont.*

				Strength		
Bemben et al. 2001	25 male college football players	9 weeks	20 g/day for 5 days 5 g/day for 8 weeks	bench press strength squat strength power clean strength Wingate cycling test isokinetic strength body composition	↑ bench press strength ↑ squat strength ↑ power clean strength ↑ anaerobic power, capacity and % decrement ↑ peak torque knee flexion ↔ peak torque knee extension ↑ lean body mass ↔ body fat	None reported
Burke et al. 2001	47 active male subjects	21 days	7.7 g/day for 21 days	bench press on a isokinetic dynamometer	↑ peak force ↑ peak power ↑ time to fatigue	None reported
Chrusch et al. 2001	30 healthy older males	12 weeks	0.03 g/kg/day for 5 days 0.07 g/kg/day for 11 weeks	lean tissue bench press strength leg press strength knee extension muscle endurance average power	↑ lean tissue ↔ bench press ↑ leg press ↑ knee torque ↑ muscle endurance ↑ average power	None reported
Wilder et al. 2001	25 male college football players	10 weeks	3 g/day for 10 week or 20 g/day for 7 days, then 5 g/day for rest of the study	squat strength body composition	↔ squat strength ↔ lean body mass ↔ fat mass	None reported

Table 4. *Cont.*

				Strength		
Burke et al. 2003	49 (20 male, 29 female) active subjects	8 weeks	0.25 g/kg lean tissue/day for 7 days 0.0625 g/kg lean tissue/day for 49 days	bench press strength leg press strength isokinetic endurance quadriceps cross-sectional area body composition	↑ bench press ↔ leg press ↑ total work ↑ body mass ↑ lean body mass ↑ cross-sectional area	None reported
Ferguson et al. 2006	26 trained females	10 weeks	0.3 g/kg for 7 days 0.03 g/kg for 9 weeks	bench press strength leg press strength body composition	↔ bench press ↔ leg press ↔ total mass ↔ lean body mass ↔ fat mass	None reported
Kerksick et al. 2009	24 trained males	4 weeks	20 g/day for 5 days 5 g/day for 23 days	bench press strength leg press strength isokinetic knee extension Wingate cycling test body composition	↑ bench press ↑ leg press ↔ peak torque ↔ peak power ↑ lean body mass ↑ fat-free mass	None reported
Camic et al. 2010	22 untrained males	28 days	5 g/day for 28 days	bench press strength leg extension strength Wingate cycling test	↑ bench press ↔ leg extension ↔ mean power ↔ peak power	None reported
Hummer et al. 2019	22 (16 males, 6 females) active subjects	6 weeks	4 g/day for 6 weeks	bench press strength bench press endurance squat strength squat endurance	↑ bench press strength ↑ bench press endurance ↑ squat strength ↔ squat endurance	None reported

Table 4. *Cont.*

				Strength and Performance Outcomes		
Kreider et al. 1998	25 college football players	28 days	15.75 g/day for 28 days	total work during sprints on a cycle ergometer bench press volume total volume	↑ total work ↑ bench press volume ↑ total volume	None reported
Stone et al. 1999	42 college football players	5 weeks	0.22 g/kg/day for 5 weeks	bench press strength squat strength counter movement vertical jump static vertical jump body composition	↑ bench press strength ↔ squat strength ↔ counter-movement vertical jump ↔ static vertical jump ↑ body mass ↑ lean body mass	None reported
Chilibeck, et al. 2007	19 male union rugby players	8 weeks	0.7 g/kg/day for 8 weeks	bench press endurance leg press endurance body composition	↔ bench press repetitions ↔ leg press repetitions ↑ when combining bench press and leg press scores ↔ total body mass ↔ lean tissue mass ↔ fat mass	None reported

BLa = blood lactate; CR or no identifier = creatine monohydrate; ↓ indicates decrease; ↑ indicates increase; ↔ indicates no difference.

3.1.3. Athletes

From a historical perspective, some of the earliest work investigating the performance benefits of creatine supplementation was done as part of strength and conditioning programs in the late 1990s [22,32,33]. Several of these studies indicated that when creatine supplementation was provided for longer durations in conjunction with a strength and conditioning program, improvements in various indices of strength, power, and body composition were reported. These findings have since been replicated across a variety of sports, as seen in Tables 4 and 5. For example, Kreider et al. [22] reported greater improvements in fat/bone-free mass, training volume, and sprint performance in NCAA

Division I collegiate football players following 28 days of creatine supplementation in conjunction with a resistance/agility training program. Similarly, following 5 weeks of creatine supplementation, freshmen and redshirt American collegiate football players experienced greater increases in bench press and squat scores, and lower body power compared to a placebo [11]. Interestingly, in a similar study involving collegiate football players participating in an 8-week strength and conditioning program, athletes were provided wither creatine with dextrose at two different dosages according to each athlete's fat-free mass (FFM) or a placebo for 8 weeks. Results revealed that both groups receiving creatine experienced significant strength gains over the training period. Still, only the creatine group (300 mg/kg) experienced significant strength gains compared to the placebo group. Thus, it appears that the dosage of creatine may be a factor when supplementing for some athletes [34]. Longer periods of creatine supplementation have also continued to show promise, with no risk of long-term benefits tapering off or individuals becoming "de-sensitized" to creatine's beneficial effects. For example, Bemben et al. [35] observed superior improvements in the bench press and back squat 1 RM strength and anaerobic capacity in collegiate redshirt football players who had been supplementing with creatine compared to training alone. In a study involving collegiate football players participating in a 9-week resistance training program, results demonstrated improvements in the maximal bench press, squat, and power clean performances in all groups, with the creatine + glucose group evincing the greatest improvement compared to creatine alone and the placebo group. The authors suggested that this may have been due to increased work volume, facilitated by enhanced recovery [35]. Similarly, following a 10-week strength and conditioning program, collegiate football players who received low-dose creatine monohydrate supplementation (5 g/day) experienced greater increases in maximal bench press, squat, and power clean performances than from training alone. This study was important in that it revealed not only that creatine supplementation may improve strength and power, but that a loading phase was not necessary to achieve these results [36].

Table 5. Summary of selected studies examining the effects of creatine supplementation on sport performance.

Author Year	Subjects	Duration	Dosing Protocol	Primary Variables	Results	Adverse Events
Grindstaff et al. 1997	18 (7 male, 11 female) junior competitive swimmers	9 days	21 g/day for 9 days	100-m sprint performance arm ergometer performance	↑ sprint swimming performance	None reported
Kreider et al. 1998	25 college football players	28 days	15.75 g/day for 28 days	total work during sprints on a cycle ergometer bench press volume total volume	↑ total work ↑ bench press volume ↑ total volume	None reported

Table 5. *Cont.*

Author Year	Subjects	Duration	Dosing Protocol	Primary Variables	Results	Adverse Events
Noonan et al. 1998	39 college football players	9 weeks	20 g/day for 5 days 100 or 300 mg/kg/fat-free mass for 8 weeks	bench press 40-yard dash % body fat fat-free mass vertical jump	↑ bench press ↑ 40-yard dash ↔ % body fat ↔ fat-free mass ↔ vertical jump	None reported
Peyrebrune et al. 1998	14 male college swimmers	5 days	9 g/day for 5 days	single 50-m sprint time repetitive 50-m sprint time	↔ single 50 m sprint time ↑ repetitive 50 m sprint time	None reported
Stout et al. 1999	24 college football players	8 weeks	21 g/day for 5 days 10 g/day thereafter	vertical jump 100-yard dash bench press strength	↑ vertical jump ↑ 100-yard dash ↑ bench press strength	None reported
Jones et al. 1999	8 elite ice hockey players	11 weeks	20 g/day for 5 days 5 g/day for 10 weeks	5 × 15 s skating sprints 6 timed 80-m skating sprints	↑ 5 × 15 s skating sprints ↑ 6 timed 80 m skating sprints	None reported
Kirksey et al. 1999	36 (16 male, 20 female) track and field athletes	6 weeks	0.3 g/kg/day	counter movement vertical jump power and total work during sprints on a cycle ergometer	↑ counter movement vertical jump ↑ peak power ↑ total work on cycle ergometer	None reported
Kreider et al. 1999	61 college football players	12 weeks	20–25 g/day	Bench press strength Bench press endurance Body composition	↑ bench press strength ↑ bench press endurance ↑ body mass ↑ soft tissue lean mass	None reported

Table 5. *Cont.*

Mujika et al. 2000	17 trained soccer players	10 weeks	20 g/day for 5 days 5 g/day for 9 weeks	counter movement jump repeated sprint ability	↔ counter-movement jump ↑ repeated sprint ability	None reported
Haff et al. 2000	36 (16 male, 20 female) track and field athletes	6 weeks	0.3 g/kg/day	counter movement vertical jump	↑ counter-movement vertical jump	None reported
Skare et al. 2001	18 male competitive sprinters	5 days	20 g/day for 5 days	100-m sprint time total sprint time (6 × 60 m)	↑ 100 m sprint time ↑ total sprint time	None reported
Romer et al. 2001	9 competitive squash players	5 days	0.075 g/kg 4 times for 5 days	single sprint repetitive sprint performance	↔ single sprint ↑ repetitive sprint performance	None reported
Izquierdo et al. 2002	19 male handball players	5 days	20 g/day for 5 days	counter movement vertical jump repetitive sprint performance	↑ counter-movement vertical jump ↑ 6 × 15 m sprints	None reported
Cox et al. 2002	12 elite female soccer players	6 days	20 g/day for 6 days	agility kick drill test agility race test repetitive sprint performance BLa	↔ kick drill test ↑ agility run ↑ repetitive sprint performance ↓ BLa	None reported
Lehmkul et al. 2003	29 (17 male, 12 female) track and field athletes	8 weeks	0.3 g/kg/day for 7 days 0.03 g/kg/day for 7 weeks.	average and peak power during repeated sprints on a cycle ergometer	↔ static vertical jump ↔ counter-movement vertical jump ↔ average power ↔ peak power	None reported
Delecluse et al. 2003	12 (7 male, 5 female competitive sprinters	7 days	0.35 g/day for 7 days	single 40-m sprint repeated 40-m sprints	↔ single 40 m sprint ↔ repeated 40 m sprints	None reported

Table 5. *Cont.*

Kocak et al. 2003	20 elite male wrestlers	5 days	20 g/day for 5 days	average and peak power during Wingate anaerobic power test	↑ average power ↑ peak power	None reported
Ostojic et al. 2004	20 young male soccer players	7 days	30 g/day for 7 days	dribbling test sprint power counter movement jump	↑ dribbling test ↑ sprint power ↑ counter-movement vertical jump	None reported
Pluim et al. 2006	36 competitive tennis players	32 days	0.3 g/day for 6 days 0.03 g/day for 28 days	serve velocity groundstroke velocity repetitive sprints	↔ serve velocity ↔ ground-stroke velocity ↔ repetitive sprints	None reported
Glaister et al. 2006	42 active males	5 days	20 g/day for 5 days	repetitive sprint performance	↔ repetitive sprint performance	None reported
Lamontagne-Lacasse et al. 2011	12 elite male volleyball players	28 days	20 g/day in days 1–4 10 g/day in days 5–6 5 g/day in days 7–28	repeated block jump spike jump	↔ repeated block jump ↔ spike jump	None reported
Ramierz-Campillo et al. 2016	30 amateur female soccer players	6 weeks	20 g/day for 7 days 5 g/day for 5 weeks	jump test repeated sprinting directional change	↑ jump test ↑ repeated sprinting ↔ directional change	None reported

BLa = blood lactate; CR or no identifier = creatine monohydrate; BLa = blood lactate; ↓ indicates decrease; ↑ indicates increase; ↔ indicates no difference.

Analogous benefits have also been observed in female collegiate athletes supplementing with creatine in conjunction with a strength and conditioning program. In a study involving female collegiate lacrosse players participating in a concurrent resistance training program as part of a 5-week preseason conditioning program, those supplementing with creatine displayed even greater strength gains than the placebo group [37]. For example, during a 13-week resistance training program, collegiate female soccer players supplementing with creatine experienced more significant strength gains in the back squat between 5 and 13 weeks than those in the placebo group [38].

3.1.4. Untrained

Creatine supplementation has even been shown to elicit ergogenic benefits to those who are previously untrained and in those who completed self-designed training programs. For example, improvements of 20%–25% in muscular strength have been reported in untrained females supplementing with creatine while undergoing 10 weeks of concurrent strength training [39]. Furthermore, Candow et al. [40] observed improvements in muscle thickness and leg press strength in men and women who were physically active but previously naïve to resistance training after 6 weeks of creatine supplementation and resistance training. Furthermore, 30 days of creatine supplementation (5 g/day) has also been shown to improve strength even without supervised concurrent strength and conditioning programs in individuals who are physically active [41].

3.2. Exercise Capacity/Prolonged High-Intensity Efforts

Harris et al. [6] were the first to show that supplementing with creatine in 4×5 g/day for 5 days resulted in a 50% increase in intramuscular total creatine content (20%–40% as PCr). Greenhaff and colleagues [8] translated these changes in total creatine content to performance outcomes when they had 12 study participants supplement with either a placebo or creatine over a 5-day supplementation period (4×5 g/day = 20 g/day) before completing five 30-s bouts of maximal muscle contractions. No performance changes were observed in the placebo group, but when creatine supplementation was employed, peak torque production increased in the final ten repetitions of the first bout and throughout the entirety of bouts 2–4. That same year, Balsom et al. [7] published data to illustrate the ergogenic potential of creatine supplementation. In this study, 16 healthy males supplemented for 6 days at a dose of 5 g/day in a double-blind format. Each participant completed 10 repeated 6-s bouts of high-intensity cycling with 30 s of rest between each bout. Work output exhibited smaller declines towards the end of each bout when creatine supplementation was provided; no such changes were observed in the placebo group. Birch and colleagues [42] supplemented individuals with either a placebo or creatine with 4×5 g/day (20 g/day) for 5 days. Before and after supplementation, study participants complete three 30-s bouts of high-intensity cycling. Although the placebo exhibited no impact on peak power, mean power or total work output, creatine supplementation significantly increased peak power after the first cycling bout and increased mean power output and total work completed during the first and second exercise bouts. No impact on power or total work was observed in the third bout. Casey et al. [43] had nine males complete two bouts of 30 s maximal cycling before and after supplementation with creatine at a dosage of 20 g/day for 5 days. Skeletal muscle biopsies confirmed an increase in intramuscular PCr content, which likely contributed to the increase in total work production of 4% that was observed. Moreover, the loss of ATP was found to be approximately 30% less when creatine was ingested, and this sustainment was found even after more work was completed. The results from this study were meaningful as they highlighted the key relationship that exists between the decay in exercise performance and intramuscular PCr status.

From there, modifications to the traditional creatine dose began to be considered, as studies had indicated some degree of individual variation (responders vs. non-responders) to supplementation [6,22,44]. As discussed previously, Kreider et al. in 1998 supplemented 25 football players for 28 days with a combination of carbohydrate (99 g) and 15.75 g of creatine and reported increases in fat-free mass, as well as improvements in bench press volume, total lifting volume, and total work performed through the first five sets of 6-s bouts of maximal cycling [22]. In accordance, Stout and investigators [45] examined the combination of creatine and carbohydrates. These authors reported that creatine supplementation (4×5 g/day for 6 days) in 26 college-aged men increased anaerobic working capacity by 9.4%, whereas no change in working capacity was noted when a 33-g dose of carbohydrate was delivered as a placebo. Notably, the combination of creatine and carbohydrates increased working capacity by 30.7%, providing additional credence to previous work that demonstrated an improvement in intramuscular creatine uptake and better sustainment of performance [6,43,44]. In 2003, Van Loon and colleagues [46] supplemented 20 college-aged men for 6 weeks in a double-blind, placebo-controlled, parallel fashion. A traditional loading phase (4×5 g/day for 5 days) was completed, which was then followed with 2 g/day for the remaining 37 days. Before and after the loading and maintenance supplementation periods, high-intensity exercise capacity was assessed by having participants perform 15 cycling sprints that were each 12 s in duration and separated by 48 s of rest. The loading phase successfully increased muscle creatine levels, which regressed back to baseline levels after 6 weeks. Additionally, peak power output after 5 days of supplementation increased performance during repeated sprints on the cycle ergometer. Moreover, these performance increases were maintained after the 6-week supplementation regimen, suggesting that continued ergogenic potential is exhibited by creatine under these parameters. Results from Barber et al. [47] illustrated in 13 healthy trained individuals that just 2 days of supplementation in a randomized, double-blind, crossover fashion with either a placebo (20 g maltodextrin + 0.5 g/kg maltodextrin) or creatine (20 g + 0.5 g/kg maltodextrin) prior to completing six repeated 10-s cycling sprints with 60 s of rest between sprints can increase relative peak power production by 4% when compared to the placebo condition. Moreover, Fukuda and colleagues reported that acute creatine supplementation significantly increased anaerobic running capacity, an intense, challenging three-minute bout of high-intensity running [48]. In summary and when viewed collectively, these results and others display a consistent pattern of research that supports the ability of creatine supplementation to increase one's capacity to perform high-intensity exercise both after acute (5–6 days) and prolonged (4–6 weeks) periods of supplementation.

3.3. Sport-Specific Performance

For many athletes and coaches, the impact of creatine supplementation on sports performance is the most important consideration. It is well established that creatine supplementation leads to increased muscle PCr content [6,14], accelerated ATP resynthesis [7,43], and enhanced performance in short-duration, predominately anaerobic intermittent exercise [13,49]. As a result of these observed benefits, it has been suggested that creatine supplementation could translate to enhanced on-field performance for competitive athletes. The ever-changing nature of sports in terms of intensity, distances covered, and duration makes the replication of sports performance difficult. Due to the challenges associated with assessing on-field performance, many experimental approaches have employed simulated play and field tests. The outcomes and designs of several studies have been summarized in Table 5.

3.3.1. Agility Performance

Numerous studies have measured performance using sport-specific drills and simulations following creatine use. Assessing agility performance is a common way to assess on-field performance during soccer. In this respect, Cox et al. [50] had 14 elite female soccer players undertake a simulated match before and after consuming either 20 g/day of creatine or placebo for 6 days. During the simulation, performance times during ten agility runs were digitally recorded and run times were significantly faster during trials 3, 5, and 8 in the creatine supplementation group. Similarly, Ramirez-Campillo et al. [51] divided 30 competitive female soccer players into three equal groups (creatine, placebo, control) with each completing a pre-test agility run. The creatine and placebo groups were also assigned to a 6-week plyometric training program. Although agility run times increased in both groups as a result of training, no performance differences between groups were identified. Other soccer studies have examined field tests, with mixed results. For example, Ostojic et al. [52] had 20 adolescent male soccer players ingest 30 g/day of creatine for 7 days, which resulted in an improvement in ball dribbling skills, jump height, and power production, whereas no improvement in kicking accuracy was observed. Similar outcomes in kicking accuracy were reported by Cox et al. [50] after creatine supplementation.

Tennis is a sport characterized by explosive, powerful actions interspersed with tremendous agility demands and notable exercise capacity demands. However, the impact of creatine ingestion on tennis performance has not been as extensively studied as other sports. Using a double-blind cross-over study design, Op't Eijnde et al. (2001) had eight well-trained tennis players consume 20 g/day of creatine or a placebo over 5 days.

Stroke quality via the Leuven Tennis Performance Test and a 70-m shuttle run were measured at baseline and after each treatment [53]. The study included a 5-week washout period between treatments and no significant differences in performance were recorded after creatine supplementation for either test. Pluim et al. (2006) had 39 male tennis players perform ball machine ground stroke drills to assess the acute (0.3 g/kg/day for 6 days) and chronic (0.03 g/kg/day for 5 weeks) effects of creatine ingestion [54]. Performance metrics included the velocity of repeated ground strokes and serving velocity. No significant differences were found in these performance measures after the acute loading phase or following 5 weeks of creatine use in any performance measure. As a result, the authors concluded that creatine should not be recommended to tennis players; however, the chosen outcome measures may not have been ideal outcome variables when considering creatine's primary ergogenic mechanism of action.

Studies that have incorporated creatine use in combat sports have yielded conflicting results. Kocak and Karli (2003) had 20 international-level wrestlers perform a 30-s Wingate test before and after consuming 20 g/day of creatine or placebo for 5 days. Average power and peak power were significantly greater when creatine supplementation was provided, whereas no change was observed in the placebo condition [55]. However, Aedma et al. [56] reported creatine supplementation (0.3 g/kg/day) for 5 days did not improve peak power, mean power, or fatigue index in repeated upper-body ergometer sprint tests in 20 amateur Brazilian Jiu Jitsu and submission wrestling athletes. Similarly, 10 taekwondo athletes failed to show improvements in Wingate anaerobic capacity tests following the use of creatine (50 mg/kg/day) or placebo for 6 weeks in a crossover design with a 6-week washout period [57]. In summary, the available research in combative sport athletes suggests that creatine supplementation may have limited potential to enhance performance in these types of sports. These results are somewhat surprising, considering the energetic nature of these events, and as a result, more research is encouraged to better determine the potential for creatine to enhance performance in sports of this nature.

3.3.2. Sprint Performance

Improvements in sprint performance after creatine supplementation have been shown in handball [58], football [59], ice hockey [60], soccer [50–52,61], swimming [62], and track athletes [63]. Briefly, Kreider in 1998 [22] and Stout in 1999 [59] reported improvements in sprint performance by collegiate American football players after supplementing their diet with combinations of creatine (5.25–15.75 g) and carbohydrates (33–99 g). Similarly, Skare et al. [63] reported faster 100-m sprint times and reduced total time of 6 × 60-m sprints in a group of well-trained male sprinters after 5 days of creatine supplementation (20 g/day). Not all studies, however, suggested that creatine supplementation improves sprint performance [54,64–66]. In this respect, Glaister and colleagues [67] reported no change in sprint running performance (30 m dash) after a standard 5-day creatine loading phase in active college-aged men who were consistently completing activities that involved sprinting. Similarly, Delecluse et al. [64] also reported that creatine supplementation did not improve 40-m dash performance after a 7-day creatine loading protocol in elite college sprinters. It was surmised from these authors that the high volume of high-intensity training already being completed by these athletes may have underpinned creatine's ability to promote further improvements in sprinting performance. This thesis, however, is refuted by Skare and investigators [63], who supplemented collegiate track and field athletes with creatine for five days (20 g/day) while also participating in a concomitant resistance training program. In this study, performance changes in 60-m and 100-m sprint performance were assessed and the authors found that 100-m sprint velocity was improved and the total time to complete multiple 60-m sprints was significantly reduced. In summary, it appears that creatine's ability to improve performance in those activities which contain a predominant gravitational component (e.g., sprinting, vertical jump, etc.) may be challenged. Thus, it has been suggested that the added body mass which occurs due to creatine supplementation may offset any ergogenic outcomes, but this conclusion has not been universally observed. As a result, people must consider that multiple factors interact (training status, type of athlete, sprint distance, the incorporation of other forms of training besides sprinting, etc.) to influence the final observed outcomes. Nevertheless, a number of studies are available that have documented the ability of creatine supplementation to improve sprinting performance.

3.3.3. Jump Performance

The ability to jump explosively both horizontally and vertically is a key attribute for many athletes. To jump effectively, an athlete must be able to generate high amounts of power relative to their body mass. Due to creatine's known ability to increase body water and subsequently body mass, concerns have been raised about the impact of supplementing with creatine on jumping performance. Lamontage-Lacasse et al. [68] compared the effect of 4 weeks of creatine to that of a placebo on 1 RM spike jump (SJ) and repeated block jump (BJ) capacity among 12 elite male volleyball players. Dosing for the study included ingesting either a placebo or creatine at 20 g/day on days 1–4, 10 g/day on days 5–6, and 5 g/day on days 7–28. Before and after the treatment, subjects performed the 1 RM SJ test, followed by the repeated BJ test (10 × 10 BJs). Following supplementation, no differences were observed in SJ between the creatine and placebo conditions. A non-significant ($p > 0.05$) 1.9% improvement in BJ performance was observed for the group supplementing with creatine compared to the placebo group. Alternatively, Izquierdo et al. [58] found that creatine supplementation (20 g/day for 5 days) attenuated a decline in countermovement jumping (CMJ) ability after a single set of half squats when compared to the placebo group. Similar findings were recorded by Mujika et al. [61], who found that 5 days of creatine loading (20 g/day) limited the observed decay in jumping performance after a maximal intermittent soccer-specific test (40 × 15-s bouts of high-intensity running interspersed with 10-s bouts of low-intensity running). Additionally, a recent meta-analysis in soccer suggested that creatine supplementation produces small but non-significant improvements in single jump performance [69]. Stone et al. [11] supplemented 42 American collegiate football players for 5 weeks and determined that creatine supplementation led to significantly greater power output and rate of force development during static vertical jumps. In addition, Haff et al. [70] reported greater rates of improvement in CMJ performance for track athletes who ingested creatine (0.3 g/kg/day over the 6-week study) compared with a placebo group. To summarize the available literature surrounding jumping performance, the majority of studies demonstrate consistent, small improvements in jumping performance, whereas other studies have reported improvements that did not reach traditional levels of statistical significance.

3.3.4. Selected Competitive Athletes
American Football

Using trained collegiate athletes, Kreider et al. [22] had 25 NCAA Division IA football players supplement in a double-blind, randomized fashion for 28 days with either 99 g of carbohydrates or an isoenergetic amount of carbohydrate + 15.75 g of creatine as part of their regular training and conditioning sessions. Increases in body mass and fat-free mass were found, along with increases in bench press volume, total volume, and total work completed during the initial bouts of a repeated sprint cycling protocol. In addition, Stout and investigators [59] supplemented 24 collegiate football players with either carbohydrates (35 g), creatine (5.25 g of creatine + 1 g of carbohydrates) or creatine + carbohydrates (5.25 g of creatine + 33 g of carbohydrates). Four doses were taken per day in all groups for the first 5 days, whereas thereafter, two doses per day were consumed. The combination of creatine + carbohydrates led to significantly greater improvements in bench press strength, vertical jump, and 100-yard dash when compared to the carbohydrate group, whereas the observed changes in the creatine only group exhibited greater mean changes, but the magnitude of these changes was not considered to be statistically significant. Stone and colleagues [11] reported increases in upper and lower body strength and peak rates of force development in American football players after 5 weeks of creatine supplementation. Finally, Kreider in 1999 [33] supplemented 51 college football players in a matched, randomized fashion to a placebo, carbohydrates + placebo, or a combination of carbohydrates + protein + creatine. The two groups which contained creatine experienced the greatest gains in lean mass in comparison to the groups that did not contain creatine.

Track and Field

Kirksey et al. [71] supplemented 36 male and female collegiate track and field athletes with either a placebo or creatine (0.30 g/kg/day) for 6 weeks. Lower body power was assessed using countermovement and static vertical jumps, whereas exercise capacity was assessed using five consecutive 10-s bouts of cycling exercise. Improvements in countermovement vertical jump height (7.0% vs. 2.3%), vertical jump power (6.8% vs. 3.1%), average cycling peak power (12.8% vs. 4.8%), average cycling power (10.8% vs. 3.1%), and cycling total work (10.8% vs. 3.5%) were all greater when athletes supplemented with creatine vs. placebo. To further extend these outcomes, Lehmkul et al. [72] supplemented 29 male and female collegiate track and field athletes for 8 weeks and assessed changes in body composition and performance. Again, supplementation with creatine after 1 week and also after an additional 7 weeks resulted in improvements in rates of power production when compared to rates observed by athletes who consumed a placebo.

Swimming

Grindstaff and investigators [73] supplemented 18 male and female junior competitive swimmers with either creatine or a placebo for 9 days during training. Before and after supplementation, all swimmers completed three 100-m freestyle swims with 60 s of recovery between each race. Those swimmers supplementing with creatine swam significantly faster than those supplementing with a placebo after the first heat. The entire distance covered by those supplementing with creatine over all three 100-m races tended to be greater. This study is significant as it was one of the first to use adolescent athletes and also incorporated female athletes. Additionally, Peyrebrune et al. [65] reported improvements in repeat swim performance among elite swimmers following creatine supplementation (9 g/day for 5 days). In terms of swimming, Papadimitriou reviewed the literature surrounding the efficacy of several ergogenic supplements, including creatine [74]. Finally, creatine has also been proposed as a potential ergogenic aid for other total body exercises with similar physiological demands, such as mixed-martial arts; however, limited evidence is currently available.

3.4. Endurance Performance

Although creatine supplementation has been quite effective in its ability to function as an ergogenic aid for anaerobic exercise, much less evidence and discussion exists surroundings its ability to benefit endurance performance [75]. When considering creatine's role from an energetic perspective, this is not surprising, as exercise bouts greater than 2 to 3 min in duration rely predominately on the oxidative system for the synthesis of ATP [76]. Currently, the efficacy of creatine supplementation appears to be more limited for endurance sports, with the magnitude of the benefit being dependent on the duration of the event, as well as the mode of exercise. As exercise duration increases, the ergogenic potential of creatine is diminished [49]. A vast majority of studies show that creatine supplementation has no appreciable effect on maximal oxygen consumption (i.e., VO_2Max or VO_2Peak) [7,58,77–82], submaximal oxygen consumption [83–85], or time trial performance [82,86–90]. A few studies demonstrated improvements in time-to-exhaustion [91,92], but a majority reported no effect [77,82,93,94].

Interestingly, a few research studies have reported small yet significant improvements in other endurance performance variables, such as blood lactate concentration at a given workload and ventilatory/lactate threshold following creatine supplementation [77,95,96]. For example, Chwalbinska-Monteta et al. [95] had elite male rowers perform an incremental exercise to exhaustion on a rowing ergometer before and after consuming 20 g/day of creatine or a placebo for 5 days. Significant reductions in blood lactate at submaximal intensities, as well as increases in lactate threshold, were reported. No differences between groups were found in maximal power or maximal blood lactate concentration after the all-out exercise. In addition to this study, a pair of studies conducted in the same laboratory [77,97] used recreationally active males completing 4 weeks of high-intensity interval training while supplementing with either 10 g per day of creatine citrate or a placebo. During an incremental cycle ergometry test, significant improvements in critical power (+6.7%) and ventilatory threshold (+6%) were observed for the creatine group compared to the placebo group. No significant differences were recorded in VO_2Peak, time to exhaustion, or total work accomplished. A 2019 study by Fenandez-Landa and colleagues [96] investigated the effects of 10 weeks of creatine on rowing performance. An incremental test to exhaustion was performed by each subject on a rowing ergometer before and after both the creatine and placebo treatments. Measurements included power at anaerobic threshold (WAT), 4 mmol (W4), and 8 mmol (W8) of blood lactate concentration. Power outputs at 8 mmol blood lactate concentration in the creatine-supplemented group increased after eight weeks, whereas no changes were observed in the placebo group, in addition to an approximate 6%–7% increase in absolute power output, which was significantly greater than the absolute power output in the placebo group.

In these studies, endurance performance variables were only slightly improved after creatine supplementation and therefore should be interpreted with caution when considering the practical benefits of using creatine for endurance performance. In addition, the mode of exercise may increase the likelihood of finding ergogenic benefits from creatine. For example, the majority of support for creatine use in endurance exercise has been observed using either an indoor rowing ergometer [95,96] or a cycle ergometer [77,97]. By contrast, the performance of endurance activities such as running, swimming, and soccer involves the propulsion of the body, which may be adversely affected by the gains in body mass typically seen with creatine supplementation, a concept also discussed previously with activities such as jumping. It is quite possible that increases in body mass may counteract any beneficial effects of creatine by increasing the energy demands of exercise during these propulsive/anti-gravity activities [98]. Finally, and as mentioned in previous sections, Mielgo-Ayuso et al. [69] conducted a systematic review and meta-analysis of the effects of creatine supplementation in soccer players. Although 90% of the energy used in soccer match play comes from aerobic metabolism, results indicated that creatine did not improve any aspect of aerobic performance.

Other outcomes of creatine supplementation have been purported to impact endurance training and performance. For example, studies have demonstrated that adding creatine to carbohydrates [44] or carbohydrates + protein supplementation [99] may help promote greater glycogen storage. In many endurance activities, the intensity and duration of training and competition cause drastic reductions in hepatic and intramuscular glycogen levels [1], thus added glycogen storage resulting from adding creatine to carbohydrate and carbohydrates + protein feeding is viewed as an essential benefit. Additionally, the volume and nature of endurance training can invoke noticeable muscle damage. Cooke and colleagues [100] have reported that creatine supplementation may reduce circulating levels of muscle damage markers and help to more quickly restore the ability of the damaged muscle to produce force, whereas other studies have also highlighted a distinct myoprotective role for creatine [101]. Beyond these findings, Santos et al. [102] previously reported that creatine supplementation (5 g/dose four times per day for 5 days) in a group of runners who were monitored for 24 h after completing a 30-km running race experienced reduced levels of soreness, muscle damage, and inflammation. Finally, athletes participating in endurance activities must deal with high ambient temperatures and humidity, which can compromise thermoregulation and subsequently performance. In this respect, studies are available that have documented creatine's ability to hyper-hydrate the cell and enhance tolerance to heat [103,104]. In considering these findings, athletes may experience a reduction in the risk of heat-related injuries when training or competing in these conditions [3,4,103–105]. To summarize, limited research is available that has directly examined the ability of creatine supplementation to impact endurance exercise. Indeed, more work is needed to explore creatine's potential to impact performance in these types of activities directly. Although some preliminary work suggests that creatine may not directly support endurance performance, other studies have provided many areas of evidence where creatine can provide support and aid the training and performance of these types of athletes (See Table 6).

Table 6. Summary of selected studies examining the effects of creatine supplementation on endurance performance.

Author-Year	Subjects	Duration	Dosing Protocol	Primary Variables	Results	Adverse Events
Rossiter et al. 1996	38 (28 male, 10 female) competitive rowers	5 days	0.25 g/kg/day for 5 days	time trial performance during rowing ergometry	↓ 2.3 s in 1000-m times	None reported
McNaughton et al. 1998	16 elite male paddlers	5 days	20 g/day for 5 days	total work, peak power, BLa during rowing ergometry	↑ in total work during 90–300 s of rowing ergometry performance	None reported
Miura et al. 1999	8 healthy males	5 days	20 g/day for 5 days	critical power test during cycle ergometry	↔ critical power ↑ anaerobic work capacity	None reported
Rico-Sanz et al. 2000	14 elite male cyclists	5 days	20 g/day for 5 days	oxygen consumption, time to exhaustion, BLa during maximal cycle ergometry	↔ VO2 max ↑ time to exhaustion ↔ BLa	None reported
Syrotuik el al. 2001	22 (12 male, 10 female) competitive rowers	6 weeks	0.3 g/kg/day for 5 days 0.03 g/kg/day for 5 weeks	time trial performance during rowing ergometry	↔ in 2000-m rowing times	None reported

Table 6. *Cont.*

Jones et al. 2002	9 active males	5 days	20 g/day for 5 days	VO2 kinetics during moderate and heavy submaximal cycle exercise	↔ VO2 kinetics ↓ VO2 during heavy cycling exercise	None reported
Chwalbinska-Moneta 2003	16 elite male rowers	5 days	20 g/day for 5 days	maximal power output, time to exhaustion, Bla during rowing ergometry	↔ maximal power output ↑ time to exhaustion ↔ BLa	None reported
Graef et al. 2009	43 active males	30 days	10 g/day for 20 days; only on training days (5 × week)	oxygen consumption, time to exhaustion, VT, total work, during maximal cycle ergometry	↔ VO2 peak ↑ time to exhaustion ↑ ventilatory threshold ↔ Total work	None reported
Kendall et al. 2009	43 active males	30 days	10 g/day for 20 days; only on training days (5 × week)	critical power and anaerobic work capacity during cycle ergometry	↑ Critical power ↔ Anaerobic work capacity	None reported
Hickner et al. 2010	12 endurance-trained males	28 days	3 g/day for 28 days	VO2peak, submaximal VO2, RER, Bla, 10 s sprints at 110% VO2peak during simulated cycling road race	↔ VO2peak ↓ submaximal VO2 ↔ RER ↔ Bla, ↔ 10-s sprints at 110% VO2peak	2 subjects reported muscle cramping at rest following supplementation
De Andrade Nemezio et al. 2015	24 male amateur cyclists	5 days	20 g/day for 5 days	time trial performance total O2 uptake, BLa during maximal cycle ergometry	↔ 1000 m time ↓ total O2 uptake ↔ BLa	None reported

Table 6. *Cont.*

| Fernandez-Landa et al. 2020 | 28 elite male rowers | 10 weeks | 0.04 g/kg/day for 10 weeks

+ 3 g HMB/day for 10 weeks | power output at AT, 4 mmol, 8 mmol Bla during rowing ergometry | ↑ power at AT for creatine-HMB and HMB only group
↑ power at 4 mmol BLa for creatine-HMB group
↑ power at 8 mmol BLa for creatine only, HMB only, and creatine-HMB groups | None reported |

BLa = blood lactate; CR or no identifier = creatine monohydrate; RER = respiratory exchange ratio; ↓ indicates decrease; ↑ indicates increase; ↔ indicates no difference.

4. Recovery

Increases in intramuscular levels of creatine phosphate secondary to creatine supplementation increase the supply of a robust, energetic substrate that can be used to resynthesize ATP. In this capacity, creatine supplementation can help increase and maintain the delivery of ATP to working muscles, allowing for an increased ability to perform work, resulting in the widespread display of ergogenic outcomes commonly reported in the literature [3,9,19,22,106]. Aside from overt improvements in the performance of single bouts of maximal efforts, creatine is able to augment performance across multiple sets of performance and subsequently demonstrates a role in enhancing recovery. The term recovery is often contextual in nature and typically pertains to either physiological, subjective, or performance-based parameters. In this respect, creatine appears to positively influence recovery in regard to physical performance following bouts of intense activity, and has been shown to enhance recovery during bouts of intermittent activity, sustaining maximal performance across multiple bouts of exercise. In addition, creatine supplementation may also reduce the post-exercise inflammatory response, thereby attenuating markers of muscle damage and soreness in the hours and days following bouts of exercise-induced muscle damage. Finally, creatine may have efficacy as a therapeutic intervention following an injury or during periods of limb immobilization.

4.1. Augmented Recovery Following Exercise
Augmented Recovery during Intermittent Activities

Creatine supplementation has consistently demonstrated the ability to augment recoverability between bouts of intermittent activity, such as repeated Wingate tests, high-intensity interval training, sprint cycling, sprinting, intermittent team sports, and certain resistance training protocols. As mentioned earlier, oral creatine supplementation is an effective means to increase the intramuscular PCr content by 10%–40% [6], a key component of the ATP-PCr energy system, which is known to contribute a high percentage of ATP yield during high-intensity bouts of activity lasting 0–30 s [7]. For these reasons, it is no surprise that creatine supplementation has been shown to improve intermittent high-intensity exercise performance [3,107]. In this respect, heightened physical performance is thought to be due to a higher initial intramuscular PCr content at the beginning of the exercise and to increased PCr resynthesis during recovery periods between intermittent anaerobic bouts. As such, PCr availability is increased throughout the completion of successive bouts, which subsequently functions to attenuate declines in power output and other indices of fatigue. Moreover, activities characterized by a high force output are more reliant on type II muscle fibers, which have a higher PCr content than the more oxidative type I fibers, and subsequently are favored during bouts of activity at or near maximal effort, with a high degree of force output [108]. Several studies involving many different types of athletes (e.g., cycling, running, swimming, hockey, handball, etc.) supplementing with creatine have documented improvements in intermittent high-intensity exercise capacity [7,42,47,60,73,78,109–115]. For example, Birch et al. [42] were one of the first to note improvements in high-intensity cycling performance (30 s) during the first two of three maximal effort bouts following five days of creatine supplementation (4×5 g/day). Kreider et al. [22] reported greater improvements in total work completed during a 12×6 s maximal-effort sprint protocol, particularly during sprints one to five, after 28 days of creatine supplementation in collegiate football players compared to a placebo. Using repeated maximal swimming efforts, Grindstaff and investigators [73] reported significantly faster swim times in their first race. The entire distance covered over swimming was set in 18 male and female junior competitive swimmers after supplementing with either creatine or a placebo for nine days during training. Jones et al. [60] examined skating performance in 16 elite ice-hockey players after having them consume 5 g of creatine monohydrate or placebo (glucose) four times per day for 5 days, after which a maintenance dose of 5 g per day for ten weeks was administered. Subjects completed six timed 80-m skating sprints, with each sprint being initiated every 30 s with a split time taken after 47 m. Mean on-ice sprint skating performance was significantly improved after 10 days, and this ergogenic outcome was sustained through ten weeks of supplementation when compared to baseline performance. Similarly, Mujika et al. [61] reported that 6 days of creatine supplementation (20 g/day) improved repeated sprint performance (6×15 m sprints with 30 s recovery) in 17 highly trained male soccer players. Other improvements in intermittent running performance have also been reported following periods of creatine supplementation. For example, Aaserud et al. [116] observed improvements in sprint times during successive ($8 \times$) 40-m sprint tests following 15 g of creatine per day over 5 days in well-trained handball players. Similarly, Deminice et al. [117] observed improvements in mean and peak power during a running-based anaerobic sprint test (RAST) consisting of six 35-m sprints following a 7-day creatine supplementation period (0.3 g/kg/day) in under 20 male soccer players.

Some contradictory findings have also been observed following creatine supplementation, whereby no improvements in performance were documented. For example, Green and colleagues [118] supplemented 19 active men with either a placebo or 20 g of creatine per day for 6 days and had them complete three upper-body and three lower-body Wingate tests. No changes in mean power or peak power production were observed for either type of test. However, the magnitude in the observed decrease in power was greater following placebo use but was lesser with creatine supplementation, suggesting that a better attainment of power may have occurred. In addition, Similar outcomes were reported by Ahmun [119] and Deutekom [120], who also failed to observe any ability of creatine loading at a dosage of 20 g/day for 5 and 6 days, respectively, to enhance repeated sprint cycling performance. Similar to repeat cycling sprint performance, Ahmum [119] reported that acute creatine supplementation did not positively impact sprint running performance when compared to a placebo, an outcome that was also reported by Glaister et al. [67]. In summary, several studies are available which have reported on the ability of creatine to positively impact repeated sprint exercise performance. In general, these studies consistently demonstrate an improvement in power production and performance times across multiple bouts of exercise. These findings are not universal, as a limited number of studies have indicated that creatine may not impact performance. Researchers are encouraged to consider the nature of the exercise and training being completed by the study participants, as well as the dosing regimen, when fully evaluating creatine's potential. Finally, in nearly all of these instances in which an ergogenic outcome was not identified, creatine supplementation, at worst, was able to maintain exercise performance.

4.2. Loss of Force Production, Muscle Damage, Soreness, and Inflammation

Preliminary evidence indicates that creatine supplementation may improve recovery from bouts of intensive exercise and subsequently improve physical performance, particularly when a high degree of exercise-induced muscle damage may have occurred. More specifically, and in addition to its recovery of force production potential, it is also possible that creatine supplementation may help to attenuate muscle damage and soreness following damaging bouts of exercise; however, the specific mechanisms underpinning this protective effect have yet to be fully elucidated. For example, Cooke et al. [100] reported greater isokinetic (10% higher) and isometric (21% higher) knee extension strength in a group that supplemented with creatine (0.3 g/kg body weight × 5 days) following a bout of eccentric-only repetitions using 120% of the subjects' 1RM on leg press, leg extension, and leg flexion exercise machines. The authors noted attenuated creatine kinase levels (a common marker of muscle damage) at 48, 72, and 96 h, and 7 days, following a bout of eccentric exercise in the creatine supplementation group. In a similar study, Rosene et al. [121] reported a higher level of maximal isometric force production in the days following a bout of eccentric leg extensions, a type of exercise known to instigate muscle damage, in those supplementing with creatine after a maintenance dosing protocol, but not after an acute loading dose protocol. A similar attenuation of post-exercise muscle damage was noted in a cohort of Ironman triathletes who completed a creatine (or placebo) loading protocol (20 g/day for 5 days) prior to an Ironman competition [122]. Those supplementing with creatine experienced blunted increases in creatine kinase, lactate dehydrogenase, and aldolase post-competition. The Ironman competitors also experienced an attenuation of glutamic oxaloacetic acid transaminase and glutamic pyruvic acid transaminase following the race, which indicates a dampened inflammatory response. Interestingly, preliminary evidence is also available to suggest that creatine supplementation (20 g/day for 5 days) may attenuate post-exercise increases in markers of muscle damage and soreness, while mitigating reductions in the range of motion above that which is normally present in instances of the "repeated bout effect" during resistance training activities [123]. For more pronounced benefits during resistance training programs, longer periods of creatine supplementation may be required to attenuate post-exercise muscle damage following bouts of resistance-based exercise. For example, Wang et al. [124] indicated that a 4-week period of creatine supplementation (20 g/day for 6 days followed by 2 g/day for 22 days) in conjunction with a complex training regimen reduced the post-exercise increase in creatine kinase compared to a placebo group. Furthermore, a review by Kin et al. [125] summarized the role of creatine supplementation in exercise-induced muscle damage and concluded that creatine might help to prevent muscle damage and facilitate recovery following high-intensity exercise. However, contradictory findings have been observed following bouts of resistance-based exercise, predominantly following shorter periods of creatine supplementation. For example, Rawson et al. [106] had subjects complete 50 maximal eccentric contractions of the elbow flexors, preceded by a creatine (or placebo) loading protocol of 20 g/day for 5 days. Both groups experienced significant losses in maximal isometric force production and range of motion, while experiencing increases in arm circumference, soreness, creatine kinase, and lactate dehydrogenase. These acute physiological responses were similar between groups, indicating that the standard creatine loading protocol used exhibited a limited potential to attenuate markers of muscle damage and soreness following the bout of resistance exercise inducing muscle damage. Similar findings have also been reported in which creatine supplementation did not appear to attenuate markers of muscle damage, soreness, strength deficits, or fatigue following exercise [121,126–129].

Various intracellular mechanisms have been discussed and proposed that outline the potential anti-oxidant and protective properties of creatine, as first proposed by Lawler et al. in the early 2000s [130]. Following this proposed model, researchers examined if creatine exerted a natural ability to act as an antioxidant against aqueous radical and reactive species ions, using an in vitro model. Shortly thereafter, Sestili et al. [131] observed creatine's ability to exert mild antioxidant activity in living cells due to creatine acting as a scavenger of reactive oxygen and nitrogen species. In this respect, Santos et al. [102] observed reductions in markers of cell damage and inflammation in runners following a 30 km race who completed a creatine loading protocol prior to the race (20 g/day for 5 days) compared to a control group. Importantly, this work was one of the first investigations to demonstrate the ability of creatine to attenuate the inflammatory response, following an endurance event that resulted in a high degree of muscle damage and pro-inflammatory signaling mechanisms. Using a repeated sprint protocol, Deminice et al. [117] reported the inhibition of TNF-α and C-reactive protein (common inflammatory markers), but not markers of oxidative stress following a 7-day creatine loading protocol (0.3 g/kg/day). As such, it appears that creatine may be more beneficial for endurance activities regarding its ability to reduce indices of post-exercise muscle damage and soreness, in addition to attenuating pro-inflammatory signaling cascades. Although potential exists for creatine supplementation to favorably influence post-exercise indices of muscle damage, soreness, and muscle function, more work is needed to identify the specific types of exercise models and sports activities that may benefit the most and how an acute attenuation of post-exercise muscle damage influences training adaptations over time. Table 7 provides a summary of selected studies examining the effects of creatine supplementation on outcomes relating to muscle damage, soreness, inflammation, and recovery.

Table 7. Summary of selected studies examining the effects of creatine supplementation on indices of muscle damage, inflammation, and recovery.

Author Year	Subjects	Duration	Dosing Protocol	Primary Variables	Results	Adverse Events
Oopik et al. 2002	5 well-trained male wrestlers	17 hours	30 g (7.5 g/serving) + 320 g glucose (80 g/serving) over 4 doses	isokinetic performance blood glucose blood lactate plasma ammonia plasma urea body mass	↑ submaximal work ↔ blood glucose ↔ Bla ↓ plasma ammonia ↔ plasma urea ↔ body mass	None reported
Hespel et al. 2001	22 (13 males, 9 females)	10 weeks	20 g/day for 3 weeks (immobilization) 15 g/day for 3 weeks (10–8) 5 g/day for 7 weeks (7–1).	quadriceps cross-sectional area (CSA) knee extension isometric force	↑ CSA ↑ knee torque ↑ isometric force	None reported

Table 7. *Cont.*

Tyler et al. 2004	60 ACL reconstruction patients (33 males, 27 females)	6 months	20 g/day for 7 days 5 g/day for 12 weeks	knee extension knee flexion hip flexion hip abduction hip adduction single leg hop	↑ knee outcome measures comparing to baseline	None reported
Rawson et al. 2007	22 trained males	10 days	0.3 g/kg/day for 5 days 0.03 g/kg/day for 5 days	maximal strength range of Motion muscle Soreness blood lactate	↔ strength ↔ ROM ↔ soreness ↔ Bla	None reported
Cooke et al. 2009	14 untrained males	20 days	0.3 g/kg/day + glucose (80 g/day) for 5 days 0.1 g/kg/day + glucose (0.4 g/day) for 14 days	isokinetic force isometric force	↑ isokinetic force ↑ isometric force	None reported
Rosene et al. 2009	20 healthy males	30 days	20 g/day for 7 days 6 g/day for 23 days	isometric force knee range of motion muscle soreness creatine kinase blood lactate	↑ isometric force	None reported
Johnston et al. 2009	7 healthy males	30 days	Maltodextrin 20 g/day for 7 days (Day 1–7) 20 g/day for 7 days (Day 15–21)	fat free mass elbow flexor strength and endurance elbow extensor strength & endurance	↑ lean tissue ↑ muscular strength ↑ muscular endurance	None reported

Table 7. *Cont.*

McKinnon et al. 2012	27 (15 male, 12 female) untrained subjects	10 days	40 g/day + CHO 40 g/day for 5 days 5 g/day + CHO 5 g/day for 5 days.	muscle force loss rate of recovery muscle soreness	↔ force loss ↔ rate of recovery ↔ muscle soreness	None reported
Boychuk et al. 2016	14 healthy males	48 hours	0.3 d/kg	maximal voluntary contraction muscle thickness electromyo graphy muscle soreness	↔ strength ↔ EMG activation ↔ muscle soreness	None reported
Backx et al. 2017	30 healthy males	12 days	20 g/day for 5 days 5 g/day for 7 days	quadriceps cross-sectional area (CSA) leg 1 RM knee extensions	↔ CSA ↔ 1 RM	None reported

1 RM = one repetition maximum; CHO = carbohydrates; CR or no identifier = creatine monohydrate; BLa = blood lactate; ↓ indicates decrease; ↑ indicates increase; ↔ indicates no difference.

4.3. Immobilization and Muscle Dysfunction

Periods of limb immobilization and muscle disuse appear to alter the metabolic functioning of skeletal muscle tissue through the downregulation of metabolic pathways, enzyme activity, and organelle function [132,133]. Furthermore, muscle disuse elicits mitochondrial-mediated apoptosis, which is thought to contribute further to muscle atrophy, while concomitantly creating additional metabolic perturbations within the cell [132,134]. Previous research has identified reductions in intracellular phosphagen and glycogen content following muscle disuse [135]. More recently, it was discovered that periods of muscle disuse also resulted in alterations in intracellular phosphagen and creatine transporter content in human skeletal muscle, demonstrating how the cell adapts to the aforementioned metabolic disruptions in cellular energy turnover [136]. These alterations in intracellular phosphate and transporter protein content, coupled with mitochondrial-mediated apoptosis, highlight a rationale for the potential therapeutic benefit of creatine supplementation to attenuate reductions in phosphagen levels associated with muscle disuse. Supplementation may also concomitantly increase the capacity of the ATP-PCr energy system to resynthesize ATP, leading to higher rates of energy turnover during tissue recovery.

Because of the greater phosphagen-mediated re-synthesis of ATP following creatine supplementation, users are likely able to achieve a greater volume load during rehabilitative activities, which may have implications in minimizing muscle atrophy and promoting other favorable adaptations [49]. However, other plausible mechanisms have been proposed for how creatine may confer anabolic properties [137,138]. For example, it has been proposed that increased intracellular osmolarity from augmented creatine storage may cause cell swelling and the concomitant stimulation of anabolic signaling pathways, independently of exercise [138]. Unfortunately, studies assessing the intrinsic capacity of creatine to stimulate hypertrophic protein remodeling are scant. Among the few studies that have been conducted assessing the utility of creatine to ameliorate disuse-mediated muscle atrophy, results are mixed in their findings and primary outcomes. One of the earlier works [139] demonstrated that creatine had no effect on the change in quadriceps cross-sectional area after two weeks of cast immobilization. However, creatine augmented the regeneration of whole muscle and fiber-type-specific cross-sectional area after a 10-week rehabilitation program following the cast immobilization. Mechanistically, the authors also reported that creatine supplementation elevated the myogenic transcription factors myogenin and myogenic factor 4 (Mrf4) [139]. Other work has demonstrated that short-term (5-day) creatine supplementation can upregulate insulin like growth factor-1 (IGF-1) mRNA at rest and augment the phosphorylation of eukaryotic translation initiation factor 4E-binding protein 1 (4E-BP1) 24 h after an acute bout of resistance exercise [140]. Both IGF-1 and 4E-BP1 have the potential to stimulate protein anabolism through PI3K-Akt-mTOR signaling, thereby positively influencing the recovery of the injured tissue [141]. Beyond its putative anabolic effects, creatine supplementation may have anti-catabolic effects as well [142]. However, these findings are not ubiquitous. Although short term oral creatine supplementation was found to attenuate the loss in muscle mass and strength following a period of upper arm immobilization in healthy young men, [143] these same benefits have not been observed following periods of lower limb immobilization [144], knee arthroplasty [145], or anterior cruciate ligament reconstruction [146]. Cumulatively, although these data suggest the potential therapeutic usage of creatine to aid in recovery, more work is needed in this area, particularly regarding whether any potential synergistic benefits may exist when creatine is combined with other nutrients or rehabilitative strategies post-injury.

5. Other Considerations

Due to the popularity associated with creatine supplementation since the first published reports in the early 1990s, a number of other questions have been evaluated and considered regarding its efficacy. For example, the majority of the published literature on creatine has been completed using male athletes, leading to much less information being available on how creatine supplementation impacts females. Previous work has highlighted gender-specific differences in creatine production and turnover, which lays the foundation for gender considerations for creatine [4]. In regard to research involving exercise performance in females, Vandenberghe et al. [39] reported that creatine supplementation increases intramuscular PCr levels, muscle mass, and strength when compared to those females who took a placebo. Other research by Hamilton [147] showed improvements in upper-body exercise capacity, and Tarnopolsky showed improvements in high-intensity exercise performance [148], whereas Kambis et al. [149] reported improvements in knee extension muscle performance. Similarly, excellent potential exists for creatine to support the health and function of older populations. Although nearly all of the original research on creatine used young, athletic populations, research in the past 10–15 years has highlighted creatine's ability to increase the ability to perform daily living activities, to delay fatigue, and to improve muscle mass in older populations [150–158].

Overwhelmingly, the majority of research that has examined the potential of creatine to impact exercise performance has been conducted using the monohydrate version. Although several other forms of creatine have been proposed and marketed as alternatives, none have been shown to offer benefits above and beyond those seen with monohydrate. In this respect, a number of studies have been completed comparing various alternative forms of creatine, and the interested reader is directed to the following papers: [3–5,30,159–163]. In this respect, one must also realize that several studies have sought to examine the impact of combining creatine with other ingredients, such as beta-alanine [164,165], beta-hydroxy-beta-methylbutyrate (HMB) [96,166–171], glutamine [72], sodium bicarbonate [47], carbohydrates [22,44,99,172,173], and protein [22,59,99] to examine the potential for any synergistic outcomes. Furthermore, the interested reader is encouraged to read the critical review on this topic by Jager et al. [159].

The level of creatine uptake is a key consideration, as it relates to the potential for health and performance outcomes. In this respect, one of the key considerations that has been identified in the literature is the presence of 'responders' and 'nonresponders'. This concept was discussed in a 1999 review by Demant and Rhodes, in which they summarized the available literature and highlighted the fact that identical supplementation regimens could lead to increases in intramuscular PCr levels, whereas the same regimen may cause limited to no changes in other people following a similar supplementation regimen [20]. To illustrate this point, Kilduff et al. [174] identified subjects in their study as responders and nonresponders based on the magnitude of change seen in intramuscular PCr. When examined together, peak force was not changed due to supplementation, but when evaluated separately, the responders significantly increased their peak force production after creatine supplementation. Later, Syrotuik et al. [175] completed an analysis aiming to build a physiological profile of responders and nonresponders. In terms of creatine uptake, a commonly discussed factor that may dictate the extent to which intramuscular PCr levels change in response to creatine supplementation is the level of vegetarianism observed by the individual. In this respect, it is well known that meat-containing foods and products contain the highest amounts of creatine, and thus people who refrain from eating meats tend to consume the lowest amounts of creatine in their diet. Several papers are available that have examined the impact of creatine supplementation on people who follow various types of vegetarianism, with initial reports showing that vegetarians do, indeed, have lower levels of intramuscular PCr [176,177]. Shomrat and colleagues [178] were some of the first to examine this question and concluded that creatine supplementation equally impacted the ability of vegetarian and non-vegetarian individuals to generate peak power after an identical regimen of supplementation. Furthermore, additional studies by Watt [179] and Lukaszuk [180] highlighted the fact that creatine supplementation in vegetarian people can impact intramuscular and plasma levels of creatine in a similar fashion. Furthering this aim, reviews by Venderley and Kaviani concluded that creatine supplementation could be an effective strategy for vegetarian individuals to increase their intramuscular levels of PCr, a key factor that may impact an individual's ability to perform high-intensity exercise [176,177]. Finally, interested readers are encouraged to review the following articles by Antonio [4], as well as the International Society of Sports Nutrition's position on creatine [3].

6. Conclusions

Augmenting intramuscular creatine stores either by creatine loading or daily supplementation over several days leads to increased concentrations of intramuscular creatine and PCr. Increases in these substrates are associated with an attenuation of ATP degradation, heightened ATP production, and an increase in energetic output during activities involving intermittent, high-intensity, short bouts of exercise. Additionally, creatine supplementation shows promise in facilitating recovery following exercise-induced muscle damage and potentially as an aid during post-injury rehabilitation. Based on the current literature, the following can be deduced involving creatine supplementation and its ergogenic potential:

1. Creatine supplementation is safe during short- and long-term intervals for healthy males and females, as well as in younger and older individuals.
2. Creatine supplementation, ingested at 0.3 g/kg/day for 3–5 consecutive days or 20 g/day for 5–7 successive days, has been shown to quickly increase intramuscular creatine, yielding immediate ergogenic benefits. Correspondingly, a regimen of 3–5 g/day over 4 weeks increased creatine stores, augmented muscle performance, mitigated recovery factors, and resulted in muscle accretion.
3. Creatine supplementation intermixed with carbohydrates or carbohydrates and protein appears to be efficacious in increasing intramuscular creatine retention, although the additional benefits in terms of performance outcomes appear to be nebulous.
4. Creatine supplementation appears to provide an ergogenic effect when assessing isolated or individual bouts of peak or maximal force production.
5. Creatine supplementation facilitates more significant improvements in strength and FFM.

6. Creatine supplementation provides benefits during single and repeated sprints and may increase agility and jumping performance.
7. Creatine supplementation appears to provide ergogenic benefits to aerobic endurance bouts with positive physiological adaptations.
8. Creatine supplementation may enhance recovery from intense exercise and possibly provide synergistic benefits during the post-injury rehabilitation period.
9. Creatine supplementation provides positive benefits to both males and females, athletes and recreational fitness enthusiasts, as well as younger and older individuals.
10. Creatine supplementation provides more significant augmentations of intramuscular creatine in vegans than omnivores, due to lower initial levels of creatine stores, with both groups receiving comparable ergogenic benefits.

Author Contributions: Conceptualization, B.W. and C.M.K.; investigation, all authors; writing—original draft preparation, all authors; writing—review and editing, all authors; visualization, all authors. All authors have read and agreed to the published version of the manuscript.

Funding: The APC was funded by Alzchem Trotsberg, GmhH.

Institutional Review Board Statement: Not applicable, due to the article being a review of existing literature and not requiring the recruitment and consent of additional study participants.

Informed Consent Statement: Not applicable.

Data Availability Statement: Data sharing not applicable as no new data were analyzed as part of this review paper.

Acknowledgments: The authors would like to thank all of the research subjects, investigators, and funding agencies contributing to the research in this paper.

Conflicts of Interest: The authors of this manuscript received financial remuneration for preparing and reviewing this paper from the sponsor of the special issue, Alzchem. B.W., J.J.M., B.L., A.R.J., R.B.K., and C.M.K. have consulted with and received external funding from companies who sell certain dietary ingredients and have received remuneration from companies for delivering scientific presentations at conferences. A.R.J. and C.M.K. also write for online and other media outlets on topics related to exercise and nutrition. None of these entities had any role in the design of the paper, collection, analyses, or interpretation of data; in the writing of the manuscript, or in the decision to publish this paper. R.B.K. has conducted industry-sponsored research on creatine, received financial support for presenting on creatine at industry-sponsored scientific conferences (including the ISSN), and served as an expert witness on cases related to creatine. Additionally, he serves as Chair of the Scientific Advisory Board for Alzchem, which manufactures creatine monohydrate.

References

1. Kerksick, C.M.; Wilborn, C.D.; Roberts, M.D.; Smith-Ryan, A.E.; Kleiner, S.M.; Jäger, R.; Collins, R.; Cooke, M.; Davis, J.N.; Galvan, E.; et al. ISSN exercise & sports nutrition review update: Research & recommendations. *J. Int. Soc. Sports Nutr.* **2018**, *15*, 38. [CrossRef]
2. Williams, M.H. Facts and fallacies of purported ergogenic amino acid supplements. *Clin. Sports Med.* **1999**, *18*, 633–649. [CrossRef]
3. Kreider, R.B.; Kalman, D.S.; Antonio, J.; Ziegenfuss, T.N.; Wildman, R.; Collins, R.; Candow, D.G.; Kleiner, S.M.; Almada, A.L.; Lopez, H.L. International Society of Sports Nutrition position stand: Safety and efficacy of creatine supplementation in exercise, sport, and medicine. *J. Int. Soc. Sports Nutr.* **2017**, *14*, 1–18. [CrossRef]
4. Antonio, J.; Candow, D.G.; Forbes, S.C.; Gualano, B.; Jagim, A.R.; Kreider, R.B.; Rawson, E.S.; Smith-Ryan, A.E.; VanDusseldorp, T.A.; Willoughby, D.S.; et al. Common questions and misconceptions about creatine supplementation: What does the scientific evidence really show? *J. Int. Soc. Sports Nutr.* **2021**, *18*, 1–17. [CrossRef]
5. Fazio, C.; Elder, C.L.; Harris, M.M. Efficacy of alternative forms of creatine supplementation on improving performance and body composition in healthy subjects: A systematic review. *J. Strength Cond. Res.* **2021**. [CrossRef]
6. Harris, R.C.; Söderlund, K.; Hultman, E. Elevation of creatine in resting and exercised muscle of normal subjects by creatine supplementation. *Clin. Sci.* **1992**, *83*, 367–374. [CrossRef]
7. Dalsom, P.D., Ekblom, B., Söerlund, K., Sjödln, B., Hultman, E. Creatine supplementation and dynamic high-intensity intermittent exercise. *Scand. J. Med. Sci. Sports* **2007**, *3*, 143–149. [CrossRef]
8. Greenhaff, P.L.; Casey, A.; Short, A.H.; Harris, R.; Soderlund, K.; Hultman, E. Influence of Oral Creatine Supplementation of Muscle Torque during Repeated Bouts of Maximal Voluntary Exercise in Man. *Clin. Sci.* **1993**, *84*, 565–571. [CrossRef] [PubMed]

9. Kreider, R.B. Effects of creatine supplementation on performance and training adaptations. *Mol. Cell. Biochem.* **2003**, *244*, 89–94. [CrossRef]
10. Rawson, E.S.; Volek, J.S. Effects of Creatine Supplementation and Resistance Training on Muscle Strength and Weightlifting Performance. *J. Strength Cond. Res.* **2003**, *17*, 822–831. [CrossRef]
11. Stone, M.H.; Sanborn, K.; Smith, L.L.; O'Bryant, H.S.; Hoke, T.; Utter, A.C.; Johnson, R.L.; Boros, R.; Hruby, J.; Pierce, K.C.; et al. Effects of In-Season (5 Weeks) Creatine and Pyruvate Supplementation on Anaerobic Performance and Body Composition in American Football Players. *Int. J. Sport Nutr.* **1999**, *9*, 146–165. [CrossRef] [PubMed]
12. Buford, T.W.; Kreider, R.B.; Stout, J.R.; Greenwood, M.; Campbell, B.; Spano, M.; Ziegenfuss, T.; Lopez, H.; Landis, J.; Antonio, J. International Society of Sports Nutrition position stand: Creatine supplementation and exercise. *J. Int. Soc. Sports Nutr.* **2007**, *4*, 6. [CrossRef]
13. Cooper, R.; Naclerio, F.; Allgrove, J.; Jimenez, A. Creatine supplementation with specific view to exercise/sports performance: An update. *J. Int. Soc. Sports Nutr.* **2012**, *9*, 33. [CrossRef]
14. Hultman, E.; Soderlund, K.; Timmons, J.A.; Cederblad, G.; Greenhaff, P.L. Muscle creatine loading in men. *J. Appl. Physiol.* **1996**, *81*, 232–237. [CrossRef]
15. Greenhaff, P.L.; Bodin, K.; Soderlund, K.; Hultman, E. Effect of oral creatine supplementation on skeletal muscle phosphocreatine resynthesis. *Am. J. Physiol. Metab.* **1994**, *266*, E725–E730. [CrossRef]
16. Law, Y.L.L.; Ong, W.S.; GillianYap, T.L.; Lim, S.C.J.; Von Chia, E. Effects of Two and Five Days of Creatine Loading on Muscular Strength and Anaerobic Power in Trained Athletes. *J. Strength Cond. Res.* **2009**, *23*, 906–914. [CrossRef]
17. Sale, C.; Harris, R.C.; Florance, J.; Kumps, A.; Sanvura, R.; Poortmans, J.R. Urinary creatine and methylamine excretion following 4×5 g·day−1or 20×1 g·day−1of creatine monohydrate for 5 days. *J. Sports Sci.* **2009**, *27*, 759–766. [CrossRef] [PubMed]
18. Lanhers, C.; Pereira, B.; Naughton, G.; Trousselard, M.; Lesage, F.-X.; Dutheil, F. Creatine Supplementation and Lower Limb Strength Performance: A Systematic Review and Meta-Analyses. *Sports Med.* **2015**, *45*, 1285–1294. [CrossRef]
19. Kraemer, W.J.; Volek, J.S. Creatine supplementation: Its Role in Human Performance. *Clin. Sports Med.* **1999**, *18*, 651–666. [CrossRef]
20. Demant, T.W.; Rhodes, E. Effects of Creatine Supplementation on Exercise Performance. *Sports Med.* **1999**, *28*, 49–60. [CrossRef]
21. Volek, J.S.; Duncan, N.D.; Mazzetti, S.A.; Staron, R.S.; Putukian, M.; Gómez, A.L.; Pearson, D.R.; Fink, W.J.; Kraemer, W.J. Performance and muscle fiber adaptations to creatine supplementation and heavy resistance training. *Med. Sci. Sports Exerc.* **1999**, *31*, 1147–1156. [CrossRef]
22. Kreider, R.B.; Ferreira, M.; Wilson, M.; Grindstaff, P.; Plisk, S.; Reinardy, J.; Cantler, E.; Almada, A.L. Effects of creatine supplementation on body composition, strength, and sprint performance. *Med. Sci. Sports Exerc.* **1998**, *30*, 73–82. [CrossRef]
23. Del Favero, S.; Roschel, H.; Artioli, G.; Ugrinowitsch, C.; Tricoli, V.; Costa, A.; Barroso, R.; Negrelli, A.L.; Otaduy, M.C.; Leite, C.D.C.; et al. Creatine but not betaine supplementation increases muscle phosphorylcreatine content and strength performance. *Amino Acids* **2011**, *42*, 2299–2305. [CrossRef]
24. Atakan, M.; Karavelioğlu, M.; Harmancı, H.; Cook, M.; Bulut, S. Short term creatine loading without weight gain improves sprint, agility and leg strength performance in female futsal players. *Sci. Sports* **2019**, *34*, 321–327. [CrossRef]
25. Stevenson, S.W.; Dudley, G.A. Creatine loading, resistance exercise performance, and muscle mechanics. *J. Strength Cond. Res.* **2001**, *15*, 413–419.
26. Zuniga, J.M.; Housh, T.J.; Camic, C.L.; Hendrix, C.R.; Mielke, M.; Johnson, G.O.; Housh, D.J.; Schmidt, R.J. The Effects of Creatine Monohydrate Loading on Anaerobic Performance and One-Repetition Maximum Strength. *J. Strength Cond. Res.* **2012**, *26*, 1651–1656. [CrossRef] [PubMed]
27. Lanhers, C.; Pereira, B.; Naughton, G.; Trousselard, M.; Lesage, F.-X.; Dutheil, F. Creatine Supplementation and Upper Limb Strength Performance: A Systematic Review and Meta-Analysis. *Sports Med.* **2016**, *47*, 163–173. [CrossRef]
28. Kelly, V.G.; Jenkins, D.G. Effect of oral creatine supplementation on near-maximal strength and repeated sets of high-intensity bench press exercise. *J. Strength Cond. Res.* **1998**, *12*, 109–115.
29. Antonio, J.; Ciccone, V. The effects of pre versus post workout supplementation of creatine monohydrate on body composition and strength. *J. Int. Soc. Sports Nutr.* **2013**, *10*, 36. [CrossRef]
30. Peeters, B.M.; Lantz, C.D.; Mayhew, J.L. Effect of oral creatine monohydrate and creatine phosphate supplementation on maximal strength indices, body composition and blood pressure. *J. Strength Cond. Res.* **1999**, *13*, 3–9.
31. Syrotuik, D.G.; Bell, G.J.; Burnham, R.; Sim, L.L.; Calvert, R.A.; Maclean, I.M. Absolute and relative strength performance following creatine monohydrate supplementation combined with periodized resistance training. *J. Strength Cond. Res.* **2000**, *14*, 182–190.
32. Kreider, R.B.; Klesges, R.; Harmon, K.; Grindstaff, P.; Ramsey, L.; Bullen, D.; Wood, L.; Li, Y.; Almada, A. Effects of Ingesting Supplements Designed to Promote Lean Tissue Accretion on Body Composition during Resistance Training. *Int. J. Sport Nutr.* **1996**, *6*, 234–246. [CrossRef] [PubMed]
33. Kreider, R.B.; Klesges, R.C.; Lotz, D.; Davis, M.; Cantler, E.; Grindstaff, P.; Ramsey, L.; Bullen, D.; Wood, L.; Almada, A.L. Effects of nutritional supplementation during off-season college football training on body composition and strength. *J. Exerc. Physiol.* **1999**, *2*, 24–39.
34. Noonan, D.; Berg, K.; Latin, R.W.; Wagner, J.C.; Reimers, K. Effects of Varying Dosages of Oral Creatine Relative to Fat Free Body Mass on Strength and Body Composition. *J. Strength Cond. Res.* **1998**, *12*, 104. [CrossRef]

35. Bemben, M.G.; Bemben, D.A.; Loftiss, D.D.; Knehans, A.W. Creatine supplementation during resistance training in college football athletes. *Med. Sci. Sports Exerc.* **2001**, *33*, 1667–1673. [CrossRef]
36. Pearson, D.R.; Hambx, W.R.; Harris, T. Long-Term Effects of Creatine Monohdrate on Strength and Power. *J. Strength Cond. Res.* **1999**, *13*, 187–192.
37. Brenner, M.; Rankin, J.W.; Sebolt, D. The effect of creatine supplementation during resistance training in women. *J. Strength Cond. Res.* **2000**, *14*, 207–213.
38. Larson-Meyer, D.E.; Hunter, G.R.; Trowbridge, C.A.; Turk, J.C.; Ernest, J.M.; Torman, S.L.; Harbin, P.A. The effect of creatine supplementation on muscle strength and composition during off-season training in female soccery players. *J. Strength Cond. Res.* **2000**, *14*, 434–442.
39. Vandenberghe, K.; Goris, M.; Van Hecke, P.; Van Leemputte, M.; Vangerven, L.; Hespel, P. Long-term creatine intake is beneficial to muscle performance during resistance training. *J. Appl. Physiol.* **1997**, *83*, 2055–2063. [CrossRef]
40. Candow, D.G.; Chilibeck, P.D.; Burke, D.G.; Mueller, K.D.; Lewis, J.D. Effect of Different Frequencies of Creatine Supplementation on Muscle Size and Strength in Young Adults. *J. Strength Cond. Res.* **2011**, *25*, 1831–1838. [CrossRef]
41. Herda, T.J.; Beck, T.W.; Smith-Ryan, A.; Smith, A.E.; Walter, A.A.; Hartman, M.J.; Stout, J.R.; Cramer, J.T. Effects of Creatine Monohydrate and Polyethylene Glycosylated Creatine Supplementation on Muscular Strength, Endurance, and Power Output. *J. Strength Cond. Res.* **2009**, *23*, 818–826. [CrossRef] [PubMed]
42. Birch, R.; Noble, D.; Greenhaff, P.L. The influence of dietary creatine supplementation on performance during repeated bouts of maximal isokinetic cycling in man. *Graefe's Arch. Clin. Exp. Ophthalmol.* **1994**, *69*, 268–270. [CrossRef]
43. Casey, A.; Constantin-Teodosiu, D.; Howell, S.; Hultman, E.; Greenhaff, P.L. Creatine ingestion favorably affects performance and muscle metabolism during maximal exercise in humans. *Am. J. Physiol. Metab.* **1996**, *271*, E31–E37. [CrossRef] [PubMed]
44. Green, A.L.; Hultman, E.; Macdonald, I.A.; Sewell, D.A.; Greenhaff, P.L. Carbohydrate ingestion augments skeletal muscle creatine accumulation during creatine supplementation in humans. *Am. J. Physiol. Content* **1996**, *271*, 821–826. [CrossRef]
45. Stout, J.R.; Eckerson, J.M.; Housh, T.J.; Ebersolevkt, K.T. The effects of creatine supplementation on anaerobic working capacity. *J. Strength Cond. Res.* **1999**, *13*, 135–138.
46. Van Loon, L.J.C.; Oosterlaar, A.M.; Hartgens, F.; Hesselink, M.K.C.; Snow, R.J.; Wagenmakers, A.J.M. Effects of creatine loading and prolonged creatine supplementation on body composition, fuel selection, sprint and endurance performance in humans. *Clin. Sci.* **2003**, *104*, 153–162. [CrossRef]
47. Barber, J.J.; McDermott, A.Y.; McGaughey, K.J.; Olmstead, J.D.; Hagobian, T.A. Effects of Combined Creatine and Sodium Bicarbonate Supplementation on Repeated Sprint Performance in Trained Men. *J. Strength Cond. Res.* **2013**, *27*, 252–258. [CrossRef]
48. Fukuda, D.H.; Smith, A.E.; Kendall, K.L.; Dwyer, T.R.; Kerksick, C.; Beck, T.W.; Cramer, J.T.; Stout, J.R. The Effects of Creatine Loading and Gender on Anaerobic Running Capacity. *J. Strength Cond. Res.* **2010**, *24*, 1826–1833. [CrossRef]
49. Branch, J.D. Effect of Creatine Supplementation on Body Composition and Performance: A Meta-analysis. *Int. J. Sport Nutr. Exerc. Metab.* **2003**, *13*, 198–226. [CrossRef] [PubMed]
50. Cox, G.; Mujika, I.; Tumilty, D.; Burke, L. Acute Creatine Supplementation and Performance during a Field Test Simulating Match Play in Elite Female Soccer Players. *Int. J. Sport Nutr. Exerc. Metab.* **2002**, *12*, 33–46. [CrossRef]
51. Ramírez-Campillo, R.; González-Jurado, J.A.; Martínez, C.; Nakamura, F.Y.; Peñailillo, L.; Meylan, C.M.; Caniuqueo, A.; Cañas-Jamet, R.; Moran, J.; Alonso-Martínez, A.M.; et al. Effects of plyometric training and creatine supplementation on maximal-intensity exercise and endurance in female soccer players. *J. Sci. Med. Sport* **2016**, *19*, 682–687. [CrossRef]
52. Ostojic, S.M. Creatine Supplementation in Young Soccer Players. *Int. J. Sport Nutr. Exerc. Metab.* **2004**, *14*, 95–103. [CrossRef] [PubMed]
53. Eijnde, B.O.; Vergauwen, L.; Hespel, P. Creatine Loading does not Impact on Stroke Performance in Tennis. *Int. J. Sports Med.* **2001**, *22*, 76–80. [CrossRef]
54. Pluim, B.M. The effects of creatine supplementation on selected factors of tennis specific training *Commentary* Commentary. *Br. J. Sports Med.* **2006**, *40*, 507–512. [CrossRef] [PubMed]
55. Koçak, S.; Karli, U. Effects of high dose oral creatine supplementation on anaerobic capacity of elite wrestlers. *J. Sports Med. Phys. Fit.* **2003**, *43*, 488–492.
56. Aedma, M.; Timpmann, S.; Lätt, E.; Ööpik, V. Short-term creatine supplementation has no impact on upper-body anaerobic power in trained wrestlers. *J. Int. Soc. Sports Nutr.* **2015**, *12*, 1–9. [CrossRef] [PubMed]
57. De Oca, R.M.-M.; Farfán-González, F.; Camarillo-Romero, S.; Tlatempa-Sotelo, P.; Francisco-Argüelles, C.; Kormanowski, A.; González-Gallego, J.; Alvear-Ordenes, I. Effects of creatine supplementation in taekwondo practitioners. *Nutr. Hosp.* **2013**, *28*, 391–399.
58. Izquierdo, M.; Ibañez, J.; González-Badillo, J.J.; Gorostiaga, E.M. Effects of creatine supplementation on muscle power, endurance, and sprint performance. *Med. Sci. Sports Exerc.* **2002**, *34*, 332–343. [CrossRef]
59. Stout, J.; Eckerson, J.; Noonan, D.; Moore, G.; Cullen, D. Effects of 8 weeks of creatine supplementation on exercise performance and fat-free weight in football players during training. *Nutr. Res.* **1999**, *19*, 217–225. [CrossRef]
60. Jones, A.M.; Atter, T.; Georg, K.P. Oral creatine supplementation improves multiple sprint performance in elite ice-hockey players. *J. Sports Med. Phys. Fit.* **1999**, *39*, 189–196. [CrossRef]
61. Mujika, I.; Padilla, S.; Ibañez, J.; Izquierdo, M.; Gorostiaga, E. Creatine supplementation and sprint performance in soccer players. *Med. Sci. Sports Exerc.* **2000**, *32*, 518–525. [CrossRef] [PubMed]

62. Theodorou, A.S.; Cooke, C.B.; King, R.F.G.J.; Hood, C.; Denison, T.; Wainwright, B.G.; Havenetidis, K. The effect of longer-term creatine supplementation on elite swimming performance after an acute creatine loading. *J. Sports Sci.* **1999**, *17*, 853–859. [CrossRef] [PubMed]

63. Skare, O.-C.; Skadberg, O.; Wisnes, A.R. Creatine supplementation improves sprint performance in male sprinters. *Scand. J. Med. Sci. Sports* **2001**, *11*, 96–102. [CrossRef] [PubMed]

64. Delecluse, C.; Diels, R.; Goris, M. Effect of Creatine Supplementation on Intermittent Sprint Running Performance in Highly Trained Athletes. *J. Strength Cond. Res.* **2003**, *17*, 446. [CrossRef] [PubMed]

65. Peyrebrune, M.C.; Nevill, M.E.; Donaldson, F.J.; Cosford, D.J. The effects of oral creatine supplementation on performance in single and repeated sprint swimming. *J. Sports Sci.* **1998**, *16*, 271–279. [CrossRef]

66. Williams, J.; Abt, G.; Kilding, A.E. Effects of Creatine Monohydrate Supplementation on Simulated Soccer Performance. *Int. J. Sports Physiol. Perform.* **2014**, *9*, 503–510. [CrossRef]

67. Glaister, M.; Lockey, R.A.; Abraham, C.S.; Staerck, A.; Goodwin, J.E.; McInnes, G. Creatine Supplementation and Multiple Sprint Running Performance. *J. Strength Cond. Res.* **2006**, *20*, 273–277. [CrossRef] [PubMed]

68. Lamontagne-Lacasse, M.; Nadon, R.; Goulet, E.D.B. Effect of Creatine Supplementation Jumping Performance in Elite Volleyball Players. *Int. J. Sports Physiol. Perform.* **2011**, *6*, 525–533. [CrossRef] [PubMed]

69. Mielgo-Ayuso, J.; Calleja-Gonzalez, J.; Marqués-Jiménez, D.; Caballero-García, A.; Córdova, A.; Fernández-Lázaro, D. Effects of Creatine Supplementation on Athletic Performance in Soccer Players: A Systematic Review and Meta-Analysis. *Nutrients* **2019**, *11*, 757. [CrossRef]

70. Haff, G.; Kirksey, K.; Stone, M.; Warren, B.; Johnson, R.; Stone, M.; O'Bryant, H.; Proulx, C. The effects of 6 weeks of creatine monohydrate supplementation on dynamic rate of force development. *J. Strength Cond. Res.* **2000**, *14*, 426–433. [CrossRef]

71. Kirksey, B.; Stone, M.; Warren, B.; Johnson, R.; Stone, M.; Haff, G.; Williams, F.; Proulx, C. The effects of 6 weeks of creatine monohydrate supplementation on performance measures and body composition in colleagiate track and field athletes. *J. Strength Cond. Res.* **1999**, *13*, 148–156.

72. Lehmkuhl, M.; Malone, M.; Justice, B.; Trone, G.; Pistilli, E.; Vinci, D.; Haff, E.E.; Kilgore, J.L.; Haff, G.G. The Effects of 8 Weeks of Creatine Monohydrate and Glutamine Supplementation on Body Composition and Performance Measures. *J. Strength Cond. Res.* **2003**, *17*, 425. [CrossRef] [PubMed]

73. Grindstaff, P.D.; Kreider, R.; Bishop, R.; Wilson, M.; Wood, L.; Alexander, C.; Almada, A. Effects of creatine supplementation on repetitive sprint performance and body composition in competitive swimmers. *Int. J. Sport Nutr.* **1997**, *7*, 330–346. [CrossRef]

74. Papadimitriou, K. Effects of legal ergogenic supplements on swimmer's performance: A bibliographic approach. *Inq. Sport Phys. Educ.* **2018**, *16*, 66–77.

75. Rothschild, J.A.; Bishop, D.J. Effects of Dietary Supplements on Adaptations to Endurance Training. *Sports Med.* **2020**, *50*, 25–53. [CrossRef]

76. Hawley, J.A.; Hargreaves, M.; Joyner, M.J.; Zierath, J.R. Integrative Biology of Exercise. *Cell* **2014**, *159*, 738–749. [CrossRef]

77. Graef, J.L.; Smith, A.E.; Kendall, K.L.; Fukuda, D.H.; Moon, J.R.; Beck, T.W.; Cramer, J.T.; Stout, J.R. The effects of four weeks of creatine supplementation and high-intensity interval training on cardiorespiratory fitness: A randomized controlled trial. *J. Int. Soc. Sports Nutr.* **2009**, *6*, 18. [CrossRef] [PubMed]

78. Barnett, C.; Hinds, M.; Jenkins, D.G. Effects of oral creatine supplementation on multiple sprint cycle performance. *Aust. J. Sci. Med. Sport* **1996**, *28*, 35–39. [PubMed]

79. Cañete, S.; Juan, A.F.S.; Pérez, M.; Gómez-Gallego, F.; López-Mojares, L.M.; Earnest, C.; Fleck, S.J.; Lucia, A. Does Creatine Supplementation Improve Functional Capacity in Elderly Women? *J. Strength Cond. Res.* **2006**, *20*, 22–28. [CrossRef]

80. Eijnde, B.O.; Van Leemputte, M.; Goris, M.; Labarque, V.; Taes, Y.; Verbessem, P.; Vanhees, L.; Ramaekers, M.; Eynde, B.V.; Van Schuylenbergh, R.; et al. Effects of creatine supplementation and exercise training on fitness in men 55–75 yr old. *J. Appl. Physiol.* **2003**, *95*, 818–828. [CrossRef] [PubMed]

81. Syrotuik, D.G.; Game, A.B.; Gillies, E.M.; Bell, G.J. Effects of Creatine Monohydrate Supplementation during Combined Strength and High Intensity Rowing Training on Performance. *Can. J. Appl. Physiol.* **2001**, *26*, 527–542. [CrossRef] [PubMed]

82. Forbes, S.C.; Sletten, N.; Durrer, C.; Myette-Côté, É.; Candow, D.; Little, J.P. Creatine Monohydrate Supplementation Does Not Augment Fitness, Performance, or Body Composition Adaptations in Response to Four Weeks of High-Intensity Interval Training in Young Females. *Int. J. Sport Nutr. Exerc. Metab.* **2017**, *27*, 285–292. [CrossRef] [PubMed]

83. Engelhardt, M.; Neumann, G.; Berbalk, A.; Reuter, I. Creatine supplementation in endurance sports. *Med. Sci. Sports Exerc.* **1998**, *30*, 1123–1129. [CrossRef]

84. Miura, A.; Kino, F.; Kajitani, S.; Sato, H.; Sato, H.; Fukuba, Y. The Effect of Oral Creatine Supplementation on the Curvature Constant Parameter of the Power-Duration Curve for Cycle Ergometry in Humans. *Jpn. J. Physiol.* **1999**, *49*, 169–174. [CrossRef] [PubMed]

85. Stroud, M.A.; Holliman, D.; Bell, D.; Green, A.L.; Macdonald, I.A.; Greenhaff, P.L. Effect of Oral Creatine Supplementation on Respiratory Gas Exchange and Blood Lactate Accumulation during Steady-State Incremental Treadmill Exercise and Recovery in Man. *Clin. Sci.* **1994**, *87*, 707–710. [CrossRef]

86. Bellinger, B.; Bold, A.; Wilson, G.; Myburgh, K.; Noakes, T. Oral creatine supplementation decreases plasma markers of adenine nucleotide degradation during a 1-h cycle test. *Acta Physiol. Scand.* **2000**, *170*, 217–224. [CrossRef]

87. Nemezio, K.M.D.A.; Bertuzzi, R.; Correia-Oliveira, C.R.; Gualano, B.; Bishop, D.; Lima-Silva, A. Effect of Creatine Loading on Oxygen Uptake during a 1-km Cycling Time Trial. *Med. Sci. Sports Exerc.* **2015**, *47*, 2660–2668. [CrossRef]

88. Jones, A.M.; Carter, H.; Pringle, J.S.M.; Campbell, I.T. Effect of creatine supplementation on oxygen uptake kinetics during submaximal cycle exercise. *J. Appl. Physiol.* **2002**, *92*, 2571–2577. [CrossRef]

89. McNaughton, L.R.; Dalton, B.; Tarr, J. The effects of creatine supplementation on high-intensity exercise performance in elite performers. *Graefe's Arch. Clin. Exp. Ophthalmol.* **1998**, *78*, 236–240. [CrossRef]

90. Rossiter, H.B.; Cannell, E.R.; Jakeman, P.M. The effect of oral creatine supplementation on the 1000-m performance of competitive rowers. *J. Sports Sci.* **1996**, *14*, 175–179. [CrossRef]

91. Maganaris, C.; Maughan, R. Creatine supplementation enhances maximum voluntary isometric force and endurance capacity in resistance trained men. *Acta Physiol. Scand.* **1998**, *163*, 279–287. [CrossRef] [PubMed]

92. Prevost, M.C.; Nelson, A.G.; Morris, G.S. Creatine Supplementation Enhances Intermittent Work Performance. *Res. Q. Exerc. Sport* **1997**, *68*, 233–240. [CrossRef]

93. Jakobi, J.M.; Rice, C.L.; Curtin, S.V.; Marsh, G.D. Contractile properties, fatigue and recovery are not influenced by short-term creatine supplementation in human muscle. *Exp. Physiol.* **2000**, *85*, 451–460. [CrossRef] [PubMed]

94. Vandebuerie, F.; Eynde, B.V.; Vandenberghe, K.; Hespel, P. Effect of Creatine Loading on Endurance Capacity and Sprint Power in Cyclists. *Int. J. Sports Med.* **1998**, *19*, 490–495. [CrossRef] [PubMed]

95. Chwalbińska-Moneta, J. Effect of creatine supplementation on aerobic performance and anaerobic capacity in elite rowers in the course of endurance training. *Int. J. Sport Nutr. Exerc. Metab.* **2003**, *13*, 173–183. [CrossRef]

96. Fernández-Landa, J.; Fernández-Lázaro, D.; Calleja-González, J.; Caballero-García, A.; Martínez, A.C.; León-Guereño, P.; Mielgo-Ayuso, J. Effect of Ten Weeks of Creatine Monohydrate Plus HMB Supplementation on Athletic Performance Tests in Elite Male Endurance Athletes. *Nutrients* **2020**, *12*, 193. [CrossRef] [PubMed]

97. Kendall, K.L.; Smith, A.E.; Graef, J.L.; Fukuda, D.H.; Moon, J.R.; Beck, T.W.; Cramer, J.T.; Stout, J.R. Effects of Four Weeks of High-Intensity Interval Training and Creatine Supplementation on Critical Power and Anaerobic Working Capacity in College-Aged Men. *J. Strength Cond. Res.* **2009**, *23*, 1663–1669. [CrossRef]

98. Mujika, I.; Padilla, S. Creatine Supplementation as an Ergogenic Aid for Sports Performance in Highly Trained Athletes: A Critical Review. *Int. J. Sports Med.* **1997**, *18*, 491–496. [CrossRef]

99. Steenge, G.R.; Simpson, E.J.; Greenhaff, P.L. Protein- and carbohydrate-induced augmentation of whole body creatine retention in humans. *J. Appl. Physiol.* **2000**, *89*, 1165–1171. [CrossRef]

100. Cooke, M.B.; Rybalka, E.; Williams, A.D.; Cribb, P.J.; Hayes, A. Creatine supplementation enhances muscle force recovery after eccentrically-induced muscle damage in healthy individuals. *J. Int. Soc. Sports Nutr.* **2009**, *6*, 13. [CrossRef] [PubMed]

101. Cooke, M.B.; Rybalka, E.; Stathis, C.G.; Hayes, A. Myoprotective Potential of Creatine Is Greater than Whey Protein after Chemically-Induced Damage in Rat Skeletal Muscle. *Nutrients* **2018**, *10*, 553. [CrossRef]

102. Santos, R.; Bassit, R.; Caperuto, E.; Rosa, L.C. The effect of creatine supplementation upon inflammatory and muscle soreness markers after a 30km race. *Life Sci.* **2004**, *75*, 1917–1924. [CrossRef]

103. Volek, J.S.; Mazzetti, S.A.; Farquhar, W.B.; Barnes, B.R.; Gómez, A.L.; Kraemer, W.J. Physiological responses to short-term exercise in the heat after creatine loading. *Med. Sci. Sports Exerc.* **2001**, *33*, 1101–1108. [CrossRef] [PubMed]

104. Weiss, B.A.; Powers, M.E. Creatine supplementation does not impair the thermoregulatory response during a bout of exercise in the heat. *J. Sports Med. Phys. Fit.* **2006**, *46*, 555–563.

105. Lopez, R.M.; Casa, D.J.; McDermott, B.P.; Ganio, M.S.; Armstrong, L.E.; Maresh, C.M. Does Creatine Supplementation Hinder Exercise Heat Tolerance or Hydration Status? A Systematic Review with Meta-Analyses. *J. Athl. Train.* **2009**, *44*, 215–223. [CrossRef] [PubMed]

106. Rawson, E.S.; Gunn, B.; Clarkson, P.M. The effects of creatine supplementation on exercise-induced muscle damage. *J. Strength Cond. Res.* **2001**, *15*, 178–184. [PubMed]

107. Yquel, R.; Arsac, L.; Thiaudière, E.; Canioni, P.; Manier, G. Effect of creatine supplementation on phosphocreatine resynthesis, inorganic phosphate accumulation and pH during intermittent maximal exercise. *J. Sports Sci.* **2002**, *20*, 427–437. [CrossRef]

108. Fitts, R. Cellular mechanisms of muscle fatigue. *Physiol. Rev.* **1994**, *74*, 49–94. [CrossRef]

109. Okudan, N.; Belviranli, M.; Pepe, H.; Gökbel, H. The effects of beta alanine plus creatine administration on performance during repeated bouts of supramaximal exercise in sedentary men. *J. Sports Med. Phys. Fit.* **2014**, *55*, 1322–1328.

110. Okudan, N.; Gokbel, H. The effects of creatine supplementation on performance during the repeated bouts of supramaximal exercise. *J. Sports Med. Phys. Fit.* **2005**, *45*, 507–511.

111. Gill, N.D.; Hall, R.D.; Blazevich, A.J. Creatine Serum Is Not as Effective as Creatine Powder for Improving Cycle Sprint Performance in Competitive Male Team-Sport Athletes. *J. Strength Cond. Res.* **2004**, *18*, 272–275. [CrossRef]

112. Griffen, C.; Rogerson, D.; Ranchordas, M.; Ruddock, A. Effects of Creatine and Sodium Bicarbonate Coingestion on Multiple Indices of Mechanical Power Output during Repeated Wingate Tests in Trained Men. *Int. J. Sport Nutr. Exerc. Metab.* **2015**, *25*, 298–306. [CrossRef]

113. Crisafulli, D.L.; Buddhadev, H.H.; Brilla, L.R.; Chalmers, G.R.; Suprak, D.N.; Juan, J.G.S. Creatine-electrolyte supplementation improves repeated sprint cycling performance: A double blind randomized control study. *J. Int. Soc. Sports Nutr.* **2018**, *15*, 1–11. [CrossRef]

114. Dawson, B.; Cutler, M.; Moody, A.; Lawrence, S.; Goodman, C.; Randall, N. Effects of oral creatine loading on single and repeated maximal short sprints. *Aust. J. Sci. Med. Sport* **1995**, *27*, 56–61.
115. Cottrell, G.T.; Coast, J.R.; Herb, R.A. Effect of recovery interval on multiple-bout sprint cycling performance after acute creatine supplementation. *J. Strength Cond. Res.* **2002**, *16*, 109–116. [PubMed]
116. Aaserud, R.; Gramvik, P.; Olsen, S.R.; Jensen, J. Creatine supplementation delays onset of fatigue during repeated bouts of sprint running. *Scand. J. Med. Sci. Sports* **1998**, *8*, 247–251. [CrossRef] [PubMed]
117. Deminice, R.; Rosa, F.T.; Franco, G.S.; Jordao, A.A.; de Freitas, E.C. Effects of creatine supplementation on oxidative stress and inflammatory markers after repeated-sprint exercise in humans. *Nutrients* **2013**, *29*, 1127–1132. [CrossRef]
118. Green, J.; McLester, J.; Smith, J.; Mansfield, E. The effects of creatine supplementation on repeated upper- and lower-body Wingate performance. *J. Strength Cond. Res.* **2001**, *15*, 36–41.
119. Ahmun, R.P.; Tong, R.J.; Grimshaw, P.N. The Effects of Acute Creatine Supplementation on Multiple Sprint Cycling and Running Performance in Rugby Players. *J. Strength Cond. Res.* **2005**, *19*, 92–97. [CrossRef]
120. Deutekom, M.; Beltman, J.G.M.; De Ruiter, C.J.; De Koning, J.J.; De Haan, A. No acute effects of short-term creatine supplementation on muscle properties and sprint performance. *Graefe's Arch. Clin. Exp. Ophthalmol.* **2000**, *82*, 223–229. [CrossRef]
121. Rosene, J.; Matthews, T.; Ryan, C.; Belmore, K.; Bergsten, A.; Blaisdell, J.; Gaylord, J.; Love, R.; Marrone, M.; Ward, K.; et al. Short and longer-term effects of creatine supplementation on exercise induced muscle damage. *J. Sports Sci. Med.* **2009**, *8*, 89–96. [PubMed]
122. Bassit, R.A.; Pinheiro, C.H.D.J.; Vitzel, K.F.; Sproesser, A.J.; Silveira, L.R.; Curi, R. Effect of short-term creatine supplementation on markers of skeletal muscle damage after strenuous contractile activity. *Graefe's Arch. Clin. Exp. Ophthalmol.* **2009**, *108*, 945–955. [CrossRef] [PubMed]
123. Veggi, K.F.T.; Machado, M.; Koch, A.J.; Santana, S.C.; Oliveira, S.S.; Stec, M.J. Oral Creatine Supplementation Augments the Repeated Bout Effect. *Int. J. Sport Nutr. Exerc. Metab.* **2013**, *23*, 378–387. [CrossRef]
124. Wang, C.-C.; Fang, C.-C.; Lee, Y.-H.; Yang, M.-T.; Chan, K.-H. Effects of 4-Week Creatine Supplementation Combined with Complex Training on Muscle Damage and Sport Performance. *Nutrients* **2018**, *10*, 1640. [CrossRef] [PubMed]
125. Kim, J.; Lee, J.; Kim, S.; Yoon, D.; Kim, J.; Sung, D.J. Role of creatine supplementation in exercise-induced muscle damage: A mini review. *J. Exerc. Rehabil.* **2015**, *11*, 244–250. [CrossRef]
126. Boychuk, K.E.; Lanovaz, J.L.; Krentz, J.R.; Lishchynsky, J.T.; Candow, D.G.; Farthing, J.P. Creatine supplementation does not alter neuromuscular recovery after eccentric exercise. *Muscle Nerve* **2016**, *54*, 487–495. [CrossRef]
127. McKinnon, N.B.; Graham, M.T.; Tiidus, P.M. Effect of Creatine Supplementation on Muscle Damage and Repair following Eccentrically-Induced Damage to the Elbow Flexor Muscles. *J. Sports Sci. Med.* **2012**, *11*, 653–659.
128. Rawson, E.S.; Conti, M.P.; Miles, M.P. Creatine Supplementation Does not Reduce Muscle Damage or Enhance Recovery from Resistance Exercise. *J. Strength Cond. Res.* **2007**, *21*, 1208–1213. [CrossRef]
129. Warren, G.L.; Fennessy, J.M.; Millard-Stafford, M.L. Strength loss after eccentric contractions is unaffected by creatine supplementation. *J. Appl. Physiol.* **2000**, *89*, 557–562. [CrossRef]
130. Lawler, J.M.; Barnes, W.S.; Wu, G.; Song, W.; Demaree, S. Direct Antioxidant Properties of Creatine. *Biochem. Biophys. Res. Commun.* **2002**, *290*, 47–52. [CrossRef]
131. Sestili, P.; Martinelli, C.; Bravi, G.; Piccoli, G.; Curci, R.; Battistelli, M.; Falcieri, E.; Agostini, D.; Gioacchini, A.M.; Stocchi, V. Creatine supplementation affords cytoprotection in oxidatively injured cultured mammalian cells via direct antioxidant activity. *Free Radic. Biol. Med.* **2006**, *40*, 837–849. [CrossRef] [PubMed]
132. Magne, H.; Savary-Auzeloux, I.; Rémond, D.; Dardevet, D. Nutritional strategies to counteract muscle atrophy caused by disuse and to improve recovery. *Nutr. Res. Rev.* **2013**, *26*, 149–165. [CrossRef]
133. Greenhaff, P.L. The Molecular Physiology of Human Limb Immobilization and Rehabilitation. *Exerc. Sport Sci. Rev.* **2006**, *34*, 159–163. [CrossRef]
134. Vazeille, E.; Codran, A.; Claustre, A.; Averous, J.; Listrat, A.; Béchet, D.; Taillandier, D.; Dardevet, D.; Attaix, D.; Combaret, L. The ubiquitin-proteasome and the mitochondria-associated apoptotic pathways are sequentially downregulated during recovery after immobilization-induced muscle atrophy. *Am. J. Physiol. Metab.* **2008**, *295*, E1181–E1190. [CrossRef]
135. MacDougall, J.D.; Ward, G.R.; Sale, D.G.; Sutton, J.R. Biochemical adaptation of human skeletal muscle to heavy resistance training and immobilization. *J. Appl. Physiol.* **1977**, *43*, 700–703. [CrossRef]
136. Luo, D.; Edwards, S.; Smeuninx, B.; McKendry, J.; Nishimura, Y.; Perkins, M.; Philp, A.; Joanisse, S.; Breen, L. Immobilization leads to alterations in intracellular phosphagen and creatine transporter content in human skeletal muscle. *Am. J. Physiol. Physiol.* **2020**, *319*, C34–C44. [CrossRef]
137. Marshall, R.N.; Smeuninx, B.; Morgan, P.T.; Breen, L. Nutritional Strategies to Offset Disuse-Induced Skeletal Muscle Atrophy and Anabolic Resistance in Older Adults: From Whole-Foods to Isolated Ingredients. *Nutrients* **2020**, *12*, 1533. [CrossRef]
138. Wall, B.T.; Morton, J.P.; Van Loon, L.J.C. Strategies to maintain skeletal muscle mass in the injured athlete: Nutritional considerations and exercise mimetics. *Eur. J. Sport Sci.* **2015**, *15*, 53–62. [CrossRef]
139. Hespel, P.; Eijnde, B.O.; Van Leemputte, M.; Ursø, B.; Greenhaff, P.L.; Labarque, V.; Dymarkowski, S.; Van Hecke, P.; Richter, E.A. Oral creatine supplementation facilitates the rehabilitation of disuse atrophy and alters the expression of muscle myogenic factors in humans. *J. Physiol.* **2001**, *536*, 625–633. [CrossRef]

140. Deldicque, L.; Louis, M.; Theisen, D.; Nielens, H.; Dehoux, M.; Thissen, J.-P.; Rennie, M.J.; Francaux, M. Increased IGF mRNA in Human Skeletal Muscle after Creatine Supplementation. *Med. Sci. Sports Exerc.* **2005**, *37*, 731–736. [CrossRef]
141. Bodine, S.C. mTOR Signaling and the Molecular Adaptation to Resistance Exercise. *Med. Sci. Sports Exerc.* **2006**, *38*, 1950–1957. [CrossRef]
142. Parise, G.; Mihic, S.; MacLennan, D.; Yarasheski, K.E.; Tarnopolsky, M.A. Effects of acute creatine monohydrate supplementation on leucine kinetics and mixed-muscle protein synthesis. *J. Appl. Physiol.* **2001**, *91*, 1041–1047. [CrossRef]
143. Johnston, A.P.W.; Burke, D.G.; MacNeil, L.G.; Candow, D.G. Effect of Creatine Supplementation during Cast-Induced Immobilization on the Preservation of Muscle Mass, Strength, and Endurance. *J. Strength Cond. Res.* **2009**, *23*, 116–120. [CrossRef]
144. Backx, E.M.P.; Hangelbroek, R.; Snijders, T.; Verscheijden, M.-L.; Verdijk, L.B.; De Groot, L.C.P.G.M.; Van Loon, L.J.C. Creatine Loading Does not Preserve Muscle Mass or Strength during Leg Immobilization in Healthy, Young Males: A Randomized Controlled Trial. *Sports Med.* **2017**, *47*, 1661–1671. [CrossRef]
145. Roy, B.D.; De Beer, J.; Harvey, D.; Tarnopolsky, M.A. Creatine Monohydrate Supplementation Does not Improve Functional Recovery after total Knee Arthroplasty. *Arch. Phys. Med. Rehabil.* **2005**, *86*, 1293–1298. [CrossRef]
146. Tyler, T.F.; Nicholas, S.J.; Hershman, E.B.; Glace, B.W.; Mullaney, M.J.; McHugh, M.P. The Effect of Creatine Supplementation on Strength Recovery after Anterior Cruciate Ligament (ACL) Reconstruction. *Am. J. Sports Med.* **2004**, *32*, 383–388. [CrossRef]
147. Hamilton, K.L.; Meyers, M.C.; Skelly, W.A.; Marley, R.J. Oral creatine supplementation and upper extremity anaerobic response in females. *Int. J. Sport Nutr. Exerc. Metab.* **2000**, *10*, 277–289. [CrossRef] [PubMed]
148. Tarnopolsky, M.A.; MacLennan, D.P. Creatine Monohydrate Supplementation Enhances High-Intensity Exercise Performance in Males and Females. *Int. J. Sport Nutr. Exerc. Metab.* **2000**, *10*, 452–463. [CrossRef] [PubMed]
149. Kambis, K.W.; Pizzedaz, S.K. Short-term creatine supplementation improves maximum quadriceps contraction in women. *Int. J. Sport Nutr. Exerc. Metab.* **2003**, *13*, 87–96. [CrossRef]
150. Candow, D.; Forbes, S.; Kirk, B.; Duque, G. Current Evidence and Possible Future Applications of Creatine Supplementation for Older Adults. *Nutrients* **2021**, *13*, 745. [CrossRef] [PubMed]
151. Forbes, S.C.; Candow, D.G.; Ferreira, L.H.B.; Souza-Junior, T.P. Effects of Creatine Supplementation on Properties of Muscle, Bone, and Brain Function in Older Adults: A Narrative Review. *J. Diet. Suppl.* **2021**, *10*, 1–18.
152. Candow, D.G.; Chilibeck, P.D.; Gordon, J.; Vogt, E.; Landeryou, T.; Kaviani, M.; Paus-Jensen, L. Effect of 12 months of creatine supplementation and whole-body resistance training on measures of bone, muscle and strength in older males. *Nutr. Health* **2021**, *27*, 151–159. [CrossRef] [PubMed]
153. Forbes, S.C.; Chilibeck, P.D.; Candow, D.G. Creatine Supplementation during Resistance Training Does not Lead to Greater Bone Mineral Density in Older Humans: A Brief Meta-Analysis. *Front. Nutr.* **2018**, *5*, 27. [CrossRef] [PubMed]
154. Chilibeck, P.D.; Kaviani, M.; Candow, D.G.; Zello, G.A. Effect of creatine supplementation during resistance training on lean tissue mass and muscular strength in older adults: A meta-analysis. *Open Access J. Sports Med.* **2017**, *8*, 213–226. [CrossRef] [PubMed]
155. Candow, D.G.; Vogt, E.; Johannsmeyer, S.; Forbes, S.C.; Farthing, J.P. Strategic creatine supplementation and resistance training in healthy older adults. *Appl. Physiol. Nutr. Metab.* **2015**, *40*, 689–694. [CrossRef] [PubMed]
156. Candow, D.G.; Zello, G.A.; Ling, B.; Farthing, J.P.; Chilibeck, P.D.; McLeod, K.; Harris, J.; Johnson, S. Comparison of Creatine Supplementation before Versus after Supervised Resistance Training in Healthy Older Adults. *Res. Sports Med.* **2014**, *22*, 61–74. [CrossRef]
157. Candow, D.G.; Little, J.P.; Chilibeck, P.D.; Abeysekara, S.; Zello, G.A.; Kazachkov, M.; Cornish, S.M.; Yu, P.H. Low-Dose Creatine Combined with Protein during Resistance Training in Older Men. *Med. Sci. Sports Exerc.* **2008**, *40*, 1645–1652. [CrossRef] [PubMed]
158. Candow, D.G.; Chilibeck, P.D.; Chad, K.E.; Chrusch, M.J.; Davison, K.S.; Burke, D.G. Effect of Ceasing Creatine Supplementation while Maintaining Resistance Training in Older Men. *J. Aging Phys. Act.* **2004**, *12*, 219–231. [CrossRef]
159. Jäger, R.; Purpura, M.; Shao, A.; Inoue, T.; Kreider, R.B. Analysis of the efficacy, safety, and regulatory status of novel forms of creatine. *Amino Acids* **2011**, *40*, 1369–1383. [CrossRef] [PubMed]
160. Kerksick, C.; Wilborn, C.D.; Campbell, W.I.; Harvey, T.M.; Marcello, B.M.; Roberts, M.D.; Parker, A.G.; Byars, A.G.; Greenwood, L.D.; Almada, A.L.; et al. The Effects of Creatine Monohydrate Supplementation With and Without D-Pinitol on Resistance Training Adaptations. *J. Strength Cond. Res.* **2009**, *23*, 2673–2682. [CrossRef] [PubMed]
161. Spillane, M.; Schoch, R.; Cooke, M.B.; Harvey, T.; Greenwood, M.; Kreider, R.B.; Willoughby, D.S. The effects of creatine ethyl ester supplementation combined with heavy resistance training on body composition, muscle performance, and serum and muscle creatine levels. *J. Int. Soc. Sports Nutr.* **2009**, *6*, 6. [CrossRef]
162. Greenwood, M. Differences in creatine retention among three nutritional formulations of oral creatine supplements. *J. Exerc. Physiol. Online* **2003**, *6*, 37–43.
163. Jagim, A.R.; Oliver, J.M.; Sanchez, A.; Galvan, E.; Fluckey, J.; Riechman, S.; Greenwood, M.; Kelly, K.; Meininger, C.; Rasmussen, C.; et al. A buffered form of creatine does not promote greater changes in muscle creatine content, body composition, or training adaptations than creatine monohydrate. *J. Int. Soc. Sports Nutr.* **2012**, *9*, 43. [CrossRef]
164. Stout, J.R.; Cramer, J.T.; Mielke, M.; O'Kroy, J.; Torok, D.J.; Zoeller, R.F. Effects of Twenty-Eight Days of Beta-Alanine and Creatine Monohydrate Supplementation on the Physical Working Capacity at Neuromuscular Fatigue Threshold. *J. Strength Cond. Res.* **2006**, *20*, 928–931. [CrossRef] [PubMed]

165. Hoffman, J.; Ratamess, N.; Kang, J.; Mangine, G.; Faigenbaum, A.; Stout, J. Effect of creatine and beta-alanine supplementation on performance and endocrine responses in strength/power athletes. *Int. J. Sport Nutr. Exerc. Metab.* **2006**, *16*, 430–446. [CrossRef]
166. Fernández-Landa, J.; Fernández-Lázaro, D.; Calleja-González, J.; Caballero-García, A.; Córdova, A.; León-Guereño, P.; Mielgo-Ayuso, J. Long-Term Effect of Combination of Creatine Monohydrate plus β-Hydroxy β-Methylbutyrate (HMB) on Exercise-Induced Muscle Damage and Anabolic/Catabolic Hormones in Elite Male Endurance Athletes. *Biomolecules* **2020**, *10*, 140. [CrossRef] [PubMed]
167. Fernández-Landa, J.; Calleja-González, J.; León-Guereño, P.; Caballero-García, A.; Córdova, A.; Mielgo-Ayuso, J. Effect of the Combination of Creatine Monohydrate Plus HMB Supplementation on Sports Performance, Body Composition, Markers of Muscle Damage and Hormone Status: A Systematic Review. *Nutrients* **2019**, *11*, 2528. [CrossRef]
168. Mobley, C.B.; Fox, C.D.; Ferguson, B.S.; Amin, R.H.; Dalbo, V.J.; Baier, S.; Rathmacher, J.A.; Wilson, J.M.; Roberts, M.D. L-leucine, beta-hydroxy-beta-methylbutyric acid (HMB) and creatine monohydrate prevent myostatin-induced Akirin-1/Mighty mRNA down-regulation and myotube atrophy. *J. Int. Soc. Sports Nutr.* **2014**, *11*, 38. [CrossRef]
169. O'Connor, D.M.; Crowe, M.J. Effects of Six Weeks of ?-Hydroxy-?-Methylbutyrate (HMB) and HMB/Creatine Supplementation on Strength, Power, and Anthropometry of Highly Trained Athletes. *J. Strength Cond. Res.* **2007**, *21*, 419–423. [CrossRef]
170. Crowe, M.J.; O'Connor, D.M.; Lukins, J.E. The Effects of ß-Hydroxy-ß-Methylbutyrate (HMB) and HMB/Creatine Supplementation on Indices of Health in Highly Trained Athletes. *Int. J. Sport Nutr. Exerc. Metab.* **2003**, *13*, 184–197. [CrossRef]
171. Jówko, E.; Ostaszewski, P.; Jank, M.; Sacharuk, J.; Zieniewicz, A.; Wilczak, J.; Nissen, S. Creatine and β-hydroxy-β-methylbutyrate (HMB) additively increase lean body mass and muscle strength during a weight-training program. *Nutrients* **2001**, *17*, 558–566. [CrossRef]
172. Green, A.L.; Simpson, E.J.; Littlewood, J.J.; Macdonald, I.A.; Greenhaff, P.L. Carbohydrate ingestion augments creatine retention during creatine feeding in humans. *Acta Physiol. Scand.* **1996**, *158*, 195–202. [CrossRef]
173. Roberts, P.A.; Fox, J.; Peirce, N.; Jones, S.; Casey, A.; Greenhaff, P.L. Creatine ingestion augments dietary carbohydrate mediated muscle glycogen supercompensation during the initial 24 h of recovery following prolonged exhaustive exercise in humans. *Amino Acids* **2016**, *48*, 1831–1842. [CrossRef] [PubMed]
174. Kilduff, L.P.; Vidakovic, P.; Cooney, G.; Twycross-Lewis, R.; Amuna, P.; Parker, M.; Paul, L.; Pitsiladis, Y.P. Effects of creatine on isometric bench-press performance in resistance-trained humans. *Med. Sci. Sports Exerc.* **2002**, *34*, 1176–1183. [CrossRef] [PubMed]
175. Syrotuik, D.G.; Bell, G.J. Acute Creatine Monohydrate Supplementation: A Descriptive Physiological Profile of Responders vs. Nonresponders. *J. Strength Cond. Res.* **2004**, *18*, 610–617. [CrossRef] [PubMed]
176. Kaviani, M.; Shaw, K.; Chilibeck, P.D. Benefits of Creatine Supplementation for Vegetarians Compared to Omnivorous Athletes: A Systematic Review. *Int. J. Environ. Res. Public Health* **2020**, *17*, 3041. [CrossRef]
177. Venderley, A.M.; Campbell, W.W. Vegetarian Diets. *Sports Med.* **2006**, *36*, 293–305. [CrossRef]
178. Shomrat, A.; Weinstein, Y.; Katz, A. Effect of creatine feeding on maximal exercise performance in vegetarians. *Graefe's Arch. Clin. Exp. Ophthalmol.* **2000**, *82*, 321–325. [CrossRef]
179. Watt, K.K.; Garnham, A.P.; Snow, R.J. Skeletal Muscle Total Creatine Content and Creatine Transporter Gene Expression in Vegetarians Prior to and Following Creatine Supplementation. *Int. J. Sport Nutr. Exerc. Metab.* **2004**, *14*, 517–531. [CrossRef]
180. Lukaszuk, J.M.; Robertson, R.J.; Arch, J.E.; Moyna, N.M. Effect of a Defined Lacto-Ovo-Vegetarian Diet and Oral Creatine Monohydrate Supplementation on Plasma Creatine Concentration. *J. Strength Cond. Res.* **2005**, *19*, 735–740. [CrossRef]

Review

The Application of Creatine Supplementation in Medical Rehabilitation

Kylie K. Harmon [1], Jeffrey R. Stout [2], David H. Fukuda [2], Patrick S. Pabian [3], Eric S. Rawson [4] and Matt S. Stock [1,*]

[1] Neuromuscular Plasticity Laboratory, Institute of Exercise Physiology and Rehabilitation Science, School of Kinesiology and Physical Therapy, University of Central Florida, Orlando, FL 32816, USA; kylie.harmon@ucf.edu

[2] Physiology of Work and Exercise Response (POWER) Laboratory, Institute of Exercise Physiology and Rehabilitation Science, School of Kinesiology and Physical Therapy, University of Central Florida, Orlando, FL 32816, USA; Jeffrey.Stout@ucf.edu (J.R.S.); david.fukuda@ucf.edu (D.H.F.)

[3] Musculoskeletal Research Laboratory, Institute of Exercise Physiology and Rehabilitation Science, School of Kinesiology and Physical Therapy, University of Central Florida, Orlando, FL 32816, USA; Patrick.pabian@ucf.edu

[4] Department of Health, Nutrition, and Exercise Science, Messiah University, Mechanicsburg, PA 17055, USA; erawson@messiah.edu

* Correspondence: matt.stock@ucf.edu

Abstract: Numerous health conditions affecting the musculoskeletal, cardiopulmonary, and nervous systems can result in physical dysfunction, impaired performance, muscle weakness, and disuse-induced atrophy. Due to its well-documented anabolic potential, creatine monohydrate has been investigated as a supplemental agent to mitigate the loss of muscle mass and function in a variety of acute and chronic conditions. A review of the literature was conducted to assess the current state of knowledge regarding the effects of creatine supplementation on rehabilitation from immobilization and injury, neurodegenerative diseases, cardiopulmonary disease, and other muscular disorders. Several of the findings are encouraging, showcasing creatine's potential efficacy as a supplemental agent via preservation of muscle mass, strength, and physical function; however, the results are not consistent. For multiple diseases, only a few creatine studies with small sample sizes have been published, making it difficult to draw definitive conclusions. Rationale for discordant findings is further complicated by differences in disease pathologies, intervention protocols, creatine dosing and duration, and patient population. While creatine supplementation demonstrates promise as a therapeutic aid, more research is needed to fill gaps in knowledge within medical rehabilitation.

Keywords: supplements; muscle damage; recovery; immobilization; atrophy; muscular dystrophy; amyotrophic lateral sclerosis; Parkinson's Disease; cardiopulmonary disease; mitochondrial cytopathy

Citation: Harmon, K.K.; Stout, J.R.; Fukuda, D.H.; Pabian, P.S.; Rawson, E.S.; Stock, M.S. The Application of Creatine Supplementation in Medical Rehabilitation. *Nutrients* **2021**, *13*, 1825. https://doi.org/10.3390/nu13061825

Academic Editor: Louise Deldicque

Received: 16 April 2021
Accepted: 26 May 2021
Published: 27 May 2021

Publisher's Note: MDPI stays neutral with regard to jurisdictional claims in published maps and institutional affiliations.

1. Introduction

Physical rehabilitation is beneficial for everyone, including children, adults, and older people with a wide range of health conditions. The goal of physical rehabilitation is to enhance, restore, or maximize functional ability to promote health and optimize quality of life. Globally, an estimated 2.4 billion people live with a health condition that would benefit from rehabilitation [1].

Numerous acute and chronic conditions can result in dysfunction and reduced health due to physical manifestations directly related to the disorder, such as orthopedic injury with subsequent disuse and post-surgical immobilization. Dysfunction also leads to inactivity, altered movement, or disuse, resulting in muscle mass loss. These symptoms can therefore also occur in association with cardiovascular disease, neurodegenerative conditions, or other disease conditions such as myopathies [2,3]. Further, in older populations,

145

significant loss of muscle and function has been associated with increased mortality risk [4]; thus, interventions focused on regaining muscle loss must be prioritized.

Muscle atrophy is related to metabolic and physiological changes, including decreased sensorimotor excitability, contractile ability of muscle fibers, increased protein degradation, and depletion of energy stores, which may promote fatigue, reduce functional capacity and hinder activities of daily living [5–7]. More recently, there has been an increased appreciation for nutritional and dietary supplement interventions, both independently and combined with exercise, to maintain or enhance clinically essential outcomes and improve rehabilitation-related adaptations in patients experiencing substantial muscle and strength reduction [8,9]. Evidence suggests that one nutritional intervention in particular, creatine monohydrate supplementation, may effectively attenuate and possibly enhance rehabilitation [7,8,10].

Even though creatine was detected in skeletal muscle 189 years ago, irrespective of its immense success as an ergogenic aid since the 1990s, much remains unknown about its biological capabilities. This is especially true for its potential of use in accordance with health and medical conditions. Chemically known as a non-protein nitrogen compound [11], approximately 95 percent of the body's creatine is stored within skeletal muscle. Moreover, small amounts of creatine are also found in the brain and testes [12,13]. Approximately two-thirds of the creatine found in skeletal muscle is stored as phosphocreatine, while the remaining quantity is stored as free creatine [13]. The standard concentration of total creatine in skeletal muscle is approximately 120 mmol·kg^{-1} (dry mass) [14,15], whereas the upper limit appears to be approximately 150–160 mmol·kg^{-1} [16–19]. It has been suggested that creatine's physiological and biochemical effects are mostly related to the functions of creatine kinase and phosphocreatine (i.e., CK/PCr system) to support the maintenance of cellular energy [17–19]. The most commonly accepted theory explaining the beneficial effects of creatine supplementation on muscle performance is the energetic theory, that the increase in skeletal muscle concentrations of creatine and phosphocreatine contribute to an increase in exercise intensity and volume, thus providing a more substantial stimulus and leading to the significantly greater adaptation to training [13].

Creatine has gained popularity as a dietary supplement because of its fundamental role in human health and physical performance. Its metabolic contributions are many and wide ranging, and it is one of the most frequently used and effective supplements available for health and performance [20]. Various research studies over the last several decades have shown that creatine supplementation has numerous positive effects on skeletal muscle structure and metabolism, including improving training volume and muscle mass (growth/hypertrophy) in conjunction with exercise [20]. Therefore, the potential of creatine to augment training-induced muscle hypertrophy is not only relevant in the context of elite sports performance but also in the context of rehabilitation of various conditions due to its ability to prevent muscle atrophy [21]. While its role as an ergogenic agent in sports is well supported by several studies [12], there is growing research interest from the medical field regarding the potential for creatine to serve as a therapeutic agent in a wide range of conditions [22,23]. Figure 1 provides an overview of the therapeutic efficacy of creatine supplementation in a variety of conditions examined within this review. Interestingly, Corn and colleagues [24] have suggested that creatine may be beneficial as an adjunct to traditional physical therapy interventions. Therefore, the purpose of this review is to examine the literature concerning the effectiveness of creatine monohydrate supplementation as a potential strategy to enhance the rehabilitation of muscle and function in physical inactivity (disuse) and disease-related conditions.

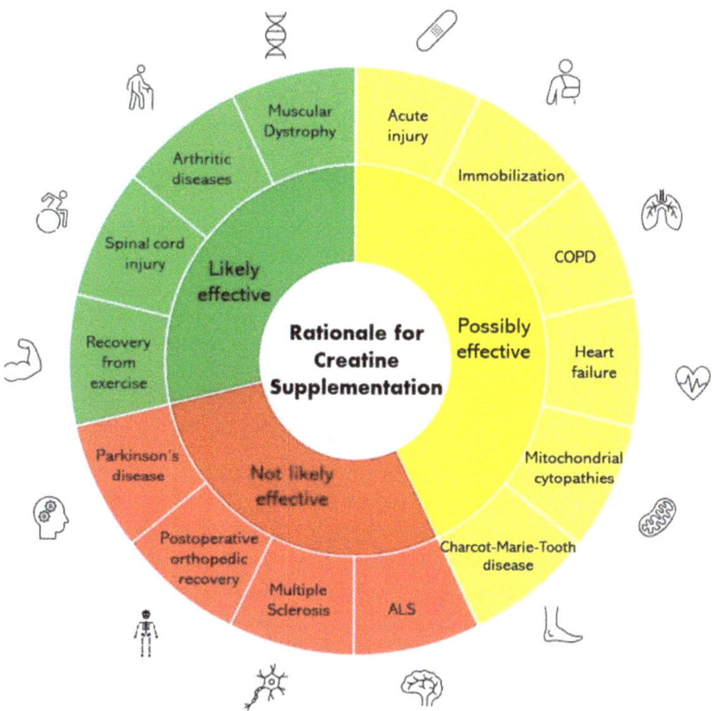

Figure 1. Rationale for examination of creatine supplementation as a rehabilitative aid. Abbreviations defined from clockwise: COPD = chronic obstructive pulmonary disease; ALS = amyotrophic lateral sclerosis.

2. Methods

A comprehensive review of the scientific and medical literature was conducted to assess the state of the science on creatine supplementation in medical rehabilitation. Specifically, the National Institutes for Health National Library of Medicine PubMed.gov and Google Scholar search engines were used to identify publications focused on the utility of creatine supplementation during rehabilitation due to immobilization, muscle disuse atrophy, neurodegenerative disorders, cardiopulmonary disease, and mitochondrial cytopathies.

3. Creatine Supplementation, Muscle Damage, and Recovery from Stressful Exercise

There have been anecdotal reports in the media that creatine supplementation is associated with increased muscle dysfunction, including increased incidence of cramps, injuries, and rhabdomyolysis [25–29]. However, this is not supported by the literature. The results of several studies show no increase or decrease in muscle cramps, tightness, strains, total injuries, missed practices, or players lost for the season in collegiate athletes ingesting creatine supplements [30–32]. Interestingly, even though anecdotal reports claim that creatine supplements exacerbate muscle dysfunction, several research teams have investigated creatine supplementation as a means to enhance recovery from intense exercise by reducing exercise-induced muscle damage [33–35]. Therefore, the potential benefit of creatine supplementation to reduce muscle damage and enhance recovery from stressful exercise is worthy of further investigation.

3.1. Mechanisms of Benefit

During stressful and particularly unaccustomed exercise, damage to muscle fibers can occur. In the days following exercise, muscle force production and range of motion decrease, while muscle soreness, swelling, muscle serum proteins, and inflammatory compounds increase [36,37]. However, creatine supplementation may offer several potentially beneficial effects in terms of recovery from damaging exercise. Increasing muscle creatine through supplementation may reduce muscle membrane fluidity and increase membrane stiffness, which could decrease damage from exercise [21,33–35]. Additionally, increased muscle creatine levels may help maintain calcium homeostasis, mitigate inflammation, and decrease free radical-induced damage following damaging exercise [21,33–35]. Beyond attenuating muscle damage, increased muscle creatine due to creatine supplementation alters the intramuscular milieu, which subsequently causes several changes beneficial to the adaptive response to resistance exercise. For example, creatine supplementation results in increased growth factor expression (e.g., myogenin, MRF-4, insulin-like growth factor I and II [IGF-I and IGF-II]) [38–41], increased satellite cell number and myonuclei concentration [42], and expression of multiple genes associated with adaptive processes related to exercise (e.g., osmosensing, cytoskeleton, remodeling, GLUT-4 translocation, glycogen and protein synthesis, satellite cell proliferation, and differentiation, DNA replication and repair, mRNA processing and transcription, and cell survival) [43]. It is possible that the increase in intramuscular water associated with creatine supplementation modulates some of the effects, as this is known to inhibit protein breakdown and RNA degradation and stimulate protein, DNA, RNA, and glycogen synthesis [33,44,45]. Collectively, these data support that creatine is not only a performance enhancing nutrient that provides greater fuel availability prior to intense exercise, but also an adaptive nutrient which, when supplemented, augments the adaptive response to training.

3.2. Specific Effects on Muscle Damage

Our review of the literature resulted in the examination of 16 clinical trials that describe the effects of creatine ingestion on markers of muscle damage in hard training individuals (e.g., resistance exercise) or those subjected to an unaccustomed exercise challenge (e.g., high-force eccentric exercise) [46–61]. The methods used in these investigations are discrepant, including differences in supplementation dosing and duration, participant training status, and exercise challenge (e.g., resistance, high-force eccentric, and endurance) [25,34,35]. Across these studies, improvements are noted in established markers of exercise-induced muscle damage, including reduced post-exercise levels of muscle serum proteins (creatine kinase, lactate dehydrogenase); reduced inflammatory compounds (prostaglandin-E_2, tumor necrosis factor-α, interferon-α, interleuikin 1-β), reduced oxidative stress markers (glutathione peroxidase, thiobarbituric acid reaction substances), increased recovery of strength, and reduced delayed onset muscle soreness. However, consistent improvements were not found across studies or across the same variables. In their systematic and meta-analytic review of creatine supplementation and muscle damage, Northeast and Clifford [35] concluded that creatine supplementation had little practical value as a recovery aid. However, they pointed out that there were less than 10 eligible studies for each outcome and time point, limiting statistical power and contributing to high heterogeneity for outcomes. Two important conclusions are worthy of consideration. First, as several studies have shown a protective effect of creatine supplementation on exercise-induced muscle damage, this area requires further exploration. As small sample size prevented previous reviews from conducting sub-analyses [35], we cannot know if creatine supplementation might benefit one population (e.g., trained vs. untrained) or be more protective in a specific type of exercise (e.g., endurance vs. eccentric-resistance). Secondly, despite lingering myths and anecdote, there are no data to support that creatine supplementation increases muscle damage following severe and/or unaccustomed exercise, which in some studies was quite severe (e.g., 50 maximal eccentric contractions [55]; Ironman triathlon [48]).

4. Rehabilitation and Creatine Use for Clinical Conditions

4.1. Immobilization and Disuse-Induced Atrophy

Skeletal muscle disuse has been repeatedly demonstrated as having deleterious effects on a variety of physiological parameters. Even short periods of disuse have caused observable decrements in muscle cross-sectional area (CSA) [62], reduced force production capabilities [63], increased muscle protein breakdown [64], and altered neuromuscular function [65]. The decrements that occur during disuse can lead to longer recovery periods, a greater chance of injury recurrence, and a decline in metabolic health [66–70]. Because of its high clinical relevance, joint immobilization has become a frequently employed experimental model to examine physiological changes in skeletal muscle as a consequence of inactivity and disuse.

Due to the well-documented ability of creatine to potentiate the anabolic effects of resistance training [71], the potential of creatine supplementation to mitigate the effects of disuse-induced maladaptation is of exceptional interest [72,73]. Indeed, seven recent studies have examined the effects of creatine supplementation during immobilization, with several having demonstrated promising results. Using a single-blind, cross-over design, Johnston et al. [74] observed the effects of creatine supplementation on the preservation of muscle mass, strength, and endurance after a 7 day immobilization period of the upper limb in creatine-naïve men. Compared with an isocaloric placebo, 20 g·day^{-1} (four doses of 5 g each) of creatine supplementation during the immobilization period better maintained lean tissue mass, elbow flexor strength, and endurance, and elbow extensor strength and endurance. Similarly, in a double-blind, placebo-controlled trial involving 2 weeks of lower-limb cast immobilization during which participants ingested either creatine or a placebo, Eijinde et al. [75] observed that 20 g·day^{-1} (four doses of 5 g each) of creatine supplementation offset the observed decreases in muscle GLUT4 protein content that occurred in the placebo group. Following a subsequent 10-week rehabilitation training period during which creatine supplementation was reduced (15 g·day^{-1} during the first 3 weeks and 5 g·day^{-1} for the remaining 7 weeks), muscle GLUT4 protein content was increased by ~40% in the creatine group while returning the baseline levels in the placebo group. Given the role of muscle GLUT4 protein in providing a mechanism for glucose to enter the cell, these results suggest that creatine supplementation may promote favorable changes in glucoregulation during immobilization.

The efficacy of creatine supplementation during periods of inactivity has also been observed in animal models. Hindlimb suspension, often used as a surrogate for cast immobilization, causes observable muscle atrophy and strength loss in rodent models [76]. Aoki et al. [77] investigated the effect of creatine supplementation (5 g·kg^{-1}·body weight·day^{-1}) both prior to and during a period of hindlimb suspension in rats. The animals underwent 7 days of hindlimb immobilization in combination with three supplemental interventions: creatine during the immobilization period, creatine for 7 days prior to and during immobilization (i.e., 14 days of creatine supplementation), and a control condition. Regardless of the treatment group, immobilization induced a decrease in muscle weight. However, creatine supplementation prior to and during immobilization appeared to mitigate muscle mass loss. This is an intriguing finding, as a period of creatine loading is a frequent practice in athletic populations [20] and could easily be incorporated into pre-treatment or preoperative protocols in patients undergoing periods of planned immobilization to attenuate muscle atrophy.

While the aforementioned findings are promising, other works provide conflicting results. Marzuca-Nassr et al. [78] observed inconclusive effects of short-term creatine supplementation during hindlimb immobilization in rats. Throughout 5 days of hindlimb immobilization, adult rats received daily creatine supplementation (5 g·kg^{-1} body weight day^{-1}) or a placebo. Despite positive changes in protein metabolism and a slight attenuation of total muscle loss in the creatine group, creatine supplementation did not prevent muscle fiber atrophy or strength loss. Three double-blind, placebo-controlled trials in humans have also demonstrated inconsistent findings of creatine supplementation throughout immobilization. Fransen et al. [79] observed no effect of 20 g·day^{-1} creatine supplementation on maintenance of work or power production following 7 days of wrist cast immobilization. Similarly, throughout 7 days of lower-limb cast immobilization during which participants received a loading dose of creatine (20 g·day^{-1} for 5 days prior to immobilization) followed by a maintenance dose (5 g·day^{1}) during the immobilization period, Backx et al. [80] found no effect of creatine supplementation on preservation of quadriceps CSA or knee-extension strength. While the loading dose of creatine was successful in increasing muscle creatine content, this was ineffective in preservation of strength or muscle mass. These results persisted upon the removal of creatine non-responders (i.e., participants who did not show an increase in muscle total creatine content exceeding 10 mmol·kg^{-1}) from the analyses and throughout a subsequent 7 day recovery period. Similarly, during 2 weeks of lower-limb cast immobilization in healthy young men, Hespel et al. [7] found no effect of 20 g·day^{-1} of daily creatine supplementation in preserving quadriceps CSA or knee-extension power. However, throughout a subsequent 10-week program of rehabilitation training during which creatine supplementation was reduced (15 g·day^{-1} the initial 3 weeks and 5 g·day^{-1} during the remaining 7 weeks), quadriceps CSA and knee-extension power recovered at a faster rate in participants supplementing with creatine versus placebo. Further, myogenic protein expression was altered in the creatine group during rehabilitation. In particular, MRF-4 expression was significantly increased in the creatine group during rehabilitation and strongly correlated to changes in muscle fiber diameter. While the anabolic potential of creatine in combination with resistance training is well documented, its efficacy in providing strategies to increase the rate of functional recovery during rehabilitation is a significant finding.

While the specific mechanisms involved in the maintenance of muscle mass during immobilization are not fully known, based on the literature, it is likely that creatine supplementation may have a protective effect on skeletal muscle at least partially due to alterations in muscle protein expression [7,41,78] and satellite cell activity [42]. Given the prolonged recovery time often observed after clinical immobilization [67], interventions to enhance the rehabilitation process are sorely needed.

4.1.1. Post-Operative Orthopedic Recovery

Despite the promising results observed during rehabilitation from immobilization [7], creatine supplementation does not appear to confer a beneficial effect during post-operative rehabilitation. In a randomized, double-blind, placebo-controlled trial, Tyler et al. [81] observed the effect of creatine supplementation on recovery of muscle strength following anterior cruciate ligament (ACL) reconstruction. Creatine supplementation began with a loading dose of 20 g·day^{-1} the day following reconstructive surgery. This dosage was decreased to 5 g·day^{-1} at day 7 when formal rehabilitation began and continued for 12 weeks. Quadriceps and hamstrings strength and power were unaffected by creatine supplementation. Utilizing a similar randomized, double-blind, placebo-controlled design, Roy et al. [82] observed no effect of creatine supplementation on functional recovery after total knee arthroplasty. In contrast to the work of Tyler et al. [81], patients in this trial received creatine pre-surgery (10 g·day^{-1} for 10 days) in addition to post-surgery (5 g·day^{-1} for 30 days). Despite the addition of a pre-surgery loading dose, creatine supplementation had no effect on preserving muscle strength or enhancing recovery, as measured by quadriceps, ankle, and handgrip strength, as well as timed walks and step climbs.

4.1.2. Acute Injury

Like periods of immobilization, muscle and nerve damage can result in muscle atrophy as a consequence of impaired function, mobility, and physical inactivity. While medical and surgical interventions are available, therapies aimed at restoring function after severe damage can still result in imperfect clinical outcomes [83]. Given the extensive recovery time from muscle and nerve damage, agents that can enhance the healing process are critically needed.

Although the literature is limited, the effects of creatine supplementation during recovery from nerve damage have demonstrated promising results. Özkan et al. [83] observed a positive effect of creatine supplementation on reinnervation of denervated muscle in the rodent model. After severing the sciatic nerve, adult rats were fed a creatine-enhanced or normal diet during subsequent recovery, with subgroups of animals undergoing surgical nerve repair and others receiving no neural anastomosis. In both those with and without surgical nerve repair, creatine supplementation significantly improved functional recovery as measured by walking analyses, pinch strength, limb circumference, and toe contracture. To the best of our knowledge, this has yet to be replicated in humans; however, this is a noteworthy finding as it indicates that supplementation with creatine may enhance recovery after denervation.

Despite the potential efficacy of creatine supplementation in recovery from nerve damage, similar effects were not observed during recovery from muscle damage in animal models [84]. After injecting rat soleus muscle with notexin to cause myotoxin-induced muscle degeneration, rats were fed a creatine-enhanced or normal diet and observed for 42 days post-injury. While creatine supplementation was able to ensure faster recovery of total muscle creatine and phosphocreatine content, it did not influence regeneration of muscle mass to pre-injury levels and had no effects on the time course of recovery. However, physical activity, which has previously been demonstrated to have a strong effect on the efficacy of both creatine supplementation and muscle recovery [85], was not manipulated. Future research should examine the effects of combined creatine supplementation and physical activity on muscle regenerative capacity.

4.1.3. Spinal Cord Injury

Individuals with spinal cord injury (SCI) frequently suffer from extreme deconditioning and muscle strength impairments due to the nature of their injuries [86]. Patients with SCI often experience diminished upper body strength and reduced work capacity which can lead to inactivity, further exacerbating these problems [87–89]. As such, patients with SCI would benefit greatly from increased levels of muscular strength, power, and endurance to improve movement for normal activities in their daily lives. Optimizing movement for activities such as transferring body weight, propelling a wheelchair, or navigating community obstacles would allow for more independence, increased activity, and improved health and quality of life.

Creatine supplementation has been investigated in individuals with SCI due to its potential as an anabolic substance. Although the literature is limited, the findings are promising. In a randomized, double-blind placebo-controlled cross-over trial of patients with cervical-level SCI, 20 g·day^{-1} of supplemental creatine enhanced upper extremity work capacity [90]. Patients with complete cervical-level SCI were supplemented with creatine or a placebo and performed incremental work capacity tests before and after 7 days of treatment. After creatine supplementation, participants had significantly improved VO$_2$, VCO$_2$, and ventilatory threshold as measured during incremental upper arm ergometry. These findings are further strengthened by the work of Amorim et al. [91], who, supplemented SCI patients with 3 g·day^{-1} creatine, vitamin D, or placebo throughout a progressive 8 week resistance training program in a randomized, double-blind, placebo-controlled fashion. In the creatine-supplemented group, upper body strength and corrected arm CSA significantly improved. This is particularly notable given the considerably lower daily dose of creatine compared to other studies. The findings from these studies indicate that creatine supplementation enhances the effects of exercise training in individuals with SCI. Similar results have also been demonstrated in rats with SCI [92]. Rats fed a creatine-enhanced diet for 4 weeks prior to surgical SCI demonstrated significantly better post-traumatic locomotor capacity than control rats. Creatine also appeared to reduce the effects of secondary neurotrauma as observed via reduction in lesion site scar tissue.

4.1.4. Arthritic Diseases

Creatine supplementation may also confer a beneficial effect on physical function in those with arthritic diseases [93]. In a randomized, double-blind, placebo-controlled fashion, post-menopausal women with knee osteoarthritis demonstrated significantly improved physical function and lower-limb lean mass when supplemented daily with both a loading dose (20 g·day^{-1} for 7 days) and maintenance dose (5 g·day^{-1} for 11 weeks) of creatine throughout a 12 week resistance training program [94]. Similar findings have been observed in those with rheumatoid arthritis [95]. Administration of oral creatine over a 3-week period (20 g·day^{-1} for 5 days followed by 2 g·day^{-1} for 16 days) improved muscle strength in patients with rheumatoid arthritis, despite no associated training protocol over the course of the intervention. It should be noted, however, that although strength did improve, total skeletal muscle creatine content did not. The authors concluded no clear clinical benefit of creatine supplementation, as functional ability and disease activity did not change. Still, positive associations have recently been observed between strength and walking self-efficacy, pain reduction, and improved function in osteoarthritis, further demonstrating the importance of maintenance of muscle strength in patients with arthritic diseases [96].

4.1.5. Rationale for Discrepant Findings throughout the Immobilization Literature

There are a variety of methodological and physiological reasons for the inconsistent observations of creatine supplementation during immobilization. Although it is difficult to compare results across studies given differing protocols, durations, and dosing, it has been suggested that the length of the immobilization protocol and muscle group involved impact the atrophic response and functional impairments [97]. It has previously been demonstrated that lower body muscle groups are more negatively impacted by disuse than upper body musculature [98,99]. It is therefore possible that creatine supplementation may have a differing effect on upper versus lower body musculature.

Further, the overall duration of creatine supplementation may significantly impact potential findings. While short-term creatine supplementation has been shown to increase muscle creatine content, longer-term creatine supplementation is associated with enhanced muscular hypertrophy and strength [100]. Given the acute nature of the bulk of the literature observing the effects of creatine on disuse-induced adaptations, it is possible that interventions utilizing creatine supplementation to counteract the effects of disuse would demonstrate greater efficacy from longer intervention protocols. When considering the positive effects of creatine loading in mitigating disuse-induced atrophy [77] as well as the increased hypertrophy [7] and strength [94] observed during longer protocols of creatine supplementation, increasing the duration of the supplementation seems promising. Further, several aforementioned studies with favorable results have incorporated exercise interventions in combination with creatine supplementation. As exercise has been demonstrated to enhance the uptake of creatine [14], the role of exercise in the efficacy of creatine supplementation is of critical importance and should be incorporated when mobility and function allow. Additionally, differences in the presence of a loading phase and maintenance dose alter the overall exposure to creatine supplementation. Loading doses of creatine are routinely administered as the greatest uptake of creatine in the muscle appears to occur in the initial phases of a heightened dose [14].

Still, it is possible that other factors are responsible for the discrepant findings in the literature. Neural impairments, such as decreased voluntary activation, have also been observed throughout immobilization [101] and can impact both muscular strength and endurance [102]. Significant alterations in neuromuscular properties, such as reduced voluntary activation [103,104], decreased H-reflex [105], and changes in motor cortex excitability [103,105] have been observed to occur with disuse, with neural factors explaining nearly 50% of initial strength loss [104,105]. If injury precedes disuse, alterations may occur in sensory stimuli, receptor activity, and signal transduction from the injured area to the central nervous system, further exacerbating functional impairments [106].

Finally, the effects of anabolic resistance, which have been repeatedly observed during periods of immobilization [107–109], must be taken into consideration. Anabolic resistance, the inability of an anabolic stimulus such as protein ingestion, hormonal response, or muscle contraction to stimulate muscle protein synthesis, is a hallmark characteristic of periods of inactivity or disuse [107–109]. Substantial inactivity or disuse decreases muscle protein synthesis with minimal change in muscle protein breakdown [110]. This can result in a dampened response to hyperaminoacidemia. This is significant, as one of the primary mechanisms postulated as responsible for creatine's efficacy in improving muscular hypertrophy and performance is its ability to enhance protein stimulus via osmotic swelling [44,45,111]. However, if disuse-induced anabolic resistance is present, its effects may interfere with the protein synthesis that occurs with creatine supplementation, thereby inhibiting preservation of muscle mass and function. Similarly, the presence of satellite cells, which have been implicated in creatine's efficacy, has been observed to decrease with disuse [112]. As satellite cells support regeneration of skeletal muscle following damage or atrophy, their role in recovery from immobilization or trauma is critical.

Promising results in immobilization studies may not be able to be extrapolated to clinical populations. While the bulk of the current disuse protocols in humans have been performed in healthy participants, patients undergoing immobilization associated with injury, surgery, or pain may experience accelerated atrophy and functional decrements. These phenomena may be partially responsible for the lack of creatine's efficacy in postoperative observations. The physiological effects of illness or injury (e.g., inflammation, immune response) may exacerbate the negative changes in whole-body protein balance that occur during immobilization. Trauma, surgery, and pain may accelerate muscle atrophy during immobilization [113], ultimately leading to longer recovery times to return to baseline. Further, orthopedic issues severe enough to warrant surgery are likely to limit physical activity in advance of surgical intervention, leading to greater overall impairment. While creatine supplementation may offer enhanced outcomes from various conditions resulting in immobilization and injury, more work is needed to determine its overall efficacy in consideration of these factors.

4.2. Neurodegenerative Diseases

Neurodegenerative diseases are debilitating conditions caused by the progressive deterioration and eventual apoptosis of alpha motor neurons and skeletal muscle fibers. Given that these diseases are incurable, adjunctive therapies are needed to slow their progression and improve quality of life by controlling symptoms. As neurodegenerative diseases result in reduced physical activity, the theoretical basis of considering creatine supplementation in these patients is well justified, given its ability to improve muscle strength, mass, and endurance [20]. In other words, creatine supplementation may serve as a general countermeasure that delays the progression of physical impairments. However, neurodegenerative diseases adversely affect the distinct neurometabolic pathways and pathological characteristics that creatine may precisely target.

There are a variety of clinical design issues that have made studying the effects of creatine supplementation in patients with neurodegenerative diseases challenging. The most obvious challenge is that these conditions are uncommon, resulting in a shallow pool of patients that might be willing to enroll in clinical studies. In addition, the timeline and severity by which patients with neurodegenerative diseases experience declines in function can vary, resulting in heterogenous responses for both within and between groups. Rapid neurodegeneration and the need for a creatine washout period [100] make implementing a within-subjects design, which offer less error variance, more difficult. Patient care for neurodegenerative diseases is also incredibly complex and requires considerable resources. As such, this type of work is likely to be limited to hospital based research centers and academic health science centers where patient care can be combined with carefully monitored clinical trials. Given these reasons, it is not surprising that prospective studies on creatine supplementation in patients with neurodegenerative diseases have featured small sample sizes and non-significant trends, highlighting the potential for type II statistical errors.

4.2.1. Muscular Dystrophies

Skeletal muscle free creatine and phosphocreatine stores are significantly reduced in patients with certain myopathies and muscular dystrophies [114]. These observations have been linked to a lower creatine transporter protein content and impaired creatine uptake/release kinetics [115]. Our review of the literature shows that the potential therapeutic benefits of creatine supplementation have been studied in patients with Duchenne muscular dystrophy and Becker's muscular dystrophy more than in other neurodegenerative diseases. These diseases are X chromosome-linked and caused by mutations with the dystrophin gene, with life expectancy in the 20's for those with Duchenne muscular dystrophy. In most patients, shortened lifespan is mainly due to respiratory or cardiac failure [116]. It is interesting to note that the magnitude of creatine's treatment effects has varied in patients with dystrophinopathies versus type 1 and type 2 myotonic dystrophy [117].

Voluntary muscle strength has been considered the primary outcome variable in most creatine studies in patients with muscular dystrophies. Three randomized, double-blind trials in boys with dystrophinopathies have reported noteworthy improvements. Louis et al. [118] assessed the effects of creatine supplementation in 12 boys with Duchenne muscular dystrophy and 3 boys (mean age = 11 years) with Becker's muscular dystrophy. They utilized a randomized, double-blind, cross-over design, with a 2-month washout period in between. Following supplementation with creatine, a significant improvement in strength was observed and time to exhaustion during a submaximal contraction nearly doubled. Louis et al. [118] also reported a 25% increase in joint stiffness during the placebo period, but no change was observed while supplementing with creatine. Similar findings were reported by Tarnopolsky and colleagues [119], who tested the hypothesis that creatine supplementation would increase muscle strength and mass in boys (mean age = 10 years) with Duchenne muscular dystrophy while utilizing a randomized, double-blind, cross-over design. Following 4 months of creatine supplementation, significant improvements in grip strength of the dominant hand and fat-free mass were noted. Utilizing a 6 month, parallel-group design, Escolar et al. [120] noted strong trends for improvements in strength and other functional tasks that may have been mitigated by the unexpected maintenance of strength in the placebo group. Of these three studies, less variability was observed in the two cross-over trials [118,119] than in the parallel-group trial [120]. In contrast to studies on dystrophinopathies, creatine supplementation has not been shown to enhance strength or other measures of physical function in patients with type 1 myotonic dystrophy [121,122] or type 2 myotonic dystrophy/proximal myotonic myopathy [123]. This would suggest that the effects of creatine supplementation in patients with a specific type of muscular dystrophy may not be replicated or extrapolated to others. However, it should be noted that studies which have demonstrated benefits of creatine supplementation have been reported in children, whereas non-significant effects have been limited to adult patients. A considerable gap in knowledge is whether benefits of creatine supplementation in patients with certain myopathies and muscular dystrophies are age specific.

Beyond improvements in muscle strength, creatine supplementation in patients with muscular dystrophies may have two other benefits. First, creatine supplementation has generally resulted in improvements in self-reported activities of daily living or subjective improvement in function [122–125]. Work by Banerjee et al. [125], for example, demonstrated that 53.8% of parents whose children had been assigned to a creatine group reported "better" outcomes. Schneider-Gold et al. [123] reported that patients assigned to a creatine treatment group reported significant improvements in the subjective assessment of activity in daily life using a visual analogue scale. This was observed even though muscle strength did not improve [123]. Overall, patients with muscular dystrophies that supplement with creatine may experience subjective improvements in activity levels or function, regardless of changes in objective measures. Second, limited evidence suggests that creatine supplementation may enhance bone density in dystrophic children. Louis and colleagues [118] reported that in a subgroup of boys able to walk throughout their trial, creatine supplementation resulted in significant improvements in bone mineral density. The potential for preservation of bone mineral density is consistent with studies demonstrating a decrease in urinary N-telopeptide excretion (a marker of bone breakdown) [118,119]. Finally, no studies have reported side effects or adverse reactions to supplementing with creatine in patients with muscular dystrophies, many of whom have been young children. While more research is needed, the overall body of literature suggests that creatine supplementation shows promise as a safe and cost-effective means of maintaining or even enhancing muscle strength, functional performance, and bone density in patients with muscular dystrophies. Smaller improvements have been observed in lean mass [119,121,122] and with assessments of muscle strength that are not highly quantitative, such as manual muscle testing [123]. Responses to these variables seem to differ among specific types of dystrophies, highlighting the need for individualized treatments. It is important to further emphasize that reductions in strength and lean mass are hallmarks of these conditions. Small improvements or even maintenance over time may have real-world clinical value to patients and their families.

4.2.2. Amyotrophic Lateral Sclerosis

Amyotrophic lateral sclerosis (ALS) is a neurodegenerative disease that results in the progressive loss of motor neurons that control voluntary muscle activity. It is the most common type of motor neuron disease and is always fatal [126]. Individuals diagnosed with ALS suffer from muscle weakness, atrophy, and difficulty with speech, among many other debilitating symptoms. With no cure available, most ALS medications are intended to minimize pain and fatigue [127].

Several authors have provided a rationale for why creatine supplementation would be a viable treatment option for individuals diagnosed with ALS [128,129]. Plausible explanations for why creatine might enhance various aspects of quality of life in patients with ALS include protection against neuron loss in the substantia nigra and motor cortex [130], as well as decreased oxidative stress [131] and mitochondrial dysfunction [132] commonly observed with this condition. Despite this initial enthusiasm and encouraging animal work [133], clinical trials in humans have reported disappointing results. Only three published trials utilizing a double-blind, randomized designed have evaluated the effects of creatine supplementation in ALS patients, and all three showed no benefits beyond placebo [134–136]. All three studies utilized doses of 5–10 g·day^{-1}, measured muscle strength, and reported that creatine was well tolerated with no major side effects. Given that these studies included patients that were late in the disease process, it remains to be determined if use of creatine as a therapeutic adjunct early after diagnosis would result in better outcomes.

4.2.3. Multiple Sclerosis

Multiple sclerosis (MS) is an immune-mediated disease caused by destruction or failure of myelin-producing cells, resulting in impaired nerve transmission. The usual onset of MS is between 20 and 50 years of age and it is twice as prevalent in women than men. The symptoms associated with MS are highly variable, but frequently include muscle weakness, difficulty with balance and vision, and fatigue. There is no known cure for MS, though physical therapy can help patients improve functionality.

There is a strong theoretical rationale as to why creatine supplementation may enhance outcomes in patients with MS [137]. For example, MS patients show compromised brain creatine metabolism [138], reduced cardiac phosphocreatine concentration [139], and elevated levels of creatine kinase in cerebrospinal fluid [140]. However, only two studies in patients with MS have evaluated the effects of creatine supplementation, with both reporting no improvements above placebo. Utilizing a randomized, double-blind, placebo-controlled trial, Lambert et al. [141] examined the effects of a loading dose (20 g·day^{-1} for

5 days) on isokinetic knee extension and flexion work. Vastus lateralis muscle biopsies were taken to measure intramuscular total creatine, phosphocreatine, and free creatine, and bioelectrical impedance analyses were used to examine body composition. The authors concluded that creatine supplementation had no influence on muscle creatine stores or high-intensity exercise capacity in patients with MS. These findings were supported by the work of Malin and colleagues [142], who, utilizing a 14 day, double-blind, cross-over trial with a three-week washout period, reported that creatine supplementation did not enhance knee joint power. Clearly, this area is ripe with opportunities for scientific discovery, but the results to date do not support the notion that creatine supplementation enhances physical function in MS.

4.2.4. Parkinson's Disease

Parkinson's Disease (PD) is a progressive neurodegenerative disease that is characterized by resting tremors, rigidity, slowness, and problems with gait and balance. In addition to motor impairments, individuals with PD frequently report anxiety, apathy, and depression [143]. The main pathological characteristic of PD is cell death in the basal ganglia, leading to a reduction in brain dopamine levels. PD is typically diagnosed in adults around the ages of 50–60 years [144]. While a variety of pharmacological drugs are used to combat symptoms associated with PD, such as dopamine agonists, levodopa, and monoamine oxidase inhibitors, there is no cure, though the disease itself is not lethal [145]. As such, innovative approaches for managing symptoms and improving quality of life in those with PD are needed.

Given creatine's well-documented ability to improve muscle function in healthy adults [17,20] some have speculated that individuals with PD may particularly benefit from supplementation [146]. Despite widespread interest, only five studies have compared creatine supplementation to placebo in a randomized, double-blind fashion in adults with PD [147–151]. Aside from one study by Bender, each of these five studies provided a dose of 10 g·day^{-1} to participants assigned to the creatine group. The most common outcome measures in these trials included the Unified Parkinson's Disease Rating Scale and its constituent parts (e.g., Mental, Motor, and Activities of Daily Living). Studies by the National Institute of Neurological Disorders and Stroke (NINDS) [149] and Kieburtz [151] utilized the Schwab and England Scale, which assesses the difficulties patients have completing chores and daily activities. The findings from these studies have shown that the impact of creatine supplementation in individuals with PD may be small. Various aspects of these studies have demonstrated conflicting results. The NINDS study [149], for example, demonstrated no change in any of the Unified Parkinson's Disease Rating Scale outcomes, but significant between-group differences for changes in the Schwab and England Scale. Collectively, the limited studies in this area have indicated that creatine supplementation may not be an effective therapeutic strategy for individuals with PD.

4.2.5. Charcot–Marie–Tooth Disease

Charcot–Marie–Tooth Disease (CMT) is a group of inherited motor and sensory neuropathies that cause muscle atrophy and weakness in the hands and feet [152]. CMT is slowly progressive and incurable. While it is typically not fatal, most patients experience difficulty with muscle stiffness and gait due to foot drop and increased foot supination [153]. Therefore, patients with CMT experience a progressive decline in strength, physical activity, and function.

Only three studies have evaluated the effects of creatine supplementation in CMT patients. Doherty and colleagues [154] evaluated the potential benefits of one month of creatine supplementation in patients (mean age = 43 years) with CMT disease type 1 ($n = 34$) and type 2 ($n = 5$) using a double-blind, placebo-controlled, cross-over design. They reported no significant differences in visual analogue activities of daily living scales, body mass, estimated percentage body fat, or fat-free mass after creatine supplementation as compared with placebo. Shortly thereafter, the same research group published two studies which aimed to test the hypothesis that creatine supplementation would enhance strength and myosin heavy chain content in patients with CMT when combined with resistance training [155,156]. Utilizing a randomized, double-blind design, Chetlin et al. [155] evaluated 20 patients (mean age = 45 years) with CMT disease who completed a 12 week resistance exercise training program. The resistance training program was divided into three 4-week phases and was completed at home with adjustable wrist and ankle weights with a therapeutic squeeze ball. The authors reported that the resistance training program resulted in improvements in most outcomes, but there were not interactive effects for creatine supplementation. Given the large effects that resistance training elicits in novices [157], it is possible that any benefits of creatine were not detectable in the treatment group. Smith and colleagues [156] further analyzed these data to determine whether the combination of resistance exercise and creatine supplementation would increase the percentage of type I myosin heavy chain content composition in the vastus lateralis and whether myosin isoform changes would correlate with improved chair rise-time in patients with CMT. Their findings indicated that, when combined with resistance training, creatine supplementation resulted in a decline in type 1 myosin heavy chain content and an increase in type II myosin heavy chain content. Moreover, these changes correlated with an increase in chair rise performance. While speculative, the work of Smith et al. [156] points to a role for creatine supplementation altering skeletal muscle protein synthesis, activating satellite cells, and modifying myosin heavy chain isoform in patients with CMT. Given that only three studies have included patients with CMT [154–156], it is unclear whether creatine supplementation is an effective treatment approach for improving clinical outcomes. Given that there are so few therapeutic treatment options available for these patients, more work in this area is needed.

4.3. Cardiopulmonary Disease

Patients with cardiovascular impairments, including those with chronic obstructive pulmonary disease (COPD) and congestive heart failure, typically display similar comorbidities and etiological factors resulting in malnutrition and muscle dysfunction [158–160] that may benefit from creatine supplementation during the process of physical rehabilitation. From a broad perspective, creatine plays a role in both cardiac function [161] and vascular health [162] with declines in endogenous creatine a likely consequence of related dysfunction that may be augmented with exogenous supplementation. While the current review is focused on physical rehabilitation rather than the influence of creatine on cardiopulmonary diseases, in-depth analysis of these topics are presented by Balestrino [161] and Clarke et al. [162]. There is evidence to support performance benefits in the literature, with improved muscular strength and endurance, maximal aerobic power, and body composition in patients suffering from COPD [163] and heart failure [164–166] following creatine supplementation without exercise interventions. However, these findings are not consistent [167,168].

When examined in combination with cardiopulmonary rehabilitation, the potential clinical benefits of creatine supplementation for patients with COPD and congestive heart failure have been limited [169]. Cardiopulmonary rehabilitation programs typically feature two to three 60–90 min exercise sessions per week, completed over a 7- to 12-week period. Each session usually includes both aerobic and resistance training components, with an intended aim of progressive overload. Minimal improvements in physical performance, health-related quality of life, and body composition with the addition of creatine supplementation [168,170–172] have been observed. However, this may be due to the robust physiological stimulus and positive outcomes for these measures provided by structured exercise programming in these individuals.

Only one study [163] reported significant increases in fat-free mass, muscular strength/endurance, and health status in COPD patients receiving supplemental creatine versus placebo during pulmonary rehabilitation, but found similar changes in pulmonary function and whole-body exercise capacity between groups. Notably, the creatine group in this study had an average body mass index (BMI) of 23.2 ± 3.6 kg·m^{-2}, which is lower than those examined elsewhere (25–28 kg·m^{-2}) and the closest to the cutoff of being classified as underweight (21 kg·m^{-2}) with the potential of protein-energy malnutrition. Further, the study design featured the most comprehensive creatine loading protocol (5.7 g, 3× daily for two weeks) of the evaluated studies, which was completed prior to commencement of the pulmonary rehabilitation program and resulted in augmented body composition (body mass and fat-free mass) and muscle function (leg extension endurance and handgrip strength/endurance). In comparison, Deacon et al. [168] also utilized a creatine loading phase (~5.5 g, 4× daily for 5 days) in COPD patients before beginning a program of rehabilitative exercise in combination with a maintenance dose of creatine. Despite this, patients in both the creatine and placebo groups showed similar improvements after the intervention. However, it should be noted that the creatine group in this investigation had the highest BMI of the evaluated studies.

From a different perspective, Hemati et al. [173] reported improvements in markers of inflammation (interleukin-6 and C-reactive protein) as well as endothelial dysfunction (P-selectin and intercellular adhesion molecule-1) in heart failure patients undergoing an 8 week aerobic training program while supplementing with a 5 g·day^{-1} maintenance dose of creatine compared to a control group receiving no treatment. Recently, Ostojic [174] proposed that along with pulmonary rehabilitation, creatine supplementation be used as adjuvant therapy as nutritional support for those experiencing "long-haul" symptoms of COVID-19. Taken together with the findings in patients with cardiopulmonary impairments (with and without exercise), additional research and larger clinical trials [175] are still needed to evaluate the efficacy of creatine supplementation in these clinical populations.

4.4. Mitochondrial Cytopathies

Mitochondrial cytopathies are a heterogenous group of genetic disorders that adversely change the electron transport chain (ETC) and its function [176]. The most common mitochondrial cytopathy is mitochondrial encephalopathy, lactic acidosis, and stroke-like episodes (MELAS). Other types include chronic progressive external opthalmoplegia (CPEO) and Kearns–Sayre syndrome [176]. Patients with mitochondrial cytopathies show poor exercise tolerance, very low VO$_2$ max values, and decreased ability to extract oxygen peripherally [177]. Interestingly, the decrease in aerobic power results in an up-regulation of the anaerobic pathways [177]. Further, mitochondria dysfunction results in an inability to meet various energy needs for proper function, notably in the nervous, cardiac, endocrine, and musculoskeletal systems [178,179]. This is especially evident within the muscular and nervous systems due to the high energy demands of muscle and nerve [179]. Since there is currently no known cure for these diseases, patients are managed and treated based upon their symptoms, which typically involve muscular weakness, ataxia, and intolerance to physical activity [180,181].

Nutritional interventions such as antioxidants have been recommended due to the high level of oxidative stress observed in patients with mitochondrial cytopathies, but to date have shown no efficacy. Interestingly, several investigators have shown a decrease in phosphocreatine/inorganic phosphate ratio and phosphocreatine concentration in patients with mitochondrial cytopathy. Thus, it appears that the functional loss of the ETC adversely alters phosphocreatine metabolism, which has been shown to delay recovery post-exercise [114,182,183]. Increasing the concentration of phosphocreatine, with creatine supplementation, in skeletal muscle has been shown to increase muscle function and accelerate recovery from exercise in healthy populations [184]. Furthermore, it has been suggested that creatine supplementation may also attenuate oxidative stress, thereby reducing free radical damage to the mitochondria [185]. Therefore, creatine supplementation would be a potentially beneficial strategy in this patient population.

Given the potential benefits of creatine supplementation in patients with mitochondrial cytopathies, several clinical studies have been conducted. Tarnopolsky and colleagues [186] administered supplemental creatine (10 g·day^{-1} for 14 days, followed by

4 g·day^{-1} for 7 days) in a double-blind fashion to primarily MELAS variant patients with severe mitochondrial cytopathy (VO$_2$ max = 10.1 mL·kg^{-1}·min^{-1}). Creatine supplementation resulted in an 11% improvement in dorsiflexion strength endurance and a 19% improvement in handgrip strength. Further, although not statistically significant, there was a 0.6 kg increase in lean mass in the creatine group.

In contrast, in a randomized, placebo-controlled cross-over trial, Klopstock and colleagues [187] provided supplemental creatine (20 g·day^{-1} for 28 days) to 16 patients with primarily CPEO variant mitochondrial cytopathy. While the large dose was well tolerated, there were no observed significant effects on exercise performance measures or activities of daily living. Using a similar design, Kornblum et al. [188] provided patients with CPEO mitochondrial cytopathies with 150 mg·kg^{-1} of body weight per day of supplemental creatine for six weeks. In agreement with Klopstock et al. [187], creatine supplementation did not improve exercise performance measures. Further, intramuscular phosphocreatine/adenosine triphosphate ratio did not increase, and there was no effect on post-exercise phosphocreatine recovery.

There are several reasons for the conflicting results in examinations of creatine's impact on mitochondrial cytopathies. Unlike the MELAS variant, CPEO patients typically have normal ETC enzyme activity and, more importantly, normal creatine and phosphocreatine concentrations in skeletal muscle. Due to the pathophysiology differences between MELAS and CPEO patients, responses to creatine supplementation may be different. For example, it is well known that patients with low skeletal muscle creatine and phosphocreatine levels respond to exercise performance measures to a greater extent with creatine supplementation [184]. Given that the CPEO patients in Kornblum et al. [188] had phosphocreatine levels that were not significantly lower than those in the control group may have resulted in a non-significant difference in response. Therefore, future studies examining the effect of creatine supplementation in this population need to identify the mitochondrial variant to determine the efficacy of the intervention. As indicated in a recent systematic review, sound research designs are challenging due to the heterogeneity in disorders and physical presentation [189]. More consistent study endpoints, design, and clinically relevant outcomes have yet to be determined, and should be considered with higher sample sizes [189].

While there is some evidence suggesting creatine supplementation combined with rehabilitation or treatments for other pathologies involving muscular dysfunction may be beneficial, there are very few clinical trials examining the effects of creatine for patients with mitochondrial cytopathies. Due to the vast array of possible physical presentations and limitations of patients with mitochondrial cytopathies, formal physical rehabilitation and physical therapy for patients should be individually-based on symptom presentation [181,190]. While treatment approaches for these patient populations demonstrate efficacy in the promotion of physical function, to the best of our knowledge no trials exist that integrate these physical interventions with the addition of creatine supplementation. Due to the potential efficacy as demonstrated individually, the combination of the two interventions may be promising for the promotion of function in patients with mitochondrial cytopathies.

5. Conclusions

Given the encouraging findings regarding the role of creatine supplementation throughout recovery from exercise, rehabilitation from immobilization or injury, and therapeutic support during various chronic conditions, creatine monohydrate demonstrates promise as a rehabilitative aid. Several notable findings have been reported. Based on the literature, creatine supplementation may:

1. Support recovery from exercise by decreasing exercise-induced damage, supporting the adaptive response to exercise, and augmenting the physiological response to training. However, further research is needed regarding whether creatine supplementation confers a benefit to a specific population (e.g., trained vs. untrained) or type of exercise (e.g., endurance vs. resistance).

2. Promote maintenance and mitigate the loss of muscle mass, muscular strength, and endurance, as well as promote healthy glucoregulation, during periods of immobilization. However, differing study protocols (e.g., muscle group involved, duration, creatine dosing) may impact the observed findings via differences in atrophic response, overall creatine exposure, alterations in neuromuscular mechanisms, and metabolic adaptations to disuse.
3. Enhance recovery after nerve damage/denervation. While promising, these findings were observed in the rodent model and have not yet been replicated in humans.
4. Improve physical function, lean mass, and muscular strength in populations with chronic arthritic diseases. However, the sample size of available trials is small, and more work is needed.
5. Improve work capacity, strength, and lean mass in individuals with SCI. However, given the small number of available studies and limited sample size of participants, more work in this area is needed.
6. Improve physical function, lean mass, muscular strength, bone density, and quality of life in patients with muscular dystrophy. Despite the promising findings in individuals with dystrophinopathies, similar findings have not been observed in patients with myotonic dystrophies, indicating that the effects of creatine supplementation are likely specific to dystrophy type.
7. Confer a beneficial effect in patients with CMT disease via an increase in type II myosin heavy chain content, alterations in skeletal muscle protein synthesis, and activation of satellite cells. However, these findings are speculative, and more work is needed.
8. Improve lean mass, muscular strength and endurance, and health status in COPD patients. However, observed responses may be complicated by the robust physiological response to exercise training in this population, and additional research and larger clinical trials are needed.
9. Improve markers of inflammation and endothelial dysfunction in patients with heart failure. However, this has only been observed in a single trial, and more work in this area is needed.
10. Improve muscular strength and endurance in certain mitochondrial cytopathies (MELAS variant). Due to the differences in pathophysiology of mitochondrial cytopathies, responses to creatine supplementation may differ between disorder, and should be further examined.

Notably, creatine supplementation appears safe and well tolerated in virtually all medical patient populations. Despite the positive impact of creatine supplementation in numerous clinical conditions, various gaps in the literature may prevent clinicians and medical professionals from strongly recommending that creatine be used to mitigate declines in physical function or for use while rehabilitating. More work in these areas is needed to gauge creatine's role as a medical and rehabilitative aid.

Author Contributions: Conceptualization, J.R.S. All authors contributed to original draft preparation, writing, review, and editing. All authors have read and agreed to the published version of the manuscript.

Funding: The APC of selected papers of this Special Issue are being funded by AlzChem, LLC. (Trostberg, Germany), which manufactures creatine monohydrate. The funders had no role in the writing of the manuscript, interpretation of the literature, or in the decision to publish the results.

Institutional Review Board Statement: Not applicable.

Informed Consent Statement: Not applicable.

Data Availability Statement: Not applicable.

Acknowledgments: The authors would like to thank all of the research participants, scholars, and funding agencies who have contributed to the research cited in this manuscript.

Conflicts of Interest: J.R.S. has conducted industry-sponsored research on creatine and other nutraceuticals over the past 25 years. Further, J.R.S. has also received financial support for presenting on the science of various nutraceuticals, except creatine, at industry-sponsored scientific conferences. E.S.R. has conducted industry-sponsored research on creatine and received financial support for presenting on creatine at industry-sponsored scientific conferences. Additionally, he serves as a member of the Scientific Advisory Board for AlzChem, who sponsored this Special Issue.

References

1. Rehabilitation. Available online: https://www.who.int/news-room/fact-sheets/detail/rehabilitation (accessed on 18 March 2021).
2. Jones, S.W.; Hill, R.J.; Krasney, P.A.; O'Conner, B.; Peirce, N.; Greenhaff, P.L. Disuse Atrophy and Exercise Rehabilitation in Humans Profoundly Affects the Expression of Genes Associated with the Regulation of Skeletal Muscle Mass. *FASEB J.* **2004**, *18*, 1025–1027. [CrossRef] [PubMed]
3. McKenna, C.F.; Fry, C.S. Altered Satellite Cell Dynamics Accompany Skeletal Muscle Atrophy during Chronic Illness, Disuse, and Aging. *Curr. Opin. Clin. Nutr. Metab. Care* **2017**, *20*, 447–452. [CrossRef]
4. Sedlmeier, A.M.; Baumeister, S.E.; Weber, A.; Fischer, B.; Thorand, B.; Ittermann, T.; Dörr, M.; Felix, S.B.; Völzke, H.; Peters, A.; et al. Relation of Body Fat Mass and Fat-Free Mass to Total Mortality: Results from 7 Prospective Cohort Studies. *Am. J. Clin. Nutr.* **2021**, *113*, 639–646. [CrossRef]
5. Nabuurs, C.I.; Choe, C.U.; Veltien, A.; Kan, H.E.; van Loon, L.J.C.; Rodenburg, R.J.T.; Matschke, J.; Wieringa, B.; Kemp, G.J.; Isbrandt, D.; et al. Disturbed Energy Metabolism and Muscular Dystrophy Caused by Pure Creatine Deficiency Are Reversible by Creatine Intake. *J. Physiol.* **2013**, *591*, 571–592. [CrossRef]
6. Reardon, T.F.; Ruell, P.A.; Fiatarone Singh, M.A.; Thompson, C.H.; Rooney, K.B. Creatine Supplementation Does Not Enhance Submaximal Aerobic Training Adaptations in Healthy Young Men and Women. *Eur. J. Appl. Physiol.* **2006**, *98*, 234–241. [CrossRef] [PubMed]
7. Hespel, P.; Op't Eijnde, B.; Leemputte, M.V.; Ursø, B.; Greenhaff, P.L.; Labarque, V.; Dymarkowski, S.; Hecke, P.V.; Richter, E.A. Oral Creatine Supplementation Facilitates the Rehabilitation of Disuse Atrophy and Alters the Expression of Muscle Myogenic Factors in Humans. *J. Physiol.* **2001**, *536*, 625–633. [CrossRef] [PubMed]
8. Fairman, C.M.; Kendall, K.L.; Hart, N.H.; Taaffe, D.R.; Galvão, D.A.; Newton, R.U. The Potential Therapeutic Effects of Creatine Supplementation on Body Composition and Muscle Function in Cancer. *Crit. Rev. Oncol. Hematol.* **2019**, *133*, 46–57. [CrossRef]
9. Wall, B.T.; van Loon, L.J. Nutritional Strategies to Attenuate Muscle Disuse Atrophy. *Nutr. Rev.* **2013**, *71*, 195–208. [CrossRef] [PubMed]
10. Pearlman, J.P.; Fielding, R.A. Creatine Monohydrate as a Therapeutic Aid in Muscular Dystrophy. *Nutr. Rev.* **2006**, *64*, 80–88. [CrossRef]
11. Bonilla, D.A.; Kreider, R.B.; Stout, J.R.; Forero, D.A.; Kerksick, C.M.; Roberts, M.D.; Rawson, R.S. Metabolic Basis of Creatine in Health and Disease: A Bioinformatics-Assisted Review. *Nutrients* **2021**, *13*, 1238. [CrossRef]
12. Buford, T.W.; Kreider, R.B.; Stout, J.R.; Greenwood, M.; Campbell, B.; Spano, M.; Ziegenfuss, T.; Lopez, H.; Landis, J.; Antonio, J. International Society of Sports Nutrition Position Stand: Creatine Supplementation and Exercise. *J. Int. Soc. Sports Nutr.* **2007**, *4*, 6. [CrossRef] [PubMed]
13. Volek, J.S.; Ballard, K.D.; Forsythe, C.E. Overview of Creatine Metabolism. In *Essentials of Creatine in Sports and Health*; Stout, J.R., Antonio, J., Kalman, D., Eds.; Humana Press: Totowa, NJ, USA, 2008; pp. 1–23. ISBN 978-1-59745-573-2.
14. Harris, R.C.; Söderlund, K.; Hultman, E. Elevation of Creatine in Resting and Exercised Muscle of Normal Subjects by Creatine Supplementation. *Clin. Sci.* **1992**, *83*, 367–374. [CrossRef] [PubMed]
15. Hultman, E.; Soderlund, K.; Timmons, J.A.; Cederblad, G.; Greenhaff, P.L. Muscle Creatine Loading in Men. *J. Appl. Physiol.* **1996**, *81*, 232–237. [CrossRef]
16. Greenhaff, P.L. The Nutritional Biochemistry of Creatine. *J. Nutr. Biochem.* **1997**, *8*, 610–618. [CrossRef]
17. Balsom, P.D.; Söderlund, K.; Ekblom, B. Creatine in Humans with Special Reference to Creatine Supplementation. *Sports Med.* **1994**, *18*, 268–280. [CrossRef]
18. Wallimann, T.; Dolder, M.; Schlattner, U.; Eder, M.; Hornemann, T.; O'Gorman, E.; Rück, A.; Brdiczka, D. Some New Aspects of Creatine Kinase (CK): Compartmentation, Structure, Function and Regulation for Cellular and Mitochondrial Bioenergetics and Physiology. *Biofactors* **1998**, *8*, 229–234. [CrossRef] [PubMed]
19. Wallimann, T.; Schlösser, T.; Eppenberger, H.M. Function of M-Line-Bound Creatine Kinase as Intramyofibrillar ATP Regenerator at the Receiving End of the Phosphorylcreatine Shuttle in Muscle. *J. Biol. Chem.* **1984**, *259*, 5238–5246. [CrossRef]
20. Kreider, R.B.; Kalman, D.S.; Antonio, J.; Ziegenfuss, T.N.; Wildman, R.; Collins, R.; Candow, D.G.; Kleiner, S.M.; Almada, A.L.; Lopez, H.L. International Society of Sports Nutrition Position Stand: Safety and Efficacy of Creatine Supplementation in Exercise, Sport, and Medicine. *J. Int. Soc. Sports Nutr.* **2017**, *14*, 18. [CrossRef]
21. Hespel, P.; Derave, W. Ergogenic Effects of Creatine in Sports and Rehabilitation. In *Creatine and Creatine Kinase in Health and Disease*; Salomons, G.S., Wyss, M., Eds.; Subcellular Biochemistry; Springer: Dordrecht, The Netherlands, 2007; pp. 246–259. ISBN 978-1-4020-6486-9.

22. Gualano, B.; Artioli, G.G.; Poortmans, J.R.; Lancha Junior, A.H. Exploring the Therapeutic Role of Creatine Supplementation. *Amino Acids* **2010**, *38*, 31–44. [CrossRef] [PubMed]

23. Weir, J.P. Clinical Applications. In *Essentials of Creatine in Sports and Health*; Stout, J.R., Antonio, J., Kalman, D., Eds.; Humana Press: Totowa, NJ, USA, 2008; pp. 173–210. ISBN 978-1-59745-573-2.

24. Corn, M.; Kallail, D.; Payne, D.; Bullock, C.; Raynes, E.A. Will the Use of Supplemental Creatine Decrease Rehabilitation Time among Young Adults Coping with the Effects of Immobilization Related Injuries? *Int. J. Soc. Health Inf. Manag.* **2011**, *4*, 89–102.

25. Rawson, E.S.; Clarkson, P.M.; Tarnopolsky, M.A. Perspectives on Exertional Rhabdomyolysis. *Sports Med* **2017**, *47*, 33–49. [CrossRef]

26. Eichner, E.R. An Outbreak of Muscle Breakdown: A Morality Play in Four Acts. *Curr. Sports Med. Rep.* **2010**, *9*, 325–326. [CrossRef]

27. Oh, J.Y.; Laidler, M.; Fiala, S.C.; Hedberg, K. Acute Exertional Rhabdomyolysis and Triceps Compartment Syndrome during a High School Football CAMP. *Sports Health* **2012**, *4*, 57–62. [CrossRef] [PubMed]

28. Smoot, M.K.; Amendola, A.; Cramer, E.; Doyle, C.; Kregel, K.C.; Chiang, H.; Cavanaugh, J.E.; Herwaldt, L.A. A Cluster of Exertional Rhabdomyolysis Affecting a Division I Football Team. *Clin. J. Sport Med.* **2013**, *23*, 365–372. [CrossRef]

29. Centers for Disease Control and Prevention (CDC). Hyperthermia and Dehydration-Related Deaths Associated with Intentional Rapid Weight Loss in Three Collegiate Wrestlers–North Carolina, Wisconsin, and Michigan, November-December 1997. *Mmwr. Morb. Mortal. Wkly. Rep.* **1998**, *47*, 105–108.

30. Greenwood, M.; Kreider, R.B.; Greenwood, L.; Byars, A. Cramping and Injury Incidence in Collegiate Football Players Are Reduced by Creatine Supplementation. *J. Athl. Train.* **2003**, *38*, 216–219. [PubMed]

31. Greenwood, M.; Kreider, R.B.; Greenwood, L.; Willoughby, D.; Byars, A. The Effects of Creatine Supplementation on Cramping and Injury Occurrence during College Baseball Training and Competition. *JEP Online* **2003**, *6*, 16–23.

32. Greenwood, M.; Kreider, R.B.; Melton, C.; Rasmussen, C.; Lancaster, S.; Cantler, E.; Milnor, P.; Almada, A. Creatine Supplementation during College Football Training Does Not Increase the Incidence of Cramping or Injury. *Mol. Cell Biochem.* **2003**, *244*, 83–88. [CrossRef] [PubMed]

33. Rawson, E.S.; Miles, M.P.; Larson-Meyer, D.E. Dietary Supplements for Health, Adaptation, and Recovery in Athletes. *Int. J. Sport Nutr. Exerc. Metab.* **2018**, *28*, 188–199. [CrossRef] [PubMed]

34. Kim, J.; Lee, J.; Kim, S.; Yoon, D.; Kim, J.; Sung, D.J. Role of Creatine Supplementation in Exercise-Induced Muscle Damage: A Mini Review. *J. Exerc. Rehabil.* **2015**, *11*, 244–250. [CrossRef]

35. Northeast, B.; Clifford, T. The Effect of Creatine Supplementation on Markers of Exercise-Induced Muscle Damage: A Systematic Review and Meta-Analysis of Human Intervention Trials. *Int. J. Sport Nutr. Exerc. Metab.* **2021**, *31*, 276–291. [CrossRef]

36. Clarkson, P.M.; Hubal, M.J. Exercise-Induced Muscle Damage in Humans. *Am. J. Phys. Med. Rehabil.* **2002**, *81*, S52–S69. [CrossRef] [PubMed]

37. Hyldahl, R.D.; Hubal, M.J. Lengthening Our Perspective: Morphological, Cellular, and Molecular Responses to Eccentric Exercise. *Muscle Nerve* **2014**, *49*, 155–170. [CrossRef] [PubMed]

38. Burke, D.G.; Candow, D.G.; Chilibeck, P.D.; MacNeil, L.G.; Roy, B.D.; Tarnopolsky, M.A.; Ziegenfuss, T. Effect of Creatine Supplementation and Resistance-Exercise Training on Muscle Insulin-like Growth Factor in Young Adults. *Int. J. Sport Nutr. Exerc. Metab.* **2008**, *18*, 389–398. [CrossRef]

39. Deldicque, L.; Louis, M.; Theisen, D.; Nielens, H.; Dehoux, M.; Thissen, J.-P.; Rennie, M.J.; Francaux, M. Increased IGF MRNA in Human Skeletal Muscle after Creatine Supplementation. *Med. Sci. Sports Exerc.* **2005**, *37*, 731–736. [CrossRef] [PubMed]

40. Willoughby, D.S.; Rosene, J. Effects of Oral Creatine and Resistance Training on Myosin Heavy Chain Expression. *Med. Sci. Sports Exerc.* **2001**, *33*, 1674–1681. [CrossRef] [PubMed]

41. Willoughby, D.S.; Rosene, J.M. Effects of Oral Creatine and Resistance Training on Myogenic Regulatory Factor Expression. *Med. Sci. Sports Exerc.* **2003**, *35*, 923–929. [CrossRef]

42. Olsen, S.; Aagaard, P.; Kadi, F.; Tufekovic, G.; Verney, J.; Olesen, J.L.; Suetta, C.; Kjaer, M. Creatine Supplementation Augments the Increase in Satellite Cell and Myonuclei Number in Human Skeletal Muscle Induced by Strength Training. *J. Physiol.* **2006**, *573*, 525–534. [CrossRef] [PubMed]

43. Safdar, A.; Yardley, N.J.; Snow, R.; Melov, S.; Tarnopolsky, M.A. Global and Targeted Gene Expression and Protein Content in Skeletal Muscle of Young Men Following Short-Term Creatine Monohydrate Supplementation. *Physiol. Genom.* **2008**, *32*, 219–228. [CrossRef]

44. Berneis, K.; Ninnis, R.; Häussinger, D.; Keller, U. Effects of Hyper- and Hypoosmolality on Whole Body Protein and Glucose Kinetics in Humans. *Am. J. Physiol.* **1999**, *276*, E188–E195. [CrossRef] [PubMed]

45. Häussinger, D.; Roth, E.; Lang, F.; Gerok, W. Cellular Hydration State: An Important Determinant of Protein Catabolism in Health and Disease. *Lancet* **1993**, *341*, 1330–1332. [CrossRef]

46. McKinnon, N.B.; Graham, M.T.; Tiidus, P.M. Effect of Creatine Supplementation on Muscle Damage and Repair Following Eccentrically-Induced Damage to the Elbow Flexor Muscles. *J. Sports Sci. Med.* **2012**, *11*, 653–659.

47. Bassit, R.A.; Curi, R.; Costa Rosa, L.F.B.P. Creatine Supplementation Reduces Plasma Levels of Pro-Inflammatory Cytokines and PGE2 after a Half-Ironman Competition. *Amino Acids* **2008**, *35*, 425–431. [CrossRef] [PubMed]

48. Bassit, R.A.; da Justa Pinheiro, C.H.; Vitzel, K.F.; Sproesser, A.J.; Silveira, L.R.; Curi, R. Effect of Short-Term Creatine Supplementation on Markers of Skeletal Muscle Damage after Strenuous Contractile Activity. *Eur. J. Appl. Physiol.* **2010**, *108*, 945–955. [CrossRef]

49. Basta, P.; Skarpańska-Stejnborn, A.; Pilaczyńska-Szcześniak, Ł. Creatine Supplementation and Parameters of Exercise-Induced Oxidative Stress after a Standard Rowing Test. *Stud. Phys. Cult. Tour.* **2006**, *6*, 17–23.
50. Boychuk, K.E.; Lanovaz, J.L.; Krentz, J.R.; Lishchynsky, J.T.; Candow, D.G.; Farthing, J.P. Creatine Supplementation Does Not Alter Neuromuscular Recovery after Eccentric Exercise. *Muscle Nerve* **2016**, *54*, 487–495. [CrossRef] [PubMed]
51. Cooke, M.B.; Rybalka, E.; Williams, A.D.; Cribb, P.J.; Hayes, A. Creatine Supplementation Enhances Muscle Force Recovery after Eccentrically-Induced Muscle Damage in Healthy Individuals. *J. Int. Soc. Sports Nutr.* **2009**, *6*, 13. [CrossRef]
52. Deminice, R.; Rosa, F.T.; Franco, G.S.; Jordao, A.A.; de Freitas, E.C. Effects of Creatine Supplementation on Oxidative Stress and Inflammatory Markers after Repeated-Sprint Exercise in Humans. *Nutrition* **2013**, *29*, 1127–1132. [CrossRef] [PubMed]
53. Machado, M.; Pereira, R.; Sampaio-Jorge, F.; Knifis, F.; Hackney, A. Creatine Supplementation: Effects on Blood Creatine Kinase Activity Responses to Resistance Exercise and Creatine Kinase Activity Measurement. *Braz. J. Pharm. Sci.* **2009**, *45*, 751–757. [CrossRef]
54. Rawson, E.S.; Conti, M.P.; Miles, M.P. Creatine Supplementation Does Not Reduce Muscle Damage or Enhance Recovery from Resistance Exercise. *J. Strength Cond. Res.* **2007**, *21*, 1208–1213. [CrossRef] [PubMed]
55. Rawson, E.S.; Gunn, B.; Clarkson, P.M. The Effects of Creatine Supplementation on Exercise-Induced Muscle Damage. *J. Strength Cond. Res.* **2001**, *15*, 178–184.
56. Rosene, J.; Matthews, T.; Ryan, C.; Belmore, K.; Bergsten, A.; Blaisdell, J.; Gaylord, J.; Love, R.; Marrone, M.; Ward, K.; et al. Short and Longer-Term Effects of Creatine Supplementation on Exercise Induced Muscle Damage. *J. Sports Sci. Med.* **2009**, *8*, 89–96.
57. Santos, R.V.T.; Bassit, R.A.; Caperuto, E.C.; Costa Rosa, L.F.B.P. The Effect of Creatine Supplementation upon Inflammatory and Muscle Soreness Markers after a 30km Race. *Life Sci.* **2004**, *75*, 1917–1924. [CrossRef]
58. Taylor, B.A.; Panza, G.; Ballard, K.D.; White, C.M.; Thompson, P.D. Creatine Supplementation Does Not Alter the Creatine Kinase Response to Eccentric Exercise in Healthy Adults on Atorvastatin. *J. Clin. Lipidol.* **2018**, *12*, 1305–1312. [CrossRef]
59. Veggi, K.F.; Machado, M.; Koch, A.J.; Santana, S.C.; Oliveira, S.S.; Stec, M.J. Oral Creatine Supplementation Augments the Repeated Bout Effect. *Int. J. Sport Nutr. Exerc. Metab.* **2013**, *23*, 378–387. [CrossRef] [PubMed]
60. Wang, C.-C.; Fang, C.-C.; Lee, Y.-H.; Yang, M.-T.; Chan, K.-H. Effects of 4-Week Creatine Supplementation Combined with Complex Training on Muscle Damage and Sport Performance. *Nutrients* **2018**, *10*, 1640. [CrossRef] [PubMed]
61. Kreider, R.B.; Melton, C.; Rasmussen, C.J.; Greenwood, M.; Lancaster, S.; Cantler, E.C.; Milnor, P.; Almada, A.L. Long-Term Creatine Supplementation Does Not Significantly Affect Clinical Markers of Health in Athletes. *Mol. Cell Biochem.* **2003**, *244*, 95–104. [CrossRef] [PubMed]
62. Hather, B.M.; Adams, G.R.; Tesch, P.A.; Dudley, G.A. Skeletal Muscle Responses to Lower Limb Suspension in Humans. *J. Appl. Physiol. (1985)* **1992**, *72*, 1493–1498. [CrossRef] [PubMed]
63. Wall, B.T.; Dirks, M.L.; Snijders, T.; Senden, J.M.G.; Dolmans, J.; van Loon, L.J.C. Substantial Skeletal Muscle Loss Occurs during Only 5 Days of Disuse. *Acta Physiol. Oxf.* **2014**, *210*, 600–611. [CrossRef]
64. Berg, H.E.; Dudley, G.A.; Häggmark, T.; Ohlsén, H.; Tesch, P.A. Effects of Lower Limb Unloading on Skeletal Muscle Mass and Function in Humans. *J. Appl. Physiol. (1985)* **1991**, *70*, 1882–1885. [CrossRef]
65. Duchateau, J.; Hainaut, K. Electrical and Mechanical Changes in Immobilized Human Muscle. *J. Appl. Physiol. (1985)* **1987**, *62*, 2168–2173. [CrossRef]
66. Stevens, J.E.; Walter, G.A.; Okereke, E.; Scarborough, M.T.; Esterhai, J.L.; George, S.Z.; Kelley, M.J.; Tillman, S.M.; Gibbs, J.D.; Elliott, M.A.; et al. Muscle Adaptations with Immobilization and Rehabilitation after Ankle Fracture. *Med. Sci. Sports Exerc.* **2004**, *36*, 1695–1701. [CrossRef] [PubMed]
67. Vandenborne, K.; Elliott, M.A.; Walter, G.A.; Abdus, S.; Okereke, E.; Shaffer, M.; Tahernia, D.; Esterhai, J.L. Longitudinal Study of Skeletal Muscle Adaptations during Immobilization and Rehabilitation. *Muscle Nerve* **1998**, *21*, 1006–1012. [CrossRef]
68. Stuart, C.A.; Shangraw, R.E.; Prince, M.J.; Peters, E.J.; Wolfe, R.R. Bed-Rest-Induced Insulin Resistance Occurs Primarily in Muscle. *Metabolism* **1988**, *37*, 802–806. [CrossRef]
69. Tzankoff, S.P.; Norris, A.H. Effect of Muscle Mass Decrease on Age-Related BMR Changes. *J. Appl. Physiol. Respir. Environ. Exerc. Physiol.* **1977**, *43*, 1001–1006. [CrossRef] [PubMed]
70. Ferrando, A.A.; Lane, H.W.; Stuart, C.A.; Davis-Street, J.; Wolfe, R.R. Prolonged Bed Rest Decreases Skeletal Muscle and Whole Body Protein Synthesis. *Am. J. Physiol.* **1996**, *270*, E627–E633. [CrossRef] [PubMed]
71. Branch, J.D. Effect of Creatine Supplementation on Body Composition and Performance: A Meta-Analysis. *Int. J. Sport Nutr. Exerc. Metab.* **2003**, *13*, 198–226. [CrossRef]
72. Fransen, J.C. Exploring the Potential of Creatine Ingestion to Maintain Muscle Function during Immobilization. *JNHFS* **2015**, *3*. [CrossRef]
73. Padilha, C.S.; Cella, P.S.; Salles, L.R.; Deminice, R. Oral Creatine Supplementation Attenuates Muscle Loss Caused by Limb Immobilization: A Systematic Review. *Fisioter. Mov.* **2017**, *30*, 831–838. [CrossRef]
74. Johnston, A.P.W.; Burke, D.G.; MacNeil, L.G.; Candow, D.G. Effect of Creatine Supplementation During Cast-Induced Immobilization on the Preservation of Muscle Mass, Strength, and Endurance. *J. Strength Cond. Res.* **2009**, *23*, 116–120. [CrossRef]
75. Eijnde, B.O.; Ursø, B.; Richter, E.A.; Greenhaff, P.L.; Hespel, P. Effect of Oral Creatine Supplementation on Human Muscle GLUT4 Protein Content After Immobilization. *Diabetes* **2001**, *50*, 18–23. [CrossRef]
76. Jaspers, S.R.; Tischler, M.E. Atrophy and Growth Failure of Rat Hindlimb Muscles in Tail-Cast Suspension | Journal of Applied Physiology. *J. Appl. Physiol.* **1984**, *57*, 1472–1479. [CrossRef] [PubMed]

77. Aoki, M. Deleteriuos Effects of Immobilization upon Rat Skeletal Muscle: Role of Creatine Supplementation. *Clin. Nutr.* **2004**, *23*, 1176–1183. [CrossRef]

78. Marzuca-Nassr, G.N.; Fortes, M.S.; Guimarães-Ferreira, L.; Murata, G.M.; Vitzel, K.F.; Vasconcelos, D.A.; Bassit, R.A.; Curi, R. Short-Term Creatine Supplementation Changes Protein Metabolism Signaling in Hindlimb Suspension. *Braz. J. Med. Biol. Res.* **2019**, *52*, e8391. [CrossRef]

79. Fransen, J.C.; Zuhl, M.; Kerksick, C.M.; Cole, N.; Altobelli, S.; Kuethe, D.O.; Schneider, S. Impact of Creatine on Muscle Performance and Phosphagen Stores after Immobilization. *Eur. J. Appl. Physiol.* **2015**, *115*, 1877–1886. [CrossRef]

80. Backx, E.M.P.; Hangelbroek, R.; Snijders, T.; Verscheijden, M.-L.; Verdijk, L.B.; de Groot, L.C.P.G.M.; van Loon, L.J.C. Creatine Loading Does Not Preserve Muscle Mass or Strength During Leg Immobilization in Healthy, Young Males: A Randomized Controlled Trial. *Sports Med.* **2017**, *47*, 1661–1671. [CrossRef]

81. Tyler, T.F.; Nicholas, S.J.; Hershman, E.B.; Glace, B.W.; Mullaney, M.J.; McHugh, M.P. The Effect of Creatine Supplementation on Strength Recovery after Anterior Cruciate Ligament (ACL) Reconstruction: A Randomized, Placebo-Controlled, Double-Blind Trial. *Am. J. Sports Med.* **2004**, *32*, 383–388. [CrossRef] [PubMed]

82. Roy, B.D.; de Beer, J.; Harvey, D.; Tarnopolsky, M.A. Creatine Monohydrate Supplementation Does Not Improve Functional Recovery after Total Knee Arthroplasty. *Arch Phys. Med. Rehabil.* **2005**, *86*, 1293–1298. [CrossRef] [PubMed]

83. Özkan, Ö.; Duman, Ö.; Haspolat, Ş.; Özgentaş, H.E.; Dikici, M.B.; Gürer, I.; Güngör, H.A.; Güzide Gökhan, A. Effect of Systemic Creatine Monohydrate Supplementation on Denervated Muscle During Reinnervation: Experimental Study in the Rat. *J. Reconstr. Microsurg.* **2005**, *21*, 573–580. [CrossRef]

84. Crassous, B.; Richard-Bulteau, H.; Deldicque, L.; Serrurier, B.; Pasdeloup, M.; Francaux, M.; Bigard, X.; Koulmann, N. Lack of Effects of Creatine on the Regeneration of Soleus Muscle after Injury in Rats. *Med. Sci. Sports Exerc.* **2009**, *41*, 1761–1769. [CrossRef]

85. Richard-Bulteau, H.; Serrurier, B.; Crassous, B.; Banzet, S.; Peinnequin, A.; Bigard, X.; Koulmann, N. Recovery of Skeletal Muscle Mass after Extensive Injury: Positive Effects of Increased Contractile Activity. *Am. J. Physiol. Cell Physiol.* **2008**, *294*, C467–C476. [CrossRef]

86. Hicks, A.L.; Martin Ginis, K.A.; Pelletier, C.A.; Ditor, D.S.; Foulon, B.; Wolfe, D.L. The Effects of Exercise Training on Physical Capacity, Strength, Body Composition and Functional Performance among Adults with Spinal Cord Injury: A Systematic Review. *Spinal Cord* **2011**, *49*, 1103–1127. [CrossRef]

87. Hopman, M.T.; Oeseburg, B.; Binkhorst, R.A. Cardiovascular Responses in Paraplegic Subjects during Arm Exercise. *Eur. J. Appl. Physiol. Occup. Physiol.* **1992**, *65*, 73–78. [CrossRef]

88. Jehl, J.L.; Gandmontagne, M.; Pastene, G.; Eyssette, M.; Flandrois, R.; Coudert, J. Cardiac Output during Exercise in Paraplegic Subjects. *Eur. J. Appl. Physiol. Occup. Physiol.* **1991**, *62*, 256–260. [CrossRef]

89. Lin, K.H.; Lai, J.S.; Kao, M.J.; Lien, I.N. Anaerobic Threshold and Maximal Oxygen Consumption during Arm Cranking Exercise in Paraplegia. *Arch. Phys. Med. Rehabil.* **1993**, *74*, 515–520. [CrossRef]

90. Jacobs, P.L.; Mahoney, E.T.; Cohn, K.A.; Sheradsky, L.F.; Green, B.A. Oral Creatine Supplementation Enhances Upper Extremity Work Capacity in Persons with Cervical-Level Spinal Cord Injury. *Arch. Phys. Med. Rehabil.* **2002**, *83*, 19–23. [CrossRef] [PubMed]

91. Amorim, S.; Teixeira, V.H.; Corredeira, R.; Cunha, M.; Maia, B.; Margalho, P.; Pires, J. Creatine or Vitamin D Supplementation in Individuals with a Spinal Cord Injury Undergoing Resistance Training: A Double-Blinded, Randomized Pilot Trial. *J. Spinal Cord Med.* **2018**, *41*, 471–478. [CrossRef] [PubMed]

92. Hausmann, O.N.; Fouad, K.; Wallimann, T.; Schwab, M.E. Protective Effects of Oral Creatine Supplementation on Spinal Cord Injury in Rats. *Spinal Cord* **2002**, *40*, 449–456. [CrossRef] [PubMed]

93. Wilkinson, T.J. Oral Creatine Supplementation: A Potential Adjunct Therapy for Rheumatoid Arthritis Patients. *WJR* **2014**, *4*, 22. [CrossRef]

94. Neves, M.; Gualano, B.; Roschel, H.; Fuller, R.; Benatti, F.B.; De Sá Pinto, A.L.; Lima, F.R.; Pereira, R.M.; Lancha, A.H.; Bonfá, E. Beneficial Effect of Creatine Supplementation in Knee Osteoarthritis. *Med. Sci. Sports Exerc.* **2011**, *43*, 1538–1543. [CrossRef]

95. Willer, B.; Stucki, G.; Hoppeler, H.; Brühlmann, P.; Krähenbühl, S. Effects of Creatine Supplementation on Muscle Weakness in Patients with Rheumatoid Arthritis. *Rheumatol. Oxf.* **2000**, *39*, 293–298. [CrossRef]

96. Lange, A.K.; Vanwanseele, B.; Fiatarone Singh, M.A. Strength Training for Treatment of Osteoarthritis of the Knee: A Systematic Review. *Arthritis Care Res. Off.* **2008**, *59*, 1488–1494. [CrossRef]

97. Clark, B.C. In Vivo Alterations in Skeletal Muscle Form and Function after Disuse Atrophy. *Med. Sci. Sports Exerc.* **2009**, *41*, 1869–1875. [CrossRef] [PubMed]

98. Candow, D.G.; Chilibeck, P.D. Differences in Size, Strength, and Power of Upper and Lower Body Muscle Groups in Young and Older Men. *J. Gerontol. A Biol. Sci. Med. Sci.* **2005**, *60*, 148–156. [CrossRef] [PubMed]

99. Lynch, N.A.; Metter, E.J.; Lindle, R.S.; Fozard, J.L.; Tobin, J.D.; Roy, T.A.; Fleg, J.L.; Hurley, B.F. Muscle Quality. I. Age-Associated Differences between Arm and Leg Muscle Groups. *J. Appl. Physiol. (1985)* **1999**, *86*, 188–194. [CrossRef] [PubMed]

100. Vandenberghe, K.; Goris, M.; Van Hecke, P.; Van Leemputte, M.; Vangerven, L.; Hespel, P. Long-Term Creatine Intake Is Beneficial to Muscle Performance during Resistance Training. *J. Appl. Physiol. (1985)* **1997**, *83*, 2055–2063. [CrossRef] [PubMed]

101. MacLennan, R.J.; Ogilvie, D.; McDorman, J.; Vargas, E.; Grusky, A.R.; Kim, Y.; Garcia, J.M.; Stock, M.S. The Time Course of Neuromuscular Impairment during Short-term Disuse in Young Women-MacLennan-2021-Physiological Reports-Wiley Online Library. *Physiol. Rep.* **2021**, *9*, e14677. [CrossRef] [PubMed]

102. McComas, A.J. Human Neuromuscular Adaptations That Accompany Changes in Activity. *Med. Sci. Sports Exerc.* **1994**, *26*, 1498–1509. [CrossRef] [PubMed]

103. Clark, B.C.; Issac, L.C.; Lane, J.L.; Damron, L.A.; Hoffman, R.L. Neuromuscular Plasticity during and Following 3 Wk of Human Forearm Cast Immobilization. *J. Appl. Physiol. (1985)* **2008**, *105*, 868–878. [CrossRef]

104. Kawakami, Y.; Akima, H.; Kubo, K.; Muraoka, Y.; Hasegawa, H.; Kouzaki, M.; Imai, M.; Suzuki, Y.; Gunji, A.; Kanehisa, H.; et al. Changes in Muscle Size, Architecture, and Neural Activation after 20 Days of Bed Rest with and without Resistance Exercise. *Eur. J. Appl. Physiol.* **2001**, *84*, 7–12. [CrossRef]

105. Clark, B.C.; Manini, T.M.; Bolanowski, S.J.; Ploutz-Snyder, L.L. Adaptations in Human Neuromuscular Function Following Prolonged Unweighting: II. Neurological Properties and Motor Imagery Efficacy. *J. Appl. Physiol. (1985)* **2006**, *101*, 264–272. [CrossRef]

106. Howard, E.E.; Pasiakos, S.M.; Fussell, M.A.; Rodriguez, N.R. Skeletal Muscle Disuse Atrophy and the Rehabilitative Role of Protein in Recovery from Musculoskeletal Injury | Advances in Nutrition | Oxford Academic. *Adv. Nutr.* **2020**, *11*, 989–1001. [CrossRef] [PubMed]

107. Morton, R.W.; Traylor, D.A.; Weijs, P.J.M.; Phillips, S.M. Defining Anabolic Resistance: Implications for Delivery of Clinical Care Nutrition. *Curr. Opin. Crit. Care* **2018**, *24*, 124–130. [CrossRef] [PubMed]

108. Breen, L.; Stokes, K.A.; Churchward-Venne, T.A.; Moore, D.R.; Baker, S.K.; Smith, K.; Atherton, P.J.; Phillips, S.M. Two Weeks of Reduced Activity Decreases Leg Lean Mass and Induces "Anabolic Resistance" of Myofibrillar Protein Synthesis in Healthy Elderly. *J. Clin. Endocrinol. Metab.* **2013**, *98*, 2604–2612. [CrossRef] [PubMed]

109. Tanner, R.E.; Brunker, L.B.; Agergaard, J.; Barrows, K.M.; Briggs, R.A.; Kwon, O.S.; Young, L.M.; Hopkins, P.N.; Volpi, E.; Marcus, R.L.; et al. Age-Related Differences in Lean Mass, Protein Synthesis and Skeletal Muscle Markers of Proteolysis after Bed Rest and Exercise Rehabilitation. *J. Physiol.* **2015**, *593*, 4259–4273. [CrossRef]

110. Symons, T.B.; Sheffield-Moore, M.; Chinkes, D.L.; Ferrando, A.A.; Paddon-Jones, D. Artificial Gravity Maintains Skeletal Muscle Protein Synthesis during 21 Days of Simulated Microgravity. *J. Appl. Physiol. (1985)* **2009**, *107*, 34–38. [CrossRef]

111. Ziegenfuss, T.N.; Rogers, M.; Lowery, L.; Mullins, N.; Mendel, R.; Antonio, J.; Lemon, P. Effect of Creatine Loading on Anaerobic Performance and Skeletal Muscle Volume in NCAA Division I Athletes. *Nutrition* **2002**, *18*, 397–402. [CrossRef]

112. Arentson-Lantz, E.J.; English, K.L.; Paddon-Jones, D.; Fry, C.S. Fourteen Days of Bed Rest Induces a Decline in Satellite Cell Content and Robust Atrophy of Skeletal Muscle Fibers in Middle-Aged Adults. *J. Appl. Physiol. (1985)* **2016**, *120*, 965–975. [CrossRef]

113. Wolfe, R.R.; Jahoor, F.; Hartl, W.H. Protein and Amino Acid Metabolism after Injury. *Diabetes Metab. Rev.* **1989**, *5*, 149–164. [CrossRef]

114. Tarnopolsky, M.A.; Parise, G. Direct Measurement of High-Energy Phosphate Compounds in Patients with Neuromuscular Disease. *Muscle Nerve* **1999**, *22*, 1228–1233. [CrossRef]

115. Fitch, C.D.; Moody, L.G. Creatine Metabolism in Skeletal Muscle V. An Intracellular Abnormality of Creatine Trapping in Dystrophic Muscle. *Proc. Soc. Exp. Biol. Med.* **1969**, *132*, 219–222. [CrossRef] [PubMed]

116. Passamano, L.; Taglia, A.; Palladino, A.; Viggiano, E.; D'AMBROSIO, P.; Scutifero, M.; Rosaria Cecio, M.; Torre, V.; De Luca, F.; Picillo, E.; et al. Improvement of Survival in Duchenne Muscular Dystrophy: Retrospective Analysis of 835 Patients. *Acta Myol.* **2012**, *31*, 121–125. [PubMed]

117. Kley, R.A.; Tarnopolsky, M.A.; Vorgerd, M. Creatine for Treating Muscle Disorders. *Cochrane Database Syst. Rev.* **2013**, CD004760. [CrossRef]

118. Louis, M.; Lebacq, J.; Poortmans, J.R.; Belpaire-Dethiou, M.-C.; Devogelaer, J.-P. Beneficial Effects of Creatine Supplementation in Dystrophic Patients. *Muscle Nerve Off. Med.* **2003**, *27*, 604–610. [CrossRef]

119. Tarnopolsky, M.A.; Mahoney, D.J.; Vajsar, J.; Rodriguez, C.; Doherty, T.J.; Roy, B.D.; Biggar, D. Creatine Monohydrate Enhances Strength and Body Composition in Duchenne Muscular Dystrophy. *Neurology* **2004**, *62*, 1771–1777. [CrossRef]

120. Escolar, D.M.; Buyse, G.; Henricson, E.; Leshner, R.; Florence, J.; Mayhew, J.; Tesi-Rocha, C.; Gorni, K.; Pasquali, L.; Patel, K.M.; et al. CINRG Randomized Controlled Trial of Creatine and Glutamine in Duchenne Muscular Dystrophy. *Ann. Neurol.* **2005**, *58*, 151–155. [CrossRef] [PubMed]

121. Tarnopolsky, M.; Mahoney, D.; Thompson, T.; Naylor, H.; Doherty, T.J. Creatine Monohydrate Supplementation Does Not Increase Muscle Strength, Lean Body Mass, or Muscle Phosphocreatine in Patients with Myotonic Dystrophy Type 1. *Muscle Nerve* **2004**, *29*, 51–58. [CrossRef] [PubMed]

122. Walter, M.C.; Reilich, P.; Lochmüller, H.; Kohnen, R.; Schlotter, B.; Hautmann, H.; Dunkl, E.; Pongratz, D.; Müller-Felber, W. Creatine Monohydrate in Myotonic Dystrophy: A Double-Blind, Placebo-Controlled Clinical Study. *J. Neurol.* **2002**, *249*, 1717–1722. [CrossRef] [PubMed]

123. Schneider-Gold, C.; Beck, M.; Wessig, C.; George, A.; Kele, H.; Reiners, K.; Toyka, K.V. Creatine Monohydrate in DM2/PROMM: A Double-Blind Placebo-Controlled Clinical Study. Proximal Myotonic Myopathy. *Neurology* **2003**, *60*, 500–502. [CrossRef]

124. Walter, M.C.; Lochmüller, H.; Reilich, P.; Klopstock, T.; Huber, R.; Hartard, M.; Hennig, M.; Pongratz, D.; Müller-Felber, W. Creatine Monohydrate In Muscular Dystrophies: A Double-Blind, Placebo-Controlled Clinical Study. *Neurology* **2000**, *54*, 1848–1850. [CrossRef]

125. Banerjee, B.; Sharma, U.; Balasubramanian, K.; Kalaivani, M.; Kalra, V.; Jagannathan, N.R. Effect of Creatine Monohydrate in Improving Cellular Energetics and Muscle Strength in Ambulatory Duchenne Muscular Dystrophy Patients: A Randomized, Placebo-Controlled 31P MRS Study. *Magn. Reson. Imaging* **2010**, *28*, 698–707. [CrossRef]

126. Zucchi, E.; Bonetto, V.; Sorarù, G.; Martinelli, I.; Parchi, P.; Liguori, R.; Mandrioli, J. Neurofilaments in Motor Neuron Disorders: Towards Promising Diagnostic and Prognostic Biomarkers. *Mol. Neurodegener.* **2020**, *15*, 1–20. [CrossRef] [PubMed]

127. Gittings, L.M.; Sattler, R. Recent Advances in Understanding Amyotrophic Lateral Sclerosis and Emerging Therapies. *Fac. Rev.* **2020**, *9*, 12. [CrossRef]

128. Tarnopolsky, M.A.; Beal, M.F. Potential for Creatine and Other Therapies Targeting Cellular Energy Dysfunction in Neurological Disorders. *Ann. Neurol.* **2001**, *49*, 561–574. [CrossRef] [PubMed]

129. Wyss, M.; Schulze, A. Health Implications of Creatine: Can Oral Creatine Supplementation Protect against Neurological and Atherosclerotic Disease? *Neuroscience* **2002**, *112*, 243–260. [CrossRef]

130. Beal, M.F. Neuroprotective Effects of Creatine. *Amino Acids* **2011**, *40*, 1305–1313. [CrossRef] [PubMed]

131. Klopstock, T.; Elstner, M.; Bender, A. Creatine in Mouse Models of Neurodegeneration and Aging. *Amino Acids* **2011**, *40*, 1297–1303. [CrossRef]

132. Hervias, I.; Beal, M.F.; Manfredi, G. Mitochondrial Dysfunction and Amyotrophic Lateral Sclerosis. *Muscle Nerve* **2006**, *33*, 598–608. [CrossRef]

133. Klivenyi, P.; Ferrante, R.J.; Matthews, R.T.; Bogdanov, M.B.; Klein, A.M.; Andreassen, O.A.; Mueller, G.; Wermer, M.; Kaddurah-Daouk, R.; Beal, M.F. Neuroprotective Effects of Creatine in a Transgenic Animal Model of Amyotrophic Lateral Sclerosis. *Nat. Med.* **1999**, *5*, 347–350. [CrossRef]

134. Groeneveld, G.J.; Veldink, J.H.; van der Tweel, I.; Kalmijn, S.; Beijer, C.; de Visser, M.; Wokke, J.H.J.; Franssen, H.; van den Berg, L.H. A Randomized Sequential Trial of Creatine in Amyotrophic Lateral Sclerosis. *Ann. Neurol.* **2003**, *53*, 437–445. [CrossRef] [PubMed]

135. Rosenfeld, J.; King, R.M.; Jackson, C.E.; Bedlack, R.S.; Barohn, R.J.; Dick, A.; Phillips, L.H.; Chapin, J.; Gelinas, D.F.; Lou, J.-S. Creatine Monohydrate in ALS: Effects on Strength, Fatigue, Respiratory Status and ALSFRS. *Amyotroph. Lateral Scler.* **2008**, *9*, 266–272. [CrossRef]

136. Shefner, J.M.; Cudkowicz, M.E.; Schoenfeld, D.; Conrad, T.; Taft, J.; Chilton, M.; Urbinelli, L.; Qureshi, M.; Zhang, H.; Pestronk, A.; et al. A Clinical Trial of Creatine in ALS. *Neurology* **2004**, *63*, 1656–1661. [CrossRef]

137. Ostojic, S.M. Creatine and Multiple Sclerosis. *Nutr. Neurosci.* **2020**, 1–8. [CrossRef]

138. Arnold, D.L.; Riess, G.T.; Matthews, P.M.; Francis, G.S.; Collins, D.L.; Wolfson, C.; Antel, J.P. Use of Proton Magnetic Resonance Spectroscopy for Monitoring Disease Progression in Multiple Sclerosis. *Ann. Neurol.* **1994**, *36*, 76–82. [CrossRef]

139. Beer, M.; Sandstede, J.; Weilbach, F.; Spindler, M.; Buchner, S.; Krug, A.; Köstler, H.; Pabst, T.; Kenn, W.; Landschütz, W.; et al. Cardiac Metabolism and Function in Patients with Multiple Sclerosis: A Combined 31P-MR-Spectroscopy and MRI Study. *RoFo* **2001**, *173*, 399–404. [CrossRef] [PubMed]

140. Joanna, L.; James, P.; Anthony, A.; Garnette, R.S. Nuclear Magnetic Resonance Study of Cerebrospinal Fluid From Patients With Multiple Sclerosis. *Can. J. Neurol. Sci.* **1993**, *20*, 194–198. [CrossRef]

141. Lambert, C.P.; Archer, R.L.; Carrithers, J.A.; Fink, W.J.; Evans, W.J.; Trappe, T.A. Influence of Creatine Monohydrate Ingestion on Muscle Metabolites and Intense Exercise Capacity in Individuals with Multiple Sclerosis. *Arch. Phys. Med. Rehabil.* **2003**, *84*, 1206–1210. [CrossRef]

142. Malin, S.K.; Cotugna, N.; Fang, C.-S. Effect of Creatine Supplementation on Muscle Capacity in Individuals with Multiple Sclerosis. *J. Diet. Suppl.* **2008**, *5*, 20–32. [CrossRef]

143. Jankovic, J. Parkinson's Disease: Clinical Features and Diagnosis. *J. Neurol. Neurosurg. Psychiatry* **2008**, *79*, 368–376. [CrossRef] [PubMed]

144. Marras, C.; Beck, J.C.; Bower, J.H.; Roberts, E.; Ritz, B.; Ross, G.W.; Abbott, R.D.; Savica, R.; Van Den Eeden, S.K.; Willis, A.W.; et al. Prevalence of Parkinson's Disease across North America. *NPJ Parkinson's Dis.* **2018**, *4*, 1–7. [CrossRef]

145. Park, A.; Stacy, M. Disease-Modifying Drugs in Parkinson's Disease. *Drugs* **2015**, *75*, 2065–2071. [CrossRef] [PubMed]

146. Klein, A.M.; Ferrante, R.J. The Neuroprotective Role of Creatine. *Subcell Biochem.* **2007**, *46*, 205–243. [CrossRef]

147. Bender, A.; Koch, W.; Elstner, M.; Schombacher, Y.; Bender, J.; Moeschl, M.; Gekeler, F.; Müller-Myhsok, B.; Gasser, T.; Tatsch, K.; et al. Creatine Supplementation in Parkinson Disease: A Placebo-Controlled Randomized Pilot Trial. *Neurology* **2006**, *67*, 1262–1264. [CrossRef] [PubMed]

148. Li, Z.; Wang, P.; Yu, Z.; Cong, Y.; Sun, H.; Zhang, J.; Zhang, J.; Sun, C.; Zhang, Y.; Ju, X. The Effect of Creatine and Coenzyme Q10 Combination Therapy on Mild Cognitive Impairment in Parkinson's Disease. *Eur. Neurol.* **2015**, *73*, 205–211. [CrossRef] [PubMed]

149. NINDS NET-PD Investigators. A Randomized, Double-Blind, Futility Clinical Trial of Creatine and Minocycline in Early Parkinson Disease. *Neurology* **2006**, *66*, 664–671. [CrossRef] [PubMed]

150. NINDS NET-PD Investigators. A Pilot Clinical Trial of Creatine and Minocycline in Early Parkinson Disease: 18-Month Results. *Clin. Neuropharmacol.* **2008**, *31*, 141–150. [CrossRef]

151. Writing Group for the NINDS Exploratory Trials in Parkinson Disease (NET-PD), Investigators; Kieburtz, K.; Tilley, B.C.; Elm, J.J.; Babcock, D.; Hauser, R.; Ross, G.W.; Augustine, A.H.; Augustine, E.U.; Aminoff, M.J.; et al. Effect of Creatine Monohydrate on Clinical Progression in Patients with Parkinson Disease: A Randomized Clinical Trial. *JAMA* **2015**, *313*, 584–593. [CrossRef]

152. McCorquodale, D.; Pucillo, E.M.; Johnson, N.E. Management of Charcot-Marie-Tooth Disease: Improving Long-Term Care with a Multidisciplinary Approach. *J. Multidiscip. Healthc.* **2016**, *9*, 7–19. [CrossRef] [PubMed]
153. Newman, C.J.; Walsh, M.; O'Sullivan, R.; Jenkinson, A.; Bennett, D.; Lynch, B.; O'Brien, T. The Characteristics of Gait in Charcot-Marie-Tooth Disease Types I and II. *Gait Posture* **2007**, *26*, 120–127. [CrossRef]
154. Doherty, T.J.; Lougheed, K.; Markez, J.; Tarnopolsky, M.A. Creatine Monohydrate Does Not Increase Strength in Patients with Hereditary Neuropathy. *Neurology* **2001**, *57*, 559–560. [CrossRef]
155. Chetlin, R.D.; Gutmann, L.; Tarnopolsky, M.A.; Ullrich, I.H.; Yeater, R.A. Resistance Training Exercise and Creatine in Patients with Charcot-Marie-Tooth Disease. *Muscle Nerve* **2004**, *30*, 69–76. [CrossRef] [PubMed]
156. Smith, C.A.; Chetlin, R.D.; Gutmann, L.; Yeater, R.A.; Alway, S.E. Effects of Exercise and Creatine on Myosin Heavy Chain Isoform Composition in Patients with Charcot–Marie–Tooth Disease. *Muscle Nerve* **2006**, *34*, 586–594. [CrossRef] [PubMed]
157. DeFreitas, J.M.; Beck, T.W.; Stock, M.S.; Dillon, M.A.; Kasishke, P.R. An Examination of the Time Course of Training-Induced Skeletal Muscle Hypertrophy. *Eur. J. Appl. Physiol.* **2011**, *111*, 2785–2790. [CrossRef]
158. Dubé, B.-P.; Laveneziana, P. Effects of Aging and Comorbidities on Nutritional Status and Muscle Dysfunction in Patients with COPD. *J. Thorac. Dis.* **2018**, *10*, S1355–S1366. [CrossRef]
159. Matsuo, H.; Yoshimura, Y.; Fujita, S.; Maeno, Y. Risk of Malnutrition Is Associated with Poor Physical Function in Patients Undergoing Cardiac Rehabilitation Following Heart Failure. *Nutr. Diet.* **2019**, *76*, 82–88. [CrossRef] [PubMed]
160. Springer, J.; Springer, J.-I.; Anker, S.D. Muscle Wasting and Sarcopenia in Heart Failure and beyond: Update 2017. *Esc Heart Fail.* **2017**, *4*, 492–498. [CrossRef] [PubMed]
161. Balestrino, M. Role of Creatine in the Heart: Health and Disease. *Nutrients* **2021**, *13*, 1215. [CrossRef]
162. Clarke, H.; Hickner, R.C.; Ormsbee, M.J. The Potential Role of Creatine in Vascular Health. *Nutrients* **2021**, *13*, 857. [CrossRef]
163. Fuld, J.P.; Kilduff, L.P.; Neder, J.A.; Pitsiladis, Y.; Lean, M.E.J.; Ward, S.A.; Cotton, M.M. Creatine Supplementation during Pulmonary Rehabilitation in Chronic Obstructive Pulmonary Disease. *Thorax* **2005**, *60*, 531–537. [CrossRef]
164. Andrews, R.; Greenhaff, P.; Curtis, S.; Perry, A.; Cowley, A.J. The Effect of Dietary Creatine Supplementation on Skeletal Muscle Metabolism in Congestive Heart Failure. *Eur. Heart. J.* **1998**, *19*, 617–622. [CrossRef] [PubMed]
165. Gordon, A.; Hultman, E.; Kaijser, L.; Kristjansson, S.; Rolf, C.J.; Nyquist, O.; Sylvén, C. Creatine Supplementation in Chronic Heart Failure Increases Skeletal Muscle Creatine Phosphate and Muscle Performance. *Cardiovasc. Res.* **1995**, *30*, 413–418. [CrossRef]
166. Kuethe, F.; Krack, A.; Richartz, B.M.; Figulla, H.R. Creatine Supplementation Improves Muscle Strength in Patients with Congestive Heart Failure. *Pharmazie* **2006**, *61*, 218–222. [PubMed]
167. Carvalho, A.P.P.F.; Rassi, S.; Fontana, K.E.; de Sousa Correa, K.; Feitosa, R.H.F. Influence of Creatine Supplementation on the Functional Capacity of Patients with Heart Failure. *Arq. Bras. Cardiol.* **2012**, *99*, 623–629. [CrossRef]
168. Deacon, S.J.; Vincent, E.E.; Greenhaff, P.L.; Fox, J.; Steiner, M.C.; Singh, S.J.; Morgan, M.D. Randomized Controlled Trial of Dietary Creatine as an Adjunct Therapy to Physical Training in Chronic Obstructive Pulmonary Disease. *Am. J. Respir. Crit. Care Med.* **2008**, *178*, 233–239. [CrossRef] [PubMed]
169. Al-Ghimlas, F.; Todd, D.C. Creatine Supplementation for Patients with COPD Receiving Pulmonary Rehabilitation: A Systematic Review and Meta-Analysis: Creatine Supplementation in COPD. *Respirology* **2010**, *15*, 785–795. [CrossRef]
170. Cornelissen, V.A.; Defoor, J.G.M.; Stevens, A.; Schepers, D.; Hespel, P.; Decramer, M.; Mortelmans, L.; Dobbels, F.; Vanhaecke, J.; Fagard, R.H.; et al. Effect of Creatine Supplementation as a Potential Adjuvant Therapy to Exercise Training in Cardiac Patients: A Randomized Controlled Trial. *Clin. Rehabil.* **2010**, *24*, 988–999. [CrossRef]
171. Faager, G.; Söderlund, K.; Sköld, C.M.; Rundgren, S.; Tollbäck, A.; Jakobsson, P. Creatine Supplementation and Physical Training in Patients with COPD: A Double Blind, Placebo-Controlled Study. *Int. J. Copd* **2006**, *1*, 445–453. [CrossRef] [PubMed]
172. Gosselink, R.; Spruit, M.A.; Troosters, T.; Klatka, D.; Sliwinski, P.; Nowinski, A.; Zielinski, J.; Decramer, M. Oral Creatine Supplementation (CR) Does Not Enhance the Effects of Exercise Training in COPD. *Eur. Respir. J.* **2003**, *22*, 2089.
173. Hemati, F.; Rahmani, A.; Asadollahi, K.; Soleimannejad, K.; Khalighi, Z. Effects of Complementary Creatine Monohydrate and Physical Training on Inflammatory and Endothelial Dysfunction Markers Among Heart Failure Patients. *Asian J. Sports Med.* **2016**, *7*, e28578. [CrossRef]
174. Ostojic, S.M. Can Creatine Help in Pulmonary Rehabilitation after COVID-19? *Adv. Respir. Dis.* **2020**, *14*, 175346662097114. [CrossRef] [PubMed]
175. Griffiths, T.L.; Proud, D. Creatine Supplementation as an Exercise Performance Enhancer for Patients with COPD? An Idea to Run With. *Thorax* **2005**, *60*, 525–526. [CrossRef] [PubMed]
176. Bourgeois, J.; Tarnopolsky, M. Creatine Supplementation in Mitochondrial Cytopathies. *Med. Sci. Symp. Ser.* **2000**, *14*, 91–100.
177. Tarnopolsky, M.A. What Can Metabolic Myopathies Teach Us about Exercise Physiology? *Appl. Physiol. Nutr. Metab.* **2006**, *31*, 21–30. [CrossRef]
178. El-Hattab, A.W.; Scaglia, F. Mitochondrial Cytopathies. *Cell Calcium.* **2016**, *60*, 199–206. [CrossRef]
179. Schmiedel, J.; Jackson, S.; Schäfer, J.; Reichmann, H. Mitochondrial Cytopathies. *J. Neurol.* **2003**, *250*, 267–277. [CrossRef]
180. DiMauro, S.; Hirano, M.; Schon, E.A. Approaches to the Treatment of Mitochondrial Diseases. *Muscle. Nerve* **2006**, *34*, 265–283. [CrossRef] [PubMed]
181. DiMauro, S.; Mancuso, M. Mitochondrial Diseases: Therapeutic Approaches. *Biosci. Rep.* **2007**, *27*, 125–137. [CrossRef] [PubMed]
182. Arnold, D.L.; Taylor, D.J.; Radda, G.K. Investigation of Human Mitochondrial Myopathies by Phosphorus Magnetic Resonance Spectroscopy. *Ann. Neurol.* **1985**, *18*, 189–196. [CrossRef]

183. Matthews, P.M.; Allaire, C.; Shoubridge, E.A.; Karpati, G.; Carpenter, S.; Arnold, D.L. In Vivo Muscle Magnetic Resonance Spectroscopy in the Clinical Investigation of Mitochondrial Disease. *Neurology* **1991**, *41*, 114–120. [CrossRef]
184. Kreider, R.B.; Stout, J.R. Creatine in Health and Disease. *Nutrients* **2021**, *13*, 447. [CrossRef]
185. Matthews, R.T.; Yang, L.; Jenkins, B.G.; Ferrante, R.J.; Rosen, B.R.; Kaddurah-Daouk, R.; Beal, M.F. Neuroprotective Effects of Creatine and Cyclocreatine in Animal Models of Huntington's Disease. *J. Neurosci.* **1998**, *18*, 156–163. [CrossRef]
186. Tarnopolsky, M.A.; Roy, B.D.; MacDonald, J.R. A Randomized, Controlled Trial of Creatine Monohydrate in Patients with Mitochondrial Cytopathies. *Muscle Nerve* **1997**, *20*, 1502–1509. [CrossRef]
187. Klopstock, T.; Querner, V.; Schmidt, F.; Gekeler, F.; Walter, M.; Hartard, M.; Henning, M.; Gasser, T.; Pongratz, D.; Straube, A.; et al. A Placebo-Controlled Crossover Trial of Creatine in Mitochondrial Diseases. *Neurology* **2000**, *55*, 1748–1751. [CrossRef] [PubMed]
188. Kornblum, C.; Schröder, R.; Müller, K.; Vorgerd, M.; Eggers, J.; Bogdanow, M.; Papassotiropoulos, A.; Fabian, K.; Klockgether, T.; Zange, J. Creatine Has No Beneficial Effect on Skeletal Muscle Energy Metabolism in Patients with Single Mitochondrial DNA Deletions: A Placebo-Controlled, Double-Blind 31P-MRS Crossover Study. *Eur. J. Neurol.* **2005**, *12*, 300–309. [CrossRef]
189. Pfeffer, G.; Majamaa, K.; Turnbull, D.M.; Thorburn, D.; Chinnery, P.F. Treatment for Mitochondrial Disorders. *Cochrane Database Syst. Rev.* **2012**, *4*, CD004426. [CrossRef] [PubMed]
190. Cup, E.H.; Pieterse, A.J.; Ten Broek-Pastoor, J.M.; Munneke, M.; van Engelen, B.G.; Hendricks, H.T.; van der Wilt, G.J.; Oostendorp, R.A. Exercise Therapy and Other Types of Physical Therapy for Patients with Neuromuscular Diseases: A Systematic Review. *Arch. Phys. Med. Rehabil.* **2007**, *88*, 1452–1464. [CrossRef] [PubMed]

nutrients

MDPI

Review

Creatine Supplementation in Women's Health: A Lifespan Perspective

Abbie E Smith-Ryan [1,2,*], Hannah E Cabre [1,2], Joan M Eckerson [3] and Darren G Candow [4]

[1] Applied Physiology Laboratory, Department of Exercise and Sport Science, University of North Carolina at Chapel Hill, Chapel Hill, NC 27713, USA; saylor16@live.unc.edu
[2] Human Movement Science Curriculum, Department of Allied Health Science, University of North Carolina at Chapel Hill, Chapel Hill, NC 27713, USA
[3] Department of Exercise Science and Pre-Health Professions, Creighton University, Omaha, NE 68178, USA; joaneckerson@creighton.edu
[4] Aging Muscle & Bone Laboratory, Faculty of Kinesiology & Healthy Studies, University of Regina, Regina, SK S4S 0A2, Canada; darren.candow@uregina.ca
* Correspondence: abbsmith@email.unc.edu; Tel.: +1-919-962-2574

Abstract: Despite extensive research on creatine, evidence for use among females is understudied. Creatine characteristics vary between males and females, with females exhibiting 70–80% lower endogenous creatine stores compared to males. Understanding creatine metabolism pre- and post-menopause yields important implications for creatine supplementation for performance and health among females. Due to the hormone-related changes to creatine kinetics and phosphocreatine resynthesis, supplementation may be particularly important during menses, pregnancy, post-partum, during and post-menopause. Creatine supplementation among pre-menopausal females appears to be effective for improving strength and exercise performance. Post-menopausal females may also experience benefits in skeletal muscle size and function when consuming high doses of creatine $(0.3 \text{ g·kg}^{-1} \cdot \text{d}^{-1})$; and favorable effects on bone when combined with resistance training. Pre-clinical and clinical evidence indicates positive effects from creatine supplementation on mood and cognition, possibly by restoring brain energy levels and homeostasis. Creatine supplementation may be even more effective for females by supporting a pro-energetic environment in the brain. The purpose of this review was to highlight the use of creatine in females across the lifespan with particular emphasis on performance, body composition, mood, and dosing strategies.

Keywords: female; dietary supplement; menstrual cycle; hormones; exercise performance; menopause; pregnancy; mood; cognition

Citation: Smith-Ryan, A.E; Cabre, H.E; Eckerson, J.M; Candow, D.G. Creatine Supplementation in Women's Health: A Lifespan Perspective. *Nutrients* **2021**, *13*, 877. https://doi.org/10.3390/nu13030877

Academic Editor: Patrick Diel

Received: 15 January 2021
Accepted: 5 March 2021
Published: 8 March 2021

Publisher's Note: MDPI stays neutral with regard to jurisdictional claims in published maps and institutional affiliations.

1. Introduction

Dietary supplement use has repeatedly been reported to be highest among educated women, and also appears to increase with age [1]. Creatine has been reported as one of the most commonly used dietary sports supplements. The ergogenic potential of creatine can be attributed to several mechanisms, and may have different effects on males and females. Innately, creatine is an essential substrate for the creatine kinase reaction to catalyze adenosine triphosphate (ATP) production from creatine and phosphocreatine (PCr). This recycling also serves as an endogenous metabolic buffer helping to maintain pH [2], and both mechanisms can support cross-bridge recycling and energy availability during exercise. Creatine concentrations in the central nervous system are also notable, which may support neural function in adaptations to exercise.

Despite widespread use and decades of research related to creatine, its effects in females are not well understood. Creatine characteristics vary between males and females. For example, females have/exhibit 70–80% lower endogenous creatine stores than males [3]. Females have also been reported to consume significantly lower amounts of dietary creatine compared to males [3], indicating that females may benefit from creatine supplementation as a strategy/means to increase endogenous stores. Interestingly, females have higher reported (~10%) resting levels of intramuscular creatine concentrations compared to males [4], which could theoretically lower their responsiveness to supplementation and require higher dosages compared to males [5]. In addition, creatine supplementation has not been shown to effectively reduce amino acid oxidation and measures of protein breakdown following exercise in females, which has been reported in males [6]. Therefore, the ergogenic potential of creatine among females has been questioned.

The current body of literature that has evaluated the effect of creatine supplementation in females suggests that the risk-to-benefit ratio is low [7], with most studies indicating that there are numerous metabolic, hormonal, and neurological benefits. The lack of discussion and exploration of creatine use among females across the lifespan is a disadvantage and missed opportunity, since a better understanding of creatine metabolism pre- and post-menopause yields important implications for improving health and exercise performance for females across their lifetime. Therefore, the aim of this review was to highlight the use of creatine in females from young adulthood to old age.

2. Creatine Homeostasis across the Lifespan

As a result of hormone-driven changes throughout various stages of female reproduction, endogenous creatine synthesis, creatine transport, creatine kinase kinetics, and creatine bioavailability are altered over time, highlighting the potential positive implications for dietary creatine supplementation for females [8]. The implications of hormone-related changes in creatine kinetics have been largely overlooked in performance-based studies [8]. Specifically, creatine supplementation may be of particular importance during menses, pregnancy, post-partum, during and post-menopause. The menstrual cycle may influence creatine homeostasis due to the cyclical nature of sex hormone regulation (Figure 1). Studies conducted in animal models have demonstrated that the expression of arginine-glycine aminotransferase (AGAT), the rate limiting step of creatine synthesis, is influenced by estrogen and testosterone levels [9]. Sex hormones, predominantly estrogen and progesterone, have been shown to effect creatine kinase activities and the expression of key enzymes for the endogenous synthesis of creatine [10].

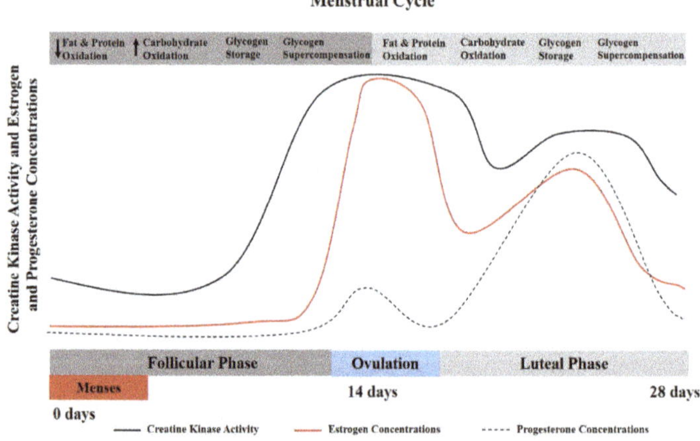

Figure 1. A theoretical model for the interplay between creatine kinase and menstrual cycle hormones [6,10,11]. Creatine metabolism and creatine kinase concentrations vary throughout the menstrual cycle and lifecycle. These alterations may also influence metabolic characteristics of protein and carbohydrate oxidation [12], which provides a physiological basis for the potential use of creatine supplementation for females.

Serum creatine kinase levels are significantly elevated during menstruation [11] compared to non-menstruating years, with creatine kinase levels decreasing with age and pregnancy. The lowest concentrations of creatine kinase values have been reported during early pregnancy (20 weeks or less), equating to about half the concentration found at peak levels (pre-menarche teenage girls) [11,13]. For a more detailed discussion related to CK in women, see Ellery et al. [8]. It has been indicated in rodent models that creatine kinase activity (and possibly creatine metabolism) synchronously increase and decrease with estrogen levels [14]. During the luteal phase when estrogen levels are at their peak, muscle damage may be reduced after eccentric exercise due to creatine kinase sparing [15]. Implications of creatine supplementation and creatine metabolism with respect to the menstrual and reproductive cycle warrant further exploration. The interplay between creatine metabolism and CK kinetics may be particularly important for females with low estrogen concentrations (follicular phase), amenorrhea, during pregnancy, and with the transition to through menopause.

Additional consideration should be given to the metabolic changes associated with a normal menstrual cycle. Estrogen is considered a master regulator of bioenergetics, with the highest levels occurring during the luteal phase of the cycle (begins just after ovulation and goes through the end of the cycle). Protein catabolism and oxidation has been shown to be elevated during this high estrogen phase (luteal); while carbohydrate storage has been shown to be reduced during the luteal phase [16]. Mechanistic support for creatine supplementation has been reported to involve muscle protein kinetics, growth factors, satellite cells, myogenic transcription factors, glycogen and calcium regulation, oxidative stress and inflammation [17,18]. Given increased protein turnover and challenges with glycogen saturation, creatine supplementation may be even more effective in the high estrogen/luteal phase.

3. Creatine Use among Pre-Menopausal Women

A considerable amount of evidence indicates that creatine is an effective ergogenic aid for increasing strength, power, and athletic performance in females without marked changes in body weight [5,19–21]. The reluctance among females to use creatine may be due to a fear of weight gain or other adverse side effects, which are largely unfounded, particularly in women [20]. This rapid weight gain is more prevalent among males; weight may rapidly and temporarily increase with a loading dose which reflects an increase in cellular hydration (i.e., water weight) [22]. This is a positive aspect for increasing hydration [22]. Weight gain may also result if creatine is consumed with a commonly recommended 1.0 g·kg^{-1} body weight of carbohydrate [17]; this is likely not the best strategy for supplementation in females (see dosing section). When reviewing the literature that has examined the effect of creatine supplementation on a variety of performance indices in females, the benefits firmly outweigh any associated risks or reported adverse events.

The potential for adverse effects from creatine supplementation are largely unfounded. An extensive recent systematic review clearly outlined the lack of adverse effect of creatine supplementation on the gastrointestinal, renal, hepatic, or cardiovascular systems among women supplementing with creatine [20]. The findings in women appear to be similar for men, supporting creatine as a safe, low risk dietary supplement when consumed in recommended doses and regimens [7,20].

Creatine supplementation is most effective for high-intensity, short duration activities or repeated bouts of high-intensity exercise with short rest periods such as jumping, sprinting, and resistance training, since increased levels PCr can more rapidly re-phosphorylate adenosine diphosphate to ATP via the creatine kinase reaction. In addition, PCr buffers hydrogen (H^{+}) ions that accumulate during high-intensity exercise and may delay the onset of fatigue. In practice, the increase in intramuscular PCr stores through creatine supplementation allows for a greater stimulus for training which results in physiological adaptations that lead to increases in muscle mass, strength, and muscle fiber hypertrophy [23]. To illustrate the effects of creatine monohydrate (CrM) supplement on exercise performance in females, its relative effects (RE) were calculated and are presented in Figures 2–4; RE was calculated using the following equation [24,25]:

$$RE = \left(\frac{\left(\frac{Post_{Cr}}{Pre_{Cr}} \right) \times 100}{\left(\frac{Post_{PL}}{Pre_{PL}} \right) \times 100} \right) \times 100$$

where Pre_{PL} is the pre-test value in the placebo group, $Post_{PL}$ is the post-test value in the placebo group, Pre_{Cr} is the pre-test value in the creating group, and $Post_{Cr}$ is the post-test value in the creatine group. A relative effect greater than 100 represents an increase or improvement in performance with creatine supplementation.

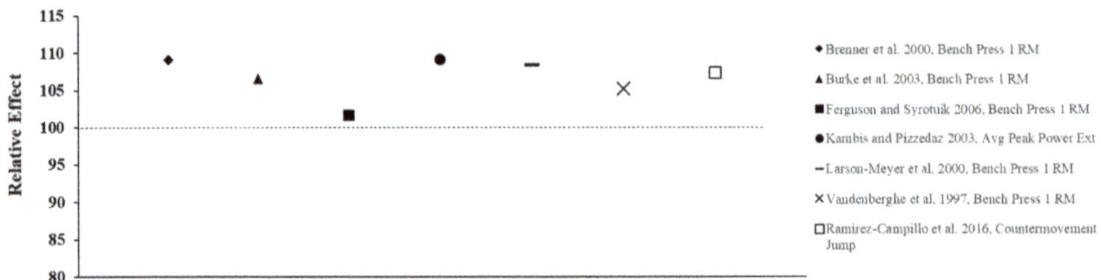

Figure 2. The relative effects of creatine supplementation in comparison to placebo for strength performance in females.

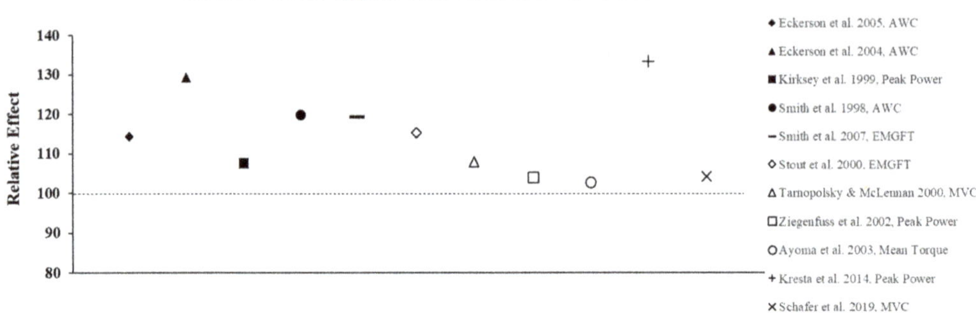

Figure 3. The relative effects of creatine supplementation in comparison to a placebo on exercise performance in females.

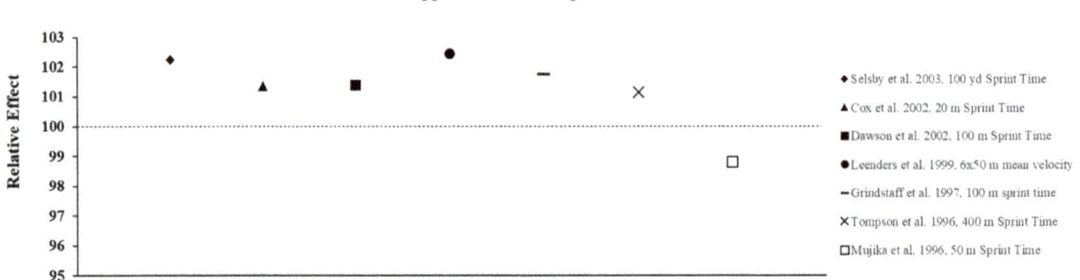

Figure 4. The relative effects of creatine supplementation in comparison to placebo on sports performance in females.

3.1. Creatine Supplementation and Strength Performance

Although studies using physically active and highly trained females as participants are lacking, both short- and long-term creatine supplementation has been shown to result in significant improvements in muscular strength and power. One of the first comprehensive studies to investigate the effects of creatine on strength performance in females was conducted by Vandenberghe et al. [26]. Healthy, sedentary females ingested CrM tablets or a maltodextrin placebo (PL) four times daily (20 $g \cdot d^{-1}$) for 4-d, followed by a maintenance dose (5 $g \cdot d^{-1}$) while participating in a 10-wk(week) resistance training program, three times per wk for 1-hr (5 sets, 12 repetitions @ 70% RM for leg press (LP), shoulder press, squat, leg extension (LE), leg curl, and bench press (BP)), followed by a 10-wk detraining and supplementation phase. The CrM group demonstrated a 6% increase ($p < 0.05$) in muscle PCr following the loading period; and the increase after 5-wk and 10-wk of training was 7% and 10% higher than baseline levels, respectively. There were no significant differences in torque output for the forearm flexors between groups following loading, however, the CrM group demonstrated significantly greater torque values at all other time points (5-wk, 10-wk, detraining) until the end of the study when supplementation had stopped for 4-wk. Results also showed that the increase in 1RM for LP, LE, and squat at 10-wk were 20–25% greater ($p < 0.05$) in the CrM group versus PL. The results for body composition showed that there were no significant differences between groups for BW and % fat; however, the change in fat free mass (FFM) was greater ($p < 0.05$) in the CrM group after both 5-wk (2.0 kg) and 10-wk (2.6 kg) of training compared to PL (1.1 kg and 1.6 kg, respectively). These findings suggest that 10-wk of CrM supplementation was effective for increasing lower-body strength and FFM beyond training alone in untrained females [26], and helped maintain strength during detraining without any significant effects on body weight or percent body fat.

Studies have also shown that CrM is effective for increasing strength performance in trained females. In a study using collegiate female soccer players ($n = 14$), 5-d CrM loading (15 $g \cdot d^{-1}$) followed by a 12-wk maintenance dose (5 $g \cdot d^{-1}$), during 13-wk of resistance training led to significant increases in 1-RM strength for the BP (18%) and squat (24%) compared to PL (9% and 12% increases for BP and squat, respectively) with both groups demonstrating similar increases in BW and FFM [27]. In a related study, Brenner et al. [28] reported significantly greater increases (mean difference: 3.4 kg) in 1-RM BP among NCAA Division I female lacrosse players completing a CrM loading phase (4×5 $g \cdot d^{-1}$ for 7-d (days)), followed by 2 $g \cdot d^{-1}$ for 4 wks. Both groups demonstrated a similar increase in BW (0.50 kg; $p < 0.05$).

Short-term CrM supplementation has also been shown to significantly increase muscular power in females. In a study by Kambis and Pizzedaz [29], 22 college-aged females randomly received a CrM loading dose relative to FFM (0.5 $g \cdot kg^{-1}$ FFM divided into four equal doses) or a PL for 5-d and were tested for isokinetic strength of the preferred quadriceps group, thigh circumference, and BW. The results showed that time to peak torque for LE significantly decreased ($p < 0.05$) and that average power in LE and leg flexion significantly increased in the CrM group compared to PL. Similar to the findings of others [26–28,30,31], there were no significant differences between groups for changes in body weight, FFM, % fat, as well as mid-quadriceps circumference or skinfold thickness of the measured thigh and suggest that CrM significantly improved muscular power without associated changes in BW or muscle volume.

Not all studies conducted in pre-menopausal women report an additive effect of CrM supplementation on strength performance in females compared to training alone. Ferguson and Syrotuik [30] reported that 10-wk of CrM supplementation (0.3 $g \cdot kg^{-1}$ for 7-d + 0.03 $g \cdot kg^{-1}$ for 9 wk) in combination with resistance training (4 $d \cdot wk^{-1}$ for 9-wk) had no additional effects on strength or body composition compared to training alone in physically active females (18–35 yrs). The results showed that both the PL and CrM groups experienced similar increases in strength and FFM without a significant change in BW over the 10-wk study. The authors suggested that the lack of non-significant findings may have been due to non-responders in the CrM group, an insufficient loading dose, insufficient training volume, or a combination of these factors [30]. Wilborn et al. [31] also recently reported that CrM and whey protein (PRO) supplementation did not enhance training adaptations to an 8-wk split-body strength training program compared to PRO alone in females. The subjects trained 4-d per week and ingested either 24 g PRO ($n = 9$) or 24 g PRO + 5 g Cr ($n = 8$) following each exercise session. The results showed that both groups demonstrated a 2.5 kg increase FFM, with no significant differences between groups for strength or lower body measures of power. The authors [31] acknowledged that the lack of a PL group was a limitation of the study, and suggested that the finding of no additive effect of CrM may have been due to higher baseline levels of intramuscular stores of creatine in their trained subjects, since it has previously been reported that individuals with lower endogenous levels have a greater capacity to increase creatine following supplementation [32].

Although not all studies show that CrM supplementation has an additive effect on adaptations to resistance training compared to training alone, there is substantial evidence to suggest that CrM supplementation is effective for increasing strength and power in both trained and untrained females, without large fluctuations in BW or FFM. Relative effects of creatine supplementation on strength performance demonstrate a consistent improvement in performance compared to placebo (Figure 2). The ergogenic effect of CrM can be attributed to an increase in intramuscular PCr stores that facilitates an increase in training intensity and enhanced recovery between successive bouts of training. A higher training stimulus over time from CrM supplementation results in greater physiological adaptations (i.e., hormonal, increased cell hydration, increased gene expression) that lead to increases in strength and hypertrophy [23].

3.2. Effect of Creatine on Exercise Performance

The ergogenic effect of Cr observed in both anerobic and aerobic exercise performance is largely attributed to its ability to prevent fatigue as a result of increased intramuscular stores of PCr, which increases ATP turnover and buffers endogenous H^+ protons to maintain pH. As a result, a number of studies have also examined the effects of CrM supplementation in females on other anaerobic indices of performance with the majority of studies showing favorable results. For example, creatine loading has been shown to improve anaerobic working capacity (AWC) estimated from the critical power test. AWC represents the maximal work potential associated with the phosphagen energy system (ATP + PCr) and, therefore, provides an estimate of anaerobic power. Eckerson et al. [33] examined the effect of 2-d and 5-d of CrM loading ($20~g \cdot d^{-1}$) on AWC in physically active females (mean age $\pm$ SD = 22 ± 5 yrs) using a double-blind, crossover design and found that 5-d of supplementation resulted in a 22% increase in AWC ($p < 0.05$), whereas the PL trial resulted in a 5% decline in performance. In a follow-up study [34] to determine if phosphate salts had a synergistic effect, AWC was increased by 13.0% and 10.8% following 6-d of loading with CrM or CrM + phosphate salts, respectively, compared to a 1.1% decline in the PL group. These findings were consistent with other studies that used physically active females as participants and reported increases in AWC ranging from 10–15% following CrM loading [35,36] (Figure 3). Tarnopolsky and MacLennan [37] also showed that short-term CrM supplementation ($20~g \cdot d^{-1} \times 4$-d) increased peak and relative peak anaerobic cycling performance (3.7%) with no gender specific responses in 24 recreationally active males and females. In a related study that used trained participants, Ziegenfuss et al. [23] found that only 3-d of CrM supplementation ($0.35~g \cdot kg^{-1}$ FFM) increased sprint cycle performance in NCAA Division I athletes and that the effect was greater in females as the sprints were repeated. In a recent study that used amateur soccer players as subjects, Ramirez-Campillo et al. [38] examined the effect of CrM supplementation and 6-wkof plyometric training on jumping, maximal and repeated sprinting, and change of direction speed performance. Females were equally and randomly assigned to one of three groups: CrM + plyometric training; PL + plyometric training; or PL only. The CrM group ingested $20~g \cdot d^{-1}$ for 1-wk in four equal doses followed by a single dose for 5-wk, and subjects in the PL groups received glucose in an identical dosing regimen. There were no changes in performance for the PL subjects; both plyometric training groups showed improvements in each of the performance indices, with the CrM group demonstrating greater improvements in jump and repeated sprint performance tests, indicating that adaptations to plyometric training were enhanced with CrM supplementation. Aoki et al. [39] examined the ergogenic effect of CrM on concurrent exercise performance in 14 females (21 $\pm$ 2 yrs) who were randomly assigned to receive either CrM (20 $g \cdot d^{-1} \times$ 5-d followed by 3 $g \cdot d^{-1} \times$ 7-d) or PL. Following the 12-d intervention, there were no differences between groups in running performance or 1RM LP following supplementation; however, there was a significant decline in the number of maximal repetitions performed during the last two sets of the repetition max test in the PL group compared to CrM. It was suggested that an increase in intramuscular stores of PCr experienced by the CrM group may have enhanced their recovery from aerobic exercise and improved resistance exercise performance during the repetition max test. The true effect of creatine on recovery has largely been unexplored. Due to sex-based differences in fatigue resistance and recovery, this would be an important area for future research.

Although creatine has not been widely investigated for its effects on endurance exercise performance, there is some evidence to suggest that it may have some ergogenic benefits. Nelson et al. [40] showed that CrM loading for 7 d resulted in a lower oxygen consumption (VO_2) at submaximal workloads, and reduced the work performed by the cardiovascular system in a study that examined the effects of CrM on cardiorespiratory responses during a graded exercise test (GXT). The results showed that CrM significantly increased total test time (20.3 $\pm$ 4 min to 21.5 $\pm$ 3.5 min) compared to PL (17.3 $\pm$ 3 min to 17.4 min $\pm$ 3 min), and that VO_2 and heart rate at the end of first five stages of the GXT were significantly lower for CrM versus no change for PL. In addition, the ventilatory threshold (VT) increased significantly from pre- to post-testing for the Cr group (66% to 78% peak VO_2), whereas the PL group demonstrated no change (70% to 68% peak VO_2). The authors [40] speculated that the decreases in sub-maximal VO_2 and heart rate were due to increased stores of PCr in muscle, which may have ultimately delayed mitochondrial respiration and lowered VO_2. In a related study, Smith et al. [41] examined the effect of CrM loading (20 $g \cdot d^{-1} \times$ 5 d) on aerobic power (VO_2 max) and critical velocity (CV), which is a theoretical velocity that can be maintained for an extended period of time using only aerobic energy stores, and was calculated in their study by having subjects complete four high speed runs to exhaustion at 90, 100, 105, and 110% of peak velocity. The results showed that CrM loading neither positively or negatively influenced VO_2 max, CV, time to exhaustion, or BW, since there were no significant differences in any of these parameters between the CrM or PL groups. It has also been shown that an increase in PCr levels following CrM supplementation delays the onset of neuromuscular fatigue (NMF), which is characterized by an increase in the electrical activity of the working muscles over time and reflects the progressive recruitment of additional motor units and/or an increase in the firing frequency of motor units that have already been recruited [42,43]. In two separate studies, Stout et al. [43] and Smith et al. [42] showed that 5-d of CrM loading significantly delayed the onset of NMF during incremental cycling exercise compared to PL using both physically active and highly trained female athletes. Both authors suggested that the delay in NMF was due to an increase in intramuscular levels of PCr levels, which may have resulted in a greater capacity to delay anaerobic glycolysis and, in turn, decreased the accumulation of lactic acid and ammonia in the working muscles and the blood. Previous studies using physically active males as subjects have shown that CrM supplementation during 4-wk of high-intensity interval training (HIIT) significantly improved VT [44] and critical power [45] compared to training alone. In contrast, Forbes et al. [46] recently reported that CrM did not augment improvements in cardiorespiratory fitness, performance, or body composition in recreationally active females following a 4-wk HIIT program. In their study [46], 17 females were randomly assigned to receive either CrM (0.3 $g \cdot kg \cdot d^{-1} \times$ 5-d followed by 0.1 $g \cdot kg \cdot d^{-1}$ for 23-d, n = 9) or PL (n = 8) and completed three HIIT sessions per week for 4-wk with 48-hr between each exercise session. HIIT improved VO$_2$peak (CrM = 10.2%; PL = 8.8%), VT (CrM = 12.7%; PL = 9.9%) and time-trial performance (CrM = −11.5%; PLA = −11.6%) with no significant differences between groups. The authors [46] suggested that the differences between their findings and those of Graef et al. [44] and Kendall et al. [45] could be associated with CrM dose, methods to assess endurance performance, and/or sex-based differences.

The studies described above suggest that females with varying levels of training and fitness may experience improvements in both anaerobic and aerobic exercise performance from both short-term and long-term creatine supplementation. Therefore, it seems likely that the ergogenic effects observed in the laboratory would carry-over to competition and allow athletes who must compete in more than one event or game on the same day, or who must compete on successive days, to recover faster and, in turn, optimize performance and improve their probability of winning. Current research suggests creatine is an effective way to improve sport performance in females (Figure 4). Given the difficulty of designing these types of studies, very few investigations have determined how CrM supplementation may influence win-loss records and the execution of skills during competition. Cox et al. [47] examined the effects of short-term CrM supplementation (4 × 5 g·d^{-1} × 6 d) on performance during a field test that simulated soccer match play using elite female soccer players from the Australian National Team and found that the athletes in the CrM group significantly improved repeated sprint performance and some agility tasks that mimicked soccer play compared to the PL group. Many studies have examined the effect of CrM supplementation on swim performance, since it is a sport that allows investigators to more closely mimic competition. While most studies show that supplementation is ineffective for improving single sprint swim performance [48–50] in a few studies that required swimmers to perform repeated sprints, improvements in the time to complete the series [51,52] and increases in work and power output [52] were reported.

4. Creatine Considerations during Pregnancy

Increased metabolic demand from growth and development during gestation, particularly from the placenta, has been associated with a reduced creatine pool [53]. Recent human data suggests a dramatic alteration in creatine homeostasis during pregnancy [54] and a reduction in creatine stores during pregnancy have been linked with low birth weight and pre-term birth [54,55].

To date, there is growing evidence in animal models that creatine supplementation during pregnancy enhances/augments neuronal cell uptake of creatine and supports mitochondrial integrity in animal offspring, thereby reducing brain injury induced by intrapartum asphyxia [55,56]. Although there are no human studies to date that have evaluated the effect of CrM supplementation during pregnancy, CrM supplementation could provide a safe, low-cost nutritional strategy for reducing intra- and post-partum complications associated with cellular energy depletion [57]. Further details for the mechanisms and implications for creatine use during pregnancy are described in this special [58].

5. Creatine Considerations for Post-Menopausal Women

The menopausal related decrease in estrogen is a main contributing factor to the age-related loss in muscle and bone mass [59] and strength (i.e., dynapenia) [60]. While the mechanisms explaining the link between estrogen levels, muscle mass, and strength remain to be determined, there is evidence to suggest that insufficient estrogen levels are associated with increased inflammation and oxidative stress [59,60] and may contribute to the blunted muscle protein synthetic and satellite cell response to anabolic stimuli (i.e., resistance training).

Creatine supplementation has been shown to act as a possible countermeasure to the menopausal related decrease in muscle, bone, and strength by reducing inflammation, oxidative stress, and serum markers of bone resorption, while also resulting in a concomitant increase in osteoblast cell activity (i.e., bone formation) [61,62]. Muscle integrity has also been upregulated with creatine use, resulting in an increase in satellite cell activity, growth factors (i.e., IGF-1), protein kinases downstream in mammalian target of rapamycin pathway, and myogenic transcription factors (for reviews see [7,18,61]). These effects have recently been explored in females, often combined with resistance training. As a result of the unique cyclical and long-term changes in estrogen across the lifespan, creatine supplementation poses an interesting therapeutic strategy for post-menopausal females.

5.1. Supplementation Only

Among post-menopausal females (65 ± 2 yrs), a short-term, high dose creatine loading period ($0.3 \text{ g·kg}^{-1} \cdot \text{d}^{-1}$ or ~20 g·d^{-1} for 7 days) augmented whole-body FFM (0.52 ± 0.05 kg), muscle strength (LP: 5.2 ± 1.8 kg, BP: 1.7 ± 0.4 kg), and sit-stand and tandem gait test performance [63]. Similar functional improvements were reported in sit-to-stand performance among post-menopausal females (60–80 years) supplementing with a similar creatine dose ($0.3 \text{ g·kg}^{-1} \cdot \text{d}^{-1}$ or ~17 g·d^{-1} for 7 days) [64]. Another related study using a similar dose (20 g·d^{-1} loading for 5 days, followed by 5 g·d^{-1} for 23 weeks) reported no differences between CrM and PL on measures of muscle mass, bone mineral, upper- and lower-body strength, or functionality in post-menopausal females (<60 years). Additionally, low-dose chronic supplementation with creatine (1 g·d^{-1} for 52 weeks) among post-menopausal females (58 ± 5 years) failed to have on effect on FFM, bone density, bone turnover, or muscle function, compared to PL [65]. Increasing the dosage of creatine to 3 g·d^{-1} (1 g dose in the morning, afternoon and evening) for an additional 52 weeks (104 weeks in total) also had no effect on the same muscle and bone measures. Furthermore, CrM had no effect on handgrip strength or the number of falls or fractures experienced among older females [66]. Collectively, it appears that a short-term high dose of creatine may have minor effects on muscle and strength among post-menopausal females.

5.2. Combined with Resistance Training

The vast majority of research involving creatine supplementation in post-menopausal females has included resistance training as part of the study design, possibly because muscle contractions (i.e., resistance training) lead to greater intramuscular creatine uptake from supplementation [67], which could augment muscle mass and performance. In post-menopausal females (>60 years), Gualano et al. [68] reported that CrM supplementation ($n = 15$; loading phase of 20 g·d^{-1} for 5 days + maintenance phase of 5 g·d^{-1} for 161 days) during supervised whole-body resistance training (7 exercises; 3 sets of 8–12 repetitions) produced greater gains (relative) in appendicular lean tissue mass and BP strength compared to PL ($n = 15$). However, CrM and resistance training had no greater effect on measures of LP strength, bone mineral density or content or serum markers of bone turnover compared to PL and resistance training. No adverse effects were reported from CrM, PL or the resistance training program. Furthermore, when compared to the creatine alone group, the combination of CrM and resistance training resulted in greater muscle accretion and strength (LP and BP) in post-menopausal females [66]. In the longest study to date, Chilibeck et al. [69] showed that CrM supplementation ($0.1 \text{ g·kg}^{-1} \cdot \text{d}^{-1}$ or ~7 g·d^{-1}) during supervised whole-body resistance training (17 exercises, 3 sets of 10 repetitions, 3 days per week for 52 weeks) reduced the rate of bone mineral density loss in the hip region and increased femoral shaft sub-periosteal width (indicator or bone strength) and upper-body strength in postmenopausal females. Additional supporting work in post-menopausal females (65 ± 5.0 yrs) showed that CrM supplementation (5 g·d^{-1}) during 12 weeks of supervised resistance training (8 exercises, 2 sets of 15 repetitions; 3 days/week) significantly increased FFM, strength (BP, leg extension, elbow flexion), and tasks of functionality (30-second chair stand, arm curl test, lying prone-to-stand test) compared to PL [70]. From a clinical perspective, Neves et al. [71] found a beneficial effect from CrM (loading phase: 20 g·d^{-1} for 7 days + maintenance phase: 5 g·d^{-1} for 79 days) during 12 weeks of lower-limb supervised resistance training on lower-limb muscle accretion and physical performance (timed-stand test) in post-menopausal females (55–65 yrs) with knee osteoarthritis compared to PL. In contrast to these studies showing some favorable effects from CrM and resistance training, Candow et al. [62,72] found no greater effect from CrM supplementation (0.1 $g \cdot kg^{-1} \cdot d^{-1}$) during 32 weeks of supervised resistance training (11 exercises, 3 sets of 10 repetitions to volitional fatigue; 3 days per week) on measures of muscle mass, bone mineral, or strength in postmenopausal females (<50 years) compared to PL. Furthermore, Pinto et al. [73] failed to show greater effects from CrM supplementation (5 $g \cdot d^{-1}$) and 12 weeks of supervised whole-body resistance training (3 sets of 13–15 repetitions) on measures of bone mineral and muscle strength. However, CrM did augment FFM more than PL over time.

Collectively, these findings suggest that post-menopausal females may experience increases in muscle mass and function when consuming high-dosage creatine (0.3 $g \cdot kg^{-1} \cdot d^{-1}$) for at least 7 consecutive days. Creatine supplementation alone or in combination with resistance training appears to provide no benefits in bone physiology in post-menopausal females. However, when combined with resistance training, the vast majority of research supports the efficacy of CrM supplementation ($\geq$5 $g \cdot d^{-1}$) for improving measures of muscle accretion, strength and tasks of physical performance in post-menopausal females. From a safety perspective, creatine poses no greater adverse effects compared to placebo. Future longer-term randomized PL controlled trials with large sample sizes are needed to fully determine whether creatine, with and without resistance training, can positively influence musculoskeletal parameters in post-menopausal females.

6. Depression and Mood

Depression rates are two times higher among females compared to males [74]. The increased prevalence of depression among females has been directly linked with hormonal milestones; major depression rates increase during puberty, during the luteal (high estrogen) phase, following pregnancy, and during perimenopause [75]. Despite the hormonal pattern, evidence suggests that this trajectory is not solely dependent on the amount of estrogen and progesterone, but rather how sensitive the brain is to these hormones [76].

Early research examining the role of dysfunctional creatine metabolism in the neurochemical foundations of depression in adults demonstrated a positive relationship between cerebral spinal fluid levels of creatine and dopamine and serotonin metabolites [77,78]. These data suggest that efficient neurotransmission of metabolites affecting mood depends upon the creatine-PCr system functioning properly. The severity of a depressive episode has been inversely linked to white matter creatine and PCr concentrations within the brain suggesting that there is a relationship between brain creatine metabolism and depression [79]. This pattern has been shown to be beneficial for anti-depressant treatment [80], suggesting dietary creatine supplementation may provide a pro-energetic effect in brain chemistry [81] through efficient regeneration of intracellular high-energy phosphates in females.

Previous research has shown that dietary creatine supplementation can promote cell survival and influence the production and usage of energy in the brain [3,82]. Clinical and pre-clinical evidence has reported positive effects of creatine supplementation on mood by restoring brain energy levels and homeostasis. Altered brain bioenergetics and mitochondrial dysfunction have been linked with depression, particularly as it relates to CK, ATP, and inorganic phosphate (P_i). In terms of energy usage, in vivo measurements of the female adult brains with a major depressive disorder, demonstrate a distinctive pattern of energy-related metabolites, specifically, a decrease in beta-nucleoside triphosphate and increased PCr level resulting from an increased use of ATP [83]. Supplementation with CrM has also been shown to significantly augment cerebral PCr and P_i [84]. Females have been reported to have lower levels of creatine in the brain, particularly the frontal lobe [85], which controls mood, cognition, memory, and emotion. As a result of sex-differences in brain creatine concentrations, supplementation may be even more effective for females for supporting a pro-energetic environment in the brain. When combined with regular antidepressant use, 8-weeks of CrM supplementation reduced depressive symptoms in female adolescents and adults with major depression [86–88]. In healthy adolescent females taking anti-depressant medication, the mean Children's Depression Rating Scale-Revised (CDRS-R) score declined from 69 to 30.6, a 56% decrease, in those consuming 4 g of CrM daily for 8 weeks [81]. Similar results were demonstrated in healthy adult females taking anti-depressant medication. There were significantly greater improvements in Hamilton Depression Rating Scale (HAM-D) score, with improvements observed after 2 weeks in females consuming $5 \text{ g} \cdot \text{d}^{-1}$ of creatine for 8 weeks [84]. This response time accelerates the effectiveness of antidepressant medications compared to the typical 4–5 weeks acclimatization to detect/identify effects with these therapeutics alone [88,89]. Dietary creatine intake is inversely proportional with depression occurrence; with a 31% greater incidence of depression in adults in the lowest quartile of creatine intake [90]. Increasing creatine concentrations in the brain as a result of increased animal protein consumption and, more effectively, through CrM supplementation, has strong evidence to support mood and depression, particularly in females. This has important relevance through various stages across the lifespan that demonstrate increased prevalence in depression as a result of cyclical hormones, including puberty, post-partum, and menopause.

7. Cognition and Sleep

Brain activity results in a rapid reduction in PCr levels to maintain ATP levels [91]. Therefore, during periods of high mental stress, which require a higher PCr demand, ATP turnover may be impaired. Creatine supplementation has been shown to support greater neural ATP resynthesis, which provides a cognitive advantage for tasks that rely on the frontal cortex (i.e., cognition, attention, memory) [92]. Brain creatine concentrations appear to be variable based/depending upon age, lifestyle choices, diet, and other factors [93], which is relevant when considering creatine supplementation for females across the lifespan.

Creatine supplementation in humans has consistently demonstrated improved cognitive performance and brain function and reduces mental fatigue during stressful mental tasks in healthy adults [94]. Greater cognitive improvements as a result of creatine supplementation have also been reported in individuals with cognitive impairments [95,96]. Creatine supplementation also appears to improve cognitive function in vegetarians due to lower brain creatine concentrations. Females process stress different than males [97], often practicing more frequent habits of multi-tasking [98] and are also more often susceptible to sleep deprivation due to pregnancy, post-partum demands, and menopausal sleep disturbances. Creatine supplementation has been shown to support these exact scenarios by augmenting mental capacity under sleep deprivation [96]. Additionally, acute and chronic sleep deprivation appear to be more detrimental to females compared to males; with lower levels of alertness and increased sleepiness-related risks [99]. Sleep deprivation has been reported to result in lower cognition, as well as a reduction in sleep quality for females during the follicular phase (low estrogen), which is also when creatine kinase levels appear to be lowest [100]. As a result of this promising research, creatine supplementation throughout the menstrual cycle may aid in attenuating its adverse effects on cognition and sleep. Cognitive and sleep benefits of creatine supplementation may be most helpful during periods of high stress and sleep deprivation.

8. Dosing Strategies

Supplementing with CrM can be accomplished using two strategies, both resulting in similar increases in intramuscular PCr levels (Table 1) [7]. A loading phase in females results in a 19% increase in total muscle creatine concentrations [67,101], which is similar to the response for males. A daily dose of 5 g is also equally effective for increasing muscle creatine stores, however this approach requires more time (~3–4 weeks) compared to a traditional loading approach (5 days) [2]. When practicing a loading phase, creatine remains elevated for about 30 days following completion of supplementation, which appears to be the same for both males and females. Based on available evidence, it appears that females can practice the same dosing strategy that is recommended for males. With creatine supplementation, there is an individual variability with the response to creatine saturation with responders, quasi-responders, and non-responders [102]. This has not yet been explored in females, but it is assumed that all individuals may respond differently.

Table 1. Dosing guidelines for creatine supplementation in females [2,67,84,101].

	Dose	Maintenance
Loading Dose	5 g 4 × daily (20 g/day; 0.3 g/kg/day) every 4 h for 5 days	3–5 g (0.03 g/kg/day) daily
Example: 150 lb female (68.2 kg)	8:00 am: 5 g 12:00 pm: 5 g 4:00 pm: 5 g 8:00 pm: 5 g	3–5 g (2.0 g/kg) daily
Routine-Consistent Dose	5 g daily	5 g daily
Brain Saturation	15–20 g daily for 3–7 days (in divided dose)	5–10 g daily

Menstrual Cycle Notes: Due to elevated protein turnover in the luteal phase, creatine supplementation/loading may support muscle protein preservation.

Skeletal muscle creatine uptake can also be influenced by insulin availability, which may enhance creatine retention [103,104]. Consuming creatine with carbohydrate (~50 g) and protein (~50 g) [105], or with $1\ g{\cdot}kg^{-1}$ of glucose, may increase total muscle creatine concentrations compared to creatine supplementation alone [106]. However, in females, the additional calories from CHO/PRO to enhance ingestion, particularly during a loading phase, may not be warranted. Specifically, for women who tend to burn fewer calories than men, if additional calories are not needed to meet training needs, the benefit from enhanced absorption, does not outweigh the potential enhanced absorptive effect; creatine monohydrate has extremely high bioavailability [107]. Additionally, due to the menstrual cycle, the lower CHO oxidation in the follicular phase may suggest the added macronutrients are not needed. One strategy is to ingest creatine with a usual meal or add it to a protein shake due to the insulin properties of amino acids.

Brain concentrations of creatine and PCr are elevated as a result of 0.13–0.80 g/kg/day for 14 days [84]. To maximize brain uptake, a loading phase (15–$20\ g{\cdot}day^{-1}$ for 3–7 days) followed by a consistent regular daily 5–10 g dose is optimal for tissue saturation. Peak absorption of creatine from supplementation is optimized when ingested as a solution vs. capsule, lozenge, or solid meat [108].

9. Conclusions and Application

As a result of changes in creatine homeostasis across the lifecycle, particularly as it relates to estrogen, creatine supplementation appears to provide many potential benefits for females. Creatine use has consistently demonstrated improvements in muscle and brain PCr levels, which has been shown to result in improvements in strength and exercise capacity. When combined with resistance training, creatine further augments body composition and bone mineral density, particularly in post-menopausal females. Creatine supplementation has also been shown to improve mood and cognition. A traditional loading dose ($0.3\ g{\cdot}day^{-1}$ 5–7 days) or a routine daily dose (5 g) for 4 weeks can be effective for females. For brain saturation, higher doses (15–$20\ g{\cdot}day^{-1}$ for 3–7 days, followed by 5–$10\ g{\cdot}day^{-1}$) of creatine are warranted. Future data evaluating more specific effects of creatine across the menstrual cycle will help to more clearly understand the varied benefits at different phases of the menstrual cycle and potential for use across the lifespan.

Author Contributions: Conceptualization, A.E.S.-R.; writing original draft preparation, A.E.S.-R., H.E.C., J.M.E., D.G.C. All authors have read and agreed to the published version of the manuscript.

Funding: The publishing fee for this manuscript was provided by Alzchem. There is no other funding to report.

Institutional Review Board Statement: No human subjects were utilized for this review paper.

Informed Consent Statement: It is assumed that all articles reviewed involving humans, also included a written informed consent.

Data Availability Statement: No new data were created or analyzed in this study. Data sharing is not applicable to this article.

Conflicts of Interest: Abbie Smith-Ryan and Darren Candow serve as scientific advisors to Alzchem, a company that makes creatine. Alzchem had no role in the content or writing of this paper.

References

1. Kantor, E.D.; Rehm, C.D.; Du, M.; White, E.; Giovannucci, E.L. Trends in Dietary Supplement Use Among US Adults From 1999–2012. *JAMA* **2016**, *316*, 1464–1474. [CrossRef]
2. Greenhaff, P.L.; Bodin, K.; Soderlund, K.; Hultman, E. Effect of oral creatine supplementation on skeletal muscle phosphocreatine resynthesis. *Am. J. Physiol.* **1994**, *266*, E725–E730. [CrossRef]
3. Brosnan, J.T.; Brosnan, M.E. Creatine: Endogenous metabolite, dietary, and therapeutic supplement. *Annu. Rev. Nutr.* **2007**, *27*, 241–261. [CrossRef]
4. Forsberg, A.M.; Nilsson, E.; Werneman, J.; Bergstrom, J.; Hultman, E. Muscle composition in relation to age and sex. *Clin. Sci. (Lond)* **1991**, *81*, 249–256. [CrossRef] [PubMed]
5. Mihic, S.; MacDonald, J.R.; McKenzie, S.; Tarnopolsky, M.A. Acute creatine loading increases fat-free mass, but does not affect blood pressure, plasma creatinine, or CK activity in men and women. *Med. Sci. Sports Exerc.* **2000**, *32*, 291–296. [CrossRef] [PubMed]
6. Parise, G.; Mihic, S.; MacLennan, D.; Yarasheski, K.E.; Tarnopolsky, M.A. Effects of acute creatine monohydrate supplementation on leucine kinetics and mixed-muscle protein synthesis. *J. Appl. Physiol.* **2001**, *91*, 1041–1047. [CrossRef] [PubMed]
7. Kreider, R.B.; Kalman, D.S.; Antonio, J.; Ziegenfuss, T.N.; Wildman, R.; Collins, R.; Candow, D.G.; Kleiner, S.M.; Almada, A.L.; Lopez, H.L. International Society of Sports Nutrition position stand: Safety and efficacy of creatine supplementation in exercise, sport, and medicine. *J. Int. Soc. Sports Nutr.* **2017**, *14*, 18. [CrossRef]
8. Ellery, S.J.; Walker, D.W.; Dickinson, H. Creatine for women: A review of the relationship between creatine and the reproductive cycle and female-specific benefits of creatine therapy. *Amino Acids* **2016**, *48*, 1807–1817. [CrossRef] [PubMed]
9. Walker, J.B. Creatine: Biosynthesis, regulation, and function. *Adv. Enzymol. Relat. Areas Mol. Biol.* **1979**, *50*, 177–242. [CrossRef] [PubMed]
10. Wyss, M.; Kaddurah-Daouk, R. Creatine and creatinine metabolism. *Physiol. Rev.* **2000**, *80*, 1107–1213. [CrossRef] [PubMed]
11. Bundey, S.; Crawley, J.M.; Edwards, J.H.; Westhead, R.A. Serum creatine kinase levels in pubertal, mature, pregnant, and postmenopausal women. *J. Med Genet.* **1979**, *16*, 117–121. [CrossRef]
12. Oosthuyse, T.; Bosch, A.N. The effect of the menstrual cycle on exercise metabolism: Implications for exercise performance in eumenorrhoeic women. *Sports Med.* **2010**, *40*, 207–227. [CrossRef] [PubMed]
13. King, B.; Spikesman, A.; Emery, A.E. The effect of pregnancy on serum levels of creatine kinase. *Clin. Chim. Acta* **1972**, *36*, 267–269. [CrossRef]
14. Somjen, D.; Weisman, Y.; Harell, A.; Berger, E.; Kaye, A.M. Direct and sex-specific stimulation by sex steroids of creatine kinase activity and DNA synthesis in rat bone. *Proc. Natl. Acad. Sci. USA* **1989**, *86*, 3361–3365. [CrossRef] [PubMed]
15. Williams, T.; Walz, E.; Lane, A.R.; Pebole, M.; Hackney, A.C. The effect of estrogen on muscle damage biomarkers following prolonged aerobic exercise in eumenorrheic women. *Biol. Sport* **2015**, *32*, 193–198. [CrossRef]
16. Draper, C.F.; Duisters, K.; Weger, B.; Chakrabarti, A.; Harms, A.C.; Brennan, L.; Hankemeier, T.; Goulet, L.; Konz, T.; Martin, F.P.; et al. Menstrual cycle rhythmicity: Metabolic patterns in healthy women. *Sci. Rep.* **2018**, *8*, 14568. [CrossRef] [PubMed]
17. Volek, J.S.; Rawson, E.S. Scientific basis and practical aspects of creatine supplementation for athletes. *Nutrition* **2004**, *20*, 609–614. [CrossRef]
18. Chilibeck, P.D.; Kaviani, M.; Candow, D.G.; Zello, G.A. Effect of creatine supplementation during resistance training on lean tissue mass and muscular strength in older adults: A meta-analysis. *Open Access J. Sports Med.* **2017**, *8*, 213–226. [CrossRef] [PubMed]
19. Branch, J.D. Effect of creatine supplementation on body composition and performance: A meta-analysis. *Int. J. Sport Nutr. Exerc. Metab.* **2003**, *13*, 198–226. [CrossRef] [PubMed]
20. de Guingand, D.L.; Palmer, K.R.; Snow, R.J.; Davies-Tuck, M.L.; Ellery, S.J. Risk of Adverse Outcomes in Females Taking Oral Creatine Monohydrate: A Systematic Review and Meta-Analysis. *Nutrients* **2020**, *12*, 1780. [CrossRef]
21. Eckerson, J. Creating as an ergogenic aid for female athletic. *Strength Cond. J.* **2016**, *38*, 53615190. [CrossRef]
22. Sobolewski, E.J.; Thompson, B.J.; Smith, A.E.; Ryan, E.D. The physiological effects of creatine supplementation on hydration: A review. *Am. J. Lifestyle Med.* **2011**, *5*, 320–327. [CrossRef]
23. Ziegenfuss, T.N.; Rogers, M.; Lowery, L.; Mullins, N.; Mendel, R.; Antonio, J.; Lemon, P. Effect of creatine loading on anaerobic performance and skeletal muscle volume in NCAA Division I athletes. *Nutrition* **2002**, *18*, 397–402. [CrossRef]
24. Cramer, J.T. Creatine supplementation in endurance sports. In *Essentials of Creatine in Sports and Health*; Stout, J., Antonio, J., Kalman, D.S., Eds.; Humana Press: Tortowa, NJ, USA, 2008; pp. 45–99.

25. Trexler, E.T.; Smith-Ryan, A.E.; Stout, J.R.; Hoffman, J.R.; Wilborn, C.D.; Sale, C.; Kreider, R.B.; Jager, R.; Earnest, C.P.; Bannock, L.; et al. International society of sports nutrition position stand: Beta-Alanine. *J. Int. Soc. Sports Nutr.* **2015**, *12*, 30. [CrossRef]

26. Vandenberghe, K.; Goris, M.; Van Hecke, P.; Van Leemputte, M.; Vangerven, L.; Hespel, P. Long-term creatine intake is beneficial to muscle performance during resistance training. *J. Appl. Physiol.* **1997**, *83*, 2055–2063. [CrossRef]

27. Larson-Meyer, D.E.; Hunter, G.R.; Trowbridge, C.A.; Turk, J.C.; Ernest, J.M.; Torman, S.; Harbin, P.A. The effect of creatine supplementation on muscle strength and body composition during off-season training in female soccer players. *J. Strength Cond. Res.* **2000**, *14*, 434–442.

28. Brenner, M.; Walberg Rankin, J.; Sebolt, D. The effect of creatine supplementation during resistance trianing in women. *J. Strength Cond. Res.* **2000**, *14*, 207–213.

29. Kambis, K.W.; Pizzedaz, S.K. Short-term creatine supplementation improves maximum quadriceps contraction in women. *Int. J. Sport Nutr. Exerc. Metab.* **2003**, *13*, 87–96. [CrossRef] [PubMed]

30. Ferguson, T.B.; Syrotuik, D.G. Effects of creatine monohydrate supplementation on body composition and strength indices in experienced resistance trained women. *J. Strength Cond. Res.* **2006**, *20*, 939–946. [CrossRef] [PubMed]

31. Wilborn, C.D.; Outlaw, J.J.; Mumford, P.W.; Urbina, S.L.; Hayward, S.; Roberts, M.D.; Taylor, L.W.; Foster, C.A. A Pilot Study Examining the Effects of 8-Week Whey Protein versus Whey Protein Plus Creatine Supplementation on Body Composition and Performance Variables in Resistance-Trained Women. *Ann. Nutr. Metab.* **2016**, *69*, 190–199. [CrossRef] [PubMed]

32. Cooper, R.; Naclerio, F.; Allgrove, J.; Jimenez, A. Creatine supplementation with specific view to exercise/sports performance: An update. *J. Int. Soc. Sports Nutr.* **2012**, *9*, 33. [CrossRef]

33. Eckerson, J.M.; Stout, J.R.; Moore, G.A.; Stone, N.J.; Nishimura, K.; Tamura, K. Effect of two and five days of creatine loading on anaerobic working capacity in women. *J. Strength Cond. Res.* **2004**, *18*, 168–173. [CrossRef] [PubMed]

34. Eckerson, J.M.; Stout, J.R.; Moore, G.A.; Stone, N.J.; Iwan, K.A.; Gebauer, A.N.; Ginsberg, R. Effect of creatine phosphate supplementation on anaerobic working capacity and body weight after two and six days of loading in men and women. *J. Strength Cond. Res.* **2005**, *19*, 756–763. [CrossRef]

35. Earnest, C.P.; Snell, P.G.; Rodriguez, R.; Almada, A.L.; Mitchell, T.L. The effect of creatine monohydrate ingestion on anaerobic power indices, muscular strength and body composition. *Acta Physiol. Scand.* **1995**, *153*, 207–209. [CrossRef] [PubMed]

36. Smith, J.C.; Stephens, D.P.; Hall, E.L.; Jackson, A.W.; Earnest, C.P. Effect of oral creatine ingestion on parameters of the work rate-time relationship and time to exhaustion in high-intensity cycling. *Eur. J. Appl. Physiol. Occup. Physiol.* **1998**, *77*, 360–365. [CrossRef]

37. Tarnopolsky, M.A.; MacLennan, D.P. Creatine monohydrate supplementation enhances high-intensity exercise performance in males and females. *Int. J. Sport Nutr. Exerc. Metab.* **2000**, *10*, 452–463. [CrossRef] [PubMed]

38. Ramirez-Campillo, R.; Gonzalez-Jurado, J.A.; Martinez, C.; Nakamura, F.Y.; Penailillo, L.; Meylan, C.M.; Caniuqueo, A.; Canas-Jamet, R.; Moran, J.; Alonso-Martinez, A.M.; et al. Effects of plyometric training and creatine supplementation on maximal-intensity exercise and endurance in female soccer players. *J. Sci. Med. Sport* **2016**, *19*, 682–687. [CrossRef] [PubMed]

39. Aoki, M.S.; Gomes, R.V.; Raso, V. Creatine supplementation attenuates the adverse effect of endurance exercise on subsequent resistance exercise performance. *Med. Sci. Sports Exerc.* **2004**, *36*, S334–S335. [CrossRef]

40. Nelson, A.G.; Day, R.; Glickman-Weiss, E.L.; Hegsted, M.; Kokkonen, J.; Sampson, B. Creatine supplementation alters the response to a graded cycle ergometer test. *Eur. J. Appl. Physiol.* **2000**, *83*, 89–94. [CrossRef] [PubMed]

41. Smith, A.E.; Fukuda, D.H.; Ryan, E.D.; Kendall, K.L.; Cramer, J.T.; Stout, J. Ergolytic/ergogenic effects of creatine on aerobic power. *Int. J. Sports Med.* **2011**, *32*, 975–981. [CrossRef] [PubMed]

42. Smith, A.E.; Walter, A.A.; Herda, T.J.; Ryan, E.D.; Moon, J.R.; Cramer, J.T.; Stout, J.R. Effects of creatine loading on electromyographic fatigue threshold during cycle ergometry in college-aged women. *J. Int. Soc. Sports Nutr.* **2007**, *4*, 20. [CrossRef] [PubMed]

43. Stout, J.; Eckerson, J.; Ebersole, K.; Moore, G.; Perry, S.; Housh, T.; Bull, A.; Cramer, J.; Batheja, A. Effect of creatine loading on neuromuscular fatigue threshold. *J. Appl. Physiol.* **2000**, *88*, 109–112. [CrossRef] [PubMed]

44. Graef, J.L.; Smith, A.E.; Kendall, K.L.; Fukuda, D.H.; Moon, J.R.; Beck, T.W.; Cramer, J.T.; Stout, J.R. The effects of four weeks of creatine supplementation and high-intensity interval training on cardiorespiratory fitness: A randomized controlled trial. *J. Int. Soc. Sports Nutr.* **2009**, *6*, 18. [CrossRef] [PubMed]

45. Kendall, K.L.; Smith, A.E.; Graef, J.L.; Fukuda, D.H.; Moon, J.R.; Beck, T.W.; Cramer, J.T.; Stout, J.R. Effects of four weeks of high-intensity interval training and creatine supplementation on critical power and anaerobic working capacity in college-aged men. *J. Strength Cond. Res.* **2009**, *23*, 1663–1669. [CrossRef]

46. Forbes, S.C.; Sletten, N.; Durrer, C.; Myette-Cote, E.; Candow, D.; Little, J.P. Creatine Monohydrate Supplementation Does Not Augment Fitness, Performance, or Body Composition Adaptations in Response to Four Weeks of High-Intensity Interval Training in Young Females. *Int. J. Sport Nutr. Exerc. Metab.* **2017**, *27*, 285–292. [CrossRef] [PubMed]

47. Cox, G.; Mujika, I.; Tumilty, D.; Burke, L. Acute creatine supplementation and performance during a field test simulating match play in elite female soccer players. *Int. J. Sport Nutr. Exerc. Metab.* **2002**, *12*, 33–46. [CrossRef]

48. Selsby, J.T.; Beckett, K.D.; Kern, M.; Devor, S.T. Swim performance following creatine supplementation in Division III athletes. *J. Strength Cond. Res.* **2003**, *17*, 421–424. [CrossRef]

49. Mujika, I.; Chatard, J.C.; Lacoste, L.; Barale, F.; Geyssant, A. Creatine supplementation does not improve sprint performance in competitive swimmers. *Med. Sci. Sports Exerc.* **1996**, *28*, 1435–1441. [CrossRef] [PubMed]

50. Dawson, B.; Vladich, T.; Blanksby, B.A. Effects of 4 weeks of creatine supplementation in junior swimmers on freestyle sprint and swim bench performance. *J. Strength Cond. Res.* **2002**, *16*, 485–490. [PubMed]
51. Peyrebrune, M.C.; Nevill, M.E.; Donaldson, F.J.; Cosford, D.J. The effects of oral creatine supplementation on performance in single and repeated sprint swimming. *J. Sports Sci.* **1998**, *16*, 271–279. [CrossRef]
52. Grindstaff, P.D.; Kreider, R.; Bishop, R.; Wilson, M.; Wood, L.; Alexander, C.; Almada, A. Effects of creatine supplementation on repetitive sprint performance and body composition in competitive swimmers. *Int. J. Sport Nutr.* **1997**, *7*, 330–346. [CrossRef] [PubMed]
53. Ellery, S.J.; LaRosa, D.A.; Kett, M.M.; Della Gatta, P.A.; Snow, R.J.; Walker, D.W.; Dickinson, H. Maternal creatine homeostasis is altered during gestation in the spiny mouse: Is this a metabolic adaptation to pregnancy? *BMC Pregnancy Childbirth* **2015**, *15*, 92. [CrossRef] [PubMed]
54. Dickinson, H.; Davies-Tuck, M.; Ellery, S.J.; Grieger, J.A.; Wallace, E.M.; Snow, R.J.; Walker, D.W.; Clifton, V.L. Maternal creatine in pregnancy: A retrospective cohort study. *BJOG Int. J. Obstet. Gynaecol.* **2016**, *123*, 1830–1838. [CrossRef]
55. Dickinson, H.; Ellery, S.; Ireland, Z.; LaRosa, D.; Snow, R.; Walker, D.W. Creatine supplementation during pregnancy: Summary of experimental studies suggesting a treatment to improve fetal and neonatal morbidity and reduce mortality in high-risk human pregnancy. *BMC Pregnancy Childbirth* **2014**, *14*, 150. [CrossRef] [PubMed]
56. Ireland, Z.; Castillo-Melendez, M.; Dickinson, H.; Snow, R.; Walker, D.W. A maternal diet supplemented with creatine from mid-pregnancy protects the newborn spiny mouse brain from birth hypoxia. *Neuroscience* **2011**, *194*, 372–379. [CrossRef] [PubMed]
57. De Guingand, D.L.; Ellery, S.J.; Davies-Tuck, M.L.; Dickinson, H. Creatine and pregnancy outcomes, a prospective cohort study in low-risk pregnant women: Study protocol. *BMJ Open* **2019**, *9*, e026756. [CrossRef] [PubMed]
58. Muccini, A.M.; Tran, N.T.; de Guingand, D.L.; Philip, M.; Della Gatta, P.A.; Galinsky, R.; Sherman, L.S.; Kelleher, M.A.; Palmer, K.R.; Berry, M.J.; et al. Creatine Metabolism in Female Reproduction, Pregnancy and Newborn Health. *Nutrients* **2021**, *13*, 490. [CrossRef]
59. Collins, B.C.; Laakkonen, E.K.; Lowe, D.A. Aging of the musculoskeletal system: How the loss of estrogen impacts muscle strength. *Bone* **2019**, *123*, 137–144. [CrossRef]
60. Messier, V.; Rabasa-Lhoret, R.; Barbat-Artigas, S.; Elisha, B.; Karelis, A.D.; Aubertin-Leheudre, M. Menopause and sarcopenia: A potential role for sex hormones. *Maturitas* **2011**, *68*, 331–336. [CrossRef]
61. Candow, D.G.; Forbes, S.C.; Chilibeck, P.D.; Cornish, S.M.; Antonio, J.; Kreider, R.B. Effectiveness of Creatine Supplementation on Aging Muscle and Bone: Focus on Falls Prevention and Inflammation. *J. Clin. Med.* **2019**, *8*, 488. [CrossRef]
62. Candow, D.G.; Forbes, S.C.; Vogt, E. Effect of pre-exercise and post-exercise creatine supplementation on bone mineral content and density in healthy aging adults. *Exp. Gerontol.* **2019**, *119*, 89–92. [CrossRef] [PubMed]
63. Gotshalk, L.A.; Kraemer, W.J.; Mendonca, M.A.; Vingren, J.L.; Kenny, A.M.; Spiering, B.A.; Hatfield, D.L.; Fragala, M.S.; Volek, J.S. Creatine supplementation improves muscular performance in older women. *Eur. J. Appl. Physiol.* **2008**, *102*, 223–231. [CrossRef] [PubMed]
64. Canete, S.; San Juan, A.F.; Perez, M.; Gomez-Gallego, F.; Lopez-Mojares, L.M.; Earnest, C.P.; Fleck, S.J.; Lucia, A. Does creatine supplementation improve functional capacity in elderly women? *J. Strength Cond. Res.* **2006**, *20*, 22–28. [CrossRef]
65. Lobo, D.M.; Tritto, A.C.; da Silva, L.R.; de Oliveira, P.B.; Benatti, F.B.; Roschel, H.; Niess, B.; Gualano, B.; Pereira, R.M. Effects of long-term low-dose dietary creatine supplementation in older women. *Exp. Gerontol.* **2015**, *70*, 97–104. [CrossRef] [PubMed]
66. Sales, L.P.; Pinto, A.J.; Rodrigues, S.F.; Alvarenga, J.C.; Goncalves, N.; Sampaio-Barros, M.M.; Benatti, F.B.; Gualano, B.; Rodrigues Pereira, R.M. Creatine Supplementation (3 g/d) and Bone Health in Older Women: A 2-Year, Randomized, Placebo-Controlled Trial. *J. Gerontol. Ser. A Boil. Sci. Med. Sci.* **2020**, *75*, 931–938. [CrossRef] [PubMed]
67. Harris, R.C.; Soderlund, K.; Hultman, E. Elevation of creatine in resting and exercised muscle of normal subjects by creatine supplementation. *Clin. Sci. (Lond)* **1992**, *83*, 367–374. [CrossRef]
68. Gualano, B.; Macedo, A.R.; Alves, C.R.; Roschel, H.; Benatti, F.B.; Takayama, L.; de Sa Pinto, A.L.; Lima, F.R.; Pereira, R.M. Creatine supplementation and resistance training in vulnerable older women: A randomized double-blind placebo-controlled clinical trial. *Exp. Gerontol.* **2014**, *53*, 7–15. [CrossRef] [PubMed]
69. Chilibeck, P.D.; Candow, D.G.; Landeryou, T.; Kaviani, M.; Paus-Jenssen, L. Effects of Creatine and Resistance Training on Bone Health in Postmenopausal Women. *Med. Sci. Sports Exerc.* **2015**, *47*, 1587–1595. [CrossRef]
70. Aguiar, A.F.; Januario, R.S.; Junior, R.P.; Gerage, A.M.; Pina, F.L.; do Nascimento, M.A.; Padovani, C.R.; Cyrino, E.S. Long-term creatine supplementation improves muscular performance during resistance training in older women. *Eur. J. Appl. Physiol.* **2013**, *113*, 987–996. [CrossRef]
71. Neves, M., Jr.; Gualano, B.; Roschel, H.; Fuller, R.; Benatti, F.B.; Pinto, A.L.; Lima, F.R.; Pereira, R.M.; Lancha, A.H., Jr.; Bonfa, E. Beneficial effect of creatine supplementation in knee osteoarthritis. *Med. Sci. Sports Exerc.* **2011**, *43*, 1538–1543. [CrossRef]
72. Candow, D.G.; Vogt, E.; Johannsmeyer, S.; Forbes, S.C.; Farthing, J.P. Strategic creatine supplementation and resistance training in healthy older adults. *Appl. Physiol. Nutr. Metab.* **2015**, *40*, 689–694. [CrossRef]
73. Pinto, C.L.; Botelho, P.B.; Carneiro, J.A.; Mota, J.F. Impact of creatine supplementation in combination with resistance training on lean mass in the elderly. *J. Cachexia Sarcopenia Muscle* **2016**, *7*, 413–421. [CrossRef]
74. Bebbington, P.; Dunn, G.; Jenkins, R.; Lewis, G.; Brugha, T.; Farrell, M.; Meltzer, H. The influence of age and sex on the prevalence of depressive conditions: Report from the National Survey of Psychiatric Morbidity. *Int. Rev. Psychiatry* **2003**, *15*, 74–83. [CrossRef] [PubMed]

75. Albert, P.R. Why is depression more prevalent in women? *J. Psychiatry Neurosci.* **2015**, *40*, 219–221. [CrossRef]
76. Eriksson, E.; Andersch, B.; Ho, H.P.; Landen, M.; Sundblad, C. Diagnosis and treatment of premenstrual dysphoria. *J. Clin. Psychiatry* **2002**, *63* (Suppl. S7), 16–23. [PubMed]
77. Agren, H.; Niklasson, F. Creatinine and creatine in CSF: Indices of brain energy metabolism in depression. Short note. *J. Neural Transm.* **1988**, *74*, 55–59. [CrossRef]
78. Allen, P.J.; D'Anci, K.E.; Kanarek, R.B.; Renshaw, P.F. Chronic creatine supplementation alters depression-like behavior in rodents in a sex-dependent manner. *Neuropsychopharmacology* **2010**, *35*, 534–546. [CrossRef]
79. Dager, S.R.; Friedman, S.D.; Parow, A.; Demopulos, C.; Stoll, A.L.; Lyoo, I.K.; Dunner, D.L.; Renshaw, P.F. Brain metabolic alterations in medication-free patients with bipolar disorder. *Arch. Gen. Psychiatry* **2004**, *61*, 450–458. [CrossRef] [PubMed]
80. Renshaw, P.F.; Parow, A.M.; Hirashima, F.; Ke, Y.; Moore, C.M.; Frederick Bde, B.; Fava, M.; Hennen, J.; Cohen, B.M. Multinuclear magnetic resonance spectroscopy studies of brain purines in major depression. *Am. J. Psychiatry* **2001**, *158*, 2048–2055. [CrossRef]
81. Kondo, D.G.; Sung, Y.H.; Hellem, T.L.; Fiedler, K.K.; Shi, X.; Jeong, E.K.; Renshaw, P.F. Open-label adjunctive creatine for female adolescents with SSRI-resistant major depressive disorder: A 31-phosphorus magnetic resonance spectroscopy study. *J. Affect. Disord.* **2011**, *135*, 354–361. [CrossRef]
82. Wallimann, T.; Wyss, M.; Brdiczka, D.; Nicolay, K.; Eppenberger, H.M. Intracellular compartmentation, structure and function of creatine kinase isoenzymes in tissues with high and fluctuating energy demands: The 'phosphocreatine circuit' for cellular energy homeostasis. *Biochem. J.* **1992**, *281 Pt 1*, 21–40. [CrossRef]
83. Iosifescu, D.V.; Bolo, N.R.; Nierenberg, A.A.; Jensen, J.E.; Fava, M.; Renshaw, P.F. Brain bioenergetics and response to triiodothyronine augmentation in major depressive disorder. *Biol. Psychiatry* **2008**, *63*, 1127–1134. [CrossRef]
84. Lyoo, I.K.; Kong, S.W.; Sung, S.M.; Hirashima, F.; Parow, A.; Hennen, J.; Cohen, B.M.; Renshaw, P.F. Multinuclear magnetic resonance spectroscopy of high-energy phosphate metabolites in human brain following oral supplementation of creatine-monohydrate. *Psychiatry Res.* **2003**, *123*, 87–100. [CrossRef]
85. Riehemann, S.; Volz, H.P.; Wenda, B.; Hubner, G.; Rossger, G.; Rzanny, R.; Sauer, H. Frontal lobe in vivo (31)P-MRS reveals gender differences in healthy controls, not in schizophrenics. *NMR Biomed.* **1999**, *12*, 483–489. [CrossRef]
86. Hellem, T.L.; Sung, Y.H.; Shi, X.F.; Pett, M.A.; Latendresse, G.; Morgan, J.; Huber, R.S.; Kuykendall, D.; Lundberg, K.J.; Renshaw, P.F. Creatine as a Novel Treatment for Depression in Females Using Methamphetamine: A Pilot Study. *J. Dual. Diagn.* **2015**, *11*, 189–202. [CrossRef] [PubMed]
87. Kondo, D.G.; Forrest, L.N.; Shi, X.; Sung, Y.H.; Hellem, T.L.; Huber, R.S.; Renshaw, P.F. Creatine target engagement with brain bioenergetics: A dose-ranging phosphorus-31 magnetic resonance spectroscopy study of adolescent females with SSRI-resistant depression. *Amino Acids* **2016**, *48*, 1941–1954. [CrossRef] [PubMed]
88. Lyoo, I.K.; Yoon, S.; Kim, T.S.; Hwang, J.; Kim, J.E.; Won, W.; Bae, S.; Renshaw, P.F. A randomized, double-blind placebo-controlled trial of oral creatine monohydrate augmentation for enhanced response to a selective serotonin reuptake inhibitor in women with major depressive disorder. *Am. J. Psychiatry* **2012**, *169*, 937–945. [CrossRef] [PubMed]
89. Roitman, S.; Green, T.; Osher, Y.; Karni, N.; Levine, J. Creatine monohydrate in resistant depression: A preliminary study. *Bipolar Disord.* **2007**, *9*, 754–758. [CrossRef] [PubMed]
90. Bakian, A.V.; Huber, R.S.; Scholl, L.; Renshaw, P.F.; Kondo, D. Dietary creatine intake and depression risk among U.S. adults. *Transl. Psychiatry* **2020**, *10*, 1–11. [CrossRef] [PubMed]
91. Rango, M.; Castelli, A.; Scarlato, G. Energetics of 3.5 s neural activation in humans: A 31P MR spectroscopy study. *Magn. Reson. Med.* **1997**, *38*, 878–883. [CrossRef] [PubMed]
92. Volz, H.P.; Rzanny, R.; Riehemann, S.; May, S.; Hegewald, H.; Preussler, B.; Hubner, G.; Kaiser, W.A.; Sauer, H. 31P magnetic resonance spectroscopy in the frontal lobe of major depressed patients. *Eur. Arch. Psychiatry Clin. Neurosci.* **1998**, *248*, 289–295. [CrossRef]
93. Rawson, E.S.; Venezia, A.C. Use of creatine in the elderly and evidence for effects on cognitive function in young and old. *Amino Acids* **2011**, *40*, 1349–1362. [CrossRef]
94. Allen, P.J. Creatine metabolism and psychiatric disorders: Does creatine supplementation have therapeutic value? *Neurosci. Biobehav. Rev.* **2012**, *36*, 1442–1462. [CrossRef] [PubMed]
95. McMorris, T.; Harris, R.C.; Howard, A.N.; Langridge, G.; Hall, B.; Corbett, J.; Dicks, M.; Hodgson, C. Creatine supplementation, sleep deprivation, cortisol, melatonin and behavior. *Physiol. Behav.* **2007**, *90*, 21–28. [CrossRef]
96. McMorris, T.; Harris, R.C.; Swain, J.; Corbett, J.; Collard, K.; Dyson, R.J.; Dye, L.; Hodgson, C.; Draper, N. Effect of creatine supplementation and sleep deprivation, with mild exercise, on cognitive and psychomotor performance, mood state, and plasma concentrations of catecholamines and cortisol. *Psychopharmacology* **2006**, *185*, 93–103. [CrossRef]
97. Matud, M.P. Gender differences in stress and coping styles. *Pers. Individ. Differ.* **2004**, *37*, 1401–1415. [CrossRef]
98. Sayer, L.C. Gender differences in the relationship between long employee hours and multitasking. *Res. Sociol. Work* **2007**, *17*, 403–435.
99. Blatter, K.; Graw, P.; Munch, M.; Knoblauch, V.; Wirz-Justice, A.; Cajochen, C. Gender and age differences in psychomotor vigilance performance under differential sleep pressure conditions. *Behav. Brain Res.* **2006**, *168*, 312–317. [CrossRef] [PubMed]
100. Vidafar, P.; Gooley, J.J.; Burns, A.C.; Rajaratnam, S.M.W.; Rueger, M.; Van Reen, E.; Czeisler, C.A.; Lockley, S.W.; Cain, S.W. Increased vulnerability to attentional failure during acute sleep deprivation in women depends on menstrual phase. *Sleep* **2018**, *41*, zsy098. [CrossRef] [PubMed]

101. McKenna, M.J.; Morton, J.; Selig, S.E.; Snow, R.J. Creatine supplementation increases muscle total creatine but not maximal intermittent exercise performance. *J. Appl. Physiol.* **1999**, *87*, 2244–2252. [CrossRef]
102. Syrotuik, D.G.; Bell, G.J. Acute creatine monohydrate supplementation: A descriptive physiological profile of responders vs. nonresponders. *J. Strength Cond. Res.* **2004**, *18*, 610–617. [CrossRef]
103. Green, A.L.; Simpson, E.J.; Littlewood, J.J.; Macdonald, I.A.; Greenhaff, P.L. Carbohydrate ingestion augments creatine retention during creatine feeding in humans. *Acta Physiol. Scand.* **1996**, *158*, 195–202. [CrossRef] [PubMed]
104. Steenge, G.R.; Lambourne, J.; Casey, A.; Macdonald, I.A.; Greenhaff, P.L. Stimulatory effect of insulin on creatine accumulation in human skeletal muscle. *Am. J. Physiol.* **1998**, *275*, E974–E979. [CrossRef]
105. Steenge, G.R.; Simpson, E.J.; Greenhaff, P.L. Protein- and carbohydrate-induced augmentation of whole body creatine retention in humans. *J. Appl. Physiol.* **2000**, *89*, 1165–1171. [CrossRef]
106. Preen, D.; Dawson, B.; Goodman, C.; Beilby, J.; Ching, S. Creatine supplementation: A comparison of loading and maintenance protocols on creatine uptake by human skeletal muscle. *Int. J. Sport Nutr. Exerc. Metab.* **2003**, *13*, 97–111. [CrossRef]
107. Jager, R.; Harris, R.C.; Purpura, M.; Francaux, M. Comparison of new forms of creatine in raising plasma creatine levels. *J. Int. Soc. Sports Nutr.* **2007**, *4*, 17. [CrossRef] [PubMed]
108. Harris, R.C.; Nevill, M.; Harris, D.B.; Fallowfield, J.L.; Bogdanis, G.C.; Wise, J.A. Absorption of creatine supplied as a drink, in meat or in solid form. *J. Sports Sci.* **2002**, *20*, 147–151. [CrossRef] [PubMed]

 *nutrients*

Review

Current Evidence and Possible Future Applications of Creatine Supplementation for Older Adults

Darren G. Candow [1,*], Scott C. Forbes [2], Ben Kirk [3,4] and Gustavo Duque [3,4]

[1] Faculty of Kinesiology and Health Studies, University of Regina, Regina, SK S4SOA2, Canada
[2] Department of Physical Education Studies, Faculty of Education, Brandon University, Brandon, MB R7A6A9, Canada; forbess@brandonu.ca
[3] Department of Medicine-Western Health, The University of Melbourne, Parkville, VIC 3021, Australia; ben.kirk@unimelb.edu.au (B.K.); Gustavo.duque@unimelb.edu.au (G.D.)
[4] Australian Institute for Musculoskeletal Science (AIMSS), The University of Melbourne and Western Health, St. Albans, VIC 3201, Australia
* Correspondence: Darren.Candow@uregina.ca; Tel.: +1-306-585-4906

Abstract: Sarcopenia, defined as age-related reduction in muscle mass, strength, and physical performance, is associated with other age-related health conditions such as osteoporosis, osteosarcopenia, sarcopenic obesity, physical frailty, and cachexia. From a healthy aging perspective, lifestyle interventions that may help overcome characteristics and associated comorbidities of sarcopenia are clinically important. One possible intervention is creatine supplementation (CR). Accumulating research over the past few decades shows that CR, primarily when combined with resistance training (RT), has favourable effects on aging muscle, bone and fat mass, muscle and bone strength, and tasks of physical performance in healthy older adults. However, research is very limited regarding the efficacy of CR in older adults with sarcopenia or osteoporosis and no research exists in older adults with osteosarcopenia, sarcopenic obesity, physical frailty, or cachexia. Therefore, the purpose of this narrative review is (1) to evaluate and summarize current research involving CR, with and without RT, on properties of muscle and bone in older adults and (2) to provide a rationale and justification for future research involving CR in older adults with osteosarcopenia, sarcopenic obesity, physical frailty, or cachexia.

Keywords: sarcopenia; osteoporosis; osteosarcopenia; frailty; cachexia

Citation: Candow, D.G.; Forbes, S.C.; Kirk, B.; Duque, G. Current Evidence and Possible Future Applications of Creatine Supplementation for Older Adults. *Nutrients* **2021**, *13*, 745. https://doi.org/10.3390/nu13030745

Academic Editor: David C. Nieman
Received: 6 January 2021
Accepted: 20 February 2021
Published: 26 February 2021

1. Introduction

Sarcopenia refers to age-related reductions in muscle strength (dynapenia), muscle mass (quantity), relative strength (strength per unit of muscle mass), muscle quality (architecture and composition), and/or physical performance (i.e., tasks of functionality) [1]. Sarcopenia typically occurs in 8–13% of adults ≥60 years of age [2] and is associated with other age-related health conditions such as osteoporosis, osteosarcopenia, sarcopenic obesity, physical frailty, and cachexia. Annually, muscle mass decreases by 0.45% in men and by 0.37% in women, but these decrements climb to 0.9% for men and to 0.7% for women starting in their seventh decade [3]. The age-related decrease in muscle strength, which is a strong predictor of poor health outcomes (mobility disability, falls, fractures, and mortality) in older adults [1], occurs more rapidly (2–5 fold faster) than the reduction in lean (muscle) mass [4].

From a global health perspective, the World Health Organization established a code (ICD-10-CM; M62.84) for sarcopenia as a means for better diagnosis, assessment, and treatment of the condition. While several definitions and subcategories of sarcopenia exist, the European Working Group on Sarcopenia in Older People (EWGSOP) defines individuals with low muscle strength (as assessed by grip-strength or chair-stand test) as having probable sarcopenia; those with low muscle strength and low muscle quantity (as assessed by dual energy X-ray absorptiometry, magnetic resonance imaging and spectroscopy, computed tomography, bioelectrical impedance, creatine dilution, and/or muscle biopsy) as having confirmed sarcopenia; and those with low muscle strength, low muscle quantity, and poor physical performance (as assessed by gait speed, short physical performance battery test, timed-up-and-go test, or 400 m walk test) as having severe sarcopenia [1]. Sarcopenia is classified as primary when its etiology is age dependent whereas secondary sarcopenia is influenced by age and/or other factors such as physical inactivity and undernutrition [1]. Contributing factors to the pathophysiology of sarcopenia include changes in neuromuscular function, skeletal muscle morphology and architecture, protein kinetics, hormonal regulation, growth factors and satellite cells, vascularization, inflammation, mitochondrial function, nutrition, and physical activity [1,3,4]. From a healthy aging perspective, interventions that may help overcome characteristics and associated comorbidities of sarcopenia (i.e., osteoporosis, osteosarcopenia, sarcopenic obesity, physical frailty, and cachexia) are clinically important.

Accumulating research over the past few decades shows that creatine supplementation (CR), primarily when combined with resistance training (RT), has some favourable effects on muscle accretion and bone mineral density, bone and muscle strength, and tasks of functionality in older adults (for reviews, see Candow et al. [5], Chilibeck et al. [6], Forbes et al. [7], Gualano et al. [8], and Kreider et al. [9]). However, research is very limited regarding the efficacy of CR in older adults with sarcopenia or osteoporosis and no research exists in older adults with osteosarcopenia, sarcopenic obesity, physical frailty, or cachexia. Therefore, the purpose of this narrative review is (1) to evaluate and summarize current research involving CR, with and without RT, on properties of muscle and bone in older adults and (2) to provide a rationale and justification for future research involving CR in older adults with osteosarcopenia, sarcopenic obesity, physical frailty, or cachexia.

2. Creatine

Creatine is an organic acid endogenously synthesized from reactions involving the amino acids arginine, glycine, and methionine in the kidneys and liver [10]. Alternatively, creatine can be exogenously consumed from meat [9] and commercially manufactured products. The vast majority ($\approx$95%) of creatine resides in skeletal muscle, with approximately 66% being stored as phosphocreatine (PCr) [9]. It is estimated that 2% of endogenous creatine stores are degraded daily to creatinine, a metabolic by-product of creatine metabolism [10]. For most individuals, excluding vegans and vegetarians, $\approx$3 g of exogenous creatine per day may help maintain creatine stores [9]. Metabolically, creatine combines with inorganic phosphate (Pi) to form PCr, which helps resynthesize and maintain adenosine triphosphate (ATP) levels [9].

3. Potential of Creatine Supplementation for Sarcopenia

The majority of aging research involving CR has focused on measures of muscle accretion and strength in response to RT. Studies published to date involving >600 older adults (>48 years of age) show divergent results, possibly because of methodological differences across studies (Table 1). We have previously reviewed the majority of these studies in detail elsewhere [5,8,11–14]. Most studies (n = 16) involved healthy older adults, whereas 4 studies involved older adults with knee osteoarthritis, osteopenia or osteoporosis, type II diabetes, or chronic obstructive pulmonary disease (COPD). The results are equivocal regarding the efficacy of CR on measures of muscle accretion and strength, with half of the studies showing greater gains from CR vs. placebo (PLA) and the other half showing similar effects between the two interventions during an RT program. Individual studies typically lack adequate statistical power to detect small changes in muscle accretion and strength from CR over time, and the responsiveness to CR in older adults may be influenced by initial resting PCr levels in different muscle regions, changes in type II muscle fibre size and quantity, and habitual dietary intake of creatine [13]. To overcome the limitations of low statistical power and high variability amongst older adult populations, three meta-analyses have been performed to determine the efficacy of CR

($\geq$3 g/day) vs. PLA during an RT program ($\geq$7 weeks) on measures of muscle accretion and strength [6,15,16]. Collectively, these meta-analyses showed that the combination of CR and RT augmented muscle accretion ($\approx$1.2 kg), and upper- and lower-body strength more than PLA and RT in older adults. Mechanistically, the greater increase in muscle accretion and strength from CR may be related to its ability to influence phosphate metabolism, calcium and glycogen regulation, cellular swelling, muscle protein signaling and breakdown, myogenic transcription factors and satellite cells, growth factors (i.e., IGF-1 and myostatin), inflammation, and oxidative stress (for reviews, see Candow et al. [5], Chilibeck et al. [6], Gualano et al. [8], and Kreider et al. [9]). Upon CR cessation, the gains in muscle accretion and strength seem to persist for up to 12 weeks when RT is maintained in older adults [17].

Table 1. Summary of studies examining creatine and resistance training on muscle outcomes in older adults.

First Author, Year	Population	Supplement Dose	Resistance Training	Duration	Outcomes
Aguiar et al. 2013 [18]	N = 18; healthy women; Mean age = 65 y	CR (5 g/day), PLA	RT = 3 x/wk	12 wks	CR ↑ gains in fat-free mass (+3.2%), muscle mass (+2.8%), 1 RM bench press, knee extension, and biceps curl compared to PLA
Alves et al. 2013 [19]	N = 47; healthy women, Mean age = 66.8 y (range: 60–80 y)	CR (20 g/day for 5 days, followed by 5 g/day thereafter), PLA with and without RT	RT = 2 x/wk	24 wks	↔1 RM strength compared to RT + PLA
Bemben et al. 2010 and Eliot et al. 2008 [20,21]	N = 42; healthy men; age = 48–72 y	CR (5 g/day), PRO (35 g/day), CR + PRO, PLA	RT = 3 x/wk	14 wks	↔lean tissue mass, 1 RM strength
Bermon et al. 1998 [22]	N = 32 (16 men, 16 women); healthy; age = 67–80 y	CR (20 g/day for 5 days followed by 3 g/day), PLA	RT = 3 x/wk	7.4 wks (52 days)	↔lower limb muscular volume, 1- and 12-repetitions maxima, and isometric intermittent endurance
Bernat et al. 2019 [23]	N = 24 healthy men; age = 59 ± 6 y	CR (0.1 g/kg/day), PLA	High-velocity RT = 2 x/wk	8 wks	↔muscle thickness, physical performance, upper-body muscle strength; CR ↑ leg press strength, total lower body strength
Brose et al. 2003 [24]	N = 28 (15 men, 13 women); healthy; age: men = 68.7, women = 70.0 y	CR (5 g/day), PLA	RT = 3 x/wk	14 wks	CR ↑ gains in lean tissue mass and isometric knee extension strength; ↔ type 1, 2 a, 2 x muscle fibre area
Candow et al. 2008 [25]	N = 35; healthy men; age = 59–77 y	CR (0.1 g/kg/day), CR + PRO (PRO: 0.3 g/kg/day), PLA	RT = 3 x/wk	10 wks	CR ↑ muscle thickness compared to PLA. CR ↑1 RM bench press ↔ 1 RM leg press
Candow et al. 2015 [26]	N = 39 (17 men, 22 women); healthy; age = 50–71 y	CR (0.1 g/kg) before RT, CR (0.1 g/kg) after RT, PLA	RT = 3 x/wk	32 wks	CR after RT ↑ lean tissue mass, 1 RM leg press, 1 RM chest press compared to PLA
Candow et al. 2020 [27]	N = 38; healthy men; age = 49–67 y	CR (On training days: 0.05 g/kg before and 0.05 g/kg after exercise) + 0.1 g/kg/day on non-train-ing days (2 equal doses) or PLA	RT = 3 x/wk	12 months	↔lean tissue mass, muscle thickness, or muscle strength
Chilibeck et al. 2015 [28]	N = 33; healthy women; Mean age = 57 y	CR (0.1 g/kg/day), PLA	RT = 3 x/wk	52 wks	↔lean tissue mass and muscle thickness gains between groups; ↑ relative bench press strength compared to PLA.
Chrusch et al. 2001 [29]	N = 30; healthy men; age = 60–84 y	CR (0.3 g/kg/d for 5 days followed by 0.07 g/kg/day), PLA	RT = 3 x/wk	12 wks	CR ↑ gains in lean tissue mass; CR ↑1 RM leg press, 1 RM knee extension, leg press endurance, and knee extension endurance; ↔ 1 RM bench press or bench press endurance.
Cooke et al. 2014 [30]	N = 20; healthy men; age = 55–70 y	CR (20 g/day for 7 days followed by 0.1 g/kg/day on training days)	RT = 3 x/wk	12 wks	↔lean tissue mass, 1 RM bench press, 1 RM leg press
Deacon et al. 2008 [31]	N = 80 (50 men, 30 women); COPD; age = 68.2 y	CR (22 g/day for 5 day followed by 3.76 g/day), PLA	RT = 3 x/wk	7 wks	↔lean tissue mass or muscle strength
Eijnde et al. 2003 [32]	N = 46; healthy men; age = 55–75 y	CR (5 g/day), PLA	Cardiorespiratory + RT = 2–3 x/wk	26 wks	↔lean tissue mass or isometric maximal strength
Gualano et al. 2011 [33]	N = 25 (9 men, 16 women); type 2 diabetes; age = 57 y	CR (5 g/day), PLA	RT = 3 x/wk	12 wks	↔lean tissue mass
Gualano et al. 2014 [34]	N = 30; "vulnera-ble" women; Mean age = 65.4 y	CR (20 g/day for 5 days; 5 g/day thereafter), PLA with and without RT	RT = 2 x/wk	24 wks	CR + RT ↑ gains in 1RM bench press and appendicular lean mass compared to PLA + RT

Table 1. *Cont.*

First Author, Year	Population	Supplement Dose	Resistance Training	Duration	Outcomes
Johannsmeyer et al. 2016 [35]	$N = 31$ (17 men, 14 women); healthy; age = 58 y	CR (0.1 g/kg/day), PLA	RT = 3 x/wk	12 wks	CR ↑ gains in lean tissue mass; ↔ 1RM strength and endurance; CR attenuated magnitude increase in time to complete balance test compared to PLA
Neves et al. 2011 [36]	$N = 24$ (postmen-opausal women with Knee osteo-arthritis); Age = 55–65 y	CR (20 g/day for 1 week, followed by 5 g/day), PLA	RT = 3 x/wk	12 wks	CR ↑ gains in limb lean mass. ↔ 1RM leg press
Pinto et al. 2016 [37]	$N = 27$ (men and women); healthy; age = 60–80 y	CR (5 g/day), PLA	RT = 3 x/wk	12 wks	CR ↑ gains in lean tissue mass; ↔ 10 RM bench press or leg press strength
Smolarek et al. 2020 [38]	$N = 26$ (5 men, 21 women); long-term care residence; age = 68.9 ± 6.8 y	CR (5 g/day), PLA	RT = 2 x/wk	16 wks	CR ↑ dominant and non-dominant handgrip strength

CR = creatine; PRO = protein; RM = repetition maximum; ↑ = significant greater; ↔ no difference between conditions; wk = weeks; y = years; g = grams; kg = kilograms.

Only three studies have determined the effects of CR and RT in older adults with different classifications of sarcopenia. Pinto et al. [37] showed that, in older adults with either probable sarcopenia ($n = 3$; skeletal muscle mass index (SMI): appendicular skeletal muscle mass/height2 <7.26 kg/m^2 for men and <5.45 kg/m^2 for women), sarcopenia ($n = 1$; SMI + handgrip strength <30 kg and <20 kg for women or gait speed <0.8 m/s), or severe sarcopenia ($n = 1$; SMI + handgrip strength <30 kg and <20 kg for women and gait speed <0.8 m/s), 12 weeks of CR (5 g/day) and supervised RT eliminated the probable and severe sarcopenia designations in 3 participants. However, creatine had no effect on the individual with sarcopenia. Furthermore, it is unknown whether creatine and RT reduced the level of severe sarcopenia to sarcopenia or probable sarcopenia. In seven older adults considered to be pre-sarcopenic (defined as relative skeletal muscle index >7.26 kg/m^2 for men and >5.5 kg/m^2 for women [39]), 8 months of CR (0.1 g/kg/day or ≈8 g/day) and supervised whole-body RT eliminated the pre-sarcopenic designation in 5 of the participants [26].

Finally, in four postmenopausal women (>60 years) who were sarcopenic (defined by appendicular lean mass, adjusted for height and weight [40]), CR (20 g/day for 5 days + 5 g/day for 23 weeks) during supervised whole-body RT (3 sets of 8–12 repetitions, 2 days per week) eliminated the sarcopenia classification in two of the women [34]. While limited by very low sample sizes, these preliminary results across studies suggest that CR (≥5 g/day) and supervised RT (>12 weeks) has some potential to mitigate sarcopenia in older adults.

Regarding physical performance (functionality), two meta-analyses of older adults demonstrated that CR in conjunction with RT resulted in greater improvements in sit-to-stand performance when compared to RT (plus PLA) alone [5,16]. These findings are of clinical relevance given that improving sit-to-stand performance may reduce the risk of falls in older adults [41].

Independent of RT, research is mixed regarding the effectiveness of CR on aging muscle, with 5 studies showing greater effects from CR vs. PLA and 5 studies showing similar effects between the two interventions (for review, see Forbes et al. [42]). While it is difficult to compare results across studies, these inconsistent findings may be related to the CR protocol and/or dosage used. The majority of studies that found beneficial effects from CR incorporated a CR loading phase (20 g/day) or used a high relative daily dosage of creatine (0.3 g/kg/day), whereas several of the studies that failed to observe beneficial effects did not use these strategies.

In summary, CR (≥3 g/day) and RT (≥7 weeks; primarily whole-body routines) can improve some measures of muscle accretion, strength, and physical performance in older adults. Independent of RT, a CR loading phase and/or high relative daily dosage of creatine (≥0.3 g/kg/day) may be required to produce some muscle benefits in older adults. It is unknown whether the combination of CR and RT provides greater fitness benefits compared to CR alone. Furthermore, the effects of CR in sarcopenic older adults is relatively unknown. No research exists regarding the efficacy of CR in older adults with inborn creatine synthesis deficiencies involving arginine–glycine amidinotransferase (AGAT), guanidinoacetate methyl transferase (GAMT), solute carrier 6 (SLC6AB), or CT1 (creatine transporter). Future research should investigate the effects of CR, with and without RT, in older clinical populations with possible musculoskeletal disorders and creatine synthesis/transporter deficiencies.

4. Potential of Creatine Supplementation for Osteoporosis

Osteoporosis refers to age-related loss of bone mineral density (BMD) and architecture [43] that increases bone fragility and the risks of falls and fractures [44]. There are 8 published studies that have examined the combined effects of CR and RT on properties of bone in older adults, with only 3 of these studies showing greater effects from creatine compared to PLA (Table 2). In healthy older men, 12 weeks of CR (loading phase: 0.3 g/kg/day for 5 days; maintenance phase: 0.07 g/kg/day for an additional 79 days) and supervised whole-body RT increased upper-limb bone mineral content (assessed by dual energy X-ray absorptiometry [DXA]) compared to PLA [45]. Additional work in healthy older men showed that 10 weeks of CR (0.1 g/kg/day) and supervised whole-body RT decreased the urinary excretion of cross-linked N-telopeptides of type I collagen (indicator of bone resorption) compared to PLA [25]). Most recently, Chilibeck et al. [28] showed that CR (0.1 g/kg/day) and supervised whole-body RT for 52 weeks attenuated the rate of bone mineral loss in the femoral neck (assessed by DXA) (Figure 1) and increased femoral shaft subperiosteal width (indicator of bone bending strength) in postmenopausal women compared to PLA.

Table 2. Study characteristics and outcomes of research examining the influence of creatine with a resistance training program on bone.

First Author, Year	Study Population	Intervention	Duration	Outcomes
Brose et al. 2003 [24]	$N = 28$; healthy (15 men, 13 women); age $\geq$ 65 y (men = 68.7 y, women = 70.8 y)	RCT; CR + RT, PLA + RT. CR = 5 g/day; RT = 3 x/wk	14 wks	$\leftrightarrow$ on osteocalcin
Candow et al. 2008 [25]	$N = 35$; older men (age: 59–77 y)	RCT; CR + PRO + RT; CR + RT, PLA + RT; CR = 0.1 g/kg/day; RT = 3 x/wk	10 wks	CR $\downarrow$ NTx
Candow et al. 2019 [5]	$N = 39$; healthy (17 men; 22 women); age $\geq$ 50 y (mean ~55 y)	RCT; CR-Before + RT, CR-After + RT, PLA + RT; CR = 0.1 g/kg/day; RT = 3 x/wk	8 mths	$\leftrightarrow$ BMD and BMC of the whole-body, limbs, femoral neck, lumbar spine, and total hip
Candow et al. 2020 [27]	$N = 38$; healthy men; age = 49–67 y	RCT; CR + RT, PLA + RT; CR = 0.1 g/kg/day; RT = 3 x/wk	12 mths	$\leftrightarrow$ BMD and geometry, bone speed of sound; CR $\uparrow$ ($p = 0.06$) section modulus of the narrow part of the femoral neck
Chilibeck et al. 2005 [45]	$N = 29$; older men (71 y).	RCT; CR + RT, PLA + RT; CR = 0.3 g/kg/day for 5 days followed by 0.07 g/kg/day for the remaining; RT = 3 x/wk	12 wks	$\uparrow$ arm BMC greater in the CR group com-pared to PLA; $\leftrightarrow$ between groups for whole-body and leg BMD
Chilibeck et al. 2015 [28]	$N = 33$; postmenopausal women; age: 57 $\pm$ 6 y	RCT; PLA + RT, CR + RT; CR = 0.1 g/kg/day (0.05 g/kg provided immediately before and 0.05 g/kg after training on training days and with two meals on non-training days); RT = 3 x/wk	12 mths	CR attenuated rate of femoral neck BMD loss compared to PLA and CR $\uparrow$ femoral shaft subperiosteal width; $\leftrightarrow$ between groups on all other outcome measures
Gualano et al. 2014 [34]	$N = 60$; older vulnerable women (age: 66 y)	RCT; PLA, CR, PLA + RT, CR + RT; CR = 20 g/day for 5 days followed by 5 g/day for the remaining; RT = 2 x/wk	24 wks	$\leftrightarrow$ bone mineral and serum bone markers between groups
Pinto et al. 2016 [37]	$N = 32$; healthy, non-athletic men and women between 60–80 y	RCT; PLA + RT, CR + RT; CR = 5 g/day; RT = 3 x/wk. Muscle groups (i.e., upper and lower body) alternated between training days, 1.5 x/wk per muscle group	12 wks	$\leftrightarrow$ BMD and BMC of all assessed sites between groups

RCT = randomized controlled trial; PLA = placebo; RT = resistance training; CR = creatine; PRO = protein; RM = repetition maximum; NTx = cross-linked N-telopeptides of type I collagen; BMD = bone mineral density; BMC = bone mineral content; $\uparrow$ = significant greater; $\leftrightarrow$ no difference be-tween conditions; wk = weeks, mth – months; y = years; g = grams; kg – kilograms

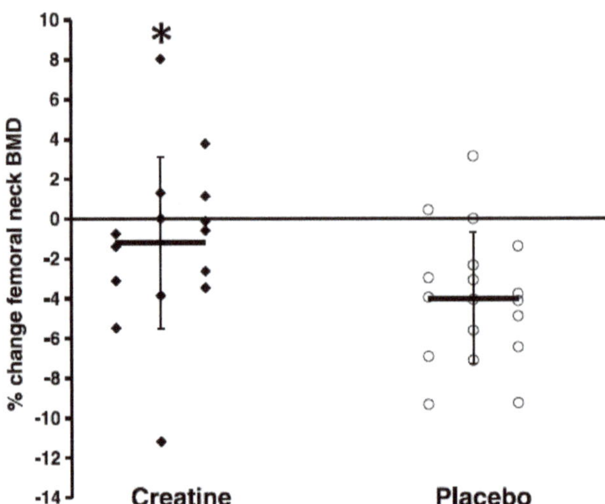

Figure 1. Relative changes in femoral neck bone mineral density (BMD). "Closed diamonds" represent changes for individual creatine group participants, and "open circles" represent placebo group participants. The "horizontal bars" represent the group means, and the "vertical bars" represent the SD. * Creatine participants lost significantly less BMD at the femoral neck compared with placebo participants ($p < 0.05$). (Reproduced with permission from Chilibeck et al. 2015 [28]).

In contrast to these studies, Brose et al. [24] was unable to find a beneficial effect from 14 weeks of CR (5 g/day) and whole-body RT on serum osteocalcin (indicator of bone formation) compared to PLA in healthy older adults. Furthermore, Gualano et al. [34] found no effect from CR (loading phase: 20 g/day for 5 days; maintenance phase: 5 g/day for an additional 24 weeks) and supervised whole-body RT on changes in bone mineral (density and content; assessed by DXA) or serum concentrations of procollagen type 1 N-propeptide (P1NP; indicator of bone formation) and type 1 collagen C-telopeptide (CTX; indicator of bone resorption) compared to PLA in older women. In addition, 12 weeks of CR (5 g/day) and supervised whole-body RT had no greater effect on measures of BMD or content (assessed by DXA) compared to PLA in healthy older adults [37]. Similarly, Candow et al. [46] was unable to find greater effects from CR (0.1 g/kg/day) and 32 weeks of supervised whole-body RT on measures of bone mineral (density and content; assessed by DXA) compared to PLA in healthy older adults. Most recently, Candow et al. [27] failed to show a beneficial effect from 52 weeks of CR (0.1 g/kg/day) and supervised whole-body RT on measures of BMD or bone geometric properties (assessed by DXA and ultrasound) in older men compared to PLA.

There are only three studies that have investigated the effects of CR alone (no exercise training stimulus) on properties of aging bone. In postmenopausal women with osteopenia or osteoporosis, 24 weeks of CR (loading phase: 20 g/day for 5 days; maintenance phase: 5 g/day for an additional 23 weeks) had no effect on measures of BMD (whole-body, lumbar, total femur, and femoral neck; assessed by DXA) or serum markers of bone turnover (CTX, P1NP) compared to PLA [34]. In two additional studies involving postmenopausal women, CR (1 g/day for 52 weeks) had no effect on measures of BMD (assessed by DXA), bone microarchitecture (assessed by high-resolution peripheral quantitative computed tomography (HR-pQCT)), CTX, or P1NP compared to PLA [47]. Increasing the dosage of creatine to 3 g/day for an additional 52 weeks (104 weeks in total) also had no effect on the same bone measures in postmenopausal women. Furthermore, creatine had no effect on the number of falls or fractures experienced [48].

Collectively, the vast majority of studies show no greater effect from CR, with and without RT, on properties of bone in older adults. In the few studies that did show beneficial effects, CR was combined with supervised whole-body RT. Importantly, no study showed any detrimental effect from CR on bone mineral or geometry. The combined effects of CR and RT on reducing the risk and incidence of falls and fractures in older adults is largely unknown. Bone tissue typically takes a long time (i.e., several months) to turnover [49], especially in older adults [50]. Future research should investigate the longer-term effects (i.e., ≥2 years) of CR, with and without RT, on properties of bone mineral and geometry and risk of falls and fractures in older adults.

5. Potential of Creatine Supplementation for Osteosarcopenia

Osteosarcopenia is a musculoskeletal syndrome characterized by low BMD (osteopenia/osteoporosis) and muscle mass and function (sarcopenia) and is predictive of functional impairments, falls, fractures, and premature mortality in older adults [51]. Despite recent commentary proposing the use of CR to combat age-related muscle and bone loss [52,53], no randomized controlled trial (RCT) has tested the effects of creatine versus PLA (control) in osteosarcopenic adults [54]. Nevertheless, there is potential for CR, with and without RT, to be used as an upstream prevention or downstream treatment strategy for this age-related debilitating syndrome.

Creatine may directly or indirectly impact components of osteosarcopenia (muscle mass, bone density, and function) via its actions on muscle and bone metabolism. In skeletal muscle, creatine is capable of upregulating anabolic signaling pathways, increasing satellite cell number/content and growth factors, as well as downregulating markers of inflammation and oxidative stress [5,6,8,9]. Creatine's role is not exclusive to muscle tissue, with preclinical studies showing that creatine may promote the differentiation of osteoblast cells involved in bone formation [55,56]. Despite this, in the absence of RT, findings from mechanistic studies involving CR have largely failed to translate into clinical improvements in musculoskeletal outcomes in healthy older adults. As highlighted previously, experimental trials have shown significant heterogeneity in study methodology, which likely relates to the inconsistent findings (see Tables 1 and 2).

Of note, 2 years of CR (3 g/day) without RT did not influence bone density/micro architecture, bone turnover markers, lean mass and muscle strength/function, or falls and fractures in postmenopausal women with osteopenia [48]. However, these participants were not sarcopenic or osteosarcopenic and the RCT was not powered to detect the effects of creatine on falls and fractures, which were considerably low throughout the trial [48]. Importantly, the authors did not rule out the possible benefits of CR in conjunction with RT.

It is somewhat surprising that no RCT exists in osteosarcopenic individuals despite the well-established biomechanical and biochemical connection between muscle and bone [57]. Indeed, skeletal muscle acts as a pulley and bone as a lever during human movement and the forces applied to myofibres during RT are transmitted to bone to initiate osteocyte-induced bone formation [57], and during inactivity, the opposite occurs, leading to degeneration of both tissues. This biomechanical interaction during activity (or lack of during inactivity) occurs alongside biochemical cross-talk via hormones and other growth factors secreted by muscle and bone cells [57]. Given that lean (muscle) mass is a major predictor of BMD [58] that and osteopenia/osteoporosis increases the risk of sarcopenia (the opposite is also true) [59], it is possible that creatine's anabolic effect on skeletal muscle may indirectly promote bone accretion and geometry. For instance, CR increases insulin-like growth factor I (IGF-1) content [60] and downregulates myostatin levels [61,62], and the former initiates osteoblastogenesis (bone formation) while the latter initiates osteoclastogenesis (bone resorption) [57]. Thus, aside from the mechanical interaction, cross-talk between muscle and bone cells represents another feasible avenue by which CR has the potential to combat osteosarcopenia. To test this hypothesis, future mechanistic studies should examine the effects of CR on hormonal factors released by the endocrine system in addition to growth factors (osteokines and myokines) secreted by muscle and bone cells. Furthermore, in order to determine the efficacy and safety of CR in older osteosarcopenic adults, future RCTs should include measures of both muscle and bone (muscle mass, strength, physical function, bone structure, and bone biomarkers) as well as clinically relevant outcomes on activities of daily living, falls, and fractures. It is also important that vital signs and adverse events are recorded in future RCTs involving creatine dosages, both of which are not consistently monitored or reported on in exercise/nutritional trials [63]. Given that adaptions in muscle mass and cortical/trabecular bone may take at least 6 months to be radiographically detected following anabolic stimuli (i.e., RT) in older osteosarcopenic adults [54], CR protocols should at least match or exceed this duration. Finally, as poor nutritional status is a risk factor for osteosarcopenia [52,53,64], the possible interaction of creatine with other essential nutrients capable of modulating muscle and bone metabolism such protein, vitamin D, and calcium [64] should be explored. Providing this information is of clinical relevance as creatine, with or without RT, may be a cost-effective strategy to treat older adults with or at risk of osteosarcopenia.

6. Potential of Creatine Supplementation for Sarcopenic Obesity

Sarcopenia has a negative effect on mobility, energy expenditure, and metabolic health, which subsequently increases adipose tissue accumulation (i.e., obesity [65]), especially in and around skeletal muscle [66]. Sarcopenic obesity (SO) occurs in approximately 20% of older adult populations [67] and increases the risk of cardiometabolic diseases, osteoporosis, disability, and premature mortality (for review, see Roh and Choi [68]). Similar to sarcopenia, there is no unanimous definition of SO. The WHO classifies obesity as a body mass index (BMI) $\geq$30 kg/m^2. However, individuals of east Asian descent have elevated body fat% compared to non-Asians with an equivalent BMI [69]; thus, east Asian's have a lower BMI cutoff point for obesity ($\geq$25 kg/m^2). Beyond BMI, body fat distribution (i.e., waist circumference) enhances the ability to predict the development of metabolic syndrome and risk of cardiovascular disease [70], with the WHO indicating cutoff values of $\geq$102 cm for men and $\geq$88 cm for women, which differ for Asian populations [71,72]. Further, the American Association of Clinical Endocrinology [73] classifies obesity using body fat thresholds of >25% in men and >35% in women. Due to the lack of a universally accepted definition, the prevalence of SO varies. For example, in a prospective study of older adults (n = 4652; >60 years of age), the prevalence of sarcopenic obesity was 18.1% in women and 42.9% in men [74].

A recent systematic review and meta-analysis of 19 studies involving older adults (n = 609 participants; $\geq$50 years of age) showed that the combination of CR and RT resulted in a greater reduction in fat mass (0.5 kg, p = 0.13) and body fat% (0.55%, p = 0.04) compared to PLA and RT [75]. Mechanistically, creatine appears to also influence adipose tissue biology. In multiple adipogenic cell culture models, creatine attenuated the accumulation of cytoplasmic triglycerides in a dose-dependent manner through inhibition of phosphatidylinositol 3-kinase activation [76]. There is also evidence that creatine can alter whole-body energetics and expenditure [77]. In rodents, diminishing creatine content impaired thermal homeostasis [78] and deletion of glycine amidinotransferase (the rate limiting enzyme of creatine synthesis) attenuated creatine content in brown adipose and impaired thermoregulation [77,79], which subsequently attenuated the capacity to activate diet-induced thermogenesis, resulting in increased adiposity [79]. Furthermore, global creatine transporter (Slc6a8) knockout mice presented with greater body fat stores compared to controls [80] possibly due to lower whole-body energy expenditure, a decrease in oxidative metabolism in beige and brown adipose tissue, and an increase in feed efficiency [81].

Collectively, CR and RT appear to be an effective intervention for decreasing body fat% in older adults. However, the effects of CR alone on adipose tissue biology in older adults are unknown. Furthermore, it remains to be determined whether CR, with and without RT or other exercise-training modalities (i.e., aerobic), can overcome SO. Based on the potential interaction of muscle and fat tissue and mechanistic actions of creatine on adipogenesis and whole-body energy expenditure, future research is warranted and may be of clinical importance for older adults.

7. Potential of Creatine Supplementation for Physical Frailty

Frailty is defined as a syndrome of physiological decline in later life, characterized by vulnerability to adverse health outcomes (i.e., hospitalization, falls, social isolation, and reduced quality of life (QoL)) [82–84]. Frailty is commonly defined according to the phenotype of physical frailty proposed by Fried et al. [82], which consists of weakness, slowness, low levels of physical activity, shrinking, and exhaustion, with one or two criteria indicating a prefrailty stage and three or more marking frailty. A recent study in a combined cohort of 8804 Australian adults aged ≥65 years (women 86%, median age 80 years) found that, while 21% of participants were frail, a staggering 48% were prefrail [85]. With the aging population and growing incidence of prefrail older adults progressing to frailty every year (at a rate of 11%), the incidence of adverse health outcomes represents a substantial burden on total healthcare costs worldwide.

Due to the predominantly musculoskeletal and physical components of the frailty phenotype, there is an unavoidable overlap between sarcopenia, osteosarcopenia, and physical frailty [86]. Regarding their common pathophysiology, these conditions share immune, endocrine, and inflammatory mechanisms, which could be targeted via nonpharmacological interventions such as exercise and nutrition (i.e., creatine) [64,87]. However, research is very limited regarding CR and physical frailty. Although some studies testing CR have involved older adults that could fulfil the clinical phenotype, those participants are usually labelled as healthy or non-sarcopenic. In the only clinical trial to directly assess the effects of CR in mildly frail older adults (defined as those with limited dependence on others for instrumental activities of daily living according to the Canadian Study of Health and Aging clinical frailty scale [88], Collins et al. [89] found no additive effect from CR (5 g/day) to whey protein (20 g) and RT (14 weeks) on measures of muscle strength and functionality compared to whey protein and training alone. However, the small sample size (n = 16) and lack of PLA and non-training control group limits the clinical application of these preliminary findings. In addition to the effects of CR on muscle and bone as previously described, there is evidence that creatine may have beneficial effects on the other components of the frailty syndrome (summarized in [90]). Creatine exhibits an anti-inflammatory effect via regulation of the cyclo-oxygenase pathway and reduction of serum levels of inflammatory cytokines (i.e., tumor necrosis factor-α (TNF-α) and IL-6), which have been associated with sarcopenia, osteoporosis, and frailty [91,92]. The small number of studies that have examined the efficacy of CR on immune system response have shown an alteration in soluble mediator production and expression of molecules involved in recognizing infections, specifically toll-like receptors. Creatine has also been proposed to be neuroprotective, an effect that could have a potential role in the treatment of the neuromuscular components of frailty [93]. Finally, creatine may act as an antioxidant, which would also benefit frail older adults susceptible to increased oxidative stress and damage [94].

In summary, despite the preclinical and clinical evidence demonstrating an effect from creatine on multiple pathophysiological mechanisms associated with frailty, no RCT has been performed examining the effects of CR (alone or in combination with exercise) in frail older adults. A major challenge in this line of research relates to the identification of frailty, which could lead to significant variability across studies. In the case of physical frailty, we propose that the adoption of Fried's criteria be used to facilitate and identify the condition through a well-accepted phenotype in which the quantification of any therapeutic effect(s) from CR can be made.

8. Potential of Creatine Supplementation for Cachexia

Cachexia can be defined as a tissue loss syndrome that involves severe weight loss and muscle wasting [95]. Cachexia is usually secondary to conditions such as cancer, COPD, chronic kidney disease, and heart failure, and therefore, therapeutic interventions should involve not only the prevention of muscle wasting but also appropriate treatment of the secondary cause.

Muscle loss in cachexia is due to both reduced protein synthesis and increased catabolism (proteolysis) due to multiple factors including reduced oral intake, high levels of inflammation, tumor-mediated effects, low physical activity, and endocrine and metabolic disturbances [96]. Some studies have demonstrated that CR can have an effect on the majority of these mechanisms; however, PLA-controlled studies have shown mixed results. Most of the studies involving CR and cachexia have been performed in cancer patients (summarized in Table 3). Overall, CR failed to produce significant effects on muscle accretion, muscle performance, or functionality. However, creatine had no detrimental effect on muscle, bone, performance, or functionality and no major adverse events were reported from those taking creatine. The inconsistent findings across studies are possibly explained by low sample sizes, multifactorial nature of the condition, deleterious effect of chemotherapy on muscle, lack of exercise intervention, length and dosage of creatine used, and heterogeneity of secondary causes of cachexia.

There may be potential for cachexia patients to experience some benefits from CR and RT. For example, some subsets of cancer are characterized by high rates of weight and muscle loss (i.e., head and neck, pancreatic, lung, colorectal, and gastric cancer) [97], which may be counteracted by creatine. Additionally, CR may also be beneficial for older adults with cachexia and/or cancer undergoing treatments that negatively affect muscle and bone mass, performance, and function (i.e., androgen deprivation therapies).

In summary, cachexia is a debilitating condition associated with multiple chronic diseases, especially cancer. Creatine has the potential to target several of the mechanisms associated with cachexia; however, research investigating the effects of creatine and cachexia is very limited. Future large-scale RCT's examining the effects of creatine, with and without exercise and pharmacological therapies, are warranted and needed.

Table 3. Studies examining Cr supplementation in a cancer context.

Authors (Year)	Patients	Treatment Modality	Dosage	Protocol Du-RATION	Compliance	Exercise Program	Results	Adverse Effects Related to Supplementation
Jatoi et al. (2017) [98]	263 cancer patients (65 ± 11 yrs.) with weight loss syndrome	210 undergoing concurrent chemotherapy	20 g/day for 5 days then 2 g/day	39 weeks	nr	n/a	↔body weight, appetite, QoL, frailty, grip strength	None reported
Bourgeois e al (2008) [99]	9 children (7.6 ± 3.8 yrs.) with ALL under-going chemotherapy	Maintenance phase of treatment on the Dann-Farber Cancer Institute protocol 2000–2001	0.1 g/kg/day	2 × 16 weeks separated by 6-week wash-out period.	nr	n/a	↓ BF% Cr, ↑ BF% NH, ↔BMD	None reported
Norman et al. (2006) [100]	31 stage III/IV colorectal cancer patients (65.10 ± 12.55 yrs.) undergoing chemotherapy	$n = 11$: fluorouracil/folic acid (5-FU FA); $n = 9$: fluorouracil/folic acid + oxaliplatin (5-FU FA + O); $n = 11$: fluorouracil/folic acid + irinotecan (5-FU FA + I)	20 g/day for 7 days then 5 g/day	8 weeks	Cr: 84.55 ± 7.77%; PLA: 87.62 ± 5.90%	n/a	↔ weight, capacitance, KE, HR, BCM, BF; ↑ HG 5-FU FA: ↑ phase angle, ECM/BCM ratio	None reported
Lonbro et al. (2013) [10]	30 Head and neck patients treated with radiotherapy	Radiotherapy according to DA-HANCA guidelines (www.da-hanca.dk) + chemotherapy ($n = 20$: cisplatin, 40 mg/m^2). N = 4 received Zalutumumab	5 g/day + 30 g Pro/day	12 weeks	69% ingested all supplementation; 19% missed ≤ 3 supplementations; 12% terminated 4 weeks early	3 days/wk., 3 × 10 total body	↑LBM ProCr group, ns↑PLA ↔ muscle strength **, ↔ Physical function **	No major adverse events reported; 2 participants stopped supplementation 4 weeks early due to muscle cramping and mucus production

ALL: acute lymphoblastic leukemia; BCM: body cell mass; BF%: body fat percentage; BMD: bone mineral density; Cr: creatine supplementation; ECM: extracellular mass; g: gram; HG: hand grip; HF: hip flexion; Kg: kilogram; KE: knee extension; LBM: lean body mass; MF: muscle function; NH: natural history group; PLA: placebo; ProCr: protein + creatine supplementation; n/a: not applicable; nr: not reported; QoL: quality of life; yrs: years old; ↑: increase; ↓: decrease; ↔: no change; ** compared to placebo. (Reproduced with permission from Fairman et al. 2019 [97]).

9. Conclusions

Sarcopenia refers to age-related reduction in muscle mass, strength, and/or physical performance and has a negative effect on the ability to perform activities of daily living and overall quality of life. Comorbidities associated with sarcopenia include osteoporosis, osteosarcopenia, sarcopenic obesity, physical frailty, and cachexia. As a possible countermeasure to sarcopenia and its age-related co-morbidities, CR (especially when combined with RT) has some favourable effects on aging muscle, bone and fat mass, muscle and bone strength, and physical performance, primarily in healthy populations (Figure 2). Independent of RT, a CR loading phase and/or high relative daily dosage of creatine ($\geq$0.3 g/kg) may be required to produce some muscle benefits in older adults. CR (independent of resistance training) for up to 2 years appears to provide no bone benefits in older females. The effects of CR alone on bone measures in older males is unknown. Despite its potential, the effects of CR in older adults with sarcopenia, osteoporosis, osteosarcopenia, sarcopenic obesity, physical frailty, and cachexia remain largely unknown and warrant future long-term clinical trials involving large sample sizes.

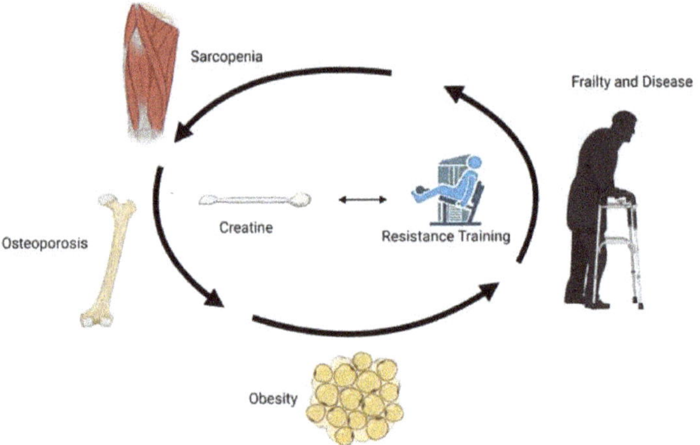

Figure 2. Potential effect of creatine, with and without resistance training.

Author Contributions: Conceptualization, D.G.C., B.K. and G.D.; writing—original draft preparation, D.G.C., S.C.F., B.K. and G.D.; writing—review and editing, D.G.C., S.C.F., B.K. and G.D. All authors have read and agreed to the published version of the manuscript.

Funding: This research received no external funding.

Conflicts of Interest: D.G.C. has conducted industry sponsored research involving creatine supplementation and received creatine donations for scientific studies and travel support for presentations involving creatine supplementation at scientific conferences. In addition, D.G.C. serves on the Scientific Advisory Board for Alzchem (a company that manufactures creatine). S.C.F. has served as a scientific advisor for a company that sells creatine products. B.K. and G.D. declare no conflicts of interest.

References

1. Cruz-Jentoft, A.J.; Bahat, G.; Bauer, J.; Boirie, Y.; Bruyere, O.; Cederholm, T.; Cooper, C.; Landi, F.; Rolland, Y.; Sayer, A.A.; et al. Writing Group for the European Working Group on Sarcopenia in Older People 2 (EWGSOP2), and the Extended Group for EWGSOP2 Sarcopenia: Revised European consensus on definition and diagnosis. *Age Ageing* **2019**, *48*, 16–31. [CrossRef]
2. Shafiee, G.; Keshtkar, A.; Soltani, A.; Ahadi, Z.; Larijani, B.; Heshmat, R. Prevalence of sarcopenia in the world: A systematic review and meta- analysis of general population studies. *J. Diabetes Metab. Disord.* **2017**, *16*, 1–10. [CrossRef]
3. Tournadre, A.; Vial, G.; Capel, F.; Soubrier, M.; Boirie, Y. Sarcopenia. *Joint Bone Spine* **2019**, *86*, 309–314. [CrossRef]

4.   Mitchell, W.K.; Williams, J.; Atherton, P.; Larvin, M.; Lund, J.; Narici, M. Sarcopenia, dynapenia, and the impact of advancing age on human skeletal muscle size and strength; a quantitative review. *Front. Physiol.* **2012**, *3*, 260. [CrossRef]

5.   Candow, D.G.; Forbes, S.C.; Chilibeck, P.D.; Cornish, S.M.; Antonio, J.; Kreider, R.B. Effectiveness of Creatine Supplementation on Aging Muscle and Bone: Focus on Falls Prevention and Inflammation. *J. Clin. Med.* **2019**, *8*, 488. [CrossRef]

6.   Chilibeck, P.D.; Kaviani, M.; Candow, D.G.; Zello, G.A. Effect of creatine supplementation during resistance training on lean tissue mass and muscular strength in older adults: A meta-analysis. *Open Access J. Sports Med.* **2017**, *8*, 213–226. [CrossRef]

7.   Forbes, S.C.; Chilibeck, P.D.; Candow, D.G. Creatine Supplementation During Resistance Training Does Not Lead to Greater Bone Mineral Density in Older Humans: A Brief Meta-Analysis. *Front. Nutr.* **2018**, *5*, 27. [CrossRef]

8.   Gualano, B.; Rawson, E.S.; Candow, D.G.; Chilibeck, P.D. Creatine supplementation in the aging population: Effects on skeletal muscle, bone and brain. *Amino Acids* **2016**, *48*, 1793–1805. [CrossRef] [PubMed]

9.   Kreider, R.B.; Kalman, D.S.; Antonio, J.; Ziegenfuss, T.N.; Wildman, R.; Collins, R.; Candow, D.G.; Kleiner, S.M.; Almada, A.L.; Lopez, H.L. International Society of Sports Nutrition position stand: Safety and efficacy of creatine supplementation in exercise, sport, and medicine. *J. Int. Soc. Sports Nutr.* **2017**, *14*, 1–18. [CrossRef] [PubMed]

10. Wyss, M.; Kaddurah-Daouk, R. Creatine and creatinine metabolism. *Physiol. Rev.* **2000**, *80*, 1107–1213. [CrossRef] [PubMed]

11. Candow, D.G. Sarcopenia: Current theories and the potential beneficial effect of creatine application strategies. *Biogerontology* **2011**, *12*, 273–281. [CrossRef]

12. Candow, D.G.; Forbes, S.C.; Little, J.P.; Cornish, S.M.; Pinkoski, C.; Chilibeck, P.D. Effect of nutritional interventions and resistance exercise on aging muscle mass and strength. *Biogerontology* **2012**, *13*, 345–358. [CrossRef]

13. Candow, D.G.; Forbes, S.C.; Chilibeck, P.D.; Cornish, S.M.; Antonio, J.; Kreider, R.B. Variables Influencing the Effectiveness of Creatine Supplementation as a Therapeutic Intervention for Sarcopenia. *Front. Nutr.* **2019**, *6*, 124. [CrossRef]

14. Forbes, S.C.; Little, J.P.; Candow, D.G. Exercise and nutritional interventions for improving aging muscle health. *Endocrine* **2012**, *42*, 29–38. [CrossRef]

15. Candow, D.G.; Chilibeck, P.D.; Forbes, S.C. Creatine supplementation and aging musculoskeletal health. *Endocrine* **2014**, *45*, 354–361. [CrossRef] [PubMed]

16. Devries, M.C.; Phillips, S.M. Creatine supplementation during resistance training in older adults-a meta-analysis. *Med. Sci. Sports Exerc.* **2014**, *46*, 1194–1203. [CrossRef] [PubMed]

17. Candow, D.G.; Chilibeck, P.D.; Chad, K.E.; Chrusch, M.J.; Davison, K.S.; Burke, D.G. Effect of ceasing creatine supplementation while maintaining resistance training in older men. *J. Aging Phys. Act.* **2004**, *12*, 219–231. [CrossRef] [PubMed]

18. Aguiar, A.F.; Januario, R.S.; Junior, R.P.; Gerage, A.M.; Pina, F.L.; do Nascimento, M.A.; Padovani, C.R.; Cyrino, E.S. Long-term creatine supplementation improves muscular performance during resistance training in older women. *Eur. J. Appl. Physiol.* **2013**, *113*, 987–996. [CrossRef]

19. Alves, C.R.; Merege Filho, C.A.; Benatti, F.B.; Brucki, S.; Pereira, R.M.; de Sa Pinto, A.L.; Lima, F.R.; Roschel, H.; Gualano, B. Creatine supplementation associated or not with strength training upon emotional and cognitive measures in older women: A randomized double-blind study. *PLoS ONE* **2013**, *8*, e76301. [CrossRef]

20. Bemben, M.G.; Witten, M.S.; Carter, J.M.; Eliot, K.A.; Knehans, A.W.; Bemben, D.A. The effects of supplementation with creatine and protein on muscle strength following a traditional resistance training program in middle-aged and older men. *J. Nutr. Health Aging* **2010**, *14*, 155–159. [CrossRef]

21. Eliot, K.A.; Knehans, A.W.; Bemben, D.A.; Witten, M.S.; Carter, J.; Bemben, M.G. The effects of creatine and whey protein supplementation on body composition in men aged 48 to 72 years during resistance training. *J. Nutr. Health Aging* **2008**, *12*, 208–212. [CrossRef]

22. Bermon, S.; Venembre, P.; Sachet, C.; Valour, S.; Dolisi, C. Effects of creatine monohydrate ingestion in sedentary and weight-trained older adults. *Acta Physiol. Scand.* **1998**, *164*, 147–155. [CrossRef]

23. Bernat, P.; Candow, D.G.; Gryzb, K.; Butchart, S.; Schoenfeld, B.J.; Bruno, P. Effects of high-velocity resistance training and creatine supplementation in untrained healthy aging males. *Appl. Physiol. Nutr. Metab.* **2019**, *44*, 1246–1253. [CrossRef] [PubMed]

24. Brose, A.; Parise, G.; Tarnopolsky, M.A. Creatine supplementation enhances isometric strength and body composition improvements following strength exercise training in older adults. *J. Gerontol. A Biol. Sci. Med. Sci.* **2003**, *58*, 11–19. [CrossRef]

25. Candow, D.G.; Little, J.P.; Chilibeck, P.D.; Abeysekara, S.; Zello, G.A.; Kazachkov, M.; Cornish, S.M.; Yu, P.H. Low-dose creatine combined with protein during resistance training in older men. *Med. Sci. Sports Exerc.* **2008**, *40*, 1645–1652. [CrossRef]

26. Candow, D.G.; Vogt, E.; Johannsmeyer, S.; Forbes, S.C.; Farthing, J.P. Strategic creatine supplementation and resistance training in healthy older adults. *Appl. Physiol. Nutr. Metab.* **2015**, *40*, 689–694. [CrossRef]

27. Candow, D.G.; Chilibeck, P.D.; Gordon, J.; Vogt, E.; Landeryou, T.; Kaviani, M.; Paus-Jensen, L. Effect of 12 months of creatine supplementation and whole-body resistance training on measures of bone, muscle and strength in older males. *Nutr. Health* **2020**, 260106020975247. [CrossRef]

28. Chilibeck, P.D.; Candow, D.G.; Landeryou, T.; Kaviani, M.; Paus-Jenssen, L. Effects of Creatine and Resistance Training on Bone Health in Postmenopausal Women. *Med. Sci. Sports Exerc.* **2015**, *47*, 1587–1595. [CrossRef] [PubMed]

29. Chrusch, M.J.; Chilibeck, P.D.; Chad, K.E.; Davison, K.S.; Burke, D.G. Creatine supplementation combined with resistance training in older men. *Med. Sci. Sports Exerc.* **2001**, *33*, 2111–2117. [CrossRef]

30. Cooke, M.B.; Brabham, B.; Buford, T.W.; Shelmadine, B.D.; McPheeters, M.; Hudson, G.M.; Stathis, C.; Greenwood, M.; Kreider, R.; Willoughby, D.S. Creatine supplementation post-exercise does not enhance training-induced adaptations in middle to older aged males. *Eur. J. Appl. Physiol.* **2014**, *114*, 1321–1332. [CrossRef] [PubMed]

31. Deacon, S.J.; Vincent, E.E.; Greenhaff, P.L.; Fox, J.; Steiner, M.C.; Singh, S.J.; Morgan, M.D. Randomized controlled trial of dietary creatine as an adjunct therapy to physical training in chronic obstructive pulmonary disease. *Am. J. Respir. Crit. Care Med.* **2008**, *178*, 233–239. [CrossRef]

32. Eijnde, B.O.; Van Leemputte, M.; Goris, M.; Labarque, V.; Taes, Y.; Verbessem, P.; Vanhees, L.; Ramaekers, M.; Vanden Eynde, B.; Van Schuylenbergh, R.; et al. Effects of creatine supplementation and exercise training on fitness in men 55-75 yr old. *J. Appl. Physiol. (1985)* **2003**, *95*, 818–828. [CrossRef]

33. Gualano, B.; de Salles Painelli, V.; Roschel, H.; Lugaresi, R.; Dorea, E.; Artioli, G.G.; Lima, F.R.; da Silva, M.E.; Cunha, M.R.; Seguro, A.C.; et al. Creatine supplementation does not impair kidney function in type 2 diabetic patients: A randomized, double-blind, placebo-controlled, clinical trial. *Eur. J. Appl. Physiol.* **2011**, *111*, 749–756. [CrossRef]

34. Gualano, B.; Macedo, A.R.; Alves, C.R.; Roschel, H.; Benatti, F.B.; Takayama, L.; de Sa Pinto, A.L.; Lima, F.R.; Pereira, R.M. Creatine supplementation and resistance training in vulnerable older women: A randomized double- blind placebo-controlled clinical trial. *Exp. Gerontol.* **2014**, *53*, 7–15. [CrossRef] [PubMed]

35. Johannsmeyer, S.; Candow, D.G.; Brahms, C.M.; Michel, D.; Zello, G.A. Effect of creatine supplementation and drop-set resistance training in untrained aging adults. *Exp. Gerontol.* **2016**, *83*, 112–119. [CrossRef]

36. Neves, M.; Gualano, B.; Roschel, H.; Fuller, R.; Benatti, F.B.; Pinto, A.L.; Lima, F.R.; Pereira, R.M.; Lancha, A.H.; Bonfa, E. Beneficial effect of creatine supplementation in knee osteoarthritis. *Med. Sci. Sports Exerc.* **2011**, *43*, 1538–1543. [CrossRef] [PubMed]

37. Pinto, C.L.; Botelho, P.B.; Carneiro, J.A.; Mota, J.F. Impact of creatine supplementation in combination with resistance training on lean mass in the elderly. *J. Cachexia Sarcopenia Muscle* **2016**, *7*, 413–421. [CrossRef]

38. Smolarek, A.C.; McAnulty, S.R.; Ferreira, L.H.; Cordeiro, G.R.; Alessi, A.; Rebesco, D.B.; Honorato, I.C.; Laat, E.F.; Mascarenhas, L.P.; Souza-Junior, T.P. Effect of 16 weeks of strength training and creatine supplementation on strength and cognition in older adults: A pilot study. *J. Exerc. Physiol. Online* **2020**, *23*, 88–94.

39. Baumgartner, R.N.; Koehler, K.M.; Gallagher, D.; Romero, L.; Heymsfield, S.B.; Ross, R.R.; Garry, P.J.; Lindeman, R.D. Epidemiology of sarcopenia among the elderly in New Mexico. *Am. J. Epidemiol.* **1998**, *147*, 755–763. [CrossRef] [PubMed]

40. Newman, A.B.; Kupelian, V.; Visser, M.; Simonsick, E.; Goodpaster, B.; Nevitt, M.; Kritchevsky, S.B.; Tylavsky, F.A.; Rubin, S.M.; Harris, T.B. Health ABC Study Investigators Sarcopenia: Alternative definitions and associations with lower extremity function. *J. Am. Geriatr. Soc.* **2003**, *51*, 1602–1609. [CrossRef] [PubMed]

41. Macrae, P.G.; Lacourse, M.; Moldavon, R. Physical performance measures that predict faller status in community- dwelling older adults. *J. Orthop. Sports Phys. Ther.* **1992**, *16*, 123–128. [CrossRef] [PubMed]

42. Forbes, S.C.; Candow, D.G.; Ferreira, L.H.B.; Souza-Junior, T.P. Effects of Creatine Supplementation on Properties of Muscle, Bone, and Brain Function in Older Adults: A Narrative Review. *J. Diet. Suppl.* **2021**, 1–18. [CrossRef]

43. Clynes, M.A.; Harvey, N.C.; Curtis, E.M.; Fuggle, N.R.; Dennison, E.M.; Cooper, C. The epidemiology of osteoporosis. *Br. Med. Bull.* **2020**, *133*, 105–117. [CrossRef]

44. Reginster, J.Y.; Beaudart, C.; Buckinx, F.; Bruyere, O. Osteoporosis and sarcopenia: Two diseases or one? *Curr. Opin. Clin. Nutr. Metab. Care* **2016**, *19*, 31–36. [CrossRef] [PubMed]

45. Chilibeck, P.D.; Chrusch, M.J.; Chad, K.E.; Shawn Davison, K.; Burke, D.G. Creatine monohydrate and resistance training increase bone mineral content and density in older men. *J. Nutr. Health Aging* **2005**, *9*, 352–353.

46. Candow, D.G.; Forbes, S.C.; Vogt, E. Effect of pre-exercise and post-exercise creatine supplementation on bone mineral content and density in healthy aging adults. *Exp. Gerontol.* **2019**, *119*, 89–92. [CrossRef]

47. Lobo, D.M.; Tritto, A.C.; da Silva, L.R.; de Oliveira, P.B.; Benatti, F.B.; Roschel, H.; Niess, B.; Gualano, B.; Pereira, M. Effects of long-term low-dose dietary creatine supplementation in older women. *Exp. Gerontol.* **2015**, *70*, 97–104. [CrossRef]

48. Sales, L.P.; Pinto, A.J.; Rodrigues, S.F.; Alvarenga, J.C.; Goncalves, N.; Sampaio-Barros, M.M.; Benatti, F.B.; Gualano, B.; Rodrigues Pereira, R.M. Creatine Supplementation (3 g/d) and Bone Health in Older Women: A 2- Year, Randomized, Placebo-Controlled Trial. *J. Gerontol. A Biol. Sci. Med. Sci.* **2020**, *75*, 931–938. [CrossRef] [PubMed]

49. Eriksen, E.F. Cellular mechanisms of bone remodeling. *Rev. Endocr. Metab. Disord.* **2010**, *11*, 219–227. [CrossRef] [PubMed]

50. Boskey, A.L.; Coleman, R. Aging and bone. *J. Dent. Res.* **2010**, *89*, 1333–1348. [CrossRef]

51. Salech, F.; Marquez, C.; Lera, L.; Angel, B.; Saguez, R.; Albala, C. Osteosarcopenia Predicts Falls, Fractures, and Mortality in Chilean Community-Dwelling Older Adults. *J. Am. Med. Dir. Assoc.* **2020**. [CrossRef] [PubMed]

52. Kirk, B.; Miller, S.; Zanker, J.; Duque, G. A clinical guide to the pathophysiology, diagnosis and treatment of osteosarcopenia. *Maturitas* **2020**, *140*, 27–33. [CrossRef] [PubMed]

53. Kirk, B.; Zanker, J.; Duque, G. Osteosarcopenia: Epidemiology, diagnosis, and treatment-facts and numbers. *J. Cachexia Sarcopenia Muscle* **2020**, *11*, 609–618. [CrossRef]

54. Atlihan, R.; Kirk, B.; Duque, G. Non-Pharmacological Interventions in Osteosarcopenia: A Systematic Review. *J. Nutr. Health Aging* **2021**, *25*, 25–32. [CrossRef]

55. Antolic, A.; Roy, B.D.; Tarnopolsky, M.A.; Zernicke, R.F.; Wohl, G.R.; Shaughnessy, S.G.; Bourgeois, J.M. Creatine monohydrate increases bone mineral density in young Sprague-Dawley rats. *Med. Sci. Sports Exerc.* **2007**, *39*, 816–820. [CrossRef] [PubMed]

56. Gerber, I.; ap Gwynn, I.; Alini, M.; Wallimann, T. Stimulatory effects of creatine on metabolic activity, differentiation and mineralization of primary osteoblast-like cells in monolayer and micromass cell cultures. *Eur. Cell. Mater.* **2005**, *10*, 8–22. [CrossRef] [PubMed]

57. Kirk, B.; Feehan, J.; Lombardi, G.; Duque, G. Muscle, Bone, and Fat Crosstalk: The Biological Role of Myokines, Osteokines, and Adipokines. *Curr. Osteoporos Rep.* **2020**, *18*, 388–400. [CrossRef]

58. Ho-Pham, L.T.; Nguyen, U.D.; Nguyen, T.V. Association between lean mass, fat mass, and bone mineral density: A meta-analysis. *J. Clin. Endocrinol. Metab.* **2014**, *99*, 30–38. [CrossRef]

59. Kirk, B.; Phu, S.; Brennan-Olsen, S.L.; Bani Hassan, E.; Duque, G. Associations between osteoporosis, the severity of sarcopenia and fragility fractures in community-dwelling older adults. *Eur. Geriatr. Med.* **2020**, *11*, 443–450. [CrossRef]

60. Burke, D.G.; Candow, D.G.; Chilibeck, P.D.; MacNeil, L.G.; Roy, B.D.; Tarnopolsky, M.A.; Ziegenfuss, T. Effect of creatine supplementation and resistance-exercise training on muscle insulin-like growth factor in young adults. *Int. J. Sport Nutr. Exerc. Metab.* **2008**, *18*, 389–398. [CrossRef]

61. Farshidfar, F.; Pinder, M.A.; Myrie, S.B. Creatine Supplementation and Skeletal Muscle Metabolism for Building Muscle Mass-Review of the Potential Mechanisms of Action. *Curr. Protein Pept. Sci.* **2017**, *18*, 1273–1287. [CrossRef]

62. Saremi, A.; Gharakhanloo, R.; Sharghi, S.; Gharaati, M.R.; Larijani, B.; Omidfar, K. Effects of oral creatine and resistance training on serum myostatin and GASP-1. *Mol. Cell. Endocrinol.* **2010**, *317*, 25–30. [CrossRef]

63. El-Kotob, R.; Ponzano, M.; Chaput, J.P.; Janssen, I.; Kho, M.E.; Poitras, V.J.; Ross, R.; Ross-White, A.; Saunders, T.J.; Giangregorio, L.M. Resistance training and health in adults: An overview of systematic reviews. *Appl. Physiol. Nutr. Metab.* **2020**, *45*, S165–S179. [CrossRef]

64. Kirk, B.; Prokopidis, K.; Duque, G. Nutrients to mitigate osteosarcopenia: The role of protein, vitamin D and calcium. *Curr. Opin. Clin. Nutr. Metab. Care* **2021**, *24*, 25–32. [CrossRef]

65. Jakicic, J.M.; Powell, K.E.; Campbell, W.W.; Dipietro, L.; Pate, R.R.; Pescatello, L.S.; Collins, K.A.; Bloodgood, B.; Piercy, K.L. 2018 physical activity guidelines advisory committee* Physical Activity and the Prevention of Weight Gain in Adults: A Systematic Review. *Med. Sci. Sports Exerc.* **2019**, *51*, 1262–1269. [CrossRef] [PubMed]

66. Goodpaster, B.H.; Carlson, C.L.; Visser, M.; Kelley, D.E.; Scherzinger, A.; Harris, T.B.; Stamm, E.; Newman, A.B. Attenuation of skeletal muscle and strength in the elderly: The Health ABC Study. *J. Appl. Physiol.* **2001**, *90*, 2157–2165. [CrossRef] [PubMed]

67. Kim, Y.S.; Lee, Y.; Chung, Y.S.; Lee, D.J.; Joo, N.S.; Hong, D.; Song, G.; Kim, H.J.; Choi, Y.J.; Kim, K.M. Prevalence of sarcopenia and sarcopenic obesity in the Korean population based on the Fourth Korean National Health and Nutritional Examination Surveys. *J. Gerontol. A Biol. Sci. Med. Sci.* **2012**, *67*, 1107–1113. [CrossRef]

68. Roh, E.; Choi, K.M. Health Consequences of Sarcopenic Obesity: A Narrative Review. *Front. Endocrinol. (Lausanne)* **2020**, *11*, 332. [CrossRef] [PubMed]

69. Jee, S.H.; Sull, J.W.; Park, J.; Lee, S.Y.; Ohrr, H.; Guallar, E.; Samet, J.M. Body-mass index and mortality in Korean men and women. *N. Engl. J. Med.* **2006**, *355*, 779–787. [CrossRef]

70. Despres, J.P. Abdominal obesity and cardiovascular disease: Is inflammation the missing link? *Can. J. Cardiol.* **2012**, *28*, 642–652. [CrossRef] [PubMed]

71. Alberti, K.G.; Zimmet, P.; Shaw, J.; IDF Epidemiology Task Forcesensus Group. The metabolic syndrome—A new worldwide definition. *Lancet* **2005**, *366*, 1059–1062. [CrossRef]

72. Lee, S.Y.; Park, H.S.; Kim, D.J.; Han, J.H.; Kim, S.M.; Cho, G.J.; Kim, D.Y.; Kwon, H.S.; Kim, S.R.; Lee, C.B.; et al. Appropriate waist circumference cutoff points for central obesity in Korean adults. *Diabetes Res. Clin. Pract.* **2007**, *75*, 72–80. [CrossRef] [PubMed]

73. Garvey, W.T.; Mechanick, J.I.; Brett, E.M.; Garber, A.J.; Hurley, D.L.; Jastreboff, A.M.; Nadolsky, K.; Pessah- Pollack, R.; Plodkowski, R. Reviewers of the AACE/ACE Obesity Clinical Practice Guidelines American Association of Clinical Endocrinologists and American College of Endocrinology Comprehensive Clinical Practice Guidelines for Medical Care of Patients with Obesity. *Endocr. Pract.* **2016**, *22*, 1–203. [CrossRef]

74. Batsis, J.A.; Mackenzie, T.A.; Lopez-Jimenez, F.; Bartels, S.J. Sarcopenia, sarcopenic obesity, and functional impairments in older adults: National Health and Nutrition Examination Surveys 1999–2004. *Nutr. Res.* **2015**, *35*, 1031–1039. [CrossRef] [PubMed]

75. Forbes, S.; Candow, D.; Krentz, J.; Roberts, M.; Young, K. Body fat changes following creatine supplementation and resistance training in adults > 50 years of age: A meta-analysis. *J. Funct. Morphol. Kinesiol.* **2019**, *4*, 62. [CrossRef]

76. Lee, N.; Kim, I.; Park, S.; Han, D.; Ha, S.; Kwon, M.; Kim, J.; Byun, S.H.; Oh, W.; Jeon, H.B.; et al. Creatine inhibits adipogenesis by downregulating insulin-induced activation of the phosphatidylinositol 3-kinase signaling pathway. *Stem Cells Dev.* **2015**, *24*, 983–994. [CrossRef]

77. Chouchani, E.T.; Kazak, L.; Spiegelman, B.M. New Advances in Adaptive Thermogenesis: UCP1 and Beyond. *Cell. Metab.* **2019**, *29*, 27–37. [CrossRef] [PubMed]

78. Wakatsuki, T.; Hirata, F.; Ohno, H.; Yamamoto, M.; Sato, Y.; Ohira, Y. Thermogenic responses to high-energy phosphate contents and/or hindlimb suspension in rats. *Jpn. J. Physiol.* **1996**, *46*, 171–175. [CrossRef]

79. Kazak, L.; Chouchani, E.T.; Lu, G.Z.; Jedrychows, P.; Bare, C.J.; Mina, A.I.; Kumari, M.; Zhang, S.; Vuckovic, I.; Laznik-Bogoslavski, D.; et al. Genetic Depletion of Adipocyte Creatine Metabolism Inhibits Diet Induced Thermogenesis and Drives Obesity. *Cell. Metab.* **2017**, *26*, 693. [CrossRef]

80. Perna, M.K.; Kokenge, A.N.; Miles, K.N.; Udobi, K.C.; Clark, J.F.; Pyne-Geithman, G.J.; Khuchua, Z.; Skelton, M.R. Creatine transporter deficiency leads to increased whole body and cellular metabolism. *Amino Acids* **2016**, *48*, 2057–2065. [CrossRef]

81. Kazak, L.; Rahbani, J.F.; Samborska, B.; Lu, G.Z.; Jedrychowski, M.P.; Lajoie, M.; Zhang, S.; Ramsay, L.C.; Dou, F.Y.; Tenen, D.; et al. Ablation of adipocyte creatine transport impairs thermogenesis and causes diet-induced obesity. *Nat. Metab.* **2019**, *1*, 360–370. [CrossRef]

82. Fried, L.P.; Tangen, C.M.; Walston, J.; Newman, A.B.; Hirsch, C.; Gottdiener, J.; Seeman, T.; Tracy, R.; Kop, W.J.; Burke, G.; et al. Cardiovascular Health Study Collaborative Research Group Frailty in older adults: Evidence for a phenotype. *J. Gerontol. A Biol. Sci. Med. Sci.* **2001**, *56*, 146. [CrossRef]

83. Song, X.; Mitnitski, A.; Rockwood, K. Prevalence and 10-year outcomes of frailty in older adults in relation to deficit accumulation. *J. Am. Geriatr. Soc.* **2010**, *58*, 681–687. [CrossRef] [PubMed]

84. Rochat, S.; Cumming, R.G.; Blyth, F.; Creasey, H.; Handelsman, D.; Le Couteur, D.G.; Naganathan, V.; Sambrook, P.N.; Seibel, M.J.; Waite, L. Frailty and use of health and community services by community-dwelling older men: The Concord Health and Ageing in Men Project. *Age Ageing* **2010**, *39*, 228–233. [CrossRef] [PubMed]

85. Thompson, M.Q.; Theou, O.; Karnon, J.; Adams, R.J.; Visvanathan, R. Frailty prevalence in Australia: Findings from four pooled Australian cohort studies. *Australas. J. Ageing* **2018**, *37*, 155–158. [CrossRef] [PubMed]

86. Martin, F.C.; Ranhoff, A.H. *Frailty and Sarcopenia*; Falaschi, P., Marsh, D., Eds.; Orthogeriatrics: The Management of Older Patients with Fragility Fractures; Springer: Cham, Switzerland, 2021; pp. 53–65.

87. Oktaviana, J.; Zanker, J.; Vogrin, S.; Duque, G. The Effect of beta-hydroxy-beta-methylbutyrate (HMB) on Sarcopenia and Functional Frailty in Older Persons: A Systematic Review. *J. Nutr. Health Aging* **2019**, *23*, 145–150. [CrossRef] [PubMed]

88. Rockwood, K.; Song, X.; MacKnight, C.; Bergman, H.; Hogan, D.B.; McDowell, I.; Mitnitski, A. A global clinical measure of fitness and frailty in elderly people. *CMAJ* **2005**, *173*, 489–495. [CrossRef]

89. Collins, J.; Longhurst, G.; Roschel, H.; Gualano, B. Resistance Training and Co-supplementation with Creatine and Protein in Older Subjects with Frailty. *J. Frailty Aging* **2016**, *5*, 126–134. [PubMed]

90. Riesberg, L.A.; Weed, S.A.; McDonald, T.L.; Eckerson, J.M.; Drescher, K.M. Beyond muscles: The untapped potential of creatine. *Int. Immunopharmacol.* **2016**, *37*, 31–42. [CrossRef]

91. Dalle, S.; Rossmeislova, L.; Koppo, K. The Role of Inflammation in Age-Related Sarcopenia. *Front. Physiol.* **2017**, *8*, 1045. [CrossRef]

92. Van Epps, P.; Oswald, D.; Higgins, P.A.; Hornick, T.R.; Aung, H.; Banks, R.E.; Wilson, B.M.; Burant, C.; Graventstein, S.; Canaday, D.H. Frailty has a stronger association with inflammation than age in older veterans. *Immun. Ageing* **2016**, *13*, 1–9. [CrossRef] [PubMed]

93. Hassan, E.B.; Imani, M.; Duque, G. Is Physical Frailty a Neuromuscular Condition? *J. Am. Med. Dir. Assoc.* **2019**, *20*, 1556–1557. [CrossRef] [PubMed]

94. Uchmanowicz, I. Oxidative Stress, Frailty and Cardiovascular Diseases: Current Evidence. *Adv. Exp. Med. Biol.* **2020**, *1216*, 65–77. [PubMed]

95. Webster, J.M.; Kempen, L.J.A.P.; Hardy, R.S.; Langen, R.C.J. Inflammation and Skeletal Muscle Wasting During Cachexia. *Front. Physiol.* **2020**, *11*, 597675. [CrossRef] [PubMed]

96. Wyart, E.; Bindels, L.B.; Mina, E.; Menga, A.; Stanga, S.; Porporato, P.E. Cachexia, a Systemic Disease beyond Muscle Atrophy. *Int. J. Mol. Sci.* **2020**, *21*, 8592. [CrossRef]

97. Fairman, C.M.; Kendall, K.L.; Hart, N.H.; Taaffe, D.R.; Galvao, D.A.; Newton, R.U. The potential therapeutic effects of creatine supplementation on body composition and muscle function in cancer. *Crit. Rev. Oncol. Hematol.* **2019**, *133*, 46–57. [CrossRef]

98. Jatoi, A.; Steen, P.D.; Atherton, P.J.; Moore, D.F.; Rowland, K.M.; Le-Lindqwister, N.A.; Adonizio, C.S.; Jaslowski, A.J.; Sloan, J.; Loprinzi, C. A double-blind, placebo-controlled randomized trial of creatine for the cancer anorexia/weight loss syndrome (N02C4): An Alliance trial. *Ann. Oncol.* **2017**, *28*, 1957–1963. [CrossRef]

99. Bourgeois, J.M.; Nagel, K.; Pearce, E.; Wright, M.; Barr, R.D.; Tarnopolsky, M.A. Creatine monohydrate attenuates body fat accumulation in children with acute lymphoblastic leukemia during maintenance chemotherapy. *Pediatr. Blood Cancer.* **2008**, *51*, 183–187. [CrossRef]

100. Norman, K.; Stubler, D.; Baier, P.; Schutz, T.; Ocran, K.; Holm, E.; Lochs, H.; Pirlich, M. Effects of creatine supplementation on nutritional status, muscle function and quality of life in patients with colorectal cancer—A double blind randomised controlled trial. *Clin. Nutr.* **2006**, *25*, 596–605. [CrossRef]

101. Lonbro, S.; Dalgas, U.; Primdahl, H.; Overgaard, J.; Overgaard, K. Feasibility and efficacy of progressive resistance training and dietary supplements in radiotherapy treated head and neck cancer patients—The DAHANCA 25A study. *Acta Oncol.* **2013**, *52*, 310–318. [CrossRef]

Review

Creatine Supplementation and Brain Health

Hamilton Roschel [1,*], Bruno Gualano [1,2], Sergej M. Ostojic [3] and Eric S. Rawson [4]

[1] Applied Physiology & Nutrition Research Group, Rheumatology Division, School of Physical Education and Sport, Faculdade de Medicina FMUSP, Universidade de Sao Paulo, Sao Paulo 01246-903, Brazil; gualano@usp.br
[2] Food Research Center, University of São Paulo, Sao Paulo 05508-080, Brazil
[3] FSPE Applied Bioenergetics Lab, University of Novi Sad, 21000 Novi Sad, Serbia; sergej.ostojic@chess.edu.rs
[4] Department of Health, Nutrition, and Exercise Science, Messiah University, Mechanicsburg, PA 17055, USA; erawson@messiah.edu
* Correspondence: hars@usp.br; Tel.: +55-11-3061-8789

Abstract: There is a robust and compelling body of evidence supporting the ergogenic and therapeutic role of creatine supplementation in muscle. Beyond these well-described effects and mechanisms, there is literature to suggest that creatine may also be beneficial to brain health (e.g., cognitive processing, brain function, and recovery from trauma). This is a growing field of research, and the purpose of this short review is to provide an update on the effects of creatine supplementation on brain health in humans. There is a potential for creatine supplementation to improve cognitive processing, especially in conditions characterized by brain creatine deficits, which could be induced by acute stressors (e.g., exercise, sleep deprivation) or chronic, pathologic conditions (e.g., creatine synthesis enzyme deficiencies, mild traumatic brain injury, aging, Alzheimer's disease, depression). Despite this, the optimal creatine protocol able to increase brain creatine levels is still to be determined. Similarly, supplementation studies concomitantly assessing brain creatine and cognitive function are needed. Collectively, data available are promising and future research in the area is warranted.

Keywords: phosphorylcreatine; dietary supplement; cognition; brain injury; concussion

Citation: Roschel, H.; Gualano, B.; Ostojic, S.M.; Rawson, E.S. Creatine Supplementation and Brain Health. *Nutrients* **2021**, *13*, 586. https://doi.org/10.3390/nu13020586

Academic Editor: Richard B. Kreider
Received: 18 January 2021
Accepted: 4 February 2021
Published: 10 February 2021

Publisher's Note: MDPI stays neutral with regard to jurisdictional claims in published maps and institutional affiliations.

1. Introduction

The ergogenic effects of creatine supplementation are well documented, with evidence supporting its efficacy in increasing muscle strength, lean mass, and exercise performance/muscle function, particularly when combined with exercise in different populations, from athletes to a wide spectrum of patient populations [1–3].

Creatine mechanisms of action involve rapid energy provision by transferring the N-phosphoryl group from phosphorylcreatine (PCr) to adenosine diphosphate (ADP), thus resynthesizing adenosine triphosphate (ATP) and spatial energy buffering, transferring energy from the mitochondria to the cytosol. These mechanisms are responsible for facilitating ATP homeostasis during high energy turnover, maintaining a low ADP concentration and reducing Ca^{2+} leakage from the sarcoplasmic reticulum and impairment of force output of the muscle [4–6]. Additionally, creatine could also attenuate the formation of reactive oxygen species by its coupling with ATP into the mitochondria or by scavenging radical species in an acellular setting [7]. Its direct and indirect antioxidant effects have been suggested to have therapeutic effects in neurodegenerative diseases [8].

Although most of the total body's creatine is found in skeletal muscle, the brain is also a very metabolically active tissue, accounting for up to 20% of the body's energy consumption [9,10]. Creatine kinase (CK), a main enzyme involved in the ATP/CK/PCr system, is also expressed in a brain-specific isoform (BB-CK) [4–6], suggesting that creatine may also be relevant for energy provision to the central nervous system (CNS). In fact, creatine-deficient syndromes involving brain creatine depletion are characterized by major mental and developmental disorders (e.g., mental retardation, learning delays, autism, and seizures), which may be partially reversed by creatine supplementation [11–14]. Cognitive processing may also be affected by creatine metabolism, as it may facilitate ATP homeostasis during periods of rapid or altered brain ATP turnover, such as during complex cognitive tasks, hypoxia, sleep deprivation, and some neurological conditions [3,15,16]. Additionally, creatine supplementation might be beneficial for mild traumatic brain injury (mTBI), which is also associated with changes in brain energy needs. The effects of creatine supplementation on brain creatine levels, cognitive processing, and mTBI have been previously reviewed [3,17,18]. As this is a growing field, the purpose of this short review is to provide an update regarding the effects of creatine supplementation on brain health in humans beyond what is discussed in Dolan et al. [3].

2. The Effects of Creatine Supplementation on Brain Creatine Levels

While muscle exclusively relies on dietary ingestion and endogenous synthesis from the liver, kidneys, and pancreas [19], the brain can synthesize creatine. The enzymatic apparatus necessary for endogenous creatine synthesis is found in the nervous system and creatine transporters are found at the blood–brain barrier, neurons and oligodendrocytes cells, indicating that brain creatine may not solely dependent on endogenous production from other organs or dietary sources [20]. Furthermore, brain creatine seems not to be influenced by habitual dietary intake from food, as similar brain PCr is found between vegetarians and omnivores [21]. Still, if the intracerebral synthesis is limited due to inherited disorders of creatine-catalyzing enzyme(s) machinery, dietary provision of the compound can positively affect brain creatine concentrations [22]. Figure 1 illustrates endogenous creatine synthesis in the brain and its transport across the blood–brain barrier.

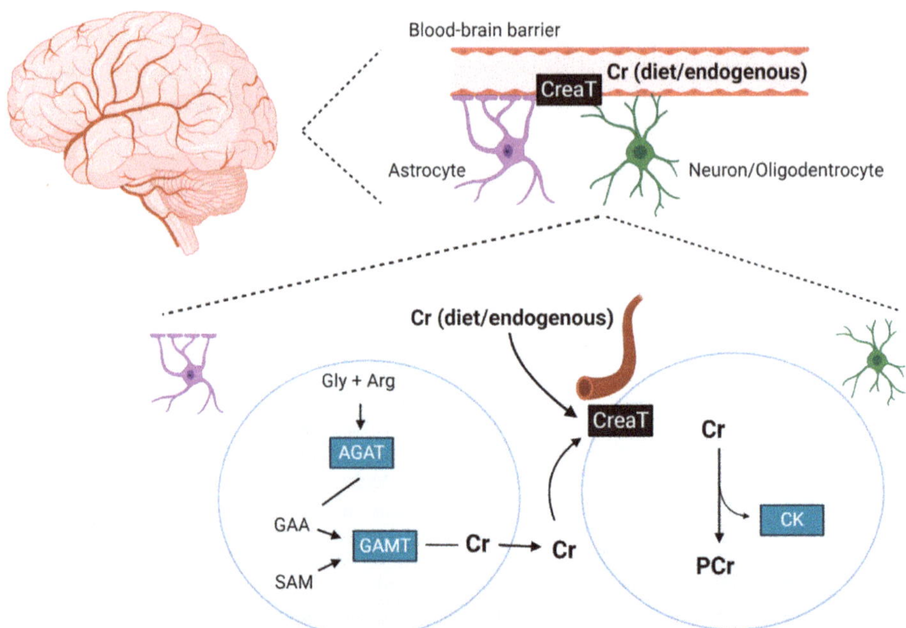

Figure 1. Dietary creatine is transported through the blood–brain barrier via a creatine transporter. Astrocytes cells can also endogenously produce creatine, which is taken up by the neurons expressing the creatine transporter. Cr: creatine; PCr: phosphocreatine; Gly: glycine; Arg: arginine; AGAT: L-ariginine: glycine amidinotransferase; GAA: guanidinoacetate; GAMT: guanidinoacetate methyltransferase, SAM: S-adenosylmethionine; CreaT: Cr transporter. Created with BioRender.com.

Brain creatine content has been suggested to be affected by other factors, such as aging [23]; however, comparable levels of brain PCr have also been found between apparently healthy elderly and young individuals [24]. Other factors related to aging that may influence brain creatine concentrations include reduced brain and/or physical activity, depression, schizophrenia, and panic disorder. The overlap between these factors may be misleading as to what might be identified as an age-related decline (reviewed in Rawson and Venezia [25]).

While consistent information is available on supplementation protocols aimed at increasing muscle creatine content [26], much less is known regarding the optimal supplementation strategy to increase brain creatine levels. A large heterogeneity in respect to brain creatine assessment technique (i.e., total brain creatine as assessed by H^1-NMR vs. brain PCr as assessed by P^{31}-NMR), supplement dose and duration (range 2 to 20 g/d for 1 to 8 weeks), and population characteristics (including habitual dietary creatine intake, health status, etc.) hampers direct comparison between the few studies on the topic. Further confusion is introduced by the fact that creatine content may differ regionally within the brain [25,27]. Nevertheless, collectively, the available literature suggests possible increases in both creatine and PCr in the brain following supplementation, though smaller than that seen in muscle (~half the increase) [3]. As reviewed in detail by Dolan et al. [3], there are currently 12 studies of the effects of creatine supplementation on brain creatine or PCr concentrations. Nine of these studies showed a significant increase in brain creatine, averaging about 5 to 10%, which is less than the increase in muscle creatine or PCr resulting from similar supplementation protocols. Some of these studies focused on patient populations who have altered brain energetics, including females with major depressive disorder, depression and amphetamine use, and selective serotonin uptake inhibitor resistant depression. Other groups investigated the effects of creatine ingestion on brain creatine levels in apparently healthy individuals. There is no clear indication why a small number of studies were ineffective at increasing brain creatine despite using similar supplementation protocols, but differences in baseline brain creatine levels, brain creatine assessment, population characteristics, and dosing strategies likely play a role.

The explanation for these differences in creatine uptake between muscle and brain remains speculative. As discussed, brain creatine content may rely less on exogenous creatine than muscle [20,21,24,28], which could theoretically involve a down-regulated response in brain creatine synthesis upon supplementation. Alternative to this hypothesis is the demonstration that the brain lacks the expression of creatine transporter in the astrocytes involved in the blood–brain barrier, thus implying a limited permeability of the brain to the circulating creatine [29], which is in line with the lack of increase in brain creatine following supplementation reported by some studies [24,28,30]. It is also plausible to speculate that if the brain is, in fact, resistant to exogenous creatine, a high-dose, long duration protocol would be needed, such as those used in the study by Dechent et al. [27] (i.e., 20 g/day for 4 weeks). The need for a higher supplementation dose in order to increase brain creatine level, as compared to the supplementation dose required for muscle, is further corroborated by data available from the only study assessing both muscle and brain creatine levels in response to supplementation, with increases found in the former, but not the latter [24]. Of interest, supplementing guanidinoacetic acid (GAA), a creatine precursor, was found superior to an equimolar dose of creatine in increasing brain creatine content [31]. While creatine is mainly transported via a specific transporter (SLC6A8 or CT1; also used for GAA transport), dietary GAA could be imported to the brain through additional delivery transporters and routes (including SLC6A6, GAT2, and passive diffusion) [32] and become readily available for methylation to creatine. Although preliminary, these data are of relevance considering the inherent capacity of the brain to synthesize creatine and its theoretical impaired ability to transport creatine through the blood–brain barrier, thus warranting further research on alternative strategies to increase brain creatine.

3. Creatine Supplementation and Cognition

The interest in the effects of creatine supplementation on cognition is not new. Despite the number of positive studies available on the subject (Summarized in Table 1), differences between investigations including study populations, cognitive function testing, and supplementation dosing and duration precludes direct comparison; however, some conclusions can be made. Although controversial [28,33,34], creatine supplementation may positively influence some aspects of cognition in different experimental paradigms [10,35–40]. Importantly, its effects seem more pronounced in stressful conditions such as hypoxia [8] and sleep deprivation combined with exercise [10,37,38]. Despite the suggestion that more complex or demanding cognitive processes are more prone to respond to supplementation (as they are more energy demanding), inconsistencies regarding cognitive test response to supplementation hampers further conclusions [37,38]), which may be attributed to differences in experimental design such as the sleep deprivation period and exercise intensity employed between studies.

Table 1. Effects of creatine supplementation on cognitive performance.

Population	Creatine Supplementation Protocol	Cognitive Tests (CT) Outcomes (O)	Reference
Healthy older women	20 g/day + 5 g/day for 24 weeks	CT: Mini-mental state examination, stroop, trail making, digit span, delay recall test and the short version of the geriatric depression scale O: No change	Alves et al. (2013) [33]
Semiprofessional, non-vegetarian, male mountain bikers	20 g/day for 7 days	CT: Simple and choice reaction time, differentiation task test, Eiksen flanker test and Corsi block test O: Creatine increased performance in choice reaction time, Eiksen flanker test and Corsi block test.	Borchio et al. (2020) [41]
Healthy young women (vegetarian and meat-eaters)	20 g/day for 5 days	CT: Word recall, simple and choice reaction time, rapid visual information processing and controlled oral word association test O: Word recall test performance was reduced in meat-eater after creatine supplementation (within-group comparison). Post supplementation performance was higher in vegetarians than in meat-eaters.	Benton and Donohoe (2011) [15]
Professional male rugby players who were sleep-deprived (3–5 h)	0.05 or 0.1 g/kg/bw for 1 day	CT: Rugby passing skill test O: Sleep deprivation reduced passing accuracy and creatine reversed this effect (trend for greater effect with larger dose).	Cook et al. (2011) [42]
Healthy young adults	20 g/day for 5 days + 5 g/day for 2 days	CT: Backward digit span test and ravens advanced progressive matrices. O: Backward digit span performance was increased after creatine.	Hammett et al. (2010) [35]
Healthy young men and women	5 g/day for 15 days	CT: Memory scanning, number-pair matching, sustained attention, arrow flankers and IQ test O: Aspect of improvement was reported in all the cognitive tests performed in the creatine group.	Ling et al. (2009) [36]
Healthy young men and women who were sleep-deprived (24 h)	20 g/day for 7 days	CT: Random number generation, forward and backward recall, visual reaction time, static balance and mood state O: Performance reduction was attenuated in the creatine group for random movement generation, choice reaction time, balance and mood.	McMorris et al. (2006) [38]
Healthy elderly men and women	20 g/day for 7 days	CT: Random number generation, forward and backward recall and long-term memory tests O: Forward number recall, forward and backward spatial recall and long-term memory performance were enhanced after creatine supplementation.	McMorris et al. (2007a) [16]
Healthy young men who were sleep-deprived (36 h)	20 g/day for 7 days	CT: Random number generation, short-term number recall, visual reaction time, cognitive effort, dynamic balance test and mood state O: Performance on the random number generation test was improved following creatine.	McMorris et al. (2007b) [37]

Table 1. *Cont.*

Population	Creatine Supplementation Protocol	Cognitive Tests (CT) Outcomes (O)	Reference
Healthy male and female children	0.3 g/kg/day for 7 days	CT: Stroop, Rey auditory verbal learning test, Raven progressive matrices and trail making test O: No change	Merege-Filho et al. (2017) [28]
Vegan and vegetarian healthy male and female young adults	5 g/day for 6 weeks	CT: Ravens advanced progressive matrices and Wechsler auditory backward digit span task O: Creatine improved performance on the Raven's test and the backward digit span task.	Rae and Broer (2015) [17]
Healthy male and female young adults	0.03 g/kg/day for 6 weeks	CT: Automated neuropsychological assessment metrics O: No change	Rawson et al. (2008) [34]
Male and female institutionalized older adults (with full physical and mental capacities preserved)	5 g/day for 16 weeks	CT: Montreal Cognitive Assessment (MoCA) questionnaire O: Creatine (plus resistance training) improved MoCA scores.	Smolarek et al. (2020) [43]
Healthy male and female young adults exposed to experimental hypoxia	20 g/day for 7 days	CT: Neuropsychological test comprising verbal and visual memory, finger tapping, symbol digit coding stroop test, test of shifting attention, continuous performance test, alertness and peripheral and corticomotor excitability O: Creatine supplementation offset hypoxia-induced decrements in a number of cognitive tests.	Turner et al. (2015) [10]
Healthy male and female young adults exposed to mental fatigue (90 min Stoop task)	20 g/day for 7 days	CT: Psychomotor performance (visuomotor task with Fitlight-hardware and software), strength endurance task, Flanker test, heart rate, blood glucose, success motivation and intrinsic motivation, mood, session ratings of perceived exertion and mental fatigue O: Accuracy throughout the 90 min Stroop task and strength endurance (in the non-dominant hand) were improved with creatine. No other effects of creatine supplementation were observed.	Van Cutsem et al. (2020) [44]
Healthy male and female young adults	8 g/day for 5 days	CT: Serial calculation task (Uchida-Kraeplin) O: Both groups increased mean performance. Mental fatigue, assessed during the second half of the test, was increased in the creatine group only.	Watanabe et al. (2002) [40]

In elderly individuals, specifically, literature is conflicting on the effects of creatine supplementation on cognitive performance. While McMorris et al. [16] showed improved cognitive performance, Alves et al. [33] found creatine (alone or associated with exercise training) ineffective. Both studies are limited by the lack of brain creatine concentration assessments, casting doubt on whether aging-related reduction in cognitive processing may arise from the presence of, for instance, neurodegenerative diseases or whether the supplementation protocol employed (designed for increasing muscle creatine content) may have been insufficient to significantly increase brain PCr. Recently, Smolarek et al. [43] found increased cognitive performance (and handgrip strength) after a 16 week intervention combining resistance training and creatine supplementation (5 g/day) in a pilot study including older adults. The results are, however, limited by the absence of an exercising control group and inconsistent cognitive performance in the control group across time, thus hampering further conclusion on the effects of supplementation alone.

It has been contended that vegetarians may differentially respond to creatine supplementation when compared to meat-eaters. In this respect, cognitive function has been shown to be improved in vegetarians after creatine supplementation [39]. Another study found greater effects on memory in vegetarians as compared to omnivores following creatine supplementation [15]. Importantly, the lack of a control group (meat-eaters) and the fact that between-group differences were due to decreased performance in the omnivores, rather than an improvement in the vegetarians, limits the conclusions of this study. Additionally, comparable brain creatine concentrations have been shown between meat-eaters and vegetarians [21], which undermines the theory that vegetarians should respond better than meat-eaters due to lower pre-supplementation brain creatine. More research should be conducted on the differential responses to creatine supplementation between vegetarians and omnivores.

Improvements in cognitive processing capability is also of interest to athletes. Several sports include motor control, decision making, coordination, reaction time, and other cognitive tasks as key aspects of performance, which may be affected by mental fatigue [45]. In this respect, creatine may play an ergogenic role, as, theoretically, it may mitigate mental fatigue, thus favouring performance. Indeed, creatine has been shown effective in attenuating the effects of sleep deprivation on throwing accuracy in rugby players [42], while no effect was observed on passing accuracy in non-stressed soccer players [46,47]. Brain creatine content was not assessed in these studies, raising uncertainty as to whether the results observed result from changes in brain creatine. Nonetheless, the discrepancy in the results may, at least partially, relate to the suggestion that creatine supplementation is most effective under stressed cognitive processes conditions such as sleep deprivation.

More recently, two studies revisited the subject, with interesting results. Borchio et al. [41] found improved performance in selected indexes of cognitive function after a time-trial track test in semi-professional mountain bikers supplemented with creatine. Interestingly, no prior cognitive deficit-inducing condition, such as sleep deprivation, was imposed, suggesting that creatine could potentially attenuate mental fatigue even in non-stressed situations. Van Cutsem et al. [44] studied the effects of creatine supplementation on mental fatigue and its negative effects on psychomotor skills in a non-athlete population and found that creatine was able to improve Stroop accuracy during a 90 min Stroop task and to increase strength endurance (assessed by a handgrip strength test) pre-to-post Stroop task. Importantly, no effects of supplementation were observed on the mental-fatigue-induced impairments in psychomotor and cognitive performance. Collectively, although these results suggest a potential role of creatine on mental fatigue, whether and to what extent this could affect specific sports performance remains to be elucidated.

4. Creatine Supplementation and Brain Injury, Concussion, and Hypoxia

One of the characteristics of traumatic brain injury is the alteration of ATP demand due to reduced blood flow and hypoxia [48]. Importantly, brain creatine is reduced following a mild traumatic brain injury (mTBI) [49], making creatine supplementation, and subsequent increase in brain creatine, a potentially valuable strategy to reduce severity of, or enhance recovery from, mTBI or concussion by offsetting negative changes in energy status. The duration of the dysregulation in brain energy metabolism is not clearly defined, but could remain for weeks if not years. Alosco et al. [50] reported on retired players from the National Football League (aged 40 to 69) who had experienced repetitive head impacts during their career and many years later had complaints of cognitive and/or behavioral/mood symptoms. In this cohort, there was a relationship between greater exposure to repetitive head impacts and decreased brain creatine in the parietal white matter. This indicates that there could be later-life derangements in brain energy metabolism subsequent to mTBI, and lends support to the concept that creatine supplementation could be valuable in enhancing recovery from mTBI, even years after the injury. In addition to its potential role in aiding the cellular energy crisis induced by injury, creatine may lessen other features of mTBI, such as membrane disruption leading to calcium influx, nerve damage, mitochondrial dysfunction, oxidative stress, and inflammation (reviewed in [48,51]).

In an experimental model mimicking the effects of mTBI, Turner et al. [10] found that supplementation was able to increase brain creatine and cognitive processing during oxygen deprivation. Animal models have also been employed to investigate the effects of creatine supplementation on traumatic brain injury. Sullivan et al. [52] found significant reduction in brain damage following traumatic brain injury in both mice (36%) and rats (50%). These large effects are compelling, but as humans only increase brain creatine about 10% in response to supplementation and some animals increase brain creatine 30 to 50%, it is difficult to generalize these data to the general population or athletes [53].

Despite its potential, experimental data in humans are still scarce; however, results from the few studies available are promising. Creatine supplementation has been shown able to improve cognition, communication, self-care, personality, and behavior, and reductions in headaches, dizziness, and fatigue in children with mTBI [54,55].

Collectively, despite limited data, creatine supplementation seems potentially beneficial in reducing severity of or enhancing recovery from mTBI, warranting further studies on its role not only as a post-injury therapy but also as a neuroprotective agent in populations at high risk of mTBI. As has been described elsewhere, creatine supplements have documented muscular performance benefits, are inexpensive, widely available, and have a strong safety profile [26,56–63]. Encouraging supplementation to reduce damage from or enhance recovery from mTBI based primarily on animal and theoretical data in lieu of clinical trials would ordinarily be considered premature. However, in this instance, given the devastating effects of mTBI, combined with the large body of safety and efficacy creatine supplementation data, encouraging supplementation for populations who are at high risk for mTBI might be considered more prudent.

5. Conclusions and Future Directions

There is a potential for creatine supplementation to improve cognitive processing, especially in conditions characterized by brain creatine deficits, which could be induced by acute stressors (e.g., exercise, sleep deprivation) or chronic, pathologic conditions (e.g., creatine synthesis enzyme deficiencies, mTBI, aging, Alzheimer's disease, depression).

However, at least three main gaps remain. First, it is important to determine the optimal creatine protocol able to increase brain creatine levels. So far, dose-response studies are lacking and protocols are heterogenous. Second, supplementation studies concomitantly assessing brain creatine levels and cognitive function are needed, as it may help establish causation for the effect of creatine supplementation on cognition. Third, the identification of novel conditions in which creatine supplementation may be more effective in improving cognitive function is warranted as creatine in a rested healthy brain has been shown to have a lessened effect.

Author Contributions: Conceptualization, H.R., B.G., S.M.O. and E.S.R.; writing—review and editing, H.R., B.G., S.M.O., E.S.R. All authors have read and agreed to the published version of the manuscript.

Funding: This research received no external funding.

Acknowledgments: The authors would like to thank all of the research participants, scholars, and funding agencies who have contributed to the research cited in this manuscript. The figure was created with BioRender.com.

Conflicts of Interest: B.G. has received research grants, creatine donation for scientific studies, travel support for participation in scientific conferences and honorarium for speaking at lectures from AlzChem (a company which manufactures creatine). Additionally, he serves as a member of the Scientific Advisory Board for Alzchem. E.S.R. has conducted industry-sponsored research on creatine and received financial support for presenting on creatine at industry-sponsored scientific conferences. Additionally, he serves as a member of the Scientific Advisory Board for AlzChem, who sponsored this special issue. S.M.O. serves as a member of the Scientific Advisory Board on creatine in health and medicine (AlzChem LLC). S.M.O. owns patent "Sports Supplements Based on Liquid Creatine" at the European Patent Office (WO2019150323 A1), and active patent application "Synergistic Creatine" at the UK Intellectual Property Office (GB2012773.4). S.M.O. has served as a speaker at Abbott Nutrition, a consultant of Allied Beverages Adriatic and IMLEK, and an advisory board member for the University of Novi Sad School of Medicine, and has received research funding related to creatine from the Serbian Ministry of Education, Science, and Technological Development, Provincial Secretariat for Higher Education and Scientific Research, AlzChem GmbH, KW Pfannenschmidt GmbH, and ThermoLife International LLC. S.M.O. is an employee of the University of Novi Sad and does not own stocks and shares in any organization.

References

1. Gualano, B.; Roschel, H.; Lancha, A.H.; Brightbill, C.E.; Rawson, E.S.; Junior, A.H.L. In sickness and in health: The widespread application of creatine supplementation. *Amino Acids* **2011**, *43*, 519–529. [CrossRef]
2. Dolan, E.; Artioli, G.G.; Pereira, R.M.R.; Gualano, B. Muscular Atrophy and Sarcopenia in the Elderly: Is There a Role for Creatine Supplementation? *Biomolecules* **2019**, *9*, 642. [CrossRef]
3. Dolan, E.; Gualano, B.; Rawson, E.S. Beyond muscle: The effects of creatine supplementation on brain creatine, cognitive processing, and traumatic brain injury. *Eur. J. Sport Sci.* **2019**, *19*, 1–14. [CrossRef]
4. Wallimann, T.; Wyss, M.; Brdiczka, D.; Nicolay, K.; Eppenberger, H.M. Intracellular compartmentation, structure and function of creatine kinase isoenzymes in tissues with high and fluctuating energy demands: The 'phosphocreatine circuit' for cellular energy homeostasis. *Biochem. J.* **1992**, *281*, 21–40. [CrossRef]
5. Wallimann, T.; Turner, D.C.; Eppenberger, H.M. Localization of creatine kinase isoenzymes in myofibrils. I. Chicken skeletal muscle. *J. Cell Biol.* **1977**, *75*, 297–317. [CrossRef]
6. Sahlin, K.; Harris, R.C. The creatine kinase reaction: A simple reaction with functional complexity. *Amino Acids* **2011**, *40*, 1363–1367. [CrossRef]
7. Sestili, P.; Martinelli, C.; Colombo, E.; Barbieri, E.; Potenza, L.; Sartini, S.; Fimognari, C. Creatine as an antioxidant. *Amino Acids* **2011**, *40*, 1385–1396. [CrossRef] [PubMed]
8. Beal, M.F. Neuroprotective effects of creatine. *Amino Acids* **2011**, *40*, 1305–1313. [CrossRef]
9. Gualano, B.; Artioli, G.G.; Poortmans, J.R.; Junior, A.H.L. Exploring the therapeutic role of creatine supplementation. *Amino Acids* **2009**, *38*, 31–44. [CrossRef] [PubMed]
10. Turner, C.E.; Byblow, W.D.; Gant, N. Creatine Supplementation Enhances Corticomotor Excitability and Cognitive Performance during Oxygen Deprivation. *J. Neurosci.* **2015**, *35*, 1773–1780. [CrossRef] [PubMed]
11. Kaldis, P.; Hemmer, W.; Zanolla, E.; Holtzman, D.; Wallimann, T. 'Hot Spots' of Creatine Kinase Localization in Brain: Cerebellum, Hippocampus and Choroid Plexus. *Dev. Neurosci.* **1996**, *18*, 542–554. [CrossRef] [PubMed]
12. Salomons, G.S.; Van Dooren, S.J.M.; Verhoeven, N.M.; Marsden, D.; Schwartz, C.; Cecil, K.M.; Degrauw, T.J.; Jakobs, C. X-linked creatine transporter defect: An overview. *J. Inherit. Metab. Dis.* **2003**, *26*, 309–318. [CrossRef]
13. Stockler, S.; Holzbach, U.; Hanefeld, F.; Marquardt, I.; Helms, G.; Requart, M.; Hanicke, W.; Frahm, J. Creatine Deficiency in the Brain: A New, Treatable Inborn Error of Metabolism. *Pediatr. Res.* **1994**, *36*, 409–413. [CrossRef]
14. Stockler, S.; Schutz, P.W.; Salomons, G.S. Cerebral creatine deficiency syndromes: Clinical aspects, treatment and pathophysiology. *Alzheimer's Dis.* **2007**, *46*, 149–166. [CrossRef]
15. Benton, D.; Donohoe, R. The influence of creatine supplementation on the cognitive functioning of vegetarians and omnivores. *Br. J. Nutr.* **2010**, *105*, 1100–1105. [CrossRef] [PubMed]
16. McMorris, T.; Mielcarz, G.; Harris, R.C.; Swain, J.P.; Howard, A.N. Creatine Supplementation and Cognitive Performance in Elderly Individuals. *Aging Neuropsychol. Cogn.* **2007**, *14*, 517–528. [CrossRef] [PubMed]
17. Rae, C.D.; Bröer, S. Creatine as a booster for human brain function. How might it work? *Neurochem. Int.* **2015**, *89*, 249–259. [CrossRef] [PubMed]
18. Avgerinos, K.I.; Spyrou, N.; Bougioukas, K.I.; Kapogiannis, D. Effects of creatine supplementation on cognitive function of healthy individuals: A systematic review of randomized controlled trials. *Exp. Gerontol.* **2018**, *108*, 166–173. [CrossRef]
19. Walker, J.B. Creatine: Biosynthesis, regulation, and function. *Adv. Enzymol. Relat. Areas Mol. Biol.* **1979**, *50*, 177–242. [PubMed]
20. Braissant, O.; Bachmann, C.; Henry, H. Expression and function of AGAT, GAMT and CT1 in the mammalian brain. *Alzheimer's Dis.* **2007**, *46*, 67–81. [CrossRef]
21. Solis, M.Y.; Painelli, V.D.S.; Artioli, G.G.; Roschel, H.; Otaduy, M.C.; Gualano, B. Brain creatine depletion in vegetarians? A cross-sectional 1H-magnetic resonance spectroscopy (1H-MRS) study. *Br. J. Nutr.* **2013**, *111*, 1272–1274. [CrossRef]
22. Stockler-Ipsiroglu, S.; Van Karnebeek, C.D.M.; Longo, N.; Korenke, G.C.; Mercimek-Mahmutoglu, S.; Marquart, I.; Barshop, B.; Grolik, C.; Schlune, A.; Angle, B.; et al. Guanidinoacetate methyltransferase (GAMT) deficiency: Outcomes in 48 individuals and recommendations for diagnosis, treatment and monitoring. *Mol. Genet. Metab.* **2014**, *111*, 16–25. [CrossRef] [PubMed]
23. Laakso, M.P.; Hiltunen, Y.; Könönen, M.; Kivipelto, M.; Koivisto, A.; Hallikainen, M.; Soininen, H. Decreased brain creatine levels in elderly apolipoprotein E epsilon 4 carriers. *J. Neural Transm.* **2003**, *110*, 267–275. [CrossRef] [PubMed]
24. Solis, M.Y.; Artioli, G.G.; Otaduy, M.C.G.; Leite, C.D.C.; Arruda, W.; Veiga, R.R.; Gualano, B. Effect of age, diet, and tissue type on PCr response to creatine supplementation. *J. Appl. Physiol.* **2017**, *123*, 407–414. [CrossRef]
25. Rawson, E.S.; Venezia, A.C. Use of creatine in the elderly and evidence for effects on cognitive function in young and old. *Amino Acids* **2011**, *40*, 1349–1362. [CrossRef] [PubMed]
26. Kreider, R.B.; Kalman, D.S.; Antonio, J.; Ziegenfuss, T.N.; Wildman, R.; Collins, R.; Candow, D.G.; Kleiner, S.M.; Almada, A.L.; Lopez, H.L. International Society of Sports Nutrition position stand: Safety and efficacy of creatine supplementation in exercise, sport, and medicine. *J. Int. Soc. Sports Nutr.* **2017**, *14*, 1–18. [CrossRef] [PubMed]
27. Dechent, P.; Pouwels, P.J.W.; Wilken, B.; Hanefeld, F.; Frahm, J. Increase of total creatine in human brain after oral supplementation of creatine-monohydrate. *Am. J. Physiol. Content* **1999**, *277*, R698–R704. [CrossRef]
28. Merege-Filho, C.A.A.; Otaduy, M.C.G.; De Sá-Pinto, A.L.; De Oliveira, M.O.; Gonçalves, L.D.S.; Hayashi, A.P.T.; Roschel, H.; Pereira, R.M.R.; Silva, C.A.; Brucki, S.M.D.; et al. Does brain creatine content rely on exogenous creatine in healthy youth? A proof-of-principle study. *Appl. Physiol. Nutr. Metab.* **2017**, *42*, 128–134. [CrossRef] [PubMed]

29. Béard, E.; Braissant, O. Synthesis and transport of creatine in the CNS: Importance for cerebral functions. *J. Neurochem.* **2010**, *115*, 297–313. [CrossRef]

30. Wilkinson, I.D.; Mitchel, N.; Breivik, S.; Greenwood, P.; Griffiths, P.D.; Winter, E.M.; Van Beek, E.J.R. Effects of Creatine Supplementation on Cerebral White Matter in Competitive Sportsmen. *Clin. J. Sport Med.* **2006**, *16*, 63–67. [CrossRef]

31. Ostojić, S.M.; Ostojić, J.; Drid, P.; Vraneš, M. Guanidinoacetic acid versus creatine for improved brain and muscle creatine levels: A superiority pilot trial in healthy men. *Appl. Physiol. Nutr. Metab.* **2016**, *41*, 1005–1007. [CrossRef]

32. Tachikawa, M.; Hosoya, K.-I. Transport characteristics of guanidino compounds at the blood-brain barrier and blood-cerebrospinal fluid barrier: Relevance to neural disorders. *Fluids Barriers CNS* **2011**, *8*, 13. [CrossRef] [PubMed]

33. Alves, C.R.R.; Filho, C.A.A.M.; Benatti, F.B.; Brucki, S.M.D.; Pereira, R.M.R.; Pinto, A.L.D.S.; Lima, F.R.; Roschel, H.; Gualano, B. Creatine Supplementation Associated or Not with Strength Training upon Emotional and Cognitive Measures in Older Women: A Randomized Double-Blind Study. *PLoS ONE* **2013**, *8*, e76301. [CrossRef]

34. Rawson, E.S.; Lieberman, H.R.; Walsh, T.M.; Zuber, S.M.; Harhart, J.M.; Matthews, T.C. Creatine supplementation does not improve cognitive function in young adults. *Physiol. Behav.* **2008**, *95*, 130–134. [CrossRef] [PubMed]

35. Hammett, S.T.; Wall, M.B.; Edwards, T.C.; Smith, A. Dietary supplementation of creatine monohydrate reduces the human fMRI BOLD signal. *Neurosci. Lett.* **2010**, *479*, 201–205. [CrossRef] [PubMed]

36. Ling, J.; Kritikos, M.; Tiplady, B. Cognitive effects of creatine ethyl ester supplementation. *Behav. Pharmacol.* **2009**, *20*, 673–679. [CrossRef]

37. McMorris, T.; Harris, R.; Howard, A.; Langridge, G.; Hall, B.; Corbett, J.; Dicks, M.; Hodgson, C. Creatine supplementation, sleep deprivation, cortisol, melatonin and behavior. *Physiol. Behav.* **2007**, *90*, 21–28. [CrossRef] [PubMed]

38. McMorris, T.; Harris, R.C.; Swain, J.; Corbett, J.; Collard, K.; Dyson, R.J.; Dye, L.; Hodgson, C.; Draper, N. Effect of creatine supplementation and sleep deprivation, with mild exercise, on cognitive and psychomotor performance, mood state, and plasma concentrations of catecholamines and cortisol. *Psychopharmacology* **2006**, *185*, 93–103. [CrossRef] [PubMed]

39. Rae, C.D.; Digney, A.L.; McEwan, S.R.; Bates, T.C. Oral creatine monohydrate supplementation improves brain performance: A double–blind, placebo–controlled, cross–over trial. *Proc. R. Soc. B Biol. Sci.* **2003**, *270*, 2147–2150. [CrossRef]

40. Watanabe, A.; Kato, N.; Kato, T. Effects of creatine on mental fatigue and cerebral hemoglobin oxygenation. *Neurosci. Res.* **2002**, *42*, 279–285. [CrossRef]

41. Borchio, L.; Machek, S.B.; Machado, M. Supplemental creatine monohydrate loading improves cognitive function in experienced mountain bikers. *J. Sports Med. Phys. Fit.* **2020**, *60*, 1168–1170. [CrossRef]

42. Cook, C.J.; Crewther, B.T.; Kilduff, L.P.; Drawer, S.; Gaviglio, C.M. Skill execution and sleep deprivation: Effects of acute caffeine or creatine supplementation—A randomized placebo-controlled trial. *J. Int. Soc. Sports Nutr.* **2011**, *8*, 2. [CrossRef] [PubMed]

43. Smolarek, A.C.; McAnulty, S.R.; Ferreira, L.H.; Cordeiro, G.R.; Alessi, A.; Rebesco, D.B.; Honorato, I.C.; Laat, E.F.; Mascarenhas, L.P.; Souza-Junior, T.P. Effect of 16 Weeks of Strength Training and Creatine Supplementation on Strength and Cognition in Older Adults: A Pilot Study. *J. Exerc. Physiol. Online* **2020**, *23*, 88–94.

44. Van Cutsem, J.; Roelands, B.; Pluym, B.; Tassignon, B.; Verschueren, J.O.; De Pauw, K.; Meeusen, R. Can Creatine Combat the Mental Fatigue-associated Decrease in Visuomotor Skills? *Med. Sci. Sports Exerc.* **2020**, *52*, 120–130. [CrossRef]

45. Meeusen, R. Exercise, Nutrition and the Brain. *Sports Med.* **2014**, *44*, 47–56. [CrossRef]

46. Cox, G.; Mujika, I.; Tumilty, D.; Burke, L. Acute Creatine Supplementation and Performance during a Field Test Simulating Match Play in Elite Female Soccer Players. *Int. J. Sport Nutr. Exerc. Metab.* **2002**, *12*, 33–46. [CrossRef]

47. Mohebbi, H.; Rahnama, N.; Moghadassi, M.; Ranjbar, K. Effect of Creatine Supplementation on Sprint and Skill Performance in Young Soccer Players. *Middle-East J. Sci. Res.* **2012**, *12*, 397–401.

48. Dean, P.J.A.; Arikan, G.; Opitz, B.; Sterr, A. Potential for use of creatine supplementation following mild traumatic brain injury. *Concussion* **2017**, *2*, CNC34. [CrossRef]

49. Vagnozzi, R.; Signoretti, S.; Floris, R.; Marziali, S.; Manara, M.; Amorini, A.M.; Belli, A.; Di Pietro, V.; D'Urso, S.; Pastore, F.S.; et al. Decrease in N-Acetylaspartate Following Concussion May Be Coupled to Decrease in Creatine. *J. Head Trauma Rehabil.* **2013**, *28*, 284–292. [CrossRef]

50. Alosco, M.L.; Tripodis, Y.; Rowland, B.; Chua, A.S.; Liao, H.; Martin, B.; Jarnagin, J.; Chaisson, C.E.; Pasternak, O.; Karmacharya, S.; et al. A magnetic resonance spectroscopy investigation in symptomatic former NFL players. *Brain Imaging Behav.* **2020**, *14*, 1419–1429. [CrossRef] [PubMed]

51. Barrett, E.C.; McBurney, M.I.; Ciappio, E.D. Omega-3 fatty acid supplementation as a potential therapeutic aid for the recovery from mild traumatic brain injury/concussion. *Adv. Nutr* **2014**, *5*, 268–277. [CrossRef]

52. Sullivan, P.G.; Geiger, J.D.; Mattson, M.P.; Scheff, S.W. Dietary supplement creatine protects against traumatic brain injury. *Ann. Neurol.* **2000**, *48*, 723–729. [CrossRef]

53. Ipsiroglu, O.S.; Stromberger, C.; Ilas, J.; Höger, H.; Mühl, A.; Stöckler-Ipsiroglu, S. Changes of tissue creatine concentrations upon oral supplementation of creatine-monohydrate in various animal species. *Life Sci.* **2001**, *69*, 1805–1815. [CrossRef]

54. Sakellaris, G.; Kotsiou, M.; Tamiolaki, M.; Kalostos, G.; Tsapaki, E.; Spanaki, M.; Spilioti, M.; Charissis, G.; Evangeliou, A. Prevention of Complications Related to Traumatic Brain Injury in Children and Adolescents with Creatine Administration: An Open Label Randomized Pilot Study. *J. Trauma: Inj. Infect. Crit. Care* **2006**, *61*, 322–329. [CrossRef]

55. Sakellaris, G.; Nasis, G.; Kotsiou, M.; Tamiolaki, M.; Charissis, G.; Evangeliou, A. Prevention of traumatic headache, dizziness and fatigue with creatine administration. A pilot study. *Acta Paediatr.* **2007**, *97*, 31–34. [CrossRef] [PubMed]

56. Gualano, B.; Painelli, V.D.S.; Roschel, H.; Lugaresi, R.; Dorea, E.; Artioli, G.G.; Lima, F.R.; Da Silva, M.E.R.; Cunha, M.R.; Seguro, A.C.; et al. Creatine supplementation does not impair kidney function in type 2 diabetic patients: A randomized, double-blind, placebo-controlled, clinical trial. *Graefe's Arch. Clin. Exp. Ophthalmol.* **2010**, *111*, 749–756. [CrossRef]
57. Gualano, B.; Ferreira, D.C.; Sapienza, M.T.; Seguro, A.C.; Lancha, A.H. Effect of Short-term High-Dose Creatine Supplementation on Measured GFR in a Young Man with a Single Kidney. *Am. J. Kidney Dis.* **2010**, *55*, e7–e9. [CrossRef] [PubMed]
58. Gualano, B.; Ugrinowitsch, C.; Novaes, R.B.; Artioli, G.G.; Shimizu, M.H.; Seguro, A.C.; Harris, R.C.; Lancha, A.H. Effects of creatine supplementation on renal function: A randomized, double-blind, placebo-controlled clinical trial. *Graefe's Arch. Clin. Exp. Ophthalmol.* **2008**, *103*, 33–40. [CrossRef]
59. Hayashi, A.P.; Solis, M.Y.; Sapienza, M.T.; Otaduy, M.C.G.; de Sa Pinto, A.L.; Silva, C.A.; Sallum, A.M.E.; Pereira, R.M.R.; Gualano, B. Efficacy and safety of creatine supplementation in childhood-onset systemic lupus erythematosus: A randomized, double-blind, placebo-controlled, crossover trial. *Lupus* **2014**, *23*, 1500–1511. [CrossRef]
60. Lugaresi, R.; Leme, M.; Painelli, V.D.S.; Murai, I.H.; Roschel, H.; Sapienza, M.T.; Lancha, A.H.J.; Gualano, B. Does long-term creatine supplementation impair kidney function in resistance-trained individuals consuming a high-protein diet? *J. Int. Soc. Sports Nutr.* **2013**, *10*, 26. [CrossRef]
61. Neves, M.; Gualano, B.; Roschel, H.; Lima, F.R.; De Sá-Pinto, A.L.; Seguro, A.C.; Shimizu, M.H.; Sapienza, M.T.; Fuller, R.; Lancha, A.H.; et al. Effect of creatine supplementation on measured glomerular filtration rate in postmenopausal women. *Appl. Physiol. Nutr. Metab.* **2011**, *36*, 419–422. [CrossRef]
62. Rawson, E.S.; Clarkson, P.M.; Tarnopolsky, M.A. Perspectives on Exertional Rhabdomyolysis. *Sports Med.* **2017**, *47*, 33–49. [CrossRef] [PubMed]
63. Solis, M.Y.; Hayashi, A.P.; Artioli, G.G.; Roschel, H.; Sapienza, M.T.; Otaduy, M.C.; De Sã Pinto, A.L.; Silva, C.A.; Sallum, A.M.E.; Pereira, R.M.R.; et al. Efficacy and safety of creatine supplementation in juvenile dermatomyositis: A randomized, double-blind, placebo-controlled crossover trial. *Muscle Nerve* **2015**, *53*, 58–66. [CrossRef] [PubMed]

 *nutrients*

Review

Potential of Creatine in Glucose Management and Diabetes

Marina Yazigi Solis, Guilherme Giannini Artioli and Bruno Gualano *

Applied Physiology & Nutrition Research Group, School of Medicine, University of Sao Paulo, Sao Paulo 01246-903, Brazil; marina.solis@gmail.com (M.Y.S.); artioli@usp.br (G.G.A.)
* Correspondence: gualano@usp.br; Tel.: +55-11-3061-8789

Abstract: Creatine is one of the most popular supplements worldwide, and it is frequently used by both athletic and non-athletic populations to improve power, strength, muscle mass and performance. A growing body of evidence has been identified potential therapeutic effects of creatine in a wide variety of clinical conditions, such as cancer, muscle dystrophy and neurodegenerative disorders. Evidence has suggested that creatine supplementation alone, and mainly in combination with exercise training, may improve glucose metabolism in health individuals and insulin-resistant individuals, such as in those with type 2 diabetes mellitus. Creatine itself may stimulate insulin secretion in vitro, improve muscle glycogen stores and ameliorate hyperglycemia in animals. In addition, exercise induces numerous metabolic benefits, including increases in insulin-independent muscle glucose uptake and insulin sensitivity. It has been speculated that creatine supplementation combined with exercise training could result in additional improvements in glucose metabolism when compared with each intervention separately. The possible mechanism underlying the effects of combined exercise and creatine supplementation is an enhanced glucose transport into muscle cell by type 4 glucose transporter (GLUT-4) translocation to sarcolemma. Although preliminary findings from small-scale trials involving patients with type 2 diabetes mellitus are promising, the efficacy of creatine for improving glycemic control is yet to be confirmed. In this review, we aim to explore the possible therapeutic role of creatine supplementation on glucose management and as a potential anti-diabetic intervention, summarizing the current knowledge and highlighting the research gaps.

Keywords: dietary supplements; exercise; skeletal muscle; glycemic control; type 2 diabetes mellitus

Citation: Solis, M.Y.; Artioli, G.G.; Gualano, B. Potential of Creatine in Glucose Management and Diabetes. *Nutrients* **2021**, *13*, 570. https://doi.org/10.3390/nu13020570

Academic Editor: Richard B. Kreider
Received: 22 January 2021
Accepted: 3 February 2021
Published: 9 February 2021

Publisher's Note: MDPI stays neutral with regard to jurisdictional claims in published maps and institutional affiliations.

1. Introduction

Type 2 diabetes mellitus (T2DM) is a major public health concern worldwide, imposing high health costs for public and private health systems. According to the Global Burden of Disease Study, diabetes incidence increased from 11.3 million in 1990 to 22.9 million in 2017, whilst prevalence increased from 211.2 million in 1990 to 476.0 million in 2017 [1]. In 2017, the International Diabetes Federation estimated that 451 million adults live with diabetes, and by 2045, this number could increase to 693 million if no preventive measures are adopted [2].

T2DM is a metabolic disorder characterized by sustained hyperglycemia resulting from impaired insulin production by pancreatic β cells, impaired insulin action (i.e., insulin resistance), or both [3]. Chronic hyperglycemia in diabetes is associated with several cardiometabolic disorders, such as hypertension, dyslipidemia, atherosclerosis and visceral obesity. Moreover, T2DM is considered one of the top 10 causes of premature deaths from noncommunicable diseases, and is associated with increased mortality from infections, cardiovascular disease, stroke, chronic kidney disease, chronic liver disease, and cancer. In fact, all-cause mortality risk increases by 2- to 3-fold in individuals with diabetes [1].

T2DM can be managed with non pharmacological treatment (i.e., weight reduction, dietary intervention, and physical activity) and/or pharmacological treatment [4]. There have been several dietary candidates to help control glycemia, so far with little or no clinical support from large, controlled trials. In the past two decades, creatine (α-methyl guanidine-acetic acid) supplementation has also been speculated as a dietary supplement potentially able to improve glucose control and insulin resistance.

Creatine is a naturally occurring amine, which is endogenously synthesized ($\sim$1 g·d^{-1}) in the liver, kidneys and pancreas from the amino acid glycine, methionine and arginine. Creatine can also be exogenously obtained from food sources ($\sim$1–5 g·d^{-1}), especially by the ingestion of beef, pork, chicken and fish. In humans, creatine is found in its free (60 to 70%) and phosphorylated (30 to 40%) forms. Approximately 95% of the total bodily creatine store is found in skeletal muscle, with the remaining 5% being found in cells with rapid energy demands, such as cardiac myocytes, retina, neurons and testicles. Creatine excretion occurs through its irreversible and non-enzymatic conversion to creatinine, which is then eliminated by the kidneys [5].

Creatine supplementation is a popular strategy to improve exercise performance in healthy individuals and athletes due to its efficacy of increasing muscle free creatine and phosphorylcreatine contents [6]. Strong evidence indicates that creatine supplementation increases muscle strength, lean mass and improve performance in high-intensity, short-duration exercise [7]. Moreover, new applications for creatine have been proposed, as creatine seems to have potential therapeutic properties in a wide variety of clinical conditions, such as muscle disorders, neurodegenerative conditions and, metabolic dysfunctions, including insulin resistance and T2DM [8].

There is preliminary evidence showing that creatine supplementation could affect glucose metabolism. Studies have demonstrated that creatine ingestion combined with carbohydrate promote greater total muscle glycogen accumulation in animals [9,10] and in humans [11]. Additionally, creatine supplementation along with carbohydrate promotes greater muscle creatine retention than creatine alone [12]. These effects may be partially explained by the fact that both muscle glucose and creatine uptake are influenced by insulin-dependent transporters. In vitro studies showed that creatine increases insulin secretion [13,14]; human studies, however, did not demonstrate the same effects [15,16]. In addition, creatine supplementation was shown to ameliorate hyperglycemia in transgenic Huntington mice and delay diabetes offset [17]. In humans, creatine supplementation associated with exercise training improved glycemia in sedentary [18] and T2DM adults [19]. Altogether, these findings provide the rationale for exploring the application of creatine as a potential anti-diabetic intervention. In this short, narrative review, we will describe the effects of creatine supplementation on glycemic control, summarizing the current knowledge and highlighting the research gaps.

2. Insulin resistance in the Context of the Interplay between Creatine and Glucose Metabolism

Under normal conditions, insulin is secreted by pancreatic β cells in response to the presence of energy substrates (e.g., glucose, fatty acids and amino acids), hormones and changes in energetic demands (e.g., fasting–feeding cycle, exercise and stress) in order to maintain normal blood glucose levels [3]. Insulin binds to its tyrosine kinase-type receptor and activates phosphorylation of a family of insulin receptor substrates (IRSs), especially IRS1 and IRS2 [20]. IRS-phosphorylated proteins bind and activate intracellular signaling molecules, such as phosphatidylinositol 3 kinase (PI3K), which in turn, promotes the translocation of the type 4 glucose transporter (GLUT-4) to the plasma membrane, ultimately resulting in the uptake of bloodstream glucose into the muscle skeletal [3]. Additionally, insulin stimulates the mitogen-activated protein kinase (MAPK) pathway, a necessary step in cell proliferation and nuclear activation [21].

Insulin resistance is a condition that precedes T2DM and, together with genetic and environmental factors, such as obesity and physical inactivity, may lead to the failure of β-cell function and, hence, a progressive decline in insulin secretion [22,23]. Insulin resistance is generally associated with suppressed PI3K pathway, with increased serine phosphorylation of IRS proteins and inhibited tyrosine phosphorylation [24]; IRS protein degradation also seems to occur in some conditions [25]. Additionally, insulin resistance in

T2DM could display a suboptimal GLUT-4 translocation, irrespective of decreased muscle GLUT-4 content [19].

Recent evidence has suggested that T2DM individuals display altered creatine metabolism [26,27]. In a prospective cohort study including more than 4700 participants, Post et al. [27] observed that higher plasma creatine concentration was associated with increased incidence of T2DM. According to the authors, higher extracellular creatine concentration and lower intracellular phosphorylcreatine/creatine content may be related to an impaired intracellular energy state that suggests mitochondrial dysfunction, a postulated mechanism involved in T2DM pathophysiology [27]. Whether creatine is either a marker or maker in this process remains to be addressed.

The potential basis for creatine supplementation to improve glycemic control involves: (1) creatine-induced increased insulin secretion; (2) creatine-induced changes in osmoregulation, and (3) creatine-induced increased glucose uptake via an augment in GLUT-4 content and/or translocation. In addition, exercise training has been suggested to have synergistic effects to creatine, leading to the assumption that these combined interventions could foster greater benefits in glycemic control vs. creatine or exercise alone [28]. The potential mechanisms that could explain the potential benefits of creatine, associated or not with exercise, on glucose control are illustrated in Figure 1.

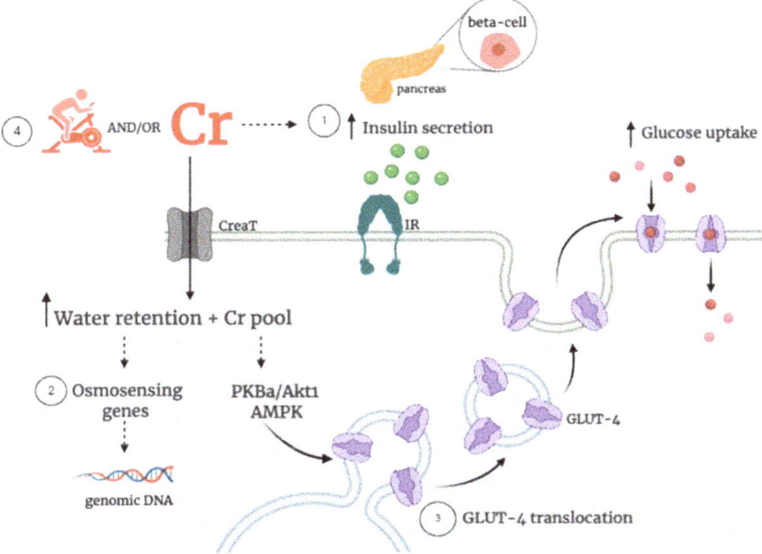

Figure 1. Possible creatine-related mechanisms on glycemic control. Potential mechanisms underpinning the role of creatine on glucose metabolism involve: (1) increased beta-cell insulin secretion; (2) improved water retention and osmosensing genes and (3) increased glucose uptake via type 4 glucose transporter (GLUT-4) content and activity. Additionally, (4) creatine supplementation could enhance the known benefits of exercise on glucose uptake/insulin sensitivity. There are currently insufficient clinical data to support all of these mechanisms. Note: Cr: creatine; CreaT: creatine transporter; IR: insulin receptor; GLUT-4: glucose transporter; PKBa/Akt1: protein kinase Bα; AMPK: Adenosine Monophosphate-activated protein kinase.

In relation to the effects of creatine on insulin secretion, Hill et al. [29] were the first to show, in dogs, that an acute dose of creatine resulted in hypoglycemia. Later, in vitro studies confirmed that supraphysiological creatine concentrations elicited a modest stimulation of insulin secretion in isolated rat pancreas [14]. Similar results were shown in ex vivo experiments with mouse islets [13] and with insulinoma cells [30]. In pre-clinical studies involving different animal models, creatine supplementation increased circulating insulin levels [31] and improved insulinogenic index in T2DM rats [32]. However, human studies involving healthy [18] and T2DM adults [19] did not show increased insulin secretion either when creatine was provided alone or in combination with exercise training [19,33].

Creatine is also able to promote muscle water retention, thereby leading to changes in cell osmolarity [34]. Increased intracellular osmolarity induces cellular swelling, which may activate cell-volume sensitive signaling cascades capable of inducing adaptive changes in intra- and extracellular osmolarity. It is well documented that cell swelling is a potent stimulus to glycogen synthesis in muscle [35] and liver [36]. Thus, creatine-induced cell swelling could improve muscle glycogen stores. In fact, 10 days of creatine supplementation in health adults was able to increase glycogen stores and modulate mRNA content of genes and protein content involved in osmosensing, which could stimulate anabolic signal transduction, such as myogenin and, consequently, satellite cell activation [34]. In addition, increased intracellular osmolarity were associated with increases in circulating IGF-1, which promotes insulin-like effects and decreases counterregulatory hormones [37], which are involved in glycogen catabolism.

Some studies have suggested that oral creatine supplementation holds potential to promote muscle glucose uptake [9,19,38]. GLUT-4 is a key protein involved in transmembrane glucose transport and glucose uptake by skeletal muscle cells [39]. Increases in membrane GLUT-4 content could ultimately result in improved insulin sensitivity and glucose tolerance [9]. In rodents, creatine has been shown to increase *SLC2A4* (GLUT-4) gene expression and to enhance GLUT-4 protein content in muscle [10]. Similar effects have also been reported in humans after creatine supplementation accompanied by an exercise training program following limb immobilization [9], although not all studies have found the same outcomes [33,34]. Interestingly, creatine-induced improvements on glycemic control in T2DM patients were linked to increased GLUT-4 translocation to the sarcolemma, but not to changes in total muscle GLUT-4 content [19]. Increased GLUT-4 translocation was also associated with increased Adenosine Monophosphate-activated protein kinase (AMPK-α) expression, a protein involved in GLUT-4 translocation [38]. Likewise, among healthy young men, short-term creatine supplementation upregulated protein kinase Bα (PKBa/Akt1), a protein that plays a role in the insulin-stimulated GLUT-4 translocation and in glycogen synthesis [34].

Exercise knowingly modulates muscle glucose uptake by increasing (1) extracellular glucose delivery, from blood to muscle, (2) muscle membrane glucose transport, and (3) intracellular glucose phosphorylation and insulin sensitivity [40]. Importantly, T2DM patients retain the capacity of GLUT4 translocation to the sarcolemma in response to exercise [41], which makes exercise a potent means to improve glucose uptake. Exercise-induced GLUT-4 translocation seems to occur via AMPK pathway in an insulin-independent manner [42]. Thus, irrespective of circulating insulin levels or peripheral insulin action (which is facilitated trough muscle contraction via improvements in insulin signaling), exercise can directly induce a substantive GLUT-4 translocation to sarcolemma, thereby improving glycemic control [43]. Of relevance, preliminary evidence suggests that creatine has the potential to enhance the well-known beneficial effects of exercise on glucose control. The role of creatine supplementation, alone or in combination with exercise, on glucose metabolism is compressively covered in the next two subsections.

3. Effects of Creatine Supplementation Alone on Glycemic Control

Creatine supplementation increases intermuscular total creatine content by 10 to 30% in children, adults and older individuals [6,44]. Clinical evidence indicates that creatine supplementation improves fat-free mass [34], delays fatigue, increases muscle strength and, particularly in older adults, improves performance in activities of daily living [45–47]. Additionally, creatine supplementation upregulates genes and protein content of kinases involved in osmosensing and signal transduction, cytoskeleton remodeling, protein and glycogen synthesis regulation, satellite cell proliferation and differentiation [34]. Therefore, even without exercise, creatine alone seems to modulate skeletal muscle signaling, leading to potential short-term, physiological adaptations.

A study with animals revealed that 3 weeks of creatine administration (2% of the diet) significantly increased muscle glycogen stores possibly due to upregulation of *GLUT-4* expression and AMPK phosphorylation in female Wistar rats [10]. Moreover, Rooney et al. [31] showed increased plasma insulin concentrations in Wistar male rats supplemented with creatine (2% of the body weight) for eight weeks. The authors attributed this result to increased pancreatic total creatine content, which may have altered insulin secretion. In contrast, Op't Eijnde et al. [9] reported that creatine intake (5 g·kg^{-1}·d^{-1} of creatine for 5 days) in Wistar male rats did not increase GLUT-4 expression or enhance the sensitivity and the responsiveness of rat muscles to insulin. Similarly, Young and Young [48] did not find changes in glucose metabolism following creatine intake (300 mg·kg^{-1}·d^{-1} of creatine over 5 weeks) in Sprague Dawley rats.

Creatine supplementation has been tested in different experimental paradigms of insulin resistance. Ferrante et al. [17] studied the effects of creatine ingestion in transgenic mice model of Huntington's disease. Different doses of creatine (1, 2, or 3%) resulted in a substantial neuroprotective effect, improvement in the rotarod test performance and a significant reduction in hyperglycemia that typically accompanies this experimental model. Interestingly, the administration of creatine also delayed the onset of diabetes in these mice. In addition, Op't Eijnde et al. [32] supplemented Goto-Kakizaki rats, a T2DM model, with creatine (2% of the diet) for eight weeks, and showed an improvement in the insulinogenic index (plasma glucose and insulin ratio), which was mostly attributed to a reduction in insulinemia. The authors concluded that creatine supplementation was able to improve insulin sensitivity in skeletal muscle of insulin-resistant rats. In contrast, using an animal model of severe muscle wasting and insulin resistance induced by high-dose dexamethasone, Nicastro et al. [49] showed that 7 days of creatine supplementation (5 g·Kg^{-1}·d^{-1} of creatine) aggravated hyperglycemia and hyperinsulinemia in male Wistar rats. These findings highlight the complexities in interpreting and generalizing data obtained from creatine studies involving animal models, since there is a large specie-specific variability in response to creatine supplementation. This makes it difficult to reconcile pre-clinical data involving creatine [50]. The studies assessing the effect of creatine supplementation alone on glycemic control in animal models are summarized in Table 1.

Table 1. Effect of creatine supplementation alone in glucose metabolism in animals.

Reference	Model	Creatine Protocol	Main Findings
Ferrante et al. [17]	Transgenic mice model of Huntington's disease	Diet supplemented with 1, 2, or 3% of Cr for 21 d	↑ glucose tolerance; ↑ neuroprotective effect; ↑ body weight; ↑ motor performance on the rotarod test.
Op't Eijnde et al. [9]	Male Wistar rats	Powdered rat chow with 5% of Cr for 5 d	↑ Cr and PCr muscle content; ↔ muscle GLUT-4 content; ↔ glucose transport rate; ↔ plasma insulin; ↔ blood glucose.
Young and Young, [48]	Male Sprague Dawley rats	300 mg·kg^{-1}·d^{-1} for 5 wk	↑ Cr and PCr muscle content; ↔ basal rates of glucose uptake; ↔ insulin-stimulated rates of glucose uptake.
Rooney et al. [31]	Male Wister rats	Chow containing 2% of Cr for 2, 4, or 8 wk	↔ fasting plasma glucose; ↔ plasma glucose after oral glucose load; ↑ fasting plasma insulin levels; ↑ pancreatic TCr content.

Table 1. *Cont.*

Reference	Model	Creatine Protocol	Main Findings
Ju et al. [10]	Female Wistar rats	Chow containing 2% of Cr for 3 wk	↑ glycogen content; ↑ muscle GLUT-4 content; ↑ *GLUT-4* mRNA; ↑ AMPK phosphorylation; ↑ Acetyl-Coa carboxylase phosphorylation.
Op't Eijnde et al. [32]	Male Goto-Kakizaki rats	Pallets enriched with 2% of Cr for 8 wk	↑ muscle Cr content only in young rats (but not in older rats); ↓ plasma insulin concentration; ↔ Blood D-glucose concentration after OGTT; ↓ insulinogenic index.
Nicastro et al. [49]	Male Wistar rats	$5 \text{ g·Kg}^{-1}\text{·d}^{-1}$ of Cr for 7 d + $5 \text{ mg·kg}^{-1}\text{·d}^{-1}$ of DXM	↑ serum glucose and insulin after Cr + DXM; ↑ HOMA-IR after Cr + DXM; ↓ GLUT-4 translocation after Cr + DXM

Notes: ↑: increase; ↓: decrease; ↔: no change; Cr: Creatine; PCr: phosphorylcreatine; TCr: total creatine; OGTT: oral glucose tolerance test; GLUT-4: glucose transporter; AMPK: AMP-activated protein kinase; HOMA-IR: Homeostatic Model Assessment for Insulin Resistance (a surrogate of insulin resistance); DXM: dexamethasone.

In humans, creatine supplementation (5 g·d^{-1} of creatine for 42 days) induced an increase in plasma glucose in response to an oral glucose load in health vegetarians adults [51]. Van Loon et al. [33] demonstrated that creatine supplementation alone (20 g·d^{-1} for 5 days followed by 2 g·d^{-1} for 6 weeks) increased glycogen content (+18%) in young, non-vegetarian adults. In contrast, Newman et al. [16] showed that creatine supplementation (20 g·d^{-1} for 5 days followed by 3 g·d^{-1} for 28 days) did not influence muscle glycogen content, plasma glucose and insulin responses during an oral glucose tolerance test in healthy, active, male adults. Additionally, insulin sensitivity surrogates (i.e., glucose-insulin index and index of insulin sensitivity) did not change with creatine ingestion, indicating that insulin action and secretion remained unaltered [16].

Van Loon et al. [33] showed no effect of creatine on *GLUT-4* mRNA and GLUT-4 protein content. Similarly, Safdar et al. [34] conducted a microarray analysis and did not observe changes in *GLUT-4* mRNA after creatine supplementation ($20 \text{ g·kg}^{-1}\text{·d}^{-1}$ for 3 days followed by $5 \text{ mg·kg}^{-1}\text{·d}^{-1}$ for 7 days) in young, healthy men. Nevertheless, they observed an increase in *PKBa/Akt1* mRNA, which participates in glycogen synthesis via increasing glycogen synthase activity. The authors also showed a 21% decrease in skeletal muscle *phosphofructokinase* and *glycogen phosphorylase* mRNA, suggesting that creatine supplementation in the absence of exercise could eventually increase muscle glycogen stores by increasing the cascade signaling leading to GLUT-4 recruitment to the sarcolemma, despite the lack of changes in GLUT-4 [34]. In a significant different model, however, creatine supplementation (20 g·d^{-1} for 2 weeks) was shown to prevent the drop in GLUT-4 protein expression induced by leg immobilization in healthy young men (+9% in the creatine group vs. −20% in the placebo group) [9]. In an open-label, cross-over study (with a 2-day washout period) involving a small sample of T2DM patients, creatine supplementation (6 g·d^{-1} for 5 days) was as effective as metformin ($2 \times 500 \text{ mg·d}^{-1}$), a widely used anti-diabetic drug, in lowering blood glucose concentrations [52]. The human studies assessing the effect of creatine supplementation alone on glycemic control are summarized in Table 2.

Table 2. Effect of creatine supplementation alone in glucose metabolism in humans.

Reference	Sample (*n*)	Study Design	Creatine Protocol	Main Findings
Newman et al. [16]	Healthy, active, untrained, male adults (17)	Sigle-blind, placebo-controlled trial	Loading phase: 20 g·d^{-1} (4 × 5 g) of Cr for 5 d + Maintenance phase: 3 g·d^{-1} for 28 d	↑ muscle TCr; ↔ muscle glycogen content; ↔ plasma glucose and insulin during OGTT; ↔ glucose-insulin index; ↔ index of insulin sensitivity.
Rooney et al. [51]	Healthy, vegetarian adults (14)	Controlled-trial	5 g·d^{-1} of Cr for 42 d	↑ plasma total Cr concentration; ↑ plasma glucose response; ↔ plasma insulin.
Van Loon et al. [33]	Young, nonvegetarians adults (20)	Double-blind placebo-controlled trial	Loading phase: 20 g·d^{-1} (4 × 5 g) of Cr for 5 d + Maintenance phase: 2 g·d^{-1} for 6 wk	↑ muscle glycogen, Cr and PCr after loading phase, with a decline in maintenance phase; ↔ *GLUT-4* mRNA; ↔ total muscle GLUT-4 protein content.
Safdar et al. [34]	Young, healthy, nonobese men (12)	Double-blind, crossover, randomized, placebo-controlled trial	Loading phase: 20 g·d^{-1} (4 × 5 g) of Cr for 3 d + Maintenance phase: 5 g·d^{-1} for 7 d	↑ muscle total Cr ↑ *PKB/Akt1* expression and protein; ↑ *MAPK* expression; ↔ *GLUT-4* mRNA
Rocic et al. [52]	Recently diagnosed T2DM patients, without anti-diabetic treatment (30)	Open-label, cross-over	6 g·d^{-1} of Cr or 1000 mg·d^{-1} of metformin for 5 d	↓ glucose concentration in both groups; ↔ insulin, C-peptide and, HbA1c.

Notes: ↑: increase; ↓: decrease; ↔: no change; Cr: Creatine; PCr: phosphorylcreatine; TCr: total creatine; GLUT-4: glucose transporter; OGTT. oral glucose tolerance test; HbA1c: glycosylated hemoglobin; Akt1: protein kinase B; MAPK: mitogen-activated protein kinases.

4. Effects of Creatine Supplementation Combined with Exercise on Glycemic Control

Exercise has been recognized as a cornerstone in diabetes management, in addition to diet and hypoglycemic/antihyperglycemic agents. Regular physical activity induces beneficial metabolic and hemodynamic changes, resulting in improvements in insulin-independent muscle glucose uptake and in insulin sensitivity, as well as increased glycogen content [39,53]. The benefits of exercise as regard to glucose metabolism are widely reported in trained and untrained healthy [54], obese [55], insulin-resistant [56] and T2DM individuals [57]. Creatine supplementation has emerged as a strategy capable of enhancing the physiological and metabolic adaptations to exercise, which could confer protection in insulin resistance conditions [19,28,58]. Data supporting this hypothesis are so far inconclusive, however.

Souza et al. [59] showed that oral creatine administration (5 g·kg^{-1}·d^{-1} for 7 days followed by 1 g·kg^{-1}·d^{-1} for 8 weeks), associated or not with high-intensity swimming training, reduced plasma glucose levels in Wistar rats. Similarly, Araujo et al. [60] demonstrated that Wistar rats receiving creatine supplementation (diet supplemented with 13% for 7 days followed by 2% for 55 days) plus high-intensity exercise training exhibited a smaller area under the curve for glucose in response to an oral glucose challenge. On the other hand, Freire et al. [61] did not show any effect of creatine (2% of creatine in drinking water for 8 weeks) plus swimming training on plasma glucose responses to an oral glucose tolerance test in Wistar rats. Likewise, Vaisy et al. [62] did not demonstrate any benefits of creatine supplementation (diet enriched with 2.5% of creatine for 12 weeks) alone or in combination with exercise training on glucose homeostasis and muscular insulin sensitivity in Wistar rats with insulin resistance induced by a sucrose-rich cafeteria diet.

These conflicting results could be partially explained by the differences in creatine supplementation regimen and exercise types. For instance, both Freire et al. [61] and

Vaisy et al. [62] used a single-phase creatine protocol throughout the entire intervention period, whereas others opted for a loading dose followed by a maintenance phase [62,63]. Training models also varied considerably with the use of treadmill [59,61,62] and swimming exercises [60], which precludes more direct comparisons concerning the exercise stimuli to which the animals were subjected. Therefore, the overall conclusions from pre-clinical studies are equivocal. The studies assessing the effects of creatine supplementation along exercise on glycemic control in animals are summarized in Table 3.

Table 3. Effect of creatine supplementation along with exercise in glucose metabolism in animals.

Reference	Model	Creatine and Training Protocol	Main Findings
Souza et al. [59]	Male Wistar rats	Loading phase: $5\ g\cdot kg^{-1}$ body weight of Cr for 7 d + Maintenance phase: $1\ g\cdot d^{-1}$ for 8 wk Training: swimming	↓ plasma glucose levels during 1–4 wk after Cr alone; ↓ plasma glucose levels during 1–8 wk after Cr and exercise protocol.
Freire et al. [61]	Male Wistar rats	Pallets enriched with 2% of Cr for 4 or 8 wk Training: swimming	↔ glucose uptake; ↔ glucose AUC during OGTT; ↔ liver and quadriceps glycogen content.
Vaisy et al. [62]	Male Wistar rats	Cafeteria diet enriched with 2.5% of Cr for 12 wk Training: swimming	↔ fasting blood glucose concentration; ↓ fasting plasma insulin level after training and training + creatine; ↓ whole body insulin level.
Araújo et al. [60]	Male Wistar rats	Loading phase: Chow containing 13% of Cr for 7d + Maintenance phase: Chow containing 2% of Cr for 55 d Training: high intensity treadmill running	↔ glucose uptake; ↓ glucose AUC during OGTT after Cr and exercise protocol.

Note: ↑: increase; ↓: decrease; ↔: no change; Cr: Creatine; PCr: phosphorylcreatine; TCr: total creatine; OGTT: oral glucose tolerance test; AUC: area under the curve.

Op't Eijnde et al. [9] provided new insights on the potential benefits of creatine plus exercise on glucose metabolism in humans. In this study, healthy individuals had one of their legs immobilized for two weeks, while they received either creatine ($20\ g\cdot d^{-1}$) or placebo supplementation. After the immobilization period, participants underwent a 10-week knee extension training (3 times a week). The immobilization period caused a significant reduction in GLUT-4 protein expression (-20%) in the control group, but not in the creatine supplemented group. The rehabilitation period promoted a normalization of the GLUT-4 content in the control group and a ~40% increase in GLUT-4 expression in the creatine group. Additionally, rehabilitation training per se did not increase muscle GLUT4 content above the baseline levels, but creatine along with training led to a substantial increase in this protein (+40%). Of relevance, the authors also reported that muscle glycogen was significantly augmented only when exercise training was undertaken in conjunction with creatine supplementation [9]. Using a similar experimental design, Derave et al. [63] showed that creatine ($15\ g\cdot d^{-1}$ during immobilization followed by $2.5\ g\cdot d^{-1}$ for 6 weeks during rehabilitation) combined or not with protein supplementation was able to increase GLUT-4 protein expression and improve oral glucose tolerance following a 6-week exercise rehabilitation program that started two weeks after a single-leg immobilization protocol.

Gualano et al. [18] demonstrated that a 12-week creatine supplementation protocol ($20\ g\cdot d^{-1}$ for 10 days followed by $10\ g\cdot d^{-1}$ throughout the remaining period) associated with moderate-intensity, aerobic exercise training (3 times a week) resulted in a greater decrease in plasma glucose in response to oral glucose tolerance test, compared to exercise alone, in sedentary but apparently healthy men. These data suggested a synergistic effect of exercise and creatine on glucose tolerance; however, in this and other studies [16,33], creatine did not change fasting insulin and Homeostatic Model Assessment for Insulin Resistance (HOMA-IR), a surrogate of insulin resistance.

Next to this preliminary study involving healthy participants, Gualano et al. [19] conducted a small-scale, double-blind, placebo-controlled trial involving T2DM patients, who underwent exercise training and received either creatine supplementation (5 g·d^{-1} for 12 weeks) or placebo. In this study, creatine supplementation along with exercise training significantly reduced glycated hemoglobin (HbA1c) and glycemia in response to a meal tolerance test. No significant differences were observed for insulin and C-peptide concentrations. The beneficial effect on glycemic control was partially explained by an improvement in GLUT-4 translocation to sarcolemma, rather than in total muscle GLUT-4 content [19]. In an ancillary analysis from this study, decreased HbA1c levels and increased GLUT-4 translocation were associated with increased AMPK protein expression [38], providing new insights on the molecular mechanisms underlying the effects of combined creatine and exercise interventions on glucose metabolism. Of clinical relevance, creatine supplementation as an adjuvant therapy appeared to be safe, as no adverse events were reported and no alterations in health-related laboratory markers were seen [19,38].

However, the efficacy of creatine supplementation was not confirmed by a subsequent randomized, double-blind, placebo-controlled, pilot trial involving community-dwelling older adults [64]. After a 12-week follow-up period, creatine supplementation (5 g·d^{-1}) and resistance training (3 times a week) did not improve insulin resistance (assessed by fasting blood glucose and insulin, and HOMA-IR). It is possible to speculate that the better glycemic control exhibited by the participants in this study, compared to those of Gualano et al.'s study, may partially explain the null findings on the basis of a "ceiling effect". The studies assessing the effect of creatine supplementation along exercise on glycemic control in humans are summarized in Table 4.

Table 4. Effect of creatine supplementation along with exercise on glucose metabolism in humans.

Reference	Sample (*n*)	Study Design	Creatine and Training Protocol	Main Findings
Op't Eijnde et al. [9]	Young, healthy subjects (22)	Double-blind placebo-controlled trial	Loading phase: 20 g·d^{-1} during immobilization period (2 wk) + Maintenance phase: 15 g·d^{-1} for 3 wk followed by 5 g·d^{-1} for 7 wk during rehabilitation training Program training: knee-extensor resistance training (3 times·wk^{-1})	Immobilization period: ↓ 20% GLUT-4 in placebo group, but not in Cr group; ↔ glycogen and Cr muscle content in both groups. Rehabilitation period: ↑ 40% GLUT-4 in Cr group; ↑ glycogen and Cr muscle content after Cr.
Derave et al. [63]	Young, healthy subjects (22)	Double-blind, placebo-controlled trial	Loading phase: 15 g·d^{-1} during immobilization period (2 wk) combined or not with protein supplementation + Maintenance phase: 2.5 g·d^{-1} for 6 wk during rehabilitation training Program training: knee-extensor resistance training (3 times·wk^{-1})	Immobilization period: ↓ GLUT-4 in placebo and Cr group, but not in Cr + protein group; ↔ glycogen and Cr muscle content in all groups. Rehabilitation period: ↑ 24% GLUT-4 in Cr group and ↑ 33% in Cr + protein group; ↑ glycogen and Cr muscle content after Cr and Cr + protein supplementation

Table 4. *Cont.*

Reference	Sample (*n*)	Study Design	Creatine and Training Protocol	Main Findings
Gualano et al. [18]	Healthy, sedentary male (22)	Double-blind, randomized-placebo-controlled trial	Loading phase: 0.3 g·kg^{-1}·d^{-1} of Cr for 1 wk + Maintenance phase: 0.15 g·kg^{-1}·d^{-1} for 11 wk Program training: aerobic training at 70% of the VO^{2}max	↓ glucose AUC after OGTT; ↔ fasting insulin; ↔ HOMA-IR.
Gualano et al. [19]	T2DM patients (25)	Double-blind, randomized-placebo-controlled trial	5 g·d^{-1} of Cr for 12 wk Program training: Aerobic training and resistance training	↓ HbA1c in Cr group; ↓ glycemia during MTT (0, 30 and 60 min) in Cr group; ↑ muscle PCr content in Cr group; ↑ muscle strength and function in Cr group
Alves et al. [38]	T2DM patients (25)	Double-blind, randomized-placebo-controlled trial	5 g·d^{-1} of Cr for 12 wk Program training: Aerobic training and resistance training	↑ AMPK protein expression; ↔ IR-β, Akt1 and MAPK.
Oliveira et al. [64]	Healthy, older adults (32)	randomized, double-blind, placebo-controlled, parallel-group clinical trial	5 g·d^{-1} of Cr for 12 wk Program training: resistance training	↔ inflammatory biomarkers ↔ fasting blood glucose; ↔ fasting insulin; ↔ HOMA-IR.

Note: ↑: increase; ↓: decrease; ↔: no change; Cr: Creatine; PCr: phosphorylcreatine; TCr: total creatine; OGTT: oral glucose tolerance test; GLUT-4: glucose transporter; T2DM: type 2 diabetes mellitus; AUC: area under the curve; HOMA-IR: Homeostatic Model Assessment for Insulin Resistance; HbA1c: glycosylated hemoglobin; IR-β: insulin receptor; Akt1: protein kinase B; MAPK: AMP-activated protein kinase.

5. Conclusions and Future Directions

Creatine supplementation has the potential to promote changes in glucose metabolism that may favor a healthier metabolic profile. This may be particularly true when exercise training is provided along with this supplement, as creatine seems to enhance the training adaptations.

Evidence supporting the role of creatine on glucose metabolism remains speculative. As discussed, pre-clinical data are difficult to interpret as creatine responses are highly dependent on the experimental model adopted. The few existing clinical trials are small-scale, short-term and, therefore, exploratory. Despite the fact that a few of them have revealed promising benefits of creatine on glucose control, especially when exercise as co-prescribed, further large, longer-term, controlled trials involving T2DM with variable disease severity and under different pharmacological treatments are necessary to drawn firm conclusions on the efficacy and safety of creatine as an anti-diabetic intervention. It is equally important to develop new investigations aimed at unraveling the mechanisms by which creatine, aligned or not with training, could regulate glucose control, as basic research is indeed useful to better target potentially most benefited populations for testing creatine in next clinical trials.

Author Contributions: Conceptualization, M.Y.S., G.G.A. and B.G.; writing—original draft preparation, M.Y.S., G.G.A. and B.G.; writing—review and editing, M.Y.S., G.G.A. and B.G.; visualization, M.Y.S., G.G.A. and B.G.; supervision, B.G. All authors have read and agreed to the published version of the manuscript.

Funding: This research received no specific funding.

Institutional Review Board Statement: Not applicable.

Informed Consent Statement: Not applicable.

Data Availability Statement: Not applicable.

Acknowledgments: Guilherme Giannini Artioli and Bruno Gualano were supported by FAPESP #2019/25032-9 and 2017/12511-0, respectively.

Conflicts of Interest: BG has received research grants, creatine donation for scientific studies, travel support for participation in scientific conferences and honorarium for speaking at lectures from AlzChem (a company which manufactures creatine). Additionally, he serves as a member of the Scientific Advisory Board for Alzchem. The others have no conflict of interests.

References

1. Lin, X.; Xu, Y.; Pan, X.; Xu, J.; Ding, Y.; Sun, X.; Song, X.; Ren, Y.; Shan, P.F. Global, regional, and national burden and trend of diabetes in 195 countries and territories: An analysis from 1990 to 2025. *Sci. Rep.* **2020**, *10*, 14790. [CrossRef]
2. Cho, N.H.; Shaw, J.E.; Karuranga, S.; Huang, Y.; da Rocha Fernandes, J.D.; Ohlrogge, A.W.; Malanda, B. IDF Diabetes Atlas: Global estimates of diabetes prevalence for 2017 and projections for 2045. *Diabetes Res. Clin. Pract.* **2018**, *138*, 271–281. [CrossRef]
3. DeFronzo, R.A.; Ferrannini, E.; Groop, L.; Henry, R.R.; Herman, W.H.; Holst, J.J.; Hu, F.B.; Kahn, C.R.; Raz, I.; Shulman, G.I.; et al. Type 2 diabetes mellitus. *Nat. Rev. Dis. Primers* **2015**, *1*, 15019. [CrossRef] [PubMed]
4. American Diabetes Association. 2. Classification and Diagnosis of Diabetes: Standards of Medical Care in Diabetes-2020. *Diabetes Care* **2020**, *43* (Suppl. 1), S14–S31. [CrossRef] [PubMed]
5. Wyss, M.; Kaddurah-Daouk, R. Creatine and creatinine metabolism. *Physiol. Rev.* **2000**, *80*, 1107–1213. [CrossRef]
6. Harris, R.C.; Soderlund, K.; Hultman, E. Elevation of creatine in resting and exercised muscle of normal subjects by creatine supplementation. *Clin. Sci.* **1992**, *83*, 367–374. [CrossRef] [PubMed]
7. Branch, J.D. Effect of creatine supplementation on body composition and performance: A meta-analysis. *Int. J. Sport Nutr. Exerc. Metab.* **2003**, *13*, 198–226. [CrossRef] [PubMed]
8. Gualano, B.; Rawson, E.S.; Candow, D.G.; Chilibeck, P.D. Creatine supplementation in the aging population: Effects on skeletal muscle, bone and brain. *Amino Acids* **2016**, *48*, 1793–1805. [CrossRef]
9. Op't Eijnde, B.; Urso, B.; Richter, E.A.; Greenhaff, P.L.; Hespel, P. Effect of oral creatine supplementation on human muscle GLUT4 protein content after immobilization. *Diabetes* **2001**, *50*, 18–23. [CrossRef] [PubMed]
10. Ju, J.S.; Smith, J.L.; Oppelt, P.J.; Fisher, J.S. Creatine feeding increases GLUT4 expression in rat skeletal muscle. *Am. J. Physiol. Endocrinol. Metab.* **2005**, *288*, E347–E352. [CrossRef] [PubMed]
11. Nelson, A.G.; Arnall, D.A.; Kokkonen, J.; Day, R.; Evans, J. Muscle glycogen supercompensation is enhanced by prior creatine supplementation. *Med. Sci. Sports Exerc.* **2001**, *33*, 1096–1100. [CrossRef]
12. Green, A.L.; Simpson, E.J.; Littlewood, J.J.; Macdonald, I.A.; Greenhaff, P.L. Carbohydrate ingestion augments creatine retention during creatine feeding in humans. *Acta Physiol. Scand.* **1996**, *158*, 195–202. [CrossRef]
13. Marco, J.; Calle, C.; Hedo, J.A.; Villanueva, M.L. Glucagon-releasing activity of guanidine compounds in mouse pancreatic islets. *FEBS Lett.* **1976**, *64*, 52–54. [CrossRef]
14. Alsever, R.N.; Georg, R.H.; Sussman, K.E. Stimulation of insulin secretion by guanidinoacetic acid and other guanidine derivatives. *Endocrinology* **1970**, *86*, 332–336. [CrossRef]
15. Steenge, G.R.; Lambourne, J.; Casey, A.; Macdonald, I.A.; Greenhaff, P.L. Stimulatory effect of insulin on creatine accumulation in human skeletal muscle. *Am. J. Physiol.* **1998**, *275*, E974–E979. [CrossRef] [PubMed]
16. Newman, J.E.; Hargreaves, M.; Garnham, A.; Snow, R.J. Effect of creatine ingestion on glucose tolerance and insulin sensitivity in men. *Med. Sci. Sports Exerc.* **2003**, *35*, 69–74. [CrossRef] [PubMed]
17. Ferrante, R.J.; Andreassen, O.A.; Jenkins, B.G.; Dedeoglu, A.; Kuemmerle, S.; Kubilus, J.K.; Kaddurah-Daouk, R.; Hersch, S.M.; Beal, M.F. Neuroprotective effects of creatine in a transgenic mouse model of Huntington's disease. *J. Neurosci.* **2000**, *20*, 4389–4397. [CrossRef]
18. Gualano, B.; Novaes, R.B.; Artioli, G.G.; Freire, T.O.; Coelho, D.F.; Scagliusi, F.B.; Rogeri, P.S.; Roschel, H.; Ugrinowitsch, C.; Lancha, A.H., Jr. Effects of creatine supplementation on glucose tolerance and insulin sensitivity in sedentary healthy males undergoing aerobic training. *Amino Acids* **2008**, *34*, 245–250. [CrossRef]
19. Gualano, B.; Painnelli, V.D.S.; Roschel, H.; Artioli, G.G.; Neves, M., Jr.; De Sa Pinto, A.L.; Da Silva, M.E.; Cunha, M.R.; Otaduy, M.C.; Leite Cda, C.; et al. Creatine in type 2 diabetes: A randomized, double-blind, placebo-controlled trial. *Med. Sci. Sports Exerc.* **2011**, *43*, 770–778. [CrossRef]
20. Kruger, M.; Kratchmarova, I.; Blagoev, B.; Tseng, Y.H.; Kahn, C.R.; Mann, M. Dissection of the insulin signaling pathway via quantitative phosphoproteomics. *Proc. Natl. Acad. Sci. USA* **2008**, *105*, 2451–2456. [CrossRef] [PubMed]
21. Ramos, P.M.; Martinez, V.B.; Granado, J.Q.; Juanatey, J.R. Advances in hypertension and diabetes in 2007. *Rev. Esp. Cardiol.* **2008**, *61* (Suppl. 1), 58–71. [PubMed]
22. DeFronzo, R.A.; Tripathy, D. Skeletal muscle insulin resistance is the primary defect in type 2 diabetes. *Diabetes Care* **2009**, *32* (Suppl. 2), S157–S163. [CrossRef]

23. Abdul-Ghani, M.A.; Tripathy, D.; DeFronzo, R.A. Contributions of beta-cell dysfunction and insulin resistance to the pathogenesis of impaired glucose tolerance and impaired fasting glucose. *Diabetes Care* **2006**, *29*, 1130–1139. [CrossRef] [PubMed]
24. Copps, K.D.; White, M.F. Regulation of insulin sensitivity by serine/threonine phosphorylation of insulin receptor substrate proteins IRS1 and IRS2. *Diabetologia* **2012**, *55*, 2565–2582. [CrossRef]
25. Bouzakri, K.; Karlsson, H.K.; Vestergaard, H.; Madsbad, S.; Christiansen, E.; Zierath, J.R. IRS-1 serine phosphorylation and insulin resistance in skeletal muscle from pancreas transplant recipients. *Diabetes* **2006**, *55*, 785–791. [CrossRef]
26. Palomino-Schatzlein, M.; Lamas-Domingo, R.; Ciudin, A.; Gutierrez-Carcedo, P.; Mares, R.; Aparicio-Gomez, C.; Hernandez, C.; Simo, R.; Herance, J.R. A Translational In Vivo and In Vitro Metabolomic Study Reveals Altered Metabolic Pathways in Red Blood Cells of Type 2 Diabetes. *J. Clin. Med.* **2020**, *9*, 1619. [CrossRef] [PubMed]
27. Post, A.; Groothof, D.; Schutten, J.C.; Flores-Guerrero, J.L.; Swarte, J.C.; Douwes, R.M.; Kema, I.P.; de Boer, R.A.; Garcia, E.; Connelly, M.A.; et al. Plasma creatine and incident type 2 diabetes in a general population-based cohort: The PREVEND study. *Clin. Endocrinol.* **2020**. [CrossRef]
28. Pinto, C.L.; Botelho, P.B.; Pimentel, G.D.; Campos-Ferraz, P.L.; Mota, J.F. Creatine supplementation and glycemic control: A systematic review. *Amino Acids* **2016**, *48*, 2103–2129. [CrossRef]
29. Hill, R. The effect of the administration of creatine on the blood sugar (abstract). *J. Biol. Chem.* **1928**, *78*, iv. (In Abstract)
30. Gempel, K.; Brdiczka, D.; Kaddurah-Daouk, R.; Wallimann, T.; Kaufhold, P.; Gerbitz, K.D. The creatine analogue cyclocreatine increases insulin secretion in INS-1 cells via a K+ channel independent mechanism. *Diabetologia* **1996**, *39*, 109.
31. Rooney, K.; Bryson, J.; Phuyal, J.; Denyer, G.; Caterson, I.; Thompson, C. Creatine supplementation alters insulin secretion and glucose homeostasis in vivo. *Metabolism* **2002**, *51*, 518–522. [CrossRef]
32. Op't Eijnde, B.; Jijakli, H.; Hespel, P.; Malaisse, W.J. Creatine supplementation increases soleus muscle creatine content and lowers the insulinogenic index in an animal model of inherited type 2 diabetes. *Int. J. Mol. Med.* **2006**, *17*, 1077–1084. [CrossRef]
33. van Loon, L.J.; Murphy, R.; Oosterlaar, A.M.; Cameron-Smith, D.; Hargreaves, M.; Wagenmakers, A.J.; Snow, R. Creatine supplementation increases glycogen storage but not GLUT-4 expression in human skeletal muscle. *Clin. Sci.* **2004**, *106*, 99–106. [CrossRef]
34. Safdar, A.; Yardley, N.J.; Snow, R.; Melov, S.; Tarnopolsky, M.A. Global and targeted gene expression and protein content in skeletal muscle of young men following short-term creatine monohydrate supplementation. *Physiol. Genom.* **2008**, *32*, 219–228. [CrossRef]
35. Low, S.Y.; Rennie, M.J.; Taylor, P.M. Modulation of glycogen synthesis in rat skeletal muscle by changes in cell volume. *J. Physiol.* **1996**, *495 Pt 2*, 299–303. [CrossRef]
36. Baquet, A.; Hue, L.; Meijer, A.J.; van Woerkom, G.M.; Plomp, P.J. Swelling of rat hepatocytes stimulates glycogen synthesis. *J. Biol. Chem.* **1990**, *265*, 955–959. [CrossRef]
37. Deldicque, L.; Louis, M.; Theisen, D.; Nielens, H.; Dehoux, M.; Thissen, J.P.; Rennie, M.J.; Francaux, M. Increased IGF mRNA in human skeletal muscle after creatine supplementation. *Med. Sci. Sports Exerc.* **2005**, *37*, 731–736. [CrossRef]
38. Alves, C.R.; Ferreira, J.C.; de Siqueira-Filho, M.A.; Carvalho, C.R.; Lancha, A.H., Jr.; Gualano, B. Creatine-induced glucose uptake in type 2 diabetes: A role for AMPK-alpha? *Amino Acids* **2012**, *43*, 1803–1807. [CrossRef]
39. Henriksen, E.J. Invited review: Effects of acute exercise and exercise training on insulin resistance. *J. Appl. Physiol.* **2002**, *93*, 788–796. [CrossRef]
40. Sigal, R.J.; Kenny, G.P.; Wasserman, D.H.; Castaneda-Sceppa, C.; White, R.D. Physical activity/exercise and type 2 diabetes: A consensus statement from the American Diabetes Association. *Diabetes Care* **2006**, *29*, 1433–1438. [CrossRef]
41. Kennedy, J.W.; Hirshman, M.F.; Gervino, E.V.; Ocel, J.V.; Forse, R.A.; Hoenig, S.J.; Aronson, D.; Goodyear, L.J.; Horton, E.S. Acute exercise induces GLUT4 translocation in skeletal muscle of normal human subjects and subjects with type 2 diabetes. *Diabetes* **1999**, *48*, 1192–1197. [CrossRef]
42. Ren, J.M.; Semenkovich, C.F.; Gulve, E.A.; Gao, J.; Holloszy, J.O. Exercise induces rapid increases in GLUT4 expression, glucose transport capacity, and insulin-stimulated glycogen storage in muscle. *J. Biol. Chem.* **1994**, *269*, 14396–14401. [CrossRef]
43. Charron, M.J.; Brosius, F.C., 3rd; Alper, S.L.; Lodish, H.F. A glucose transport protein expressed predominately in insulin-responsive tissues. *Proc. Natl. Acad. Sci. USA* **1989**, *86*, 2535–2539. [CrossRef] [PubMed]
44. Solis, M.Y.; Artioli, G.G.; Otaduy, M.C.G.; Leite, C.D.C.; Arruda, W.; Veiga, R.R.; Gualano, B. Effect of age, diet, and tissue type on PCr response to creatine supplementation. *J. Appl. Physiol.* **2017**, *123*, 407–414. [CrossRef]
45. Gotshalk, L.A.; Kraemer, W.J.; Mendonca, M.A.; Vingren, J.L.; Kenny, A.M.; Spiering, B.A.; Hatfield, D.L.; Fragala, M.S.; Volek, J.S. Creatine supplementation improves muscular performance in older women. *Eur. J. Appl. Physiol.* **2008**, *102*, 223–231. [CrossRef]
46. Rawson, E.S.; Clarkson, P.M. Acute creatine supplementation in older men. *Int. J. Sports Med.* **2000**, *21*, 71–75. [CrossRef] [PubMed]
47. Rawson, E.S.; Wehnert, M.L.; Clarkson, P.M. Effects of 30 days of creatine ingestion in older men. *Eur. J. Appl. Physiol. Occup. Physiol.* **1999**, *80*, 139–144. [CrossRef]
48. Young, J.C.; Young, R.E. The effect of creatine supplementation on glucose uptake in rat skeletal muscle. *Life Sci.* **2002**, *71*, 1731–1737. [CrossRef]
49. Nicastro, H.; Gualano, B.; de Moraes, W.M.; de Salles Painelli, V.; da Luz, C.R.; dos Santos Costa, A.; de Salvi Guimaraes, F.; Medeiros, A.; Brum, P.C.; Lancha, A.H., Jr. Effects of creatine supplementation on muscle wasting and glucose homeostasis in rats treated with dexamethasone. *Amino Acids* **2012**, *42*, 1695–1701. [CrossRef]

50. Tarnopolsky, M.A.; Bourgeois, J.M.; Snow, R.; Keys, S.; Roy, B.D.; Kwiecien, J.M.; Turnbull, J. Histological assessment of intermediate- and long-term creatine monohydrate supplementation in mice and rats. *Am. J. Physiol. Regul. Integr. Comp. Physiol.* **2003**, *285*, R762–R769. [CrossRef]
51. Rooney, K.B.; Bryson, J.M.; Digney, A.L.; Rae, C.D.; Thompson, C.H. Creatine supplementation affects glucose homeostasis but not insulin secretion in humans. *Ann. Nutr. Metab.* **2003**, *47*, 11–15. [CrossRef]
52. Rocic, B.; Bajuk, N.B.; Rocic, P.; Weber, D.S.; Boras, J.; Lovrencic, M.V. Comparison of antihyperglycemic effects of creatine and metformin in type II diabetic patients. *Clin. Investig. Med.* **2009**, *32*, E322. [CrossRef] [PubMed]
53. Robinson, T.M.; Sewell, D.A.; Hultman, E.; Greenhaff, P.L. Role of submaximal exercise in promoting creatine and glycogen accumulation in human skeletal muscle. *J. Appl. Physiol.* **1999**, *87*, 598–604. [CrossRef]
54. King, D.S.; Dalsky, G.P.; Staten, M.A.; Clutter, W.E.; Van Houten, D.R.; Holloszy, J.O. Insulin action and secretion in endurance-trained and untrained humans. *J. Appl. Physiol.* **1987**, *63*, 2247–2252. [CrossRef]
55. Gan, S.K.; Kriketos, A.D.; Ellis, B.A.; Thompson, C.H.; Kraegen, E.W.; Chisholm, D.J. Changes in aerobic capacity and visceral fat but not myocyte lipid levels predict increased insulin action after exercise in overweight and obese men. *Diabetes Care* **2003**, *26*, 1706–1713. [CrossRef]
56. Devlin, J.T.; Hirshman, M.; Horton, E.D.; Horton, E.S. Enhanced peripheral and splanchnic insulin sensitivity in NIDDM men after single bout of exercise. *Diabetes* **1987**, *36*, 434–439. [CrossRef]
57. Bruce, C.R.; Kriketos, A.D.; Cooney, G.J.; Hawley, J.A. Disassociation of muscle triglyceride content and insulin sensitivity after exercise training in patients with Type 2 diabetes. *Diabetologia* **2004**, *47*, 23–30. [CrossRef]
58. Rocic, B.; Vucic, M.; Mesic, R.; Rocic, P.; Coce, F. Hypoglycemic effect of creatine in insulin dependent diabetic patients. *Diabetol. Croatica* **1995**, *24*, 117–120.
59. Souza, R.A.; Santos, R.M.; Osório, R.A.L.; Cogo, J.C.; Prianti-Júnior, A.C.G.; Martins, R.A.B.L.; Ribeiro, W. Influence of the short and long term supplementation of creatine on the plasmatic concentrations of glucose and lactate in Wistar rats. *Rev. Bras. Med. Esporte* **2006**, *12*, 361–365. [CrossRef]
60. Araújo, M.B.D.; Vieira Junior, R.C.; Moura, L.P.D.; Costa Junior, M.; Dalia, R.A.; Sponton, A.C.D.S.; Ribeiro, C.; Mello, M.A.R.D. Influence of creatine supplementation on indicators of glucose metabolism in skeletal muscle of exercised rats. *Motriz* **2013**, *19*, 709–716. [CrossRef]
61. Freire, T.O.; Gualano, B.; Leme, M.D.; Polacow, V.O.; Lancha, A.H., Jr. Effects of creatine supplementation on glucose uptake in rats submitted to exercise training. *Rev. Bras. Med. Esporte* **2008**, *14*, 431–435. [CrossRef]
62. Vaisy, M.; Szlufcik, K.; De Bock, K.; Eijnde, B.O.; Van Proeyen, K.; Verbeke, K.; Van Veldhoven, P.; Hespel, P. Exercise-induced, but not creatine-induced, decrease in intramyocellular lipid content improves insulin sensitivity in rats. *J. Nutr. Biochem.* **2011**, *22*, 1178–1185. [CrossRef] [PubMed]
63. Derave, W.; Eijnde, B.O.; Verbessem, P.; Ramaekers, M.; Van Leemputte, M.; Richter, E.A.; Hespel, P. Combined creatine and protein supplementation in conjunction with resistance training promotes muscle GLUT-4 content and glucose tolerance in humans. *J. Appl. Physiol.* **2003**, *94*, 1910–1916. [CrossRef] [PubMed]
64. Oliveira, C.L.P.; Antunes, B.M.M.; Gomes, A.C.; Lira, F.S.; Pimentel, G.D.; Boule, N.G.; Mota, J.F. Creatine supplementation does not promote additional effects on inflammation and insulin resistance in older adults: A pilot randomized, double-blind, placebo-controlled trial. *Clin. Nutr. ESPEN* **2020**, *38*, 94–98. [CrossRef]

Review

The Role of Creatine in the Development and Activation of Immune Responses

Eric C. Bredahl [1], Joan M. Eckerson [1], Steven M. Tracy [2], Thomas L. McDonald [2] and Kristen M. Drescher [3,*]

[1] Department of Exercise Science and Pre-Health Professions, Creighton University, Omaha, NE 68178, USA; ericbredahl@creighton.edu (E.C.B.); joaneckerson@creighton.edu (J.M.E.)
[2] Department of Pathology and Microbiology, University of Nebraska Medical Center, Omaha, NE 68198, USA; stracy@unmc.edu (S.M.T.); tmcdonal@unmc.edu (T.L.M.)
[3] Department of Medical Microbiology and Immunology, Creighton University, Omaha, NE 68178, USA
* Correspondence: kristendrescher@creighton.edu; Tel.: +001-(402)-280-2725

Abstract: The use of dietary supplements has become increasingly common over the past 20 years. Whereas supplements were formerly used mainly by elite athletes, age and fitness status no longer dictates who uses these substances. Indeed, many nutritional supplements are recommended by health care professionals to their patients. Creatine (CR) is a widely used dietary supplement that has been well-studied for its effects on performance and health. CR also aids in recovery from strenuous bouts of exercise by reducing inflammation. Although CR is considered to be very safe in recommended doses, a caveat is that a preponderance of the studies have focused upon young athletic individuals; thus there is limited knowledge regarding the effects of CR on children or the elderly. In this review, we examine the potential of CR to impact the host outside of the musculoskeletal system, specifically, the immune system, and discuss the available data demonstrating that CR can impact both innate and adaptive immune responses, together with how the effects on the immune system might be exploited to enhance human health.

Keywords: innate immunity; adaptive immunity; nutritional supplements; inflammation; macrophage polarization; cytotoxic T cells; toll-like receptors

Citation: Bredahl, E.C.; Eckerson, J.M.; Tracy, S.M.; McDonald, T.L.; Drescher, K.M. The Role of Creatine in the Development and Activation of Immune Responses. *Nutrients* **2021**, *13*, 751. https://doi.org/10.3390/nu13030751

Academic Editor: Elena Barbieri
Received: 16 January 2021
Accepted: 13 February 2021
Published: 26 February 2021

Publisher's Note: MDPI stays neutral with regard to jurisdictional claims in published maps and institutional affiliations.

1. Introduction

Creatine (CR) synthesis occurs in vertebrates in the liver, kidney, and pancreas, requiring arginine, methionine, and glycine as its building blocks via the arginine biosynthesis pathway. Creatine eventually metabolizes to form creatinine (CRN) [1,2], classically considered to be an inert waste product that is excreted in the urine [2,3]. However, there is evidence that suggests that CRN can have similar activity to CR in vitro [4,5]. Creatine, derived from the Greek word for flesh (*kreas*), was discovered by the French chemist Michel-Eugene Chevreul in 1832 as an integral component of meat, a finding subsequently confirmed by von Liebig in 1847. The first CR supplementation studies began in animals and humans in the early 1900s, but it was not until the 1990s that CR use became mainstream, gaining widespread research attention after two gold medalists from the 1992 Barcelona Olympics credited CR as part of their success [6]. Creatine is primarily stored in skeletal muscle as either free CR (~40%) or as phosphocreatine (~60%), which plays a critical role in the phosphagen energy system. Because of this, CR supplementation is most effective for high-intensity, short-duration activities, or repeated bouts of high-intensity exercise with short rest periods, since increased levels of phosphocreatine can rapidly re-phosphorylate adenosine diphosphate (ADP) to adenosine triphosphate (ATP) through the CR kinase reaction [2,7,8]. The increase in ATP turnover is achieved when CR is transported into the muscles via the CR transporter (Slc6a8), which is both sodium and chloride-dependent [2,9,10]. Upon entry into the muscles, creatine kinase is responsible for catalyzing CR into phosphocreatine, which provides an available phosphate group to donate to ADP to form ATP in a reversible reaction [1,2] (Figure 1).

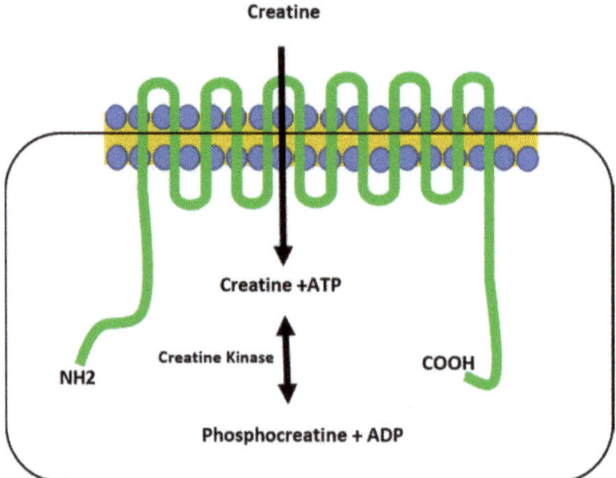

Figure 1. The creatine transporter. The creatine transporter (green) shuttles creatine from the extracellular space into the cytoplasm of the cell and is comprised of 12 transmembrane domains. Creatine kinase catalyzes a reversible reaction between creatine and phosphocreatine, resulting in the generation of ATP. NH2 = amine terminus, COOH = carboxy terminus.

While CR is naturally synthesized by vertebrates in the liver, pancreas, and kidneys, it is also consumed in a diet containing meat, fish, and other animal products [7]. For example, beef, tuna, salmon, and pork all contain approximately 4–5 g of CR per kilogram. However, the average CR pool for a 70 kg individual ranges from 120 to 40 g, and approximately $2\ \mathrm{g\ d^{-1}}$ is lost in the form of CRN [8]. This loss is replaced by both dietary and endogenous CR synthesis, which is approximately $1\ \mathrm{g\ d^{-1}}$. Therefore, many athletes utilize CR supplementation (most often in the form of CR monohydrate [11–13]) to increase intramuscular stores of CR and phosphocreatine [8,11–23]. Many different CR loading paradigms have been used [24], but the most commonly used dosing strategy occurs in two phases. The first phase is a loading phase in which an individual ingests $20\ \mathrm{g\ d^{-1}}$ in four 5 g doses for five to seven days, followed by a maintenance dose of $3–5\ \mathrm{g\ d^{-1}}$ of CR for at least a month, and often much longer [2,17,24–26]. During the loading phase, total muscular CR stores have been reported to increase between 20 and 40% [27,28], with reported side effects limited to cramping and bloating during the loading phase [29]. It is important to note that approximately 20–30% of individuals are non-responders to CR supplementation, which has been defined by Greenhaff et al. as individuals with changes in CR stores of $<10\ \mathrm{mmol\ kg^{-1}}$ dry weight (dw) following the loading phase [30]. Syrotuik and Bell [31] reported that responders (increased total CR by $>20\ \mathrm{mmol\ kg^{-1}}$ dw) had lower initial levels of free CR and phosphocreatine, a greater percentage and cross-sectional area of type II muscle fibers, and a larger fat-free mass compared to non-responders [31]. Other factors influencing an individual's response to creatine supplementation include hydration status, dietary factors, caffeine use, and the dose of creatine ingested [24,26,32–35].Until recently, most studies on the effects of CR supplementation have focused upon its ability to enhance athletic performance and recovery [8,14–19]. In this review, we will examine evidence to determine how CR impacts both the innate and adaptive arms of the immune system, and what effect this may have on individuals using CR as an ergogenic aid. In each section, we will first discuss the general immunological processes that occur in the host, and then delve into the available evidence regarding the influence of CR on these events.

2. Creatine and the Innate Immune System

2.1. Toll-Like Receptors Are Downregulated in Response to Exposure to Creatine

The innate immune system represents the first line of defense against microbial and viral assault for the host. It is apparent that to function properly, the host must first be able to differentiate between self and non-self molecules. The host accomplishes this by utilizing a class of molecules known as pattern recognition receptors (PRRs) that bind with moieties that are unique to classes of pathogens, known as pathogen-associated molecular patterns (PAMPs). Found only in pathogens, PAMPs are stable (not subject to genetic drift), and are expressed at all stages of the pathogen's life cycle [36]. Examples of PAMPs include substances such as bacterial lipopolysaccharide (LPS), flagellin, and single-stranded RNA [37,38].

A major class of PRRs include the toll-like receptor (TLR) family. TLRs are widely expressed in the host and, upon interaction with a PAMP ligand, initiate a cascade of events inducing an inflammatory response that results in the generation of cytokines, reactive oxygen species, and ultimately the recruitment of cells of both the innate and adaptive immune system to the site of infection [37]. Some TLRs (TLR-3, -7, -8, and -9) will also induce the production of type I interferons (IFN α/β), which are critical in inducing an anti-viral state in the cells, thereby aiding in containing the spread of a viral infection [38].

To initiate investigation into how CR may impact the host immune system, Leland and colleagues examined the effects of treating a mouse macrophage cell line with either CR or CRN hydrochloride (CRN-HCl) on the expression of four TLRs (TLR-2, TLR-3, TLR-4, and TLR-7), using both qRT-PCR and immunostaining [5]. Because PRR/PAMP interactions rapidly occur after infection [37], we examined the expression of the TLRs over the course of an hour. These TLRs were chosen for examination as they are the PRRs that represent distinct classes of pathogens based on their ligands, *viz.* Gram-positive bacteria (TLR-2; lipoteichoic acid), double-stranded (ds) RNA viruses (TLR-3; dsRNA), Gram-negative bacteria (TLR-4; LPS), and single-stranded (ss) RNA viruses (TLR-7; ssRNA) [37]. The authors observed that both CR and CRN-HCl downregulated the expression of all the TLRs studied, although the kinetics of the downregulation varied over the time course [5] (Figure 2). We postulate that the downregulation is related either directly to the reduced expression of proinflammatory mediators [4,39], or indirectly to the altered microRNA expression. Despite being carried out in vitro, these studies suggested the intriguing possibility that CR supplementation might serve to decrease the ability of the host to detect infections.

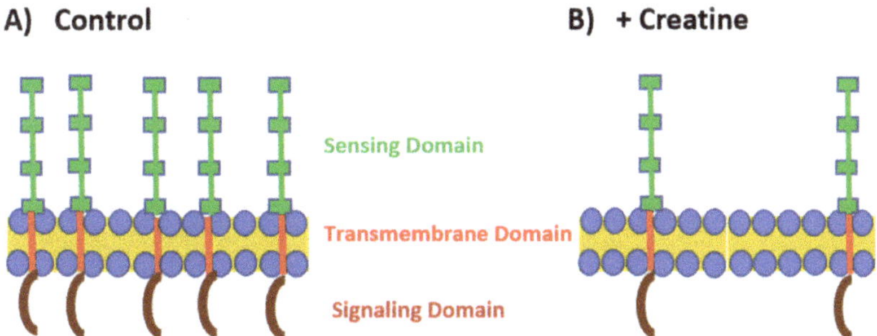

Figure 2. TLR expression under control and creatine-stimulated conditions. (**A**) Under normal conditions, TLRs are highly expressed on the cell surface or within the endosome. The sensing domain (green) of the TLR is located outside the cell or within the endosome. There is a transmembrane domain that spans the cell membrane (red) and the signaling domain (maroon) is located within the cytoplasm. (**B**) In the presence of creatine, TLR expression is downregulated by the cells.

A recent study [40] examined the impact of oral CR supplementation prior to lung transplantation on ischemia-reperfusion injury in rats. In this model, pathologic inflammation, perivascular edema, and alveolar damage were assessed [41]. Rats with lungs from donors which had been pretreated with CR had higher levels of serum CRN than control-treated animals [40], indicating the uptake of CR and shedding of the end product CRN into blood. Consistent with our in vitro studies [5], lungs from rats with higher serum CRN levels also had lower levels of TLR-4 expressed in the lung parenchyma, in addition to having fewer infiltrating mononuclear cells compared to control rats [40]. Unlike what was observed in vitro [5], however, no changes in TLR-7 expression were observed following CR treatment [40]. The reduction in pathology observed in the rats is likely due to two distinct mechanisms: the antioxidant properties of CR and a reduced level of inflammatory mediators [40]. The finding that CR administration reduced pathologic damage resulting from ischemia-reperfusion injury is potentially of clinical significance, as there is currently no treatment for this condition in humans [41].

Although more work is needed to fully understand this topic, the aforementioned studies demonstrate that the CR-induced modulation of TLR expression can be observed both in vitro and in vivo. The downregulation of these sensing molecules of the innate immune system by CR could have implications in a small subset of individuals who use CR as a supplement, by potentially slowing the host response to certain infections. It is unlikely that this downregulation in TLRs would negatively affect relatively healthy individuals. On the other hand, CR-induced TLR downregulation might potentially be exploited for a patient's health. For example, consider a disease such as septic shock [42] in which bacterial LPS activates the innate immune system and induces a suite of symptoms rapidly leading to severe illness and death. In such cases, CR supplementation to suppress TLR-4 activation might be a potential clinical adjunct to be used as an aid in controlling a particular disease state.

2.2. Macrophages Undergo Changes in Phenotype Following Exposure to Creatine In Vitro and In Vivo

Macrophages are multi-faceted white blood cells that, among other things, engulf pathogens and cellular debris, aid in activating the adaptive immune system, and present antigens to T cells. Macrophages can develop into one of two forms, termed M1 and M2 [43]. M1 macrophages are critical in anti-microbial defenses and inflammatory responses, while M2 macrophages are involved in tissue remodeling and anti-worm defenses [44–46]. M2 macrophages can be further subdivided [47–51], but this is beyond the scope of this review. These phenotypic divisions are akin to those observed in CD4+ T helper 1 (Th1) and T helper 2 (Th2) cells [52]. Cytokines, including tumor necrosis factor-α (TNF-α), IFN-γ and granulocyte macrophage stimulating factor (GM-CSF), and bacterial components such as LPS can drive the development of the macrophages toward an M1 cell phenotype, while interleukin (IL)-4 can drive development toward an M2 cell phenotype [44–46,53]. Somewhat like Janus, the Roman god with two faces looking in opposite directions, the phenotype of these macrophages is fluid, depending on the microenvironment [53], and this phenotypic plasticity is important for occasions when an infection occurs: it is desirable to have macrophages that produce large amounts of proinflammatory mediators such as IL-12, tumor necrosis factor alpha (TNF-α), IL-23, IL-1β, and IL-6 [54,55], and then, as the infection begins to resolve, it is advantageous to have macrophages produce mediators that help to repair the tissue damage caused by the acute or chronic inflammatory process [54,55]. The polarization towards the M2 phenotype will downregulate the inflammatory response and the production of the above-mentioned proinflammatory mediators [55–57]. The disruption of the balance of M1 versus M2 macrophages in the host has been implicated in the pathogenesis of both infectious and autoimmune diseases, including rheumatoid arthritis, irritable bowel disease, Sjogren's syndrome, and systemic lupus erythematosus [58–63].

The differential metabolism of L-arginine is observed in M1 and M2 macrophages [64,65]. Due to macrophages' dependence on transcription factors associated with the macrophage phenotype, inducible nitric oxide synthase (iNOS) is a feature of M1 macrophages, while arginase 1 is associated with M2 macrophages; these associations provide valuable experimental markers. iNOS production requires the signal transducer and activator of transcription 1 (STAT1), while arginase 1 is dependent on the activation of STAT6 [64,65].

Recent research by the Chen and Hu groups [66] studied the impact of CR metabolism on the polarization of macrophages. In a series of studies, the authors cultured peritoneal macrophages obtained from wild type mice treated with CR, and demonstrated an increase in the intracellular concentration of CR concordant with an inhibition of M1 development and a polarization towards the M2 phenotype. This finding was consistent with our earlier work, which found in both murine and human macrophage lines, co-cultured with CRN, a downregulated product of M1 macrophages, TNF-α [4], although to date, the authors have not examined whether there is a polarization towards the M2 phenotype. These groups performed a similar experiment, culturing murine macrophages containing a defect in the CR transporter (Slc6a8$^{-/y}$ mice) with CR or CRN. There was a polarization of the cells towards an M1 phenotype, dominated by the production of iNOS and the chemokine CXCL9 [60]. Creatine hydrolyzes to CRN in an aqueous environment and CRN does not require the CR transporter (it diffuses through the cell membrane) to enter the cell [2,7]. Because the Slc6a8$^{-/y}$ mice did not acquire a M2 phenotype, it supports the hypothesis that CR drove the polarization of the cells toward the M2 phenotype, not CRN.

Using the fact that *Listeria monocytogenes* infections are controlled by highly activated macrophages [67], Ji et al. [66] designed an elegant experiment using cre-lox technology [68] to specifically knock out the CR transporter gene Slc6a8 in macrophages in order to examine the outcome of an *L. monocytogenes* infection in a mouse model of infection. When mice with the macrophage-specific deletion of Slc6a8 were infected with *L. monocytogenes*, they showed enhanced survival and increased body weights relative to infected wild type (Slc6a8 normal) control mice. The authors showed that the knocked-out Slc6a8 gene did not interfere with other functions that may be relevant to host defense by administering CR to *L. monocytogenes*-infected wild type mice. The result showed that the infected CR-treated mice experienced a poorer outcome compared to infected control mice, which were not provided CR [66].

To demonstrate that polarization towards a M2 phenotype was functionally relevant in pathogenesis, an experiment was performed using in an in vivo model of carbon tetrachloride-induced liver fibrosis [66]. In this model, carbon-tetrachloride was injected intraperitoneally into mice twice weekly for four weeks, a procedure which induces fibrosis and increased levels of liver enzymes, specifically aspartate aminotransferase (AST) and alanine aminotransferase (ALT) [69]. These results showed that mice with a macrophage-specific deficiency in the creatine transporter gene Slc6a8 (and which thus would be CR-depleted) exhibited increased levels of fibrosis compared to wild type mice [66].

While this series of experiments strongly supports the concept that CR can influence the development of activated macrophage function (Figure 3), other factors such as microRNA (miRNA) expression [70–72] may also be involved in macrophage polarization, particularly in complex physiological settings. Though the mechanisms by which CR or CRN influence macrophage plasticity remain as yet incompletely defined, the further definition of miRNA expression may provide insight into macrophage polarization. Several comprehensive reviews, however, have discussed the specific miRNAs that are hypothesized to impact macrophage polarization [70,71,73].

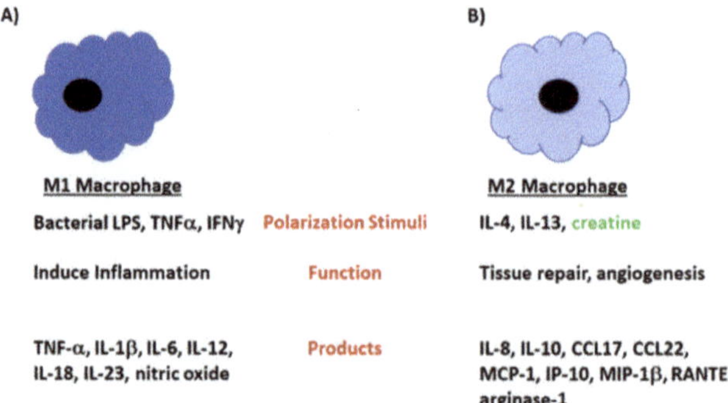

Figure 3. M1 and M2 macrophages develop under different stimuli and perform unique functions in the host. (**A**) Macrophages polarize to the M1 phenotype under conditions that highly activate the cells, such as when LPS or IFN-γ is present in the microenvironment. M1 macrophages are highly phagocytic and produce large amounts of proinflammatory mediators including TNF-α, IL-12, and IL-6. (**B**) M2 macrophages develop when IL-4 is present in the microenvironment. Cells of the M2 subtype produce mediators that are involved in tissue repair and angiogenesis. These cells produce large amounts IL-10, CCL17, CCL22 and arginase-1. In the presence of creatine, the polarization shifts towards an M2 phenotype.

2.3. Creatine Treatment Can Alter the Inflammatory Response

It is intuitively apparent that inflammation must be finely controlled by the host to prevent immunopathologic changes. The inflammatory response is beneficial when it is controlled, is appropriate for the pathogen, and is resolved when the antigen is cleared. However, in the absence of these factors, the host is at risk for the development of immunopathologic damage and autoimmune disease.

The role of CR in ameliorating inflammation was first probed in the late 1970s by Madan and Khanna by injecting carrageenan into the foot pads of rats, and then injecting the animals intraperitoneally (i.p.) with either CR or vehicle [74]. Carrageenan-induced inflammation is acute (within hours) and is now known to be mediated by high levels of cytokines, including TNF-α, and IL-1β, as well as other proinflammatory substances including nitric oxide, inducible nitric oxide synthase (iNOS), prostaglandins, and cyclooxygenase. In addition, there is considerable neutrophil infiltration of the injection site [75]. The inflammation is caused by the activation of TLR-4 due to the activation of the nuclear factor -κβ (NF-κβ) pathway [76]. They noted that the intraperitoneal injection of CR significantly reduced paw edema compared to vehicle-treated animals [74]. A subsequent study from the same authors compared the efficacy of CR to that of the nonsteroidal anti-inflammatory drug (NSAID) phenylbutazone [77], and found CR to be similar in efficacy to phenylbutazone [74], results that suggested a similar mechanism at work. This group also tested the effects of CR in an additional model of foot pad swelling to ensure that the results were not a model-specific artifact [78]. Injection of the anti-fungal agent nystatin into the paw induces several highly proinflammatory mediators, including IL-1β, IL-8, and TNF-α, by triggering TLR-2 signaling and activating the NF-κβ pathway [79]. Oral feeding of CR in the rat model of nystatin-induced paw edema also resulted in a significant reduction in swelling [78]. Cumulatively, these studies demonstrated that CR was an effective anti-inflammatory agent in both acute (carrageenan-induced) and chronic (nystatin-induced) models of inflammation.

The first phase of an inflammatory response involves the secretion of mediators such as IL-8, TNF-α, and IL-1β; these cytokines will alter the surface of endothelial cells to permit the recruitment of cells from the periphery so that they can eventually enter the tissue [80]. For this to successfully occur, endothelial cells must be induced to express intercellular adhesion molecule-1 (ICAM-1) and E-selectin. There also needs to be an increase in vascular permeability for diapedesis to occur [81]. To determine whether CR affected these processes, human pulmonary endothelial cells were co-cultured with TNF-α to induce high levels of these adhesion molecules on the surfaces of the cells [82]. Following the addition of CR to the cultures, decreased levels of in ICAM-1 and E-selectin were observed on the endothelial cells [82]. The inference to be drawn from this finding in vitro is that neutrophils would be less able to adhere to the endothelium and thereby less likely to be efficiently recruited to the site where the inflammatory response has been triggered. To determine the functional significance of this finding, the authors indirectly measured the ability of ⁵¹Cr-labeled neutrophils to adhere to TNF-α-treated endothelial cells [83] in the presence or absence of CR by measuring ⁵¹Cr released from lysed endothelial cells incubated with labeled neutrophils. The authors determined that the level of neutrophil adhesion to the TNF-α and CR-treated endothelial cells, as measured by the release of ⁵¹Cr, was similar to that observed in control cells that had not been treated with TNF-α. Although the authors did not specifically quantitate the endothelial levels of E-selectin and ICAM-1 expression (either at the level of mRNA or protein) in the target cultures in the labeling experiment, the functional evidence supported the hypothesis that CR lowered the adhesion molecule levels in the TNF-α treated cells to those levels found in untreated target cells.

While adhesion molecule expression is required for cellular recruitment, it is not the sole determinant of whether diapedesis will occur; endothelial cell junctions must also be loosened [80,81]. As both serotonin [84] and hydrogen peroxide [85] are well-described inducers of endothelial cell permeability, in a study by Nomura and colleagues, endothelial cells cultured either with serotonin (0.1 μM) or hydrogen peroxide (0.1 μM) were treated as well with CR [82], and endothelial permeability was measured using fluorescein isothiocyanate-dextran in a standard cellular permeability assay [86,87]. The results of this experiment demonstrated that cultures containing either serotonin or hydrogen peroxide treated cultures containing 5 mM CR and showed similar levels of endothelial cell permeability to control (no serotonin or hydrogen peroxide)-treated cultures [82]. These studies are consistent with the hypothesis that the inflammatory processes involving cellular recruitment may be downregulated in individuals supplemented with CR (Figure 4). If valid, this result could be desirable in individuals with certain autoimmune conditions, including those driven by proinflammatory mediators [62].

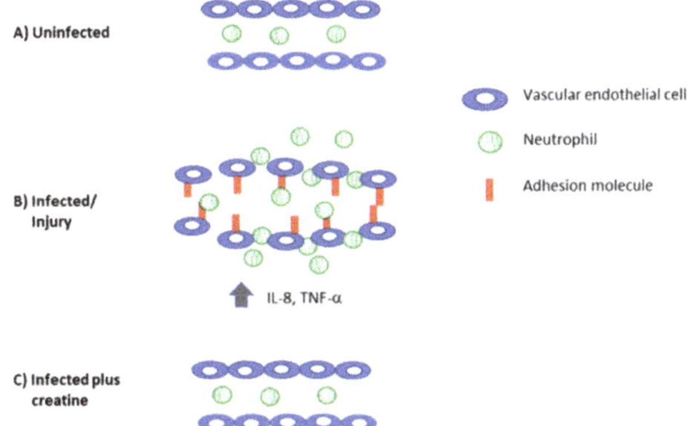

Figure 4. Creatine reduces inflammation to control levels. (**A**) In the absence of infection, there is no adhesion molecule expression on the vascular endothelium and neutrophils are not recruited into the tissue. (**B**) When the host is injured or infected, there is increased production of proinflammatory mediators that recruit immune cells (IL-8) and induce the expression of adhesion molecules on the vascular endothelium (TNF-α, IL-1β). Together, these mediators also induce swelling and loosen the interactions between the endothelial cells. These changes permit neutrophils (and eventually other immune cells) to be recruited to the site of damage/infection. (**C**) Creatine reduces inflammation, downregulates adhesion molecule and cytokine expression, and preserves the tight junctions in the endothelial cells to reduce inflammation.

Studies examining the immune response in humans following CR supplementation have largely focused on the inflammatory response, and the utility of CR supplementation in human inflammatory conditions has been mixed [88–91]. Santos et al. (2004) measured the levels of TNF-α and prostaglandin E2 (PGE2) in individuals following completion of a 30 km race in participants who were supplemented with 4×5 g d^{-1} CR prior to the competition compared to control participants. CR treatment reduced TNF-α and PGE2 levels, indicating a reduction in inflammation [91]. A similar study performed in half-ironman participants demonstrated that CR-supplemented athletes experienced decreased levels of TNF-α, PGE2, and IL-1β compared to the levels observed in control participants [90]. Deminice et al (2013) also demonstrated that TNF-α and CRP levels were reduced following acute exercise (acute sprint test) in individuals supplemented with CR [89]. In contrast, in a study focusing on patients with osteoarthritis in the knee, patients were supplemented with creatine monohydrate for one week (20 g d^{-1}) and then entered a maintenance phase where they were supplemented with 5 g d^{-1}. To determine whether CR could alter known markers of inflammation, TNF-α, IL-6, IL-1β, and C-reactive protein were measured in the sera of the study subjects. No significant differences in these markers were noted between CR-supplemented and control patients [86]. Together, these human studies raise a number of questions. Does CR supplementation work best in healthy individuals? Furthermore, in individuals with ongoing inflammatory conditions, does the CR dose need to be increased to reduce preexisting inflammation?

3. Creatine and the Adaptive Immune System

3.1. Creatine Kinase B (CKB) Is Required for T Cell Development

T cell develop occurs within the thymus where cells undergo both positive and negative selection. Positive selection is defined as the process by which the T cell receptor (TCR) on the developing T cell (thymocyte) interacts with the host major histocompatibility complex (MHC), which determines whether the interactions are appropriate. This process occurs in the thymic cortex and is referred to as self-restriction [92]. If the thymocyte fails this process, it undergoes apoptotic cell death, but if the interaction is successful, the thymocyte migrates to the medulla where those cells which are self-reactive undergo negative selection [92]. The cells that survive this second selection process are said to be self-tolerant. The process of T cell development is dependent upon the triggering of a signaling cascade that involves a series of phosphorylation events [93–95]. Cells that successfully navigate the selection processes leave the thymus and become mature CD4+ or CD8+ T cells. Once CD4+ T cells interact with antigens in the periphery, CD4+ T cells differentiate further. The two main subsets of CD4+ cells are helper T (Th) cells, the main function of which is to produce soluble mediators that activate macrophages (termed Th1 cells) or induce class switching in B cells (termed Th2 cells). CD8+ cells are cytotoxic T cells (CTL) that function to kill virus-infected host cells and to control tumors [96]. The ratio of CD4+ to CD8+ T cells in the periphery is approximately 2.5:1 [75].

Signaling via the TCR requires ATP due to a series of phosphorylation events which must occur when the TCR is stimulated [97]. Creatine kinase B (CKB) is a key mediator in ATP generation in developing thymocytes and mature T cells [98]. During development in the thymus, thymocytes can be either double positive (expressing both CD4 and CD8 in the cortex) or single positive (expressing CD4 or CD8 in the medulla). CKB modulates thymocyte population sizes: CD4+CD8+ thymocytes have been shown to express low levels of CKB, while single positive CD4+ or CD8+ thymocytes each express high levels of CKB [98]. A CKB transgenic mouse was created and the CKB gene was placed under the control of the CD2 protein, whose expression is found on cells of the T cell and natural killer cell lineages [99]. In these animals, a reduction in the overall number of thymocytes was observed [98]. The transgenic expression of CKB under the control of the CD2 promoter lowered the numbers of CD4+CD8+ cells in the cortex due to increased levels of apoptosis. In the periphery, however, the single positive mature CKB transgenic T cells underwent intense proliferation and produced high levels of IL-2, a T cell proliferation and survival factor [100], and IFN-γ, a strong activator of macrophages [101]. The inhibition of CKB in mature T cells using RNA interference resulted in reduced levels of activation, indicating the requirement of CKB in T cell function. Higher levels of CKB were found in single positive CD4+ thymocytes compared to single positive CD8+ thymocytes [89], suggesting that the levels of CKB in a single positive thymocyte could also impact whether the thymocyte becomes a mature CD4+ or CD8+ cell in the periphery (Figure 5).

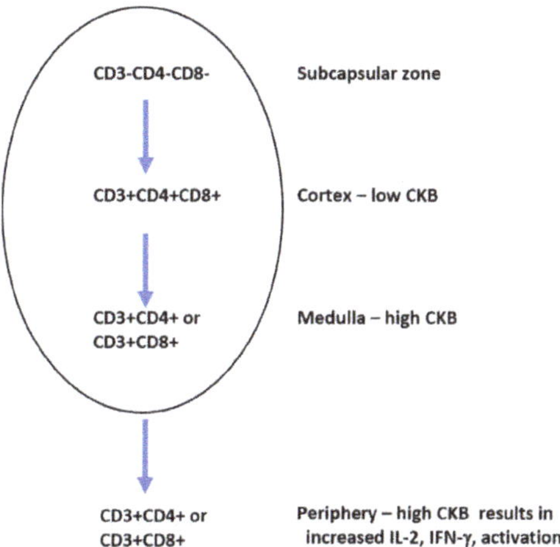

Figure 5. Creatine kinase B (CKB) levels vary during T cell development. Triple positive T cells in the cortex express low levels of CKB.

The overexpression of CKB in developing T cells results in a reduction in the number of total T cells, indicating that CKB plays an integral role in T cell development. When T cells acquire the CD4 or CD8 phenotype in the medulla, there are high levels of CKB. High levels of CKB in T cells in the periphery result in high levels of IL-2 and IFN-γ production.

Together, this work provides strong evidence that the creatine kinase system is required for certain stages of T cell development and may play a significant role in determining whether a T cell acquires the CD4+ or CD8+ phenotype. Given that the effector phase of the T cell response (that is, the time when the CD4+ or CD8+ cell is responding to an assault in the periphery, either by proliferating and producing cytokines (CD4+ T cells) or killing infected targets (CD8+ T cells)) requires significant energy, it is reasonable that the highest levels of CKB would be found in these cells.

3.2. Creatine and CD8+ T Cell Function

To understand the role of CR in CD8+ T cell function in tumor control, Di Biase and colleagues examined the expression of the CR transporter, Slc6a8 [102]. Using a well-described model [103] that has been used to understand the tumor microenvironment and test potential therapeutic options for melanoma treatment, melanomas were induced by injection of B16-OVA cells into wild type mice [103,104], then tumor-infiltrating lymphocytes (TILs) were isolated from the tumors. TILs isolated from tumors in many model systems, as well as humans with certain types of tumors, show skewed CD4+:CD8+ cell ratios with increased numbers of CD8+ T cells within the tumor, but not in the periphery [105–107] (recall that normal peripheral CD4+:CD8+ ratios are about 2.5:1). Because of their ability to lyse cells, CD8+ T cells are an integral component of the host defense against tumors [96]. It is critical that activated CD8+ T cells require increased levels of ATP to function properly [108]. Higher levels of the creatine transporter protein Slc6a8 were expressed in TILs compared to T cells isolated from tumor-free spleens of the tumor-bearing animals [102]. Following this observation, B16-OVA melanoma cells were then injected into wild type mice as before, or into Slc6a8-deficient mice that had been treated with CR (delivered i.p.). The tumor burden was then assessed to address whether the creatine transporter is relevant to anti-tumor immunity. Interestingly, it appears so: Slc6a8-deficent mice were less able to control tumor burden than wild type mice [102], and wild type mice treated with CR (either i.p. or orally) had smaller tumors than those found in control mice.

Could the creatine transporter Slc6a8 play a role in T cell activation? To explore this, CD8+ T cells (CTLs) were isolated from wild type control and Slc6a8-deficient mice, and then were nonspecifically stimulated using an anti-CD3 antibody to activate the T cells by cross-linking the CD3 molecules on the surface of the T cells [109]. Wild type (control) mouse CTLs showed superior activation compared to those from creatine transporter-deficient mice in all parameters measured, *viz.* proliferation, IL-2 and IFN-γ production, CD25 expression, and the production of granzyme, the molecule responsible for the cytotoxic function of CD8+ T cells [102] (Figure 6). These results indicated that Slc6a8 plays a role in T cell activation, supporting the requirement for CTLs to have the capacity to take up CR in order to efficiently perform their cytotoxic function. While the available data support a role for CR in the CD8+ T cell-mediated control of the tumors, it is important to note that the tumors induced by B16-OVA cells also express the creatine transporter [102], leaving open the (yet untested) possibility that CR uptake could have an undefined effect on the tumor cells.

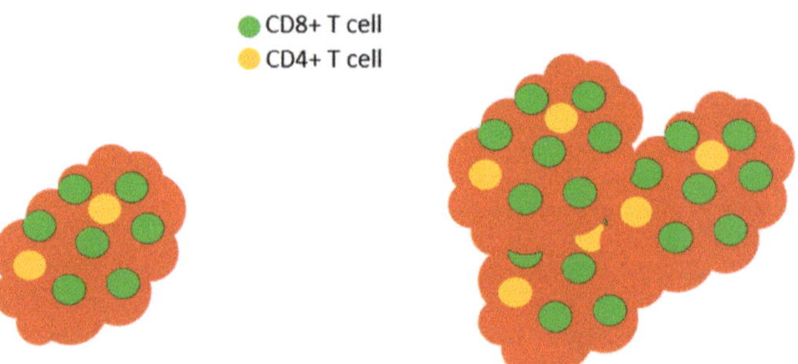

Figure 6. Slc6a8 is required to control tumors. In normal mice with tumors, the tumor (red) is infiltrated by high levels of CD8+ T cells (left). The isolation and in vitro activation of these CD8+ T cells demonstrate that the cells produce large amounts of IL-2 and IFN-γ and express high levels of CD25 (the high affinity IL-2 receptor) and granzyme. In Slc6a8 knockout mice (right), the tumor is poorly controlled and the CD8+ T cells produce low levels of IL-2, IFN-γ, CD25, and granzyme following in vitro activation.

To determine whether the results were model-specific and not generally applicable, and to determine the efficacy of CR in combination with another well-described cancer therapy, studies were performed using the MC38 cell mouse model of colon cancer [110]. In this model, tumors respond to anti-programmed cell death protein 1 (PD1) treatment (a therapeutic option for several human cancers [111]). It is hypothesized that anti-PD1 treatment alters the tumor microenvironment to tip the balance of energy usage in favor of the T cells. The hypothesis tested was that daily CR supplementation in conjunction with anti-PD1 treatment would reduce tumor burden in the mice. Twenty-one days after tumor induction, tumor size was quantitated. Animals that received the anti-PD1 antibody or CR demonstrated reduced tumor size compared to control mice, although anti-PD1 treatment was significantly more effective than CR alone in reducing tumor size. However, when used in conjunction with anti-PD1 therapy, the combination was significantly more effective at controlling tumor growth than anti-PD1 treatment alone; indeed, four of five mice showed no evidence of remaining tumor. When the surviving tumor-free animals received yet another injection of MC38 cells 3 months after the end of the study, no tumors were detected 6 months later, indicating that this protection was long-lived [102]. As the

MC38 cells do not express the creatine transporter Slc6a8 [110], the action of CR was not directly on the tumor. The mechanism by which the anti-PD1 treatment works is that PD1–PDL1 interactions inhibit T cell function [112], while CR is proposed to increase the levels of granzyme in CD8+ T cells responsible for tumor killing [102]. Anti-PD1 treatment combined with CR supplementation represent two distinct mechanisms of tumor control that, to date, appear to be beneficial in an animal model of cancer.

Recently, a commentary based upon studies treating various pulmonary conditions proposed that CR supplementation might be beneficial to patients undergoing pulmonary rehabilitation during and following SARS-CoV-2 infection (also known as COVID-19) [113]. COVID-19 patients have been described as having T cells that are "functionally exhausted" based on reduced cytokine expression and increased levels of PD-1 [114]. Given the general lack of treatment modalities available for COVID-19 and the established role of CD8+ T cells in helping to clear virus infections, this intriguing proposal merits further investigation. While there have been a limited number of studies that have examined the impact of CR supplementation on T cell function, the studies described herein support the need for further investigation in order to gain a better understanding of how CR impacts T cell development and function. Because CR supplementation represents a safe, inexpensive adjunct therapy that, based upon animal studies, appears to have a significant potential to augment anti-tumor responses, its clinical significance deserves exploration.

3.3. Creatine Influences CD4+ Th2 Cell-Mediated Disease

Studies from the laboratory of Vieira have examined the effects of CR on the pulmonary system in a murine asthma model [39,115]. Using the well-accepted ovalbumin (OVA)-induced model of asthma, the laboratory examined the impact of CR supplementation on the characteristic airway inflammation and remodeling in mice that is induced by the strong Th2 response dominated by IL-4, IL-5, and insulin-like growth factor-1 (IGF-1) [116]. Extensive eosinophil infiltration is observed in these animals in the absence of any immunomodulatory intervention, as well as increased smooth muscle thickness and collagen deposition [115]. Notably, the CR treatment of mice with OVA-induced airway disease resulted in significantly increased pathologic changes compared to control-treated mice, as well as increased levels of IL-4, IL-5, and IGF-1 in the bronchoalveolar lavage fluid [115]. These data are consistent with our studies demonstrating that TNF-α levels were reduced in macrophages treated with CRN [4]. Collectively, these results suggest that while in this instance CR treatment was not beneficial, it should be noted that the results showed that CR may skew the immune response towards a strong Th2-like response, which would be desirable when the damaging pathological change is driven by a strong Th1 response (Figure 7).

Creatine's effects are exerted beyond the specific immune cells involved in airway remodeling and allergic asthma. Consistent with the stronger Th2-mediated response observed in the OVA-sensitized mice, the CR treatment of the animals in this model also demonstrated reduced levels of NF-$\kappa\beta$ activation in endothelial cells compared to control-treated animals [75]. The levels of chemokine (C-C motif) ligand 5 (CCL5), a chemokine involved in eosinophil recruitment [117], and CCL2, a chemokine that recruits monocytes, T cells, and dendritic cells [118], were increased. The levels of tissue inhibitor of metalloproteinase (TIMP)-1 and -2, matrix metalloproteinase-9 (MMP-9) and -12, transforming growth factor-β1 (TGF-β1), IGF-1, IL-5, and the epidermal growth factor receptor (EGFR) were also increased in the epithelial cells of OVA-sensitized CR-treated mice compared to OVA-sensitized control mice [119].

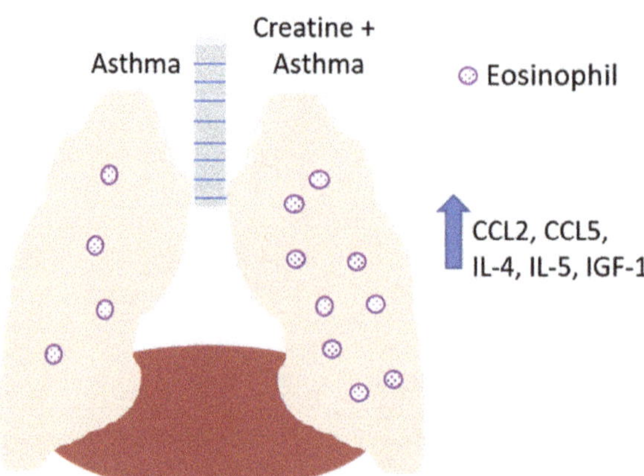

Figure 7. Effect of creatine on the development of disease in a mouse model of asthma. In the absence of creatine supplementation (left lobe of lung), asthma induction results in the recruitment of eosinophils to the airways, with IL-4, IL-5, and IGF-1 detected in the bronchoalveolar lavage fluid. Creatine supplementation (right lobe of lung) resulted in increased levels of these soluble mediators, as well as the increased production of CCL-2 and CCL-5 by endothelial cells, resulting in the increased recruitment of monocytes, lymphocytes, dendritic cells (CCL2) and eosinophils (CCL5).

4. Future Directions

Creatine, used originally as an ergogenic aid by elite athletes to enhance performance, has found its way into the lives of "ordinary people". Individuals of all ages and fitness levels use CR on a regular basis [120–128], as demonstrated by sales in 2019 that surpassed USD 360 million [129]. The question then becomes: who should use CR and what types of benefits can various subsets of individuals gain from its use? Additionally, perhaps more importantly, are there people who should not use CR? The studies discussed in this review indicate that CR has diverse effects on components of the innate and adaptive immune repertoire. In turn, these results suggest that these immune effects are not trivial, and under some circumstances, might negatively impact the CR user. As with nearly any nutrient, when used to excess or abused, health effects can occur. Besides sounding a warning, these studies also suggest instances wherein CR supplementation may be actually beneficial by reducing pathologic changes in the host. In instances where the mechanism of immune-mediated protection or immunology is understood, it may be possible to make an educated guess as to whether CR supplementation will help the clinical situation. For example, if an individual has a condition exacerbated by proinflammatory mediators, then CR administration should be considered as an adjuvant therapy since it appears to ameliorate proinflammatory processes characteristic of an M1 phenotype, and all available data attest to its safety. Based on the current literature, there is clearly a path to justify the continued investigation of the potential influence that CR has upon the immune response, particularly in the realm of autoimmune and infectious diseases.

Author Contributions: Conceptualization K.M.D.; writing—original draft preparation E.C.B., J.M.E., S.M.T., T.L.M., K.M.D.; writing—review and editing, E.C.B., J.M.E., S.M.T., T.L.M., K.M.D. All authors have read and agreed to the published version of the manuscript.

Funding: The APC was funded by AlzChem L.L.C.

Institutional Review Board Statement: Not applicable.

Informed Consent Statement: Not applicable.

Data Availability Statement: No novel data was generated for this review article.

Conflicts of Interest: K.M.D. is a paid member of the AlzChem Scientific Advisory Board. AlzChem L.L.C. had no role in the writing of this manuscript or in the review process.

References

1. Brosnan, J.T.; Brosnan, M.E. Creatine: Endogenous Metabolite, Dietary, and Therapeutic Supplement. *Annu. Rev. Nutr.* **2007**, *27*, 241–261. [CrossRef] [PubMed]
2. Wyss, M.; Kaddurah-Daouk, R. Creatine and Creatinine Metabolism. *Physiol. Rev.* **2000**, *80*, 1107–1213. [CrossRef]
3. Dröge, W. Free Radicals in the Physiological Control of Cell Function. *Physiol. Rev.* **2002**, *82*, 47–95. [CrossRef] [PubMed]
4. Riesberg, L.A.; McDonald, T.L.; Wang, Y.; Chen, X.-M.; Holzmer, S.W.; Tracy, S.M.; Drescher, K.M. Creatinine downregulates TNF-α in macrophage and T cell lines. *Cytokine* **2018**, *110*, 29–38. [CrossRef] [PubMed]
5. Leland, K.M.; McDonald, T.L.; Drescher, K.M. Effect of creatine, creatinine, and creatine ethyl ester on TLR expression in macrophages. *Int. Immunopharmacol.* **2011**, *11*, 1341–1347. [CrossRef] [PubMed]
6. Rawson, E.S.; Clarkson, P.M. Creatine Supplementation: The Athlete's Friend or Foe? *Int. Sports Med. J.* **2000**, *1*, 1–4.
7. Greenhaff, P.L. The nutritional biochemistry of creatine. *J. Nutr. Biochem.* **1997**, *8*, 610–618. [CrossRef]
8. Bemben, M.G.; Lamont, H.S. Creatine Supplementation and Exercise Performance. *Sports Med.* **2005**, *35*, 107–125. [CrossRef] [PubMed]
9. Snow, R.J.; Murphy, R.M. Creatine and the creatine transporter: A review. *Mol. Cell. Biochem.* **2001**, *224*, 169–181. [CrossRef]
10. Speer, O.; Neukomm, L.J.; Murphy, R.M.; Zanolla, E.; Schlattner, U.; Henry, H.; Snow, R.J.; Wallimann, T. Creatine transporters: A reappraisal. *Mol. Cell. Biochem.* **2004**, *256*, 407–424. [CrossRef]
11. Volek, J.S.; Kraemer, W.J. Creatine Supplementation: Its Effect on Human Muscular Performance and Body Composition. *J. Strength Cond. Res.* **1996**, *10*, 200. [CrossRef]
12. Calfee, R.; Fadale, P. Popular Ergogenic Drugs and Supplements in Young Athletes. *Pediatrics* **2006**, *117*, e577–e589. [CrossRef] [PubMed]
13. Metzl, J.D.; Small, E.; Levine, S.R.; Gershel, J.C. Popular ergogenic drugs and supplements in young athletes. *Pediatrics* **2001**, *108*, 421–425. [CrossRef] [PubMed]
14. Bosco, C.; Tihanyi, J.; Pucspk, J.; Kovacs, I.; Gabossy, A.; Colli, R.; Pulvirenti, G.; Tranquilli, C.; Foti, C.; Viru, M.; et al. Effect of Oral Creatine Supplementation on Jumping and Running Performance. *Int. J. Sports Med.* **1997**, *18*, 369–372. [CrossRef] [PubMed]
15. Dawson, B.; Cutler, M.; Moody, A.; Lawrence, S.; Goodman, C.; Randall, N. Effects of oral creatine loading on single and repeated maximal short sprints. *Aust. J. Sci. Med. Sport* **1995**, *27*, 56–61. [PubMed]
16. Volek, J.S.; Duncan, N.D.; Mazzetti, S.A.; Staron, R.S.; Putukian, M.; Gómez, A.L.; Pearson, D.R.; Fink, W.J.; Kraemer, W.J. Performance and muscle fiber adaptations to creatine supplementation and heavy resistance training. *Med. Sci. Sports Exerc.* **1999**, *31*, 1147–1156. [CrossRef] [PubMed]
17. Kreider, R.B. Effects of creatine supplementation on performance and training adaptations. *Mol. Cell. Biochem.* **2003**, *244*, 89–94. [CrossRef] [PubMed]
18. Karimian, J.; Esfahani, P.S. Supplement consumption in body builder athletes. *J. Res. Med. Sci.* **2011**, *16*, 1347–1353. [PubMed]
19. Terjung, R.L.; Clarkson, P.; Eichner, E.R.; Greenhaff, P.L.; Hespel, P.J.; Israel, R.G.; Kraemer, W.J.; Meyer, R.A.; Spriet, L.L.; Tarnopolsky, M.A.; et al. Physiological and Health Effects of Oral Creatine Supplementation. *Med. Sci. Sports Exerc.* **2000**, *32*, 706–717. [CrossRef] [PubMed]
20. Mihic, S.; Macdonald, J.R.; McKenzie, S.; Tarnopolsky, M.A. Acute creatine loading increases fat-free mass, but does not affect blood pressure, plasma creatinine, or CK activity in men and women. *Med. Sci. Sports Exerc.* **2000**, *32*, 291–296. [CrossRef]
21. Grindstaff, P.D.; Kreider, R.; Bishop, R.; Wilson, M.; Wood, L.; Alexander, C.; Almada, A. Effects of creatine supplementation on repetitive sprint performance and body composition in competitive swimmers. *Int. J. Sport Nutr.* **1997**, *7*, 330–346. [CrossRef] [PubMed]
22. Burns, R.D.; Schiller, M.; Merrick, M.A.; Wolf, K.N. Intercollegiate student athlete use of nutritional supplements and the role of athletic trainers and dietitians in nutrition counseling. *J. Am. Diet. Assoc.* **2004**, *104*, 246–249. [CrossRef] [PubMed]
23. Poortmans, J.R.; Francaux, M. Adverse Effects of Creatine Supplementation. *Sports Med.* **2000**, *30*, 155–170. [CrossRef] [PubMed]
24. Juhn, M.; O'Kane, J.W.; Vinci, D.M. Oral Creatine Supplementation in Mate Collegiate Athletes. *J. Am. Diet. Assoc.* **1999**, *99*, 593–595. [CrossRef]
25. Jones, A.M.; Atter, T.; Georg, K.P. Oral creatine supplementation improves multiple sprint performance in elite ice-hockey players. *J. Sports Med. Phys. Fit.* **1999**, *39*, 189–196. [CrossRef]
26. Greenwood, M.; Farris, J.; Kreider, R.; Greenwood, L.; Byars, A. Creatine Supplementation Patterns and Perceived Effects in Select Division I Collegiate Athletes. *Clin. J. Sport Med.* **2000**, *10*, 191–194. [CrossRef]
27. Kreider, R.B. Creatine supplementation in exercise and sport. In *Energy-Yeilding Macronutrients and Energy Metabolism in Sports Nutrition*; Driskell, J., Wolinsky, J., Eds.; CRC Press LLC: Boca Raton, FL, USA, 1999; pp. 213–242.
28. Kreider, R. Creatine supplementation: Analysis of ergogenic value, medical safety, and concerns. *J. Exerc. Physiol. Online* **1998**, *1*, 7–18.
29. Juhn, M.S.; Tarnopolsky, M. Potential Side Effects of Oral Creatine Supplementation. *Clin. J. Sport Med.* **1998**, *8*, 298–304. [CrossRef] [PubMed]

30. Greenhaff, P.L.; Bodin, K.; Söderlund, K.; Hultman, E. Effect of oral creatine supplementation on skeletal muscle phophocreatine resynthesis. *Am. J. Physiol.* **1994**, *266*, E725–E730. [CrossRef] [PubMed]

31. Syrotuik, D.G.; Bell, G.J. Acute Creatine Monohydrate Supplementation: A Descriptive Physiological Profile of Responders vs. Nonresponders. *J. Strength Cond. Res.* **2004**, *18*, 610–617. [CrossRef] [PubMed]

32. Lee, C.-L.; Lin, J.-C.; Cheng, C.-F. Effect of caffeine ingestion after creatine supplementation on intermittent high-intensity sprint performance. *Graefe's Arch. Clin. Exp. Ophthalmol.* **2011**, *111*, 1669–1677. [CrossRef] [PubMed]

33. Kaviani, M.; Shaw, K.; Chilibeck, P.D. Benefits of Creatine Supplementation for Vegetarians Compared to Omnivorous Athletes: A Systematic Review. *Int. J. Environ. Res. Public Health* **2020**, *17*, 3041. [CrossRef] [PubMed]

34. Ferretti, R.; Moura, E.G.; Dos Santos, V.C.; Caldeira, E.J.; Conte, M.; Matsumura, C.Y.; Pertille, A.; Mosqueira, M. High-fat diet suppresses the positive effect of creatine supplementation on skeletal muscle function by reducing protein expression of IGF-PI3K-AKT-mTOR pathway. *PLoS ONE* **2018**, *13*, e0199728. [CrossRef]

35. Ostojic, S.M.; Ahmetovic, Z. Gastrointestinal Distress after Creatine Supplementation in Athletes: Are Side Effects Dose Dependent? *Res. Sports Med.* **2008**, *16*, 15–22. [CrossRef] [PubMed]

36. Li, Y.; Li, Y.; Cao, X.; Jin, X.; Jin, T. Pattern recognition receptors in zebrafish provide functional and evolutionary insight into innate immune signaling pathways. *Cell. Mol. Immunol.* **2017**, *14*, 80–89. [CrossRef] [PubMed]

37. Beutler, B.A. TLRs and innate immunity. *Blood* **2009**, *113*, 1399–1407. [CrossRef]

38. Browne, E.P. The Role of Toll-like Receptors in Retroviral Infection. *Microorganisms* **2020**, *8*, 1787. [CrossRef] [PubMed]

39. Vieira, R.P.; Duarte, A.C.S.; Claudino, R.C.; Perini, A.; Santos Ângela, B.G.; Moriya, H.T.; Arantes-Costa, F.M.; Martins, M.A.; Carvalho, C.R.F.; Dolhnikoff, M. Creatine Supplementation Exacerbates Allergic Lung Inflammation and Airway Remodeling in Mice. *Am. J. Respir. Cell Mol. Biol.* **2007**, *37*, 660–667. [CrossRef]

40. Almeida, F.M.; Battochio, A.S.; Napoli, J.P.; Alves, K.A.; Balbin, G.S.; Oliveira-Junior, M.; Moriya, H.T.; Pego-Fernandes, P.M.; Vieira, R.P.; Pazetti, R. Creatine Supply Attenuates Ischemia-Reperfusion Injury in Lung Transplantation in Rats. *Nutrients* **2020**, *12*, 2765. [CrossRef]

41. Laubach, V.E.; Sharma, A.K. Mechanisms of lung ischemia-reperfusion injury. *Curr. Opin. Organ. Transplant.* **2016**, *21*, 246–252. [CrossRef] [PubMed]

42. Opal, S.M.; E Huber, C. Bench-to-bedside review: Toll-like receptors and their role in septic shock. *Crit. Care* **2002**, *6*, 125–136. [CrossRef]

43. Tarique, A.A.; Logan, J.; Thomas, E.; Holt, P.G.; Sly, P.D.; Fantino, E. Phenotypic, Functional, and Plasticity Features of Classical and Alternatively Activated Human Macrophages. *Am. J. Respir. Cell Mol. Biol.* **2015**, *53*, 676–688. [CrossRef] [PubMed]

44. Ivashkiv, L.B. Epigenetic regulation of macrophage polarization and function. *Trends Immunol.* **2013**, *34*, 216–223. [CrossRef] [PubMed]

45. Sica, A.; Mantovani, A. Macrophage plasticity and polarization: In vivo veritas. *J. Clin. Investig.* **2012**, *122*, 787–795. [CrossRef]

46. Murray, P.J.; Allen, J.E.; Biswas, S.K.; Fisher, E.A.; Gilroy, D.W.; Goerdt, S.; Gordon, S.; Hamilton, J.A.; Ivashkiv, L.B.; Lawrence, T.; et al. Macrophage Activation and Polarization: Nomenclature and Experimental Guidelines. *Immunity* **2014**, *41*, 14–20. [CrossRef] [PubMed]

47. Atri, C.; Guerfali, F.Z.; Laouini, D. Role of Human Macrophage Polarization in Inflammation during Infectious Diseases. *Int. J. Mol. Sci.* **2018**, *19*, 1801. [CrossRef] [PubMed]

48. Gordon, S.; Martinez, F.O. Alternative Activation of Macrophages: Mechanism and Functions. *Immunity* **2010**, *32*, 593–604. [CrossRef] [PubMed]

49. Gordon, S.; Taylor, P.R. Monocyte and macrophage heterogeneity. *Nat. Rev. Immunol.* **2005**, *5*, 953–964. [CrossRef]

50. Guilliams, M.; Ginhoux, F.; Jakubzick, C.; Naik, S.H.; Onai, N.; Schraml, B.U.; Segura, E.; Tussiwand, R.; Yona, S. Dendritic cells, monocytes and macrophages: A unified nomenclature based on ontogeny. *Nat. Rev. Immunol.* **2014**, *14*, 571–578. [CrossRef] [PubMed]

51. Junttila, I.S.; Mizukami, K.; Dickensheets, H.; Meier-Schellersheim, M.; Yamane, H.; Donnelly, R.P.; Paul, W.E. Tuning sensitivity to IL-4 and IL-13: Differential expression of IL-4Rα, IL-13Rα1, and γc regulates relative cytokine sensitivity. *J. Exp. Med.* **2008**, *205*, 2595–2608. [CrossRef]

52. Mosmann, T.R.; Cherwinski, H.; Bond, M.W.; Giedlin, M.A.; Coffman, R.L. Two types of murine helper T cell clone. I. Definition according to profiles of lymphokine activities and secreted proteins. *J. Immunol.* **1986**, *136*, 2348–2357. [PubMed]

53. Martinez, F.O.; Gordon, S. The M1 and M2 paradigm of macrophage activation: Time for reassessment. *F1000Prime Rep.* **2014**, *6*, 13. [CrossRef] [PubMed]

54. Mantovani, A.; Sica, A.; Sozzani, S.; Allavena, P.; Vecchi, A.; Locati, M. The chemokine system in diverse forms of macrophage activation and polarization. *Trends Immunol.* **2004**, *25*, 677–686. [CrossRef]

55. Martinez, F.O. Macrophage activation and polarization. *Front. Biosci.* **2008**, *13*, 453–461. [CrossRef]

56. Sunderkötter, C.; Nikolic, T.; Dillon, M.J.; Van Rooijen, N.; Stehling, M.; Drevets, D.A.; Leenen, P.J.M. Subpopulations of Mouse Blood Monocytes Differ in Maturation Stage and Inflammatory Response. *J. Immunol.* **2004**, *172*, 4410–4417. [CrossRef] [PubMed]

57. Auffray, C.; Fogg, D.; Garfa, M.; Elain, G.; Join-Lambert, O.; Kayal, S.; Sarnacki, S.; Cumano, A.; Lauvau, G.; Geissmann, F. Monitoring of Blood Vessels and Tissues by a Population of Monocytes with Patrolling Behavior. *Science* **2007**, *317*, 666–670. [CrossRef] [PubMed]

58. Ruscitti, P.; Cipriani, P.; Di Benedetto, P.; Liakouli, V.; Berardicurti, O.; Carubbi, F.; Ciccia, F.; Alvaro, S.; Triolo, G.; Giacomelli, R. Monocytes from patients with rheumatoid arthritis and type 2 diabetes mellitus display an increased production of interleukin (IL)-1β via the nucleotide-binding domain and leucine-rich repeat containing family pyrin 3(NLRP3)-inflammasome activation: A pos. *Clin. Exp. Immunol.* **2015**, *182*, 35–44. [CrossRef] [PubMed]

59. Liu, Y.-C.; Zou, X.-B.; Chai, Y.-F.; Yao, Y.-M. Macrophage Polarization in Inflammatory Diseases. *Int. J. Biol. Sci.* **2014**, *10*, 520–529. [CrossRef] [PubMed]

60. Jansen, A.; Homo-Delarche, F.; Hooijkaas, H.; Leenen, P.J.; Dardenne, M.; Drexhage, H.A. Immunohistochemical Characterization of Monocytes-Macrophages and Dendritic Cells Involved in the Initiation of the Insulitis and -Cell Destruction in NOD Mice. *Diabetes* **1994**, *43*, 667–675. [CrossRef] [PubMed]

61. Furlan, R.; Cuomo, C.; Martino, G. *Animal Models of Multiple Sclerosis*; Walker, K.M., Ed.; Springer: London, UK, 2009; Volume 549, pp. 157–173.

62. Murphy, C.A.; Langrish, C.L.; Chen, Y.; Blumenschein, W.; McClanahan, T.; Kastelein, R.A.; Sedgwick, J.D.; Cua, D.J. Divergent Pro- and Antiinflammatory Roles for IL-23 and IL-12 in Joint Autoimmune Inflammation. *J. Exp. Med.* **2003**, *198*, 1951–1957. [CrossRef]

63. Smith, A.M.; Rahman, F.Z.; Hayee, B.; Graham, S.J.; Marks, D.J.; Sewell, G.W.; Palmer, C.D.; Wilde, J.; Foxwell, B.M.; Gloger, I.S.; et al. Disordered macrophage cytokine secretion underlies impaired acute inflammation and bacterial clearance in Crohn's disease. *J. Exp. Med.* **2009**, *206*, 1883–1897. [CrossRef] [PubMed]

64. Italiani, P.; Mazza, E.M.C.; Lucchesi, D.; Cifola, I.; Gemelli, C.; Grande, A.; Battaglia, C.; Bicciato, S.; Boraschi, D. Transcriptomic Profiling of the Development of the Inflammatory Response in Human Monocytes In Vitro. *PLoS ONE* **2014**, *9*, e87680. [CrossRef] [PubMed]

65. Rackov, G.; Hernández-Jiménez, E.; Shokri, R.; Carmona-Rodríguez, L.; Mañes, S.; Álvarez-Mon, M.; López-Collazo, E.; Martínez-A, C.; Balomenos, D. p21 mediates macrophage reprogramming through regulation of p50-p50 NF-κB and IFN-β. *J. Clin. Investig.* **2016**, *126*, 3089–3103. [CrossRef] [PubMed]

66. Ji, L.; Zhao, X.; Zhang, B.; Kang, L.; Song, W.; Zhao, B.; Xie, W.; Chen, L.; Hu, X. Slc6a8-Mediated Creatine Uptake and Accumulation Reprogram Macrophage Polarization via Regulating Cytokine Responses. *Immunity* **2019**, *51*, 272–284.e7. [CrossRef] [PubMed]

67. Wang, G.; Zhao, H.; Zheng, B.; Li, D.; Yuan, Y.; Han, Q.; Tian, Z.; Zhang, J. TLR2 Promotes Monocyte/Macrophage Recruitment into the Liver and Microabscess Formation to Limit the Spread of Listeria Monocytogenes. *Front. Immunol.* **2019**, *10*, 1388. [CrossRef] [PubMed]

68. Song, A.J.; Palmiter, R.D. Detecting and Avoiding Problems When Using the Cre–lox System. *Trends Genet.* **2018**, *34*, 333–340. [CrossRef] [PubMed]

69. Ramachandran, P.; Pellicoro, A.; Vernon, M.A.; Boulter, L.; Aucott, R.L.; Ali, A.; Hartland, S.N.; Snowdon, V.K.; Cappon, A.; Gordon-Walker, T.T.; et al. Differential Ly-6C expression identifies the recruited macrophage phenotype, which orchestrates the regression of murine liver fibrosis. *Proc. Natl. Acad. Sci. USA* **2012**, *109*, E3186–E3195. [CrossRef] [PubMed]

70. Li, H.; Jiang, T.; Li, M.-Q.; Zheng, X.-L.; Zhao, G.-J. Transcriptional Regulation of Macrophages Polarization by MicroRNAs. *Front. Immunol.* **2018**, *9*, 1175. [CrossRef] [PubMed]

71. Essandoh, K.; Li, Y.; Huo, J.; Fan, G.-C. MiRNA-Mediated Macrophage Polarization and its Potential Role in the Regulation of Inflammatory Response. *Shock* **2016**, *46*, 122–131. [CrossRef]

72. Self-Fordham, J.B.; Naqvi, A.R.; Uttamani, J.R.; Kulkarni, V.; Nares, S. MicroRNA: Dynamic Regulators of Macrophage Polarization and Plasticity. *Front. Immunol.* **2017**, *8*, 1062. [CrossRef]

73. Vergadi, E.; Ieronymaki, E.; Lyroni, K.; Vaporidi, K.; Tsatsanis, C. Akt Signaling Pathway in Macrophage Activation and M1/M2 Polarization. *J. Immunol.* **2017**, *198*, 1006–1014. [CrossRef]

74. Madan, B.R.; Khanna, N.K. Effects of amino acids on the carrageenan-induced paw oedema in rats: A preliminary report. *Ind. J. Physiol. Pharmacol.* **1976**, *8*, 227–229.

75. Ghorbanzadeh, B.; Mansouri, M.T.; Hemmati, A.A.; NaghiZadeh, B.; Mard, S.A.; Rezaie, A. A study of the mechanisms underlying the anti-inflammatory effect of ellagic acid in carrageenan-induced paw edema in rats. *Indian J. Pharmacol.* **2015**, *47*, 292–298. [CrossRef] [PubMed]

76. Bhattacharyya, S.; Gill, R.K.; Chen, M.L.; Zhang, F.; Linhardt, R.J.; Dudeja, P.K.; Tobacman, J.K. Toll-like Receptor 4 Mediates Induction of the Bcl10-NFκB-Interleukin-8 Inflammatory Pathway by Carrageenan in Human Intestinal Epithelial Cells. *J. Biol. Chem.* **2008**, *283*, 10550–10558. [CrossRef] [PubMed]

77. Van Hoogmoed, L.M.; Snyder, J.R.; Harmon, F. In vitro investigation of the effect of prostaglandins and nonsteroidal anti-inflammatory drugs on contractile activity of the equine smooth muscle of the dorsal colon, ventral colon, and pelvic flexure. *Am. J. Veter. Res.* **2000**, *61*, 1259–1266. [CrossRef] [PubMed]

78. Khanna, N.K.; Madan, B.R. Studies on the anti-inflammatory activity of creatine. *Arch. Int. Pharmacodyn. Ther.* **1978**, *231*, 340–350. [PubMed]

79. Razonable, R.R.; Henault, M.; Watson, H.L.; Paya, C.V. Nystatin Induces Secretion of Interleukin (IL)-1β, IL-8, and Tumor Necrosis Factor Alpha by a Toll-Like Receptor-Dependent Mechanism. *Antimicrob. Agents Chemother.* **2005**, *49*, 3546–3549. [CrossRef] [PubMed]

80. Shaw, S.K.; Ma, S.; Kim, M.B.; Rao, R.M.; Hartman, C.U.; Froio, R.M.; Yang, L.; Jones, T.; Liu, Y.; Nusrat, A.; et al. Coordinated Redistribution of Leukocyte LFA-1 and Endothelial Cell ICAM-1 Accompany Neutrophil Transmigration. *J. Exp. Med.* **2004**, *200*, 1571–1580. [CrossRef]

81. Muller, W.A. Getting Leukocytes to the Site of Inflammation. *Veter- Pathol.* **2013**, *50*, 7–22. [CrossRef] [PubMed]

82. Nomura, A.; Zhang, M.; Sakamoto, T.; Ishii, Y.; Morishima, Y.; Mochizuki, M.; Kimura, T.; Uchida, Y.; Sekizawa, K. Anti-inflammatory activity of creatine supplementation in endothelial cells in vitro. *Br. J. Pharmacol.* **2003**, *139*, 715–720. [CrossRef] [PubMed]

83. Ishii, Y.; Kimura, T.; Morishima, Y.; Mochizuki, M.; Nomura, A.; Sakamoto, T.; Uchida, Y.; Sekizawa, K. S-carboxymethylcysteine inhibits neutrophil activation mediated by N-formyl-methionyl-leucyl-phenylalanine. *Eur. J. Pharmacol.* **2002**, *449*, 183–189. [CrossRef]

84. Dvorak, A.M. Mast Cell-Derived Mediators of Enhanced Microvascular Permeability, Vascular Permeability Factor/Vascular Endothelial Growth Factor, Histamine, and Serotonin, Cause Leakage of Macromolecules through a New Endothelial Cell Permeability Organelle, the Vesiculo-Vacuolar Organelle. *Chem. Immunol. Allergy* **2005**, *85*, 185–204. [CrossRef] [PubMed]

85. Lee, K.S.; Kim, S.R.; Park, S.J.; Park, H.S.; Min, K.H.; Lee, M.H.; Jin, S.M.; Jin, G.Y.; Yoo, W.H.; Lee, Y.C. Hydrogen Peroxide Induces Vascular Permeability via Regulation of Vascular Endothelial Growth Factor. *Am. J. Respir. Cell Mol. Biol.* **2006**, *35*, 190–197. [CrossRef] [PubMed]

86. Hashida, R.; Anamizu, C.; Yagyu-Mizuno, Y.; Ohkuma, S.; Takano, T. Transcellular transport of fluorescein dextran through an arterial endothelial cell monolayer. *Cell Struct. Funct.* **1986**, *11*, 343–349. [CrossRef] [PubMed]

87. Simionescu, N.; Palade, G.E. Dextrans and glycogens as particulate tracers for studying capillary permeability. *J. Cell Biol.* **1971**, *50*, 616–624. [CrossRef] [PubMed]

88. Cornish, S.M.; Peeler, J.D. No effect of creatine monohydrate supplementation on inflammatory and cartilage degradation biomarkers in individuals with knee osteoarthritis. *Nutr. Res.* **2018**, *51*, 57–66. [CrossRef]

89. Deminice, R.; Rosa, F.T.; Franco, G.S.; Jordao, A.A.; De Freitas, E.C. Effects of creatine supplementation on oxidative stress and inflammatory markers after repeated-sprint exercise in humans. *Nutrients* **2013**, *29*, 1127–1132. [CrossRef]

90. Bassit, R.A.; Curi, R.; Rosa, L.F.B.P.C. Creatine supplementation reduces plasma levels of pro-inflammatory cytokines and PGE2 after a half-ironman competition. *Amino Acids* **2007**, *35*, 425–431. [CrossRef] [PubMed]

91. Santos, R.; Bassit, R.; Caperuto, E.; Rosa, L.C. The effect of creatine supplementation upon inflammatory and muscle soreness markers after a 30km race. *Life Sci.* **2004**, *75*, 1917–1924. [CrossRef] [PubMed]

92. Kondo, K.; Ohigashi, I.; Takahama, Y. Thymus machinery for T-cell selection. *Int. Immunol.* **2019**, *31*, 119–125. [CrossRef] [PubMed]

93. Mustelin, T.; Taskén, K. Positive and negative regulation of T-cell activation through kinases and phosphatases. *Biochem. J.* **2003**, *371*, 15–27. [CrossRef]

94. Qian, D.; Weiss, A. T cell antigen receptor signal transduction. *Curr. Opin. Cell Biol.* **1997**, *9*, 205–212. [CrossRef]

95. Owen, M.J.; Venkitaraman, A.R. Signalling in lymphocyte development. *Curr. Opin. Immunol.* **1996**, *8*, 191–198. [CrossRef]

96. Kägi, D.; Ledermann, B.; Bürki, K.; Seiler, P.; Odermatt, B.; Olsen, K.J.; Podack, E.R.; Zinkernagel, R.M.; Hengartner, H. Cytotoxicity mediated by T cells and natural killer cells is greatly impaired in perforin-deficient mice. *Nat. Cell Biol.* **1994**, *369*, 31–37. [CrossRef] [PubMed]

97. Li, Y.; Yin, Y.; Mariuzza, R.A. Structural and Biophysical Insights into the Role of CD4 and CD8 in T Cell Activation. *Front. Immunol.* **2013**, *4*, 206. [CrossRef] [PubMed]

98. Zhang, Y.; Li, H.; Wang, X.; Gao, X.; Liu, X. Regulation of T Cell Development and Activation by Creatine Kinase B. *PLoS ONE* **2009**, *4*, e5000. [CrossRef] [PubMed]

99. Bell, G.M.; E Seaman, W.; Niemi, E.C.; Imboden, J.B. The OX-44 molecule couples to signaling pathways and is associated with CD2 on rat T lymphocytes and a natural killer cell line. *J. Exp. Med.* **1992**, *175*, 527–536. [CrossRef] [PubMed]

100. Kelly, E.; Won, A.; Refaeli, Y.; Van Parijs, L. IL-2 and Related Cytokines Can Promote T Cell Survival by Activating AKT. *J. Immunol.* **2002**, *168*, 597–603. [CrossRef] [PubMed]

101. Valledor, A.F.; Comalada, M.; Santamaría-Babí, L.F.; Lloberas, J.; Celada, A. Macrophage Proinflammatory Activation and Deactivation. *Adv. Immunol.* **2010**, *108*, 1–20. [CrossRef] [PubMed]

102. Di Biase, S.; Ma, X.; Wang, X.; Yu, J.; Wang, Y.-C.; Smith, D.J.; Zhou, Y.; Li, Z.; Kim, Y.J.; Clarke, N.; et al. Creatine uptake regulates CD8 T cell antitumor immunity. *J. Exp. Med.* **2019**, *216*, 2869–2882. [CrossRef] [PubMed]

103. Gough, M.; Crittenden, M.; Thanarajasingam, U.; Sanchez-Perez, L.; Thompson, J.; Jevremovic, D.; Vile, R. Gene Therapy to Manipulate Effector T Cell Trafficking to Tumors for Immunotherapy. *J. Immunol.* **2005**, *174*, 5766–5773. [CrossRef] [PubMed]

104. Overwijk, W.W.; Restifo, N.P. B16 as a Mouse Model for Human Melanoma. *Curr. Protoc. Immunol.* **2000**, *39*, 20.1.1–20.1.29. [CrossRef] [PubMed]

105. Smyrk, T.C.; Watson, P.; Kaul, K.; Lynch, H.L. Tumor infiltrating lymphocytes are a marker for microsatellite instability in colo-rectal cancer. *Cancer* **2001**, *91*, 2417. [CrossRef]

106. Michaelrobinson, J.M.; Biemer-Hüttmann, A.-E.; Purdie, D.M.; Walsh, M.D.; Simms, L.A.; Biden, K.G.; Young, J.P.; Leggett, B.A.; Jass, J.R.; Radford-Smith, G.L. Tumour infiltrating lymphocytes and apoptosis are independent features in colorectal cancer stratified according to microsatellite instability status. *Gut* **2001**, *48*, 360–366. [CrossRef] [PubMed]

107. Drescher, K.M.; Sharma, P.; Watson, P.; Gatalica, Z.; Thibodeau, S.N.; Lynch, H.T. Lymphocyte recruitment into the tumor site is altered in patients with MSI-H colon cancer. *Fam. Cancer* **2009**, *8*, 231–239. [CrossRef] [PubMed]
108. Patel, C.H.; Powell, J.D. Targeting T cell metabolism to regulate T cell activation, differentiation and function in disease. *Curr. Opin. Immunol.* **2017**, *46*, 82–88. [CrossRef] [PubMed]
109. Tsoukas, C.D.; Landgraf, B.; Bentin, J.; Valentine, M.; Lotz, M.; Vaughan, J.H.; A Carson, D. Activation of resting T lymphocytes by anti-CD3 (T3) antibodies in the absence of monocytes. *J. Immunol.* **1985**, *135*, 1719–1723.
110. Kazak, L.; Cohen, P. Creatine metabolism: Energy homeostasis, immunity and cancer biology. *Nat. Rev. Endocrinol.* **2020**, *16*, 421–436. [CrossRef] [PubMed]
111. Ohaegbulam, K.C.; Assal, A.; Lazar-Molnar, E.; Yao, Y.; Zang, X. Human cancer immunotherapy with antibodies to the PD-1 and PD-L1 pathway. *Trends Mol. Med.* **2015**, *21*, 24–33. [CrossRef] [PubMed]
112. Sunshine, J.; Taube, J.M. PD-1/PD-L1 inhibitors. *Curr. Opin. Pharmacol.* **2015**, *23*, 32–38. [CrossRef] [PubMed]
113. Ostojic, S.M. Can creatine help in pulmonary rehabilitation after COVID-19? *Ther. Adv. Respir. Dis.* **2020**, *14*, 1–2. [CrossRef] [PubMed]
114. Diao, B.; Wang, C.; Tan, Y.; Chen, X.; Liu, Y.; Ning, L.; Chen, L.; Li, M.; Liu, Y.; Wang, G.; et al. Reduction and Functional Exhaustion of T Cells in Patients with Coronavirus Disease 2019 (COVID-19). *Front. Immunol.* **2020**, *11*, 827. [CrossRef] [PubMed]
115. Vieira, R.P.; Claudino, R.C.; Duarte, A.C.S.; Santos Ângela, B.G.; Perini, A.; Neto, H.C.C.F.; Mauad, T.; Martins, M.A.; Dolhnikoff, M.; Carvalho, C.R.F. Aerobic Exercise Decreases Chronic Allergic Lung Inflammation and Airway Remodeling in Mice. *Am. J. Respir. Crit. Care Med.* **2007**, *176*, 871–877. [CrossRef]
116. MacKenzie, J.R.; Mattes, J.; Dent, L.A.; Foster, P.S. Eosinophils Promote Allergic Disease of the Lung by Regulating CD4+Th2 Lymphocyte Function. *J. Immunol.* **2001**, *167*, 3146–3155. [CrossRef] [PubMed]
117. Emad, A.; Emad, Y. Relationship between Eosinophilia and Levels of Chemokines (CCL5 and CCL11) and IL-5 in Bronchoalveolar Lavage Fluid of Patients with Mustard Gas-Induced Pulmonary Fibrosis. *J. Clin. Immunol.* **2007**, *27*, 605–612. [CrossRef] [PubMed]
118. Raghu, H.; Lepus, C.M.; Wang, Q.; Wong, H.H.; Lingampalli, N.; Oliviero, F.; Punzi, L.; Giori, N.J.; Goodman, S.B.; Chu, C.R.; et al. CCL2/CCR2, but not CCL5/CCR5, mediates monocyte recruitment, inflammation and cartilage destruction in osteoarthritis. *Ann. Rheum. Dis.* **2017**, *76*, 914–922. [CrossRef] [PubMed]
119. Ferreira, S.C.; Toledo, A.C.; Hage, M.; Santos, A.B.G.; Medeiros, M.C.R.; Martins, M.A.; Carvalho, C.; Dolhnikoff, M.; Vieira, R.P. Creatine Activates Airway Epithelium in Asthma. *Endoscopy* **2010**, *31*, 906–912. [CrossRef] [PubMed]
120. Pearlman, J.P.; Fielding, R.A. Creatine Monohydrate as a Therapeutic Aid in Muscular Dystrophy. *Nutr. Rev.* **2006**, *64*, 80–88. [CrossRef] [PubMed]
121. McGuine, T.A.; Sullivan, J.C.; Bernhardt, D.T. Creatine Supplementation in High School Football Players. *Clin. J. Sport Med.* **2001**, *11*, 247–253. [CrossRef] [PubMed]
122. McGuine, T.A.; Sullivan, J.C.; A Bernhardt, D. Creatine supplementation in Wisconsin high school athletes. *WMJ Off. Publ. State Med. Soc. Wis.* **2002**, *101*, 25–30.
123. McGuine, T.A.; Sullivan, J.C.; Bernhardt, D.A. Use of nutritional supplements by high school football and volleyball players. *Iowa Orthop. J.* **2001**, *21*, 43–48.
124. Smith, J.; Dahm, D.L. Creatine Use among a Select Population of High School Athletes. *Mayo Clin. Proc.* **2000**, *75*, 1257–1263. [CrossRef] [PubMed]
125. Neves, M.; Gualano, B.; Roschel, H.; Lima, F.R.; De Sá-Pinto, A.L.; Seguro, A.C.; Shimizu, M.H.; Sapienza, M.T.; Fuller, R.; Lancha, A.H.; et al. Effect of creatine supplementation on measured glomerular filtration rate in postmenopausal women. *Appl. Physiol. Nutr. Metab.* **2011**, *36*, 419–422. [CrossRef] [PubMed]
126. Rawson, E.S.; Venezia, A.C. Use of creatine in the elderly and evidence for effects on cognitive function in young and old. *Amino Acids* **2011**, *40*, 1349–1362. [CrossRef] [PubMed]
127. Rawson, E.S.; Clarkson, P.M. Acute Creatine Supplementation in Older Men. *Int. J. Sports Med.* **2000**, *21*, 71–75. [CrossRef] [PubMed]
128. Rawson, E.S.; Wehnert, M.L.; Clarkson, P.M. Effects of 30 days of creatine ingestion in older men. *Graefe's Arch. Clin. Exp. Ophthalmol.* **1999**, *80*, 139–144. [CrossRef] [PubMed]
129. Have a News Tip for 21-WFMJ? Available online: https://www.wfmj.com/story/43004281/global-creatine-market-size-worth-around-usd-520-million-by-2024-from-usd-360-million-in-2020-at-a-carg-of-65-during-2020-2024-with-top-countries-data (accessed on 23 February 2021).

Review

Creatine in T Cell Antitumor Immunity and Cancer Immunotherapy

Bo Li [1,*] and Lili Yang [1,2,3,4,*]

[1] Department of Microbiology, Immunology & Molecular Genetics, University of California, Los Angeles, Los Angeles, CA 90095, USA
[2] Eli and Edythe Broad Center of Regenerative Medicine and Stem Cell Research, University of California, Los Angeles, Los Angeles, CA 90095, USA
[3] Jonsson Comprehensive Cancer Center, The David Geffen School of Medicine, University of California, Los Angeles, Los Angeles, CA 90095, USA
[4] Molecular Biology Institute, University of California, Los Angeles, Los Angeles, CA 90095, USA
* Correspondence: bo.li@ucla.edu (B.L.); liliyang@ucla.edu (L.Y.)

Abstract: Creatine is a broadly used dietary supplement that has been extensively studied for its benefit on the musculoskeletal system. Yet, there is limited knowledge regarding the metabolic regulation of creatine in cells beyond the muscle. New insights concerning various regulatory functions for creatine in other physiological systems are developing. Here, we highlight the latest advances in understanding creatine regulation of T cell antitumor immunity, a topic that has previously gained little attention in the creatine research field. Creatine has been identified as an important metabolic regulator conserving bioenergy to power CD8 T cell antitumor reactivity in a tumor microenvironment; creatine supplementation has been shown to enhance antitumor T cell immunity in multiple preclinical mouse tumor models and, importantly, to synergize with other cancer immunotherapy modalities, such as the PD-1/PD-L1 blockade therapy, to improve antitumor efficacy. The potential application of creatine supplementation for cancer immunotherapy and the relevant considerations are discussed.

Keywords: creatine; T cell antitumor immunity; metabolic regulator; cancer immunotherapy

Citation: Li, B.; Yang, L. Creatine in T Cell Antitumor Immunity and Cancer Immunotherapy. *Nutrients* **2021**, *13*, 1633. https://doi.org/10.3390/nu13051633

Academic Editors: Richard B. Kreider and Jeffrey R. Stout

Received: 28 March 2021
Accepted: 11 May 2021
Published: 13 May 2021

Publisher's Note: MDPI stays neutral with regard to jurisdictional claims in published maps and institutional affiliations.

1. Introduction

Creatine (Cr) is a nitrogenous organic acid naturally occurring in vertebrates. Endogenous creatine is synthesized from arginine and glycine mainly in the kidney and liver by two enzymes, L-arginine:glycine amidinotransferase (AGAT; also known as GATM) and guanidinoacetate *N*-methyltransferase (GAMT). Synthesized creatine is released into the circulation and specifically taken up by the creatine transporter (CrT or SLC6A8) expressing cells, where creatine is further phosphorylated by creatine kinase (CK) using adenosine triphosphate (ATP) to form phosphocreatine (PCr) [1]. Four CK isoforms have been identified: two cytosolic isoforms, the muscle-type (CKM) and the brain-type (CKB); and two mitochondrial isoforms, the ubiquitous-type (CKMT1) and the sarcomeric-type (CKMT2) [2]. The CK/PCr/Cr system is coupled with ATP-adenosine diphosphate (ADP) transition to buffer intracellular ATP levels. PCr can be break down by CK to resynthesize ATP that supplies cellular energy needs in an acute manner [1].

Besides de novo synthesis, diet is also a primary source of natural creatine for humans. A person who eats red meat, poultry, and/or fish obtains approximately 50% of daily creatine need (about 2 g per day) from food intake [3]. Compared to non-vegetarians, vegetarians have significantly lower levels of creatine and phosphocreatine in the muscle [4,5]. The average creatine stores for a 70 kg adult are between 120 and 140 g [6]. However, approximately 2 g per day of creatine in the muscle are degraded into creatinine that is excreted in the urine [6]. Degradation of creatine is even greater in individuals with higher physical activity and/or larger muscle mass. Wallimann et al. have reported that normal omnivore diets generally provide 0.75–1.5 g/day of creatine and are not sufficient to fully saturate creatine stores in the body [7]. Therefore, creatine supplementation is an effective way to increase the creatine reservoir. Creatine is predominantly stored in skeletal muscle as free creatine (~40%) or phosphocreatine (~60%); the latter is a major source of bioenergy to the body [1].

Creatine was discovered as a constituent of meat by the French chemist Michel-Eugene Chevreul more than 180 years ago. However, creatine did not gain wide attention until the 1990s when two gold medalists from the 1992 Barcelona Olympics credited it with helping them enhance performance. The phosphagen energy metabolic system produces ATP more rapidly than other metabolic systems, such as the glycolysis system and the aerobic system [8]. Concentrations of cellular PCr in some tissues such as muscles or the brain can reach up to 30–40 mM, resulting in rapidly replenishing ATP stores that can be immediately used during high-energy demand states [9]. Over the past three decades, oral creatine supplements have been broadly utilized by athletes to improve performance during high-intensity exercise with repeated bouts and short rest periods, such as running, swimming, sprinting, jumping, and strength training [9–13]. The improvement in performance can reach up to 20% on various high-intensity exercise tasks [10]. A major mechanism is that increasing the stores of PCr in skeletal muscle leads to the rephosphorylation of ADP to ATP during bursts of high intensity exercise, thereby increasing the availability of bioenergy [14,15]. Hydrogen ions resulting from increased lactic acid accumulation during high-intensity exercise is a key contributor to muscle fatigue. Given that the breakdown of PCr to creatine and phosphate consumes a hydrogen ion, creatine can also buffer the pH changes caused by increased hydrogen ion concentration and maximize performance [16].

In addition to its use by athletes, creatine is the most popular nutrition supplement used by body builders to gain muscle mass [14]. The popular form of creatine used by athletes and body builders is creatine monohydrate [17]. The dosing strategy typically consists of two phases: the first phase is a loading phase in which athletes ingest 20 g per day of creatine over four doses for five to seven days, followed by a maintenance phase where 1–10 g per day of creatine is administered for a month or much longer [1,10]. The loading phase has been reported to increase muscle stores of PCr between 20% and 40% [9]. Importantly, the creatine-loading phase results in limited side effects, such as cramping, nausea, and fluid retention [18]. Alternatively, a line of evidence has indicated that the ingestion of 3 g per day of creatine for a minimum 28 days is as effective as the higher dose-loading regimen at raising total skeletal muscle creatine stores [19]. Gradually increasing the muscle stores of creatine may alleviate side effects that are commonly associated with the higher dose-loading regimen.

Despite 95% of creatine stores in skeletal muscle, a significant amount of creatine exists in the brain [20], where creatine plays key roles in maintaining normal neurological functions. Human genetic deficiencies in creatine pathway components involved in creatine synthesis or transport result in decreased levels of ATP in the brain, which is associated with various clinical symptoms including developmental delays, speech impairment, and mental retardation [21]. Several lines of evidence have revealed that oral creatine supplements protect against neurological disorders such as traumatic brain injury [22], amyotrophic lateral sclerosis [23], Huntington's Disease [24], and Parkinson's Disease [25].

Clinically, patients with deficiency in creatine synthesis can benefit from creatine supplementation; however, no treatment is available to patients with deficiency in the creatine transporter gene *CrT* [26,27]. The gene *CrT* is located on the X chromosome. This gene encodes the solute carrier family 6 member 8 (SLC6A8) that is a plasma membrane protein whose function is to transport creatine into and out of cells in a sodium- and chloride-dependent manner. A high expression of *CrT* gene is required for normal physiological functions of high-energy demanding cells and organs such as muscle and the brain [1]. Patients harboring human *CrT* mutations have been associated with a group of muscle and brain disorders [1]. In line with the clinical data, mice with systemic CrT deficiency have smaller body weight and exhibit deficits in memory and spatial learning [28]. However, the function and regulation of CrT/creatine system outside of the muscle and the brain remain largely elusive. New concepts of creatine biology in other systems and cell types have just begun to be developed in the past few years.

In this review, we aim to summarize recent advances in understanding creatine regulation of antitumor immunity, a topic that has previously gained little attention in the creatine research field. We also discuss the potential application of creatine supplementation for cancer immunotherapy and relevant considerations.

2. Creatine Regulation of T Cell Antitumor Immunity

T cells play a central role in mediating immune responses against cancer. In a tumor, CD8 cytotoxic T cells are activated through their T cell receptor (TCR) recognition of tumor antigenic peptides presented by major histocompatibility complex (MHC) class I molecules on tumor cells. Activated CD8 T cells can efficiently kill tumor cells by releasing cytotoxic molecules (e.g., granzymes and perforin) or using death ligands (e.g., TNF-related apoptosis-inducing ligand and CD95 ligand) [29,30]. Therefore, CD8 T cells have emerged as attractive therapeutic targets for cancer treatment [31–34].

The activation and proliferation of T cells are energy-demanding activities, which require large amounts of energy in the form of ATP [35]. Distinct metabolic programs such as glycolysis and tricarboxylic acid (TCA) cycle are utilized to convert nutrients (e.g., glucose, amino acids, and lipids) into ATP to support CD8 T cell homeostasis and effector function [35–37]. However, in the tumor microenvironment, T cells face fierce competition with fast-growing tumor cells for the limited availability of nutrients [38]. Therefore, an economical and efficient energy metabolism is required for tumor-infiltrating T cells to sustain effective antitumor responses [39]. However, metabolic regulators of T cell antitumor immunity remains poorly understood [40–43]. Studies of the metabolic regulation of T cell antitumor immunity thus may identify new therapeutic targets for treating cancer.

By using a B16-OVA mouse melanoma model, Di Biase and colleagues observed an upregulation of *CrT* gene expression in tumor-infiltrating CD8 T cells compared to CD8 T cells isolated from tumor-free mice. Further study of the tumor-infiltrating CD8 T cell subsets revealed an upregulation of *CrT* gene expression that was much more significant in the PD-1^{hi} cells compared to the PD-1^{lo} cells; the highest levels of *CrT* gene expression were detected in the PD-1^{hi}Tim-3^{hi}LAG-3^{hi} subset that is considered to be the most "exhausted" [44]. These findings suggest a possible feedback loop in tumor-fighting CD8 T cells that compensates for bioenergy-insufficiency by increasing creatine uptake; in particular, the "exhausted" CD8 T cells are highly capable of uptaking creatine and may benefit the most from creatine supplementation treatment [44].

To study if the CrT/creatine system regulates the energy metabolism of tumor-fighting CD8 cytotoxic T cells, systemic *CrT*-knockout (KO) mice were used for the B16-OVA tumor challenge. Notably, *CrT*-KO mice with creatine supplementation exhibited accelerated tumor growth compared to their *CrT*-wild-type (WT) littermates with creatine supplementation [44]. In line with these results, *CrT*-KO tumor-infiltrating CD8 T cells expressed higher levels of PD-1 compared to their wildtype counterparts [44], suggesting CrT deficiency may lead to bioenergy insufficiency and exhaustion of antitumor T cells.

To study if creatine directly or indirectly regulates antitumor immunity, WT CD45.1 congenic mice were reconstituted with bone marrow cells from either *CrT*-WT or *CrT*-KO donor mice and then challenged with B16-OVA melanoma cells. Indeed, CrT deficiency impeded the capacity of the reconstituted immune system to control tumor growth. Furthermore, OT1 transgenic (Tg)/*CrT*-KO mice were generated to produce OVA-specific CD8 T cells deficient in CrT. After transferring these T cells into WT CD45.1 congenic mice bearing pre-established B16-OVA tumors, they were less effective in suppressing tumor growth compared to control WT T cells [44]. Similarly, the OT1 Tg/*CrT*-KO CD8 T cells upregulated PD-1 levels and produced a smaller amount of effector T cell cytokines (e.g., IL-2 and IFN-γ) [44]. After in vitro antigen stimulation, *CrT*-KO CD8 T cells exhibited a dramatic reduction in cell proliferation, surface activation marker production (e.g., CD25), and effector cytokine (e.g., IL-2 and IFN-γ) and cytotoxic molecule (e.g., Granzyme B) production, whereas overexpression of *CrT* gene rescued the proliferation and function of *CrT*-KO CD8 T cells. Study of tumor antigen-specific CD8 T cells deficient in CrT exhibited similar in vitro results [44].

Taken together, these studies strongly support a new role of the creatine/CrT pathway in positively regulating an effector CD8 T cell response against tumor.

3. Molecular Mechanisms Mediating Creatine Regulation of CD8 T Cell Responses

Muscle cells and brain cells power cellular activities by using creatine to buffer intracellular ATP levels via the CK/PCr/Cr system [1]. Recent evidence has indicated that the expression levels of the *CrT* gene and *Ckb* gene are markedly upregulated in CD8 T cells after TCR stimulation [44], suggesting activated CD8 T cells may use a similar mechanism to power T cell activities, in particular antitumor reactivity. Compared to *CrT*-WT CD8 T cells, *CrT*-KO CD8 T cells contained significantly reduced levels of ATP after antigen stimulation. Supplementing ATP to T cell culture not only rescued hypoactivation of *CrT*-KO CD8 T cells, but also further activated *CrT*-WT CD8 T cells [44]. Another research group also reported that overexpression of the *Ckb* gene enhanced the cytokine production and proliferation of mature T cells, while blockade of CKB using specific CK inhibitor or *Ckb* short hairpin RNA (shRNA) resulted in severe impairment of T cell function [45]. Collectively, these results indicate that the CK/PCr/Cr ATP-buffering system is essential for a productive CD8 T cell response to antigen stimulation.

Besides the kidney and liver cells that are responsible for the classical pathway of creatine synthesis, other cell types, such as the muscle cells [46], adipocytes [47], and pancreatic acinar cells [48], have been indicated to synthesize creatine. Could *de novo* synthesis be another source to feed the CK/PCr/Cr system in T cells? To address this, expression of two genes encoding the key enzymes controlling creatine synthesis, *Agat* and *Gamt*, was examined in T cells. Both genes were expressed at low levels in CD8 T cells, and the expression of *Gamt* gene was further downregulated after TCR stimulation. Intracellular creatine levels were undetectable in activated *CrT*-KO CD8 T cells [44]. Hence, activated CD8 T cells may have a very limited capacity for *de novo* creatine synthesis and mainly rely on uptaking extracellular creatine through CrT to sustain the CK/PCr/Cr ATP-buffering system.

TCR stimulation leads to a cascade of phosphorylation events in a sequential manner in T cells [49]. Given that ATP supplies the phosphate group for the phosphorylation reactions catalyzed by the protein kinases [40], the intracellular ATP levels may have a strong impact on T cell activation signaling pathways. Creatine uptake deficiency in CD8 T cells dampened activation of the TCR proximal signaling molecule zeta chain of TCR-associated protein kinase 70 (Zap-70) and the downstream transcription factors, NFAT and AP-1. Supplementing ATP to T cell culture rescued the impaired TCR signaling in *CrT*-KO CD8 T cells [44]. Creatine supplementation further enhanced the phosphorylation of Zap-70 in *CrT*-WT CD8 T cells. Interestingly, the activation of another key downstream transcription factor, NF-κB, was not effected by creatine uptake deficiency [44], suggesting that the NF-κB signaling pathway may better resist ATP fluctuation during T cell response. Together, these data suggest that a CD8 T cell requires the CK/PCr/Cr ATP buffering system to properly activate TCR signaling pathways in response to antigen stimulation.

5′ Adenosine monophosphate-activated protein kinase (AMPK) can function as a nutrient and energy sensor to maintain cell energy homeostasis by detecting shifts in the AMP:ATP ratio. AMPK is activated in cells with low energy status. In order to restore cell energy homeostasis, activated AMPK stimulates ATP-producing catabolic pathways and inhibits ATP-consuming anabolic pathways. While AMPK as a key metabolic regulator plays critical role in T cell metabolism and function [50–53], excessive AMPK activation impairs the function of antigen-specific CD8 T cells [54]. In *CrT*-KO CD8 T cells, the decreased ATP levels were associated with increased activation of AMPK. In addition, AICAR, an AMPK activator, significantly suppressed activation of AP-1 transcription factor as well as production of effector T cell cytokines and surface activation markers in *CrT*-WT CD8 T cells [44]. Therefore, creatine-mediated ATP buffering system may enhance effector CD8 T cell activities in part through AMPK regulation of TCR signaling pathways. Further studies are necessary to understand whether creatine-mediated ATP/energy buffering system cross-regulates other signaling pathways and transcriptional events to augment antitumor T cell function.

4. Creatine Supplementation for Cancer Therapy—Potential Application for Cancer Immunotherapy

Many cancer immunotherapies have been designed to target immune cell metabolism in the tumor microenvironment [38,40–42,55]. For example, immune checkpoint blockade (ICB) therapies, such as PD-1/PD-L1 therapies, reduce tumor glycolysis and switch the energy metabolism to favor T cells [32,43,56–58]. By linking creatine to antitumor T cell activities, recent findings update the picture of the metabolic regulatory network that controls T cell antitumor immunity. A "hybrid-engine" model has been proposed (Figure 1A,B), in which a tumor antigen-specific CD8 T cell uses glucose, amino acids, and lipids as fuels for glycolysis and tricarboxylic acid cycle to generate ATP, while using creatine-mediated ATP buffering system, a "molecular battery", to store energy. This efficient "hybrid-engine" system enables tumor antigen-specific CD8 T cells to make maximal use of limited nutrients and mount their effector function in a metabolically challenging environment [35,39,59]. CD8 T cells rely heavily on the uptake of creatine from extracellular resources (Figure 1B). Therefore, manipulating creatine-mediated energy buffering system to reinvigorate tumor-fighting CD8 T cells could open up a new avenue in cancer immunotherapy. Although creatine can be taken from a creatine-rich diet as well as from dietary supplements (Figure 1C), creatine administration as a therapeutic intervention would generate the best antitumor benefits (Figure 1D).

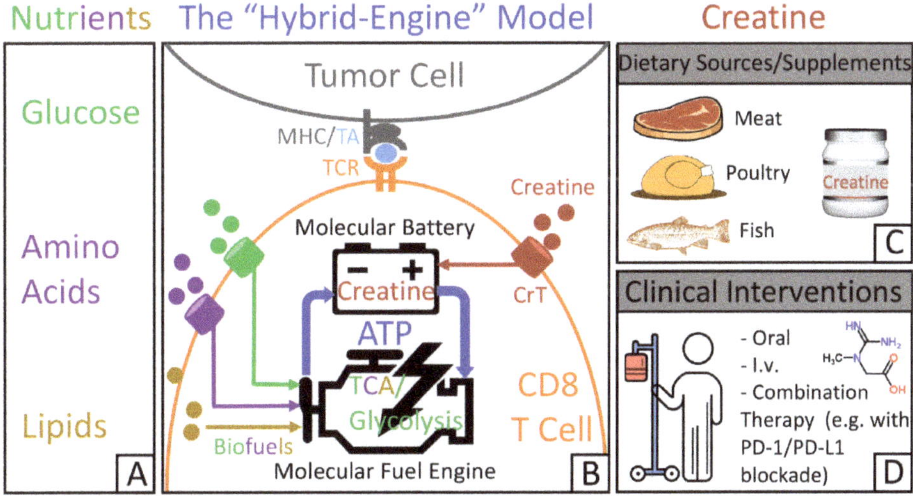

Figure 1. The "hybrid-engine" model in T cell antitumor immunity and potential application of creatine supplementation for cancer immunotherapy. (**A**) Limited nutrients for immune cells in the tumor microenvironment. (**B**) The "hybrid-engine" model. Analogous to the hybrid car, a tumor antigen-specific CD8 T cell uses a "molecular fuel engine", including tricarboxylic acid cycle and glycolysis, to convert nutrients into adenosine triphosphate (ATP), while utilizing creatine mediated energy buffering system, a "molecular battery", to store ATP and power T cell activities. (**C**) Creatine can be obtained from creatine-rich diet (e.g., red meat, poultry, and fish) and dietary supplements. (**D**) Creatine administration as a therapeutic intervention would result in the best therapeutic effects. ©2019 Di Biase et al. Originally published in *J. Exp. Med.* doi: 10.1084/jem.20182044.

In a mouse B16-OVA melanoma model, creatine supplementation either through i.p. injection or through oral administration effectively suppressed tumor growth, which was associated with a dramatic reduction in the number of PD-1^{hi}CD62L^{lo} cells (an "exhaustion-prone" T cell phenotype) among the tumor-infiltrating CD8 T cells [44]. In another mouse syngeneic MC38 colon cancer model, animals receiving the creatine supplementation also had significantly reduced tumor growth compared to control animals [44], suggesting the creatine-induced tumor suppression effect is not tumor model specific and may be applicable to many different types of cancer. As the expression of CrT in the MC38 colon cancer cells is undetectable, the action of creatine supplementation is not on the tumor. In addition, creatine supplementation was not able to suppress B16-OVA tumor growth in immunodeficient NOD/SCID/IL-2R$\gamma^{-/-}$ (NSG) mice or in C57BL/6J WT mice with T cell depletion [44], suggesting that immune cells, especially T cells, may mediate the tumor suppression effect of creatine supplementation. Therefore, creatine supplementation has the translational potential as a new means of improving T cell antitumor activities for cancer immunotherapy.

If creatine provides a potent energy benefit for antitumor CD8 T cells, which is non-redundant to major immune checkpoint regulatory pathways, creatine supplementation may be a valuable component of combination immunotherapy to further enhance efficacy of current ICB treatments. Previous study has shown that the MC38 colon cancer model is responsive to PD-1/PD-L1 blockade therapy [60]. In this model, creatine supplementation in combination with anti-PD-1 therapy generated a more powerful antitumor effect compared to that of each monotherapy [44]. Of note, all the surviving mice from their primary tumors were protected against a second tumor challenge for another 6 months. This attractive antitumor effect was associated with significantly increased number of memory CD8$^+$CD44$^+$ T cells in the surviving mice [44]. Collectively, these encouraging findings point to a translational potential of creatine supplementation for combination cancer immunotherapy.

5. Creatine Supplementation for Cancer Therapy—Other Potential Benefits

In addition to its potential application for cancer immunotherapy, creatine supplementation has been indicated to augment the efficacy of the anticancer drug methylglyoxal (MG) [61]. Pal et al. reported that administration of creatine enhanced the antitumor effects of MG and ascorbic acid in sarcoma animal model in vivo and tumor burden was completely eradicated [62]. These results suggest that creatine supplementation can function as an adjunctive therapeutic intervention with other anticancer agents. Moreover, given the beneficial effects of creatine on muscle mass and physical function, emerging evidence has suggested that creatine supplementation may also have therapeutic potential for attenuating cancer-related weight loss and maintaining muscle function for cancer patients [63,64]. However, application in various cancer contexts has just began and further studies are needed to fully understand the impact of creatine supplementation on clinical outcomes in the cancer patient population at a risk of muscle wasting.

6. Creatine Supplementation for Cancer Therapy—Possible Influences on Cancer Cells

One of the hallmarks of cancer is the reprogramming of cellular metabolism [65]. In contrast to normal cells, cancer cells utilize cellular metabolites to support the high proliferation through distinct mechanisms in their local microenvironment [66]. Creatine-mediated ATP buffering system efficiently provides energy when cells demand high levels of ATP. Prior studies have shown that PCr can be used to transiently increase energy metabolism to promote cancer cell growth by buffering the ATP stores [67–69]. However, a number of studies suggest that creatine and its analogues can suppress tumor growth [1,61,62,70,71]. It has been initially demonstrated that creatine analogue cyclocreatine has antitumor properties in vitro [72]. A possible mechanism of antitumor effect is that phosphocyclocreatine generated from cyclocreatine by CK has a poor substrate activity in the CK reverse reaction and results in energy depletion in cancer cells [73]. However, cyclocreatine also suppressed tumor growth in colon adenocarcinoma without indications of energy depletion in cancer cells [70]. In line with this study, creatine has been shown to inhibit the growth and progression of mammary tumors, sarcoma and neuroblastoma tumors in both rats and mice models [71]. Creatine combined with MG, an anticancer drug, induced higher cytotoxicity and apoptosis in the human breast cancer MCF-7 cell line and chemically transformed the mouse C2C12 muscle cell line, compared to MG alone [62]. In contrast, no detrimental effects were observed in normal C2C12 muscle cells treated with MG plus creatine [62], suggesting that enhanced cytotoxic effects of MG plus creatine are specifically limited within cancer cells. Several lines of evidence have indicated that tumor creatine content is low in multiple types of cancer tissue [61,62,74,75]. Creatine content was restored to almost normal levels with the concomitant regression of tumor cells after creatine treatment [61,75]. Both creatine and cyclocreatine exhibit antitumor effects under certain tumor conditions, suggesting that creatine may employ additional mechanisms, which are independent of sustaining cellular energy charge, to mediate its antitumor effects.

A number of studies have linked cancer to creatine by investigating CK expression and their association with prognosis of cancer patients [76–81]. The expression of CKB was found to be upregulated in different types of cancer [76,78–81]. In addition, elevated expression of CKMT1 was correlated with poor prognosis in patients with breast cancer [76] or liver cancer [78]. Of note, the proliferation of cancer cells expressing high levels of CKB can be greatly impaired by cyclocreatine; however, the cancer cells with low levels of CK were resistant to the antitumor effect of cyclocreatine [72]. Overexpression of CKB in the cancer cells that expressed low levels of CK increased their sensitivity to cyclocreatine inhibition [72]. In the liver microenvironment, CKB released by metastatic colorectal cancer cells could use extracellular ATP to phosphorylate hepatocyte-secreted creatine to produce PCr. Metastatic cells then use PCr to regenerate ATP to support their survival in the liver [82]. Although the source for extracellular ATP is not clear, these data suggest that a crosstalk between normal cells and cancer cells is necessary to support PCr-mediated cancer cell survival under certain conditions. Supplementing PCr to colorectal cancer cells with ATP depletion rescued their ATP levels [82]. The mechanism of PCr atypically imported by CrT on cancer cells remains to be determined.

Abnormal gene expression of ecotropic virus integration site-1 (EVI1), an oncogenic transcription factor, was observed in some patients with acute myeloid leukemia (AML) and was correlated with poor survival of patients [83]. The EVI1-positive AML subtype has a poor response to current treatment regimens [84]. By using a screen of pooled shRNAs, Fenouille and colleagues have demonstrated that CKMT1 contributes to survival of EVI1-expressing cells in EVI1-positive AML patients [68]. EVI1 has been suggested to promote CKMT1 expression by suppressing the myeloid differentiation regulator RUNX1. Inhibition of creatine metabolism by either CKMT1 depletion or using creatine analogue cyclocreatine specifically enhanced the cell cycle arrest and apoptotic cell death of EVI1-positive cancer cells, and increased animal survival in AML mouse syngeneic and xenograft models [68]. The blockade of CKMT1 impaired proliferative capacity, mitochondrial function, and ATP production in cancer cells, which could be rescued by exogenous PCr in vitro and in vivo [68]. Thus, these findings point to the therapeutic potential of targeting CKMT1 pathway for treating EVI1-positive AML.

Kurmi and colleagues have reported that oncogenic HER2 signaling induces CKMT1 phosphorylation via ABL tyrosine kinase in HER2 positive breast cancer cells [67]. CKMT1 phosphorylation enhanced the stabilization of CKMT1, promoting the PCr energy shuttle and breast cancer cell proliferation [67]. Blockade of the PCr-mediated metabolism by using either CKMT1 shRNA or the creatine analogue cyclocreatine decreased proliferation of HER2 positive cell lines in vitro and *in vivo*. The impaired proliferation could be rescued by PCr supplementation [67]. Although the HER2-directed monoclonal antibody trastuzumab offers significant clinical benefit selectively in HER2 positive breast cancer patients, intrinsic and acquired resistance to this therapy leads to no response for many patients [85–89]. Cyclocreatine combined with the HER2 kinase inhibitor lapatinib efficiently suppressed tumor growth in a HER2 positive trastuzumab-resistant patient-derived xenograft model [67]. Collectively, cyclocreatine can inhibit trastuzumab-resistant HER2 positive breast cancer cells via disrupting the PCr energy shuttle and improve the efficacy of existing breast cancer treatments.

Recently, Maguire et al. investigated the crosstalk between adipocytes and neoplastic cells in the tumor microenvironment and identified the upregulation of *Gatm* and the fatty acyl-CoA synthetase gene *Acsbg1* in adipocytes and in breast cancer cells, respectively [90]. Genetic inhibition of either *Gatm* in adipocytes or *CrT* in breast cancer cells attenuated tumor growth in obesity [90], suggesting that adipocyte-derived creatine is required for obesity-driven tumor progression. High *Acsbg1* expression in cancer cells enhanced ATP production through oxidative phosphorylation and uptake of adipocyte-derived creatine through CrT, which supported the production of PCr and drove tumor cell proliferation [90]. Although these findings revealed a protumoral role of creatine in regulating obesity-accelerated breast cancer cell proliferation, creatine supplementation did not promote tumor growth in lean or obese animals [90]. Thus, only adipocyte-derived creatine, not systemic creatine, is essential for tumor cell growth in this specific breast cancer model. Meanwhile, another group reported that creatine supplementation or GATM-mediated *de novo* synthesis of creatine promoted colorectal and breast cancer metastasis in orthotopic mouse models by increasing Snail and Slug expression through monopolar spindle 1 (MPS1)-activated Smad2 and Smad3 phosphorylation [91]. Moreover, GATM knockdown or MPS1 inhibition attenuated cancer metastasis and lower survival in mice. Notably, creatine supplementation suppressed primary tumor growth in mouse colon cancer (CT26) and mouse breast cancer (4T1) models [91]. Several notes of caution should be made in interpreting these results of cancer metastasis: (1) some studies were performed using severely immunocompromised mice (i.e., the NSG mice) and therefore failed to take into account the creatine regulation of antitumor immunity effect [44]; (2) creatine supplementation doses used in some studies were exceedingly high, way above the recommended safe dose range for humans [10,92]; and (3) long-term creatine supplementation has been indicated to induce species-specific liver inflammation in mice, which could contribute to promoting cancer metastasis in liver observed in these studies [93–95].

7. Conclusions

Recent studies identified creatine as an important metabolic regulator conserving bioenergy to power CD8 T cell antitumor immunity and suggested a potential application of creatine supplementation for cancer immunotherapy. The safety of long-term creatine supplementation in both healthy individuals and patients has been well documented, which provides a clear and expedient path forward for utilizing creatine supplementation to treat cancer [13,64,96–99]. Additionally, creatine supplementation can augment muscle and enhance strength, which may also benefit cancer patients suffering from cachexia at their late stages [63,64,100,101]. Both oral and intravenous administration routes can be effective in animal tumor models [44]. However, bioavailability of creatine through oral administration in humans is low because creatine is rapidly converted into creatinine in the high acidic environment of the stomach [102,103]. Thus, for the best cancer therapy benefits, optimal administration routes and dosing strategies for creatine clinical intervention still need to be further investigated.

Creatine supplementation synergizes with the anti-PD-1 therapy to yield superior antitumor efficacy [44], because creatine may activate tumor infiltrating CD8 T cells via an energy-buffering mechanism that is non-redundant to the mechanisms utilized by ICB therapies. Therefore, creatine supplementation has a potential to become an economical and effective treatment for enhancing ICB therapies. In addition, many other cancer therapies, including new T cell immunotherapies, traditional chemotherapy, targeted therapy, and radiation therapy, may also have improved therapeutic efficacy when combined with creatine supplementation treatment [31–34,67,104–106]. For the full translation to clinical applications, a speedy clinical development of creatine supplementation for combination therapies treating different types of cancer is urgently needed in the future.

Creatine-mediated energy buffering system is efficient and supports CD8 T cell antitumor activities in a metabolically challenging microenvironment via ATP/AMPK-mediated regulation of TCR signaling pathways [35,39,44,59]. The hyporesponsiveness of other immune cells in *CrT*-KO mice bearing tumors has been observed, indicating that the immune regulatory function of this energy system may go beyond modulating CD8 T cells in a tumor microenvironment. Creatine has been indicated to have anti-inflammatory properties in acute and chronic animal models of inflammation [107]. Creatine uptake can reprogram macrophage M1/M2 polarization by regulating IFN-γ and IL-4 cytokine responses partly in an ATP-dependent manner under infection conditions [108]. Whether and how other immune cells, such as regulatory T cells and tumor-associated macrophages, in the tumor microenvironment mediate antitumor effects of creatine will be interesting topics and certainly merit investigation. Additionally, creatine and creatine analogues can inhibit tumor cell survival likely through additional mechanisms that are independent of energy metabolism disruption. The regulatory mechanisms of creatine and creatine analogues in cancer cells warrant further investigation, especially in light of the recent studies on creatine promotion of cancer metastasis.

Author Contributions: Conceptualization, B.L. and L.Y.; writing—original draft preparation, B.L.; writing—review and editing, B.L. and L.Y.; funding acquisition, L.Y. Both authors have read and agreed to the published version of the manuscript.

Funding: The APC was funded by AlzChem, LLC. (Trostberg, Germany), a company that makes creatine. The funder had no role in the writing of the manuscript, interpretation of the literature, or in the decision to publish the results.

Institutional Review Board Statement: Not applicable.

Informed Consent Statement: Not applicable.

Data Availability Statement: Not applicable.

Acknowledgments: We thank Samuel Zeng for critical reading of this manuscript.

Conflicts of Interest: L.Y. is a scientific advisor to AlzChem, and is also an inventor on patents relating to creatine for immunotherapy filed by the University of California, Los Angeles (UCLA).

References

1. Wyss, M.; Kaddurah-Daouk, R. Creatine and creatinine metabolism. *Physiol. Rev.* **2000**, *80*, 1107–1213. [CrossRef]
2. Kazak, L.; Cohen, P. Creatine metabolism: Energy homeostasis, immunity and cancer biology. *Nat. Rev. Endocrinol.* **2020**, *16*, 421–436. [CrossRef] [PubMed]
3. Candow, D.G. Sarcopenia: Current theories and the potential beneficial effect of creatine application strategies. *Biogerontology* **2011**, *12*, 273–281. [CrossRef]
4. Shomrat, A.; Weinstein, Y.; Katz, A. Effect of creatine feeding on maximal exercise performance in vegetarians. *Eur. J. Appl. Physiol.* **2000**, *82*, 321–325. [CrossRef]
5. Delanghe, J.; De Slypere, J.P.; De Buyzere, M.; Robbrecht, J.; Wieme, R.; Vermeulen, A. Normal reference values for creatine, creatinine, and carnitine are lower in vegetarians. *Clin. Chem.* **1989**, *35*, 1802–1803. [CrossRef]
6. Bemben, M.G.; Witten, M.S.; Carter, J.M.; Eliot, K.A.; Knehans, A.W.; Bemben, D.A. The effects of supplementation with creatine and protein on muscle strength following a traditional resistance training program in middle-aged and older men. *J. Nutr. Health Aging* **2010**, *14*, 155–159. [CrossRef]
7. Wallimann, T.; Tokarska-Schlattner, M.; Schlattner, U. The creatine kinase system and pleiotropic effects of creatine. *Amino Acids* **2011**, *40*, 1271–1296. [CrossRef] [PubMed]
8. Gastin, P.B. Energy system interaction and relative contribution during maximal exercise. *Sports Med.* **2001**, *31*, 725–741. [CrossRef] [PubMed]
9. Riesberg, L.A.; Weed, S.A.; McDonald, T.L.; Eckerson, J.M.; Drescher, K.M. Beyond muscles: The untapped potential of creatine. *Int. Immunopharmacol.* **2016**, *37*, 31–42. [CrossRef]
10. Kreider, R.B. Effects of creatine supplementation on performance and training adaptations. *Mol. Cell. Biochem.* **2003**, *244*, 89–94. [CrossRef] [PubMed]

11. Peyrebrune, M.C.; Nevill, M.E.; Donaldson, F.J.; Cosford, D.J. The effects of oral creatine supplementation on performance in single and repeated sprint swimming. *J. Sports Sci.* **1998**, *16*, 271–279. [CrossRef]
12. Bemben, M.G.; Lamont, H.S. Creatine supplementation and exercise performance: Recent findings. *Sports Med.* **2005**, *35*, 107–125. [CrossRef]
13. Kreider, R.B.; Kalman, D.S.; Antonio, J.; Ziegenfuss, T.N.; Wildman, R.; Collins, R.; Candow, D.G.; Kleiner, S.M.; Almada, A.L.; Lopez, H.L. International Society of Sports Nutrition position stand: Safety and efficacy of creatine supplementation in exercise, sport, and medicine. *J. Int. Soc. Sports Nutr.* **2017**, *14*, 18. [CrossRef]
14. Karimian, J.; Esfahani, P.S. Supplement consumption in body builder athletes. *J. Res. Med Sci. Off. J. Isfahan Univ. Med. Sci.* **2011**, *16*, 1347–1353.
15. Burke, D.G.; Candow, D.G.; Chilibeck, P.D.; MacNeil, L.G.; Roy, B.D.; Tarnopolsky, M.A.; Ziegenfuss, T. Effect of creatine supplementation and resistance-exercise training on muscle insulin-like growth factor in young adults. *Int. J. Sport Nutr. Exerc. Metab.* **2008**, *18*, 389–398. [CrossRef] [PubMed]
16. Demant, T.W.; Rhodes, E.C. Effects of creatine supplementation on exercise performance. *Sports Med.* **1999**, *28*, 49–60. [CrossRef]
17. Calfee, R.; Fadale, P. Popular ergogenic drugs and supplements in young athletes. *Pediatrics* **2006**, *117*, e577–e589. [CrossRef] [PubMed]
18. Juhn, M.S.; Tarnopolsky, M. Potential side effects of oral creatine supplementation: A critical review. *Clin. J. Sport Med. Off. J. Can. Acad. Sport Med.* **1998**, *8*, 298–304. [CrossRef] [PubMed]
19. Hultman, E.; Soderlund, K.; Timmons, J.A.; Cederblad, G.; Greenhaff, P.L. Muscle creatine loading in men. *J. Appl. Physiol.* **1996**, *81*, 232–237. [CrossRef] [PubMed]
20. Braissant, O.; Henry, H.; Loup, M.; Eilers, B.; Bachmann, C. Endogenous synthesis and transport of creatine in the rat brain: An in situ hybridization study. *Brain Res. Mol. Brain Res.* **2001**, *86*, 193–201. [CrossRef]
21. Joncquel-Chevalier Curt, M.; Voicu, P.M.; Fontaine, M.; Dessein, A.F.; Porchet, N.; Mention-Mulliez, K.; Dobbelaere, D.; Soto-Ares, G.; Cheillan, D.; Vamecq, J. Creatine biosynthesis and transport in health and disease. *Biochimie* **2015**, *119*, 146–165. [CrossRef] [PubMed]
22. Sullivan, P.G.; Geiger, J.D.; Mattson, M.P.; Scheff, S.W. Dietary supplement creatine protects against traumatic brain injury. *Ann. Neurol.* **2000**, *48*, 723–729. [CrossRef]
23. Klivenyi, P.; Ferrante, R.J.; Matthews, R.T.; Bogdanov, M.B.; Klein, A.M.; Andreassen, O.A.; Mueller, G.; Wermer, M.; Kaddurah-Daouk, R.; Beal, M.F. Neuroprotective effects of creatine in a transgenic animal model of amyotrophic lateral sclerosis. *Nat. Med.* **1999**, *5*, 347–350. [CrossRef] [PubMed]
24. Ferrante, R.J.; Andreassen, O.A.; Jenkins, B.G.; Dedeoglu, A.; Kuemmerle, S.; Kubilus, J.K.; Kaddurah-Daouk, R.; Hersch, S.M.; Beal, M.F. Neuroprotective effects of creatine in a transgenic mouse model of Huntington's disease. *J. Neurosci. Off. J. Soc. Neurosci.* **2000**, *20*, 4389–4397. [CrossRef]
25. Matthews, R.T.; Ferrante, R.J.; Klivenyi, P.; Yang, L.; Klein, A.M.; Mueller, G.; Kaddurah-Daouk, R.; Beal, M.F. Creatine and cyclocreatine attenuate MPTP neurotoxicity. *Exp. Neurol.* **1999**, *157*, 142–149. [CrossRef] [PubMed]
26. Schulze, A. Creatine deficiency syndromes. *Mol. Cell. Biochem.* **2003**, *244*, 143–150. [CrossRef]
27. Nasrallah, F.; Feki, M.; Kaabachi, N. Creatine and creatine deficiency syndromes: Biochemical and clinical aspects. *Pediatric Neurol.* **2010**, *42*, 163–171. [CrossRef]
28. Skelton, M.R.; Schaefer, T.L.; Graham, D.L.; Degrauw, T.J.; Clark, J.F.; Williams, M.T.; Vorhees, C.V. Creatine transporter (CrT.; Slc6a8) knockout mice as a model of human CrT deficiency. *PLoS ONE* **2011**, *6*, e16187. [CrossRef]
29. Rousalova, I.; Krepela, E. Granzyme B-induced apoptosis in cancer cells and its regulation (review). *Int. J. Oncol.* **2010**, *37*, 1361–1378. [CrossRef] [PubMed]
30. Kayagaki, N.; Yamaguchi, N.; Nakayama, M.; Eto, H.; Okumura, K.; Yagita, H. Type I interferons (IFNs) regulate tumor necrosis factor-related apoptosis-inducing ligand (TRAIL) expression on human T cells: A novel mechanism for the antitumor effects of type I IFNs. *J. Exp. Med.* **1999**, *189*, 1451–1460. [CrossRef]
31. Lim, W.A.; June, C.H. The Principles of Engineering Immune Cells to Treat Cancer. *Cell* **2017**, *168*, 724–740. [CrossRef] [PubMed]
32. Baumeister, S.H.; Freeman, G.J.; Dranoff, G.; Sharpe, A.H. Coinhibitory Pathways in Immunotherapy for Cancer. *Annu. Rev. Immunol.* **2016**, *34*, 539–573. [CrossRef]
33. Rosenberg, S.A.; Restifo, N.P. Adoptive cell transfer as personalized immunotherapy for human cancer. *Science* **2015**, *348*, 62–68. [CrossRef] [PubMed]
34. Ribas, A. Releasing the Brakes on Cancer Immunotherapy. *N. Engl. J. Med.* **2015**, *373*, 1490–1492. [CrossRef]
35. Fox, C.J.; Hammerman, P.S.; Thompson, C.B. Fuel feeds function: Energy metabolism and the T-cell response. *Nat. Rev. Immunol.* **2005**, *5*, 844–852. [CrossRef]
36. Zeng, H.; Chi, H. mTOR signaling in the differentiation and function of regulatory and effector T cells. *Curr. Opin. Immunol.* **2017**, *46*, 103–111. [CrossRef]
37. O'Neill, L.A.; Kishton, R.J.; Rathmell, J. A guide to immunometabolism for immunologists. *Nat. Rev. Immunol.* **2016**, *16*, 553–565. [CrossRef] [PubMed]
38. McCarthy, S.A.; Mufson, R.A.; Pearce, E.J.; Rathmell, J.C.; Howcroft, T.K. Metabolic reprogramming of the immune response in the tumor microenvironment. *Cancer Biol. Ther.* **2013**, *14*, 315–318. [CrossRef]
39. Siska, P.J.; Rathmell, J.C. T cell metabolic fitness in antitumor immunity. *Trends Immunol.* **2015**, *36*, 257–264. [CrossRef]

40. Patel, C.H.; Powell, J.D. Targeting T cell metabolism to regulate T cell activation, differentiation and function in disease. *Curr. Opin. Immunol.* **2017**, *46*, 82–88. [CrossRef]

41. Kishton, R.J.; Sukumar, M.; Restifo, N.P. Metabolic Regulation of T Cell Longevity and Function in Tumor Immunotherapy. *Cell Metab.* **2017**, *26*, 94–109. [CrossRef]

42. Ho, P.C.; Kaech, S.M. Reenergizing T cell anti-tumor immunity by harnessing immunometabolic checkpoints and machineries. *Curr. Opin. Immunol.* **2017**, *46*, 38–44. [CrossRef]

43. Chang, C.H.; Qiu, J.; O'Sullivan, D.; Buck, M.D.; Noguchi, T.; Curtis, J.D.; Chen, Q.; Gindin, M.; Gubin, M.M.; van der Windt, G.J.; et al. Metabolic Competition in the Tumor Microenvironment Is a Driver of Cancer Progression. *Cell* **2015**, *162*, 1229–1241. [CrossRef]

44. Di Biase, S.; Ma, X.; Wang, X.; Yu, J.; Wang, Y.C.; Smith, D.J.; Zhou, Y.; Li, Z.; Kim, Y.J.; Clarke, N.; et al. Creatine uptake regulates CD8 T cell antitumor immunity. *J. Exp. Med.* **2019**, *216*, 2869–2882. [CrossRef]

45. Zhang, Y.; Li, H.; Wang, X.; Gao, X.; Liu, X. Regulation of T cell development and activation by creatine kinase B. *PLoS ONE* **2009**, *4*, e5000. [CrossRef] [PubMed]

46. Russell, A.P.; Ghobrial, L.; Wright, C.R.; Lamon, S.; Brown, E.L.; Kon, M.; Skelton, M.R.; Snow, R.J. Creatine transporter (SLC6A8) knockout mice display an increased capacity for in vitro creatine biosynthesis in skeletal muscle. *Front. Physiol.* **2014**, *5*, 314. [CrossRef]

47. Kazak, L.; Chouchani, E.T.; Lu, G.Z.; Jedrychowski, M.P.; Bare, C.J.; Mina, A.I.; Kumari, M.; Zhang, S.; Vuckovic, I.; Laznik-Bogoslavski, D.; et al. Genetic Depletion of Adipocyte Creatine Metabolism Inhibits Diet-Induced Thermogenesis and Drives Obesity. *Cell Metab.* **2017**, *26*, 693. [CrossRef] [PubMed]

48. da Silva, R.P.; Clow, K.; Brosnan, J.T.; Brosnan, M.E. Synthesis of guanidinoacetate and creatine from amino acids by rat pancreas. *Br. J. Nutr.* **2014**, *111*, 571–577. [CrossRef]

49. Li, Y.; Yin, Y.; Mariuzza, R.A. Structural and biophysical insights into the role of CD4 and CD8 in T cell activation. *Front. Immunol.* **2013**, *4*, 206. [CrossRef] [PubMed]

50. Ma, E.H.; Poffenberger, M.C.; Wong, A.H.; Jones, R.G. The role of AMPK in T cell metabolism and function. *Curr. Opin. Immunol.* **2017**, *46*, 45–52. [CrossRef] [PubMed]

51. Rao, E.; Zhang, Y.; Li, Q.; Hao, J.; Egilmez, N.K.; Suttles, J.; Li, B. AMPK-Dependent and independent effects of AICAR and compound C on T-cell responses. *Oncotarget* **2016**, *7*, 33783–33795. [CrossRef] [PubMed]

52. Hardie, D.G.; Ross, F.A.; Hawley, S.A. AMPK: A nutrient and energy sensor that maintains energy homeostasis. *Nat. Rev. Mol. Cell Biol.* **2012**, *13*, 251–262. [CrossRef]

53. Tamas, P.; Hawley, S.A.; Clarke, R.G.; Mustard, K.J.; Green, K.; Hardie, D.G.; Cantrell, D.A. Regulation of the energy sensor AMP activated protein kinase by antigen receptor and Ca2+ in T lymphocytes. *J. Exp. Med.* **2006**, *203*, 1665–1670. [CrossRef]

54. Son, J.; Cho, Y.W.; Woo, Y.J.; Baek, Y.A.; Kim, E.J.; Cho, Y.; Kim, J.Y.; Kim, B.S.; Song, J.J.; Ha, S.J. Metabolic Reprogramming by the Excessive AMPK Activation Exacerbates Antigen-Specific Memory CD8(+) T Cell Differentiation after Acute Lymphocytic Choriomeningitis Virus Infection. *Immune Netw.* **2019**, *19*, e11. [CrossRef] [PubMed]

55. Guo, C.; Chen, S.; Liu, W.; Ma, Y.; Li, J.; Fisher, P.B.; Fang, X.; Wang, X.Y. Immunometabolism: A new target for improving cancer immunotherapy. *Adv. Cancer Res.* **2019**, *143*, 195–253. [CrossRef] [PubMed]

56. Scharping, N.E.; Menk, A.V.; Moreci, R.S.; Whetstone, R.D.; Dadey, R.E.; Watkins, S.C.; Ferris, R.L.; Delgoffe, G.M. The Tumor Microenvironment Represses T Cell Mitochondrial Biogenesis to Drive Intratumoral T Cell Metabolic Insufficiency and Dysfunction. *Immunity* **2016**, *45*, 701–703. [CrossRef] [PubMed]

57. Bengsch, B.; Johnson, A.L.; Kurachi, M.; Odorizzi, P.M.; Pauken, K.E.; Attanasio, J.; Stelekati, E.; McLane, L.M.; Paley, M.A.; Delgoffe, G.M.; et al. Bioenergetic Insufficiencies Due to Metabolic Alterations Regulated by the Inhibitory Receptor PD-1 Are an Early Driver of CD8(+) T Cell Exhaustion. *Immunity* **2016**, *45*, 358–373. [CrossRef] [PubMed]

58. Gubin, M.M.; Zhang, X.; Schuster, H.; Caron, E.; Ward, J.P.; Noguchi, T.; Ivanova, Y.; Hundal, J.; Arthur, C.D.; Krebber, W.J.; et al. Checkpoint blockade cancer immunotherapy targets tumour-specific mutant antigens. *Nature* **2014**, *515*, 577–581. [CrossRef] [PubMed]

59. Wherry, E.J.; Kurachi, M. Molecular and cellular insights into T cell exhaustion. *Nat. Rev. Immunol.* **2015**, *15*, 486–499. [CrossRef] [PubMed]

60. Homet Moreno, B.; Zaretsky, J.M.; Garcia-Diaz, A.; Tsoi, J.; Parisi, G.; Robert, L.; Meeth, K.; Ndoye, A.; Bosenberg, M.; Weeraratna, A.T.; et al. Response to Programmed Cell Death-1 Blockade in a Murine Melanoma Syngeneic Model Requires Costimulation, CD4, and CD8 T Cells. *Cancer Immunol. Res.* **2016**, *4*, 845–857. [CrossRef]

61. Patra, S.; Ghosh, A.; Roy, S.S.; Bera, S.; Das, M.; Talukdar, D.; Ray, S.; Wallimann, T.; Ray, M. A short review on creatine-creatine kinase system in relation to cancer and some experimental results on creatine as adjuvant in cancer therapy. *Amino Acids* **2012**, *42*, 2319–2330. [CrossRef]

62. Pal, A.; Roy, A.; Ray, M. Creatine supplementation with methylglyoxal: A potent therapy for cancer in experimental models. *Amino Acids* **2016**, *48*, 2003–2013. [CrossRef]

63. van de Worp, W.; Schols, A.; Theys, J.; van Helvoort, A.; Langen, R.C.J. Nutritional Interventions in Cancer Cachexia: Evidence and Perspectives From Experimental Models. *Front. Nutr.* **2020**, *7*, 601329. [CrossRef]

64. Fairman, C.M.; Kendall, K.L.; Hart, N.H.; Taaffe, D.R.; Galvao, D.A.; Newton, R.U. The potential therapeutic effects of creatine supplementation on body composition and muscle function in cancer. *Crit. Rev. Oncol. Hematol.* **2019**, *133*, 46–57. [CrossRef]

65. Yoshida, G.J. Metabolic reprogramming: The emerging concept and associated therapeutic strategies. *J. Exp. Clin. Cancer Res.* **2015**, *34*, 111. [CrossRef]

66. Pavlova, N.N.; Thompson, C.B. The Emerging Hallmarks of Cancer Metabolism. *Cell Metab.* **2016**, *23*, 27–47. [CrossRef]

67. Kurmi, K.; Hitosugi, S.; Yu, J.; Boakye-Agyeman, F.; Wiese, E.K.; Larson, T.R.; Dai, Q.; Machida, Y.J.; Lou, Z.; Wang, L.; et al. Tyrosine Phosphorylation of Mitochondrial Creatine Kinase 1 Enhances a Druggable Tumor Energy Shuttle Pathway. *Cell Metab.* **2018**, *28*, 833–847.e838. [CrossRef] [PubMed]

68. Fenouille, N.; Bassil, C.F.; Ben-Sahra, I.; Benajiba, L.; Alexe, G.; Ramos, A.; Pikman, Y.; Conway, A.S.; Burgess, M.R.; Li, Q.; et al. The creatine kinase pathway is a metabolic vulnerability in EVI1-positive acute myeloid leukemia. *Nat. Med.* **2017**, *23*, 301–313. [CrossRef] [PubMed]

69. MacPherson, R.E.; Gamu, D.; Frendo-Cumbo, S.; Castellani, L.; Kwon, F.; Tupling, A.R.; Wright, D.C. Sarcolipin knockout mice fed a high-fat diet exhibit altered indices of adipose tissue inflammation and remodeling. *Obesity* **2016**, *24*, 1499–1505. [CrossRef]

70. Kristensen, C.A.; Askenasy, N.; Jain, R.K.; Koretsky, A.P. Creatine and cyclocreatine treatment of human colon adenocarcinoma xenografts: 31P and 1H magnetic resonance spectroscopic studies. *Br. J. Cancer* **1999**, *79*, 278–285. [CrossRef]

71. Miller, E.E.; Evans, A.E.; Cohn, M. Inhibition of rate of tumor growth by creatine and cyclocreatine. *Proc. Natl. Acad. Sci. USA* **1993**, *90*, 3304–3308. [CrossRef]

72. Lillie, J.W.; O'Keefe, M.; Valinski, H.; Hamlin, H.A., Jr.; Varban, M.L.; Kaddurah-Daouk, R. Cyclocreatine (1-carboxymethyl-2-iminoimidazolidine) inhibits growth of a broad spectrum of cancer cells derived from solid tumors. *Cancer Res.* **1993**, *53*, 3172–3178. [PubMed]

73. Annesley, T.M.; Walker, J.B. Cyclocreatine phosphate as a substitute for creatine phosphate in vertebrate tissues. Energistic considerations. *Biochem. Biophys. Res. Commun.* **1977**, *74*, 185–190. [CrossRef]

74. Soares, J.D.P.; Howell, S.L.; Teixeira, F.J.; Pimentel, G.D. Dietary Amino Acids and Immunonutrition Supplementation in Cancer-Induced Skeletal Muscle Mass Depletion: A Mini-Review. *Curr. Pharm. Des.* **2020**, *26*, 970–978. [CrossRef] [PubMed]

75. Campos-Ferraz, P.L.; Gualano, B.; das Neves, W.; Andrade, I.T.; Hangai, I.; Pereira, R.T.; Bezerra, R.N.; Deminice, R.; Seelaender, M.; Lancha, A.H. Exploratory studies of the potential anti-cancer effects of creatine. *Amino Acids* **2016**, *48*, 1993–2001. [CrossRef]

76. Qian, X.L.; Li, Y.Q.; Gu, F.; Liu, F.F.; Li, W.D.; Zhang, X.M.; Fu, L. Overexpression of ubiquitous mitochondrial creatine kinase (uMtCK) accelerates tumor growth by inhibiting apoptosis of breast cancer cells and is associated with a poor prognosis in breast cancer patients. *Biochem. Biophys. Res. Commun.* **2012**, *427*, 60–66. [CrossRef]

77. Amamoto, R.; Uchiumi, T.; Yagi, M.; Monji, K.; Song, Y.; Oda, Y.; Shiota, M.; Yokomizo, A.; Naito, S.; Kang, D. The Expression of Ubiquitous Mitochondrial Creatine Kinase Is Downregulated as Prostate Cancer Progression. *J. Cancer* **2016**, *7*, 50–59. [CrossRef]

78. Uranbileg, B.; Enooku, K.; Soroida, Y.; Ohkawa, R.; Kudo, Y.; Nakagawa, H.; Tateishi, R.; Yoshida, H.; Shinzawa, S.; Moriya, K.; et al. High ubiquitous mitochondrial creatine kinase expression in hepatocellular carcinoma denotes a poor prognosis with highly malignant potential. *Int. J. Cancer* **2014**, *134*, 2189–2198. [CrossRef]

79. Li, X.H.; Chen, X.J.; Ou, W.B.; Zhang, Q.; Lv, Z.R.; Zhan, Y.; Ma, L.; Huang, T.; Yan, Y.B.; Zhou, H.M. Knockdown of creatine kinase B inhibits ovarian cancer progression by decreasing glycolysis. *Int. J. Biochem. Cell Biol.* **2013**, *45*, 979–986. [CrossRef] [PubMed]

80. Gazdar, A.F.; Zweig, M.H.; Carney, D.N.; Van Steirteghen, A.C.; Baylin, S.B.; Minna, J.D. Levels of creatine kinase and its BB isoenzyme in lung cancer specimens and cultures. *Cancer Res.* **1981**, *41*, 2773–2777.

81. Feld, R.D.; Witte, D.L. Presence of creatine kinase BB isoenzyme in some patients with prostatic carcinoma. *Clin. Chem.* **1977**, *23*, 1930–1932. [CrossRef] [PubMed]

82. Loo, J.M.; Scherl, A.; Nguyen, A.; Man, F.Y.; Weinberg, E.; Zeng, Z.; Saltz, L.; Paty, P.B.; Tavazoie, S.F. Extracellular metabolic energetics can promote cancer progression. *Cell* **2015**, *160*, 393–406. [CrossRef]

83. Glass, C.; Wilson, M.; Gonzalez, R.; Zhang, Y.; Perkins, A.S. The role of EVI1 in myeloid malignancies. *Blood Cells Mol. Dis.* **2014**, *53*, 67–76. [CrossRef]

84. Groschel, S.; Lugthart, S.; Schlenk, R.F.; Valk, P.J.; Eiwen, K.; Goudswaard, C.; van Putten, W.J.; Kayser, S.; Verdonck, L.F.; Lubbert, M.; et al. High EVI1 expression predicts outcome in younger adult patients with acute myeloid leukemia and is associated with distinct cytogenetic abnormalities. *J. Clin. Oncol.* **2010**, *28*, 2101–2107. [CrossRef] [PubMed]

85. Rexer, B.N.; Arteaga, C.L. Intrinsic and acquired resistance to HER2-targeted therapies in HER2 gene-amplified breast cancer: Mechanisms and clinical implications. *Crit. Rev. Oncog.* **2012**, *17*, 1–16. [CrossRef]

86. Gale, M.; Li, Y.; Cao, J.; Liu, Z.Z.; Holmbeck, M.A.; Zhang, M.; Lang, S.M.; Wu, L.; Do Carmo, M.; Gupta, S.; et al. Acquired Resistance to HER2-Targeted Therapies Creates Vulnerability to ATP Synthase Inhibition. *Cancer Res.* **2020**, *80*, 524–535. [CrossRef] [PubMed]

87. Hunter, F.W.; Barker, H.R.; Lipert, B.; Rothe, F.; Gebhart, G.; Piccart-Gebhart, M.J.; Sotiriou, C.; Jamieson, S.M.F. Mechanisms of resistance to trastuzumab emtansine (T-DM1) in HER2-positive breast cancer. *Br. J. Cancer* **2020**, *122*, 603–612. [CrossRef]

88. Costa, R.L.B.; Czerniecki, B.J. Clinical development of immunotherapies for HER2(+) breast cancer: A review of HER2-directed monoclonal antibodies and beyond. *NPJ Breast Cancer* **2020**, *6*, 10. [CrossRef]

89. Gajria, D.; Chandarlapaty, S. HER2-Amplified breast cancer: Mechanisms of trastuzumab resistance and novel targeted therapies. *Expert Rev. Anticancer Ther.* **2011**, *11*, 263–275. [CrossRef] [PubMed]

90. Maguire, O.A.; Ackerman, S.E.; Szwed, S.K.; Maganti, A.V.; Marchildon, F.; Huang, X.; Kramer, D.J.; Rosas-Villegas, A.; Gelfer, R.G.; Turner, L.E.; et al. Creatine-Mediated crosstalk between adipocytes and cancer cells regulates obesity-driven breast cancer. *Cell Metab.* **2021**, *33*, 499–512.e496. [CrossRef]
91. Zhang, L.; Zhu, Z.; Yan, H.; Wang, W.; Wu, Z.; Zhang, F.; Zhang, Q.; Shi, G.; Du, J.; Cai, H.; et al. Creatine promotes cancer metastasis through activation of Smad2/3. *Cell Metab.* **2021**. [CrossRef]
92. Kreider, R.B.; Stout, J.R. Creatine in Health and Disease. *Nutrients* **2021**, *13*, 447. [CrossRef]
93. Potikha, T.; Stoyanov, E.; Pappo, O.; Frolov, A.; Mizrahi, L.; Olam, D.; Shnitzer-Perlman, T.; Weiss, I.; Barashi, N.; Peled, A.; et al. Interstrain differences in chronic hepatitis and tumor development in a murine model of inflammation-mediated hepatocarcinogenesis. *Hepatology* **2013**, *58*, 192–204. [CrossRef] [PubMed]
94. Brodt, P. Role of the Microenvironment in Liver Metastasis: From Pre- to Prometastatic Niches. *Clin. Cancer Res. Off. J. Am. Assoc. Cancer Res.* **2016**, *22*, 5971–5982. [CrossRef]
95. Tarnopolsky, M.A.; Bourgeois, J.M.; Snow, R.; Keys, S.; Roy, B.D.; Kwiecien, J.M.; Turnbull, J. Histological assessment of intermediate- and long-term creatine monohydrate supplementation in mice and rats. *Am. J. Physiol. Regul. Integr. Comp. Physiol.* **2003**, *285*, R762–R769. [CrossRef] [PubMed]
96. Smith, R.N.; Agharkar, A.S.; Gonzales, E.B. A review of creatine supplementation in age-related diseases: More than a supplement for athletes. *F1000Research* **2014**, *3*, 222. [CrossRef] [PubMed]
97. Kreider, R.B.; Melton, C.; Rasmussen, C.J.; Greenwood, M.; Lancaster, S.; Cantler, E.C.; Milnor, P.; Almada, A.L. Long-Term creatine supplementation does not significantly affect clinical markers of health in athletes. *Mol. Cell. Biochem.* **2003**, *244*, 95–104. [CrossRef]
98. Kim, H.J.; Kim, C.K.; Carpentier, A.; Poortmans, J.R. Studies on the safety of creatine supplementation. *Amino Acids* **2011**, *40*, 1409–1418. [CrossRef]
99. Jatoi, A.; Steen, P.D.; Atherton, P.J.; Moore, D.F.; Rowland, K.M.; Le-Lindqwister, N.A.; Adonizio, C.S.; Jaslowski, A.J.; Sloan, J.; Loprinzi, C. A double-blind, placebo-controlled randomized trial of creatine for the cancer anorexia/weight loss syndrome (N02C4): An Alliance trial. *Ann. Oncol.* **2017**, *28*, 1957–1963. [CrossRef]
100. Sakkas, G.K.; Schambelan, M.; Mulligan, K. Can the use of creatine supplementation attenuate muscle loss in cachexia and wasting? *Curr. Opin. Clin. Nutr. Metab. Care* **2009**, *12*, 623–627. [CrossRef]
101. de Campos-Ferraz, P.L.; Andrade, I.; das Neves, W.; Hangai, I.; Alves, C.R.; Lancha, A.H., Jr. An overview of amines as nutritional supplements to counteract cancer cachexia. *J. Cachexia Sarcopenia Muscle* **2014**, *5*, 105–110. [CrossRef] [PubMed]
102. Gufford, B.T.; Ezell, E.L.; Robinson, D.H.; Miller, D.W.; Miller, N.J.; Gu, X.; Vennerstrom, J.L. pH-Dependent stability of creatine ethyl ester: Relevance to oral absorption. *J. Diet. Suppl.* **2013**, *10*, 241–251. [CrossRef] [PubMed]
103. Gufford, D.T.; Sriraghavan, K.; Miller, N.J.; Miller, D.W.; Gu, X.; Vennerstrom, J.L.; Robinson, D.H. Physicochemical characterization of creatine N-methylguanidinium salts. *J. Diet. Suppl.* **2010**, *7*, 240–252. [CrossRef]
104. Page, D.B.; Postow, M.A.; Callahan, M.K.; Allison, J.P.; Wolchok, J.D. Immune modulation in cancer with antibodies. *Annu. Rev. Med.* **2014**, *65*, 185–202. [CrossRef]
105. Couzin-Frankel, J. Breakthrough of the year 2013. Cancer immunotherapy. *Science* **2013**, *342*, 1432–1433. [CrossRef]
106. Pardoll, D.M. The blockade of immune checkpoints in cancer immunotherapy. *Nat. Rev. Cancer* **2012**, *12*, 252–264. [CrossRef]
107. Bredahl, E.C.; Eckerson, J.M.; Tracy, S.M.; McDonald, T.L.; Drescher, K.M. The Role of Creatine in the Development and Activation of Immune Responses. *Nutrients* **2021**, *13*, 751. [CrossRef]
108. Ji, L.; Zhao, X.; Zhang, B.; Kang, L.; Song, W.; Zhao, B.; Xie, W.; Chen, L.; Hu, X. Slc6a8-Mediated Creatine Uptake and Accumulation Reprogram Macrophage Polarization via Regulating Cytokine Responses. *Immunity* **2019**, *51*, 272–284.e277. [CrossRef]

 *nutrients*

Review

Role of Creatine in the Heart: Health and Disease

Maurizio Balestrino [1,2]

1 Dipartimento di Neuroscienze, Riabilitazione, Oftalmologia, Genetica e Scienze Materno-Infantili (DINOGMI), University of Genoa, Largo Daneo 3, 16132 Genova, Italy; mbalestrino@neurologia.unige.it; Tel.: +39-010-353-7040

2 IRCCS Ospedale Policlinico San Martino, Largo Rosanna Benzi 10, 16132 Genova, Italy

Abstract: Creatine is a key player in heart contraction and energy metabolism. Creatine supplementation (throughout the paper, only supplementation with creatine monohydrate will be reviewed, as this is by far the most used and best-known way of supplementing creatine) increases creatine content even in the normal heart, and it is generally safe. In heart failure, creatine and phosphocreatine decrease because of decreased expression of the creatine transporter, and because phosphocreatine degrades to prevent adenosine triphosphate (ATP) exhaustion. This causes decreased contractility reserve of the myocardium and correlates with left ventricular ejection fraction, and it is a predictor of mortality. Thus, there is a strong rationale to supplement with creatine the failing heart. Pending additional trials, creatine supplementation in heart failure may be useful given data showing its effectiveness (1) against specific parameters of heart failure, and (2) against the decrease in muscle strength and endurance of heart failure patients. In heart ischemia, the majority of trials used phosphocreatine, whose mechanism of action is mostly unrelated to changes in the ergogenic creatine-phosphocreatine system. Nevertheless, preliminary data with creatine supplementation are encouraging, and warrant additional studies. Prevention of cardiac toxicity of the chemotherapy compounds anthracyclines is a novel field where creatine supplementation may also be useful. Creatine effectiveness in this case may be because anthracyclines reduce expression of the creatine transporter, and because of the pleiotropic antioxidant properties of creatine. Moreover, creatine may also reduce concomitant muscle damage by anthracyclines.

Keywords: phosphocreatine; creatine transporter; supplementation; treatment; heart; heart failure; ischemia; myocardial infarction; anthracycline; cardiac toxicity

Citation: Balestrino, M. Role of Creatine in the Heart: Health and Disease. *Nutrients* **2021**, *13*, 1215. https://doi.org/10.3390/nu13041215

Academic Editors: Richard B. Kreider and Jeffrey R. Stout

Received: 9 March 2021
Accepted: 3 April 2021
Published: 7 April 2021

Publisher's Note: MDPI stays neutral with regard to jurisdictional claims in published maps and institutional affiliations.

1. Metabolism and Role of Creatine

1.1. Functions of Creatine

Creatine plays a key role in cellular energy metabolism. The creatine kinase enzyme reversibly phosphorylates it to phosphocreatine. Then, when phosphocreatine is reverted to creatine, its phosphate bond breaks, and such a break provides enough energy to allow phosphorylating a molecule of adenosine diphosphate (ADP) to adenosine triphosphate (ATP). Thus, phosphocreatine acts as an energy reserve to synthesize ATP rapidly, with no need for oxygen. The reaction is the following one:

$$\text{Creatine} + \text{ATP} \rightleftharpoons \text{Phosphocreatine} + \text{ADP} + \text{H}^+$$

This reaction plays a crucial role in heart contraction [1]. Its roles have been reviewed elsewhere [2] and are, in summary:

(1) Transfer of ATP from its production site (mitochondria) to its place of exploitation (neuronal membrane or cytoplasm). This function is often called "the ATP shuttle". To carry out this transport ("shuttle") of ATP energy, creatine first receives the phosphate from ATP near the mitochondria, becoming phosphocreatine. It then diffuses through its concentration gradient to the periphery of the cell. Near cytoplasmic ATPase, it donates its phosphate

to ADP, effectively forming ATP far away from the mitochondria and delivering it precisely where and when it is required. In doing so, it reverts to creatine and diffuses back, again along its concentration gradient, to the mitochondrion to start the cycle again. The reason why the cell needs this complex mechanism to transport energy between mitochondria and cytoplasmic ATPase is that ATP is a very large molecule, therefore its diffusion through the organelle-filled cytoplasm is slow and cumbersome. By contrast, phosphocreatine is a much smaller molecule, thus it diffuses more quickly through the cytoplasm.

(2) Restoration by phosphocreatine of ATP concentration in conditions of increased energy demand and in diseases involving a reduced supply of blood or oxygen. In the first scenario, the consumption of ATP is excessive compared to the ability of the cell to synthesize it. For instance, a muscle exposed to a particularly intense effort quickly uses more ATP than it can produce, thus exhausting its reserve. In the second scenario, an organ cannot produce enough ATP because of a blood deficiency (ischemia) or an oxygen deficiency (anoxia). For example, in case of a myocardial infarction, phosphocreatine intervenes by transferring its phosphoric group to ADP, to provide ATP at a time when the heart cannot synthesize it due to ischemia.

Among all the biochemical reactions that our cells use to synthesize ATP, the one that starts from the creatine/phosphocreatine system is the quickest in buffering ATP levels at times of increased energy expenditure [3]. This explains the researchers' interest in this molecule, whose administration has been proposed in various conditions, both physiological and pathological [4,5].

1.2. Procurement of Creatine by the Organism, with Specific Reference to the Heart

Creatine is normally degraded to creatinine [6], leading to a steady depletion of the body creatine store. The creatine store is replenished partly from endogenous synthesis and partly by ingesting creatine with food [7].

1.2.1. Endogenous Synthesis of Creatine

In the body as a whole, creatine is synthesized in the kidney and in the liver [6]. Specifically, the kidney accomplishes the first step of the synthesis, forming guanidinoacetic acid from arginine and glycine. Guanidinoacetic acid is then transported to the liver, where it is converted into creatine with the intervention of the methyl donor S-adenosyl-methionine [6]. Specific organs can additionally synthesize creatine for their own consumption, such as the brain [8] and the testis [9,10]. Concerning the heart, it is generally believed that cardiomyocytes cannot synthesize creatine [11,12]. However, there is evidence that such synthesis may actually occur, i.e., cardiomyocytes may indeed synthesize creatine just as other organs do. In fact, the heart expresses the first enzyme of creatine synthesis, arginine-glycine amidino transferase (AGAT). Its expression in the heart is comparable to that of other tissues under basal conditions and increases several fold in pathological states [13]. Moreover, experiments showed that the addition of arginine (a precursor of creatine) to the incubation medium of toad hearts mitigated the decrease in creatine upon in vitro incubation, as if arginine was metabolized to creatine [14]. Furthermore, the addition of arginine to isolated rabbit hearts caused an increase in their content of creatinine (the product of creatine cyclization) [15]. Both these latter experiments strongly suggested that in the isolated heart arginine is indeed converted into creatine, as it is in other organs [16,17]. However, the possible creatine synthesis by the heart has received so far very little attention, and additional research is definitely warranted [18].

1.2.2. Uptake of Creatine from Dietary or Supplement Sources

About half of the creatine that the organism needs is normally ingested and taken up from dietary sources [19]. Creatine is not present in vegetables, but it is only present in foods of animal origin [4]. Thus, subjects who do not regularly consume meat or fish tend to have some degree of creatine deficiency, and should supplement their diet with it [4].

Moreover, exogenous supplementation of creatine permits administering high amounts of this compound, and increasing its content above normal levels [20–22]. Throughout the paper, only supplementation with creatine monohydrate will be reviewed, as this is by far the most used and best-known way of supplementing creatine. When administered in this way at adequate doses, creatine is stored in the tissues, where it increases the intracellular pool of both creatine and phosphocreatine. Such an increase is especially relevant for the muscular tissue. Creatine supplementation allows the muscles to contract more powerfully and to a longer extent [23], an effect that is exploited by athletes to improve their performance [24,25].

The metabolism and functions of creatine in the heart is similar to that in the muscle and in other tissues. Figure 1 provides a summary of the metabolism of creatine in the myocardium.

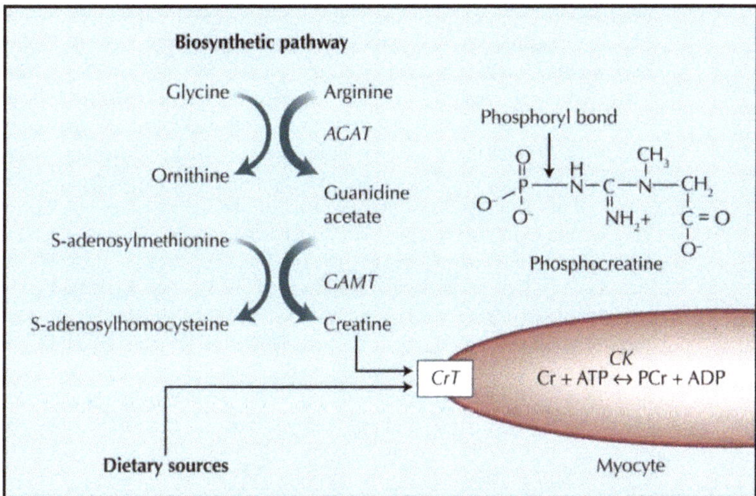

Figure 1. The primary sources for blood-borne creatine (Cr) are diet (meat) and a two-step biosynthesis that occurs primarily in the kidney, liver, and pancreas. Cr, a β-amino acid, is made by the transfer of glycine onto the arginine side chain catalyzed by arginine:glycine amidinotransferase (AGAT) to form guanidinoacetate. The methyl group is transferred to the guanidino group via guanidinoacetate methyltransferase (GAMT). Cr accumulates in muscles and brain through the action of the Cr transporter (CrT) in the sarcolemma. Cr is trapped by phosphorylation to phosphocreatine (PCr, see structure) by creatine kinase (CK). ADP—adenosine diphosphate; ATP—adenosine triphosphate; CrP—creatine phosphate. Reprinted by permission from Springer Nature Customer Service Centre GmbH:Springer Nature, Current Hypertension Reports (On the hypothesis that the failing heart is energy starved: Lessons learned from the metabolism of ATP and creatine, Joanne S. Ingwall) Copyright Springer Nature Customer Service Centre GmbH, 2006.

Specific mechanisms of the benefit provided by creatine supplementation include:

1. Restoration of normal creatine content when it is lower than normal due to lifestyle (e.g., vegetarian or vegan subjects [26]) or to disease (e.g., heart failure, see below).
2. Increase in energy availability (obtained by increasing phosphocreatine concentration in the tissue) in cases where the balance between energy availability and requirement is limited by decreased energy production (as is the case in hypoxia or ischemia), or by increased demand (e.g., the muscle of athletes during athletic performance).

Finally, we should note that creatine by itself does not enter cells, instead it needs a specific transporter to cross plasma membranes [16]. The same happens in the heart, where the creatine transporter is present on the plasma membrane of the myocytes and is necessary for creatine to enter myocardial cells [27,28].

2. Cardiac Effects of Creatine Supplementation in Healthy Subjects

2.1. Cardiac Effects of Creatine Supplementation in the Normal Heart

2.1.1. In Vitro Studies

Before discussing the effects of creatine supplementation in normal subjects, we shall mention some effects of creatine supplementation in in vitro normal heart preparations.

In heart strips prepared from frog ventricles, the force of contraction was increased when the preparation was perfused with a rather high concentration of creatine (9.2 mM) [29]. In an in vitro frog ventricular preparation, perfusion with rather high creatine concentrations (10 and 20 mM) reversibly caused, among others, increased force of contraction [30]. Interestingly, in both papers force of contraction was, by contrast, decreased when the in vitro preparation was perfused with much higher creatine concentrations (20–70 mM) [29,30], thus suggesting harmful effects by very high intracellular creatine increase. Santacruz et al. [31] reported increased content of phosphocreatine in in vitro cardiomyocytes upon creatine supplementation. Interestingly, these authors also reported such an increase in hypoxic conditions as well, although hypoxia decreased creatine uptake. Kilian et al. [32] used creatine supplementation in in vitro isolated hearts. They found that such a supplementation decreased heart rate, increased left ventricular systolic pressure, increased coronary flow, increased ATP content and decreased isocitrate dehydrogenase, a marker of cell death. Besides these positive findings, the authors add a puzzling statement (contained in the abstract and in the discussion sections, not in the methods nor in the results ones) saying that "when glucose supply was limited, conduction abnormalities occurred at a greater frequency in creatine-supplemented hearts as compared with the control group". Surprisingly, there are no data presented to support this sentence, so the value of this statement is unclear. Summing up, these in vitro data suggest that supplementation with creatine of the normal heart may improve some physiologically relevant parameters, like force of contraction and coronary flow. However, they suggest that very high increase in creatine content may cause decrease in the force of contraction. We should emphasize that such harmful effects occurred only when the in vitro hearts were perfused with creatine concentrations extremely high, probably impossible to obtain in vivo (20–70 mM).

2.1.2. In Vivo Studies

The normal heart has a high creatine content in comparison to other organs, as all excitable tissues have. For example, Horn et al. [33] reported in rats an average content of creatine about 80 nmol/mg protein in heart and brain, compared to about 20 nmol/mg protein in the liver and 25 nmol/mg protein in the kidney. Ipsiroglu et al. [21] found, in guinea pigs, a total creatine content (mean $\pm$ standard deviation) of 12.9 ± 0.10 μmol/g weight in the heart, 10.1 ± 0.45 μmol/g weight in brain, 7.5 ± 0.47 μmol/g weight in the liver and 5 ± 0.60 μmol/g weight in the kidney. Corresponding levels varied slightly in mouse and rat, but the proportions remained unchanged [21]. Human heart has an even higher creatine content. Neubauer et al. [34] confirmed a creatine content in the rat heart of 87.5 ± 4.2 nmol/mg protein, but found a content of 136.4 ± 6.1 nmol/mg protein in humans. The latter finding was confirmed by Nascimben et al. [35], who reported a creatine content of 131 ± 28 nmol/mg protein in human hearts. Such a high creatine content leads to downregulation of the creatine transporter and, it was stated, to little possibility of further increasing creatine content [12]. In agreement with this view, two rodent studies by a single laboratory concluded that creatine supplementation is unable to increase heart creatine content in the normal heart [33,36]. These studied concluded that a different strategy is needed to augment heart creatine, namely use of creatine derivatives that may enter cardiac myocytes with no need of the creatine transporter. It should be noted, however, that in neither of the two papers [33,36] creatine was increased in the muscles, a tissue where, by contrast, creatine is usually reported to increase upon adequate supplementation in both humans and animals [20,37,38]. Thus, the fact that in those two papers creatine did not increase not only in the heart but also in the muscle strongly suggests that the amount of creatine used in those studies [33,36] may have been insufficient. Accordingly, they reported creatine supplementation not as absolute daily amount (grams/day) but as percent of creatine in the total feed. Since the authors did not quantify the amount of food consumed daily, it is impossible to judge how large (or how insufficient) such supplementation might have been. By contrast, other authors reported increase in creatine content in normal hearts supplemented with creatine. Boehm et al. [39] found an 11% increased total creatine (i.e., creatine + phosphocreatine) in the hearts of animals whose feed was supplemented with 3% creatine. In this latter paper, ATP content was not affected by creatine supplementation, strongly suggesting that the creatine-phosphocreatine system created a reserve energy that might have possibly been used at times of increased energy demand or insufficient energy production. Ipsiroglu et al. [21] also found an increase in the creatine content of normal rodents' heart upon creatine supplementation, and emphasized that such increase required a 4-week supplementation to become statistically significant. The increase reported by Ipsiroglu et al. [21] was, after 4 weeks of supplementation, 15% in the guinea pig, 17% in the mouse, 28% in the rat. Summing up, these studies suggest that increase in creatine upon its supplementation in the normal heart is possible, although one should administer high doses for several weeks, possibly because the normal heart already contains a large amount of creatine.

A different approach to experimentally increasing creatine in the normal heart involved genetically modified rodents where the creatine transporter was overexpressed, thus increasing intracellular creatine. The research group that first used this approach to increase myocardial creatine reported that increasing creatine transporter expression did cause increased intracellular creatine up to four times its basal value. However, they unexpectedly found that such increase in creatine content in normal hearts was attended by left ventricular hypertrophy and heart failure [40,41]. Those authors tentatively explained such apparent paradox with the fact that a very large increase in intracellular creatine required large amounts of ATP to phosphorylate creatine, and that this process caused a large decrease in intracellular ATP that, those authors argued, ended up by being detrimental to the heart [40,41]. Moreover, they found that such an increase in intracellular creatine caused downregulation of the cellular enzyme enolase, which supposedly decreased glycolytic activity of the cardiomyocytes [40]. However, a different group later used the same transporter over-expression to obtain a large increase in intracellular creatine, but did not find any adverse effect of this increase [42]. In an interesting study, Zervou et al. [43] compared wildtype mice, mice overexpressing the creatine transporter that showed a moderate increase in creatine, and mice overexpressing the transporter that showed a large creatine increase (average creatine content was 81 nmol/mg protein in wildtype animals, 123 and 220 in the other two groups, respectively). They found harmful effects only in the group with the maximal creatine increase, not in the group with the moderate creatine increase [43]. Furthermore, Lygate et al. [11] reported no adverse effect by overexpressing creatine transporter with an only moderate increase in creatine content. These data suggest that harmful effects on cardiac performance may unexpectedly accompany a very large creatine increase, but that this risk seems not to exist at all when creatine increases only by about 60% of its baseline level [43]. It is worth noting that in the human skeletal muscle creatine supplementation causes an increase in creatine content that does not exceed 50% of baseline value [20], and that a "ceiling" is eventually reached preventing further creatine uptake [7].

2.2. Considerations on the Effects and Safety of Creatine Supplementation in Healthy Subjects

The above data suggest that increase in creatine content of normal hearts is possible upon creatine supplementation. As reported above, in vitro data suggest improvement in cardiac function by creatine supplementation even in normal cardiac preparations. Such improvement has not been reported in vivo in healthy subjects, however the above reviewed in vitro data suggest that creatine supplementation might possibly be useful to healthy subjects as well. Moreover, in vivo data suggest that creatine supplementation is safe in normal subjects. In fact, although animal data suggest that harmful effects on cardiac performance may unexpectedly accompany a very large creatine increase, this risk seems not to exist at all when creatine increases only by about 60% of its baseline level [43]. Therefore, there is little reason for concern about dietary supplementation of creatine. In fact, although the extent of creatine increase in normal humans upon supplementation has not been measured, in rodent models creatine supplementation increased creatine content in the heart by 11% in the study of Boehm et al. [39] and by 15–28% in diverse species in the study by Ipsiroglu et al. [21].

Some confirmation of the cardiac safety of creatine supplementation in normal subjects comes from an interesting paper that studied the cardiac effects of creatine supplementation in bodybuilders [44]. Those authors found that creatine supplementation only had the effect of slightly reducing the bradycardia that bodybuilders experienced because of their training. Specifically, they studied 16 controls (not body-builders nor creatine-supplementing subjects), 16 body-builders who did not use creatine supplementation and 16 body builders who regularly supplemented their diet with creatine (range 3.5–15.0 mg/day). They found that the resting heart rate was (beats/min, mean ± standard deviation) 71.5 ± 12.6 in controls; 61.8 ± 6.8 in body-builders who did not use creatine supplementation; and 69.63 ± 14.1 in body-builders who used to supplement their diet with creatine. This difference was statistically significant ($p = 0.048$). By contrast, systolic and diastolic blood pressure, interval of the QT segment of the electrocardiogram (both raw and corrected using the Bazett's formula) did not differ between groups. Although the mechanism of the different heart rate is unclear, it is certainly not a harmful effect, given that the typical resting heart rate for adults is between 50 and 90 beats per minute [45]. Although the value of such observation is limited to bodybuilders, this finding further suggests that creatine supplementation in normal subjects is safe from the cardiac point of view. Furthermore, a review of the literature [4] found no risk of, among else, cardiac adverse events by creatine supplementation.

3. Heart Diseases Where Creatine Supplementation May Be Useful

Heart failure, heart ischemia and anthracycline cardiotoxicity are the disease conditions of the heart where creatine supplementation has been proposed and investigated [12]. Individual reviews of the creatine role in each of these conditions follow.

4. Creatine Supplementation in Heart Failure

The European Society of Cardiology [46] defines heart failure as "a clinical syndrome characterized by typical symptoms (e.g., breathlessness, ankle swelling and fatigue) that may be accompanied by signs (e.g., elevated jugular venous pressure, pulmonary crackles and peripheral oedema) caused by a structural and/or functional cardiac abnormality, resulting in a reduced cardiac output and/or elevated intracardiac pressures at rest or during stress". One of the main consequences of heart failure is thus the inability of the heart to pump blood to an extent that is adequate to support body functions. Therefore, heart failure is a serious condition, for which usually there is no cure.

4.1. Decrease in Creatine in Heart Failure

This section will report numerous data across a long period showing that creatine, phosphocreatine, or both decrease in heart failure.

In preclinical research, Feinstein [47] demonstrated decrease in phosphocreatine, total creatine (i.e., creatine + phosphocreatine) and adenosine triphosphate (ATP) in the hearts of guinea pigs subjected to various experimental conditions which affect the performance of the heart in vivo. The latter conditions included experimental congestive heart failure, acute asphyxia, and ouabain treatment. This author concluded that decreased rate of synthesis of high-energy compounds is the most important determinant of heart failure. In agreement with this hypothesis, it was much later discovered that one of the main roles of creatine is allowing fast re-synthesis of ATP at the sites of its utilization [2]. A few years later, Fox et al. [48] showed that in dogs affected by chronic heart failure due to experimental pulmonary arterial stenosis, phosphocreatine and total creatine were decreased by approximately 33–43% compared to controls, while ATP was reduced to a lesser extent (about 12%). Still in in vivo dogs, Shen et al. [49] studied experimental heart failure due to chronic pacing of the right ventricle, finding that both ATP and creatine were decreased, and that creatine decreased at an earlier time compared to ATP. Those authors quite reasonably hypothesized that in the early stages of heart failure phosphocreatine was used to replenish the ATP stock, thus slowing ATP decrease. More recently, Ten Hove et al. [50] confirmed a decreased creatine content in experimental rat heart failure and attributed it to the concurrent decrease in creatine transporter that they also found.

As said, the decrease in creatine content of the failing heart is thought to be due to the down regulation of the creatine transporter that was demonstrated in the failing hearts of both rodents and humans [34,50]. However, it is noteworthy that while in the failing heart ATP levels are stable or moderately decreased, phosphocreatine levels are usually decreased to a much larger extent, with a corresponding reduction in the phosphocreatine/ATP ratio [12]. Apparently, this indicates a discrepancy between ATP synthesis in the mitochondrion and ATP requirements in the cytoplasm. Thus, it is very likely that phosphocreatine is used to rapidly re-synthesize ATP, thus slowing ATP decrease in the failing heart. This is an additional mechanism of phosphocreatine depletion in heart failure, besides decreased creatine transporter expression.

Summarizing these preclinical studies, they suggest that in the failing heart the content of creatine decreases because of (1) a decrease in its uptake due to down-regulation of the creatine transporter and (2) a consumption of phosphocreatine, which is used to prevent or delay exhaustion of ATP.

Moving to studies in humans affected by heart failure, Nascimben et al. [35] found that both creatine kinase activity and creatine decreased in heart failure patients, and they suggested that this decrease impaired the ability of cardiomyocytes to rapidly provide energy to the systems that required it. Neubauer et al. [34] confirmed the decrease in creatine content in both human patients and in an experimental rodent model of heart failure due to coronary artery ligation. Furthermore, they found in humans a concurrent decrease in creatine transporter, which they concluded was the cause of the creatine decrease. Winter et al. [51] found with magnetic resonance spectroscopy a decrease in total creatine in patients affected by cardiac failure not due to ischemia. Neubauer et al. [52] used in vivo 31P-MR spectroscopy to investigate the phosphocreatine/ATP ratio in human volunteers and in heart failure patients. They found not only that heart failure patients had a lower average ratio than controls, but that in individual patients such decrease was a statistically significant predictor of mortality. These findings were later confirmed by a different group using in vivo proton magnetic resonance spectroscopy [53]. In a later research, the same group not only further confirmed the decreased creatine content of the failing human heart due to a large variety of causes, but reported a positive correlation between myocardial creatine content and left ventricular ejection fraction [54]. These data confirmed the preclinical ones, and led to the theory that (1) heart failure is caused by decreased energy availability, and that (2) one of the possible strategies to counter it might have been reversing and normalizing the decreased phosphocreatine content of failing hearts [40,55].

4.2. Effects of Decreasing Creatine on Cardiac Function

Indeed, decreasing heart creatine per se has harmful effects on contractility. Saks et al. [29] showed in frog hearts that decreasing cardiac creatine content caused decreased force of contraction. Ten Hove et al. [56] developed a strategy to decrease the intracellular content of creatine in the rodents' hearts. They found that these hearts did not show significant anomalies at rest, but they had decreased contractile capacity when challenged with a sympathomimetic compound. In other words, they had a decreased contractility reserve, and they could not efficiently increase cardiac output when stimulated. Moreover, they proved more vulnerable to ischemic damage. Kapelko et al. [57] showed that isolated rat hearts which had been depleted of creatine by treatment with guanidinopropionic acid (an antagonist of the creatine transporter) had near-normal cardiac output when subjected to a submaximal pressure load, but showed a 43% decrease in pressure-volume work at maximal pressure load. Both these latter papers suggested that decreased heart creatine content did not have major effects at rest or at low levels of stimulation, but prevented increased cardiac output at times of higher need for contractility.

Field [58] proposed the interesting observation that, since available evidence does not show that in heart failure creatine decreases below the Km of the creatine kinase enzyme, its content is usually sufficient to maintain creatine metabolism at efficiency level. Although interesting, this opinion conflicts with the above mentioned findings that in later years would have demonstrated that the decrease in creatine in heart failure is indeed clinically relevant, being a predictor of mortality in individual patients [52,53] and correlating with left ventricular ejection fraction [54].

The above data show that (1) decreased creatine content of failing hearts and (2) its correlation with decreased contractility strength are both robust findings, and provide a rationale to investigate the effects of creatine supplementation in heart failure patients.

4.3. Effects of Creatine Supplementation in Heart Failure Patients

Creatine supplementation of the failing heart aims to normalize the decreased creatine content that, as reported above, is known to occur in this condition. The effects of creatine supplementation in heart failure have been investigated in a few trials, either preclinical or clinical.

At the preclinical level, Faller et al. [59] used, within a study on the effects of ribose supplementation in the failing heart, mice that overexpressed the creatine transporter. They selected those mice that showed only a moderate increase in creatine content, subjected them to coronary artery ligation surgery to induce chronic myocardial infarction and supplemented their diet with ribose. They found that this treatment did not improve cardiac function. Although relevant, this study did not use creatine supplementation but over-expression of the creatine transporter. Furthermore, it involved the administration of ribose, the effects of which might possibly have interfered with the creatine increase.

At the clinical level, Fumagalli et al. [60] investigated the effects of supplementation with both coenzyme Q10 and creatine (320 and 340 mg daily, respectively, for 8 weeks) in a randomized, placebo-controlled trial [61] on heart failure patients in the New York Heart Association functional class II to III. They found in the treated group a higher peak oxygen consumption, with no adverse effects. In this trial, the rather low dose of creatine that was administered (340 mg daily) is noteworthy. This finding raised interest, and it was suggested that the effect of the treatment might have been improving the function not of the myocardium but of the skeletal muscle. Carvalho et al. [62] investigated the effects of creatine supplementation (5 g/day for 6 months) in humans' heart failure (New York Heart Association functional class II to IV) using a randomized, placebo-controlled design. They did not find any effect on the various parameters they explored, but, interestingly, they found only in the creatine-treated group a significant positive correlation between peak oxygen consumption and the distance covered in the six-minute walk test. In principle, this result indicates a more efficient oxygen utilization only in the creatine-treated group.

Moreover, some studies showed that in heart failure patients creatine supplementation is able to improve muscle performance, leading to a global functional improvement. Gordon et al. [63] found in a double-blind, placebo-controlled study that creatine supplementation improved muscle strength and endurance in heart failure patients. Andrews et al. [64] demonstrated in a placebo-controlled study that creatine supplementation (20 g/day for 5 days) significantly improves muscle function. Specifically, they reported that creatine increased muscle endurance, defined as the number of contractions until exhaustion at 75% of maximum voluntary strength, and reduced lactate and ammonia production under the same conditions. These findings raised interest, but an accompanying editorial [65] led to a misunderstanding when stating "only patients with low muscle creatine levels benefit from the therapy". Actually, Andrews et al. did not measure creatine content in their patients; they referred instead to previous studies showing "a reduction in total creatine content in skeletal muscle in patients with severe chronic heart failure" and demonstrating that "dietary creatine supplementation in chronic heart failure produces a significant increase in skeletal muscle creatine and phosphocreatine content". Thus, the correct interpretation of those authors' data should be that all patients with heart failure as a population should benefit from creatine supplementation. Moreover, the same editorial [65] raised concern about the safety of long-term creatine supplementation, concern that later research would have dispelled [4]. Finally, Kuethe et al. [66] found in a double blind, placebo-controlled and crossover-designed study that in patients with severe heart failure creatine supplementation (4 g 5 times a day) was able to improve muscle strength. Summing up, these studies demonstrate that although creatine supplementation may have some positive effects on cardiac function in heart failure patients, its more robust effect lies in improving the endurance and strength of skeletal muscle, an effect that is anyway theoretically able to improve the quality of life of these mostly incurable patients.

5. Creatine Supplementation in Heart Ischemia

Ischemia is probably the condition where creatine supplementation has the strongest rationale, given its energy-boosting properties. At times of ischemia all organs, including the heart, decrease or lose their ability to synthesize ATP because of the decreased supply of nutrients such as glucose and oxygen. In the case of the heart, ischemia leads to life-threatening conditions like angina pectoris and myocardial infarction. There is a strong rationale for increasing the intracellular level of creatine, phosphocreatine or both in conditions of ischemia, because as already said (see above) phosphocreatine acts as an extra energy source, effectively synthesizing ATP even in the absence of oxygen and glucose [17], thus improving tissue resistance to ischemic damage. The effects of creatine or phosphocreatine supplementation in heart ischemia have been previously reviewed [5,67].

5.1. Preclinical Studies

5.1.1. Effects of Decreasing Heart Creatine on Vulnerability to Ischemia

Ten Hove et al. [56] found that decreasing creatine levels in the rodents' hearts made the latter more vulnerable to ischemic damage. By contrast, Lygate et al. [68] reported that mortality from coronary artery ligation was not different in transgenic mice lacking the first enzyme of creatine synthesis and in controls. The authors explained the latter finding with the fact that those transgenic mice accumulated guanidinoacetate, the precursor of creatine that may itself be phosphorylated and vicariate to an extent the role of phosphocreatine [69]. However, those authors did not measure creatine in the hearts of their transgenic mice; therefore, we cannot rule out the possibility that those hearts did actually contain some creatine from the diet. Incidentally, the same group reported that a moderate increase in myocardial creatine obtained by over-expressing the creatine transporter was protective against ischemia damage [11]. Horn et al. [70] reported that decreasing creatine content in the heart by blocking the creatine transporter with beta-guanidinopropionate was accompanied by increased mortality and ATP reduction upon infarction due to coronary artery ligation.

5.1.2. Effects of Creatine Supplementation on Ischemic Damage

In a preclinical study, Webster et al. [71] found that creatine supplementation before ischemia improved heart contraction during ischemia in rats, although this effect was limited to rats that were sedentary before ischemia and was not observed in rats that routinely exercised before ischemia. Lygate et al. [11] found that a moderate increase in myocardial creatine obtained by over-expression of the creatine transporter protected against ischemic damage due to experimental infarction in rats.

5.2. Lack of Clinical Studies

No trial has investigated the effects of creatine supplementation in human patients with myocardial infarction. Accordingly, a 2011 Cochrane meta-analysis [72] reached the conclusion that "The trials in patients with acute myocardial infarction only evaluated intravenous creatine phosphate".

5.3. The Use of Creatine Phosphate in Human Myocardial Infarction

While creatine has not been studied in human myocardial infarction (see above), numerous investigations have studied the effects of phosphocreatine. These effects have been reviewed elsewhere [5,67] and in a Cochrane review [72]. Generally speaking, they reported some encouraging results, although the Cochrane review found it still insufficient for recommending its routine use in clinical practice [72]. The studies using phosphocreatine should, however, be kept distinct from those using creatine. In fact, it is questionable that phosphocreatine enters cells upon its systemic (oral or parenteral) administration. Phosphocreatine does not cross biological membranes, as we demonstrated in our laboratory by showing that phosphocreatine did not increase creatine nor phosphocreatine content in in vitro brain slices [73]. Moreover, there is no evidence that phosphocreatine has a transporter, nor that it can use the creatine transporter or another one. Soboll et al. [74] found that phosphocreatine was taken up by both isolated rat heart mitochondria and liposomes, an observation that is obviously not relevant to uptake by whole cells or organs. Preobrazhenskiĭ et al. [75] reported that isolated perfused rat hearts took up 32P-phosphocreatine, especially after they were made ischemic. This latter observation is at variance with what we found in brain slices (see above), and it is unclear what the mechanism for such uptake may be. Accordingly, the protective cardiac effects by phosphocreatine in heart ischemia are usually explained by mechanisms other than increase in creatine or phosphocreatine in the cardiac cells. These mechanisms include insertion of phosphocreatine into the sarcolemma to modify its physical properties [76,77] and inhibition of platelet aggregation [76,78]. These mechanisms have been reviewed by Saks et al. [76,79], who correctly mention phosphocreatine penetration into cells only as a possibility waiting for demonstration [76]. Thus, phosphocreatine administration seems to act as a cardio protectant in a substantially different way compared to creatine.

6. Creatine Supplementation in Anthracycline Toxicity

6.1. Use and Adverse Effects of Anthracyclines

Anthracyclines are antitumor agents used in many types of cancers, including breast cancer and hematological malignancies [80]. The two most used compounds are doxorubicin (also called adriamycin) [81] and daunorubicin [82]. Epirubicin and idarubicin also belong to this class and are used in chemotherapy [80]. The mechanism of action of these molecules is multifactorial, however two of their effects are considered the most important ones: mitochondrial damage due to the production of reactive oxygen species (ROS) and inhibition of DNA replication due to binding of anthracyclines to topoisomerase, an enzyme that intervenes in DNA duplication [80].

Production by anthracyclines of ROS leads to mitochondrial damage, and is a prominent mechanism of the antitumor activity of anthracyclines. It occurs because of reduction in the anthracycline molecule by cellular oxido-reductases (including, in the heart, the NADH dehydrogenase). In the presence of molecular oxygen, the molecule resulting from this reduction spontaneously auto-oxidizes to generate again the parent anthracycline and a superoxide anion, starting a self-perpetuating loop [80,83]. Moreover, mitochondrial dysfunction is worsened by the generation of toxic radical and reactive nitrogen species resulting from the interaction of anthracyclines with cellular iron [80,84].

Although anthracyclines have widespread toxic effects, cardiac adverse effects are especially important for their high frequency of occurrence and their ability to limit the clinical use of these compounds. They range from mild cases showing only elevation of markers of cardiac damage with no or few symptoms to life-threatening heart failure [80].

Some cardioprotective strategies have been investigated, including the co-administration of the iron-chelating compound dexrazoxane, inhibitors of the angiotensin-converting enzyme (ACE-inhibitors), angiotensin II receptor blockers, and beta-blockers, but none of them is currently in routine use [80].

6.2. Studies Linking Anthracyclines Toxicity and Creatine Metabolism

Several preclinical studies link anthracyclines toxicity with the creatine metabolism.

Darrabie et al. [85] found that anthracyclines administration reduced the expression of the creatine transporter, and consequently reduced creatine uptake by cardiomyocytes [86]. DeAtley et al. reported that anthracycline administration reduced the activity of creatine kinase in vitro. Tokarska-Schlattner et al. [87,88] found that anthracycline administration damaged the mitochondrial creatine kinase and reduced the capability of creatine to stimulate respiration of in vitro isolated mitochondria.

Although the above studies provide a specific framework for cardiac protection by creatine, we should not forget that besides its ergogenic effects, creatine is also an antioxidant [89], thus it may have a non-specific effect against anthracycline toxicity by reducing the formation of reactive oxygen species.

Very importantly, Gupta et al. [90] found that in in vivo rodents, overexpression of creatine kinase decreased anthracycline damage to cardiac energy metabolism and contraction force, and improved survival. These findings were later challenged by Aksentijevic et al. [91]. These authors confirmed in isolated hearts that oxidative stress by hydrogen peroxide (H_2O_2) as well as by doxorubicin treatment caused a decrease in the force of contraction. However, they did not find any protection by increasing creatine content through the over-expression of the creatine transporter, nor by addition of creatine to the heart perfusion medium.

6.3. Effects of Creatine Supplementation in Animal Models

Several papers investigated in animal models the effects of creatine supplementation on cardiac anthracycline toxicity.

The first one [92] investigated the distinct effects of administering either a high dose of creatine (0.2 g/Kg/day, which would correspond to 14 g/day for a human weighing 70 Kg) or a mixture of vitamins C and E for 30 days before one single dose of doxorubicin. Those authors found that both treatment groups showed an approximately double survival time compared to controls and improved several parameters of damage that were increased by doxorubicin. Based on the effect on these parameters, the authors concluded that the two vitamins had a more noticeable effect than that of creatine. Vitamins C and E are both powerful antioxidants [93,94]. Thus, while this paper is important because it demonstrates a protective effect by creatine (average survival time was 3 days in the control group, 6 days in the creatine-treated one), it suggests that its antioxidant properties [89] may be more important in this context than its ergogenic ones.

Later, Santacruz et al. [95] demonstrated in in vitro cultured cardiomyocytes that perfusion with 5 mM creatine significantly improved markers of damage, apoptosis and ROS generation after doxorubicin treatment.

Of particular interest are two papers that demonstrate creatine efficacy in countering the toxic effects of anthracyclines on the skeletal muscle. In in vitro isolated muscles [96], doxorubicin treatment decreased the force of contraction and increased latency to fatigue in both type I and type II muscle fibers, and pre-treatment with creatine prevented these effects. These findings were very recently confirmed in an experiment [97] involving both in vivo rats and in vitro isolated muscles, where doxorubicin worsened grip strength and latency to fatigue while pre-treatment with creatine fully prevented both harmful effects.

Thus, preclinical papers investigating the effects of creatine supplementation on cardiac or muscular toxicity by anthracyclines reported improvement, both on cardiac toxicity and on the associated muscle damage. We should emphasize that all of them administered creatine before, not after, the challenge with anthracyclines.

6.4. Effects of Phosphocreatine

Researchers investigated phosphocreatine, too, in anthracycline toxicity.

In in vivo rodents, adriamycin (the alternate name of doxorubicin) decreased the expression of a micro-RNA (miRNA 378/378*) and of the calcium-binding protein calumenin, and both these effects were reversed by phosphocreatine [98]. Another group from the same university later reported [99] that in in vitro rat cultured cardiomyocytes incubation with phosphocreatine for 12 or 24 h before administration of doxorubicin prevented the large mortality that the drug caused, an effect that the authors attributed to a phosphocreatine-induced increase in the concentration of the calcium-binding protein calumenin. Interestingly, in the latter paper the effect of phosphocreatine increased with the duration of its perfusion before doxorubicin administration, and was not statistically significant for an infusion time as short as 6 h.

As for clinical research on human patients, Parve et al. [100] reported the case of a 52 y.o. woman who had developed cardiomyopathy after doxorubicin and radiotherapy, and whose cardiomyopathy improved after phosphocreatine treatment. A Cochrane review of medical interventions for treating anthracycline-induced cardiotoxicity in childhood cancer [101] quoted a paper, which I was not able to retrieve, that compared in 68 patients treatment with phosphocreatine for 2 weeks with what was called a "control treatment" (consisting of vitamin C, adenosine triphosphate, vitamin E and oral coenzyme Q10) in anthracycline-induced cardiotoxicity. The Cochrane paper reported that this trial "found no differences in overall survival, mortality due to heart failure, echocardiographic cardiac function, and adverse events between treatment and control groups". However, we should note that both arms consisted of active treatments, and that apparently a control, untreated group was not included. Thus, we cannot rule out the possibility that both treatments (including phosphocreatine) were equally effective in reducing anthracycline-induced cardiotoxicity.

Thus, treatment with phosphocreatine showed some encouraging preliminary data that may warrant further research. However, we repeat that, as discussed above, it is doubtful that phosphocreatine enters cells, thus its mechanisms of action are probably unrelated to changes in the creatine-phosphocreatine complex (see above).

7. Concluding Remarks

The above data allow some conclusions to be drawn. Since such conclusions descend strictly from all the above-discussed data, individual references will not be quoted again in the ensuing paragraphs. Instead, the reader who looks for more details and references is referred to the preceding sections of this paper.

In vitro studies suggest some improvement by creatine supplementation in the function even of the healthy heart; however, such improvements could not be confirmed in in vivo studies of healthy subjects.

Creatine supplementation in the healthy heart is safe. Some harmful effects that were reported in preclinical experiments on transgenic animals were consequent to very high increases in creatine content, so high that they are not possible with creatine supplementation alone. In fact, in vivo creatine supplementation increases cardiomyocyte content by about 11–28%, and no adverse events were observed when creatine increase was limited to 60% of its basal value. Furthermore, extensive use of creatine in human placebo-controlled trials have concluded that creatine supplementation is safe even at high doses and for extended time, with the possible exception of subjects affected by kidney damage.

In heart failure creatine, phosphocreatine or both decrease in both animal and human studies. The mechanisms for this decrease are both a decrease in the creatine transporter that decreases uptake, and a consumption of phosphocreatine that attempts to prevent or delay exhaustion of ATP.

The decrease in phosphocreatine in heart failure is highly clinically relevant, because it causes a decreased contractility reserve of the myocardium and correlates with left ventricular ejection fraction. Phosphocreatine decrease is so clinically relevant that in individual patients it is a predictor of mortality.

Thus, the above data provide a strong rationale to supplement with creatine the failing heart, in an attempt to reverse the above-described harmful effects.

So far, studies of creatine supplementation in heart failure demonstrated encouraging results, which warrant further investigations. Especially, it would be useful to understand in which patients (presumably, those with the most marked decrease in creatine or phosphocreatine) creatine supplementation is useful. Such a study would require dosing cardiac creatine in vivo using, for example, magnetic resonance spectroscopy, but it would be very valuable and useful. Despite the described paucity of current evidence, and pending additional trials, creatine supplementation in heart failure may nevertheless be useful, at least in selected patients, given (1) the additional evidence showing that in heart failure patients creatine improves muscle strength and endurance, and (2) the evidence that demonstrates that such creatine supplementation is feasible and safe. Pending additional trials, at least heart failure patients in whom weakness and fatigue are prominent symptoms, and in whom kidney function is normal, should trial creatine supplementation.

There is a strong rationale to use creatine supplementation in heart ischemia; however, the majority of the studies so far have used phosphocreatine, not creatine, supplementation. Although the two molecules are obviously strictly related to each other, there is currently no clear evidence that phosphocreatine can enter cells as creatine does. For this reason, its mechanism of action is mostly unrelated to changes in the ergogenic creatine-phosphocreatine system.

Antagonism of cardiac toxicity of the anti-tumor compounds belonging to the anthracycline family is a novel, clinically relevant field where creatine supplementation may be useful. Several studies suggest that anthracyclines damage the creatine-phosphocreatine system by reducing the expression of the creatine transporter and by impairing the activity of creatine kinase. Moreover, anthracycline cardiac toxicity is largely due to generation of reacting oxygen and nitrogen species, therefore creatine may counter it because of its anti-oxidant properties, too.

Studies on genetically modified rodents over-expressing the creatine transporter gave conflicting results in the prevention of anthracycline toxicity. However, supplementation with creatine before anthracycline administration has proved very effective in preventing cardiac toxicity in preclinical studies, both in vitro and in vivo. Furthermore, supplementation with creatine before anthracycline challenge fully prevents the damage by these compounds on skeletal muscle, both in vitro and in vivo. Thus, we should definitely carry out clinical studies to investigate whether or not administration of creatine before a planned anthracycline treatment is able to prevent or decrease the cardiac toxicity that currently limits in a significant way the use of these important chemotherapy agents. In designing such studies, we should bear in mind that creatine may be superior to other antioxidants in clinical contexts because of its additional effects improving muscle anthracycline toxicity. Phosphocreatine should also be further investigated; however, its mechanism of action is probably unrelated to the ergogenic creatine-phosphocreatine system.

8. Scientific Significance and Translational Opportunities

From the scientific point of view, the value of a literature review consists of delivering a comprehensive view of a clinical issue, in such a way as to put individual papers in a global perspective that may suggest translational opportunities. The above-reported review on creatine supplementation in the heart makes it possible to formulate the following suggestions for clinical translation. Again, the reader should refer to the above paragraphs for details and quotes. Please note that, as already stated, we discuss only supplementation with creatine monohydrate, not with other forms of creatine.

1. In healthy hearts, there is currently no demonstration that creatine supplementation may improve cardiac function. However, creatine supplementation is safe, with the possible exception of subjects with renal failure (elevated plasma creati-

      nine), thus fear of adverse events should not prevent willing subjects from trialing creatine supplementation.

2. In heart failure, there is a decrease in the creatine content of the myocytes, and such a decrease is highly relevant from the clinical point of view. Moreover, creatine supplementation improves muscle function in these patients. Thus:

   a. Creatine supplementation should be trialed in heart failure patients, especially when weakness and fatigue are prominent symptoms.

   b. Further research should correlate, in individual patients, creatine and phosphocreatine content of the myocardium with the clinical benefits obtained from supplementation.

3. Further research should be carried out on the effects of creatine supplementation in heart ischemia.

4. Mitigation of anthracyclines toxicity is an unmet clinical need. Thus, treatment of oncological patients with anthracyclines might even now be preceded by an adequate period of creatine supplementation, possibly together with vitamins C and E, to prevent chemotherapy toxicity both to the heart and to the muscle. Moreover, research should be carried out in ample clinical cohorts to definitively determine the usefulness of this supplementation in anthracyclines chemotherapy.

Funding: This work was supported by grants from Italian Ministry of Health (Ricerca Corrente). AlzChem Trostberg GmbH (Trostberg, Germany) kindly paid the article processing charges for publication of this paper.

Institutional Review Board Statement: Not applicable.

Informed Consent Statement: Not applicable.

Data Availability Statement: Not applicable.

Conflicts of Interest: The author is founding member and president of NovaNeuro Srl, an academic spinoff that ideates and commercializes dietary supplements based on creatine. He received an honorarium by AlzChem Trostberg GmbH (Trostberg, Germany) to write this article. The funders had no role in the design of the study; in the collection, analyses, or interpretation of data; in the writing of the manuscript, or in the decision to publish the results.

References

1. Ventura-Clapier, R.; Vassort, G. The hypodynamic state of the frog heart. Further evidence for a phosphocreatine—Creatine pathway. *J. Physiol.* **1980**, *76*, 583–589.
2. Wallimann, T.; Tokarska-Schlattner, M.; Schlattner, U. The Creatine kinase system and pleiotropic Effects of creatine. *Amino Acids* **2011**, *40*, 1271–1296. [CrossRef] [PubMed]
3. Sahlin, K.; Harris, R.C. The Creatine Kinase Reaction: A Simple reaction with functional complexity. *Amino Acids* **2011**, *40*, 1363–1367. [CrossRef]
4. Balestrino, M.; Adriano, E. Beyond sports: Efficacy and Safety of creatine supplementation in pathological or paraphysiological conditions of brain and muscle. *Med. Res. Rev.* **2019**, *39*. [CrossRef] [PubMed]
5. Balestrino, M.; Sarocchi, M.; Adriano, E.; Spallarossa, P. Potential of Creatine or phosphocreatine supplementation in cerebrovascular disease and in ischemic heart disease. *Amino Acids* **2016**, *48*, 1955–1967. [CrossRef]
6. Wyss, M.; Kaddurah-Daouk, R. Creatine and creatinine metabolism. *Physiol. Rev.* **2000**, *80*, 1107–1213. [CrossRef] [PubMed]
7. Casey, A.; Greenhaff, P.L. Does Dietary Creatine Supplementation Play a Role in Skeletal Muscle Metabolism and Performance? *Am. J. Clin. Nutr.* **2000**, *72*, 607S–617S. [CrossRef] [PubMed]
8. Hanna-El-Daher, L.; Braissant, O. Creatine synthesis and exchanges between brain cells: What can be learned from human creatine deficiencies and various experimental models? *Amino Acids* **2016**, *48*, 1877–1895. [CrossRef]
9. Koszalka, T.R. Creatine synthesis in the testis. *Proc. Soc. Exp. Biol. Med.* **1968**, *128*, 1130–1137. [CrossRef]
10. Lee, H.; Kim, J.H.; Chae, Y.J.; Ogawa, H.; Lee, M.H.; Gerton, G.L. Creatine synthesis and transport systems in the male rat reproductive tract. *Biol. Reprod.* **1998**, *58*, 1437–1444. [CrossRef]
11. Lygate, C.A.; Bohl, S.; ten Hove, M.; Faller, K.M.E.; Ostrowski, P.J.; Zervou, S.; Medway, D.J.; Aksentijevic, D.; Sebag-Montefiore, L.; Wallis, J.; et al. Moderate elevation of intracellular creatine by targeting the creatine transporter protects mice from acute myocardial infarction. *Cardiovasc. Res.* **2012**, *96*, 466–475. [CrossRef]

12. Zervou, S.; Whittington, H.J.; Russell, A.J.; Lygate, C.A. Augmentation of creatine in the heart. *Mini Rev. Med. Chem.* **2016**, *16*, 19–28. [CrossRef] [PubMed]
13. Cullen, M.E.; Yuen, A.H.Y.; Felkin, L.E.; Smolenski, R.T.; Hall, J.L.; Grindle, S.; Miller, L.W.; Birks, E.J.; Yacoub, M.H.; Barton, P.J.R. Myocardial Expression of the arginine: Glycine Amidinotransferase gene is elevated in heart failure and normalized after recovery: Potential implications for local Creatine synthesis. *Circulation* **2006**, *114* (Suppl. 1), I16–I20. [CrossRef]
14. Nekhorocheff, J. Degradation and synthesis of creatine in isolated toad heart. *C. R. Hebd. Séance Acad. Sci.* **1955**, *240*, 1284–1285.
15. Fisher, R.B.; Wilhelmi, A.E. The Metabolism of creatine: The conversion of arginine into creatine in the isolated rabbit heart. *Biochem. J.* **1937**, *31*, 1136–1156. [CrossRef] [PubMed]
16. Snow, R.J.; Murphy, R.M. Creatine and the creatine transporter: A review. *Mol. Cell Biochem.* **2001**, *224*, 169–181. [CrossRef] [PubMed]
17. Marques, E.P.; Wyse, A.T.S. Creatine as a Neuroprotector: An Actor that Can Play Many Parts. *Neurotox Res.* **2019**, *423*, 411–423. [CrossRef] [PubMed]
18. Ingwall, J.S. On the hypothesis that the failing heart is energy starved: Lessons learned from the metabolism of ATP and creatine. *Curr. Hypertens. Rep.* **2006**, *8*, 457–464. [CrossRef]
19. Brosnan, M.E.; Brosnan, J.T. The role of dietary creatine. *Amino Acids* **2016**, *48*, 1785–1791. [CrossRef]
20. Harris, R.C.; Söderlund, K.; Hultman, E. Elevation of creatine in resting and exercised muscle of normal subjects by Creatine supplementation. *Clin. Sci.* **1992**, *83*, 367–374. [CrossRef]
21. Ipsiroglu, O.S.; Stromberger, C.; Ilas, J.; Höger, H.; Mühl, A.; Stöckler-Ipsiroglu, S. Changes of tissue creatine concentrations upon oral supplementation of Creatine-monohydrate in various animal species. *Life Sci.* **2001**, *69*, 1805–1815. [CrossRef]
22. Dechent, P.; Pouwels, P.J.W.; Wilken, B.; Hanefeld, F.; Frahm, J. Increase of total creatine in human brain after oral supplementation of Creatine-monohydrate. *Am. J. Physiol. Regul. Integr. Comp. Physiol.* **1999**, *277*, R698–R704. [CrossRef]
23. Sweeney, H.L. The Importance of the Creatine Kinase Reaction: The Concept of Metabolic Capacitance. *Med. Sci. Sports Exerc.* **1994**, *26*, 30–36. [CrossRef]
24. Peeling, P.; Binnie, M.J.; Goods, P.S.R.; Sim, M.; Burke, L.M. Evidence-based supplements for the enhancement of athletic performance. *Int. J. Sport Nutr. Exerc. Metab.* **2018**, *28*, 178–187. [CrossRef]
25. Kreider, R.B.; Kalman, D.S.; Antonio, J.; Ziegenfuss, T.N.; Wildman, R.; Collins, R.; Candow, D.G.; Kleiner, S.M.; Almada, A.L.; Lopez, H.L. International Society of Sports Nutrition Position Stand: Safety and efficacy of creatine supplementation in exercise, sport, and medicine. *J. Int. Soc. Sports Nutr.* **2017**, *14*. [CrossRef]
26. Blancquaert, L.; Baguet, A.; Bex, T.; Volkaert, A.; Everaert, I.; Delanghe, J.; Petrovic, M.; Vervaet, C.; De Henauw, S.; Constantin-Teodosiu, D.; et al. Changing to a vegetarian diet reduces the body creatine pool in omnivorous women, but appears not to affect carnitine and carnosine homeostasis: A randomised trial. *Br. J. Nutr.* **2018**, *119*, 759–770. [CrossRef]
27. Guimbal, C.; Kilimann, M.W. A Na(+)-dependent creatine transporter in rabbit brain, muscle, heart, and kidney. CDNA cloning and functional expression. *J. Biol. Chem.* **1993**, *268*, 8418–8421. [CrossRef]
28. Fischer, A.; Ten Hove, M.; Sebag-Montefiore, L.; Wagner, H.; Clarke, K.; Watkins, H.; Lygate, C.A.; Neubauer, S. Changes in creatine transporter function during cardiac maturation in the rat. *BMC Dev. Biol.* **2010**, *10*, 70. [CrossRef] [PubMed]
29. Saks, V.A.; Rosenshtraukh, L.V.; Undrovinas, A.I.; Smirnov, V.N.; Chazov, E.I. Studies of energy transport in heart cells. intracellular creatine content as a regulatory factor of frog heart energetics and force of contraction. *Biochem. Med.* **1976**, *16*, 21–36. [CrossRef]
30. Rosenshtraukh, L.V.; Saks, V.A.; Undrovinas, A.I.; Chazov, E.I.; Smirnov, V.N.; Sharov, V.G. Studies of energy transport in heart cells. The effect of creatine phosphate on the frog ventricular contractile force and action potential duration. *Biochem. Med.* **1978**, *19*, 148–164. [CrossRef]
31. Santacruz, L.; Arciniegas, A.J.L.; Darrabie, M.; Mantilla, J.G.; Baron, R.M.; Bowles, D.E.; Mishra, R.; Jacobs, D.O. Hypoxia decreases creatine uptake in cardiomyocytes, while creatine supplementation enhances HIF activation. *Physiol. Rep.* **2017**, *5*. [CrossRef]
32. Kilian, G.; Jana, A.K.; Grant, G.D.; Milne, P.J. The Effects of creatine on the retrogradely perfused isolated rat heart. *J. Pharm.* **2002**, *54*, 105–109. [CrossRef]
33. Horn, M.; Frantz, S.; Remkes, H.; Laser, A.; Urban, B.; Mettenleiter, A.; Schnackerz, K.; Neubauer, S. Effects of chronic dietary Creatine feeding on cardiac energy metabolism and on creatine content in heart, skeletal muscle, brain, liver and kidney. *J. Mol. Cell Cardiol.* **1998**, *30*, 277–284. [CrossRef]
34. Neubauer, S.; Remkes, H.; Spindler, M.; Horn, M.; Wiesmann, F.; Prestle, J.; Walzel, B.; Ertl, G.; Hasenfuss, G.; Wallimann, T. Downregulation of the Na(+)-creatine cotransporter in failing human myocardium and in experimental heart failure. *Circulation* **1999**, *100*, 1847–1850. [CrossRef] [PubMed]
35. Nascimben, L.; Ingwall, J.S.; Pauletto, P.; Friedrich, J.; Gwathmey, J.K.; Saks, V.; Pessina, A.C.; Allen, P.D. Creatine kinase system in failing and nonfailing human myocardium. *Circulation* **1996**, *94*, 1894–1901. [CrossRef] [PubMed]
36. Horn, M.; Remkes, H.; Dienesch, C.; Hu, K.; Ertl, G.; Neubauer, S. Chronic high-dose creatine feeding does not attenuate left ventricular remodeling in rat hearts post-myocardial infarction. *Cardiovasc. Res.* **1999**, *43*, 117–124. [CrossRef]
37. del Favero, S.; Roschel, H.; Artioli, G.; Ugrinowitsch, C.; Tricoli, V.; Costa, A.; Barroso, R.; Negrelli, A.L.; Otaduy, M.C.; da Costa Leite, C.; et al. Creatine but not Betaine supplementation increases muscle Phosphorylcreatine content and strength performance. *Amino Acids* **2012**, *42*, 2299–2305. [CrossRef] [PubMed]

38. Op't Eijnde, B.; Jijakli, H.; Hespel, P.; Malaisse, W.J. Creatine supplementation increases soleus muscle creatine content and lowers the insulinogenic index in an animal model of inherited Type 2 Diabetes. *Int. J. Mol. Med.* **2006**, *17*, 1077–1084. [CrossRef] [PubMed]

39. Boehm, E.; Chan, S.; Monfared, M.; Wallimann, T.; Clarke, K.; Neubauer, S. Creatine transporter activity and content in the rat heart supplemented by and depleted of creatine. *Am. J. Physiol. Endocrinol. Metab.* **2003**, *284*, E399–E406. [CrossRef] [PubMed]

40. Phillips, D.; Ten Hove, M.; Schneider, J.E.; Wu, C.O.; Sebag-Montefiore, L.; Aponte, A.M.; Lygate, C.A.; Wallis, J.; Clarke, K.; Watkins, H.; et al. Mice over-expressing the myocardial creatine transporter develop progressive heart failure and show decreased glycolytic capacity. *J. Mol. Cell Cardiol.* **2010**, *48*, 582–590. [CrossRef]

41. Wallis, J.; Lygate, C.A.; Fischer, A.; ten Hove, M.; Schneider, J.E.; Sebag-Montefiore, L.; Dawson, D.; Hulbert, K.; Zhang, W.; Zhang, M.H.; et al. Supranormal myocardial creatine and phosphocreatine concentrations lead to cardiac hypertrophy and heart failure: Insights from creatine transporter-overexpressing transgenic mice. *Circulation* **2005**, *112*, 3131–3139. [CrossRef] [PubMed]

42. Santacruz, L.; Hernandez, A.; Nienaber, J.; Mishra, R.; Pinilla, M.; Burchette, J.; Mao, L.; Rockman, H.A.; Jacobs, D.O. Normal cardiac function in mice with supraphysiological cardiac creatine levels. *Am. J. Physiol. Heart Circ. Physiol.* **2014**, *306*, H373–H381. [CrossRef] [PubMed]

43. Zervou, S.; Yin, X.; Nabeebaccus, A.A.; O'Brien, B.A.; Cross, R.L.; McAndrew, D.J.; Atkinson, R.A.; Eykyn, T.R.; Mayr, M.; Neubauer, S.; et al. Proteomic and metabolomic changes driven by elevating myocardial creatine suggest novel metabolic feedback mechanisms. *Amino Acids* **2016**, *48*, 1969–1981. [CrossRef] [PubMed]

44. Mert, K.; Ilgüy, S.; Dural, M.; Mert, G.; Ozakin, E. Effects of creatine supplementation on cardiac autonomic functions in bodybuilders. *Pacing Clin. Electrophysiol.* **2017**, *40*. [CrossRef]

45. Nanchen, D. Resting heart rate: What is normal? *Heart* **2018**, *104*, 1048–1049. [CrossRef]

46. Ponikowski, P.; Voors, A.A.; Anker, S.D.; Bueno, H.; Cleland, J.G.F.; Coats, A.J.S.; Falk, V.; González-Juanatey, J.R.; Harjola, V.-P.; Jankowska, E.A.; et al. 2016 ESC Guidelines for the diagnosis and treatment of acute and chronic heart failure: The task force for the diagnosis and treatment of acute and chronic heart failure of the european society of cardiology (ESC)developed with the special contribution of the heart failure association (HFA) of the ESC. *Eur. Heart. J.* **2016**, *37*, 2129–2200. [CrossRef]

47. Feinstein, M.B. Effects of experimental congestive heart failure, ouabain, and asphyxia on the high-energy phosphate and creatine content of the guinea pig heart. *Circ. Res.* **1962**, *10*, 333–346. [CrossRef]

48. Fox, A.C.; Wikler, N.S.; Reed, G.E. High energy phosphate compounds in the myocardium during experimental congestive heart failure. purine and pyrimidine nucleotides, creatine, and creatine phosphate in normal and in failing hearts. *J. Clin. Investig.* **1965**, *44*, 202–218. [CrossRef]

49. Shen, W.; Asai, K.; Uechi, M.; Mathier, M.A.; Shannon, R.P.; Vatner, S.F.; Ingwall, J.S. Progressive loss of myocardial ATP due to a loss of total purines during the development of heart failure in dogs: A compensatory role for the parallel loss of creatine. *Circulation* **1999**, *100*, 2113–2118. [CrossRef]

50. Ten Hove, M.; Chan, S.; Lygate, C.; Monfared, M.; Boehm, E.; Hulbert, K.; Watkins, H.; Clarke, K.; Neubauer, S. Mechanisms of Creatine Depletion in Chronically Failing Rat Heart. *J. Mol. Cell Cardiol.* **2005**, *38*, 309–313. [CrossRef] [PubMed]

51. Winter, J.L.; Castro, P.; Meneses, L.; Chalhub, M.; Verdejo, H.; Greig, D.; Gabrielli, L.; Chiong, M.; Concepción, R.; Mellado, R.; et al. Myocardial Lipids and Creatine Measured by Magnetic Resonance Spectroscopy among Patients with Heart Failure. *Rev. Méd. Chile* **2010**, *138*, 1475–1479. [CrossRef]

52. Neubauer, S.; Horn, M.; Cramer, M.; Harre, K.; Newell, J.B.; Peters, W.; Pabst, T.; Ertl, G.; Hahn, D.; Ingwall, J.S.; et al. Myocardial Phosphocreatine-to-ATP Ratio Is a Predictor of Mortality in Patients with Dilated Cardiomyopathy. *Circulation* **1997**, *96*, 2190–2196. [CrossRef] [PubMed]

53. Nakae, I.; Mitsunami, K.; Matsuo, S.; Matsumoto, T.; Morikawa, S.; Inubushi, T.; Koh, T.; Horie, M. Assessment of Myocardial Creatine Concentration in Dysfunctional Human Heart by Proton Magnetic Resonance Spectroscopy. *Magn. Reason. Med. Sci.* **2004**, *3*, 19–25. [CrossRef] [PubMed]

54. Nakae, I.; Mitsunami, K.; Matsuo, S.; Inubushi, T.; Morikawa, S.; Tsutamoto, T.; Koh, T.; Horie, M. Myocardial Creatine Concentration in Various Nonischemic Heart Diseases Assessed by 1H Magnetic Resonance Spectroscopy. *Circ. J.* **2005**, *69*, 711–716. [CrossRef] [PubMed]

55. Neubauer, S. The Failing Heart—An engine out of fuel. *N. Engl. J. Med.* **2007**, *356*, 1140–1151. [CrossRef]

56. Ten Hove, M.; Lygate, C.A.; Fischer, A.; Schneider, J.E.; Sang, A.E.; Hulbert, K.; Sebag-Montefiore, L.; Watkins, H.; Clarke, K.; Isbrandt, D.; et al. Reduced inotropic reserve and increased susceptibility to cardiac ischemia/reperfusion injury in phosphocreatine-deficient Guanidinoacetate-N-methyltransferase–knockout mice. *Circulation* **2005**, *111*, 2477–2485. [CrossRef] [PubMed]

57. Kapelko, V.I.; Saks, V.A.; Novikova, N.A.; Golikov, M.A.; Kupriyanov, V.V.; Popovich, M.I. Adaptation of Cardiac Contractile Function to Conditions of Chronic Energy Deficiency. *J. Mol. Cell Cardiol.* **1989**, *21* (Suppl. 1), 79–83. [CrossRef]

58. Field, M.L. Creatine Supplementation in congestive heart failure. *Cardiovasc. Res.* **1996**, *31*, 174–176. [CrossRef]

59. Faller, K.M.E.; Medway, D.J.; Aksentijevic, D.; Sebag-Montefiore, L.; Schneider, J.E.; Lygate, C.A.; Neubauer, S. Ribose supplementation alone or with elevated creatine does not preserve high energy nucleotides or cardiac function in the failing mouse heart. *PLoS ONE* **2013**, *8*, e66461. [CrossRef]

60. Ahmed, M.; Anderson, S.D.; Schofield, R.S. Coenzyme Q10 and creatine in heart failure: Micronutrients, macrobenefit? *Clin. Cardiol.* **2011**, *34*, 196–197. [CrossRef] [PubMed]

61. Fumagalli, S.; Fattirolli, F.; Guarducci, L.; Cellai, T.; Baldasseroni, S.; Tarantini, F.; Di Bari, M.; Masotti, G.; Marchionni, N. Coenzyme Q10 Terclatrate and creatine in chronic heart failure: A randomized, placebo-controlled, double-blind study. *Clin. Cardiol.* **2011**, *34*, 211–217. [CrossRef]

62. Carvalho, A.P.P.F.; Rassi, S.; Fontana, K.E.; Correa, K.S.; Feitosa, R.H.F. Influence of creatine supplementation on the functional capacity of patients with heart failure. *Arq. Bras. Cardiol.* **2012**, *99*, 623–629. [CrossRef] [PubMed]

63. Gordon, A.; Hultman, E.; Kaijser, L.; Kristjansson, S.; Rolf, C.J.; Nyquist, O.; Sylvén, C. Creatine supplementation in chronic heart failure increases skeletal muscle creatine phosphate and muscle performance. *Cardiovasc. Res.* **1995**, *30*, 413–418. [CrossRef]

64. Andrews, R.; Greenhaff, P.; Curtis, S.; Perry, A.; Cowley, A.J. The effect of dietary creatine supplementation on skeletal muscle metabolism in congestive heart failure. *Eur. Heart J.* **1998**, *19*, 617–622. [CrossRef] [PubMed]

65. Schaufelberger, M.; Swedberg, K. Is Creatine supplementation helpful for patients with chronic heart failure? *Eur. Heart J.* **1998**, *19*, 533–534. [CrossRef]

66. Kuethe, F.; Krack, A.; Richartz, B.M.; Figulla, H.R. Creatine supplementation improves muscle strength in patients with congestive heart failure. *Pharmazie* **2006**, *61*, 218–222. [PubMed]

67. Perasso, L.; Spallarossa, P.; Gandolfo, C.; Ruggeri, P.; Balestrino, M. Therapeutic use of creatine in brain or heart ischemia: Available data and future perspectives. *Med. Res. Rev.* **2013**, *33*, 336–363. [CrossRef] [PubMed]

68. Lygate, C.A.; Aksentijevic, D.; Dawson, D.; ten Hove, M.; Phillips, D.; de Bono, J.P.; Medway, D.J.; Sebag-Montefiore, L.; Hunyor, I.; Channon, K.M.; et al. Living without creatine: Unchanged exercise capacity and response to chronic myocardial infarction in creatine-deficient mice. *Circ. Res.* **2013**, *112*, 945–955. [CrossRef]

69. Kan, H.E.; Renema, W.K.J.; Isbrandt, D.; Heerschap, A. Phosphorylated guanidinoacetate partly compensates for the lack of phosphocreatine in skeletal muscle of mice lacking guanidinoacetate methyltransferase. *J. Physiol.* **2004**, *560*, 219–229. [CrossRef] [PubMed]

70. Horn, M.; Remkes, H.; Strömer, H.; Dienesch, C.; Neubauer, S. Chronic phosphocreatine depletion by the creatine analogue beta-guanidinopropionate is associated with increased mortality and loss of ATP in rats after myocardial infarction. *Circulation* **2001**, *104*, 1844–1849. [CrossRef]

71. Webster, I.; Toit, E.F.D.; Huisamen, B.; Lochner, A. The effect of creatine supplementation on myocardial function, mitochondrial respiration and susceptibility to ischaemia/reperfusion injury in sedentary and exercised rats. *Acta Physiol.* **2012**, *206*, 6–19. [CrossRef]

72. Horjus, D.L.; Oudman, I.; van Montfrans, G.A.; Brewster, L.M. Creatine and creatine analogues in hypertension and cardiovascular disease. *Cochrane Database Syst. Rev.* **2011**. [CrossRef]

73. Perasso, L.; Lunardi, G.L.; Risso, F.; Pohvozcheva, A.V.; Leko, M.V.; Gandolfo, C.; Florio, T.; Cupello, A.; Burov, S.V.; Balestrino, M. Protective effects of some creatine derivatives in brain tissue anoxia. *Neurochem. Res.* **2008**, *33*, 765–775. [CrossRef] [PubMed]

74. Soboll, S.; Conrad, A.; Eistert, A.; Herick, K.; Krämer, R. Uptake of Creatine Phosphate into heart mitochondria: A leak in the creatine shuttle. *Biochim. Biophys. Acta* **1997**, *1320*, 27–33. [CrossRef]

75. Preobrazhenskiĭ, A.N.; Dzhavadov, S.A.; Saks, V.A. Possible mechanism of the protective effect of phosphocreatine on the ischemic myocardium. *Biokhimiia* **1986**, *51*, 675–683. [PubMed]

76. Saks, V.A.; Dzhaliashvili, I.V.; Konorev, E.A.; Strumia, E. Molecular and cellular aspects of the cardioprotective mechanism of phosphocreatine. *Biokhimiia* **1992**, *57*, 1763–1784. [CrossRef]

77. Tokarska-Schlattner, M.; Epand, R.F.; Meiler, F.; Zandomeneghi, G.; Neumann, D.; Widmer, H.R.; Meier, B.H.; Epand, R.M.; Saks, V.; Wallimann, T.; et al. Phosphocreatine interacts with phospholipids, affects membrane properties and exerts membrane-protective Effects. *PLoS ONE* **2012**, *7*, e43178. [CrossRef] [PubMed]

78. Panchenko, E.; Dobrovolsky, A.; Rogoza, A.; Sorokin, E.; Ageeva, N.; Markova, L.; Titaeva, E.; Anuchin, V.; Karpov, Y.; Saks, V. The effect of exogenous phosphocreatine on maximal walking distance, blood rheology, platelet aggregation, and fibrinolysis in patients with intermittent claudication. *Int. Angiol.* **1994**, *13*, 59–64.

79. Saks, V.A.; Strumia, E. Phosphocreatine: Molecular and cellular aspects of the mechanism of cardioprotective action. *Curr. Ther. Res.* **1993**, *53*, 565–598. [CrossRef]

80. McGowan, J.V.; Chung, R.; Maulik, A.; Piotrowska, I.; Walker, J.M.; Yellon, D.M. Anthracycline chemotherapy and cardiotoxicity. *Cardiovasc. Drugs* **2017**, *31*, 63–75. [CrossRef]

81. Rivankar, S. An Overview of doxorubicin formulations in cancer therapy. *J. Cancer Res.* **2014**, *10*, 853–858. [CrossRef]

82. Saleem, T.; Kasi, A. Daunorubicin. In *StatPearls*; StatPearls Publishing: Treasure Island, FL, USA, 2021.

83. Berthiaume, J.M.; Wallace, K.B. Adriamycin-induced oxidative mitochondrial cardiotoxicity. *Cell Biol. Toxicol.* **2007**, *23*, 15–25. [CrossRef] [PubMed]

84. Hahn, V.S.; Lenihan, D.J.; Ky, B. Cancer therapy-induced cardiotoxicity: Basic mechanisms and potential cardioprotective therapies. *J. Am. Heart Assoc.* **2014**, *3*, e000665. [CrossRef] [PubMed]

85. DeAtley, S.M.; Aksenov, M.Y.; Aksenova, M.V.; Jordan, B.; Carney, J.M.; Butterfield, D.A. Adriamycin-induced changes of creatine kinase activity in vivo and in cardiomyocyte culture. *Toxicology* **1999**, *134*, 51–62. [CrossRef]

86. Darrabie, M.D.; Arciniegas, A.J.L.; Mantilla, J.G.; Mishra, R.; Vera, M.P.; Santacruz, L.; Jacobs, D.O. Exposing cardiomyocytes to subclinical concentrations of doxorubicin rapidly reduces their creatine transport. *Am. J. Physiol. Heart Circ. Physiol.* **2012**, *303*, H539–H548. [CrossRef] [PubMed]

87. Tokarska-Schlattner, M.; Wallimann, T.; Schlattner, U. Multiple interference of anthracyclines with mitochondrial creatine kinases: Preferential damage of the cardiac isoenzyme and its implications for drug cardiotoxicity. *Mol. Pharm.* **2002**, *61*, 516–523. [CrossRef]
88. Tokarska-Schlattner, M.; Dolder, M.; Gerber, I.; Speer, O.; Wallimann, T.; Schlattner, U. Reduced creatine-stimulated respiration in doxorubicin challenged mitochondria: Particular sensitivity of the heart. *Biochim. Biophys. Acta* **2007**, *1767*, 1276–1284. [CrossRef]
89. Sestili, P.; Martinelli, C.; Colombo, E.; Barbieri, E.; Potenza, L.; Sartini, S.; Fimognari, C. Creatine as an antioxidant. *Amino Acids* **2011**, *40*, 1385–1396. [CrossRef]
90. Gupta, A.; Rohlfsen, C.; Leppo, M.K.; Chacko, V.P.; Wang, Y.; Steenbergen, C.; Weiss, R.G. Creatine kinase-overexpression improves myocardial energetics, contractile dysfunction and survival in murine doxorubicin cardiotoxicity. *PLoS ONE* **2013**, *8*, e74675. [CrossRef]
91. Aksentijević, D.; Zervou, S.; Faller, K.M.E.; McAndrew, D.J.; Schneider, J.E.; Neubauer, S.; Lygate, C.A. Myocardial creatine levels do not influence response to acute oxidative stress in isolated perfused heart. *PLoS ONE* **2014**, *9*, e109021. [CrossRef]
92. Santos, R.V.T.; Batista, M.L.; Caperuto, E.C.; Costa Rosa, L.F. Chronic supplementation of creatine and vitamins C and E increases survival and improves biochemical parameters after doxorubicin treatment in rats. *Clin. Exp. Pharm. Physiol.* **2007**, *34*, 1294–1299. [CrossRef]
93. Padayatty, S.J.; Katz, A.; Wang, Y.; Eck, P.; Kwon, O.; Lee, J.-H.; Chen, S.; Corpe, C.; Dutta, A.; Dutta, S.K.; et al. Vitamin C as an antioxidant: Evaluation of its role in disease prevention. *J. Am. Coll. Nutr.* **2003**, *22*, 18–35. [CrossRef] [PubMed]
94. Mustacich, D.J.; Bruno, R.S.; Traber, M.G. Vitamin E. *Vitam. Horm.* **2007**, *76*, 1–21. [CrossRef] [PubMed]
95. Santacruz, L.; Darrabie, M.D.; Mantilla, J.G.; Mishra, R.; Feger, B.J.; Jacobs, D.O. Creatine supplementation reduces doxorubicin-induced cardiomyocellular injury. *Cardiovasc. Toxicol.* **2015**, *15*, 180–188. [CrossRef] [PubMed]
96. Bredahl, E.C.; Hydock, D.S. Creatine supplementation and doxorubicin-induced skeletal muscle dysfunction: An ex vivo investigation. *Nutr. Cancer* **2017**, *69*, 607–615. [CrossRef] [PubMed]
97. Torok, Z.A.; Busekrus, R.B.; Hydock, D.S. Effects of creatine supplementation on muscle fatigue in rats receiving doxorubicin treatment. *Nutr. Cancer* **2020**, *72*, 252–259. [CrossRef] [PubMed]
98. Miao, Y.U.; Zhi-Hui, H.E.; Li-Ying, X.; Xiao-Tong, S.; Jie, L.; Jing-Yi, F.; Yu, W.; Ming, Z. Effect of creatine phosphate sodium on miRNA378, miRNA378* and calumenin mRNA in adriamycin-injured cardiomyocytes. *Zhongguo Ying Yong Sheng Li Xue Za Zhi* **2016**, *32*, 514–518. [CrossRef] [PubMed]
99. Wang, Y.; Sun, Y.; Guo, X.; Fu, Y.; Long, J.; Wei, C.-X.; Zhao, M. Creatine phosphate disodium salt protects against dox-induced cardiotoxicity by increasing calumenin. *Med. Mol. Morphol.* **2018**, *51*, 96–101. [CrossRef]
100. Parve, S.; Aliakberova, G.I.; Gylmanov, A.A.; Abdulganieva, D.I. Role of exogenous phosphocreatine in chemotherapy-induced cardiomyopathy. *Rev. Cardiovasc. Med.* **2017**, *18*, 82–87.
101. Cheuk, D.K.L.; Sieswerda, E.; van Dalen, E.C.; Postma, A.; Kremer, L.C.M. medical interventions for treating anthracycline-induced symptomatic and asymptomatic cardiotoxicity during and after treatment for childhood cancer. *Cochrane Database Syst. Rev.* **2016**, CD008011. [CrossRef]

Review

The Potential Role of Creatine in Vascular Health

Holly Clarke [1], Robert C. Hickner [1,2,3] and Michael J. Ormsbee [1,2,3,*]

[1] Department of Nutrition, Food and Exercise Sciences, Florida State University, Tallahassee, FL 32306, USA; hec17e@my.fsu.edu (H.C.); rhickner@fsu.edu (R.C.H.)
[2] Department of Biokenetics, Exercise and Leisure Sciences, School of Health Science, University of KwaZulu-Natal, Westville 4041, South Africa
[3] Institute of Sports Sciences and Medicine, Florida State University, 1104 Spirit Way, Tallahassee, FL 32306, USA
* Correspondence: mormsbee@fsu.edu

Abstract: Creatine is an organic compound, consumed exogenously in the diet and synthesized endogenously via an intricate inter-organ process. Functioning in conjunction with creatine kinase, creatine has long been known for its pivotal role in cellular energy provision and energy shuttling. In addition to the abundance of evidence supporting the ergogenic benefits of creatine supplementation, recent evidence suggests a far broader application for creatine within various myopathies, neurodegenerative diseases, and other pathologies. Furthermore, creatine has been found to exhibit non-energy related properties, contributing as a possible direct and in-direct antioxidant and eliciting anti-inflammatory effects. In spite of the new clinical success of supplemental creatine, there is little scientific insight into the potential effects of creatine on cardiovascular disease (CVD), the leading cause of mortality. Taking into consideration the non-energy related actions of creatine, highlighted in this review, it can be speculated that creatine supplementation may serve as an adjuvant therapy for the management of vascular health in at-risk populations. This review, therefore, not only aims to summarize the current literature surrounding creatine and vascular health, but to also shed light onto the potential mechanisms in which creatine may be able to serve as a beneficial supplement capable of imparting vascular-protective properties and promoting vascular health.

Keywords: creatine; vascular pathology; cardiovascular disease; oxidative stress; vascular health

Citation: Clarke, H.; Hickner, R.C.; Ormsbee, M.J. The Potential Role of Creatine in Vascular Health. *Nutrients* **2021**, *13*, 857. https://doi.org/10.3390/nu13030857

Academic Editor: Roberto Cangemi

Received: 14 January 2021
Accepted: 4 March 2021
Published: 5 March 2021

1. Introduction

According to the World Health Organization (WHO), since the 1970s, cardiovascular disease (CVD) has steadily remained the leading cause of mortality in developed countries around the world, taking an estimated 17.9 million lives every year. Within the United States (US) alone, it has been estimated that around 655,000 deaths occur annually due to CVDs [1]. The American Heart Association (AHA) reported that more than 43.7 million adults aged >60 years suffered from at least one or more CVD in 2016, with around two thirds of CVD deaths occurring in those aged >75 years [2]. Furthermore, the AHA predicts that without effective intervention 40% of US adults will have one or more CVDs by 2030 [3], placing a substantial financial strain on the healthcare system. Not forgoing the economic impact imparted by the 2020 worldwide coronavirus pandemic, the national healthcare expenditure projections to account for CVDs alone has been estimated to further rise to ~USD 5.7 trillion by 2026 [4]. Furthermore, CVD additionally has a major impact upon wellbeing and individual quality of life, with the incidence of CVD being closely associated with an increased risk of depression [5,6] and the reduction in ability to perform activities of daily living [7]. Considering the multidisciplinary impact of CVD, effective interventions and therapies are clearly warranted. In an attempt to avoid the reliance upon pharmaceuticals and often invasive surgical procedures, naturally derived therapies have grown in popularity for the prevention and management of CVD. Common examples can include nutritional choices such as the ingestion of nutrient dense, antioxidant rich "super foods" or dietary supplements. While multivitamins, fruits and vegetables are commonly praised for their contributions to health, there is growing evidence that suggests creatine supplementation may also serve as a potential nutritional adjuvant therapy [8,9]. This naturally occurring amino-acid derivative, often taken for its benefits to skeletal muscle, has been shown to serve as an anti-inflammatory and an antioxidant, among other advantageous properties [9]. These properties suggest creatine has potential to attenuate the detrimental characteristics of CVD.

CVD is an umbrella term used to denote a variety of pathological disorders of the heart and or the vessels that stem through the body and vascularize all tissues and organs. Although both the heart and vessels work collectively to function as the cardiovascular system, pathophysiological changes in either the heart or vasculature can independently contribute to the development of CVD. Common vascular-specific pathologies can include coronary artery disease, stroke, atherosclerosis, hypertension, and peripheral artery disease. Despite all being classified under a common umbrella term, the pathological etiology of each is multifactorial, with numerous risk factors contributing to the development and severity of each. Major examples of well recognized risk factors include dyslipidemia, hypertension, systemic inflammation, obesity, diabetes, tobacco use, lack of physical activity, and alcohol abuse; all of which represent more than 90% of the CVD risks in current epidemiological studies [10]. Physiologically speaking, vascular pathologies are commonly characterized by deteriorations in vascular integrity or alterations in vessel structure. Vessel walls can be described as consisting of three structurally distinct layers, or tunicas [11]. The tunica intima is the inner most layer lining the entire vascular system, consisting of flattened endothelial cells (ECs) arranged in a longitudinal manner in the direction of blood flow [11]. The tunica media provides vasotone, mechanical strength and contractile power, by which these properties are provided by the presence of vascular smooth muscle cells (VSMCs), embedded in an intricate network of elastin and collagen [12]. Finally, the tunica adventitia is the outermost layer and provides structure whilst anchoring vessels to surrounding tissues [11]. All vessels, with the exception of the capillaries found in the microvasculature, have varying degrees of each layer in accordance to their function and location. Any disturbance in these layers however can lead to cellular dysfunction and consequential challenges to vessel health and the development of vascular pathologies [13]. Some of the most common deteriorations seen in vessel structure include arterial wall thickening, wall enlargement, arterial stiffening, endothelial dysfunction, and inflammation [14–17]; all of which are further augmented by the risk factors previously mentioned.

It can be confidently concluded that the maintenance of healthy vasculature is paramount for longevity, which is in part why vascular pathology remains the leading cause of mortality in the world. Maintenance of the inner most ECs, found lining the vessels of the vascular system, has been found to be integral to overall vascular health and reduced risk of future adverse cardiovascular events [18–20]. ECs contribute to the delicate control and homeostasis of the vascular system, adapting to hormonal [21], mechanical [22], and neural stimuli [23] to sustain vascular function. ECs additionally contribute to fluid filtration, control of vasomotor tone via the release and synthesis of vasoactive factors such as nitric oxide (NO), regulation of both local and global blood flow, regulation of blood pressure, hemostasis, hormone trafficking, angiogenesis, immune response and inflammation [17,24]. Despite their complex contribution to vascular health, ECs are delicate and can be easily damaged by varying factors such as free radicals or reaction oxygen species (ROS) [25] and chronic inflammation [26], both of which can be augmented by, or a result of, risk factors such as dyslipidemia [27], hyperglycemia [28], tobacco use [29], alcohol abuse [30], obesity [31], and unfortunately the inevitable process of aging [32]. Any damage suffered can result in adverse alterations to endothelial physiology, consequentially leading to endothelial dysfunction (ED). ED can be characterized by a reduction in the bioavailability of NO, a potent vasoactive compound, which leads to the impairment of endothelium-dependent vasodilation often seen underlining many, if not all, CVDs [15]. Furthermore, ED often leads to ECs frequently expressing and releasing more procoagulation factors, shifting what would be a hemostatic balance towards a more prothrombotic and proinflammatory state [17], all of which similarly underly a multitude of CVDs [33].

2. Combatting the Development of Vascular Pathology

While modern advances in medicine and pharmaceuticals such as blood thinners (anticoagulants), angiotensin-converting enzyme (ACE) inhibitors, beta-blockers, and diuretics have been shown to be effective for the management of vascular pathologies and CVDs, these are not without their limitations. Many medications come with inherent risk of side effects such as dizziness, constipation or diarrhea, skin irritations and excessive bleeding. These side effects, in addition to the financial costs commonly associated with prescriptions, often hinders the medical adherence of many consumers. A Keiser Family Foundation data note published in 2019 regarding prescription drugs in older adults, reported that 76% of older adults believe prescription medication cost is unreasonable, with 23% reporting difficulty in affording their prescribed medications [34]. Perhaps more shockingly, about one in five older adults (21%) reported not taking their medication as prescribed due to costs, with over half of these individuals not even informing their doctor or health care provider. Taking this data into consideration, despite the ability of pharmaceuticals to help manage the pathological characteristics of vascular diseases, it could be argued that other, or adjuvant, interventional strategies are comparably important to offer those at who are at risk of, or suffering from, vascular diseases or CVDs in general.

With the noted limitations of pharmaceuticals, the consumption of dietary nutritional supplements, or the use of nutraceuticals or functional "super" foods, has increasingly grown in popularity and is utilized by many individuals to help promote health and wellbeing [35]. In the Council for Responsible Nutrition 2018 consumer survey, 75% of US adults reported taking some form of dietary supplement to either benefit overall wellness, fill nutrient gaps, promote bone health, benefit heart health, support healthy aging, and or to aid in joint health [35]. Some common "superfoods" reporting significant benefits to cardiovascular health include fruits and vegetables high in vitamin C, polyphenols, carotenoids and lycopene (citrus fruits, blueberries, red peppers, melon, strawberries, carrots, tomatoes), nuts and peanuts (walnuts, almonds), and whole grains [36]. Other dietary supplements applauded for their vascular benefits include l-arginine [37,38], co-enzyme Q10 [39,40], nitrites [41], tetrahydrobiopterin (BH4) [42], and l-citrulline [43]. Literature suggests that it is the vitamin richness, nutrient density, and antioxidant capacity of many of these that contribute to the benefits often reported [36,44–46].

Creatine is known as an efficacious and widely popular ergogenic supplement, often taken to help enhance energy stores and the buffering capacity of skeletal muscle during high-intensity exercises. Although a popular supplement, creatine can be found naturally in meat and fish sources, with creatine content ranging between 3–5 g/kg of raw meat [47]. Despite being found naturally, to achieve the typical 20 g/day "loading" amount of creatine necessary to quickly increase skeletal muscle stores [48,49], one would have to consume approximately 4 kg of meat per day. This high meat intake, for some, could be difficult to achieve due to dietary concerns, high calorie intake, cost, or other concerns such as the inherent high fat intake. Due to these factors, creatine is now synthetically produced and remains one of the most commonly consumed supplements [50]. Creatine monohydrate was one of the earliest forms of synthetic creatine, first marketed in 1990s, and remains the most commonly consumed and utilized form of creatine in scientific literature [50]. Other forms of creatine however have been marketed with varying claims of superior absorption, chemical and physiological properties, although these claims are often unsupported [50]. Examples of other marketed creatine analogs include creatine ethyl ester, creatine malate, creatine pyruvate, and sodium creatine phosphate; all of which have a lower creatine content percent than creatine monohydrate [50]. For the sake of this review, when discussing varying studies, the term "supplemental creatine" will denote the investigator's use of creatine monohydrate. The use of another form, if used, will be noted where necessary. In addition to the known ergogenic benefits of creatine monohydrate [49,51], recent evidence also supports the ability of creatine to exhibit other non-energy related benefits, aiding in the attenuation of risk factors that are associated with vascular pathology. As aforementioned earlier, the accumulation of ROS is closely associated with a deterioration in EC health and vascular integrity. Although primarily in vitro data, there is evidence to suggest that creatine may serve as both a direct- and indirect-antioxidant [52,53], which could benefit vascular and EC health through ROS reduction. Furthermore, creatine has also been shown to serve as an anti-inflammatory agent [54], reducing EC damage induced by chronic and exercise induced inflammation. In addition to these, there is also evidence to suggest that creatine can help manage dyslipidemia [55,56], improve glycemic control [57], and improve mitochondrial function [58], all of which have been found to characterize vascular pathologies. It is due to, in part, these non-energy related properties that researchers believe is why creatine has shown such promise in other clinical populations such as those suffering from metabolic [57,59], muscular [60,61] and neurological diseases [62–64]. In addition to all the reported beneficial and pleiotropic properties of creatine, creatine supplementation also exhibits an excellent safety profile, with minimal reported side effects after acute or chronic supplementation, in moderate or large doses, in a variety of populations from young to old [49,65].

Despite the evidence supporting creatine's application as an ergogenic aid, as a potential adjuvant role in the management of varying pathologies, and its ability to influence other disease risk factors, there is scarce scientific insight into the possible application of creatine for the improvement of vascular health and CVD. Taking into consideration that creatine has been found to elicit non-energy related benefits, such as reducing damaging free radicals and ROS, reducing inflammation, reducing mitochondria-specific ROS, and possibly reducing circulating levels of homocysteine [56,66], all of which have been scientifically linked closely to vascular disease [67–69], there may be mechanisms by which creatine could potentially impact vascular health. Furthermore, the lack of clinical trials in this area highlights a major gap in the creatine literature. Although the exact mechanisms by which creatine exerts these non-energy-related benefits are still relatively speculative, with CVD remaining the leading cause of mortality and the clear untapped potential of creatine, there is a need for further research and insight into the more novel applications and functions of creatine supplementation.

The following sections of this review aim to outline the current literature available regarding creatine's application for vascular health and function, in addition to potential mechanisms by which creatine may contribute towards vascular health.

3. Existing Research on the Effects of Creatine Supplementation and Vascular Health

Since its isolation and extraction from skeletal muscle in 1832 by French chemist Michel Eugène Chevreul, the metabolic, ergogenic and physiological application of creatine has been extensively researched. Despite the recently found exciting potential of creatine to serve as an adjuvant therapy in varying clinical applications, in addition to the variety of non-energy-related properties that collectively highlight the expansive value of creatine, there have been few studies investigating the role of creatine for vascular health. Following an extensive literature search, we found only four clinical studies that reported investigating the direct impact of creatine supplementation on vascular health and function. Two of these studies [70,71] primarily looked at creatine's impact upon the macrovasculature, specifically pertaining to the larger blood vessels such as the aorta and sizeable arteries in the brain and limbs; whereas, the remaining two studies [55,56] looked primarily at the impact of creatine on the microvasculature, the portion of the vascular system that is composed of the smallest vessels such as the arterioles and capillaries.

3.1. Creatine and the Macrovasculature

Arciero et al. [70] was among the first to investigate the impact of creatine, taken either alone or in combination with resistance training, on blood flow of the lower leg (calf) and forearm. This randomized, double-blind, placebo-controlled study allocated

30 healthy male participants into one of three major groups: creatine only, creatine + resistance training, and placebo + resistance training. Both creatine and placebo (dextrose) supplementation protocols consisted of 20 g/day for the first 5 days, followed by 23 days of 10 g/day. To determine the resulting impact of supplement and/or training on blood flow, Arciero utilized the method of venous occlusion plethysmography, in which changes in limb circumference in response to rapid occlusion and reperfusion are indicative of limb blood flow, calculated as $mL \cdot 100\ mL^{-1} \cdot min^{-1}$. Following creatine supplementation, the authors reported a significant increase in both calf and forearm blood flow; however, these changes were seen only in the creatine + resistance training group and not in the creatine alone or placebo group. Considering that effects were only significant when supplementation was taken in combination with resistance training, the authors concluded that these novel results indicate a "synergistic" interplay, or "additive" effect. Although these findings support those of other studies, indicating that the benefits of supplements are often augmented when used in combination with other physical interventions [72,73], Arciero's reported findings when taken alone provide no definitive evidence to suggest that creatine could independently benefit vascular health.

Similarly, in a double-blind placebo-controlled study, Sanchez-Gonzalez et al. [71] looked to determine the effect of three weeks of creatine supplementation (2×5 g/day), in comparison to a maltodextrin placebo, on hemodynamic and arterial stiffness responses after acute isokinetic exercise in 16 healthy young males. To determine hemodynamic responses, heart rate was monitored continuously using bipolar electrocardiogram (ECG), and arterial stiffness was assessed using the current non-invasive gold standard method, pulse wave velocity (PWV) [74]. Authors reported that following the 21-day supplement period, creatine attenuated the increase in systolic blood pressure (SBP) 5 min post-exercise, and heart rate response at both 5- and 15-min post-exercise. It was also reported that creatine suppressed the increase in brachial-ankle PWV (baPWV) following the fatiguing bout of isokinetic exercise, and that those in the creatine group displayed a faster return of heart rate to resting values in comparison to those in the placebo group. The authors concluded that creatine supplementation contributed to improved hemodynamic and vascular responses to acute isokinetic bouts of exercise. Although no direct mechanisms were assessed, this effect was speculated to be due to a reduction in left ventricle afterload and reduced muscle ammonia and lactic acid production, which would have otherwise led to sympathetic-mediated increases in heart rate and blood pressures [75]. As previously mentioned, increases in arterial stiffness often indicate the presence of impaired arterial wall health and is a major contributor to the development of CVD; therefore, these results suggest a potential benefit of creatine supplementation for vascular health. Furthermore, heart rate recovery time following exercise is a powerful indicator of overall mortality [76], again suggesting a benefit of creatine to cardiovascular health. These findings are in contrast to that of Arciero et al.

3.2. Creatine and the Microvasculature

Moraes et al. [55] was the first to investigate the impact of creatine supplementation on the systemic microcirculation, rather than macrocirculation. Moraes investigated the effect of one week of 20 g/day micronized creatine, a commonly utilized "loading" protocol [49,77], on systemic microcirculation, microvascular reactivity, and skin capillary density in young healthy males. In addition to microvascular assessments, Moraes also reported on the impact of creatine on blood lipids such as low-density lipoprotein cholesterol (LDL-C), high-density lipoprotein cholesterol (HDL-C) and total cholesterol, all of which have documented associations with CVD risk [78]. Furthermore, circulating homocysteine levels were assessed, a sulfhydryl-containing amino acid known to adversely impact vascular health and increase CVD risk [79]. Following supplementation, authors reported no significant impact upon circulating homocysteine levels; however, a significant reduction was seen in both total cholesterol and low-density lipoprotein cholesterol (LDL-C). It was further reported that creatine significantly increased skin functional capillary density and recruitment post-occlusive reactive hyperemia, and that cutaneous microvascular vasodilation induced by hyperemia also increased. Despite being extremely novel at the time, this study was not without its limitations. For example, this study lacked a true control group or placebo, was open label in that all participants were aware of the supplement they were receiving, and assessed variables following a relatively short time frame of creatine supplementation. Despite these noted limitations however, the improvements found by Moraes hold great promise and suggest potential vascular-protective properties for creatine. As mentioned previously in this review, the functional ability of ECs to synthesize and release potent vasoactive compounds to control vasomotor tone is vital for the management of blood flow, blood distribution, and blood pressures. The inability of ECs to release vasoactive agents such as NO, or to generate endothelium-derived hyperpolarization factors (EDHFs) to induce vasodilation, is a common underlining characteristic of a multitude of vascular pathologies [80,81]. Moraes et al. demonstrated here that following even only an acute period of creatine supplementation, microvascular reactivity and recruitment significantly improved, indicating improvements in EC function and possible increases in upstream contribution. Not to forget as well, creatine demonstrated the potential to lower both LDL-C and total cholesterol, both of which positively impact vascular health and reduce CVD risk [82]. Although no direct mechanisms by which creatine may have imparted these vascular benefits was assessed or reported, these findings still remain promising and bolster the need for further investigation into new novel mechanisms by which creatine may operate.

Van Bavel et al. [56] further studied the effect of creatine supplementation on the microvasculature, but chose to only include participants that followed a strict vegan diet that was rich in fruits and vegetables and void of any animal-derived foods. Considering that creatine is naturally found in meats and fish, it has been shown that those following a vegetarian or vegan diet present with substantially lower creatine stores than those ingesting an omnivorous diet [83]. Therefore, there is speculation that vegetarians or vegans may benefit from creatine to a greater extent, due to the potential to store more creatine during supplementation [83]. For this single-blinded, randomized study by Van Bavel et al., forty-nine vegan subjects aged between 20 and 45 years were separated into one of two major groups: creatine group (5 g/day for three weeks) or placebo group (5 g/day maltodextrin for three weeks). Similar to Moraes et al., blood specimens were collected before and after supplementation to determine any alterations in blood lipids (total cholesterol, LDL-C and HDL-C) and homocysteine. To determine the impact of supplementation on the microvasculature, laser speckle imaging with acetylcholine (ACh) skin iontophoresis was used to assess cutaneous microvascular reactivity, and intravital video-microscopy was used to measure skin capillary density and reactivity at both rest and following post-occlusive reactive hyperemia. The authors reported a reduction in homocysteine following creatine, however this was only reported in those previously presenting with elevated levels of homocysteine (hyperhomocysteinemia). Although regarded as statistically insignificant, there was also a trend towards the reduction in LDL-C and total cholesterol in those who ingested the creatine supplementation. In regard to the microvasculature, the authors reported that the basal capillary density of the creatine group was significantly increased in comparison to the placebo group. Furthermore, the authors reported a significant increase in capillary recruitment following post-occlusive reactive hyperemia for those in the creatine group but not those in the placebo group. These findings further support those reported by Moraes. Despite these promising results, the authors did not assess specific mechanisms by which creatine may have yielded these microvascular benefits. However, it was proposed in the concluding remarks that reductions in vascular oxidative stress (reduction in ROS) may have resulted in these vascular benefits; however, no biomarkers of oxidative stress were measured. Despite the lack of knowledge regarding creatine's mechanistic contribution to vascular improvements, these findings further support the proposition that creatine may possess vascular-protective properties.

The four above-mentioned clinical trials are the only studies to directly explore the effect of creatine on the vasculature, utilizing laboratory methods that are vascular-specific, such as PWV, venous occlusion plethysmography, laser speckle imaging and microscopy. Despite the scarcity of studies, Sanchez-Gonzalez et al., Moraes et al., and Van Bavel et al. all reported significant improvements in vascular variables following creatine supplementation regimens that ranged from 7–21 days, during which time participants consumed 100–210 g of creatine in total over the entire 7–21 days. Furthermore, even though benefits were not reported by Arciero et al. following creatine supplementation alone, there was a noticeable and significant synergistic improvement when this supplement was combined with resistance training. It is also important to highlight the primary commonality, and possible limitations, shared between these studies, in that all studies utilized relatively young, healthy individuals as study subjects. Although the development of vascular pathology and CVDs are multifactorial and can be influenced by a variety of lifestyle factors, the primary unmodifiable risk factor for CVD is age [84,85], hence the vast proportion of CVD deaths occurring in those aged 75 years and above [2]. Therefore, when considering the integrity of the cardiovascular system of these young, healthy individuals, it seems justified to assume that no vascular dysfunction, or need for improvement, was even present prior to study involvement. Thus, if these authors reported benefits to the vasculature, albeit minor but statistically significant, in this young demographic, one could argue that the benefits of creatine may be even greater when applied in a population already at risk of vascular dysfunction (chronic smokers, elderly, post-menopausal women), or those suffering from CVDs already. Hence, there is a clear need for further, more population-specific, investigation into the potential of creatine to benefit vascular health.

4. The Presence and Function of Creatine within the Vascular Endothelium

Prior to proposing potential mechanisms by which creatine may be able to serve as a therapeutic supplement for vascular health, it is important to note the presence of creatine and its functional constituents within the vascular endothelium specifically.

Creatine that has either been consumed naturally via the diet or through supplementation is actively absorbed across cell membranes into the intracellular compartments via a creatine transporter (CRT), also known as SLC6A8. This transporter is sodium-(Na^+) and chloride-(Cl^-) dependent, requiring at least two Na^+ ions and one Cl^- ion for the transport of one creatine molecule [86]. Although the vast majority of creatine stores can be found within human skeletal muscle (~95%) [49], the presence of the CRT has been identified on enterocytes [87], kidney epithelial cells [88], the blood–brain barrier [89], and the ECs that line the vascular system [90]. Once located intracellularly, creatine can exist either in a free form or in a phosphorylated form, phosphocreatine (PCr). Both creatine and PCr, together with globally present creatine kinase (CK) isoenzymes, function as high-energy compounds crucial for cellular metabolism, working as the known creatine-phosphocreatine system (Cr-PCr system) [9]. In short, during cases of low cellular adenosine triphosphate (ATP) levels or high energy demand, CK will catalyze the transfer of the *N*-phosphoryl group from PCr to adenosine diphosphate (ADP) to resynthesize ATP and replenish the cells intracellular ATP pool. Conversely, when ATP production from either glycolytic or oxidative pathways are greater than ATP utilization, CK can function in reverse to capture and store this cellular energy by replenishing PCr stores. For further, more in-depth information regarding the cellular intricacies of the creatine system, readers are encouraged to read reviews by Wallimann et al. [91] and Persky and Brazeau [92]. In addition to its role as a temporal high-energy phosphate buffer, considering the complexity and variety of CK isoenzymes and their subcellular compartmentalization and distribution, the Cr-PCr system is also believed to function as a spatial high-energy shuttle. This shuttle serves to quickly, and efficiently, shuttle high-energy phosphates (potential energy) between sites of ATP production (such as the mitochondrial electron transport chain) and sites of ATP utilization (such as ATP-gated ion channels, ATP-regulated receptors, ATP-regulated ion pumps; contractile processes, cell motility, cell signaling, or organelle transport [93]).

Taking into consideration the complexity and vast functionality of the Cr-PCr system, it is clear that creatine plays a vital role in cellular function.

Decking et al. [90] investigated the importance and functional aspects of CK isoenzymes in a variety of ECs throughout the vascular system. Using both the ECs of the aorta (AECs) and the microvasculature (MVECs) of pigs and rats, Decking looked to assess the presence of varying CK isoenzymes in each subtype of ECs, the intracellular concentration of energy phosphates, and the function of CK isoenzymes during substrate depletion. First and foremost, supporting that of other studies [94,95], Decking reported that each variety of EC demonstrated the ability to take up creatine from the medium, thus indicating the presence and function of an EC-CRT. Through the use of phosphorus-31 nuclear magnetic resonance (31^P-NMR) spectroscopy to monitor cellular energy status, Decking also illustrated that porcine AECs contained a considerable amount of PCr, which suggested the presence of cytosolic CKs. Furthermore, when incubated in a creatine-rich medium, AEC concentrations of PCr increased, thus confirming the presence of a CRT and Cr-PCr system. Interestingly, when the medium was devoid of creatine, PCr accumulation rate was reduced by >90%. Decking had ultimately demonstrated not only the presence of PCr in both types of ECs, but that these concentrations were reversible, thereby indicating intricate intracellular control of energy substrates via a CK pathway. In regard to specific isoenzymes, Decking reported only the brain-brain CK (BB-CK) isoenzyme being present in the cytosol of cultured macro- and micro-ECs. However, both rat and pig ECs presented with mitochondria-specific CKs (Mt-CKs). Taking into consideration the finding of both a cytosolic CK and mitochondrial CK, Decking concluded that this strongly suggested the presence of an active Cr-PCr system, or energy shuttle between compartments. Although the expression and activity of CK and the Cr-PCr system varies between vascular beds, this pioneering research supports the presence and function of creatine and the Cr-PCr system within the vasculature. Other notable studies that complement the findings of Decking et al., supporting the presence and function of creatine, CRT and CK in ECs, include those by Loike et al. [96], Nomura et al. [54], and Sestili et al. [97].

5. Possible Application of Creatine for the Promotion of Vascular Health

Taking into consideration the additional non-energy related properties of creatine, and the beneficial application of this supplement within other pathologies [98,99], there is evidence to suggest that creatine may contribute to other novel mechanisms which could therapeutically benefit vascular health. However, scientific evidence of potential novel mechanisms is sparse. The following sections provide a unique perspective on creatine, highlighting some promising properties of creatine that may contribute to vascular health in a novel way, such as the impact of creatine on oxidative stress, nitric oxide (NO) and endothelial nitric oxide synthase (eNOS), endothelium-derived hyperpolarization factors (EDHFs), endothelial cell integrity, and the protection of cellular deoxyribonucleic acid (DNA) and ribonucleic acid (RNA).

5.1. Creatine, Oxidative Stress and Nitric Oxide Bioavailability

Oxidative stress is a pathological state of imbalance between the production of damaging free radicals and their removal by antioxidant defenses [100]. The highly reactive, unstable free radicals that characterize oxidative stress can be formed from many compounds including oxygen and nitrogen, forming reactive oxygen species (ROS) and reactive nitrogen species (RNS), respectively; or, reactive oxygen and nitrogen species (RONS) collectively [101]. Unsurprisingly, the presence or progressive accumulation of oxidative stress has been found to underline a multitude of diseases, including CVDs such as hypertension, atherosclerosis, and stroke [67]. Furthermore, other notable lifestyle risk factors that are closely associated with CVD such as smoking, alcohol abuse, lack of physical activity, and poor diet (high fat intake) [10], similarly contribute to the production of harmful free radicals [102]. Free radicals, being inherently unstable, often steal electrons from other biomolecules to satisfy their need for valent shell completion. As a consequence of this, lipids, proteins, carbohydrates, RNA and even DNA can become oxidized (loss of an electron), which alters their structure and often renders them dysfunctional. Fortunately, biologically we possess an intricate antioxidant system poised to protect, quench and to reduce the accumulation of free radicals before they can impose their harmful effects. The antioxidant system is diverse in nature, expressing both enzymatic and non-enzymatic processes which function to either directly or in indirectly reduce free radicals. However, despite these natural defenses, the antioxidant system has been shown to slowly diminish and become overwhelmed as a result of biological aging [103], in addition to being significantly impaired by certain factors such as smoking [104], poor nutrition [105] and lack of physical activity [106]. Despite the lifelong pressure put upon our antioxidant defenses, fortunately antioxidants also exist in a variety of natural sources that we can ingest through our diet. For example, foods rich in polyphenols [107], carotenoids [108], lycopene [109], vitamins (A, C and E) [110], and flavonoids [111], have all been shown to support the body's natural antioxidant defenses. It is due to the benefit of these naturally occurring antioxidants that antioxidant-containing foods and supplements have been well researched [44,112,113] and used as a therapeutic strategy for the treatment of many diseases. It is important to note that antioxidant supplementation, as with medications and other therapies, is not without limitations. Despite promising in vitro data and the emphasis put upon positive clinical trials, other clinical research has produced few promising results, reporting a variation in effectiveness between diseases and populations, with further issues surrounding individual antioxidant bioavailability and metabolism [114,115]. Therefore, although supplemental antioxidant protection is theoretically sound and physiological promising, further clinical research is still required to expand this field.

Considering the above findings, the question that remains is where does creatine supplementation fit into the realm of antioxidant defenses against oxidative stress, and ultimately improvement in vascular health? Despite creatine being structurally different from other natural antioxidants, and rarely being marketed as a potential antioxidant, there is evidence to suggest that creatine may possess both in-direct and direct-antioxidant properties [9,52,53]. For an in-depth perspective on creatine and its antioxidant potential readers are suggested to read that by Sestili et al. [53]. Matthews et al. [62] was one of the first to investigate the potential of creatine to protect against oxidative stress and neurotoxicity induced by intrastriatal injections of malonate or intraperitoneal injections of nitropropionic acid (3-NP), an animal model of Huntington's disease. It was first reported that following two weeks of 1% dietary creatine supplementation, PCr concentrations within striatal slices significantly increased, suggesting the presence of a CRT and the ability of creatine supplementation to increase brain stores of energy metabolites. Furthermore, the authors reported that those animals consuming creatine for just two weeks had significantly reduced striatal lesions following both malonate and 3-NP neurotoxicity, suggesting significant neuroprotection. Finally, Matthews reported that creatine reduced markers of hydroxyl ($\cdot$OH) free radical generation. These findings led to the first postulation that creatine may infer antioxidant-like properties. Following this study, Lawler et al. [52] investigated the direct antioxidant properties of creatine. Lawler, using a highly controlled acellular setting, looked to determine the impact of varying doses of creatine on five ROS systems: xanthine oxidase for superoxide anions ($O_2{}^-$), hydrogen peroxide (H_2O_2), peroxynitrite ($ONOO^-$), lipid peroxidation, removal of 2,2'-azino-bis3-ethylbenzothiazoline-6-sulphonic acid ($ABTS^+$) cation radical, and tert-butyl-hydroperoxide (tBOOH). Antioxidant scavenging capacity (ASC) was also assessed. Following these experiments, authors reported that creatine exhibited significant oxidant scavenging potential of ionized radicals such as $ABTS^+$, $O_2{}^-$, and $ONOO^-$. Furthermore, there was a direct dose–response relationship found between creatine and total ASC. All in all, despite being in a controlled acellular setting, Lawler was the first to show the direct antioxidant potential of creatine, thereby sparking the need and interest of further research both in vivo and in vitro. To approach the issue regarding the need for further in vitro data, Sestili et al. [97] investigated the ability of creatine to serve as an antioxidant within animal (murine myoblasts–C2C12) and human (promonocytic–U937; umbilical vein endothelial cells–HUVECs) cultured cell lines that were oxidatively injured with H_2O_2, tBOOH, and $ONOO^-$. Sestili reported that creatine treatment significantly reduced the cytotoxic effects of all oxidative stressors, consequently increasing cell vitality. Supporting that previously found by Lawler, Sestili also reported a significant dose-dependent effect. Although no exact mechanism was offered by Sestili, it is interesting to note that cellular concentrations of creatine increased following supplementation, but had no impact upon antioxidant enzymes. Furthermore, the antioxidant effects of creatine were abolished following the addition of a CRT inhibitor, β-guanidinopropionic acid. This excitingly suggests that creatine offered these antioxidant effects directly and were dependent upon intracellular creatine concentrations. In addition to the promising findings outlined above, other studies that similarly support the ability of creatine to protect and serve as an antioxidant include that by Fimognari et al. [116], Rambo et al. [117], Sestili et al. [118], and Hosamani et al. [119].

Although the above aforementioned studies all elude to the ability of creatine to serve as an antioxidant, most studies utilized oxidative stressors that were externally derived (i.e., controlled in the experiment, added to the medium, added to the cell lines). Meyer et al. [58] however, looked to determine whether creatine could directly serve as an antioxidant against one of the largest contributing sources of ROS in the body, the mitochondria. Mitochondria are complex organelles whose primary function is respiration, or oxidative phosphorylation, promoting energy production in the form of ATP through a series of intermittent stages termed the electron transport chain (ETC). Both O_2^- and H_2O_2 are produced following the leakage of electrons from varying redox centers of the ETC and other associated metabolic enzymes. Eleven distinct mitochondrial sites of O_2^- and/or H_2O_2 production have been identified, each with unique properties and contributing to mitochondrial ROS (mtROS) formation in varying amounts [120]. Although, in small amounts, mtROS is necessary for certain physiological processes such as cell signaling [120], high production or accumulation of mtROS has been closely associated with cell death (apoptosis) and cellular energy dysfunction, and has been shown to be associated with a multitude of vascular pathologies such as atherosclerosis, stroke, and hypertension [67]. The rate of mtROS production has been found to be dependent upon mitochondrial membrane potential, cellular PCr/Cr ratios, and adenine concentrations [121]. Adequate creatine availability and functional mitochondrial creatine kinases (mt-CK) have similarly been found to be necessary for the maintenance of ADP/ATP ratios that are favorable for the respiratory chain to proceed at sustainable, low ROS producing rates [58]. Meyer et al. was the first to show that exogenous creatine supplementation was capable of stimulating efficient mt-CK function, through the augmentation of cellular PCr/Cr ratios and stabilization of cellular ADP/ATP ratios. These collectively led to a reduction in H_2O_2 production through mt-CK-dependent ADP recycling in adult male Wistar rats. An additional study by Barbieri et al. [122] also demonstrated the protective effect of creatine supplementation upon the mitochondrial membrane potential, thereby sustaining the efficiency and integrity of the mitochondria in oxidatively injured (H_2O_2 treated) C2C12 mouse myoblasts. All in all, these findings support the ability of creatine to not only serve as an antioxidant, but to also successfully reduce ROS production by facilitating healthy mitochondrial function, which could therefore help reduce the risk of developing vascular pathologies.

Despite the growing body of knowledge around creatine as an antioxidant, very few studies have looked at the ability of creatine to exhibit these same properties in humans. Rahimi et al. [123] investigated the effect of creatine supplementation (4×5 g/day for 7 days) on exercise-induced oxidative stress following an acute bout of resistance exercise. This double-blind, placebo-controlled study utilized twenty-seven healthy young men, and assessed oxidative stress through plasma malondialdehyde (MDA) and urinary 8-hydroxy-2-deoxyguanosine (8-OHdG) immediately and 24-h post exercise. Rahimi reported that those supplemented with creatine presented with significantly lower markers of 8-OHdG immediately and 24-h post exercise, in comparison to placebo. Furthermore, plasma

MDA concentrations were significantly higher in the placebo group post exercise. Rahimi concluded that these results supported the ability of creatine to reduce oxidative damage induced by acute resistance exercise. Despite these promising findings, however, further research is still needed to fully determine the application of creatine as an antioxidant in humans specifically.

Taking into consideration that many, if not all, CVDs are characterized by oxidative stress and the subsequent vascular dysfunction [67,113,124–126], the ability of supplemental creatine to potentially serve as an antioxidant is just one of the novel ways in which it may be able to benefit vascular health. Further expanding upon the promising implication that creatine may serve against oxidative stress, it could be proposed that creatine could benefit vascular health through its impact on nitric oxide (NO) and endothelial nitric oxide synthase (eNOS).

One of the primary functions of healthy vascular ECs is to synthesize and release vasoactive factors to aid in the control of vasomotor tone, stimulating vasodilation or vasoconstriction to sustain healthy blood flow and regulation. NO, first described by Furchgott & Zawadzki in 1980 [127], is one of the most crucial endothelium-dependent vasodilators in the vascular system [128]. NO is synthesized in vascular ECs by the specific eNOS enzyme, in which L-arginine is converted into NO and citrulline [129]. NO exerts its vasodilatory properties in a paracrine manner by diffusing from ECs into adjacent VSMCs. The inhibition of NO via the perfusion of NG monomethyl-L-arginine (L-NMMA) has been shown to result in a dose-dependent increase in blood pressure due to increased global vasoconstriction, which can then be reversed by reintroducing NO [130]. Furthermore, even though NO is technically a free radical by definition, there is a clear association between NO-deficiency and CVD risk; with a decrease in NO bioavailability or eNOS dysfunction being linked to atherosclerosis [131], hypertension [132], Type II diabetes [133], arterial stiffness [134], stroke [135], heart disease [80], and overall mortality [136]. Clearly, sufficient NO is necessary for vascular health. Unfortunately, due to the short half-life and diffusion distance of around 1–10 s and 50–1000 μm, respectively, NO is often targeted and rendered biologically futile by the presence of other ROS such as O_2^- [137] In fact, the rate at which NO and O_2^- react occurs at the near-diffusion-limited rate of $6.7 \times 109\ M^{-1}{\cdot}s^{-1}$ [138]; therefore, nearly every collision between O_2^- and NO results in the irreversible formation of $ONOO^-$, another biologically relevant cytotoxic free radical [139]. Thus, in situations in which oxidative stress is high, overwhelming the antioxidant system, the bioavailability of NO can be detrimentally reduced, thereby impairing vascular function. Evidence of this can be seen in populations characterized by heightened levels of oxidative stress, in which NO bioavailability is low and vascular health and function is impaired, such as the elderly [140] and those with pathologies such as chronic kidney disease [141,142]. In addition to this, oxidative stress has also been shown to target important co-factors necessary for eNOS to efficiently synthesize NO. For example, free radicals have been shown to oxidize tetrahydrobiopterin (BH_4) into dihydrobiopterin (BH_2), consequentially leading to eNOS uncoupling, the reduction in NO production, and the potentiating of preexisting oxidative stress via the production of RNS [143,144]. The uncoupling and oxidative damage to eNOS has also been linked closely with a variety of CVDs [144–146].

Taking into consideration that oxidative stress can hinder NO's synthesis, function and bioavailability, if creatine exhibits antioxidant properties it could be hypothesized that creatine may mechanistically aid in the reduction and scavenging of ROS, thereby increasing NO bioavailability and contributing to improved vascular health. Ahsan et al. [147] investigated this novel theory and assessed the role of PCr in HUVECs injured by oxidized low-density lipoproteins (OxLDLs), a consequence of ROS, and its influence on the eNOS pathway. Ahsan reported that following damaged induced by OxLDL insult, HUVECs pre-treated with 10–30 mM of PCr had significantly reduced signs of endothelial apoptosis (cell death), reduced ROS generation and improved NO content. Interestingly, Ahsan also reported that those HUVECs treated with PCr had sustained eNOS signaling via the phosphatidylinositol 3-kinase/protein kinase B/endothelial nitric oxide synthase

(PI3K/Akt/eNOS) pathway. Ahsan therefore hypothesized that these PCr-mediated benefits were due to, in part, the antioxidant activity of PCr and the ability of PCr supplementation to modulate the PI3K/Akt/eNOS pathway. In addition to this, there is further evidence that demonstrates the ability of other supplemental antioxidants to improve NO bioavailability and eNOS coupling through ROS reduction, thereby supporting vascular health [148–153]. Taking all of this into consideration, if creatine can truly serve as an antioxidant and potential eNOS-stimulating supplement, creatine could serve as a vascular-protective supplement. However, more clinical trials are needed to quantify and to truly measure whether creatine supplementation could provide vascular benefits through this novel antioxidant, eNOS stimulating, NO increasing mechanism, especially in humans.

5.2. Creatine and Endothelium-Derived Hyperpolarization Factors

In addition to the important endothelium-derived relaxing factor NO, the endothelium is known to stimulate two additional relaxing factors, prostacyclin (PGI_2—via the cyclooxygenase pathway [154]) and endothelium-derived hyperpolarization factor (EDHF) [155]. Although the contribution of NO and PGI_2 to endothelium-dependent vasodilation and control of both local and systemic blood pressures is well established [156], the exact contribution of EDHFs are less known. Nevertheless, evidence suggests that EDHFs play a vital role in the control of vasomotor tone, working in synergy with other vasoactive factors to similarly contribute to blood pressure regulation [157]. For example, Scotland et al. reported that following the inhibition of both NO and PGI_2 pathways, vasodilation still occurred following stimulation, highlighting the key compensatory role of EDHFs [158]. Furthermore, it has been shown that a reduction in the potential of ECs to stimulate EDHFs is closely associated with increased CVD risk [159–162].

As made apparent by its given name, EDHFs stimulate vessel vasodilation through the hyperpolarization of neighboring VSMCs. The hyperpolarization initialized within ECs can either be propagated to neighboring VSMCs directly, via specialized myo-endothelial gap junctions, or trigger an increase of potassium (K^+) ion efflux into the subendothelial space. The resulting increase in K^+ within the interstitium can then activate calcium-dependent potassium channels (KCa^+), inwardly rectifying potassium channels (KIR^+), or the sodium-potassium pump (Na^+-K^+ pump) on VSMCs, causing VSMC hyperpolarization. This hyperpolarization then causes the closure of smooth muscle voltage gated calcium (Ca^{2+}) channels, resulting in the reduction of intracellular Ca^{2+} concentrations and ultimately vessel relaxation [155]. The initial EC derived hyperpolarization can be stimulated by a variety of stimuli, including acetylcholine (ACh), mechanical factors such as shear stress, and bradykinin. Upon stimulation by varying agonists, intracellular levels of Ca^{2+} within the ECs increase, thereby activating endothelial-specific KCa^+ channels that initialize the original hyperpolarization. Currently, both myo-endothelial gap junction coupling and fluctuations in K^+ specifically (intracellularly and intercellularly), remain comparably important and appropriate explanations for EDHF. However, it is clear by the vast involvement of multiple channels, pumps and specialized gap junctions, that the mechanism by which EDHFs function is multifactorial. It is evidential that the pathological disruption or dysfunction of any of these structures would lead to the impairment of EDHF stimulation, thereby reducing healthy vascular function.

Among the variety of influential K^+ pumps that contribute to EDHFs and vascular tone regulation, is the ATP-sensitive potassium pump (KATP). The KATP has been found to be widely distributed in a variety of tissues, including cardiac tissue (cardiac myocytes [163]), pancreatic β cells [164], vascular ECs and VSMCs [165–168]. The KATP channel consists of four pore-forming inward rectifier K^+ channel subunits, in addition to four sulfonylurea receptors (SUR) [168]. This specific KATP channel has been demonstrated to play a crucial role in vascular tone regulation in a variety of studies utilizing pharmacological approaches [169,170], transgenic mouse models [171,172], and in human patients [173,174]. KATP channels, hence their name, are subjected to regulation by intracellular ATP and ADP levels. High amounts of intracellular ATP reduces the vascular KATP channel activity (closure), whereas ADP concentrations between 0.1–3 mmol/L causes stimulatory (opening) effects [175]. The inhibition, or blockage, of KATP channels has been shown to decrease the extent of vasodilation in cerebral arterioles, basil arteries, coronary arteries, mesenteric arteries and internal mammary arteries [168]. Taking into consideration the evidence that supports the role in which KATP channels play in vascular tone, any dysfunction or deterioration of these channels is likely to contribute to the pathogenesis of many CVDs.

Therefore, how may creatine play a potential role in the support of EDHFs, thereby supporting and potentially improving vascular health and function? As stated previously, the Cr-PCr system functions as both a high-energy phosphate buffer and a spatial high-energy phosphate shuttle [93]. Considering that KATP channels are regulated by ADP and ATP, and that the Cr-PCr-system helps regulate metabolites within the cytosol and locally around ATP-consuming proteins [93]; one could claim that through creatine supplementation, KATP channels could be benefitted. In a well-written review by Dzeja & Terzic [176], the authors highlighted how the microenvironment of these specific channels harbors several phosphotransfer enzymes, including CKs, that enable the local transfer and control of ADP and ATP. It was concluded that through the delivery and removal of adenine nucleotides at these KATP channels, that the ATP/ADP ratio could be controlled, thereby influencing the opening or closing of the channel and vascular tone regulation. Other studies, such as that by Selivanov et al. [177], further support the presence of CK enzymes coupled to these specific KATP channels. Selivanov suggested that it is the collaboration between CKs and adenylate kinases (AKs) that serves as the metabolic sensor of KATP channels, in addition to controlling intracellular phosphotransfer fluxes, the regulation of channel function and vascular response. In other words, increased intracellular levels of creatine in tissues such as the endothelium, could help regulate KATP channels, hyperpolarizing the neighboring VSMCs, and contributing to the enhancement of EDHF mediated vasodilation and ultimately vascular health. It is this hypothesis that Moraes et al. claimed could have led to the improvement in microvascular reactivity and density he reported following creatine supplementation [55].

Another study that could potentially support the hypothesis of creatine supporting the function of EDHFs, is that conducted by Guerrero and colleagues [178]. Guerrero et al. looked to investigate the impact of varying ATP-generating systems upon the function of the Na^+-K^+ pump, also known as the Na^+-K^+-ATPase. As noted previously, the Na^+-K^+ pump also contributes to the generation of VSMC hyperpolarization, and quite like the described KATP channel, is ATP-dependent. Following the inducing and inhibition of varying ATP-generating systems (glycolysis, oxidative phosphorylation, and the CK-system), Guerrero concluded that even in the absence of glycolytic and oxidative ATP, the CK system and 3 mM of supplemental PCr was able to support the ATP supply required for Na^+-K^+ pump activity. Although this study was conducted in kidney cell epithelia, these results suggest a role of the CK system and supplemental PCr in the function of the Na^+-K^+ pump; a pump that has been evidentially shown to also contribute greatly to the stimulation of EDHFs. Therefore, this could justify the need for similar experiments investigating the role of creatine and ATP-dependent pumps in vascular cells specifically.

Despite the lack of direct evidence surrounding the impact or contribution of creatine supplementation towards the function of pumps, channels and structural junctions involved in the development of EDHFs, the evidence provided here outlines another novel mechanism by which creatine could potentially benefit vascular health. This evidence further emphasizes the need for future clinical trials, both in vivo and in vitro, to identify the role creatine and the Cr-PCr system in EDHFs.

5.3. Creatine, Endothelial Cell Integrity, and Inflammation

As stated in previous sections, the ability of ECs to synthesize and release a variety of vasoactive factors is paramount for overall vascular health [17], with the manifestation of ED substantially increasing the risk of CVD development [15]. In addition to ED however, endothelial structural integrity and permeability is also closely associated with EC health and CVD risk [179]. ECs line the entirety of the vascular system, directly facing the vessel lumen. Therefore, ECs serve as a semi-permeable barrier between the blood and its circulating components (water, nutrients, hormones, proteins) and the underlining tissues, allowing for the selective movement and diffusion of both small and large molecules [180]. Evidence has shown that any destabilization in the junctions that connect each EC, or disruption in the endothelial barrier itself can result in increased permeability (also known as "endothelial leakiness") and an increased risk of CVDs such as atherosclerosis, coronary artery disease, stroke, and thrombosis [179,181].

Vascular EC barrier function has been shown to be critically supported by intercellular junctions located between neighboring ECs [182]. Two major subtypes of intercellular junctions include tight junctions (TJ) and adherens junctions (AJ); however, it is the TJs that are majorly responsible for barrier function and the control of EC permeability [183]. In healthy physiological conditions, vascular ECs are regulated, and permeability is intricately controlled. In vascular pathologies however, proinflammatory signals activate the expression of varying adhesion molecules and the attraction of damaging leukocytes that collectively destabilize and debilitate the endothelial barrier [179]. It is evident in diseased vessels that barrier function of the ECs has been weakened, resulting in increased permeability and pathological alterations in EC structure [179]. Risk factors that are associated with these pathological changes include dyslipidemia [184], diabetes [185], obesity [186], and smoking [187], all of which lead to chronic inflammation and the accumulation of ROS (oxidative stress), which are also risk factors to barrier dysfunction [179]. Considering the destructive consequences of such risk factors, therapeutically reducing these risk factors and increasing the stability of TJs and the EC plasma membrane, could collectively help prevent the manifestation of endothelial barrier dysfunction.

Nomura et al. [54] was among the first to report on the impact of creatine on inflammation and EC cell membrane permeability. Using pulmonary ECs in culture, Nomura demonstrated that upon incubation with a creatine, EC intracellular stores of creatine and PCr significantly increased, signifying the presence of a CRT and CK system. When looking closer at the impact of this increased creatine and PCr, Nomura reported that following 5 mM of supplemental creatine, neutrophil adhesion to ECs had significantly reduced. Additionally, following only 0.5 mM of creatine, the expression of inflammatory markers such as intercellular adhesion molecule-1 (ICAM-1) and E-selectin was similarly reduced. Taking into consideration the known impact these inflammatory markers have upon endothelial permeability and CVD [188,189]; these results suggest potential benefits for creatine and EC health. Nomura's reported benefits, however, did not stop at a reduction in inflammatory markers. Nomura also reported that following cytotoxicity induced by serotonin and H_2O_2, 5 mM of supplemental creatine significantly reduced EC permeability, improved EC membrane stability, and reduced overall EC leakiness. Collectively these findings support the hypothesis that creatine may benefit EC stability, vascular protection, and health.

Similarly investigating the role of creatine in EC stability, Tokarska-Schlattner et al. [190] proposed a novel mechanism in which creatine, due to its zwitterionic structure, may directly interact with membrane phospholipids. Schlattner directly tested the lipid interaction of both creatine and PCr, in addition to the cyclic analogues cyclo-creatine (cCr) and phospho-cyclo creatine (PcCr), using liposome model systems. Following surface plasmon resonance spectroscopy, authors reported a low affinity PCr/phospholipid interaction which additionally induced changes in liposome shape, indicating alterations in the lipid bilayers. In addition to this, authors reported that PCr efficiently protected against induced membrane permeabilization in two separate models: induced liposome-permeabilization by membrane-active peptide melittin, and erythrocyte hemolysis induced by doxorubicin (oxidative drug), hypoosmotic stress or saponin (mild detergent). These excitingly novel results were the first to demonstrate the potential of guanidino compounds, such as PCr, to interact directly with membrane phospholipids, resulting in the modulation of membrane properties and membrane stabilization. The authors thereby concluded that PCr, which could be augmented by supplemental creatine, could exert protective effects upon cell membranes under pathological conditions. Therefore, creatine supplementation may be able to stabilize EC membranes, attenuate EC leakiness, increase EC health and protect against vascular pathologies. Furthermore, taking into consideration the presence of varying cell signaling molecules embedded within the lipid bilayer, such as G-protein coupled receptors, mechanotransducers and bioactive lipids, it could be speculated that the insertion of PCr within the lipid bilayer could have stimulatory or inhibitory effects on these molecules, thereon impacting downstream pathways and protein signaling. This speculative proposal, however, requires further investigation.

As previously mentioned, varying disorders such as diabetes have been shown to both lead to and be characterized by endothelial leakiness. Considering the pleiotropic benefits of creatine, Rahmani et al. [185] looked to determine the effect of creatine supplementation on serum biochemical markers associated with diabetes [triglycerides (TGs), total cholesterol, LDL-C, and HDL-C], and the endothelial permeability of coronary arteries within diabetic rats. Thirty-two Wistar rats were allocated randomly into one of four groups: control; supplemented creatine; diabetic; diabetic and creatine. Supplementary creatine monohydrate was given to those allocated at a dose of 400 mg/kg/daily for five months. Rahmani reported that following the supplementation period, those rats undergoing creatine treatment had significantly reduced serum levels of TGs, cholesterol, and LDL-C, with an additional positive increase in the more vascular-protective HDL-C. Furthermore, coronary permeability was significantly reduced in the diabetic group treated with creatine supplementation, in comparison to other un-treated groups. The authors therefore concluded that not only did creatine serve as a lipid-lowering supplement but, similar to that shown by Nomura et al. and Schlattner et al., creatine also reduced EC permeability and leakiness.

Recently, Hall et al. [191] identified a unique requirement for creatine and the Cr/PCr system within intestinal epithelial cells (IECs). Those suffering from irritable bowel disease (IBD) have been shown to suffer from intestinal barrier dysfunction, or intestinal epithelial leakiness. In a previous animal model of colitis, evidence suggests that creatine regulates energy distribution within IECs and further reduces the severity of colitis [192]. Hall and colleagues therefore looked to determine the importance of the CRT and Cr/PCr system within human patients suffering from IBD, and in an animal model in which mice presented with or without the CRT gene. Interestingly it was reported that the CRT was found localized around the TJs of IECs, and that the CRT closely regulated the intracellular creatine concentration and resultant barrier formation. Furthermore, in the absence of creatine, the stability of TJs was significantly altered, resulting in increased intestinal leakiness. Finally, in human biopsies, those suffering from IBD had reduced levels of messenger-RNA encoding for the CRT. It was concluded by the authors that the CRT regulates the energy balance in IECs needed for the sufficient function of TJs; therefore, it is the CRT and intracellular creatine that regulates epithelial integrity and barrier function. Despite this study being conducted in IECs specifically, what is interesting here is that Hall et al. provides evidence that illustrates the need of operational CRTs and sufficient intracellular creatine to maintain the efficient function of TJs. As alluded to earlier, in ECs specifically, the healthy function of TJs is paramount for EC integrity and barrier health. Therefore, one could hypothesize that through creatine supplementation and an increase in EC intracellular creatine concentrations, TJ energy requirements could be sufficiently managed, leading to efficient function of TJs, improved integrity and reduced risk of EC leakiness.

EC leakiness underlines a multitude of CVDs and can be augmented by a variety of risk factors such as inflammation, oxidative stress, diabetes and dyslipidemia. Thus far, we have already outlined the potential for creatine to act as an antioxidant, which in itself could have positive benefits upon vascular EC membrane stability. In addition to this however, there is evidence to suggest that creatine supplementation could reduce circulating lipids, reduce EC leakiness caused by inflammation and diabetes, and aid in the maintenance of TJ function. Despite the novelty of these properties, and the lack of true clinical trials, again we are left with what could be another benefit of creatine supplementation, one that could benefit vascular health through the improvement in EC integrity and reduction in EC leakiness.

5.4. Creatine and Vascular DNA/RNA Protection

Without the protection, maintenance, and accurate repair of DNA in response to disruption or damage, the intricate process of DNA replication and genetic coding can fail. Conclusively, the integrity of DNA is vital for cellular health, and unless precisely repaired in the event of corruption, nuclear DNA and mtDNA damage can lead to devastating mutations or other deleterious cellular consequences [193]. Damage to both DNA and ribonucleic acid (RNA) has been found to underline a plethora of vascular pathologies such as hypertension [194], coronary artery disease [195], atherosclerosis [196], and peripheral artery disease [197]. Furthermore, damage to mtDNA specifically has been closely associated with a decline in mitochondrial function, increase in mtROS production, mitochondrial apoptosis, and the manifestation of CVDs [198]. Although DNA and RNA damage can be a natural consequence of the aging [193,199], damage can also be elicited by factors such as ROS [200], radiation (such as ultraviolet rays [201]), and chemotherapeutic drugs [202]. Taking into consideration that lifestyle factors such as obesity [203], poor nutrition [204], smoking [29], and alcohol abuse [205] all lead to an increase in ROS, one could connect these lifestyle factors to an increased risk of DNA and RNA damage also. Despite DNA being a complex target for therapeutics, some promise has been found within the use of RNA therapeutics for the treatment of CVDs [206] and vascular pathologies specifically [207,208]. However, these therapeutic strategies are still relatively new; therefore, reducing inducers of DNA and RNA damage, such as ROS, could serve as a more realistic target for immediate treatment strategies.

In addition to the plethora of non-energy related properties exhibited by creatine supplementation, creatine has also been shown to protect against a variety of DNA and RNA damage. Berneburg et al. [209] investigated the impact of creatine on mitochondrial mutagenesis, function and mtDNA damage in cultured human skin fibroblasts, exposed to intense UV-radiation. Following UV-radiation, mtDNA was extracted and analyzed using real-time polymerase chain reaction (PCR) to determine DNA damage, and complete mitochondria were analyzed for oxygen consumption, ATP content, and mitochondrial membrane potential, used as markers for mitochondrial function. Following UV exposure, mtDNA was significantly damaged, resulting in a decrease in mitochondrial function. The authors further reported however, that mtDNA damage and mutagenesis, in addition to functional assessments, were all normalized and improved by increasing intracellular creatine levels. Furthermore, quite like that of other studies, creatine supplementation resulted in a dose-dependent increase in intracellular creatine concentration. Berneburg concluded here that increases in intracellular creatine played a significant protective role, protecting fibroblasts from functional-deteriorations and mtDNA damage induced by radiation.

Barbieri et al. [122] further investigated the impact of creatine supplementation on mitochondrial biosynthesis and mtDNA in differentiating C2C12 cells under oxidative conditions. Prior to analysis, creatine was preloaded to C2C12 cells by adding either 3 mM or 10 mM over the first 24 h of differentiation. Cultures were then exposed to 0 or 0.3 mM H_2O_2 for 1-h to induced oxidative stress. In regard to cell vitality, authors reported that creatine supplementation of just 3 mM significantly improved cell vitality in response to oxidative challenge. Barbieri further reported that creatine supplementation significantly increased mitochondrial markers of biosynthesis, and aided in the protection of mtDNA. Barbieri concluded that these results further supported the notion that creatine was capable of protecting mitochondria and mtDNA against oxidative insult. Considering the association between mtDNA damage, mitochondrial apoptosis and vascular pathologies [210–212], these findings suggest a potential benefit of creatine to support vascular health. On another interesting note, Barbieri found that creatine was capable of activating adenosine monophosphate-activated kinase (AMPK) within C2C12 cells. In ECs specifically, AMPK can phosphorylate and stimulate eNOS, thus the synthesis of NO. Therefore, this could be another unique and novel molecular pathway to investigate in future studies regarding creatine and vascular health.

In addition to those noted above, other studies that support the role of creatine supplementation in the protection of DNA damage include that by Rahimi et al. [123] and Mirzaei et al. [213], who both report the ability of creatine to protect DNA against oxidative stress induced by exercise. Furthermore, Qasim & Mahmood [214] report findings that suggest a DNA-protective role of creatine in human erythrocytes and lymphocytes oxidatively challenged by H_2O_2.

In an attempt to explore creatine's role in DNA protection within the vasculature specifically, Guidi et al. [215] looked to evaluate the protective effects of creatine supplementation on oxidatively injured nuclear DNA (nDNA) and mtDNA, in both an acellular system using plasmid DNA and in vascular specific cultured HUVECs. To induce oxidative damage, HUVECs pretreated with varying doses of creatine were exposed to 200 μM of H_2O_2. A PCR-based assay was then used to assess the level of DNA damage in both systems, and HUVEC cell viability was further determined 72 h post-oxidative damage. Authors reported that creatine, in a dose-dependent manner, demonstrated the ability to protect both nDNA and mtDNA against oxidative damage; however, these protective benefits were more pronounced in HUVEC mtDNA than nDNA. Furthermore, HUVECs pretreated with creatine for 24 h showed increased cell viability in comparison to controls, and presented with significantly less mtDNA breaks and mutations. Guidi also concluded that creatine could play an important role in protecting vascular EC mtDNA against oxidative insults. Furthermore, Guidi proposed that creatine could aid in mitochondrial stability, contributing to the stabilization of oxidative phosphorylation and the reduction of mtROS production.

Although evidence suggests that creatine may protect DNA from detrimental damage, less is known regarding creatine's impact upon RNA which is comparably susceptible to damage and similarly associated with vascular pathology risk. To address this, Fimognari et al. [116] aimed to investigate whether cultured Jurkat T-leukemia cell RNA could be protected against a variety of chemical insults, following either 1, 3 or 10 mM of creatine. The RNA toxic compounds utilized in this study included ethyl methanesulfonate (EMS), H_2O_2, doxorubicin (chemotherapy drug), spermine NONOate (a NO donor), and S-nitro-N-acetylpenicillamine (SNAP). The authors reported that all chemical insults significantly impaired cell vitality and damaged RNA. Creatine supplementation was found to significantly reduce RNA-damaging activity; however, was found to only do so with H_2O_2 and doxorubicin. Taking into consideration that H_2O_2 is a biologically prominent free radical, despite the inability to counteract other chemical insults, these results can still be considered promising. Fimognari concluded that creatine supplementation does indeed exhibit RNA-protective properties and hypothesized that this could have been partially due to its capacity to directly scavenge free radicals, and/or its ability to maintain cellular energy stores. Taking into consideration the detrimental role DNA and RNA damage plays in vascular pathologies [196,206], and the evidence that demonstrates the ability of creatine to protect and attenuate damage elicited by varying toxic stimuli, one could suggest that creatine may be a suitable therapeutic for the support of vascular health through the promotion of DNA and RNA maintenance and health.

6. Discussion

Creatine's popularity as an ergogenic aid exponentially increased following the success of two notable Olympic gold medalists who consumed creatine supplementation during the 1992 Barcelona Olympic games [216,217]. Since its success on a global platform, the popularity of creatine has continued to grow, generating a substantial body of literature supporting its beneficial ergogenic effects, including improvements in lean mass [218], fatigue resistance [219], and intermittent high-intensity exercise capacity [220]. Furthermore, considering the physiological impairments such as intellectual disability, seizures, muscle weakness and gastrointestinal issues that symptomatically characterize genetic diseases of creatine deficiency (such as arginine: glycine amidinotransferase deficiency, guanidinoacetate methyltransferase deficiency, and creatine transporter deficiency); it is clear that creatine plays a vital role in bioenergetics, metabolism and overall cellular health. Due to its broad application and cellular importance, many have hypothesized the potential therapeutic role of creatine. Creatine supplementation has since been shown to benefit pathologies such as myopathies [60,61], chronic kidney disease [221], diabetes [57], and neurodegenerative disorders [63]. Research has additionally highlighted other non-energy related potentials of creatine, such as its ability to serve as both a direct- and indirect-antioxidant [52,53].

Due to the widespread potential of creatine, some have argued this supplement as one of the most promising, pleiotropic, nutritional supplement therapeutics currently available. Creatine can also be sourced, however, in its natural form. Arguably more beneficial in its whole food form due to the additional nutritional value, creatine concentrations can range from 3–5 g/kg of raw meat [47]. Despite these natural sources of creatine, to successfully ingest the recommended "loading" dose of 20 g/day required to rapidly increase skeletal muscle stores [48,49], one would have to consume approximately 4 kg of meat per day. Consuming this large amount of meat could be difficult for those with dietary restrictions, those conscious of calorie intake, those wanting cheaper alternatives, or those who are cautious of other macronutrients such as fat and protein. While skeletal muscle creatine stores can similarly be increased through the ingestion of smaller amounts (3–5 g/day) over longer durations of time [48], this would still require the consumption of 1 kg of meat a day, which may be unachievable for many. Thus, creatine supplementation in the zero-calorie (if unflavored), powdered monohydrate form specifically, has remained the most efficacious method of increasing dietary creatine intake. Furthermore, creatine supplementation has been found to exhibit an excellent safety profile. As covered by Kreider et al., clinical populations supplemented with high levels of creatine (0.3–0.8 g/kg/day or 21–56 g/day for a 70 kg individual) for up to 5 years, produced no reports of significant or severe adverse events [49]. This tolerance and safety profile has been reported in both young and older populations, over both acute and long-term periods [222]. The few common side effects mentioned in creatine literature include bloating and weight gain from increases in water retention [49], while serious adverse events such as kidney dysfunction are extremely rare and are often only a consequence of poor adherence or compliance with dosing recommendations.

Although creatine supplementation appears to have high therapeutic potential, supporting published scientific literature is still limited in a variety of areas. For example, although creatine has shown great promise and has benefited certain clinical populations, there is scarce information on the role creatine plays within vascular health. Therefore, throughout this review we have discussed the few studies that have directly shown the potential of creatine supplementation to benefit vascular health. Furthermore, we have elucidated the novel ways in which creatine may serve to alleviate various risk factors that contribute to the development of vascular pathologies and CVDs. For example, we have touched upon the ability of creatine to (see Figure 1):

1. Increase natural EC stores of high-energy metabolites [54,90].
2. Serve as both a direct- and indirect- antioxidant [52,58,62,97], scavenging free radicals which could thereby improve eNOS efficiency, NO synthesis, and NO bioavailability.
3. Improve the integrity and efficiency of the mitochondria resulting in reduced mtROS production [58,215].
4. Increase microvascular density, recruitment, and vasomotor function [55,56].
5. Improve EC membrane stability and decrease EC leakiness [54,185,190,191].
6. Aid in the function of EC and VSMC energy-dependent ion pumps [176–178], thereby benefiting the propagation of EDHFs.

7. Reduce circulating amounts of damaging lipids such as LDL-C and total cholesterol [55,123].
8. Protect both DNA and RNA from cytotoxic stimuli such as oxidative stress [116,122,209,215].

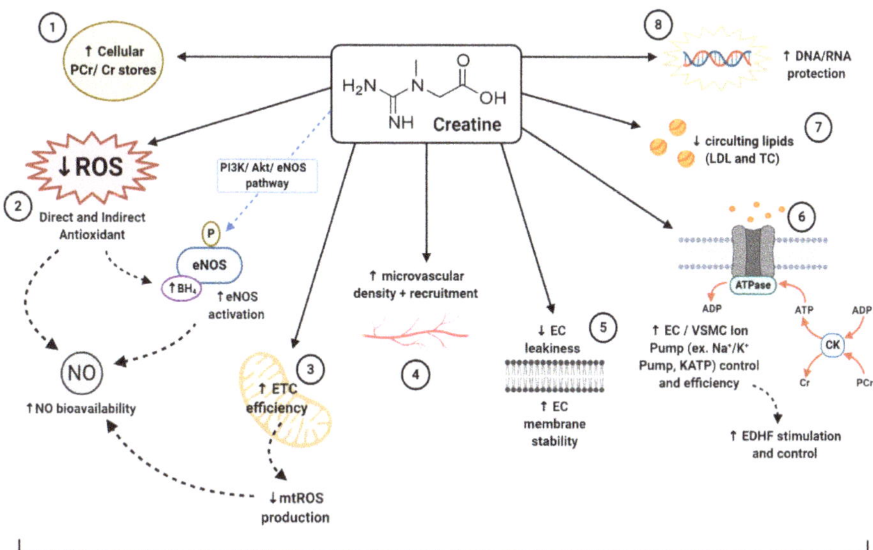

Figure 1. Novel mechanisms by which creatine supplementation may benefit vascular health and function; PCr = phosphocreatine, Cr = creatine, EC = endothelial cell, DNA = deoxyribonucleic acid, RNA = ribonucleic acid, ETC = mitochondria electron transport chain, mtROS = mitochondrial-specific reactive oxygen species, ROS = reactive oxygen species, LDL = low density lipoprotein, TC = total cholesterol, PI3K/Akt/eNOS pathway = Phosphatidylinositol 3-kinase/Protein kinase B/endothelial nitric oxide synthase pathway, eNOS = endothelial nitric oxide synthase, P = phosphorylation, BH₄ = tetrahydrobiopterin, NO = nitric oxide, ADP = adenosine diphosphate, ATP = adenosine triphosphate, CK = creatine kinase, VSMC = vascular smooth muscle cell, Na⁺/K⁺ Pump = sodium/potassium pump, KATP = ATP-sensitive potassium pump, EDHF = endothelium-derived hyperpolarization factor, ↓ = decrease/reduction in, ↑ = increase in, "blue dashed line" = proposed pathway by Ahsan et al. [147], phosphorylation and stimulation of eNOS.

In addition to these novel ways by which creatine may serve a protective role in vascular health, creatine has also been shown to help reduce inflammation and homocysteine; both of which are closely associated with incidence of CVD [68,69]. For more information regarding the impact of creatine on inflammation and homocysteine, readers are advised to reference the review by Clarke et al. [223]. Furthermore, evidence also suggests the potential of creatine and/or PCr to directly stimulate the synthesis of NO through the PI3K/Akt/eNOS pathway in endothelial cells [147]; however, this interesting mechanism requires further investigation.

7. Conclusions

To conclude, the current body of literature surrounding creatine's more clinical applications, serving as a therapeutic or adjuvant therapy in human populations, is still relatively new and requires further investigation. Despite the availability of numerous in vitro and in vivo data supporting the potential of creatine to function in these new and novel ways, there is very little translational science investigating the impact of creatine supplementation on human vasculature, specifically. To date, only four studies have directly investigated the effects of creatine on the vasculature in humans. Therefore, as highlighted throughout this review, although there is evidence to suggest that creatine may possess unique properties that may impart novel benefits upon the vasculature, further clinical research is needed to truly determine the impact of creatine supplementation on vascular health. Furthermore, population specific clinical trials involving those at risk of, or presenting with, vascular pathology would further contribute to the body of knowledge, helping to uncover the potential applications of creatine supplementation within vascular health.

Author Contributions: Conceptualization, H.C.; writing—original draft preparation, H.C.; writing—review and editing, H.C., M.J.O., R.C.H.; funding acquisition, R.C.H. All authors have read and agreed to the published version of the manuscript.

Funding: R.C.H. and H.C. are supported by the National Heart, Lung, and Blood Institute (NHLBI) or NIH: 2R15HL113854-02.

Institutional Review Board Statement: Not applicable, as this review incorporates all previously published studies. No new study was conducted by the authors.

Informed Consent Statement: Not applicable.

Data Availability Statement: No new data were created or analyzed by the authors and incorporated in this review. Data sharing is not applicable to this article.

Acknowledgments: The authors would like to acknowledge and thank BioRender.com from which Figure 1 was created.

Conflicts of Interest: The authors declare no conflict of interest.

References

1. Virani, S.S.; Alonso, A.; Aparicio, H.J.; Benjamin, E.J.; Bittencourt, M.S.; Callaway, C.W.; Carson, A.P.; Chamberlain, A.M.; Cheng, S.; Delling, F.N.; et al. Heart Disease and Stroke Statistics—2021 Update: A Report from the American Heart Association. *Circulation* **2020**, E139–E596. [CrossRef]

2. Mozaffarian, D.; Benjamin, E.J.; Go, A.S.; Arnett, D.K.; Blaha, M.J.; Cushman, M.; Das, S.R.; De Ferranti, S.; Després, J.-P.; Fullerton, H.J.; et al. Heart Disease and Stroke Statistics—2016 Update: A Report from the American Heart Association. American Heart Association Statistics Committee, Stroke Statistics Subcommittee. *Circulation* **2016**, *133*, e38–e360. [CrossRef] [PubMed]

3. Heidenreich, P.A.; Trogdon, J.G.; Khavjou, O.A.; Butler, J.; Dracup, K.; Ezekowitz, M.D.; Finkelstein, E.A.; Hong, Y.; Johnston, S.C.; Khera, A.; et al. Forecasting the Future of Cardiovascular Disease in the United States. *Circulation* **2011**, *123*, 933–944. [CrossRef] [PubMed]

4. Cuckler, G.A.; Sisko, A.M.; Poisal, J.A.; Keehan, S.P.; Smith, S.D.; Madison, A.J.; Wolfe, C.J.; Hardesty, J.C. National Health Expenditure Projections, 2017–2026: Despite Uncertainty, Fundamentals Primarily Drive Spending Growth. *Health Aff.* **2018**, *37*, 482–492. [CrossRef] [PubMed]

5. Celano, C.M.; Huffman, J.C. Depression and Cardiac Disease. *Cardiol. Rev.* **2011**, *19*, 130–142. [CrossRef]

6. Stewart, J.C.; Rollman, B.L. Optimizing Approaches to Addressing Depression in Cardiac Patients: A Comment on O'Neil et al. *Ann. Behav. Med.* **2014**, *48*, 142–144. [CrossRef]

7. De Leon, C.F.M.; Krumholz, H.M.; Vaccarino, V.; Williams, C.S.; Glass, T.A.; Berkman, L.F.; Kas, S.V. A population-based perspective of changes in health-related quality of life after myocardial infarction in older men and women. *J. Clin. Epidemiol.* **1998**, *51*, 609–616. [CrossRef]

8. Smith, R.N.; Agharkar, A.S.; Gonzales, E.B. A review of creatine supplementation in age-related diseases: More than a supplement for athletes. *F1000Research* **2014**, *3*, 222. [CrossRef]

9. Wallimann, T.; Tokarska-Schlattner, M.; Schlattner, U. The creatine kinase system and pleiotropic effects of creatine. *Amino Acids* **2011**, *40*, 1271–1296. [CrossRef] [PubMed]

10. Flora, G.D. A Brief Review of Cardiovascular Diseases, Associated Risk Factors and Current Treatment Regimes. *Curr. Pharm. Des.* **2019**, *25*, 4063–4084. [CrossRef] [PubMed]

11. Martinez-Lemus, L.A. The Dynamic Structure of Arterioles. *Basic Clin. Pharmacol. Toxicol.* **2011**, *110*, 5–11. [CrossRef] [PubMed]

12. Herring, N.; Paterson, D.J. *Levick's Introduction to Cardiovascular Physiology*, 6th ed.; CRC Press: Boca Raton, FL, USA, 2018.

13. Bennett, M.R.; Sinha, S.; Owens, G.K. Vascular Smooth Muscle Cells in Atherosclerosis. *Circ. Res.* **2016**, *118*, 692–702. [CrossRef] [PubMed]

14. Seals, D.R.; Jablonski, K.L.; Donato, A.J. Aging and vascular endothelial function in humans. *Clin. Sci.* **2011**, *120*, 357–375. [CrossRef] [PubMed]

15. Widmer, R.J.; Lerman, A. Endothelial dysfunction and cardiovascular disease. *Glob. Cardiol. Sci. Pract.* **2014**, *2014*, 291–308. [CrossRef]

16. Kohn, J.C.; Lampi, M.C.; Reinhart-King, C.A. Age-related vascular stiffening: Causes and consequences. *Front. Genet.* **2015**, *6*, 112. [CrossRef] [PubMed]

17. Rajendran, P.; Rengarajan, T.; Thangavel, J.; Nishigaki, Y.; Sakthisekaran, D.; Sethi, G.; Nishigaki, I. The Vascular Endothelium and Human Diseases. *Int. J. Biol. Sci.* **2013**, *9*, 1057–1069. [CrossRef]

18. Schwartz, B.G.; Economides, C.; Mayeda, G.S.; Burstein, S.; A Kloner, R. The endothelial cell in health and disease: Its function, dysfunction, measurement and therapy. *Int. J. Impot. Res.* **2009**, *22*, 77–90. [CrossRef]

19. Barthelmes, J.; Nägele, M.P.; Ludovici, V.; Ruschitzka, F.; Sudano, I.; Flammer, A.J. Endothelial dysfunction in cardiovascular disease and Flammer syndrome—Similarities and differences. *EPMA J.* **2017**, *8*, 99–109. [CrossRef]

20. Favero, G.; Paganelli, C.; Buffoli, B.; Rodella, L.F.; Rezzani, R. Endothelium and Its Alterations in Cardiovascular Diseases: Life Style Intervention. *BioMed Res. Int.* **2014**, *2014*, 801896 . [CrossRef]

21. Henderson, K.K.; Byron, K.L. Vasopressin-induced vasoconstriction: Two concentration-dependent signaling pathways. *J. Appl. Physiol.* **2007**, *102*, 1402–1409. [CrossRef]

22. Ye, G.J.; Nesmith, A.P.; Parker, K.K. The Role of Mechanotransduction on Vascular Smooth Muscle Myocytes Cytoskeleton and Contractile Function. *Anat. Rec. Adv. Integr. Anat. Evol. Biol.* **2014**, *297*, 1758–1769. [CrossRef]

23. Thomas, G.D. Neural control of the circulation. *Adv. Physiol. Educ.* **2011**, *35*, 28–32. [CrossRef]

24. Pugsley, M.K.; Tabrizchi, R. The vascular system. An overview of structure and function. *J. Pharmacol. Toxicol. Methods* **2000**, *44*, 333–340. [CrossRef]

25. Incalza, M.A.; D'Oria, R.; Natalicchio, A.; Perrini, S.; Laviola, L.; Giorgino, F. Oxidative stress and reactive oxygen species in endothelial dysfunction associated with cardiovascular and metabolic diseases. *Vasc. Pharmacol.* **2018**, *100*, 1–19. [CrossRef]

26. Castellon, X.; Bogdanova, V. Chronic Inflammatory Diseases and Endothelial Dysfunction. *Aging Dis.* **2016**, *7*, 81–89. [CrossRef]

27. Kim, J.-A.; Montagnani, M.; Chandrasekran, S.; Quon, M.J. Role of Lipotoxicity in Endothelial Dysfunction. *Heart Fail. Clin.* **2012**, *8*, 589–607. [CrossRef]

28. Popov, D. Endothelial cell dysfunction in hyperglycemia: Phenotypic change, intracellular signaling modification, ultrastructural alteration, and potential clinical outcomes. *Int. J. Diabetes Mellit.* **2010**, *2*, 189–195. [CrossRef]

29. Pittilo, R.M. Cigarette smoking, endothelial injury and cardiovascular disease. *Int. J. Exp. Pathol.* **2001**, *81*, 219–230. [CrossRef] [PubMed]

30. Tanaka, A.; Cui, R.; Kitamura, A.; Liu, K.; Imano, H.; Yamagishi, K.; Kiyama, M.; Okada, T.; Iso, H.; CIRCS Investigators. Heavy Alcohol Consumption is Associated with Impaired Endothelial Function: The Circulatory Risk in Communities Study (CIRCS). *J. Atheroscler. Thromb.* **2016**, *23*, 1047–1054. [CrossRef] [PubMed]

31. Lobato, N.; Filgueira, F.P.; Akamine, E.H.; Tostes, R.; De Carvalho, M.H.C.; Fortes, Z.B. Mechanisms of endothelial dysfunction in obesity-associated hypertension. *Braz. J. Med. Biol. Res.* **2012**, *45*, 392–400. [CrossRef]

32. Fleg, J.L.; Strait, J. Age-associated changes in cardiovascular structure and function: A fertile milieu for future disease. *Heart Fail. Rev.* **2012**, *17*, 545–554. [CrossRef]

33. Widlansky, M.E.; Gokce, N.; Keaney, J.F.; Vita, J.A. The clinical implications of endothelial dysfunction. *J. Am. Coll. Cardiol.* **2003**, *42*, 1149–1160. [CrossRef]

34. Kaiser Family Foundation. Data Note Prescription Drugs and Older Adults. Available online: https://www.kff.org/health-reform/issue-brief/data-note-prescription-drugs-and-older-adults/ (accessed on 26 December 2020).

35. Council for Responsible Nutrition. 2018 Consumer Survey on Dietary Supplements. Available online: https://www.crnusa.org/sites/default/files/images/2018-survey/CRN-ConsumerSurvey-Infographic-2019f.pdf (accessed on 26 December 2020).

36. Johnston, C. Functional Foods as Modifiers of Cardiovascular Disease. *Am. J. Lifestyle Med.* **2009**, *3*, 39S–43S. [CrossRef]

37. Bode-Böger, S.M.; Muke, J.; Surdacki, A.; Brabant, G.; Böger, R.H.; Frölich, J.C. Oral L-arginine improves endothelial function in healthy individuals older than 70 years. *Vasc. Med.* **2003**, *8*, 77–81. [CrossRef] [PubMed]

38. Yin, W.-H.; Chen, J.-W.; Tsai, C.; Chiang, M.-C.; Young, M.S.; Lin, S.-J. L-arginine improves endothelial function and reduces LDL oxidation in patients with stable coronary artery disease. *Clin. Nutr.* **2005**, *24*, 988–997. [CrossRef] [PubMed]

39. Dai, Y.-L.; Luk, T.-H.; Yiu, K.-H.; Wang, M.; Yip, P.M.; Lee, S.W.; Li, S.-W.; Tam, S.; Fong, B.; Lau, C.-P.; et al. Reversal of mitochondrial dysfunction by coenzyme Q10 supplement improves endothelial function in patients with ischaemic left ventricular systolic dysfunction: A randomized controlled trial. *Atherosclerosis* **2011**, *216*, 395–401. [CrossRef] [PubMed]

40. Zozina, V.I.; Covantev, S.; Goroshko, O.A.; Krasnykh, L.M.; Kukes, V.G. Coenzyme Q10 in Cardiovascular and Metabolic Diseases: Current State of the Problem. *Curr. Cardiol. Rev.* **2018**, *14*, 164–174. [CrossRef]

41. Walker, M.A.; Bailey, T.G.; McIlvenna, L.; Allen, J.D.; Green, D.J.; Askew, C.D. Acute Dietary Nitrate Supplementation Improves Flow Mediated Dilatation of the Superficial Femoral Artery in Healthy Older Males. *Nutrients* **2019**, *11*, 954. [CrossRef]

42. Eskurza, I.; Myerburgh, L.A.; Kahn, Z.D.; Seals, D.R. Tetrahydrobiopterin augments endothelium-dependent dilatation in sedentary but not in habitually exercising older adults. *J. Physiol.* **2005**, *568*, 1057–1065. [CrossRef]

43. Romero, M.J.; Platt, D.H.; Caldwell, R.W. Therapeutic Use of Citrulline in Cardiovascular Disease. *Cardiovasc. Drug Rev.* **2006**, *24*, 275–290. [CrossRef]

44. Fusco, D.; Colloca, G.; Monaco, M.R.L.; Cesari, M. Effects of antioxidant supplementation on the aging process. *Clin. Interv. Aging* **2007**, *2*, 377 387. [PubMed]

45. Durante, A.; Bronzato, S. Dietary supplements and cardiovascular diseases. *Int. J. Prev. Med.* **2018**, *9*, 80. [CrossRef]

46. Sosnowska, B.; Penson, P.; Banach, M. The role of nutraceuticals in the prevention of cardiovascular disease. *Cardiovasc. Diagn. Ther.* **2017**, *67*, S21–S31. [CrossRef] [PubMed]

47. Rasmussen, C.; Greenwood, M.; Kalman, D.; Antonio, J. Nutritional Supplements for Endurance Athletes. In *Nutritional Supplements in Sports and Exercise*; Humana Press: Totowa, NJ, USA, 2008; pp. 369–407.
48. Hultman, E.; Soderlund, K.; Timmons, J.A.; Cederblad, G.; Greenhaff, P.L. Muscle creatine loading in men. *J. Appl. Physiol.* **1996**, *81*, 232–237. [CrossRef]
49. Kreider, R.B.; Kalman, D.S.; Antonio, J.; Ziegenfuss, T.N.; Wildman, R.; Collins, R.; Candow, D.G.; Kleiner, S.M.; Almada, A.L.; Lopez, H.L. International Society of Sports Nutrition position stand: Safety and efficacy of creatine supplementation in exercise, sport, and medicine. *J. Int. Soc. Sports Nutr.* **2017**, *14*, 1–18. [CrossRef]
50. Jäger, R.; Purpura, M.; Shao, A.; Inoue, T.; Kreider, R.B. Analysis of the efficacy, safety, and regulatory status of novel forms of creatine. *Amino Acids* **2011**, *40*, 1369–1383. [CrossRef]
51. Cooper, R.; Naclerio, F.; Allgrove, J.; Jimenez, A. Creatine supplementation with specific view to exercise/sports performance: An update. *J. Int. Soc. Sports Nutr.* **2012**, *9*, 33. [CrossRef]
52. Lawler, J.M.; Barnes, W.S.; Wu, G.; Song, W.; Demaree, S. Direct Antioxidant Properties of Creatine. *Biochem. Biophys. Res. Commun.* **2002**, *290*, 47–52. [CrossRef] [PubMed]
53. Sestili, P.; Martinelli, C.; Colombo, E.; Barbieri, E.; Potenza, L.; Sartini, S.; Fimognari, C. Creatine as an antioxidant. *Amino Acids* **2011**, *40*, 1385–1396. [CrossRef] [PubMed]
54. Nomura, A.; Zhang, M.; Sakamoto, T.; Ishii, Y.; Morishima, Y.; Mochizuki, M.; Kimura, T.; Uchida, Y.; Sekizawa, K. Anti-inflammatory activity of creatine supplementation in endothelial cells in vitro. *Br. J. Pharmacol.* **2003**, *139*, 715–720. [CrossRef]
55. De Moraes, R.; Van Bavel, D.; De Moraes, B.S.; Tibiriçá, E. Effects of dietary creatine supplementation on systemic microvascular density and reactivity in healthy young adults. *Nutr. J.* **2014**, *13*, 115. [CrossRef]
56. Van Bavel, D.; De Moraes, R.; Tibirica, E. Effects of dietary supplementation with creatine on homocysteinemia and systemic microvascular endothelial function in individuals adhering to vegan diets. *Fundam. Clin. Pharmacol.* **2019**, *33*, 428–440. [CrossRef]
57. Pinto, C.L.; Botelho, P.B.; Pimentel, G.D.; Campos-Ferraz, P.L.; Mota, J.F. Creatine supplementation and glycemic control: A systematic review. *Amino Acids* **2016**, *48*, 2103–2129. [CrossRef]
58. Meyer, L.E.; Machado, L.B.; Santiago, A.P.S.; Da-Silva, W.S.; De Felice, F.G.; Holub, O.; Oliveira, M.F.; Galina, A. Mitochondrial Creatine Kinase Activity Prevents Reactive Oxygen Species Generation. *J. Biol. Chem.* **2006**, *281*, 37361–37371. [CrossRef]
59. Gualano, B.; Painelli, V.D.S.; Roschel, H.; Lugaresi, R.; Dorea, E.; Artioli, G.G.; Lima, F.R.; Da Silva, M.E.R.; Cunha, M.R.; Seguro, A.C.; et al. Creatine supplementation does not impair kidney function in type 2 diabetic patients: A randomized, double-blind, placebo-controlled, clinical trial. *Graefe Arch. Clin. Exp. Ophthalmol.* **2010**, *111*, 749–756. [CrossRef]
60. Tarnopolsky, M.; Martin, J. Creatine monohydrate increases strength in patients with neuromuscular disease. *Neurology* **1999**, *52*, 854. [CrossRef]
61. Tarnopolsky, M.A.; Mahoney, D.J.; Vajsar, J.; Rodriguez, C.; Doherty, T.J.; Roy, B.D.; Biggar, D. Creatine monohydrate enhances strength and body composition in Duchenne muscular dystrophy. *Neurology* **2004**, *62*, 1771–1777. [CrossRef]
62. Matthews, R.T.; Yang, L.; Jenkins, B.G.; Ferrante, R.J.; Rosen, B.R.; Kaddurah-Daouk, R.; Beal, M.F. Neuroprotective Effects of Creatine and Cyclocreatine in Animal Models of Huntington's Disease. *J. Neurosci.* **1998**, *18*, 156–163. [CrossRef] [PubMed]
63. Matthews, R.T.; Ferrante, R.J.; Klivenyia, P.; Yanga, L.; Klein, A.M.; Muellera, G.; Daoukc, R.K.; Beala, M.F. Creatine and Cyclocreatine Attenuate MPTP Neurotoxicity. *Exp. Neurol.* **1999**, *157*, 142–149. [CrossRef] [PubMed]
64. Li, Z.; Wang, P.; Yu, Z.; Cong, Y.; Sun, H.; Zhang, J.; Zhang, J.; Sun, C.; Zhang, Y.; Ju, X. The effect of creatine and coenzyme q10 combination therapy on mild cognitive impairment in Parkinson's disease. *Eur. Neurol.* **2015**, *73*, 205–211. [CrossRef] [PubMed]
65. Persky, A.M.; Rawson, E.S. Safety of Creatine Supplementation. *Cholest. Bind. Cholest. Transp. Proteins* **2007**, *46*, 275–289. [CrossRef]
66. Korzun, W.J. Oral creatine supplements lower plasma homocysteine concentrations in humans. *Am. Soc. Clin. Lab. Sci.* **2004**, *17*, 102–106.
67. Dhalla, N.S.; Temsah, R.M.; Netticadan, T. Role of oxidative stress in cardiovascular diseases. *J. Hypertens.* **2000**, *18*, 655–673. [CrossRef]
68. Willerson, J.T. Inflammation as a Cardiovascular Risk Factor. *Circulation* **2004**, *109*, II-2. [CrossRef] [PubMed]
69. Wald, D.S.; Law, M.; Morris, J.K. Homocysteine and cardiovascular disease: Evidence on causality from a meta-analysis. *BMJ* **2002**, *325*, 1202–1206. [CrossRef] [PubMed]
70. Arciero, P.J.; Hannibal, N.S.; Nindl, B.C.; Gentile, C.L.; Hamed, J.; Vukovich, M.D. Comparison of creatine ingestion and resistance training on energy expenditure and limb blood flow. *Metabolism* **2001**, *50*, 1429–1434. [CrossRef] [PubMed]
71. Sanchez-Gonzalez, M.A.; Wieder, R.; Kim, J.-S.; Vicil, F.; Figueroa, A. Creatine supplementation attenuates hemodynamic and arterial stiffness responses following an acute bout of isokinetic exercise. *Graefe Arch. Clin. Exp. Ophthalmol.* **2011**, *111*, 1965–1971. [CrossRef]
72. Englund, D.A.; Kirn, D.R.; Koochek, A.; Zhu, H.; Travison, T.G.; Reid, K.F.; Von Berens, Å.; Melin, M.; Cederholm, T.; Gustafsson, T.; et al. Nutritional Supplementation With Physical Activity Improves Muscle Composition in Mobility-Limited Older Adults, The VIVE2 Study: A Randomized, Double-Blind, Placebo-Controlled Trial. *J. Gerontol. Ser. A Boil. Sci. Med. Sci.* **2018**, *73*, 95–101. [CrossRef] [PubMed]
73. Beaudart, C.; Dawson, A.; Shaw, S.C.; Harvey, N.C.; Kanis, J.A.; Binkley, N.; Reginster, J.Y.; Chapurlat, R.; Chan, D.C.; Bruyère, O.; et al. Nutrition and physical activity in the prevention and treatment of sarcopenia: Systematic review. *Osteoporos. Int.* **2017**, *28*, 1817–1833. [CrossRef]

74. Laurent, S.; Cockcroft, J.; Van Bortel, L.; Boutouyrie, P.; Giannattasio, C.; Hayoz, D.; Pannier, B.; Vlachopoulos, C.; Wilkinson, I.; Struijker-Boudier, H.; et al. Expert consensus document on arterial stiffness: Methodological issues and clinical applications. *Eur. Heart J.* **2006**, *27*, 2588–2605. [CrossRef]

75. Cui, J.; Mascarenhas, V.; Moradkhan, R.; Blaha, C.; Sinoway, L.I. Effects of muscle metabolites on responses of muscle sympathetic nerve activity to mechanoreceptor(s) stimulation in healthy humans. *Am. J. Physiol. Integr. Comp. Physiol.* **2008**, *294*, R458–R466. [CrossRef] [PubMed]

76. Cole, C.R.; Blackstone, E.H.; Pashkow, F.J.; Snader, C.E.; Lauer, M.S. Heart-Rate Recovery Immediately after Exercise as a Predictor of Mortality. *N. Engl. J. Med.* **1999**, *341*, 1351–1357. [CrossRef] [PubMed]

77. Harris, R.C.; Söderlund, K.; Hultman, E. Elevation of creatine in resting and exercised muscle of normal subjects by creatine supplementation. *Clin. Sci.* **1992**, *83*, 367–374. [CrossRef] [PubMed]

78. Nelson, R.H. Hyperlipidemia as a Risk Factor for Cardiovascular Disease. *Prim. Care Clin. Off. Pract.* **2013**, *40*, 195–211. [CrossRef]

79. Kumar, A.; Palfrey, H.A.; Pathak, R.; Kadowitz, P.J.; Gettys, T.W.; Murthy, S.N. The metabolism and significance of homocysteine in nutrition and health. *Nutr. Metab.* **2017**, *14*, 1–12. [CrossRef] [PubMed]

80. Raddino, R.; Carretta, G.; Teli, M.; Bonadei, I.; Robba, D.; Zanini, G.; Madureri, A.; Nodari, S.; Cas, L.D. Nitric oxide and cardiovascular risk factors. *Heart Int.* **2007**, *3*, 18. [CrossRef] [PubMed]

81. Sverdlov, A.L.; Ngo, D.T.; Chan, W.P.; Chirkov, Y.Y.; Horowitz, J.D. Aging of the Nitric Oxide System: Are We as Old as Our NO? *J. Am. Heart Assoc.* **2014**, *3*, e000973. [CrossRef]

82. Silverman, M.G.; Ference, B.A.; Im, K.; Wiviott, S.D.; Giugliano, R.P.; Grundy, S.M.; Braunwald, E.; Sabatine, M.S. Association Between Lowering LDL-C and Cardiovascular Risk Reduction Among Different Therapeutic Interventions. *JAMA* **2016**, *316*, 1289–1297. [CrossRef] [PubMed]

83. Kaviani, M.; Shaw, K.; Chilibeck, P.D. Benefits of Creatine Supplementation for Vegetarians Compared to Omnivorous Athletes: A Systematic Review. *Int. J. Environ. Res. Public Health* **2020**, *17*, 3041. [CrossRef] [PubMed]

84. Dhingra, R.; Vasan, R.S. Age as a Risk Factor. *Med. Clin. N. Am.* **2012**, *96*, 87–91. [CrossRef] [PubMed]

85. Harman, D. The aging process: Major risk factor for disease and death. *Proc. Natl. Acad. Sci. USA* **1991**, *88*, 5360–5363. [CrossRef] [PubMed]

86. Snow, R.J.; Murphy, R.M. Creatine and the creatine transporter: A review. *Mol. Cell. Biochem.* **2001**, *224*, 169–181. [CrossRef] [PubMed]

87. Peral, M.J.; García-Delgado, M.; Calonge, M.L.; Durán, J.M.; De La Horra, M.C.; Wallimann, T.; Speer, O.; Ilundáin, A.A. Human, rat and chicken small intestinal Na^+-Cl^--creatine transporter: Functional, molecular characterization and localization. *J. Physiol.* **2002**, *545*, 133–144. [CrossRef] [PubMed]

88. Li, H.; Thali, R.F.; Smolak, C.; Gong, F.; Alzamora, R.; Wallimann, T.; Scholz, R.; Pastor-Soler, N.M.; Neumann, D.; Hallows, K.R. Regulation of the creatine transporter by AMP-activated protein kinase in kidney epithelial cells. *Am. J. Physiol. Physiol.* **2010**, *299*, F167–F177. [CrossRef] [PubMed]

89. Braissant, O. Creatine and guanidinoacetate transport at blood-brain and blood-cerebrospinal fluid barriers. *J. Inherit. Metab. Dis.* **2012**, *35*, 655–664. [CrossRef]

90. Decking, U.K.M.; Alves, C.; Wallimann, T.; Wyss, M.; Schrader, J. Functional aspects of creatine kinase isoenzymes in endothelial cells. *Am. J. Physiol. Physiol.* **2001**, *281*, C320–C328. [CrossRef]

91. Wallimann, T.; Dolder, M.; Schlattner, U.; Eder, M.; Hornemann, T.; O'Gorman, E.; Rück, A.; Brdiczka, D. Some new aspects of creatine kinase (CK): Compartmentation, structure, function and regulation for cellular and mitochondrial bioenergetics and physiology. *BioFactors* **1998**, *8*, 229–234. [CrossRef] [PubMed]

92. Persky, A.M.; A Brazeau, G. Clinical pharmacology of the dietary supplement creatine monohydrate. *Pharmacol. Rev.* **2001**, *53*, 161–176. [PubMed]

93. Wallimann, T.; Wyss, M.; Brdiczka, D.; Nicolay, K.; Eppenberger, H.M. Intracellular compartmentation, structure and function of creatine kinase isoenzymes in tissues with high and fluctuating energy demands: The 'phosphocreatine circuit' for cellular energy homeostasis. *Biochem. J.* **1992**, *281*, 21–40. [CrossRef] [PubMed]

94. Fitch, C.D.; Shields, R.P. Creatine metabolism in skeletal muscle. I. Creatine movement across muscle membranes. *J. Biol. Chem.* **1966**, *241*, 3611–3614. [CrossRef]

95. Guerrero-Ontiveros, M.L.; Wallimann, T. Creatine supplementation in health and disease. Effects of chronic creatine ingestion in vivo: Down-regulation of the expression of creatine transporter isoforms in skeletal muscle. *Mol. Cell. Biochem.* **1998**, *184*, 427–437. [CrossRef]

96. Loike, J.D.; Cao, L.; Brett, J.; Ogawa, S.; Silverstein, S.C.; Stern, D. Hypoxia induces glucose transporter expression in endothelial cells. *Am. J. Physiol. Physiol.* **1992**, *263*, C326–C333. [CrossRef]

97. Sestili, P.; Martinelli, C.; Bravi, G.; Piccoli, G.; Curci, R.; Battistelli, M.; Falcieri, E.; Agostini, D.; Gioacchini, A.M.; Stocchi, V. Creatine supplementation affords cytoprotection in oxidatively injured cultured mammalian cells via direct antioxidant activity. *Free Radic. Biol. Med.* **2006**, *40*, 837–849. [CrossRef]

98. Gualano, B.; Artioli, G.G.; Poortmans, J.R.; Junior, A.H.L. Exploring the therapeutic role of creatine supplementation. *Amino Acids* **2009**, *38*, 31–44. [CrossRef] [PubMed]

99. Gualano, B.; Roschel, H.; Lancha, A.H., Jr.; Brightbill, C.E.; Rawson, E.S. In sickness and in health: The widespread application of creatine supplementation. *Amino Acids* **2011**, *43*, 519–529. [CrossRef] [PubMed]

100. Betteridge, D.J. What is oxidative stress? *Metabolism* **2000**, *49*, 3–8. [CrossRef]
101. Burton, G.J.; Jauniaux, E. Oxidative stress. *Best Pract. Res. Clin. Obstet. Gynaecol.* **2011**, *25*, 287–299. [CrossRef] [PubMed]
102. Kandola, K.; Bowman, A.; Birch-Machin, M.A. Oxidative stress—A key emerging impact factor in health, ageing, lifestyle and aesthetics. *Int. J. Cosmet. Sci.* **2015**, *37*, 1–8. [CrossRef]
103. Liguori, I.; Russo, G.; Curcio, F.; Bulli, G.; Aran, L.; Della-Morte, D.; Gargiulo, G.; Testa, G.; Cacciatore, F.; Bonaduce, D.; et al. Oxidative stress, aging, and diseases. *Clin. Interv. Aging* **2018**, *13*, 757–772. [CrossRef] [PubMed]
104. Bloomer, R.J. Decreased blood antioxidant capacity and increased lipid peroxidation in young cigarette smokers compared to nonsmokers: Impact of dietary intake. *Nutr. J.* **2007**, *6*, 39. [CrossRef]
105. Liu, Z.; Ren, Z.; Zhang, J.; Chuang, C.-C.; Kandaswamy, E.; Zhou, T.; Zuo, L. Role of ROS and Nutritional Antioxidants in Human Diseases. *Front. Physiol.* **2018**, *9*, 477. [CrossRef]
106. Simioni, C.; Zauli, G.; Martelli, A.M.; Vitale, M.; Sacchetti, G.; Gonelli, A.; Neri, L.M. Oxidative stress: Role of physical exercise and antioxidant nutraceuticals in adulthood and aging. *Oncotarget* **2018**, *9*, 17181–17198. [CrossRef] [PubMed]
107. Álvarez, R.; Araya, H.; Navarro-Lisboa, R.; De Dicastillo, C.L. Evaluation of Polyphenol Content and Antioxidant Capacity of Fruits and Vegetables Using a Modified Enzymatic Extraction. *Food Technol. Biotechnol.* **2016**, *54*, 462–467. [CrossRef] [PubMed]
108. Young, A.J.; Lowe, G.L. Carotenoids—Antioxidant Properties. *Antioxidants* **2018**, *7*, 28. [CrossRef] [PubMed]
109. Djuric, Z.; Powell, L. Antioxidant capacity of lycopene-containing foods. *Int. J. Food Sci. Nutr.* **2001**, *52*, 143–149. [CrossRef]
110. Lara-Padilla, E.; Kormanovski, A.; Grave, P.A.; Olivares-Corichi, I.M.; Santillan, R.M.; Hicks, J.J. Increased antioxidant capacity in healthy volunteers taking a mixture of oral antioxidants versus vitamin C or E supplementation. *Adv. Ther.* **2007**, *24*, 50–59. [CrossRef]
111. Pietta, P.-G. Flavonoids as Antioxidants. *J. Nat. Prod.* **2000**, *63*, 1035–1042. [CrossRef]
112. Pham-Huy, L.A.; He, H.; Pham-Huy, C. Free Radicals, Antioxidants in Disease and Health. *Int. J. Biomed. Sci. IJBS* **2008**, *4*, 89–96.
113. Senoner, T.; Dichtl, W. Oxidative Stress in Cardiovascular Diseases: Still a Therapeutic Target? *Nutrients* **2019**, *11*, 2090. [CrossRef]
114. Frei, B. Efficacy of Dietary Antioxidants to Prevent Oxidative Damage and Inhibit Chronic Disease. *J. Nutr.* **2004**, *134*, 3196S–3198S. [CrossRef]
115. Temple, N.J. Antioxidants and disease: More questions than answers. *Nutr. Res.* **2000**, *20*, 449–459. [CrossRef]
116. Fimognari, C.; Sestili, P.; Lenzi, M.; Cantelli-Forti, G.; Hrelia, P. Protective effect of creatine against RNA damage. *Mutat. Res. Mol. Mech. Mutagen.* **2009**, *670*, 59–67. [CrossRef] [PubMed]
117. Rambo, L.M.; Ribeiro, L.R.; Oliveira, M.S.; Furian, A.F.; Lima, F.D.; Souza, M.A.; Silva, L.F.A.; Retamoso, L.T.; Corte, C.L.D.; Puntel, G.O.; et al. Additive anticonvulsant effects of creatine supplementation and physical exercise against pentylenetetrazol-induced seizures. *Neurochem. Int.* **2009**, *55*, 333–340. [CrossRef]
118. Sestili, P.; Barbieri, E.; Martinelli, C.; Battistelli, M.; Guescini, M.; Vallorani, L.; Casadei, L.; D'Emilio, A.; Falcieri, E.; Piccoli, G.; et al. Creatine supplementation prevents the inhibition of myogenic differentiation in oxidatively injured C2C12 murine myoblasts. *Mol. Nutr. Food Res.* **2009**, *53*, 1187–1204. [CrossRef]
119. Hosamani, R.; Ramesh, S.R.; Muralidhara. Attenuation of Rotenone-Induced Mitochondrial Oxidative Damage and Neurotoxicty in Drosophila melanogaster Supplemented with Creatine. *Neurochem. Res.* **2010**, *35*, 1402–1412. [CrossRef]
120. Wong, H.-S.; Dighe, P.A.; Mezera, V.; Monternier, P.-A.; Brand, M.D. Production of superoxide and hydrogen peroxide from specific mitochondrial sites under different bioenergetic conditions. *J. Biol. Chem.* **2017**, *292*, 16804–16809. [CrossRef]
121. Korshunov, S.S.; Skulachev, V.P.; Starkov, A.A. High protonic potential actuates a mechanism of production of reactive oxygen species in mitochondria. *FEBS Lett.* **1997**, *416*, 15–18. [CrossRef]
122. Barbieri, E.; Guescini, M.; Calcabrini, C.; Vallorani, L.; Diaz, A.R.; Fimognari, C.; Canonico, B.; Luchetti, F.; Papa, S.; Battistelli, M.; et al. Creatine Prevents the Structural and Functional Damage to Mitochondria in Myogenic, Oxidatively Stressed C2C12 Cells and Restores Their Differentiation Capacity. *Oxidative Med. Cell. Longev.* **2016**, *2016*, 1–12. [CrossRef]
123. Rahimi, R. Creatine Supplementation Decreases Oxidative DNA Damage and Lipid Peroxidation Induced by a Single Bout of Resistance Exercise. *J. Strength Cond. Res.* **2011**, *25*, 3448–3455. [CrossRef] [PubMed]
124. Gracia, K.C.; Llanas-Cornejo, D.; Husi, H. CVD and Oxidative Stress. *J. Clin. Med.* **2017**, *6*, 22. [CrossRef] [PubMed]
125. He, F.; Zuo, L. Redox Roles of Reactive Oxygen Species in Cardiovascular Diseases. *Int. J. Mol. Sci.* **2015**, *16*, 27770–27780. [CrossRef]
126. Madamanchi, N.R.; Vendrov, A.; Runge, M.S. Oxidative Stress and Vascular Disease. *Arter. Thromb. Vasc. Biol.* **2005**, *25*, 29–38. [CrossRef]
127. Furchgott, R.F.; Zawadzki, J.V. The obligatory role of endothelial cells in the relaxation of arterial smooth muscle by acetylcholine. *Nat. Cell Biol.* **1980**, *288*, 373–376. [CrossRef]
128. Tousoulis, D.; Kampoli, A.-M.; Papageorgiou, C.T.N.; Stefanadis, C. The Role of Nitric Oxide on Endothelial Function. *Curr. Vasc. Pharmacol.* **2012**, *10*, 4–18. [CrossRef]
129. Forstermann, U.; Sessa, W.C. Nitric oxide synthases: Regulation and function. *Eur. Heart J.* **2012**, *33*, 829–837. [CrossRef]
130. Rees, D.D.; Palmer, R.M.; Moncada, S. Role of endothelium-derived nitric oxide in the regulation of blood pressure. *Proc. Natl. Acad. Sci. USA* **1989**, *86*, 3375–3378. [CrossRef] [PubMed]
131. Chen, J.-Y.; Ye, Z.-X.; Wang, X.-F.; Chang, J.; Yang, M.-W.; Zhong, H.-H.; Hong, F.-F.; Yang, S.-L. Nitric oxide bioavailability dysfunction involves in atherosclerosis. *Biomed. Pharmacother.* **2018**, *97*, 423–428. [CrossRef]

132. Hadi, H.A.R.; Carr, C.S.; Al Suwaidi, J. Endothelial Dysfunction: Cardiovascular Risk Factors, Therapy, and Outcome. *Vasc. Health Risk Manag.* **2005**, *1*, 183–198.

133. Kingwell, B.A.; Formosa, M.; Muhlmann, M.; Bradley, S.J.; McConell, G.K. Nitric oxide synthase inhibition reduces glucose uptake during exercise in individuals with type 2 diabetes more than in control subjects. *Diabetes* **2002**, *51*, 2572–2580. [CrossRef]

134. Wilkinson, I.B.; Maccallum, H.; Cockcroft, J.R.; Webb, D.J. Inhibition of basal nitric oxide synthesis increases aortic augmentation index and pulse wave velocity in vivo. *Br. J. Clin. Pharmacol.* **2002**, *53*, 189–192. [CrossRef]

135. Wang, Z.; Chen, G.; Chen, Z.-Q.; Mou, R.-T.; Feng, D.-X. The role of nitric oxide in stroke. *Med. Gas Res.* **2017**, *7*, 194–203. [CrossRef] [PubMed]

136. Giannitsi, S.; Bougiakli, M.; Bechlioulis, A.; Naka, K.; Mpougiaklh, M. Endothelial dysfunction and heart failure: A review of the existing bibliography with emphasis on flow mediated dilation. *JRSM Cardiovasc. Dis.* **2019**, *8*, 2048004019843047. [CrossRef]

137. Radi, R. Oxygen radicals, nitric oxide, and peroxynitrite: Redox pathways in molecular medicine. *Proc. Natl. Acad. Sci. USA* **2018**, *115*, 5839–5848. [CrossRef]

138. Huie, R.E.; Padmaja, S. The Reaction of no With Superoxide. *Free Radic. Res. Commun.* **1993**, *18*, 195–199. [CrossRef]

139. Beckman, J.S.; Beckman, T.W.; Chen, J.; Marshall, P.A.; Freeman, B.A. Apparent hydroxyl radical production by peroxynitrite: Implications for endothelial injury from nitric oxide and superoxide. *Proc. Natl. Acad. Sci. USA* **1990**, *87*, 1620–1624. [CrossRef]

140. Wadley, A.J.; Van Zanten, J.J.C.S.V.; Aldred, S. The interactions of oxidative stress and inflammation with vascular dysfunction in ageing: The vascular health triad. *AGE* **2012**, *35*, 705–718. [CrossRef]

141. Dupont, J.J.; Ramick, M.G.; Farquhar, W.B.; Townsend, R.R.; Edwards, D.G. NADPH oxidase-derived reactive oxygen species contribute to impaired cutaneous microvascular function in chronic kidney disease. *Am. J. Physiol. Physiol.* **2014**, *306*, F1499–F1506. [CrossRef]

142. Baylis, C. Nitric oxide deficiency in chronic kidney disease. *Am. J. Physiol. Physiol.* **2008**, *294*, F1–F9. [CrossRef]

143. Łuczak, A.; Madej, M.; Kasprzyk, A.; Doroszko, A. Role of the eNOS Uncoupling and the Nitric Oxide Metabolic Pathway in the Pathogenesis of Autoimmune Rheumatic Diseases. *Oxidative Med. Cell. Longev.* **2020**, *2020*, 1417981. [CrossRef]

144. Crabtree, M.J.; Channon, K.M. Synthesis and recycling of tetrahydrobiopterin in endothelial function and vascular disease. *Nitric Oxide* **2011**, *25*, 81–88. [CrossRef]

145. Karbach, S.; Wenzel, P.; Waisman, A.; Munzel, T.; Daiber, A. eNOS Uncoupling in Cardiovascular Diseases—The Role of Oxidative Stress and Inflammation. *Curr. Pharm. Des.* **2014**, *20*, 3579–3594. [CrossRef] [PubMed]

146. Ning, X.; Ulrich, F.; Huige, L. Implication of eNOS Uncoupling in Cardiovascular Disease. *React. Oxyg. Species* **2017**, *3*, 38–46.

147. Ahsan, A.; Han, G.; Pan, J.; Liu, S.; Padhiar, A.A.; Chu, P.; Sun, Z.; Zhang, Z.; Sun, B.; Wu, J.; et al. Phosphocreatine protects endothelial cells from oxidized low-density lipoprotein-induced apoptosis by modulating the PI3K/Akt/eNOS pathway. *Apoptosis* **2015**, *20*, 1563–1576. [CrossRef] [PubMed]

148. Varadharaj, S.; Kelly, O.J.; Khayat, R.N.; Kumar, P.S.; Ahmed, N.; Zweier, J.L. Role of Dietary Antioxidants in the Preservation of Vascular Function and the Modulation of Health and Disease. *Front. Cardiovasc. Med.* **2017**, *4*, 64. [CrossRef] [PubMed]

149. Gokce, N.; Keaney, J.F.; Frei, B.; Holbrook, M.; Olesiak, M.; Zachariah, B.J.; Leeuwenburgh, C.; Heinecke, J.W.; Vita, J.A. Long-Term Ascorbic Acid Administration Reverses Endothelial Vasomotor Dysfunction in Patients With Coronary Artery Disease. *Circulation* **1999**, *99*, 3234–3240. [CrossRef]

150. Wang-Polagruto, J.F.; Villablanca, A.C.; Polagruto, J.A.; Lee, L.; Holt, R.R.; Schrader, H.R.; Ensunsa, J.L.; Steinberg, F.M.; Schmitz, H.H.; Keen, C.L. Chronic Consumption of Flavanol-rich Cocoa Improves Endothelial Function and Decreases Vascular Cell Adhesion Molecule in Hypercholesterolemic Postmenopausal Women. *J. Cardiovasc. Pharmacol.* **2006**, *47*, S177–S186. [CrossRef]

151. Tomasian, D.; Keaney, J.F.; Vita, J.A. Antioxidants and the bioactivity of endothelium-derived nitric oxide. *Cardiovasc. Res.* **2000**, *47*, 426–435. [CrossRef]

152. Malinski, T.; Dawoud, H. Vitamin D3, L-Arginine, L-Citrulline, and antioxidant supplementation enhances nitric oxide bioavailability and reduces oxidative stress in the vascular endothelium—Clinical implications for cardiovascular system. *Pharmacogn. Res.* **2020**, *12*, 17. [CrossRef]

153. Ting, H.H.; Timimi, F.K.; Boles, K.S.; Creager, S.J.; Ganz, P.; A Creager, M. Vitamin C improves endothelium-dependent vasodilation in patients with non-insulin-dependent diabetes mellitus. *J. Clin. Investig.* **1996**, *97*, 22–28. [CrossRef]

154. Kirkby, N.S.; Lundberg, M.H.; Harrington, L.S.; Leadbeater, P.D.M.; Milne, G.L.; Potter, C.M.F.; Al-Yamani, M.; Adeyemi, O.; Warner, T.D.; Mitchell, J.A. Cyclooxygenase-1, not cyclooxygenase-2, is responsible for physiological production of prostacyclin in the cardiovascular system. *Proc. Natl. Acad. Sci. USA* **2012**, *109*, 17597–17602. [CrossRef]

155. Ozkor, M.A.; Quyyumi, A.A. Endothelium-Derived Hyperpolarizing Factor and Vascular Function. *Cardiol. Res. Pract.* **2011**, *2011*, 156146. [CrossRef] [PubMed]

156. Sandoo, A.; Van Zanten, J.J.V.; Metsios, G.S.; Carroll, D.; Kitas, G.D. The Endothelium and Its Role in Regulating Vascular Tone. *Open Cardiovasc. Med. J.* **2010**, *4*, 302–312. [CrossRef]

157. Ozkor, M.A.; Murrow, J.R.; Rahman, A.M.; Kavtaradze, N.; Lin, J.; Manatunga, A.; Quyyumi, A.A. Endothelium-Derived Hyperpolarizing Factor Determines Resting and Stimulated Forearm Vasodilator Tone in Health and in Disease. *Circulation* **2011**, *123*, 2244–2253. [CrossRef]

158. Scotland, R.S.; Madhani, M.; Chauhan, S.; Moncada, S.; Andresen, J.; Nilsson, H.; Hobbs, A.J.; Ahluwalia, A. Investigation of Vascular Responses in Endothelial Nitric Oxide Synthase/Cyclooxygenase-1 Double-Knockout Mice. *Circulation* **2005**, *111*, 796–803. [CrossRef]

159. Jia, G.; Durante, W.; Sowers, J.R. Endothelium-Derived Hyperpolarizing Factors: A Potential Therapeutic Target for Vascular Dysfunction in Obesity and Insulin Resistance. *Diabetes* **2016**, *65*, 2118–2120. [CrossRef]

160. Kang, K.-T. Endothelium-derived Relaxing Factors of Small Resistance Arteries in Hypertension. *Toxicol. Res.* **2014**, *30*, 141–148. [CrossRef]

161. Godo, S.; Sawada, A.; Saito, H.; Ikeda, S.; Enkhjargal, B.; Suzuki, K.; Tanaka, S.; Shimokawa, H. Disruption of Physiological Balance Between Nitric Oxide and Endothelium-Dependent Hyperpolarization Impairs Cardiovascular Homeostasis in Mice. *Arter. Thromb. Vasc. Biol.* **2016**, *36*, 97–107. [CrossRef]

162. Luksha, L.; Agewall, S.; Kublickiene, K. Endothelium-derived hyperpolarizing factor in vascular physiology and cardiovascular disease. *Atherosclerosis* **2009**, *202*, 330–344. [CrossRef]

163. Spruce, A.E.; Standen, N.B.; Stanfield, P.R. Voltage-dependent ATP-sensitive potassium channels of skeletal muscle membrane. *Nat. Cell Biol.* **1985**, *316*, 736–738. [CrossRef]

164. Trube, G.; Rorsman, P.; Ohno-Shosaku, T. Opposite effects of tolbutamide and diazoxide on the ATP-dependent K+ channel in mouse pancreatic β-cells. *Pflügers Archiv.* **1986**, *407*, 493–499. [CrossRef]

165. Standen, N.B.; Quayle, J.M.; Davies, N.W.; Brayden, J.E.; Huang, Y.; Nelson, M.T. Hyperpolarizing vasodilators activate ATP-sensitive K+ channels in arterial smooth muscle. *Science* **1989**, *245*, 177–180. [CrossRef] [PubMed]

166. Janigro, D.; West, G.A.; Gordon, E.L.; Winn, H.R. ATP-sensitive K+ channels in rat aorta and brain microvascular endothelial cells. *Am. J. Physiol. Physiol.* **1993**, *265*, C812–C821. [CrossRef]

167. Schnitzler, M.M.Y.; Derst, C.; Daut, J.; Preisig-Müller, R. ATP-sensitive potassium channels in capillaries isolated from guinea-pig heart. *J. Physiol.* **2000**, *525*, 307–317. [CrossRef]

168. Shi, W.-W.; Yang, Y.; Shi, Y.; Jiang, C.; Wei-Wei, S.; Yun, S.; Chun, J. K(ATP) channel action in vascular tone regulation: From genetics to diseases. *Sheng Li Xue Bao Acta Physiol. Sin.* **2012**, *64*, 1–13.

169. Zhang, Y.-L.; Chen, Y.-P.; Wang, H. Targeting Small Arteries of Hypertensive Status with Novel ATP-Sensitive Potassium Channel Openers. *Curr. Vasc. Pharmacol.* **2005**, *3*, 119–124. [CrossRef]

170. Wang, H. Cardiovascular ATP-sensitive K+ channel as a new molecular target for development of antihypertensive drugs. *Zhongguo Yao Li Xue Bao Acta Pharmacol. Sin.* **1998**, *19*, 397.

171. Chutkow, W.A.; Pu, J.; Wheeler, M.T.; Wada, T.; Makielski, J.C.; Burant, C.F.; McNally, E.M. Episodic coronary artery vasospasm and hypertension develop in the absence of Sur2 KATP channels. *J. Clin. Investig.* **2002**, *110*, 203–208. [CrossRef]

172. Miki, T.; Suzuki, M.; Shibasaki, T.; Uemura, H.; Sato, T.; Yamaguchi, K.; Koseki, H.; Iwanaga, T.; Nakaya, H.; Seino, S. Mouse model of Prinzmetal angina by disruption of the inward rectifier Kir6.1. *Nat. Med.* **2002**, *8*, 466–472. [CrossRef] [PubMed]

173. Medeiros-Domingo, A.; Tan, B.-H.; Crotti, L.; Tester, D.J.; Eckhardt, L.; Cuoretti, A.; Kroboth, S.L.; Song, C.; Zhou, Q.; Kopp, D.; et al. Gain-of-function mutation S422L in the KCNJ8-encoded cardiac KATP channel Kir6.1 as a pathogenic substrate for J-wave syndromes. *Heart Rhythm.* **2010**, *7*, 1466–1471. [CrossRef] [PubMed]

174. Tester, D.J.; Tan, B.-H.; Medeiros-Domingo, A.; Song, C.; Makielski, J.C.; Ackerman, M.J. Loss-of-Function Mutations in the KCNJ8 -Encoded Kir6.1 K ATP Channel and Sudden Infant Death Syndrome. *Circ. Cardiovasc. Genet.* **2011**, *4*, 510–515. [CrossRef] [PubMed]

175. Randak, C.O.; Welsh, M.J. ADP inhibits function of the ABC transporter cystic fibrosis transmembrane conductance regulator via its adenylate kinase activity. *Proc. Natl. Acad. Sci. USA* **2005**, *102*, 2216–2220. [CrossRef]

176. Dzeja, P.P.; Terzic, A. Phosphotransfer reactions in the regulation of ATP-sensitive K + channels. *FASEB J.* **1998**, *12*, 523–529. [CrossRef]

177. Selivanov, V.A.; Alekseev, A.E.; Hodgson, D.M.; Dzeja, P.P.; Terzic, A. Nucleotide-gated KATP channels integrated with creatine and adenylate kinases: Amplification, tuning and sensing of energetic signals in the compartmentalized cellular environment. *Mol. Cell Biochem.* **2004**, *256–257*, 243–256. [CrossRef] [PubMed]

178. Guerrero, M.L.; Beron, J.; Spindler, B.; Groscurth, P.; Wallimann, T.; Verrey, F. Metabolic support of Na+ pump in apically permeabilized A6 kidney cell epithelia: Role of creatine kinase. *Am. J. Physiol. Physiol.* **1997**, *272*, C697–C706. [CrossRef]

179. Chistiakov, D.A.; Orekhov, A.N.; Bobryshev, Y.V. Endothelial Barrier and Its Abnormalities in Cardiovascular Disease. *Front. Physiol.* **2015**, *6*, 365. [CrossRef] [PubMed]

180. Galley, H.F.; Webster, N.R. Physiology of the endothelium. *Br. J. Anaesth.* **2004**, *93*, 105–113. [CrossRef]

181. Park-Windhol, C.; D'Amore, P.A. Disorders of Vascular Permeability. *Annu. Rev. Pathol. Mech. Dis.* **2016**, *11*, 251–281. [CrossRef]

182. Vestweber, D. Relevance of endothelial junctions in leukocyte extravasation and vascular permeability. *Ann. N. Y. Acad. Sci.* **2012**, *1257*, 184–192. [CrossRef] [PubMed]

183. Bazzoni, G.; Dejana, E. Endothelial Cell-to-Cell Junctions: Molecular Organization and Role in Vascular Homeostasis. *Physiol. Rev.* **2004**, *84*, 869–901. [CrossRef] [PubMed]

184. Mundi, S.; Massaro, M.; Scoditti, E.; Carluccio, M.A.; Van Hinsbergh, V.W.M.; Iruela-Arispe, M.L.; De Caterina, R. Endothelial permeability, LDL deposition, and cardiovascular risk factors—A review. *Cardiovasc. Res.* **2018**, *114*, 35–52. [CrossRef] [PubMed]

185. Rahmani, A.; Asadollahi, K.; Soleimannejad, K.; Khalighi, Z.; Mohsenzadeh, Y.; Hemati, R.; Moradkhani, A.; Abangah, G. The Effects of Creatine Monohydrate on Permeability of Coronary Artery Endothelium and Level of Blood Lipoprotein in Diabetic Rats. *Ann. Clin. Lab. Sci.* **2016**, *46*, 495–501. [PubMed]

186. Wang, L.; Chen, Y.; Li, X.; Zhang, Y.; Gulbins, E.; Zhang, Y. Enhancement of endothelial permeability by free fatty acid through lysosomal cathepsin B-mediated Nlrp3 inflammasome activation. *Oncotarget* **2016**, *7*, 73229–73241. [CrossRef] [PubMed]

187. Lu, Q.; Gottlieb, E.; Rounds, S. Effects of cigarette smoke on pulmonary endothelial cells. *Am. J. Physiol. Cell. Mol. Physiol.* **2018**, *314*, L743–L756. [CrossRef]
188. Luc, G.; Arveiler, D.; Evans, A.; Amouyel, P.; Ferrieres, J.; Bard, J.-M.; Elkhalil, L.; Fruchart, J.-C.; Ducimetiere, P. Circulating soluble adhesion molecules ICAM-1 and VCAM-1 and incident coronary heart disease: The PRIME Study. *Atherosclerosis* **2003**, *170*, 169–176. [CrossRef]
189. Roldán, V.; Marín, F.; Lip, G.Y.H.; Blann, A.D. Soluble E-selectin in cardiovascular disease and its risk factors. *Thromb. Haemost.* **2003**, *90*, 1007–1020. [CrossRef] [PubMed]
190. Tokarska-Schlattner, M.; Epand, R.F.; Meiler, F.; Zandomeneghi, G.; Neumann, D.; Widmer, H.R.; Meier, B.H.; Epand, R.M.; Saks, V.; Wallimann, T.; et al. Phosphocreatine Interacts with Phospholipids, Affects Membrane Properties and Exerts Membrane-Protective Effects. *PLoS ONE* **2012**, *7*, e43178. [CrossRef] [PubMed]
191. Hall, C.H.; Lee, J.S.; Murphy, E.M.; Gerich, M.E.; Dran, R.; Glover, L.E.; Abdulla, Z.I.; Skelton, M.R.; Colgan, S.P. Creatine Transporter, Reduced in Colon Tissues from Patients with Inflammatory Bowel Diseases, Regulates Energy Balance in Intestinal Epithelial Cells, Epithelial Integrity, and Barrier Function. *Gastroenterology* **2020**, *159*, 984–998. [CrossRef] [PubMed]
192. Glover, L.E.; Bowers, B.E.; Saeedi, B.; Ehrentraut, S.F.; Campbell, E.L.; Bayless, A.J.; Dobrinskikh, E.; Kendrick, A.A.; Kelly, C.J.; Burgess, A.; et al. Control of creatine metabolism by HIF is an endogenous mechanism of barrier regulation in colitis. *Proc. Natl. Acad. Sci. USA* **2013**, *110*, 19820–19825. [CrossRef] [PubMed]
193. Lombard, D.B.; Chua, K.F.; Mostoslavsky, R.; Franco, S.; Gostissa, M.; Alt, F.W. DNA Repair, Genome Stability, and Aging. *Cell* **2005**, *120*, 497–512. [CrossRef]
194. Negishi, H.; Ikeda, K.; Kuga, S.; Noguchi, T.; Kanda, T.; Njelekela, M.; Liu, L.; Miki, T.; Nara, Y.; Sato, T.; et al. The relation of oxidative DNA damage to hypertension and other cardiovascular risk factors in Tanzania. *J. Hypertens.* **2001**, *19*, 529–533. [CrossRef]
195. Bhat, M.A.; Mahajan, N.; Gandhi, G. DNA and chromosomal damage in coronary artery disease patients. *EXCLI J.* **2013**, *12*, 872–884.
196. Shah, N.R.; Mahmoudi, M. The role of DNA damage and repair in atherosclerosis: A review. *J. Mol. Cell. Cardiol.* **2015**, *86*, 147–157. [CrossRef]
197. Brass, E.P.; Wang, H.; Hiatt, W.R. Multiple skeletal muscle mitochondrial DNA deletions in patients with unilateral peripheral arterial disease. *Vasc. Med.* **2000**, *5*, 225–230. [CrossRef] [PubMed]
198. Poznyak, A.; Ivanova, E.; Sobenin, I.; Yet, S.-F.; Orekhov, A. The Role of Mitochondria in Cardiovascular Diseases. *Biology* **2020**, *9*, 137. [CrossRef] [PubMed]
199. Gorbunova, V.; Seluanov, A.; Mao, Z.; Hine, C. Changes in DNA repair during aging. *Nucleic Acids Res.* **2007**, *35*, 7466–7474. [CrossRef]
200. Cooke, M.S.; Evans, M.D.; Dizdaroglu, M.; Lunec, J. Oxidative DNA damage: Mechanisms, mutation, and disease. *FASEB J.* **2003**, *17*, 1195–1214. [CrossRef]
201. Rastogi, R.P.; Kumar, A.; Tyagi, M.B.; Sinha, R.P. Molecular Mechanisms of Ultraviolet Radiation-Induced DNA Damage and Repair. *J. Nucleic Acids* **2010**, *2010*, 1–32. [CrossRef]
202. Woods, D.; Turchi, J.J. Chemotherapy induced DNA damage response. *Cancer Biol. Ther.* **2013**, *14*, 379–389. [CrossRef]
203. Marseglia, L.; Manti, S.; D'Angelo, G.; Nicotera, A.; Parisi, E.; Di Rosa, G.; Gitto, E.; Arrigo, T. Oxidative Stress in Obesity: A Critical Component in Human Diseases. *Int. J. Mol. Sci.* **2014**, *16*, 378–400. [CrossRef]
204. Vetrani, C.; Costabile, G.; Di Marino, L.; Rivellese, A.A. Nutrition and oxidative stress: A systematic review of human studies. *Int. J. Food Sci. Nutr.* **2012**, *64*, 312–326. [CrossRef] [PubMed]
205. Albano, E. Alcohol, oxidative stress and free radical damage. *Proc. Nutr. Soc.* **2006**, *65*, 278–290. [CrossRef] [PubMed]
206. Lucas, T.; Bonauer, A.; Dimmeler, S. RNA Therapeutics in Cardiovascular Disease. *Circ. Res.* **2018**, *123*, 205–220. [CrossRef] [PubMed]
207. Pradhan-Nabzdyk, L.; Huang, C.; LoGerfo, F.W.; Nabzdyk, C.S. Current siRNA targets in atherosclerosis and aortic aneurysm. *Discov. Med.* **2014**, *17*, 233–246.
208. Nabzdyk, C.S.; Pradhan-Nabzdyk, L.; LoGerfo, F.W. RNAi therapy to the wall of arteries and veins: Anatomical, physiologic, and pharmacological considerations. *J. Transl. Med.* **2017**, *15*, 164. [CrossRef] [PubMed]
209. Berneburg, M.; Gremmel, T.; Kürten, V.; Schroeder, P.; Hertel, I.; Von Mikecz, A.; Wild, S.; Chen, M.; Declercq, L.; Matsui, M.; et al. Creatine Supplementation Normalizes Mutagenesis of Mitochondrial DNA as Well as Functional Consequences. *J. Investig. Dermatol.* **2005**, *125*, 213–220. [CrossRef] [PubMed]
210. Fetterman, J.L.; Holbrook, M.; Westbrook, D.G.; Brown, J.A.; Feeley, K.P.; Bretón-Romero, R.; Linder, E.A.; Berk, B.D.; Weisbrod, R.M.; Widlansky, M.E.; et al. Mitochondrial DNA damage and vascular function in patients with diabetes mellitus and atherosclerotic cardiovascular disease. *Cardiovasc. Diabetol.* **2016**, *15*, 53. [CrossRef] [PubMed]
211. Yu, E.P.; Bennett, M.R. The role of mitochondrial DNA damage in the development of atherosclerosis. *Free. Radic. Biol. Med.* **2016**, *100*, 223–230. [CrossRef]
212. Uryga, A.; Gray, K.; Bennett, M. DNA Damage and Repair in Vascular Disease. *Annu. Rev. Physiol.* **2016**, *78*, 45–66. [CrossRef]
213. Mirzaei, D.; Rahmani-Nia, F.; Salehi, Z.; Rahimi, R. Effects of creatine monohydrate supplementation on oxidative DNA damage and lipid peroxidation induced by acute incremental exercise to exhaustion in wrestlers. *Kinesiol. Int. J. Fundam. Appl. Kinesiol.* **2013**, *45*, 30–40.

214. Qasim, N.; Mahmood, R. Diminution of Oxidative Damage to Human Erythrocytes and Lymphocytes by Creatine: Possible Role of Creatine in Blood. *PLoS ONE* **2015**, *10*, e0141975. [CrossRef]
215. Guidi, C.; Potenza, L.; Sestili, P.; Martinelli, C.; Guescini, M.; Stocchi, L.; Zeppa, S.; Polidori, E.; Annibalini, G.; Stocchi, V. Differential effect of creatine on oxidatively-injured mitochondrial and nuclear DNA. *Biochim. Biophys. Acta (BBA) Gen. Subj.* **2008**, *1780*, 16–26. [CrossRef]
216. Anderson, O. Creatine propels British athletes to Olympic gold medals: Is creatine the one true ergogenic aid. *Run. Res. News* **1993**, *9*, 1–5.
217. Bird, S.P. Creatine Supplementation and Exercise Performance: A Brief Review. *J. Sports Sci. Med.* **2003**, *2*, 123–132.
218. Chilibeck, P.D.; Kaviani, M.; Candow, D.G.; A Zello, G. Effect of creatine supplementation during resistance training on lean tissue mass and muscular strength in older adults: A meta-analysis. *Open Access J. Sports Med.* **2017**, *8*, 213–226. [CrossRef]
219. Rawson, E.S.; Stec, M.J.; Frederickson, S.J.; Miles, M.P. Low-dose creatine supplementation enhances fatigue resistance in the absence of weight gain. *Nutrition* **2011**, *27*, 451–455. [CrossRef] [PubMed]
220. Kreider, R.B.; Ferreira, M.; Wilson, M.; Grindstaff, P.; Plisk, S.; Reinardy, J.; Cantler, E.; Almada, A.L. Effects of creatine supplementation on body composition, strength, and sprint performance. *Med. Sci. Sports Exerc.* **1998**, *30*, 73–82. [CrossRef] [PubMed]
221. Post, A.; Tsikas, D.; Bakker, S.J. Creatine is a Conditionally Essential Nutrient in Chronic Kidney Disease: A Hypothesis and Narrative Literature Review. *Nutritions* **2019**, *11*, 1044. [CrossRef]
222. Kreider, R.B.; Melton, C.; Rasmussen, C.J.; Greenwood, M.; Lancaster, S.; Cantler, E.C.; Milnor, P.; Almada, A.L. Long-term creatine supplementation does not significantly affect clinical markers of health in athletes. *Mol. Cell. Biochem.* **2003**, *244*, 95–104. [CrossRef]
223. Clarke, H.; Kim, D.-H.; Meza, C.A.; Ormsbee, M.J.; Hickner, R.C. The Evolving Applications of Creatine Supplementation: Could Creatine Improve Vascular Health? *Nutrients* **2020**, *12*, 2834. [CrossRef]

 *nutrients*

Review

Creatine Supplementation for Patients with Inflammatory Bowel Diseases: A Scientific Rationale for a Clinical Trial

Theo Wallimann [1,*], Caroline H. T. Hall [2], Sean P. Colgan [3] and Louise E. Glover [4,*]

[1] Department of Biology, ETH-Zurich, Emeritus, 8962 Bergdietikon, Switzerland
[2] Mucosal Inflammation Program and Department of Pediatrics, Division of Gastroenterology, Hepatology and Nutrition, Children's Hospital Colorado, University of Colorado, 12700 E. 19th Ave, Aurora, CO 80045, USA; caroline.hall@childrenscolorado.org
[3] Mucosal Inflammation Program and Department of Medicine, Division of Gastroenterology and Hepatology, University of Colorado, 12700 E 19th Ave, Aurora, CO 80045, USA; sean.colgan@cuanschutz.edu
[4] Comparative Immunology Group, School of Biochemistry and Immunology, Trinity Biomedical Sciences Institute, Trinity College Dublin, D2 Dublin, Ireland
* Correspondence: theo.wallimann@cell.biol.ethz.ch (T.W.); gloverl@tcd.ie (L.E.G.)

Abstract: Based on theoretical considerations, experimental data with cells in vitro, animal studies in vivo, as well as a single case pilot study with one colitis patient, a consolidated hypothesis can be put forward, stating that "oral supplementation with creatine monohydrate (Cr), a pleiotropic cellular energy precursor, is likely to be effective in inducing a favorable response and/or remission in patients with inflammatory bowel diseases (IBD), like ulcerative colitis and/or Crohn's disease". A current pilot clinical trial that incorporates the use of oral Cr at a dose of 2×7 g per day, over an initial period of 2 months in conjunction with ongoing therapies (NCT02463305) will be informative for the proposed larger, more long-term Cr supplementation study of 2×3–5 g of Cr per day for a time of 3–6 months. This strategy should be insightful to the potential for Cr in reducing or alleviating the symptoms of IBD. Supplementation with chemically pure Cr, a natural nutritional supplement, is well tolerated not only by healthy subjects, but also by patients with diverse neuromuscular diseases. If the outcome of such a clinical pilot study with Cr as monotherapy or in conjunction with metformin were positive, oral Cr supplementation could then be used in the future as potentially useful adjuvant therapeutic intervention for patients with IBD, preferably together with standard medication used for treating patients with chronic ulcerative colitis and/or Crohn's disease.

Keywords: pleiotropic effects of creatine (Cr) supplementation; inflammatory bowel diseases (IBD); ulcerative colitis; Crohn's disease; creatine kinase (CK); phosphocreatine (PCr); creatine transporter (CrT); intestinal epithelial cell protection; intestinal tissue protection; creatine perfusion; organ transplantation; Adenosine mono-phosphate (AMP); activated protein kinase (AMPK); liver kinase B1 (LKB1); mitochondrial permeability transition pore (mPTP); reactive oxygen species (ROS); glucose transporter (GLUT)

Citation: Wallimann, T.; Hall, C.H.T.; Colgan, S.P.; Glover, L.E. Creatine Supplementation for Patients with Inflammatory Bowel Diseases: A Scientific Rationale for a Clinical Trial. *Nutrients* **2021**, *13*, 1429. https://doi.org/10.3390/nu13051429

Academic Editor: Richard B. Kreider

Received: 2 March 2021
Accepted: 22 April 2021
Published: 23 April 2021

Publisher's Note: MDPI stays neutral with regard to jurisdictional claims in published maps and institutional affiliations.

1. Introduction

Inflammatory bowel disease (IBD), including ulcerative colitis (UC) and Crohn's disease, are chronic and relapsing-remitting inflammatory disorders of the gastrointestinal tract that may develop in genetically susceptible individuals in response to unknown antigenic triggers. Although the etiology of IBD remains a conundrum, it seems definitive that multiple factors, such as genetic predisposition, environment, malfunction of the immune system and changes in the intestinal gut microbiota are involved in the onset and progression of IBD (for review see [1]). Since, in most cases, IBD cannot be cured completely, adjuvant therapies for IBD that may alleviate symptoms and thus improve quality of life parameters of afflicted patients, is a poorly characterized area of study [2].

Oral supplementation with chemically pure creatine (Cr) monohydrate, a natural nutritional supplement with pleiotropic beneficial influences, may fill such a gap in treatment options [3]. Cr may be a good candidate for an adjuvant treatment of IBD, since it has been shown to generally improve the energy state of cells, to enhance resilience of cells against several cell stressors, to modulate the immune system, to display anti-inflammatory influences and to dampen nociceptive pain, as will be elucidated below.

Besides water, Cr is one of the most abundant single molecular compound in the human body, with a propensity of 120–140 g in a 70 kg person and a concentration of 40–50 mM in fast-twitch skeletal muscle. Cr monohydrate has become a very popular and the most effective ergogenic aid for athletic and recreational sports [4] and is used also as clinical adjuvant therapeutic intervention for patients with neuromuscular diseases [5]. Millions of persons worldwide consume Cr and, as one of the most intensively studied nutritional supplements, it is recommended as effective and safe by international sports societies [6] and national associations for food safety, such as EFSA and others. A dosage of 3–5 g of chemically pure Cr monohydrate per day as a long-term maintenance dose, without interruption needed, is generally recommended. For athletes in weight lifting and high-intensity performance disciplines, a short-term loading phase over 5–7 days with 20 g per day is recommended to quickly load the endogenous Cr pool in the muscles before switching to the above maintenance dosage [4]. Based on 30 years of practical experience with Cr supplementation and a very large number publications on this issue, involving athletes and regular people, as well as patients (children and elderly), oral supplementation with chemically pure Cr monohydrate if taken within the officially recommended dosages is safe for humans with no significant side effects (see the series of publications in this issue: https://www.mdpi.com/journal/nutrients/special_issues/creatine_supplementation, accessed on 2 March 2021, and in an earlier issue (https://link.springer.com/journal/726/48 /8/page/1, accessed on 2 March 2021). As a matter of fact, Cr is necessary for optimal cell and body physiology, for genetic defects in either one of the two enzymes for endogenous Cr synthesis (AGAT or GAMT) or in the creatine transporter (CrT), the latter facilitating Cr uptake into target cells, lead to more or less severe Cr-deficiency syndromes in transgenic animal models, as well as in humans (for review see [7]).

1.1. The Phospho-Creatine Creatine Kinase System in Intestinal Epithelial Cells

The creatine kinase (CK)/phospho-creatine (PCr) system, with Cr as an energy precursor, plays a crucial physiological role for cells and tissues with high and fluctuating energy requirements, including skeletal, heart and smooth muscles, brain and nervous tissues, as well as other tissues and cells [3,8,9]. This also holds true for intestinal smooth muscle and intestinal epithelial cells, where cytosolic brain-type BB-CK and mitochondrial mtCK isoenzymes are prominently co-expressed [10] (see Figure 1). In addition, a specific creatine transporter (CrT1), belonging to the X-linked gene SLC6A8, as a member of a solute carrier family, is present in the apical cell membrane of intestinal epithelial cells [11]. By this electrogenic Na^+- and Cl^--dependent Cr-cotransporter (CrT1), with a high affinity for Cr (Km for Cr of 30 µM), epithelial cells of the intestine take up Cr ingested with the diet, e.g., from meat and fish, as the most significant alimentary sources for Cr [12]. As CrT1 function is dependent on Na^+, inhibition of the Na^+/Ka^+-ATPase, e.g., by the action of Lyn kinase, also inhibits Cr uptake into cells, as shown recently [13].

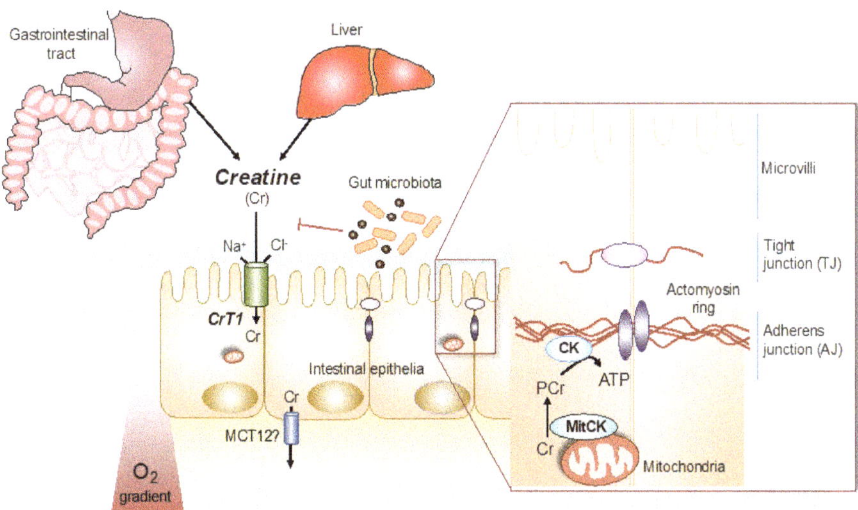

Figure 1. Cr/CK shuttle and the intestinal mucosal barrier. Cr is derived from dietary sources in the gastrointestinal tract, or by de novo synthesis primarily in the kidney and in the liver [12]. The Na$^+$ and Cl$^-$ dependent creatine transporter (CrT1), expressed in the apical membrane of intestinal epithelial cells, facilitates Cr uptake from the gut lumen [11,14]. Potential routes for Cr absorption into systemic circulation include paracellular movement by solvent drag transport, or via basolateral Cr transport by the monocarboxylate transporter 12 (MCT12) [15]. Gut microbiota express specific enzymes that can mediate Cr and creatinine (Crn) breakdown. In hypoxic intestinal epithelial cells, cytosolic CK localizes to apical adherens junctions in complex with the actomyosin cytoskeletal network, providing a conduit for rapid ATP generation during the energy-dependent processes of epithelial junction assembly and barrier restitution [16]. Adapted from [17].

In humans, a significant proportion of the Cr taken-up via CrT1 into intestinal epithelial cells, is released into the blood stream by a basolateral monocarboxylate transporter, MCT12, that works as a novel facilitative CrT2 transporter [15]. From the blood, Cr is taken up by the target organs, such as skeletal and cardiac muscle, as well as neuronal tissues and other cells [12]. The uptake of Cr by intestinal epithelial cells, followed by trans-epithelial release into the blood stream leads to a systemic exposure of the body by Cr that is then taken up by those target organs, which depend on Cr, via their own CrT. Intestinal epithelial cells themselves also depend on the CK/PCr system for optimal physiological function. In these cells, PCr works in a similar way, as has been shown to be the case in other cells with high energy requirements [3], as an immediate high-energy buffer and as an energy transport vehicle to guarantee the maintenance of locally high PCr/ATP and ATP/ADP ratios in the vicinity of ATP-dependent processes, such as ion pumps and metabolite transporters [3,8,9], thus increasing the thermodynamic efficiency of intestinal epithelial cells [17] in a similar way as had been shown with other cells [3,8,10].

1.2. Creatine for Cytoprotection against Ischemia, Hypoxia, Oxidative Stress and Acidosis

Experimental and clinical data strongly support profound cell-protective properties of creatine in neuronal cells and tissues in vitro and in vivo [18–20]. Similar beneficial effects have also been observed with direct intra-venous phospho-creatine (PCr) injections against hypoxic cardiovascular stress [21]. Hypoxia also seems to play a key role in the pathogenesis of intestinal mucosal epithelial diseases [22], compromising cellular energy metabolism by lowering the cellular energy charge, i.e., the PCr/ATP ratio. Hypoxia renders cells more vulnerable to cellular stressors, such as ROS, inflammation and toxins [20,23,24]. In addition, hypoxia has been shown to diminish Cr uptake into cardiac cells [25]. Although the above positive influences of Cr on cell metabolism and cell integrity have largely been studied in other tissues and cells compared to in intestinal epithelial cells, the physiological workings of the CK/PCr system, the CrT and of creatine as such turned out to be very similar in the different cells and, thus, they are likely to also hold true for intestinal epithelial cells. Indeed, it was recently shown that hypoxia profoundly decreases levels of Cr, PCr and total available energy (PCr + ATP + (0.5 × ADP) in intestinal epithelial cells [26]. Thus, creatine supplementation is likely to normalize the intracellular Cr and PCr levels and thus also the energy charge, exemplified as the PCr/ATP ratio, in intestinal epithelial cells. At the same time, creatine supplementation is also highly cytoprotective against oxidative stress by ROS [23,27]. Creatine also leads to a significant increase in cell survival after hypoxic insult in vivo [19,20,28]. Recent work using transgenic mice that overexpress mtCK in cardiac muscle by only 25% over the normal level, shows that hearts from these mice functionally recover much better from ischemia reperfusion damage, with a significant reduction in cardiac infarct size [29]. This complements nicely with earlier data showing that transgenic mice overexpressing the creatine transporter (CrT) in heart and thus importing more creatine into this tissue, are also more resistant to ischemia reperfusion-related cardiac tissue damage. In addition, the extent of tissue damage is significantly lower in the mtCK overexpressing mice compared to normal control mice without elevated creatine concentrations in their heart muscle [30]. Given the cytoprotective role of creatine in the setting of hypoxia and ischemia referenced above and the evidence that hypoxia occurs in the setting of IBD [31] with associated intestinal cell barrier dysfunction [32], it follows that creatine is likely to provide a protective influence on the concomitant hypoxia observed in IBD.

Finally, as acidosis promotes lipid peroxidation and other manifestations of oxidant-mediated damage in various cell types, this condition is also relevant for intestinal epithelium. Acidosis, associated with inflammatory conditions, produces oxidative stress and amplifies these effects, e.g., at an acidotic pH, the response of the intestine to an oxidative insult is magnified [33]. In a rat model of chronic acidosis, creatine supplementation was shown to exert direct anti-oxidant properties by directly scavenging ROS and creatine abolished the chronic reduction in the expression levels of glucose transporter (GLUT2) [34]. Moreover, the administration of creatine under chronic acidosis led to functional strengthening of this jejunal acidotic phenotype, making the tissue more resistant to acidosis [34]. Thus, the beneficial influence and alleviation by creatine with respect to ischemic, oxidative and acidotic insults is certainly relevant for intestinal epithelial tissue, as well as for the entire intestine.

1.3. Creatine Stimulates Mitochondrial Respiration and Serves as an Anti-Apoptotic Effector

Mitochondrial creatine kinase (mtCK) and creatine stimulate mitochondrial respiration [35] and thus contribute significantly to maintain a healthy energy state of cells, especially under metabolic stress or toxic insults. Within this context, MtCK and creatine also play a crucial role in an early event of apoptosis; that is, in controlling the opening of the so-called mitochondrial permeability transition pore (mPTP) [36] that is sensitive to cyclosporine A. The addition of creatine to liver mitochondria from transgenic mice expressing mtCK in their livers, after a challenge by 40 mM calcium plus 5 mM atractyloside, prevents swelling of mitochondria and the release of apoptotic factors and reactive oxygen species (ROS) in a similar fashion as the bona fide anti-apoptotic agent cyclosporine A. In liver mitochondria from normal mice, which do not express mtCK in their liver, no effect of creatine on mitochondrial swelling could be seen, indicating that the action of mtCK, located in the mitochondrial inter-membrane space and present in all cells except for liver, is necessary together with creatine to prevent challenged mitochondria from swelling and mitochondrial permeability transition pore (mPTP) from opening [37]. Similar anti-apoptotic protection by creatine or phospho-creatine could be observed with intact cardiomyocytes [38], or with human umbilical vein endothelial cells that both were protected by creatine from lipopolysaccharide (LPS)-induced apoptosis [39]. This anti-apoptotic effect of creatine could also be demonstrated in vivo in hyper-cholesterolemic mice, where pravastatin-induced mitochondrial mPTP opening in skeletal muscles was minimized by creatine [40]. A significant factor for cell protection by creatine is mediated by the action of octameric mtCK that stabilizes mitochondrial contact sites and protects mitochondrial and cell integrity [41]. Thus, creatine, together with mtCK, regulate mitochondrial oxidative phosphorylation and exert a significant anti-apoptotic effect on a variety of cells by protecting them from different cytotoxic insults (for review [42]). In line with this notion, the anti-apoptotic effects and preservation of the function and structural integrity of mitochondria under metabolic stress was demonstrated directly in vivo in murine cardiac muscle [30]. Intestinal epithelial cell apoptosis significantly contributes to the development of ulcerative colitis and IBD in humans and mice, and therapies that target the inflammatory cytokine TNF, as well as the p53-upregulated modulator of apoptosis (PUMA) that are both upregulated in colitis tissues, have been found to inhibit apoptosis in intestinal epithelial cells and to promote mucosal healing [43]. Taking the evidence for the role of creatine in inhibition of apoptosis in other tissues and the finding that apoptosis contributes to disease activity in IBD [44], it is thus very reasonable to postulate that creatine would be anti-apoptotic in the mucosa of IBD patients as well.

1.4. Creatine as Anti-Inflammatory, Nociceptive and Immune Modulatory Compound

There is solid evidence in sports medicine, e.g., from ironman competitions, that creatine supplementation reduces the plasma levels of pro-inflammatory cytokines and prostaglandin E2 (PGE2) [45]. Very recent data have provided evidence that creatine may also have the potential to lower pain sensitivity associated with inflammation by antagonizing the acid-sensing ion channel (ASIC3) [46]. This effect is most likely based on the structural similarity of creatine, itself a guanidino compound, to other guanidino compound ligands of the ASIC3 pain receptor, such as GMQ and amiloride, that modulated this ion sensing channel [46]. Thus, one may expect from creatine supplementation, as an additional beneficial effect, an improvement of the abdominal pain associated with intestinal inflammation. This is an important determinant of quality of life for patients with IBD both in the setting of active inflammation as well as in IBD patients with irritable bowel syndrome symptoms in the absence of inflammation [47].

Immune cells themselves express CK and seem to depend on the CK/PCr system in a similar way to muscle and brain cells [3,10], as pointed out in a recent review [48]. For example, leucocytes express the CRT-1 transporter for the import of extracellular creatine [49]. Additionally, in macrophages, CK has a functional impact on these cells by supporting the formation of actin-based protrusions needed for macrophage motility and phagocytosis [50]. In addition, the uptake and accumulation of creatine into macrophages leads to the reprogramming and polarization of macrophages by modulating cellular responses to cytokines, such as IFN-γ and Il-4, thus enhancing the ability of macrophages to sense viral and bacterial antigens [51].

With regard to T cells, creatine kinase is involved in T cell development and activation [52] and creatine uptake regulates CD8 T cell immunity [53]. Thus, it seems obvious that the CK/PCr system has a profound impact both on the innate and adaptive immune response, exhibiting significant immune modulatory effects [54] and, therefore, it may be inferred that patients with IBD, who often suffer from intestinal infections, may benefit by creatine supplementation as a general activator of immune responses [54].

1.5. Creatine Affords Anti-Depressant Effects

There is growing evidence from human genetics, epidemiology, neuroimaging, as well as from animal studies that disruptions in brain energy metabolism, e.g., brain energy production, storage and utilization in the form of PCr and Cr [55] are implicated in the development and maintenance of depression, and that creatine has the potential to improve these disruptions in some depressive patients [56].

With respect to the clinical fact that many chronic colitis patients not only suffer from abdominal pain but also from depression, it is important to note that creatine has been shown to exert beneficial effects in the clinical management of depression, even in patients resistant to conventional anti-depression treatment [57,58]. Exciting new experiments with animal models show that creatine is able to afford its anti-depressant-like effects in a similar way as ketamine does [59,60]. Depression is a common finding in patients with chronic diseases, including those with IBD in whom depression significantly worsens their quality of life [61]. Thus, it will certainly be worthwhile to test the anti-depressant effects of creatine supplementation in our cohort of chronic colitis patients.

2. Scientific Rationale Specifically for Intestinal Tissue

2.1. HIF Controls Creatine Kinase (CK) Expression and CK Together with Creatine Are Involved in the Energetics of Mucosal Barrier Regulation

Fully charged cellular energy batteries are a prerequisite for optimal body function not only for muscle and brain cells, but also for intestinal smooth muscle and epithelial cells. Long-term decay or failure of cellular energetics, e.g., by chronic ischemia, inflammation and progressive dysfunction of mitochondria, as well as deterioration of mucosal barrier functions, are important aspects of IBD that are accompanied by a state of chronic inflammation [17]. Mucosal surfaces of the lower gastrointestinal tract are subject to pronounced fluctuations in oxygen (O_2) tension, particularly during inflammation (Figure 2). As an adaptive response to hypoxia, the hypoxia-induced transcription factors (HIF-1 and HIIF-2) become stabilized [62]. An unbiased analysis of HIF target genes identified creatine kinases (both cytosolic BB-CK and mtCK) and the major Cr transporter SLC6A8 that are all coordinately regulated by HIF [16]. Further analysis revealed that cytosolic BB-CK is expressed in a HIF-2 dependent manner and that this enzyme localizes to apical intestinal epithelium cell adherence junctions, where it is critically involved in the ATP-dependent junction assembly, epithelial integrity and mucosal barrier function (see Figure 1). This same study revealed that tissue transcripts from 30 IBD patients (including both Crohn's disease and ulcerative colitis) showed a marked reduction in the expression of all three isoforms of CK compared to non-IBD controls. In light of this observation, it is notable that the interaction of epithelial junctions with the actin cytoskeleton is a significant energy sink within the mucosa. Energy deficiencies associated with IBD, including those associated with microbial dysbiosis, likely contribute to barrier dysfunction during active inflammation [63].

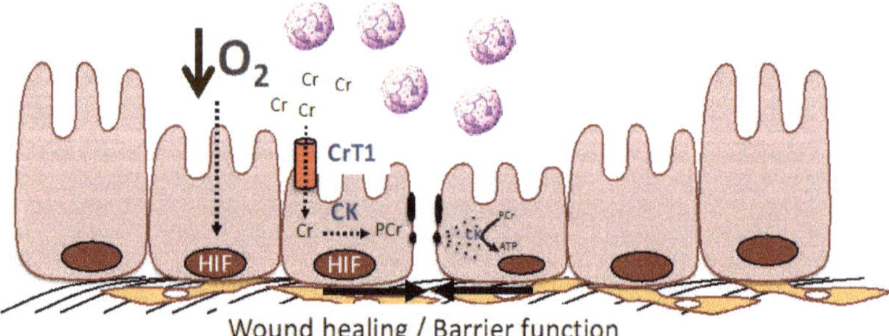

Figure 2. The PCr/Cr shuttle promotes mucosal barrier and wound healing by improving cellular energetics and by stabilization of the adherence junctions. During active inflammation, such as that seen in IBD, low O_2 levels result in the stabilization of HIF and resultant induction of creatine kinase (CK) isoenzymes and the Cr transporter (CrT1) within intestinal epithelial cells (see [16,62,63] for further details). CK localizes to epithelial junctions that are stabilized by interactions with the actin cytoskeleton. In response to epithelial disruption during inflammation, large amounts of ATP are necessary to accommodate the demand for cytoskeletal reorganization, including the acto-myosin ATPase at epithelial cellular junctions. Under such conditions, CK and CrT1 coordinately promote wound healing and barrier function by generating ATP from PCr to efficiently promote homeostasis.

2.2. Creatine Supplementation Regulates the Energy Balance of Intestinal Epithelial Cells, Epithelial Integrity and Barrier Function

Patients with IBD present with intestinal barrier dysfunction that is likely related to disturbed cellular energetics and dysbiosis. Since the CK/PCr system, as well as the creatine transporter (CrT1) are involved in a plethora of processes that are important for cellular energetics [3,8], also in intestinal epithelial cells [16,17,62] it is interesting that the investigation of mucosal biopsied from 30 patients with Crohn's disease and 27 patients with ulcerative colitis both showed lower expression levels of CrT1, which might contribute to the reduced barrier function of intestinal epithelium [14] (see Figure 3). Notably, in intestinal epithelial cells (IECs), CrT1 localized specifically around tight junctions and knockdown or overexpression of CrT1 in these cells corroborated the idea that CrT1, besides regulating the intracellular creatine concentration in IECs, was also modulating epithelial barrier formation and wound healing [14]. In CrT1 knockdown IECs—that is, in the absence of adequate creatine transport—these cells transformed to a stressed, glycolysis-predominant energy metabolism, resulting in leaky tight junctions and mislocalization of actin and tight junction proteins [14]. Despite the significant impacts of CrT1 loss, proliferation was not altered in CrT1 knockdown intestinal epithelial cells [14]. It is noteworthy that metabolomic analysis has revealed that the actin cytoskeleton demands nearly 20% of total available energy within the epithelium [26]. Taken together, these data support the fact that CrT1, together with CK, phosphocreatine (PCr) and creatine, regulates the energy balance of IECs and enforces the structural and functional integrity of the tight-junction-actin cytoskeleton. These are excellent arguments that speak for a clinical trial, using creatine supplementation directly on patients with IBD.

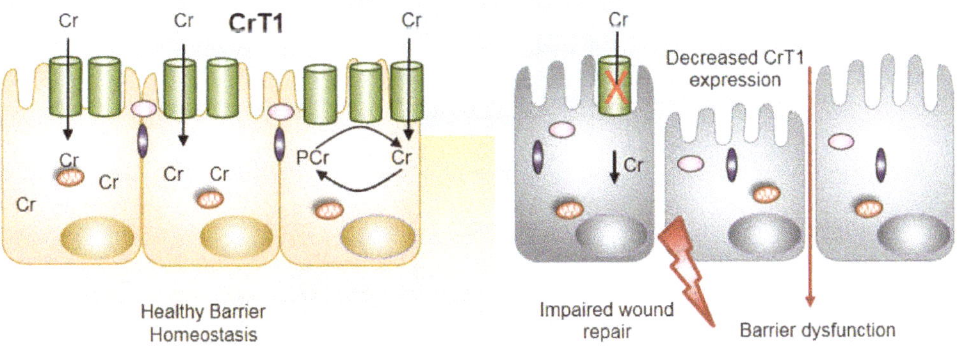

Figure 3. Decreased expression of CrT1 in IBD promotes barrier dysfunction. Normal expression of CrT1 on the apical surface of intestinal epithelia (left panel) results in adequate supplies of Cr via dietary sources to promote healthy barrier function and intestinal homeostasis. Patients with IBD express lower levels of CrT1 (right panel) and disrupt the Cr-PCr energy shuttle to the extent that wound healing potential and barrier are dysfunctional (see [14,16] for further details).

2.3. Creatine Supplementation Maintains Intestinal Epithelial Energy Homeostasis and Protects against Colitis in Animal Models

Oral creatine supplementation in mice with experimentally induced colitis markedly ameliorated both disease severity and inflammation in TNBS and DSS mouse colitis models [16]. Furthermore, as an indicator for pathology, mucosal CK expression was lowered in patients with ulcerative colitis and Crohn's disease. Thus, a role for HIF-regulated CK expression in intestinal epithelial homeostasis was established and this revealed a fundamental link between cellular bioenergetics and mucosal barrier [16,64]; see schematic representation of CK function for maintaining intestinal mucosal barrier function in [17]. This model (see Figure 1) is representative of the well-documented PCr shuttling function from mitochondria, via mitochondrial mtCK to intracellular sites of ATP consumption, where cytosolic CK is specifically localized for in-situ ATP-regeneration [3,8,9,12].

2.4. Creatine-Loading Preserves Intestinal Barrier Function during Intestine Organ Preservation by Static Cold Storage

Very recent data demonstrate that, in two rodent models, a single flush of intestines with intraluminal preservation solution supplemented with 50 mM Cr significantly improved intestinal barrier function and electrophysiology, reflecting superior mucosal integrity after 10 h of cold storage, in comparison to the placebo group without Cr [65]. Permeability and trans-epithelia resistance measurements remained at fresh tissue values and oxidative injury was controlled in the Cr group by the preferential utilization of glutathione [65]. Thus, Cr supplementation of intestinal tissue improved graft quality, among others, by preserving the cellular energy state in this tissue, which is reflected in a greater PCr energy charge of 324% in Cr treated versus control intestinal tissue. The significant improvement in cellular energy state, reflected by an increased PCr/ATP, as well as ATP/AMP ratios in the Cr group, was most likely responsible for alleviating tissue damage due to the ischemic storage of intestine. Thus, augmenting the cellular energy reserves by Cr supplementation of intestinal tissue led to improved tissue integrity and physiological function, such that a fully energized state is facilitated upon the reperfusion of transplanted intestinal tissue [65]. Similar cell-protecting effects against ischemia or ischemia-reperfusion tissue damage were already described for cardiac muscle and brain tissues [3,7,19–21,23,55]. The above data show that perfusion of intestinal tissue of the mouse, before transplantation, has a tissue protective effect and improves the results of intestine organ transplantation. If these data are confirmed with human patients, the perfusion of intestinal tissue with Cr could become a standard procedure in intestinal tissue transplantation.

2.5. A Genetic Screen with Mice Susceptible for Colitis Reveals a Link to Creatine Metabolism

The data presented above are supported by a full-fletched genetic study involving the screening of 36'530 third generation germ-line mutant mice, derived from N-ethyl-N-nitroso-urea-mutagenized grandsires for abnormalities in intestinal homeostasis and abnormalities after oral administration of dextran sodium sulfate (DSS) to induce colitis [66]. The team around Dr. Bruce Beutler, Nobelist in Physiology, 2011, of Texas Southwestern Medical Center in Dallas, identified, among 27 mice susceptible to the colitis phenotype, one mutant mouse that was strongly correlated with a missense mutation in one of the two enzymes that are important for endogenous creatine biosynthesis; that is, arginine-glycine-aminotransferase (AGAT) catalyzing the rate-limiting first step of creatine synthesis, mainly taking place in the kidney [12]. The intestinal epithelium of the AGAT mutant mice displayed significantly increased cell death and decreased proliferation during DSS treatment, compared to control mice under the same challenge. Then, the supplementation of homozygous AGAT mutants with exogenous creatine ameliorated and significantly improved the histological parameters and the colitis phenotype of these mice, respectively [66]. These findings establish an in vivo requirement for the rapid replenishment of cytoplasmic ATP by the CK/PCr-system within colonic epithelial cells for the maintenance of the mucosal barrier after injury [66].

2.6. Involvement of AMPK Activation to Restore Adherence Junction Assembly in Intestinal Epithelium

Adenosine monophosphate-activated protein kinase (AMPK), an evolutionarily conserved serine/threonine protein kinase, plays a central role in the maintenance of the cellular energy balance [67]. AMPK works as a cellular energy sensor, responding to cellular energy stress situation, e.g., oxidative stress, ischemia and anoxia, as indicated by an elevated AMP/ATP ratio. Under such conditions, specifically arising in pathological cases of IBD, AMPK is activated both by allosteric binding of AMP [68], as well as by upstream kinases. For example, AMPK is phosphorylated and activated by liver kinase B1 (LKB1) [69], a protein kinase that plays a conserved role in epithelial polarity and regulation [70]. Several lines of experimental evidence point towards an important role of AMPK in the regulation of epithelial adherence junction assembly and disassembly and demonstrates an intriguing link between cellular energy status and adherence junction function [71]. For example, the activation and phosphorylation of AMPK increases during calcium-induced adherence junction assembly and this increase depends on the activity of LKB1. Conversely, a kinase-dead mutant of AMPK inhibits adherence junction assembly and barrier function, as measured by the localization of specific adherence junction proteins and by trans-epithelial resistance, respectively [71]. In fact, metformin a widely prescribed, clinically safe anti-diabetes drug and well-known activator of AMPK [72] promotes the expression and assembly of adherence junctions via an AMPK-dependent way [73]. This is corroborated by the fact that a reduced risk of IBD is consistently observed in patients with type 2 diabetes mellitus, who have been treated with metformin [74].

Therefore, not only CK and the PCr/Cr energy support system, but also AMPK, as a regulator of cellular energetics, play a direct role in the structural and functional integrity or adherence junctions and thus for epithelial barrier function. These data would argue that one could combine and thus enforce oral creatine supplementation with low-dose metformin treatment as adjuvant therapy for IBD (see below).

2.7. Creatine Supplementation in One Single Case of Crohn's Disease Improved Both Symptomatic and Endoscopic Characteristics of Ulcerative Colitis

Fully in line with these pioneering data is a first single case study with a 33-year-old patient with a two-year history of Crohn's ileitis, who responded very well to creatine supplementation (1.5 g per day, given as monotherapy for a time period of 6 months) with both symptomatic and endoscopic improvement in disease activity [75]. Specifically, before creatine supplementation, the colonoscopy of the patient showed large ulcers of 0.5–2.0 cm in diameter with >30% ulcerative surface, 50–70% affected surface and no narrowing (SDS-CD:7), whereas after creatine supplementation, the same patient presented with aphthous ulcers <0.5 cm in diameter, <10% ulcerated surface, <50% affected surface and no-narrowing (SES-CD:3) (see Figures 1–3 in [75]).

3. Proposal, Methodology and Clinical End-Points

Based on the theoretical concept of CK function and experimental evidence from the DSS colitis mouse models, as well as on the single case clinical trial, plus the extensive genetic screen that revealed a correlation of colitis to a failure in creatine metabolism (mutated AGAT), a pilot clinical trial is ongoing. Indeed, a randomized, placebo-controlled clinical pilot trial with creatine supplementation (2 × 7 g of creatine monohydrate per day, ingested for 8 weeks, 6 patients per arm) is underway in 12 patients, aged 18–70, with mild to moderate ulcerative colitis at the University of Colorado (NCT02463305). Note that this study is not a monotherapy trial and patients taking mesalamine or thiopurines will be allowed concomitant use of these drugs during the trial. The primary outcome is improvement in the endoscopic score as an assessment of mucosal inflammation from biopsy samples obtained before and after the 8-week treatment course. Multiple secondary outcomes include intestinal permeability, as measured by urinary saccharide excretion testing, symptom severity, colonic inflammatory biomarkers, CK and Cr levels from biopsies and blood, intestinal microbiome as well as clinical remission status. This ongoing trial should give insight into the tolerability, safety and efficacy of creatine monohydrate for patients with ulcerative colitis.

(A) Creatine monotherapy trial: In addition to the current study mentioned above, we believe that there is utility in conducting a detailed investigation into the long term effects of creatine supplementation in IBD which expands on the above study. Such a Cr monotherapy trial could include a careful evaluation of 40 patients treated with placebo (control group) and 40 patients treated with creatine monotherapy (2 × 3–5 g of

Cr per day for a time of 3–6 months (verum group) and should be carried out by using standard clinical questionnaires for ulcerative colitis and Crohn's disease, (e.g., IBDQ). Intestinal inflammation should be quantitatively assessed using established inflammation biomarkers (e.g., fecal calprotectin, serum C-reactive protein) and intestinal permeability should be quantified using urinary saccharide excretion. Finally, patients should undergo pre- and post-treatment coloscopy, currently the gold standard of assessment for mucosal inflammation. Patients with diagnosed colorectal cancers will be excluded from the study. The reason for this precaution is that, in an animal model with orthotopically implanted colorectal cancer cell lines, very high dosage of external Cr fed to these cancer mice promoted cancer metastasis, resulting in a higher incidence of liver metastases compare to control animals without externally added Cr [76]. However, the amounts of orally fed Cr to the cancer mice corresponded to a calculated amount of 50–150 g of Cr/day for a human of 70 kg body weight. On the other hand, in accordance with earlier studies [48,77,78], the same authors found that Cr instead had an anti-proliferative effect on primary tumor growth. As a matter of fact, Cr has convincingly been shown to represent an important metabolic regulator controlling anti-tumor T cell immunity [53], which also represents a key player in the etiology of IBD's. Based on these results, Cr supplementation has been proposed as a valid strategy in cancer protection and/or management [79], as well as to prevent cancer-related loss of weight and muscle mass [80]. Questionnaires, measurement of clinical parameters, as well as coloscopy, should be performed at the beginning and after three and six months of the clinical study.

(B) Creatine treatment in combination with metformin: As stated above, metformin has shown a great potential in experimental cell and animal models of IBD [71–73] and offered interesting retrospective insights from patients with type 2 diabetes, who took metformin [74]. Therefore, a combination treatment with 2 × 3–5 g Cr per day plus a low dose of metformin (1 × 0.5–1.0 g/day) may be suggested, for the two are likely to work synergistically and may provide better clinical results compared to creatine monotherapy alone. Since the safety profile of both, creatine and metformin, is excellent, such a clinical study should be straight-forward, considering that millions of people take either creatine or metformin for other purposes.

4. Anticipated Outcome

The anticipated outcome will be a proof of concept from the ongoing Cr adjuvant clinical pilot study in mild to moderate ulcerative colitis. It is also proposed that a Cr monotherapy pilot using 2 × 3–5 g of chemically pure creatine monohydrate per day, given for 3–6 months, either as a monotherapy or in combination with low dose metformin (1 × 0.5–1.0 g per day) would potentially be able to improve several quality-of-life parameters of patients with chronic colitis, e.g., improvement in muscle strength, mobility and vital functions, with less fatigue and depression. Most importantly, one would expect a possibly significant reduction or even alleviation of colitis symptoms, such as abdominal pain, clinical symptoms and inflammation of the colon, as judged by coloscopy. Almost certainly, since IBDs are multifactorial diseases with phenotypical and genetic subtypes, Cr is unlikely solving all the problems associated with these serious diseases, but if Cr should indeed turn out to be helpful in inducing such favorable responses or even the remission of the ulcerative pathology in colitis patients, an inclusion of creatine supplementation, either alone or in combination with metformin, as standard adjuvant therapeutic intervention for ulcerative colitis and/or Crohn's disease, could be envisaged, preferably together with established medical treatment options. As a prerequisite for this, however, a multicenter study with significant patient numbers will be needed. Excluding the animal experiments and one human case, described above, no other clinical study with Cr supplementation has been performed with IBD patients so far. Therefore, it is to be expected that, based on the results of the proposed studies, either with creatine and metformin alone or in combination, some potentially unique information concerning potentially beneficial effects of Cr on colitis and Crohn's disease should be revealed by such a study as a clinical endpoint.

5. Conclusions

Based on the available experimental data and the scientific rationale, it seems appropriate and timely to propose a full-fletched clinical trial with oral creatine monohydrate supplementation alone or in combination with metformin on patients suffering from IBD. Depending on the outcome of the results proposed herein with Cr supplementation and/or metformin alone or in combination, such treatments could in the future become a standard adjuvant therapeutic intervention for ulcerative colitis and/or Crohn's disease that could be combined with established medical treatments for these pathological situations, hopefully for the benefit of IBD patients.

Author Contributions: Original conceptualization and draft preparation T.W.; discussion, reviewing, editing and visualization C.H.T.H., S.P.C. and L.E.G. All authors have read and agreed to the published version of the manuscript.

Funding: Funded by a National Institute of Health grant DK05189 and the Veterans Administration grant BX002182 (to SPC); and by an National Institute of Health grant NIH KL2-TR002534 (to C.H.T.H.).

Institutional Review Board Statement: The ongoing pilot clinical is underway with 12 patients, aged 18–70, with mild to moderate ulcerative colitis at the University of Colorado (NCT02463305) under protocol 13-3054 approved 4 June 2015.

Informed Consent Statement: Informed consent was obtained from all subjects in the above study.

Data Availability Statement: Data is contained within the article.

Conflicts of Interest: The authors declare no conflict of interest.

References

1. Jin, L.; Li, L.; Hu, C.; Paez-Cortez, J.; Bi, Y.; Macoritto, M.; Cao, S.; Tian, Y. Integrative Analysis of Transcriptomic and Proteomic Profiling in Inflammatory Bowel Disease Colon Biopsies. *Inflamm. Bowel Dis.* **2019**, *25*, 1906–1918. [CrossRef]
2. Wang, Q.; Mi, S.; Yu, Z.; Li, Q.; Lei, J. Opening a Window on Attention: Adjuvant Therapies for Inflammatory Bowel Disease. *Can. J. Gastroenterol. Hepatol.* **2020**, 7397523. [CrossRef]
3. Wallimann, T.; Tokarska-Schlattner, M.; Schlattner, U. The creatine kinase system and pleiotropic effects of creatine. *Amino Acids* **2011**, *40*, 1271–1296. [CrossRef]
4. Kreider, R.B.; Stout, J.R. Creatine in Health and Disease. *Nutrients* **2021**, *13*, 447. [CrossRef]
5. Tarnopolsky, M.A. Clinical use of creatine in neuromuscular and neurometabolic disorders. *Subcell Biochem.* **2007**, *46*, 183–204. [CrossRef]
6. Kreider, R.B.; Kalman, D.S.; Antonio, J.; Ziegenfuss, T.N.; Wildman, R.; Collins, R.; Candow, D.G.; Kleiner, S.M.; Almada, A.L.; Lopez, H.L. International Society of Sports Nutrition position stand: Safety and efficacy of creatine supplementation in exercise, sport, and medicine. *J. Int. Soc. Sports Nutr.* **2017**, *14*, 18. [CrossRef] [PubMed]
7. Wallimann, T.; Harris, R. Creatine: A miserable life without it. *Amino Acids.* **2016**, *48*, 1739–1750. [CrossRef]
8. Wallimann, T.; Wyss, M.; Brdiczka, D.; Nicolay, K.; Eppenberger, H.M. Intracellular compartmentation, structure and function of creatine kinase isoenzymes in tissues with high and fluctuation energy demands: The phosphocreatine circuit for cellular energy homeostasis. *Biochem. J.* **1992**, *281*, 21–40. [CrossRef] [PubMed]
9. Bessman, S.P.; Carpenter, C.L. The creatine-creatine phosphate energy shuttle. *Annu. Rev. Biochem.* **1985**, *54*, 831–862. [CrossRef]
10. Wallimann, T.; Hemmer, W. Creatine kinase in non-muscle tissues and cells. *Mol. Cell Biochem.* **1994**, *133–134*, 193–220. [CrossRef]
11. Peral, M.J.; García-Delgado, M.; Calonge, M.L.; Durán, J.M.; De La Horra, M.C.; Wallimann, T.; Speer, O.; Ilundáin, A. Human, rat and chicken small intestinal Na^+-Cl^--creatine transporter: Functional, molecular characterization and localization. *J. Physiol.* **2002**, *545*, 133–144. [CrossRef]
12. Wyss, M.; Kaddurah-Daouk, R. Creatine and creatinine metabolism. *Physiol Rev.* **2000**, *80*, 1107–1121. [CrossRef] [PubMed]
13. Okuma, D.O.; Aponte-Collazo, L.J.; Dewar, B.J.; Cox, N.J.; East, M.P.; Tech, K.; McDonald, I.M.; Tikunov, A.P.; Holmuhamedov, E.; Macdonald, J.M.; et al. Lyn regulates creatine uptake in an imatinib-resistant CML cell line. *Biochim. Biophys. Acta Gen. Subj.* **2020**, *1864*, 129507. [CrossRef] [PubMed]
14. Hall, C.H.T.; Lee, J.S.; Murphy, E.M.; Gerich, M.E.; Dran, R.; Glover, L.E.; Abdulla, Z.I.; Skelton, M.R.; Colgan, S.P. Creatine Transporter, Reduced in Colon Tissues from Patients With Inflammatory Bowel Diseases, Regulates Energy Balance in In-testinal Epithelial Cells, Epithelial Integrity, and Barrier Function. *Gastroenterology* **2020**, *159*, 984–998. [CrossRef]
15. Takahashi, M.; Kishimoto, H.; Shirasaka, Y.; Inoue, K. Functional characterization of monocarboxylate transporter 12 (SLC16A12/MCT12) as a facilitative creatine transporter. *Drug Metab. Pharmacokinet.* **2020**, *35*, 281–287. [CrossRef]

16. Glover, L.E.; Bowers, B.E.; Saeedi, B.; Ehrentraut, S.F.; Campbell, E.L.; Bayless, A.J.; Dobrinskikh, E.; Kendrick, A.A.; Kelly, C.J.; Burgess, A.; et al. Control of creatine metabolism by HIF is an endogenous mechanism of barrier regulation in colitis. *Proc. Natl. Acad. Sci. USA* **2013**, *110*, 19820–19825. [CrossRef]

17. Kitzenberg, D.; Colgan, S.P.; Glover, L.E. Creatine kinase in ischemic and inflammatory disorders. *Clin. Transl. Med.* **2016**, *5*, 31. [CrossRef]

18. Brewer, G.J.; Wallimann, T. Protective effect of the energy precursor creatine against toxicity of glutamate and beta-amyloid in rat hippocampal neurons. *J. Neurochem.* **2000**, *74*, 1968–1978. [CrossRef]

19. Adcock, K.H.; Nedelcu, J.; Loenneker, T.; Martin, E.; Wallimann, T.; Wagner, B.P. Neuroprotection of creatine supplementation in neonatal rats with transient cerebral hypoxia-ischemia. *Dev. Neurosci.* **2002**, *24*, 382–388. [CrossRef]

20. Prass, K.; Royl, G.; Lindauer, U.; Freyer, D.; Megow, D.; Dirnagl, U.; Stöckler-Ipsiroglu, G.; Wallimann, T.; Priller, J. Improved reperfustion and neuroprotection by creatine in a mouse model of stroke. *J. Cereb. Blood Flow Metab.* **2007**, *27*, 452–459. [CrossRef]

21. Balestrino, M.; Sarocchi, M.; Adriano, E.; Spallarossa, P. Potential of creatine or phosphocreatine supplementation in cerebrovascular disease and in ischemic heart disease. *Amino Acids* **2016**, *48*, 1955–1967. [CrossRef]

22. Vavricka, S.; Ruiz, P.A.; Scharl, S.; Biedermann, L.; Scharl, M.; de Vallière, C.; Lundby, C.; Wenger, R.H.; Held, L.; Merz, T.M.; et al. Protocol for a prospective, controlled, observational study to evaluate the influence of hypoxia on healthy volunteers and patients with inflammatory bowel disease: The Altitude IBD-Study. *BMJ Open* **2017**, *7*, e013477. [CrossRef]

23. Barbieri, E.; Guescini, M.; Calcabrini, C.; Vallorani, L.; Diaz, A.R.; Fimognari, C.; Canonico, B.; Luchetti, F.; Papa, S.; Battistelli, M.; et al. Creatine prevents the structural and functional damage to mitochondria in myogenic oxidatively stressed C2C12 cells and restores their differentiation capacity. *Oxid. Med. Cell Longev.* **2016**, *2016*. [CrossRef] [PubMed]

24. Hemati, F.; Rahmani, A.; Asadollahi, K.; Soleimannejad, K.; Khalighi, Z. Effects of Complimentary Creatine Monohydrate and Physical Training on Inflammatory and Endothelial Dysfunction Markers Among Heart Failure Patients. *Asian J. Sports Med.* **2016**, *7*, e28578. [CrossRef] [PubMed]

25. Santacuz, L.; Arciniegas, A.J.L.; Darrabie, M.; Mantilla, J.G.; Baron, R.M.; Bowles, D.E.; Mishra, R.; Jacobs, D.O. Hypoxia decreases creatine uptake in cardiomyocytes, while creatine supplementation enhances HIF activation. *Physiol. Rep.* **2017**, *5*, e13382. [CrossRef] [PubMed]

26. Lee, J.S.; Wang, R.X.; Alexeev, E.E.; Lanis, J.M.; Battista, K.D.; Glover, L.E.; Colgan, S.P. Hypoxanthine is a checkpoint stress metabolite in colonic epithelial energy modulation and barrier function. *J. Biol. Chem.* **2018**, *293*, 6039–6051. [CrossRef] [PubMed]

27. Sestili, P.; Martinelli, C.; Colombo, E.; Barbieri, E.; Potenza, L.; Sartini, S.; Fimognari, C. Creatine as an antioxidant. *Amino Acids* **2011**, *40*, 1385–1396. [CrossRef]

28. Ellery, S.J.; Dickinson, H.; McKenzie, M.; Walker, D.W. Dietary interventions designed to protect the perinatal brain from hypoxic-ischemic encephalopathy—Creatine prophylaxis and the need for multi-organ protection. *Neurochem. Int.* **2016**, *95*, 15–23. [CrossRef]

29. Whittington, H.J.; Ostrowski, P.J.; McAndrew, D.J.; Cao, F.; Shaw, A.; Eykyn, T.R.; Lake, H.; Tyler, J.; Schneider, J.E.; Neubauer, S.; et al. Over-expression of mitochondrial creatine kinase in the murine heart improves functional recovery and protects against injury following ischemia-reperfusion. *Cardiovasc. Res.* **2018**, *114*, 858–869. [CrossRef]

30. Whittington, H.J.; McAndrew, D.J.; Cross, R.L.; Neubauer, S.; Lygate, C.A. Protective effect of creatine elevation against ischaemia reperfusion injury is retained in the presence of co-morbidities and during cardioplegia. *PLoS ONE* **2016**, *11*, e0146429. [CrossRef]

31. Giatromanolaki, A.; Sivridis, E.; Maltezos, E.; Papazoglou, D.; Simopoulos, C.; Gatter, K.C.; Harris, A.L.; Koukourakis, M.I. Hypoxia inducible factor 1alpha and 2alpha overexpression in inflammatory bowel disease. *J. Clin. Pathol.* **2003**, *56*, 209–213. [CrossRef] [PubMed]

32. Taylor, C.T.; Dzus, A.L.; Colgan, S.P. Autocrine regulation of epithelial permeability by hypoxia: Role for polarized release of tumor necrosis factor alpha. *Gastroenterology* **1998**, *114*, 657–668. [CrossRef]

33. Pedoto, A.; Nandi, J.; Oler, A.; Camporesi, E.M.; Hakim, T.S.; Levine, R.A. Role of Nitric Oxide in Acidosis-Induced Intestinal Injury in Anesthetized Rats. *J. Lab. Clin. Med.* **2001**, *138*, 270–276. [CrossRef] [PubMed]

34. Sironi, C.; Bodega, F.; Zocchi, L.; Porta, C. Effects of creatine treatment on jejunial phenotypes in a rat model of acidosis. *Antioxidants* **2019**, *8*, 225. [CrossRef] [PubMed]

35. Kay, L.; Nicolay, K.; Wieringa, B.; Saks, V.; Wallimann, T. Direct evidence for the control of mitochondrial respiration by mitochondrial creatine kinase in oxidative muscle cells in situ. *J. Biol. Chem.* **2000**, *275*, 6937–6944. [CrossRef] [PubMed]

36. O'Gorman, E.; Beutner, G.; Dolder, M.; Koretsky, A.P.; Brdiczka, D.; Wallimann, T. The role of creatine kinase and creatine in inhibition of mitochondrial permeability transition. *FEBS Lett.* **1997**, *414*, 253–257. [CrossRef]

37. Dolder, M.; Walzel, B.; Speer, O.; Schlattner, U.; Wallimann, T. Inhibition of the mitochondrial permeability transition by creatine kinase substrates. Requirement for micro-compartmentation. *J. Biol. Chem.* **2003**, *278*, 17760–17766. [CrossRef]

38. Caretti, A.; Bianciardi, P.; Sala, G.; Terruzzi, C.; Lucchina, F.; Samaja, M. Supplementation of creatine and ribose prevents apoptosis in ischemic cardiomyocytes. *Cell Physiol. Biochem.* **2010**, *26*, 831–838. [CrossRef]

39. Sun, Z.; Lan, X.; Ahsan, A.; Xi, Y.; Liu, S.; Zhang, Z.; Chu, P.; Song, Y.; Piao, F.; Peng, J.; et al. Phosphocreatine protects against LPS-induced human umbilical vein endothelial cell apoptosis by regulating mitochondrial oxidative phosphorylation. *Apoptosis* **2016**, *21*, 283–297. [CrossRef]

40. Busanello, E.N.B.; Marques, A.C.; Lander, N.; de Oliveira, D.N.; Catharino, R.R.; Oliveira, H.C.F.; Vercesi, A.E. Pravastatin chronic treatment sensitizes hypercholesterolemic mice muscle to mitochondrial permeability transition: Protection by creatine or coenzyme Q10. *Front. Pharmacol.* **2017**, *8*, 185. [CrossRef]

41. Speer, O.; Bäck, N.; Buerklen, T.; Brdiczka, D.; Koretsky, A.; Wallimann, T.; Eriksson, O. Octameric mitochondrial creatine kinase induces and stabilizes contact sites between the inner and outer mitochondrial membrane. *Biochem. J.* **2005**, *385*, 445–450. [CrossRef]

42. Wallimann, T.; Riek, U.; Möddel, M. Intradialytic creatine supplementation: A scientific rationale for improving the health and quality of life of dialysis patients. *Med. Hypotheses* **2017**, *99*, 1–14. [CrossRef]

43. Qiu, W.; Wu, B.; Wang, X.; Buchanan, M.E.; Regueiro, M.D.; Hartman, D.J.; Schoen, R.E.; Yu, J.; Zhang, L. PUMA-mediated intestinal epithelial apoptosis contributes to ulcerative colitis in humans and mice. *J. Clin. Investig.* **2011**, *121*, 1722–1732. [CrossRef] [PubMed]

44. Günther, C.; Martini, E.; Nadine Wittkopf, N.; Amann, K.; Weigmann, B.; Neumann, H.; Waldner, M.; Hedrick, S.M.; Tenzer, S.; Neurath, M.F.; et al. Caspase-8 regulates TNF-alpha induced epithelial necroptosis and terminal ileitis. *Nature* **2011**, *477*, 335–339. [CrossRef] [PubMed]

45. Bassit, R.A.; Pinheiro, C.H.; Vitzel, K.F.; Sproesser, A.J.; Silveira, L.R.; Curi, R. Effect of short-term creatine supplementation on markers of skeletal muscle damage after strenuous contractile activity. *Eur. J. Appl. Physiol.* **2010**, *108*, 945–955. [CrossRef]

46. Izurieta Munoz, H.; Gonzales, E.B.; Sumien, N. Effects of creatine supplementation on nociception in young male and female mice. *Pharmacol. Rep.* **2018**, *70*, 316–321. [CrossRef] [PubMed]

47. Gajula, P.; Quigley, E.M. Overlapping irritable bowel syndrome and inflammatory bowel disease. *Minerva Gastrolenterologica Diabetol.* **2019**, *65*, 107–115. [CrossRef]

48. Kazak, L.; Cohen, P. Creatine metabolism: Energy homeostasis, immunity and cancer biology. *Nat. Rev. Endocrinol.* **2020**, *16*, 421–436. [CrossRef] [PubMed]

49. Taii, A.; Tachikawa, M.; Ohta, Y.; Hosoya, K.I.; Terasaki, T. Determination of Intrinsic Creatine Transporter (Slc6a8) Activity and Creatine Transport Function of Leukocytes in Rats. *Biol. Pharm. Bull.* **2020**, *43*, 474–479. [CrossRef]

50. Venter, G.; Polling, S.; Pluk, H.; Venselaar, H.; Wijers, M.; Willemse, M.; Fransen, J.A.M.; Wieringa, B. Submembranous recruitment of creatine kinase B supports formation of dynamic actin-based protrusions of macrophages and relies on its C-terminal flexible loop. *Eur. J. Cell Biol.* **2015**, *94*, 114–127. [CrossRef]

51. Ji, L.; Zhao, X.; Zhang, B.; Kang, L.; Song, W.; Zhao, B.; Xie, W.; Chen, L.; Hu, X. Slc6a8-Mediated Creatine Uptake and Accumulation Reprogram Macrophage Polarization via Regulating Cytokine Responses. *Immunity* **2019**, *51*, 272–284.e7. [CrossRef]

52. Zhang, Y.; Li, H.; Wang, X.; Gao, X.; Liu, X. Regulation of T Cell Development and Activation by Creatine Kinase B. *PLoS ONE* **2009**, *4*, e5000. [CrossRef]

53. Di Biase, S.; Ma, X.; Wang, X.; Yu, J.; Wang, Y.C.; Smith, D.J.; Zhou, Y.; Li, Z.; Kim, Y.J.; Clarke, N.; et al. Creatine uptake regulates CD8 T cell antitumor immunity. *J. Exp. Med.* **2019**, *216*, 2869–2882. [CrossRef]

54. Bredahl, E.C.; Eckerson, J.M.; Tracy, S.M.; McDonald, T.L.; Drescher, K.M. The Role of Creatine in the Development and Activation of Immune Responses. *Nutrients* **2021**, *13*, 751. [CrossRef] [PubMed]

55. Andres, D.; Ducray, A.D.; Schlattner, U.; Wallimann, T.; Widmer, H.R. Functions and effects of creatine in the central nervous system. *Brain Res. Bull.* **2008**, *76*, 329–343. [CrossRef]

56. Kious, B.M.; Kondo, D.G.; Renshaw, P.E. Creatine for the treatment of depression. *Biomolecules* **2019**, *9*, 406. [CrossRef] [PubMed]

57. Kondo, D.G.; Forrest, L.N.; Shi, X.; Sung, Y.H.; Hellem, T.L.; Huber, R.S.; Renshaw, P.F. Creatine target engagement with brain bioenergetics: A dose-range phosphorus-31-magnetic resonance spectroscopy study of adolescent females with SSRI-resistant depression. *Amino Acids* **2016**, *48*, 1941–1954. [CrossRef] [PubMed]

58. Kious, B.M.; Sabic, H.; Sung, Y.H.; Kondo, D.G.; Renshaw, P. An open-label pilot study of combined augmentation with creatine monohydrate and 5-hydroxytryptophan for selective serotonin reuptake inhibitor- or serotonin-norepinephrine reuptake inhibitor-resistant depression in adult women. *J. Clin. Psychopharmacol.* **2017**, *37*, 578–583. [CrossRef]

59. Cunha, M.P.; Pazini, F.L.; Rosa, J.M.; Ramos-Hryb, A.B.; Oliveira, A.; Kaster, M.P.; Rodrigues, A.L. Creatine, similarly to ketamine, affords anti-depressant-like effects in the tail suspension test via adenosine A1 and A2A receptor activities. *Purinergic Signal.* **2015**, *11*, 215–227. [CrossRef] [PubMed]

60. Pazini, F.L.; Cunha, M.P.; Rosa, J.M.; Colla, A.R.; Lieberknecht, V.; Oliveira, Á.; Rodrigues, A.L. Creatine, similar to ketamine, counteracts depressive-like behavior induced by corticosterone via PI3K/Akt/mTOR pathway. *Mol. Neurobiol.* **2016**, *53*, 6818–6834. [CrossRef]

61. Zhang, C.K.; Hewett, J.; Hemming, J.; Grant, T.; Zhao, H.; Abraham, C.; Oikonomou, I.; Berkowitz, M.; Cho, J.H.; Proctor, D.D. The Influence of Depression on Quality of Life in Patients with Inflammatory Bowel Disease. *Inflamm. Bowel Dis.* **2013**, *19*, 1732–1739. [CrossRef]

62. Glover, L.E.; Colgan, S.P. Epithelial barrier regulation by hypoxia inducible factor. *Ann. Am. Thorac. Soc.* **2017**, *14*, S233–S236. [CrossRef] [PubMed]

63. Lee, J.S.; Wang, R.X.; Alexeev, E.E.; Colgan, S.P. Intestinal Inflammation as a Dysbiosis of Energy Procurement: New Insights into an Old Topic. *Gut Microbes* **2021**, *13*, 1–20. [CrossRef] [PubMed]

64. Colgan, S.P.; Curtis, V.F.; Lanis, J.M.; Glover, L.E. Metabolic regulation of intestinal epithelial barrier during inflammation. *Tissue Barriers* **2015**, *3*, e970936. [CrossRef] [PubMed]

65. Mueller, K.; Kokotilo, M.S.; Carter, J.; Thiesen, A.; Madsen, K.L.; Studzinski, J.; Khadaroo, R.G.; Churchill, T.A. Creatine-loading preserves intestinal barrier function during organ preservation. *Cryobiology* **2018**, *84*, 69–76. [CrossRef] [PubMed]

66. Turer, E.; McAlpine, W.; Wang, K.W.; Lu, T.; Li, X.; Tang, M.; Zhan, X.; Wang, T.; Zhan, X.; Bu, C.H.; et al. Creatine maintains intestinal homeostasis and protects against colitis. *Proc. Natl. Acad. Sci. USA* **2017**, *114*, E1273–E1281. [CrossRef]

67. Ross, F.A.; MacKintosh, C.; Hardie, D.G. AMP-activated protein kinase: A cellular energy sensor that comes in 12 flavours. *FEBS J.* **2016**, *283*, 2987–3001. [CrossRef] [PubMed]

68. Suter, M.; Riek, U.; Tuerk, R.; Schlattner, U.; Wallimann, T.; Neumann, D. Dissecting the role of 5′-AMPK for allosteric stimulation, activation and deactivation of AMP-activated protein kinase. *J. Biol. Chem.* **2006**, *281*, 32207–32216. [CrossRef]

69. Woods, A.; Johnstone, S.R.; Dickerson, K.; Leiper, F.C.; Fryer, L.G.; Neumann, D.; Schlattner, U.; Wallimann, T.; Carlson, M.; Carling, D. LKB1 is the upstream kinase in the AMP-activated protein kinase cascade. *Curr. Biol.* **2003**, *13*, 2004–2008. [CrossRef]

70. Alessi, D.R.; Sakamoto, K.; Bayascas, J.R. LKB1-dependent signaling pathways. *Annu. Rev. Biochem.* **2006**, *75*, 137–163. [CrossRef]

71. Zheng, B.; Cantley, L.C. Regulation of epithelial tight junction assembly and disassembly by AMP-activated protein kinase. *Proc. Natl. Acad. Sci. USA* **2007**, *104*, 819–822. [CrossRef] [PubMed]

72. Rena, G.; Hardie, D.G.; Pearson, E.R. The mechanisms of action of metformin. *Diabetologia* **2017**, *60*, 1577–1585. [CrossRef]

73. Chen, L.; Wang, J.; You, Q.; He, S.; Meng, Q.; Gao, J.; Wu, X.; Shen, Y.; Sun, Y.; Wu, X.; et al. Activating AMPK to Restore Tight Junction Assembly in Intestinal Epithelium and to Attenuate Experimental Colitis by Metformin. *Front. Pharmacol.* **2018**, *9*, 761. [CrossRef] [PubMed]

74. Tseng, C.H. Metformin use is associated with a lower risk of inflammatory bowel disease in patients with type 2 diabetes mellitus. *J. Crohns Colitis* **2021**, *15*, 64–73. [CrossRef]

75. Roy, A.; Lee, D. Dietary Creatine as a possible novel treatment for Crohn's ileitis. *ACG Case Rep. J.* **2016**, *3*, e173. [CrossRef]

76. Zhang, L.; Zhu, Z.; Yan, H.; Wang, W.; Wu, Z.; Zhang, F.; Zhang, Q.; Shi, G.; Du, J.; Cai, H.; et al. Creatine promotes cancer metastasis through activation of Smad2/3. *Cell Metab.* **2021**, S1550-413100116-9. [CrossRef]

77. Cella, P.S.; Marinello, P.C.; Padilha, C.S.; Testa, M.T.; Guirro, P.B.; Cecchini, R.; Duarte, J.A.; Guarnier, F.A.; Deminice, R. Creatine supplementation does not promote tumor growth or enhance tumor aggressiveness in Walker-256 tumor-bearing rats. *Nutrition* **2020**, *79*, 110958. [CrossRef]

78. Campos-Ferraz, P.L.; Gualano, B.; das Neves, W.; Andrade, I.T.; Hangai, I.; Pereira, R.T.; Bezerra, R.N.; Deminice, R.; Seelaender, M.; Lancha, A.H. Exploratory studies of the potential anti-cancer effects of creatine. *Amino Acids* **2016**, *48*, 1993–2001. [CrossRef]

79. Li, B.; Yang, L. Creatine in T cell antitumor immunity and cancer therapy. *Nutrients* **2021**, (in press).

80. Fairman, C.M.; Kendall, K.L.; Hart, N.H.; Taaffe, D.R.; Galvão, D.A.; Newton, R.U. The potential therapeutic effects of creatine supplementation on body composition and muscle function in cancer. *Crit. Rev. Oncol. Hematol.* **2019**, *133*, 46–57. [CrossRef]

Review

Chronic Dialysis Patients Are Depleted of Creatine: Review and Rationale for Intradialytic Creatine Supplementation

Yvonne van der Veen [1,*,†], Adrian Post [1,*,†], Daan Kremer [1], Christa A. Koops [2], Erik Marsman [3], Theo Y. Jerôme Appeldoorn [4], Daan J. Touw [4], Ralf Westerhuis [3], Margaretha Rebecca Heiner-Fokkema [2], Casper F. M. Franssen [1], Theo Wallimann [5] and Stephan J. L. Bakker [1,*]

[1] Department of Internal Medicine, University Medical Center Groningen, University of Groningen, 9713 GZ Groningen, The Netherlands; d.kremer@umcg.nl (D.K.); c.f.m.franssen@umcg.nl (C.F.M.F.)
[2] Department of Laboratory Medicine, University Medical Center Groningen, University of Groningen, 9713 GZ Groningen, The Netherlands; c.a.koops@umcg.nl (C.A.K.); m.r.heiner@umcg.nl (M.R.H.-F.)
[3] Dialysis Center Groningen, 9713 GZ Groningen, The Netherlands; e.marsman@dcg.nl (E.M.); r.westerhuis@dcg.nl (R.W.)
[4] Department of Clinical Pharmacy and Pharmacology, University Medical Center Groningen, University of Groningen, 9713 GZ Groningen, The Netherlands; t.y.j.appeldoorn@umcg.nl (T.Y.J.A.); d.j.touw@umcg.nl (D.J.T.)
[5] Department of Biology, ETH Zurich, 8092 Zurich, Switzerland; theo.wallimann@cell.biol.ethz.ch
*   Correspondence: y.van.der.veen@umcg.nl (Y.v.d.V.); a.post01@umcg.nl (A.P.); s.j.l.bakker@umcg.nl (S.J.L.B.)
†   Both authors contributed equally.

Citation: van der Veen, Y.; Post, A.; Kremer, D.; Koops, C.A.; Marsman, E.; Appeldoorn, T.Y.J.; Touw, D.J.; Westerhuis, R.; Heiner-Fokkema, M.R.; Franssen, C.F.M.; et al. Chronic Dialysis Patients Are Depleted of Creatine: Review and Rationale for Intradialytic Creatine Supplementation. *Nutrients* **2021**, *13*, 2709. https://doi.org/10.3390/nu13082709

Academic Editor: Elena Barbieri

Received: 7 June 2021
Accepted: 3 August 2021
Published: 6 August 2021

Publisher's Note: MDPI stays neutral with regard to jurisdictional claims in published maps and institutional affiliations.

Abstract: There is great need for the identification of new, potentially modifiable risk factors for the poor health-related quality of life (HRQoL) and of the excess risk of mortality in dialysis-dependent chronic kidney disease patients. Creatine is an essential contributor to cellular energy homeostasis, yet, on a daily basis, 1.6–1.7% of the total creatine pool is non-enzymatically degraded to creatinine and subsequently lost via urinary excretion, thereby necessitating a continuous supply of new creatine in order to remain in steady-state. Because of an insufficient ability to synthesize creatine, unopposed losses to the dialysis fluid, and insufficient intake due to dietary recommendations that are increasingly steered towards more plant-based diets, hemodialysis patients are prone to creatine deficiency, and may benefit from creatine supplementation. To avoid problems with compliance and fluid balance, and, furthermore, to prevent intradialytic losses of creatine to the dialysate, we aim to investigate the potential of intradialytic creatine supplementation in improving outcomes. Given the known physiological effects of creatine, intradialytic creatine supplementation may help to maintain creatine homeostasis among dialysis-dependent chronic kidney disease patients, and consequently improve muscle status, nutritional status, neurocognitive status, HRQoL. Additionally, we describe the rationale and design for a block-randomized, double-blind, placebo-controlled pilot study. The aim of the pilot study is to explore the creatine uptake in the circulation and tissues following different creatine supplementation dosages.

Keywords: creatine; intradialytic creatine supplementation; hemodialysis; muscle; protein energy wasting; clinical trial; muscle weakness; chronic fatigue; cognitive impairment; depression; anemia

1. Introduction

End-stage kidney disease (ESKD) is one of the world's leading causes of morbidity and mortality. It is estimated that more than 2.5 million people received kidney replacement therapy worldwide in 2010, and this number is projected to increase to over 5 million by 2030 [1]. Dialysis is a life-saving treatment, but unfortunately, the health-related quality of life (HRQoL) of dialysis patients is poor and mortality risks are high, as compared with the general population [2]. Although several potentially modifiable risk factors (e.g., pre-dialysis care and nutritional status) and unmodifiable risk factors (e.g., age and genetics) for excess risk of mortality and poor HRQoL have been identified in dialysis-dependent CKD patients, there is great need for the identification of new, potentially modifiable risk factors [2–7]. We hypothesize that creatine deficiency is such a modifiable risk factor, which underlies several important causes for impaired HRQoL in patients with dialysis-dependent chronic kidney disease (CKD), including protein energy wasting (PEW), sarcopenia, fatigue, muscle weakness, depression, cognitive impairment, and increased susceptibility, as well as a higher risk of an adverse course of infectious diseases.

Creatine is a natural nitrogenous organic acid that is integral to energy metabolism, and is crucial for proper cell functioning [8,9]. Creatine can be charged to the high-energy product phosphocreatine by creatine kinase and ATP [4,10]. In the human body, the majority of creatine (>90%) is present in skeletal muscle, cardiac muscle, smooth muscle, the brain, and in the nervous tissue [8,9]. Furthermore, smaller amounts are present in other tissues and cell types, including the kidney, erythrocytes, and leucocytes [8,9]. In all these tissues and cells, the phosphocreatine–creatine circuit serves as an energy buffer, facilitating quick transitions in energy requirements [9]. Importantly, on a daily basis, approximately 1.6–1.7% of the total, mainly intracellular, creatine pool is non-enzymatically degraded to creatinine, which subsequently leaves the cells and is excreted by the kidney as a waste product in urine [11,12]. This loss necessitates a continuous replenishment of the total creatine pool, with new creatine to remain in balance [12].

Generally, a common omnivorous diet can provide up to 50% of the daily requirement of creatine replenishment from alimentary sources, like meat, fish, and dairy products [7,12]. Coverage of the other 50% requires endogenous synthesis, and the requirements for endogenous synthesis increase if diets become more plant-based. The endogenous synthesis of creatine involves a metabolic pathway, of which the first and rate-limiting step is primarily situated in the kidney [13], where the enzyme arginine:glycine amidino-transferase (AGAT) converts arginine plus glycine into the creatine precursor guanidinoacetate (GAA) [14], which is exported from the kidney into circulation [8,12]. From there, GAA is taken up into the liver where it is subsequently converted by the enzyme guanidinoacetate methyl-transferase (GAMT) into creatine, and is released into the blood stream (Figure 1) [4,8,12]. Because the rate-limiting step of endogenous creatine synthesis takes place in the kidney, and the capacity for the endogenous synthesis of metabolites by AGAT has been shown to be almost halved after the donation of a kidney [7], it may be expected that patients with dialysis-dependent CKD with heavily impaired kidney function would also have a virtually absent capacity for endogenous creatine synthesis [4,7].

It should also be realized that, apart from decreased or absent endogenous synthesis, creatine losses are likely higher in patients with dialysis-dependent CKD than in healthy subjects. In healthy subjects, creatine losses will mainly be limited to the non-enzymatic conversion of creatine to creatinine, as the majority of creatine is reabsorbed after glomerular filtration [15–17]. In patients with dialysis-dependent CKD, creatine losses will be the consequence of the non-enzymatic conversion of creatine to creatinine and—on top of that—of unopposed losses into the dialysate, because the dialysis filter cannot reabsorb creatine after filtration from circulation [18]. The same holds true for amino acids, including the AGAT substrates glycine and arginine, which are essential for the synthesis of the creatine precursor GAA, and other valuable small water soluble molecules, like GAA itself [17,18].

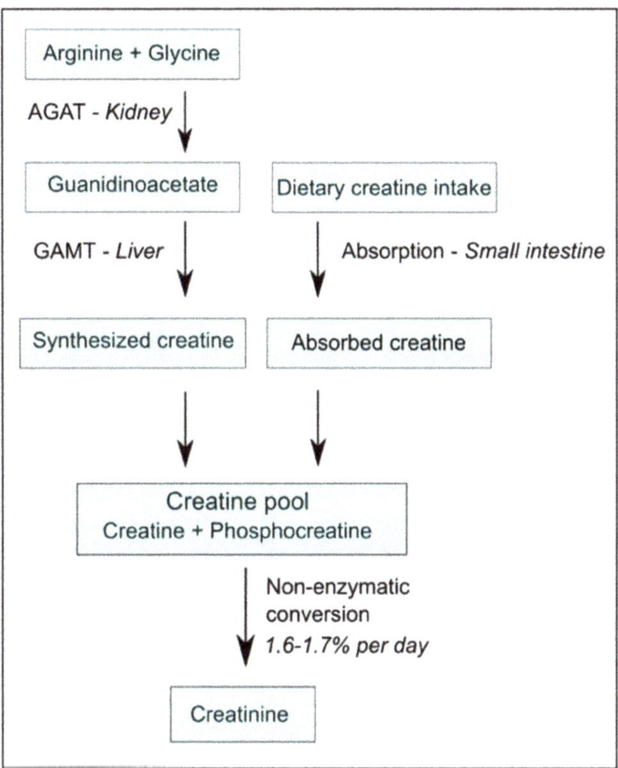

Figure 1. Simplified schematic overview of creatine homeostasis. AGAT-arginine:glycine amidino-transferase; GAMT-guanidinoacetate methyltransferase.

The existing tendency towards a negative creatine balance in chronic dialysis patients is further exacerbated by the current dietary recommendations for patients with CKD, which are increasingly steered towards more plant-based diets [19,20]. The reasons for this are that plant-based diets provide lower loads of phosphorus and acid than animal-based diets, which may seem helpful for controlling hyperphosphatemia and acidosis. However, a plant-based diet might put patients with dialysis-dependent CKD at an even higher risk for a negative creatine balance and developing creatine deficiency, because plant-based diets lack naturally derived creatine [21]. It should be noted that amino acids necessary for the endogenous synthesis of creatine are present in plant-based foods, but because of the absence of kidney function in patients with dialysis-dependent CKD, the enzymatic capacity for the endogenous synthesis of creatine from these amino acids is severely impaired [4,7]. This is underscored by the fact that even without dietary recommendations toward a plant-based diet, roughly 43% of the general U.S. population has an average intake of creatine below the recommended daily allowance of 1.0 g of dietary creatine per day [22].

Together, the poor endogenous synthesis of creatine, unopposed loss of creatine and creatine precursors during dialysis, and low dietary intake of creatine may add up to creatine deficiency in patients with dialysis-dependent CKD. In general, the plasma concentrations of small molecules tend to be higher in hemodialysis patients compared with healthy individuals, because of a lower glomerular filtration rate. However, the fasting plasma creatine and GAA levels are lower in dialysis patients compared with healthy individuals [18]. This indicates that dialysis-dependent CKD, in comparison with normal subjects, have a generally lower level of creatine and its metabolites. This is supported by the fact that skeletal muscle biopsies in CKD showed significantly low ATP and phosphocreatine levels [23]. This notion is further supported by studies showing lower phosphocreatine concentrations and a decreased phosphocreatine/ATP energy-charge ratio in the hearts and skeletal muscle of patients on either hemodialysis or peritoneal dialysis treatment, as shown by means of non-invasive in vivo 31P-NMR imaging [24–26].

2. Rationale

Creatine supplementation is frequently used among athletes, during rehabilitation, and in patients with neuromuscular diseases [27–29]. We propose that creatine supplementation is particularly important for patients with dialysis-dependent CKD. The reasons why this is likely to be of particular importance in these patients are because (1) the endogenous synthesis of creatine in these patients is severely impaired because of the virtual absence of kidney function accompanied by the virtual absence of the first enzymatic step required for the endogenous absence of creatine from the amino acids arginine and glycine [4,7]; (2) unopposed losses of creatine to the dialysis fluid during dialysis sessions [17,18]; and (3) inadequate intake of creatine due to advice towards a primary plant-based diet in these patients [21]. All of this comes on top of the normally existing continuous non-enzymatic conversion of approximately 1.6–1.7% of the endogenous creatine pool to creatinine, which necessitates the continuous replenishment of creatine by the combination of endogenous synthesis and dietary intake in order to remain in steady-state [11,12]. This is novel because, until recently, it was not recognized that kidney function is an important contributor to endogenous creatine synthesis, so the capacity of patients with dialysis-dependent CKD to maintain creatine homeostasis in the light of ongoing conversion of creatine into creatinine is severely impaired, and there are additional unopposed losses of creatine to the dialysis fluid and an inadequate dietary intake. Patients with dialysis-dependent CKD could benefit from creatine supplementation by allowing for the maintenance of their endogenous creatine pools, which would help them to sustain bodily functions that depend on creatine availability, including normal functioning of the muscles, heart, immune system, and brain [24,30,31].

Thus far, only a few creatine supplementation trials have been performed in patients with dialysis-dependent CKD. In a double-blind, randomized, placebo-controlled trial with 10 patients, presenting with frequent muscle cramps during hemodialysis, an oral dose of 12 g of creatine monohydrate or placebo was given before each dialysis session for 4 weeks [32]. The authors found a 60% reduction in muscle cramps after four weeks of creatine supplementation, while no difference was found in the placebo group. After a wash-out period, the frequency of muscle cramps returned to the previous level. In another double-blind, randomized, placebo-controlled trial with 30 hemodialysis patients, an initial loading phase of 1 week, in which an oral dose of 40 g of creatine monohydrate or placebo was given per day, was followed by a period of 3 weeks, in which an oral dose of 10 g of creatine monohydrate or placebo was given per day [33]. It was evaluated whether oral creatine supplementation could attenuate the loss of muscle mass and malnutrition-inflammation score (MIS). MIS was developed to examine PEW in relation to inflammation, and is made up of 10 components, namely: weight change, dietary intake, gastro-intestinal symptoms, functional capacity, comorbidity, subcutaneous fat, signs of muscle wasting, body mass index (BMI), serum albumin level, and total iron-binding capacity (TIBC). Each of these 10 components of MIS has four levels for severity, ranging from 0 (normal) to 3 (severely abnormal). The sum of all 10 components can therefore range from 0 (normal) to 30 (severely malnourished and inflamed) [34]. Compared with the placebo arm, the creatine treated arm demonstrated a significant increase in lean body mass after 4 weeks of supplementation. Additionally, the creatine treated arm demonstrated a sizable reduction of MIS. However, to date, no studies have investigated whether creatine supplementation in patients with dialysis-dependent CKD is able to improve muscle strength, cognitive impairment, HRQoL, ability to perform daily tasks, infectious diseases, fatigue, or depression, individually or altogether. Furthermore, the aforementioned studies each used oral creatine supplementation with either single or multiple doses throughout the day, dissolved in relatively large volumes of water, which may negatively affect both patient compliance and fluid balance.

To avoid problems with compliance and fluid balance, and, furthermore, to prevent intradialytic losses of creatine to the dialysate, we aim to investigate the potential of intradialytic creatine supplementation in improving muscle status and HRQoL. To do so, the optimal creatine concentration in the dialysate in relation to maximal creatine uptake, as well as the tolerability of the intradialytic creatine application, must be determined first. The second goal of the study is to determine the effects of intradialytic creatine supplementation of chronic hemodialysis patients on muscle mass, muscle strength, cognitive functions, HRQoL, frailty, fatigue, and depression parameters.

3. A Clinical Pilot Study

3.1. Study Design and Setting

In this randomized, double-blind, placebo-controlled pilot study, a total of 16 hemodialysis patients will be included, which will be divided into four groups (0.5 mM, 1.0 mM, 1.5 mM, and 2.0 mM) each consisting of three patients receiving creatine and one receiving placebo. The study is a pilot study to generate data to allow for sample size calculations for a future larger study to be performed, as no data on the standard deviations of changes of intra-erythrocyte creatine concentrations over time and other outcome parameters over time are currently available to allow for an adequate sample size calculation. So, no sample size was calculated for the current study. The study will be conducted at the University Medical Center Groningen (UMCG) and the Dialysis Center Groningen (DCG). The protocol was written in accordance with the Standard Protocol Items: Recommendations for Interventional Trials (SPIRIT) guideline. Ethics committee approval will be obtained from the UMCG, the Netherlands.

The study population comprises of patients with dialysis-dependent CKD being treated with hemodialysis. The targeted study population of dialysis-dependent kidney disease patients in the local study centers is primarily >55 years old, predominantly Caucasian, with similar proportions of men and women, with approximately 50% of patients having residual diuresis. Only subjects who provide written informed consent are eligible. Additional inclusion criteria for this study are as follows:

- Age $\geq$ 18 years;
- Hemodialysis treatment in the UMCG or DCG;
- Dialysis vintage $\geq$2 months;
- Conventional hemodialysis, thrice weekly treatment with three to five hours per dialysis treatment;
- Hemoglobin at previous routine monthly assessment greater than or equal to 6.5 mmol/L;
- Signed informed consent.
- Exclusion criteria for the current study are:
- Pregnancy;
- Presence of clinical signs of infection;
- Confirmed diagnosis of malignancies;
- Incapacity of the Dutch language;
- Inability to complete questionnaires.

3.2. Interventions, Blinding, and Randomization

In this study, creatine will be added to the dialysis fluid. Adverse events are defined as any untoward medical occurrence in patients participating in the study. These adverse events do not necessarily have to have a causal relationship with the treatment. No major side-effects are expected. As the principle of intradialytic creatine supplementation is new, this pilot study will take a step-up approach with block randomization. The hemodialysis patients are divided into four different dosage groups in which patients will continuously receive either 0.5 mM, 1.0 mM, 1.5 mM, or 2.0 mM of creatine dissolved into the dialysis fluid during the entirety of a dialysis session. Each group consisted of four patients, in which three patients will be randomized to receive creatine and one patient received the placebo. Randomization will be performed according to the subsequent blocks, each of which will consiste of three patients receiving creatine and one receiving placebo. The first randomization block will be assigned to a creatine dosage of 0.5 mM. If after one week no adverse effects are observed, the next, higher dosage group will be started (Figure 2).

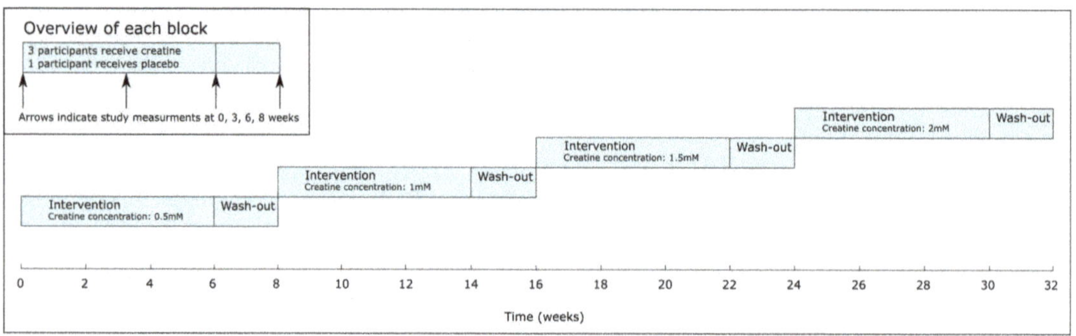

Figure 2. Timeline overview.

Oral supplementation has been extensively studied in the last thirty years, both in athletes as well as in the general population, in various age groups ranging from children to the elderly [35,36]. Oral supplementation with chemically pure creatine monohydrate, as used in this study, if taken within the officially recommended dosages, is safe with no significant side effects, except for moderate weight gain [33,37]. This weight gain is initially due to water retention, and in the long term is due to an increase in fat free muscle mass [33]. In dialysis patients, the initial water retention can be compensated by the dialysis itself. The increase in fat free muscle mass would, in fact, be beneficial for dialysis-dependent CKD, as they often suffer from PEW and/or sarcopenia. On top of that, as the gastrointestinal tract is spared when using intradialytic supplementation, complaints such as gastrointestinal complaints and bloating are unlikely to occur.

The patients will receive creatine supplementation or placebo (hemodialysis quality water) during each hemodialysis session for a total length of 6 weeks. Creatine-monohydrate, Creapure® "Pharma Quality" (not GMP), produced by AlzChem Trostberg, Germany will be used for the preparation of a 50 mmol/L stock solution of creatine which will be added to the dialysis fluid to reach the projected dialysate concentrations, as indicated above.

To ensure the double-blind design, optically similar solutions will be packed and coded by Ace Pharmaceuticals, the Netherlands. The creatine solution will be added with an infusion pump at a sampling tip on the afferent dialysis fluid hose for the artificial kidney. Depending on the setting of the infusion pump and the dialysis flow, different creatine concentrations (between 0.5 and 2.0 mM) will be reached. The total dialysis treatment duration for the study was six weeks, with three dialysis sessions per week. The total creatine uptake at the different dosages (0.5 mM, 1.0 mM, 1.5 mM, or 2.0 mM) will be calculated as the added amount of creatine minus the creatine measured in the collected hemodialysis fluid (the creatine that did not diffuse to the blood).

3.3. Measurements

Study measurements will be performed at baseline, week 3, week 6, and after a washout period of 2 weeks. The primary outcome parameters of the study will be the assessment of the plasma creatine and intra-erythrocyte creatine concentration. At each study visit, the blood, interdialytic urine, and dialysate will be collected from each participant. Blood will be drawn from the hemodialysis lines both prior to and after the hemodialysis treatment. Participants with residual diuresis will be instructed to collect all of the urine in the interdialytic interval prior to the hemodialysis session of the visit, according to a strict protocol. At the end of the hemodialysis session prior to the hemodialysis session of the study visit, participants will be instructed to start collecting all of the urine until the start of the hemodialysis session of the study visit. For participants dialyzing thrice weekly, this will be roughly 48 h. During the hemodialysis session, all dialysate will be collected in a 200-L tank. The total dialysate volume will be measured by calculating the weight difference of the tank before and after the hemodialysis session. At the end of hemodialysis, all dialysate will be homogenized, and samples will be taken for analyses and storage.

Plasma, intra-erythrocyte, and urinary creatine concentrations will be measured with liquid chromatography mass spectrometry (LC-MS/MS), validated according to ISO 151 89 guidelines, using an LC30AD UPLC (Shimadzu, Kyoto, Japan) and an API-4500 mass spectrometer (SCIEX, Framingham, MA, USA). Creatine will be separated using a Phenomenex Kinetex Evo™ C18 (2.6 μm, 100 Å, 150 × 4.6 mm) column, with a Phenomenex SecurityGuard™ ULTRA cartridges for C18 (2.6 μm, 2 × 4.6 mm) guard column. Creatine is detected using positive-ion electrospray ionization in the multiple reaction monitoring mode using the following transitions: m/z 132→90 for creatine and m/z 135→93 for the internal standard (D3-creatine). Data will be analyzed using Analyst MD 1.6.2 (Sciex). The intra-assay and inter-assay coefficients of variation are 3.9% and 6.5% at 17 μmol/L and 1.7% and 1.7% at 1065 μmol/L, respectively.

Anthropometry measurements will include body weight in kilograms (kg), body height in centimeters (cm), and hip and waist circumference both in centimeters (cm). Body weight will be measured both before and after the hemodialysis session in lightweight clothing without shoes using a calibrated digital measuring scale (seca 877, seca, Hamburg, Germany). Body height will be measured before the start of the hemodialysis session using a wall-secured stadiometer (seca 222). Waist and hip circumference will be measured in centimeters (cm) twice after the hemodialysis session, using a measuring tape roll with a standardized retraction mechanism (seca 201). Blood pressure (mmHg) will be measured according to a standard clinical protocol, using an automatic device (Phillips Suresign VS2+, Andover, MA, USA). Blood pressure (mmHg) will be measured before the start of the hemodialysis session, each hour during the dialysis period, and at the end of the hemodialysis session. Blood pressure will be measured in standing position before and after the dialysis session. Patients will be asked to stand up straight for 1 min, after which blood pressure and heart rate measurements will be performed. Body composition will be determined using a multifrequency bioelectrical impedance analyzer (BIA, Quadscan 4000, Bodystat, Douglas, UK) at 5, 50, 100, and 200 Hz, which allowes for distinguishing between lean body mass and fat body mass, taking into account the differences in volume status [38]. According to recommendations [39], the BIA measurement will be performed before the hemodialysis session, with the participant in a supine position with arms and legs abducted from the body.

Hand grip strength will be assessed in kilograms (kg) in both arms with the Jamar Hydraulic Hand Dynamometer (Patterson Medical JAMAR 5030J1, Warenville, IL, USA) [40]. The hand grip strength will be assessed before the start of the hemodialysis session, in a seated position, with shoulders in adduction, arms rotated into a neutral position, elbows flexed to 90°, and forearms and wrists held in a neutral position. After the participant is positioned in the right way, they will be instructed to perform a maximal isometric contraction. To avoid the occurrence of motivation bias, all patients will be encouraged through vocal commands at high volume [41]. Hand grip strength will be tested three times with an interval of 30 s between each attempt. As it has been shown that the difference between the arm with the fistulae compared with the contralateral is greater than the difference between the dominant compared with the non-dominant hand [42], the measurements will be performed on the arm in which no fistula is present, and when no fistula is present in either of the arms, measurements will be performed on the hand of the dominant arm.

The Short Physical Performance Battery (SPPB) is a relatively simple test that can be used to gain insight into balance, gait, strength, and endurance [43], and will be performed before the start of the hemodialysis session. The SPPB consists of three separate tests, namely: a balance test, a walking test, and a repeated chair–stand test. For the balance test, participants need to stand still for at least 10 s in three progressively difficult positions, with the feet together in side-by-side, semi-tandem, and tandem positions. Lower limb strength will be tested with the repeated chair–stand test. Participants will be asked to fold their arm across their chests while standing up and sitting down five times as quickly as possible, and the time will be recorded in seconds (s).

The 4 m Gait Speed Test (4MWT) will be used to measure locomotion [44–46]. For the 4MWT, two pylons will be placed 4 m apart and the subject will be instructed to walk at their usual place. Time from start to end of the 4 m will be recorded in seconds (s). The 4MWT will be measured twice, before the start of the hemodialysis session.

Forced expiratory volume for 1 s (FEV_1) will be measured in liters (L), both before the start of the dialysis session as well as at the end of the hemodialysis session, by means of an Asma-1 handheld spirometer (Vitalograph, Buckingham, UK) [47].

The 9-Hole Peg Test (9-HPT, Sammsons Preston Rolyan, Chicaco, IL, USA) requires participants to place and remove nine pegs into nine holes, one at a time, as quickly as possible, and is therefore a measure of hand dexterity [48]. Participants will be asked to perform the 9-HPT with the dominant hand first and then with the non-dominant hand.

The following questionnaires will be administered: Six-Dimensional EuroQol Instrument (EQ-6D) [49] and the Short Form 36 Health Survey (SF36) [50] to assess HRQoL; Checklist Individual Strength (CIS) [51] to assess fatigue; Dialysis Symptom Indicator (DSI) [16] to assess the prevalence, severity, and impact of symptoms in hemodialysis patients; Groninger Frailty Indicator (GFI) [52] to identify frailty; Cognitive Failure Questionnaire (CFQ) [53] to assess cognitive failure; Patient Health Questionnaire-9 (PHQ-9) [54] to measure depressive disorders and depression severity; and the Patient-Generated Subjective Global Assessment (PG-SGA) [55–57] to assess nutritional status. Patients can fill in the questionnaires at home before the start of the hemodialysis session, or during the dialysis session with the help of a researcher.

3.4. Outcomes

The aim of the pilot study is to explore the creatine uptake in the circulation and tissues following different creatine supplementation dosages. Therefore, the main parameters for the pilot study are the plasma creatine concentration and intra-erythrocytic creatine concentration of both the pre- and post-hemodialysis samples. Intra-erythrocytic creatine concentration will be used as a non-invasive proxy for creatine tissue uptake. The secondary study parameters are hand grip strength as a measure of muscle strength, the combined interdialytic urinary and intradialytic dialysate excretion of creatinine as a measure of muscle mass [58], and bioelectrical impedance analysis (BIA) as a measure of body composition. Other study parameters are N-terminal pro-brain natriuretic peptide (NT-proBNP) as a cardiac function marker; high sensitivity troponin T (hs-TNT) as a cardiac ischaemia marker; C-reactive protein as an inflammation marker; self-reported physical health using the EQ-6D, SF36, and DSI; fatigue using the CIS; and cognitive functions using the CFQ.

3.5. Statistical Analyses

Statistical analyses will be performed using R statistical software (Vienna, Austria) (http://cran.r-project.org/ (accessed on 1 August 2021)). The results will be expressed as mean $\pm$ standard deviation (SD), median (interquartile range), or number (percentage) for normally distributed, skewed, and categorical data, respectively. For all of the analyses, *p*-value < 0.05 will be considered statistically significant. To determine the effect of creatine supplementation on the primary and secondary outcomes, linear mixed-effect models will be used to compare the change in outcome over time between the different treatment groups. Linear mixed-effect models will be performed using the "lmer" function of the "lme4" package. Treatment group, time, and the interaction between treatment group and time will be used as fixed factors, and random intercepts and slopes will be added for the subjects. To minimize the false positive rate, as recommended by Barr et al., we aim to use the maximal random effect structure, even if a random effect does not improve the overall model fit [59].

4. Discussion

Creatine supplementation as a nutritional intervention has been extensively studied in several populations, showing promising results [60–62]. To date, almost all creatine supplementation studies have been performed with oral supplementation. In dialysis-dependent kidney disease patients, the oral supplementation of creatine is less suitable as it requires dissolving the creatine in large volumes of water, which may negatively affect the fluid balance. In addition, the causes for creatine deficiency in dialysis-dependent kidney disease patients, being impaired endogenous synthesis [4,7], losses to the dialysis fluid [17,18], and insufficient intake [19,20], are chronic in nature and therefore require long-term supplementation of creatine. As many CKD patients are dependent on dialysis for years, sometimes even lifelong, treating creatine deficiency with oral supplementation requires a high level of compliance. In addition, oral supplementation does not prevent losses of creatine to the dialysis fluid. In contrast to oral supplementation, intradialytic supplementation offers the possibility to supplement creatine in a controlled manner, while preventing unopposed losses of creatine to the dialysis fluid, volume overload due to the necessary ingestion of large volumes of water, and potential problems with compliance.

A creatine deficient state is not without consequences. It has become increasingly clear that low creatine levels play an important role in many different causes for impaired HRQoL and have higher mortality rates in hemodialysis patients. For example, a recent study showed that higher plasma creatine concentrations are associated with lower odds of low muscle mass, low protein intake, hypoalbuminemia, and severe fatigue, indicating a potential role for creatine supplementation in hemodialysis patients [18].

Protein-energy-wasting (PEW), a progressive depletion of protein and energy stores, is highly prevalent in hemodialysis patients (up to 50–75%) and is associated with both increased morbidity and mortality, and impaired quality of life. Potential causes of PEW in patients with dialysis-dependent CKD include reduced protein and energy intake, reduced physical activity, increased catabolism, reduced anabolism, and comorbidities (e.g., diabetes), as well as the dialysis treatment itself, causing, among others, a loss of amino acids to the dialysate [63,64]. Additionally, the situation is exacerbated by inflammatory processes, and loss of residual renal function [63]. These causes often occur simultaneously and exacerbate the general pathological state of dialysis patients.

It has been shown that oral creatine supplementation increases the intramuscular total creatine (creatine plus phosphocreatine) concentrations [62] and, in parallel, significantly increases muscle mass [65] and muscle performance for high-intensity and endurance performance [62,66]. In addition, there is a growing body of evidence that creatine supplementation, combined with moderate resistance training, can, at least partially, counteract the loss of muscle mass caused by ageing or immobilization [67,68].

Furthermore, hemodialysis patients frequently suffer from fatigue [2,69]. The burden of fatigue is underscored by the results of a cross-sectional study, in which 94% of the patients would accept more frequent hemodialysis if it would increase their energy level, but only 19% would do so for an increase in survival time of up to 3 years [70]. It has been hypothesized previously that creatine administration plays an important role in reducing both mental and muscular fatigue by increasing the brain content of phosphocreatine [71]. Therefore, intradialytic creatine supplementation may benefit hemodialysis patients by potentially attenuating the fatigue often experienced with kidney disease.

Another contributor to increased mortality and impaired HRQoL in hemodialysis patients is cognitive impairment [72]. Cognitive impairment is associated with lower compliance concerning nutritional restrictions, fluid restrictions, and medication [73]. The underlying pathogenesis is not fully understood, but it has been indicated that creatine supplementation improves brain health, improves cognition, and is effective at alleviating brain ischemia and hypoxia [30]. Importantly, creatine has also been shown to alleviate treatment-resistant depression, especially in women [74–76], and recent study results indicate a significant negative relationship between dietary creatine intake and depression in a nationally representative adult cohort in the USA, indicating that a low creatine intake may enhance the incidence of depression [75]. Therefore, intradialytic creatine supplementation may benefit hemodialysis patients by potentially improving the mental complaints often experienced with kidney disease.

Additionally, as creatine has a positive impact on both the development and activation of the innate and adaptive immune response [31], dialysis patients who are supplemented with creatine may have a lower incidence of infections during the time of chronic dialysis treatments.

Furthermore, as cardiovascular diseases are prominent pathologies in dialysis patients, creatine, with its reported benefit for vascular health, such as alleviating oxidative stress and inflammation [71,77,78], may be helpful in these respects in this vulnerable population.

During hemodialysis treatment, erythrocytes are subject to mechanical and oxidative stress, potentially leading to anemia. Studies suggest that creatine, by its ability to inhibit erythrocyte lipid peroxidation, may contribute to the maintenance of normal cell deformability [79–81]. Intradialytic creatine supplementation could potentially lead to reduced losses of erythrocytes, and may therefore reduce the erythropoietin (EPO) requirements of the dialysis patients. The reduced administration of EPO will circumvent the possible EPO-supported progression of tumors and should markedly reduce costs to the healthcare system [82,83].

5. Conclusions

Patients with CKD relying on dialysis treatments likely suffer from creatine deficiency due to a decreased endogenous production of creatine and unopposed losses of creatine from the blood into the dialysate. In the current study, we have provided the rationale for intradialytic creatine supplementation and described the study protocol for a pilot study in preparation for a large double-blind, placebo-controlled supplementation trial.

Intradialytic creatine supplementation may help to maintain creatine homeostasis among dialysis-dependent CKD patients, and consequently improve important causes for impaired HRQoL, including protein energy wasting (PEW), fatigue, muscle weakness, depression, and cognitive impairment.

Author Contributions: Conceptualization, T.W., S.J.L.B., C.F.M.F., Y.v.d.V. and A.P.; investigation, Y.v.d.V., A.P.; methodology, C.A.K., M.R.H.-F., E.M. and R.W.; resources, T.Y.J.A. and D.J.T.; supervision, T.W. and S.J.L.B.; writing, Y.v.d.V., A.P., D.K. and C.A.K. All authors have read and agreed to the published version of the manuscript.

Funding: This research received no external funding.

Institutional Review Board Statement: The study will be conducted according to the guidelines of the Declaration of Helsinki, after approval by the Ethics Committee of the University Medical Center Groningen.

Informed Consent Statement: Informed consent will be obtained from all subjects involved in the study.

Conflicts of Interest: T.W. is involved as a scientific consultant at Crearene AG, Switzerland.

Abbreviations

Abbreviations

AGA—arginine:glycine amidino-transferase; AMPK—AMP-activated protein kinase; ATP—adenosine-tri-phosphate; Cr—creatine; PCr—Phosphocreatine; CK—creatine kinase; ESKD—end-stage kidney disease; GAA—guanidinoacetate; GAMT—guanidino-acetate-amino-transferase; HRQoL—health-related quality of life.

References

1. Al Ismaili, F.; Al Salmi, I.; Al Maimani, Y.; Metry, A.M.; Al Marhoobi, H.; Hola, A.; Pisoni, R.L. Epidemiological Transition of End-Stage Kidney Disease in Oman. *Kidney Int. Rep.* **2017**, *2*, 27–35. [CrossRef]
2. Kraus, M.A.; Fluck, R.J.; Weinhandl, E.D.; Kansal, S.; Copland, M.; Komenda, P.; Finkelstein, F.O. Intensive Hemodialysis and Health-Related Quality of Life. *Am. J. Kidney Dis.* **2016**, *68*, S33–S42. [CrossRef]
3. Ma, L.; Zhao, S. Risk factors for mortality in patients undergoing hemodialysis: A systematic review and meta-analysis. *Int. J. Cardiol.* **2017**, *238*, 151–158. [CrossRef]
4. Wallimann, T.; Riek, U.; Möddel, M. Intradialytic creatine supplementation: A scientific rationale for improving the health and quality of life of dialysis patients. *Med. Hypotheses* **2017**, *99*, 1–14. [CrossRef]
5. Levin, A. Identification of patients and risk factors in chronic kidney disease-evaluating risk factors and therapeutic strategies. *Nephrol. Dial. Transplant.* **2001**, *16*, 57–60. [CrossRef] [PubMed]
6. McQuillan, R.; Trpeski, L.; Fenton, S.; Lok, C.E. Modifiable risk factors for early mortality on hemodialysis. *Int. J. Nephrol.* **2012**, *2012*, 435736. [CrossRef]
7. Post, A.; Tsikas, D.; Bakker, S.J.L. Creatine is a conditionally essential nutrient in chronic kidney disease: A hypothesis and narrative literature review. *Nutrients* **2019**, *11*, 1044. [CrossRef] [PubMed]
8. Wyss, M.; Kaddurah-Daouk, R. Creatine and Creatinine Metabolism. *Physiol. Rev.* **2000**, *80*, 1107–1213. [CrossRef]
9. Wallimann, T.; Tokarska-Schlattner, M.; Schlattner, U. The creatine kinase system and pleiotropic effects of creatine. *Amino Acids* **2011**, *40*, 1271–1296. [CrossRef]
10. Clark, J.F. Creatine and Phosphocreatine: A Review of Their Use in Exercise and Sport. *J. Athl. Train.* **1997**, *32*, 45. [PubMed]
11. Wilson, F.P.; Xie, D.; Anderson, A.H.; Leonard, M.B.; Reese, P.P.; Delafontaine, P.; Horwitz, E.; Kallem, R.; Navaneethan, S.; Ojo, A.; et al. Urinary Creatinine Excretion, Bioelectrical Impedance Analysis, and Clinical Outcomes in Patients with CKD: The CRIC Study. *Clin. J. Am. Soc. Nephrol.* **2014**, *9*, 2095–2103. [CrossRef] [PubMed]
12. Brosnan, J.T.; da Silva, R.P.; Brosnan, M.E. The metabolic burden of creatine synthesis. *Amino Acids* **2011**, *40*, 1325–1331. [CrossRef] [PubMed]
13. Da Silva, R.P.; Clow, K.; Brosnan, J.T.; Brosnan, M.E. Synthesis of guanidinoacetate and creatine from amino acids by rat pancreas. *Br. J. Nutr.* **2014**, *111*, 571–577. [CrossRef]
14. Edison, E.E.; Brosnan, M.E.; Meyer, C.; Brosnan, J.T. Creatine synthesis: Production of guanidinoacetate by the rat and human kidney in vivo. *Am. J. Physiol. Ren. Physiol.* **2007**, *293*, 1799–1804. [CrossRef]
15. Verrey, F.; Singer, D.; Ramadan, T.; Vuille-Dit-Bille, R.N.; Mariotta, L.; Camargo, S.M.R. Kidney amino acid transport. *Pflug. Arch. Eur. J. Physiol.* **2009**, *458*, 53–60. [CrossRef] [PubMed]
16. Weisbord, S.D.; Fried, L.F.; Arnold, R.M.; Rotondi, A.J.; Fine, M.J.; Levenson, D.J.; Switzer, G.E. Development of a symptom assessment instrument for chronic hemodialysis patients: The dialysis symptom index. *J. Pain Symptom Manag.* **2004**, *27*, 226–240. [CrossRef] [PubMed]
17. Hendriks, F.K.; Smeets, J.S.J.; Broers, N.J.H.; van Kranenburg, J.M.X.; van der Sande, F.M.; Kooman, J.P.; Van Loon, L.J.C. End-stage renal disease patients lose a substantial amount of amino acids during hemodialysis. *J. Nutr.* **2020**, *150*, 1160–1166. [CrossRef] [PubMed]
18. Post, A.; Schutten, J.C.; Kremer, D.; van der Veen, Y.; Groothof, D.; Sotomayor, C.G.; Koops, C.A.; de Blaauw, P.; Kema, I.P.; Westerhuis, R.; et al. Creatine homeostasis and protein energy wasting in hemodialysis patients. *J. Transl. Med.* **2021**, *19*, 115. [CrossRef]
19. Joshi, S.; McMacken, M.; Kalantar-Zadeh, K. Plant-Based Diets for Kidney Disease: A Guide for Clinicians. *Am. J. Kidney Dis.* **2021**, *77*, 287–296. [CrossRef] [PubMed]
20. Carrero, J.J.; González-Ortiz, A.; Avesani, C.M.; Bakker, S.J.L.; Bellizzi, V.; Chauveau, P.; Clase, C.M.; Cupisti, A.; Espinosa-Cuevas, A.; Molina, P.; et al. Plant-based diets to manage the risks and complications of chronic kidney disease. *Nat. Rev. Nephrol.* **2020**, *16*, 525–542. [CrossRef]
21. Balestrino, M.; Adriano, E. Beyond sports: Efficacy and safety of creatine supplementation in pathological or paraphysiological conditions of brain and muscle. *Med. Res. Rev.* **2019**, *39*, 2427–2459. [CrossRef]
22. Ostojic, S.M. Dietary creatine intake in U.S. population: NHANES 2017–2018. *Nutrition* **2021**, *87–88*, 111207. [CrossRef]
23. Pastoris, O.; Aquilani, R.; Foppa, P.; Bovio, G.; Segagni, S.; Baiardi, P.; Catapano, M.; Maccario, M.; Salvadeo, A.; Dossena, M. Altered muscle energy metabolism in post-absorptive patients with chronic renal failure. *Scand. J. Urol. Nephrol.* **1997**, *31*, 281–287. [CrossRef] [PubMed]

24. Durozard, D.; Pimmel, P.; Baretto, S.; Caillette, A.; Labeeuw, M.; Baverel, G.; Zech, P. 31P NMR spectroscopy investigation of muscle metabolism in hemodialysis patients. *Kidney Int.* **1993**, *43*, 885–892. [CrossRef] [PubMed]

25. Tagami, T.; Sakuma, H.; Matsumura, K.; Takeda, K.; Mori, S.; Takeuchi, T.; Nakano, T. Evaluation of altered myocardial high energy phosphate metabolism patients on maintenance dialysis using phosphorus-31 magnetic resonance spectroscopy. *Investig. Radiol.* **1998**, *33*, 171–176. [CrossRef]

26. Ogimoto, G.; Sakurada, T.; Imamura, K.; Kuboshima, S.; Maeba, T.; Kimura, K.; Owada, S. Alteration of energy production by the heart in CRF patients undergoing peritoneal dialysis. *Mol. Cell. Biochem.* **2003**, *244*, 135–138. [CrossRef]

27. MA, T. Clinical use of creatine in neuromuscular and neurometabolic disorders. *Subcell. Biochem.* **2007**, *46*, 183–204. [CrossRef]

28. Harmon, K.K.; Stout, J.R.; Fukuda, D.H.; Pabian, P.S.; Rawson, E.S.; Stock, M.S. The Application of Creatine Supplementation in Medical Rehabilitation. *Nutrients* **2021**, *13*, 1825. [CrossRef] [PubMed]

29. Racette, S.B. Creatine Supplementation and Athletic Performance. *J. Orthop. Sports Phys. Ther.* **2003**, *33*, 615–621. [CrossRef]

30. Roschel, H.; Gualano, B.; Ostojic, S.M.; Rawson, E.S. Creatine Supplementation and Brain Health. *Nutrients* **2021**, *13*, 586. [CrossRef]

31. Bredahl, E.C.; Eckerson, J.M.; Tracy, S.M.; McDonald, T.L.; Drescher, K.M. The Role of Creatine in the Development and Activation of Immune Responses. *Nutrients* **2021**, *13*, 751. [CrossRef]

32. Chang, C.T.; Wu, C.H.; Yang, C.W.; Huang, J.Y.; Wu, M.S. Creatine monohydrate treatment alleviates muscle cramps associated with haemodialysis. *Nephrol. Dial. Transplant.* **2002**, *17*, 1978–1981. [CrossRef] [PubMed]

33. Marini, A.C.B.; Motobu, R.D.; Freitas, A.T.V.; Mota, J.F.; Wall, B.T.; Pichard, C.; Laviano, A.; Pimentel, G.D. Short-Term Creatine Supplementation May Alleviate the Malnutrition-Inflammation Score and Lean Body Mass Loss in Hemodialysis Patients: A Pilot Randomized Placebo-Controlled Trial. *J. Parenter. Enter. Nutr.* **2020**, *44*, 815–822. [CrossRef] [PubMed]

34. Rambod, M.; Bross, R.; Zitterkoph, J.; Benner, D.; Pithia, J.; Colman, S.; Kovesdy, C.; Kopple, J.D.; Kalantar-Zadeh, K. Association of Malnutrition-Inflammation Score with Quality of Life and Mortality in Maintenance Hemodialysis Patients: A 5-Year Prospective Cohort Study. *Am. J. Kidney Dis.* **2009**, *53*, 298. [CrossRef] [PubMed]

35. Dalbo, V.J.; Roberts, M.D.; Lockwood, C.M.; Tucker, P.S.; Kreider, R.B.; Kerksick, C.M. The effects of age on skeletal muscle and the phosphocreatine energy system: Can creatine supplementation help older adults. *Dyn. Med.* **2009**, *8*, 2009. [CrossRef]

36. Kreider, R.B. Effects of creatine supplementation on performance and training adaptations. *Mol. Cell. Biochem.* **2003**, *244*, 89–94. [CrossRef] [PubMed]

37. Twycross-Lewis, R.; Kilduff, L.P.; Wang, G.; Pitsiladis, Y.P. The effects of creatine supplementation on thermoregulation and physical (cognitive) performance: A review and future prospects. *Amin. Acids* **2016**, *48*, 1843–1855. [CrossRef]

38. Kyle, U.G.; Bosaeus, I.; De Lorenzo, A.D.; Deurenberg, P.; Elia, M.; Gómez, J.M.; Heitmann, B.L.; Kent-Smith, L.; Melchior, J.C.; Pirlich, M.; et al. Bioelectrical impedance analysis—Part I: Review of principles and methods. *Clin. Nutr.* **2004**, *23*, 1226–1243. [CrossRef] [PubMed]

39. Tattersall, J. Bioimpedance Analysis in Dialysis: State of the Art and What We Can Expect. *Blood Purif.* **2009**, *27*, 70–74. [CrossRef]

40. Angst, F.; Drerup, S.; Werle, S.; Herren, D.B.; Simmen, B.R.; Goldhahn, J. Prediction of grip and key pinch strength in 978 healthy subjects. *BMC Musculoskelet. Disord.* **2010**, *11*, 94. [CrossRef]

41. Johansson, C.A.; Kent, B.E.; Shepard, K.F. Relationship between verbal command volume and magnitude of muscle contraction. *Phys. Ther.* **1983**, *63*, 1260–1265. [CrossRef]

42. Pajek, M.B.; Čuk, I.; Pajek, J. Vascular Access Effects on Motor Performance and Anthropometric Indices of Upper Extremities. *Ther. Apher. Dial.* **2016**, *20*, 295–301. [CrossRef]

43. Guralnik, J.M.; Simonsick, E.M.; Ferrucci, L.; Glynn, R.J.; Berkman, L.F.; Blazer, D.G.; Scherr, P.A.; Wallace, R.B. A short physical performance battery assessing lower extremity function: Association with self-reported disability and prediction of mortality and nursing home admission. *J. Gerontol.* **1994**, *49*, M85–M94. [CrossRef] [PubMed]

44. Kutner, N.G.; Zhang, R.; Huang, Y.; Painter, P. Gait Speed and Mortality, Hospitalization, and Functional Status Change among Hemodialysis Patients: A US Renal Data System Special Study. *Am. J. Kidney Dis.* **2015**, *66*, 297–304. [CrossRef]

45. Perera, S.; Patel, K.V.; Rosano, C.; Rubin, S.M.; Satterfield, S.; Harris, T.; Ensrud, K.; Orwoll, E.; Lee, C.G.; Chandler, J.M.; et al. Gait Speed Predicts Incident Disability: A Pooled Analysis. *J. Gerontol. Ser. A Biol. Sci. Med. Sci.* **2015**, *71*, 63–71. [CrossRef]

46. Studenski, S.; Perera, S.; Patel, K.; Rosano, C.; Faulkner, K.; Inzitari, M.; Brach, J.; Chandler, J.; Cawthon, P.; Connor, E.B.; et al. Gait speed and survival in older adults. *JAMA J. Am. Med. Assoc.* **2011**, *305*, 50–58. [CrossRef] [PubMed]

47. Drew, C.D.; Hughes, D.T. Characteristics of the Vitalograph spirometer. *Thorax* **1969**, *24*, 703–706. [CrossRef] [PubMed]

48. Wang, Y.C.; Bohannon, R.W.; Kapellusch, J.; Garg, A.; Gershon, R.C. Dexterity as measured with the 9-Hole Peg Test (9-HPT) across the age span. *J. Hand Ther.* **2015**, *28*, 53–60. [CrossRef] [PubMed]

49. Hoeymans, N.; Van Lindert, H.; Westert, G.P. The health status of the Dutch population as assessed by the EQ-6D. *Qual. Life Res.* **2005**, *14*, 655–663. [CrossRef] [PubMed]

50. Larson, J.S. The MOS 36-Item Short form Health Survey. *Eval. Health Prof.* **1997**, *20*, 14–27. [CrossRef]

51. Beurskens, A.J.H.M.; Bültmann, U.; Kant, I.; Vercoulen, J.H.M.M.; Bleijenberg, G.; Swaen, G.M.H. Fatigue among working people: Validity of a questionnaire measure. *Occup. Environ. Med.* **2000**, *57*, 353–357. [CrossRef]

52. Meulendijks, F.G.; Hamaker, M.E.; Boereboom, F.T.J.; Kalf, A.; Vögtlander, N.P.J.; van Munster, B.C. Groningen frailty indicator in older patients with end-stage renal disease. *Ren. Fail.* **2015**, *37*, 1419–1424. [CrossRef] [PubMed]

53. Gillanders, D.T.; Bolderston, H.; Bond, F.W.; Dempster, M.; Flaxman, P.E.; Campbell, L.; Kerr, S.; Tansey, L.; Noel, P.; Ferenbach, C.; et al. The Development and Initial Validation of the Cognitive Fusion Questionnaire. *Behav. Ther.* **2014**, *45*, 83–101. [CrossRef] [PubMed]

54. Kroenke, K.; Spitzer, R.L.; Williams, J.B.W. The PHQ-9: Validity of a brief depression severity measure. *J. Gen. Intern. Med.* **2001**, *16*, 606–613. [CrossRef]

55. van Vliet, I.M.Y.; Gomes-Neto, A.W.; de Jong, M.F.C.; Bakker, S.J.L.; Jager-Wittenaar, H.; Navis, G.J. Malnutrition screening on hospital admission: Impact of overweight and obesity on comparative performance of MUST and PG-SGA SF. *Eur. J. Clin. Nutr.* **2021**, 1–9. [CrossRef]

56. Desbrow, B.; Bauer, J.; Blum, C.; Kandasamy, A.; McDonald, A.; Montgomery, K. Assessment of nutritional status in hemodialysis patients using patient-generated subjective global assessment. *J. Ren. Nutr.* **2005**, *15*, 211–216. [CrossRef]

57. Jager-Wittenaar, H.; Ottery, F.D. Assessing nutritional status in cancer. *Curr. Opin. Clin. Nutr. Metab. Care* **2017**, *20*, 322–329. [CrossRef] [PubMed]

58. Post, A.; Ozyilmaz, A.; Westerhuis, R.; Ipema, K.J.R.; Bakker, S.J.L.; Franssen, C.F.M. Complementary biomarker assessment of components absorbed from diet and creatinine excretion rate reflecting muscle mass in dialysis patients. *Nutrients* **2018**, *10*, 1827. [CrossRef]

59. Barr, D.J.; Levy, R.; Scheepers, C.; Tily, H.J. Random effects structure for confirmatory hypothesis testing: Keep it maximal. *J. Mem. Lang.* **2013**, *68*, 255–278. [CrossRef] [PubMed]

60. Devries, M.C.; Phillips, S.M. Creatine supplementation during resistance training in older adults-a meta-analysis. *Med. Sci. Sports Exerc.* **2014**, *46*, 1194–1203. [CrossRef]

61. Chilibeck, P.D.; Kaviani, M.; Candow, D.G.; Zello, G.A. Effect of creatine supplementation during resistance training on lean tissue mass and muscular strength in older adults: A meta-analysis. *Open Access J. Sports Med.* **2017**, *8*, 213. [CrossRef]

62. Kreider, R.B.; Kalman, D.S.; Antonio, J.; Ziegenfuss, T.N.; Wildman, R.; Collins, R.; Candow, D.G.; Kleiner, S.M.; Almada, A.L.; Lopez, H.L. International Society of Sports Nutrition position stand: Safety and efficacy of creatine supplementation in exercise, sport, and medicine. *J. Int. Soc. Sports Nutr.* **2017**, *14*, 18. [CrossRef] [PubMed]

63. Sabatino, A.; Regolisti, G.; Karupaiah, T.; Sahathevan, S.; Sadu Singh, B.K.; Khor, B.H.; Salhab, N.; Karavetian, M.; Cupisti, A.; Fiaccadori, E. Protein-energy wasting and nutritional supplementation in patients with end-stage renal disease on hemodialysis. *Clin. Nutr.* **2017**, *36*, 663–671. [CrossRef]

64. Jadeja, Y.P.; Kher, V. Protein energy wasting in chronic kidney disease: An update with focus on nutritional interventions to improve outcomes. *Indian J. Endocrinol. Metab.* **2012**, *16*, 246. [CrossRef] [PubMed]

65. Farshidfar, F.; Pinder, M.A.; Myrie, S.B. Creatine Supplementation and Skeletal Muscle Metabolism for Building Muscle Mass-Review of the Potential Mechanisms of Action. *Curr. Protein Pept. Sci.* **2017**, *18*, 1273–1287. [CrossRef] [PubMed]

66. Kreider, R.B.; Stout, J.R. Creatine in health and disease. *Nutrients* **2021**, *13*, 447. [CrossRef]

67. Tamopolsky, M.; Zimmer, A.; Paikin, J.; Safdar, A.; Aboud, A.; Pearce, E.; Roy, B.; Doherty, T. Creatine monohydrate and conjugated linoleic acid improve strength and body composition following resistance exercise in older adults. *PLoS ONE* **2007**, *2*, e991. [CrossRef]

68. Dolan, E.; Artioli, G.G.; Pereira, R.M.R.; Gualano, B. Muscular Atrophy and Sarcopenia in the Elderly: Is There a Role for Creatine Supplementation? *Biomolecules* **2019**, *9*, 642. [CrossRef]

69. Zyga, S.; Alikari, V.; Sachlas, A.; Fradelos, E.C.; Stathoulis, J.; Panoutsopoulos, G.; Georgopoulou, M.; Theophilou, P.; Lavdaniti, M. Assessment of Fatigue in End Stage Renal Disease Patients Undergoing Hemodialysis: Prevalence and Associated Factors. *Med. Arch.* **2015**, *69*, 376–380. [CrossRef]

70. Ramkumar, N.; Beddhu, S.; Eggers, P.; Pappas, L.M.; Cheung, A.K. Patient preferences for in-center intense hemodialysis. *Hemodial. Int.* **2005**, *9*, 281–295. [CrossRef]

71. Balestrino, M.; Adriano, E. Creatine as a Candidate to Prevent Statin Myopathy. *Biomolecules* **2019**, *9*, 496. [CrossRef]

72. Angermann, S.; Schier, J.; Baumann, M.; Steubl, D.; Hauser, C.; Lorenz, G.; Günthner, R.; Braunisch, M.C.; Kemmner, S.; Satanovskij, R.; et al. Cognitive Impairment is Associated with Mortality in Hemodialysis Patients. *J. Alzheimer's Dis.* **2018**, *66*, 1529–1537. [CrossRef] [PubMed]

73. Sehgal, A.R.; Grey, S.F.; DeOreo, P.B.; Whitehouse, P.J. Prevalence, recognition, and implications of mental impairment among hemodialysis patients. *Am. J. Kidney Dis.* **1997**, *30*, 41–49. [CrossRef]

74. Kious, B.M.; Kondo, D.G.; Renshaw, P.F. Creatine for the treatment of depression. *Biomolecules* **2019**, *9*, 406. [CrossRef] [PubMed]

75. Bakian, A.V.; Huber, R.S.; Scholl, L.; Renshaw, P.F.; Kondo, D. Dietary creatine intake and depression risk among U.S. adults. *Transl. Psychiatry* **2020**, *10*, 1–11. [CrossRef]

76. Seliger, S.L.; Weiner, D.E. Cognitive Impairment in Dialysis Patients: Focus on the Blood Vessels? *Am. J. Kidney Dis.* **2013**, *61*, 187. [CrossRef]

77. Sestili, P.; Martinelli, C.; Colombo, E.; Barbieri, E.; Potenza, L.; Sartini, S.; Fimognari, C. Creatine as an antioxidant. *Amino Acids* **2011**, *40*, 1385–1396. [CrossRef] [PubMed]

78. Clarke, H.; Kim, D. H.; Meza, C.A.; Ormsbee, M.J.; Hickner, R.C. The Evolving Applications of Creatine Supplementation: Could Creatine Improve Vascular Health? *Nutrients* **2020**, *12*, 2834. [CrossRef]

79. Lipovac, V.; Gavella, M.; Vučić, M.V.; Mrzljak, V.; Ročić, B.R. Effect of creatine on erythrocyte rheology in vitro. *Clin. Hemorheol. Microcirc.* **2000**, *22*, 45–52.

80. Tokarska-Schlattner, M.; Epand, R.F.; Meiler, F.; Zandomeneghi, G.; Neumann, D.; Widmer, H.R.; Meier, B.H.; Epand, R.M.; Saks, V.; Wallimann, T.; et al. Phosphocreatine Interacts with Phospholipids, Affects Membrane Properties and Exerts Membrane-Protective Effects. *PLoS ONE* **2012**, *7*, e43178. [CrossRef]
81. Sestili, P.; Ambrogini, P.; Barbieri, E.; Sartini, S.; Fimognari, C.; Calcabrini, C.; Diaz, A.R.; Guescini, M.; Polidori, E.; Luchetti, F.; et al. New insights into the trophic and cytoprotective effects of creatine in in vitro and in vivo models of cell maturation. *Amino Acids* **2016**, *48*, 1897–1911. [CrossRef] [PubMed]
82. Zhou, Y.; Xu, H.; Ding, Y.; Lu, Q.; Zou, M.H.; Song, P. AMPKα1 deletion in fibroblasts promotes tumorigenesis in athymic nude mice by p52-mediated elevation of erythropoietin and CDK2. *Oncotarget* **2016**, *7*, 53654–53667. [CrossRef] [PubMed]
83. Pradeep, S.; Huang, J.; Mora, E.M.; Nick, A.M.; Cho, M.S.; Wu, S.Y.; Noh, K.; Pecot, C.V.; Rupaimoole, R.; Stein, M.A.; et al. Erythropoietin Stimulates Tumor Growth via EphB4. *Cancer Cell* **2015**, *28*, 610–622. [CrossRef] [PubMed]

Review

Diagnostic and Pharmacological Potency of Creatine in Post-Viral Fatigue Syndrome

Sergej M. Ostojic [1,2]

1   FSPE Applied Bioenergetics Lab, University of Novi Sad, 21000 Novi Sad, Serbia; sergej.ostojic@chess.edu.rs
2   Faculty of Health Sciences, University of Pecs, H-7621 Pecs, Hungary

Abstract: Post-viral fatigue syndrome (PVFS) is a widespread chronic neurological disease with no definite etiological factor(s), no actual diagnostic test, and no approved pharmacological treatment, therapy, or cure. Among other features, PVFS could be accompanied by various irregularities in creatine metabolism, perturbing either tissue levels of creatine in the brain, the rates of phospho-creatine resynthesis in the skeletal muscle, or the concentrations of the enzyme creatine kinase in the blood. Furthermore, supplemental creatine and related guanidino compounds appear to impact both patient- and clinician-reported outcomes in syndromes and maladies with chronic fatigue. This paper critically overviews the most common disturbances in creatine metabolism in various PVFS populations, summarizes human trials on dietary creatine and creatine analogs in the syndrome, and discusses new frontiers and open questions for using creatine in a post-COVID-19 world.

Keywords: post-viral fatigue syndrome; chronic fatigue syndrome; creatine; GAA; creatine kinase

Citation: Ostojic, S.M. Diagnostic and Pharmacological Potency of Creatine in Post-Viral Fatigue Syndrome. *Nutrients* **2021**, *13*, 503. https://doi.org/10.3390/nu13020503

Academic Editors: Richard B. Kreider and Jeffrey R. Stout
Received: 15 December 2020
Accepted: 28 January 2021
Published: 4 February 2021

Publisher's Note: MDPI stays neutral with regard to jurisdictional claims in published maps and institutional affiliations.

1. Introduction

Post-viral fatigue syndrome (PVFS) is a medical condition that is categorized among other disorders of the nervous system (Code: 8E49) in the eleventh revision of the International Classification of Diseases [1]. According to the current codification system published by the World Health Organization in 2019, PVFS covers chronic fatigue syndrome (CFS) and benign myalgic encephalomyelitis (ME), puzzling conditions previously designated as individual entities, and now systematized under the PVFS umbrella. Although some CFS/ME cases are not preceded by a viral infection [2], all conditions share clinical features that allow for a mutual medical/scientific exploration. Mainly characterized by a prolonged severe post-exercise malaise, an impairment in various cognitive functions, non-inflammatory myalgia and joint pain, and unrefreshing sleep, PVFS has an unknown cause, no pathognomonic diagnostic criteria, and no approved medical treatment (for a detailed review see [3–5]). Various viral infections have often been reported before the first appearance of PVFS, including Epstein–Barr virus, cytomegalovirus, coxsackieviruses, and coronaviruses, and the onset might be sudden or gradual [6]. Besides viruses, many physiological and psychological factors appear to work together to predispose an individual to PVFS and to precipitate and perpetuate the illness [7], making this ailment even more baffling and hard to tackle. Beyond other risk factors, creatine shortfall may be one of the hallmarks of PVFS pathology, with compensating for the lack of creatine perhaps seen as an adjunct management strategy in this mysterious disease [8]. This review paper outlines the irregularities of creatine metabolism in PVFS, summarizes studies on creatine supplementation in PVFS and similar syndromes, and discusses new frontiers of using creatine by emphasizing COVID-19 pandemics and post-COVID-19 convalescence and nutritional care.

2. Biomarkers of Creatine Metabolism in Post-Viral Fatigue Syndrome

A pioneering biochemical and muscle study from the late 1980s and early 1990s on patients with PVFS revealed minimal changes in surrogate markers of creatine turnover/muscle physiology. A mildly elevated creatine kinase (CK) and indistinguishable muscle biopsies were found in 96 patients who had suffered from the PVFS, although enterovirus RNA was present in the skeletal muscle of some patients up to 20 years after the onset of disease [9]. Preedy and co-workers [10] reported only minor abnormalities or expected outcomes after a quantitative morphometric analysis of skeletal muscle fibers in 22 patients with PVFS. Mean muscle RNA composition (mg RNA/mg DNA) was reduced by 15% in acute onset PVFS, implying lower muscle protein synthetic potential (but not muscle bulk), while plasma carnosinase and CK levels were within normal ranges. Another study demonstrated mostly regular electromyographic and muscle histopathology studies, and normal plasma CK levels in 35 patients with chronic fatigue [11], yet abnormal fiber density was found in several patients who did not have acute-onset PVFS. Wassif and colleagues [12] confirmed that the conventional biochemical markers (e.g., albumin, liver enzymes, CK, and carnosinase) are insensitive to discriminate patients with CFS and other myopathies, while histological examination revealed relatively mild abnormalities in 3 out of 10 patients with PVFS (e.g., macrophage infiltration and muscle fiber atrophy). An exciting trial found that stress-induced neutrophil mobilization might be disrupted in CFS, with healthy women demonstrating a strong correlation between exercise-induced neutrophilia and plasma CK while this link was not observed in the CFS patients [13].

Studies following those seminal trials brought a somewhat better understanding of tissue metabolism of creatine in PVFS by using magnetic resonance spectroscopy (MRS). Wong and co-workers [14] were arguably the first who evaluated skeletal muscle metabolism in CFS during rest and exercise using ^{31}P MRS to reflect minute-to-minute intracellular high-energy phosphate metabolism. The authors found that CFS patients and normal controls have similar skeletal muscle metabolic patterns during and after exercise. However, CFS patients reach exhaustion much more rapidly than normal subjects, at which point they also have relatively reduced intracellular concentrations of ATP (adenosine triphosphate), while no intergroup differences were found for muscle phosphocreatine levels. A follow-up study confirmed that no significant metabolic abnormalities are associated with fatigue in CFS patients, although abnormalities may be present in a minority of patients [15]. However, other ^{31}P MRS trials reported significantly reduced the maximal rate of post-exercise phosphocreatine resynthesis in CFS patients compared to sedentary controls [16], or decreased resting values of phosphocreatine-to-phosphocreatine plus phosphate and increased pH levels during exercise in the CFS population [17]. This implies CFS-driven perturbation in energy metabolism, although not all studies reported an impaired rate of phosphocreatine resynthesis after exercise [18,19]. Finally, Brooks and colleagues [20] demonstrated a trend of reduced levels of hippocampal creatine in CFS patients as compared with controls (8.6 mM vs. 10.9 mM), suggesting an impaired creatine metabolism in the brain as well.

Chronic fatigue syndrome in childhood also appears to be characterized by vascular and metabolic alterations in the brain [21], with lower blood flow in the temporal and occipital lobes and markedly higher blood flow in the basal ganglia and thalamus in patients with CFS as compared to healthy children. This was accompanied by a notable elevation of the choline-to-creatine ratio in children with CFS, which was arguably the first time that a possible creatine alteration in CFS in the young brain has been described, although no individual levels for brain metabolites were reported. A choline-to-creatine ratio is a well-known surrogate ^{1}H MRS biomarker of altered brain metabolism, with elevated levels perhaps demonstrating a reduction in brain creatine (or an elevation in brain choline). Soon afterward, an elevation in the choline-to-creatine ratio in the basal ganglia and white matter was found in patients with histologically mild hepatitis C suffering from CFS [22]. Altered cerebral metabolism was also found in patients with CFS who demonstrated higher *N*-acetyl aspartate-to-choline and choline-to-creatine ratios in the occipital cortex, as compared to healthy controls [23,24]. Various cardiac bioenergetic abnormalities were found in 12 CFS patients, with the mean phosphocreatine-to-ATP ratio in the CFS group tending to be lower than that seen in the control group, with values consistent with significant cardiac impairment [25]. In addition, the half-time for phosphocreatine recovery from end-exercise to baseline levels was prolonged in CFS patients. Van der Schaaf and co-workers [26] used functional brain imaging in 89 women with CFS, evaluating the possible link between the brain metabolism and clinical features of CFS. They found that more pain in CFS was associated with reduced gray matter volume and decreased *N*-acetyl aspartate-to-creatine ratio in the dorsolateral prefrontal cortex. Nevertheless, most of the studies provided no absolute levels of creatine in relevant tissues, making the case for creatine alterations in CFS incomplete.

Widespread metabolic abnormalities in CFS were corroborated in a recent whole-brain magnetic resonance spectroscopy trial [27]. A notably elevated choline-to-creatine ratio was found in the left anterior cingulate, with metabolite ratios correlated with fatigue in seven brain regions. Specifically, creatine levels in the parietal cortex were lower in CFS patients than in the control group (6.4 mM vs. 7.3 mM, $p = 0.03$), while in the putamen, creatine was higher in patients than in controls (6.3 mM vs. 5.7 mM; $p = 0.01$), suggesting location-specific variation in brain creatine in CFS. Patients with CFS also showed higher brain temperatures than healthy controls in several brain regions, suggesting that neuroinflammation, mitochondrial dysfunction, and aberrant neuronal communication may contribute to metabolic perturbations in the CFS brain. An elevated creatine excretion via urine has been identified as a metabolomic signature of fibromyalgia syndrome, a chronic condition similar to PVFS, with creatine urinary loss correlating well with fatigue and pain severity [28]. The increased utilization of creatine to form ATP in CFS has been suggested in a metabolomics trial [29], as illustrated by elevated urinary creatinine (an end product of creatine metabolism) and decreased serum glycine (a precursor of creatine) in blood and urine samples from 34 women with ME/CFS.

In the search for valid diagnostic/prognostic biomarkers of CFS, Nacul and co-workers [30] recently reviewed lab tests from 272 people with CFS and 136 healthy controls participating in the UK ME/CFS Biobank. The authors reported that patients with severe CFS actually have lower CK levels compared to healthy controls and non-severe CFS patients, with differences persisting after adjusting for sex, age, body mass index, muscle mass, disease duration, and activity levels. This interesting discovery was corroborated in a consecutive trial where among the 30 clinical parameters evaluated at the UK ME/CFS Biobank, only blood CK levels showed statistically significant differences between groups, with levels lower in CFS patients than in healthy controls (59.93 U/L vs. 88.67 U/L; $p = 0.006$) [31]. This was accompanied by a CFS-driven dysregulation of microRNA profiles, which represent genes interconnected with neuronal and endocrine-metabolic system pathways, including an upregulation of *NTRK1*, which is essential for the development and survival of neurons, and downregulation of *MECP2* and *AGO2* genes, which provide instructions for modifying chromatin and RNA-mediated silencing, respectively. Low CK activity in CFS is perhaps another indicator of an inadequate turnover of a key enzyme involved in creatine utilization and may be a symptom of the low availability of cellular energy that might involve both mitochondrial and cytosolic pathways [32,33].

3. Dietary Creatine and Alternatives in Syndromes with Prolonged Fatigue

Keeping in mind the fact that supplemental creatine has been investigated in a plethora of neurological, neuromuscular, and immune disorders characterized by creatine deficit or perturbation, it is odd that PVFS mainly remained outside of the scope of the creatine research community. A single clinical trial on creatine supplementation in CFS was registered at ClinicalTrials.gov in 2015 (NCT02374112), and the study is still on going, with no results published so far. Besides, only a handful of trials evaluated the effects of supplemental creatine and/or other guanidino compounds in the PVFS population or similar disorders with prolonged fatigue of unknown source.

Almost 20 years ago, Brouwers and co-workers [34] assessed the effect of a polynutrient supplement on the fatigue severity, clinical symptoms, and physical activity of patients with CFS. The authors conducted a prospective randomized placebo-controlled, double-blind trial in 53 CFS patients who received a multi-component product specifically developed to have a high antioxidative capacity for ten weeks. The product contained protein, carbohydrates, fat, trace elements, minerals, vitamins, and other components, including creatine (1200 mg per 100 mL). The authors found no significant differences between the placebo and the treated group on any of the outcome measures. Besides other methodological limitations of this study (e.g., no assessment of the nutritional status of CFS patients prior to the treatment, some components of the mixture having potential contrasting effects), the creatine dosage used here appears to be insufficient as compared to traditional supplementation regimens.

A research group from the University of Sao Paolo evaluated the effects of creatine supplementation in fibromyalgia, a condition similar (if not equivalent) to CFS and characterized by widespread musculoskeletal pain accompanied by fatigue, sleep, memory, and mood issues [35]. The authors supplemented creatine (20 g of creatine monohydrate for five days followed by 5 g per day throughout the trial) to 43 fibromyalgia patients in a randomized, double-blind, placebo-controlled, parallel-group design, and monitored muscle performance, cognitive function, sleep, and tissue metabolism at baseline and 16-week follow up. Creatine intervention provoked higher muscle phosphoryl creatine levels when compared with the placebo group, accompanied by greater dynamic and isometric muscular strength. The mental health domain from the 36-item Short-Form Health Survey was also improved following creatine supplementation, along with incidental memory from the Delay Recall Test, while the other markers of cognition, quality of life, or sleep remained unchanged. Another randomized controlled crossover trial evaluated the effects of supplemental guanidinoacetic acid (GAA), a natural precursor of creatine, in 21 mid-age women with CFS [36]. Three months of oral GAA (2.4 g/day) induced a significant elevation of total muscle creatine levels compared with the placebo group (36.3% vs. 2.4%; $p < 0.01$), complemented by a superior rise in quadriceps isometric strength and maximal oxygen uptake. GAA also attenuated several aspects of fatigue, such as activity, motivation, and mental fatigue, and improved both physical and mental domains of health-related quality of life assessed through the 36-item Short-Form Health Survey. However, GAA demonstrated no effects on the main clinical outcomes, such as general fatigue and musculoskeletal soreness at rest and during activity.

An interesting cross-sectional study assessed the dietary habits and food avoidance-behaviors in women with CFS [37]. Although no homogeneous pattern of food habits was established in this trial, CFS patients appear to often avoid many foods rich in creatine (e.g., meat, milk, and dairy products). A connection between dietary intake of creatine and clinical features of CFS has not been established so far; however, low creatine consumption from food sources may play an essential role in the creatine metabolism irregularities associated with CFS, and perhaps calls for creatine compensation through the prescription of creatine-rich foods and/or creatine supplementation. Jenkins and Raymen also reported intakes below the reference values for animal-based nutrients (e.g., vitamin D, vitamin A, calcium, zinc, and iron) in CFS patients [38]. To understand this under-investigated disorder will require careful nutritional approaches.

4. Alternative Mechanisms of Creatine Action

Both animal studies and human trials indicate the beneficial effects of supplemental creatine in domains beyond cellular bioenergetics, including neuroprotection, immunomodulation, and antioxidant activity (for a detailed review see [39]), domains often compromised in syndromes with chronic fatigue. That being said, creatine may help individuals cope with PVFS through several auxiliary means (Figure 1).

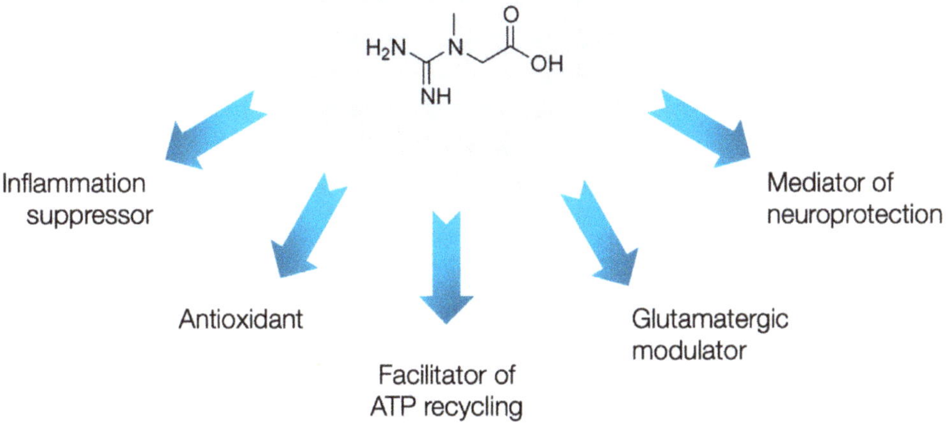

Figure 1. Possible mechanisms of creatine action in post-viral fatigue syndrome.

For instance, creatine can act as a mediator of neuroprotection in a variety of neurological conditions, including traumatic brain injury [40], neurodegenerative diseases [41], and cerebral ischemia [42]. This perhaps happens due to a creatine-mediated inhibition of the mitochondrial permeability transition pore (MPTP), an inner membrane protein linked to neuronal cell death, by stabilizing mitochondrial CK and stimulating the production of phosphocreatine, which stabilizes ATP levels in the neuron [43]. Since neuronal mitochondrial dysfunction and the opening of the MPTP play a vital role in PVFS development [44], creatine may renew mitochondrial viability and perhaps alleviate the cognitive dysfunction seen in the syndrome. In addition, creatine can act as a suppressor of acute and chronic inflammation by downregulating membrane proteins (such as toll-like receptors) that play a role in innate immunity [45]. Toll-like receptor upregulation seems to trigger an inflammatory signaling cascade leading to neuroinflammation and neurodegeneration in CFS/ME [46], and supplemental creatine may limit this hyperimmune response. Furthermore, the antioxidant activity of creatine emerges as an additional mechanism that is likely to play a supportive role in the creatine-cytoprotection paradigm [47]. For instance, creatine possesses direct antioxidant activity and can protect mitochondrial DNA from ROS-mediated damage [48]. Owing to the fact that oxidative stress is a major contributor to debilitating chronic fatigue [49], using supplemental creatine to alleviate oxidative damage might be a promising practice in PVFS. Finally, creatine acts as a partial agonist of central $GABA_A$ receptors [50], and a modulator of the NMDA receptor [51], both employed in regulating glutamatergic function. An impaired glutamatergic regulation and transport, accompanied by endotoxin intolerance, have been suggested in fatigue syndromes [52], and creatine may play a putative role as a fine-tuning agent of glutamate conveyance in the PVFS brain.

5. Open Questions and Creatine in the Post-COVID-19 World

Before considering of whether dietary creatine (and creatine analogs) has any therapeutic value in PVFS, a number of issues have to be addressed in more detail. Firstly, a possible creatine deficiency in the syndrome should be described in terms of demographics that include age, gender, ethnicity, co-morbidities, disease severity and duration, and tissue specificity, among others. Secondly, the intervention protocols should be scrutinized for both prospective and retrospective approaches, with creatine administered being throughout the acute phase of viral infection to induce PVFS, or during an episode of persistent

PVFS. We also have to investigate the effects of dietary creatine in tissues besides the skeletal muscle, evaluating brain creatine in PVFS in a region-specific manner, and alternative mechanisms of creatine supplementation via both patient- and clinician-reported outcomes and biomarkers. Creatine intervention specifics in PVFS should also be probed for the chemical formulation that is most effective for sufficient brain uptake, the acceptable dosages in terms of clinical efficacy and pharmacovigilance, and the optimal time span of creatine intervention due to the rather prolonged character of the disease. Specifically, recent data suggest that creatine can cross the blood–brain barrier but only with poor efficiency [53]. The creatine transporter (CT1 or SLC6A8) is not present in the astrocytic feet, which covers ~ 98% of the blood–brain barrier [54], implying somewhat limited uptake of creatine by the brain from the circulation. This problem might be overcome by modifying the creatine molecule to allow it to cross biological membranes by designing lipophilic creatine derivatives that could more adequately enrich the brain creatine pool [55]. Robust research designs that incorporate well-sampled long-term randomized controlled multicenter trials are highly warranted in various PVFS cohorts, including survivors of the coronavirus disease 2019 (COVID-19).

The COVID-19 pandemic caused by the severe acute respiratory syndrome coronavirus 2 (SARS-CoV-2) may bring PVFS into focus since coronaviruses are already recognized for their role in PVFS etiopathogenesis [56]. The medical community has started to see a large number of patients experiencing post-viral persistent fatigue following SARS-CoV-2 infection [57], calling for an effective and affordable treatment for COVID-19 convalescents. Dietary creatine has been recently suggested as a possible adjuvant therapeutic agent for use in COVID-19 recovery [58], due to the beneficial effects demonstrated during rehabilitation in various pulmonary conditions. Being a safe and inexpensive dietary compound, creatine should be investigated post-haste as a possible component of nutritional care for post-COVID-19 fatigue syndrome, along with other nutraceuticals, to reduce post-viral fatigue, promote a swift recovery, and fortify future resistance in often poorly nourished patients [59]. However, like other promising therapeutics for post-COVID-19 subjects, creatine requires accelerated yet attentive research and approval pathways, with sufficient efficacy and safety guarantees. In particular, the effects of creatine on renal function in elderly COVID-19 convalescents should be carefully monitored due to age- and disease-specific kidney impairments that might be exacerbated by creatine intervention.

6. Conclusions

Currently, there is not enough evidence to unequivocally endorse supplemental creatine for PVFS. However, the findings from initial trials on the metabolic substrate of PVFS, along with promising results from interventional studies, emphasize the need to explore creatine and similar compounds in this ever-prevalent yet baffling disorder. The need for an effective, low-risk, and affordable dietary intervention to tackle post-COVID-19 fatigue, which is going to remain an issue for years to come, perhaps provides a unique research opportunity to explore creatine in PVFS using expedited yet diligent approaches.

Funding: This research received no external funding.

Institutional Review Board Statement: Not applicable.

Informed Consent Statement: Not applicable.

Data Availability Statement: Not applicable.

Conflicts of Interest: S.M.O. serves as a member of the Scientific Advisory Board on creatine in health and medicine (AlzChem LLC). S.M.O. owns patent "Sports Supplements Based on Liquid Creatine" at European Patent Office (WO2019150323 A1), and active patent application "Synergistic Creatine" at UK Intellectual Property Office (GB2012773.4). S.M.O. has served as a speaker at Abbott Nutrition, a consultant of Allied Beverages Adriatic and IMLEK, and an advisory board member for the University of Novi Sad School of Medicine, and has received research funding related to creatine from the Serbian Ministry of Education, Science, and Technological Development, Provincial

Secretariat for Higher Education and Scientific Research, AlzChem GmbH, KW Pfannenschmidt GmbH, and ThermoLife International LLC. SMO is an employee of the University of Novi Sad. SMO does not own stocks and shares in any organization. The funders had no role in the design of the study; in the collection, analyses, or interpretation of data; in the writing of the manuscript.

References

1. World Health Organization. International Classification of Diseases for Mortality and Morbidity Statistics (11th Revision). 2019. Available online: https://icd.who.int/browse11/l-m/en (accessed on 15 December 2020).
2. Unger, E.R.; Lin, J.S.; Brimmer, D.J.; Lapp, C.W.; Komaroff, A.L.; Nath, A.; Laird, S.; Iskander, J. CDC grand rounds: Chronic fatigue syndrome-advancing research and clinical education. *MMWR Morb. Mortal. Wkly. Rep.* **2016**, *65*, 1434–1438. [CrossRef]
3. Lim, E.J.; Son, C.G. Review of case definitions for myalgic encephalomyelitis/chronic fatigue syndrome (ME/CFS). *J. Transl. Med.* **2020**, *18*, 289. [CrossRef]
4. Sandler, C.X.; Lloyd, A.R. Chronic fatigue syndrome: Progress and possibilities. *Med. J. Aust.* **2020**, *212*, 428–433. [CrossRef]
5. Kim, D.Y.; Lee, J.S.; Park, S.Y.; Kim, S.J.; Son, C.G. Systematic review of randomized controlled trials for chronic fatigue syndrome/myalgic encephalomyelitis (CFS/ME). *J. Transl. Med.* **2020**, *18*, 7. [CrossRef]
6. Morris, G.; Berk, M.; Walder, K.; Maes, M. The putative role of viruses, bacteria, and chronic fungal biotoxin exposure in the genesis of intractable fatigue accompanied by cognitive and physical disability. *Mol. Neurobiol.* **2016**, *53*, 2550–2571. [CrossRef]
7. Afari, N.; Buchwald, D. Chronic fatigue syndrome: A review. *Am. J. Psychiatry* **2003**, *160*, 221–236. [CrossRef] [PubMed]
8. Ostojic, S.M. Postviral fatigue syndrome and creatine: A piece of the puzzle? *Nutr. Neurosci* **2020**, in press. [CrossRef] [PubMed]
9. Archard, L.C.; Bowles, N.E.; Behan, P.O.; Bell, E.J.; Doyle, D. Postviral fatigue syndrome: Persistence of enterovirus RNA in muscle and elevated creatine kinase. *J. R. Soc. Med.* **1988**, *81*, 326–329. [CrossRef] [PubMed]
10. Preedy, V.R.; Smith, D.G.; Salisbury, J.R.; Peters, T.J. Biochemical and muscle studies in patients with acute onset post-viral fatigue syndrome. *J. Clin. Pathol.* **1993**, *46*, 722–726. [CrossRef] [PubMed]
11. Connolly, S.; Smith, D.G.; Doyle, D.; Fowler, C.J. Chronic fatigue: Electromyographic and neuropathological evaluation. *J. Neurol.* **1993**, *240*, 435–438. [CrossRef]
12. Wassif, W.S.; Sherman, D.; Salisbury, J.R.; Peters, T.J. Use of dynamic tests of muscle function and histomorphometry of quadriceps muscle biopsies in the investigation of patients with chronic alcohol misuse and chronic fatigue syndrome. *Ann. Clin. Biochem.* **1994**, *31*, 462–468. [CrossRef]
13. Cannon, J.G.; Angel, J.B.; Abad, L.W.; O'Grady, J.; Lundgren, N.; Fagioli, L.; Komaroff, A.L. Hormonal influences on stress-induced neutrophil mobilization in health and chronic fatigue syndrome. *J. Clin. Immunol.* **1998**, *18*, 291–298. [CrossRef] [PubMed]
14. Wong, R.; Lopaschuk, G.; Zhu, G.; Walker, D.; Catellier, D.; Burton, D.; Teo, K.; Collins Nakai, R.; Montague, T. Skeletal muscle metabolism in the chronic fatigue syndrome. In vivo assessment by 31P nuclear magnetic resonance spectroscopy. *Chest* **1992**, *102*, 1716–1722. [CrossRef] [PubMed]
15. Barnes, P.R.; Taylor, D.J.; Kemp, G.J.; Radda, G.K. Skeletal muscle bioenergetics in the chronic fatigue syndrome. *J. Neurol. Neurosurg. Psychiatry* **1993**, *56*, 679–683. [CrossRef] [PubMed]
16. McCully, K.K.; Natelson, B.H.; Iotti, S.; Sisto, S.; Leigh, J.S., Jr. Reduced oxidative muscle metabolism in chronic fatigue syndrome. *Muscle Nerve* **1996**, *19*, 621–625. [CrossRef]
17. Block, W.; Träber, F.; Kuhl, C.K.; Keller, E.; Lamerichs, R.; Karitzky, J.; Rink, H.; Schild, H.H. 31P-MR- [31P-mr spectroscopy of peripheral skeletal musculature under load: Demonstration of normal energy metabolites compared with metabolic muscle diseases]. *Rofo* **1998**, *168*, 250–257. [CrossRef]
18. McCully, K.K.; Smith, S.; Rajaei, S.; Leigh, J.S., Jr.; Natelson, B.H. Blood flow and muscle metabolism in chronic fatigue syndrome. *Clin. Sci.* **2003**, *104*, 641–647. [CrossRef]
19. McCully, K.K.; Smith, S.; Rajaei, S.; Leigh, J.S., Jr.; Natelson, B.H. Muscle metabolism with blood flow restriction in chronic fatigue syndrome. *J. Appl. Physiol.* **2004**, *96*, 871–878. [CrossRef]
20. Brooks, J.C.; Roberts, N.; Whitehouse, G.; Majeed, T. Proton magnetic resonance spectroscopy and morphometry of the hippocampus in chronic fatigue syndrome. *Br. J. Radiol.* **2000**, *73*, 1206–1208. [CrossRef]
21. Tomoda, A.; Miike, T.; Yamada, E.; Honda, H.; Moroi, T.; Ogawa, M.; Ohtani, Y.; Morishita, S. Chronic fatigue syndrome in childhood. *Brain Dev.* **2000**, *22*, 60–64. [CrossRef]
22. Forton, D.M.; Allsop, J.M.; Main, J.; Foster, G.R.; Thomas, H.C.; Taylor-Robinson, S.D. Evidence for a cerebral effect of the hepatitis C virus. *Lancet* **2001**, *358*, 38–39. [CrossRef]
23. Puri, B.K.; Counsell, S.J.; Zaman, R.; Main, J.; Collins, A.G.; Hajnal, J.V.; Davey, N.J. Relative increase in choline in the occipital cortex in chronic fatigue syndrome. *Acta Psychiatr. Scand.* **2002**, *106*, 224–226. [CrossRef] [PubMed]
24. Chaudhuri, A.; Condon, B.R.; Gow, J.W.; Brennan, D.; Hadley, D.M. Proton magnetic resonance spectroscopy of basal ganglia in chronic fatigue syndrome. *Neuroreport* **2003**, *14*, 225–228. [CrossRef]
25. Hollingsworth, K.G.; Jones, D.E.; Taylor, R.; Blamire, A.M.; Newton, J.L. Impaired cardiovascular response to standing in chronic fatigue syndrome. *Eur. J. Clin. Investig.* **2010**, *40*, 608–615. [CrossRef] [PubMed]
26. Van der Schaaf, M.E.; De Lange, F.P.; Schmits, I.C.; Geurts, D.E.M.; Roelofs, K.; van der Meer, J.W.M.; Toni, I.; Knoop, H. Prefrontal structure varies as a function of pain symptoms in chronic fatigue syndrome. *Biol. Psychiatry* **2017**, *81*, 358–365. [CrossRef] [PubMed]

27. Mueller, C.; Lin, J.C.; Sheriff, S.; Maudsley, A.A.; Younger, J.W. Evidence of widespread metabolite abnormalities in Myalgic encephalomyelitis/chronic fatigue syndrome: Assessment with whole-brain magnetic resonance spectroscopy. *Brain Imaging Behav.* **2020**, *14*, 562–572. [CrossRef] [PubMed]

28. Malatji, B.G.; Meyer, H.; Mason, S.; Engelke, U.F.H.; Wevers, R.A.; van Reenen, M.; Reinecke, C.J. A diagnostic biomarker profile for fibromyalgia syndrome based on an NMR metabolomics study of selected patients and controls. *BMC Neurol.* **2017**, *17*, 88. [CrossRef]

29. Armstrong, C.W.; McGregor, N.R.; Lewis, D.P.; Butt, H.L.; Gooley, P.R. Metabolic profiling reveals anomalous energy metabolism and oxidative stress pathways in chronic fatigue syndrome patients. *Metabolomics* **2015**, *11*, 1626–1639. [CrossRef]

30. Nacul, L.; de Barros, B.; Kingdon, C.C.; Cliff, J.M.; Clark, T.G.; Mudie, K.; Dockrell, H.M.; Lacerda, E.M. Evidence of clinical pathology abnormalities in people with myalgic encephalomyelitis/chronic fatigue syndrome (me/cfs) from an analytic cross-sectional study. *Diagnostics* **2019**, *9*, 41. [CrossRef]

31. Almenar-Pérez, E.; Sarría, L.; Nathanson, L.; Oltra, E. Assessing diagnostic value of microRNAs from peripheral blood mononu-clear cells and extracellular vesicles in myalgic encephalomyelitis/chronic fatigue syndrome. *Sci. Rep.* **2020**, *10*, 2064. [CrossRef]

32. Germain, A.; Ruppert, D.; Levine, S.M.; Hanson, M.R. Metabolic profiling of a myalgic encephalomyelitis/chronic fatigue syndrome discovery cohort reveals disturbances in fatty acid and lipid metabolism. *Mol. Biosyst.* **2017**, *13*, 371–379. [CrossRef] [PubMed]

33. Tomas, C.; Brown, A.; Strassheim, V.; Elson, J.L.; Newton, J.; Manning, P. Cellular bioenergetics is impaired in patients with chronic fatigue syndrome. *PLoS ONE* **2017**, *12*, e0186802. [CrossRef] [PubMed]

34. Brouwers, F.M.; Van Der Werf, S.; Bleijenberg, G.; Van Der Zee, L.; Van Der Meer, J.W. The effect of a polynutrient supplement on fatigue and physical activity of patients with chronic fatigue syndrome: A double-blind randomized controlled trial. *QJM* **2002**, *95*, 677–683. [CrossRef]

35. Alves, C.R.; Santiago, B.M.; Lima, F.R.; Otaduy, M.C.; Calich, A.L.; Tritto, A.C.; de Sá Pinto, A.L.; Roschel, H.; Leite, C.C.; Benatti, F.B.; et al. Creatine supplementation in fibromyalgia: A randomized, double-blind, placebo-controlled trial. *Arthritis Care Res.* **2013**, *65*, 1449–1459. [CrossRef] [PubMed]

36. Ostojic, S.M.; Stojanovic, M.; Drid, P.; Hoffman, J.R.; Sekulic, D.; Zenic, N. Supplementation with guanidinoacetic acid in women with chronic fatigue syndrome. *Nutrients* **2016**, *8*, 72. [CrossRef] [PubMed]

37. Trabal, J.; Leyes, P.; Fernández-Solá, J.; Forga, M.; Fernández-Huerta, J. Patterns of food avoidance in chronic fatigue syndrome: Is there a case for dietary recommendations? *Nutr. Hosp.* **2012**, *27*, 659–662. [PubMed]

38. Jenkins, M.; Rayman, M. Nutrient intake is unrelated to nutrient status in patients with chronic fatigue syndrome. *J. Nutr. Environ. Med.* **2005**, *15*, 177–189. [CrossRef]

39. Riesberg, L.A.; Weed, S.A.; McDonald, T.L.; Eckerson, J.M.; Drescher, K.M. Beyond muscles: The untapped potential of creatine. *Int. Immunopharmacol.* **2016**, *37*, 31–42. [CrossRef]

40. Dolan, E.; Gualano, B.; Rawson, E.S. Beyond muscle: The effects of creatine supplementation on brain creatine, cognitive processing, and traumatic brain injury. *Eur. J. Sport Sci.* **2019**, *19*, 1–14. [CrossRef]

41. Marques, E.P.; Wyse, A.T.S. Creatine as a neuroprotector: An actor that can play many parts. *Neurotox. Res.* **2019**, *36*, 411–423. [CrossRef]

42. Balestrino, M.; Sarocchi, M.; Adriano, E.; Spallarossa, P. Potential of creatine or phosphocreatine supplementation in cerebrovas-cular disease and in ischemic heart disease. *Amino Acids* **2016**, *48*, 1955–1967. [CrossRef] [PubMed]

43. Beal, M.F. Neuroprotective effects of creatine. *Amino Acids* **2011**, *40*, 1305–1313. [CrossRef] [PubMed]

44. Morris, G.; Maes, M. Mitochondrial dysfunctions in myalgic encephalomyelitis/chronic fatigue syndrome explained by activated immuno-inflammatory, oxidative and nitrosative stress pathways. *Metab. Brain Dis.* **2014**, *29*, 19–36. [CrossRef] [PubMed]

45. Leland, K.M.; McDonald, T.L.; Drescher, K.M. Effect of creatine, creatinine, and creatine ethyl ester on TLR expression in macrophages. *Int. Immunopharmacol.* **2011**, *11*, 1341–1347. [CrossRef] [PubMed]

46. Gambuzza, M.E.; Salmeri, F.M.; Soraci, L.; Soraci, G.; Sofo, V.; Marino, S.; Bramanti, P. The role of toll-like receptors in chronic fatigue syndrome/myalgic encephalomyelitis: A new promising therapeutic approach? *CNS Neurol. Disord. Drug Targets* **2015**, *14*, 903–914. [CrossRef] [PubMed]

47. Sestili, P.; Martinelli, C.; Colombo, E.; Barbieri, E.; Potenza, L.; Sartini, S.; Fimognari, C. Creatine as an antioxidant. *Amino Acids* **2011**, *40*, 1385–1396. [CrossRef]

48. Guidi, C.; Potenza, L.; Sestili, P.; Martinelli, C.; Guescini, M.; Stocchi, L.; Zeppa, S.; Polidori, E.; Annibalini, G.; Stocchi, V. Differential effect of creatine on oxidatively-injured mitochondrial and nuclear DNA. *Biochim. Biophys. Acta* **2008**, *1780*, 16–26. [CrossRef]

49. Lee, J.S.; Kim, H.G.; Lee, D.S.; Son, C.G. Oxidative stress is a convincing contributor to idiopathic chronic fatigue. *Sci. Rep.* **2018**, *8*, 12890. [CrossRef]

50. Koga, Y.; Takahashi, H.; Oikawa, D.; Tachibana, T.; Denbow, D.M.; Furuse, M. Brain creatine functions to attenuate acute stress responses through GABAergic system in chicks. *Neuroscience* **2005**, *132*, 65–71. [CrossRef]

51. Royes, L.F.; Fighera, M.R.; Furian, A.F.; Oliveira, M.S.; Fiorenza, N.G.; Ferreira, J.; da Silva, A.C.; Priel, M.R.; Ueda, E.S.; Calixto, J.B.; et al. Neuromodulatory effect of creatine on extracellular action potentials in rat hippocampus: Role of NMDA receptors. *Neurochem. Int.* **2008**, *53*, 33–37. [CrossRef]

52. Rönnbäck, L.; Hansson, E. On the potential role of glutamate transport in mental fatigue. *J. Neuroinflamm.* **2004**, *1*, 22. [CrossRef]

53. Béard, E.; Braissant, O. Synthesis and transport of creatine in the CNS: Importance for cerebral functions. *J. Neurochem.* **2010**, *115*, 297–313.

54. Ainsley Dean, P.J.; Arikan, G.; Opitz, B.; Sterr, A. Potential for use of creatine supplementation following mild traumatic brain injury. *Concussion* **2017**, *2*, CNC34. [CrossRef] [PubMed]
55. Adriano, E.; Gulino, M.; Arkel, M.; Salis, A.; Damonte, G.; Liessi, N.; Millo, E.; Garbati, P.; Balestrino, M. Di-acetyl creatine ethyl ester, a new creatine derivative for the possible treatment of creatine transporter deficiency. *Neurosci. Lett.* **2018**, *665*, 217–223. [CrossRef] [PubMed]
56. Lam, M.H.; Wing, Y.K.; Yu, M.W.; Leung, C.M.; Ma, R.C.; Kong, A.P.; So, W.Y.; Fong, S.Y.; Lam, S.P. Mental morbidities and chronic fatigue in severe acute respiratory syndrome survivors: Long-term follow-up. *Arch. Intern. Med.* **2009**, *169*, 2142–2147. [CrossRef] [PubMed]
57. Townsend, L.; Dyer, A.H.; Jones, K.; Dunne, J.; Mooney, A.; Gaffney, F.; O'Connor, L.; Leavy, D.; O'Brien, K.; Dowds, J.; et al. Persistent fatigue following SARS-CoV-2 infection is common and independent of severity of initial infection. *PLoS ONE* **2020**, *15*, e0240784. [CrossRef]
58. Ostojic, S.M. Can creatine help in pulmonary rehabilitation after COVID-19? *Ther. Adv. Respir. Dis.* **2020**, in press. [CrossRef] [PubMed]
59. Butters, D.; Whitehouse, M. COVID-19 and nutriceutical therapies, especially using zinc to supplement antimicrobials. *Inflammopharmacology* **2020**, in press. [CrossRef] [PubMed]

Review

Role of Creatine Supplementation in Conditions Involving Mitochondrial Dysfunction: A Narrative Review

Robert Percy Marshall [1,*], Jan-Niklas Droste [1], Jürgen Giessing [2] and Richard B. Kreider [3]

1 Medical Department, RasenBallsport Leipzig GmbH, 04177 Leipzig, Germany; jan-niklas.droste@redbulls.com
2 Faculty of Natural and Environmental Sciences, Institute of Sports Science, Universität Koblenz-Landau, 76829 Landau, Germany; giessing@uni-landau.de
3 Exercise & Sport Nutrition Lab, Human Clinical Research Facility, Department of Health & Kinesiology, Texas A&M University, College Station, TX 77843, USA; rbkreider@tamu.edu
* Correspondence: Robert.Marshall@redbulls.com

Abstract: Creatine monohydrate (CrM) is one of the most widely used nutritional supplements among active individuals and athletes to improve high-intensity exercise performance and training adaptations. However, research suggests that CrM supplementation may also serve as a therapeutic tool in the management of some chronic and traumatic diseases. Creatine supplementation has been reported to improve high-energy phosphate availability as well as have antioxidative, neuroprotective, anti-lactatic, and calcium-homoeostatic effects. These characteristics may have a direct impact on mitochondrion's survival and health particularly during stressful conditions such as ischemia and injury. This narrative review discusses current scientific evidence for use or supplemental CrM as a therapeutic agent during conditions associated with mitochondrial dysfunction. Based on this analysis, it appears that CrM supplementation may have a role in improving cellular bioenergetics in several mitochondrial dysfunction-related diseases, ischemic conditions, and injury pathology and thereby could provide therapeutic benefit in the management of these conditions. However, larger clinical trials are needed to explore these potential therapeutic applications before definitive conclusions can be drawn.

Keywords: mitochondriopathia; cardiac infarction; chronic fatigue syndrome; long COVID; ischemia; hypoxia; stroke; neurodegenerative diseases; oxidative stress; noncommunicable disease

Citation: Marshall, R.P.; Droste, J.-N.; Giessing, J.; Kreider, R.B. Role of Creatine Supplementation in Conditions Involving Mitochondrial Dysfunction: A Narrative Review. *Nutrients* **2022**, *14*, 529. https://doi.org/10.3390/nu14030529

Academic Editor: Theo Wallimann

Received: 14 December 2021
Accepted: 24 January 2022
Published: 26 January 2022

Publisher's Note: MDPI stays neutral with regard to jurisdictional claims in published maps and institutional affiliations.

1. Introduction

Creatine (*N*-aminoiminomethyl-*N*-methyl glycine) is a naturally occurring and nitrogen containing compound synthesized from the amino acids glycine, methionine that is classified within the family of guanidine phosphagens [1,2]. About one half the daily need for creatine is obtained from endogenous synthesis while the remaining is obtained from the diet, primarily red meat, fish, or dietary supplements [3,4]. Creatine is mainly stored in the muscle (95%) with the remaining found in the heart, brain, and testes [3–6], with about 2/3 in the form of PCr and the remaining as free creatine [4,5,7]. The metabolic basis of creatine in health and disease has been recently reviewed in detail by Bonilla and colleagues [1] (see Figure 1). Briefly, adenosine triphosphate (ATP) serves as the primary source of energy in most living cells. Enzymatic degradation of ATP into adenosine diphosphate (ADP) and inorganic phosphate (Pi) liberates free energy to fuel metabolic activity. However, only a small amount of ATP is stored in the cell. Energy derived from the degradation of phosphocreatine (PCr) serves to resynthesize ADP and Pi back to ATP to maintain cellular function until glycolysis in the cytosol and oxidative phosphorylation in the mitochondria can produce enough ATP to meet metabolic demands. Creatine also plays an important role in shuttling Pi from the mitochondria into the cytosol to form PCr to help maintain cellular bioenergetics (i.e., Creatine Phosphate Shuttle) [8]. In this way, PCr can donate its phosphate to ADP, thereby restoring ATP for cellular needs leaving creatine in the cytosol

343

to diffuse back into the mitochondria to shuttle the next phosphate to locations far from its production site [8]. The ATP stored in the cells is usually sufficient for energy depletion that lasts less than two seconds. However, another two to seven seconds of muscle contractions are fueled by depleting available PCr stores [9]. Together, the ATP–PCr energy system provides energy to fuel short-term explosive exercise. Increasing PCr and creatine in muscle provides an energy reserve to meet anaerobic energy needs, thereby providing a critical source of energy particularly during ischemia, injury, and/or in response to impaired mitochondrial function [8,10].

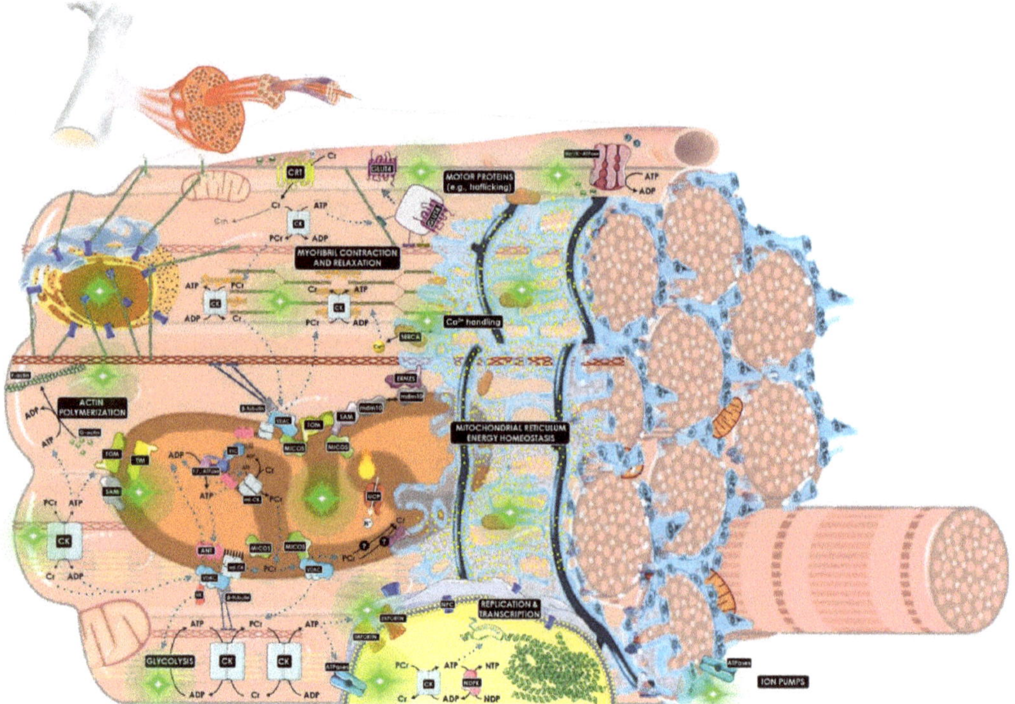

Figure 1. General overview of the metabolic role of creatine in the creatine kinase/phosphocreatine (CK/PCr) system [1]. The diagram depicts connected subcellular energy production and cellular mechanics of creatine metabolism. This chemo-mechanical energy transduction network involves structural and functional coupling of the mitochondrial reticulum (mitochondrial interactosome and oxidative metabolism), phosphagen and glycolytic system (extramitochondrial ATP production), the linker of nucleoskeleton and cytoskeleton complex (nesprins interaction with microtubules, actin polymerization, β-tubulins), motor proteins (e.g., myofibrillar ATPase machinery, vesicles transport), and ion pumps (e.g., SERCA, Na⁺/K⁺-ATPase). The cardiolipin-rich domain is represented by parallel black lines. Green sparkled circles represent the subcellular processes where the CK/PCr system is important for functionality. Several proteins of the endoplasmic reticulum–mitochondria organizing network (ERMIONE), the SERCA complex, the TIM/TOM complex, the MICOS complex, the linker of nucleoskeleton and cytoskeleton complex, and the architecture of sarcomere and cytoskeleton are not depicted for readability. ANT: adenine nucleotide translocase; CK: creatine kinase; Cr: creatine; Crn: creatinine; CRT: Na⁺/Cl⁻-dependent creatine transporter; ERMES: endoplasmic reticulum-mitochondria encounter structure; ETC: electron transport chain; GLUT-4: glucose transporter type 4;

HK: hexokinase; mdm10: mitochondrial distribution and morphology protein 10; MICOS: mitochondrial contact site and cristae organizing system; NDPK: nucleoside-diphosphate kinase; NPC: nuclear pore complex; PCr: phosphocreatine; SAM: sorting and assembly machinery; SERCA: Sarco/Endoplasmic Reticulum Ca^{2+} ATPase; TIM: translocase of the inner membrane complex; TOM: translocase of the outer membrane complex; UCP: uncoupling protein; VDAC: voltage-dependent anion channel. Reprinted with permission. See Bonilla et al. [1] for more details about the metabolic basis of creatine in energy production and disease.

Numerous studies over the last three decades have shown that creatine monohydrate (CrM) supplementation (e.g., 4 × 5 g/day for 5–7 days or 3–6 g/day for 4–12 weeks) increases muscle creatine and PCr content by 20–40% [5,11–15] and brain creatine content by 5–15% [16–21]. Creatine monohydrate supplementation has been reported to safely improve high-intensity exercise performance by 10–20% leading to greater training adaptations in adolescents [22–26], young adults [27–38], and older individuals [21,39–48]. No clinically significant side effects have been reported other than a desired weight gain [49]. Additionally, there is little to no evidence that CrM causes anecdotal reports of bloating, gastrointestinal distress, disproportionate increase in water retention, increased stress on the kidneys, increased susceptibility to injury, etc. [49,50]. In fact, studies directly assessing whether creatine causes some of those issues found no or opposite effects. As a result, there has been interest in assessing whether CrM supplementation may benefit a number of clinical populations including conditions that impair mitochondrial function [6]. The rationale is that since CrM supplementation can increase high-energy phosphate availability and also has antioxidant, neuroprotective, anti-lactic, and calcium-homoeostatic effects, increasing phosphagen availability may help improve cell survival and/or health outcomes in conditions in which mitochondrial function is compromised (e.g., ischemia, injury, and/or non-communicable chronic diseases). The purpose of this review is to examine the literature related to the role of CrM supplementation in the management of various conditions characterized by mitochondrial dysfunction and make recommendations about further work needed in this area.

2. Methods

The methodological basis of this narrative review is a selective literature search in the PubMed database, supplemented by a free Internet search (German and English). In a first explorative step, the search terms "creatine supplementation" and/or "mitochondrial dysfunction" and "creatine" and/or "mitochondrial disease" were used. After a first analysis of the searched literature identifying 68 articles, a new selective literature search was performed in the sources described above using the terms mentioned above, adding relevant cited sources and cross-references. Subsequently, titles, abstracts and finally full-text articles were examined by the scientific team with regard to the suitability of the articles in terms of content and, in a subsequent step, in terms of quality. After the qualitative criteria had been verified, the content exploration was carried out following thematic questions related to the role of creatine in context: (1) Ergogenic role in mitochondrial dysfunction; (2) Noncommunicable chronic diseases (NCD); (3) Cardiovascular disease and ischemic heart failure; (4) Traumatic and ischemic CNS injuries; (5) Neurodegenerative disorders; (6) Psychological disorders; and (7) Chronic Fatigue Syndrome, Post Viral Fatigue Syndrome and Long COVID.

3. Creatine's Ergogenic Role in Mitochondrial Dysfunction

Although there is not clear definition of mitochondrial dysfunction, it generally refers to conditions that reduce the ability of the mitochondria to contribute to production of energy in the form of ATP. However, any alteration of normal mitochondrial function could be called "mitochondrial dysfunction" as well [51]. Mitochondrial dysfunction can be of primary origin through inheriting pathological altered mitochondrial DNA (mtDNA) or acquiring secondary dysfunction through aging and exposure to mtDNA damaging processes [52,53]. This can be due to traumatic ischemic (blood deficient) or anoxic (oxygen deficient) as well as chronic conditions. Most common reasons for mitochondrial dysfunction are hypoxia, overexpression of reactive oxygen species (ROS), and an alteration of the intracellular calcium homoeostasis. Since creatine supplementation increases the availability of PCr, it may help cells withstand ischemic challenges and/or offset energy deficits associated with mitochondrial dysfunction

3.1. Acute, Traumatic Mitochondrial Dysfunction

Figure 2 shows the schematic sequence of an acute traumatic mitochondrial dysfunction with possible subsequent ischemia. The mechanical forces of injury result in an influx of calcium, potassium, and sodium. A calcium gradient is created, which reduces mitochondrial function [54,55]. In addition, an injury can lead to short-term ischemia (hypoxia) due to swelling, edema formation, development of neuroinflammation, obstruction of vessels, or hemorrhage [56]. The resulting oxygen deficiency interrupts the respiratory chain in the mitochondria. In both cases, the cell must switch to the energetic emergency plan and produce energy glycolytically, thereby increasing lactate production [57–61]. Oxygen radicals are generated, causing oxidative stress. This leads to cell damage and ultimately to cell death (apoptosis) [62–64]. If sufficient creatine phosphate reserves are present, the cell can compensate short-term energy deficits. ATP-dependent calcium transporters can counteract the calcium gradient under consumption of ATP and PCr, maintain the cell milieu, and thus normalize mitochondrial function [65,66]. Oxygen radicals can be intercepted [67]. Even transient hypoxia of a few seconds can be counteracted by the body in this way [68]. There is evidence that creatine and cyclocreatine inhibit the mitochondrial–creatine kinase–adenine nucleotide translocator (Mi-Cr-ANT) complex and the mitochondrial permeability transition that is associated with ischemic injury and apoptosis [69]. Additionally, creatine enhances the ability of Mi-CK to shuttle ADP for oxidative phosphorylation and PCr formation, thereby decreasing mitochondrial membrane and production of reactive oxygen species (ROS) [70]. Since impairment in cellular energy production and increased oxidative stress are common features in several neuromuscular degenerative diseases, creatine supplementation may provide some therapeutic benefit [69,70]. In support of this premise, Sakellaris et al. [71,72] reported that oral administered creatine can be used as an additional supplement in treatment of acute mitochondrial dysfunction after brain injury. These studies showed clear improvement in clinical outcomes of patients with additional creatine-supplementation in comparison to no creatine-intake. Table 1 shows the level of evidence in humans that creatine supplementation may have a positive effect on treatment outcomes in patients with traumatic brain injury.

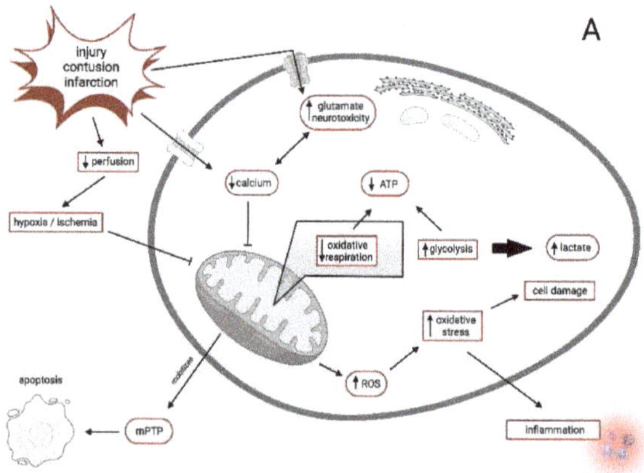

Figure 2. *Cont.*

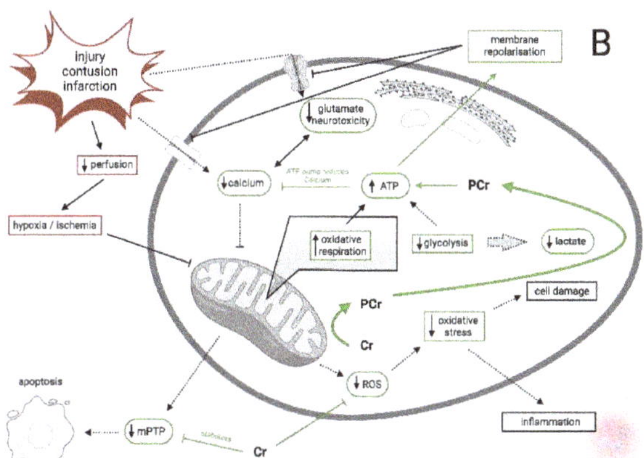

Figure 2. Panel **A**: Intracellular cascade after injury, infarction or contusion leads to mitochondrial dysfunction. Panel **B**: Impact of creatine on mitochondrial dysfunction. Green shows direct increase/stimulation of Cr/PCr, red shows direct decrease/inhibition of Cr/PCr, dotted line represents indirect impact of Cr/PCr on cellular pathways. ATP is adenosine triphosphate; Cr is creatine; PCr is phosphocreatine; ROS is reactive oxygen species; mPTP is mitochondrial permeability transition pore. Adapted from Dean et al. [55].

Table 1. Level of evidence for creatine supplementation in acute traumatic mitochondrial dysfunction.

Study	Disease	Subject	Treatment	Randomized	Subjects	Efficacy	Effect Role
Sakellaris et al [71]	Traumatic brain injury	Human	0.4 g/kg per day for 6 months	Yes	39	Improved self-care, cognition, behavior functions and communication	Direct effect on disease
Sakellaris et al. [72]	Traumatic brain injury	Human	0.4 g/kg per day for 6 months	Yes	39	Reduced fatigue, headache and dizziness	Direct effect on disease

3.2. Chronic, Atraumatic Mitochondrial Dysfunction

Many chronic diseases such as cancer and age-related pathological conditions have been related to an altered mitochondrial function [73–101]. Chronic mitochondrial dysfunction is usually caused by slow changes in mitochondrial homeostasis eventually leading to an increase in ROS/NOS, glycolysis, and hyper-acidosis. There are multiple factors that directly damage mitochondrial function (Figure 3). Hypoxia is a common factor in conditions such as solid tumor, ischemia, or inflammation that leads to a depletion of oxygen and eventually through production of ROS to an alteration of intracellular proteins, lipids and DNA [89]. On the other hand, research was able to prove that malignant cells tend to create energy under glycolytic conditions although sufficient oxygen is provided. This pathological mechanism is called "Warburg Effect" [102,103]. This leads to an increase in cell acidity and an increase in ROS with damaging of DNA. Other factors leading to chronic mitochondrial dysfunction are toxic metals or reactive nitrogen species (NOS) [104]. An increase in ingested carbohydrates bigger than the individual needs leads to hyperinsulinemia. As a chronic condition, this will lead to an increase in receptor for advanced glycation end products (RAGE). Thus, nitrosative stress increases, manipulating mitochondrial function [105–109]. Increasing stress will lead to an intracellular accumulation of ammonium [110–112], ROS [113], lactate [114], ultimately inhibiting the Krebs cycle and oxidative metabolism.

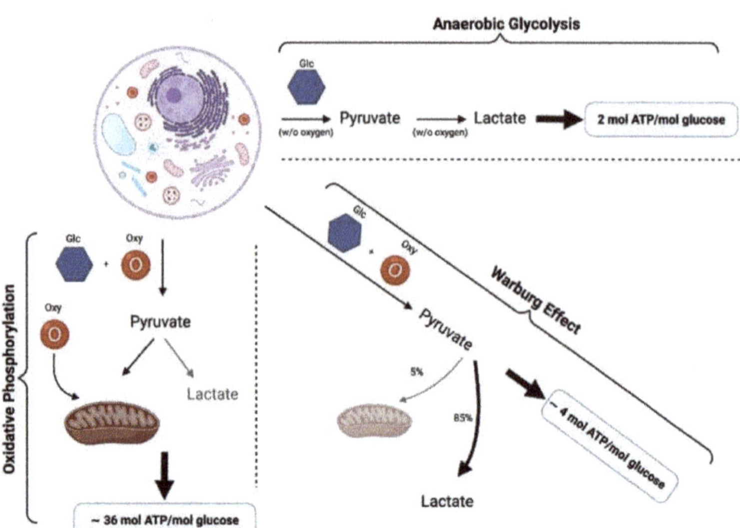

Figure 3. Warburg Effect: glycolysis produces 2 ATP instead of 36 ATP, in pathological tissues even despite aerobic conditions. Glc is glucose, Oxy is oxygen, ATP is adenosine triphosphate. Adapted from Vander Heiden et al. [91].

Typical factors that lead to a disturbance in the cellular respiration are hypoxia, inflammation, viruses, mutations, oncogenes, age, radiation, and carcinogens [115]. The ultimate, most common denominators are reactive species which damage mtDNA. As soon as cellular defense systems such as antioxidants, intracellular energetic buffer, and enzymatic reactions are worn down, chronic alteration of cellular organelles begins [116]. As mentioned above, it is hard to differentiate in chronic mitochondrial dysfunction whether pathological conditions lead to hypoxia that produces ROS/NOS which eventually harms mtDNA or whether an altered mtDNA leads to an overexpression of ROS/NOS damaging itself [117]. It is widely accepted, however, that this chronic status is a vicious circle leading to a lethal cellular condition harming the host.

Magnetic resonance spectroscopy (MRS) is an analytical tool that detects electromagnetical signals that are produced by the atomic nuclei within the molecules. Thus, it can be used to (non-invasively) measure concentrations for specific molecules in tissue. This technique has extensively been used in neurological research to identify phosphorus and proton metabolites in tissue in vivo [118–121]. Using this, research was able to prove mitochondrial dysfunction in patients with bipolar disorders. These patients also suffered from an impaired energy production [122], increased levels of lactate (hyperacidotic state) [123] and PCr concentration [114,124,125]. Therefore, it was assumed that creatine supplementation could improve clinical outcome in cases of mitochondrial dysfunction. Creatine is able to buffer lactate accumulation by reducing the need for glycolysis [126], reducing ROS [127] and restoring calcium homeostasis. Table 2 presents an overview of the level of evidence for creatine supplementation for chronic, atraumatic mitochondrial dysfunction.

Table 2. Level of evidence for creatine supplementation for chronic, atraumatic mitochondrial dysfunction.

Study	Disease	Subject	Treatment	Randomized	Subjects	Efficacy	Effect Role
Guimarães-Ferreira et al. [128]	-	Animal/vitro	5 g/kg per day for 6 days	no	39	Decrease in ROS in muscle tissue	Anima model
Kato et al. [124]	Bipolar disorder	Humans	None	No	25 (disease) vs. 21 (control)	Abnormal energy phosphate metabolism in bipolar disorder	No intervention, only descriptive, observational findings

4. Noncommunicable Chronic Diseases (NCD)

Modern ways of (unhealthy) living like over nutrition, exposure to toxic substances, and sedentarism combined with an individual's genetic background led to the development of NCD [90]. Four disease clusters are associated with NCD such as cardiovascular diseases, cancers, chronic pulmonary diseases, and diabetes mellitus [129]. NCD are associated with low-grade inflammation and an increase in oxidative stress [130]. Through the past decades, they have become the biggest health threat of modern society [131–133]. Lately, there has been a link established between NCD and mitochondrial dysfunction. Reduced oxygen consumption rates have been shown in cardiovascular diseases such as hypertension and atherosclerosis. Additionally, they suffer from calcium overload due to mitochondrial calcium mishandling and ROS overproduction [134–137]. Obesity [138–141] as well as diabetes mellitus [142–149] are associated with an increased mitochondrial fragmentation rate, impaired ATP production, as well as ROS overproduction and calcium mishandling. In regards to creatine and its connection to mitochondrial dysfunction, reduced levels were detected in human myocytes in diabetes mellitus [150], obesity [151], and hypertension [152]. Not surprisingly, NCD are the most common factors contributing to the development of an acute ischemic heart attack or acute ischemic brain disease (Figure 1).

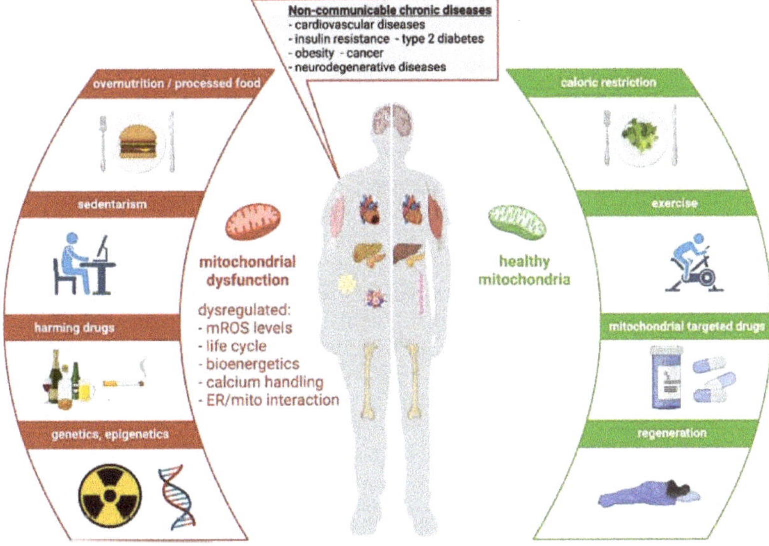

Figure 4. Mitochondrial dysfunction and non-communicable diseases. Adapted from Díaz-Vegas et al. [90].

Table 3 shows some of the studies that have been conducted on creatine supplementation in noncommunicable chronic diseases. Creatine's benefits in physical activity and thus counteracting NCD development have been widely explained [20,153–163]. There is, however, substantial evidence for the beneficial effects of supplementation even without combining it with sports. The sole intake of creatine has been able to significantly lower blood lipids such as cholesterol and triglycerides, slow down the development of fatty liver, and lower the HbA1C in human and animal studies, thus improving the clinical outcome and progression of the metabolic syndrome [164–166].

Table 3. Level of evidence of creatine's role in noncommunicable chronic disease.

Study	Disease	Subject	Treatment	Randomized	Subjects	Efficacy	Effect Role
Rider et al. [151]	Obesity	Human	None	None	64	Deranged cardiac energetics and diastolic dysfunction in obesity group	Observational, disease related changes in metabolism
Scheuermann-Freestone et al. [150]	Diabetes Type 2	Human	None	None	36	Impaired myocardial and skeletal muscle metabolism (reduced PCR/ATP ratio)	Observational disease related changes in metabolism
Lamb et al. [152]	Hypertension	Human	None	None	24	Altered high-energy phosphate metabolism in hypertension. Cardiac dysfunction correlates with metabolic alterations	Observational, disease related changes in metabolism
Gualano et al. [164]	Diabetes Type 2	Human	5 g creatine for 12 weeks + physical activity program	Yes	25	Improved glycemic control in supplementation group (by GLUT-4 recruitment)	Direct effect on disease related metabolic effects
Earnest et al. [165]	Hyper-cholester-inaemia	Human	4 × 5 g creatine for 5 days and afterwards 2 times per day for 51 days (orally)	Yes	34	Minor reduction of total cholesterol during supplementation. Reduction of triacylglycerol's and very-low-density-lipoprotein c 4 weeks after finishing supplementation	Direct effect of supplementation on metabolism.
Deminice et al. [166]	Fatty liver	Animal	Control vs. 0.25% choline diet vs. 0.25% choline + 2% creatine diet	None	24	Prevention of fat liver accumulation and hepatic events in creatine-fed group	Animal model

5. Cardiovascular Disease and Ischemic Heart Failure

Optimal replenishment of creatine reserves was able (in experimental studies) to slow down disease progression of the other above mentioned NCD and cardiomyopathy. Therefore, creatine supplementation has been identified to be of special therapeutic interest in treatment of cardiovascular diseases and their course [167,168]. The heart has its own four creatine kinase (CK) isozymes, proving the importance of ensuring filled energy depots [169]. A gradual reduction of myocardial total creatine content has been shown on chronic heart failure in human as well as animal studies [170–173]. The ratio of PCr/ATP has been defined to better judge myocardial creatine metabolism [174]. Low ratios have been positively correlated with low contractile function, more severe heart failure symptoms, and a higher risk of mortality [175–177].

Creatine supplementation in patients with chronic heart failure and similar animal studies have not shown any beneficial effect on clinical outcome, neither on myocardial creatine concentrations [178–180]. The transmembrane Creatine-Transporter (CrT) seems to be the limiting factor in this matter [181]. Question remains if other creatine-analogues that pass intracellular without the need of CrT might prove of better help in cardiovascular diseases. The energy deficiency resulting from local hypoxia during an ischemic heart attack leads to mitochondrial dysfunction, which in turn can have arrhythmogenic consequences and lead to sudden cardiac death [182–184]. Therefore, it is not surprising that creatine plays a critical role during a cardiac ischemic event [185,186]. First in vitro studies allow the hypothesis that saturation of myocardial creatine stores may lead to protection in the event of a transient ischemic attack [49]. In this context, in animal studies, filled ATP stores have a positive inotropic, apoptosis-protective effect and counteract a post-ischemic inflammatory cascade [187].

Intravenous in vivo administration of phosphocreatine was able to confer significant myocardial protection after bypass surgery [188], resulting in a reduction in the incidence of ventricular fibrillation and myocardial infarction as well as arrhythmias [189]. The newly developed special form of creatine, cyclo-creatine, deserves special attention. After an oral loading phase prior to elective cardiac interventions (PCI, ACVB, HTX), cyclo-creatine has a similar protective effect against lethal events [183,187,190,191]. However, large-scale human studies have yet to confirm the initial promising results. Table 4 summarizes the level of evidence available on the role of creatine in cardiovascular disease and ischemic heart failure [187–191].

Table 4. Level of evidence for creatine supplementation for chronic, atraumatic mitochondrial dysfunction.

Study	Disease	Subject	Treatment	Randomized	Subjects	Efficacy	Effect Role
Elgebaly et al. [187]	-	Animal/ vitro	500 mg/kg BW	no	6	Better aortic flow, coronary flow, cardiac output, stroke volume, and stroke work	Animal model
Cisowski et al. [188]	Cardiac surgery	Humans	6 g 3 days pre-surgery, intra-surgical and two days post- surgery i.v.	yes	40	Reduced arrhythmic events, reduced need of ionotropic medication	Direct effect on surgical procedure
Ruda et al. [189]	Ischemic myocardial infarct	human	2 g bolus + 4 g/h over 2 h	Yes	60	Reduced arrhythmic events	Direct effect on short term outcome
Chida et al. [192]	Dilated Cardio- myopathy	Human	None	None	13	Plasma BNP level was correlated negatively with the myocardial phosphocrea- tine/adenosine triphosphate	Observational finding
Roberts et al. [191].	None	Animal	Oral creatine- feeding	None	Not clear	Higher cellular ATP during ischemia in creatine-fed rat hearts	Animal model

6. Traumatic and Ischemic Central Nervous System Injuries

Mitochondrial function and ATP production are crucial for the neuronal survival and excitability [193]. At the same time, however, mitochondrial dysfunction leads to the overproduction of ROS and neuronal apoptosis which is closely related to neurodegenerative diseases and cerebral ischemia [193–197]. Whereas earlier research mainly focused on mitochondrial bioenergetic roles, new studies have shown the importance of apoptotic signaling, mitochondrial biogenesis, and mitophagy in the development of cerebrovascular disease and stroke. Mitochondrial health is therefore essential for neurological survival and rehabilitation [198,199]. Reperfusion injury is another acute complication feared by medical doctors involving mitochondria and clinical outcomes [200,201]. Following reperfusion of the injured brain tissue, excessive ROS and calcium produced under hypoxic conditions are washed in the body's periphery, causing damage on cellular and molecular level [202]. Intracellular calcium deregulation enhances neuronal cell death after stroke, giving the stability of the mitochondrial (calcium) permeability transition pore (mPTP) a special predictive measure [203].

The acute protective effects of creatine on the central nervous system (CNS) have long been known. Similar to the effect in the myocardium, energy buffering for short-term hypoxic conditions can be achieved by saturating intracellular PCr. This may lead to protection against ischemia and cell death, as well as calcium gradients created by mechanical stimuli [204–206]. In animal experiments, researchers were able to show that idiopathically caused brain damage and spinal cord injuries developed to a lesser extent after creatine oral administration [207,208]. Creatine supplementation also had a positive effect on infarct sizes after insult in ischemic mouse models [209]. These results suggest that creatine administration may lead to preventive CNS protection against concussions, traumatic brain injury, spinal cord injury, and insults [210].

Adding to the above-mentioned protective effects of Creatine during a hypoxic situation, special advantages of creatine on the CNS have been proven. The term excitotoxicity describes the destruction of neuronal cells due to pathological activation of its excitatory receptors [202]. Research was able to show that excitatory amino acids, such as Glutamate, become more neurotoxic when the cell's energy levels are reduced by hypoxia [211]. Activation of the glutamate NMDA receptor correlates with reduced ATP and PCr levels [212]. Creatine was able to protect animal brain tissues from the apoptotic effects of excitatory amino acids [213,214]. Lastly, it was shown that Creatine stabilizes mPTP in rodent studies, thus protecting brain tissue from apoptosis and cell death [67]. Table 5 presents a summary of the level of evidence related to creatine supplementation for traumatic and ischemic CNS injuries [205–207].

Table 5. Level of evidence for the role of creatine supplementation in individuals with traumatic and ischemic CNS injuries.

Study	Disease	Subject	Treatment	Randomized	Subjects	Efficacy	Effect Role
Zhu et al. [206]	None/induced ischemia	Animal	2% creatine-supplemented diet for 4 weeks	None	6 per group	Reduction in ischemia induced infarct size	Animal model
Turner et al. [205]	None/induced hypoxia	Human	7-ds oral creatine-supplementation	Yes	15	Less decrease in cognitive performance, attentional capacity, corticomotor excitability for creatine-group	Human brain metabolism
Hausmann et al. [207]	None/induced spinal cord injury	Animal	4 weeks oral creatine-supplementation	none	20	Better locomotor scores after 1 week for creatine-group. Less scar tissue for creatine-group after 2 weeks	Animal model
Sullivan et al. [208]	None/induced traumatic brain injury	Animal	Mice: 0.1 mL/10 g/BW creatine monohydrate injection for 1, 3 or 5 days Rats: 1% creatine diet for 4 weeks.	none	40 mice/ 24 rats	Reduction of brain tissue damage size by 36% mice and 50% rats	Animal model
Prass et al. [209]	None/induced experimental stroke	Animal	Creatine-rich diet (0%, 0.5%, 1%, 2% for 3 weeks	None	Unclear	Reduction of infarct size by 40% in 2% creatine-fed group	Animal model

7. Neurodegenerative Disorders

Ageing has been defined as a "progressive accumulation of changes with time that are associated with or responsible for the ever-increasing susceptibility to disease and death" [215]. Brain tissue is due to its high-energy demands especially vulnerable to mitochondrial deficits, ROS, hypoxia, and energy depletion [216,217]. Although ROS are of special need to neurons and brain tissue needed for synaptic plasticity, learning and memory function, their overproduction is closely related to nitration of proteins, mtDNA impairment and the development of neurodegenerative diseases, ageing, and cognitive deficits [218–220]. Insulin resistance and diabetes mellitus deteriorate these conditions and accelerate cognitive decline as well as incidence of neurogenerative diseases [221–223]. RAGE and ammonium level up the documented damage to mitochondria, neuronal cells, and brain tissue [224–226]. Alzheimer's disease has already been named "type 3 diabetes" [227]. Pathologically altered mitochondria have been shown to be swollen, have altered membrane potential, and reductions of ATP levels [228]. Therefore, mitochondrial protection and reduction of oxidative stress have been suggested to be of high therapeutic importance for the treatment of neurodegenerative disorders [229]. Anti-inflammatory nutrition, caloric restriction, as well as the use of supplements have been discussed to be improve mitochondrial functioning and cognition [230–233]. Various studies have also shown that creatine supplementation has a positive effect on cognition and brain function [234,235]. The effect was greater the more the participant was exposed to external stressors (e.g., hypoxia, sleep deprivation, etc.) [45,205] or the more complex the tasks were performed [236]. In this context, intake led to a lower need for sleep, earlier wake-up times, and improved sleep behavior [237].

Neurodegenerative diseases are usually characterized by the destruction or dysfunction of neurons in a specific brain area. Depending on the affected brain area, course, and severity, the forms of the disease differ. These include Alzheimer's disease (MA), amyotrophic lateral sclerosis (ALS), multiple sclerosis (MS), Huntington's disease (MH), and Parkinson's disease (MP). Impaired energy balance with mitochondrial dysfunction and oxidative stress are common to all diseases [238]. Similar findings have been made with intellectual disability-related diseases [239]. This bioenergetic deficit is thought to lead to apoptosis and necrosis and ultimately to neuronal degeneration [240]. Therefore, it is reasonable to assume that an improvement in mitochondrial health could enable a positive influence on the course of the disease. Table 6 provides a summary of the level of evidence related to the role of creatine supplementation for neurodegenerative disorders [45,234,236]. Initial studies suggest that creatine supplementation may be neuroprotective. For example, in 2013, Kley and coworkers [241] conducted a Cochrane review on the role of creatine monohydrate supplementation for treating muscle disorders. The researchers found sound evidence from randomized clinical trials that creatine supplementation increased strength and functional capacity in muscular dystrophy and idiopathic inflammatory myopathy while having no effect in patients with metabolic-related myopathies and McArdle disease. More long-term research is needed to evaluate the long-term effects of creatine in neurodegenerative diseases that impair muscle function.

Table 6. Level of evidence for the role of creatine supplementation in individuals with neurodegenerative disorders.

Study	Disease	Subject	Treatment	Randomized	Subjects	Efficacy	Effect Role
Hammett et al. [234]	None	Human	20 g/d creatine for 5 days + 5 g/d for 2-days	Yes	22	Reduction of stress related blood oxygen level dependent in fMRI in creatine-group	Human metabolic response
Watanabe et al. [235]	None	Human	8 g/d for 5-days	Yes	24	Reduction of mental fatigue and increased brain oxygen consumption in creatine-group	Human metabolic response
McMorris et al. [236]	None	Human	4 × 5 g/d	yes	20	Better in central complex executive tasks with creatine while sleep deprivation	Human metabolic response
McMorris et al. [45]	None	Human	4 × 5 g/d	Yes	15	random number generation, forward number and spatial recall, and long-term memory	Human metabolism

8. Psychological Disorders

In the 1980s, a link was established between bioenergetic deficits and depression [190, 242–244], bipolar disorders [114,245,246], and obsessive–compulsive disorders [247,248]. It is believed that there is an increase in energy demand with depletion of PCr stores at the onset of disease [124,249]. In clinical trials with depressed patients [250–252], a positive effect on subjective impairment after adjuvant creatine supplementation could be demonstrated. The higher the increase in cerebral PCr after creatine supplementation, the lower the depressive or manic symptoms [253]. The combination of antidepressants and creatine was more effective than simple pharmacological medication [254]. Creatine administration was even effective when drug therapy with SSRIs proved to be ineffective [255]. In this content, creatine has also been discussed as a potential therapeutic agent in the treatment of drug addiction and its psychic related disorders [256]. Positive effects of creatine supplementation have also been reported in post-traumatic stress disorders [257]. Schizophrenic and stress patients seem to gain no benefit from creatine intake. There is, however, ongoing debate on higher dosage for a needed benefit in these sub-groups [258]. Table 7 presents a summary of the literature related to the effects of creatine supplementation on individuals with psychological disorders [251,252,255].

Table 7. Level of evidence for the role of creatine supplementation in individuals with psychological disorders.

Study	Disease	Subject	Treatment	Randomized	Subjects	Efficacy	Effect Role
Kondo et al. [250]	Adolescent major depressive disorder	Human	4 g/d creatine for 8 weeks	None	15	Reduction in children-depression symptom scores. Significant increase in brain phosphocreatine level. Development of hypomania/mania in bipolar patients.	Direct effect on disease (no RCT)
Roitman et al. [251]	Treatment resistant depression	Human	3–5 g/d creatine for 4 weeks	None	8 unipolar depressed patients and two bipolar patients	Improved Hamilton Depression Rating Scale, Hamilton Anxiety Scale, and Clinical Global Impression for 7 of 8 unipolar depressed patients	Direct effect on disease (no RCT)

Table 7. *Cont.*

Study	Disease	Subject	Treatment	Randomized	Subjects	Efficacy	Effect Role
Toniolo et al. [252]	Depressive episode of Bipolar Type 1 and Type 2	Human	6 g/d creatine for 6 weeks	Yes	35	No significant difference in Montgomery-Åsberg Depression Rating Scale by intervention but higher remission rate in creatine supplemented group	Direct effect on disease
Kondo et al. [255]	Adolescent with SSRI resistant major depressive disorder	Human	0 g vs. 2 g vs. 4 g vs. 10 g creatine supplementation for 8 weeks	Yes	34	Clinical depression scores correlated inversely with brain phosphocreatine (PCR) levels. PCR level improved with higher dose.	Potential direct effect on disease

9. Chronic Fatigue Syndrome, Post Viral Fatigue Syndrome, and Long COVID

Fatigue is the most characteristic symptom of an energy deficit. There does not, however, exist a proper definition of the fatigue syndrome [259]. Fibromyalgia is a similar pathological entity closely related to CFS. Initially thought to be purely a psychological problem, linking fatigue to depression or other psychiatric diseases, newer research has been able to prove a metabolic dysfunction causing the symptoms [99,260,261]. Linking this clinical state with mitochondrial dysfunction was first able when lowered mitochondrial ATP levels were shown using MRS on patients with fatigue syndrome [262]. Later muscle biopsies and serum biomarkers have been able to show reduced mitochondrial biomarkers [263,264]. These markers have been Carnitine and CoQ10 [265]. On a mitochondrial level fatty acid metabolism was altered, electron transport chain was disrupted, there was a greater need in glucose concentrations and higher levels of lactate were shown [266]. Higher creatinine excretion via urine was shown to correlate positively with fatigue and pain severity. Being the end product of creatine, this urine marker could imply a higher turnover and depletion of the body's creatine storage [267]. More recent hypotheses state that these alterations have been caused by an activation of immune–inflammatory pathways due to viral infections (e.g., Epstein Barr, Q Fever, Ross River Infection) [268].

Long COVID is a persistent fatigue state after Sars-2-CoV-2 infection [269,270]. Interestingly, even asymptomatic patients exhibited raised biomarkers involved in inflammation and stress response [271]. Long COVID, Chronic Fatigue Syndrome, and Post Viral Fatigue Syndrome are believed to be the same entity [248,272]. Supplementation of guadinioacteic acid, a precursor of creatine, was able to attenuate several aspects of fatigue in fibromyalgia patients [273]. In combination of experimental findings as well as these first promising clinical outcomes, creatine might be an important key in the rehabilitation process of CFS and Long COVID patients [274]. Table 8 summarizes the available literature on the effects of the creatine precursor GAA on chronic fatigue and Post-COVID syndrome [274].

Table 8. Summary of literature on the effects of creatine precursors on chronic fatigue and Post-COVID syndrome.

Study	Disease	Subject	Treatment	Randomized	Subjects	Efficacy	Effect Role
Ostojic et al. [264]	Chronic Fatigue syndrome	Human	2 g, 4 g oral Guanidinoacetic Acid for 3 months vs. placebo	Yes	21	Higher muscle creatine-phosphate level and better oxidative capacity. However, no significant improvement of fatigue symptoms	Direct effect on disease related metabolism

10. Conclusions

This review summarizes creatine's impact on mitochondrial function besides restoring ATP-storage. Creatine monohydrate is one of the best-known nutrient supplements mainly being used for improvement of athletic performance. However, there is growing evidence for a broader therapeutic spectrum of this nitrogen–amino-compound. Various health-promoting effects on cell-metabolism after the intake of creatine have been shown. Its impact on mitochondrial integrity has become of special interest. Mitochondrial dysfunction has become a central pathological hallmark of non-communicable diseases. The supplementation of creatine monohydrate may have some synergistic effects in the treatment of CND. This seems to be directly related to its protective effects on mitochondria. Different from pharmaceutical products, the intake of creatine is safe age- and gender-independent with nearly no side-effects [49,50]. Although these findings are promising, much of the available data has been generated with in vitro or animal studies. Therefore, there is a need to conduct more clinical trials in humans to assess the potential therapeutic effects of creatine monohydrate supplementation on conditions influencing mitochondrial function.

Author Contributions: Conceptualization, R.P.M.; writing—original draft preparation, R.P.M.; writing—review and editing, R.P.M., J.-N.D., J.G., and R.B.K.; funding acquisition, R.P.M. All authors have read and agreed to the published version of the manuscript.

Funding: The APC of selected papers of this special issue are being funded by AlzChem, LLC. (Trostberg, Germany) who manufactures creatine monohydrate. The funders had no role in the writing of the manuscript, interpretation of the literature, or in the decision to publish the results.

Institutional Review Board Statement: Not applicable.

Informed Consent Statement: Not applicable.

Data Availability Statement: Not applicable.

Acknowledgments: The authors would like to thank all of the research participants, scholars, and funding agencies who have contributed to the research cited in this manuscript.

Conflicts of Interest: R.P.M. received financial support for presenting on creatine at industry sponsored scientific conferences. J.N.D. declares there is no financial and no non-financial conflict of interest. J.G. reports no conflict of interest. R.B.K. has conducted industry sponsored research on creatine, received financial support for presenting on creatine at industry sponsored scientific conferences, and has served as an expert witness on cases related to creatine. Additionally, he serves as Chair of the Scientific Advisory Board for AlzChem who sponsored this special issue.

References

1. Bonilla, D.A.; Kreider, R.B.; Stout, J.R.; Forero, D.A.; Kerksick, C.M.; Roberts, M.D.; Rawson, E.S. Metabolic Basis of Creatine in Health and Disease: A Bioinformatics-Assisted Review. *Nutrients* **2021**, *13*, 1238. [CrossRef]
2. Candow, D.G.; Forbes, S.C.; Chilibeck, P.D.; Cornish, S.M.; Antonio, J.; Kreider, R.B. Effectiveness of Creatine Supplementation on Aging Muscle and Bone: Focus on Falls Prevention and Inflammation. *J. Clin. Med.* **2019**, *8*, 488. [CrossRef]
3. Brosnan, M.E.; Brosnan, J.T. The role of dietary creatine. *Amino Acids* **2016**, *48*, 1785–1791. [CrossRef]
4. Harris, R. Creatine in health, medicine and sport: An introduction to a meeting held at Downing College, University of Cambridge, July 2010. *Amino Acids* **2011**, *40*, 1267. [CrossRef]
5. Harris, R.C.; Soderlund, K.; Hultman, E. Elevation of creatine in resting and exercised muscle of normal subjects by creatine supplementation. *Clin. Sci.* **1992**, *83*, 367–374. [CrossRef]
6. Kreider, R.B.; Stout, J.R. Creatine in Health and Disease. *Nutrients* **2021**, *13*, 447. [CrossRef]
7. Hultman, E.; Soderlund, K.; Timmons, J.A.; Cederblad, G.; Greenhaff, P.L. Muscle creatine loading in men. *J. Appl. Physiol.* **1996**, *81*, 232–237. [CrossRef]
8. Wallimann, T.; Tokarska-Schlattner, M.; Schlattner, U. The creatine kinase system and pleiotropic effects of creatine. *Amino Acids* **2011**, *40*, 1271–1296. [CrossRef]
9. Huertas, J.R.; Casuso, R.A.; Agustín, P.H.; Cogliati, S. Stay Fit, Stay Young: Mitochondria in Movement: The Role of Exercise in the New Mitochondrial Paradigm. *Oxidative Med. Cell. Longev.* **2019**, *2019*, e7058350. [CrossRef]
10. Negro, M.; Avanzato, I.; D'Antona, G. Chapter 2.7—Creatine in Skeletal Muscle Physiology. In *Nonvitamin and Nonmineral Nutritional Supplements*; Nabavi, S.M., Silva, A.S., Eds.; Academic Press: Cambridge, MA, USA, 2019; pp. 59–68.

11. Nelson, A.G.; Arnall, D.A.; Kokkonen, J.; Day, R.; Evans, J. Muscle glycogen supercompensation is enhanced by prior creatine supplementation. *Med. Sci. Sports Exerc.* **2001**, *33*, 1096–1100. [CrossRef]

12. Tarnopolsky, M.A.; Parise, G. Direct measurement of high-energy phosphate compounds in patients with neuromuscular disease. *Muscle Nerve* **1999**, *22*, 1228–1233. [CrossRef]

13. McKenna, M.J.; Morton, J.; Selig, S.E.; Snow, R.J. Creatine supplementation increases muscle total creatine but not maximal intermittent exercise performance. *J. Appl. Physiol.* **1999**, *87*, 2244–2252. [CrossRef]

14. Greenhaff, P.L.; Bodin, K.; Soderlund, K.; Hultman, E. Effect of oral creatine supplementation on skeletal muscle phosphocreatine resynthesis. *Am. J. Physiol.* **1994**, *266*, E725–E730. [CrossRef]

15. Greenwood, M.; Kreider, R.B.; Earnest, C.P.; Rasmussen, C.; Almada, A. Differences in creatine retention among three nutritional formulations of oral creatine supplements. *J. Exerc. Physiol. Online* **2003**, *6*, 37–43.

16. Choi, J.K.; Kustermann, E.; Dedeoglu, A.; Jenkins, B.G. Magnetic resonance spectroscopy of regional brain metabolite markers in FALS mice and the effects of dietary creatine supplementation. *Eur. J. Neurosci.* **2009**, *30*, 2143–2150. [CrossRef]

17. Lyoo, I.K.; Kong, S.W.; Sung, S.M.; Hirashima, F.; Parow, A.; Hennen, J.; Cohen, B.M.; Renshaw, P.F. Multinuclear magnetic resonance spectroscopy of high-energy phosphate metabolites in human brain following oral supplementation of creatine-monohydrate. *Psychiatry Res.* **2003**, *123*, 87–100. [CrossRef]

18. Roschel, H.; Gualano, B.; Ostojic, S.M.; Rawson, E.S. Creatine Supplementation and Brain Health. *Nutrients* **2021**, *13*, 586. [CrossRef]

19. Dolan, E.; Gualano, B.; Rawson, E.S. Beyond muscle: The effects of creatine supplementation on brain creatine, cognitive processing, and traumatic brain injury. *Eur. J. Sport Sci.* **2019**, *19*, 1–14. [CrossRef]

20. Gualano, B.; Rawson, E.S.; Candow, D.G.; Chilibeck, P.D. Creatine supplementation in the aging population: Effects on skeletal muscle, bone and brain. *Amino Acids* **2016**, *48*, 1793–1805. [CrossRef]

21. Rawson, E.S.; Venezia, A.C. Use of creatine in the elderly and evidence for effects on cognitive function in young and old. *Amino Acids* **2011**, *40*, 1349–1362. [CrossRef]

22. Cornish, S.M.; Chilibeck, P.D.; Burke, D.G. The effect of creatine monohydrate supplementation on sprint skating in ice-hockey players. *J. Sports Med. Phys. Fitness* **2006**, *46*, 90–98.

23. Dawson, B.; Vladich, T.; Blanksby, B.A. Effects of 4 weeks of creatine supplementation in junior swimmers on freestyle sprint and swim bench performance. *J. Strength Cond. Res.* **2002**, *16*, 485–490.

24. Grindstaff, P.D.; Kreider, R.; Bishop, R.; Wilson, M.; Wood, L.; Alexander, C.; Almada, A. Effects of creatine supplementation on repetitive sprint performance and body composition in competitive swimmers. *Int. J. Sport Nutr.* **1997**, *7*, 330–346. [CrossRef]

25. Juhasz, I.; Gyore, I.; Csende, Z.; Racz, L.; Tihanyi, J. Creatine supplementation improves the anaerobic performance of elite junior fin swimmers. *Acta Physiol. Hung.* **2009**, *96*, 325–336. [CrossRef]

26. Silva, A.J.; Machado Reis, V.; Guidetti, L.; Bessone Alves, F.; Mota, P.; Freitas, J.; Baldari, C. Effect of creatine on swimming velocity, body composition and hydrodynamic variables. *J. Sports Med. Phys. Fitness* **2007**, *47*, 58–64.

27. Kreider, R.B.; Ferreira, M.; Wilson, M.; Grindstaff, P.; Plisk, S.; Reinardy, J.; Cantler, E.; Almada, A.L. Effects of creatine supplementation on body composition, strength, and sprint performance. *Med. Sci. Sports Exerc.* **1998**, *30*, 73–82. [CrossRef]

28. Stone, M.H.; Sanborn, K.; Smith, L.L.; O'Bryant, H.S.; Hoke, T.; Utter, A.C.; Johnson, R.L.; Boros, R.; Hruby, J.; Pierce, K.C.; et al. Effects of in-season (5 weeks) creatine and pyruvate supplementation on anaerobic performance and body composition in American football players. *Int. J. Sport Nutr.* **1999**, *9*, 146–165. [CrossRef]

29. Bemben, M.G.; Bemben, D.A.; Loftiss, D.D.; Knehans, A.W. Creatine supplementation during resistance training in college football athletes. *Med. Sci. Sports Exerc.* **2001**, *33*, 1667–1673. [CrossRef]

30. Hoffman, J.; Ratamess, N.; Kang, J.; Mangine, G.; Faigenbaum, A.; Stout, J. Effect of creatine and beta-alanine supplementation on performance and endocrine responses in strength/power athletes. *Int. J. Sport Nutr. Exerc. Metab.* **2006**, *16*, 430–446. [CrossRef]

31. Chilibeck, P.D.; Magnus, C.; Anderson, M. Effect of in-season creatine supplementation on body composition and performance in rugby union football players. *Appl. Physiol. Nutr. Metab.* **2007**, *32*, 1052–1057. [CrossRef]

32. Claudino, J.G.; Mezencio, B.; Amaral, S.; Zanetti, V.; Benatti, F.; Roschel, H.; Gualano, B.; Amadio, A.C.; Serrao, J.C. Creatine monohydrate supplementation on lower-limb muscle power in Brazilian elite soccer players. *J. Int. Soc. Sports Nutr.* **2014**, *11*, 32. [CrossRef]

33. Kerksick, C.M.; Rasmussen, C.; Lancaster, S.; Starks, M.; Smith, P.; Melton, C.; Greenwood, M.; Almada, A.; Kreider, R. Impact of differing protein sources and a creatine containing nutritional formula after 12 weeks of resistance training. *Nutrition* **2007**, *23*, 647–656. [CrossRef]

34. Kerksick, C.M.; Wilborn, C.D.; Campbell, W.I.; Harvey, T.M.; Marcello, B.M.; Roberts, M.D.; Parker, A.G.; Byars, A.G.; Greenwood, L.D.; Almada, A.L.; et al. The effects of creatine monohydrate supplementation with and without D-pinitol on resistance training adaptations. *J. Strength Cond. Res.* **2009**, *23*, 2673–2682. [CrossRef]

35. Galvan, E.; Walker, D.K.; Simbo, S.Y.; Dalton, R.; Levers, K.; O'Connor, A.; Goodenough, C.; Barringer, N.D.; Greenwood, M.; Rasmussen, C.; et al. Acute and chronic safety and efficacy of dose dependent creatine nitrate supplementation and exercise performance. *J. Int. Soc. Sports Nutr.* **2016**, *13*, 12. [CrossRef]

36. Volek, J.S.; Kraemer, W.J.; Bush, J.A.; Boetes, M.; Incledon, T.; Clark, K.L.; Lynch, J.M. Creatine supplementation enhances muscular performance during high-intensity resistance exercise. *J. Am. Diet. Assoc.* **1997**, *97*, 765–770. [CrossRef]

37. Volek, J.S.; Mazzetti, S.A.; Farquhar, W.B.; Barnes, B.R.; Gomez, A.L.; Kraemer, W.J. Physiological responses to short-term exercise in the heat after creatine loading. *Med. Sci. Sports Exerc.* **2001**, *33*, 1101–1108. [CrossRef]

38. Volek, J.S.; Ratamess, N.A.; Rubin, M.R.; Gomez, A.L.; French, D.N.; McGuigan, M.M.; Scheett, T.P.; Sharman, M.J.; Hakkinen, K.; Kraemer, W.J. The effects of creatine supplementation on muscular performance and body composition responses to short-term resistance training overreaching. *Eur. J. Appl. Physiol.* **2004**, *91*, 628–637. [CrossRef]

39. Buford, T.W.; Kreider, R.B.; Stout, J.R.; Greenwood, M.; Campbell, B.; Spano, M.; Ziegenfuss, T.; Lopez, H.; Landis, J.; Antonio, J. International Society of Sports Nutrition position stand: Creatine supplementation and exercise. *J. Int. Soc. Sports Nutr.* **2007**, *4*, 6. [CrossRef]

40. Kreider, R.B.; Wilborn, C.D.; Taylor, L.; Campbell, B.; Almada, A.L.; Collins, R.; Cooke, M.; Earnest, C.P.; Greenwood, M.; Kalman, D.S.; et al. ISSN exercise & sport nutrition review: Research & recommendations. *J. Int. Soc. Sports Nutr.* **2010**, *7*, 7. [CrossRef]

41. Branch, J.D. Effect of creatine supplementation on body composition and performance: A meta-analysis. *Int. J. Sport Nutr. Exerc. Metab.* **2003**, *13*, 198–226. [CrossRef]

42. Devries, M.C.; Phillips, S.M. Creatine supplementation during resistance training in older adults-a meta-analysis. *Med. Sci. Sports Exerc.* **2014**, *46*, 1194–1203. [CrossRef]

43. Lanhers, C.; Pereira, B.; Naughton, G.; Trousselard, M.; Lesage, F.X.; Dutheil, F. Creatine Supplementation and Lower Limb Strength Performance: A Systematic Review and Meta-Analyses. *Sports Med.* **2015**, *45*, 1285–1294. [CrossRef]

44. Wiroth, J.B.; Bermon, S.; Andrei, S.; Dalloz, E.; Hebuterne, X.; Dolisi, C. Effects of oral creatine supplementation on maximal pedalling performance in older adults. *Eur. J. Appl. Physiol.* **2001**, *84*, 533–539. [CrossRef]

45. McMorris, T.; Mielcarz, G.; Harris, R.C.; Swain, J.P.; Howard, A. Creatine supplementation and cognitive performance in elderly individuals. *Neuropsychol. Dev. Cogn. B Aging Neuropsychol. Cogn.* **2007**, *14*, 517–528. [CrossRef]

46. Rawson, E.S.; Clarkson, P.M. Acute creatine supplementation in older men. *Int. J. Sports Med.* **2000**, *21*, 71–75. [CrossRef]

47. Tarnopolsky, M.A. Potential benefits of creatine monohydrate supplementation in the elderly. *Curr. Opin. Clin. Nutr. Metab. Care* **2000**, *3*, 497–502. [CrossRef]

48. Aguiar, A.F.; Januario, R.S.; Junior, R.P.; Gerage, A.M.; Pina, F.L.; do Nascimento, M.A.; Padovani, C.R.; Cyrino, E.S. Long-term creatine supplementation improves muscular performance during resistance training in older women. *Eur. J. Appl. Physiol.* **2013**, *113*, 987–996. [CrossRef]

49. Kreider, R.B.; Kalman, D.S.; Antonio, J.; Ziegenfuss, T.N.; Wildman, R.; Collins, R.; Candow, D.G.; Kleiner, S.M.; Almada, A.L.; Lopez, H.L. International Society of Sports Nutrition position stand: Safety and efficacy of creatine supplementation in exercise, sport, and medicine. *J. Int. Soc. Sports Nutr.* **2017**, *14*, 18. [CrossRef]

50. Antonio, J.; Candow, D.G.; Forbes, S.C.; Gualano, B.; Jagim, A.R.; Kreider, R.B.; Rawson, E.S.; Smith-Ryan, A.E.; VanDusseldorp, T.A.; Willoughby, D.S.; et al. Common questions and misconceptions about creatine supplementation: What does the scientific evidence really show? *J. Int. Soc. Sports Nutr.* **2021**, *18*, 13. [CrossRef]

51. Brand, M.D.; Nicholls, D.G. Assessing mitochondrial dysfunction in cells. *Biochem. J.* **2011**, *435*, 297–312. [CrossRef]

52. Read, C.Y.; Calnan, R.J. Mitochondrial disease: Beyond etiology unknown. *J. Pediatr. Nurs.* **2000**, *15*, 232–241. [CrossRef]

53. Cohen, B.H.; Gold, D.R. Mitochondrial cytopathy in adults: What we know so far. *Clev. Clin. J. Med.* **2001**, *68*, 625–642. [CrossRef]

54. Giza, C.C.; Hovda, D.A. The Neurometabolic Cascade of Concussion. *J. Athl. Train* **2001**, *36*, 228–235. [CrossRef]

55. Dean, A.; Philip, J.; Arikan, G.; Opitz, B.; Sterr, A. Potential for use of creatine supplementation following mild traumatic brain injury. *Concussion* **2017**, *2*, CNC34. [CrossRef]

56. Gaetz, M. The neurophysiology of brain injury. *Clin. Neurophysiol.* **2004**, *115*, 4–18. [CrossRef]

57. Brooke, N.S.; Ouwerkerk, R.; Adams, C.B.; Radda, G.K.; Ledingham, J.G.; Rajagopalan, B. Phosphorus-31 magnetic resonance spectra reveal prolonged intracellular acidosis in the brain following subarachnoid hemorrhage. *Proc. Natl. Acad. Sci. USA* **1994**, *91*, 1903–1907. [CrossRef]

58. Abe, K.; Aoki, M.; Kawagoe, J.; Yoshida, T.; Hattori, A.; Kogure, K.; Itoyama, Y. Ischemic Delayed Neuronal Death. *Stroke* **1995**, *26*, 1478–1489. [CrossRef]

59. Ankarcrona, M.; Dypbukt, J.M.; Bonfoco, E.; Zhivotovsky, B.; Orrenius, S.; Lipton, S.A.; Nicotera, P. Glutamate-induced neuronal death: A succession of necrosis or apoptosis depending on mitochondrial function. *Neuron* **1995**, *15*, 961–973. [CrossRef]

60. Fiskum, G.; Murphy, A.N.; Beal, M.F. Mitochondria in Neurodegeneration: Acute Ischemia and Chronic Neurodegenerative Diseases. *J. Cereb. Blood Flow Metab.* **1999**, *19*, 351–369. [CrossRef]

61. Schinder, A.F.; Olson, E.C.; Spitzer, N.C.; Montal, M. Mitochondrial Dysfunction Is a Primary Event in Glutamate Neurotoxicity. *J. Neurosci.* **1996**, *16*, 6125–6133. [CrossRef]

62. Béard, E.; Braissant, O. Synthesis and transport of creatine in the CNS: Importance for cerebral functions. *J. Neurochem.* **2010**, *115*, 297–313. [CrossRef]

63. Rabinowitz, A.R.; Li, X.; Levin, H.S. Sport and nonsport etiologies of mild traumatic brain injury: Similarities and differences. *Annu. Rev. Psychol.* **2014**, *65*, 301–331. [CrossRef]

64. Signoretti, S.; Lazzarino, G.; Tavazzi, B.; Vagnozzi, R. The pathophysiology of concussion. *PM R* **2011**, *3*, S359–S368. [CrossRef]

65. Andres, R.H.; Ducray, A.D.; Schlattner, U.; Wallimann, T.; Widmer, H.R. Functions and effects of creatine in the central nervous system. *Brain Res. Bull.* **2008**, *76*, 329–343. [CrossRef]

66. Gualano, B.; Roschel, H.; Lancha, A.H.; Brightbill, C.E.; Rawson, E.S. In sickness and in health: The widespread application of creatine supplementation. *Amino Acids* **2012**, *43*, 519–529. [CrossRef]

67. Rae, C.D.; Bröer, S. Creatine as a booster for human brain function. How might it work? *Neurochem. Int.* **2015**, *89*, 249–259. [CrossRef]
68. Perasso, L.; Spallarossa, P.; Gandolfo, C.; Ruggeri, P.; Balestrino, M. Therapeutic Use of Creatine in Brain or Heart Ischemia: Available Data and Future Perspectives. *Med. Res. Rev.* **2013**, *33*, 336–363. [CrossRef]
69. O'Gorman, E.; Beutner, G.; Dolder, M.; Koretsky, A.P.; Brdiczka, D.; Wallimann, T. The role of creatine kinase in inhibition of mitochondrial permeability transition. *FEBS Lett.* **1997**, *414*, 253–257. [CrossRef]
70. Meyer, L.E.; Machado, L.B.; Santiago, A.P.; Da-Silva, W.S.; De Felice, F.G.; Holub, O.; Oliveira, M.F.; Galina, A. Mitochondrial creatine kinase activity prevents reactive oxygen species generation: Antioxidant role of mitochondrial kinase-dependent ADP re-cycling activity. *J. Biol. Chem.* **2006**, *281*, 37361–37371. [CrossRef]
71. Sakellaris, G.; Kotsiou, M.; Tamiolaki, M.; Kalostos, G.; Tsapaki, E.; Spanaki, M.; Spilioti, M.; Charissis, G.; Evangeliou, A. Prevention of complications related to traumatic brain injury in children and adolescents with creatine administration: An open label randomized pilot study. *J. Trauma* **2006**, *61*, 322–329. [CrossRef]
72. Sakellaris, G.; Nasis, G.; Kotsiou, M.; Tamiolaki, M.; Charissis, G.; Evangeliou, A. Prevention of traumatic headache, dizziness and fatigue with creatine administration. A pilot study. *Acta Paediatr.* **2008**, *97*, 31–34. [CrossRef]
73. Sosa, V.; Moliné, T.; Somoza, R.; Paciucci, R.; Kondoh, H.; Lleonart, M.E. Oxidative stress and cancer: An overview. *Ageing Res. Rev.* **2013**, *12*, 376–390. [CrossRef]
74. Nicolson, G.L.; Ferreira, G.; Settineri, R.; Ellithorpe, R.R.; Breeding, P.; Ash, M.E. Mitochondrial Dysfunction and Chronic Disease: Treatment with Membrane Lipid Replacement and Other Natural Supplements. In *Mitochondrial Biology and Experimental Therapeutics*; Oliveira, P.J., Ed.; Springer International Publishing: Berlin/Heidelberg, Germany, 2018; pp. 499–522.
75. Newell, C.; Leduc-Pessah, H.; Khan, A.; Shearer, J. Mitochondrial Dysfunction in Chronic Disease. In *The Routledge Handbook on Biochemistry of Exercise*; Routledge: Abingdon, UK, 2020.
76. Victor, M.V.; Rocha, M.; Herance, R.; Hernandez-Mijares, A. Oxidative Stress and Mitochondrial Dysfunction in Type 2 Diabetes. *Curr. Pharm. Des.* **2011**, *17*, 3947–3958. [CrossRef]
77. Picard, M.; Turnbull, D.M. Linking the Metabolic State and Mitochondrial DNA in Chronic Disease, Health, and Aging. *Diabetes* **2013**, *62*, 672–678. [CrossRef]
78. Pieczenik, S.R.; Neustadt, J. Mitochondrial dysfunction and molecular pathways of disease. *Exp. Mol. Pathol.* **2007**, *83*, 84–92. [CrossRef]
79. Madamanchi, N.R.; Runge, M.S. Mitochondrial Dysfunction in Atherosclerosis. *Circ. Res.* **2007**, *100*, 460–473. [CrossRef]
80. Galvan, D.L.; Green, N.H.; Danesh, F.R. The hallmarks of mitochondrial dysfunction in chronic kidney disease. *Kidney Int.* **2017**, *92*, 1051–1057. [CrossRef]
81. Cloonan, S.M.; Kim, K.; Esteves, P.; Trian, T.; Barnes, P.J. Mitochondrial dysfunction in lung ageing and disease. *Eur. Respir. Rev.* **2020**, *29*, 157. [CrossRef]
82. Wei, P.Z.; Szeto, C.C. Mitochondrial dysfunction in diabetic kidney disease. *Clin. Chim. Acta* **2019**, *496*, 108–116. [CrossRef]
83. Mansouri, A.; Gattolliat, C.-H.; Asselah, T. Mitochondrial Dysfunction and Signaling in Chronic Liver Diseases. *Gastroenterology* **2018**, *155*, 629–647. [CrossRef]
84. Fang, T.; Wang, M.; Xiao, H.; Wei, X. Mitochondrial dysfunction and chronic lung disease. *Cell Biol. Toxicol.* **2019**, *35*, 493–502. [CrossRef] [PubMed]
85. López-Armada, M.J.; Riveiro-Naveira, R.R.; Vaamonde-García, C.; Valcárcel-Ares, M.N. Mitochondrial dysfunction and the inflammatory response. *Mitochondrion* **2013**, *13*, 106–118. [CrossRef] [PubMed]
86. Castellani, R.; Hirai, K.; Aliev, G.; Drew, K.L.; Nunomura, A.; Takeda, A.; Cash, A.D.; Obrenovich, M.E.; Perry, G.; Smith, M.A. Role of mitochondrial dysfunction in Alzheimer's disease. *J. Neurosci. Res.* **2002**, *70*, 357–360. [CrossRef] [PubMed]
87. Sorrentino, V.; Menzies, K.J.; Auwerx, J. Repairing Mitochondrial Dysfunction in Disease. *Annu. Rev. Pharmacol. Toxicol.* **2018**, *58*, 353–389. [CrossRef]
88. Abrigo, J.; Simon, F.; Cabrera, D.; Vilos, C.; Cabello-Verrugio, C. Mitochondrial Dysfunction in Skeletal Muscle Pathologies. *Curr. Protein Pept. Sci.* **2019**, *20*, 536–546. [CrossRef] [PubMed]
89. Prakash, Y.S.; Pabelick, C.M.; Sieck, G.C. Mitochondrial Dysfunction in Airway Disease. *Chest* **2017**, *152*, 618–626. [CrossRef]
90. Diaz-Vegas, A.; Sanchez-Aguilera, P.; Krycer, J.R.; Morales, P.E.; Monsalves-Alvarez, M.; Cifuentes, M.; Rothermel, B.A.; Lavandero, S. Is Mitochondrial Dysfunction a Common Root of Noncommunicable Chronic Diseases? *Endocr. Rev.* **2020**, *41*, 491–517. [CrossRef]
91. Novak, E.A.; Mollen, K.P. Mitochondrial dysfunction in inflammatory bowel disease. *Front. Cell Dev. Biol.* **2015**, *3*, 62. [CrossRef]
92. Ballinger, S.W. Mitochondrial dysfunction in cardiovascular disease. *Free Radic. Biol. Med.* **2005**, *38*, 1278–1295. [CrossRef]
93. Rosca, M.G.; Hoppel, C.L. Mitochondrial dysfunction in heart failure. *Heart Fail. Rev.* **2013**, *18*, 607–622. [CrossRef]
94. Chistiakov, D.A.; Shkurat, T.P.; Melnichenko, A.A.; Grechko, A.V.; Orekhov, A.N. The role of mitochondrial dysfunction in cardiovascular disease: A brief review. *Ann. Med.* **2018**, *50*, 121–127. [CrossRef] [PubMed]
95. Li, X.; Zhang, W.; Cao, Q.; Wang, Z.; Zhao, M.; Xu, L.; Zhuang, Q. Mitochondrial dysfunction in fibrotic diseases. *Cell Death Discov.* **2020**, *6*, 1–14. [CrossRef] [PubMed]
96. Johri, A.; Beal, M.F. Mitochondrial Dysfunction in Neurodegenerative Diseases. *J. Pharmacol. Exp. Ther.* **2012**, *342*, 619–630. [CrossRef] [PubMed]

97. Barot, M.; Gokulgandhi, M.R.; Mitra, A.K. Mitochondrial Dysfunction in Retinal Diseases. *Curr. Eye Res.* **2011**, *36*, 1069–1077. [CrossRef] [PubMed]
98. Hu, F.; Liu, F. Mitochondrial stress: A bridge between mitochondrial dysfunction and metabolic diseases? *Cell. Signal.* **2011**, *23*, 1528–1533. [CrossRef] [PubMed]
99. Myhill, S.; Booth, N.E.; McLaren-Howard, J. Chronic fatigue syndrome and mitochondrial dysfunction. *Int. J. Clin. Exp. Med.* **2009**, *2*, 1–16.
100. Haas, R.H. Mitochondrial Dysfunction in Aging and Diseases of Aging. *Biology* **2019**, *8*, 48. [CrossRef]
101. Kemp, G.J. Mitochondrial dysfunction in chronic ischemia and peripheral vascular disease. *Mitochondrion* **2004**, *4*, 629–640. [CrossRef]
102. Liberti, M.V.; Locasale, J.W. The Warburg Effect: How Does it Benefit Cancer Cells? *Trends Biochem. Sci.* **2016**, *41*, 211–218. [CrossRef]
103. Vander Heiden, M.G.; Cantley, L.C.; Thompson, C.B. Understanding the Warburg Effect: The Metabolic Requirements of Cell Proliferation. *Science* **2009**, *324*, 1029–1033. [CrossRef]
104. Zapelini, P.H.; Rezin, G.T.; Cardoso, M.R.; Ritter, C.; Klamt, F.; Moreira, J.C.F.; Streck, E.L.; Dal-Pizzol, F. Antioxidant treatment reverses mitochondrial dysfunction in a sepsis animal model. *Mitochondrion* **2008**, *8*, 211–218. [CrossRef] [PubMed]
105. Molnár, A.G.; Kun, S.; Sélley, E.; Kertész, M.; Szélig, L.; Csontos, C.; Böddi, K.; Bogár, L.; Miseta, A.; Wittmann, I. Role of Tyrosine Isomers in Acute and Chronic Diseases Leading to Oxidative Stress—A Review. *Curr. Med. Chem.* **2016**, *23*, 667–685. [CrossRef] [PubMed]
106. Ortiz, G.G.; Pacheco Moisés, F.P.; Mireles-Ramírez, M.; Flores-Alvarado, L.J.; González-Usigli, H.; Sánchez-González, V.J.; Sánchez-López, A.L.; Sánchez-Romero, L.; Díaz-Barba, E.I.; Santoscoy-Gutiérrez, J.F.; et al. Chapter One-Oxidative Stress: Love and Hate History in Central Nervous System. In *Advances in Protein Chemistry and Structural Biology*; Stress and Inflammation in Disorders; Donev, R., Ed.; Academic Press: Cambridge, MA, USA, 2017; Volume 108, pp. 1–31.
107. Incalza, M.A.; D'Oria, R.; Natalicchio, A.; Perrini, S.; Laviola, L.; Giorgino, F. Oxidative stress and reactive oxygen species in endothelial dysfunction associated with cardiovascular and metabolic diseases. *Vasc. Pharmacol.* **2018**, *100*, 1–19. [CrossRef] [PubMed]
108. Pall, M.L.; Levine, S. Nrf2, a master regulator of detoxification and also antioxidant, anti-inflammatory and other cytoprotective mechanisms, is raised by health promoting factors. *Sheng Li Xue Bao* **2015**, *67*, 1–18.
109. Moldogazieva, N.T.; Mokhosoev, I.M.; Mel'nikova, T.I.; Porozov, Y.B.; Terentiev, A.A. Oxidative Stress and Advanced Lipoxidation and Glycation End Products (ALEs and AGEs) in Aging and Age-Related Diseases. *Oxidative Med. Cell. Longev.* **2019**, *2019*, e3085756. [CrossRef] [PubMed]
110. Bangsbo, J. Energy demands in competitive soccer. *J. Sports Sci.* **1994**, *12*, S5–S12. [CrossRef]
111. Adeva, M.M.; Souto, G.; Blanco, N.; Donapetry, C. Ammonium metabolism in humans. *Metabolism* **2012**, *61*, 1495–1511. [CrossRef]
112. Mutch, B.J.; Banister, E.W. Ammonia metabolism in exercise and fatigue: A review. *Med. Sci. Sports Exerc.* **1983**, *15*, 41–50. [CrossRef]
113. Srinivasan, S.; Guha, M.; Kashina, A.; Avadhani, N.G. Mitochondrial Dysfunction and Mitochondrial Dynamics-The Cancer Connection. *Biochim. Biophys. Acta* **2017**, *1858*, 602–614. [CrossRef]
114. Stork, C.; Renshaw, P.F. Mitochondrial dysfunction in bipolar disorder: Evidence from magnetic resonance spectroscopy research. *Mol. Psychiatry* **2005**, *10*, 900–919. [CrossRef]
115. Devic, S. Warburg Effect-a Consequence or the Cause of Carcinogenesis? *J. Cancer* **2016**, *7*, 817–822. [CrossRef] [PubMed]
116. Rani, V.; Deep, G.; Singh, R.K.; Palle, K.; Yadav, U.C.S. Oxidative stress and metabolic disorders: Pathogenesis and therapeutic strategies. *Life Sci.* **2016**, *148*, 183–193. [CrossRef] [PubMed]
117. Stepien, K.M.; Heaton, R.; Rankin, S.; Murphy, A.; Bentley, J.; Sexton, D.; Hargreaves, I.P. Evidence of Oxidative Stress and Secondary Mitochondrial Dysfunction in Metabolic and Non-Metabolic Disorders. *J. Clin. Med.* **2017**, *6*, 71. [CrossRef] [PubMed]
118. Mlynárik, V. Introduction to nuclear magnetic resonance. *Anal. Biochem.* **2017**, *529*, 4–9. [CrossRef] [PubMed]
119. Prost, R.W. Magnetic resonance spectroscopy. *Med. Phys.* **2008**, *35*, 4530–4544. [CrossRef] [PubMed]
120. Henning, A. Proton and multinuclear magnetic resonance spectroscopy in the human brain at ultra-high field strength: A review. *Neuroimage* **2018**, *168*, 181–198. [CrossRef]
121. Porter, D.A.; Smith, M.A. Magnetic resonance spectroscopy in vivo. *J. Biomed. Eng.* **1988**, *10*, 562–568. [CrossRef]
122. Bertolino, A.; Frye, M.; Callicott, J.H.; Mattay, V.S.; Rakow, R.; Shelton-Repella, J.; Post, R.; Weinberger, D.R. Neuronal pathology in the hippocampal area of patients with bipolar disorder: A study with proton magnetic resonance spectroscopic imaging. *Biol. Psychiatry* **2003**, *53*, 906–913. [CrossRef]
123. Dager, S.R.; Friedman, S.D.; Parow, A.; Demopulos, C.; Stoll, A.L.; Lyoo, I.K.; Dunner, D.L.; Renshaw, P.F. Brain Metabolic Alterations in Medication-Free Patients with BipolarDisorder. *Arch. Gen. Psychiatry* **2004**, *61*, 450–458. [CrossRef]
124. Kato, T.; Takahashi, S.; Shioiri, T.; Inubushi, T. Brain phosphorous metabolism in depressive disorders detected by phosphorus-31 magnetic resonance spectroscopy. *J. Affect. Disord.* **1992**, *26*, 223–230. [CrossRef]
125. Kato, T.; Shioiri, T.; Murashita, J.; Hamakawa, H.; Takahashi, Y.; Inubushi, T.; Takahashi, S. Lateralized abnormality of high energy phosphate metabolism in the frontal lobes of patients with bipolar disorder detected by phase-encoded 31P-MRS. *Psychol. Med.* **1995**, *25*, 557–566. [CrossRef] [PubMed]

126. Riesberg, L.A.; Weed, S.A.; McDonald, T.L.; Eckerson, J.M.; Drescher, K.M. Beyond muscles: The untapped potential of creatine. *Int. Immunopharmacol.* **2016**, *37*, 31–42. [CrossRef] [PubMed]

127. Wyss, M.; Schulze, A. Health implications of creatine: Can oral creatine supplementation protect against neurological and atherosclerotic disease? *Neuroscience* **2002**, *112*, 243–260. [CrossRef]

128. Guimarães-Ferreira, L.; Pinheiro, C.H.J.; Gerlinger-Romero, F.; Vitzel, K.F.; Nachbar, R.T.; Curi, R.; Nunes, M.T. Short-term creatine supplementation decreases reactive oxygen species content with no changes in expression and activity of antioxidant enzymes in skeletal muscle. *Eur. J. Appl. Physiol.* **2012**, *112*, 3905–3911. [CrossRef]

129. Hunter, D.J.; Reddy, K.S. Noncommunicable Diseases. *N. Engl. J. Med.* **2013**, *369*, 1336–1343. [CrossRef]

130. Peña-Oyarzun, D.; Bravo-Sagua, R.; Diaz-Vega, A.; Aleman, L.; Chiong, M.; Garcia, L.; Bambs, C.; Troncoso, R.; Cifuentes, M.; Morselli, E.; et al. Autophagy and oxidative stress in non-communicable diseases: A matter of the inflammatory state? *Free Radic. Biol. Med.* **2018**, *124*, 61–78. [CrossRef]

131. Lozano, R.; Naghavi, M.; Foreman, K.; Lim, S.; Shibuya, K.; Aboyans, V.; Abraham, J.; Adair, T.; Aggarwal, R.; Ahn, S.Y.; et al. Global and regional mortality from 235 causes of death for 20 age groups in 1990 and 2010: A systematic analysis for the Global Burden of Disease Study 2010. *Lancet* **2012**, *380*, 2095–2128. [CrossRef]

132. Murray, C.J.L.; Lopez, A.D. Measuring the Global Burden of Disease. *N. Engl. J. Med.* **2013**, *369*, 448–457. [CrossRef]

133. World Health Organization. *Noncommunicable Diseases: Progress Monitor 2020*; World Health Organization: Geneva, Switzerland, 2020.

134. Yu, E.P.K.; Reinhold, J.; Yu, H.; Starks, L.; Uryga, A.K.; Foote, K.; Finigan, A.; Figg, N.; Pung, Y.-F.; Logan, A.; et al. Mitochondrial Respiration Is Reduced in Atherosclerosis, Promoting Necrotic Core Formation and Reducing Relative Fibrous Cap Thickness. *Arterioscler. Thromb. Vasc. Biol.* **2017**, *37*, 2322–2332. [CrossRef]

135. Shimizu, S.; Ishibashi, M.; Kumagai, S.; Wajima, T.; Hiroi, T.; Kurihara, T.; Ishii, M.; Kiuchi, Y. Decreased cardiac mitochondrial tetrahydrobiopterin in a rat model of pressure overload. *Int. J. Mol. Med.* **2013**, *31*, 589–596. [CrossRef]

136. Tang, Y.; Mi, C.; Liu, J.; Gao, F.; Long, J. Compromised mitochondrial remodeling in compensatory hypertrophied myocardium of spontaneously hypertensive rat. *Cardiovasc. Pathol.* **2014**, *23*, 101–106. [CrossRef] [PubMed]

137. Walther, T.; Tschöpe, C.; Sterner-Kock, A.; Westermann, D.; Heringer-Walther, S.; Riad, A.; Nikolic, A.; Wang, Y.; Ebermann, L.; Siems, W.-E.; et al. Accelerated Mitochondrial Adenosine Diphosphate/Adenosine Triphosphate Transport Improves Hypertension-Induced Heart Disease. *Circulation* **2007**, *115*, 333–344. [CrossRef] [PubMed]

138. Yu, T.; Robotham, J.L.; Yoon, Y. Increased production of reactive oxygen species in hyperglycemic conditions requires dynamic change of mitochondrial morphology. *Proc. Natl. Acad. Sci. USA* **2006**, *103*, 2653–2658. [CrossRef] [PubMed]

139. Tormos, K.V.; Anso, E.; Hamanaka, R.B.; Eisenbart, J.; Joseph, J.; Kalyanaraman, B.; Chandel, N.S. Mitochondrial Complex III ROS Regulate Adipocyte Differentiation. *Cell Metab.* **2011**, *11*, 537–544. [CrossRef]

140. Teodoro, J.S.; Rolo, A.P.; Duarte, F.V.; Simões, A.M.; Palmeira, C.M. Differential alterations in mitochondrial function induced by a choline-deficient diet: Understanding fatty liver disease progression. *Mitochondrion* **2008**, *8*, 367–376. [CrossRef]

141. Galloway, C.A.; Lee, H.; Brookes, P.S.; Yoon, Y. Decreasing mitochondrial fission alleviates hepatic steatosis in a murine model of nonalcoholic fatty liver disease. *Am. J. Physiol. Gastrointest. Liver Physiol.* **2014**, *307*, G632–G641. [CrossRef]

142. Tubbs, E.; Chanon, S.; Robert, M.; Bendridi, N.; Bidaux, G.; Chauvin, M.-A.; Ji-Cao, J.; Durand, C.; Gauvrit-Ramette, D.; Vidal, H.; et al. Disruption of Mitochondria-Associated Endoplasmic Reticulum Membrane (MAM) Integrity Contributes to Muscle Insulin Resistance in Mice and Humans. *Diabetes* **2018**, *67*, 636–650. [CrossRef]

143. Fazakerley, D.J.; Minard, A.Y.; Krycer, J.R.; Thomas, K.C.; Stöckli, J.; Harney, D.J.; Burchfield, J.G.; Maghzal, G.J.; Caldwell, S.T.; Hartley, R.C.; et al. Mitochondrial oxidative stress causes insulin resistance without disrupting oxidative phosphorylation. *J. Biol. Chem.* **2018**, *293*, 7315–7328. [CrossRef]

144. Anderson, E.J.; Lustig, M.E.; Boyle, K.E.; Woodlief, T.L.; Kane, D.A.; Lin, C.-T.; Price, J.W.; Kang, L.; Rabinovitch, P.S.; Szeto, H.H.; et al. Mitochondrial H_2O_2 emission and cellular redox state link excess fat intake to insulin resistance in both rodents and humans. *J. Clin. Investig.* **2009**, *119*, 573–581. [CrossRef]

145. Gutiérrez, T.; Parra, V.; Troncoso, R.; Pennanen, C.; Contreras-Ferrat, A.; Vasquez-Trincado, C.; Morales, P.E.; Lopez-Crisosto, C.; Sotomayor-Flores, C.; Chiong, M.; et al. Alteration in mitochondrial Ca^{2+} uptake disrupts insulin signaling in hypertrophic cardiomyocytes. *Cell Commun. Signal.* **2014**, *12*, 68. [CrossRef]

146. Tubbs, E.; Theurey, P.; Vial, G.; Bendridi, N.; Bravard, A.; Chauvin, M.-A.; Ji-Cao, J.; Zoulim, F.; Bartosch, B.; Ovize, M.; et al. Mitochondria-Associated Endoplasmic Reticulum Membrane (MAM) Integrity Is Required for Insulin Signaling and Is Implicated in Hepatic Insulin Resistance. *Diabetes* **2014**, *63*, 3279–3294. [CrossRef] [PubMed]

147. Zhang, Z.; Wakabayashi, N.; Wakabayashi, J.; Tamura, Y.; Song, W.-J.; Sereda, S.; Clerc, P.; Polster, B.M.; Aja, S.M.; Pletnikov, M.V.; et al. The dynamin-related GTPase Opa1 is required for glucose-stimulated ATP production in pancreatic beta cells. *Mol. Biol. Cell* **2011**, *22*, 2235–2245. [CrossRef] [PubMed]

148. Reinhardt, F.; Schultz, J.; Waterstradt, R.; Baltrusch, S. Drp1 guarding of the mitochondrial network is important for glucose-stimulated insulin secretion in pancreatic beta cells. *Biochem. Biophys. Res. Commun.* **2016**, *474*, 646–651. [CrossRef] [PubMed]

149. Anello, M.; Lupi, R.; Spampinato, D.; Piro, S.; Masini, M.; Boggi, U.; Del Prato, S.; Rabuazzo, A.M.; Purrello, F.; Marchetti, P. Functional and morphological alterations of mitochondria in pancreatic beta cells from type 2 diabetic patients. *Diabetologia* **2005**, *48*, 282–289. [CrossRef]

150. Scheuermann-Freestone, M.; Madsen, P.L.; Manners, D.; Blamire, A.M.; Buckingham, R.E.; Styles, P.; Radda, G.K.; Neubauer, S.; Clarke, K. Abnormal cardiac and skeletal muscle energy metabolism in patients with type 2 diabetes. *Circulation* **2003**, *107*, 3040–3046. [CrossRef]

151. Rider, O.J.; Francis, J.M.; Ali, M.K.; Holloway, C.; Pegg, T.; Robson, M.D.; Tyler, D.; Byrne, J.; Clarke, K.; Neubauer, S. Effects of catecholamine stress on diastolic function and myocardial energetics in obesity. *Circulation* **2012**, *125*, 1511–1519. [CrossRef]

152. Lamb, H.J.; Beyerbacht, H.P.; Van der Laarse, A.; Stoel, B.C.; Doornbos, J.; Van der Wall, E.E.; De Roos, A. Diastolic dysfunction in hypertensive heart disease is associated with altered myocardial metabolism. *Circulation* **1999**, *99*, 2261–2267. [CrossRef]

153. Guescini, M.; Tiano, L.; Genova, M.L.; Polidori, E.; Silvestri, S.; Orlando, P.; Fimognari, C.; Calcabrini, C.; Stocchi, V.; Sestili, P. The Combination of Physical Exercise with Muscle-Directed Antioxidants to Counteract Sarcopenia: A Biomedical Rationale for Pleiotropic Treatment with Creatine and Coenzyme Q10. *Oxidative Med. Cell. Longev.* **2017**, *2017*, e7083049. [CrossRef]

154. Alves, C.R.R.; Filho, C.A.A.M.; Benatti, F.B.; Brucki, S.; Pereira, R.M.R.; Pinto, A.L.d.S.; Lima, F.R.; Roschel, H.; Gualano, B. Creatine Supplementation Associated or Not with Strength Training upon Emotional and Cognitive Measures in Older Women: A Randomized Double-Blind Study. *PLoS ONE* **2013**, *8*, e76301. [CrossRef]

155. Candow, D.G.; Forbes, S.C.; Chilibeck, P.D.; Cornish, S.M.; Antonio, J.; Kreider, R.B. Variables Influencing the Effectiveness of Creatine Supplementation as a Therapeutic Intervention for Sarcopenia. *Front. Nutr.* **2019**, *6*, 124. [CrossRef]

156. Gualano, B.; Macedo, A.R.; Alves, C.R.; Roschel, H.; Benatti, F.B.; Takayama, L.; De Sa Pinto, A.L.; Lima, F.R.; Pereira, R.M. Creatine supplementation and resistance training in vulnerable older women: A randomized double-blind placebo-controlled clinical trial. *Exp. Gerontol.* **2014**, *53*, 7–15. [CrossRef] [PubMed]

157. Lobo, D.M.; Tritto, A.C.; Da Silva, L.R.; De Oliveira, P.B.; Benatti, F.B.; Roschel, H.; Nieß, B.; Gualano, B.; Pereira, R.M.R. Effects of long-term low-dose dietary creatine supplementation in older women. *Exp. Gerontol.* **2015**, *70*, 97–104. [CrossRef] [PubMed]

158. Pinto, C.L.; Botelho, P.B.; Carneiro, J.A.; Mota, J.F. Impact of creatine supplementation in combination with resistance training on lean mass in the elderly. *J. Cach. Sarc. Muscle* **2016**, *7*, 413–421. [CrossRef] [PubMed]

159. Gualano, B.; Artioli, G.G.; Poortmans, J.R.; Lancha Junior, A.H. Exploring the therapeutic role of creatine supplementation. *Amino Acids* **2010**, *38*, 31–44. [CrossRef]

160. Candow, D.G.; Vogt, E.; Johannsmeyer, S.; Forbes, S.C.; Farthing, J.P. Strategic creatine supplementation and resistance training in healthy older adults. *Appl. Physiol. Nutr. Metab.* **2015**, *40*, 689–694. [CrossRef]

161. De Sousa, M.V.; Da Silva Soares, D.B.; Caraça, E.R.; Cardoso, R. Dietary protein and exercise for preservation of lean mass and perspectives on type 2 diabetes prevention. *Exp. Biol. Med.* **2019**, *244*, 992–1004. [CrossRef]

162. Barney, B.; Beck, G.R. Nutrition Interventions in Heart Failure. In *Manual of Heart Failure Management*; Bisognano, J.D., Earley, M.B., Baker, M.L., Eds.; Springer: Berlin/Heidelberg, Germany, 2009; pp. 207–217.

163. Solis, M.Y.; Artioli, G.G.; Gualano, B. Potential of Creatine in Glucose Management and Diabetes. *Nutrients* **2021**, *13*, 570. [CrossRef]

164. Gualano, B.; De Salles Painneli, V.; Roschel, H.; Artioli, G.G.; Neves, M.; De Sá Pinto, A.L.; Da Silva, M.E.R.; Cunha, M.R.; Otaduy, M.C.G.; Leite, C.D.C.; et al. Creatine in type 2 diabetes: A randomized, double-blind, placebo-controlled trial. *Med. Sci. Sports Exerc.* **2011**, *43*, 770–778. [CrossRef]

165. Earnest, C.P.; Almada, A.L.; Mitchell, T.L. High-performance capillary electrophoresis-pure creatine monohydrate reduces blood lipids in men and women. *Clin. Sci.* **1996**, *91*, 113–118. [CrossRef]

166. Deminice, R.; De Castro, G.S.F.; Francisco, L.V.; Da Silva, L.E.C.M.; Cardoso, J.F.R.; Frajacomo, F.T.T.; Teodoro, B.G.; Dos Reis Silveira, L.; Jordao, A.A. Creatine supplementation prevents fatty liver in rats fed choline-deficient diet: A burden of one-carbon and fatty acid metabolism. *J. Nutr. Biochem.* **2015**, *26*, 391–397. [CrossRef]

167. Gupta, A.; Akki, A.; Wang, Y.; Leppo, M.K.; Chacko, V.P.; Foster, D.B.; Caceres, V.; Shi, S.; Kirk, J.A.; Su, J.; et al. Creatine kinase-mediated improvement of function in failing mouse hearts provides causal evidence the failing heart is energy starved. *J. Clin. Investig.* **2012**, *122*, 291–302. [CrossRef] [PubMed]

168. Wallis, J.; Lygate, C.A.; Fischer, A.; ten Hove, M.; Schneider, J.E.; Sebag-Montefiore, L.; Dawson, D.; Hulbert, K.; Zhang, W.; Zhang, M.H.; et al. Supranormal Myocardial Creatine and Phosphocreatine Concentrations Lead to Cardiac Hypertrophy and Heart Failure. *Circulation* **2005**, *112*, 3131–3139. [CrossRef] [PubMed]

169. Nascimben, L.; Ingwall, J.S.; Pauletto, P.; Friedrich, J.; Gwathmey, J.K.; Saks, V.; Pessina, A.C.; Allen, P.d. Creatine Kinase System in Failing and Nonfailing Human Myocardium. *Circulation* **1996**, *94*, 1894–1901. [CrossRef] [PubMed]

170. Neubauer, S. The Failing Heart—An Engine Out of Fuel. *N. Engl. J. Med.* **2007**, *356*, 1140–1151. [CrossRef] [PubMed]

171. Shen, W.; Spindler, M.; Higgins, M.A.; Jin, N.; Gill, R.M.; Bloem, L.J.; Ryan, T.P.; Ingwall, J.S. The fall in creatine levels and creatine kinase isozyme changes in the failing heart are reversible: Complex post-transcriptional regulation of the components of the CK system. *J. Mol. Cell. Cardiol.* **2005**, *39*, 537–544. [CrossRef] [PubMed]

172. Lygate, C.A.; Fischer, A.; Sebag-Montefiore, L.; Wallis, J.; ten Hove, M.; Neubauer, S. The creatine kinase energy transport system in the failing mouse heart. *J. Mol. Cell. Cardiol.* **2007**, *42*, 1129–1136. [CrossRef]

173. Liao, R.; Nascimben, L.; Friedrich, J.; Gwathmey, J.K.; Ingwall, J.S. Decreased Energy Reserve in an Animal Model of Dilated Cardiomyopathy. *Circ. Res.* **1996**, *78*, 893–902. [CrossRef]

174. Zervou, S.; Whittington, H.J.; Russell, A.J.; Lygate, C.A. Augmentation of Creatine in the Heart. *Mini Rev. Med. Chem.* **2016**, *16*, 19–28. [CrossRef]

175. Neubauer, S.; Krahe, T.; Schindler, R.; Horn, M.; Hillenbrand, H.; Entzeroth, C.; Mader, H.; Kromer, E.P.; Riegger, G.A.; Lackner, K. 31P magnetic resonance spectroscopy in dilated cardiomyopathy and coronary artery disease. Altered cardiac high-energy phosphate metabolism in heart failure. *Circulation* **1992**, *86*, 1810–1818. [CrossRef]
176. Neubauer, S.; Horn, M.; Pabst, T.; Gödde, M.; Lübke, D.; Jilling, B.; Hahn, D.; Ertl, G. Contributions of 31P-magnetic resonance spectroscopy to the understanding of dilated heart muscle disease. *Eur. Heart J.* **1995**, *16*, 115–118. [CrossRef]
177. Neubauer, S.; Horn, M.; Cramer, M.; Harre, K.; Newell, J.B.; Peters, W.; Pabst, T.; Ertl, G.; Hahn, D.; Ingwall, J.S.; et al. Myocardial Phosphocreatine-to-ATP Ratio Is a Predictor of Mortality in Patients with Dilated Cardiomyopathy. *Circulation* **1997**, *96*, 2190–2196. [CrossRef] [PubMed]
178. Horn, M.; Remkes, H.; Dienesch, C.; Hu, K.; Ertl, G.; Neubauer, S. Chronic high-dose creatine feeding does not attenuate left ventricular remodeling in rat hearts post-myocardial infarction. *Cardiovasc. Res.* **1999**, *43*, 117–124. [CrossRef]
179. McClung, J.; Hand, G.; Davis, J.; Carson, J. Effect of creatine supplementation on cardiac muscle of exercise-stressed rats. *Eur. J. Appl. Physiol.* **2003**, *89*, 26–33. [CrossRef] [PubMed]
180. Bo, H.; Jiang, N.; Ma, G.; Qu, J.; Zhang, G.; Cao, D.; Wen, L.; Liu, S.; Ji, L.L.; Zhang, Y. Regulation of mitochondrial uncoupling respiration during exercise in rat heart: Role of reactive oxygen species (ROS) and uncoupling protein 2. *Free Radic. Biol. Med.* **2008**, *44*, 1373–1381. [CrossRef]
181. Cao, F.; Zervou, S.; Lygate, C.A. The creatine kinase system as a therapeutic target for myocardial ischaemia–reperfusion injury. *Biochem. Soc. Trans.* **2018**, *46*, 1119–1127. [CrossRef] [PubMed]
182. Hultman, J.; Ronquist, G.; Forsberg, J.O.; Hansson, H.E. Myocardial energy restoration of ischemic damage by administration of phosphoenolpyruvate during reperfusion. A study in a paracorporeal rat heart model. *Eur. Surg. Res.* **1983**, *15*, 200–207. [CrossRef] [PubMed]
183. Osbakken, M.; Ito, K.; Zhang, D.; Ponomarenko, I.; Ivanics, T.; Jahngen, E.G.; Cohn, M. Creatine and cyclocreatine effects on ischemic myocardium: 31P nuclear magnetic resonance evaluation of intact heart. *Cardiology* **1992**, *80*, 184–195. [CrossRef]
184. Sharov, V.G.; Saks, V.A.; Kupriyanov, V.V.; Lakomkin, V.L.; Kapelko, V.I.; Steinschneider, A.Y.; Javadov, S.A. Protection of ischemic myocardium by exogenous phosphocreatine. I. Morphologic and phosphorus 31-nuclear magnetic resonance studies. *J. Thorac. Cardiovasc. Surg.* **1987**, *94*, 749–761. [CrossRef]
185. Balestrino, M.; Sarocchi, M.; Adriano, E.; Spallarossa, P. Potential of creatine or phosphocreatine supplementation in cerebrovascular disease and in ischemic heart disease. *Amino Acids* **2016**, *48*, 1955–1967. [CrossRef]
186. ten Hove, M.; Lygate, C.A.; Fischer, A.; Schneider, J.E.; Sang, A.E.; Hulbert, K.; Sebag-Montefiore, L.; Watkins, H.; Clarke, K.; Isbrandt, D.; et al. Reduced Inotropic Reserve and Increased Susceptibility to Cardiac Ischemia/Reperfusion Injury in Phosphocreatine-Deficient Guanidinoacetate-N-Methyltransferase–Knockout Mice. *Circulation* **2005**, *111*, 2477–2485. [CrossRef]
187. Elgebaly, S.A.; Wei, Z.; Tyles, E.; Elkerm, A.F.; Houser, S.L.; Gillies, C.; Kaddurah-Daouk, R. Enhancement of the recovery of rat hearts after prolonged cold storage by cyclocreatine phosphate. *Transplantation* **1994**, *57*, 803–806. [CrossRef] [PubMed]
188. Cisowski, M.; Bochenek, A.; Kucewicz, E.; Wnuk-Wojnar, A.M.; Morawski, W.; Skalski, J.; Grzybek, H. The use of exogenous creatine phosphate for myocardial protection in patients undergoing coronary artery bypass surgery. *J. Cardiovasc. Surg.* **1996**, *37*, 75–80.
189. Ruda, M.Y.; Samarenko, M.B.; Afonskaya, N.I.; Saks, V.A. Reduction of ventricular arrhythmias by phosphocreatine (Neoton) in patients with acute myocardial infarction. *Am. Heart J.* **1988**, *116*, 393–397. [CrossRef]
190. Elgebaly, S.A.; Poston, R.; Todd, R.; Helmy, T.; Almaghraby, A.M.; Elbayoumi, T.; Kreutzer, D.L. Cyclocreatine protects against ischemic injury and enhances cardiac recovery during early reperfusion. *Expert Rev. Cardiovasc. Ther.* **2019**, *17*, 683–697. [CrossRef]
191. Roberts, J.J.; Walker, J.B. Feeding a creatine analogue delays ATP depletion and onset of rigor in ischemic heart. *Am. J. Physiol. Heart Circ. Physiol.* **1982**, *243*, H911–H916. [CrossRef]
192. Chida, K.; Otani, H.; Kohzuki, M.; Saito, H.; Kagaya, Y.; Takai, Y.; Takahashi, S.; Yamada, S.; Zuguchi, M. The Relationship between Plasma BNP Level and the Myocardial Phosphocreatine/Adenosine Triphosphate Ratio Determined by Phosphorus-31 Magnetic Resonance Spectroscopy in Patients with Dilated Cardiomyopathy. *Cardiology* **2006**, *106*, 132–136. [CrossRef]
193. Russo, E.; Nguyen, H.; Lippert, T.; Tuazon, J.; Borlongan, C.V.; Napoli, E. Mitochondrial targeting as a novel therapy for stroke. *Brain Circ.* **2018**, *4*, 84–94. [CrossRef]
194. Soustiel, J.F.; Zaaroor, M. Mitochondrial targeting for development of novel drug strategies in brain injury. *Cent. Nerv. Syst. Agents Med. Chem.* **2012**, *12*, 131–145. [CrossRef]
195. Niizuma, K.; Endo, H.; Chan, P.H. Oxidative stress and mitochondrial dysfunction as determinants of ischemic neuronal death and survival. *J. Neurochem.* **2009**, *109*, 133–138. [CrossRef]
196. Niizuma, K.; Yoshioka, H.; Chen, H.; Kim, G.S.; Jung, J.E.; Katsu, M.; Okami, N.; Chan, P.H. Mitochondrial and apoptotic neuronal death signaling pathways in cerebral ischemia. *Biochim. Biophys. Acta (BBA) Mol. Basis Dis.* **2010**, *1802*, 92–99. [CrossRef]
197. Lin, M.T.; Beal, M.F. Mitochondrial dysfunction and oxidative stress in neurodegenerative diseases. *Nature* **2006**, *443*, 787–795. [CrossRef] [PubMed]
198. He, Z.; Ning, N.; Zhou, Q.; Khoshnam, S.E.; Farzaneh, M. Mitochondria as a therapeutic target for ischemic stroke. *Free Radic. Biol. Med.* **2020**, *146*, 45–58. [CrossRef] [PubMed]
199. Nguyen, H.; Zarriello, S.; Rajani, M.; Tuazon, J.; Napoli, E.; Borlongan, C.V. Understanding the Role of Dysfunctional and Healthy Mitochondria in Stroke Pathology and Its Treatment. *Int. J. Mol. Sci.* **2018**, *19*, 2127. [CrossRef] [PubMed]

200. Sanderson, T.H.; Reynolds, C.A.; Kumar, R.; Przyklenk, K.; Hüttemann, M. Molecular Mechanisms of Ischemia–Reperfusion Injury in Brain: Pivotal Role of the Mitochondrial Membrane Potential in Reactive Oxygen Species Generation. *Mol. Neurobiol.* **2013**, *47*, 9–23. [CrossRef]

201. Chouchani, E.T.; Pell, V.R.; James, A.M.; Work, L.M.; Saeb-Parsy, K.; Frezza, C.; Krieg, T.; Murphy, M.P. A Unifying Mechanism for Mitochondrial Superoxide Production during Ischemia-Reperfusion Injury. *Cell Metab.* **2016**, *23*, 254–263. [CrossRef]

202. Andrabi, S.S.; Parvez, S.; Tabassum, H. Ischemic stroke and mitochondria: Mechanisms and targets. *Protoplasma* **2020**, *257*, 335–343. [CrossRef]

203. Nicholls, D.G. Mitochondrial calcium function and dysfunction in the central nervous system. *Biochim. Biophys. Acta* **2009**, *1787*, 1416–1424. [CrossRef]

204. Blennow, K.; Hardy, J.; Zetterberg, H. The neuropathology and neurobiology of traumatic brain injury. *Neuron* **2012**, *76*, 886–899. [CrossRef]

205. Turner, C.E.; Byblow, W.D.; Gant, N. Creatine supplementation enhances corticomotor excitability and cognitive performance during oxygen deprivation. *J. Neurosci.* **2015**, *35*, 1773–1780. [CrossRef]

206. Zhu, S.; Li, M.; Figueroa, B.E.; Liu, A.; Stavrovskaya, I.G.; Pasinelli, P.; Beal, M.F.; Brown, R.H.; Kristal, B.S.; Ferrante, R.J.; et al. Prophylactic creatine administration mediates neuroprotection in cerebral ischemia in mice. *J. Neurosci* **2004**, *24*, 5909–5912. [CrossRef]

207. Hausmann, O.N.; Fouad, K.; Wallimann, T.; Schwab, M.E. Protective effects of oral creatine supplementation on spinal cord injury in rats. *Spin. Cord* **2002**, *40*, 449–456. [CrossRef] [PubMed]

208. Sullivan, P.G.; Geiger, J.D.; Mattson, M.P.; Scheff, S.W. Dietary supplement creatine protects against traumatic brain injury. *Ann. Neurol.* **2000**, *48*, 723–729. [CrossRef]

209. Prass, K.; Royl, G.; Lindauer, U.; Freyer, D.; Megow, D.; Dirnagl, U.; Stöckler-Ipsiroglu, G.; Wallimann, T.; Priller, J. Improved reperfusion and neuroprotection by creatine in a mouse model of stroke. *J. Cereb. Blood Flow Metab.* **2007**, *27*, 452–459. [CrossRef] [PubMed]

210. Freire Royes, L.F.; Cassol, G. The Effects of Creatine Supplementation and Physical Exercise on Traumatic Brain Injury. *Mini Rev. Med. Chem.* **2016**, *16*, 29–39. [CrossRef] [PubMed]

211. Novelli, A.; Reilly, J.A.; Lysko, P.G.; Henneberry, R.C. Glutamate becomes neurotoxic via the N-methyl-d-aspartate receptor when intracellular energy levels are reduced. *Brain Res.* **1988**, *451*, 205–212. [CrossRef]

212. Tsuji, K.; Nakamura, Y.; Ogata, T.; Shibata, T.; Kataoka, K. Rapid decrease in ATP content without recovery phase during glutamate-induced cell death in cultured spinal neurons. *Brain Res.* **1994**, *662*, 289–292. [CrossRef]

213. Carter, A.J.; Müller, R.E.; Pschorn, U.; Stransky, W. Preincubation with Creatine Enhances Levels of Creatine Phosphate and Prevents Anoxic Damage in Rat Hippocampal Slices. *J. Neurochem.* **1995**, *64*, 2691–2699. [CrossRef]

214. Brustovetsky, N.; Brustovetsky, T.; Dubinsky, J.M. On the mechanisms of neuroprotection by creatine and phosphocreatine. *J. Neurochem.* **2001**, *76*, 425–434. [CrossRef]

215. Harman, D. The aging process. *Proc. Natl. Acad. Sci. USA* **1981**, *78*, 7124–7128. [CrossRef]

216. Grimm, A.; Friedland, K.; Eckert, A. Mitochondrial dysfunction: The missing link between aging and sporadic Alzheimer's disease. *Biogerontology* **2016**, *17*, 281–296. [CrossRef]

217. Leuner, K.; Hauptmann, S.; Abdel-Kader, R.; Scherping, I.; Keil, U.; Strosznajder, J.B.; Eckert, A.; Müller, W.E. Mitochondrial dysfunction: The first domino in brain aging and Alzheimer's disease? *Antiox. Redox Signal.* **2007**, *9*, 1659–1675. [CrossRef] [PubMed]

218. Bishop, N.A.; Lu, T.; Yankner, B.A. Neural mechanisms of ageing and cognitive decline. *Nature* **2010**, *464*, 529–535. [CrossRef] [PubMed]

219. Grimm, A.; Eckert, A. Brain aging and neurodegeneration: From a mitochondrial point of view. *J. Neurochem.* **2017**, *143*, 418–431. [CrossRef] [PubMed]

220. Geary, D.C. Mitochondrial Functioning and the Relations among Health, Cognition, and Aging: Where Cell Biology Meets Cognitive Science. *Int. J. Mol. Sci.* **2021**, *22*, 3562. [CrossRef] [PubMed]

221. González-Reyes, R.E.; Aliev, G.; Ávila-Rodrigues, M.; Barreto, G.E. Alterations in Glucose Metabolism on Cognition: A Possible Link Between Diabetes and Dementia. *Curr. Pharm. Des.* **2016**, *22*, 812–818. [CrossRef] [PubMed]

222. Shieh, J.C.-C.; Huang, P.-T.; Lin, Y.-F. Alzheimer's Disease and Diabetes: Insulin Signaling as the Bridge Linking Two Pathologies. *Mol. Neurobiol.* **2020**, *57*, 1966–1977. [CrossRef] [PubMed]

223. Cao, Y.; Yan, Z.; Zhou, T.; Wang, G. SIRT1 Regulates Cognitive Performance and Ability of Learning and Memory in Diabetic and Nondiabetic Models. *J. Diabetes. Res.* **2017**, *2017*, 7121827. [CrossRef]

224. Jo, D.; Kim, B.C.; Cho, K.A.; Song, J. The Cerebral Effect of Ammonia in Brain Aging: Blood-Brain Barrier Breakdown, Mitochondrial Dysfunction, and Neuroinflammation. *J. Clin. Med.* **2021**, *10*, 2773. [CrossRef]

225. Bustamante, J.; Czerniczyniec, A.; Lores-Arnaiz, S. Brain nitric oxide synthases and mitochondrial function. *Front. Biosci.* **2007**, *12*, 1034–1040. [CrossRef]

226. Felipo, V.; Butterworth, R.F. Mitochondrial dysfunction in acute hyperammonemia. *Neurochem. Int.* **2002**, *40*, 487–491. [CrossRef]

227. De la Monte, S.M.; Wands, J.R. Alzheimer's Disease is Type 3 Diabetes—Evidence Reviewed. *J. Diabetes Sci. Technol.* **2008**, *2*, 1101–1113. [CrossRef] [PubMed]

228. Sripetchwandee, J.; Chattipakorn, N.; Chattipakorn, S.C. Links Between Obesity-Induced Brain Insulin Resistance, Brain Mitochondrial Dysfunction, and Dementia. *Front. Endocrinol.* **2018**, *9*, 496. [CrossRef] [PubMed]
229. Müller, W.E.; Eckert, A.; Kurz, C.; Eckert, G.P.; Leuner, K. Mitochondrial dysfunction: Common final pathway in brain aging and Alzheimer's disease—Therapeutic aspects. *Mol. Neurobiol.* **2010**, *41*, 159–171. [CrossRef]
230. Kaliszewska, A.; Allison, J.; Martini, M.; Arias, N. The Interaction of Diet and Mitochondrial Dysfunction in Aging and Cognition. *Int. J. Mol. Sci.* **2021**, *22*, 3574. [CrossRef] [PubMed]
231. Francis, H.M.; Stevenson, R.J. Potential for diet to prevent and remediate cognitive deficits in neurological disorders. *Nutr. Rev.* **2018**, *76*, 204–217. [CrossRef] [PubMed]
232. Head, E. Oxidative damage and cognitive dysfunction: Antioxidant treatments to promote healthy brain aging. *Neurochem. Res.* **2009**, *34*, 670–678. [CrossRef]
233. Poddar, J.; Pradhan, M.; Ganguly, G.; Chakrabarti, S. Biochemical deficits and cognitive decline in brain aging: Intervention by dietary supplements. *J. Chem. Neuroanat.* **2019**, *95*, 70–80. [CrossRef]
234. Hammett, S.T.; Wall, M.B.; Edwards, T.C.; Smith, A.T. Dietary supplementation of creatine monohydrate reduces the human fMRI BOLD signal. *Neurosci. Lett.* **2010**, *479*, 201–205. [CrossRef]
235. Watanabe, A.; Kato, N.; Kato, T. Effects of creatine on mental fatigue and cerebral hemoglobin oxygenation. *Neurosci. Res.* **2002**, *42*, 279–285. [CrossRef]
236. McMorris, T.; Harris, R.C.; Howard, A.N.; Langridge, G.; Hall, B.; Corbett, J.; Dicks, M.; Hodgson, C. Creatine supplementation, sleep deprivation, cortisol, melatonin and behavior. *Physiol. Behav.* **2007**, *90*, 21–28. [CrossRef]
237. Dworak, M.; Kim, T.; McCarley, R.W.; Basheer, R. Creatine supplementation reduces sleep need and homeostatic sleep pressure in rats. *J. Sleep Res.* **2017**, *26*, 377–385. [CrossRef] [PubMed]
238. Gibson, G.E.; Starkov, A.; Blass, J.P.; Ratan, R.R.; Beal, M.F. Cause and consequence: Mitochondrial dysfunction initiates and propagates neuronal dysfunction, neuronal death and behavioral abnormalities in age-associated neurodegenerative diseases. *Biochim. Biophys. Acta* **2010**, *1802*, 122–134. [CrossRef] [PubMed]
239. Valenti, D.; De Bari, L.; De Filippis, B.; Henrion-Caude, A.; Vacca, R.A. Mitochondrial dysfunction as a central actor in intellectual disability-related diseases: An overview of Down syndrome, autism, Fragile X and Rett syndrome. *Neurosci. Biobehav. Rev.* **2014**, *46*, 202–217. [CrossRef] [PubMed]
240. Adhihetty, P.J.; Beal, M.F. Creatine and its potential therapeutic value for targeting cellular energy impairment in neurodegenerative diseases. *Neuromol. Med.* **2008**, *10*, 275–290. [CrossRef]
241. Kley, R.A.; Tarnopolsky, M.A.; Vorgerd, M. Creatine for treating muscle disorders. *Cochrane Database Syst. Rev.* **2013**, *2013*, CD004760. [CrossRef]
242. Shao, A.; Lin, D.; Wang, L.; Tu, S.; Lenahan, C.; Zhang, J. Oxidative Stress at the Crossroads of Aging, Stroke and Depression. *Aging Dis.* **2020**, *11*, 1537–1566. [CrossRef]
243. Martin, E.I.; Ressler, K.J.; Binder, E.; Nemeroff, C.B. The Neurobiology of Anxiety Disorders: Brain Imaging, Genetics, and Psychoneuroendocrinology. *Psychiatr. Clin.* **2009**, *32*, 549–575. [CrossRef]
244. Yildiz-Yesiloglu, A.; Ankerst, D.P. Review of 1H magnetic resonance spectroscopy findings in major depressive disorder: A meta-analysis. *Psychiatry Res. Neuroimag.* **2006**, *147*, 1–25. [CrossRef]
245. Scaglia, F. The role of mitochondrial dysfunction in psychiatric disease. *Dev. Disabil. Res. Rev.* **2010**, *16*, 136–143. [CrossRef]
246. Agren, H.; Niklasson, F. Creatinine and creatine in CSF: Indices of brain energy metabolism in depression. Short note. *J. Neural. Transm.* **1988**, *74*, 55–59. [CrossRef]
247. Mirza, Y.; O'Neill, J.; Smith, E.A.; Russell, A.; Smith, J.M.; Banerjee, S.P.; Bhandari, R.; Boyd, C.; Rose, M.; Ivey, J.; et al. Increased medial thalamic creatine-phosphocreatine found by proton magnetic resonance spectroscopy in children with obsessive-compulsive disorder versus major depression and healthy controls. *J. Child Neurol.* **2006**, *21*, 106–111. [CrossRef] [PubMed]
248. Wood, E.; Hall, K.H.; Tate, W. Role of mitochondria, oxidative stress and the response to antioxidants in myalgic encephalomyelitis/chronic fatigue syndrome: A possible approach to SARS-CoV-2 "long-haulers"? *Chronic Dis. Transl. Med.* **2021**, *7*, 14–26. [CrossRef] [PubMed]
249. Frye, M.A.; Watzl, J.; Banakar, S.; O'Neill, J.; Mintz, J.; Davanzo, P.; Fischer, J.; Chirichigno, J.W.; Ventura, J.; Elman, S.; et al. Increased anterior cingulate/medial prefrontal cortical glutamate and creatine in bipolar depression. *Neuropsychopharmacology* **2007**, *32*, 2490–2499. [CrossRef] [PubMed]
250. Kondo, D.G.; Sung, Y.-H.; Hellem, T.L.; Fiedler, K.K.; Shi, X.; Jeong, E.-K.; Renshaw, P.F. Open-label adjunctive creatine for female adolescents with SSRI-resistant major depressive disorder: A 31-phosphorus magnetic resonance spectroscopy study. *J. Affect. Disord.* **2011**, *135*, 354–361. [CrossRef]
251. Roitman, S.; Green, T.; Osher, Y.; Karni, N.; Levine, J. Creatine monohydrate in resistant depression: A preliminary study. *Bipolar Disord.* **2007**, *9*, 754–758. [CrossRef]
252. Toniolo, R.A.; Silva, M.; Fernandes, F.d.B.F.; Amaral, J.A.d.M.S.; Dias, R.d.S.; Lafer, B. A randomized, double-blind, placebo-controlled, proof-of-concept trial of creatine monohydrate as adjunctive treatment for bipolar depression. *J. Neural. Transm.* **2018**, *125*, 247–257. [CrossRef]
253. Kious, B.M.; Kondo, D.G.; Renshaw, P.F. Creatine for the Treatment of Depression. *Biomolecules* **2019**, *9*, 406. [CrossRef]
254. Pazini, F.L.; Cunha, M.P.; Rodrigues, A.L.S. The possible beneficial effects of creatine for the management of depression. *Prog. Neuropsychopharmacol. Biol. Psychiatry* **2019**, *89*, 193–206. [CrossRef]

255. Kondo, D.G.; Forrest, L.N.; Shi, X.; Sung, Y.-H.; Hellem, T.L.; Huber, R.S.; Renshaw, P.F. Creatine target engagement with brain bioenergetics: A dose-ranging phosphorus-31 magnetic resonance spectroscopy study of adolescent females with SSRI-resistant depression. *Amino Acids* **2016**, *48*, 1941–1954. [CrossRef]

256. D'Anci, K.E.; Allen, P.J.; Kanarek, R.B. A potential role for creatine in drug abuse? *Mol. Neurobiol.* **2011**, *44*, 136–141. [CrossRef]

257. Amital, D.; Vishne, T.; Roitman, S.; Kotler, M.; Levine, J. Open Study of Creatine Monohydrate in Treatment-Resistant Posttraumatic Stress Disorder. *J. Clin. Psychiatry* **2006**, *67*, 836–837. [CrossRef] [PubMed]

258. Allen, P.J. Creatine metabolism and psychiatric disorders: Does creatine supplementation have therapeutic value? *Neurosci. Biobehav. Rev.* **2012**, *36*, 1442–1462. [CrossRef] [PubMed]

259. Rosenthal, T.C.; Majeroni, B.A.; Pretorius, R.; Malik, K. Fatigue: An overview. *Am. Fam. Physician* **2008**, *78*, 1173–1179.

260. Jamal, G.A.; Hansen, S. Post-Viral Fatigue Syndrome: Evidence for Underlying Organic Disturbance in the Muscle Fibre. *Eur. Neurol.* **1989**, *29*, 273–276. [CrossRef] [PubMed]

261. Edwards, R.H.T.; Newham, D.J.; Peters, T.J. Muscle biochemistry and pathophysiology in postviral fatigue syndrome. *Br. Med. Bull.* **1991**, *47*, 826–837. [CrossRef] [PubMed]

262. Lane, R.J.M.; Barrett, M.C.; Taylor, D.J.; Kemp, G.J.; Lodi, R. Heterogeneity in chronic fatigue syndrome: Evidence from magnetic resonance spectroscopy of muscle. *Neuromuscul. Disord.* **1998**, *8*, 204–209. [CrossRef]

263. Behan, W.M.H.; More, I.A.R.; Behan, P.O. Mitochondrial abnormalities in the postviral fatigue syndrome. *Acta Neuropathol.* **1991**, *83*, 61–65. [CrossRef]

264. Zhang, C.; Baumer, A.; Mackay, I.R.; Linnane, A.W.; Nagley, P. Unusual pattern of mitochondrial DNA deletions in skeletal muscle of an adult human with chronic fatigue syndrome. *Hum. Mol. Genet.* **1995**, *4*, 751–754. [CrossRef]

265. Filler, K.; Lyon, D.; Bennett, J.; McCain, N.; Elswick, R.; Lukkahatai, N.; Saligan, L.N. Association of mitochondrial dysfunction and fatigue: A review of the literature. *BBA Clin.* **2014**, *1*, 12–23. [CrossRef]

266. Morris, G.; Maes, M. Mitochondrial dysfunctions in Myalgic Encephalomyelitis/chronic fatigue syndrome explained by activated immuno-inflammatory, oxidative and nitrosative stress pathways. *Metab. Brain Dis.* **2014**, *29*, 19–36. [CrossRef]

267. Malatji, B.G.; Meyer, H.; Mason, S.; Engelke, U.F.H.; Wevers, R.A.; Van Reenen, M.; Reinecke, C.J. A diagnostic biomarker profile for fibromyalgia syndrome based on an NMR metabolomics study of selected patients and controls. *BMC Neurol.* **2017**, *17*, 88. [CrossRef] [PubMed]

268. Derakhshan, M. Viral infection, a suggestive hypothesis for aetiology of chronic fatigue syndrome. *J. Med. Hypotheses Ideas* **2008**, *2*, 10–11.

269. Smith, A.P. Post-viral Fatigue: Implications for Long Covid. *Asian J. Res. Infect. Dis.* **2021**, 17–23. [CrossRef]

270. Carfi, A.; Bernabei, R.; Landi, F. Persistent Symptoms in Patients After Acute COVID-19. *JAMA* **2020**, *324*, 603–605. [CrossRef]

271. Doykov, I.; Hällqvist, J.; Gilmour, K.C.; Grandjean, L.; Mills, K.; Heywood, W.E. 'The long tail of Covid-19'-The detection of a prolonged inflammatory response after a SARS-CoV-2 infection in asymptomatic and mildly affected patients. *F1000Research* **2021**, *9*, 1349. [CrossRef]

272. Poenaru, S.; Abdallah, S.J.; Corrales-Medina, V.; Cowan, J. COVID-19 and post-infectious myalgic encephalomyelitis/chronic fatigue syndrome: A narrative review. *Ther. Adv. Infect.* **2021**, *8*, 20499361211009385. [CrossRef]

273. Ostojic, S.M.; Stojanovic, M.; Drid, P.; Hoffman, J.R.; Sekulic, D.; Zenic, N. Supplementation with Guanidinoacetic Acid in Women with Chronic Fatigue Syndrome. *Nutrients* **2016**, *8*, 72. [CrossRef]

274. Ostojic, S.M. Diagnostic and Pharmacological Potency of Creatine in Post-Viral Fatigue Syndrome. *Nutrients* **2021**, *13*, 503. [CrossRef]

Article

Bioavailability, Efficacy, Safety, and Regulatory Status of Creatine and Related Compounds: A Critical Review

Richard B. Kreider [1,*], **Ralf Jäger** [2] and **Martin Purpura** [2]

[1] Exercise & Sport Nutrition Lab, Human Clinical Research Facility, Department of Health & Kinesiology, Texas A&M University, College Station, TX 77843, USA

[2] Increnovo LLC, Milwaukee, WI 53202, USA; ralf.jaeger@increnovo.com (R.J.); martin.purpura@increnovo.com (M.P.)

* Correspondence: rbkreider@tamu.edu; Tel.: +1-972-458-1498

Abstract: In 2011, we published a paper providing an overview about the bioavailability, efficacy, and regulatory status of creatine monohydrate (CrM), as well as other "novel forms" of creatine that were being marketed at the time. This paper concluded that no other purported form of creatine had been shown to be a more effective source of creatine than CrM, and that CrM was recognized by international regulatory authorities as safe for use in dietary supplements. Moreover, that most purported "forms" of creatine that were being marketed at the time were either less bioavailable, less effective, more expensive, and/or not sufficiently studied in terms of safety and/or efficacy. We also provided examples of several "forms" of creatine that were being marketed that were not bioavailable sources of creatine or less effective than CrM in comparative effectiveness trials. We had hoped that this paper would encourage supplement manufacturers to use CrM in dietary supplements given the overwhelming efficacy and safety profile. Alternatively, encourage them to conduct research to show their purported "form" of creatine was a bioavailable, effective, and safe source of creatine before making unsubstantiated claims of greater efficacy and/or safety than CrM. Unfortunately, unsupported misrepresentations about the effectiveness and safety of various "forms" of creatine have continued. The purpose of this critical review is to: (1) provide an overview of the physiochemical properties, bioavailability, and safety of CrM; (2) describe the data needed to substantiate claims that a "novel form" of creatine is a bioavailable, effective, and safe source of creatine; (3) examine whether other marketed sources of creatine are more effective sources of creatine than CrM; (4) provide an update about the regulatory status of CrM and other purported sources of creatine sold as dietary supplements; and (5) provide guidance regarding the type of research needed to validate that a purported "new form" of creatine is a bioavailable, effective and safe source of creatine for dietary supplements. Based on this analysis, we categorized forms of creatine that are being sold as dietary supplements as either having strong, some, or no evidence of bioavailability and safety. As will be seen, CrM continues to be the only source of creatine that has substantial evidence to support bioavailability, efficacy, and safety. Additionally, CrM is the source of creatine recommended explicitly by professional societies and organizations and approved for use in global markets as a dietary ingredient or food additive.

Keywords: dietary ingredients; ergogenic aids; exercise; performance

Citation: Kreider, R.B.; Jäger, R.; Purpura, M. Bioavailability, Efficacy, Safety, and Regulatory Status of Creatine and Related Compounds: A Critical Review. *Nutrients* **2022**, *14*, 1035. https://doi.org/10.3390/nu14051035

Academic Editor: David C. Nieman

Received: 8 February 2022
Accepted: 25 February 2022
Published: 28 February 2022

Publisher's Note: MDPI stays neutral with regard to jurisdictional claims in published maps and institutional affiliations.

1. Introduction

Creatine (N-(aminoiminomethyl)-N-methyl glycine) is a naturally occurring nitrogen-containing compound that plays an integral role in cellular metabolism. While creatine is commonly referred to as an amino acid, it is not actually an amino acid in the traditional sense. It is not incorporated into proteins or an essential, conditionally essential, and non-essential amino acid that serves as building blocks of protein. Instead, creatine is an amino acid derivative that is endogenously synthesized from the amino acids arginine and glycine by L-arginine: glycine amidinotransferase (AGAT) to guanidinoacetate (GAA). The GAA is then methylated (i.e., CH_3 group added) by the enzyme guanidinoacetate N-methyltransferase (GAMT) with S-adenosyl methionine (SAMe) to form creatine [1]. The kidney, liver, pancreas, and some areas within the brain contain AGAT. Most GAA is formed in the kidney and converted by GMAT to creatine in the liver [2–4]. The primary role of creatine is to bind with inorganic phosphate (Pi) in the cell to form phosphocreatine (PCr), and thereby serve as a high-energy phosphate source of energy to resynthesize adenosine triphosphate (ATP) that has been degraded to adenosine diphosphate (ADP) + Pi as a source of energy to fuel cellular metabolism [5]. Creatine also plays a critical role in translocating energy-related intermediates from the electron transport system in the mitochondria to the cytosol [6,7].

About 95% of creatine is stored in the muscle, with the remaining amount found in other tissues like the heart, brain, and testes [8,9]. About two-thirds of creatine is bound with Pi and stored as PCr with the remaining one-third stored as free creatine (Cr). The total creatine pool (Cr + PCr) is about 120 mmol/kg of dry muscle mass for an individual who consumes a diet with red meat and fish [10]. The body breaks down about 1–2% of the intramuscular creatine pool into creatinine, which is excreted in the urine [10–12]. Daily degradation of creatine to creatinine is greater in individuals with larger muscle mass and individuals with higher levels of physical activity [13]. Creatine synthesis provides about half of the daily need for creatine [2]. The remaining creatine needed to maintain normal tissue levels is obtained in the diet primarily from red meat and fish [14–17] or dietary supplements containing a bioavailable source of creatine [14–16]. Since creatine stores are not fully saturated on vegan or omnivorous diets that typically provide 0 to 1.5 g/day of creatine, daily dietary creatine needs have been estimated to be 2–4 g/day [2,6,15]. For this reason, dietary supplementation of creatine has been recommended to optimize creatine stores [5,6]. The most extensively studied and effective form of creatine found in nutritional supplements that professional organizations recommend for use is creatine monohydrate (CrM) [14,15,17,18]. Over the last 30 years, several studies have shown that CrM supplementation (e.g., 5 g/day for 5–7 days or 3–5 g/day for 30 days) increases blood, muscle, and tissue levels of creatine and PCr by 20–40% [11,12,19,20]. Co-ingesting CrM with carbohydrates [20–22] and carbohydrate and protein [23] promotes more consistent and greater creatine retention. The increased creatine levels have been reported to enhance high-intensity exercise performance and exercise training adaptations [14,15,24]. Furthermore, there is accumulating research that CrM supplementation may have health and clinical benefits in populations that may benefit from increasing creatine availability to the cell [5].

In 2011, we published a paper overviewing the known pharmacokinetics, bioavailability, efficacy, and regulatory status of various purported "forms" of creatine "marketed with claims of improved physical, chemical, and physiological properties in comparison to CrM" [25]. The article concluded that "the efficacy, safety, and regulatory status of most of the newer forms of creatine found in dietary supplements have not been well-established" and "there is little to no evidence supporting marketing claims that these newer forms of creatine are more stable, digested faster, more effective in increasing muscle creatine levels, and/or associated with fewer side effects than CrM" [25]. Similarly, in its most recent position stand on creatine supplementation, the International Society of Sport Nutrition (ISSN) concluded: (1) "Creatine monohydrate is the most extensively studied and clinically effective form of creatine for use in nutritional supplements in terms of muscle uptake and ability to increase high-intensity exercise capacity"; (2) "Claims that different forms of creatine are degraded to a lesser degree than creatine monohydrate in vivo or result in a greater uptake to muscle are currently unfounded [25]"; and, (3) "Clinical evidence has not demonstrated that different forms of creatine such as creatine citrate [22], creatine serum [26], creatine ethyl ester [27], buffered forms of creatine [28], or creatine nitrate [29] promote greater creatine retention than creatine monohydrate [25]." These conclusions were reiterated more recently in a review related to myths and misconceptions about creatine supplementation noting "there are no peer-reviewed published papers showing that the ingestion of equal amounts of creatine salts [30–33] or other forms of creatine like effervescent creatine [22], creatine ethyl ester [27,34,35], buffered creatine [28], creatine nitrate [29,36], creatine dipeptides, or the micro amounts of creatine contained in creatine serum [26,37,38] and beverages (e.g., 25–50 mg) increases creatine storage in muscle to a greater degree than creatine monohydrate [25]. Most studies show that ingestion of these other forms have less physiological impact than creatine monohydrate on intramuscular creatine stores and/or performance and that any performance differences were more related to other nutrients that creatine is bound to or co-ingested within supplement formulations."

Despite this clear scientific evidence, guidance from professional organizations, and regulatory challenges of selling other forms of creatine in the global marketplace, companies continue to market "new forms" of creatine that are purportedly more stable, bioavailable, and/or effective than CrM. Some have argued that mentioning some of the forms of creatine described in our 2011 paper served as validation or an endorsement that these forms of creatine were valid sources of creatine rather than the intended conclusion that these purported forms of creatine were not scientifically validated, safe, effective, and regulatory approved sources of creatine for dietary supplements. The intent of this comprehensive review is to provide an update regarding (1) how creatine is absorbed from food and/or dietary supplements into the body; (2) whether sources of creatine currently marketed and/or used in dietary supplements are bioavailable sources of creatine; and (3) whether any of these purportedly alternate forms of creatine are as effective in increasing creatine stores in the body to a greater degree than CrM. Based on this assessment, compounds commonly marketed as sources of creatine will be categorized as: (1) Strong evidence to support bioavailability, efficacy, and safety; (2) limited evidence to support bioavailability, efficacy, and safety; or (3) no evidence to support bioavailability, efficacy, and safety.

2. Methods

This paper was conducted as a systematic review of the literature related to sources of creatine marketed as ingredients for dietary supplements or found in dietary supplements, food products, and/or beverages marketed as containing creatine. This was accomplished by performing a PubMed search of the US National Library of Medicine database of different sources of creatine found in dietary supplements related to solubility, stability, bioavailability, supplementation, and performance. In addition, we conducted patent searches of the United States Patent and Trademark Office, the European Patent Office, the Japan Patent Office, and World Intellectual Property Organization (WIPO) mainly when no articles were found from a PubMed search on the purported form of creatine. We also reviewed company websites to assess claims and studies they cited to support claims, as well as publicly available reports of studies submitted as evidence to the court in cases related to these purported forms of creatine. To assess the current regulatory status, we conducted a search of key markets around the world, including the US Food and Drug Administration (FDA); the Therapeutic Goods Administration in the Department of Health of Australia; Health Canada; the State Administration for Market Regulation of China; the European Union Commission; the Japan Ministry of Health, Labor and Welfare; and the Korea Food and Drug Administration. The information obtained from this search was used to assess the legal and regulatory status related to the sale of purported forms of creatine in these marketplaces. Figure 1 shows a Preferred Reporting Items for Systematic Reviews and Meta-Analyses (PRISMA) flow chart [39].

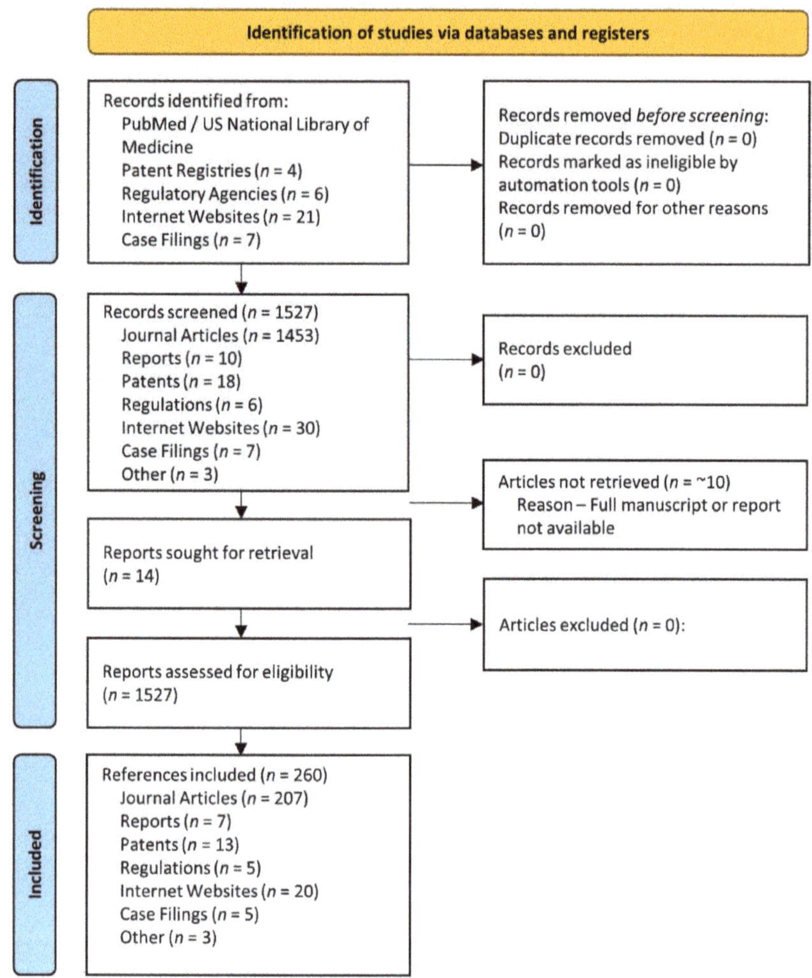

Figure 1. PRISMA flow chart.

3. Bioavailability

Bioavailability refers to the degree or rate at which a drug or substance is absorbed into the body, reaches the intended target site, and is available to influence physiological activity [40]. In terms of nutrients, bioavailability refers to the amount of the nutrient contained in the food or supplement that is delivered to the target tissue and available in the intended tissue for metabolic activity [40]. If food or supplement contains a large number of nutrients, but only a small percentage of the nutrient is liberated from the food or supplement, transported through the blood to the tissue, and ultimately taken up by the tissue, it is not very bioavailable. Similarly, if a similar quantity of food or supplement delivers less of the active nutrient to the target tissue than another food or supplement of equal quantity, it is comparatively less bioavailable. In the case of creatine, individuals who consume meat and fish in their diet typically have a plasma creatine level of around 25 µmol/L (about 3.75 µg/mL), and a muscle creatine content of about 120 mmol/kg dry muscle mass (DMM) [11,12,19,20]. Muscle creatine levels are typically lower in individuals following a vegan diet [41,42] and the elderly, who may not consume as much protein in their diet or have more difficulty digesting dietary protein [43,44]. Plasma creatine levels increase after consuming creatine-containing food and dietary supplements containing a bioavailable source of creatine in proportion to the amount creatine ingested and digestion rate [2,41,42,45,46].

For a dietary source of creatine to be bioavailable, creatine must be: (1) absorbed as creatine into the blood and transported to tissues [2,25]; (2) transported into tissue via tissue-specific creatine transporter genes (e.g., CRT1 or SLC6A8 in muscle) [6,7]; and (3) increase tissue and cellular creatine and PCr content by a physiologically meaningful amount to influence metabolic activity (e.g., 20–40% in muscle and 10–20% in brain) [5,12,15,20,24,47]. In creatine research, the efficacy of creatine supplementation is determined by assessing the magnitude in which creatine supplementation protocol increases muscle creatine content as typically measured from muscle biopsy samples and/or muscle and brain creatine content as determined from magnetic resonance spectroscopy (MRS) [15]. Since oral ingestion of CrM is nearly 100% bioavailable (i.e., it's either absorbed by tissue or excreted in urine), whole-body creatine retention with CrM supplementation can also be estimated as the difference between daily intake of CrM and urinary creatine output [22]. Purported forms of creatine that do not increase blood creatine concentrations, do not increase uptake of creatine through tissue-specific creatine transporters, and ultimately do not increase tissue creatine levels by physiologically meaningful amounts would not affect creatine-related metabolic function. This is regardless of whether the purported form of creatine is more solubility in water, is more stable under various temperature and pH conditions outside of the body, or is delivered in food, gel, and liquid sources, as will be discussed below.

In support of this contention, classic experiments by Harris et al. [12] indicated that a dose of orally ingested creatine should ideally increase plasma creatine levels to greater than 500 μmol/L (75 μg/mL) to optimize tissue uptake. They reported that oral ingestion of 1 g or less of CrM had negligible effects on blood creatine content (i.e., rarely exceeding 100 μmol/L (15 μg/mL)). However, ingestion of one oral dose of 5 g of CrM (equivalent to about 1.1 kg of uncooked beef) resulted in plasma creatine levels of about 800 μmol/L (120 μg/mL) after 1 h of ingestion. It sustained plasma creatine above 200 μmol/L (30 μg/mL) for over 4–5 h (see Figure 2A). Supplementation of 5 g of CrM every 2 h maintained peak plasma creatine to levels exceeding 1000 μmol/L (150 μg/mL). Moreover, ingesting 5 g of CrM, 4 to 6 times a daily for 2-days or more significantly increased muscle creatine content of about 35%. Creatine uptake into the muscle was greatest during the first two days of CrM supplementation and declined over the next few days as muscle creatine levels became fully saturated. Subsequent studies from Hultman and colleagues [11] investigated the effects of various CrM supplementation strategies on changes in muscle creatine content. These experiments revealed that: (1) consuming 4×5 g of CrM per day for 6-days (i.e., creatine loading strategy) significantly increased muscle free creatine content by 33% and returned to baseline within 4-weeks after supplementation; (2) ingesting 4×5-g doses of CrM for 6-days followed by ingestion of 2 g/day of CrM for 28 days maintained a 36% increase in muscle creatine levels; and (3) ingesting 3 g/day of CrM for 35 days (i.e., low dose supplementation strategy) resulted in a gradual 16.7% increase in muscle creatine content. About 70% of the increase in the total creatine pool was observed in changes in free creatine content in the muscle.

Harris and associates [45] also conducted two experiments assessing the bioavailability of CrM in solution compared to creatine obtained from meat, crushed lozenges, and suspended in gel (see Figure 2B,C). In the first study, the researchers reported that ingestion of 2.5 g of CrM in solution (providing about 2.2 g of creatine) promoted a more rapid and greater increase in peak plasma creatine (287 ± 115 μmol/L or 42 μg/mL) than ingesting 408 g of lightly cooked steak containing 5.4 g (182 ± 52 μmol/L or 27 μg/mL). However, ingesting 5.4 g of creatine in lightly cooked meat promoted a more sustained increase in plasma creatine. Nevertheless, both strategies resulted in similar increases in area under the curve (AUC) values (507 ± 205 and 518 ± 153 μmol/h/L, respectively). In the second study, the researchers reported that ingestion of 2 g of CrM in solution resulted in peak plasma creatine levels of 386 ± 88 μmol/L (57 μg/mL) with an AUC of

622 ± 193 µmol/h/L. This compared to the peak value of 269 ± 67 µmol/L (41 µg/mL) with an AUC of 399 ± 196 µmol/h/L when CrM was administered in suspended gel and a peak creatine level of 277 ± 53 µmol/L and an AUC of 438 ± 131 µmol/h/L when CrM was administered as crushed lozenges. Collectively, these findings and are important because they demonstrate: (1) ingestion of 5 g of creatine from meat or 2–2.5 g of CrM administered in fluid, gels, and solids increased plasma creatine content by physiologically significant amounts needed to promote creatine uptake into the muscle; (2) the optimal single dose of CrM to increase plasma creatine levels is 5 g, but that ingesting 2–3 g will increase plasma creatine to sufficient levels to promote creatine uptake to tissue; (3) ingesting 5 g of CrM, 4 to 6 times a day for as little as 2 days was sufficient to significantly increase muscle creatine levels; (4) ingesting 5 g of CrM, 4 times a day for 6 days (i.e., 120 g total) increased muscle free creatine by about 35%; (5) consuming 3 g/day of CrM for 35 days (i.e., 105 g total) promoted a gradual 16.7% increase in muscle creatine; and (6) they provided a scientific basis for CrM supplementation recommendations and data for comparison of the efficacy of other purported forms of creatine marketed in dietary supplements [15].

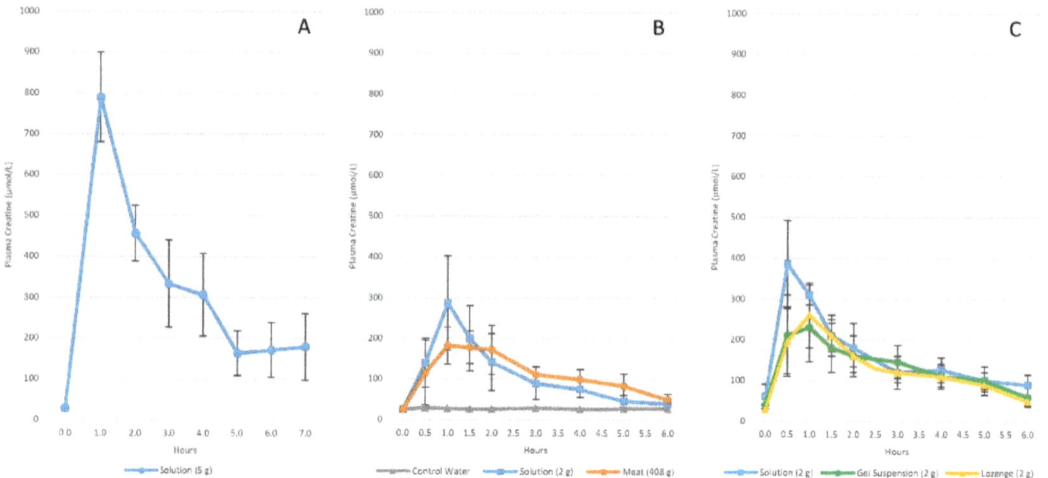

Figure 2. Changes in plasma creatine concentrations after administration of 5 g of creatine monohydrate (CrM) in solution (**A**) [12]; water, 2 g of CrM administered in solution, or 408 g of slightly cooked meat containing 5.4 g of creatine (**B**) [45]; or 2 g of CrM provided in solution, gel suspension, or in a hard candy lozenge (**C**) [45].

3.1. Methods to Assess Bioavailability

3.1.1. Assess Chemical Structure

The first step in assessing the bioavailability of a purported novel form of creatine is to determine whether the purported source of creatine contains a creatine molecule. Although this seems obvious, as will be seen below, some purported sources of creatine do not contain the complete creatine molecule. Instead, they only contain a portion of the creatine molecule structure or rearrange the chemical structure such that the compound is not really creatine. Moreover, some purported sources of creatine bind or complex compounds to creatine (or a portion of its structure) with bonds so strong (e.g., amide bonds) they would not likely break down into creatine through normal digestion, and therefore would not likely increase creatine levels in the blood or tissues. In other words, they are simply not bioavailable sources of creatine. Consequently, assessing the chemical structure of a purported form of creatine and whether associated bonds would easily disassociate into a creatine molecule is the first question that should be asked when evaluating claims about a purported "novel form" of creatine.

3.1.2. Assess Changes in Blood Creatine Content

The next step is to determine if the purported source of creatine is absorbed through normal digestive processes into the blood and increases plasma creatine levels in a physiologically significant manner. Purported forms of creatine that do not significantly increase plasma creatine above normal fasting levels (i.e., about 25 μmol/L or 3.75 μg/mL) would have no effect on muscle creatine content because it would not deliver any creatine to tissues for uptake by creatine transporters. Likewise, sources of creatine that increase plasma creatine levels by less than 100 μmol/L (15 μg/mL) would not be considered viable sources of creatine for dietary supplements because they would not deliver enough creatine to target tissue to significantly increase creatine content [12,45]. Viable sources of creatine in dietary supplements should increase plasma creatine levels above 200–500 μmol/L (30–75 μg/mL) within the first hour of oral ingestion and promote a large increase in the AUC of creatine over a 4-to-5 h period [11,12,45]. However, an increase in plasma creatine alone does not provide definitive proof that a purported source of creatine is bioavailable or effective. Differences in plasma creatine after ingestion of a bioavailable source of creatine only suggests that absorption rates may differ. Higher blood creatine could mean that the source is not taken up as quickly into tissue, while lower levels could mean that less appears in the blood or creatine absorption into tissue is faster [48]. Ultimately, the source of creatine must be transported into tissues by creatine-specific transporters and increase tissue levels of creatine in physiologically significant amounts to affect creatine-related metabolism (i.e., 20–40%). Thus, it cannot be assumed that a purported source of creatine will be effectively transported into muscle based on solubility properties in fluid and/or changes in blood concentrations alone because sources of creatine that are not effectively transported into tissue could have higher plasma levels than those that promote a more rapid transport into the tissue. Consequently, it is important to assess the difference between arterial content (amount of creatine delivered to tissue) and venous creatine content or measure the amount of creatine retained in tissues directly to determine the amount of orally ingested creatine taken up by tissue.

3.1.3. Assess Changes in Tissue Creatine Content

Thus, the third step in verifying the bioavailability of a purportedly novel source of creatine is to directly assess the effects of oral ingestion at recommended dosages on tissue creatine content. This is most often done by determining changes in muscle creatine content since 95% of creatine is stored in skeletal muscle. So-called forms of creatine that have been marketed as creatine but do not have any data showing the source of creatine increases muscle creatine content in humans should not be considered a viable source of creatine until such data are available. Purported sources of creatine that increase blood levels of creatine but do not increase tissue levels of creatine in physiologically effective amounts (i.e., 20–40% in muscle and 10–20% in brain) are not bioavailable sources of creatine. Purported sources of creatine that do not deliver similar increases in muscle creatine content than equivalent doses of CrM are less bioavailable sources of creatine than CrM. Moreover, purported derivatives or analogs of creatine that have no measurable effects on plasma creatine levels and do not increase muscle creatine content are not bioavailable sources of creatine, and therefore could have no physiological effects that have been reported in the literature from CrM supplementation. Likewise, if a form of creatine does not significantly increase muscle and/or brain creatine content in humans at the recommended dosages, it should not be considered a viable source of creatine for a dietary supplement. This includes making claims that lower doses of a purportedly more bioavailable source of creatine (e.g., 1–2 g) or "sprinkling" physiologically insignificant amounts of CrM or other purported creatine sources, derivatives, or analogs of creatine in supplements or beverages (e.g., 25–250 mg) promote similar or better benefits as CrM loading (e.g., 4 × 5 g/day for 5–7 days) or long-term supplementation (e.g., 3 g/day).

4. Physio-Chemical Properties

Figure 3 shows the structure of creatine and CrM. Creatine monohydrate was the first source of creatine marketed as a dietary supplement and remains the most common source of creatine found in dietary supplements [25]. Creatine monohydrate is considered the gold standard to compare other purported sources of creatine because of its well-known physiochemical properties, high bioavailability, stability, low cost, and a large number of studies that have demonstrated efficacy and safety [15,25]. CrM has been so extensively studied compared to other purported forms of creatine that when creatine supplementation is discussed in the literature, it is understood the authors are referring to CrM unless otherwise specified [15,17,18,49]. Nevertheless, CrM is formed by crystallization with water forming monoclinic prisms that hold one molecule of water per molecule of creatine [25]. This provides a powder containing 87.9% creatine that readily dissociates into creatine and water upon oral ingestion. The water in CrM can also be removed when exposed to heat at about 100 °C yielding anhydrous creatine that is 100% creatine [25]. However, due to the increased temperature used during the drying, anhydrous creatine contains higher amounts of the impurity creatinine. Creatine appears as internal salt and is considered a fairly weak base (pKb 11.02 at 25 °C) that forms salts with strong acids (i.e., pKa < 3.98) [25]. Creatine can also complex with other compounds via ionic binding (i.e., the attraction of positive cation and negative anion charges).

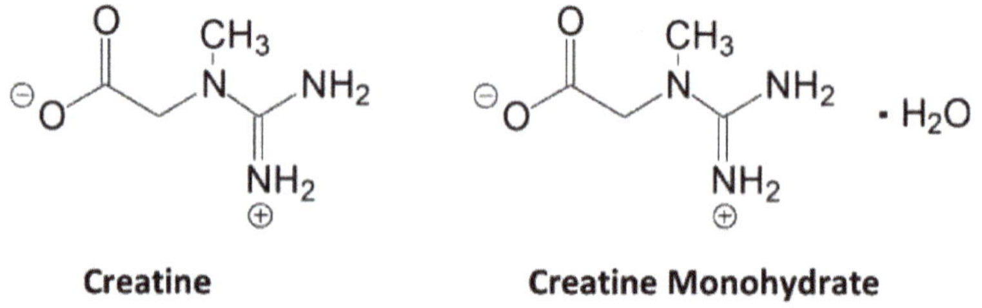

Creatine   **Creatine Monohydrate**

Figure 3. Chemical structure of creatine and creatine monohydrate.

Pischel and Gastner [50] described the basic process of industrial synthesis of CrM. The process involves adding acetic acid to an aqueous sodium sarconsinate solution and stirring to a pH value of about 10 and a temperature of about 80 °C. An aqueous cyanamide solution is then added to the medium and agitated to facilitate the reaction. After cooling, the crystalline CrM is filtered, separated, and then dried [50]. Creatine monohydrate is manufactured by using water as solvent in Germany has produced 99.9% pure CrM with no contaminants under the brand name Creapure®. Other sources of CrM, particularly from China, have been reported to contain contaminates like dicyandiamide, dihydrotriazine, dimethyl sulphate, thiourea, creatinine, and/or higher concentrations of heavy metals like mercury and lead due to the use of different chemical precursors, poorly controlled synthesis processes, using organic solvents, and/or less than adequate filtration methods that increase production of these contaminants [50]. For this reason, creatine monohydrate manufactured by AlzChem in Germany is considered the gold standard of creatine and has been the primary source of creatine used in hundreds of clinical trials conducted on CrM to establish safety and efficacy [15,25].

5. Stability

Creatine monohydrate is very stable in powder form, showing no signs of degradation to creatinine over years, even at elevated temperatures [25]. For example, Jäger [51] reported that CrM powder showed no signs of degrading to creatinine even with temperatures up to 40 °C (104 °F) for more than three years. Additionally, even when CrM was stored at 60 °C (140 °F), creatinine could only be detected in trace amounts after 44 months of storage [51]. However, creatine is not as stable in solution due to intramolecular cyclization that converts creatine to creatinine (see Figure 4A). Generally, creatine degrades to creatinine in solution at a faster rate as pH decreases and temperature increases [25,52–54]. For example, as seen in Figure 4B, Harris and coworkers [55] reported that creatine is relatively stable for 3 days in solution at neutral pH (6.5 to 7.5) However, the rate of degradation to creatinine increased when stored at 25 °C when pH decreased (e.g., 4% at 5.5 pH; 12% at 4.5 pH; and 21% at 3.5 pH) [55]. However, as described below, the conversion of creatine to creatinine is halted at pH levels < 2.5. For this reason, it is recommended that CrM should be consumed immediately after it is mixed in an acidic beverage or refrigerated to slow the degradation to creatinine and consumed within a couple of days. However, recent reports presented shelf-life stability data of CrM suspended in a solution of 70% for 13-months at neutral pH and 100% for 12 months at a pH of 2.8 [56,57].

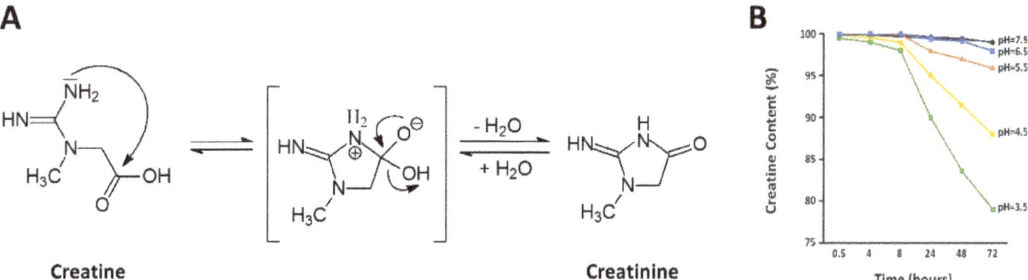

Figure 4. Conversion of creatine to creatinine (**A**) and influence of pH on creatine stability in solution (**B**). Creatine stability figure adapted from Howard and Harris [55].

As mentioned above, the degradation of creatine can be limited or prevented when creatine is in very low or very high pH environments [25]. In this regard, a pH > 12.1 promotes the deprotonation of the acid group. This makes it more difficult for intramolecular cyclization to form creatinine [25]. On the other hand, when pH is <2.5, the amide functional group on the creatine molecule is protonated and prevents the intramolecular cyclization (see Figure 5) [25]. Since stomach acid is generally less than 2.5, less than 1% of CrM is degraded to creatinine during digestion and 99% of creatine is taken up by tissue or excreted in urine after ingestion [12,25,58,59].

Figure 5. Influence of pH on creatine stability in solution.

6. Solubility

One of the limitations in terms of developing consumer products containing CrM is that CrM powder is not highly soluble. For example, when mixing CrM in solution, some CrM residue remains at the bottom of the glass requiring consumers to add more fluid, swirl, and quickly ingest to ensure they consumed all the creatine. While this has no effect on creatine bioavailability as CrM is nearly 100% bioavailable [12,25,58,59], there has been interest in finding ways to improve the solubility of creatine. The solubility of creatine in water increases linearly with increasing temperature. In this regard, about 6 g of creatine dissolves in one liter of water at 4 °C (39.2 °F) while 14 g/L are dissolved at 20 °C (68 °F),

34 g/L are dissolved at 50 °C (122 °F); and, 45 g/L are dissolved at 60 °C (140 °F) [25]. This is the reason that some researchers initially administered CrM to participants in warm to hot water [12] or hot tea [60]. Creatine solubility can also be improved by administering CrM in lower pH solutions like juices and sport drinks that generally have pH levels ranging from 2.5–3.5 [61] and/or blending CrM with carbohydrate and/or protein powders or in juice which helps suspend CrM in solution, reduce sedimentation, and enhance creatine uptake into muscle [20–23,62].

Dissolving creatine in more acidic environment is also the rationale in providing creatine in the form of easily disassociated creatine salts. Adding an acidic moiety to a creatine molecule lowers the pH of the water. For example, adding tri-creatine citrate to water yields a pH of 3.2 and increases solubility to 29 g/L whereas adding creatine pyruvate to water yields a pH of 2.6 and increases solubility to 54 g/L [63]. Creatine hydrochloride (HCl) has also been marketed as a pH lowering source of creatine with greater solubility than CrM [64,65]. While lowering pH and/or mixing creatine salts into solution enhances solubility, the amount of creatine contained in these forms of creatine salt must be equalized to CrM to deliver the same amount of creatine to the blood and tissues. In this regard, CrM contains 87.9% creatine whereas creatine citrate (40.6%), di-creatine citrate (57.7%), creatine pyruvate (59.8%), and creatine HCl (78.2%) contain less creatine by weight compared to CrM. Therefore, one would need to mix 1.54, 1.34, 1.32, and 1.11 times more of these forms of creatine in solution to deliver the same amount of creatine than CrM. Additionally, while mixing CrM in common juices and beverages with pHs ranging from 2.5–3.5 [61] would enhance solubility, it would also promote the conversion of creatine to creatinine over time [66]. Therefore, it is best to consume creatine salts or CrM with acid beverages soon after it is mixed so that the conversion of creatine to creatinine would be halted upon entering a more acidic environment in the stomach.

With that said, the only real advantage of mixing CrM in an acidic beverage is that it would leave less CrM in crystalized form at the bottom of a cup to swirl and consume during the last drink. If an individual consumes all the CrM (soluble or not), it will be bioavailability in terms of intestinal absorption, transport of creatine in the blood, transport of creatine through tissue-specific creatine transporters, and uptake of creatine into tissues. Similarly, if a bioavailable source of creatine is consumed at physiologically effective doses, it is not degraded during digestion, and it increases blood and tissue creatine content by physiologically meaningful amounts (i.e., 20–40%), it does not matter whether a form of creatine has better mixing characteristics and/or is more soluble. Research has clearly shown that CrM is not degraded into creatinine during normal digestion, it is nearly 100% bioavailable [12,25,58,59], and markedly increases blood and tissue creatine content. There are no data showing that any other purported form of creatine is more effective in increasing tissue creatine content than CrM [62]. Therefore, claims that a given form of creatine is more effectively absorbed than CrM because it is more soluble in water is unsupported marketing hyperbole.

7. Purported Creatine Related Compounds

Table 1 provides a list of creatine-related compounds that are listed in PubChem when doing a search on the word creatine. As of this writing, 88 creatine-related compounds are listed in the PubChem database. A few others that have been mentioned in patents, published literature, or on international regulatory authorities' lists are also mentioned at the bottom of Table 1. While not all these compounds have been used in dietary supplements or assessed for bioavailability, it provides an overview of the types of creatine-related compounds that have been developed. Interestingly, some creatine derivatives and analogs that have been marketed as bioavailable sources of creatine in dietary supplements are not listed in this database because the molecular structures have been altered such that the compound does not contain the creatine molecule. Consequently, they should not be considered a bioavailable source of creatine unless studies show that it increases creatine levels in the blood and target tissue (i.e., muscle and brain). This view is consistent with the United States FDA definition that a dietary supplement is considered a new dietary ingredient (NDI) if the ingredient has been chemically altered from its natural form [67].

Table 1. Theoretical creatine content of purported sources of creatine-related compounds in the PubChem database based on molecular weight.

Compound	Molecular Formula	Molecular Weight (g/mol)	Theoretical Percent Creatine by MW [†]	Difference from CrM (%)
Creatine (Creatine Anhydrous)	$C_4H_9N_3O_2$	131.13	100.0	13.8
Creatine Monohydrate	$C_4H_{11}N_3O_3$	149.15	87.9	0.0
Magnesium Creatine	$C_4H_9MgN_3O_2$	155.44	84.4	−4.0
Creatine Ethyl Ester	$C_6H_{13}N_3O_2$	159.19	82.4	−6.3
Methyl-Amino-Creatine	$C_5H_{12}N_4O_2$	160.17	81.9	−6.9
Creatine Hydrochloride	$C_4H_{10}ClN_3O$	167.59	78.2	−11.0
Creatine Methyl Ester Hydrochloride	$C_5H_{12}ClN_3O_2$	181.62	72.2	−17.9
Creatine Nitrate	$C_4H_{10}N_4O_5$	194.15	67.5	−23.2
Creatinol-O-Phosphate	$C_4H_{12}N_3O_4P$	197.13	66.5	−24.3
Tri-Creatine Citrate	$C_{14}H_{26}N_6O_{11}$	585.50	67.2	−23.5
Phospho-Creatine	$C_4H_{10}N_3O_5P$	211.11	62.1	−29.3
Creatine Pyruvate	$C_7H_{13}N_3O_5$	219.20	59.8	−31.9
Creatine Beta-Alaninate	$C_7H_{16}N_4O_4$	220.23	59.5	−32.3
Creatine Lactate	$C_7H_{15}N_3O_5$	221.21	59.3	−32.6
Creatine Benzyl Ester	$C_{11}H_{15}N_3O_2$	221.26	59.3	−32.6
Di-Creatine Citrate	$C_{14}H_{26}N_6O_{11}$	454.39	57.7	−34.3
Creatine Sulfate	$C_4H_{11}N_3O_6S$	229.21	57.2	−34.9
Creatine Pyruvate Monohydrate	$C_7H_{15}N_3O_6$	237.21	55.3	−37.1
Di-Acetyl Creatine Ethyl Ester	$C_{10}H_{17}N_3O_4$	243.26	53.9	−38.7
Creatine Sulfate Monohydrate	$C_4H_{13}N_3O_7S$	247.23	53.0	−39.7
Creatine Ethyl Ester Pyruvate	$C_9H_{17}N_3O_5$	247.25	53.0	−39.7
Sodium Creatine Phosphate	$C_4H_8N_3Na_2O_5P$	255.08	51.4	−41.5
Creatine Taurinate	$C_6H_{16}N_1O_5S$	256.28	51.2	−41.8
Creatine Pyroglutamate	$C_9H_{16}N_4O_5$	260.25	50.4	−42.7
Creatine Malate	$C_8H_{15}N_3O_7$	265.22	49.4	−43.8
Creatine Glutamate	$C_9H_{16}N_4O_6$	276.25	47.5	−46.0
Creatine Orotate	$C_9H_{13}N_5O_6$	287.23	45.7	−48.1
Creatine Carnitine	$C_{11}H_{24}N_4O_5$	292.33	44.9	−49.0
Creatine Ethyl Ester Malate	$C_{10}H_{19}N_3O_7$	293.27	44.7	−49.1
5-Hydroxytryptamine Creatine	$C_{14}H_{21}N_5O_3$	307.35	42.7	−51.5
Creatine Trinitrate	$C_4H_{12}N_6O_{11}$	320.17	41.0	−53.4
Creatine α-ketoglutarate	$C_{11}H_{20}N_4O_7$	320.30	40.9	−53.4
Creatine Citrate	$C_{10}H_{17}N_3O_9$	323.26	40.6	−53.9
D-Gluconic Acid Creatine Salt	$C_{10}H_{21}N_3O_9$	327.29	40.1	−54.4
Creatine Monohydrate Dextrose	$C_{10}H_{23}N_3O_9$	329.30	39.8	−54.7
Creatine Hydroxycitrate	$C_{10}H_{17}N_3O_{10}$	339.26	38.7	−56.0
Disodium Creatine Phosphate Tetrahydrate	$C_4H_{18}N_3Na_2O_{10}P$	345.15	76.0	−13.6
Creatine Phosphate Lactate	$C_{13}H_{22}N_3O_{15}P$	491.30	26.7	−69.6
Creatine-CoA	$C_{25}H_{43}N_{10}O_{17}P_3S$	880.70	14.9	−83.1

Adapted from Jaeger et al. [25]. MW represents molecular weight. [†] represents theoretical creatine content based on the MW of creatine assuming a full molecule of creatine is in the compound, and it is liberated as creatine from the compound. Many sources have not been studied to verify ingestion increases blood levels of creatine or tissue creatine content. Therefore, listing a compound in this table does not validate if the marketed source contains creatine, is bioavailable as creatine, or is an effective source of creatine. Not listed on PubChem as a creatine-containing compound: creatine maleate, creatine fumarate, creatylglycine ethyl ester fumarate, polyethylene glycosylated creatine, polyethylene glycosylated creatine HCL, Creatine Serum, creatyl-L-leucine.

It should be noted that different creatine-related compounds shown in Table 1 contain less creatine by molecular weight than CrM, assuming that a full creatine molecule is contained in the compound and would be liberated as creatine in circulation. Consequently, it would take a greater amount of most of these sources of creatine in a dietary supplement to provide equivalent amounts of creatine delivered from CrM if the other sources were indeed bioavailable. For example, creatine citrate (CC) contains 53.9% less creatine by molecular weight than CrM. Therefore, a supplement would need to provide 9.3 g of CC to equal a typical supplementation dose of 5 g of CrM if it had similar bioavailability. Creatine ethyl ester (CEE) has only 6.3% less creatine by molecular weight than CrM. However, as will be discussed below, some of the creatine in CEE converts to creatinine during digestion and therefore it is less bioavailable than CrM. Therefore, more CEE would have to be provided in a supplement to provide an equivalent amount of creatine to tissue and it may increase serum creatinine levels to a greater degree. Additionally, if a purported source or derivative of creatine does not break down and increase blood creatine levels and creatine content in muscle, it would not be a bioavailable source of creatine no matter how much was in the supplement. These are important points to consider when developing nutritional formulations or conducting research with these other sources of creatine.

8. Strong Evidence to Support Bioavailability, Efficacy, and Safety

Creatine Monohydrate

As noted above, CrM is the gold standard to compare other purported forms of creatine due to its known bioavailability, pharmacokinetics, efficacy, and safety [5,14,15,25,62]. Prior studies indicate that CrM loading (i.e., 4 × 5 g/day for 5–7 days) or low-dose long-term intake (e.g., 3–6 g/day for 4–12 weeks) increases muscle creatine retention typically by 20–40% depending on initial creatine content in the muscle [12,22,68–71] and brain creatine content by 5–15% [72–77]. CrM supplementation has been reported to improve acute exercise performance particularly in intermittent high-intensity exercise bouts as well as enhance training adaptations in adolescents [78–82], young adults [29,55,83–92], and older individuals [8,77,93–101]. High-intensity exercise performance is generally increased by 10–20% with greater improvements seen in individuals starting the supplementation protocol with lower muscle creatine and PCr content [102]. Improvements in performance have been reported in individuals participating in weight training [55,89,95,103–113], running [114–118], soccer [87,119,120], swimming [79,80,121–124], volleyball [125], softball [126], ice hockey [127], golf [128], among others [24]. Men and women have been reported to benefit from CrM supplementation in populations ranging from children to elderly populations [47,80,119,126,129–133].

Uptake of creatine into muscle with CrM supplementation has been reported to be enhanced when CrM is consumed with carbohydrate [20–22] and carbohydrate and protein [23]. Co-ingesting CrM with nutrients that improve insulin sensitivity like D-pinitol [134], Russian Tarragon [135], and Fenugreek extract [136] have been reported to enhance creatine retention with limited to no additive effects on exercise performance or training adaptations [89,136,137]. There is also no evidence that supplementation of micronized versions of CrM (i.e., CrM with smaller mesh size particles) are more bioavailable than CrM with normal mesh sized particles [138–140] or consuming CrM in effervescent fluid promotes greater creatine retention or performance benefits [22,141]. This makes sense in that while small differences in creatine retention may enhance the rate of saturating creatine stores (e.g., in 4 versus 5 days) or ensure more consistent response to CrM supplementation (e.g., consuming creatine with carbohydrate and/or carbohydrate and protein), once creatine stores are fully saturated, there would be no ergogenic or training advantage. No side effects have been reported with CrM supplementation other than a desired weight gain [15]. Additionally, there is no convincing evidence that CrM causes common anecdotal myths like bloating, gastrointestinal distress, disproportionate increase in water retention, increased stress on the kidneys, increased susceptibility to injury, etc. [62,142–144]. Many of these claims have been described in marketing materials by companies attempting to gain market share for their purported creatine-containing products [25,62].

Based on a large body of evidence, the ISSN concluded that CrM is the most effective ergogenic nutrient currently available to athletes in terms of increasing high-intensity exercise capacity and lean body mass during training [8,15]. Position stands by the American Academy of Nutrition, Dietitians of Canada, and the American College of Sports Medicine on nutrition and athletic performance [17,49]; the International Olympic Committee consensus statement on dietary supplements and the high-performance athlete [18]; and, the Office of Dietary Supplements, Dietary Supplement Fact Sheets on Dietary Supplements for Exercise and Athletic Performance [145] provided similar conclusions. Thus, there is consensus among professional organizations that CrM is an effective nutritional ergogenic aid that may benefit athletes involved in numerous sports and individuals initiating exercise training to promote health and fitness.

Given the metabolic and ergogenic properties of CrM [7,14,15,24], there has also been interest in assessing the effects of CrM supplementation in various clinical populations that may benefit from increasing high-energy phosphate availability and/or increasing strength and muscle mass [5]. A recent Special Issue on creatine supplementation for health and clinical diseases overviewed the metabolic basis of creatine in health and disease [5] and potential health and/or therapeutic benefits of CrM supplementation for pregnancy and newborn health [146], children and adolescents [147–151], physically active young adults [24], rehabilitation [152], women's health [133], older adults [44], brain health and cognitive function [74], glucose management and diabetes mellitus [153], immunity [154], T cell antitumor immunity and cancer therapy [155], heart health [156], vascular health [157], inflammatory bowel disease [158], chronic renal disease management [159], and post-viral fatigue [160]. From what we can see, all studies in these populations used CrM as the source of creatine. Thus, there is substantial evidence to support the safety and efficacy of CrM supplementation (see Table 2). Additionally, this body of evidence provides the basis to compare the efficacy and safety of other purported sources of creatine. For this reason, CrM is classified as having strong evidence from pharmacokinetic and tissue bioavailability studies, numerous randomized and controlled clinical trials, and a long-history of safety assessed in clinical trials and historical widespread use supporting bioavailability, efficacy, and safety.

Table 2. Example of studies showing bioavailability, efficacy, and safety of CrM supplementation.

Reference	Participants	Design	Duration	Dosing Protocol	Findings	Side Effects
Short-term Studies (<14 Days)						
Greenhaff et al. [71]	8 healthy males	SB	5 days	4 × 5 g CrM	CrM ↑ TCr by 25% and PCr resynthesis following electrically evoked isometric contractions.	None reported
Balsom et al. [161]	7 males	SB	6 days	4 × 5 g CrM	↑ in total muscle total creatine (18%), weight (1.1 kg), and 5 × 6 s cycling sprint performance and PCr recovery	None reported
Green et al. [20]	24 healthy men	RDBP	5 days	4 × 5 g CrM followed by 93 g CHO or CHO	Ingesting CrM with CHO ↑ muscle TCr and glycogen	None reported
Vandenberghe et al. [162]	9 healthy non-vegetarian males	RDBPC	5 days with 5 week washout	25 g/day CrM or PLA	CrM ↑ muscle PCr by 11% and 16% after 2 and 5 days. PCr resynthesis rate was not affected.	None reported
Bellinger et al. [163]	20 endurance cyclists	RDBP	7 days	20 g/day CrM or PLA	CrM ↑ muscle creatine content by 30% and decreased TAN contribution to sprint	None reported

Table 2. *Cont.*

Reference	Participants	Design	Duration	Dosing Protocol	Findings	Side Effects
Francaux et al. [164]	14 physically active males	RDBP	14 days	3 × 7 g of CrM or PLA	CrM ↑ MRS PCr by ~20% and PCr repletion by 15% and 10% during 40% and 70% MVCs.	None reported
Preen et al. [116]	14 physically active men	RDBP	5 days	20 g/day CrM or PLA	CrM increase TCr stores and work during 80-min of repeated cycling sprint exercise.	None reported
Burke et al. [165]	20 male resistance-trained athletes (18–32 years)	RDBP	5 days	4 × 5 g CrM, 4 × 5 g CrM + 25 g Sucrose, or 4 × 5 g CrM + 25 g Sucrose + 250 mg α-LA or PLA	CrM ↑ body weight (2.1 kg) with no differences among groups, TCr was ↑ more in the CrM + sucrose + α-LA group.	None reported
Longer-Term Studies (>14 days)						
Vandenberghe et al. [47]	19 young female volunteers	RDBP	10 weeks phase I (n = 19); 10 weeks phase II (n = 13)	4 × 5 g CrM for 4 days, 5 g/day thereafter or PLA	CrM ↑ muscle PCr, strength, and exercise capacity	None reported
Kreider et al. [55]	25 American college football players during offseason resistance and agility training	RDBP	28 days	CrM 15.75 g/day with glucose or glucose PLA	↑ FFM, ↑ strength, ↑ muscular endurance, ↑ 6 × 6-s cycling sprint performance with 30-s rest	None reported
Volek et al. [113]	19 healthy resistance-trained males	RDBP	12 weeks	CrM 5 × 5 g for 7 days, 5 g/day for 11 weeks or PLA	↑ FFM, strength, and muscle morphology	No differences
Kreider et al. [166]	51 American college football players during offseason resistance and agility training and spring football	RDBP	12 weeks	20 g/day and 25 g/day of CrM with CHO and PRO; CHO only; or CHO + PRO only	CrM groups ↑ FFM, ↑ strength, ↑ muscular endurance. No changes in blood chemistry panels.	CrM groups had less GI complaints than those ingesting CHO and CHO + PRO.
Tarnopolsky et al. [167]	23 young healthy but untrained males	RDBP	8 weeks	10 g/day CrM with 75 g CHO or PLA	CrM with CHO promoted greater ↑ in body mass and FFM during training.	None reported
Willoughby et al. [168]	22 untrained males during resistance-training	RDBP	12 weeks	CrM 6 g/day or PLA	CrM promoted > increases in body mass, FFM, thigh volume, muscle strength, myofibrillar protein content, and myosin heavy chain mRNA expression for Type I, IIa, and IIx fibers	None reported

Table 2. *Cont.*

Reference	Participants	Design	Duration	Dosing Protocol	Findings	Side Effects
Burke et al. [108]	18 vegan and 24 non-vegan (20 men, 22 female)	RDBP	56 days	0.25 g/kg FFM/d of CrM for 7 days, 0.0625 g/kg FFM/d for 49 days or PLA	TCr content was lower in vegans. CrM ↑ PCr, TCr, and gains in bench press strength, isotonic work, Type II fiber area, and FFM during resistance training.	None reported
Lyoo et al. [73]	15 males (23–35 years)	RDBP	56 days	2 × 0.15 g/kg CrM for 7 days, 2 × 0.015 g/kg CrM for 49 days or PLA	CrM ↑ brain PCr (3.4%), Pi (9.8%), and Cr (8.1%) while decreasing β-nucleoside triphosphate (NTP) by 7.8%.	None reported
Newman et al. [169]	17 healthy active but untrained men	RDBP	33 days	4 × 5 g CrM + 3.75 glucose for 5-days, 3 g CrM + 3 g glucose thereafter or PLA	CrM ↑ muscle TCr after loading and maintenance doses. CrM had no effects on muscle glycogen, glucose tolerance or insulin sensitivity.	None reported
Tarnopolsky et al. [170]	Moderately active younger (13 men, 14 women; 19 resistance-trained men; Older resistance-trained men (15) and women (15)	RDBP	5 days; 8 weeks; 14 weeks	4 × 5 g CrM for 5 days; 10 g/day CrM with 75 g dextrose for 8 weeks during training; 5 g/day CrM + 2 g/day dextrose for 14 weeks during training or PLA	CrM ↑ muscle TCr in each study compared to placebo. CrM nor training influenced creatine transporter protein content. Citrate synthase was increased in older participants.	None reported
Willoughby et al. [171]	22 untrained males during resistance-training	RDBP	12 weeks	6 g/day CrM or PLA	CrM promoted > ↑ in muscle CK, myogenin, and MRF-4.	None reported

R = randomized; DB = double-blind; *p* = placebo; SB = single blind; C = crossover, CrM = creatine monohydrate; PCr = phosphocreatine; TCr = total creatine; Pi = inorganic phosphate; CHO = carbohydrate; PRO = protein; FFM = Fat-Free Mass, TAN = total adenine nucleotide pool; MVC = maximal voluntary contractions; α-LA = alpha lipoic acid; MRS = magnetic resonance spectroscopy; GI = gastrointestinal.

9. Some Evidence to Support Bioavailability, Efficacy, and Safety

9.1. Creatine Salts

Creatine salts were introduced into the marketplace in the early 1990s and are formed by adding an acid moiety to creatine, complexing an acid to the creatine molecule, or adding an acid to a complexation product [25]. The rationale was to combine creatine with acids that could easily dissociate (e.g., ionic bond) upon ingestion, thereby not only serving as a viable way to deliver creatine to tissue, but also deliver other nutrients that may have ergogenic properties and/or promote a synergistic metabolic effect with creatine. Additionally, to find ways to improve physical characteristics like solubility of creatine. To do so, the creatine salt must deliver physiologically effective doses of creatine (i.e., 3–5 g per serving) to the blood and tissue in an equivalent manner as CrM to be comparatively effective. Additionally, the acid added to creatine would have to provide a more synergistic effect than simply co-ingesting CrM with the acid independently in a nutritional formulation. Finally, the theoretically added benefit must justify the additional expense in producing the creatine salt and including it in a nutritional formulation.

A number of creatine salts have been marketed as sources of creatine for dietary supplements including creatine citrate (including di- and tri- forms) [172–174]; creatine maleate, creatine fumarate, creatine tartrate [31]; creatine pyruvate [32,174–176]; creatine ascorbate [33]; and creatine orotate [30,176,177], among others. Some creatine salts are less stable when compared to CrM. For example, storage of tri-creatine citrate at 40 °C (104 °F) for 28 days results in formation of 770 ppm of creatinine compared to no measurable amount with CrM powder [63]. Adding carbohydrate to the formulation has been reported to improve the stability of some creatine salts [178]. However, creatine salts would also have less stability than CrM in solution since adding the acid to creatine decreases pH to ranges that would promote greater formation of creatine to creatinine in solution over time. The following summarizes results of studies that provide some evidence of bioavailability, efficacy, and safety of creatine salts.

9.1.1. Creatine Citrate

Figure 6 shows the chemical structure of two common creatine salts. Jäger et al. [63] compared the effects of oral ingestion of 5 g of CrM to 6.7 g of tri-creatine citrate (CC) and 7.7 g of creatine pyruvate (CPY) that provided equimolar amounts of creatine. Tri-creatine citrate is a 1:1 salt of creatine and citric acid, with two additional creatines forming a complex with the 1:1 salt. The second and third acid moiety of citric acid are not strong enough acids to form salts with creatine. Peak concentrations of creatine were significantly higher with CPY (CrM 761.9 ± 107.7, CC 855.3 ± 165.1, CPY 972.2 ± 184.1 µmol/L) while AUC values did not significantly differ among treatments (2384 ± 376.5, 2627 ± 506.8, 2985 ± 540.6 mM/h, respectively). Results support contentions that provision of equimolar amounts of CC and CPY can serve as a bioavailable source of creatine. Conversely, Gufford et al. [179] reported that CC and CPY had less permeability than CrM across caco-2 monolayers cells, which is used as a model to assess intestinal absorption. However, it is unlikely that the small differences in blood creatine levels observed would promote greater creatine retention and/or a bioenergetic advantage. To date, we are unaware of any study that has evaluated the effects of CC or CPY on muscle or brain creatine content.

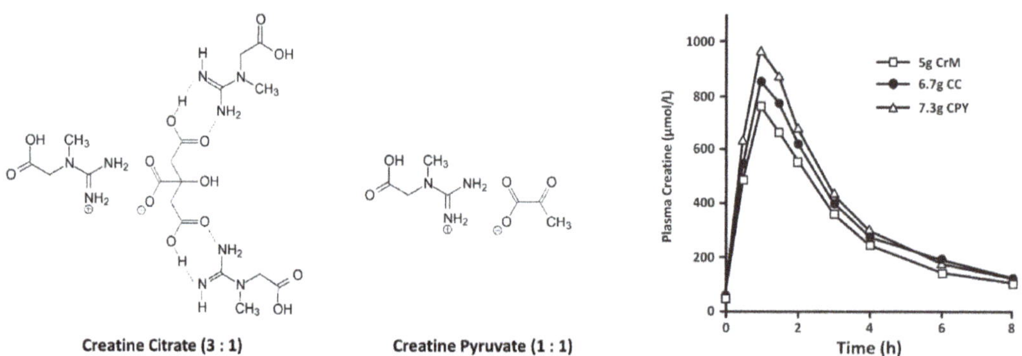

Figure 6. Chemical structure of tri-creatine citrate and creatine pyruvate with plasma creatine changes after oral administration of equal molar doses of creatine monohydrate (CrM), creatine citrate (CC) and creatine pyruvate (CPY). Adapted from Jäger et al. [63].

In terms of efficacy, several studies have evaluated whether creatine citrate (CC) supplementation can affect exercise capacity. For example, Smith and colleagues [180] found ingesting 20 g/day of di-creatine citrate for 5 days delayed neuromuscular fatigue in women. Jäger et al. [174] reported that ingestion of 5 g/day of CC (providing 3.25 g/day of creatine) for 28 days significantly increased intermittent maximal effort handgrip force compared to placebo. Graef and coworkers [181] reported that 10 g/day of CC supplementation for 30 days during high-intensity interval training significantly increased the ventilatory threshold, but did not enhance maximal aerobic capacity. Smith and colleagues [172] reported that 20 g/day of creatine di-citrate supplementation for 5 days had no detrimental or ergogenic effects on running critical velocity, aerobic capacity, or time to exhaustion. Finally, Fukuda and assistants [182] reported that CC supplementation (4 × 5 g with 18 g dextrose for 5 days) improved anaerobic run capacity in men, but not women. While these studies provide evidence that CC can serve as a viable source of creatine by increasing blood creatine level in a similar manner as CrM and there is some data supporting the ergogenic benefit compared to placebo, the impact of CC supplementation has not been assessed on muscle or brain creatine content. Therefore, there are no studies indicating that CC is more bioavailable, more effective, or a safer source of creatine than CrM. Given this, CC is categorized as having limited evidence to support bioavailability, efficacy, and safety.

9.1.2. Creatine Pyruvate

Calcium pyruvate supplementation (e.g., 6–25 g/day) has been reported to affect exercise performance and promote fat loss [183–186]. Stone et al. [83] also reported that CrM and the combination of CrM and calcium pyruvate supplementation during 5 weeks of off-season training improved training and body composition adaptations. In contrast, calcium pyruvate supplementation alone had no effects. Therefore, there was some rationale in developing a creatine salt with pyruvate [32]. As seen in Figure 7, ingestion of 7.3 g of CPY promoted significant increases in plasma creatine levels in a similar manner as ingesting 5 g of CrM [63]. This group also reported that ingestion of 5 g/day of CPY (providing 3 g/day of creatine) for 28 days significantly increased intermittent maximal effort handgrip force [174]. Another study found that ingesting 7 g/day of CPY for 7 days had no effects on endurance capacity or repeated sprint performance in cyclists [175]. However, ingesting 7.5 g/day of CPY improved paddle rate and lowered blood lactate in Olympic canoeists suggesting an improvement in aerobic exercise efficiency [187]. These studies indicate that CPY can serve as effective source of creatine to increase blood creatine content and a few short-term studies suggest there may be some ergogenic values. However, since there are no data assessing the effect of CPY supplementation on muscle or brain creatine content and only a limited number of studies have evaluated efficacy and safety, CPY is classified in the limited bioavailability, efficacy, and safety category. With that said, there is no evidence that CPY is more effective than CrM in increasing muscle creatine content and/or performance.

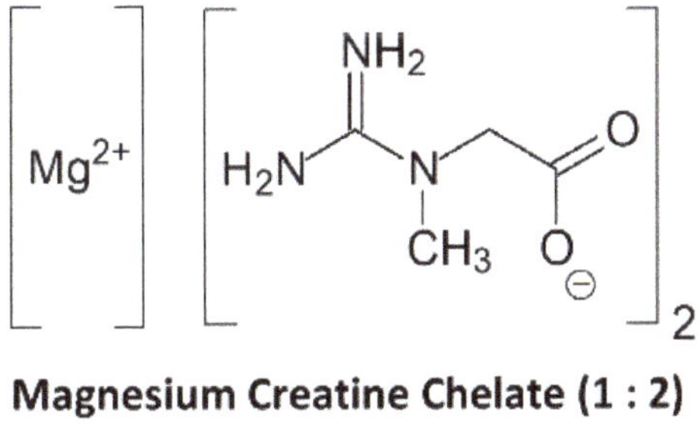

Magnesium Creatine Chelate (1 : 2)

Figure 7. Chemical structure of magnesium creatine.

9.2. Magnesium Creatine Chelate

Magnesium creatine chelate (MgCr-C) has been marketed as a more bioavailable source of creatine (see Figure 7). The rationale is that since magnesium is a cofactor in ATP reactions and the only mineral that decreases during exercise, there may be additive benefit in combining creatine with magnesium. There are also claims that MgCr-C supplementation can improve muscle protein synthesis. A patent described bioavailable chelates of creatine and essential metals [188]. Since MgCr-C contains about 84.4% of creatine by molecular weight (see Table 1), it could theoretically serve as a good source of creatine if creatine easily dissociates from MgCr-C and equimolar amounts of creatine were consumed compared to CrM. However, it is marketed as a much more bioavailable source of creatine than CrM with recommended doses of only 1 g per 18.2 kg (40 lbs.) of body weight per day (about 3.8 g/day for a 70 kg individual). We are aware of no data showing that MgCr-C increases blood creatine levels or promotes greater creatine retention in skeletal muscle. Thus, there are no data supporting that MgCr-C is more bioavailable to tissue than CrM.

Several studies have evaluated the effects of MgCr-C on performance related variables. For example, Brilla and coworkers [189] evaluated the effects of ingesting 5 g of CrM with 800 mg of magnesium oxide or magnesium plus 5 g of MgCr-C for 2 weeks compared to placebo on body water and isokinetic strength performance in recreationally active participants. Results revealed that body water increased with MgCr-C, but not CrM while torque and power increased similarly with CrM and MgCr-C. Selsby and colleagues [190] evaluated the effects of supplementing 2.5 g/day of CrM, MgCr-C, or a placebo for 10 days on strength and muscle endurance. Results revealed that CrM and MgCr-C were both effective in increasing performance with no differences observed between types of creatine ingested. Finally, Zajac et al. [191] reported that 5.5 g/day of MgCr-C supplementation during 16 weeks of soccer training improved repeated sprint ability performance compared to placebo. Interestingly, creatinine levels were also significantly increased throughout training with MgCr-C (0.83 to 1.87 mmol/L) compared to placebo (0.92 to 0.82 mmol/L), which is higher than reported in other long-term CrM studies during training in hot and humid environments that administered 5–10 g/day of CrM for 21 months in well-trained athletes [144]. However, this study did not compare the effects of consuming MgCr-C to CrM, and performance changes were consistent with other studies conducted on CrM supplementation. Consequently, there is no data showing that MgCr-C increases blood or muscle creatine and there is only limited data showing potential ergogenic effect. There is also no evidence that MgCr-C is more bioavailable, efficacious, and/or a safer source of creatine than CrM. For this reason, MgCr-C is listed in some evidence to support bioavailability, efficacy, and safety category.

9.3. Creatine Ethyl Ester

Another source of creatine that claims to have better solubility, bioavailability, and efficacy than CrM is creatine ethyl ester (CEE). CEE is basically a creatine molecule with a H+ removed from the second N position (i.e., NH versus NH_2) and a methyl group (CH_2-CH_3) added to the terminal O position through an esterification reaction (see Figure 8). Thus, CEE is a chemical alteration of creatine and is not actually creatine. For CEE to act like creatine it would have to be de-ethylated and an H+ added back to the NH of the molecule at nearly 100% efficiency to deliver 94% of an equivalent CrM dose. Marketing claims suggest that CEE is absorbed faster and more efficiently than CrM, so no loading dose is needed. Additionally, CEE is claimed to have less anecdotal side effects than CrM like bloating and dehydration. For this reason, the recommended dosages of CEE are typically 2–6 g/day.

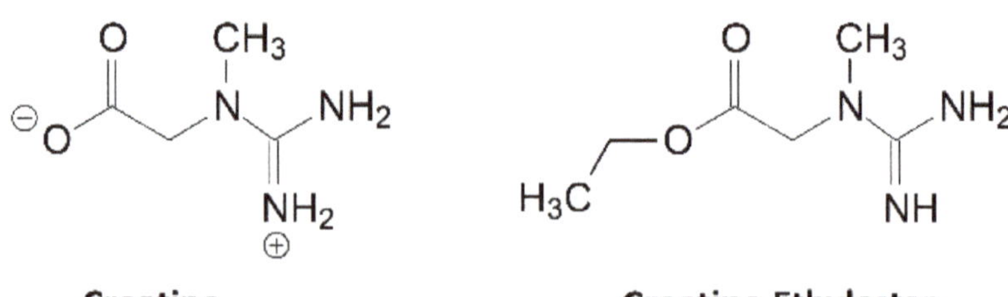

Figure 8. Chemical structure of creatine and creatine ethyl ester.

While proponents of CEE have assumed that all orally ingested CEE is converted to creatine in vivo, available studies suggest it is less efficient. For example, Childs and Tallon reported that CEE rapidly degrades to creatinine when exposed to stomach acid [34]. Giese et al. [35,192] reported that under physiological conditions, CEE non-enzymatically converted to creatinine with no measurable conversion to creatine (see Figure 9). Likewise, Katseres and colleagues [193] reported that the half-life of CEE was in the order of one minute suggesting CEE may hydrolyze too quickly to reach muscle cells in its ester form. On the other hand, Gufford et al. [194] reported that CEE converted to creatinine in a linear manner as pH levels dropped below 8.0 and that CEE was mostly stable at a pH of 1.0. Since acidity in the stomach generally ranges from 1.5 to 3.5, it is likely that some CEE is degraded into creatinine during normal digestion while delivering some level of creatine to blood.

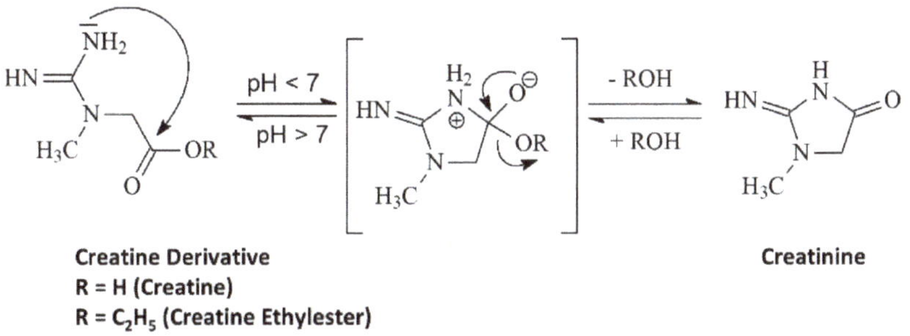

Figure 9. Degradation of creatine ethyl ester to creatinine. Adapted from Jäger et al. [25].

Spillane and colleagues [27] compared the effects of supplementing the diet with a 0.30 g/kg of fat-free body mass (approximately 20 g/day) for 5 days followed by ingestion of 0.075 g/kg of fat free mass (approximately 5 g/day) for 42 days of a placebo, CrM, or CEE on muscle creatine content and performance adaptations. If CEE is bioavailable, those taking CEE should have increased muscle creatine content better than those taking a placebo. Likewise, if CEE was more bioavailable than CrM, greater changes would be seen in the CEE group compared to those in the CrM group. As seen in Figure 10A, fasting serum creatine levels significantly increased only in the CrM group. While this was not a pharmacokinetic oral dose study, it is interesting that CEE supplementation had no effect on fasting serum creatine levels compared to a placebo. Conversely, CEE significantly increased serum creatinine levels by more than two-fold after 6, 27, and 48 days of supplementation in comparison to the placebo and CrM groups (Figure 10B). The values observed exceeded normal creatine values even for highly trained athletes training in hot and humid environments [13]. In addition, while CEE supplementation promoted a significant increase in muscle total creatine content after 27 days of supplementation compared to those ingesting a placebo, those taking CrM observed significantly greater increases compared to the placebo and CEE groups (Figure 10C). These findings suggest a large amount of CEE is converted to creatinine and CEE is less effective in increasing muscle creatine content than CrM. This was despite including a 20 g/day loading dose of CEE that manufacturers claimed is unnecessary due to greater bioavailability. Moreover, CEE supplementation did not promote greater changes in body composition, strength, or anaerobic power during training compared to CrM supplementation. These findings directly refute claims that CEE is more bioavailable source of creatine than CrM and that CEE promotes greater training adaptations than CrM. Further, the clinically significant increase in creatinine levels observed should raise some concerns about potential safety of CEE as has reported two case studies [195,196].

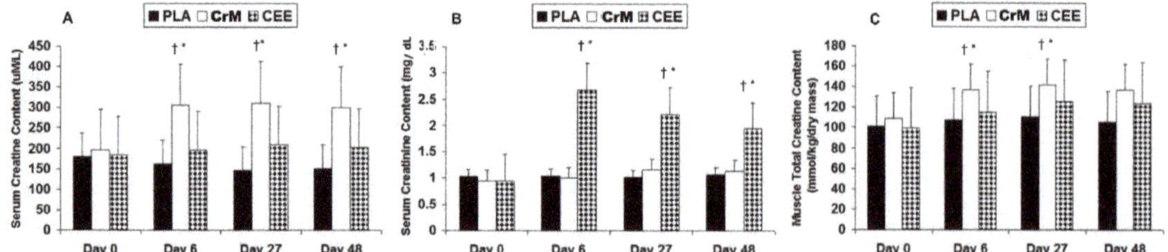

Figure 10. Serum creatine (**A**), serum creatinine (**B**), and muscle creatine content (**C**). * Represents significant change from baseline. † in panel A indicates significantly higher serum creatine concentrations in CrM when compared to PLA ($p = 0.007$) and CEE ($p = 0.005$). † in panel B indicates serum creatinine in the CEE group was greater than PLA ($p = 0.001$) and CrM ($p = 0.001$). † in panel C shows the PLA group was significantly less than the CrM ($p = 0.026$) and CEE ($p = 0.041$) groups. Adapted from Spillane et al. [27].

With that said, some have pointed to the results of a recent study conducted by Arazi and associates [197] to support claims about the efficacy of CEE. The researchers evaluated the effects of 6 weeks of CEE supplementation (20 g/day for 5 days and 5 g/day for 37 days) compared to consuming a placebo during resistance training (3 sets of 8–10 repetitions at 60–80% or one repetition maximum, 3 times per week) in untrained, younger, and underweight men. The investigators reported that CEE supplementation during resistance-training promoted significant increases in body weight and leg press strength, while percent body fat decreased to a greater degree with placebo ingestion. In addition, some differences were reported in anabolic and catabolic hormones that prior have not been previously studies conducted on creatine had not reported [62]. While this study indicates that higher than recommended doses of CEE can positively affect training adaptations, this study did not assess the effects of CEE supplementation on blood or muscle creatine content or compare the efficacy of CEE supplementation to CrM. The results observed are consistent with those found in the Spillane et al. study [27] in that CEE had some benefit over placebo ingestion, but results were not better than CrM. Given that CEE promoted a modest but less effective increase in muscle creatine content in that study, one would expect some benefit of CEE supplementation during training if higher than recommended doses are ingested. However, there is no evidence that ingesting recommended doses of CEE is effective or that ingesting typical CrM loading, and maintenance doses of CEE is more effective than CrM. We are also not aware of any studies that have evaluated other marketed forms of CEE (i.e., creatine methyl ester hydrochloride, di-acetyl creatine ethyl ester, creatine ethyl ester pyruvate, creatine ethyl ester malate, or creatylglycine ethyl ester fumarate). Nevertheless, since there is some evidence that ingesting high doses of CEE can increase muscle creatine content and performance compared to placebo, we have categorized CEE in some evidence category. However, we recommend that additional research evaluate safety given the increased creatinine levels observed.

9.4. Creatine HCl

Creatine hydrochloride (Cr-HCl) has been marketed as a more bioavailable source of creatine than CrM. As shown in Figure 11, Cr-HCl is a salt of HCL and creatine molecule. Like other creatine salts, adding hydrochloric acid to creatine would be expected to decrease pH and improve solubility. Marketing claims indicate that Cr-HCl has a 38 times greater bioavailability than CrM [198]. The basis for this claim appears to come from a report from Gufford and colleagues [179] who conducted physiochemical characterization studies on several N-methylguanidinium salts, including creatine Cr-HCl. They reported that Cr-HCl contains about 78% creatine by molecular weight and that Cr-HCl was 37.9 times more soluble in water than CrM at 25 °C. However, CrM was assessed at a saturation pH of 8.6 while Cr-HCL was measured at a saturation pH of 0.3. While mixing creatine in acidic solutions may improve solubility when mixed in water, as noted above, it would have no effect on bioavailability. These claims are also apparently based on a report from Alraddadi et al. [199] who conducted a bioavailability study in rats with labeled creatine (creatine-^{13}C) at a low (10 mg/kg) and high (70 mg/kg) oral doses. They then assessed the amount of creatine-^{13}C incorporated into plasma, muscle and brain tissue and used a simulated prediction model to estimate how Cr-HCl would theoretically affect tissue creatine retention based on differences in solubility. While this is an interesting approach, there are several problems with using these findings to make claims about Cr-HCl. First, this is only a theoretical modeling study. The researchers did not directly compare Cr-HCl to CrM intake on plasma or tissue creatine content. Second, it is well-known that there are species specific differences in creatine metabolism and storage [200]. Therefore, you cannot directly extrapolate results from mice or rat data to human creatine oral dosing studies. Pharmacokinetic and creatine retention studies need to be conducted in humans to assess whether Cr-HCl promotes greater creatine retention in tissue to assess the validity of this claim.

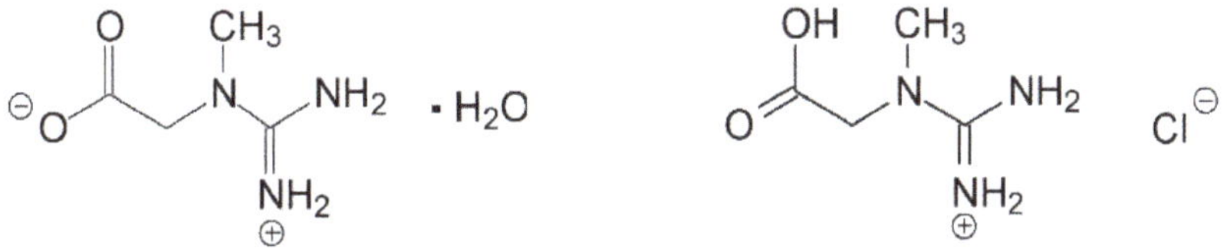

Creatine Monohydrate   **Creatine Hydrochloride**

Figure 11. Comparison of creatine monohydrate and creatine HCl structures.

As of this writing, no PubMed indexed articles have been published on Cr-HCl and muscle creatine retention or performance. However, several articles have been published in non-indexed journals from Brazil that have been cited in marketing materials. In the first study, de França and colleagues [65] evaluated the effects of supplementing the diet with 1.5 g/day of Cr-HCl, 5 g/day of Cr-HCl, and 5 g/day of CrM compared to controls during 4 weeks of resistance-training in 40 young recreational weightlifters on strength gains and skinfold caliper determined body composition. The researchers reported some benefits of Cr-HCl and CrM supplementation on leg press and skinfold determine body composition. However, the use of skinfold calipers to estimate body composition and statistical analysis methods employed make it difficult to draw any conclusions. In fact, gains in fat-free mass were greatest in the CrM group (+1.7 kg), but supposedly not significantly different than observed with 5 g/day of Cr-HCL (+1.6 kg) that were reported to be significantly different than controls (+1.1 kg) and those ingesting 1.5 g/day of Cr-HCl (+1.1 kg). In a follow-up study [201], this research group administered 5 g/day of CrM or 1.5 g/day of Cr-HCL with 3.5 g/day of resistant starch for 30-days to Brazilian Olympic level athletes. Results revealed both groups increased skinfold caliper determined fat-free mass and strength, although bioelectric impedance determined total body water was increased to a greater degree in the CrM group (CrM + 1.81 L vs. Cr-HCL + 0.24 L). This would be expected, given the creatine content based on molecular weight in these dosages was 35.1 g in the Cr-HCl group compared to 131.9 g in the CrM group over the 30-day period. Finally, a study conducted by Tayebi and Arazi [202] evaluated the effects of ingesting 3 g/day of Cr-HCL, 3 g/day of CrM, and 20 g/day of CrM, or a placebo for 7 days on anaerobic power and hormone levels. Results revealed that ingestion of 3 g/day of Cr-HCl did not promote greater gains in performance or hormonal responses than 3 or 20 g/day of CrM as claimed. The authors concluded that Cr-HCl does not appear to be a more effective source of creatine than CrM. Thus, while Cr-HCl is a simple salt that should readily disassociate into creatine and HCL, there is no evidence that Cr-HCl is absorbed more effectively than CrM in humans; Cr-HCl promotes greater muscle creatine retention than CrM at equivalent doses; or, that lower doses of Cr-HCl are as effective as standard supplementation protocols with CrM. Given this analysis, Cr-HCl is classified in some evidence category to support efficacy compared to placebo. However, claims that Cr-HCl is more bioavailable, effective, and/or a safer source of creatine than CrM are not supported.

9.5. Creatine Nitrate

A number of studies have indicated that dietary nitrates, typically ingested in the form of beet root juice or nitrate powder, can improve endurance [203] and explosive exercise capacity [204]. Recommended dosages generally range between 300–600 mg ingested 1–2 h prior to exercise [204–206]. Since nitrate can ionically bond to creatine and form a salt (see Figure 12), creatine nitrate (CrN) has been developed and marketed as a bioavailable source of creatine for dietary supplements [207]. By molecular weight, CrN contains 67.5% creatine. Therefore, ingesting 1 g of CrN would theoretically provide 0.675 g of creatine and 0.325 g of nitrate. Marketing claims suggest greater bioavailability and therefore recommended doses are typically 1–2 g of CrN per/day [208]. In terms of bioavailability, there is limited data available. However, Galvan and colleagues [29] conducted a pharmacokinetic study evaluating the effects of acute oral ingestion of a placebo, 1.5 g of CrN (CrN-Low), 3 g of CrN (CrN-High), and 5 g CrM on blood creatine and nitrate levels. Results revealed that the plasma creatine AUC over a 5 h period for CrM (5634.4 $\pm$ 1949.8 μmol/L) was significantly greater than the placebo (1012.4 $\pm$ 1882.2 μmol/L), CrN-Low (2342.0 $\pm$ 3133.3 μmol/L, $p = 0.004$), and CrN-High (1761.7 $\pm$ 3408.8 μmol/L, $p = 0.007$) treatments with no differences seen between the CrN dosages. Conversely, the nitrate AUCs in the CrN groups were significantly greater than the placebo and CrM treatments in a dose related manner. These investigators also evaluated the effects of ingesting four doses a day of either a placebo (5 g dextrose), CrM (3 g CRM with 2 g dextrose, CrN-Low (1.5 g CrN, 3.5 g dextrose), and CrN-High (3 g CrN, 2 g dextrose) for 7 days followed by ingesting one dose per day for 21 days as a maintenance dose. Muscle biopsies were obtained at 0, 7, and 28 days to assess muscle creatine content. Results revealed that 7 days of creatine loading (12 g/day of CrM and CrN) significantly increased muscle creatine content in the CrM (7.1 mmol/kg DW) and CrN-High (4.6 mmol/kg DW). However, no difference was seen compared to placebo was observed when ingesting 4×1.5 g/day of CrN for 7 days or after 28 days of taking 1.5 or 3 g/day of CrN. These findings suggest that CrN can be a bioavailable source of creatine proportional to the amount of creatine delivered during the loading phase (i.e., 54.6 g for CrN-High versus 73.8 g for CrM), but not more bioavailable than CrM when equivalent doses are ingested. On the other hand, Ostojic et al. [209] conducted a study evaluating the effects of CrM and CrN supplementation on MRS determined skeletal muscle creatine content and markers of health. In a randomized and crossover manner with a 7 day washout period, participants ingested a placebo, 3 g/day of CrN, 3 g/day of CrM, or 3 g/day of CrN + 3 g/day of CrM for 5 days. This theoretically provided a total of 0, 9.75, 13.2, and 22.9 g of creatine during the 5-day period. The researchers found that peak serum creatine increased to a greater degree with CrN + CrM supplementation (CrM 118.6 $\pm$ 12.9, CrN 163.8 $\pm$ 12.9; CrN + CrM 183.7 $\pm$ 15.5 μmol/L) while muscle creatine increased to a greater degree with CrN ingestion (CrM 2.1%, CrN 8.0%, CrN + CrM 9.6%). However, a limitation to this study is that only a 7-day washout period was observed between treatments. It is well known that it takes about 4 weeks for muscle creatine to return to normal after creatine supplementation [47]. Thus, it is possible that the testing order may have confounded results. Nevertheless, results are conflicting on whether short- or long-term CrN supplementation (3 g/day) significantly increase muscle creatine levels.

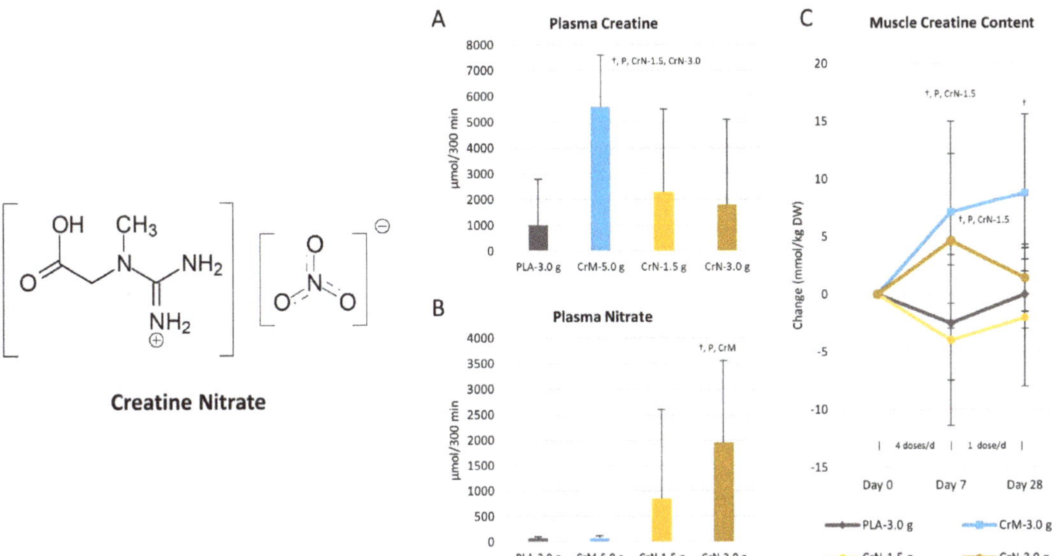

Figure 12. Chemical structure of creatine nitrate. (**A,B**) show 5 h area under the curve data after plasma creatine and plasma nitrate, respectively. (**C**) shows mean changes in muscle creatine content with 95% confidence intervals after 7 days of loading 4 doses/day and 21 days of ingesting 1 those/day. † Represents significant change from baseline, PLA and P represent placebo, CrM represents creatine monohydrate, and CrN represents creatine nitrate.

In terms of performance, Gavan and colleagues found some ergogenic benefit of CrN supplementation on muscular endurance, but they were unrelated to changes in muscle creatine content, suggesting the ergogenic benefit was primarily due to providing ergogenic levels of nitrate and not creatine at the dosages studied. Dalton and colleagues [36] reported that ingesting 3 and 6 g/day of CrN for 5 days significantly improved some measures of strength and muscle endurance compared to placebo. However, it is unclear whether these changes were primarily due to creatine and/or nitrate. Some improvement in exercise performance was also reported with acute [210] and 8-weeks [211] supplementation of a pre-workout supplement containing 2 g/day of CrN. However, since the supplement contained caffeine and other ergogenic nutrients, the benefits cannot be attributed to CrN. Further, it remains to be determined whether CrN supplementation has any additional benefit than simply co-ingesting CrM another source of nitrate (e.g., beet powder).

In terms of safety, since nitrates may lower blood pressure, there has been some concern that CrN may promote hypotension, particularly around intense exercise and/or if individuals take higher than recommended doses. Several studies have assessed safety of acute and chronic CrN supplementation. Dalton et al. [36] reported that ingestion of up to 6 g of CrN for 6 days does not negatively affect resting hemodynamics, response to a postural challenge, the ability to perform high-intensity exercise, or clinical chemistry profiles. Joy and colleagues [212] reported that 28 days of CrN supplementation (1 and 2 g/day) during training had no adverse effects on clinical blood chemistries compared to a non-supplemented group. Galvan and coworkers [29] also found no adverse effects after 28 days of supplementation (3 g/day). Finally, Jung and associates reported no adverse effects of participants consuming a pre-workout supplement containing 2 g/day of CrN for 8 weeks. Thus, CrN appears to be safe when taken in these amounts and timeframes. However, CrN has only been approved as a dietary supplement by the U.S. FDA at levels of 750 mg per day, which is below any meaningful level expected to increase muscular creatine levels and performance. Based on this analysis, there is some evidence showing

CrN may serve as a bioavailable and effective source of creatine. However, most studies have used higher than recommended doses, and those studies show that CrN is not more effective than CrM supplementation.

9.6. Buffered Creatine Monohydrate

In the early 2010s, a "buffered" or "pH-correct" form of CrM was heavily marketed as a more bioavailable source of creatine than CrM [213]. According to the patent [214], CrM was better stabilized by adding an alkaline powder (e.g., soda ash, magnesium glycerol phosphate, bicarbonate) to CrM (or other purported forms of creatine) in order to increase pH between 7–14. Consequently, they developed CrM that was "synthesized to a pH of 12" (CrM-Alk) and claimed that due to greater stability in preventing the conversion of CrM to creatinine, CrM-Alk was up to 10 times more bioavailable than CrM [213]. Therefore, 1.5 g of CrM-Alk was purported to be equivalent to ingesting 10–15 g of CrM. Additionally, the company theorized that since less CrM-Alk was needed to be ingested, there would be fewer side effects than CrM [213]. To support these claims, the manufacturers cite a non-peer reviewed report from Bulgaria on their website [215]. In this report, 24 healthy Olympic level soccer players were administered increasing doses of CrM-Alk or CrM at one-month intervals (i.e., 0, 1.5, 4.5, and 6 g/day). The authors reported that CrM-Alk promoted less of an increase in urine creatinine than CrM despite changes being <0.2% different between groups at each time point; that urine pH increased by 0.65 in the CrM-Alk group, but only 0.1 in the CrM group (CrM-Alk 5.27 to 5.92; CrM 5.5 to 5.6); and, peak oxygen uptake increased in the CrM-Alk group (<30 mL/min over time or <1.0% for a trained individual). No differences in body weight were reported. These investigators concluded the CrM-Alk group "outperformed creatine monohydrate as a creatine product" despite not performing any statistical analysis to determine if these minimal differences were statistically significant. Thus, the report does not validate claims than CrM-Alk supplementation is a more bioavailable, efficacious, or safer form of creatine than CrM.

Conversely, in a very well-controlled clinical trial, Jagim and colleagues [28] compared the effects of CrM-Alk supplementation at recommended and equivalent doses to CrM during 28 days of training in resistance-trained athletes with no recent history of creatine supplementation. In a double-blind manner, 36 resistance-trained participants were randomly assigned to ingest CrM (4×5 g/day for 7-days, 5 g/day for 21-days), CrM-Alk at recommended doses (1.5 g/day for 28-days), or CrM-Alk with equivalent doses to CrM (4×5 g/day for 7-days, 5 g/day for 21-days). Muscle biopsies, dual-energy x-ray absorptiometry (DXA) determined body composition, and performance measures were obtained after 0, 7, and 28 days of supplementation. Results revealed that neither recommended doses of CrM-Alk or loading and maintenance equivalent doses of CrM-Alk to CrM promoted greater changes in muscle creatine content, body composition, strength, or anaerobic capacity than CrM (see Figure 13). In fact, muscle creatine content was not significantly increased after 7 or 28 days of supplementation at recommended doses (-6.4 ± 37.8; 13.7 ± 42.2 %, respectively). There was some evidence that ingesting higher doses of CrM-Alk increased muscle creatine content after 28 days (6.2 ± 29.2; $27.3 \pm 49.1\%$, respectively), but these values were less than observed with CrM (23.5 ± 49.0; $50.4 \pm 44.8\%$, respectively). Thus, while high doses of CrM-Alk may increase muscle creatine content to some degree over time, there is no evidence that CrM-Alk is up to 10 times more bioavailable than CrM and/or recommended doses are efficacious. Additionally, was no evidence that CrM-Alk promoted greater training adaptations than those taking CrM or that participants taking CrM-Alk experienced fewer side effects than those taking CrM. Therefore, buffered creatine monohydrate is classified in some evidence to the support bioavailability, efficacy, and safety categories, but there is no evidence that buffered creatine is better.

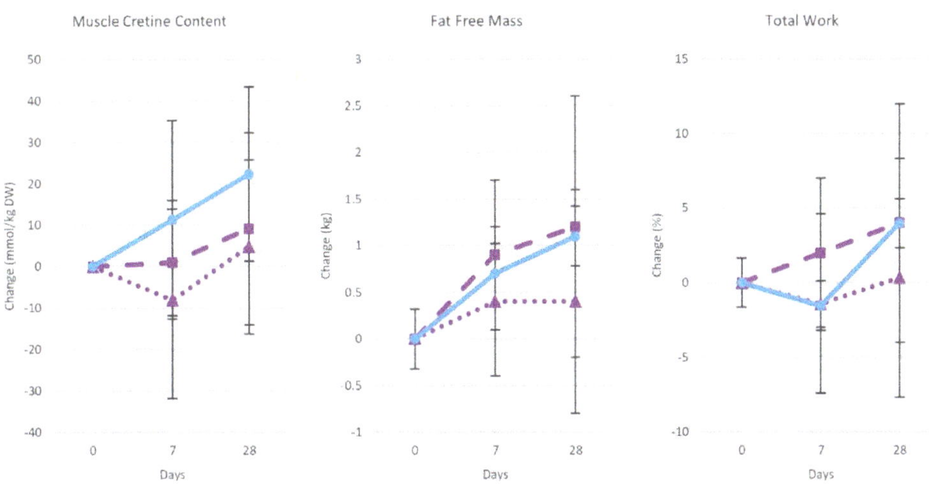

Figure 13. Changes in muscle creatine content, fat-free mass, and 30 s cycling sprint performance after 7 and 28 days of CrM-Alk supplementation at 1.5 g/day recommended doses (▲), CrM-Alk supplementation of 20 g/day for 7 days and 5 g/day for 21 days (■), or CrM supplementation of 20 g/day for 7 days and 5 g/day for 21 days (●). Adapted from Jagim et al. [28].

Table 3 summarizes results of studies assessing creatine-containing compounds in which there is some evidence supporting bioavailability, efficacy, and/or safety. As described above, while some effects were observed compared to placebo ingestion and in some instances have comparable effects on performance as CrM, none of these forms have been shown to promote greater creatine retention in muscle than CrM.

Table 3. Creatine containing compounds other than CrM with some evidence supporting bioavailability, efficacy, and/or safety.

Reference	Participants	Design	Duration	Dosing Protocol	Findings	Side Effects
Creatine Salts						
Jäger et al. [63]	3 females and 3 males	RDBPC	1 oral dose with 7 day washout	5 g CrM 6.7 g CC 7.3 g CPY	Creatine peak AUC was higher with CPY with no differences in absorption kinetics	None reported
Smith et al. [180]	15 recreationally active women (22.3 ± 0.6 yrs)	RDBP	5 days	20 g/day of CC	CC loading delayed the onset of neuromuscular fatigue during cycle ergometry.	None reported
Jäger et al. [174]	49 healthy males (26.5 ± 4 yrs)	RDBP	28 days	5 g/day of CC, CPY, or PLA	CPY and CC ↑ intermittent handgrip exercise of maximal intensity. Some evidence CPY might benefit endurance exercise.	None reported
Graef and coworkers [181]	43 recreationally active men (22.6 ± 5 yrs)	RDBP	5 days/week for 6-weeks	2 × 5 g/day of PLA or CC on training days	CC increases ventilatory anaerobic threshold (PLA 10%, CC 16%). No differences in time to exhaustion or total work.	None reported

Table 3. *Cont.*

Reference	Participants	Design	Duration	Dosing Protocol	Findings	Side Effects
Smith et al. [172]	55 active men (27) and women (28)	RDBP	5 days	4 × 5 g/day of CC or PLA	CC did not positively or negatively affect maximal aerobic capacity, critical velocity, time to exhaustion, or body mass.	None reported
Fukuda et al. [182]	50 recreationally active men (24) and women (26) 22 ± 3 yrs	RDBP	5 days	4 × 5 g/day of CC or PLA	CC loading ↑ anaerobic running capacity (+23%) with no effect in PLA group in men but not women.	None reported
Stone et al. [83]	42 American football players	RDBP	5 weeks	0.22 g/kg/day of PLA, CrM, caPYR, or CrM + caPYR	CrM and CrM + caPYR ↑ strength, FFM, and power output. No difference from PLA or caPYR alone.	GI issues with caPYR. None reported with CrM
Van Schuylenbergh et al. [175]	14 well-trained male endurance athletes (4 cyclists, 10 triathletes)	RDBP	7 days	2 × 3.5 g of CPY with 8 g CHO or PLA	CYP had no effects on 1-h time trial steady-state power output, interval sprints, total work lactate, or heart rate.	None reported
Nuuttilla et al. [187]	Olympic canoeists	RDBP	7 days	7.5 g/day of CPY or PLA	CPY improved paddle rate and lowered blood lactate suggesting an improvement in aerobic exercise efficiency.	None reported
Magnesium Creatine Chelate						
Brilla et al. [189]	35 recreationally active men	RDBP	14 days	800 mg/day magnesium (Mg) and 5 g/day Cr as Mg oxide plus Cr or MgCr-C	Body mass and power ↑ in both Cr groups while intracellular and extracellular water and peak torque only increased in the MgCr-C group	None reported
Selsby et al. [190]	31 resistance-trained men	RDBP	10 days	2.5 g/day of PLA, Cr or Mg-Cr	Both Cr groups improved bench press total work compared to PLA. No differences between groups.	None reported
Zajac et al. [191]	20 elite soccer players	RDBP	16 weeks	5.5 g/day of r MgCr-C or PLA	MgCr-C ↑ 35 m repeated sprint performance, total time, average power, and peak power with no changes in PLA group.	MgCr-C ↑ serum creatinine compared to PLA
Creatine Ethyl Ester						
Spillane et al. [27]	30 healthy males (20.4 ± 1.7 yrs)	RDBP	47 days	0.30 g/kg FFM for 5-days, 0.075 g/kg FFM for 42 days of PLA, CrM, or CEE 392	CEE ↑ in muscle TCr after 27-days compared to PLA. However, CrM observed significantly greater ↑ in TCr compared to PLA and CEE. CEE did not promote > training adaptations.	CEE ↑ serum creatinine twofold > than PLA and CrM. None reported with CrM.

Table 3. *Cont.*

Reference	Participants	Design	Duration	Dosing Protocol	Findings	Side Effects
Arazi et al. [197]	16 resistance trained males	RDBP	42 days	4 × 5 g/day of PLA or CEE for 5 days, 5 g/day for 37 days	CEE during resistance-training ↑ body weight and leg press strength while percent body fat ↓ with some evidence of an ↑ in testosterone and growth hormone.	None reported.
Creatine HCl						
de França et al. [65]	40 healthy males and females	RDBP	28 days	5 g/day PLA, 1.5 g/day of Cr-HCl, 5 g/day of Cr-HCl, or 5 g/day CrM	Reported some effects on skinfold determined fat mass and FFM and leg press strength but gains in CrM were greater than Cr-HCl	None reported.
Yoshioka et al. [201]	11 healthy elite Brazilian gymnasts	RDBP	30 days	5 g/dayay of CrM or 1.5 g/dayay of Cr-HCL with 3.5 g/dayay of resistant starch	Skinfold caliper determined FFM, strength, and BIA determined total body water was increased to a greater degree in the CrM group (CrM + 1.81 L vs. Cr-HCL +0.24 L).	None reported.
Tayebi et al. [202]	36 resistance trained men	RDBP	7 days	20 g/day CrM, 3 g/day CrM, 3 g/day Cr-HCL, or PLA	3 g/day of Cr-HCl did not promote greater gains in performance or hormonal responses than 3 or 20 g/day of CrM.	None reported.
Creatine Nitrate						
Ostojic et al. [209]	10 healthy men	RDBPC	1 oral dose	3 g CrN + 3 g CNN, 3 g CrN, 3 g CrM	CrN + CNN ingestion promoted a greater increase in serum creatine AUC levels (183.7 ± 15.5, 163.8 ± 12.9, and 118.6 ± 12.9 μmol/L, respectively).	None reported.
Ostojic et al. [209]	10 healthy men	RDBPC	5 days	3 g/day CrN + 3 g/day CNN, 3 g/day CrN, 3 g/day CrM	MRS determined muscle creatine content increased to a greater degree with CrN + CNN (9.6%, 8.0%, 2.1%, respectively)	Irregular bowel movement (1 CrN and CrN + CNN), Excessive sleepiness (1 CrN), Seldom stomach bloating (1 CrM). CrN + CNN decrease eGFR determined kidney function.

Table 3. *Cont.*

Reference	Participants	Design	Duration	Dosing Protocol	Findings	Side Effects
Galvan et al. [29]	13 males	RDBPC	1 oral dose with 7 day washout	1.5 g CrN (CrN-Low), 3 g CrN (CrN-High), 5 g CrM or a placebo	CrM ↑ plasma Cr AUC to a greater degree than PLA, CrN-Low, and CrN-High while plasma nitrate ↑ in CrN treatments.	None reported.
Galvan et al. [29]	48 active males	RDBP	28 days	4 × 5 g PLA, 4 × 1.5 g/day of CrN (CrN-Low), 4 × 3 g/day CrN (CrN-High), 4 × 3 g/day CrM for 7 days and 1 dose/d for 21 days	Creatine loading (12 g/day of CrM and CrN) ↑ muscle TCr in the CrM (7.1 mmol/kg DW) and CrN-High (4.6 mmol/kg DW) groups. CrM maintained ↑ muscle TCr. CrN-Low had no effects on TCr compared to PLA after 7 and 28 days. 3 g/day of CrN was not sufficient to maintain elevated muscle TCr after 28 days.	None reported.
Dalton et al. [36]	28 participants (18 men, 10 women)	RDBPC	6 days	3 g/day of PLA. 3 g/day CrN, 6 g/day CrN.	Up to 6 g of CrN for 6-days does not negatively affect resting hemodynamics, response to a postural challenge, the ability to perform high-intensity exercise, or clinical chemistry profiles.	None reported.
Joy et al. [212]	58 young males and females (24.3 ± 4 yrs)	R	28 days	Control group, 1 g/day CrN, 2 g/day CrN with other nutrients	1–2 g/day of CrN supplementation during training had no adverse effects on clinical blood chemistries compared to a non-supplemented group.	None reported.
Buffered Creatine						
Jagim et al. [28]	36 resistance trained males	RDBP	28 days	CrM (4 × 5 g/day for 7 days, 5 g/day for 21 days); CrM-Alk at recommended doses (1.5 g/day for 28 days); or CrM-Alk with equivalent doses to CrM (4 × 5 g/day for 7 days, 5 g/day for 21 days).	Neither recommended doses nor loading and maintenance equivalent doses of CrM-Alk promoted greater changes in muscle TCr, body composition, strength, or anaerobic capacity compared to CrM. Recommended doses did not ↑ TCr.	None reported.

R = randomized; DB = double-blind; *p* = placebo; SB = single blind; C = crossover; yrs = years; PLA = placebo; CHO = carbohydrate; PRO = protein; Cr = creatine; CrM = creatine monohydrate; CC = creatine citrate; CPY = creatine pyruvate; caPYR = calcium pyruvate, MgCr-C = magnesium creatine chelate, CEE = creatine ethyl ester; Cr-HCl = creatine hydrochloride; CrN = creatine nitrate; CNN = creatinine; CrM = Alk = buffered creatine; PCr = phosphocreatine; TCr = total creatine; AUC = area under the curve; FFM = Fat-Free Mass; BIA = bioelectrical impedance; GI = gastrointestinal; MRS = magnetic resonance spectroscopy; DW = dry weight.

10. No Evidence to Support Bioavailability, Efficacy, and Safety

10.1. Other Creatine Salts

While there is some bioavailability, efficacy, and safety data on CC and CPY, little to no data are available on several other creatine salts listed in Table 1. A patent from Negrisoli and Del Corona [31] disclosed several hydrosoluble organic salts of creatine including creatine maleate, creatine fumarate, creatine tartrate, and creatine malate. The inventors claimed that these creatine salts had solubilities of 10, 19, 3, 8.5, and 4.5 g/100 mL (%), respectively. Since salts are relatively weak bonds, it is likely that these creatine salts would increase creatine in the blood at equimolar doses that are generally about 1.3–1.6 times greater than CrM. However, there are no data indicating that these creatine salts increase blood creatine content, increase tissue creatine content, have any ergogenic value, are safe for long-term supplementation, or are more effective sources of creatine than CrM. A patent disclosing creatine ascorbate was also filed in the late 1990s [33]. The rationale was to provide a means of increasing creatine and ascorbic acid availability to improve exercise capacity while supporting the immune system [33]. However, we are not aware of any pharmacokinetic or exercise related studies to test this hypothesis and benefits would seemingly be similar and more effectively dosed by co-ingestion of CrM and vitamin C. Finally, there has been interest in tri-creatine orotate (CO) as a creatine salt (71% creatine by molecular weight) and a few raw material suppliers offer CO as a source of creatine to manufacturers [30,176,177]. Supplement companies who sell creatine orotate claim it provides creatine and orotic acid that is purported to aid in the production of carnosine in the muscle, and therefore improves muscle-buffering capacity. However, as of this writing, we are aware of no data showing bioavailability and/or efficacy of CO supplementation [216]. The European Food and Safety Authority has also expressed concerns about the potential cancerogenic effects of orotic acid [176,177]. Therefore, CO does not seem to be a good alternative for CrM in dietary supplements particularly when co-ingestion of effective doses of CrM and beta alanine would seeming be more effective. Based on this analysis, these creatine salts are classified in the no evidence to support bioavailability, efficacy, and safety category. Therefore, they cannot be considered more effective than CrM.

10.2. Creatine Serum

As noted above, there has been interest in developing shelf-life stable liquid, gels, and/or beverages containing creatine. The theoretical rationale has been that these types of products may be more convenient to consume, absorbed faster into the blood, and/or promote a greater efficiency in transport of creatine to the muscle. One product that was heavily marked in the late 1990s and early 2000s is "creatine serum" (CS). This product claimed to deliver 2.5 g of creatine per 5 mL oral dose by providing a "creatine phosphate complex" that was designed to be absorbed via mucosal thereby bypassing the supposed degradation of creatine to creatinine through digestion [217]. Their rationale was based on general pharmacokinetic absorption studies indicating that drugs and/or nutrients are absorbed faster through the mucosal lining in the mouth. However, when researchers evaluated the creatine content of CS, they found that CS contained <10 mg of creatine and 69 mg of creatinine per 5 mL dose in multiple samples and lot numbers [38]. Additionally, they found that one 5 mL oral dose of CS purportedly providing 2.5 g of creatine had no effect on plasma creatine levels (same as water) whereas ingestion of 2.5 g of CrM increased plasma creatine levels to about 300 μmol/L after one hour of ingestion and declining in a classical manner throughout the next 8 h (see Figure 14A) [38]. No changes in creatinine levels were seen among participants ingesting CS, CrM or water. Consequently, this study shows that CS does not contain creatine and has no bioavailability in the blood [38]. To further assess the bioavailability of serum creatine, Kreider and colleagues [26] evaluated the effects of ingesting 5 mL of a flavored placebo; 5 mL of CS (purportedly providing 2.5 g of CrM); 8 × 5 mL doses of CS per day (purportedly providing 20 g/day of CrM); and 4 × 5 g doses of CrM (20 g/day) for 5 days on muscle creatine, phosphocreatine and content. Results revealed that CrM loading significantly increased total muscle creatine (+31%) and phosphocreatine (+16%) (see Figure 13). However, CS ingestion at recommended and equivalent doses had no effects on muscle free creatine, phosphocreatine, total creatine content, or ATP concentrations. Collectively, these studies show that CS is not a bioavailable source of creatine, and therefore can have no creatine-related efficacy. Therefore, it is classified in the no evidence to support category, and there is no evidence that CS outperforms CrM. Unfortunately, this product remains in the marketplace despite data showing it is a completely ineffective source of creatine.

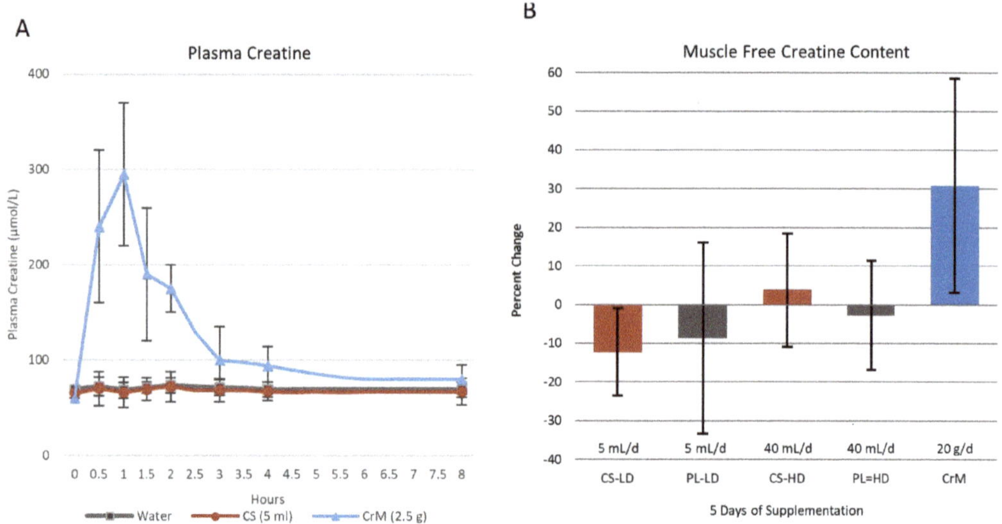

Figure 14. Changes in plasma creatine levels after oral ingestion of water, 5 mL of creatine serum (CS) purportedly providing 2.5 g of CrM, and 2.5 g of CrM in solution (**A**) [26] and 5 days of 5 mL of CS purportedly providing 2.5 g of creatine (CS-LD), 5 mL of a flavored placebo (PL-LD), 8 × 5 mL of CS (CS-HD) purportedly providing 20 g/day of creatine, 8 × 5 mL of flavored placebo (PL-HD), or 4 × 5 g/day of CrM (**B**) [26], Adapted from Harris et al. [38] and Kreider et al. [26].

10.3. Creatyl-L-Leucine

Creatyl-L-Leucine (CLL) has been marketed as "super creatine" [218]. As described in a patent [219], CLL is claimed to be "stable aqueous composition" of an "amide-protected, biologically-active form of creatine (creatyl-amide) molecule" that is "stable across a wide range pH's and temperatures" and "can provide a wide range of physiological benefits including, for example, regeneration of ADP to ATP in muscle tissue, increasing the serum concentration of creatine, increasing muscle fiber size/cross-sectional area and lean body mass, activating satellite cells, enhancing memory and cognitive function, enhancing the functional capacity of a mammal having a neuromuscular disease, increasing muscular strength, endurance and/or power, enhancing cognitive function in infants with inborn errors of creatine metabolism, and/or alleviating the deleterious effects of sleep deprivation". Analysis of the structure of CLL (see Figure 15) indicates that CLL does not contain a creatine molecule. Additionally, amide bonds are generally very strong, so pharmacokinetic data would need to show that CLL breaks down into creatine, increases creatine in the blood, and increases tissue creatine content to establish that CLL is a bioavailable source of creatine.

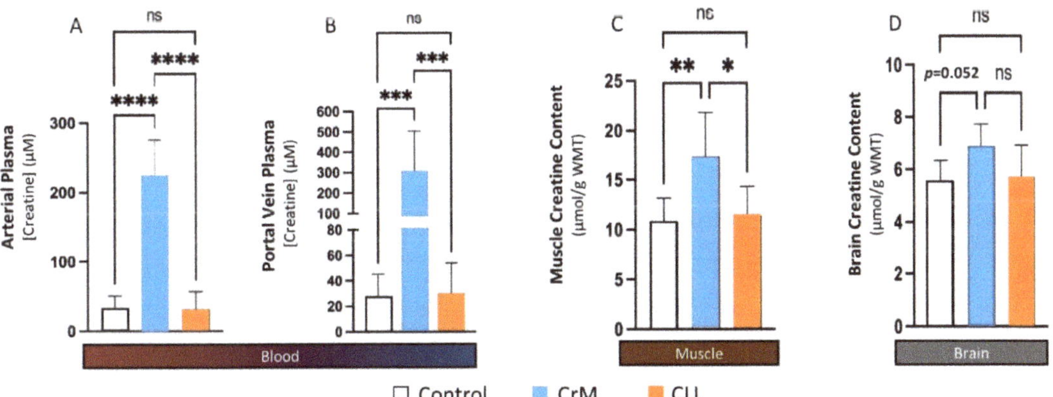

Figure 15. Comparison of creatine, creatine monohydrate, L-leucine, and creatyl-L-leucine chemical structures.

As of this writing, only two published articles have assessed the safety and/or efficacy of CLL. The first study was toxicology assessment of the administration of large doses of CLL in rats [220]. This study found that CLL did not cause mortality, toxic effects, or adverse effects in rats administered CLL for 90-days by oral gavage at doses of 1250, 2500, and 5000 mg/kg/day. More recently, da Silva [221] conducted an elegant study assessing the effects of feeding 24 rats either a control diet, a diet containing 4.0 g/kg/day of CrM, or a diet containing 6.56 g/kg/day of CLL for 7 days on arterial delivery of creatine, tissue uptake, and storage. According to the researchers, for a 70 kg individual, this would equate to a dose of 17.6 g/day of CrM and 28.9 g/day of CLL providing equimolar amounts of creatine if CLL based on the molecular weight of creatine if CLL degraded into creatine. As shown in Figure 16, rats fed CrM experienced significant increases in creatine concentrations in arterial plasma (+7-fold), portal vein plasma (+10-fold), muscle creatine content (+1.63-fold from control, and +1.53-fold from CLL) while tending to increase brain creatine content ($p = 0.052$) compared to controls. These changes were significantly greater than rats fed a control or CLL containing diet. Additionally, rats fed CLL did not increase blood, muscle, and brain creatine content above rats fed a control diet. The researcher concluded that provision of large doses of CLL to rats did not increase creatine bioaccumulation indicating that CLL is poorly absorbed by the intestine and is not a bioavailable source of creatine.

Figure 16. Blood, muscle, and brain creatine content in response to rats fed a control diet, 4.0 g/kg/day of creatine monohydrate (CrM), or 6.56 g/kg/day of creatyl-L-Leucine (CLL) for 7 days. (**A**) presents arterial plasma creatine concentration, (**B**) presents portal vein creatine concentration, (**C**) presents muscle creatine content, and (**D**) presents brain creatine content data for each group. Data are means ± standard deviations. **** = $p < 0.0001$, *** = $p < 0.001$, ** = $p < 0.01$, * $p < 0.05$, ns = not statistically significant between groups identified in brackets. Adapted from da Silva [221].

Several other studies have been recently conducted on CLL supplementation with human participants by experienced researchers at respected institutions with reports of results submitted in ongoing lawsuits [222–226]. These reports provide additional data showing CLL is not degraded into creatine upon oral ingestion [222,223,226], CLL does not increase blood creatine content [222,223,226], and CLL does not increase muscle [224,225] or brain creatine content [224] even when administered at doses much higher than found in marketed products containing CLL in humans [222–226]. Additionally, several of these studies reported that ingesting equivalent doses of CrM promoted significantly greater increases in blood [222,223,226] and tissue creatine content [225] than those ingesting CLL and CLL ingestion was no different than placebo controls [225,226]. Thus, available evidence indicates that CLL is not creatine, CLL is not a bioavailable source of creatine, and CLL is not "super creatine" compared to CrM. Therefore, CLL is listed in the no evidence category.

10.4. Creatinol-O-Phosphate

Figure 17 shows the chemical structure of creatinol-O-phosphate (COP). Creatinol in the form of COP is not creatine, nor was it intended to increase muscle creatine content. Rather, it was initially studied in the 1970s to intravenously deliver phosphate to improve myocardial function and reduce arrhythmias during ischemic conditions [227–230]. For example, Melloni and colleagues [228] investigated the effects of intravenous administration of 1020 mg, 2040 mg, and 3060 mg of COP compared to placebo on arterial blood pressure, heart rate and arrhythmias. The researchers found that COP administration increased blood phosphate levels as well as urinary excretion of phosphate and creatinine. Phosphate loading has been found to increase myocardial ejection fraction during exercise and maximal aerobic capacity [231,232] and is considered an ergogenic aid for endurance athletes [14]. While this is unrelated to creatine supplementation, the increase in urinary creatinine excretion led some to speculate that creatinol may act as a precursor of creatine and thereby serve as a source of creatine the body [228]. However, pharmacokinetic studies indicated that absorption of COP was complete when administered intramuscularly and distributed primarily to the kidney, liver, and heart and that COP could cross myocardial cell membranes [230]. One study from 1975 has been reported by others to show that intramuscular and intravenous administration of COP increased handgrip performance [233]. However, it is difficult to find details about this study. We are also not aware of any study that has evaluated whether oral COP has any effect on muscle creatine levels or exercise performance. Nevertheless, some companies have included COP as a source of creatine in dietary supplements and energy drink beverages. There is no evidence that oral COP ingestion has any effect on muscle creatine content or creatine-related metabolism. Claims that oral COP is a source of creatine and/or is more bioavailable than CrM are not supported.

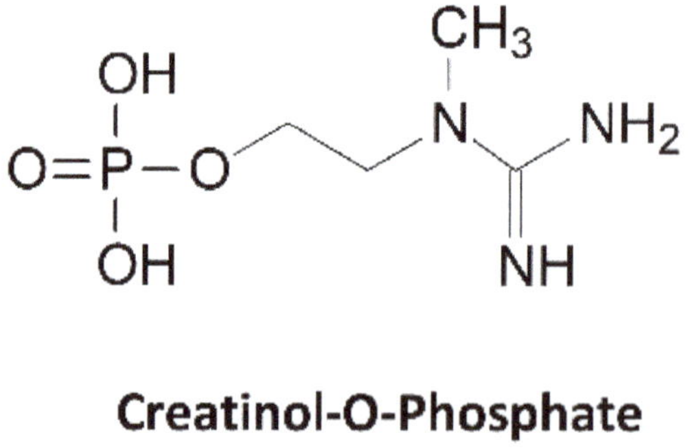

Figure 17. Chemical structure of Creatinol-O-Phosphate.

Table 4 summarizes the results of studies that have evaluated sources of creatine that currently have no evidence supporting of bioavailability, efficacy, and/or safety. As can be seen, there are limited published data on these purported sources of creatine, and the available evidence indicates that they are not bioavailable sources of creatine.

Table 4. Creatine containing compounds that have no evidence of bioavailability, efficacy, and/or safety.

Reference	Participants	Design	Duration	Dosing Protocol	Findings	Side Effects
Creatine Serum						
Harris et al. [38]	6 males	R	1 oral dose with 7 day washout	Water Control, 5 mL CS (purportedly delivering 2.5 g CrM), 2.5 g CrM	CrM ↑ plasma Cr while CS had no effects and was similar to water. Analytic chemistry analysis showed < 10 mg of creatine and 90 mg creatinine in CS sample.	None reported.
Kreider et al. [27]	40 males (18–30 years)	RDBP	5 days	5 mL PLA, 5 mL of CS (recommended dose purportedly providing 2.5 g of CrM); 8 × 5 mL/day CS (purportedly providing 20 g/day of CrM); and 4 × 5 g doses of CrM (20 g/day)	CrM increased muscle creatine stores. Consumption of CS at recommended and 8× recommended levels had no effect.	None reported.
Creatyl-L-Leucine						
Reddeman et al. [189]	Rats	Open Label	90 days	Repeated-dose oral gavage toxicity study at doses of 1250, 2500, and 5000 mg/kg body weight per day.	There was no genotoxic activity observed in an in vivo mammalian micronucleus test at concentrations up to the limit dose of 2000 mg/kg body weight per day. The no observed adverse effect level from the 90-day study was determined to be 5000 mg/kg body weight per day, which was the highest dose tested for male and female rats.	None reported.
da Silva [221]	24 rats	R	7 days	Control diet, a diet containing 4.0 g/kg/day CrM, or a diet containing 6.56 g/kg/day CLL	CrM ↑ [creatine] in arterial plasma (+7-fold), portal vein plasma (+10-fold), muscle TCr (+1.63-fold from control, and +1.53-fold from CLL) while tending to increase brain creatine content compared to controls. CLL did not increase blood, muscle, and brain creatine content above rats fed a control diet with values lower than CrM.	None reported.
Creatinol-O-Phosphate						
Nicaise et al. [233]	-	-	-	Intramuscular and intravenous injection	COP↑ handgrip performance.	None reported.

R = randomized, *p* = placebo, DB = double-blind, SB = single blind; C = crossover, PLA = placebo, Cr = creatine; TCr = total creatine; CrM = creatine monohydrate, CS = creatine serum; CLL = creatyl-L-leucine; COP = creatinol-O-phosphate; ↑ = increase.

11. Regulatory Status

11.1. United States

In the United States (US), congress enacted the Dietary Supplement Health and Education Act (DSHEA) of 1994 that placed dietary supplements in a special category of foods under the jurisdiction of the FDA. According to DSHEA, a dietary supplement is a product intended to supplement the diet, ingested orally, and contains a "dietary ingredient". Dietary ingredients include vitamins, minerals, amino acids, herbs, botanicals, and other substances such as extracts, metabolites, or concentrates of those substances [14,234]. Dietary supplements can be delivered in powders, pills, capsules, hard and chewable tablets, soft gels, gummies, liquids, and even properly labeled energy bars that are intended for oral ingestion. However, they cannot include products promoted for sublingual, intranasal, transdermal, injected, or in any other route of administration [14]. DSHEA also established laws for FDA oversight over "new dietary ingredients" (NDI), which are ingredients that introduced to the marketplace after DSHEA was enacted [235]. A dietary supplement containing an NDI is deemed adulterated by the FDA, and therefore may not be lawfully distributed, unless (1) the NDI has "been present in the food supply as an article used for food in a form in which the food has not been chemically altered" or (2) there is a "history of use or other evidence of safety" that is submitted to the FDA for at least 75 days before selling the product (i.e., an "NDI Notification"). Dietary ingredients that were sold in the U.S. prior to October 15, 1994, were considered "grandfathered," and therefore not NDIs subject to these requirements. This included CrM since it was introduced into the U.S. market in 1993 [236].

Since all of the other purported sources of creatine described above were introduced into the U.S. marketplace after 15 October 1994, they are considered NDI's and manufacturers and distributors were expected to notify the FDA about these ingredients (See Section 413(d) of the Federal Food, Drug, and Cosmetic Act (the FD&C Act), 21 U.S.C. 350b(d)) [235] unless they meet the "present in the food supply" exemption noted above. An NDI Notification should include documentation of how the product containing the NDI is "reasonably expected to be safe" along with (1) the name of the new dietary ingredient (or Latin binomial name if it is an herb or botanical); and (2) "a description of the dietary supplement that contains the new dietary ingredient, including (a) the level of the new dietary ingredient in the product, (b) conditions of use of the product stated in the labeling, or if no conditions of use are stated, the ordinary conditions of use, and (c) a history of use or other evidence of safety establishing that the dietary ingredient, when used under the conditions recommended or suggested in the labeling of the dietary supplement, is reasonably expected to be safe" [14]. Once submitted, the FDA has 75 days to object to the notification. If the FDA does not respond within this timeframe, the NDI can be included in dietary supplements and legally sold in the U.S. market. However, it is important to understand that an NDI Notification only indicates that the FDA considers the NDI to be reasonably be considered as safe for human consumption. It does not affirm efficacy and/or validate any claims made about the NDI.

Since DSHEA and FDA regulations do not provide sufficient clarification on many issues, there has been a lot of confusion in the dietary supplement industry on what is an NDI, what manufacturing or other changes made to an ingredient cause it to be a "new" ingredient, when an NDI Notification is required, and what information it should contain. To provide clarification, the FDA released a "Draft Guidance for Industry" entitled "Dietary Supplements: New Dietary Ingredient Notifications and Related Issues" in July of 2011. However, that draft guidance prompted even more confusion and controversy, so the FDA released a revised draft guidance in 2016. While a guidance does not carry enforcement authority like a law or regulation, it provides the FDA's perspective of how they interpret the laws and regulations related to NDI's to help dietary supplement manufacturers know whether they are required to submit an NDI Notification to FDA, how to prepare NDI Notifications consistent with FDA review expectations, and how to improve the quality of submissions [235]. The 2016 Draft Guidance has also been criticized for a lack of clarity concerning what was considered a grandfathered ingredient and whether an NDI Notification was required if another manufacturer had already submitted an NDI Notification among other issues. This led to ingredients that should have been considered to be NDIs entering the marketplace with a notification, and several NDI Notifications being rejected by the FDA for lack of adequate safety data and/or other issues.

As an alternative to submitting an NDI Notification, some companies have pursued obtaining "Self-Affirmed" "Generally Recognized as Safe" (GRAS) status by conducting toxicology studies on animals and having scientific experts review the safety data and affirm the ingredient was reasonably expected to be safe. In this case, the company commissions studies to assess safety (e.g., toxicology studies in animals) and maintains internal documents to show safety if requested by the FDA. Once an ingredient is self-affirmed as GRAS, it can be introduced into the food supply, and is then not required to submit an NDI Notification to FDA to be included in dietary supplement under the "present in the food supply" exemption noted above. Companies can voluntarily submit their ingredient's GRAS determination to FDA, however, there is no requirement to do so, and therefore a company's GRAS self-affirmation can remain private. While lawful, the FDA has expressed some concern about this approach and discourages dietary supplement manufacturers from self-affirming GRAS to avoid submitting an NDI Notification.

After a dietary supplement is entered into the U.S. market, the FDA can restrict or ban its sale if it is deemed adulterated (e.g., unsafe). The FDA works with the Federal Trade Commission (FTC), who has jurisdiction over marketing claims that companies make about dietary supplements in advertising to regulate the dietary supplement industry. The FTC can act against companies for disseminating false, misleading, or unsubstantiated claims about dietary supplements. Marketing claims made for dietary supplements can also be challenged in litigation brought by consumers or competitors. This background provides the basis for understanding how various purported forms of creatine have entered the US market. The legal and regulatory status of CrM in dietary supplements is indisputable because it appeared on the US market in 1993 [236], and there was a large body of evidence showing it was reasonably expected to be safe. Since then, AlzChem Trostberg GmbH (the Germany manufacturer of CrM) voluntary submitted a GRAS application to the FDA that was not acted upon, meaning it could claim that CrM has FDA approved GRAS status [237,238]. Consequently, CrM is considered GRAS for inclusion as a dietary ingredient in dietary supplements, energy drinks, protein bars and powders, milkshakes, meal replacement powders and bars, meat replacement products, powdered drink mixes, and functional foods. As of this writing, CrM is the only form of creatine that is listed on the FDA's inventory of GRAS notices [239] (see Table 5). As noted above, a company is not required to submit its GRAS self-affirmation to FDA, and therefore it may not be public information.

Table 5. Regulatory status of nutrients marketed as creatine supplements in the United States of America.

FDA Generally Recognized As Safe (GRAS)				
Purported Creatine Source Submitted	FDA Report Number	Submission Year	Intended Dosage	FDA Response
Creatine Monohydrate (Creapure®)	GRN 931	2020	1 g creatine (1.12 g creatine monohydrate) as an ingredient in "energy" drinks, protein bars and powders, milk shakes, meal replacement powders and bars, meat analogs, and powdered drink mixes (excluding infant formula.	FDA has no questions at this time.
Self-Affirmed Generally Recognized As Safe (GRAS)				

Table 5. *Cont.*

FDA Generally Recognized As Safe (GRAS)				
Purported Creatine Source Submitted	**FDA Report Number**	**Submission Year**	**Intended Dosage**	**FDA Response**
Creatine chelated with Mg (Creatine MagnaPower®)		2013		
New Dietary Ingredient Notifications (NDIN) *				
Purported Creatine Source Submitted	**FDA Report Number**	**Submission Year**	**Intended Dosage**	**FDA Response**
Creatine Pyruvate	RPT28	1998	5–10 g/day in 2 equal doses	Filed by FDA without substantive comments.
Creatine Ethylesters [Brand: Cre-Ester™]	RPT154	2002	Maximum daily dose of 30 g	Objected by the FDA.
Creatine Ethylesters [Brand: Cre-Ester™]	RPT190	2002	0.5–5.0 g/day	Objected by the FDA.
Tricreatine Orotate	RTP201	2003	1–2 g 3 ×/day (3–6 g/day)	Objected by the FDA.
Creatine ethyl ester HCL [Brand: CE2™]	RTP249	2004	500 mg–5 g/day	Objected by the FDA.
Creatine from creatine ethyl ester HCL [Brand: CE2™]	RTP264	2004	500 mg–3 g/day	Objected by the FDA.
Beta Creatine	RPT660	2010	4.5–7.5 g/day creatine 3–6 g/day beta-alanine	Objected by the FDA.
Creatine Nitrate	RTP696	2011	1.5 g serving, maximum dose 3 g/day	Objected by the FDA.
Creatine Nitrate	RPT993	2017	750 mg per day	Acknowledged with no objections by FDA.
Creatine acesulfame	RPT1064	2018	10 g per day	Objected by the FDA.

FDA = Food and Drug Administration. * Data retrieved from AHPA NDI Database http://ndi.ahpa.org/ (accessed on 1 January 2022).

The legal status of other purported sources of creatine is less clear. According to the FDA's NDI Notification database, since 1995, NDI Notifications for creatine pyruvate (1998), creatine ethyl ester (2003), creatine ethyl ester HCL (2004), tri-creatine orotate (2003), β-creatine (2010), creatine nitrate (2011 and 2017), and creatine acesulfame (2018) have been submitted. Several of these notifications were initially objected to by the FDA citing: (1) the form of creatine may not be a legal dietary ingredient as defined by the FD&C Act §201(ff); (2) inadequate safety information to conclude that the form of creatine is reasonably expected to be safe; and/or (3) a lack of information about the chemical identity of the creatine form [25]. However, creatine pyruvate at doses of 5–10 g/day and creatine nitrate at 750 mg per day have not been objected to by the FDA and can be sold as dietary ingredients. The AHPA NDIN database does not include notifications for creatine maleate, creatine fumarate, creatine tartrate, creatine ascorbate, creatine citrate, magnesium creatine chelate, creatine HCL, alkaline creatine (although it is a buffered form of CrM), creatine serum, CLL, or COP. Since some of these sources of creatine are ingredients in dietary supplements, any company selling these nutrients would have to have documentation that the ingredient was on the market in the US before 15 October 1994 or is present in the food supply in a form that is not chemically altered. While FDA GRAS notifications are published and are accessible to the public, self-affirmed GRAS files are not published, which makes them difficult to search and to validate its content. Based on a press release, we identified one self-affirmed GRAS affirmation for a creatine chelated with Mg (Creatine MagnaPower®).

11.2. International Regulation

Every country has independent laws and regulations governing dietary supplements. Some follow FDA guidance while others classify dietary supplements and drugs and have additional oversight, approval, and/or limitations about dosages [240]. Creatine monohydrate can be legally sold throughout the world although some countries limit the amount to of CrM that can be included per dose (e.g., no more than 3–5 g/serving). Some of the other forms of creatine marketed as ingredients for dietary supplements are not permitted to be included in dietary supplements in their country due to a lack of safety data. Therefore, except for CrM, one cannot assume that of the purported other sources of creatine described above can be legally sold as a dietary supplement in all countries. The following provides a brief overview in major markets that creatine is sold and regulatory oversight. Table 6 describes the responsible agencies and regulatory status of creatine containing dietary supplements in various countries. As can be seen, CrM remains the only source of creatine that is approved for sale in Australia, Canada, China, the European Union, Japan, and South Korea. Additionally, it is the only source of creatine that has approved health claims in the European Union, Canada, Japan, and South Korea.

Table 6. Regulatory status of nutrients marketed as creatine supplements primary international markets.

Country	Responsible Agency	Primary Regulations/Statutes	Regulatory Status of Creatine
Australia	Department of Health, Therapeutic Goods Administration (TGA).	Dietary supplements are considered complimentary medicines and regulated under the Therapeutic Goods Act (TGA) of 1989 [241] and 1990 TGA regulations [198]. Medicinal products are categorized as lower risk medicines that can be listed on the Australian Register of Therapeutic Goods (ARTG) [198] while higher risks medicines must be registered with the ARTG [241].	As of this writing, of the 90,988 products listed in the ARTG database, only 25 products contain creatine. Of these, CrM is the only source of creatine listed as an ingredient.
Canada	Health Canada [242]	Natural and Non-prescription Health Products Directorate (NNHPD) of Health Canada [242]. The NNHPD maintains a compendium of articles that reviews the safety and efficacy of licensed NHP's [243]. CrM was assigned a monograph by the NNHPD that overviews research on CrM to substantiate safety and efficacy. Only products containing CrM can benefit from an abbreviated licensing process by referencing the monograph. Applicants using all other creatine forms are required to submit their own evidence of safety and efficacy for review as part of the pre-market licensing process.	As of this writing, 20 compounds purported to contain creatine are included in the NHP Ingredient Database [244] including creatine, creatine-alpha-ketoglutarate, creatine ethyl ester, creatine ethyl ester HCl, creatine gluconate, creatine HCl, creatine hydroxycitrate, creatine monohydrate, creatine nitrate, creatine orotate, creatine phosphate, creatine pyroglutamate, creatine pyruvate, creatine taurinate, dicreatine malate, disodium creatine phosphate, magnesium creatine chelate, polyethylene glycosylated creatine, polyethylene glycosylated creatine HCl, and tricreatine citrate. Creatinol-O-phosphate is listed as a medicinal product in the NHP database [244].

Table 6. *Cont.*

Country	Responsible Agency	Primary Regulations/Statutes	Regulatory Status of Creatine
China	New Food Safety Law of the People's Republic of China and the Administrative Permission Law of the People's Republic of China, CFDA [198].	Nutritional supplements in China must be orally ingested, have at least one of 22 preventive functions as recognized by the Ministry of Health, and cannot be a curative drug [198]. Imported supplements must be approved by the National Medical Product Administration while foods are supervised by the State Administration for Market Regulation) [198].	Importers of dietary supplements and foods containing creatine must submit notification materials for review and approval before being allowed to be sold in China. Since CrM and other forms of creatine are produced in China, they would seemingly be legal to consume. However, it is unclear which forms of creatine are allowed to be imported into China.
European Union (EU)	European Commission Directive on Food Supplements [245–247]	The European Food Safety Authority (EFSA) evaluates scientific health claims. Creatine is considered a substance that may be added for specific nutritional purposes in foods for particular nutritional uses (FPNU) [246].	In 2004, EFSA indicated that the use of creatine in foods for nutritional use was not a matter of concern provided that the source had high purity (99.95%), did not contain impurities, and that dose of up to 3 g/day of supplemental creatine which is similar to the normal daily turnover rate of creatine was unlikely to pose any risk [248]. EFSA substantiated scientific health claims of CrM include: (1) CrM increases physical performance during short-term, high intensity, repeated exercise bouts, endurance capacity, and endurance performance [249], (2) CrM increases attention and improves memory [250], and, (3) CrM (at least 3 g/day) in combination with resistance training and improved muscle strength. All studies cited were performance on pure CrM so the regulatory status of other "forms" of creatine in the EU are less clear.
Japan	Ministry of Health, Labor and Welfare (MHLW)	Dietary substances in Japan are legally classified as food, food additives or "non-drug" (food). The Consumer Affairs Agency started "Foods with Function Claims" which reviews and approves health claims related to dietary supplements.	CrM is considered a "non-drug" [251] that is allowed to be sold as a food ingredient and additive under the Food Sanitation Law [252]. Health claims of CrM for muscle maintenance with exercise was accepted in 2019. Thus, CrM can be imported, distributed, and produced in Japan. CEE has been included on the "non-drug" list. In order for other forms to be imported, distributed, and/or produced in Japan, safety data and similarity of the proposed form to CrM must be submitted and approved by the MHLW [253]. In addition to CrM, creatine citrate and creatine pyruvate have been approved to be imported into Japan. It is unclear whether other forms of creatine can be imported into Japan and sold as dietary supplements.
South Korea	Ministry of Food and Drug Safety (MFDS) [254].	Similar to the U.S., new dietary ingredients must have sufficient safety data including toxicology studies in animals and supporting safety and efficacy data from human clinical trials to support efficacy at the recommended daily doses marketed.	An application to register CrM as a dietary supplement was filed in 2005 and approved by the MFDA for use as a dietary supplement 2008 with an accompanying health claim [255]. Given these requirements, forms of creatine reviewed above that have bioavailability data at recommended doses substantiating efficacy and safety seemingly be eligible for approval while those that do not have that data would likely experience more difficulty obtaining approval to sell their form of creatine in South Korea.

CrM = creatine monohydrate.

11.3. Assessment and Guidance for Industry

The regulatory status of CrM is unequivocal in the global markets as a dietary or food supplement [25]. However, the scientific basis and regulatory status of other forms of creatine continues to be less clear. Since our last review in 2011, more data is available about several forms of creatine and several now have data providing some support as to efficacy and safety while others do not. No alternative form of creatine has shown superior bioavailability, efficacy or safety compared to CrM. Consequently, despite marketing hyperbole, CrM remains the gold standard to compare other forms of creatine with the strongest bioavailability, efficacy, and safety portfolios. Alternatives to CrM continue to be prevalent in the marketplace, including several that do not appear to meet regulatory requirements in several countries.

In the US, a major factor in determining whether an NDI notification is required to be submitted to the FDA is whether the NDI has been present in the food supply and/or has been chemically altered from its original form [26]. Any change in the chemical structure of an ingredient is assumed to alter the biological activity of the ingredient thereby requiring toxicology studies to establish that the NDI is reasonably expected to be safe as altered and that the biological behavior is comparable to the native ingredient. Most of the forms of creatine listed in Table 1 have been chemically altered in some way (e.g., covalently binding or complexing) to the creatine molecule. Some have clearly rearranged the creatine molecule. Therefore, they should have all been submitted as and NDI notification to the FDA prior to marketing. Yet, as described above, only nine of these newer forms are listed in the IND notification inventory. Of these, the FDA initially rejected some of the IND notification applications, yet the forms were sold for years before finally submitting an acceptable IND notification application or Self-Affirming GRAS. Even then, many have little to no data supporting bioavailability and/or efficacy despite making bold claims that the source is more bioavailable, effective, and/or a safer form of creatine than CrM.

The reason why alternate marketed forms of creatine are in the marketplace without pre-market IND notification is likely due to confusion over legal definitions of dietary supplements, natural health products, and/or food additives in different countries as well as what is meant by chemical alteration in a nutrient. Regardless, confusion of over laws regulating dietary supplements combined with inadequate enforcement by regulators has created an environment where there are often little consequences of non-compliance. For example, studies published in 2003 clearly showed that creatine serum was not a bioavailable source of creatine, yet it continues to be sold as a creatine-containing product. While alternative forms of creatine are unlikely to pose a health risk, they are typically more expensive than CrM. Additionally, misleading claims that lower doses of an alternate form of creatine are as effective as CrM may limit the benefits consumers may achieve from creatine supplementation. Given the health benefits of creatine, availability of ineffective sources of creatine or recommendations to take less creatine than needed to increase creatine stores in the muscle and/or brain can limit the benefits theses populations may derive from creatine supplementation [5]. Thus, we give the following recommendations as guidance to researchers and industry as they consider developing new dietary ingredients containing creatine.

(1) Only consider developing creatine supplements that contain a creatine molecule. Alteration of the chemical structure of creatine in any way is assumed to change the chemical activity and biological function and may negate any benefit of creatine supplementation. Additionally, binding creatine to other compounds may prevent creatine from being liberated in vivo, thereby making the form of creatine non-bioavailable or less bioavailable source of creatine.

(2) Companies who develop new forms of creatine should conduct toxicology studies in animals to establish that high dose ingestion is safe and conduct clinical trials in humans to validate safety. We then recommend obtaining FDA GRAS status or Self-Affirming GRAS status.

(3)   Pharmacokinetic studies must be performed to show that the novel form of creatine is degraded into creatine and increases blood creatine levels to physiological levels necessary to promote creatine uptake into tissue (e.g., >200–500 µmol/L or 25–65 µg/L).

(4)   Bioavailability studies should be conducted to show recommended doses increase muscle and/or brain creatine content.

(5)   Placebo, double blind, and randomized clinical trials should be performed to substantiate that the form of creatine provides ergogenic benefit and does not cause any untoward side effects.

(6)   Comparative effectiveness trials at recommended and equivalent doses must be performed to show a new form of creatine increases muscle and/or brain creatine content to a greater degree than CrM to substantiate those claims.

(7)   Comparative effectiveness trials at recommended and equivalent doses must also be performed to determine if a new form of creatine is more effective and/or a safer alternative to CrM to substantiate those types of claims.

(8)   Supplement companies should clearly declare the source and amount of creatine contained in their products so consumers can know if they are taking effective doses.

(9)   Claims made about a form of creatine should be based on research conducted on that form of creatine at recommended doses, not untested hypotheses, speculation, assumptions, and/or marketing hyperbole. Such practices only undermine the scientific validity and consumer confidence about creatine supplementation.

(10)   Pure CrM is the only source of creatine with strong evidence of bioavailability, efficacy, and safety and considered as GRAS by the FDA, approved for use in the EU and Australia, and evaluated for safety by Health Canada.

(11)   Consumers should only consider taking supplements that contain sources of creatine that research has shown is bioavailable, effective, safe, and devoid of impurities.

12. Summary

CrM supplementation increases muscle phosphagen levels, improves repetitive high-intensity exercise performance, and promotes greater training adaptations [15]. No significant side effects other than weight gain have been reported from CrM supplementation despite widespread use throughout the world. Research on CrM has served as the basis to establish professional guidelines, recommendations, and establish regulation. CrM remains the only source of creatine that has substantial evidence of bioavailability, efficacy and safety and is considered GRAS by the U.S. FDA, is approved for use with accompanying health claims in the EU, has been extensively reviewed and approved by Health Canada, and is approved to be sold in major global markets. The bioavailability, efficacy, safety, and regulatory status of other purported sources of creatine are less clear, with only a few having some data supporting efficacy compared to placebo (see Table 7). However, there is no evidence that other "forms" of creatine are more bioavailable, effective, or safer forms of creatine compared to CrM. We recommend that companies interested in developing and marketing novel forms of creatine ensure the purported source contains the creatine molecule and conduct high-dose safety data in animals, pharmacokinetic studies to show the source of creatine increases blood and tissue concentrations of creatine, and comparative effectiveness studies to support structure and function claims. Additionally, the should company clearly list the amount of creatine contained in the supplement on supplement facts labels so consumers can make an informed decision about whether that purported source of creatine may deliver enough creatine to increase tissue creatine content by physiological levels needed to effect exercise and/or health.

Table 7. Categorization of purported sources of creatine based on bioavailability, efficacy, and safety.

Strong Evidence	Some Evidence	No Evidence
Creatine Monohydrate	Creatine Citrate	5-Hydroxytryptamine Creatine
	Creatine Pyruvate	Creatine Benzyl Ester
	Magnesium Creatine Chelate	Creatine Beta-Alaninate
	Creatine Ethyl Ester	Creatine Carnitine
	Creatine HCl	Creatine Ethyl Ester Malate
	Creatine Nitrate	Creatine Ethyl Ester Pyruvate
	Buffered Creatine Monohydrate	Creatine Fumarate
		Creatine Gluconate
		Creatine Glutamate
		Creatine Hydroxycitrate
		Creatine Lactate
		Creatine Malate
		Creatine Maleate
		Creatine Methyl Ester HCL
		Creatine Monohydrate Dextrose
		Creatine Orotate
		Creatine Phosphate Lactate
		Creatine Pyroglutamate
		Creatine Pyruvate Monohydrate
		Creatine Serum
		Creatine Sulfate Monohydrate
		Creatine Taurinate
		Creatine Trinitrate
		Creatine α-ketoglutarate
		Creatine-CoA
		Creatinol-0-Phosphate
		Creatyl-L-Leucine
		Di-Acetyl Creatine Ethyl Ester
		Disodium Creatine Phosphate
		Methyl-Amino-Creatine
		Phospho-Creatine
		Polyethylene Glycosylated Creatine

Author Contributions: Conceptualization, R.B.K., R.J. and M.P.; writing—original draft preparation, R.B.K.; writing—review and editing, R.J. and M.P.; funding acquisition, R.B.K. All authors have read and agreed to the published version of the manuscript.

Funding: The APC of selected papers of this Special Issue are being funded by Alzchem, LLC. (Trostberg, Germany), which manufactures creatine monohydrate. The funders had no role in the writing of the manuscript, interpretation of the literature, or in the decision to publish the results.

Institutional Review Board Statement: Not applicable.

Informed Consent Statement: Not applicable.

Data Availability Statement: Not applicable.

Acknowledgments: We would like to thank all the study participants, scholars, and funding agencies who have contributed to the research cited in this manuscript. We would also like to thank Ivan Wasserman (U.S.), Toshitada Inoue (Japan/South Korea), and John Doherty (Canada) for reviewing the local regulations.

Conflicts of Interest: R.J. and M.P. are researchers and principals for Increnovo, LLC, which conducts research, develops intellectual property, and consults with industry about raw ingredients and product formulations. They have filed patents for creatine while being employed by SKW Trostberg/Degussa AG (now Alzchem), from 1999 to 2007, with all the patents being expired or abandoned (WO2006015774A1, US20110123654A1, WO2003071884A1, US20050096392A1, WO2006122809A1, WO2002052957A1, WO2003047367A1, US20040006139A1, US20020072541A1, DE10244281A1, DE10119608A1), and conducted research on various forms of creatine. R.B.K. has conducted sponsored research on dietary supplements including creatine through grants awarded to the universities he has been affiliated with, received honorarium for presenting research related to dietary supplements and creatine at industry-sponsored scientific conferences, has served as an expert witness on cases related to dietary supplements, including past and current cases related to creatine, and is acting Chair of the Scientific Advisory Board on Creatine for Alzchem. He has also presented research related to creatine at a number of international conferences.

References

1. Paddon-Jones, D.; Borsheim, E.; Wolfe, R.R. Potential ergogenic effects of arginine and creatine supplementation. *J. Nutr.* **2004**, *134*, 2888S–2894S, discussion 2895S. [CrossRef] [PubMed]
2. Brosnan, M.E.; Brosnan, J.T. The role of dietary creatine. *Amino Acids* **2016**, *48*, 1785–1791. [CrossRef]
3. Da Silva, R.P.; Clow, K.; Brosnan, J.T.; Brosnan, M.E. Synthesis of guanidinoacetate and creatine from amino acids by rat pancreas. *Br. J. Nutr.* **2014**, *111*, 571–577. [CrossRef] [PubMed]
4. Da Silva, R.P.; Nissim, I.; Brosnan, M.E.; Brosnan, J.T. Creatine synthesis: Hepatic metabolism of guanidinoacetate and creatine in the rat in vitro and in vivo. *Am. J. Physiol. Endocrinol. Metab.* **2009**, *296*, E256–E261. [CrossRef]
5. Kreider, R.B.; Stout, J.R. Creatine in Health and Disease. *Nutrients* **2021**, *13*, 447. [CrossRef]
6. Wallimann, T.; Tokarska-Schlattner, M.; Schlattner, U. The creatine kinase system and pleiotropic effects of creatine. *Amino Acids* **2011**, *40*, 1271–1296. [CrossRef]
7. Bonilla, D.A.; Kreider, R.B.; Stout, J.R.; Forero, D.A.; Kerksick, C.M.; Roberts, M.D.; Rawson, E.S. Metabolic Basis of Creatine in Health and Disease: A Bioinformatics-Assisted Review. *Nutrients* **2021**, *13*, 1238. [CrossRef] [PubMed]
8. Buford, T.W.; Kreider, R.B.; Stout, J.R.; Greenwood, M.; Campbell, B.; Spano, M.; Ziegenfuss, T.; Lopez, H.; Landis, J.; Antonio, J. International Society of Sports Nutrition position stand: Creatine supplementation and exercise. *J. Int. Soc. Sports Nutr.* **2007**, *4*, 6. [CrossRef]
9. Kreider, R.B.; Jung, Y.P. Creatine supplementation in exercise, sport, and medicine. *J. Exerc. Nutr. Biochem.* **2011**, *15*, 53–69. [CrossRef]
10. Balsom, P.D.; Soderlund, K.; Ekblom, B. Creatine in humans with special reference to creatine supplementation. *Sports Med.* **1994**, *18*, 268–280. [CrossRef]
11. Hultman, E.; Soderlund, K.; Timmons, J.A.; Cederblad, G.; Greenhaff, P.L. Muscle creatine loading in men. *J. Appl. Physiol.* **1996**, *81*, 232–237. [CrossRef] [PubMed]
12. Harris, R.C.; Soderlund, K.; Hultman, E. Elevation of creatine in resting and exercised muscle of normal subjects by creatine supplementation. *Clin. Sci.* **1992**, *83*, 367–374. [CrossRef]
13. Kreider, R.B.; Melton, C.; Rasmussen, C.J.; Greenwood, M.; Lancaster, S.; Cantler, E.C.; Milnor, P.; Almada, A.L. Long-term creatine supplementation does not significantly affect clinical markers of health in athletes. *Mol. Cell. Biochem.* **2003**, *244*, 95–104. [CrossRef] [PubMed]
14. Kerksick, C.M.; Wilborn, C.D.; Roberts, M.D.; Smith-Ryan, A.; Kleiner, S.M.; Jager, R.; Collins, R.; Cooke, M.; Davis, J.N.; Galvan, E.; et al. ISSN exercise & sports nutrition review update: Research & recommendations. *J. Int. Soc. Sports Nutr.* **2018**, *15*, 38. [CrossRef] [PubMed]
15. Kreider, R.B.; Kalman, D.S.; Antonio, J.; Ziegenfuss, T.N.; Wildman, R.; Collins, R.; Candow, D.G.; Kleiner, S.M.; Almada, A.L.; Lopez, H.L. International Society of Sports Nutrition position stand: Safety and efficacy of creatine supplementation in exercise, sport, and medicine. *J. Int. Soc. Sports Nutr.* **2017**, *14*, 18. [CrossRef] [PubMed]
16. Meyers, S. Sports nutrition market growth watch. In *Natural Products Insider*; Informa Exhibitions: Irving, TX, USA, 2016.
17. Thomas, D.T.; Erdman, K.A.; Burke, L.M. Position of the Academy of Nutrition and Dietetics, Dietitians of Canada, and the American College of Sports Medicine: Nutrition and Athletic Performance. *J. Acad. Nutr. Diet* **2016**, *116*, 501–528. [CrossRef] [PubMed]
18. Maughan, R.J.; Burke, L.M.; Dvorak, J.; Larson-Meyer, D.E.; Peeling, P.; Phillips, S.M.; Rawson, E.S.; Walsh, N.P.; Garthe, I.; Geyer, H.; et al. IOC Consensus Statement: Dietary Supplements and the High-Performance Athlete. *Int. J. Sport Nutr. Exerc. Metab.* **2018**, *28*, 104–125. [CrossRef] [PubMed]
19. Casey, A.; Constantin-Teodosiu, D.; Howell, S.; Hultman, E.; Greenhaff, P.L. Creatine ingestion favorably affects performance and muscle metabolism during maximal exercise in humans. *Am. J. Physiol.* **1996**, *271*, E31–E37. [CrossRef] [PubMed]
20. Green, A.L.; Hultman, E.; Macdonald, I.A.; Sewell, D.A.; Greenhaff, P.L. Carbohydrate ingestion augments skeletal muscle creatine accumulation during creatine supplementation in humans. *Am. J. Physiol.* **1996**, *271*, E821–E826. [CrossRef]
21. Green, A.L.; Simpson, E.J.; Littlewood, J.J.; Macdonald, I.A.; Greenhaff, P. Carbohydrate ingestion augments creatine retention during creatine feeding in humans. *Acta Physiol. Scand.* **1996**, *158*, 195–202. [CrossRef]
22. Greenwood, M.; Kreider, R.B.; Earnest, C.P.; Rasmussen, C.; Almada, A. Differences in creatine retention among three nutritional formulations of oral creatine supplements. *J. Exerc. Physiol. Online* **2003**, *6*, 37–43.
23. Steenge, G.R.; Simpson, E.J.; Greenhaff, P.L. Protein- and carbohydrate-induced augmentation of whole body creatine retention in humans. *J. Appl. Physiol.* **2000**, *89*, 1165–1171. [CrossRef] [PubMed]
24. Wax, B.; Kerksick, C.M.; Jagim, A.R.; Mayo, J.J.; Lyons, B.C.; Kreider, R.B. Creatine for Exercise and Sports Performance, with Recovery Considerations for Healthy Populations. *Nutrients* **2021**, *13*, 1915. [CrossRef]

25. Jager, R.; Purpura, M.; Shao, A.; Inoue, T.; Kreider, R.B. Analysis of the efficacy, safety, and regulatory status of novel forms of creatine. *Amino Acids* **2011**, *40*, 1369–1383. [CrossRef]
26. Kreider, R.B.; Willoughby, D.S.; Greenwood, M.; Parise, G.; Payne, E.; Tarnopolsky, M.A. Effects of serum creatine supplementation on muscle creatine content. *J. Exerc. Physiol. Online* **2003**, *6*, 24–33.
27. Spillane, M.; Schoch, R.; Cooke, M.; Harvey, T.; Greenwood, M.; Kreider, R.; Willoughby, D.S. The effects of creatine ethyl ester supplementation combined with heavy resistance training on body composition, muscle performance, and serum and muscle creatine levels. *J. Int. Soc. Sports Nutr.* **2009**, *6*, 6. [CrossRef]
28. Jagim, A.R.; Oliver, J.M.; Sanchez, A.; Galvan, E.; Fluckey, J.; Riechman, S.; Greenwood, M.; Kelly, K.; Meininger, C.; Rasmussen, C.; et al. A buffered form of creatine does not promote greater changes in muscle creatine content, body composition, or training adaptations than creatine monohydrate. *J. Int. Soc. Sports Nutr.* **2012**, *9*, 43. [CrossRef] [PubMed]
29. Galvan, E.; Walker, D.K.; Simbo, S.Y.; Dalton, R.; Levers, K.; O'Connor, A.; Goodenough, C.; Barringer, N.D.; Greenwood, M.; Rasmussen, C.; et al. Acute and chronic safety and efficacy of dose dependent creatine nitrate supplementation and exercise performance. *J. Int. Soc. Sports Nutr.* **2016**, *13*, 12. [CrossRef] [PubMed]
30. Abraham, S.; Jiang, S. Process for Preparing a Creatine Heterocyclic Acid Salt and Method of Use. U.S. Patent No. 6,838,562, 4 January 2005.
31. Negrisoli, G.; Del Corona, L. Hydrosoluble Organic Salts of Creatine. U.S. Patent No. 5,973,199, Application No. 08/649,620, 26 October 1999.
32. Pischel, I.; Weiss, S. New Creatine Pyruvate Derivatives from Crystallisation in Polar Solvents. Germany Patent WO1998028263A1, 20 December 1996.
33. Pischel, I.; Weiss, S.; Gloxhuber, C.; Mertschenk, B. Creatine Ascorbates and a Method of Producing Them. U.S. Patent No. 5,863,939, 26 January 1999.
34. Child, R.; Tallon, M.J. Creatine ethyl ester rapidly degrades to creatinine in stomach acid. In Proceedings of the International Society of Sports Nutrition 4th Annual Meeting, Las Vegas, NV, USA, 12 June 2007.
35. Giese, M.W.; Lecher, C.S. Non-enzymatic cyclization of creatine ethyl ester to creatinine. *Biochem. Biophys. Res. Commun.* **2009**, *388*, 252–255. [CrossRef]
36. Dalton, R.L.; Sowinski, R.J.; Grubic, T.J.; Collins, P.B.; Coletta, A.M.; Reyes, A.G.; Sanchez, B.; Koozehchian, M.; Jung, Y.P.; Rasmussen, C.; et al. Hematological and Hemodynamic Responses to Acute and Short-Term Creatine Nitrate Supplementation. *Nutrients* **2017**, *9*, 1359. [CrossRef] [PubMed]
37. Gill, N.D.; Hall, R.D.; Blazevich, A.J. Creatine serum is not as effective as creatine powder for improving cycle sprint performance in competitive male team-sport athletes. *J. Strength Cond. Res.* **2004**, *18*, 272–275. [CrossRef] [PubMed]
38. Harris, R.C.; Almada, A.L.; Harris, D.B.; Dunnett, M.; Hespel, P. The creatine content of Creatine Serum and the change in the plasma concentration with ingestion of a single dose. *J. Sports Sci.* **2004**, *22*, 851–857. [CrossRef]
39. Page, M.J.; McKenzie, J.E.; Bossuyt, P.M.; Boutron, I.; Hoffmann, T.C.; Mulrow, C.D.; Shamseer, L.; Tetzlaff, J.M.; Akl, E.A.; Brennan, S.E.; et al. The PRISMA 2020 statement: An updated guideline for reporting systematic reviews. *BMJ* **2021**, *372*, n71. [CrossRef] [PubMed]
40. Schumann, K.; Classen, H.G.; Hages, M.; Prinz-Langenohl, R.; Pietrzik, K.; Biesalski, H.K. Bioavailability of oral vitamins, minerals, and trace elements in perspective. *Arzneimittelforschung* **1997**, *47*, 369–380. [PubMed]
41. Blancquaert, L.; Baguet, A.; Bex, T.; Volkaert, A.; Everaert, I.; Delanghe, J.; Petrovic, M.; Vervaet, C.; De Henauw, S.; Constantin-Teodosiu, D.; et al. Changing to a vegetarian diet reduces the body creatine pool in omnivorous women, but appears not to affect carnitine and carnosine homeostasis: A randomised trial. *Br. J. Nutr.* **2018**, *119*, 759–770. [CrossRef]
42. Watt, K.K.; Garnham, A.P.; Snow, R.J. Skeletal muscle total creatine content and creatine transporter gene expression in vegetarians prior to and following creatine supplementation. *Int. J. Sport Nutr. Exerc. Metab.* **2004**, *14*, 517–531. [CrossRef] [PubMed]
43. Forbes, S.C.; Candow, D.G.; Ostojic, S.M.; Roberts, M.D.; Chilibeck, P.D. Meta-Analysis Examining the Importance of Creatine Ingestion Strategies on Lean Tissue Mass and Strength in Older Adults. *Nutrients* **2021**, *13*, 1912. [CrossRef]
44. Candow, D.G.; Forbes, S.C.; Kirk, B.; Duque, G. Current Evidence and Possible Future Applications of Creatine Supplementation for Older Adults. *Nutrients* **2021**, *13*, 745. [CrossRef] [PubMed]
45. Harris, R.C.; Nevill, M.; Harris, D.B.; Fallowfield, J.L.; Bogdanis, G.C.; Wise, J.A. Absorption of creatine supplied as a drink, in meat or in solid form. *J. Sports Sci.* **2002**, *20*, 147–151. [CrossRef]
46. Kaviani, M.; Shaw, K.; Chilibeck, P.D. Benefits of Creatine Supplementation for Vegetarians Compared to Omnivorous Athletes: A Systematic Review. *Int. J. Environ. Res. Public Health* **2020**, *17*, 3041. [CrossRef]
47. Vandenberghe, K.; Goris, M.; Van Hecke, P.; Van Leemputte, M.; Vangerven, L.; Hespel, P. Long-term creatine intake is beneficial to muscle performance during resistance training. *J. Appl. Physiol.* **1997**, *83*, 2055–2063. [CrossRef]
48. Jäger, R.; Purpura, M.; Harris, R.C. Reduction of plasma creatine concentrations as an indicator of improved bioavailability. *J. Int. Soc. Sports Nutr.* **2016**, *13*, 3–4. [CrossRef]
49. Rodriguez, N.R.; DiMarco, N.M.; Langley, S.; American Dietetic Association; Dietitians of Canada; American College of Sports Medicine: Nutrition and Athletic Performance. Position of the American Dietetic Association, Dietitians of Canada, and the American College of Sports Medicine: Nutrition and athletic performance. *J. Am. Diet. Assoc.* **2009**, *109*, 509–527. [PubMed]
50. Pischel, I.; Gastner, T. Creatine–its chemical synthesis, chemistry, and legal status. *Subcell. Biochem.* **2007**, *46*, 291–307. [CrossRef]
51. Jäger, R. The Use of Creatine Monohydrate in Sports Nutrition. Degussa BioActives Publications: Freising, Germany, 2003.

52. Howard, A.N.; Harris, R.C. Compositions Containing Creatine. U.S. Patent No. 5,969,544, Application No. 08/866,517, 19 October 1999.

53. Edgar, G.; Shiver, H.E. The equilibrium between creatine and creatinine, in aqueous solution: The effect of hydrogen ion. *J. Am. Chem. Soc.* **1925**, *47*, 1179–1188. [CrossRef]

54. Cannon, J.G.; Orencole, S.F.; Fielding, R.A.; Meydani, M.; Meydani, S.N.; Fiatarone, M.A.; Blumberg, J.B.; Evans, W.J. Acute phase reponse in exercise: Interaction of age and vitamin E on neutrophils and muscle enzyme release. *Am. J. Physiol.* **1990**, *259*, R1214–R1219.

55. Kreider, R.B.; Ferreira, M.; Wilson, M.; Grindstaff, P.; Plisk, S.; Reinardy, J.; Cantler, E.; Almada, A.L. Effects of creatine supplementation on body composition, strength, and sprint performance. *Med. Sci. Sports Exerc.* **1998**, *30*, 73–82. [CrossRef] [PubMed]

56. Parandak, K.; Petersen, B.L. Shelf-Stable Nitrogenous Organic Acid Compositions. U.S. Patent Application No. 16/164,762, 5 January 2019.

57. CreaBev™: Soluble and Stable Creatine Monohydrate for Superior Performance Beverages. Available online: https://www.glanbianutritionals.com/en/nutri-knowledge-center/nutritional-resources/creabevtm (accessed on 5 January 2022).

58. Persky, A.M.; Brazeau, G.A.; Hochhaus, G. Pharmacokinetics of the dietary supplement creatine. *Clin. Pharmacokinet.* **2003**, *42*, 557–574. [CrossRef] [PubMed]

59. Deldicque, L.; Decombaz, J.; Zbinden Foncea, H.; Vuichoud, J.; Poortmans, J.R.; Francaux, M. Kinetics of creatine ingested as a food ingredient. *Eur. J. Appl. Physiol.* **2008**, *102*, 133–143. [CrossRef] [PubMed]

60. Greenhaff, P.; Bodin, K.; Harris, R.; Hultman, E.; Jones, D.G.; McIntyre, D.; Soderlund, K.; Turner, D.L. The influence of oral creatine supplementation on muscle phosphocreatine resynthesis following intenst contraction in man. *J. Physiol.* **1993**, *467*, 75P.

61. Reddy, A.; Norris, D.F.; Momeni, S.S.; Waldo, B.; Ruby, J.D. The pH of beverages in the United States. *J. Am. Dent. Assoc.* **2016**, *147*, 255–263. [CrossRef] [PubMed]

62. Antonio, J.; Candow, D.G.; Forbes, S.C.; Gualano, B.; Jagim, A.R.; Kreider, R.B.; Rawson, E.S.; Smith-Ryan, A.E.; VanDusseldorp, T.A.; Willoughby, D.S.; et al. Common questions and misconceptions about creatine supplementation: What does the scientific evidence really show? *J. Int. Soc. Sports Nutr.* **2021**, *18*, 13. [CrossRef] [PubMed]

63. Jäger, R.; Harris, R.C.; Purpura, M.; Francaux, M. Comparison of new forms of creatine in raising plasma creatine levels. *J. Int. Soc. Sports Nutr.* **2007**, *4*, 17. [CrossRef]

64. Miller, D.W.; Vennerstrom, J.L.; Faulkner, M.C. Creatine Oral Supplementation Using Creatine Hydrochloride Salt. U.S. Patent No. 7,608,641 B2, Application No. 10/846,782, 27 October 2009.

65. França, E.; Avelar, B.; Yoshioka, C.; Santana, J.; Madureira, D.; Rocha, L.; Zocoler, C.; Rossi, F.; Lira, F.; Rodrigues, B.; et al. Creatine HCl and Creatine Monohydrate Improve Strength but Only Creatine HCl Induced Changes on Body Composition in Recreational Weightlifters. *Food Nutr. Sci.* **2015**, *6*, 1624–1630. [CrossRef]

66. Howard, A.N.; Harris, R. Compositions Containing Creatine and Aloe Vera Extract. U.S. Patent No. 6,168,802, 2 June 1999.

67. Federal Food, Drug, and Cosmetic Act, 413(a) [21 U.S.C. 350b], United States Congress, 15 October 1994. Available online: https://uscode.house.gov/view.xhtml?req=(title:21%20section:350b%20edition:prelim (accessed on 7 January 2022).

68. Nelson, A.G.; Arnall, D.A.; Kokkonen, J.; Day, R.; Evans, J. Muscle glycogen supercompensation is enhanced by prior creatine supplementation. *Med. Sci. Sports Exerc.* **2001**, *33*, 1096–1100. [CrossRef]

69. Tarnopolsky, M.A.; Parise, G. Direct measurement of high-energy phosphate compounds in patients with neuromuscular disease. *Muscle Nerve* **1999**, *22*, 1228–1233. [CrossRef]

70. McKenna, M.J.; Morton, J.; Selig, S.E.; Snow, R.J. Creatine supplementation increases muscle total creatine but not maximal intermittent exercise performance. *J. Appl. Physiol.* **1999**, *87*, 2244–2252. [CrossRef] [PubMed]

71. Greenhaff, P.L.; Bodin, K.; Soderlund, K.; Hultman, E. Effect of oral creatine supplementation on skeletal muscle phosphocreatine resynthesis. *Am. J. Physiol.* **1994**, *266*, E725–E730. [CrossRef]

72. Choi, J.K.; Kustermann, E.; Dedeoglu, A.; Jenkins, B.G. Magnetic resonance spectroscopy of regional brain metabolite markers in FALS mice and the effects of dietary creatine supplementation. *Eur. J. Neurosci.* **2009**, *30*, 2143–2150. [CrossRef] [PubMed]

73. Lyoo, I.K.; Kong, S.W.; Sung, S.M.; Hirashima, F.; Parow, A.; Hennen, J.; Cohen, B.M.; Renshaw, P.F. Multinuclear magnetic resonance spectroscopy of high-energy phosphate metabolites in human brain following oral supplementation of creatine-monohydrate. *Psychiatry Res.* **2003**, *123*, 87–100. [CrossRef]

74. Roschel, H.; Gualano, B.; Ostojic, S.M.; Rawson, E.S. Creatine Supplementation and Brain Health. *Nutrients* **2021**, *13*, 586. [CrossRef]

75. Dolan, E.; Gualano, B.; Rawson, E.S. Beyond muscle: The effects of creatine supplementation on brain creatine, cognitive processing, and traumatic brain injury. *Eur. J. Sport Sci.* **2019**, *19*, 1–14. [CrossRef] [PubMed]

76. Gualano, B.; Rawson, E.S.; Candow, D.G.; Chilibeck, P.D. Creatine supplementation in the aging population: Effects on skeletal muscle, bone and brain. *Amino Acids* **2016**, *48*, 1793–1805. [CrossRef] [PubMed]

77. Rawson, E.S.; Venezia, A.C. Use of creatine in the elderly and evidence for effects on cognitive function in young and old. *Amino Acids* **2011**, *40*, 1349–1362. [CrossRef] [PubMed]

78. Cornish, S.M.; Chilibeck, P.D.; Burke, D.G. The effect of creatine monohydrate supplementation on sprint skating in ice-hockey players. *J. Sports Med. Phys. Fit.* **2006**, *46*, 90–98.

79. Dawson, B.; Vladich, T.; Blanksby, B.A. Effects of 4 weeks of creatine supplementation in junior swimmers on freestyle sprint and swim bench performance. *J. Strength Cond. Res.* **2002**, *16*, 485–490.

80. Grindstaff, P.D.; Kreider, R.; Bishop, R.; Wilson, M.; Wood, L.; Alexander, C.; Almada, A. Effects of creatine supplementation on repetitive sprint performance and body composition in competitive swimmers. *Int. J. Sport Nutr.* **1997**, *7*, 330–346. [CrossRef]

81. Juhasz, I.; Gyore, I.; Csende, Z.; Racz, L.; Tihanyi, J. Creatine supplementation improves the anaerobic performance of elite junior fin swimmers. *Acta Physiol. Hung.* **2009**, *96*, 325–336. [CrossRef] [PubMed]

82. Silva, A.J.; Machado Reis, V.; Guidetti, L.; Bessone Alves, F.; Mota, P.; Freitas, J.; Baldari, C. Effect of creatine on swimming velocity, body composition and hydrodynamic variables. *J. Sports Med. Phys. Fit.* **2007**, *47*, 58–64.

83. Stone, M.H.; Sanborn, K.; Smith, L.L.; O'Bryant, H.S.; Hoke, T.; Utter, A.C.; Johnson, R.L.; Boros, R.; Hruby, J.; Pierce, K.C.; et al. Effects of in-season (5 weeks) creatine and pyruvate supplementation on anaerobic performance and body composition in American football players. *Int. J. Sport Nutr.* **1999**, *9*, 146–165. [CrossRef] [PubMed]

84. Bemben, M.G.; Bemben, D.A.; Loftiss, D.D.; Knehans, A.W. Creatine supplementation during resistance training in college football athletes. *Med. Sci. Sports Exerc.* **2001**, *33*, 1667–1673. [CrossRef] [PubMed]

85. Hoffman, J.; Ratamess, N.; Kang, J.; Mangine, G.; Faigenbaum, A.; Stout, J. Effect of creatine and beta-alanine supplementation on performance and endocrine responses in strength/power athletes. *Int. J. Sport Nutr. Exerc. Metab.* **2006**, *16*, 430–446. [CrossRef] [PubMed]

86. Chilibeck, P.D.; Magnus, C.; Anderson, M. Effect of in-season creatine supplementation on body composition and performance in rugby union football players. *Appl. Physiol. Nutr. Metab.* **2007**, *32*, 1052–1057. [CrossRef] [PubMed]

87. Claudino, J.G.; Mezencio, B.; Amaral, S.; Zanetti, V.; Benatti, F.; Roschel, H.; Gualano, B.; Amadio, A.C.; Serrao, J.C. Creatine monohydrate supplementation on lower-limb muscle power in Brazilian elite soccer players. *J. Int. Soc. Sports Nutr.* **2014**, *11*, 32. [CrossRef] [PubMed]

88. Kerksick, C.M.; Rasmussen, C.; Lancaster, S.; Starks, M.; Smith, P.; Melton, C.; Greenwood, M.; Almada, A.; Kreider, R. Impact of differing protein sources and a creatine containing nutritional formula after 12 weeks of resistance training. *Nutrition* **2007**, *23*, 647–656. [CrossRef]

89. Kerksick, C.M.; Wilborn, C.D.; Campbell, W.I.; Harvey, T.M.; Marcello, B.M.; Roberts, M.D.; Parker, A.G.; Byars, A.G.; Greenwood, L.D.; Almada, A.L.; et al. The effects of creatine monohydrate supplementation with and without D-pinitol on resistance training adaptations. *J. Strength Cond. Res.* **2009**, *23*, 2673–2682. [CrossRef] [PubMed]

90. Volek, J.S.; Kraemer, W.J.; Bush, J.A.; Boetes, M.; Incledon, T.; Clark, K.L.; Lynch, J.M. Creatine supplementation enhances muscular performance during high-intensity resistance exercise. *J. Am. Diet. Assoc.* **1997**, *97*, 765–770. [CrossRef]

91. Volek, J.S.; Mazzetti, S.A.; Farquhar, W.B.; Barnes, B.R.; Gomez, A.L.; Kraemer, W.J. Physiological responses to short-term exercise in the heat after creatine loading. *Med. Sci. Sports Exerc.* **2001**, *33*, 1101–1108. [CrossRef] [PubMed]

92. Volek, J.S.; Ratamess, N.A.; Rubin, M.R.; Gomez, A.L.; French, D.N.; McGuigan, M.M.; Scheett, T.P.; Sharman, M.J.; Hakkinen, K.; Kraemer, W.J. The effects of creatine supplementation on muscular performance and body composition responses to short-term resistance training overreaching. *Eur. J. Appl. Physiol.* **2004**, *91*, 628–637. [CrossRef]

93. Kreider, R.B.; Wilborn, C.D.; Taylor, L.; Campbell, B.; Almada, A.L.; Collins, R.; Cooke, M.; Earnest, C.P.; Greenwood, M.; Kalman, D.S.; et al. ISSN exercise & sport nutrition review: Research & recommendations. *J. Int. Soc. Sports Nutr.* **2010**, *7*, 7. [CrossRef]

94. Branch, J.D. Effect of creatine supplementation on body composition and performance: A meta-analysis. *Int. J. Sport Nutr. Exerc. Metab.* **2003**, *13*, 198–226. [CrossRef] [PubMed]

95. Devries, M.C.; Phillips, S.M. Creatine supplementation during resistance training in older adults-a meta-analysis. *Med. Sci. Sports Exerc.* **2014**, *46*, 1194–1203. [CrossRef]

96. Lanhers, C.; Pereira, B.; Naughton, G.; Trousselard, M.; Lesage, F.X.; Dutheil, F. Creatine Supplementation and Lower Limb Strength Performance: A Systematic Review and Meta-Analyses. *Sports Med.* **2015**, *45*, 1285–1294. [CrossRef] [PubMed]

97. Wiroth, J.B.; Bermon, S.; Andrei, S.; Dalloz, E.; Hebuterne, X.; Dolisi, C. Effects of oral creatine supplementation on maximal pedalling performance in older adults. *Eur. J. Appl. Physiol.* **2001**, *84*, 533–539. [CrossRef] [PubMed]

98. McMorris, T.; Mielcarz, G.; Harris, R.C.; Swain, J.P.; Howard, A. Creatine supplementation and cognitive performance in elderly individuals. *Neuropsychol. Dev. Cogn. B Aging Neuropsychol. Cogn.* **2007**, *14*, 517–528. [CrossRef] [PubMed]

99. Rawson, E.S.; Clarkson, P.M. Acute creatine supplementation in older men. *Int. J. Sports Med.* **2000**, *21*, 71–75. [CrossRef]

100. Tarnopolsky, M.A. Potential benefits of creatine monohydrate supplementation in the elderly. *Curr. Opin. Clin. Nutr. Metab. Care* **2000**, *3*, 497–502. [CrossRef] [PubMed]

101. Aguiar, A.F.; Januario, R.S.; Junior, R.P.; Gerage, A.M.; Pina, F.L.; do Nascimento, M.A.; Padovani, C.R.; Cyrino, E.S. Long-term creatine supplementation improves muscular performance during resistance training in older women. *Eur. J. Appl. Physiol.* **2013**, *113*, 987–996. [CrossRef] [PubMed]

102. Kreider, R.B. Effects of creatine supplementation on performance and training adaptations. *Mol. Cell. Biochem.* **2003**, *244*, 89–94. [CrossRef] [PubMed]

103. Gualano, B.; Macedo, A.R.; Alves, C.R.; Roschel, H.; Benatti, F.B.; Takayama, L.; de Sa Pinto, A.L.; Lima, F.R.; Pereira, R.M. Creatine supplementation and resistance training in vulnerable older women: A randomized double-blind placebo-controlled clinical trial. *Exp. Gerontol.* **2014**, *53*, 7–15. [CrossRef] [PubMed]

104. Candow, D.G.; Little, J.P.; Chilibeck, P.D.; Abeysekara, S.; Zello, G.A.; Kazachkov, M.; Cornish, S.M.; Yu, P.H. Low-dose creatine combined with protein during resistance training in older men. *Med. Sci. Sports Exerc.* **2008**, *40*, 1645–1652. [CrossRef]

105. Hass, C.J.; Collins, M.A.; Juncos, J.L. Resistance training with creatine monohydrate improves upper-body strength in patients with Parkinson disease: A randomized trial. *Neurorehabilit. Neural Repair* **2007**, *21*, 107–115. [CrossRef]

106. Candow, D.G.; Chilibeck, P.D. Effect of creatine supplementation during resistance training on muscle accretion in the elderly. *J. Nutr. Health Aging* **2007**, *11*, 185–188.

107. Chilibeck, P.D.; Chrusch, M.J.; Chad, K.E.; Shawn Davison, K.; Burke, D.G. Creatine monohydrate and resistance training increase bone mineral content and density in older men. *J. Nutr. Health Aging* **2005**, *9*, 352–353.

108. Burke, D.G.; Chilibeck, P.D.; Parise, G.; Candow, D.G.; Mahoney, D.; Tarnopolsky, M. Effect of creatine and weight training on muscle creatine and performance in vegetarians. *Med. Sci. Sports Exerc.* **2003**, *35*, 1946–1955. [CrossRef] [PubMed]

109. Wilder, N.; Gilders, R.; Hagerman, F.; Deivert, R.G. The effects of a 10-week, periodized, off-season resistance-training program and creatine supplementation among collegiate football players. *J. Strength Cond. Res.* **2002**, *16*, 343–352. [PubMed]

110. Izquierdo, M.; Ibanez, J.; Gonzalez-Badillo, J.J.; Gorostiaga, E.M. Effects of creatine supplementation on muscle power, endurance, and sprint performance. *Med. Sci. Sports Exerc.* **2002**, *34*, 332–343. [CrossRef] [PubMed]

111. Chrusch, M.J.; Chilibeck, P.D.; Chad, K.E.; Davison, K.S.; Burke, D.G. Creatine supplementation combined with resistance training in older men. *Med. Sci. Sports Exerc.* **2001**, *33*, 2111–2117. [CrossRef] [PubMed]

112. Becque, M.D.; Lochmann, J.D.; Melrose, D.R. Effects of oral creatine supplementation on muscular strength and body composition. *Med. Sci. Sports Exerc.* **2000**, *32*, 654–658. [CrossRef] [PubMed]

113. Volek, J.S.; Duncan, N.D.; Mazzetti, S.A.; Staron, R.S.; Putukian, M.; Gomez, A.L.; Pearson, D.R.; Fink, W.J.; Kraemer, W.J. Performance and muscle fiber adaptations to creatine supplementation and heavy resistance training. *Med. Sci. Sports Exerc.* **1999**, *31*, 1147–1156. [CrossRef] [PubMed]

114. Ahmun, R.P.; Tong, R.J.; Grimshaw, P.N. The effects of acute creatine supplementation on multiple sprint cycling and running performance in rugby players. *J. Strength Cond. Res.* **2005**, *19*, 92–97. [CrossRef]

115. Cox, G.; Mujika, I.; Tumilty, D.; Burke, L. Acute creatine supplementation and performance during a field test simulating match play in elite female soccer players. *Int. J. Sport Nutr. Exerc. Metab.* **2002**, *12*, 33–46. [CrossRef]

116. Preen, D.; Dawson, B.; Goodman, C.; Lawrence, S.; Beilby, J.; Ching, S. Effect of creatine loading on long-term sprint exercise performance and metabolism. *Med. Sci. Sports Exerc.* **2001**, *33*, 814–821. [CrossRef]

117. Aaserud, R.; Gramvik, P.; Olsen, S.R.; Jensen, J. Creatine supplementation delays onset of fatigue during repeated bouts of sprint running. *Scand. J. Med. Sci. Sports* **1998**, *8*, 247–251. [CrossRef]

118. Bosco, C.; Tihanyi, J.; Pucspk, J.; Kovacs, I.; Gabossy, A.; Colli, R.; Pulvirenti, G.; Tranquilli, C.; Foti, C.; Viru, M.; et al. Effect of oral creatine supplementation on jumping and running performance. *Int. J. Sports Med.* **1997**, *18*, 369–372. [CrossRef] [PubMed]

119. Ramirez-Campillo, R.; Gonzalez-Jurado, J.A.; Martinez, C.; Nakamura, F.Y.; Penailillo, L.; Meylan, C.M.; Caniuqueo, A.; Canas-Jamet, R.; Moran, J.; Alonso-Martinez, A.M.; et al. Effects of plyometric training and creatine supplementation on maximal-intensity exercise and endurance in female soccer players. *J. Sci. Med. Sport* **2016**, *19*, 682–687. [CrossRef] [PubMed]

120. Yanez-Silva, A.; Buzzachera, C.F.; Picarro, I.D.; Januario, R.S.; Ferreira, L.H.; McAnulty, S.R.; Utter, A.C.; Souza-Junior, T.P. Effect of low dose, short-term creatine supplementation on muscle power output in elite youth soccer players. *J. Int. Soc. Sports Nutr.* **2017**, *14*, 5. [CrossRef] [PubMed]

121. Dabidi Roshan, V.; Babaei, H.; Hosseinzadeh, M.; Arendt-Nielsen, L. The effect of creatine supplementation on muscle fatigue and physiological indices following intermittent swimming bouts. *J. Sports Med. Phys. Fit.* **2013**, *53*, 232–239.

122. Selsby, J.T.; Beckett, K.D.; Kern, M.; Devor, S.T. Swim performance following creatine supplementation in Division III athletes. *J. Strength Cond. Res.* **2003**, *17*, 421–424. [PubMed]

123. Leenders, N.M.; Lamb, D.R.; Nelson, T.E. Creatine supplementation and swimming performance. *Int. J. Sport Nutr.* **1999**, *9*, 251–262. [CrossRef] [PubMed]

124. Peyrebrune, M.C.; Nevill, M.E.; Donaldson, F.J.; Cosford, D.J. The effects of oral creatine supplementation on performance in single and repeated sprint swimming. *J. Sports Sci.* **1998**, *16*, 271–279. [CrossRef]

125. Lamontagne-Lacasse, M.; Nadon, R.; Goulet, E.D. Effect of creatine supplementation on jumping performance in elite volleyball players. *Int. J. Sports Physiol. Perform.* **2011**, *6*, 525–533. [CrossRef]

126. Ayoama, R.; Hiruma, E.; Sasaki, H. Effects of creatine loading on muscular strength and endurance of female softball players. *J. Sports Med. Phys. Fit.* **2003**, *43*, 481–487.

127. Jones, A.M.; Atter, T.; Georg, K.P. Oral creatine supplementation improves multiple sprint performance in elite ice-hockey players. *J. Sports Med. Phys. Fit.* **1999**, *39*, 189–196. [CrossRef]

128. Ziegenfuss, T.N.; Habowski, S.M.; Lemieux, R.; Sandrock, J.E.; Kedia, A.W.; Kerksick, C.M.; Lopez, H.L. Effects of a dietary supplement on golf drive distance and functional indices of golf performance. *J. Int. Soc. Sports Nutr.* **2015**, *12*, 4. [CrossRef]

129. Tarnopolsky, M.A.; MacLennan, D.P. Creatine monohydrate supplementation enhances high-intensity exercise performance in males and females. *Int. J. Sport Nutr. Exerc. Metab.* **2000**, *10*, 452–463. [CrossRef]

130. Ziegenfuss, T.N.; Rogers, M.; Lowery, L.; Mullins, N.; Mendel, R.; Antonio, J.; Lemon, P. Effect of creatine loading on anaerobic performance and skeletal muscle volume in NCAA Division I athletes. *Nutrition* **2002**, *18*, 397–402. [CrossRef]

131. Benton, D.; Donohoe, R. The influence of creatine supplementation on the cognitive functioning of vegetarians and omnivores. *Br. J. Nutr.* **2011**, *105*, 1100–1105. [CrossRef]

132. Johannsmeyer, S.; Candow, D.G.; Brahms, C.M.; Michel, D.; Zello, G.A. Effect of creatine supplementation and drop-set resistance training in untrained aging adults. *Exp. Gerontol.* **2016**, *83*, 112–119. [CrossRef] [PubMed]

133. Smith-Ryan, A.E.; Cabre, H.E.; Eckerson, J.M.; Candow, D.G. Creatine Supplementation in Women's Health: A Lifespan Perspective. *Nutrients* **2021**, *13*, 877. [CrossRef]

134. Greenwood, M.; Kreider, R.B.; Rasmussen, C.; Almada, A.L.; Earnest, C.P. D-Pinitol augments whole body creatine retention in man. *J. Exerc. Physiol. Online* **2001**, *4*, 41–47.

135. Jäger, R.; Kendrick, I.; Purpura, M.; Harris, R.; Ribnicky, D.; Pischel, I. The effect of Russian Tarragon (*Artemisia dracunculus* L.) on the plasma creatine concentration with creatine monohydrate administration. *J. Int. Soc. Sports Nutr.* **2008**, *5*, P4. [CrossRef]

136. Taylor, L.; Poole, C.; Pena, E.; Lewing, M.; Kreider, R.; Foster, C.; Wilborn, C. Effects of Combined Creatine Plus Fenugreek Extract vs. Creatine Plus Carbohydrate Supplementation on Resistance Training Adaptations. *J. Sports Sci. Med.* **2011**, *10*, 254–260. [PubMed]

137. Oliver, J.M.; Jagim, A.R.; Pischel, I.; Jager, R.; Purpura, M.; Sanchez, A.; Fluckey, J.; Riechman, S.; Greenwood, M.; Kelly, K.; et al. Effects of short-term ingestion of Russian Tarragon prior to creatine monohydrate supplementation on whole body and muscle creatine retention and anaerobic sprint capacity: A preliminary investigation. *J. Int. Soc. Sports Nutr.* **2014**, *11*, 6. [CrossRef] [PubMed]

138. Pakulak, A.; Candow, D.G.; Totosy de Zepetnek, J.; Forbes, S.C.; Basta, D. Effects of Creatine and Caffeine Supplementation During Resistance Training on Body Composition, Strength, Endurance, Rating of Perceived Exertion and Fatigue in Trained Young Adults. *J. Diet. Suppl.* **2021**, 1–16. [CrossRef] [PubMed]

139. Van Bavel, D.; de Moraes, R.; Tibirica, E. Effects of dietary supplementation with creatine on homocysteinemia and systemic microvascular endothelial function in individuals adhering to vegan diets. *Fundam. Clin. Pharmacol.* **2019**, *33*, 428–440. [CrossRef]

140. Moraes, R.; Van Bavel, D.; Moraes, B.S.; Tibirica, E. Effects of dietary creatine supplementation on systemic microvascular density and reactivity in healthy young adults. *Nutr. J.* **2014**, *13*, 115. [CrossRef] [PubMed]

141. Falk, D.J.; Heelan, K.A.; Thyfault, J.P.; Koch, A.J. Effects of effervescent creatine, ribose, and glutamine supplementation on muscular strength, muscular endurance, and body composition. *J. Strength Cond. Res.* **2003**, *17*, 810–816. [CrossRef] [PubMed]

142. Greenwood, M.; Farris, J.; Kreider, R.; Greenwood, L.; Byars, A. Creatine supplementation patterns and perceived effects in select division I collegiate athletes. *Clin. J. Sport Med.* **2000**, *10*, 191–194. [CrossRef]

143. Greenwood, M.; Kreider, R.B.; Greenwood, L.; Byars, A. Cramping and Injury Incidence in Collegiate Football Players Are Reduced by Creatine Supplementation. *J. Athl. Train.* **2003**, *38*, 216–219. [PubMed]

144. Greenwood, M.; Kreider, R.B.; Melton, C.; Rasmussen, C.; Lancaster, S.; Cantler, E.; Milnor, P.; Almada, A. Creatine supplementation during college football training does not increase the incidence of cramping or injury. *Mol. Cell. Biochem.* **2003**, *244*, 83–88. [CrossRef]

145. National Institutes of Health; Office of Dietary Supplements. Dietary Supplement Fact Sheets: Dietary Supplements for Exercise and Athletic Performance. Available online. https://ods.od.nih.gov/factsheets/ExerciseAndAthleticPerformance-HealthProfessional/ (accessed on 5 January 2022).

146. Muccini, A.M.; Tran, N.T.; de Guingand, D.L.; Philip, M.; Della Gatta, P.A.; Galinsky, R.; Sherman, L.S.; Kelleher, M.A.; Palmer, K.R.; Berry, M.J.; et al. Creatine Metabolism in Female Reproduction, Pregnancy and Newborn Health. *Nutrients* **2021**, *13*, 490. [CrossRef]

147. Jagim, A.R.; Kerksick, C.M. Creatine Supplementation in Children and Adolescents. *Nutrients* **2021**, *13*, 664. [CrossRef]

148. Dover, S.; Stephens, S.; Schneiderman, J.E.; Pullenayegum, E.; Wells, G.D.; Levy, D.M.; Marcuz, J.A.; Whitney, K.; Schulze, A.; Tein, I.; et al. The Effect of Creatine Supplementation on Muscle Function in Childhood Myositis: A Randomized, Double-blind, Placebo-controlled Feasibility Study. *J. Rheumatol.* **2021**, *48*, 434–441. [CrossRef] [PubMed]

149. Korovljev, D.; Todorovic, N.; Stajer, V.; Ostojic, S.M. Dietary Intake of Creatine in Children Aged 0-24 Months. *Ann. Nutr. Metab.* **2021**, *77*, 185–188. [CrossRef]

150. Korovljev, D.; Todorovic, N.; Stajer, V.; Ostojic, S.M. Food Creatine and DXA-Derived Body Composition in Boys and Girls Aged 8 to 19 Years. *Nutr. Metab. Insights* **2021**, *14*, 11786388211059368. [CrossRef] [PubMed]

151. Korovljev, D.; Stajer, V.; Ostojic, S.M. Relationship between Dietary Creatine and Growth Indicators in Children and Adolescents Aged 2-19 Years: A Cross-Sectional Study. *Nutrients* **2021**, *13*, 1027. [CrossRef]

152. Harmon, K.K.; Stout, J.R.; Fukuda, D.H.; Pabian, P.S.; Rawson, E.S.; Stock, M.S. The Application of Creatine Supplementation in Medical Rehabilitation. *Nutrients* **2021**, *13*, 1825. [CrossRef] [PubMed]

153. Solis, M.Y.; Artioli, G.G.; Gualano, B. Potential of Creatine in Glucose Management and Diabetes. *Nutrients* **2021**, *13*, 570. [CrossRef]

154. Bredahl, E.C.; Eckerson, J.M.; Tracy, S.M.; McDonald, T.L.; Drescher, K.M. The Role of Creatine in the Development and Activation of Immune Responses. *Nutrients* **2021**, *13*, 751. [CrossRef]

155. Li, B.; Yang, L. Creatine in T Cell Antitumor Immunity and Cancer Immunotherapy. *Nutrients* **2021**, *13*, 1633. [CrossRef] [PubMed]

156. Balestrino, M. Role of Creatine in the Heart: Health and Disease. *Nutrients* **2021**, *13*, 1215. [CrossRef] [PubMed]

157. Clarke, H.; Hickner, R.C.; Ormsbee, M.J. The Potential Role of Creatine in Vascular Health. *Nutrients* **2021**, *13*, 857. [CrossRef]

158. Wallimann, T.; Hall, C.H.T.; Colgan, S.P.; Glover, L.E. Creatine Supplementation for Patients with Inflammatory Bowel Diseases: A Scientific Rationale for a Clinical Trial. *Nutrients* **2021**, *13*, 1429. [CrossRef]

159. van der Veen, Y.; Post, A.; Kremer, D.; Koops, C.A.; Marsman, E.; Appeldoorn, T.Y.J.; Touw, D.J.; Westerhuis, R.; Heiner-Fokkema, M.R.; Franssen, C.F.M.; et al. Chronic Dialysis Patients Are Depleted of Creatine: Review and Rationale for Intradialytic Creatine Supplementation. *Nutrients* **2021**, *13*, 2709. [CrossRef]

160. Ostojic, S.M. Diagnostic and Pharmacological Potency of Creatine in Post-Viral Fatigue Syndrome. *Nutrients* **2021**, *13*, 503. [CrossRef]

161. Balsom, P.D.; Soderlund, K.; Sjodin, B.; Ekblom, B. Skeletal muscle metabolism during short duration high-intensity exercise: Influence of creatine supplementation. *Acta Physiol. Scand.* **1995**, *154*, 303–310. [CrossRef]

162. Vandenberghe, K.; Van Hecke, P.; Van Leemputte, M.; Vanstapel, F.; Hespel, P. Phosphocreatine resynthesis is not affected by creatine loading. *Med. Sci. Sports Exerc.* **1999**, *31*, 236–242. [CrossRef]

163. Bellinger, B.M.; Bold, A.; Wilson, G.R.; Noakes, T.D.; Myburgh, K.H. Oral creatine supplementation decreases plasma markers of adenine nucleotide degradation during a 1-h cycle test. *Acta Physiol. Scand.* **2000**, *170*, 217–224. [CrossRef]

164. Francaux, M.; Demeure, R.; Goudemant, J.F.; Poortmans, J.R. Effect of exogenous creatine supplementation on muscle PCr metabolism. *Int. J. Sports Med.* **2000**, *21*, 139–145. [CrossRef] [PubMed]

165. Burke, D.G.; Chilibeck, P.D.; Parise, G.; Tarnopolsky, M.A.; Candow, D.G. Effect of alpha-lipoic acid combined with creatine monohydrate on human skeletal muscle creatine and phosphagen concentration. *Int. J. Sport Nutr. Exerc. Metab.* **2003**, *13*, 294–302. [CrossRef] [PubMed]

166. Kreider, R.B.; Klesges, R.; Lotz, D.; Davis, M.; Cantler, E.; Grindstaff, P.; Ramsey, L.; Bullen, D.; Wood, L.; Almada, A. Effects of nutritional supplementation during off-season college football training on body composition and strength. *J. Exerc. Physiol. Online* **1999**, *2*, 24–39.

167. Tarnopolsky, M.A.; Parise, G.; Yardley, N.J.; Ballantyne, C.S.; Olatinji, S.; Phillips, S.M. Creatine-dextrose and protein-dextrose induce similar strength gains during training. *Med. Sci. Sports Exerc.* **2001**, *33*, 2044–2052. [CrossRef] [PubMed]

168. Willoughby, D.S.; Rosene, J. Effects of oral creatine and resistance training on myosin heavy chain expression. *Med. Sci. Sports Exerc.* **2001**, *33*, 1674–1681. [CrossRef] [PubMed]

169. Newman, J.E.; Hargreaves, M.; Garnham, A.; Snow, R.J. Effect of creatine ingestion on glucose tolerance and insulin sensitivity in men. *Med. Sci. Sports Exerc.* **2003**, *35*, 69–74. [CrossRef]

170. Tarnopolsky, M.; Parise, G.; Fu, M.H.; Brose, A.; Parshad, A.; Speer, O.; Wallimann, T. Acute and moderate-term creatine monohydrate supplementation does not affect creatine transporter mRNA or protein content in either young or elderly humans. *Mol. Cell. Biochem.* **2003**, *244*, 159–166. [CrossRef] [PubMed]

171. Willoughby, D.S.; Rosene, J.M. Effects of oral creatine and resistance training on myogenic regulatory factor expression. *Med. Sci. Sports Exerc.* **2003**, *35*, 923–929. [CrossRef] [PubMed]

172. Smith, A.E.; Fukuda, D.H.; Ryan, E.D.; Kendall, K.L.; Cramer, J.T.; Stout, J. Ergolytic/ergogenic effects of creatine on aerobic power. *Int. J. Sports Med.* **2011**, *32*, 975–981. [CrossRef] [PubMed]

173. Horecky, J.; Gvozdjakova, A.; Kucharska, J.; Obrenovich, M.E.; Palacios, H.H.; Li, Y.; Vancova, O.; Aliev, G. Effects of coenzyme Q and creatine supplementation on brain energy metabolism in rats exposed to chronic cerebral hypoperfusion. *Curr. Alzheimer Res.* **2011**, *8*, 868–875. [CrossRef] [PubMed]

174. Jäger, R.; Metzger, J.; Lautmann, K.; Shushakov, V.; Purpura, M.; Geiss, K.R.; Maassen, N. The effects of creatine pyruvate and creatine citrate on performance during high intensity exercise. *J. Int. Soc. Sports Nutr.* **2008**, *5*, 4. [CrossRef] [PubMed]

175. Van Schuylenbergh, R.; Van Leemputte, M.; Hespel, P. Effects of oral creatine-pyruvate supplementation in cycling performance. *Int. J. Sports Med.* **2003**, *24*, 144–150. [CrossRef] [PubMed]

176. Andres, S.; Ziegenhagen, R.; Trefflich, I.; Pevny, S.; Schultrich, K.; Braun, H.; Schanzer, W.; Hirsch-Ernst, K.I.; Schafer, B.; Lampen, A. Creatine and creatine forms intended for sports nutrition. *Mol. Nutr. Food Res.* **2017**, *61*, 1600772. [CrossRef] [PubMed]

177. European Food Safety Authority. Orotic Acid Salts as Sources of Orotic Acid and Various Minerals Added for Nutritional Purposes to Food Supplements, Scientific Opinion of the Panel on Food Additives and Nutrient Sources Added to Food (ANS). *EFSA J.* **2009**, *7*, 1187. [CrossRef]

178. Purpura, M.; Pischel, I.; Jäger, R.; Ortenburger, G. Solid and Stable Creatine/Citric Acid Composition(s) and Compositions Carbohydrate(s) or Hydrates Thereof, Method for the Production and Use Thereof. U.S. Patent Application No. 10/495,827, 17 February 2005.

179. Gufford, B.T.; Sriraghavan, K.; Miller, N.J.; Miller, D.W.; Gu, X.; Vennerstrom, J.L.; Robinson, D.H. Physicochemical characterization of creatine N-methylguanidinium salts. *J. Diet. Suppl.* **2010**, *7*, 240–252. [CrossRef]

180. Smith, A.E.; Walter, A.A.; Herda, T.J.; Ryan, E.D.; Moon, J.R.; Cramer, J.T.; Stout, J.R. Effects of creatine loading on electromyographic fatigue threshold during cycle ergometry in college-aged women. *J. Int. Soc. Sports Nutr.* **2007**, *4*, 20. [CrossRef] [PubMed]

181. Graef, J.L.; Smith, A.E.; Kendall, K.L.; Fukuda, D.H.; Moon, J.R.; Beck, T.W.; Cramer, J.T.; Stout, J.R. The effects of four weeks of creatine supplementation and high-intensity interval training on cardiorespiratory fitness: A randomized controlled trial. *J. Int. Soc. Sports Nutr.* **2009**, *6*, 18. [CrossRef]

182. Fukuda, D.H.; Smith, A.E.; Kendall, K.L.; Dwyer, T.R.; Kerksick, C.M.; Beck, T.W.; Cramer, J.T.; Stout, J.R. The effects of creatine loading and gender on anaerobic running capacity. *J. Strength Cond. Res.* **2010**, *24*, 1826–1833. [CrossRef] [PubMed]

183. Koh-Banerjee, P.K.; Ferreira, M.P.; Greenwood, M.; Bowden, R.G.; Cowan, P.N.; Almada, A.L.; Kreider, R.B. Effects of calcium pyruvate supplementation during training on body composition, exercise capacity, and metabolic responses to exercise. *Nutrition* **2005**, *21*, 312–319. [CrossRef]

184. Constantin-Teodosiu, D.; Peirce, N.S.; Fox, J.; Greenhaff, P.L. Muscle pyruvate availability can limit the flux, but not activation, of the pyruvate dehydrogenase complex during submaximal exercise in humans. *J. Physiol.* **2004**, *561*, 647–655. [CrossRef] [PubMed]

185. Ostojic, S.M.; Ahmetovic, Z. The effect of 4 weeks treatment with a 2-gram daily dose of pyruvate on body composition in healthy trained men. *Int. J. Vitam. Nutr. Res.* **2009**, *79*, 173–179. [CrossRef]
186. Ivy, J.L. Effect of pyruvate and dihydroxyacetone on metabolism and aerobic endurance capacity. *Med. Sci. Sports Exerc.* **1998**, *30*, 837–843. [PubMed]
187. Nuuttilla, S. Edustusmelojat testasivat kreatiinipyruvaatin. *Suom. Urheilulehti* **2000**, *23*.
188. Wheelwright, D.C.; Ashmead, S.D. Bioavailable Chelates of Creatine and Essential Metals. U.S. Patent No. 6,114,379, 5 September 2000.
189. Brilla, L.R.; Giroux, M.S.; Taylor, A.; Knutzen, K.M. Magnesium-creatine supplementation effects on body water. *Metabolism* **2003**, *52*, 1136–1140. [CrossRef]
190. Selsby, J.T.; DiSilvestro, R.A.; Devor, S.T. Mg^{2+}-creatine chelate and a low-dose creatine supplementation regimen improve exercise performance. *J. Strength Cond. Res.* **2004**, *18*, 311–315. [CrossRef] [PubMed]
191. Zajac, A.; Golas, A.; Chycki, J.; Halz, M.; Michalczyk, M.M. The Effects of Long-Term Magnesium Creatine Chelate Supplementation on Repeated Sprint Ability (RAST) in Elite Soccer Players. *Nutrients* **2020**, *12*, 2961. [CrossRef] [PubMed]
192. Giese, M.W.; Lecher, C.S. Qualitative in vitro NMR analysis of creatine ethyl ester pronutrient in human plasma. *Int. J. Sports Med.* **2009**, *30*, 766–770. [CrossRef] [PubMed]
193. Katseres, N.S.; Reading, D.W.; Shayya, L.; Dicesare, J.C.; Purser, G.H. Non-enzymatic hydrolysis of creatine ethyl ester. *Biochem. Biophys. Res. Commun.* **2009**, *386*, 363–367. [CrossRef] [PubMed]
194. Gufford, B.T.; Ezell, E.L.; Robinson, D.H.; Miller, D.W.; Miller, N.J.; Gu, X.; Vennerstrom, J.L. pH-dependent stability of creatine ethyl ester: Relevance to oral absorption. *J. Diet. Suppl.* **2013**, *10*, 241–251. [CrossRef]
195. Law, J.P.; Di Gerlando, S.; Pankhurst, T.; Kamesh, L. Elevation of serum creatinine in a renal transplant patient following oral creatine supplementation. *Clin. Kidney J.* **2019**, *12*, 600–601. [CrossRef]
196. Velema, M.S.; de Ronde, W. Elevated plasma creatinine due to creatine ethyl ester use. *Neth. J. Med.* **2011**, *69*, 79–81.
197. Arazi, H.; Eghbali, E.; Karimifard, M. Effect of creatine ethyl ester supplementation and resistance training on hormonal changes, body composition and muscle strength in underweight non-athlete men. *Biomed. Hum. Kinet.* **2019**, *11*, 158–166. [CrossRef]
198. Stoppani, J. *Supplement Breakdown: Creatine HCL*; JS Stoppani Blog: Westlake Village, CA, USA, 24 September 2021.
199. Alraddadi, E.A.; Lillico, R.; Vennerstrom, J.L.; Lakowski, T.M.; Miller, D.W. Absolute Oral Bioavailability of Creatine Monohydrate in Rats: Debunking a Myth. *Pharmaceutics* **2018**, *10*, 31. [CrossRef] [PubMed]
200. Kreider, R.B. Species-specific responses to creatine supplementation. *Am. J. Physiol. Regul. Integr. Comp. Physiol.* **2003**, *285*, R725–R726. [CrossRef]
201. Yoshioka, C.A.F.; Madureira, D.; Carrara, P.; Gusmão, N.; Ressureição, K.S.; Santana, J.O.; Lamolha, M.A.; Viebig, R.F.; Sanches, I.C.; de Lira, F.S.; et al. Comparison between creatine monohydrate and creatine HCl on body composition and performance of the Brazilian Olympic team. *Int. J. Food Nutr. Res.* **2019**, *3*, 1–12.
202. Tayebi, M.; Arazi, H. Is creatine hydrochloride better than creatine monohydrate for the improvement of physical performance and hormonal changes in young trained men? *Sci. Sports* **2020**, *35*, e135–e141. [CrossRef]
203. Hlinský, T.; Kumstát, M.; Vajda, P. Effects of Dietary Nitrates on Time Trial Performance in Athletes with Different Training Status: Systematic Review. *Nutrients* **2020**, *12*, 2734. [CrossRef] [PubMed]
204. Tan, R.; Cano, L.; Lago-Rodríguez, Á.; Domínguez, R. The Effects of Dietary Nitrate Supplementation on Explosive Exercise Performance: A Systematic Review. *Int. J. Environ. Res. Public Health* **2022**, *19*, 762. [CrossRef]
205. Gao, C.; Gupta, S.; Adli, T.; Hou, W.; Coolsaet, R.; Hayes, A.; Kim, K.; Pandey, A.; Gordon, J.; Chahil, G.; et al. The effects of dietary nitrate supplementation on endurance exercise performance and cardiorespiratory measures in healthy adults: A systematic review and meta-analysis. *J. Int. Soc. Sports Nutr.* **2021**, *18*, 55. [CrossRef]
206. Macuh, M.; Knap, B. Effects of Nitrate Supplementation on Exercise Performance in Humans: A Narrative Review. *Nutrients* **2021**, *13*, 3183. [CrossRef]
207. Kramer, R.; Nikolaidis, A. Amino Acid Compositions. U.S. Patent No. 11155524 B2, Application No. 16/893,319, 26 October 2021.
208. Yazdi, P. *Creatine Nitrate: Benefits, Side Effects, Dosage, Reviews*; SelfDecode: Miami, FL, USA, 9 September 2021.
209. Ostojic, S.M.; Stajer, V.; Vranes, M.; Ostojic, J. Searching for a better formulation to enhance muscle bioenergetics: A randomized controlled trial of creatine nitrate plus creatinine vs. creatine nitrate vs. creatine monohydrate in healthy men. *Food Sci. Nutr.* **2019**, *7*, 3766–3773. [CrossRef] [PubMed]
210. Jung, Y.P.; Earnest, C.P.; Koozehchian, M.; Galvan, E.; Dalton, R.; Walker, D.; Rasmussen, C.; Murano, P.S.; Greenwood, M.; Kreider, R.B. Effects of acute ingestion of a pre-workout dietary supplement with and without p-synephrine on resting energy expenditure, cognitive function and exercise performance. *J. Int. Soc. Sports Nutr.* **2017**, *14*, 3. [CrossRef] [PubMed]
211. Jung, Y.P.; Earnest, C.P.; Koozehchian, M.; Cho, M.; Barringer, N.; Walker, D.; Rasmussen, C.; Greenwood, M.; Murano, P.S.; Kreider, R.B. Effects of ingesting a pre-workout dietary supplement with and without synephrine for 8 weeks on training adaptations in resistance-trained males. *J. Int. Soc. Sports Nutr.* **2017**, *14*, 1. [CrossRef]
212. Joy, J.M.; Lowery, R.P.; Falcone, P.H.; Mosman, M.M.; Vogel, R.M.; Carson, L.R.; Tai, C.Y.; Choate, D.; Kimber, D.; Ormes, J.A.; et al. 28 days of creatine nitrate supplementation is apparently safe in healthy individuals. *J. Int. Soc. Sports Nutr.* **2014**, *11*, 60. [CrossRef] [PubMed]
213. All American EFX. Kre-Alkalyn–The World's Most Potent Creatine. Available online: http://krealkalyn.com/ (accessed on 5 January 2022).

214. Gollini, J.M. Oral Creatine Supplement and Method for Making Same. U.S. Patent 6,399,661 B1, 4 June 2002.

215. Affouras, A.; Vodenicharova, K.; Shishmanova, D.; Goranov, K.; Stroychev, K. *Clinical Trial Comparing Kre-Alkalyn to Creatine Monohydrate*; Medical Center: Sofia, Bulgaria, 2006; p. 3.

216. Creatine Orotate. Available online: https://www.exercise.com/supplements/creatine-orotate/ (accessed on 5 January 2022).

217. Muscle Marketing USA: XXTRA Powerlifter's Creatine Serum. Available online: https://mmusa.com/ (accessed on 5 January 2022).

218. Owac, J.H. Bang Revolution: Super Creatine, a Highly Significant Patent in History. 4 September 2019. Available online: https://www.youtube.com/watch?v=IRj2F-wsz5U&feature=youtu.be (accessed on 13 January 2022).

219. Owac, J.H. Stable aqueous compositions comprising amide-protected bioactive creatine species and uses thereof. U.S. Patent 8,445,466 B2, 21 May 2013.

220. Reddeman, R.A.; Glavits, R.; Endres, J.R.; Murbach, T.S.; Hirka, G.; Vertesi, A.; Beres, E.; Szakonyine, I.P. A Toxicological Assessment of Creatyl-l-Leucine. *Int. J. Toxicol.* **2018**, *37*, 171–187. [CrossRef] [PubMed]

221. da Silva, R.P. The Dietary Supplement Creatyl-l-Leucine Does Not Bioaccumulate in Muscle, Brain or Plasma and Is Not a Significant Bioavailable Source of Creatine. *Nutrients* **2022**, *14*, 701. [CrossRef]

222. Monster Energy Co., v. Vital Pharmaceuticals, Inc. et al. Case No. 5:18-cv-1882-JGB-SHK, Dkt. 434-46. C.D. Cal. Available online: https://www.law360.com/cases/5b8f06c4a5fd16342f7dd1b6 (accessed on 15 January 2022).

223. Monster Energy Co., v. Vital Pharmaceuticals, Inc. et al. Case No. 5:18-cv-1882-JGB-SHK, Dkt. 434-53. C.D. Cal. Available online: https://www.law360.com/cases/5b8f06c4a5fd16342f7dd1b6 (accessed on 15 January 2022).

224. Monster Energy Co., v. Vital Pharmaceuticals, Inc. et al. Case No. 5:18-cv-1882-JGB-SHK, Dkt. 434-50 (C.D. Cal.). Available online: https://www.law360.com/cases/5b8f06c4a5fd16342f7dd1b6 (accessed on 15 January 2022).

225. Monster Energy Co., v. Vital Pharmaceuticals, Inc. et al. Case No. 5:18-cv-1882-JGB-SHK, Dkt. 434-51. (C.D. Cal.). Available online: https://www.law360.com/cases/5b8f06c4a5fd16342f7dd1b6 (accessed on 15 January 2022).

226. Monster Energy Co., v. Vital Pharmaceuticals, Inc. et al. Case No. 5:18-cv-1882-JGB-SHK, Dkt. 434-133 (C.D. Cal.). Available online: https://www.law360.com/cases/5b8f06c4a5fd16342f7dd1b6 (accessed on 15 January 2022).

227. Guglielmi, G.; Mammarella, A. A controlled clinical study of the use of creatinol-O-phosphate in subjects with deficient myocardial circulation. *Clin. Ter.* **1979**, *91*, 355–382. [PubMed]

228. Melloni, G.F.; Minoja, G.M.; Lureti, G.F.; Merlo, L.; Pamparana, F.; Brusoni, B. Acute clinical tolerance of creatinol-O-phosphate. *Arzneimittelforschung* **1979**, *29*, 1477–1479.

229. Godfraind, T.; Ghiradi, P.; Ferrari, G.; Cassagrande, C. Creatinol-O-Phopshate Having Therapeutical Action. U.S. Patent No. 4,376,117, Application No. 198,263, 8 March 1983.

230. Marzo, A.; Ghirardi, P. Pharmacological and toxicological properties of creatinol O-phosphate. A review. *Arzneimittelforschung* **1979**, *29*, 1449–1452.

231. Kreider, R.B.; Miller, G.W.; Williams, M.H.; Somma, C.T.; Nasser, T.A. Effects of phosphate loading on oxygen uptake, ventilatory anaerobic threshold, and run performance. *Med. Sci. Sports Exerc.* **1990**, *22*, 250–256. [PubMed]

232. Kreider, R.B.; Miller, G.W.; Schenck, D.; Cortes, C.W.; Miriel, V.; Somma, C.T.; Rowland, P.; Turner, C.; Hill, D. Effects of phosphate loading on metabolic and myocardial responses to maximal and endurance exercise. *Int. J. Sport Nutr.* **1992**, *2*, 20–47. [CrossRef] [PubMed]

233. Nicaise, J. Creatinol-O-phosphate (COP) and muscular performance: A controlled clinical trial. *Curr. Ther. Res. Clin. Exp.* **1975**, *17*, 531–534. [PubMed]

234. New Dietary Ingredients in Dietary Supplements-Background for Industry. Available online: https://www.fda.gov/food/new-dietary-ingredients-ndi-notification-process/new-dietary-ingredients-dietary-supplements-background-industry (accessed on 26 October 2020).

235. EAS. Product Phosphagen™ Marketed by Experimental and Applied Sciences (EAS™). Available online: https://www.eas.com (accessed on 7 February 2022).

236. *GRN, No. 931 Creatine Monohydrate*; U.S. Food and Drug Administation. U.S. Department of Health and Human Services: Washington, DC USA, 2020.

237. Carlson, S.J. *GRAS Notice No. GRN 000931*; Division of Food Ingredients; Office of Food Additive Safety; Center for Food Safety and Applied Nutrition: Washington, DC USA, 2021.

238. U.S. Food & Drug Administration. GRAS Notices. Available online: https://www.cfsanappsexternal.fda.gov/scripts/fdcc/index.cfm?set=GRASNotices&sort=GRN_No&order=DESC&startrow=1&type=basic&search=CLL (accessed on 5 January 2022).

239. Dwyer, J.T.; Coates, P.M.; Smith, M.J. Dietary Supplements: Regulatory Challenges and Research Resources. *Nutrients* **2018**, *10*, 41. [CrossRef] [PubMed]

240. An Overview of the Regulation of Complementary Medicines in Australia. Available online: https://www.tga.gov.au/overview-regulation-complementary-medicines-australia (accessed on 5 January 2022).

241. NNHPD. *Natural Health Products Directorate*; Health Canada: Ottawa, ON, Canada, 2013.

242. NHPD. *Natural Health Products Directorate*; Health Canada: Ottawa, ON, Canada, 2018.

243. Natural Health Products Ingredients Database. Available online: http://webprod.hc-sc.gc.ca/nhpid-bdipsn/ingredsReq.do?srchRchTxt=creatine&srchRchRole=-1&mthd=Search&lang=eng (accessed on 20 January 2022).

244. European Commission on Food Safety: Food Supplements. Available online: https://ec.europa.eu/food/safety/labelling-and-nutrition/food-supplements_en (accessed on 20 January 2022).
245. EPC. Commission Directive 2001/15/EC of 15 February 2001 on Substances That May be Added for Specific Nutritional Purposes in Foods for Particular Nutritional Uses. 2011. Available online: http://eur-lex.europa.eu/LexUriServ/LexUriServ.do?uri=CELEX:32001L0015:en:NOT (accessed on 20 January 2022).
246. EPC. Directive 2002/46/EC of the European Parliament and of the Council of 10 June 2002 on the Approximation of the Laws of the Member States Relating to Food Supplements. 2002. Available online: https://eur-lex.europa.eu/eli/dir/2002/46/oj (accessed on 20 January 2022).
247. EFSA. *Opinion of the Scientific Panel on Food Additives, Flavourings, Processing Aids and Materials in Contact with Food (AFC) on a Request from the Commission Related to Creatine Monohydrate for Use in Foods for Particular Nutritional Uses*; European Food Safey Authority: Parma, Italy, 2004.
248. EFSA. Scientific Opinion on the substantiation of health claims related to creatine and increase in physical performance during short-term, high intensity, repeated exercise bouts (ID 739, 1520, *1521*, 1522, 1523, *1525*, 1526, 1531, *1532*, 1533, 1534, *1922*, 1923, 1924), increase in endurance capacity (ID 1527, 1535), and increase in endurance performance (ID 1521, 1963) pursuant to Article 13(1) of Regulation (EC) No 1924/2006. *EFSA J.* **2011**, *9*, 2303.
249. EFSA. Scientific Opinion on the substantiation of health claims related to creatine and increased attention (ID 1524) and improvement of memory (ID 1528) pursuant to Article 13(1) of Regulation (EC) No 1924/2006. *EFSA J.* **2011**, *9*, 2216. [CrossRef]
250. MHLW. List of the Ingredients (Raw Materials) Not Deemed to be Drugs Unless Medicinal Indications/Efficacies Are Noted (Translated). 2009. Available online: http://www.fukushihoken.metro.tokyo.jp/kenkou/kenko_shokuhin/ken_syoku/kanshi/seibun/index.html (accessed on 20 January 2022).
251. MHLW. *Director of Standards Division, Dept of Food Safety, Pharmaceutical and Food Safety Bureau*; Ministry of Health, Labor and Welfare: Tokyo, Japan, 2001.
252. MMHLW. Food for Specified Health Uses (FOSHU). 2022. Available online: https://www.mhlw.go.jp/english/topics/foodsafety/fhc/02.html (accessed on 20 January 2022).
253. Shimizu, T. Korean latest conditions of health food labeling. *Food Style* **2008**, *12*, 54–56.
254. KFDA. Functional Food Approval Status (Translated). 2009. Available online: http://hfoodi.kfda.go.kr/index.jsp (accessed on 20 January 2022).
255. Federal Food, Drug and Cosmetic Act. In *§413(a)*; United States Congress: Washington, DC, USA, 1938.

Review

Timing of Creatine Supplementation around Exercise: A Real Concern?

Felipe Ribeiro [1,2], Igor Longobardi [1], Pedro Perim [1], Breno Duarte [1], Pedro Ferreira [1,2], Bruno Gualano [1,3], Hamilton Roschel [1] and Bryan Saunders [1,4,*]

1   Applied Physiology and Nutrition Research Group, School of Physical Education and Sport, Rheumatology Division, Faculdade de Medicina FMUSP, University of São Paulo, São Paulo 01246-903, SP, Brazil; felipe.is2@hotmail.com (F.R.); i.long@usp.br (I.L.); pedroperim13@gmail.com (P.P.); brenoduarte@usp.br (B.D.); pedro.h.a.ferreira95@gmail.com (P.F.); gualano@usp.br (B.G.); hars@usp.br (H.R.)
2   Centro Universitário São Camilo, São Paulo 04263-200, SP, Brazil
3   Food Research Center (FoRC), University of São Paulo, São Paulo 05508-080, SP, Brazil
4   Institute of Orthopaedics and Traumatology, Faculty of Medicine FMUSP, University of São Paulo, São Paulo 01246-903, SP, Brazil
*   Correspondence: drbryansaunders@outlook.com

Abstract: Creatine has been considered an effective ergogenic aid for several decades; it can help athletes engaged in a variety of sports and obtain performance gains. Creatine supplementation increases muscle creatine stores; several factors have been identified that may modify the intramuscular increase and subsequent performance benefits, including baseline muscle Cr content, type II muscle fibre content and size, habitual dietary intake of Cr, aging, and exercise. Timing of creatine supplementation in relation to exercise has recently been proposed as an important consideration to optimise muscle loading and performance gains, although current consensus is lacking regarding the ideal ingestion time. Research has shifted towards comparing creatine supplementation strategies pre-, during-, or post-exercise. Emerging evidence suggests greater benefits when creatine is consumed after exercise compared to pre-exercise, although methodological limitations currently preclude solid conclusions. Furthermore, physiological and mechanistic data are lacking, in regard to claims that the timing of creatine supplementation around exercise moderates gains in muscle creatine and exercise performance. This review discusses novel scientific evidence on the timing of creatine intake, the possible mechanisms that may be involved, and whether the timing of creatine supplementation around exercise is truly a real concern.

Keywords: dietary supplements; ergogenic aid; hypertrophy; resistance training; sports nutrition; strength; supplementation

Citation: Ribeiro, F.; Longobardi, I.; Perim, P.; Duarte, B.; Ferreira, P.; Gualano, B.; Roschel, H.; Saunders, B. Timing of Creatine Supplementation around Exercise: A Real Concern? *Nutrients* **2021**, *13*, 2844. https://doi.org/10.3390/nu13082844

Academic Editor: Ajmol Ali

Received: 26 June 2021
Accepted: 10 August 2021
Published: 19 August 2021

Publisher's Note: MDPI stays neutral with regard to jurisdictional claims in published maps and institutional affiliations.

1. Introduction

Athletes (and physically active individuals) are interested in nutritional strategies that are aimed at enhancing exercise performance. Creatine (Cr) deserves a special place among the plethora of ergogenic supplements, as it is one of the most studied and scientifically supported supplements on the market [1,2]. Creatine is a naturally occurring non-protein nitrogen compound synthesised in the liver and kidney from precursor amino acids, arginine, glycine, and methionine. Most of the body's Cr is found in muscle (95%), of which two-thirds are stored as phosphorylcreatine (PCr), the remaining third as free Cr [3], with less than 5% found in other tissues, such as the brain and testes [4]. In a seminal study by Harris et al. (1992), it was demonstrated for the first time in humans that Cr supplementation, at varying doses of 20–30 g/day, ingested over several individual 5 g doses throughout the day, could increase total intramuscular Cr content (TCr = PCr + Cr) by as much as 20% [5]. Numerous subsequent studies have shown the efficacy of Cr supplementation in increasing muscle Cr content, including using more gradual loading protocols [6–8].

Several factors could influence the individual intramuscular increase in TCr and subsequent performance benefits as a consequence of Cr supplementation, including baseline muscle Cr content, type II muscle fibre content and size, and habitual dietary intake of Cr and aging [9,10]. Interestingly, it has been known for some time that exercise can enhance Cr loading in muscle [5], and the specific timing of Cr supplementation in relation to exercise has more recently been touted as an important consideration, in order to optimise training gains, although the current consensus on its importance is lacking. Emerging evidence suggests that post-exercise Cr ingestion may provide superior benefits compared to pre-exercise consumption [11,12], although several methodological limitations presented by these investigations currently preclude definitive interpretations of these results.

The purpose of this narrative review is to summarise and discuss current evidence and new emerging questions on the influence of Cr supplementation timing, in regard to exercise, on muscle Cr content and physical performance.

2. Creatine Supplementation

Studies show that creatine supplementation in doses of 5–20 g/day for >5 days can increase intramuscular Cr and PCr to the point of saturation [8,13]. This increase in PCr is associated with the main mechanism of action, regarding the ergogenic effect of Cr supplementation [14]. Phosphorylcreatine can provide an inorganic phosphate (Pi) molecule for the resynthesis of ATP via the Cr kinase reaction, in which Pi donation from PCr degradation is used by adenosine diphosphate (ADP) and, consequently, increases ATP resynthesis. Creatine phosphokinase is the enzyme that catalyses this reaction and is limited only by the concentration of its substrates and products, namely Cr and PCr [15]. This phosphagen system, also termed the ATP-CP system, is the fastest way to supply ATP for skeletal muscle metabolism [16]. Precisely for this reason, the ATP-CP system is related to high-intensity and short-duration exercises [17] and, thus, is associated with greater total work capacity [1]. In this respect, the ATP-CP system serves as an important regulator of muscle metabolism, which explains the ergogenic benefits of Cr supplementation throughout training. Enhancing the capacity of ATP resynthesis should increase available energy during exercise, prolonging the work capacity of the skeletal muscles, delaying the onset of muscle fatigue, and improving performance.

Strong scientific evidence suggests that Cr can lead to beneficial improvements in exercise performance; however, there also appears to be some variations in the response to Cr supplementation due to a number of factors, which will be presented in the next section.

Factors Modifying the Effect of Creatine Supplementation on Muscle Creatine Content

Several factors have been shown to modify the effects of Cr supplementation on muscle Cr content. Daily dose and duration play important roles in how quickly and how much Cr stores are increased. Five to seven days of supplementation with a daily dose of 20 g·day^{-1} is sufficient to saturate muscle creatine stores [5], which is approximately 140 to 160 mmol·kg^{-1} of dry muscle. This has become a commonly employed dose in the literature and termed the "loading phase". Nonetheless, a more gradual dosing strategy of 3 g·day^{-1} leads to similar increases, but over a longer period (~28 days; [8]). Greater increases in muscle Cr are shown in those with lower initial muscle Cr content [5], while carbohydrate co-supplementation may increase Cr uptake via insulin-mediated stimulation [6,18] of the Cr transporter, CreaT. Although this mechanism of insulin stimulated Cr uptake remains to be mechanistically confirmed, if it holds true, this will only be relevant within the first few days at high doses (e.g., days 1 to 3 at 20 g·day^{-1}) of supplementation prior to saturation of muscle Cr stores [19], but may be more relevant at lower doses (e.g., 3–5 g·day^{-1}, which takes up to 28 days to saturate). Indeed, the upper threshold of saturation across individuals appears remarkably consistent [1], meaning the dosing protocol will be important.

Exercise has also been shown to enhance Cr accrual in muscles. In the seminal study from Harris et al. (1992), various doses of Cr were given to healthy participants aged 20 to

62 years, with varying levels of fitness. An additional aim of this study was to determine the effect of exercise upon Cr uptake into muscles using a unilateral leg exercise model. Throughout supplementation, participants performed 1 h of cycling exercises in one leg, while the other leg rested. Results showed that exercise potentiated the resultant increase in intramuscular Cr, with greater increases in the exercised versus non-exercised leg. These were the first data to suggest that exercise may influence the Cr loading of muscles with supplementation. These data were subsequently supported by a further study that showed a 68% greater increase in total creatine content following supplementation when a single-leg exercise (cycling at 60–70% of maximal heart rate until exhaustion) was performed [20] (Robinson et al., 1999). Thus, exercise appears to enhance the accrual of intracellular TCr with Cr supplementation, although, again, the importance thereof will likely be linked to the daily dose employed. Since high doses (20 g·day^{-1}) lead to saturation in as little as 5 days, it seems unlikely that quicker loading will have much impact. However, should supplementation occur more gradually, as with doses of 3–5 g·day^{-1} over 28 days, then faster loading might incur earlier and greater benefits. It must be acknowledged that it is unclear if these studies showed an increased uptake of Cr into muscle, or an increased Cr retention. Logic would suggest that it is likely reflective of an increased muscle uptake although this should be mechanistically confirmed.

The influence of exercise on Cr loading is apparent; however, more recent investigations have suggested that the timing of Cr supplementation in relation to the exercise bout may be important, too [11,12,21,22]. To better understand why timing of supplementation in relation to exercise might be important, it is important to appreciate how exercise might enhance Cr uptake, which will be discussed in the following section.

3. How Creatine Timing around Exercise May Influence Subsequent Loading

3.1. Creatine Concentration in the Bloodstream and Training Duration

The mechanisms via which exercise may increase Cr uptake into muscles are not entirely understood and are hypothetical, as no study has experimentally demonstrated the mechanism behind this phenomenon. Nonetheless, one proposed mechanism via which timing of Cr ingestion in relation to exercise may modify the efficacy of supplementation is through exercise hyperaemia, namely increased blood perfusion to the working muscle (Figure 1A). Blood flow increases within one second of the onset of muscular contraction, and exercise can increase skeletal muscle blood flow by 100-fold compared to values seen at rest [23]; it is important to maintain adequate oxygen and nutrient delivery, in order to support the energetic demands of the skeletal muscles during exercise. The extent to which blood flow increases during and after exercise is influenced by factors such as the duration, type, intensity, and volume of exercise. This is important to note because muscle blood flow is closely matched to the metabolic demands of contractions induced by the exercise [24]. Theoretically, greater blood flow to the muscle could lead to greater delivery of Cr and, thus, enhance its uptake and retention, although this would primarily be restricted to the exercised muscles. An increase in blood flow kinetics as well as Cr transport to exercised muscles can result in greater delivery, retention, and metabolization of the nutrients to the exercised muscles [25]. Thus, if supplementation is provided around exercise, then circulating Cr could coincide with increased blood flow to the muscle (Figure 1B).

An important factor to consider, in regard to the timing of Cr supplementation in relation to exercise, is the time it takes for the Cr concentration to become elevated in the bloodstream. This is relevant to determine if ingestion of Cr pre- or post-exercise would provide distinct elevations of intramuscular Cr. In humans, Cr is actively absorbed from the gastrointestinal tract before entering the bloodstream to be delivered to various tissues throughout the body [26]. Creatine monohydrate absorption is close to 100% [27], and when 2 g of Cr is consumed in an aqueous solution, it reaches peak plasma concentration in approximately 1 h. This is similar to other protocols in which maximum plasma concentration of Cr occurred in <2 h when the dose administered was <10 g [18,28,29]. Although higher doses >10 g can take up to 2.5 to reach peak concentration in the blood [30], the most employed single-dose of 5 g should peak around 1–2 h following ingestion, remaining elevated for a further 4 h [5,30]. This information could be crucial when optimizing supplement timing; if Cr-uptake is maximised when there is increased muscle blood flow, then individuals should look to coincide peak circulating Cr levels with hyperaemia.

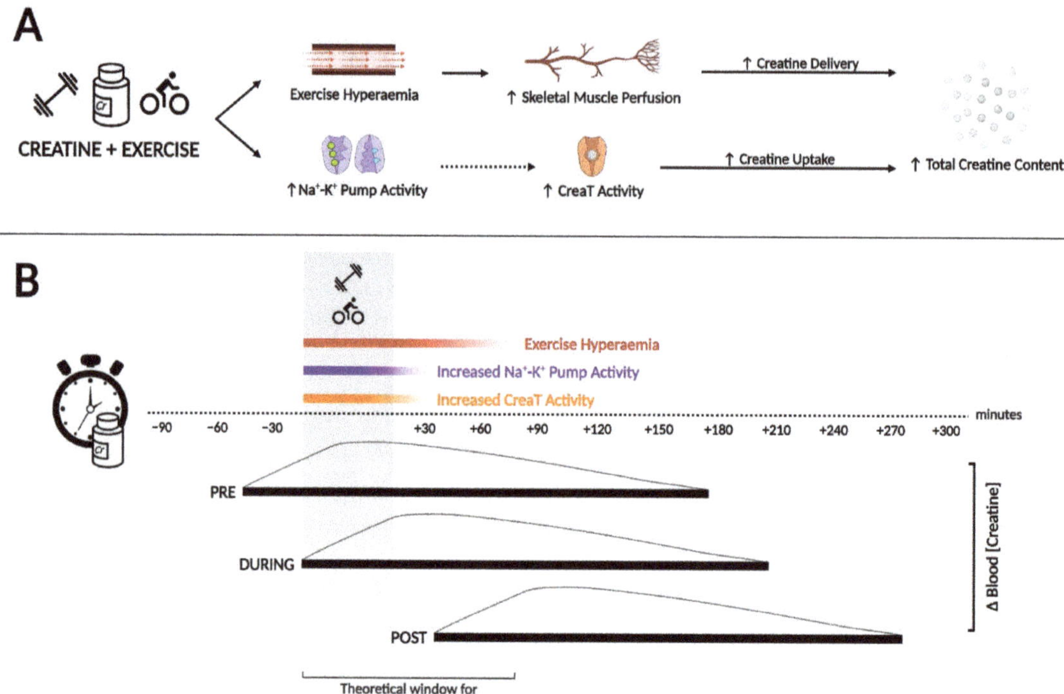

Figure 1. The hypothetical mechanisms behind an exercise-mediated increase in total creatine content with creatine supplementation. (**Panel A**): exercise hyperaemia increases tissue perfusion, enhancing creatine delivery. Additionally, increased Na+/K+ pump activity during exercise supports the $[Na^+]$ gradient favouring creatine uptake by CreaT. Together, these effects may acutely potentiate the uptake and increase in total muscle creatine content. (**Panel B**): theoretical overlap of events according to the timing of creatine supplementation, in relation to exercise and its potential benefits regarding the delivery and uptake of creatine to the muscle. Created with BioRender.com.

As an example, the duration of a typical resistance training session varies between 40 and 90 min, and elite bodybuilders reported an average of 60–70 min per training session [31]. Thus, should an individual supplement immediately pre-exercise or even during exercise, Cr would begin to accumulate throughout the training session, and it is possible that peak Cr concentration in the blood would still occur during exercise. Due to the exercise undertaken, this would lead to increased blood flow to the working muscles, which may lead to increased delivery and subsequent uptake of Cr, explaining, at least partially, the greater increases in Cr loading shown previously [5,30]. This also suggests that pre-exercise Cr supplementation may be more effective at increasing muscle Cr content than post-exercise supplementation. Increased blood flow to the muscles can decrease within the following 30 min after exercise [32], although the magnitude of the post-exercise hyperaemia is proportional to the strength of the contraction and its duration [33]. Depending on the modality, intensity, and duration of the exercise (i.e., muscle contraction), vasodilation may continue for up to 120 min post-exercise. Taken together, post-exercise supplementation may not benefit from exercise-induced muscle blood flow to the same extent as pre-exercise supplementation due to a shorter overlap between circulating Cr and exercise-induced hyperaemia (Figure 1, Panel B).

Thus, if the primary mechanism by which exercise induces an increased Cr loading of muscle is via an exercise-stimulated increase in blood flow to the working muscles, then pre-exercise Cr supplementation would be expected to be the most effective supplementation strategy compared to supplementation intra- or post-workout, or at any other moment of the day.

3.2. Na^+-K^+ Pump Activity and Exercise

Creatine transport into muscle cells is performed by a specific Cr transporter, CreaT [34]. This transport occurs against a concentration gradient and is dependent on the presence of extracellular Na^+ [35], meaning Cr uptake is achieved via a Na^+-Cr cotransport system, which makes use of the sarcolemmal Na^+-K^+ pumps [36]. Thus, one other mechanism that might optimise Cr supplementation is an upregulation of the kinetics involved in the Cr transport from the bloodstream to the skeletal muscle, via an increase in Na^+-K^+ pump activity during and following exercise [37]. Indeed, exercise training involving a 2-h exercise cycle per day, for 6 consecutive days at 65% of maximal aerobic power, induced upregulation in sarcolemmal Na^+-K^+-ATPase concentration in humans, after only one week of training, in the exercised muscle [38].

Studies have shown that the Na^+-K^+ pump regulates transsarcolemmal $[Na^+]$ and $[K^+]$ gradients in skeletal muscles and is critical for the maintenance of membrane excitability and contractility [20,39]. Odoom et al. [36] showed that the pharmacological activation and inhibition of Na^+-K^+ pump activity in mouse myoblast cells were paralleled by up- and downregulation of cellular Cr accumulation, demonstrating the relationship between Na^+-K^+ pump activity and Cr uptake. This mechanistic evidence suggests that increased Na^+-K^+ pump activity, as occurs with exercise, might lead to increased Cr uptake (Figure 1A), although it is currently speculative as to whether it occurs in humans.

Since the upregulation of muscle Na^+-K^+ pump function in the exercised limb facilitates muscle Cr transport, if this mechanism holds true for humans, the timing of Cr supplementation around exercise could alter the uptake into the muscle. Specifically, pre-exercise supplementation might ensure that high circulating levels of Cr coincide with peak activation of the Na^+-K^+ pump during exercise-induced muscle contraction, leading to greater intramuscular Cr accumulation (Figure 1B). However, there is a residual effect of exercise that could last from several minutes up to 48 h post-exercise, depending on the action of interest (e.g., insulin sensitivity; [40]), meaning that post-exercise Cr supplementation might also benefit from a contraction-induced potentiation of Na^+-K^+ pump activity, although this is highly speculative. Furthermore, most individuals taking Cr supplements undertake regular exercise training, which chronically upregulates Na^+-K^+ pump activity [38]. Thus, it is possible that timing in relation to each exercise session may not be important, but that exercise training in general leads to greater Cr accumulation in muscles due to chronic adaptations in Na^+-K^+ pump activity. This is somewhat in contrast to evidence suggesting that ingestion of Cr close to an exercise session leads to greater increases in TCr than supplementation that is not close to the exercise session (>5 h) [41].

Skeletal muscles appear to be highly amenable to Cr supplementation, while chronic exercise appears to further increase the response to supplementation. Nonetheless, what is unclear is whether timing around exercise also generates differential responses in the muscle Cr loading response, and subsequent performance gains. The hypothetical mechanisms discussed provide some support to suggest that exact timing, in relation to exercise, may exert differential effects. The following section will detail studies that have investigated the effects of pre-, during-, or post-workout Cr supplementation on several outcomes.

4. Creatine Supplementation Pre-, During- or Post-Workout: The Evidence

Cribb and Hayes [41] investigated the effects of Cr supplement timing during 10-weeks of resistance exercise training on intramuscular TCr content, muscle-fibre hypertrophy, strength, and body composition. Recreational male bodybuilders were allocated into two groups: one that consumed their supplements immediately pre- and post-workout on training days, and the other that consumed their supplements in the morning before breakfast and late evening before sleep. The supplements contained 40 g glucose, 40 g protein, <0.5 g of fat, and 7 g of Cr monohydrate per 100 g; participants consumed 1 $g \cdot kg^{-1} \cdot day^{-1}$ twice on training days only. The 10-week training program was performed 4 times a week and was designed specifically to increase strength and muscle size, with a progressive overload consisting of three compound exercises with free weights based on repetition maximum (RM) of the participants. The group who ingested the supplements around their workouts had greater increases in intramuscular TCr and greater gains in maximum dynamic strength, lean mass, and cross-sectional area type II fibres compared to the group who consumed Cr at alternate times of the day. These findings suggest that supplement timing can play an important role in strength and muscle gains, although the strength and muscle gains cannot be limited to Cr supplementation only, since the supplement contained various other ingredients, including a substantial amount of protein. It is known that the timing of protein ingestion around exercise may influence hypertrophy and strength gains [42]. Furthermore, carbohydrates were included in the supplement, which was shown to enhance Cr uptake into the muscle [18]; thus, the isolated effect of exercise on muscle TCr loading was not determined. Finally, supplementation was provided, both pre- and post-exercise, as well as "not close" to the exercise session, meaning no inferences can be made regarding whether supplementation pre-, during-, or post-exercise influences these responses.

Timing of Cr supplementation in relation to exercise has been suggested to influence the accrual of muscle Cr [43], which may impact subsequent performance gains. We have discussed the physiological mechanisms through which Cr timing around exercise might modify its loading effects, but it is important to determine the true impact of timing experimentally. A few studies have investigated the influence of supplement timing with Cr in relation to exercise on a number of different outcomes.

The first study that specifically investigated whether Cr supplementation around exercise modified its effects was performed by Antonio and Ciccone (2013). They investigated the effects of Cr supplementation, either immediately pre- or post-exercise, throughout resistance exercise training on body composition and muscle strength (Table 1). Nineteen healthy recreational bodybuilders were randomly assigned to one of two groups, ingesting either 5 g of Cr immediately pre-workout or 5 g of Cr immediately post-workout. Supplements were ingested according to the volunteer's convenience on non-training days. Training consisted of resistance training 5 days a week for four weeks. Results showed greater muscle hypertrophy and strength gains when Cr was ingested post- versus pre-exercise. Specifically, post-exercise ingestion led to a 3% gain in fat-free mass and 7.5% gain in 1-RM bench press, compared to a 1.3% increase in fat-free mass and 6.8% 1-RM bench press improvement with pre-exercise ingestion. The authors concluded that consuming Cr immediately post-workout is superior to pre-workout on body composition and strength. These results somewhat contrast what might be expected, since increased plasma Cr levels will not coincide with increased blood flow that occurs during the exercise. However, it is important to note that no significant interactions were shown, and that magnitude-based inferences were used to determine possible and likely beneficial effects of timing on outcomes. Unfortunately, this analysis method has come under substantial criticism [44,45], while the absolute difference in fat-free mass and bench press gains were small, with overlapping confidence intervals. Thus, the true importance of these differences is somewhat unclear.

This first study to directly investigate the influence of timing of Cr supplementation around exercise has some important strengths, such as the dose administered, which is commonly employed by bodybuilders [46], and provides favourable outcomes to increase lean muscle mass [47]. Training frequency and volume per week were in the range frequently used by this population [48], while protein intake was high, namely 1.9 $g \cdot kg^{-1} \cdot day^{-1}$, which is expected to contribute to muscle strength and hypertrophy when combined with resistance training [49]. Despite these strengths, this study also has limitations, which must be considered when extrapolating the findings. First, the researchers did not analyse intramuscular Cr content, limiting the interpretation of the potential mechanisms related to the benefits observed with the timing of Cr ingestion. There was no direct and reliable measurement of skeletal muscle hypertrophy [50], which could have strengthened the findings. There was also no placebo-control group, which makes it impossible to consider the isolated influence of the resistance training session on body composition and strength, and the variations thereof. The differences shown here may well be within normal variations expected with a resistance training program. Volunteers also knew their timing of Cr supplementation, which could have created certain expectations in the participants [51]. Although protein intake was high, it is unclear when protein was ingested around exercise, which may have influenced hypertrophy and strength gains [42], while the small sample size was not conducive to clear conclusions. Finally, the authors did not report whether the athletes were familiarised with the 1-RM test used as the primary outcome measure. Since there may be a learning effect for such outcomes, and the athletes may have had different levels of familiarity with the exercise, this may have influenced the results to some extent [52]. Although the results by Antonio and Ciccone [11] suggest that Cr supplementation post-workout may provide greater gains in muscle strength and fat-free mass compared to pre-exercise Cr, the aforementioned limitations preclude solid recommendations without further supporting evidence.

Candow et al. [22], following the pioneering work by Antonio and Ciccone [11], compared the effects of Cr supplementation ingested immediately before vs. immediately after supervised resistance training in healthy older adults (Table 1). The 22 participants, who were not previously engaged in resistance training, were randomised in a double-blind design to one of two supplementation groups: one received Cr before (0.1 g·kg^{-1} Cr + 0.1 g·kg^{-1} placebo after exercise) and the other after (0.1 g·kg^{-1} placebo before + 0.1 g·kg^{-1} Cr after exercise) exercise. Resistance training was performed 3 days/week, on non-consecutive days, for 12 weeks. Results showed there was no difference between groups for gains in maximum strength, increases in muscle thickness, and changes in body composition, suggesting that supplement timing of Cr does not affect these measures.

This study has some important strengths, including a 12-week intervention, which is more than enough time to verify the effects of Cr supplementation [8]. Muscle thickness was measured, an important and relevant measure to observe resistance training-induced hypertrophy. However, similar to the study by Antonio and Ciccone (2013), the protocol did not include a placebo-only group to determine the true effects of the training alone. The study also lacked measures on muscle TCr changes with supplementation and training, an important consideration, since the supplementation protocol differed somewhat to those commonly employed in the literature. Cr ingested at a dose of 20 g for 5–7 days is sufficient to saturate muscle Cr stores [5], while ~3–5 g/day of Cr for 4 weeks similarly saturates skeletal muscle Cr levels [8,53]. The Cr supplementation protocol used by Candow et al. (2014) was not assessed in this sense. Although the ingested dose on training days was similar to the latter dosing strategy (~7 g of Cr for a 70 kg individual), supplementation was only performed three times per week. The extent to which this strategy increases TCr in the initial days/weeks, or when it would saturate muscles, is unknown. Nonetheless, it is likely that this dosing strategy would lead to a slow, transient increase in muscle Cr stores, meaning the timing of Cr supplementation in relation to exercise—there might be differences during such a supplementation protocol if it truly has an impact. It is also currently unclear if these same results apply to young healthy adults since elderly adults appear to respond differently to Cr supplementation [9]. Despite these limitations, the results of this study suggest that when intermittent low doses of Cr are consumed during chronic resistance training for 12 weeks, then the timing of supplementation pre- or post-exercise does not exert differential effects on strength, hypertrophy, and body composition in healthy older adults.

The largest study to date on the topic of Cr timing involved a 32-week resistance exercise training program [12] (Table 1). Thirty-nine healthy older adults completed the double-blind placebo-controlled design, and were randomised into three groups: "Cr-Before" (0.1 g·kg^{-1} Cr immediately before +0.1 g·kg^{-1} placebo immediately after RT); "Cr-After" (placebo immediately before + Cr immediately after RT); or placebo (corn starch maltodextrin immediately before and immediately after RT). Creatine was ingested only on training days and the resistance training intervention consisted of a supervised whole-body program performed 3-days per week in which the participants completed 3 sets of 10 repetitions at an intensity corresponding to their 10-RM for each exercise. Following the 32-week intervention, both Cr groups exhibited similar strength gains, with changes greater than the placebo control group.

It is interesting to note that the group that ingested Cr immediately after the training sessions showed greater increases in lean tissue mass compared to the control group, although not compared to the group that consumed Cr pre-training. There were also no differences in lean tissue mass changes between individuals in the control group and those supplementing Cr pre-training. The authors attributed the greater improvements obtained from post-workout supplementation to a better Cr absorption kinetics [54] and an increase in skeletal muscle blood flow during resistance training, which would result in greater Cr transport and accumulation in the exercised muscles, although this appears contradictory to expectation. Pre-exercise supplementation would mean that increased plasma Cr would coincide with an increased exercise-induced blood flow, potentially enhancing Cr uptake into muscle. Therefore, the reasons for these results are unclear. However, this study is similarly limited by a lack of muscle Cr determination, which might have helped to explain, at least in part, some of the differences shown between groups. A lack of direct measurement of the changes in muscle size also limits the evaluation of the strength training-induced local hypertrophy. Again, any extrapolation of these results must be limited to the elderly population due to distinct changes in muscle Cr content with supplementation compared to younger adults [9]. Finally, the lack of a statistical difference between the pre- and post-training Cr groups does not strongly support a differential effect of Cr timing.

More recently, a study examined the effects of Cr supplementation ingested throughout exercise training. Specifically, Mills et al. [55] analysed the effects of intra-set Cr supplementation during resistance training sessions on skeletal muscle mass and exercise performance in physically active young adults, which engaged in a structured resistance training program (Table 1). Twenty-two participants were randomised in a double-blind placebo-controlled design to a group supplemented with Cr (0.0055 g·kg^{-1} following each training set, totalling 18 sets per session) or a group supplemented with a placebo (maltodextrin, 0.0055 g·kg^{-1} following each training set, totalling 18 sets per session) during six weeks of RT performed five days per week. Muscle thickness, power, 1-RM, and muscular endurance were determined pre- and post-intervention. Leg press, chest press, total body strength, and leg press endurance improved in the Cr group, with no significant changes in the placebo group. Although these results interestingly demonstrate that chronic Cr throughout resistance exercise workouts can improve strength gains, the lack of a supplementation group that ingested Cr either pre- or post-workout, or "not close" to the workout session, hinders any solid conclusions other than the fact that Cr supplementation improves strength gains with training, something that was already well-established. It would be interesting to determine whether intra-set Cr supplementation is superior to supplementation at any other moment of the day, whether it be outside of training hours, immediately pre-exercise, or immediately post-exercise.

Table 1. Study protocols that investigated the timing effects of creatine supplementation before, during, and after exercise.

Reference	Population	Intervention	Outcomes
Antonio and Ciccone, [11]	19 recreational male bodybuilders.	4 weeks of 5 g·day^{-1} Cr: Group 1: Cr pre-exercise Group 2: Cr post-exercise. RT consisted of 5 d·wk^{-1} sessions.	$\leftrightarrow$ ΔBM, ΔFFM, ΔFM and Δ1-RM BP. *Possibly* (FFM, FM) and *likely* (1-RM BP) beneficial for Cr post vs. Cr pre.
Candow et al. [22]	9 men and 13 women, non-RT healthy older adults.	12 weeks of 0.1 g·kg^{-1} Cr and 0.1 g·kg^{-1} PL: Group 1: Cr before + PL after Group 2: PL before + Cr after. Cr ingested only on training days: 3 d·wk^{-1} RT session.	$\leftrightarrow$ ΔFFM $\leftrightarrow$ ΔLimb muscle thickness $\leftrightarrow$ Δ1-RM BP $\leftrightarrow$ Δ1-RM LP
Candow et al. [12]	22 women and 17 men, non-RT healthy older adults.	32 weeks of 0.1 g·kg^{-1} Cr and/or 0.1 g·kg^{-1} PL: Group 1: Cr pre + PL post Group 2: Cr post + PL pre Group 3: Placebo control. Cr ingested only on training days: 3 d·wk^{-1} RT session.	ΔLBM: ↑ Cr post PL; $\leftrightarrow$ Cr pre vs. Cr post and PL. $\leftrightarrow$ Cr groups for 1-RM BP and LP ↑ Strength for both Cr groups compared to PL.
Mills et al. [55]	22 Physically active men and women.	6 weeks of Cr or PL post each set (intra-workout). Group 1: 0.0055 g·kg^{-1} Cr post each set Group 2: 0.0055 g·kg^{-1} Pl post each set. Cr was ingested only on training days: 5 d·wk^{-1} RT session.	↑ 1-RM BP and LP for Cr vs. PL Cr pre to post-intervention: ↑ 1-RM ↑ Muscle endurance

Abbreviations: Cr = creatine; RT = resistance training; FFM = fat free mass; FM = Fat mass; BM = body mass; RM = repetition maximum; BP = bench press; LP = leg press; LBM = lean body mass; TCr = total creatine content; $\leftrightarrow$ = No difference; ↑ = increased; Δ = pre- to post-change.

Currently, evidence is unclear as to whether timing of Cr supplementation around exercise modifies its efficacy (Figure 2). There is some weak evidence to support post-exercise Cr supplementation compared to pre-exercise supplementation, though the physiological mechanism underpinning these superior gains were not determined. This is of great importance since the importance of timing is only likely to have an effect during the initial phase of muscle Cr loading, and will probably be irrelevant once muscle TCr is saturated. Certainly, the limitations in the protocols highlighted herein should be considered when we try to apply the outcomes of these studies to real life, and determine whether the timing of Cr supplementation in relation to exercise is an important factor to optimize subsequent gains.

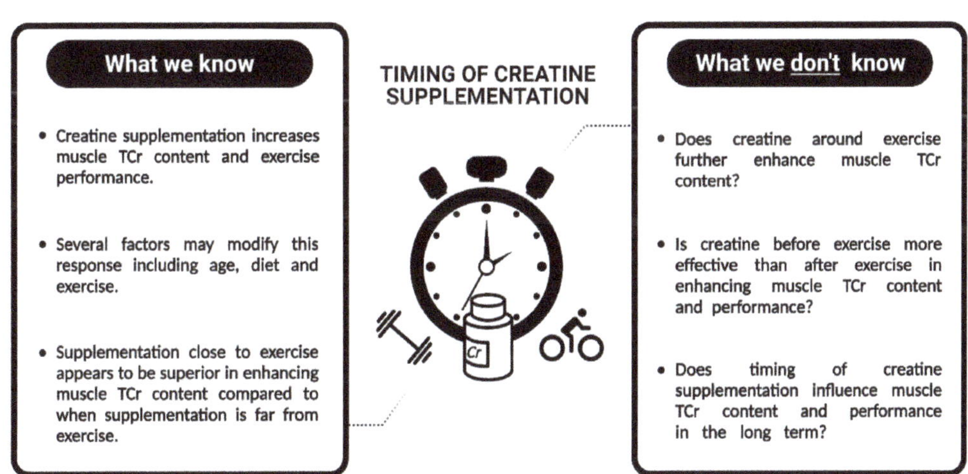

Figure 2. Overview of what is known about the timing of creatine (Cr) supplementation and what is yet to be determined. TCr = total creatine. Created with BioRender.com.

5. Future Directions for Research

There is evidence suggesting that Cr loading can be enhanced by exercise, with very limited data showing that Cr consumption close to exercise sessions can be more effective than ingestion in other moments of the day, at least in respect to muscle Cr loading. However, whether the timing of supplementation pre-, during-, or post-exercise has an impact is less clear. Recently, evidence suggested that post-exercise supplementation can increase muscle mass, but not strength, to a superior magnitude, when compared to pre-exercise Cr supplementation. However, considering the theoretical variables and mechanisms that influence Cr uptake into the muscles (discussed in the present review), and a series of methodological limitations presented by the available investigations, the interpretation of these results is limited. Thus, here we provide guidelines for future studies investigating this topic to ensure clarity in results and interpretation to advance our knowledge in the area.

Firstly, the lack of muscle Cr measurements in previous studies preclude solid conclusions on the efficacy of supplement timing. Without this information, it is only possible to speculate as to the effect of Cr timing in relation to exercise on changes in muscle Cr content (Figure 2). This is particularly relevant regarding the low doses of Cr typically used in these studies. It is known that a high daily dose of Cr (e.g., 20 g·day^{-1}) will saturate muscle Cr content in 5–7 days [5], meaning that timing of supplementation with such a high dose will likely be unimportant. However, since lower daily doses (e.g., 5 g·day^{-1}) will only saturate muscle Cr in up to 28 days [8,53], it is more than possible that enhanced loading, perhaps due to timing, might lead to greater benefits with these lower doses. Analyses during the early phases of supplementation, for example, the first 1–3 weeks with low doses (e.g., 5–7 g·day^{-1}) may be most important, since measurements (once muscle Cr is already saturated) are unlikely to show any differences. Further work should ensure this measurement is included to confirm increases in muscle Cr and the extent to which it differs between timing protocols. Any subsequent changes in muscle strength, hypertrophy, or exercise performance can then be associated with these changes.

Another limitation in the current literature is related to clinical differences in the populations studied, with one study employing recreational male bodybuilders and two studies recruiting elderly non-trained populations. This is a potentially important consideration, since TCr increases in response to a standardised Cr supplementation protocol may be affected by factors, such as age, physical exercise, and diet [9]. Furthermore, small sample sizes currently hinder the strength of evidence since several studies that suggest greater improvements with one timing versus another are not supported by statistical strength [11,12]. The clinical relevance of these small differences due to timing for different populations (e.g., young vs. elderly; competitive vs. non-competitive) may be worthwhile or irrelevant. Larger population samples are required to tease out any benefits of one specific timing over another, should differences truly exist.

Further, well-designed confirmatory research is necessary to determine the magnitude of effect that Cr timing around resistance training might incur on muscle strength and hypertrophy changes with Cr supplementation. To that end, in addition to the points already raised, it would be wise to employ exercise protocols and strength measures that have previously shown robust and clear effects following Cr supplementation (e.g., 1-RM test and/or resistance training program) to quantify the size of the contribution of timing to outcomes. The use of muscular endurance tests may also be of interest since many bodybuilders frequently train until volitional fatigue, and this may be a sensitive measure to Cr supplementation [56]. Several studies appeared not to familiarise their participants to the strength tests [11,55], which is a recommendable practice for diminishing the influence of the learning effect on the strength test assessments. Including familiarisation to the exercise protocols is vital to ensure more accurate measurements. If the goal is to verify skeletal muscle hypertrophy in response to exercise and Cr supplementation, as routinely occurs with resistance exercise studies, more direct and reliable measurements of skeletal muscle hypertrophy are important assessments of exercise and resistance training-induced local hypertrophy. For example, the measurement of muscle volume or muscle cross-sectional area via ultrasound imaging, magnetic resonance imaging, or computed tomography scans would be relevant protocols to check strength training-induced local hypertrophy [57] and whether it differs according to Cr timing.

Current studies on Cr supplement timing have focussed only on timing around resistance exercise. Creatine supplementation timing should be analysed in relation to other types of exercise (e.g., repeated sprints, endurance), not only on resistance training, due to the diversity of athletes who consume this supplement regularly to improve sports performance [1]. This would allow one to determine whether the type of exercise matters to induce the contraction-stimulated uptake of Cr into the muscles, and if this is modified by timing around the activity. Again, employing exercise tests to measure performances that were previously shown susceptible to improvements with Cr supplementation might be of particular interest in this scenario. Mechanistic studies should also strive to determine whether any changes in muscle Cr content, due to timing, is due to increased uptake or increased retention via infusion and microdialysis techniques. In addition to the effects of pre- versus post-exercise timing, another possibility of ingestion of Cr is during exercise [55], so a comparison of Cr ingested during workouts and Cr consumed pre- and post-workout, and/or in other moments of the day, may be of interest. This will provide important information as to the necessity of this small-dose multi-moment ingestion strategy.

6. Conclusions

Although exercise appears to enhance Cr accrual in muscles with Cr supplementation, evidence supporting the importance of timing of Cr supplementation around exercise (i.e., pre- versus post- versus during-exercise) is currently limited to only a few studies. Existing data are somewhat contradictory, likely due to differing supplementation protocols, sample populations, and training protocols. As it stands, adapting Cr timing specifically, according to when training is performed, is not currently supported by solid evidence and should not be considered a real concern for now. More well-controlled studies determining whether the timing of Cr supplementation around training truly influences the increases in intramuscular Cr content and its ergogenic effects are required to substantiate any such claims.

Author Contributions: Conceptualization, F.R., I.L., and B.S.; writing—review and editing, F.R., I.L., P.P., B.D., P.F., B.G., H.R., and B.S. All authors have read and agreed to the published version of the manuscript.

Funding: The authors received no specific funding for this review. F.R. (2018/19981-5), P.P. (2018/01594-5), B.D. (2019/06140-5), and B.S. (2016/50438-0) have been financially supported by Fundação de Amparo à Pesquisa do Estado de São Paulo. P.F. received a grant from Programa Institucional de Bolsas de Iniciação Científica, CNPq (165737/2020-4). H.R. has been financially supported by Conselho Nacional de Desenvolvimento Científico e Tecnológico—CNPq (301571/2017-1). Bryan Saunders received a grant from Faculdade de Medicina da Universidade de São Paulo (2020.1.362.5.2).

Acknowledgments: The authors would like to thank all of the research participants, scholars, and funding agencies who contributed to the research cited in this manuscript.

Conflicts of Interest: B.G. received research grants, creatine donation for scientific studies, travel support for participation in scientific conferences, and honorarium for speaking at lectures from AlzChem (a company that manufactures creatine). Additionally, he serves as a member of the Scientific Advisory Board for AlzChem. B.S. previously received creatine supplements free of charge from AlzChem to perform research on supplementation and exercise. AlzChem did not provide any input in regard to the intellectual content of the present review.

References

1. Kreider, R.B.; Kalman, D.S.; Antonio, J.; Ziegenfuss, T.N.; Wildman, R.; Collins, R.; Candow, D.G.; Kleiner, S.M.; Almada, A.L.; Lopez, H.L. International Society of Sports Nutrition position stand: Safety and efficacy of creatine supplementation in exercise, sport, and medicine. *J. Int. Soc. Sports Nutr.* **2017**, *14*, 18. [CrossRef]
2. Maughan, R.J.; Burke, L.M.; Dvorak, J.; Larson-Meyer, D.E.; Peeling, P.; Phillips, S.M.; Rawson, E.S.; Walsh, N.P.; Garthe, I.; Geyer, H.; et al. IOC consensus statement: Dietary supplements and the high-performance athlete. *Br. J. Sports Med.* **2018**, *52*, 439–455. [CrossRef]
3. Kreider, R.B.; Jung, Y.P. Creatine supplementation in exercise, sport, and medicine. *J. Exerc. Nutr. Biochem.* **2011**, *15*, 53–69. [CrossRef]
4. Buford, T.W.; Kreider, R.B.; Stout, J.R.; Greenwood, M.; Campbell, B.; Spano, M.; Ziegenfuss, T.; Lopez, H.; Landis, J.; Antonio, J. International Society of Sports Nutrition position stand: Creatine supplementation and exercise. *J. Int. Soc. Sports Nutr.* **2007**, *4*, 6. [CrossRef]
5. Harris, R.C.; Soderlund, K.; Hultman, E. Elevation of creatine in resting and exercised muscle of normal subjects by creatine supplementation. *Clin. Sci. (Lond.)* **1992**, *83*, 367–374. [CrossRef]
6. Green, A.L.; Simpson, E.J.; Littlewood, J.J.; Macdonald, I.A.; Greenhaff, P.L. Carbohydrate ingestion augments creatine retention during creatine feeding in humans. *Acta Physiol. Scand.* **1996**, *158*, 195–202. [CrossRef]
7. Greenhaff, P.L.; Casey, A.; Short, A.H.; Harris, R.; Soderlund, K.; Hultman, E. Influence of oral creatine supplementation of muscle torque during repeated bouts of maximal voluntary exercise in man. *Clin. Sci. (Lond.)* **1993**, *84*, 565–571. [CrossRef]
8. Hultman, E.; Soderlund, K.; Timmons, J.A.; Cederblad, G.; Greenhaff, P.L. Muscle creatine loading in men. *J. Appl. Physiol.* **1996**, *81*, 232–237. [CrossRef]
9. Solis, M.Y.; Artioli, G.G.; Otaduy, M.C.G.; Leite, C.D.C.; Arruda, W.; Veiga, R.R.; Gualano, B. Effect of age, diet, and tissue type on PCr response to creatine supplementation. *J. Appl. Physiol. (1985)* **2017**, *123*, 407–414. [CrossRef] [PubMed]
10. Candow, D.G.; Forbes, S.C.; Chilibeck, P.D.; Cornish, S.M.; Antonio, J.; Kreider, R.B. Variables Influencing the Effectiveness of Creatine Supplementation as a Therapeutic Intervention for Sarcopenia. *Front. Nutr.* **2019**, *6*, 124. [CrossRef]
11. Antonio, J.; Ciccone, V. The effects of pre versus post workout supplementation of creatine monohydrate on body composition and strength. *J. Int. Soc. Sports Nutr.* **2013**, *10*, 36. [CrossRef] [PubMed]
12. Candow, D.G.; Vogt, E.; Johannsmeyer, S.; Forbes, S.C.; Farthing, J.P. Strategic creatine supplementation and resistance training in healthy older adults. *Appl. Physiol. Nutr. Metab.* **2015**, *40*, 689–694. [CrossRef]
13. Kreider, R.B.; Melton, C.; Rasmussen, C.J.; Greenwood, M.; Lancaster, S.; Cantler, E.C.; Milnor, P.; Almada, A.L. Long-term creatine supplementation does not significantly affect clinical markers of health in athletes. *Mol. Cell. Biochem.* **2003**, *244*, 95–104. [CrossRef] [PubMed]
14. Mujika, I.; Padilla, S. Creatine supplementation as an ergogenic aid for sports performance in highly trained athletes: A critical review. *Int. J. Sports Med.* **1997**, *18*, 491–496. [CrossRef]
15. Wallimann, T.; Wyss, M.; Brdiczka, D.; Nicolay, K.; Eppenberger, H.M. Intracellular compartmentation, structure and function of creatine kinase isoenzymes in tissues with high and fluctuating energy demands: The 'phosphocreatine circuit' for cellular energy homeostasis. *Biochem. J.* **1992**, *281 Pt 1*, 21–40. [CrossRef]
16. Hargreaves, M.; Spriet, L.L. Skeletal muscle energy metabolism during exercise. *Nat. Metab.* **2020**, *2*, 817–828. [CrossRef]
17. Sahlin, K. Muscle Energetics During Explosive Activities and Potential Effects of Nutrition and Training. *Sports Med.* **2014**, *44*, S167–S173. [CrossRef]
18. Green, A.L.; Hultman, E.; Macdonald, I.A.; Sewell, D.A.; Greenhaff, P.L. Carbohydrate ingestion augments skeletal muscle creatine accumulation during creatine supplementation in humans. *Am. J. Physiol.* **1996**, *271*, E821–E826. [CrossRef]
19. Greenwood, M.; Kreider, R.; Earnest, C.; Rasmussen, C.; Almada, A. Differences in creatine retention among three nutritional formulations of oral creatine supplements. *J. Exerc. Physiol. Online* **2003**, *6*, 37–43.
20. Robinson, T.M.; Sewell, D.A.; Hultman, E.; Greenhaff, P.L. Role of submaximal exercise in promoting creatine and glycogen accumulation in human skeletal muscle. *J. Appl. Physiol. (1985)* **1999**, *87*, 598–604. [CrossRef]
21. Candow, D.G.; Chilibeck, P.D. Timing of creatine or protein supplementation and resistance training in the elderly. *Appl. Physiol. Nutr. Metab.* **2008**, *33*, 184–190. [CrossRef]
22. Candow, D.G.; Zello, G.A.; Ling, B.; Farthing, J.P.; Chilibeck, P.D.; McLeod, K.; Harris, J.; Johnson, S. Comparison of creatine supplementation before versus after supervised resistance training in healthy older adults. *Res. Sports Med.* **2014**, *22*, 61–74. [CrossRef]
23. Joyner, M.J.; Wilkins, B.W. Exercise hyperaemia: Is anything obligatory but the hyperaemia? *J. Physiol.* **2007**, *583*, 855–860. [CrossRef]
24. Joyner, M.J.; Casey, D.P. Regulation of increased blood flow (hyperemia) to muscles during exercise: A hierarchy of competing physiological needs. *Physiol. Rev.* **2015**, *95*, 549–601. [CrossRef]
25. Roberts, P.A.; Fox, J.; Peirce, N.; Jones, S.W.; Casey, A.; Greenhaff, P.L. Creatine ingestion augments dietary carbohydrate mediated muscle glycogen supercompensation during the initial 24 h of recovery following prolonged exhaustive exercise in humans. *Amino Acids* **2016**, *48*, 1831–1842. [CrossRef]
26. Persky, A.M.; Brazeau, G.A.; Hochhaus, G. Pharmacokinetics of the dietary supplement creatine. *Clin. Pharmacokinet.* **2003**, *42*, 557–574. [CrossRef]

27. Jäger, R.; Harris, R.C.; Purpura, M.; Francaux, M. Comparison of new forms of creatine in raising plasma creatine levels. *J. Int. Soc. Sports Nutr.* **2007**, *4*, 17. [CrossRef] [PubMed]

28. Rawson, E.S.; Clarkson, P.M.; Price, T.B.; Miles, M.P. Differential response of muscle phosphocreatine to creatine supplementation in young and old subjects. *Acta Physiol. Scand.* **2002**, *174*, 57–65. [CrossRef]

29. Steenge, G.R.; Simpson, E.J.; Greenhaff, P.L. Protein- and carbohydrate-induced augmentation of whole body creatine retention in humans. *J. Appl. Physiol. (1985)* **2000**, *89*, 1165–1171. [CrossRef]

30. Schedel, J.M.; Tanaka, H.; Kiyonaga, A.; Shindo, M.; Schutz, Y. Acute creatine ingestion in human: Consequences on serum creatine and creatinine concentrations. *Life Sci.* **1999**, *65*, 2463–2470. [CrossRef]

31. Hackett, D.A.; Johnson, N.A.; Chow, C.M. Training practices and ergogenic aids used by male bodybuilders. *J. Strength Cond. Res.* **2013**, *27*, 1609–1617. [CrossRef]

32. Bangsbo, J.; Hellsten, Y. Muscle blood flow and oxygen uptake in recovery from exercise. *Acta Physiol. Scand.* **1998**, *162*, 305–312. [CrossRef] [PubMed]

33. Korthuis, R.J. Chapter 4, Exercise Hyperemia and Regulation of Tissue Oxygenation During Muscular Activity. In *Skeletal Muscle Circulation*; Korthuis, R.J., Ed.; Morgan & Claypool Life Sciences: San Rafael, CA, USA, 2011.

34. Christie, D.L. Functional insights into the creatine transporter. *Sub Cell. Biochem.* **2007**, *46*, 99–118. [CrossRef]

35. Daly, M.M.; Seifter, S. Uptake of creatine by cultured cells. *Arch. Biochem. Biophys.* **1980**, *203*, 317–324. [CrossRef]

36. Odoom, J.E.; Kemp, G.J.; Radda, G.K. The regulation of total creatine content in a myoblast cell line. *Mol. Cell. Biochem.* **1996**, *158*, 179–188. [CrossRef] [PubMed]

37. Forbes, S.C.; Candow, D.G. Timing of Creatine Supplementation and Resistance Training: A Brief Review. *J. Exerc. Nutr.* **2018**, *1*, 1–6.

38. Green, H.J.; Chin, E.R.; Ball-Burnett, M.; Ranney, D. Increases in human skeletal muscle Na(+)-K(+)-ATPase concentration with short-term training. *Am. J. Physiol.* **1993**, *264*, C1538–C1541. [CrossRef] [PubMed]

39. Aughey, R.J.; Murphy, K.T.; Clark, S.A.; Garnham, A.P.; Snow, R.J.; Cameron-Smith, D.; Hawley, J.A.; McKenna, M.J. Muscle Na+-K+-ATPase activity and isoform adaptations to intense interval exercise and training in well-trained athletes. *J. Appl. Physiol. (1985)* **2007**, *103*, 39–47. [CrossRef]

40. Holloszy, J.O. Exercise-induced increase in muscle insulin sensitivity. *J. Appl. Physiol. (1985)* **2005**, *99*, 338–343. [CrossRef]

41. Cribb, P.J.; Hayes, A. Effects of supplement timing and resistance exercise on skeletal muscle hypertrophy. *Med. Sci. Sports Exerc.* **2006**, *38*, 1918–1925. [CrossRef] [PubMed]

42. Tipton, K.D.; Rasmussen, B.B.; Miller, S.L.; Wolf, S.E.; Owens-Stovall, S.K.; Petrini, B.E.; Wolfe, R.R. Timing of amino acid-carbohydrate ingestion alters anabolic response of muscle to resistance exercise. *Am. J. Physiol. Endocrinol. Metab.* **2001**, *281*, E197–E206. [CrossRef]

43. Stecker, R.A.; Harty, P.S.; Jagim, A.R.; Candow, D.G.; Kerksick, C.M. Timing of ergogenic aids and micronutrients on muscle and exercise performance. *J. Int. Soc. Sports Nutr.* **2019**, *16*, 37. [CrossRef] [PubMed]

44. Sainani, K.L. The Problem with "Magnitude-based Inference". *Med. Sci. Sports Exerc.* **2018**, *50*, 2166–2176. [CrossRef]

45. Sainani, K.L.; Lohse, K.R.; Jones, P.R.; Vickers, A. Magnitude-based Inference is not Bayesian and is not a valid method of inference. *Scand. J. Med. Sci. Sports* **2019**, *29*, 1428–1436. [CrossRef]

46. Chappell, A.J.; Simper, T.; Helms, E. Nutritional strategies of British professional and amateur natural bodybuilders during competition preparation. *J. Int. Soc. Sports Nutr.* **2019**, *16*, 35. [CrossRef] [PubMed]

47. Devries, M.C.; Phillips, S.M. Creatine supplementation during resistance training in older adults-a meta-analysis. *Med. Sci. Sports Exerc.* **2014**, *46*, 1194–1203. [CrossRef]

48. Alves, R.C.; Prestes, J.; Enes, A. Training Programs Designed for Muscle Hypertrophy in Bodybuilders: A Narrative Review. *Sports* **2020**, *8*, 149. [CrossRef]

49. Morton, R.W.; Murphy, K.T.; McKellar, S.R.; Schoenfeld, B.J.; Henselmans, M.; Helms, E.; Aragon, A.A.; Devries, M.C.; Banfield, L.; Krieger, J.W.; et al. A systematic review, meta-analysis and meta-regression of the effect of protein supplementation on resistance training-induced gains in muscle mass and strength in healthy adults. *Br. J. Sports Med.* **2018**, *52*, 376–384. [CrossRef]

50. Franchi, M.V.; Longo, S.; Mallinson, J.; Quinlan, J.I.; Taylor, T.; Greenhaff, P.L.; Narici, M.V. Muscle thickness correlates to muscle cross-sectional area in the assessment of strength training-induced hypertrophy. *Scand. J. Med. Sci. Sports* **2018**, *28*, 846–853. [CrossRef]

51. Raglin, J.; Szabo, A. Understanding placebo and nocebo effects in the context of sport: A psychological perspective. *Eur. J. Sport Sci.* **2020**, *20*, 293–301. [CrossRef] [PubMed]

52. Soares-Caldeira, L.F.; Ritti-Dias, R.M.; Okuno, N.M.; Cyrino, E.S.; Gurjão, A.L.; Ploutz-Snyder, L.L. Familiarization indexes in sessions of 1-RM tests in adult women. *J. Strength Cond. Res.* **2009**, *23*, 2039–2045. [CrossRef]

53. Antonio, J.; Candow, D.G.; Forbes, S.C.; Gualano, B.; Jagim, A.R.; Kreider, R.B.; Rawson, E.S.; Smith-Ryan, A.E.; VanDusseldorp, T.A.; Willoughby, D.S.; et al. Common questions and misconceptions about creatine supplementation: What does the scientific evidence really show? *J. Int. Soc. Sports Nutr.* **2021**, *18*, 13. [CrossRef]

54. Preen, D.; Dawson, B.; Goodman, C.; Lawrence, S.; Beilby, J.; Ching, S. Pre-exercise oral creatine ingestion does not improve prolonged intermittent sprint exercise in humans. *J. Sports Med. Phys. Fit.* **2002**, *42*, 320–329.

55. Mills, S.; Candow, D.G.; Forbes, S.C.; Neary, J.P. Effects of Creatine Supplementation during Resistance Training Sessions in Physically Active Young Adults. *Nutrients* **2020**, *12*, 1880. [CrossRef] [PubMed]

56. Dankel, S.J.; Jessee, M.B.; Mattocks, K.T.; Mouser, J.G.; Counts, B.R.; Buckner, S.L.; Loenneke, J.P. Training to Fatigue: The Answer for Standardization When Assessing Muscle Hypertrophy? *Sports Med.* **2017**, *47*, 1021–1027. [CrossRef]
57. Haun, C.T.; Vann, C.G.; Roberts, B.M.; Vigotsky, A.D.; Schoenfeld, B.J.; Roberts, M.D. A Critical Evaluation of the Biological Construct Skeletal Muscle Hypertrophy: Size Matters but So does the Measurement. *Front. Physiol.* **2019**, *10*, 247. [CrossRef] [PubMed]

Review

Safety of Dietary Guanidinoacetic Acid: A Villain of a Good Guy?

Sergej M. Ostojic [1,2]

1   Department of Nutrition and Public Health, University of Agder, 4604 Kristiansand, Norway; sergej.ostojic@uia.no; Tel.: +47-38-14-13-64
2   FSPE Applied Bioenergetics Lab, University of Novi Sad, 21000 Novi Sad, Serbia

Abstract: Guanidinoacetic acid (GAA) is a natural amino acid derivative that is well-recognized for its central role in the biosynthesis of creatine, an essential compound involved in cellular energy metabolism. GAA (also known as glycocyamine or betacyamine) has been investigated as an energy-boosting dietary supplement in humans for more than 70 years. GAA is suggested to effectively increase low levels of tissue creatine and improve clinical features of cardiometabolic and neurological diseases, with GAA often outcompeting traditional bioenergetics agents in maintaining ATP status during stress. This perhaps happens due to a favorable delivery of GAA through specific membrane transporters (such as SLC6A6 and SLC6A13), previously dismissed as un-targetable carriers by other therapeutics, including creatine. The promising effects of dietary GAA might be countered by side-effects and possible toxicity. Animal studies reported neurotoxic and pro-oxidant effects of GAA accumulation, with exogenous GAA also appearing to increase methylation demand and circulating homocysteine, implying a possible metabolic burden of GAA intervention. This mini-review summarizes GAA toxicity evidence in human nutrition and outlines functional GAA safety through benefit-risk assessment and multi-criteria decision analysis.

Keywords: toxicity; methylation; hyperhomocysteinemia; creatine; neuromodulation; MCDA

Citation: Ostojic, S.M. Safety of Dietary Guanidinoacetic Acid: A Villain of a Good Guy? *Nutrients* **2022**, *14*, 75. https://doi.org/10.3390/nu14010075

Academic Editor: David C. Nieman

Received: 6 December 2021
Accepted: 23 December 2021
Published: 24 December 2021

Publisher's Note: MDPI stays neutral with regard to jurisdictional claims in published maps and institutional affiliations.

1. GAA Physiology, Biomolecular Interactions and Pathways

Guanidinoacetic acid (GAA, also known as glycocyamine, betacyamine or *N*-amidinoglycine) belongs to the class of organic compounds known as alpha amino acids and derivatives. GAA (chemical formula $C_3H_7N_3O_2$) is produced endogenously in the human body from non-essential amino acids glycine and arginine, in a reaction controlled by an enzyme *L*-arginine:glycine amidinotransferase (AGAT) (Figure 1). AGAT catalyzes the transfer of an amidino group ($-C(=NH)NH_2$) from arginine to glycine to synthesize GAA, with ornithine as a byproduct. The reaction mainly takes place in the kidney, liver, and pancreas; however, GAA is also produced in the skeletal muscle, brain, and across the gut [1–3]. In the next step, GAA is combined with *S*-adenosyl-L-methionine, a reaction catalyzed by guanidinoacetate N-methyltransferase (GAMT), to produce creatine and *S*-adenosyl-L-homocysteine. The formation of creatine is a major metabolic fate of GAA, and AGAT-driven reaction is considered a rate-limiting step of creatine biosynthesis [4]. Since creatine is recognized as a critical molecular facilitator of cellular bioenergetics [5], the GAA synthesis-breakdown cycle thus remains of utmost importance for energy homeostasis. The role of GAA in the control and provision of cellular energy could also be highlighted by its interaction with cellular transporters for taurine (SLC6A6) and γ-aminobutyric acid (SLC6A13) [6,7], previously dismissed as un-targetable carriers by other bioenergetics therapeutics, including creatine (for a review, see Ref. [8]); monocarboxylate transporter 12 (SLC16A12) is involved in GAA efflux [9].

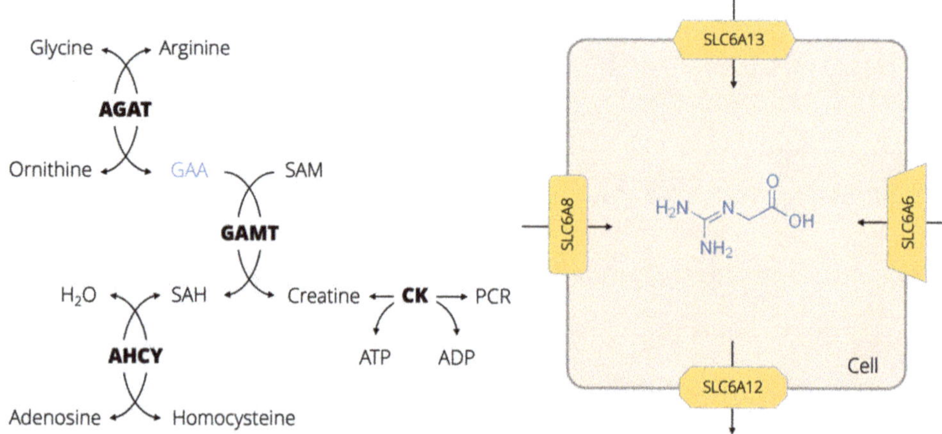

Figure 1. Metabolism and transport channels of guanidinoacetic acid (GAA). Abbreviations: AGAT, L-arginine:glycine amidinotransferase; SAM, *S*-adenosyl-L-methionine; GAMT, guanidinoacetate N-methyltransferase; SAH, *S*-adenosyl-L-homocysteine; CK, creatine kinase; PCR, phosphocreatine; AHCY, adenosylhomocysteinase; ATP, adenosine triphosphate; ADP, adenosine diphosphate. Yellow shapes depict different influx and efflux cellular transporters for GAA, including creatine transporter (SLC6A8), GABA transporter (SLC6A13), taurine transporter (SLC6A6), and 5-monocarboxylate transporter 12 (SLC16A12).

Plasma and urine GAA concentrations of 2.3 ± 0.8 μmol/L and 31.2 ± 21.7 mmol/mol of creatinine likely illustrate natural equilibrium in GAA metabolism [10]. However, several pathological conditions can affect GAA production-utilization circle, including kidney dysfunction, neurological diseases, or inherited metabolic disorders (for a detailed review, see Ref. [11]). Besides serving as an immediate precursor of creatine, GAA can also have several non-creatine-related metabolic roles, including the stimulation of hormonal release and neuromodulation, alteration of metabolic utilization of amino acids, vasodilation, and oxidant–antioxidant tuning (for a detailed review, see Ref. [12]). In addition, GAA could be obtained by a regular diet [13,14] and gut microbiota [15], yet these pathways contribute marginally to the total daily turnover of GAA.

2. GAA as a Dietary Agent in Human Nutrition

The first documented report of GAA utilization as an experimental nutritional intervention in humans arguaby dates back approximately 70 years ago. The group of Henry Borsook from Caltech University demonstrated the beneficial effects of GAA (combined with betaine) in treating cardiac decompensation [16]. The authors treated cardiac patients with a daily dosage of ~70 mg of GAA per kg body weight for up to 12 months, and many patients reported the so-called 'sthenic effect', comprising of an improved sense of wellbeing, less fatigue, and enhanced mental and physical performance. This was attributed to a GAA-driven recovery of phosphocreatine, the main reservoir of immediately available energy in energy-demanding tissues. The promising effects were soon corroborated in both hospital and ambulatory patients with heart disease, who reported feeling better after taking GAA, while the treatment produced no harmful effects, even when ingested over a long period of time [17]. A few patients with congestive heart failure were able to discontinue pharmacological treatment entirely while consuming GAA, without any unfavorable sequels [18]. Following these pioneering trials, GAA was intensively studied during the 1950's in patients with arthritis and concurrent disease [19], acute and chronic poliomyelitis [20–23], myopathic muscular dystrophy [24], anxiety and depression [25], coronary arteriosclerosis [26], myasthenia gravis [27], motor-neuron disease [28], and neuromuscular disease [29]. Those historical studies were characterized by several methodological constraints and yielded equivocal results with regard to GAA efficacy, yet the supplementation with GAA was found harmless and non-toxic.

After this inception, subsequent decades prompted a relatively limited interest in studying dietary GAA until 1999, when a Japanese group put forward GAA as a nutritional supplement to compensate for a disease-driven GAA shortage in patients with chronic renal failure [30]. During the past decade, studies in healthy humans evaluated the effectiveness and safety of supplemental GAA when administered solely or combined with other nutrients [31–33], the dose-response effects of GAA [34,35], and the impact of dietary GAA on neuromodulation [36], exercise performance [37], oxidant–antioxidant capacity [38], skeletal muscle and brain bioenergetics [39,40], and epigenetic pathways [41]. Dietary GAA was also administered in women with chronic fatigue syndrome [42], or older adults [43], and put forward as a possible treatment in AGAT deficiency [44], and skeletal muscle medicine [45]. The contemporary studies mainly paralleled findings of the seminal trials from the early 1950's, implying the advantageous effects of supplemental GAA on clinician- and patient-reported outcomes, now complemented by more robust study designs and an extensive list of pertinent biomarkers. In addition, recent trials suggested that GAA might be superior to creatine for improving bioenergetics in energy-demanding tissues [46], probably due to better transportability to target organs [8], and fewer non-responders compared to creatine [47]. This might be a rationale for its possible application in human nutrition, as an alternative or substitute of creatine, at least in some circumstances. Even so, GAA is still deemed as an experimental dietary additive since its utilization is not completely described in terms of efficacy, approval, labeling, and pharmacovigilance. At this moment, GAA can be found in several commercial formulations available in the U.S. and European markets, although no recorded standards of identity, quality, and corresponding analytical methods for GAA are currently available in the U.S., European, or Japanese pharmacopeias. The end-consumers might be, therefore, exposed to supplemental GAA while being unaware of possible safety issues.

3. Dietary GAA Safety and Toxicity

3.1. Methyl Group Depletion

Even the seminal paper that described the biochemical basis of GAA treatment recognized the possible risk of methyl group depletion following GAA consumption [16]. Since the transformation of GAA to creatine requires a donation of a methyl group (-CH3) from *S*-adenosyl-L-methionine, an excessive GAA intake can hypothetically drain the stores of methyl donors in the human body (e.g., methionine, choline, folic acid, B vitamins). The metabolic burden of methyl donor deficiency can perturb many cellular functions, including DNA methylation, neurotransmission, antioxidant defense, and protein synthesis [48]. Several human studies so far evaluated the risk of methyl group depletion after GAA intake. Our group investigated whether three different dosages of GAA (up to 4.8 g per day) administered for six weeks in healthy volunteers affect various serum and urinary variables related to GAA metabolism, including B vitamins [34]. We found that serum concentrations of folic acid, vitamin B6, B12, and holo-transcobalamin (carrier protein which binds the active form of vitamin B12) were not affected by the placebo or GAA intervention, implying that GAA dosages administered in this trial are probably insufficient to significantly impact circulating biomarkers of methyl donor micronutrients. Another trial evaluated the effects of supplemental GAA on DNA methylation [41], a critical epigenetic process for genome regulation. In this open-label, repeated-measure interventional trial, the authors evaluated the impact of 12 weeks of GAA supplementation (3 g per day) on global DNA methylation in healthy men and women. Dietary provision of GAA had no effect on DNA methylation, with 5-methylcytosine (a methylated form of the DNA base cytosine) non significantly increased at post-administration, while a non-significant DNA hypomethylation was found in 3 of 14 participants. However, it remains unknown whether methyl group depletion that might be caused by dietary GAA affects other biological methylation pathways, including amino acid and protein methylations or polysaccharide methylation.

3.2. Hyperhomocysteinemia

S-adenosyl-L-homocysteine is a byproduct of GAA utilization that is further converted to homocysteine in a simple one-step reaction catalyzed by adenosylhomocysteinase: *S*-adenosyl-L-homocysteine + H_2O <=> L-homocysteine + adenosine. Since hyperhomocysteinemia has been recognized as an independent risk factor for various cardiometabolic diseases [49], a possible augmentation of circulating homocysteine after GAA intake could thus be considered as a possible adverse effect of the intervention. Indeed, healthy young men and women who received 2.4 g of GAA per day for six weeks experienced a significant rise in serum homocysteine [31]. Elevated homocysteine levels were found in 55.6% of participants supplemented with GAA in another interventional trial [32], with a distinct dose-response effect of dietary GAA demonstrated for elevated serum homocysteine concentrations [34]. In the longest human study so far, supplemental GAA (3 g per day) significantly elevated serum homocysteine at 10-week follow-up (73.5% corresponding to 5.0 μmol/L), with 4 out of 20 participants (20.0%) experiencing clinically relevant hyperhomocysteinemia (>15.0 μmol/L) at post-administration [50]. This trial also revealed no effects of GAA on traditional biomarkers of cardiometabolic risk and inflammation, including HDL cholesterol, triglycerides, high-sensitive C-reactive protein, insulin, and ferritin. An atherogenic profile remained essentially unaffected by the intervention, indicating no major cardiometabolic burden of medium-term GAA intervention in healthy humans. Interestingly, co-administration of GAA and homocysteine-lowering agents can suppress or prevent a rise in homocysteine. No cases of elevated serum homocysteine were found in healthy human subjects supplemented with GAA and betaine (also vitamin B12, vitamin B6, and folic acid) [32]. Similarly, adding creatine to GAA largely prevented a GAA-driven rise in homocysteine in metabolically healthy men and women [33,51]. Having this in mind, dietary GAA should be carefully scrutinized as an experimental dietary additive due to its proven capacity to drive increased homocysteine production, which encourages its future utilization in human nutrition along with homocysteine-reducing agents.

3.3. Neurotoxicity

Several in vitro and animal studies documented neurotoxic effects of exogenously administered GAA. The group of Angela Wyse reported that the intrastriatal injection of GAA induced inhibition of Na+/K+-ATPase activity [52], glutamate uptake [53], and antioxidant defense in the rat brain [54]. GAA can affect brain cell development in rat brain cell cultures by causing axonal hyper-sprouting and decrease in natural apoptosis, followed by an induction of non-apoptotic cell death [55], suggesting that GAA may have different toxicity in the developing brain than in adults. Neu et al. [56] suggested that activating gamma-aminobutyric acid (GABA) receptors A might represent a candidate mechanism explaining neurological dysfunction induced by GAA. The accumulation of GAA in the brain was also found in children with inborn errors of creatine metabolism [57], suggesting that extra GAA might contribute to neurological complications in humans, such as epilepsy and seizures [58]. However, a recent study found that dietary GAA (up to 60 mg per kg body weight) does not accumulate in the brain of healthy men [59], with GAA levels remained essentially unchanged at eight-week follow-up when averaged across twelve white and grey matter locations. This study suggests that the GAA-driven neurotoxicity might be referenced to the level of GAA exposure, with the threshold of toxicological concern (although currently unidentified) highly unlikely to be acquired after dietary intake. For instance, brain GAA levels are 100–300 times higher than normal in experimental models with intrastriatal administration of GAA and/or inborn errors of creatine metabolism [57], which is substantially below GAA concentrations after dietary supplementation. Interestingly, GAA loading appears to affect peripheral GABA metabolism in healthy men and potentially down-regulates GABA synthesis in peripheral tissues [36]. Although safety sequels of this study are not elaborated further, GABA modulation should be considered as a possible neuromodulatory effect of dietary GAA.

3.4. Other Adverse Effects

An exposure to GAA in animal studies exacerbated ethanol-induced liver injury [60,61], stimulated osteoclastogenesis [62], generated reactive oxygen species [63], and modulated cerebral cortex potentials [64]. However, dietary GAA produced no effects on hepatic panel and cumulative action of antioxidants present in plasma after medium-term intake in healthy men and women [31,38]. Supplemental GAA can provoke a minor transitory gastrointestinal distress (e.g., intestinal cramping, bloating, abdominal pain) in healthy adults; the proportion of participants who reported gut-related side effects was not different from placebo [31]. GAA also induced a rise in serum creatinine due to an increase in creatine and subsequent degradation to creatinine. Creatinine is a surrogate marker of kidney damage, yet the other biomarkers of kidney damage remained unaffected by GAA intake [31].

4. Blueprint for GAA Risk-Benefit Assessment

The risk-benefit assessment is an emerging concept in the area of food safety [65,66], with the multi-criteria decision analysis (MCDA) advances as a new tool to estimate the hazards and advantages associated with the use of food interventions and dietary choices. MCDA combines heterogeneous research data into a compendious numeric that can be used to guide the selection of various health interventions in the context of risk-benefit profile [67]. In the case of GAA, the risk-benefit analysis could help in the pragmatic weighing of probabilities of side effect(s) against the benefit(s) as a consequence of GAA dietary exposure. Since GAA is an experimental dietary supplement, with only a handful of human studies available at this moment, MCDA could be created as a preliminary framework at best. In the first step (defining the decision problem), health professionals should establish the issue of interest, such as selecting the appropriate amount of supplemental GAA that safely alleviates the features of impaired energy metabolism. This step also involves nominating the target group, perhaps the clinical and general population, who can profit from energy-uplifting agents. The second step includes identifying the evaluation criteria (e.g., the prevalence of harmful health effects, improved brain function, reduced fatigue) against which various GAA dosages will be appraised. This step requires building a risk-benefit tree, based on feedback from experts and/or published trials (Figure 2). The third step of MCDA entails collecting data (e.g., qualitative, semi-quantitative, quantitative) on criteria selected above, and building the performance matrix describing how varying amounts of supplemental GAA perform when evaluated against each criterion. Only the effects with the highest level of evidence should be included in the final assessment, perhaps extracted from randomized controlled trials in the case of GAA. The fourth step defines the weights of the criteria, which enables setting up relative importance for each criterion depending on clinical significance using the 'swing weighing' approach [68]. For example, criteria weight for GAA-driven rise in serum homocysteine might include either quantitative scale (e.g., an absolute increase in µmol per liter) or qualitative method (e.g., using 'yes' or 'no' for inducing hyperhomocysteinemia); alternative scenarios representing different criteria-weighing schemes are also possible. Step five involves analyzing and synthesizing all risks and benefits while producing a single metric that can be used in grading GAA overall performance for efficacy-safety decisions; this requires accounting for uncertainty using statistical modeling [69]. The final sixth step includes reporting narrative and graphical results of an aggregated measure, enabling a decision-maker to rank alternative GAA dosages in terms of safety-efficacy profile.

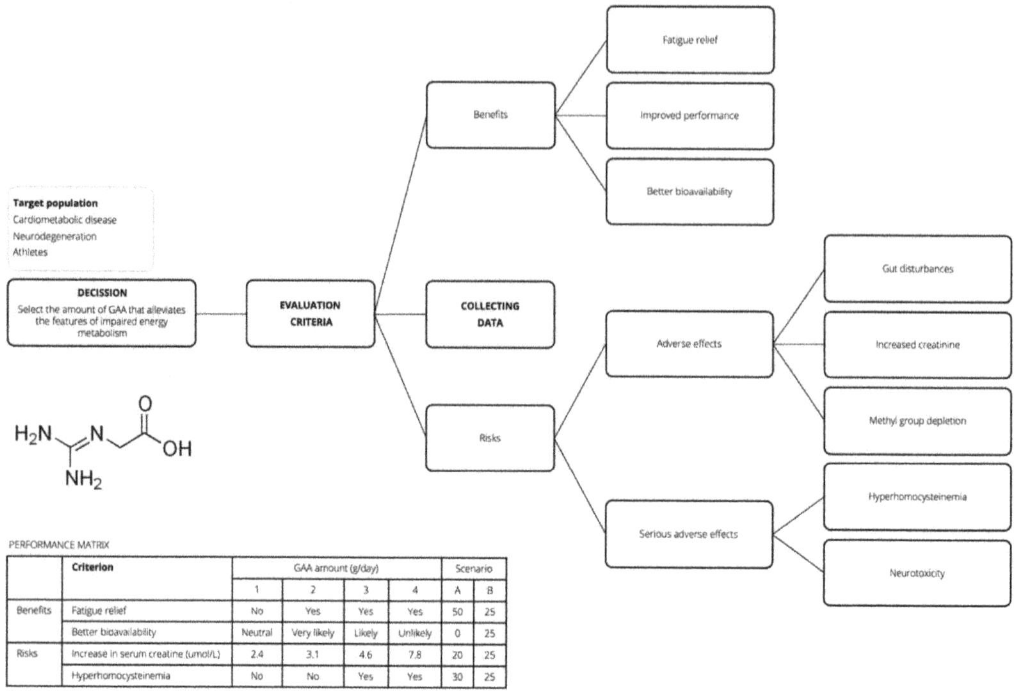

PERFORMANCE MATRIX

	Criterion	GAA amount (g/day)				Scenario	
		1	2	3	4	A	B
Benefits	Fatigue relief	No	Yes	Yes	Yes	50	25
	Better bioavailability	Neutral	Very likely	Likely	Unlikely	0	25
Risks	Increase in serum creatine (umol/L)	2.4	3.1	4.6	7.8	20	25
	Hyperhomocysteinemia	No	No	Yes	Yes	30	25

Figure 2. Risk-benefit tree and performance matrix for a hypothetical guanidinoacetic acid (GAA) case study.

5. Conclusions

GAA is an investigational dietary supplement. Preliminary human studies suggest that dietary GAA has a relatively acceptable safety profile, yet medium-term intake appears to provoke unfavorable biochemical abnormalities, such as the rise in serum homocysteine (which could be attanuated by GAA co-ingested with creatine). Other adverse events demonstrated in animal studies with non-enteral administration of GAA are not confirmed in human trials with supplemental GAA thus far. Still, the possible toxic effects of GAA in the nervous system reported in pre-clinical research (e.g., modulation of GABA-ergic neurotransmission, impairment of brain cell development, epileptogenic activity) remain of high concern, with additional human studies required before advancing GAA for human use. Whether beneficial effects outcompete side effects of GAA currently remains unknown in terms of evidence-based efficacy and safety data. A comparative analysis of supplemental GAA safety using multi-criteria decision analysis remains highly warranted to optimize treatment selection within the settings of nutrition and clinical bioenergetics.

Funding: This work was not funded by any agency in the public, commercial, or not-for-profit sectors.

Conflicts of Interest: SMO serves as a member of the Scientific Advisory Board on creatine in health and medicine (AlzChem LLC, Trostberg, Germany). SMO owns patent "Sports Supplements Based on Liquid Creatine" at European Patent Office (WO2019150323 A1), and active patent application "Synergistic Creatine" at UK Intellectual Property Office (GB2012773.4). SMO has served as a speaker at Abbott Nutrition, a consultant of Allied Beverages Adriatic and IMLEK, and has received research funding related to creatine and/or guanidinoacetic acid from the Serbian Ministry of Education, Science, and Technological Development, Provincial Secretariat for Higher Education and Scientific Research, AlzChem GmbH, KW Pfannenschmidt GmbH, Hueston Hennigan LLP, and ThermoLife International LLC. SMO does not own stocks and shares in any organization.

References

1. Edison, E.E.; Brosnan, M.E.; Meyer, C.; Brosnan, J.T. Creatine synthesis: Production of guanidinoacetate by the rat and human kidney in vivo. *Am. J. Physiol. Renal Physiol.* **2007**, *293*, F1799–F1804. [CrossRef] [PubMed]
2. Baker, S.A.; Gajera, C.R.; Wawro, A.M.; Corces, M.R.; Montine, T.J. GATM and GAMT synthesize creatine locally throughout the mammalian body and within oligodendrocytes of the brain. *Brain Res.* **2021**, *1770*, 147627. [CrossRef]
3. Ostojic, S.M. Creatine synthesis in the skeletal muscle: The times they are a-changin'. *Am. J. Physiol. Endocrinol. Metab.* **2021**, *320*, E390–E391. [CrossRef] [PubMed]
4. Walker, J.B. Creatine: Biosynthesis, Regulation, and Function. *Adv. Enzymol. Relat. Areas Mol. Biol.* **1979**, *50*, 177–242. [CrossRef] [PubMed]
5. Wallimann, T.; Tokarska-Schlattner, M.; Schlattner, U. The creatine kinase system and pleiotropic effects of creatine. *Amino Acids* **2011**, *40*, 1271–1296. [CrossRef] [PubMed]
6. Tachikawa, M.; Kasai, Y.; Yokoyama, R.; Fujinawa, J.; Ganapathy, V.; Terasaki, T.; Hosoya, K.-I. The blood-brain barrier transport and cerebral distribution of guanidinoacetate in rats: Involvement of creatine and taurine transporters. *J. Neurochem.* **2009**, *111*, 499–509. [CrossRef] [PubMed]
7. Tachikawa, M.; Yashiki, A.; Akanuma, S.-I.; Matsukawa, H.; Ide, S.; Minami, M.; Hosoya, K.-I. Astrocytic γ-aminobutyric acid (GABA) transporters mediate guanidinoacetate transport in rat brain. *Neurochem. Int.* **2018**, *113*, 1–7. [CrossRef]
8. Ostojic, S.M. Tackling guanidinoacetic acid for advanced cellular bioenergetics. *Nutrition* **2017**, *34*, 55–57. [CrossRef]
9. Jomura, R.; Tanno, Y.; Akanuma, S.-I.; Kubo, Y.; Tachikawa, M.; Hosoya, K.-I. Monocarboxylate transporter 12 as a guanidinoacetate efflux transporter in renal proximal tubular epithelial cells. *Biochim. Biophys. Acta BBA Biomembr.* **2020**, *1862*, 183434. [CrossRef]
10. Curt, M.J.-C.; Cheillan, D.; Briand, G.; Salomons, G.S.; Mention-Mulliez, K.; Dobbelaere, D.; Cuisset, J.-M.; Lion-François, L.; Portes, V.D.; Chabli, A.; et al. Creatine and guanidinoacetate reference values in a French population. *Mol. Genet. Metab.* **2013**, *110*, 263–267. [CrossRef]
11. Ostojic, S.M.; Ratgeber, L.; Olah, A.; Betlehem, J.; Acs, P. Guanidinoacetic acid deficiency: A new entity in clinical medicine? *Int. J. Med. Sci.* **2020**, *17*, 2544–2550. [CrossRef]
12. Ostojic, S.M. Advanced physiological roles of guanidinoacetic acid. *Eur. J. Nutr.* **2015**, *54*, 1211–1215. [CrossRef]
13. European Food Safety Authority. Panel on Additives and Products or Substances used in Animal Feed (FEEDAP). Safety and efficacy of guanidinoacetic acid for chickens for fattening, breeder hens and roosters, and pigs. *EFSA J.* **2016**, *14*, 4394. [CrossRef]
14. Ostojic, S.M.; Stajer, V.; Ratgeber, L.; Betlehem, J.; Acs, P. Guanidinoacetic Acid Consumption via Regular Diet in Adults. *Ann. Nutr. Metab.* **2021**, 1–2. [CrossRef] [PubMed]
15. Ostojic, S.M. Human gut microbiota as a source of guanidinoacetic acid. *Med. Hypotheses* **2020**, *142*, 109745. [CrossRef] [PubMed]
16. Borsook, H.; Borsook, M.E. The biochemical basis of betaine-glycocyamine therapy. *Ann. West. Med. Surg.* **1951**, *5*, 825–829.
17. Graybiel, A.; Patterson, C.A. Use of betaine and glycocyamine in the treatment of patients with heart disease: Preliminary report. *Ann. West. Med. Surg.* **1951**, *5*, 863–875. [PubMed]
18. Van Zandt, V.; Borsook, H. New biochemical approach to the treatment of congestive heart failure. *Ann. West. Med. Surg.* **1951**, *5*, 856–862.
19. Higgins, A.R.; Harper, H.A.; Kline, E.F.; Merrill, R.S.; Jones, R.E.; Smith, T.W.D.; Kimmel, J.R. Effects of creatine precursors in arthritis; clinical and metabolic study of glycocyamine and betaine. *Calif. Med.* **1952**, *77*, 14–18. [PubMed]
20. Borsook, M.E.; Billig, H.K.; Golseth, J.G. Betaine and glycocyamine in the treatment of disability resulting from acute anterior poliomyelitis. *Ann. West. Med. Surg.* **1952**, *6*, 423–427.
21. Fallis, B.D.; Lam, R.L. Betaine and glycocyamine therapy for the chronic residuals of poliomyelitis. *J. Am. Med. Assoc.* **1952**, *150*, 851–853. [CrossRef] [PubMed]
22. Watkins, A.L. Betaine and Glycocyamine in Treatment of Poliomyelitis. *N. Engl. J. Med.* **1953**, *248*, 621–623. [CrossRef] [PubMed]
23. Basom, W.C.; Breck, L.W.; Leonard, M.H. The effect of betaine and glycocyamine in the management of chronic anterior polio-myelitis. *Int. Rec. Med. Gen. Pract. Clin.* **1955**, *168*, 70–71.
24. Benassi, P. Effects of guanidinoacetic acid on the metabolism of creatine and creatinine in myopathic muscular dystrophy. *Boll. Soc. Ital. Biol. Sper.* **1954**, *30*, 365–368.
25. Dixon, H.H.; Dickel, H.A.; Shanklin, J.G.; Peterson, R.D.; West, E.S. Therapy in anxiety states and anxiety complicated by depression. *West. J. Surg. Obstet. Gynecol.* **1954**, *62*, 338–341.
26. Mune, N. Betasyamine in therapy of coronary arteriosclerosis. *Ugeskr. Laeger* **1954**, *116*, 1505.
27. Billig, H.E., Jr.; Morehouse, L.E. Performance and metabolic alterations during betaine glycocyamine feeding in myasthenia gravis. *Arch. Phys. Med. Rehabil.* **1955**, *36*, 233–236.
28. Liveksedge, L. Glycocyamine and betaine in motor-neurone disease. *Lancet* **1956**, *268*, 1136–1138. [CrossRef]
29. Aldes, J.H. Glycocyamine betaine as an adjunct in the treatment of neuromuscular disease patients. *J. Ark. Med. Soc.* **1957**, *54*, 186–194.
30. Tsubakihara, Y.; Suzuki, A.; Hayashi, T.; Shoji, T.; Togawa, M.; Okada, N. The effect of guanidinoacetic acid supplementation in patients with chronic renal failure. In *Guanidino Compounds in Biology and Medicine*; Mori, A., Ishida, M., Clark, J.F., Eds.; Blackwell Science: Japan, Tokyo, 1999; Volume 5, pp. 139–144.

31. Ostojic, S.M.; Niess, B.; Stojanovic, M.; Obrenovic, M. Creatine Metabolism and Safety Profiles after Six-Week Oral Guanidinoacetic Acid Administration in Healthy Humans. *Int. J. Med. Sci.* **2013**, *10*, 141–147. [CrossRef]
32. Ostojic, S.M.; Niess, B.; Stojanovic, M.; Obrenovic, M. Co-administration of methyl donors along with guanidinoacetic acid reduces the incidence of hyperhomocysteinaemia compared with guanidinoacetic acid administration alone. *Br. J. Nutr.* **2013**, *110*, 865–870. [CrossRef]
33. Semeredi, S.; Stajer, V.; Ostojic, J.; Vranes, M.; Ostojic, S.M. Guanidinoacetic acid with creatine compared with creatine alone for tissue creatine content, hyperhomocysteinemia, and exercise performance: A randomized, double-blind superiority trial. *Nutrition* **2019**, *57*, 162–166. [CrossRef]
34. Ostojic, S.M.; Stojanovic, M.; Drid, P.; Hoffman, J. Dose–response effects of oral guanidinoacetic acid on serum creatine, homocysteine and B vitamins levels. *Eur. J. Nutr.* **2014**, *53*, 1637–1643. [CrossRef] [PubMed]
35. Ostojic, S.M.; Vojvodic-Ostojic, A. Single-dose oral guanidinoacetic acid exhibits dose-dependent pharmacokinetics in healthy volunteers. *Nutr. Res.* **2015**, *35*, 198–205. [CrossRef]
36. Ostojic, S.M.; Stojanovic, M. Guanidinoacetic acid loading affects plasma γ-aminobutyric acid in healthy men. *Eur. J. Nutr.* **2015**, *54*, 855–858. [CrossRef]
37. Ostojic, S.; Stojanovic, M.D.; Hoffman, J. Six-Week Oral Guanidinoacetic Acid Administration Improves Muscular Performance in Healthy Volunteers. *J. Investig. Med.* **2015**, *63*, 942–946. [CrossRef] [PubMed]
38. Ostojic, S.M.; Stojanovic, M.D.; Olcina, G. Oxidant-Antioxidant Capacity of Dietary Guanidinoacetic Acid. *Ann. Nutr. Metab.* **2015**, *67*, 243–246. [CrossRef]
39. Ostojic, S.; Drid, P.; Ostojic, J. Guanidinoacetic acid increases skeletal muscle creatine stores in healthy men. *Nutrition* **2016**, *32*, 723–724. [CrossRef]
40. Ostojic, S.M.; Ostojic, J.; Drid, P.; Vranes, M.; Jovanov, P. Dietary guanidinoacetic acid increases brain creatine levels in healthy men. *Nutrition* **2017**, *33*, 149–156. [CrossRef]
41. Ostojic, S.M.; Mojsin, M.; Drid, P.; Vranes, M. Does Dietary Provision of Guanidinoacetic Acid Induce Global DNA Hypomethylation in Healthy Men and Women? *Lifestyle Genom.* **2018**, *11*, 16–18. [CrossRef]
42. Ostojic, S.M.; Stojanovic, M.; Drid, P.; Hoffman, J.R.; Sekulic, D.; Zenic, N. Supplementation with Guanidinoacetic Acid in Women with Chronic Fatigue Syndrome. *Nutrients* **2016**, *8*, 72. [CrossRef]
43. Seper, V.; Korovljev, D.; Todorovic, N.; Stajer, V.; Ostojic, J.; Nesic, N.; Ostojic, S.M. Guanidinoacetate-Creatine Supplementation Improves Functional Performance and Muscle and Brain Bioenergetics in the Elderly: A Pilot Study. *Ann. Nutr. Metab.* **2021**, *77*, 244–247. [CrossRef]
44. Ostojic, S.M. Benefits and drawbacks of guanidinoacetic acid as a possible treatment to replenish cerebral creatine in AGAT deficiency. *Nutr. Neurosci.* **2019**, *22*, 302–305. [CrossRef] [PubMed]
45. Ostojic, S.M.; Premusz, V.; Nagy, D.; Acs, P. Guanidinoacetic acid as a novel food for skeletal muscle health. *J. Funct. Foods* **2020**, *73*, 104129. [CrossRef]
46. Ostojic, S.M.; Ostojic, J.; Drid, P.; Vranes, M. Guanidinoacetic acid versus creatine for improved brain and muscle creatine levels: A superiority pilot trial in healthy men. *Appl. Physiol. Nutr. Metab.* **2016**, *41*, 1005–1007. [CrossRef] [PubMed]
47. Ostojic, S.M. Short-term GAA loading: Responders versus nonresponders analysis. *Food Sci. Nutr.* **2020**, *8*, 4446–4448. [CrossRef]
48. Obeid, R. The Metabolic Burden of Methyl Donor Deficiency with Focus on the Betaine Homocysteine Methyltransferase Pathway. *Nutrients* **2013**, *5*, 3481–3495. [CrossRef]
49. Ganguly, P.; Alam, S.F. Role of homocysteine in the development of cardiovascular disease. *Nutr. J.* **2015**, *14*, 6. [CrossRef]
50. Ostojic, S.M.; Trivic, T.; Drid, P.; Stajer, V.; Vranes, M. Effects of Guanidinoacetic Acid Loading on Biomarkers of Cardiometabolic Risk and Inflammation. *Ann. Nutr. Metab.* **2018**, *72*, 18–20. [CrossRef]
51. Ostojic, S.M.; Todorovic, N.; Stajer, V. Effect of Creatine and Guanidinoacetate Supplementation on Plasma Homocysteine in Metabolically Healthy Men and Women. *Ann. Nutr. Metab.* **2021**, *77*, 307–308. [CrossRef]
52. Zugno, A.I.; Stefanello, F.M.; Streck, E.L.; Calcagnotto, T.; Wannmacher, C.M.; Wajner, M.; Wyse, A.T. Inhibition of Na+, K+ -ATPase activity in rat striatum by guanidinoacetate. *Int. J. Dev. Neurosci.* **2003**, *21*, 183–189. [CrossRef]
53. Zugno, A.I.; Oliveira, D.L.; Scherer, E.B.S.; Wajner, M.; Wofchuk, S.; Wyse, A.T.S. Guanidinoacetate Inhibits Glutamate Uptake in Rat Striatum of Rats at Different Ages. *Neurochem. Res.* **2007**, *32*, 959–964. [CrossRef]
54. Zugno, A.I.; Stefanello, F.M.; Scherer, E.B.S.; Mattos, C.; Pederzolli, C.D.; Andrade, V.M.; Wannmacher, C.M.D.; Wajner, M.; Dutra-Filho, C.S.; Wyse, A.T.S. Guanidinoacetate Decreases Antioxidant Defenses and Total Protein Sulfhydryl Content in Striatum of Rats. *Neurochem. Res.* **2008**, *33*, 1804–1810. [CrossRef] [PubMed]
55. Hanna-El-Daher, L.; Béard, E.; Henry, H.; Tenenbaum, L.; Braissant, O. Mild guanidinoacetate increase under partial guanidinoacetate methyltransferase deficiency strongly affects brain cell development. *Neurobiol. Dis.* **2015**, *79*, 14–27. [CrossRef]
56. Neuab, A.; Neuhoffa, H.; Trubec, G.; Fehra, S.; Ullrichb, K.; Roeper, J.; Isbrandta, D. Activation of GABAA Receptors by Guanidinoacetate: A Novel Pathophysiological Mechanism. *Neurobiol. Dis.* **2002**, *11*, 298–307. [CrossRef]
57. Stromberger, C.; Bodamer, O.A.; Stöckler-Ipsiroglu, S. Clinical characteristics and diagnostic clues in inborn errors of creatine metabolism. *J. Inherit. Metab. Dis.* **2003**, *26*, 299–308. [CrossRef]
58. Almeida, L.S.; Verhoeven, N.M.; Roos, B.; Valongo, C.; Cardoso, M.L.; Vilarinho, L.; Salomons, G.S.; Jakobs, C. Creatine and guanidinoacetate: Diagnostic markers for inborn errors in creatine biosynthesis and transport. *Mol. Genet. Metab.* **2004**, *82*, 214–219. [CrossRef]

59. Ostojic, S.M.; Ostojic, J. Dietary guanidinoacetic acid does not accumulate in the brain of healthy men. *Eur. J. Nutr.* **2018**, *57*, 3003–3005. [CrossRef] [PubMed]
60. Kharbanda, K.K.; Todero, S.L.; Thomes, P.G.; Orlicky, D.J.; Osna, N.A.; French, S.W.; Tuma, D.J. Increased methylation demand exacerbates ethanol-induced liver injury. *Exp. Mol. Pathol.* **2014**, *97*, 49–56. [CrossRef]
61. Osna, N.A.; Feng, D.; Ganesan, M.; Maillacheruvu, P.F.; Orlicky, D.J.; French, S.W.; Tuma, D.J.; Kharbanda, K.K. Prolonged feeding with guanidinoacetate, a methyl group consumer, exacerbates ethanol-induced liver injury. *World J. Gastroenterol.* **2016**, *22*, 8497–8508. [CrossRef]
62. Schepers, E.; Glorieux, G.; Dou, L.; Cerini, C.; Gayrard, N.; Louvet, L.; Maugard, C.; Preus, P.; Rodriguez-Ortiz, M.; Argiles, A.; et al. Guanidino Compounds as Cause of Cardiovascular Damage in Chronic Kidney Disease: An in vitro Evaluation. *Blood Purif.* **2010**, *30*, 277–287. [CrossRef] [PubMed]
63. Mori, A.; Kohno, M.; Masumizu, T.; Noda, Y.; Packer, L. Guanidino compounds generate reactive oxygen species. *IUBMB Life* **1996**, *40*, 135–143. [CrossRef] [PubMed]
64. Takahashi, H.; Arai, B.; Koshino, C. Effects of guanidinoacetic acid, γ-guanidinobutyric acid and γ-guanidinobutyrylmethylester on the mammalian cerebral cortex. *Jpn. J. Physiol.* **1961**, *11*, 403–409. [CrossRef]
65. Van Der Voet, H.; De Mul, A.; Van Klaveren, J.D. A probabilistic model for simultaneous exposure to multiple compounds from food and its use for risk–benefit assessment. *Food Chem. Toxicol.* **2007**, *45*, 1496–1506. [CrossRef]
66. Ruzante, J.M.; Grieger, K.; Woodward, K.; Lambertini, E.; Kowalcyk, B. The use of multi-criteria decision analysis in food safety risk-benefit assessment. *Food Protect Trends* **2017**, *37*, 132–139.
67. Stratil, J.M.; Baltussen, R.; Scheel, I.; Nacken, A.; Rehfuess, E.A. Development of the WHO-INTEGRATE evidence-to-decision framework: An overview of systematic reviews of decision criteria for health decision-making. *Cost Eff. Resour. Alloc.* **2020**, *18*, 1–15. [CrossRef]
68. Schug, S.; Pogatzki-Zahn, E.; Phillips, L.D.; Essex, M.N.; Xia, F.; Reader, A.J.; Pawinski, R. Multi-Criteria Decision Analysis to Develop an Efficacy-Safety Profile of Parenteral Analgesics Used in the Treatment of Postoperative Pain. *J. Pain Res.* **2020**, *13*, 1969–1977. [CrossRef]
69. Durbach, I.N.; Stewart, T.J. Modeling uncertainty in multi-criteria decision analysis. *Eur. J. Oper. Res.* **2012**, *223*, 1–14. [CrossRef]

Article

Relationship between Dietary Creatine and Growth Indicators in Children and Adolescents Aged 2–19 Years: A Cross-Sectional Study

Darinka Korovljev [1], Valdemar Stajer [1] and Sergej M. Ostojic [1,2,*]

[1]   Applied Bioenergetics Lab, Faculty of Sport and Physical Education, University of Novi Sad, 21000 Novi Sad, Serbia; korovljev.darinka@gmail.com (D.K.); stajervaldemar@yahoo.com (V.S.)

[2]   Faculty of Health Sciences, University of Pecs, H-7621 Pecs, Hungary

*   Correspondence: sergej.ostojic@chess.edu.rs; Tel.: +381-21-450-188

Citation: Korovljev, D.; Stajer, V.; Ostojic, S.M. Relationship between Dietary Creatine and Growth Indicators in Children and Adolescents Aged 2–19 Years: A Cross-Sectional Study. *Nutrients* **2021**, *13*, 1027. https://doi.org/10.3390/nu13031027

Academic Editor: David C. Nieman and Richard B. Kreider

Received: 12 February 2021
Accepted: 16 March 2021
Published: 23 March 2021

Publisher's Note: MDPI stays neutral with regard to jurisdictional claims in published maps and institutional affiliations.

Abstract: A possible role of dietary creatine for ensuring proper growth and development remains unknown. The main aim of this cross-sectional study was to quantify the amount of creatine consumed through regular diet among U.S. children and adolescents aged 2 to 19 years and investigate the relationship between creatine intake and growth indicators, using data from the 2001–2002 National Health and Nutrition Examination Survey (NHANES). We included data for NHANES 2001–2002 respondents (4291 participants, 2133 boys and 2158 girls) aged 2 to 19 years at the time of screening, who provided valid dietary information and examination measures (standing height and weight). Individual values for total grams of creatine consumed per day for each participant were computed using the average amount of creatine (3.88 g/kg) across all sources of meat-based foods. All participants were categorized for height-for-age and BMI-for-age categories. The average daily intake of creatine across the whole sample was 1.07 ± 1.07 g (95% CI, from 1.04 to 1.10). Height, weight, and BMI were significantly different across creatine quartiles ($p < 0.001$), with all measures significantly higher in the 4th quartile of creatine intake (≥ 1.5 g/day) than those in other quartiles ($p < 0.05$). The participants from the 3rd quartile of creatine intake (0.84–1.49 g/day) were significantly different from others with respect to having lower rates of normal stature and higher rates of tall stature ($p < 0.05$). Each additional 0.1 g of creatine consumed per day increases height by 0.60 cm (simple model) or 0.30 cm (adjusted model). The daily intake of creatine from a regular diet in taller children and adolescents was higher than in shorter peers aged 2–19 years. Future research has to monitor temporal changes in growth and dietary creatine and validate our findings in interventional studies across pediatric populations.

Keywords: creatine; children; height; BMI-for-age; stature-for-age; growth

1. Introduction

Providing a diet with a variety of food components plays a crucial role in supporting normal growth in children and adolescents [1]. Besides meeting the high energy requirements and supplying essential macro and micronutrients, increasing evidence has shown that humans do have dietary needs of non-essential amino acids and by-products to fulfill their genetic potential for maximum growth, as well as optimal health and well-being [2,3]. Of particular note, adequate provision of dietary creatine may be necessary for optimizing human growth, development, and health. Creatine is a non-proteinogenic amino acid derivative that occurs naturally in the human body. About half of its daily turnover comes from a carnivorous diet while another half is synthesized endogenously in the liver, kidney, and pancreas, maintaining a creatine homeostatic load of ~2 g per day for an average person [4]. Creatine is a pleiotropic nutraceutical that plays an essential role in several metabolic pathways, including tissue bioenergetics and cellular growth. Its contribution to early growth and development is apparent throughout the literature, with de novo and dietary creatine necessary to uphold optimal placental function, maintenance of pregnancy, as well as fetal growth and maturation [5]. Dietary creatine is also proving to effectively tackle creatine deficiency syndrome, a genetic malfunction of creatine biosynthetic enzymatic machinery characterized by developmental delay and neuromuscular manifestations, allowing neonates and youngest children with the condition to thrive [6]. Still, whether dietary creatine affects growth in older children and adolescents at the populational level has been to date mostly unaddressed. A handful of previous small-scale studies demonstrated the efficacy of supplemental creatine among athletic adolescents and pediatric patients [7,8]; however, research in this area omitted to provide any data about dietary creatine intake and growth. Therefore, the purpose of this cross-sectional study was to quantify the amount of creatine consumed through regular diet among U.S. children and adolescents aged 2 to 19 years and investigate the relationship between creatine intake and growth indicators, using data from the 2001–2002 National Health and Nutrition Examination Survey (NHANES).

2. Materials and Methods

2.1. Study Population

Data for this study were obtained from the NHANES 2001–2002 round. NHANES is an annual survey research program operated through the U.S. National Center for Health Statistics, Division of Health Examination Statistics, a part of the Centers for Disease Control and Prevention (CDC). The program is created in 1959 to assess children and adults' health and nutritional status in the United States. The principal objective of NHANES is to estimate the number and percent of persons in the U.S. population and designated subgroups, with selected diseases and risk factors, and monitor trends in the prevalence, awareness, treatment, and control of selected conditions. The NHANES 2001–2002 round included 11,039 civilian, non-institutionalized male and female individuals aged 0 to 85 years. For this report, we sorted out data for respondents aged 2 to 19 years at the time of screening, who provided valid dietary information and examination measures (see below). The data collection for NHANES 2001–2002 was carried out between January 2001 and December 2002, with informed consent obtained from all participants or parents/guardians. The ethical approval to conduct the NHANES 2001–2002 was granted by the NHANES Institutional Review Board (Protocol #98-12).

2.2. Dietary Data

Dietary intake information was acquired from the NHANES 2001–2002 Dietary Data component. The dietary intake data were used to estimate the types and amounts of foods and beverages consumed during the 24 h before the in-person interview (midnight to midnight) and to calculate intakes of energy, relevant nutrients, and other food components from foods and beverages consumed; proxy interviews were conducted for survey participants less than six years of age, and assisted interviews were conducted with survey participants 6 to 11 years of age. To calculate creatine intake, we first identified meat-based protein foods using 8-digit food codes from the U.S. Department of Agriculture (USDA) using dietary interview entries for individual foods; red meat, poultry, fish, and seafood are recognized as the primary sources of dietary creatine [9,10]. We subsequently recorded the gram weight of each food component containing meat (USDA codes from 20000000 to 28522000) and calculated the net intake of those foods for each individual by merging all relevant food items on a daily basis. Individual values for total grams of creatine consumed per day for each participant were computed using the average amount of creatine (3.88 g/kg) across all meat sources, as previously described [11]. Nutrient intakes reported did not include those obtained from dietary supplements, medications, or plain drinking water. The intakes were calculated using the USDA Food and Nutrient Database for Dietary Studies, which contains the most up-to-date food composition values available for this time frame. The primary exposure used in this report was the mean grams of creatine consumed per day, and the secondary exposure included the average daily dietary intake of creatine categorized into quartiles.

2.3. Body Measurements

Data for this domain were acquired from NHANES 2001–2002 Examination Data component for body measures. At a minimum, body weight and standing height were measured for all participants. Body weight was taken in underwear using a digital scale (Mettler Toledo, Columbus, OH, USA), with readings recorded to the nearest 0.1 kg by the automated system. Standing height was measured with a stadiometer in the Frankfort horizontal plane to the nearest 0.1 cm (Seca, Chino, CA, USA). All measures were obtained by trained health technicians using a standardized protocol with calibrated equipment, and measurement components performed in a specially equipped room in the mobile examination center. Detailed NHANES body measurements component procedures are presented elsewhere [12]. Body mass index (BMI) was calculated for each participant as weight in kilograms divided by the square of height in meters (kg/m^2). For height and BMI, we also calculated individual percentiles for each participant, using CDC growth charts for ages 2–20 years [13], based on CDC LMS tables tabulated at half-month intervals. The participants were further categorized for height-for-age percentiles as short stature (height-for-age less than the 5th percentile), normal stature (5th to less than the 95th percentile), and tall stature (equal to or greater than the 95th percentile). The weight status categories and the corresponding percentiles were as follows: underweight (BMI-for-age less than the 5th percentile), normal weight (5th percentile to less than the 85th percentile), overweight (85th to less than the 95th percentile), and obese (equal to or greater than the 95th percentile). The primary outcome used in the analyses was height; the secondary outcomes include weight, BMI, height-for-age, and BMI-for-age. Height-for-age and BMI-for-age are recognized as common growth indicators in children and adolescents by the World Health Organization [14].

2.4. Statistical Analyses

Descriptive statistics were used to describe the characteristics of the study population. Data series were analyzed by Kolmogorov–Smirnov test for normality of distribution. Kruskal–Wallis non-parametric one-way ANOVA was used to compare body measures across creatine quartiles, and dietary creatine intake across different growth categories, with post hoc pairwise comparison tests employed to identify the differences between individual sample pairs. Single and multiple regression analyses with entering procedures were conducted to assess the association between creatine intake and growth indicators. The regression models were adjusted for an a priori defined set of covariates, including demographic variables (gender, race/ethnicity, annual household income), and nutritional variables (energy, total protein). Finally, chi-square cross-tabulation analysis was conducted for comparing observed frequencies for growth indicators (stature-for-age and BMI-for-age categories) across different creatine quartiles, with individual proportions compared with Z-test adjusted for standardized residuals. Data were analyzed using SPSS Statistics for Mac (Version 24.0) (IBM, Armonk, NY, USA), with the significance level set at $p < 0.05$, and all statistical tests were two-sided.

3. Results

A total of 4291 participants aged 2 to 19 years (2133 boys (49.7%) and 2158 girls (50.3%)) provided individual dietary data and were assessed for body weight and standing height. Table 1 displays the basic demographic and nutritional characteristics of the study sample. The mean age was approximately 11 years of age, and the mean caloric intake was ~2100 kcal per day. The average daily intake of creatine across the whole sample was 1.07 ± 1.07 g (95% confidence interval, from 1.04 to 1.10). The average creatine intake was categorized into quartiles, ranging from 0.00–0.28 g (1st quartile;

mean $\pm$ SD = 0.06 $\pm$ 0.09 g), 0.28–0.83 g (2nd quartile; 0.56 $\pm$ 0.16 g), 0.84–1.49 g (3rd quartile; 1.13 $\pm$ 0.19 g), and 1.50–8.28 g (4th quartile; 2.52 $\pm$ 1.06 g).

Table 1. Sample demographic and nutritional characteristics. BMI—body mass index.

Variable	
Participants, *n*	4291
Gender (%)	
Female	50.3
Age (years), mean $\pm$ SD	11.1 $\pm$ 5.3
Generation (%)	
Preschooler (2–5 years)	20.0
School-aged children (6–12 years)	33.6
Adolescent (aged 13–19)	46.4
Ethnicity (%)	
Non-Hispanic White	29.5
Non-Hispanic Black	4.7
Mexican American	30.9
Other Race	30.6
Other Hispanic	4.3
Annual household income (%)	
Less than $24,999	35.2
$25,000–$54,999	31.5
$55,000–$74,999	11.9
$75,000 and over	21.0
Body measures, mean $\pm$ SD	
Weight (kg)	48.1 $\pm$ 24.9
Height (cm)	144.7 $\pm$ 27.2
BMI (kg/m^2)	21.0 $\pm$ 5.8
Overweight (%)	32.7
Dietary intake	
Energy (kcal) mean $\pm$ SD	2104 $\pm$ 965
Protein (g) mean $\pm$ SD	103.8 $\pm$ 65.1

Height, weight, and BMI were significantly different across creatine quartiles ($p < 0.001$), with differences between individual sample pairs for each body measure depicted in Table 2. All measures were significantly higher in the 4th quartile of creatine intake than those in other quartiles ($p < 0.05$), with differences for other interquartile comparisons displayed a significant downward trend towards the first quartile of creatine intake (Q3 > Q2 > Q1) for most variables ($p < 0.05$). No significant differences were found in mean dietary creatine intake across different stature-for-age categories ($p = 0.154$), and BMI-for-age categories ($p = 0.138$) (Figure 1), although daily creatine consumption tended to be higher in children and adolescents with tall stature and obese individuals.

Table 2. Body measures across creatine quartiles.

Variable	Q1	Q2	Q3	Q4	*p*	Post-Hoc *
Height (cm)	140.7 $\pm$ 28.0	137.2 $\pm$ 27.9	145.4 $\pm$ 26.5	155.3 $\pm$ 22.4	<0.001	a b c d e f
Weight (kg)	44.7 $\pm$ 24.5	42.1 $\pm$ 23.8	49.0 $\pm$ 25.6	56.1 $\pm$ 23.4	<0.001	b c d e f
BMI (kg/m^2)	20.4 $\pm$ 5.5	20.1 $\pm$ 5.5	21.2 $\pm$ 6.0	22.1 $\pm$ 5.8	<0.001	b c d e f

Abbreviations: BMI—body mass index. * Superscript letters indicate a significant difference at $p < 0.05$ between individual sample pairs after post-hoc analysis, as follows: [a] Q1 vs. Q2, [b] Q1 vs. Q3, [c] Q1 vs. Q4, [d] Q2 vs. Q3, [e] Q2 vs. Q4, [f] Q3 vs. Q4.

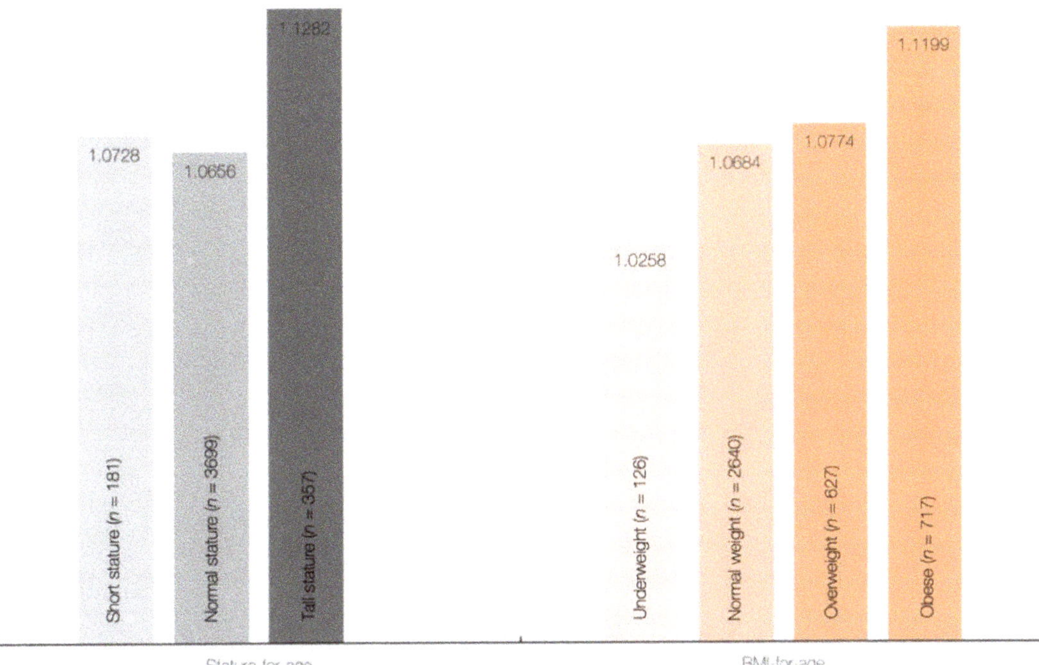

Figure 1. Dietary intake of creatine (g/day) across stature-for-age and BMI-for-age categories.

The results of regression analysis with simple and multivariable models (adjusted for gender, ethnicity, annual income, energy intake, and total protein) showed a significant association between primary exposure (dietary creatine intake) and most primary and secondary outcomes across the whole sample (Table 3), except for BMI-for-age in the crude model ($p = 0.227$), and height-for-age in the adjusted model ($p = 0.460$). Each additional 0.1 g of creatine consumed per day increases height by 0.60 cm (simple model) or 0.30 cm (adjusted model). Finally, the cross-tabulation analysis revealed a significant difference across creatine quartiles and stature-for-age categories ($p = 0.039$), with participants from Q3 are significantly different from others with respect to having lower rates of normal stature and higher rates of tall stature ($p < 0.05$) (Figure 2). No significant difference was found across creatine quartiles and BMI-for-age categories ($p = 0.483$).

Table 3. Simple and adjusted multiple regression results of the relationship between dietary creatine intake and growth indicators. BMI—body mass index.

	Crude Model			Adjusted Model		
	B	**SE**	*p*	*B*	**SE**	*p*
Height	0.24	0.57	<0.001	0.117	0.44	<0.001
Weight	0.20	0.35	<0.001	0.131	0.41	<0.001
BMI	0.12	0.08	<0.001	0.115	0.10	<0.001
Height-for-age	0.01	0.01	<0.001	−0.013	0.01	0.460
BMI-for-age	0.02	0.01	0.227	0.053	0.014	0.004

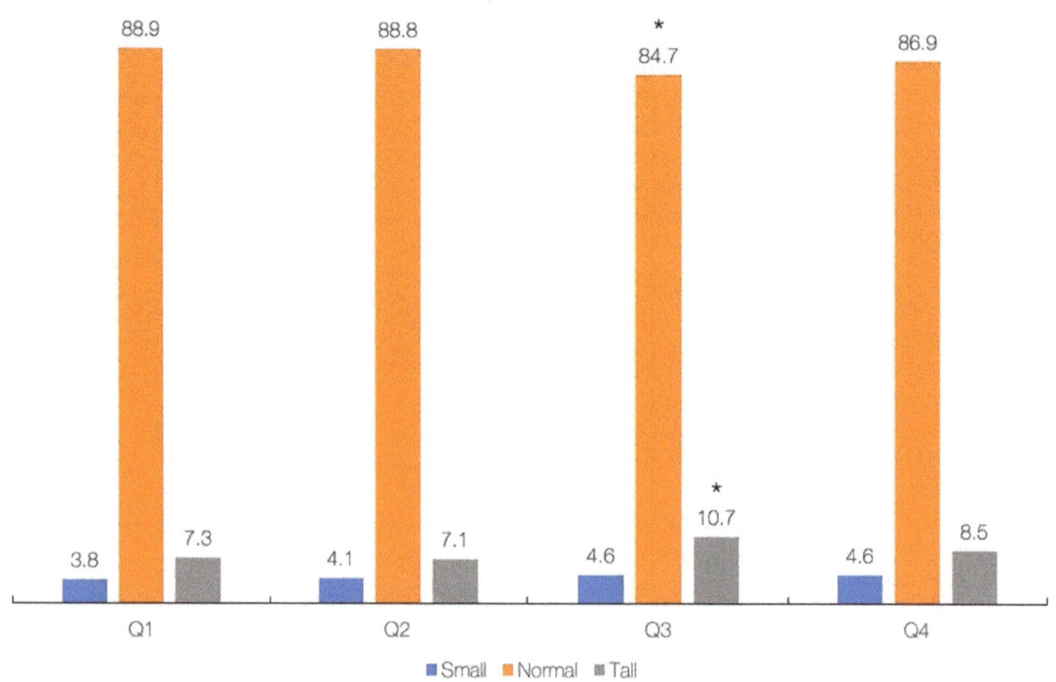

Figure 2. Dietary intake of creatine (g/day) across stature-for-age. Creatine intake across quartiles was 0.00–0.28 g (Q1), 0.28–0.83 g (Q2), 0.84–1.49 g (Q3), and 1.50–8.28 g (Q4). Asterisk (*) indicates a significant difference at $p < 0.05$ between quartiles.

4. Discussion

To our knowledge, this is the first cross-sectional populational study that evaluated the association between daily intake of creatine from regular diet and growth indicators in U.S. children and adolescents aged two years and over. We found that children and adolescents with a higher intake of creatine have higher stature, weight, and BMI compared to lower intake peers, following a stepwise rise corresponding to an incremental increase in dietary creatine intake. The mean dietary creatine intake appeared similar across different stature-for-age and BMI-for-age categories; however, the participants in the 3rd quartile of creatine intake (0.84–1.49 g/day) were significantly different from others with respect to having higher rates of tall stature. After controlling for demographic and nutritional variables, dietary creatine intake appeared positively associated with most growth indicators.

Creatine (methyl guanidino acetic acid) is a nitrogen-containing metabolite of arginine, glycine, and methionine synthesized endogenously in the human body and acquired through diet. Growing evidence suggests that creatine, along with other amino acid derivatives originating from animal-based foods, may play an essential role in human growth and development, from intrauterine growth onwards [10]. A preliminary metabolomics study advances normal maternal creatine levels during the third trimester of pregnancy as a protective factor against poor perinatal outcomes, including a small-for-gestational-age infant, preterm birth, and neonatal intensive care admission [15]. A retrospective cohort study shows that fetal growth is positively associated with creatine levels, with each unit increase in maternal creatine goes with a 1.23 unit increase in birthweight centile and a 0.11-cm increase in birth length [16]. These studies perhaps point to increased requirements for creatine due to the fetus's rapid growth and increased metabolic needs throughout pregnancy. Another prospective study demonstrated an increase in fetal brain creatine levels in healthy pregnant women between 18- and 40-weeks gestational age who underwent proton-MRS [17]. This continues for the postnatal period, with cerebral creatine increment occurs during the first months of life and growth [18]. Creatine appears to be related to the normal growth of older children as well. In a small-scale trial, 19 undernourished boys aged 8–11 years demonstrated low creatine levels due to diet deficient in creatine-building components, with nutritional supplementation provides compensation for creatine levels towards weight gain [19]. Dietary creatine normalizes creatine concentrations, makes substantial developmental progress, and attenuates clinical features in children suffering from creatine deficiency syndromes [6]. In addition, supplemental creatine elicited a significant increase in height in children with acute lymphoblastic leukemia [20], implying the importance of creatine in the growth and maturation across various pediatric settings.

Our results corroborate previous findings concerning the positive association between dietary creatine and growth in a nationally representative cohort of children and adolescents. The link between food creatine and growth remained robust even after controlling for energy and total protein, suggesting a distinctive role of dietary creatine during growth. We hypothesized that creatine might favorably affect growth by several means that involve augmented energy metabolism [21], bone mass accretion [20], and fat-free mass augmentation [22]. Additional mechanistic research is highly warranted to explore what underpins the possible benefits of dietary creatine for growth and development. Establishing a threshold of dietary creatine linked with advanced growth in the youth population remains complex. We found that children and adolescents who consume extra creatine are more likely to have tall stature, with those individuals who took 0.84 to 1.49 g of creatine per day (3rd quartile) more often having tall stature than other subgroups. Nevertheless, the proportion of tall stature among participants from the highest quartile (4th) did not differ from the lowest (1st) and lower (2nd) quartiles. This perhaps suggests that consuming additional dietary creatine (>1.5 g/day) was not necessarily accompanied by a higher prevalence of tall stature and advanced growth. The daily amount of creatine consumed by children and adolescents in the 3rd quartile (mean 1.13 g) might be appropriate to facilitate growth, keeping in mind the fact that the uptake of creatine from the diet of about 1 g per day is required to achieve steady state in the adult population [4], with youth perhaps needing more dietary creatine to sustain growth and maturation.

Since meat-based foods are the primary source of creatine [10], our findings may have implications for children and adolescents who limit the intake of those foods in their diet. Previous studies reported a higher risk of growth retardation among young children with low regular meat intake, while meat consumption was associated with a reduced likelihood of stunting [23,24]. In light of our findings, the protective effect of meat consumption against stunting should emphasize the possible role of creatine, along with other food components abundant in meat, to improve nutritional practices in the pediatric population. With this in mind, creatine intake might be recognized as another nutritional factor that positively affects growth and well-being in infants and children [10]. Addressing optimal creatine consumption could be thus considered among public measures for pediatric nutrition, either via fostering diets rich in creatine-containing foods, creatine supplementation, and/or food fortification with creatine [25]. The reference intervals of dietary creatine requirements for children and adolescents remain to be elucidated.

Study strengths include using a relatively large NHANES sample, complemented with growth indicators (e.g., stature-for-age and BMI-for-age) calculated and labeled for each participant, while the correlation between creatine intake and growth indicators controlled for main demographic and nutritional variables, including gender and total protein intake. Nevertheless, several limitations have to be considered when the study results are interpreted. The cross-sectional design of the current study prevents any conclusions about a cause and effect between creatine intake and body measures and analyzing temporal changes in those variables occurring with growth. The mean dietary creatine intake has been calculated using single-day self- or proxy-reported 24-h interviews, which could be susceptible to recall bias and cannot account for a day-to-day variation. The creatine calculation method used here omitted to consider variability in creatine content across various meat-based foods and non-meat sources; the amount of creatine may differ within animal protein subgroups [11]. In addition, NHANES data provide no measure or estimation of endogenous creatine production, a possible modifying variable that could account for a total daily creatine load (a sum of creatine synthesized de novo along with creatine consumed from a diet). Finally, nutritional conditions evaluated in this cohort (NHANES 2001–2002) might be very different from the contemporary diet, and future trials should evaluate possible time-related changes in dietary creatine intake among present-day children and adolescents, also across various countries.

5. Conclusions

Our study demonstrated that the daily intake of creatine from a regular diet in taller children and adolescents was higher than those in shorter peers aged 2–19 years. A positive association between creatine consumption and growth remained robust after adjusting for main demographic and nutritional variables. Therefore, taking enough creatine with regular food should be considered to ensure advanced growth in U.S. children and adolescents. Future research has to monitor temporal changes in growth and dietary creatine using objective biomarkers of creatine consumption and validate our findings in interventional studies across pediatric populations.

Author Contributions: D.K.: conducted research, analyzed data and performed statistical analysis, wrote paper draft, revised the paper. V.S.: conducted research, analyzed data and performed statistical analysis, revised the paper. S.M.O.: designed research (project conception, development of overall research plan, and study oversight), analyzed data and performed statistical analysis, wrote paper draft, and had primary responsibility for final content. All authors have read and agreed to the published version of the manuscript.

Funding: This research received no external funding.

Institutional Review Board Statement: The study was conducted according to the guidelines of the Declaration of Helsinki. The ethical approval to conduct the NHANES 2001–2002 was granted by the NHANES Institutional Review Board (Protocol #98-12).

Informed Consent Statement: Informed consent was obtained from all subjects involved in the study.

Data Availability Statement: Data described in the manuscript will be made available upon request.

Conflicts of Interest: D.K. and V.S. declare no conflicts of interest. S.M.O. serves as a member of the Scientific Advisory Board on creatine in health and medicine (AlzChem LLC). S.M.O. owns patent "Sports Supplements Based on Liquid Creatine" at European Patent Office (WO2019150323 A1) and active patent application "Synergistic Creatine" at UK Intellectual Property Office (GB2012773.4). S.M.O. has served as a speaker at Abbott Nutrition, a consultant of Allied Beverages Adriatic and IMLEK, and an advisory board member for the University of Novi Sad School of Medicine and has received research funding related to creatine from the Serbian Ministry of Education, Science, and Technological Development, Provincial Secretariat for Higher Education and Scientific Research, AlzChem GmbH, KW Pfannenschmidt GmbH, and ThermoLife International LLC. S.M.O. is an employee of the University of Novi Sad and does not own stocks and shares in any organization. The funders had no role in the design of the study; in the collection, analyses, or interpretation of data; in the writing of the manuscript; or in the decision to publish the results.

References

1. Corkins, M.R.; Daniels, S.R.; de Ferranti, S.D.; Golden, N.H.; Kim, J.H.; Magge, S.N.; Schwarzenberg, S.J. Nutrition in children and adolescents. *Med. Clin. N. Am.* **2016**, *100*, 1217–1235. [CrossRef]
2. Hou, Y.; Yin, Y.; Wu, G. Dietary essentiality of "nutritionally non-essential amino acids" for animals and humans. *Exp. Biol. Med.* **2015**, *240*, 997–1007. [CrossRef]
3. Hou, Y.; Wu, G. Nutritionally nonessential amino acids: A misnomer in nutritional sciences. *Adv. Nutr.* **2017**, *8*, 137–139. [CrossRef] [PubMed]
4. Brosnan, J.T.; Brosnan, M.E. Creatine: Endogenous metabolite, dietary, and therapeutic supplement. *Annu. Rev. Nutr.* **2007**, *27*, 241–261. [CrossRef]

5. Muccini, A.M.; Tran, N.T.; de Guingand, D.L.; Philip, M.; Della Gatta, P.A.; Galinsky, R.; Sherman, L.S.; Kelleher, M.A.; Palmer, K.R.; Berry, M.J.; et al. Creatine metabolism in female reproduction, pregnancy and newborn health. *Nutrients* **2021**, *13*, 490. [CrossRef] [PubMed]

6. Stöckler, S.; Hanefeld, F.; Frahm, J. Creatine replacement therapy in guanidinoacetate methyltransferase deficiency, a novel inborn error of metabolism. *Lancet* **1996**, *348*, 789–790. [CrossRef]

7. Evangeliou, A.; Vasilaki, K.; Karagianni, P.; Nikolaidis, N. Clinical applications of creatine supplementation on paediatrics. *Curr. Pharm. Biotechnol.* **2009**, *10*, 683–690. [CrossRef]

8. Jagim, A.R.; Stecker, R.A.; Harty, P.S.; Erickson, J.L.; Kerksick, C.M. Safety of creatine supplementation in active adolescents and youth: A brief review. *Front. Nutr.* **2018**, *5*, 115. [CrossRef]

9. Brosnan, J.T.; da Silva, R.P.; Brosnan, M.E. The metabolic burden of creatine synthesis. *Amino Acids* **2011**, *40*, 1325–1331. [CrossRef] [PubMed]

10. Wu, G. Important roles of dietary taurine, creatine, carnosine, anserine and 4-hydroxyproline in human nutrition and health. *Amino Acids* **2020**, *52*, 329–360. [CrossRef]

11. Bakian, A.V.; Huber, R.S.; Scholl, L.; Renshaw, P.F.; Kondo, D. Dietary creatine intake and depression risk among U.S. adults. *Transl. Psychiatry* **2020**, *10*, 52. [CrossRef] [PubMed]

12. NHANES. *Anthropometry Procedures Manual*; CDC: Atlanta, GA, USA, 2000.

13. Kuczmarski, R.J.; Ogden, C.L.; Guo, S.S.; Grummer-Strawn, L.M.; Flegal, K.M.; Mei, Z.; Wei, R.; Curtin, L.R.; Roche, A.F.; Johnson, C.L. 2000 CDC Growth Charts for the United States: Methods and development. *Vital Health Stat.* **2002**, *11*, 1–190.

14. de Onis, M.; Garza, C.; Victora, C.G.; Onyango, A.W.; Frongillo, E.A.; Martines, J. The WHO Multicentre Growth Reference Study: Planning, study design, and methodology. *Food Nutr. Bull.* **2004**, *25*, S15–S26. [CrossRef] [PubMed]

15. Heazell, A.E.; Bernatavicius, G.; Warrander, L.; Brown, M.C.; Dunn, W.B. A metabolomic approach identifies differences in maternal serum in third trimester pregnancies that end in poor perinatal outcome. *Reprod. Sci.* **2012**, *19*, 863–875. [CrossRef] [PubMed]

16. Dickinson, H.; Davies-Tuck, M.; Ellery, S.J.; Grieger, J.A.; Wallace, E.M.; Snow, R.J.; Walker, D.W.; Clifton, V.L. Maternal creatine in pregnancy: A retrospective cohort study. *BJOG* **2016**, *123*, 1830–1838. [CrossRef] [PubMed]

17. Evangelou, I.E.; du Plessis, A.J.; Vezina, G.; Noeske, R.; Limperopoulos, C. Elucidating metabolic maturation in the healthy fetal brain using 1H-MR spectroscopy. *Am. J. Neuroradiol.* **2016**, *37*, 360–366. [CrossRef] [PubMed]

18. Blüml, S.; Wisnowski, J.L.; Nelson, M.D., Jr.; Paquette, L.; Panigrahy, A. Metabolic maturation of white matter is altered in preterm infants. *PLoS ONE* **2014**, *9*, e85829. [CrossRef]

19. Koyanagi, T.; Hareyama, S.; Takanohashi, T. Effect of supplementation of vitamin, phosphorus, methionine or skim milk on the cystine content of hair, dark adaptation, creatine-creatinine excretion and growth of undernourished children. *Tohoku J. Exp. Med.* **1965**, *85*, 108–114. [CrossRef]

20. Bourgeois, J.M.; Nagel, K.; Pearce, E.; Wright, M.; Barr, R.D.; Tarnopolsky, M.A. Creatine monohydrate attenuates body fat accumulation in children with acute lymphoblastic leukemia during maintenance chemotherapy. *Pediatr. Blood Cancer* **2008**, *51*, 183–187. [CrossRef]

21. Banerjee, B.; Sharma, U.; Balasubramanian, K.; Kalaivani, M.; Kalra, V.; Jagannathan, N.R. Effect of creatine monohydrate in improving cellular energetics and muscle strength in ambulatory Duchenne muscular dystrophy patients: A randomized, placebo-controlled 31P MRS study. *Magn. Reson. Imaging* **2010**, *28*, 698–707. [CrossRef]

22. Tarnopolsky, M.A.; Mahoney, D.J.; Vajsar, J.; Rodriguez, C.; Doherty, T.J.; Roy, B.D.; Biggar, D. Creatine monohydrate enhances strength and body composition in Duchenne muscular dystrophy. *Neurology* **2004**, *62*, 1771–1777. [CrossRef] [PubMed]

23. Marquis, G.S.; Habicht, J.P.; Lanata, C.F.; Black, R.E.; Rasmussen, K.M. Breast milk or animal-product foods improve linear growth of Peruvian toddlers consuming marginal diets. *Am. J. Clin. Nutr.* **1997**, *66*, 1102–1109. [CrossRef] [PubMed]

24. Krebs, N.F.; Mazariegos, M.; Tshefu, A.; Bose, C.; Sami, N.; Chomba, E.; Carlo, W.; Goco, N.; Kindem, M.; Wright, L.L.; et al. Meat consumption is associated with less stunting among toddlers in four diverse low-income settings. *Food Nutr. Bull.* **2011**, *32*, 185–191. [CrossRef] [PubMed]

25. Ostojic, S.M. Eat less meat: Fortifying food with creatine to tackle climate change. *Clin. Nutr.* **2020**, *39*, 2320. [CrossRef]

Article

A Convergent Functional Genomics Analysis to Identify Biological Regulators Mediating Effects of Creatine Supplementation

Diego A. Bonilla [1,2,3,4,*], Yurany Moreno [1,2], Eric S. Rawson [5], Diego A. Forero [6], Jeffrey R. Stout [7], Chad M. Kerksick [8], Michael D. Roberts [9,10] and Richard B. Kreider [11]

[1] Research Division, Dynamical Business & Science Society—DBSS International SAS, Bogotá 110861, Colombia; luzyuranymoreno@gmail.com
[2] Research Group in Biochemistry and Molecular Biology, Universidad Distrital Francisco José de Caldas, Bogotá 110311, Colombia
[3] Research Group in Physical Activity, Sports and Health Sciences (GICAFS), Universidad de Córdoba, Montería 230002, Colombia
[4] kDNA Genomics®, Joxe Mari Korta Research Center, University of the Basque Country UPV/EHU, 20018 Donostia-San Sebastián, Spain
[5] Department of Health, Nutrition and Exercise Science, Messiah University, Mechanicsburg, PA 17055, USA; erawson@messiah.edu
[6] Professional Program in Sport Training, School of Health and Sport Sciences, Fundación Universitaria del Área Andina, Bogotá 111221, Colombia; dforero41@areandina.edu.co
[7] Physiology of Work and Exercise Response (POWER) Laboratory, Institute of Exercise Physiology and Rehabilitation Science, University of Central Florida, Orlando, FL 32816, USA; jeffrey.stout@ucf.edu
[8] Exercise and Performance Nutrition Laboratory, School of Health Sciences, Lindenwood University, Saint Charles, MO 63301, USA; ckerksick@lindenwood.edu
[9] School of Kinesiology, Auburn University, Auburn, AL 36849, USA; mdr0024@auburn.edu
[10] Edward via College of Osteopathic Medicine, Auburn, AL 36849, USA
[11] Exercise & Sport Nutrition Laboratory, Human Clinical Research Facility, Texas A&M University, College Station, TX 77843, USA; rbkreider@tamu.edu
[*] Correspondence: dabonilla@dbss.pro; Tel.: +57 320 335 2050

Citation: Bonilla, D.A.; Moreno, Y.; Rawson, E.S.; Forero, D.A.; Stout, J.R.; Kerksick, C.M.; Roberts, M.D.; Kreider, R.B. A Convergent Functional Genomics Analysis to Identify Biological Regulators Mediating Effects of Creatine Supplementation. *Nutrients* **2021**, *13*, 2521. https://doi.org/10.3390/nu13082521

Academic Editors: Maria Luz Fernandez and David C. Nieman

Received: 24 June 2021
Accepted: 21 July 2021
Published: 23 July 2021

Publisher's Note: MDPI stays neutral with regard to jurisdictional claims in published maps and institutional affiliations.

Abstract: Creatine (Cr) and phosphocreatine (PCr) are physiologically essential molecules for life, given they serve as rapid and localized support of energy- and mechanical-dependent processes. This evolutionary advantage is based on the action of creatine kinase (CK) isozymes that connect places of ATP synthesis with sites of ATP consumption (the CK/PCr system). Supplementation with creatine monohydrate (CrM) can enhance this system, resulting in well-known ergogenic effects and potential health or therapeutic benefits. In spite of our vast knowledge about these molecules, no integrative analysis of molecular mechanisms under a systems biology approach has been performed to date; thus, we aimed to perform for the first time a convergent functional genomics analysis to identify biological regulators mediating the effects of Cr supplementation in health and disease. A total of 35 differentially expressed genes were analyzed. We identified top-ranked pathways and biological processes mediating the effects of Cr supplementation. The impact of CrM on miRNAs merits more research. We also cautiously suggest two dose–response functional pathways (kinase- and ubiquitin-driven) for the regulation of the Cr uptake. Our functional enrichment analysis, the knowledge-based pathway reconstruction, and the identification of hub nodes provide meaningful information for future studies. This work contributes to a better understanding of the well-reported benefits of Cr in sports and its potential in health and disease conditions, although further clinical research is needed to validate the proposed mechanisms.

Keywords: creatine kinase; systems biology; bioinformatics; MAP kinase signaling system; sodium-chloride-dependent neurotransmitter symporters; signal transduction

1. Introduction

Creatine (Cr), or alpha-methylguanidinoacetic acid (PubChem CID: 586), and its phosphorylated form, phosphocreatine (PCr), are essential molecules for the optimal functioning of tissues with high and fluctuating energy demands [1]. They provide an evolutionary advantage via several creatine kinase (CK) isozymes that functionally connect places of adenosine triphosphate (ATP) synthesis with sites of ATP consumption (the CK/PCr system) [2]; therefore, Cr and PCr are physiologically essential for life through a rapid and localized support of energy- and mechanical-dependent processes (i.e., cell survival, growth, proliferation, differentiation, and migration or motility) [3,4]. For a recent and comprehensive review of the Cr metabolism, please refer to [3].

The CK/PCr system can be enhanced through supplementation with creatine monohydrate (CrM), which is the most studied, safe, and effective nutritional supplement to optimize physical performance [5–8], with potential benefits in health and disease [9–19]. It seems that the elevation of intracellular PCr concentration causes a greater capacity for phosphagens to contribute to energy metabolism, while working to reduce the accumulation of Pi and H^+ and improving Ca^{2+} handling as important mediators of fatigability in young and older adults [20–23]. This has previously been reported in vivo and in vitro after CrM supplementation [24–29]. Notwithstanding, a cellular environment rich in high-energy phosphates might also trigger downstream signaling pathways that are sensitive to energy changes by activating secondary messengers and protein kinases [30,31]. In this sense, several clinical trials [32–35] and research in animal models [36–38] have revealed that CrM administration regulates the expression of particular genes or proteins using low-throughput screening. In accordance with Kontou et al. [39], individual experiments can only identify targeted regulators (based on prior knowledge), with limited comprehension of how the biological system works. As a result, data integration from multiple experimental studies and public repositories is necessary to understand the function of biological entities (e.g., genes, proteins) and their expression patterns under certain conditions. Additionally, 'omics' technologies in conjunction with the advance of bioinformatics tools allow for data integration and the extraction of biologically relevant information, such as identifying biomarkers and regulatory components within a network [40]. Consequently, the use of a systems biology approach guarantees identification, at a systems scale, of the molecular signatures of cellular processes, molecular interactions, and relevant metabolic pathways present in the complex physiological responses or diseases with multi-factorial underpinnings [41,42]. Among the different approaches, convergent functional genomics (CFG) is an interesting methodological approach for the integrative analysis of molecular mechanisms by combining multiple lines of genomic evidence from different species [43]. CFG takes advantage of the conserved nature of metabolic circuits between several species (e.g., rodents and humans) [44] to provide relevant information about the structural and functional changes within the cell when there is not enough available data or the experiments are difficult, if not impossible, to conduct in humans (e.g., those in brain tissue) [45]. While human data increase the clinical relevance (specificity), animal model data increase the ability to identify the signal (sensitivity). Combined together, we enhance our ability to distinguish signals from noise, even with limited cohorts and datasets [46]; therefore, CFG is useful for identifying novel candidate genes and pathways for specific phenotypes [47–50] and compound-mediated gene regulation [51,52]. It is necessary to point out the complementary features of low- and high-throughput analysis, given that subsequent validation of the identified metabolic hubs requires the high sensitivity, lower noise, and reproducibility of low-throughput techniques (e.g., post-transcriptional regulation assessment, targeted-molecules expression, protein–protein interactions) [53].

In the context of novel models of physiological regulation, the concept of allostasis was developed. This highlights the importance of the anticipation of needs (such as the timely provision of energy and adequate environmental conditions) for the functional and structural stability of cells through adaptive changes [54]. Although efforts have been made to integrate the different points of metabolic regulation to explain the positive effects of CrM supplementation on physical performance [55–57] and health or therapeutic benefits [58–60], no systems biology analysis has been performed to date. Readers are encouraged to refer to the comprehensive reviews in the Special Issue on "Creatine Supplementation for Health and Clinical Diseases" to learn more about the effects of CrM supplementation [9]. The use of a systems biology approach might contribute to better comprehension of the molecular, cellular, tissue, and systemic effects of CrM and its applications to health and disease; thus, the aim of this study was to perform for the first time a CFG analysis to identify relevant pathways and biological processes mediating the effects of Cr supplementation in health and disease. This secondary analysis of the available data on differentially expressed genes after CrM administration in humans and mice will provide meaningful information for future studies.

2. Methods

2.1. Functional Genomic Analysis

2.1.1. Search and Sources of Evidence

In order to collect the gene expression data, two public repositories (NCBI Gene Expression Omnibus (GEO: http://www.ncbi.nlm.nih.gov/geo (accessed on 14 January 2021)) and the ArrayExpress Archive (https://www.ebi.ac.uk/arrayexpress/) (accessed on 14 January 2021) were searched following the PRISMA statement guidelines [61] and international recommendations [62]. The following Boolean algorithm was used in the GEO repository: *("creatine"[MeSH Terms] OR creatine monohydrate[All Fields]) AND supplementation[All Fields]*, while the free term "creatine" was used in the ArrayExpress Archive. Repositories were accessed on 14 January 2021, although an updated search was conducted prior to manuscript submission. We also contacted corresponding authors (e-mail communication) to obtain missing raw data when no record was found in the repositories.

2.1.2. Eligibility Criteria

The inclusion of gene expression data for the CFG analysis was restricted to the following: (1) original experimental studies that screened for genes differing between Cr and controls; (2) raw data deposited in the NCBI GEO or the ArrayExpress Archive; (3) expression data obtained from any tissue or cell; (4) human or mouse experimental models. The search excluded data obtained from the combination of Cr with other compounds. Information was extracted from each identified record and reported in a table, including the organism, reference, GEO accession number, sample type, numbers of cases and controls, and platform.

2.1.3. Analysis of Differentially Expressed Genes and Convergence

The GEO2R web application [63] was used to identify differentially expressed (DE) genes in the datasets for human and mouse models (freely available at http://www.ncbi.nlm.nih.gov/geo/geo2r/, accessed on 18 January 2021). For the analysis with the GEO2R tool, samples from Cr treatment were taken as the experimental group, while untreated samples were taken as the control group. GEO2R provides the following summary statistics as generated by 'limma', which performs the top table computation to extract a table of the top-ranked genes, including the adjusted p-values and raw p-values; moderated t-statistics, B-statistics, or log-odds that the gene is differentially expressed; the log2-fold changes between pairs of experimental conditions; and moderated F-statistics (which combines the t-statistics for all the pair-wise comparisons into an overall test of significance for that gene) [64].

A CFG approach was used to identify a list of candidate genes with multiple lines of evidence from humans and mouse models. After the GEO2R analysis, the gene names were extracted from the differentially expressed genes based on the adjusted p-values using the Benjamini–Hochberg method. An online tool developed by Bioinformatics and Evolutionary Genomics at VIB/UGent (Gent, Belgium) (http://bioinformatics.psb.

ugent.be/webtools/Venn) was used to identify convergence in the list of genes from these datasets (accessed on 29 January 2021).

2.1.4. Functional Enrichment Analysis

Gene Ontology (GO) enrichment analysis of the list of genes from the convergent analysis was performed with the gene set enrichment analysis tool for mammalian gene sets from Enrichr, which is a comprehensive resource for curated gene sets that was published in 2013 [65] (http://amp.pharm.mssm.edu/Enrichr, accessed on 12 May 2021). An exploratory analysis with the DAVID tool (https://david.ncifcrf.gov/, accessed on 12 May 2021) showed similar results to top terms sorted by p-value ranking using the Enrichr platform, an outcome that was somewhat expected given the use a scoring method similar to both the hypergeometric test and Fisher's exact test; however, the advantages of Enrichr over other tools are its comprehensiveness, ease of use, interactive visualization of the results, and the calculation of a combined score, which has been demonstrated to recover more of the 'correct' terms compared with the other methods (e.g., the over-representation analysis, the Jaccard distance, or the number of overlapping genes) [66]. Enrichr also retrieves the computational predictions of interactions between the list of genes from the convergent analysis and small non-coding microRNAs (known as miRNAs) using the miRTarBase library [67], which contains experimentally validated miRNA–mRNA interactions. We reported the categories ranked as statistically significant based on adjusted *p*-values using an inherent z-score permutation background correction on Fisher's exact test. The bioinformatics tools were accessed from 20 February to 12 May 2021.

2.1.5. Upstream Regulatory Pathway Analysis

We utilized the differentially expressed genes from the CFG analysis to infer the upstream regulatory networks using the computational pipeline of the eXpression2Kinases (X2K) Web (freely available at https://amp.pharm.mssm.edu/X2K/, accessed on 12 May 2021) [68]. X2K Web is an enhanced algorithm that performs an enrichment analysis to prioritize transcription factors that most likely regulate the observed changes in mRNA expression (ChEA and PWM), which was previously validated by Chen et al. [42]. It then utilizes known protein–protein interactions (PPIs) to connect the identified transcription factors to form a subnetwork. Finally, kinase enrichment analysis (KEA) is performed to prioritize protein kinases known to phosphorylate substrates within the subnetwork of transcription factors and the intermediate proteins that connect them. Top kinases and regulatory proteins (ranked by hypergeometric *p*-values, which indicate unusual differential expression in the database) were contrasted to individual low-throughput experimental reports available in the literature.

3. Findings

3.1. Selection of Gene Expression Datasets

After retrieving the records using the Boolean algorithm, we obtained 33 datasets from the GEO database, 20 from the ArrayExpress Archive, and one from the hand searching process in Google Scholar. The expression data for one study that assessed changes in muscle transcriptome after Cr supplementation during knee immobilization in healthy young men [69] was not published within a public repository (unsuccessful e-mail communication). After the assessment for inclusion criteria, only four records were suitable for eligibility, although one study was excluded from the analysis because a treatment combination with Cr, tacrine, and moclobemide was carried out. The overall procedure for data extraction is shown in the PRISMA flow chart (Figure 1).

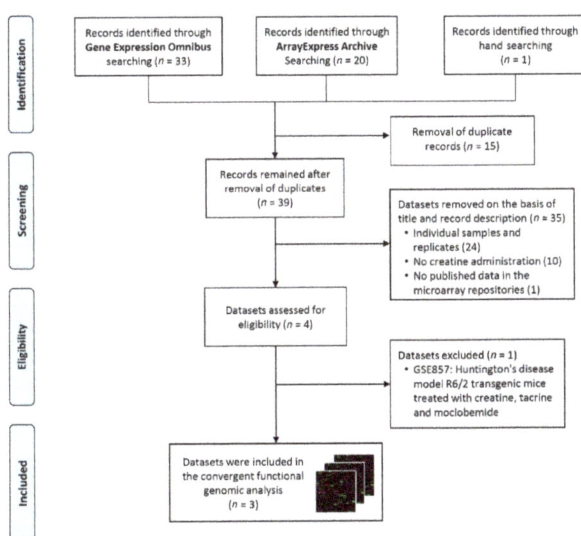

Figure 1. PRISMA flow diagram.

We downloaded the gene expression datasets GSE7877 [70] (human), GSE5140 [71] and GSE42356 [72] (mouse) from the GEO. Table 1 shows the characteristics of the studies and datasets that were included in our CFG analysis.

Table 1. Datasets used in the convergent functional genomic analysis.

Organism	Reference	GEO Number	Design	Creatine	Control	Platform
Human	[70]	GSE7877	Expression profiling of *vastus lateralis* muscle in a randomized, placebo-controlled, crossover, double-blind design in young, healthy, non-obese men supplemented with CrM vs. placebo (dextrose) for ten days	12	12	Buck Institute_Homo sapiens_25K_verC
Mouse	[71]	GSE5140	Analysis of brains of C57Bl/6J animals fed a Cr-supplemented diet for six months	6	7	Affymetrix Mouse Genome 430 2.0 Array
Mouse	[72]	GSE42356	3T3 fibroblasts overexpressing CRT were treated with 5mM CrM	3	3	Illumina MouseWG-6 v2.0 expression beadchip

3.2. Analysis of Differentially Expressed Genes

After the analysis with the GEO2R tool, the differentially expressed genes were selected using the adjusted $p < 0.05$ as the cut-off. Subsequently, these lists of genes were filtered for duplicates and empty gene names to retrieve 112 genes from GSE5140 (Table S1) and 11 genes from GSE42356 (Table S2). The available expression profile analyzed by Safdar et al. [70] with 34 genes from GSE7877 was used for subsequent analysis (Table S3)

3.3. Analysis of Convergence between Datasets

The results of the convergent analysis are shown in Figure 2. The human and mice datasets converged in polyglutamine-binding protein 1 (*PQBP1*). This gene encodes a scaffold protein involved in the regulation of transcription, alternative mRNA splicing (via spliceosome), innate immunity, and neuron projection development (UniProtKB-O60828). The list of genes used for the subsequent enrichment analysis encompassed two genes from the converged results and 33 differentially expressed genes from the GSE7877 dataset (*Homo sapiens*).

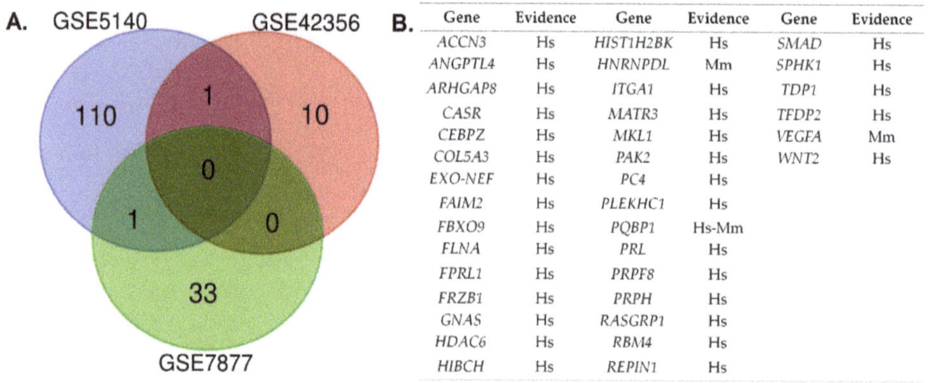

Figure 2. Convergent analysis of the differentially expressed genes obtained from the three datasets: (**A**) Venn diagram comparing differentially expressed genes identified in the included datasets (adjusted $p < 0.05$); (**B**) list of genes derived from the convergent functional genomics analysis in human and mouse models. Hs: *Homo sapiens*; Mm: *Mus musculus*.

3.4. Functional Enrichment Analysis

Functional annotation revealed that the 35 selected genes are involved in the regulation of apoptosis, proliferation–differentiation, and replication–transcription processes. Table 2 shows the functional annotation GO terms (biological process, molecular function, and cellular component) and the prediction analysis of interactions between the input list and miRNAs.

Table 2. Functional annotation and miRNA enrichment analysis of the list of genes from the convergent analysis.

Category	GO ID	Term	Adjusted *p*-Value
	GO:0030879	Mammary gland development	0.00538
	GO:0043069	Negative regulation of programmed cell death	0.01899
Biological Process	GO:0045765	Regulation of angiogenesis	0.03483
	GO:0071542	Dopaminergic neuron differentiation	0.05505
	GO:0001934	Positive regulation of protein phosphorylation	0.05505
	GO:0005664	Nuclear origin of replication recognition complex	0.2386
	GO:0005682	U5 snRNP	0.2386
Cellular Component	GO:0031904	Endosome lumen	0.2386
	GO:0046540	U4/U6 x U5 tri-snRNP complex	0.2386
	GO:0005637	Nuclear inner membrane	0.2386

Table 2. *Cont.*

Category	GO ID	Term	Adjusted *p*-Value
	GO:0035198	miRNA binding	0.05290
	GO:0036002	Pre-mRNA binding	0.05290
	GO:0048365	Rac GTPase binding	0.05290
Molecular Function	GO:0031593	Polyubiquitin-modification-dependent protein binding	0.05290
	GO:0001664	G-protein-coupled receptor binding	0.05290

Database	miRBase Accession	Description	Adjusted *p*-Value
	MIMAT0000416	Mature sequence *Homo sapiens* miR-1-3p	0.0942
	MIMAT0000275	Mature sequence *Homo sapiens* miR-218-5p	0.1405
	MIMAT0000447	Mature sequence *Homo sapiens* miR-134-5p	0.1513
miRTarBase	MIMAT0022487	Mature sequence *Homo sapiens* miR-5694	0.1961
	MI0000542	Stem loop sequence *Homo sapiens* miR-320a	0.2492

3.5. Upstream Regulatory Pathway Analysis

The upstream pathway analysis resulted in the proto-oncogene c-Fos (FOS), specificity protein 1 (SP1), nuclear transcription factor Y subunit alpha (NFYA), the enhancer of zeste homolog 2 (EZH2) and suppressor of zeste 12 (SUZ12) complex, zinc finger MIZ-type containing 1 (ZMIZ1), Myc proto-oncogene protein (MYC), protein max (MAX, also known as Myc-associated factor X), and erythroid transcription factor (GATA1) as the top transcription factors that most likely regulate the observed changes in gene expression after Cr administration. Our kinase enrichment analysis showed that mitogen-activated protein kinase 14 (MAPK14, also known as p38), cyclin-dependent kinases (CDKs), casein kinase II subunit alpha (CSNK2A1, also known as CK2A1), extracellular signal-regulated kinases (ERKs), protein kinase B (Akt/PKB), and c-Jun N-terminal kinases (JNKs) are the top signaling pathways that activate kinases known to phosphorylate substrates within our subnetwork of transcription factors and intermediate proteins (Figure 3).

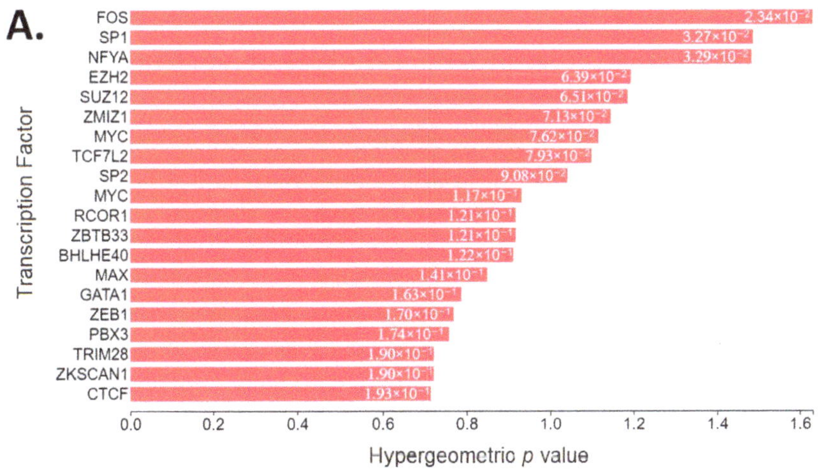

Figure 3. *Cont.*

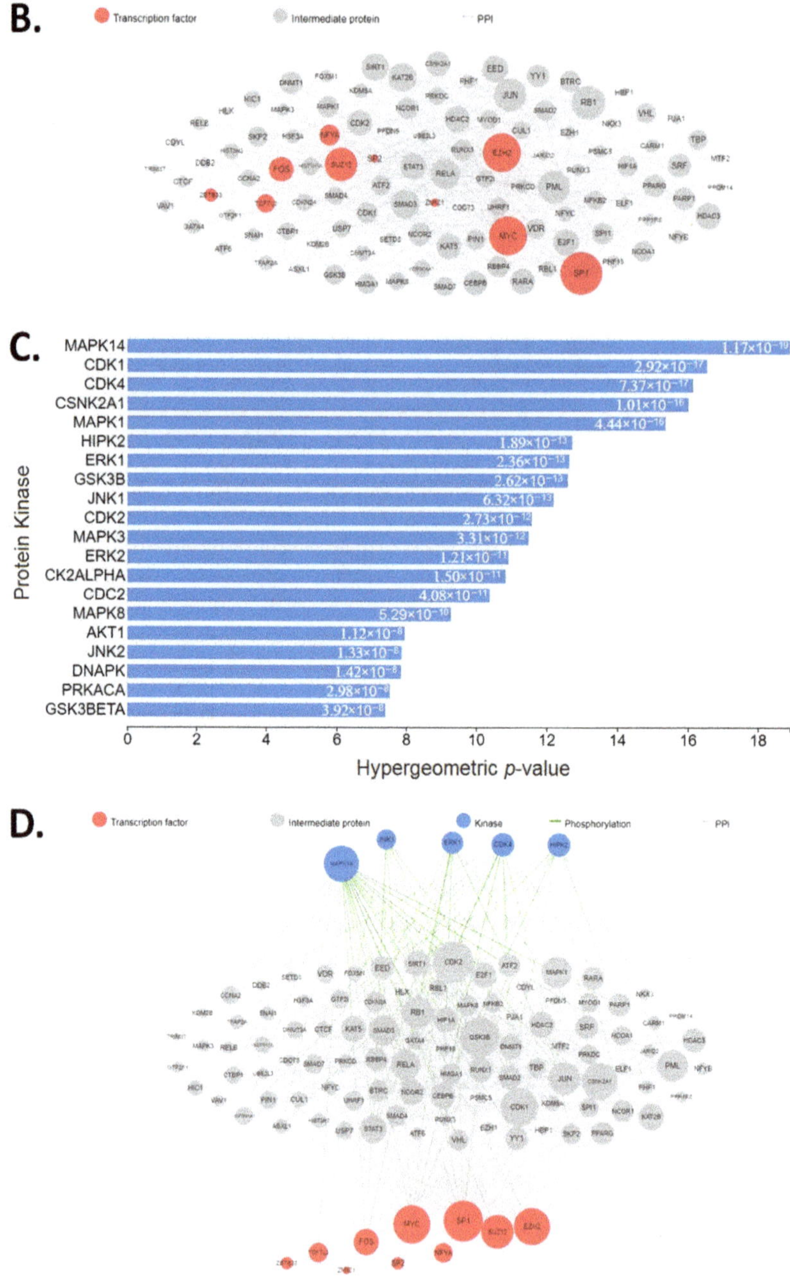

Figure 3. Upstream regulatory pathway analysis of the 35 selected differentially expressed genes: (**A**). transcription factor enrichment analysis; (**B**) protein–protein interaction expansion; (**C**) kinase enrichment analysis; (**D**) eXpression2Kinases network. Figures were obtained from the X2K Web (https://amp.pharm.mssm.edu/X2K/, accessed on 12 May 2021) after running the upstream pathway analysis of the selected genes as the input gene list.

4. Discussion

From a systems biology perspective, the analysis of interactions and characteristics of the system's components (e.g., molecular system bioenergetics, genome-wide gene-expression profiling, and pathway identification) facilitates the deciphering of the action mechanisms of various system-level properties and biological functions [73]. The CFG approach has been described as a Bayesian way of cross-validating biological findings while reducing uncertainty. In addition, the bioinformatics enrichment of groups of converged genes leads to insights into pathways and mechanisms that may be involved in different phenotypes [74]. This methodology has been successfully utilized, even with limited size cohorts and datasets, in the study of gene associations with chronic fatigue syndrome [50]; biomarker identification for suicidality [48], attention-deficit–hyperactivity disorder [49], stress-related psychiatric disorders [75], and mood disorders [52]; and the discovery of novel candidate genes and signaling pathways for epileptogenesis [47] and retinol or retinoic acid exposure [51]. In order to understand the underlying biological processes and mechanisms mediating the effects of Cr administration, we performed a CFG analysis on differentially expressed genes in humans and mice for the first time. Under the allostasis paradigm (that is, adaptation to changes through mechanisms that alter the set point of metabolic or physiological variables) [76], the CK/PCr system should be seen as an essential mechanism for life (cell survival, growth, proliferation, differentiation, and migration or motility). We previously suggested that the CK/PCr system works as a dynamic biosensor of chemomechanical energy transduction [3] with 'concurrent reactions with exchange motifs' [77] that might account for the wide range of diseases after alterations in intracellular Cr concentrations. In agreement with the established regulation of cellular allostasis through a complex balance of subcellular energy production and cellular mechanics [78], our CFG and enrichment analysis showed several biological regulators related to energy metabolism (extra- and intramitochondrial pathways) and cytoskeletal machinery (motor and cytolinker proteins). This makes the CK/PCr system a fractal model that can be used to exemplify the cytoskeleton-mediated, energy-driven, mechanoadaptative processes of the cells. This is the first time an integrative approach has been implemented to elucidate how the enhanced Cr metabolism (via CrM administration) is directly involved in the cellular adaptations through a complex balance of subcellular energy production and cellular mechanics.

4.1. Biological Pathways Mediating Effects

The functional annotation of the 35 selected differentially expressed genes after CrM administration showed that these are involved in anti-apoptotic processes, cell differentiation, and positive regulation of protein phosphorylation cascades. It must be noted that these genes perform functions throughout the cell (nuclear origin of replication recognition complex (nucleus), U5 snRNP (transcription complex), and endosome lumen (vesicles)) through either receptor activation and secondary messengers (G-protein-coupled receptor binding, Rac GTPase binding, and polyubiquitin modification-dependent protein binding) or direct regulation of de novo biosynthesis (miRNA binding and pre-mRNA binding) (Table 2). After transcription factor enrichment analysis, the proteins that most likely regulate the observed changes in gene expression after Cr administration are involved in cell survival, apoptosis, proliferation, differentiation, migration, and the cytoskeletal structure. To highlight this, among the top ranked transcription factors (Figure 3A) of our analysis we found: (i) FOS is a G0/G1 switch regulatory protein that heterodimerizes with members of the JUN family of transcription factors to form AP-1 complexes and regulate signal transduction, cell proliferation, and differentiation [79]; (ii) SP1 was the first to be cloned and characterized of the specificity protein–Krüppel-like factor (Sp/KLF) family of transcription factors, which are involved in multitude of cellular pathways and processes by regulating tissue- and developmental-stage-specific gene expression [80]; (iii) NFYA is one of the three subunits of a highly conserved and ubiquitously expressed heterotrimeric transcription factor that regulate gene expression at promoter regions [81]; (iv) EZH2 and

SUZ12 are essential for the integrity of the polycomb group complexes, the expression of which rises at the G1/S phase transition [82,83]; (v) ZMIZ1 regulates the activity of various transcription factors, including the androgen receptor, Smad3/4, and p53 [84]; and (vi) MYC is implicated in cell growth [85], proliferation, differentiation [86], and cellular adhesion and migration [87,88], and may dimerize with MAX to bind DNA and exert its effects [89].

The analysis of upstream regulatory signaling pathways resulted in MAPKs, CDKs, CSNK2A1, and Akt/PKB being the most representative after Cr administration. Although no clinical research has been conducted to study the changes in CSNK2A1 or CDK expression after Cr supplementation, one would expect a possible regulatory effect given that both proteins play an essential role during cell cycle progression and differentiation [90,91]. It is interesting to note that CK and CSNK2A1 activities vary similarly during muscle cell differentiation, with CSNK2A1 being dispensable for (i) the maintenance of the myogenic identity, (ii) the expression of early myogenic markers and late muscle-specific gene expression, and (iii) the control of myoblast fusion [92]. CSNK2 has been reported to control the Janus kinase–signal transducer and activator of transcription (JAK/STAT) signaling pathway [93], which is the principal mechanism for a wide range of growth factors and cytokines where other pathways such as MAPKs and PI3K/Akt are involved [94]. This JAK/STAT pathway activation might regulate somatic growth via binding transcriptional enhancers in the *IGF1* locus [95]. In addition, CSNK2A and CDKs share a common target, the integrator of neurotransmitters called dopamine- and cAMP-regulated phosphoprotein-32 (DARPP-32) [96]. DARPP-32 interacts with β-adducin at the cytoskeleton to mediate rapid environmental effects on neurons [97], which might explain its upregulation after Cr treatment to promote differentiation and maturation of neurons [98]. Again, we highlight the dynamic biosensor role of the CK/PCr system in this type of chemomechanical energy transduction process. Recently, CSNK2A has been reported to be a critical element of the Th17/Treg cell balance and differentiation [99], meaning more research is needed to unravel the mechanisms that explain the effects of Cr on chromatin remodeling in immune cells. Since immunity regulation goes beyond the scope of this review, please refer to [15] for a recent comprehensive summary of the current findings and future directions in this regard.

The phosphorylation cascades of MAPKs are key components of extracellular signal transduction, with important roles in cell survival, proliferation, and differentiation. The signal begins at the activation of receptor tyrosine kinases (RTKs) and other transmembrane receptors via guanine nucleotide exchange factors (GEFs, such as the son of sevenless [SOS]) that lead to the active form of the small G-proteins Ras [100]. After their activation, the Ras variants interact with MAPK kinase kinases (MAPKKK, also called MAP3Ks or MEKKs, with Raf being the most representative member) to activate MAPK kinase (MAPKK, also known as MEK), then finally MAPKs through scaffold and adaptor proteins such as KSR, JIP, and OSM [101]. There are three main MAPK families with different isoforms, namely the ERKs, the JNKs, and the p38 MAPKs (p38 $\alpha/\beta/\gamma/\delta$) [102]. Besides the hormone regulation of RTKs (e.g., growth hormone, IGF-1), the MAPK pathway may be activated by energy-driven mechanosensing [103,104] and osmosensing [105] mechanisms. Alternatively, activation of the IGF-1 receptor may stimulate the PI3K/Akt/mTORC1 pathway, which has a crucial role in protein synthesis via RPS6K and 4E-BP1 [106]. Our functional annotation analysis indicated that G-protein-coupled receptor binding ($p = 0.002921$) and Rac GTPase binding (a subgroup of the Ras superfamily of GTP hydrolases) ($p = 0.003855$) are among the top molecular functions of the differentially expressed genes that change after Cr administration. It is possible that the experimentally reported activation of MAPKs after Cr administration might be due to mechano- and osmosensing mechanisms driven by the optimization of the CK/PCr system. For instance, CrM supplementation for ten days has resulted in the mRNA overexpression of *SPHK1* (osmosensing gene) and various MAPKs in healthy men [70] (Appendix A Figure A1).

Interestingly, the results of our kinase enrichment analysis and the identification of hub nodes (downstream effectors of the MAPK and IGF-1/PI3K/Akt pathways) (Figure 3) are in high agreement with the available low-throughput, high-sensitivity experimental in vitro and in vivo evidence after Cr administration, such as qRT-PCR, Western blotting, and electrophoretic mobility shift assay. Several human and animal studies have shown that Cr brings higher growth hormone concentrations [107]; overexpression of IGF-1 [34,36,108,109]; upregulation and higher activity of Akt/PKB [70,110,111]; downregulation of myostatin and increase in GASP-1 [35]; overexpression of p38α (also called MAPK14) [69,112,113]; overexpression and higher activity of RPS6K and 4E-BP1 [37,38,114]; upregulation of myocyte enhancer factor isoforms [111,115]; overexpression of myogenic regulatory factors, such as MyoD, Myogenin, Myf5, and MRF4/Myf6/Herculin [32,33,36]; and overexpression of myosin heavy chain (MHC) isoforms [116]. Remarkably, our kinase enrichment analysis (Figure 3C–D) showed high agreement with the protein kinase content after CrM supplementation, as reported by Safdar et al. (Figure A1) [70]. All aforementioned experimental evidence validates the power of the computational prediction of the multi-species convergent analysis, which highlights the need to include this before performing low-throughput experimental analysis. It must be noted that many of the signaling pathways that might be activated after Cr administration (Figure 4) follow 'concurrent reaction with exchange motifs', which are characterized by the high level of enzymes transferring phosphorus-containing groups (EC 2.7) [77].

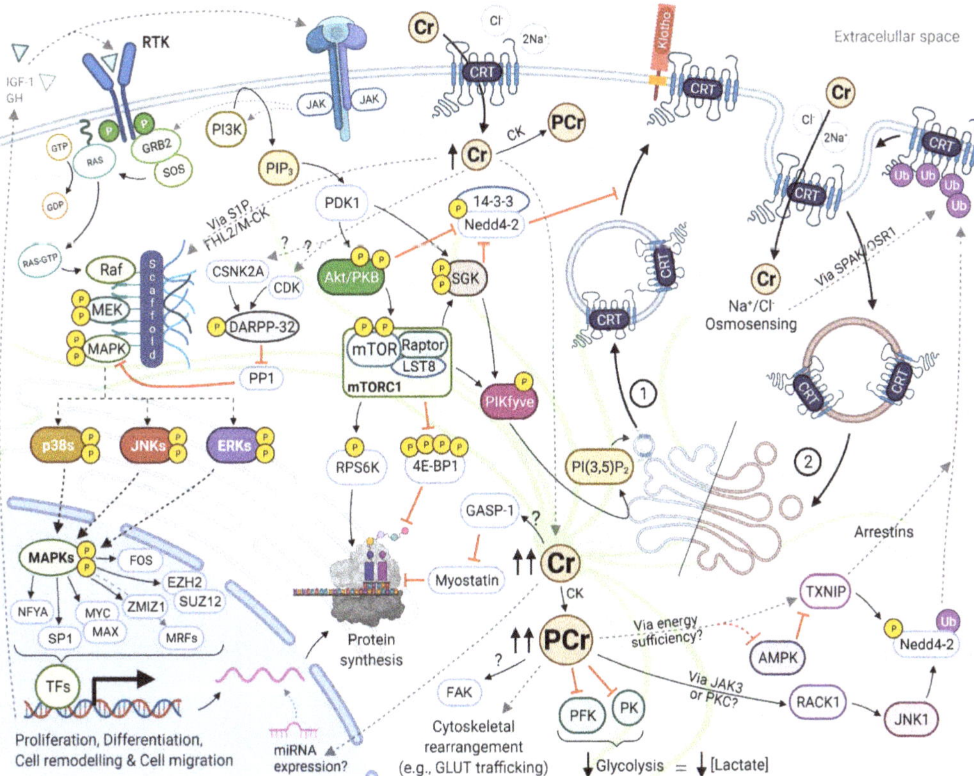

Figure 4. Bioinformatics- and knowledge-based pathway reconstruction after Cr supplementation. This is a representation of pathways interactions based on the results of our enrichment analysis of differentially expressed genes after increasing cellular Cr concentration and the available experimental evidence. This functional network follows the 'bio-logic' (integration of bottom-up and top-down directions) of the genotype–outcome interaction. MAPK activation can occur via osmosensing pathways that activate Ras/Raf (e.g., S1P/SPHK1) and mechanosensing pathways that involve mechanical and energy

optimization of the cytoskeleton (e.g., Four-and-a-Half Lim 2 is an important mechanosensor that triggers hypertrophy in response to strain and also docks key metabolic enzymes involved in the energy transduction process, such as M-CK, adenylate kinase, and phosphofructokinase). Several subunits of the protein complexes and the architecture of the cytoskeleton are not depicted for readability. We cautiously suggest two dose–response functional pathways for the regulation of the Cr uptake: a kinase-driven mechanism as a result of the initial Cr-enriched environment, which is more related to the anterograde trafficking via endolysosome-specific phosphoinositide compounds (1); and a ubiquitin-driven mechanism that controls the excessive Cr uptake, which is more related to the retrograde trafficking via clathrin-dependent and clathrin-independent processes (2). Interlinking protein filaments of the cytoskeleton are represented with lighter-colored lines in the background. See the sections of the manuscript for rationale, citations, and more abbreviations. Dashed arrows represent multiple steps. AMPK: AMP-activated protein kinase; CK: creatine kinase; GASP-1: growth and differentiation factor (GDF)-associated serum protein-1; GDP: guanosine diphosphate; GH: growth hormone; GLUT: glucose transporter; GRB2: growth factor receptor-bound protein 2; GTP: guanosine triphosphate; IGF-1: insulin-like growth factor-1; LST8: target of rapamycin complex subunit LST8; MRFs: myogenic regulatory factors; Nedd4-2: E3 ubiquitin–protein ligase NEDD4-like; OSR1: oxidative-stress-responsive kinase 1; PDK1: phosphoinositide-dependent kinase-1; PFK: phosphofructokinase; PI(3,5)P$_2$: phosphatidylinositol 3,5-bisphosphate; PIKfyve: 1-phosphatidylinositol 3-phosphate 5-kinase; PI3K: phosphoinositide 3-kinase; PIP$_3$: phosphatidylinositol (3,4,5)-trisphosphate; PK: pyruvate kinase; SGK: serum- and glucocorticoid-regulated kinase; SPAK: SPS1-related proline–alanine-rich kinase; RACK1: receptor for activated C kinase 1; RTK: receptor tyrosine kinases; TFs: transcription factors. Source: created by the authors (D.A.B.) with BioRender—https://biorender.com/ (accessed on 10 May 2021).

4.2. Creatine and miRNAs

Interesting results were obtained from Enrichr using the miRTarBase data library to analyze the miRNAs interactions. Regarding cancer, recent evidence has suggested that Cr supplementation might have a carcinogenic effect [117]; nevertheless, contrary to this hypothesis, the formation of carcinogenic heterocyclic amines is unrelated to CrM supplementation [118], and even clinical research has demonstrated a potential anti-tumor progression [119]. In fact, downregulation of the CK isozymes and low levels of PCr and Cr are associated with sarcoma and adenocarcinoma progression [120]. Moreover, Cr has been reported to enhance the anti-cancer effects of methylglyoxal in chemically induced muscle cancerous cells in vitro and in sarcoma mouse cells in vivo [121]. In our bioinformatics-assisted review of the Cr metabolism [3], we discussed the observed latency towards reliance on glycolysis at high physical work rates after Cr administration, which might explain the observed reduction in lactate accumulation (possibly via inhibition of phosphofructokinase [122] and pyruvate kinase [123]) and the potential anti-tumor progression of Cr and its derivatives [124–126]. In this sense, the human muscle transcriptome analysis performed by Safdar et al. [70] demonstrated that a 10-day CrM supplementation period decreased the phosphofructokinase mRNA content by 21% versus placebo. For a recent review on the role of Cr in T cell anti-tumor immunity and cancer immunotherapy, please refer to Li and Yang [127].

Our enrichment analysis revealed for the first time that Cr administration might impact certain miRNAs that control cancer progression and muscle function. We are aware that more experimental evidence is needed to the identify clinical effects, therapeutic targets, and potential biomarkers in health and disease states, especially in certain cancer phenotypes where a Cr-dependent tumor progression has been proposed based on preclinical data [128]. Table 3 briefly describes the functions of several of the top-ranked miRNAs that were computationally predicted to interact with the list of selected differentially expressed genes altered after CrM administration.

Table 3. Information of predicted miRNAs interacting with the list of genes from the CFG analysis.

MicroRNA	Relevant Information
miR-1-3p	Suppresses the proliferation of hepatocellular carcinoma [129] and slows the proliferation and invasion of gastric [130] and lung adenocarcinoma [131].
miR-218-5p	Significantly upregulated during myogenic differentiation after activating the IGF-1 and MAPK/ERK pathways [132].
miR-134-5p	Lower levels are found in prostate cancer compared to benign prostatic hyperplasia [133]. In addition, it might have neuroprotective effects by regulating the miR-134-5p/CREB pathway in both humans and mice [134].
miR-5694	Mediates downregulation of AF9 (a subunit of the super elongation complex and associates with the histone methyltransferases) and provides metastatic advantages in basal-like breast cancer cells [135].
miR-320a	Although associated with certain types of cancer, it has been shown to inhibit the proliferation and progression of melanoma [136] and gastric adenocarcinoma [137].
miR-200b-3p	Higher expression is found in prostate cancer compared to benign prostatic hyperplasia [133].
miR-126a-3p	It targets low-density lipoprotein-receptor-related protein 1 and blocks WNT signaling, which partially explain the anti-tumor effects of curcumin [138].
miR-378a-3p	Exhibits tumor-suppressive and anti-metastatic effects in esophageal squamous cell carcinoma [139] and glioblastoma multiforme [140]; however, miR-378a might also have a pro-angiogenic effect on myoblasts and control vascularization of skeletal muscle [141].

Since most of the identified miRNAs repress cancer progression, it would be appropriate to investigate the effects of CrM administration on these biological elements to evaluate at which point Cr might regulate excessive proliferation. These post-transcriptional gene regulators open a new field of research regarding the therapeutic role of CrM supplementation on health-related conditions. It needs to be noted that the osmosensing activation of MAPK via Sphk1/S1P is related to several miRNAs [142], as well as the notion that the effect of Cr on focal adhesion kinase [38,70] (Figure A1) might control muscle cell differentiation through a small set of miRNAs that are connected to the focal adhesion signaling during muscle regeneration, as was reported recently [143]. Further experimental validation is warranted in the future.

4.3. The Regulation of the Creatine Transporter

In agreement with Santacruz et al. (2015), the elucidation for creatine transporter (CRT, also known as SLC6A8) regulation merits further study given its importance to the optimal function of several human tissues [144]. It needs to be noted that Cr belongs to a set of seven putative systems biomarkers for Alzheimer's disease, Parkinson's disease, and amyotrophic lateral sclerosis [145], which highlights its remarkable influence on neuron survival and function. Since it is not able to increase itself to the same extent as skeletal muscle, most of the research on enhancing Cr uptake using intermediate compounds (e.g., guanidinoacetate) or derivatives (e.g., Cr ethyl ester, dodecyl Cr ester, cyclocreatine, and Cr gluconate) is focused on the brain [18] and the heart [146]. Additionally, even if intracellular Cr levels increase after CrM (or Cr analogous) supplementation, the uptake is limited by CRT downregulation due to mechanisms that are not fully understood. CrM supplementation has been shown to reduce the maximum rate of CRT activity (V_{max}) with no changes in the CRT expression [144,147,148], reinforcing the hypothesis of endosomal internalization.

Based on the results of our CFG and enrichment analysis, we performed a pathway reconstruction using well-characterized and experimentally validated PPIs to identify the possible mechanistic progression for the trafficking regulation of the CRT after CrM supplementation. By combining a knowledge-based approach [149] and the results of our upstream regulatory analysis to build the pathway, we obtained two dose–response and complementary functional networks: (i) a kinase-driven mechanism as a result of the initial Cr-enriched environment (more related to the anterograde trafficking); (ii) a ubiquitin-driven mechanism that controls the excessive Cr uptake (more related to the retrograde trafficking) (Figure 4). A third possible mechanism encompasses the initial upregulation of the full-length CRT by splice variants (SLC6A8C and SLC6A8D), considering that our functional annotation revealed that the differentially expressed genes perform a function in the U5–snRNP complex, which is involved in the pre-mRNA splicing events; however, this will not be further discussed because of the lack of experimental support beyond the study by Ndika et al. [150].

The enhancement of the CK/PCr system (Cr-enriched environment) via CrM supplementation leads to the previously discussed activation of mechano- and energy-sensing pathways, such as MAPKs and IGF-1/PI3K/Akt. The CRT function may be susceptible to regulation by several kinases of these cascades [151]. Two-electrode voltage clamp recordings have revealed that mTOR [152], SGK1/3 [153], and 1-phosphatidylinositol 3-phosphate 5-kinase (PIKfyve) [154] enhance Cr uptake. It has been suggested that this process might be mediated by the production of phosphatidylinositol-3,5-bisphosphate (PI(3,5)P$_2$), which is implicated in cytoskeleton rearrangement and cellular motility and provides spatial and temporal control for membrane trafficking [155]. For example, GLUT-4 translocation is regulated by PI(3,5)P$_2$ concentrations, given that PIKfyve is localized in a subpopulation of highly dynamic vesicles containing this transporter and may be activated by Akt/PKB after triggering the insulin receptor pathway [156]. Additionally, SGK can phosphorylate the E3 ubiquitin–protein ligase NEDD4-like (Nedd4-2), which reduces the ability of Nedd4-2 to interact with target proteins due to the interaction of the phosphorylated form with its scaffolding protein 14-3-3 [157]. This positive feedforward mechanism has been reported in the trafficking regulation of other membrane transporters [158,159]. Interestingly, Klotho protein, a transmembrane protein determinant of aging and life span, upregulates the activity of CRT by stabilizing the carrier protein in the cell membrane [160]. Klotho serves as a powerful regulator of cellular transport across the plasma membrane [161] and is associated with the activation of ERK 1/2 and SGK1 signaling cascades [162]. In recent years, Klotho has been linked to glycolysis inhibition and anti-cancer activity [163], which deserves more research due to its effect on Cr uptake. Further research is also needed to understand the molecular processes that account for the increased CRT expression and Cr uptake after activation of the estrogen-related receptor α [164].

As part of the hormetic dose–response, it is also plausible that the continuous increase in Cr concentration activates cellular responses that negatively regulate Cr uptake via possible ubiquitin-related mechanisms. Zervou et al. (2016) showed that very high Cr concentrations ($>$160 nmol·mg^{-1} protein) might lead to impaired energy metabolism in cardiomyocytes of transgenic mice overexpressing CRT, in what the authors defined as a 'substrate-rich but energy-poor heart' [165]; thus, similar to other protein carriers, downregulation of the CRT might result from nutrient sensitivity, energy sufficiency, and osmotic changes [166]. Thioredoxin-interacting protein (Txnip), a member of a novel family of proteins termed α-arrestin or arrestin-domain-containing proteins that possess homology to β-arrestins [167], has been identified as the only gene upregulated after saturating cells with Cr in vitro (Figure 2). This seems to be relevant in vivo, since higher mRNA (57.6%) and protein (28.7%) levels of Txnip were found in animal models overexpressing CRT, while Cr-deficient mouse hearts showed lower mRNA (39.71%) expression in comparison to wild-type animals [72]. Thioredoxin (*Txn1*) is a small and ubiquitously expressed protein, which in conjunction with thioredoxin reductase, reduces free radical oxygen species, protein disulfides, and other oxidants [168]. The Txnip binds to Txn1 to exert critical functions in terms of energy metabolism (e.g., increase redox stress, inhibit cellular glucose uptake, among others) [169]. Notably, the arrestin domains are the crucial structural elements in the metabolic functions of proteins such as Txnip [170], given that they operate as multifaceted protein trafficking adaptors that serve as signaling scaffolds of multiple protein kinases. They bind to membrane cargo proteins and interact with the adaptor protein complex 2, which is the second most abundant component of clathrin-coated vesicles, in order to promote endocytic turnover of their cargos [171]. As the archetypal β-arrestin, Tnxip has two major structural domains: the NH_2 domain for protein–protein interaction, with SH_3-binding proteins and MAP3Ks [172]; and the COOH domain, with proline-rich motifs that not only bind to both adaptin and clathrin heavy chains but also interact and recruit WW-domain-containing E3 ubiquitin ligases, such as Nedd4-2, to ubiquitinate proteins and promote internalization to endosomes [173]. In addition, α-arrestins are likely to utilize other mechanisms to mark cargo for internalization by clathrin-independent endocytosis [174]. It must be noted that JNK1, a top-ranked protein kinase from our enrichment analysis, may phosphorylate and activate Nedd4-2 [175,176]. Interestingly, upstream activators of JNK1 such as JAK2 [177], JAK3 [178], and PKC [179] have been reported to be negative regulators of the CRT; therefore, endosomal trafficking of CRT might be highly regulated by Cr concentration and energy sufficiency. Likewise, regulation of GLUT proteins by Txnip depends on glucose and energy-sensing pathways, taking into account that if the AMP/ATP ratio increases, AMPK becomes activated and phosphorylates Txnip to induce its degradation. This results in the repression of GLUT protein endocytosis and promotes glucose uptake to relieve energy stress [180]. Regarding Cr uptake, contradictory findings have been found regarding the regulation of CRT by AMPK [181,182], which deserves more research; the biologically conserved response to the mechanical stress induced by altered osmolarity [166] might also contribute to the control of Cr uptake (as a cellular hyperhydrating agent). Besides the kinase activity and the possible ubiquitination of CRT via Txnip/Nedd4-2, the JNK is considered the main mechanism for osmosensing signal transduction [183]. Furthermore, it has been demonstrated that SPAK and OSR1 are negative regulators of the CRT [184]. These kinases are part of the osmosensing WNK-SPAK/OSR1 pathway, considered the master regulator of cation-chloride cotransporters [185], such as the CRT. This osmosensing regulation of clathrin-mediated endocytosis is preserved among several species [186–189].

It is worth mentioning the successful strategies carried out in recent years to rescue misfolded and endoplasmic-reticulum-trapped CRT variants with the use of pharmacochaperones such as the FDA-approved 4-phenylbutyrate [190]. Several mutations of the CRT may result in transporter malfunction due to misfolding followed by impaired expression or reduced trafficking to the plasmalemma surface. Conformational changes of the CRT might trigger quality control mechanisms involving N-glycosylation (e.g., unfolded protein response) [191]. For a recent comprehensive review on this topic and novel therapeutic strategies related to Cr deficiency syndrome, please refer to [192].

5. Limitations, Strengths, and Future Directions

This study should be interpreted in light of the following limitations and strengths: (i) A very low number of 'omics' studies have evaluated the effects of CrM supplementation in humans, meaning future (epi)genomics, transcriptomics, proteomics, and metabolomics studies in this area are needed. Specifically, only one human dataset was deposited in the repositories (GEO), showing again that more publicly available datasets are needed to strengthen the current findings; however, the CFG approach has been highlighted for its ability to distinguish signals from noise, even with limited cohorts and datasets [74,193]. (ii) The findings of the CFG, the bioinformatics enrichment analysis, as well as the conclusions from in vitro and in vivo animal models should be interpreted with caution, given they might not fully reflect cellular changes in humans after CrM supplementation but rather represent a mechanistic insight into cellular dynamics and proof-of-concept evidence to develop novel therapeutic strategies through the assessment of pharmacological activators and inhibitors. (iii) Despite the limitations in the number of expression datasets, the results obtained in the enrichment analysis (Enrichr and X2K) were contrasted and supported by low-throughput, high-sensitivity experimental evidence that has identified targeted genes and proteins related to the activated pathways after CrM supplementation in humans. However, experimental validation of candidate genes, protein, and miRNA hubs from our analysis is warranted in the future. (iv) This study is a clear example of the powerful features of the 'omics' high-throughput technologies and bioinformatics tools and may represent a workflow for future studies that analyze emergent nutrients with potential applications in sports and health or disease. In fact, understanding the interactions between system components and their regulatory aspects allow following a 'biologic' interpretation that is different and much more valuable than the common top-down or bottom-up approaches.

Future studies on identifying biological regulators of CrM supplementation on health and disease include: (i) the changes in miRNAs content (and other regulator non-coding RNAs); for instance, the use of a small-interfering RNA against Txnip resulted in increased Cr uptake [72], meaning further work will contribute to elucidating the mechanisms of the Txnip–CRT interaction and its potential therapeutic use as a next-generation medicine [194]; (ii) the pharmacochaperones (e.g., 4-phenylbutyrate) and their safety and efficacy to treat pathologies associated with the Cr deficiency syndrome; (iii) the implementation of a systems biology approach as a necessary and unavoidable process to study other metabolic networks of high complexity, such as the Cr metabolism; (iv) the integration of (epi)genomics, transcriptomics, phosphoproteomics, and metabolomics analyses (multi-omic analysis). In this sense, a very recent tool called Causal Oriented Search of Multi-Omic Space (COSMOS) [195] was developed to extract mechanistic insights in a more consistent and robust manner. This opens up an exciting field of research with multiple applications in several human conditions.

6. Conclusions

The CK/PCr system acts as a hub for chemomechanical energy transduction (i.e., dynamic biosensor) during the cellular allodynamic states. For the first time, a CFG with enrichment analysis was performed to identify relevant pathways and biological processes mediating the effects of Cr in health and disease. The results of our secondary analysis of available gene expression data showed that several cytoskeleton-mediated, energy-driven, mechanoadaptative processes possibly account for the wide range of effects and diseases after alterations of intracellular Cr concentrations. Additionally, we cautiously suggest two dose–response and complementary functional networks for the negative regulation of CRT after the continuous increase in Cr concentration: (i) a kinase-driven mechanism responsible for anterograde trafficking during the initial Cr-enriched environment; (ii) a ubiquitin-driven mechanism that controls the excessive Cr uptake. In general, Cr metabolism encompasses one of the most complex and dynamic networks, meaning further studies will be expected to outline evidence detailing the positive roles of CrM supplementation in other uncovered areas of health and disease. This work might contribute to a better understanding of the well-reported benefits of Cr in sports and its potential in health and disease, although more research is warranted to validate some of the proposed mechanisms.

Supplementary Materials: The following are available online at https://www.mdpi.com/article/10.3390/nu13082521/s1, Table S1: GSE5140 Top genes, Table S2: GSE42356 Top genes, Table S3: GSE7877 Top genes.

Author Contributions: Conceptualization, D.A.B. and D.A.F.; methodology, formal analysis and visualization, D.A.B.; writing—original draft preparation, D.A.B.; writing—critical review and editing, Y.M., E.S.R., D.A.F., J.R.S., C.M.K., M.D.R. and R.B.K.; project administration, D.A.B. All authors have read and agreed to the published version of the manuscript.

Funding: The APC was funded by the Exercise and Sport Nutrition Laboratory (Texas A&M).

Institutional Review Board Statement: Not applicable.

Informed Consent Statement: Not applicable.

Data Availability Statement: The datasets supporting the reported results in this article are available in the GEO (Gene Expression Omnibus) repository (https://www.ncbi.nlm.nih.gov/geo/): GSE7877 (https://www.ncbi.nlm.nih.gov/geo/query/acc.cgi?acc=GSE7877), GSE5140 (https://www.ncbi.nlm.nih.gov/geo/query/acc.cgi?acc=GSE5140), GSE42356 (https://www.ncbi.nlm.nih.gov/geo/query/acc.cgi?acc=GSE42356). All datasets were accessed on 14 January 2021.

Acknowledgments: The authors would like to thank Theo Wallimann for providing comments and suggestions for this manuscript.

Conflicts of Interest: D.A.B. serves as science product manager for MTX Corporation®, a company that produces, distributes, sells, and does research on dietary supplements (including creatine) in Europe, has acted as a scientific consultant for MET-Rx and Healthy Sports in Colombia, and has received honoraria for speaking about creatine at international conferences. Y.M. declares no conflicts of interest. E.S.R. has conducted industry-sponsored research on creatine and received financial support for presenting on creatine at industry-sponsored scientific conferences. D.A.F. has been previously supported by grants from MinCiencias but not related to creatine. J.R.S. has conducted industry-sponsored research on creatine and other nutraceuticals over the past 25 years. Further, J.R.S has also received financial support for presenting on the science of various nutraceuticals, except creatine, at industry-sponsored scientific conferences. C.M.K. has consulted with and received external funding from companies who sell certain dietary ingredients and has received remuneration from companies for delivering scientific presentations at conferences. M.D.R. has received academic and industry funding related to dietary supplements, served as a non-paid consultant for the industry, and received honoraria for speaking at various conferences. R.B.K. has conducted industry-sponsored research on creatine, received financial support for presenting on creatine at industry-sponsored scientific conferences, and has served as an expert witness on cases related to creatine. Additionally, R.B.K. serves as chair of the "Creatine in Health" scientific advisory board for AlzChem Tostberg GmbH, while D.A.B., E.S.R., J.R.S., C.M.K. and M.D.R. serve as members of this board.

Appendix A

Kinase	Abbreviation	Type	Fold change
Hematopoietic progenitor kinase 1	HPK1	PSTK	6.87
Focal adhesion kinase	FAK	PYK	6.64
ZIP kinase	ZIP	PSTK	4.65
Janus kinase 1	JAK1	PTK	4.43
Protein kinase B alpha	PKBa	PSTK	4.21
G protein-coupled receptor kinase 2	GRK2	PSTK	4.03
Protein kinase G1 (cGMP-dependent protein kinase)	PKG1	PSTK	3.65
Protein kinase C epsilon	PKCe	PSTK	3.51
Protein kinase A (cAMP-dependent protein kinase)	PKA	PSTK	2.82
Protein kinase C alpha	PKCa	PSTK	2.82
3-phosphoinositide dependent protein kinase 1	PDK1	PSTK	2.78
Cyclin-dependent kinase 7	CDK7	PSTK	2.72
MAP kinase kinase 2	MEK2	PTYK	2.67
Extracellular-regulated MAP kinase 6	ERK6	PSTK	2.50
Extracellular regulated kinase 2	ERK2	PSTK	2.02
Extracellular regulated kinase 1	ERK1	PSTK	1.97
Cyclin-dependent kinase 9	CDK9	PSTK	1.89
p38 MAP kinase	p38 MAPK	PSTK	1.81
Oncogene Raf 1	RAF1	PSTK	1.79
Cancer Osaka thyroid oncogene	COT	PSTK	1.73
MAP Kinase Kinase 1	MEK1	PTYK	1.67
v-mos Moloney murine sarcoma viral oncogene homolog 1	MOS1	PSTK	1.52
MAP Kinase Kinase 6	MEK6	PTYK	1.52
Protein kinase C zeta	PKCz	PSTK	1.30
Casein kinase 2	CK2	PSTK	1.22
MAP Kinase Kinase 7	MEK7	PTYK	1.18
Germinal centre kinase	GCK	PSTK	0.96
Casein kinase 1 delta	CK1d	PSTK	0.96
Yamaguchi sarcoma viral oncogene homolog 1	YES1	PYK	0.76
Ribosomal S6 kinase 1	RSK1	PSTK	0.54

Figure A1. Effects of ten days of CrM supplementation in healthy young men (with no changes in their dietary intake) on protein kinase content using the Kinetworks KPKS 1.0 Protein Kinase screen. PSTK = protein–serine/threonine kinase; PTYK = protein–threonine/tyrosine kinase; PYK = protein–tyrosine kinase; PTK = protein–threonine kinase. Significant increases (red; $p < 0.05$), trends for increases (green; $p < 0.09$), and no change (grey) in protein content with CrM supplementation vs. placebo are indicated. Reproduced from Safdar et al. (2008).

References

1. Wallimann, T. The extended, dynamic mitochondrial reticulum in skeletal muscle and the creatine kinase (CK)/phosphocreatine (PCr) shuttle are working hand in hand for optimal energy provision. *J. Muscle Res. Cell Motil.* **2015**, *36*, 297–300. [CrossRef] [PubMed]
2. Wallimann, T.; Wyss, M.; Brdiczka, D.; Nicolay, K.; Eppenberger, H.M. Intracellular compartmentation, structure and function of creatine kinase isoenzymes in tissues with high and fluctuating energy demands: The 'phosphocreatine circuit' for cellular energy homeostasis. *Biochem. J.* **1992**, *281*, 21–40. [CrossRef] [PubMed]
3. Bonilla, D.A.; Kreider, R.B.; Stout, J.R.; Forero, D.A.; Kerksick, C.M.; Roberts, M.D.; Rawson, E.S. Metabolic Basis of Creatine in Health and Disease: A Bioinformatics-Assisted Review. *Nutrients* **2021**, *13*, 1238. [CrossRef] [PubMed]
4. Wallimann, T.; Harris, R. Creatine: A miserable life without it. *Amino Acids* **2016**, *48*, 1739–1750. [CrossRef]
5. Kreider, R.B.; Kalman, D.S.; Antonio, J.; Ziegenfuss, T.N.; Wildman, R.; Collins, R.; Candow, D.G.; Kleiner, S.M.; Almada, A.L.; Lopez, H.L. International Society of Sports Nutrition position stand: Safety and efficacy of creatine supplementation in exercise, sport, and medicine. *J. Int. Soc. Sports Nutr.* **2017**, *14*, 18. [CrossRef]
6. Maughan, R.J.; Burke, L.M.; Dvorak, J.; Larson-Meyer, D.E.; Peeling, P.; Phillips, S.M.; Rawson, E.S.; Walsh, N.P.; Garthe, I.; Geyer, H.; et al. IOC consensus statement: Dietary supplements and the high-performance athlete. *Br. J. Sports Med.* **2018**, *52*, 439–455. [CrossRef]
7. Bonilla, D.A.; Kreider, R.B.; Petro, J.L.; Romance, R.; García-Sillero, M.; Benítez-Porres, J.; Vargas-Molina, S. Creatine Enhances the Effects of Cluster-Set Resistance Training on Lower-Limb Body Composition and Strength in Resistance-Trained Men: A Pilot Study. *Nutrients* **2021**, *13*, 2303. [CrossRef]
8. Antonio, J.; Candow, D.G.; Forbes, S.C.; Gualano, B.; Jagim, A.R.; Kreider, R.B.; Rawson, E.S.; Smith-Ryan, A.E.; VanDusseldorp, T.A.; Willoughby, D.S.; et al. Common questions and misconceptions about creatine supplementation: What does the scientific evidence really show? *J. Int. Soc. Sports Nutr.* **2021**, *18*, 13. [CrossRef] [PubMed]
9. Kreider, R.B.; Stout, J.R. Creatine in Health and Disease. *Nutrients* **2021**, *13*, 447. [CrossRef]
10. Wallimann, T.; Tokarska-Schlattner, M.; Schlattner, U. The creatine kinase system and pleiotropic effects of creatine. *Amino Acids* **2011**, *40*, 1271–1296. [CrossRef]
11. Wallimann, T.; Hall, C.H.T.; Colgan, S.P.; Glover, L.E. Creatine Supplementation for Patients with Inflammatory Bowel Diseases: A Scientific Rationale for a Clinical Trial. *Nutrients* **2021**, *13*, 1429. [CrossRef] [PubMed]
12. Balestrino, M. Role of Creatine in the Heart: Health and Disease. *Nutrients* **2021**, *13*, 1215. [CrossRef] [PubMed]
13. Smith-Ryan, A.E.; Cabre, H.E.; Eckerson, J.M.; Candow, D.G. Creatine Supplementation in Women's Health: A Lifespan Perspective. *Nutrients* **2021**, *13*, 877. [CrossRef] [PubMed]
14. Clarke, H.; Hickner, R.C.; Ormsbee, M.J. The Potential Role of Creatine in Vascular Health. *Nutrients* **2021**, *13*, 857. [CrossRef] [PubMed]
15. Bredahl, E.C.; Eckerson, J.M.; Tracy, S.M.; McDonald, T.L.; Drescher, K.M. The Role of Creatine in the Development and Activation of Immune Responses. *Nutrients* **2021**, *13*, 751. [CrossRef] [PubMed]
16. Jagim, A.R.; Kerksick, C.M. Creatine Supplementation in Children and Adolescents. *Nutrients* **2021**, *13*, 664. [CrossRef]
17. Solis, M.Y.; Artioli, G.G.; Gualano, B. Potential of Creatine in Glucose Management and Diabetes. *Nutrients* **2021**, *13*, 570. [CrossRef]
18. Roschel, H.; Gualano, B.; Ostojic, S.M.; Rawson, E.S. Creatine Supplementation and Brain Health. *Nutrients* **2021**, *13*, 586. [CrossRef]
19. Muccini, A.M.; Tran, N.T.; de Guingand, D.L.; Philip, M.; Della Gatta, P.A.; Galinsky, R.; Sherman, L.S.; Kelleher, M.A.; Palmer, K.R.; Berry, M.J.; et al. Creatine Metabolism in Female Reproduction, Pregnancy and Newborn Health. *Nutrients* **2021**, *13*, 490. [CrossRef]
20. Li, J.L.; Wang, X.N.; Fraser, S.F.; Carey, M.F.; Wrigley, T.V.; McKenna, M.J. Effects of fatigue and training on sarcoplasmic reticulum Ca^{2+} regulation in human skeletal muscle. *J. Appl. Physiol.* **2002**, *92*, 912–922. [CrossRef]
21. Sundberg, C.W.; Hunter, S.K.; Trappe, S.W.; Smith, C.S.; Fitts, R.H. Effects of elevated H+ and Pi on the contractile mechanics of skeletal muscle fibres from young and old men: Implications for muscle fatigue in humans. *J. Physiol.* **2018**, *596*, 3993–4015. [CrossRef] [PubMed]
22. Cheng, A.J.; Place, N.; Westerblad, H. Molecular Basis for Exercise-Induced Fatigue: The Importance of Strictly Controlled Cellular Ca2+ Handling. *Cold Spring Harb. Perspect. Med.* **2018**, *8*. [CrossRef] [PubMed]
23. Sundberg, C.W.; Fitts, R.H. Bioenergetic basis of skeletal muscle fatigue. *Curr. Opin. Physiol.* **2019**, *10*, 118–127. [CrossRef] [PubMed]
24. Mesa, J.L.M.; Ruiz, J.R.; Gonzalez-Gross, M.M.; Gutierrez Sainz, A.; Castillo Garzon, M.J. Oral Creatine Supplementation and Skeletal Muscle Metabolism in Physical Exercise. *Sports Med.* **2002**, *32*, 903–944. [CrossRef] [PubMed]
25. Jones, A.M.; Wilkerson, D.P.; Fulford, J. Influence of dietary creatine supplementation on muscle phosphocreatine kinetics during knee-extensor exercise in humans. *Am. J. Physiol. Regul. Integr. Comp. Physiol.* **2009**, *296*, R1078–R1087. [CrossRef]
26. Rico-Sanz, J. Creatine reduces human muscle PCr and pH decrements and Pi accumulation during low-intensity exercise. *J. Appl. Physiol.* **2000**, *88*, 1181–1191. [CrossRef]

27. Santos, M.G.; López de Viñaspre, P.; González de Suso, J.M.; Moreno, A.; Alonso, J.; Cabañas, M.; Pons, V.; Porta, J.; Arús, C. Efecto de la suplementación oral con monohidrato de creatina en el metabolismo energértico muscular y en la composición corporal de sujetos que practican actividad física. *Revista Chilena de Nutrición* **2003**, *30*, 58–63. [CrossRef]
28. Pulido, S.M.; Passaquin, A.C.; Leijendekker, W.J.; Challet, C.; Wallimann, T.; Rüegg, U.T. Creatine supplementation improves intracellular Ca2+ handling and survival in mdx skeletal muscle cells. *FEBS Lett.* **1998**, *439*, 357–362. [CrossRef]
29. Gallo, M.; MacLean, I.; Tyreman, N.; Martins, K.J.B.; Syrotuik, D.; Gordon, T.; Putman, C.T. Adaptive responses to creatine loading and exercise in fast-twitch rat skeletal muscle. *Am. J. Physiol. Regul. Integr. Comp. Physiol.* **2008**, *294*, R1319–R1328. [CrossRef]
30. Pignatti, C.; D'Adamo, S.; Flamigni, F.; Cetrulllo, S. Molecular Mechanisms Linking Nutrition to Metabolic Homeostasis: An Overview Picture of Current Understanding. *Crit. Rev. Eukaryot. Gene Expr.* **2020**, *30*, 543–564. [CrossRef]
31. Hurtado-Carneiro, V.; Pérez-García, A.; Alvarez, E.; Sanz, C. PAS Kinase: A Nutrient and Energy Sensor "Master Key" in the Response to Fasting/Feeding Conditions. *Front. Endocrinol.* **2020**, *11*, 999. [CrossRef]
32. Hespel, P.; Op't Eijnde, B.; Leemputte, M.V.; Ursø, B.; Greenhaff, P.L.; Labarque, V.; Dymarkowski, S.; Hecke, P.V.; Richter, E.A. Oral creatine supplementation facilitates the rehabilitation of disuse atrophy and alters the expression of muscle myogenic factors in humans. *J. Physiol.* **2001**, *536*, 625–633. [CrossRef]
33. Willoughby, D.S.; Rosene, J.M. Effects of Oral Creatine and Resistance Training on Myogenic Regulatory Factor Expression. *Med. Sci. Sports Exerc.* **2003**, *35*, 923–929. [CrossRef]
34. Deldicque, L.; Louis, M.; Theisen, D.; Nielens, H.; Dehoux, M.L.; Thissen, J.-P.; Rennie, M.J.; Francaux, M. Increased IGF mRNA in Human Skeletal Muscle after Creatine Supplementation. *Med. Sci. Sports Exerc.* **2005**, *37*, 731–736. [CrossRef]
35. Saremi, A.; Gharakhanloo, R.; Sharghi, S.; Gharaati, M.R.; Larijani, B.; Omidfar, K. Effects of oral creatine and resistance training on serum myostatin and GASP-1. *Mol. Cell. Endocrinol.* **2010**, *317*, 25–30. [CrossRef] [PubMed]
36. Louis, M.; Van Beneden, R.; Dehoux, M.; Thissen, J.P.; Francaux, M. Creatine increases IGF-I and myogenic regulatory factor mRNA in C2C12 cells. *FEBS Lett.* **2004**, *557*, 243–247. [CrossRef]
37. Cunha, M.P.; Budni, J.; Ludka, F.K.; Pazini, F.L.; Rosa, J.M.; Oliveira, Á.; Lopes, M.W.; Tasca, C.I.; Leal, R.B.; Rodrigues, A.L.S. Involvement of PI3K/Akt Signaling Pathway and Its Downstream Intracellular Targets in the Antidepressant-Like Effect of Creatine. *Mol. Neurobiol.* **2015**, *53*, 2954–2968. [CrossRef] [PubMed]
38. Marzuca-Nassr, G.N.; Fortes, M.A.S.; Guimarães-Ferreira, L.; Murata, G.M.; Vitzel, K.F.; Vasconcelos, D.A.A.; Bassit, R.A.; Curi, R. Short-term creatine supplementation changes protein metabolism signaling in hindlimb suspension. *Braz. J. Med. Biol. Res.* **2019**, *52*. [CrossRef] [PubMed]
39. Kontou, P.I.; Pavlopoulou, A.; Bagos, P.G. Methods of Analysis and Meta-Analysis for Identifying Differentially Expressed Genes. In *Genet. Epidemiol*; Humana Press: New York, NY, USA, 2018; pp. 183–210.
40. Van Laere, S.; Dirix, L.; Vermeulen, P. Molecular profiles to biology and pathways: A systems biology approach. *Chin. J. Cancer* **2016**, *35*, 53. [CrossRef] [PubMed]
41. Angione, C. Human Systems Biology and Metabolic Modelling: A Review—From Disease Metabolism to Precision Medicine. *BioMed Res. Int.* **2019**, *2019*, 8304260. [CrossRef]
42. Chen, E.Y.; Xu, H.; Gordonov, S.; Lim, M.P.; Perkins, M.H.; Ma'ayan, A. Expression2Kinases: mRNA profiling linked to multiple upstream regulatory layers. *Bioinformatics* **2012**, *28*, 105–111. [CrossRef]
43. Forero, D.A.; González-Giraldo, Y. Convergent functional genomics of cocaine misuse in humans and animal models. *Am. J. Drug Alcohol Abus.* **2019**, *46*, 22–30. [CrossRef] [PubMed]
44. Monaco, G.; van Dam, S.; Casal Novo Ribeiro, J.L.; Larbi, A.; de Magalhães, J.P. A comparison of human and mouse gene co-expression networks reveals conservation and divergence at the tissue, pathway and disease levels. *BMC Evol. Biol.* **2015**, *15*, 259. [CrossRef] [PubMed]
45. Wang, D.; Levine, J.L.S.; Avila-Quintero, V.; Bloch, M.; Kaffman, A. Systematic review and meta-analysis: Effects of maternal separation on anxiety-like behavior in rodents. *Transl. Psychiatry* **2020**, *10*, 174. [CrossRef] [PubMed]
46. Niculescu, A.B.; Le-Niculescu, H. Convergent Functional Genomics: What we have learned and can learn about genes, pathways, and mechanisms. *Neuropsychopharmacology* **2009**, *35*, 355–356. [CrossRef]
47. Forero, D.A. Functional Genomics of Epileptogenesis in Animal Models and Humans. *Cell. Mol. Neurobiol.* **2020**. [CrossRef] [PubMed]
48. Niculescu, A.B.; Levey, D.F.; Phalen, P.L.; Le-Niculescu, H.; Dainton, H.D.; Jain, N.; Belanger, E.; James, A.; George, S.; Weber, H.; et al. Understanding and predicting suicidality using a combined genomic and clinical risk assessment approach. *Mol. Psychiatry* **2015**, *20*, 1266–1285. [CrossRef]
49. Bonvicini, C.; Faraone, S.V.; Scassellati, C. Common and specific genes and peripheral biomarkers in children and adults with attention-deficit/hyperactivity disorder. *World J. Biol. Psychiatry* **2018**, *19*, 80–100. [CrossRef] [PubMed]
50. Smith, A.K.; Fang, H.; Whistler, T.; Unger, E.R.; Rajeevan, M.S. Convergent genomic studies identify association of GRIK2 and NPAS2 with chronic fatigue syndrome. *Neuropsychobiology* **2011**, *64*, 183–194. [CrossRef] [PubMed]
51. Falker-Gieske, C.; Mott, A.; Franzenburg, S.; Tetens, J. Multi-species transcriptome meta-analysis of the response to retinoic acid in vertebrates and comparative analysis of the effects of retinol and retinoic acid on gene expression in LMH cells. *BMC Genom.* **2021**, *22*, 146. [CrossRef]

52. Le-Niculescu, H.; Roseberry, K.; Gill, S.S.; Levey, D.F.; Phalen, P.L.; Mullen, J.; Williams, A.; Bhairo, S.; Voegtline, T.; Davis, H.; et al. Precision medicine for mood disorders: Objective assessment, risk prediction, pharmacogenomics, and repurposed drugs. *Mol. Psychiatry* **2021**. [CrossRef]

53. Lévesque, A.; Gagnon-Carignan, S.; Lachance, S. From low- to high-throughput analysis. *Bioanalysis* **2016**, *8*, 135–141. [CrossRef]

54. Ramsay, D.S.; Woods, S.C. Physiological Regulation: How It Really Works. *Cell Metab.* **2016**, *24*, 361–364. [CrossRef] [PubMed]

55. Rawson, E.S.; Persky, A.M. Mechanisms of muscular adaptations to creatine supplementation: Review article. *Int. SportMed J.* **2007**, *8*, 43–53. [CrossRef]

56. Bonilla, D.A.; Moreno, Y. Molecular and metabolic insights of creatine supplementation on resistance training. *Revista Colombiana de Química* **2015**, *44*, 11–18. [CrossRef]

57. Farshidfar, F.; Pinder, M.A.; Myrie, S.B. Creatine Supplementation and Skeletal Muscle Metabolism for Building Muscle Mass-Review of the Potential Mechanisms of Action. *Curr. Protein Pept. Sci.* **2017**, *18*, 1273–1287. [CrossRef] [PubMed]

58. Bender, A.; Klopstock, T. Creatine for neuroprotection in neurodegenerative disease: End of story? *Amino Acids* **2016**, *48*, 1929–1940. [CrossRef]

59. Rae, C.D.; Bröer, S. Creatine as a booster for human brain function. How might it work? *Neurochem. Int.* **2015**, *89*, 249–259. [CrossRef] [PubMed]

60. Riesberg, L.A.; Weed, S.A.; McDonald, T.L.; Eckerson, J.M.; Drescher, K.M. Beyond muscles: The untapped potential of creatine. *Int. Immunopharmacol.* **2016**, *37*, 31–42. [CrossRef] [PubMed]

61. Page, M.J.; McKenzie, J.E.; Bossuyt, P.M.; Boutron, I.; Hoffmann, T.C.; Mulrow, C.D.; Shamseer, L.; Tetzlaff, J.M.; Akl, E.A.; Brennan, S.E.; et al. The PRISMA 2020 statement: An updated guideline for reporting systematic reviews. *BMJ* **2021**. [CrossRef]

62. Ramasamy, A.; Mondry, A.; Holmes, C.C.; Altman, D.G. Key Issues in Conducting a Meta-Analysis of Gene Expression Microarray Datasets. *PLoS Med.* **2008**, *5*, e184. [CrossRef] [PubMed]

63. Barrett, T.; Wilhite, S.E.; Ledoux, P.; Evangelista, C.; Kim, I.F.; Tomashevsky, M.; Marshall, K.A.; Phillippy, K.H.; Sherman, P.M.; Holko, M.; et al. NCBI GEO: Archive for functional genomics data sets—Update. *Nucleic Acids Res.* **2012**, *41*, D991–D995. [CrossRef] [PubMed]

64. Ritchie, M.E.; Phipson, B.; Wu, D.; Hu, Y.; Law, C.W.; Shi, W.; Smyth, G.K. limma powers differential expression analyses for RNA-sequencing and microarray studies. *Nucleic Acids Res.* **2015**, *43*, e47. [CrossRef]

65. Chen, E.Y.; Tan, C.M.; Kou, Y.; Duan, Q.; Wang, Z.; Meirelles, G.; Clark, N.R.; Ma'ayan, A. Enrichr: Interactive and collaborative HTML5 gene list enrichment analysis tool. *BMC Bioinform.* **2013**, *14*, 128. [CrossRef] [PubMed]

66. Kuleshov, M.V.; Jones, M.R.; Rouillard, A.D.; Fernandez, N.F.; Duan, Q.; Wang, Z.; Koplev, S.; Jenkins, S.L.; Jagodnik, K.M.; Lachmann, A.; et al. Enrichr: A comprehensive gene set enrichment analysis web server 2016 update. *Nucleic Acids Res.* **2016**, *44*, W90–W97. [CrossRef] [PubMed]

67. Huang, H.-Y.; Lin, Y.-C.-D.; Li, J.; Huang, K.-Y.; Shrestha, S.; Hong, H.-C.; Tang, Y.; Chen, Y.-G.; Jin, C.-N.; Yu, Y.; et al. miRTarBase 2020: Updates to the experimentally validated microRNA–target interaction database. *Nucleic Acids Res.* **2019**. [CrossRef]

68. Clarke, D.J.B.; Kuleshov, M.V.; Schilder, B.M.; Torre, D.; Duffy, M.E.; Keenan, A.B.; Lachmann, A.; Feldmann, A.S.; Gundersen, G.W.; Silverstein, M.C.; et al. eXpression2Kinases (X2K) Web: Linking expression signatures to upstream cell signaling networks. *Nucleic Acids Res.* **2018**, *46*, W171–W179. [CrossRef] [PubMed]

69. Hangelbroek, R.; Backx, E.; Kersten, S.; van Duynhoven, J.; Verdijk, L.; van Loon, L.; de Groot, L.; Boekschoten, M. Creatine supplementation during knee immobilization attenuates changes in muscle transcriptome. In Proceedings of the Phenotypes and Prevention: The Interplay of Genes, Life-Style and Gut Environment, Copenhagen, Denmark, 5–8 September 2016; p. 59.

70. Safdar, A.; Yardley, N.J.; Snow, R.; Melov, S.; Tarnopolsky, M.A. Global and targeted gene expression and protein content in skeletal muscle of young men following short-term creatine monohydrate supplementation. *Physiol. Genom.* **2008**, *32*, 219–228. [CrossRef]

71. Bender, A.; Beckers, J.; Schneider, I.; Holter, S.M.; Haack, T.; Ruthsatz, T.; Vogt-Weisenhorn, D.M.; Becker, L.; Genius, J.; Rujescu, D.; et al. Creatine improves health and survival of mice. *Neurobiol. Aging* **2008**, *29*, 1404–1411. [CrossRef]

72. Zervou, S.; Ray, T.; Sahgal, N.; Sebag-Montefiore, L.; Cross, R.; Medway, D.J.; Ostrowski, P.J.; Neubauer, S.; Lygate, C.A. A role for thioredoxin-interacting protein (Txnip) in cellular creatine homeostasis. *Am. J. Physiol. Endocrinol. Metab.* **2013**, *305*, E263–E270. [CrossRef]

73. Saks, V.; Monge, C.; Guzun, R. Philosophical Basis and Some Historical Aspects of Systems Biology: From Hegel to Noble—Applications for Bioenergetic Research. *Int. J. Mol. Sci.* **2009**, *10*, 1161–1192. [CrossRef] [PubMed]

74. Bertsch, B.; Ogden, C.A.; Sidhu, K.; Le-Niculescu, H.; Kuczenski, R.; Niculescu, A.B. Convergent functional genomics: A Bayesian candidate gene identification approach for complex disorders. *Methods* **2005**, *37*, 274–279. [CrossRef]

75. Cattaneo, A.; Cattane, N.; Scassellati, C.; D'Aprile, I.; Riva, M.A.; Pariante, C.M. Convergent Functional Genomics approach to prioritize molecular targets of risk in early life stress-related psychiatric disorders. *Brain Behav. Immun. Health* **2020**, *8*. [CrossRef]

76. Sterling, P. Allostasis: A model of predictive regulation. *Physiol. Behav.* **2012**, *106*, 5–15. [CrossRef] [PubMed]

77. Shellman, E.R.; Burant, C.F.; Schnell, S. Network motifs provide signatures that characterize metabolism. *Mol. BioSyst.* **2013**, *9*, 352–360. [CrossRef] [PubMed]

78. Wang, Q.; Qian, W.; Xu, X.; Bajpai, A.; Guan, K.; Zhang, Z.; Chen, R.; Flamini, V.; Chen, W. Energy-Mediated Machinery Drives Cellular Mechanical Allostasis. *Adv. Mater.* **2019**, *31*, 1900453. [CrossRef] [PubMed]

79. Shaulian, E.; Karin, M. AP-1 as a regulator of cell life and death. *Nat. Cell Biol.* **2002**, *4*, E131–E136. [CrossRef]

80. O'Connor, L.; Gilmour, J.; Bonifer, C. The Role of the Ubiquitously Expressed Transcription Factor Sp1 in Tissue-specific Transcriptional Regulation and in Disease. *Yale J. Biol. Med.* **2016**, *89*, 513–525.

81. Oldfield, A.J.; Henriques, T.; Kumar, D.; Burkholder, A.B.; Cinghu, S.; Paulet, D.; Bennett, B.D.; Yang, P.; Scruggs, B.S.; Lavender, C.A.; et al. NF-Y controls fidelity of transcription initiation at gene promoters through maintenance of the nucleosome-depleted region. *Nat. Commun.* **2019**, *10*, 3072. [CrossRef]

82. Cao, R.; Zhang, Y. SUZ12 is required for both the histone methyltransferase activity and the silencing function of the EED-EZH2 complex. *Mol. Cell* **2004**, *15*, 57–67. [CrossRef]

83. Fitieh, A.; Locke, A.J.; Motamedi, M.; Ismail, I.H. The Role of Polycomb Group Protein BMI1 in DNA Repair and Genomic Stability. *Int. J. Mol. Sci.* **2021**, *22*, 2976. [CrossRef]

84. Li, X.; Thyssen, G.; Beliakoff, J.; Sun, Z. The Novel PIAS-like Protein hZimp10 Enhances Smad Transcriptional Activity. *J. Biol. Chem.* **2006**, *281*, 23748–23756. [CrossRef]

85. Carroll, P.A.; Freie, B.W.; Mathsyaraja, H.; Eisenman, R.N. The MYC transcription factor network: Balancing metabolism, proliferation and oncogenesis. *Front. Med.* **2018**, *12*, 412–425. [CrossRef] [PubMed]

86. Smith, K.; Dalton, S. Myc transcription factors: Key regulators behind establishment and maintenance of pluripotency. *Regen. Med.* **2010**, *5*, 947–959. [CrossRef] [PubMed]

87. Hong, J.-H.; Kwak, Y.; Woo, Y.; Park, C.; Lee, S.-A.; Lee, H.; Park, S.J.; Suh, Y.; Suh, B.K.; Goo, B.S.; et al. Regulation of the actin cytoskeleton by the Ndel1-Tara complex is critical for cell migration. *Sci. Rep.* **2016**, *6*, 31827. [CrossRef]

88. Anderson, S.; Poudel, K.R.; Roh-Johnson, M.; Brabletz, T.; Yu, M.; Borenstein-Auerbach, N.; Grady, W.N.; Bai, J.; Moens, C.B.; Eisenman, R.N.; et al. MYC-nick promotes cell migration by inducing fascin expression and Cdc42 activation. *Proc. Natl. Acad. Sci. USA* **2016**, *113*, E5481–E5490. [CrossRef]

89. Hurlin, P.J.; Huang, J. The MAX-interacting transcription factor network. *Semin. Cancer Biol.* **2006**, *16*, 265–274. [CrossRef] [PubMed]

90. Volodina Iu, L.; Shtil, A.A. Casein kinase 2, the versatile regulator of cell survival. *Mol. Biol.* **2012**, *46*, 423–433. [CrossRef]

91. Ding, L.; Cao, J.; Lin, W.; Chen, H.; Xiong, X.; Ao, H.; Yu, M.; Lin, J.; Cui, Q. The Roles of Cyclin-Dependent Kinases in Cell-Cycle Progression and Therapeutic Strategies in Human Breast Cancer. *Int. J. Mol. Sci.* **2020**, *21*, 1960. [CrossRef]

92. Salizzato, V.; Zanin, S.; Borgo, C.; Lidron, E.; Salvi, M.; Rizzuto, R.; Pallafacchina, G.; Donella-Deana, A. Protein kinase CK2 subunits exert specific and coordinated functions in skeletal muscle differentiation and fusogenic activity. *FASEB J.* **2019**, *33*, 10648–10667. [CrossRef]

93. Zheng, Y.; Qin, H.; Frank, S.J.; Deng, L.; Litchfield, D.W.; Tefferi, A.; Pardanani, A.; Lin, F.T.; Li, J.; Sha, B.; et al. A CK2-dependent mechanism for activation of the JAK-STAT signaling pathway. *Blood* **2011**, *118*, 156–166. [CrossRef]

94. Bousoik, E.; Montazeri Aliabadi, H. "Do We Know Jack" About JAK? A Closer Look at JAK/STAT Signaling Pathway. *Front. Oncol.* **2018**, *8*, 287. [CrossRef]

95. Rotwein, P. Regulation of gene expression by growth hormone. *Mol. Cell. Endocrinol.* **2020**, *507*, 110788. [CrossRef] [PubMed]

96. Reis, H.J.; Rosa, D.V.F.; Guimarães, M.M.; Souza, B.R.; Barros, A.G.A.; Pimenta, F.J.; Souza, R.P.; Torres, K.C.L.; Romano-Silva, M.A. Is DARPP-32 a potential therapeutic target? *Expert Opin. Ther. Targets* **2007**, *11*, 1649–1661. [CrossRef] [PubMed]

97. Engmann, O.; Giralt, A.; Gervasi, N.; Marion-Poll, L.; Gasmi, L.; Filhol, O.; Picciotto, M.R.; Gilligan, D.; Greengard, P.; Nairn, A.C.; et al. DARPP-32 interaction with adducin may mediate rapid environmental effects on striatal neurons. *Nat. Commun.* **2015**, *6*, 10099. [CrossRef]

98. Andres, R.H.; Ducray, A.D.; Huber, A.W.; Pérez-Bouza, A.; Krebs, S.H.; Schlattner, U.; Seiler, R.W.; Wallimann, T.; Widmer, H.R. Effects of creatine treatment on survival and differentiation of GABA-ergic neurons in cultured striatal tissue. *J. Neurochem.* **2005**, *95*, 33–45. [CrossRef] [PubMed]

99. Jang, S.W.; Hwang, S.S.; Kim, H.S.; Lee, K.O.; Kim, M.K.; Lee, W.; Kim, K.; Lee, G.R. Casein kinase 2 is a critical determinant of the balance of Th17 and Treg cell differentiation. *Exp. Mol. Med.* **2017**, *49*, e375. [CrossRef]

100. Braicu, C.; Buse, M.; Busuioc, C.; Drula, R.; Gulei, D.; Raduly, L.; Berindan-Neagoe, I. A Comprehensive Review on MAPK: A Promising Therapeutic Target in Cancer. *Cancers* **2019**, *11*, 1618. [CrossRef]

101. Dhanasekaran, D.N.; Kashef, K.; Lee, C.M.; Xu, H.; Reddy, E.P. Scaffold proteins of MAP-kinase modules. *Oncogene* **2007**, *26*, 3185–3202. [CrossRef]

102. Krishna, M.; Narang, H. The complexity of mitogen-activated protein kinases (MAPKs) made simple. *Cell. Mol. Life Sci.* **2008**, *65*, 3525–3544. [CrossRef]

103. Gyoeva, F.K. The role of motor proteins in signal propagation. *Biochemistry* **2014**, *79*, 849–855. [CrossRef]

104. Hoffman, L.; Jensen, C.C.; Yoshigi, M.; Beckerle, M.; Weaver, V.M. Mechanical signals activate p38 MAPK pathway-dependent reinforcement of actin via mechanosensitive HspB1. *Mol. Biol. Cell* **2017**, *28*, 2661–2675. [CrossRef]

105. Munnik, T.; Ligterink, W.; Meskiene, I.I.; Calderini, O.; Beyerly, J.; Musgrave, A.; Hirt, H. Distinct osmo-sensing protein kinase pathways are involved in signalling moderate and severe hyper-osmotic stress. *Plant J.* **1999**, *20*, 381–388. [CrossRef] [PubMed]

106. Drummond, M.J.; Dreyer, H.C.; Fry, C.S.; Glynn, E.L.; Rasmussen, B.B. Nutritional and contractile regulation of human skeletal muscle protein synthesis and mTORC1 signaling. *J. Appl. Physiol.* **2009**, *106*, 1374–1384. [CrossRef]

107. Schedel, J.M.; Tanaka, H.; Kiyonaga, A.; Shindo, M.; Schutz, Y. Acute creatine loading enhances human growth hormone secretion. *J. Sports Med. Phys. Fit.* **2000**, *40*, 336–342.

108. Poprzecki, S.; Zając, A.; Czuba, M.; Waskiewicz, Z. The Effects of Terminating Creatine Supplementation and Resistance Training on Anaerobic Power and Chosen Biochemical Variables in Male Subjects. *J. Hum. Kinet.* **2008**, *20*, 99–110. [CrossRef]

109. Burke, D.G.; Candow, D.G.; Chilibeck, P.D.; MacNeil, L.G.; Roy, B.D.; Tarnopolsky, M.A.; Ziegenfuss, T. Effect of creatine supplementation and resistance-exercise training on muscle insulin-like growth factor in young adults. *Int. J. Sport Nutr. Exerc. Metab.* **2008**, *18*, 389–398. [CrossRef]

110. Snow, R.; Wright, C.; Quick, M.; Garnham, A.; Watt, K.; Russell, A. The effects of acute exercise and creatine supplementation on Akt signalling in human skeletal muscle. *Mol. Cell* **2004**, *14*, 395–403.

111. Al-Khalili, L.; Kramer, D.; Wretenberg, P.; Krook, A. Human skeletal muscle cell differentiation is associated with changes in myogenic markers and enhanced insulin-mediated MAPK and PKB phosphorylation. *Acta Physiol. Scand.* **2004**, *180*, 395–403. [CrossRef]

112. Deldicque, L.; Theisen, D.; Bertrand, L.; Hespel, P.; Hue, L.; Francaux, M. Creatine enhances differentiation of myogenic C2C12 cells by activating both p38 and Akt/PKB pathways. *Am. J. Physiol. Cell Physiol.* **2007**, *293*, C1263–C1271. [CrossRef]

113. Deldicque, L.; Atherton, P.; Patel, R.; Theisen, D.; Nielens, H.; Rennie, M.J.; Francaux, M. Effects of resistance exercise with and without creatine supplementation on gene expression and cell signaling in human skeletal muscle. *J. Appl. Physiol.* **2008**, *104*, 371–378. [CrossRef] [PubMed]

114. Pazini, F.L.; Cunha, M.P.; Rosa, J.M.; Colla, A.R.; Lieberknecht, V.; Oliveira, Á.; Rodrigues, A.L. Creatine, similar to ketamine, counteracts depressive-like behavior induced by corticosterone via PI3K/Akt/mTOR pathway. *Mol Neurobiol.* **2016**, *53*, 6818–6834. [CrossRef]

115. Ju, J.S.; Smith, J.L.; Oppelt, P.J.; Fisher, J.S. Creatine feeding increases GLUT4 expression in rat skeletal muscle. *Am. J. Physiol. Endocrinol. Metab.* **2005**, *288*, E347–E352. [CrossRef] [PubMed]

116. Willoughby, D.S.; Rosene, J. Effects of oral creatine and resistance training on myosin heavy chain expression. *Med. Sci. Sports Exerc.* **2001**, *33*, 1674–1681. [CrossRef]

117. Zhang, L.; Zhu, Z.; Yan, H.; Wang, W.; Wu, Z.; Zhang, F.; Zhang, Q.; Shi, G.; Du, J.; Cai, H.; et al. Creatine promotes cancer metastasis through activation of Smad2/3. *Cell Metab.* **2021**. [CrossRef] [PubMed]

118. Pereira, R.T.d.S.; Dörr, F.A.; Pinto, E.; Solis, M.Y.; Artioli, G.G.; Fernandes, A.L.; Murai, I.H.; Dantas, W.S.; Seguro, A.C.; Santinho, M.A.R.; et al. Can creatine supplementation form carcinogenic heterocyclic amines in humans? *J. Physiol.* **2015**, *593*, 3959–3971. [CrossRef] [PubMed]

119. Kazak, L.; Cohen, P. Creatine metabolism: Energy homeostasis, immunity and cancer biology. *Nat. Rev. Endocrinol.* **2020**, *16*, 421–436. [CrossRef]

120. Patra, S.; Bera, S.; SinhaRoy, S.; Ghoshal, S.; Ray, S.; Basu, A.; Schlattner, U.; Wallimann, T.; Ray, M. Progressive decrease of phosphocreatine, creatine and creatine kinase in skeletal muscle upon transformation to sarcoma. *FEBS J.* **2008**, *275*, 3236–3247. [CrossRef]

121. Pal, A.; Roy, A.; Ray, M. Creatine supplementation with methylglyoxal: A potent therapy for cancer in experimental models. *Amino Acids* **2016**, *48*, 2003–2013. [CrossRef]

122. Storey, K.B.; Hochachka, P.W. Activation of muscle glycolysis: A role for creatine phosphate in phosphofructokinase regulation. *FEBS Lett.* **1974**, *46*, 337–339. [CrossRef]

123. Kemp, R.G. Inhibition of muscle pyruvate kinase by creatine phosphate. *J. Biol. Chem.* **1973**, *248*, 3963–3967. [CrossRef]

124. Bergnes, G.; Yuan, W.; Khandekar, V.S.; O'Keefe, M.M.; Martin, K.J.; Teicher, B.A.; Kaddurah-Daouk, R. Creatine and phosphocreatine analogs: Anticancer activity and enzymatic analysis. *Oncol. Res.* **1996**, *8*, 121–130.

125. Campos-Ferraz, P.L.; Gualano, B.; das Neves, W.; Andrade, I.T.; Hangai, I.; Pereira, R.T.; Bezerra, R.N.; Deminice, R.; Seelaender, M.; Lancha, A.H. Exploratory studies of the potential anti-cancer effects of creatine. *Amino Acids* **2016**, *48*, 1993–2001. [CrossRef] [PubMed]

126. Di Biase, S.; Ma, X.; Wang, X.; Yu, J.; Wang, Y.C.; Smith, D.J.; Zhou, Y.; Li, Z.; Kim, Y.J.; Clarke, N.; et al. Creatine uptake regulates CD8 T cell antitumor immunity. *J. Exp. Med.* **2019**, *216*, 2869–2882. [CrossRef]

127. Li, B.; Yang, L. Creatine in T Cell Antitumor Immunity and Cancer Immunotherapy. *Nutrients* **2021**, *13*, 1633. [CrossRef] [PubMed]

128. Maguire, O.A.; Ackerman, S.E.; Szwed, S.K.; Maganti, A.V.; Marchildon, F.; Huang, X.; Kramer, D.J.; Rosas-Villegas, A.; Gelfer, R.G.; Turner, L.E.; et al. Creatine-mediated crosstalk between adipocytes and cancer cells regulates obesity-driven breast cancer. *Cell Metab.* **2021**, *33*, 499–512.e496. [CrossRef]

129. Zhang, H.; Zhang, Z.; Gao, L.; Qiao, Z.; Yu, M.; Yu, B.; Yang, T. miR-1-3p suppresses proliferation of hepatocellular carcinoma through targeting SOX9. *Onco Targets Ther.* **2019**, *12*, 2149–2157. [CrossRef] [PubMed]

130. Ke, J.; Zhang, B.H.; Li, Y.Y.; Zhong, M.; Ma, W.; Xue, H.; Wen, Y.D.; Cai, Y.D. MiR-1-3p suppresses cell proliferation and invasion and targets STC2 in gastric cancer. *Eur. Rev. Med. Pharmacol. Sci.* **2019**, *23*, 8870–8877. [CrossRef] [PubMed]

131. Li, T.; Wang, X.; Jing, L.; Li, Y. MiR-1-3p Inhibits Lung Adenocarcinoma Cell Tumorigenesis via Targeting Protein Regulator of Cytokinesis 1. *Front. Oncol.* **2019**, *9*, 120. [CrossRef]

132. Meyer, S.U.; Thirion, C.; Polesskaya, A.; Bauersachs, S.; Kaiser, S.; Krause, S.; Pfaffl, M.W. TNF-α and IGF1 modify the microRNA signature in skeletal muscle cell differentiation. *Cell Commun. Signal.* **2015**, *13*, 4. [CrossRef]

133. Pelka, K.; Klicka, K.; Grzywa, T.M.; Gondek, A.; Marczewska, J.M.; Garbicz, F.; Szczepaniak, K.; Paskal, W.; Wlodarski, P.K. miR-96-5p, miR-134-5p, miR-181b-5p and miR-200b-3p heterogenous expression in sites of prostate cancer versus benign prostate hyperplasia-archival samples study. *Histochem. Cell Biol.* **2020**. [CrossRef]

134. Feng, Z.; Zhang, L.; Wang, S.; Hong, Q. Circular RNA circDLGAP4 exerts neuroprotective effects via modulating miR-134-5p/CREB pathway in Parkinson's disease. *Biochem. Biophys. Res. Commun.* **2020**, *522*, 388–394. [CrossRef]
135. Tian, X.; Yu, H.; Li, D.; Jin, G.; Dai, S.; Gong, P.; Kong, C.; Wang, X. The miR-5694/AF9/Snail Axis Provides Metastatic Advantages and a Therapeutic Target in Basal-like Breast Cancer. *Mol. Ther.* **2021**, *29*, 1239–1257. [CrossRef] [PubMed]
136. Fu, G.; Lu, J.; Zheng, Y.; Wang, P.; Shen, Q. MiR-320a inhibits malignant phenotype of melanoma cells via targeting PBX3. *J. BUON* **2020**, *25*, 2071–2077.
137. Li, Y.; Liu, H.; Shao, J.; Xing, G. miR-320a serves as a negative regulator in the progression of gastric cancer by targeting RAB14. *Mol. Med. Rep.* **2017**, *16*, 2652–2658. [CrossRef]
138. Li, H.; Yue, L.; Xu, H.; Li, N.; Li, J.; Zhang, Z.; Zhao, R.C. Curcumin suppresses osteogenesis by inducing miR-126a-3p and subsequently suppressing the WNT/LRP6 pathway. *Aging* **2019**, *11*, 6983–6998. [CrossRef]
139. Ding, N.; Sun, X.; Wang, T.; Huang, L.; Wen, J.; Zhou, Y. miR378a3p exerts tumor suppressive function on the tumorigenesis of esophageal squamous cell carcinoma by targeting Rab10. *Int. J. Mol. Med.* **2018**, *42*, 381–391. [CrossRef] [PubMed]
140. Guo, X.B.; Zhang, X.C.; Chen, P.; Ma, L.M.; Shen, Z.Q. miR378a3p inhibits cellular proliferation and migration in glioblastoma multiforme by targeting tetraspanin 17. *Oncol. Rep.* **2019**, *42*, 1957–1971. [CrossRef]
141. Krist, B.; Podkalicka, P.; Mucha, O.; Mendel, M.; Sepiol, A.; Rusiecka, O.M.; Jozefczuk, E.; Bukowska-Strakova, K.; Grochot-Przeczek, A.; Tomczyk, M.; et al. miR-378a influences vascularization in skeletal muscles. *Cardiovasc. Res.* **2020**, *116*, 1386–1397. [CrossRef] [PubMed]
142. Khoei, S.G.; Sadeghi, H.; Samadi, P.; Najafi, R.; Saidijam, M. Relationship between Sphk1/S1P and microRNAs in human cancers. *Biotechnol. Appl. Biochem.* **2020**. [CrossRef]
143. Luca, E.; Turcekova, K.; Hartung, A.; Mathes, S.; Rehrauer, H.; Krützfeldt, J. Genetic deletion of microRNA biogenesis in muscle cells reveals a hierarchical non-clustered network that controls focal adhesion signaling during muscle regeneration. *Mol. Metab.* **2020**, *36*, 100967. [CrossRef] [PubMed]
144. Santacruz, L.; Darrabie, M.D.; Mishra, R.; Jacobs, D.O. Removal of Potential Phosphorylation Sites does not Alter Creatine Transporter Response to PKC or Substrate Availability. *Cell. Physiol. Biochem.* **2015**, *37*, 353–360. [CrossRef] [PubMed]
145. Kori, M.; Aydın, B.; Unal, S.; Arga, K.Y.; Kazan, D. Metabolic Biomarkers and Neurodegeneration: A Pathway Enrichment Analysis of Alzheimer's Disease, Parkinson's Disease, and Amyotrophic Lateral Sclerosis. *OMICS* **2016**, *20*, 645–661. [CrossRef] [PubMed]
146. Zervou, S.; Whittington, H.J.; Russell, A.J.; Lygate, C.A. Augmentation of Creatine in the Heart. *Mini Rev. Med. Chem.* **2016**, *16*, 19–28. [CrossRef] [PubMed]
147. Brault, J.J.; Abraham, K.A.; Terjung, R.L. Muscle creatine uptake and creatine transporter expression in response to creatine supplementation and depletion. *J. Appl. Physiol.* **2003**, *94*, 2173–2180. [CrossRef]
148. Tarnopolsky, M.; Parise, G.; Fu, M.H.; Brose, A.; Parshad, A.; Speer, O.; Wallimann, T. Acute and moderate-term creatine monohydrate supplementation does not affect creatine transporter mRNA or protein content in either young or elderly humans. *Mol. Cell. Biochem.* **2003**, *244*, 159–166. [CrossRef] [PubMed]
149. Qi, Q.; Li, J.; Cheng, J. Reconstruction of metabolic pathways by combining probabilistic graphical model-based and knowledge-based methods. *BMC Proc.* **2014**, *8*, S5. [CrossRef] [PubMed]
150. Ndika, J.D.T.; Martinez-Munoz, C.; Anand, N.; van Dooren, S.J.M.; Kanhai, W.; Smith, D.E.C.; Jakobs, C.; Salomons, G.S. Post-transcriptional regulation of the creatine transporter gene: Functional relevance of alternative splicing. *Biochim. Et Biophys. Acta (BBA) Gen. Subj.* **2014**, *1840*, 2070–2079. [CrossRef]
151. Santacruz, L.; Jacobs, D.O. Structural correlates of the creatine transporter function regulation: The undiscovered country. *Amino Acids* **2016**, *48*, 2049–2055. [CrossRef]
152. Shojaiefard, M.; Christie, D.L.; Lang, F. Stimulation of the creatine transporter SLC6A8 by the protein kinase mTOR. *Biochem. Biophys. Res. Commun.* **2006**, *341*, 945–949. [CrossRef]
153. Shojaiefard, M.; Christie, D.L.; Lang, F. Stimulation of the creatine transporter SLC6A8 by the protein kinases SGK1 and SGK3. *Biochem. Biophys. Res. Commun.* **2005**, *334*, 742–746. [CrossRef] [PubMed]
154. Strutz-Seebohm, N.; Shojaiefard, M.; Christie, D.; Tavare, J.; Seebohm, G.; Lang, F. PIKfyve in the SGK1 Mediated Regulation of the Creatine Transporter SLC6A8. *Cell. Physiol. Biochem.* **2007**, *20*, 729–734. [CrossRef] [PubMed]
155. Jin, N.; Lang, M.J.; Weisman, L.S. Phosphatidylinositol 3,5-bisphosphate: Regulation of cellular events in space and time. *Biochem. Soc. Trans.* **2016**, *44*, 177–184. [CrossRef]
156. Berwick, D.C.; Dell, G.C.; Welsh, G.I.; Heesom, K.J.; Hers, I.; Fletcher, L.M.; Cooke, F.T.; Tavare, J.M. Protein kinase B phosphorylation of PIKfyve regulates the trafficking of GLUT4 vesicles. *J. Cell Sci.* **2004**, *117*, 5985–5993. [CrossRef]
157. Kreindler, J.L.; Wiemuth, D.; Lott, J.S.; Ly, K.; Ke, Y.; Teesdale-Spittle, P.; Snyder, P.M.; McDonald, F.J. Interaction of Serum- and Glucocorticoid Regulated Kinase 1 (SGK1) with the WW-Domains of Nedd4-2 Is Required for Epithelial Sodium Channel Regulation. *PLoS ONE* **2010**, *5*, e12163. [CrossRef]
158. Dieter, M.; Palmada, M.; Rajamanickam, J.; Aydin, A.; Busjahn, A.; Boehmer, C.; Luft, F.C.; Lang, F. Regulation of Glucose Transporter SGLT1 by Ubiquitin Ligase Nedd4-2 and Kinases SGK1, SGK3, and PKB. *Obes. Res.* **2004**, *12*, 862–870. [CrossRef]
159. Pakladok, T.; Almilaji, A.; Munoz, C.; Alesutan, I.; Lang, F. PIKfyve Sensitivity of hERG Channels. *Cell. Physiol. Biochem.* **2013**, *31*, 785–794. [CrossRef]

160. Almilaji, A.; Sopjani, M.; Elvira, B.; Borras, J.; Dërmaku-Sopjani, M.; Munoz, C.; Warsi, J.; Lang, U.E.; Lang, F. Upregulation of the Creatine Transporter Slc6A8 by Klotho. *Kidney Blood Press. Res.* **2014**, *39*, 516–525. [CrossRef] [PubMed]
161. Sopjani, M.; Dermaku-Sopjani, M. Klotho-Dependent Cellular Transport Regulation. *Vitam. Horm.* **2016**, *101*, 59–84. [CrossRef]
162. Buchanan, S.; Combet, E.; Stenvinkel, P.; Shiels, P.G. Klotho, Aging, and the Failing Kidney. *Front. Endocrinol.* **2020**, *11*. [CrossRef]
163. Li, Q.; Li, Y.; Liang, L.; Li, J.; Luo, D.; Liu, Q.; Cai, S.; Li, X. Klotho negatively regulated aerobic glycolysis in colorectal cancer via ERK/HIF1alpha axis. *Cell Commun. Signal.* **2018**, *16*, 26. [CrossRef] [PubMed]
164. Brown, E.L.; Snow, R.J.; Wright, C.R.; Cho, Y.; Wallace, M.A.; Kralli, A.; Russell, A.P. PGC-1alpha and PGC-1beta increase CrT expression and creatine uptake in myotubes via ERRalpha. *Biochim. Biophys. Acta* **2014**, *1843*, 2937–2943. [CrossRef] [PubMed]
165. Zervou, S.; Yin, X.; Nabeebaccus, A.A.; O'Brien, B.A.; Cross, R.L.; McAndrew, D.J.; Atkinson, R.A.; Eykyn, T.R.; Mayr, M.; Neubauer, S.; et al. Proteomic and metabolomic changes driven by elevating myocardial creatine suggest novel metabolic feedback mechanisms. *Amino Acids* **2016**, *48*, 1969–1981. [CrossRef] [PubMed]
166. López-Hernández, T.; Haucke, V.; Maritzen, T. Endocytosis in the adaptation to cellular stress. *Cell Stress* **2020**, *4*, 230–247. [CrossRef] [PubMed]
167. Kang, D.S.; Tian, X.; Benovic, J.L. Role of beta-arrestins and arrestin domain-containing proteins in G protein-coupled receptor trafficking. *Curr. Opin. Cell Biol.* **2014**, *27*, 63–71. [CrossRef]
168. Lee, S.; Kim, S.M.; Lee, R.T. Thioredoxin and Thioredoxin Target Proteins: From Molecular Mechanisms to Functional Significance. *Antioxid. Redox Signal.* **2013**, *18*, 1165–1207. [CrossRef]
169. Patwari, P.; Lee, R.T. An expanded family of arrestins regulate metabolism. *Trends Endocrinol. Metab.* **2012**, *23*, 216–222. [CrossRef]
170. Patwari, P.; Chutkow, W.A.; Cummings, K.; Verstraeten, V.L.; Lammerding, J.; Schreiter, E.R.; Lee, R.T. Thioredoxin-independent regulation of metabolism by the alpha-arrestin proteins. *J. Biol. Chem.* **2009**, *284*, 24996–25003. [CrossRef]
171. O'Donnell, A.F.; Schmidt, M.C. AMPK-Mediated Regulation of Alpha-Arrestins and Protein Trafficking. *Int. J. Mol. Sci.* **2019**, *20*, 515. [CrossRef]
172. Spindel, O.N.; World, C.; Berk, B.C. Thioredoxin Interacting Protein: Redox Dependent and Independent Regulatory Mechanisms. *Antioxid. Redox Signal.* **2012**, *16*, 587–596. [CrossRef]
173. Kommaddi, R.P.; Shenoy, S.K. Arrestins and Protein Ubiquitination. *Prog. Mol. Biol. Transl. Sci.* **2013**, *118*, 175–204. [PubMed]
174. Prosser, D.C.; Pannunzio, A.E.; Brodsky, J.L.; Thorner, J.; Wendland, B.; O'Donnell, A.F. α-Arrestins participate in cargo selection for both clathrin-independent and clathrin-mediated endocytosis. *J. Cell Sci.* **2015**, *128*, 4220–4234. [CrossRef] [PubMed]
175. Hallows, K.R.; Bhalla, V.; Oyster, N.M.; Wijngaarden, M.A.; Lee, J.K.; Li, H.; Chandran, S.; Xia, X.; Huang, Z.; Chalkley, R.J.; et al. Phosphopeptide Screen Uncovers Novel Phosphorylation Sites of Nedd4-2 That Potentiate Its Inhibition of the Epithelial Na+ Channel. *J. Biol. Chem.* **2010**, *285*, 21671–21678. [CrossRef] [PubMed]
176. An, H.; Krist, D.T.; Statsyuk, A.V. Crosstalk between kinases and Nedd4 family ubiquitin ligases. *Mol. BioSyst.* **2014**, *10*, 1643–1657. [CrossRef]
177. Shojaiefard, M.; Hosseinzadeh, Z.; Bhavsar, S.K.; Lang, F. Downregulation of the creatine transporter SLC6A8 by JAK2. *J. Membr. Biol.* **2012**, *245*, 157–163. [CrossRef]
178. Fezai, M.; Warsi, J.; Lang, F. Regulation of the Na+,Cl- Coupled Creatine Transporter CreaT (SLC6A8) by the Janus Kinase JAK3. *Neurosignals* **2015**, *23*, 11–19. [CrossRef] [PubMed]
179. Dai, W.; Vinnakota, S.; Qian, X.; Kunze, D.L.; Sarkar, H.K. Molecular Characterization of the Human CRT-1 Creatine Transporter Expressed in Xenopus Oocytes. *Arch. Biochem. Biophys.* **1999**, *361*, 75–84. [CrossRef]
180. Wu, N.; Zheng, B.; Shaywitz, A.; Dagon, Y.; Tower, C.; Bellinger, G.; Shen, C.-H.; Wen, J.; Asara, J.; McGraw, T.E.; et al. AMPK-Dependent Degradation of TXNIP upon Energy Stress Leads to Enhanced Glucose Uptake via GLUT1. *Mol. Cell* **2013**, *49*, 1167–1175. [CrossRef]
181. Darrabie, M.D.; Arciniegas, A.J.; Mishra, R.; Bowles, D.E.; Jacobs, D.O.; Santacruz, L. AMPK and substrate availability regulate creatine transport in cultured cardiomyocytes. *Am. J. Physiol. Endocrinol. Metab.* **2011**, *300*, E870–E876. [CrossRef]
182. Li, H.; Thali, R.F.; Smolak, C.; Gong, F.; Alzamora, R.; Wallimann, T.; Scholz, R.; Pastor-Soler, N.M.; Neumann, D.; Hallows, K.R. Regulation of the creatine transporter by AMP-activated protein kinase in kidney epithelial cells. *Am. J. Physiol. Ren. Physiol.* **2010**, *299*, F167–F177. [CrossRef]
183. Galcheva-Gargova, Z.; Derijard, B.; Wu, I.H.; Davis, R.J. An osmosensing signal transduction pathway in mammalian cells. *Science* **1994**, *265*, 806–808. [CrossRef] [PubMed]
184. Fezai, M.; Elvira, B.; Borras, J.; Ben-Attia, M.; Hoseinzadeh, Z.; Lang, F. Negative regulation of the creatine transporter SLC6A8 by SPAK and OSR1. *Kidney Blood Press. Res.* **2014**, *39*, 546–554. [CrossRef] [PubMed]
185. Alessi, D.R.; Zhang, J.; Khanna, A.; Hochdorfer, T.; Shang, Y.; Kahle, K.T. The WNK-SPAK/OSR1 pathway: Master regulator of cation-chloride cotransporters. *Sci. Signal.* **2014**, *7*, re3. [CrossRef] [PubMed]
186. Zwiewka, M.; Nodzyński, T.; Robert, S.; Vanneste, S.; Friml, J. Osmotic Stress Modulates the Balance between Exocytosis and Clathrin-Mediated Endocytosis in Arabidopsis thaliana. *Mol. Plant* **2015**, *8*, 1175–1187. [CrossRef] [PubMed]
187. Pedersen, S.F.; Kapus, A.; Hoffmann, E.K. Osmosensory Mechanisms in Cellular and Systemic Volume Regulation. *J. Am. Soc. Nephrol.* **2011**, *22*, 1587–1597. [CrossRef]
188. Morbach, S.; Krämer, R. Osmoregulation and osmosensing by uptake carriers for compatible solutes in bacteria. In *Molecular Mechanisms Controlling Transmembrane Transport*; Springer: Berlin/Heidelberg, Germany, 2004; pp. 121–153.
189. Wood, J.M. Bacterial Osmosensing Transporters. *Methods Enzymol.* **2007**, *428*, 77–107.

190. El-Kasaby, A.; Kasture, A.; Koban, F.; Hotka, M.; Asjad, H.M.M.; Kubista, H.; Freissmuth, M.; Sucic, S. Rescue by 4-phenylbutyrate of several misfolded creatine transporter-1 variants linked to the creatine transporter deficiency syndrome. *Neuropharmacology* **2019**, *161*, 107572. [CrossRef] [PubMed]
191. Wang, Q.; Groenendyk, J.; Michalak, M. Glycoprotein Quality Control and Endoplasmic Reticulum Stress. *Molecules* **2015**, *20*, 13689–13704. [CrossRef] [PubMed]
192. Farr, C.V.; El-Kasaby, A.; Freissmuth, M.; Sucic, S. The Creatine Transporter Unfolded: A Knotty Premise in the Cerebral Creatine Deficiency Syndrome. *Front. Synaptic Neurosci.* **2020**, *12*, 588954. [CrossRef]
193. Niculescu, A.B. Convergent functional genomics of stem cell-derived cells. *Transl. Psychiatry* **2013**, *3*, e305. [CrossRef]
194. Chakraborty, C.; Sharma, A.R.; Sharma, G.; Doss, C.G.P.; Lee, S.S. Therapeutic miRNA and siRNA: Moving from Bench to Clinic as Next Generation Medicine. *Mol. Nucleic Acids* **2017**, *8*, 132–143. [CrossRef] [PubMed]
195. Dugourd, A.; Kuppe, C.; Sciacovelli, M.; Gjerga, E.; Gabor, A.; Emdal, K.B.; Vieira, V.; Bekker-Jensen, D.B.; Kranz, J.; Bindels, E.M.J.; et al. Causal integration of multi-omics data with prior knowledge to generate mechanistic hypotheses. *Mol. Syst. Biol.* **2021**, *17*, e9730. [CrossRef] [PubMed]

 *nutrients*

Review

Meta-Analysis Examining the Importance of Creatine Ingestion Strategies on Lean Tissue Mass and Strength in Older Adults

Scott C. Forbes [1,*], Darren G. Candow [2], Sergej M. Ostojic [3], Michael D. Roberts [4] and Philip D. Chilibeck [5]

[1] Department of Physical Education Studies, Faculty of Education, Brandon University, Brandon, MB R7A6A9, Canada
[2] Faculty of Kinesiology and Health Studies, University of Regina, Regina, SK S4SOA2, Canada; Darren.candow@uregina.ca
[3] Applied Bioenergetics Lab, Faculty of Sport and Physical Education, University of Novi Sad, Lovcenska 16, 21000 Novi Sad, Serbia; sergej.ostojic@chess.edu.rs
[4] School of Kinesiology, Auburn University, Auburn, AL 36849, USA; mdr0024@auburn.edu
[5] College of Kinesiology, University of Saskatchewan, Saskatoon, SK S7N 5B2, Canada; phil.chilibeck@usask.ca
* Correspondence: forbess@brandonu.ca; Tel.: +1-204-727-9637

Citation: Forbes, S.C.; Candow, D.G.; Ostojic, S.M.; Roberts, M.D.; Chilibeck, P.D. Meta-Analysis Examining the Importance of Creatine Ingestion Strategies on Lean Tissue Mass and Strength in Older Adults. *Nutrients* **2021**, *13*, 1912. https://doi.org/10.3390/nu13061912

Academic Editor: Elena Barbieri

Received: 1 May 2021
Accepted: 30 May 2021
Published: 2 June 2021

Abstract: Creatine supplementation in conjunction with resistance training (RT) augments gains in lean tissue mass and strength in aging adults; however, there is a large amount of heterogeneity between individual studies that may be related to creatine ingestion strategies. Therefore, the purpose of this review was to (1) perform updated meta-analyses comparing creatine vs. placebo (independent of dosage and frequency of ingestion) during a resistance training program on measures of lean tissue mass and strength, (2) perform meta-analyses examining the effects of different creatine dosing strategies (lower: ≤ 5 g/day and higher: >5 g/day), with and without a creatine-loading phase (≥ 20 g/day for 5–7 days), and (3) perform meta-analyses determining whether creatine supplementation only on resistance training days influences measures of lean tissue mass and strength. Overall, creatine (independent of dosing strategy) augments lean tissue mass and strength increase from RT vs. placebo. Subanalyses showed that creatine-loading followed by lower-dose creatine (≤ 5 g/day) increased chest press strength vs. placebo. Higher-dose creatine (>5 g/day), with and without a creatine-loading phase, produced significant gains in leg press strength vs. placebo. However, when studies involving a creatine-loading phase were excluded from the analyses, creatine had no greater effect on chest press or leg press strength vs. placebo. Finally, creatine supplementation only on resistance training days significantly increased measures of lean tissue mass and strength vs. placebo.

Keywords: supplements; hypertrophy; sarcopenia

1. Introduction

The age-related decrease in lean tissue mass and strength are two main factors that contribute to the development of sarcopenia [1]. Approximately 10% of the adult population ≥ 60 years of age has sarcopenia [2], which has a profound negative effect on functional independence and overall quality of life [3]. Furthermore, sarcopenia is associated with other age-related diseases and health conditions such as osteoporosis and physical frailty [3,4]. Several lines of research suggest that sarcopenia is caused by age-related changes in muscle protein kinetics, neuromuscular function and physiology, skeletal muscle morphology, inflammation, and mitochondrial dysregulation [1,5,6]. In addition to these cellular and mechanistic changes, insufficient physical activity and nutritional intake also contribute to sarcopenia [3,7]. Interestingly, dietary intake of creatine, a key component for muscular bioenergetics, decreases with age [8].

The combination of creatine supplementation and resistance training has the potential to serve as an effective countermeasure to the age-related loss in lean tissue mass and

479

strength, possibly by influencing anaerobic energy metabolism, calcium and glycogen regulation, muscle protein kinetics, inflammation and oxidative stress [3,9,10]. However, results from individual studies (n = 20) are mixed, with 10 studies showing beneficial effects on measures of lean tissue mass and/or strength (leg press, chest press) while 10 studies found no greater benefit from creatine vs. placebo (Table 1). While numerous methodological variables may explain these inconsistent findings, differences in creatine dosage and frequency of ingestion during the resistance training program is likely involved [10]. For example, of the 20 studies performed, 6 studies used a lower-dose creatine strategy ($\leq$5 g/day for 12–26 weeks), 6 studies used a creatine-loading phase ($\geq$20 g/day for 5–7 days) followed by a lower-dose creatine strategy ($\leq$5 g/day) while 2 studies used a creatine-loading phase ($\geq$20 g/day for 5–7 days), followed by a higher-dose creatine strategy (>5 g/day for 11 weeks). Furthermore, 6 studies used a higher-dose creatine strategy (>5 g/day) for 8–52 weeks. Finally, 4 of the 20 studies had participants ingest creatine only on resistance training days. The average sample size across studies was only 34 participants. Therefore, these studies were likely unpowered to detect small differences in lean tissue mass and strength (leg press, chest press). To overcome low statistical power across studies, meta-analyses are often performed.

Table 1. Study characteristics, dosing strategy, and outcomes of research examining the influence of creatine in older adults with a resistance training program.

First Author, Year	Population	Supplement Protocol		Resistance Training	Duration	Outcomes
		Loading Protocol	Maintenance Dose			
		Lower-Dose/Absolute Studies ($\leq$5 g/day)				
Alves et al. [11]	N = 47; healthy women, Mean age = 66.8 years (range: 60–80 years)	CR 20 g/day for 5 days	CR (5 g/day) or PLA	RT = 2 days/wk	24 wks	$\leftrightarrow$ 1RM strength compared to RT + PLA
Aguiar et al. [12]	N = 18; healthy women; Mean age = 65 years	None	CR (5 g/day) or PLA	RT = 3 days/wk	12 wks	CR $\uparrow$ gains in fat-free mass (+3.2%), muscle mass (+2.8%), 1RM bench press, knee extension, and biceps curl compared to PLA
Bemben et al. and Eliot et al. [13,14]	N = 42; healthy men; age = 48–72 years	None	CR (5 g/day)	RT = 3 days/wk	14 wks	$\leftrightarrow$ lean tissue mass, 1RM strength
Bermon et al. [15]	N = 32 (16 men, 16 women); healthy; age = 67–80 years	CR 20 g/day for 5 days	CR (3 g/day) or PLA	RT = 3 days/wk	7.4 wks (52 days)	$\leftrightarrow$ lower limb muscular volume, 1-, 12-repetitions maxima, and the isometric intermittent endurance
Brose et al. [16]	N = 28 (15 men, 13 women); healthy; age: men = 68.7, women = 70.8 years	None	CR (5 g/day) or PLA	RT = 3 days/wk	14 wks	CR $\uparrow$ gains in lean tissue mass and isometric knee extension strength; $\leftrightarrow$ type 1, 2a, 2x muscle fiber area
Deacon et al. [17]	N = 80 (50 men, 30 women); COPD; age = 68.2 years	CR 22 g/day for 5 days	CR (3.76 g/day) or PLA	RT = 3 days/wk	7 wks	$\leftrightarrow$ lean tissue mass or muscle strength
Eijnde et al. [18]	N = 46; healthy men; age = 55–75 years	None	CR (5 g/day) or PLA	Cardiorespiratory + RT = 2–3 days/wk	26 wks	$\leftrightarrow$ lean tissue mass or isometric maximal strength
Gualano et al. [19]	N = 25 (9 men, 16 women); type 2 diabetes; age = 57 years	None	CR (5 g/day) or PLA	RT = 3 days/wk	12 wks	$\leftrightarrow$ lean tissue mass
Gualano et al. [20]	N = 30; "vulnerable" women; Mean age = 65.4 years	CR 20 g/day for 5 days	CR (5 g/day) or PLA	RT = 2 days/wk	24 wks	CR + RT $\uparrow$ gains in 1RM bench press and appendicular lean mass compared to PLA + RT
Hass et al. [21]	N = 20 (17 men, 3 women with idiopathetic Parkinson's disease); Mean age = 62 years	CR 20 g/day for 5 days	CR (5 g/day) or PLA	RT = 2 days/wk	12 wks	CR $\uparrow$ chest press strength, chair rise performance; $\leftrightarrow$ Leg extension 1RM, muscular endurance

Table 1. *Cont.*

First Author, Year	Population	Supplement Protocol Loading Protocol	Maintenance Dose	Resistance Training	Duration	Outcomes
Neves et al. [22]	N = 24 (postmenopausal women with knee osteoarthritis); Age = 55–65 years	CR 20 g/day for 1 week	CR 5 (g/day) or PLA	RT=3 days/wk	12 wks	CR ↑ gains in limb lean mass. ↔ 1RM leg press
Pinto et al. [23]	N = 27 (men and women); healthy; age = 60–80 years	None	CR (5 g/day) or PLA	RT = 3 days/wk	12 wks	CR ↑ gains in lean tissue mass. ↔ 10 RM bench press or leg press strength
			Higher-Dose/Relative Studies (>5 g/day)			
Bernat et al. [24]	N = 24 healthy men; age = 59 ± 6 years	None	CR (0.1 g/kg/day; ~9.5 g/day) or PLA	High-velocity RT = 2 days/wk	8 wks	↔ muscle thickness, physical performance, upper body muscle strength. CR ↑ leg press strength, total lower body strength
Candow et al. [25]	N = 35; healthy men; age = 59–77 years	None	CR (0.1 g/kg/day; ~8.6 g/day) or PLA	RT = 3 days/wk	10 wks	CR ↑ muscle thickness compared to PLA. CR ↑ 1RM bench press ↔ 1RM leg press
Candow et al. [26]	N = 39 (17 men, 22 women); healthy; age = 50–71 years	None	CR (0.1 g/kg; ~7.7 g/day) before RT, CR (0.1 g/kg; ~8.8 g/day) after RT, or PLA	RT = 3 days/wk	32 wks	CR after RT ↑ lean tissue mass, 1RM leg press, 1RM chest press compared to PLA
Candow et al. [27]	N = 38; healthy men; age = 49–67 years	None	CR (On training days: 0.05 g/kg before and 0.05 g/kg after exercise; total ~9.3 g/day) + 0.1 g/kg/day on non-training days (2 equal doses)	RT = 3 days/wk	12 months	↔ lean tissue mass, muscle thickness, or muscle strength
Chilibeck et al. [28]	N = 33; healthy women; Mean age = 57 years	None	CR (0.1 g/kg/day; ~6.9 g/day) or PLA	RT = 3 days/wk	52 wks	↔ lean tissue mass and muscle thickness gains between groups. ↑ relative bench press strength compared to PLA.
Chrusch et al. [29]	N = 30; healthy men; age = 60–84 years	CR 0.3 g/kg/d for 5 days	CR 0.07 g/kg/day; ~6.2 g/day or PLA	RT = 3 days/wk	12 wks	CR ↑ gains in lean tissue mass. CR ↑ 1RM leg press, 1RM knee extension, leg press endurance, and knee extension endurance. ↔ 1RM bench press or bench press endurance.
Cooke et al. [30]	N = 20; healthy men; age = 55–70 years	CR 20 g/day for 7 days	CR 0.1 g/kg/day or ~8.8 g/day on training days	RT = 3 days/wk	12 wks	↔ lean tissue mass, 1RM bench press, 1RM leg press
Johannsmeyer et al. [31]	N = 31 (17 men, 14 women); healthy; age = 58 years	None	CR 0.1 g/kg/day; ~7.8 g/day or PLA	RT = 3 days/wk	12 wks	CR ↑ gains in lean tissue mass and 1RM strength in men only

To date, three meta-analyses have been performed involving creatine supplementation and resistance training in older adults [9,32,33]. Collectively, results showed that creatine and resistance training increased measures of lean tissue mass by ~1.2 kg and strength (leg press, chest press) more than placebo and resistance training. However, no sub-analyses were performed to determine whether the dosage of creatine used or the frequency of ingestion (i.e., only on resistance training days) influenced measures of lean tissue mass and/or strength. Since the publication of these meta-analyses, two additional studies involving creatine supplementation and resistance training in older adults have been published. Therefore, the purpose of this review was to (1) perform updated meta-analyses comparing creatine vs. placebo (independent of dosage and frequency of ingestion) during a resistance training program on measures of lean tissue mass and strength, (2) perform meta-analyses examining the effects of different creatine dosing strategies (lower: ≤5 g/day vs. higher: >5 g/day), with and without a creatine-loading phase (20 g/day for 5–7 days, and (3) perform meta-analyses determining whether creatine supplementation only on resistance training days influences measures of lean tissue mass and strength. Results from these meta-analyses may provide important information for the design of optimal creatine supplementation strategies for older adults.

2. Materials and Methods

We have previously published two meta-analyses in 2014 [9] and 2017 [32]. Based on our expertise in the literature, we updated these meta-analyses with recently published studies since the date of the 2017 publication [32]. PubMed and SPORTDiscus databases were searched. Similar to our previous meta-analysis [32] key terms and similar phrases were used (creatine OR creatine monohydrate OR creatine supplementation OR creatine-loading) AND (weight lifting OR weight training OR resistance training, OR resistance exercise OR strength training) AND (age OR middle-age OR older adults OR elderly). Studies with the following criteria were included: (1) healthy and chronic disease participants with a mean age >50 years of age; (2) must be a randomized control trial (RCT) where participants were randomized to an intervention group consisting of creatine monohydrate with resistance training or placebo with resistance training; (3) included outcome measures of whole-body lean tissue mass (determined with dual-energy X-ray absorptiometry [DEXA], hydrostatic weighing, air displacement plethysmography, bioelectrical impedance, or multi-site ultrasound), or upper-(chest press) or lower-body (leg press) muscular strength. Studies were excluded if they were <5 weeks in duration.

Two researchers (S.C.F. and D.G.C.) determined whether the relevant articles were to be included, and any disagreements were resolved by consensus. Databases were searched up until February 2021. Means and standard deviations for baseline and post-training measurements were extracted from each study for estimation of mean changes and the standard deviation of mean changes across the interventions. Change scores were calculated as the pre-training mean subtracted from the post-training mean. Standard deviations (SD) for the change scores were estimated from pre and post-training standard deviations (SD-pre and SD-post) using the following equation derived from the *Cochrane Handbook for Systematic Reviews of Interventions*:

$$\text{SD change score} = [(\text{SD pre})^2 + (\text{SD post})^2 - 2 * (\text{correlation between pre and post scores}) * \text{SD pre} * \text{SD post}]^{1/2}$$

We used 0.8 as the assumed correlation between pre- and post-scores. Heterogeneity was evaluated using χ^2 and I^2 tests where heterogeneity was indicated by either χ^2 p-value ≤ 0.1 or I^2 test value > 75%. We used a fixed-effects model for our meta-analysis. Weighted mean differences were calculated for lean tissue mass, along with the 95% CI. As units of measurement differed across studies for measurements of strength, calculated standardized mean differences (SMDs) and 95% CIs for leg press and chest press strength were used. Forest plots were generated using Review Manager 5.3 Software (Cochrane Community, London, UK). Significance was established at $p \leq 0.05$. Funnel plots were generated and inspected for publication bias. Adverse events were also extracted.

Sub-Analyses

To examine the influence of creatine dosage, dosing strategy was extracted and classified as either higher (>5 g/day) or lower (≤ 5 g/day), as well as whether the study included a "loading phase" (≥ 20 g/day for 5–7 days) and whether creatine was only consumed on resistance training days. Only two studies [15,17] used a creatine dosage <5 g/day. Both absolute and relative (based on body mass) dosing strategy studies were included. We estimated an absolute dose of creatine ingested per day from the product of the average body mass and the relative dose. Several sub-analyses were performed to examine the effects of creatine within each classification. Furthermore, sensitivity analysis was conducted to explore whether the overall effects depended on a single specific study.

3. Results

3.1. Lean Tissue Mass

The analysis of 16 RCTs with 18 treatment arms (n = 509) revealed that creatine supplementation and resistance training increased measures of lean tissue mass vs. placebo and resistance training (Figure 1: mean difference = 1.32 kg [95% CI: 0.93, 1.72] $p < 0.000001$).

Sub-analyses showed that higher-dose creatine, with and without a creatine-loading phase, produced significant gains in lean tissue mass vs. placebo (Figure 1: mean difference = 1.21 kg [95% CI: 0.57, 1.85] $p = 0.0002$). Even when studies incorporating a creatine-loading phase were excluded, higher-dose creatine remained effective (Figure S1: mean difference = 1.16 kg [95% CI: 0.49, 1.82] $p = 0.0006$).

Lower-dose creatine, with and without a creatine-loading phase, increased lean tissue mass vs. placebo (Figure 1: mean difference = 1.40 kg [95% CI: 0.89, 1.91] $p < 0.00001$). When studies incorporating a creatine-loading phase were excluded, lower-dose creatine was still more beneficial than placebo (Figure S2: mean difference = 1.81 kg [95% CI: 1.20, 2.42] $p < 0.00001$).

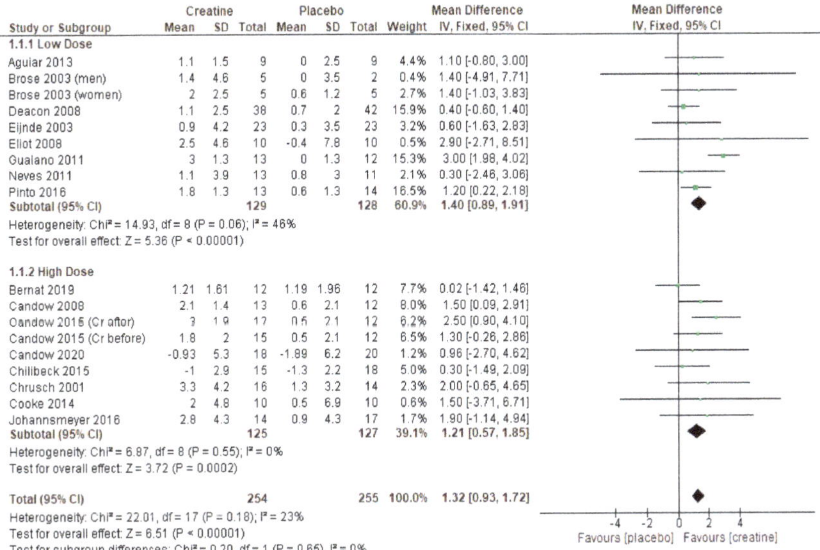

Study or Subgroup	Creatine Mean	SD	Total	Placebo Mean	SD	Total	Weight	Mean Difference IV, Fixed, 95% CI
1.1.1 Low Dose								
Aguiar 2013	1.1	1.5	9	0	2.5	9	4.4%	1.10 [-0.80, 3.00]
Brose 2003 (men)	1.4	4.6	5	0	3.5	2	0.4%	1.40 [-4.91, 7.71]
Brose 2003 (women)	2	2.5	5	0.6	1.2	5	2.7%	1.40 [-1.03, 3.83]
Deacon 2008	1.1	2.5	38	0.7	2	42	15.9%	0.40 [-0.60, 1.40]
Eijnde 2003	0.9	4.2	23	0.3	3.5	23	3.2%	0.60 [-1.63, 2.83]
Eliot 2008	2.5	4.6	10	-0.4	7.8	10	0.5%	2.90 [-2.71, 8.51]
Gualano 2011	3	1.3	13	0	1.3	12	15.3%	3.00 [1.98, 4.02]
Neves 2011	1.1	3.9	13	0.8	3	11	2.1%	0.30 [-2.46, 3.06]
Pinto 2016	1.8	1.3	13	0.6	1.3	14	16.5%	1.20 [0.22, 2.18]
Subtotal (95% CI)			129			128	60.9%	**1.40 [0.89, 1.91]**
Heterogeneity: Chi² = 14.93, df = 8 (P = 0.06); I² = 46%								
Test for overall effect: Z = 5.36 (P < 0.00001)								
1.1.2 High Dose								
Bernat 2019	1.21	1.61	12	1.19	1.96	12	7.7%	0.02 [-1.42, 1.46]
Candow 2008	2.1	1.4	13	0.6	2.1	12	8.0%	1.50 [0.09, 2.91]
Candow 2015 (Cr after)	3	1.9	12	0.5	2.1	12	6.2%	2.50 [0.90, 4.10]
Candow 2015 (Cr before)	1.8	2	15	0.5	2.1	12	6.5%	1.30 [-0.26, 2.86]
Candow 2020	-0.93	5.3	18	-1.89	6.2	20	1.2%	0.96 [-2.70, 4.62]
Chilibeck 2015	-1	2.9	15	-1.3	2.2	18	5.0%	0.30 [-1.49, 2.09]
Chrusch 2001	3.3	4.2	16	1.3	3.2	14	2.3%	2.00 [-0.65, 4.65]
Cooke 2014	2	4.8	10	0.5	6.9	10	0.6%	1.50 [-3.71, 6.71]
Johannsmeyer 2016	2.8	4.3	14	0.9	4.3	17	1.7%	1.90 [-1.14, 4.94]
Subtotal (95% CI)			125			127	39.1%	**1.21 [0.57, 1.85]**
Heterogeneity: Chi² = 6.87, df = 8 (P = 0.55); I² = 0%								
Test for overall effect: Z = 3.72 (P = 0.0002)								
Total (95% CI)			254			255	100.0%	**1.32 [0.93, 1.72]**
Heterogeneity: Chi² = 22.01, df = 17 (P = 0.18); I² = 23%								
Test for overall effect: Z = 6.51 (P < 0.00001)								
Test for subgroup differences: Chi² = 0.20, df = 1 (P = 0.65), I² = 0%								

Figure 1. Forest plot of studies on lean tissue mass with sub-analyses using lower-dose creatine studies ($\leq$5 g/day) and of higher-dose creatine studies (>5 g/day) on lean tissue mass.

3.2. Chest Press Strength

The analysis of 17 RCTs with 19 treatment arms (n = 456) revealed that creatine supplementation and resistance training significantly increased chest press strength vs. placebo and resistance training (Figure 2: standard mean difference = 0.28 [95% CI: 0.09, 0.47] $p = 0.004$).

Subanalyses showed that studies using higher-dose creatine, with and without a creatine-loading phase, found similar effects compared to the placebo (Figure 2 and Figure S3; $p > 0.05$). However, sensitivity analysis indicated that omitting the Candow et al. [27] study changed the overall effect to significantly favor creatine (Figures S4 and S5; $p = 0.008$).

Studies using a creatine-loading phase followed by lower-dose creatine revealed a significant benefit in favor of creatine (Figure 2: standard mean difference = 0.33 [95% CI: 0.05, 0.61] $p = 0.02$). However, when studies incorporating a creatine-loading phase were excluded from the analysis, lower-dose creatine was similar to the placebo (Figure 3; $p = 0.12$).

Study or Subgroup	Creatine Mean	SD	Total	Placebo Mean	SD	Total	Weight	Std. Mean Difference IV, Fixed, 95% CI
1.2.1 Low Dose								
Aguiar 2013	14	17	9	9	5	9	4.0%	0.38 [-0.55, 1.31]
Alves 2013	3.4	4	12	2.7	2.2	10	5.0%	0.20 [-0.64, 1.05]
Bemben 2010	42	13	10	33	6	10	4.1%	0.85 [-0.07, 1.78]
Bermon 1998	1	1	8	0.6	0.4	8	3.5%	0.50 [-0.50, 1.50]
Brose 2003 (men)	30	20	7	22	12	7	3.1%	0.45 [-0.61, 1.52]
Brose 2003 (women)	15	14	6	13	7	6	2.7%	0.17 [-0.97, 1.30]
Gualano 2011	8	16	13	13	8	12	5.6%	-0.38 [-1.17, 0.42]
Gualano 2014	2.6	4.3	15	1.8	4.9	15	6.8%	0.17 [-0.55, 0.89]
Hass 2007	0.2	0.07	10	0.1	0.06	10	3.4%	1.47 [0.46, 2.48]
Pinto 2016	12	5	13	11	6	14	6.1%	0.17 [-0.58, 0.93]
Subtotal (95% CI)			103			101	44.5%	0.33 [0.05, 0.61]

Heterogeneity: Chi² = 9.83, df = 9 (P = 0.36); I² = 8%
Test for overall effect: Z = 2.32 (P = 0.02)

Study or Subgroup	Creatine Mean	SD	Total	Placebo Mean	SD	Total	Weight	Std. Mean Difference IV, Fixed, 95% CI
1.2.2 High Dose								
Bernat 2019	19	12.9	12	17.58	7	12	5.5%	0.13 [-0.67, 0.93]
Candow 2008	11	11	13	9	14	12	5.7%	0.15 [-0.63, 0.94]
Candow 2015 (Cr after)	16	13	12	2	15	12	4.8%	0.96 [0.11, 1.82]
Candow 2015 (Cr before)	15	13	15	2	15	12	5.5%	0.91 [0.10, 1.71]
Candow 2020	11	20.9	18	22	21	20	8.4%	-0.51 [-1.16, 0.13]
Chilibeck 2015	18	15	15	11	14	18	7.3%	0.47 [-0.22, 1.17]
Chrusch 2001	19	13	16	16	11	14	6.8%	0.24 [-0.48, 0.96]
Cooke 2014	8	9	10	7	13	10	4.6%	0.09 [-0.79, 0.96]
Johannsmeyer 2016	15	20	14	13	22	17	7.0%	0.09 [-0.62, 0.80]
Subtotal (95% CI)			125			127	55.5%	0.23 [-0.02, 0.49]

Heterogeneity: Chi² = 11.42, df = 8 (P = 0.18); I² = 30%
Test for overall effect: Z = 1.82 (P = 0.07)

Study or Subgroup			Total			Total	Weight	Std. Mean Difference IV, Fixed, 95% CI
Total (95% CI)			228			228	100.0%	0.28 [0.09, 0.47]

Heterogeneity: Chi² = 21.51, df = 18 (P = 0.25); I² = 16%
Test for overall effect: Z = 2.91 (P = 0.004)
Test for subgroup differences: Chi² = 0.26, df = 1 (P = 0.61), I² = 0%

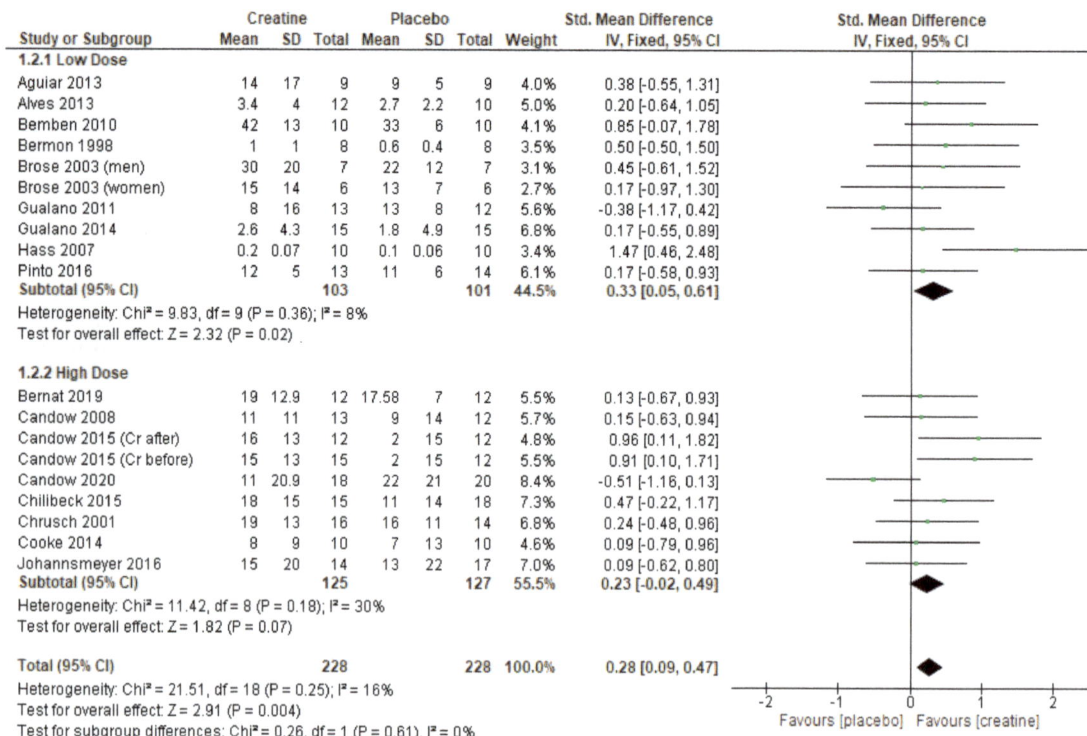

Figure 2. Forest plot of studies on chest press strength.

Study or Subgroup	Creatine Mean	SD	Total	Placebo Mean	SD	Total	Weight	Std. Mean Difference IV, Fixed, 95% CI
Aguiar 2013	14	17	9	9	5	9	16.1%	0.38 [-0.55, 1.31]
Alves 2013	3.4	4	12	2.7	2.2	10	0.0%	0.20 [-0.64, 1.05]
Bemben 2010	42	13	10	33	6	10	0.0%	0.85 [-0.07, 1.78]
Bermon 1998	1	1	8	0.6	0.4	8	0.0%	0.50 [-0.50, 1.50]
Bernat 2019	19	12.9	12	17.58	7	12	0.0%	0.13 [-0.67, 0.93]
Brose 2003 (men)	30	20	7	22	12	7	12.4%	0.45 [-0.61, 1.52]
Brose 2003 (women)	15	14	6	13	7	6	10.9%	0.17 [-0.97, 1.30]
Candow 2008	11	11	13	9	14	12	0.0%	0.15 [-0.63, 0.94]
Candow 2015 (Cr after)	16	13	12	2	15	12	0.0%	0.96 [0.11, 1.82]
Candow 2015 (Cr before)	15	13	15	2	15	12	0.0%	0.91 [0.10, 1.71]
Candow 2020	11	20.9	18	22	21	20	0.0%	-0.51 [-1.16, 0.13]
Chilibeck 2015	18	15	15	11	14	18	0.0%	0.47 [-0.22, 1.17]
Chrusch 2001	19	13	16	16	11	14	0.0%	0.24 [-0.48, 0.96]
Cooke 2014	8	9	10	7	13	10	0.0%	0.09 [-0.79, 0.96]
Gualano 2011	8	16	13	13	8	12	22.4%	-0.38 [-1.17, 0.42]
Gualano 2014	2.6	4.3	15	1.8	4.9	15	0.0%	0.17 [-0.55, 0.89]
Hass 2007	0.2	0.07	10	0.1	0.06	10	13.7%	1.47 [0.46, 2.48]
Johannsmeyer 2016	15	20	14	13	22	17	0.0%	0.09 [-0.62, 0.80]
Pinto 2016	12	5	13	11	6	14	24.6%	0.17 [-0.58, 0.93]
Total (95% CI)			58			58	100.0%	0.30 [-0.08, 0.67]

Heterogeneity: Chi² = 8.19, df = 5 (P = 0.15); I² = 39%
Test for overall effect: Z = 1.54 (P = 0.12)

Figure 3. Forest plot of lower-dose creatine studies (≤5 g/day) on chest press strength with exclusion of creatine loading studies.

3.3. Leg Press Strength

The analysis of 15 RCTs with 17 treatment arms (n = 426) revealed that creatine supplementation and resistance training significantly increased leg press strength vs. placebo and resistance training (Figure 4: standard mean difference = 0.20 [95% CI: 0.00, 0.39] p = 0.05).

Sub-analyses showed that higher-dose creatine, with and without a creatine-loading phase, produced greater gains in leg press strength vs. placebo (Figure 4: mean difference = 0.29 [95% CI: 0.04, 0.54] p = 0.02). However, when studies incorporating a creatine-loading phase were excluded, higher-dose creatine was similar to the placebo (Figure 5: p = 0.12).

Studies using lower-dose creatine, with and without a creatine-loading phase, had no greater effect on leg press strength vs. placebo (Figure 4; p = 0.69 and Figure S6; p = 0.88).

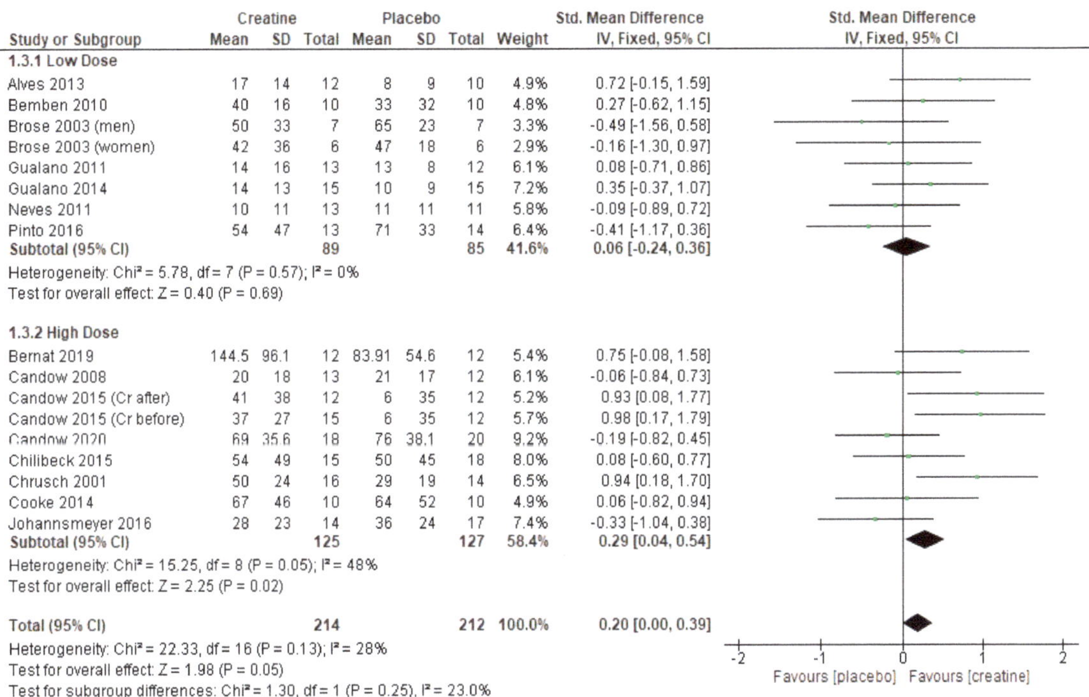

Figure 4. Forest plot of studies on leg press strength.

3.4. Creatine Only on Training Days

When only including studies that provided creatine on resistance training days, there were significant overall effects for favoring creatine on measures of lean tissue mass (Figure 6: mean difference = 1.73 kg [95% CI: 0.87, 2.89] p < 0.0001), chest press strength (Figure 7: standard mean difference = 0.58 [95% CI: 0.20, 0.96] p = 0.003), and leg press strength (Figure 8: standard mean difference = 0.44 [95% CI: 0.06, 0.81] p = 0.02). Of note, Cooke et al. [30] incorporated a creatine-loading phase followed by lower-dose creatine ($\leq$5 g/day) whereas the studies by Candow et al. [25,26] used higher-dose creatine (>5 g/day).

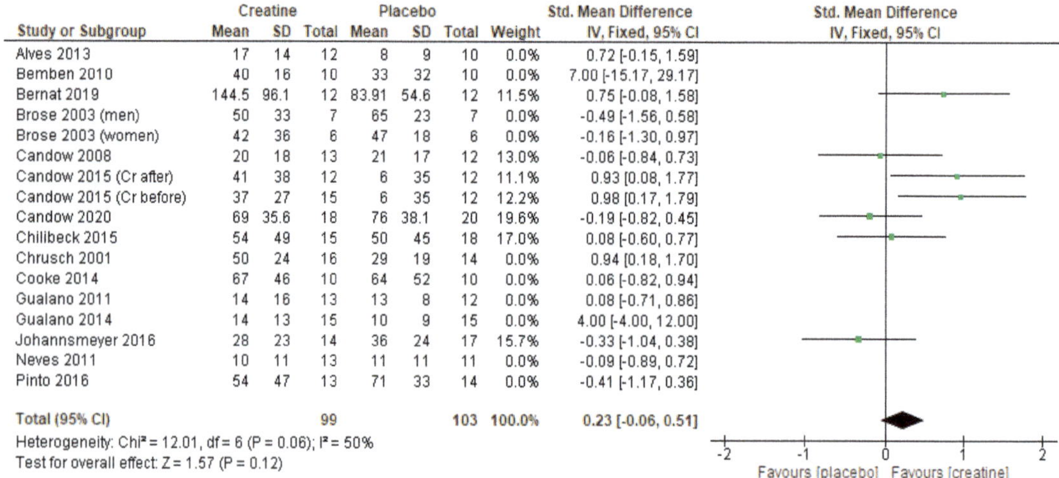

Study or Subgroup	Creatine Mean	SD	Total	Placebo Mean	SD	Total	Weight	Std. Mean Difference IV, Fixed, 95% CI
Alves 2013	17	14	12	8	9	10	0.0%	0.72 [-0.15, 1.59]
Bemben 2010	40	16	10	33	32	10	0.0%	7.00 [-15.17, 29.17]
Bernat 2019	144.5	96.1	12	83.91	54.6	12	11.5%	0.75 [-0.08, 1.58]
Brose 2003 (men)	50	33	7	65	23	7	0.0%	-0.49 [-1.56, 0.58]
Brose 2003 (women)	42	36	6	47	18	6	0.0%	-0.16 [-1.30, 0.97]
Candow 2008	20	18	13	21	17	12	13.0%	-0.06 [-0.84, 0.73]
Candow 2015 (Cr after)	41	38	12	6	35	12	11.1%	0.93 [0.08, 1.77]
Candow 2015 (Cr before)	37	27	15	6	35	12	12.2%	0.98 [0.17, 1.79]
Candow 2020	69	35.6	18	76	38.1	20	19.6%	-0.19 [-0.82, 0.45]
Chilibeck 2015	54	49	15	50	45	18	17.0%	0.08 [-0.60, 0.77]
Chrusch 2001	50	24	16	29	19	14	0.0%	0.94 [0.18, 1.70]
Cooke 2014	67	46	10	64	52	10	0.0%	0.06 [-0.82, 0.94]
Gualano 2011	14	16	13	13	8	12	0.0%	0.08 [-0.71, 0.86]
Gualano 2014	14	13	15	10	9	15	0.0%	4.00 [-4.00, 12.00]
Johannsmeyer 2016	28	23	14	36	24	17	15.7%	-0.33 [-1.04, 0.38]
Neves 2011	10	11	13	11	11	11	0.0%	-0.09 [-0.89, 0.72]
Pinto 2016	54	47	13	71	33	14	0.0%	-0.41 [-1.17, 0.36]
Total (95% CI)			99			103	100.0%	0.23 [-0.06, 0.51]

Heterogeneity: Chi² = 12.01, df = 6 (P = 0.06); I² = 50%
Test for overall effect: Z = 1.57 (P = 0.12)

Favours [placebo] Favours [creatine]

Figure 5. Forest plot of higher-dose creatine studies (>5 g/day) on leg press strength with exclusion of creatine-loading studies.

Study or Subgroup	Creatine Mean	SD	Total	Placebo Mean	SD	Total	Weight	Mean Difference IV, Fixed, 95% CI
Candow 2008	2.1	1.4	13	0.6	2.1	12	37.5%	1.50 [0.09, 2.91]
Candow 2015 (Cr after)	3	1.9	12	0.5	2.1	12	29.1%	2.50 [0.90, 4.10]
Candow 2015 (Cr before)	1.8	2	15	0.5	2.1	12	30.7%	1.30 [-0.26, 2.86]
Cooke 2014	2	4.8	10	0.5	6.9	10	2.8%	1.50 [-3.71, 6.71]
Total (95% CI)			50			46	100.0%	1.73 [0.87, 2.59]

Heterogeneity: Chi² = 1.29, df = 3 (P = 0.73); I² = 0%
Test for overall effect: Z = 3.92 (P < 0.0001)

Favours [placebo] Favours [creatine]

Figure 6. Forest plot of studies on lean tissue mass.

Study or Subgroup	Creatine Mean	SD	Total	Placebo Mean	SD	Total	Weight	Std. Mean Difference IV, Fixed, 95% CI
Bemben 2010	42	13	10	33	6	10	16.7%	0.85 [-0.07, 1.78]
Candow 2008	11	11	13	9	14	12	23.1%	0.15 [-0.63, 0.94]
Candow 2015 (Cr after)	16	13	12	2	15	12	19.6%	0.96 [0.11, 1.82]
Candow 2015 (Cr before)	15	13	15	2	15	12	22.1%	0.91 [0.10, 1.71]
Cooke 2014	8	9	10	7	13	10	18.5%	0.09 [-0.79, 0.96]
Total (95% CI)			60			56	100.0%	0.58 [0.20, 0.96]

Heterogeneity: Chi² = 4.08, df = 4 (P = 0.39); I² = 2%
Test for overall effect: Z = 3.02 (P = 0.003)

Favours [placebo] Favours [creatine]

Figure 7. Forest plot of studies on chest press strength.

Study or Subgroup	Creatine Mean	SD	Total	Placebo Mean	SD	Total	Weight	Std. Mean Difference IV, Fixed, 95% CI
Bemben 2010	40	16	10	33	32	10	18.1%	0.27 [-0.62, 1.15]
Candow 2008	20	18	13	21	17	12	22.8%	-0.06 [-0.84, 0.73]
Candow 2015 (Cr after)	41	38	12	6	35	12	19.4%	0.93 [0.08, 1.77]
Candow 2015 (Cr before)	37	27	15	6	35	12	21.4%	0.98 [0.17, 1.79]
Cooke 2014	67	46	10	64	52	10	18.3%	0.06 [-0.82, 0.94]
Total (95% CI)			60			56	100.0%	0.44 [0.06, 0.81]

Heterogeneity: Chi² = 5.35, df = 4 (P = 0.25); I² = 25%
Test for overall effect: Z = 2.28 (P = 0.02)

Favours [placebo] Favours [creatine]

Figure 8. Forest plot of studies on leg press strength.

3.5. Publication Bias

Funnel plots for each meta-analysis were visually inspected and showed no evidence of publication bias.

3.6. Adverse Events

In the lower-dose studies ($\leq$5 g/day), 10 studies reported no adverse events. One study reported a single mild bout of gastro-intestinal distress from creatine [16] and one study reported an overuse shoulder injury following creatine supplementation [18]. Neither of these studies used a loading phase.

In the higher-dose studies (>5 g/day), five studies reported no adverse events. Two studies similarly reported five incidences of gastrointestinal distress from creatine and two incidences from placebo and two incidences of muscle cramps from both the creatine and placebo group [27,28]. One of the two studies utilizing a loading phase reported an increase in GI distress during the loading phase [29].

4. Discussion

The most important results from these meta-analyses were: (1) creatine supplementation (independent of creatine-loading, maintenance dosage and frequency of ingestion) during a resistance training program increased measures of lean tissue mass and strength compared to the placebo and resistance training in older adults, (2) the combination of creatine-loading followed by lower-dose creatine ($\leq$5 g/day) was effective for increasing chest press strength, (3) the combination of creatine-loading and higher-dose creatine (>5 g/day) was effective for increasing leg press strength, (4) creatine supplementation only on resistance training days significantly increased measures of lean tissue mass and strength compared to the placebo. These results have application for the design of effective creatine supplementation strategies for older adults. For example, older adults wanting to improve whole-body lean tissue mass and strength may expect these benefits from creatine supplementation (i.e., $\geq$5 g) either daily or only on training days during a resistance training program.

Increasing whole-body lean tissue mass and strength is fundamental for mitigating sarcopenia and associated conditions of osteoporosis and physical frailty *(3)*. Older adults specifically looking to improve upper-body strength (perhaps to improve functionality, posture and/or the ability to perform upper-body activities of daily living such as carrying groceries) may need to load with creatine before proceeding to a lower daily dosage ($\leq$5 g) during their resistance training program. To specifically increase lower-body strength (perhaps to improve balance, reduce the risk of falls and/or the ability to perform lower-body activities of daily living such as climbing stairs), older adults may need to load with creatine before proceeding to a higher daily dosage (>5 g) during their resistance training program. While some have hypothesized creatine may have harmful effects [34], a plethora of evidence shows no adverse events (compared to the placebo) with long-term supplementation [35–37].

Previous meta-analyses have shown greater gains in measures of lean tissue mass (~1.2–1.3 kg) and strength from creatine supplementation and resistance training in older adults compared to the placebo [9,32,33]. Since the date of these publications, two additional studies [24,27] have been performed. When these studies were included in the current meta-analyses, creatine supplementation and resistance training still increased measures of lean tissue mass (~1.32 kg) and strength compared to the placebo. Collectively, results across meta-analyses suggest that the combination of creatine supplementation and resistance training has the potential to mitigate sarcopenia. Although none of the studies included in any of the meta-analyses were powdered to directly examine the effects of creatine vs. placebo in older adults diagnosed with sarcopenia, sub-analyses from three studies showed that the combination of creatine and resistance training eliminated the classification of sarcopenia in 11 older adults [20,23,26]. Creatine supplementation may augment lean tissue mass and strength through various mechanisms [3,4,10,32,37].

First, supplementation increases intramuscular PCr resulting in greater resynthesis of ATP during and following muscle contractions. Supplementation also increases muscle GLUT-4 content and translocation to the sarcolemma which may increase glucose uptake and subsequent glycogen resynthesis [38,39]. Creatine supplementation facilitates calcium re-uptake via creatine kinase into the sarcoplasmic reticulum, and this may increase myofibrillar cross-bride cycling, cell swelling, the expression of myogenic transcription factors (i.e., Mrf4, myogenin), satellite cell proliferation, and the expression of growth factors (i.e., insulin-like growth factor-1) [40,41]. Creatine supplementation enhances the activation of protein kinases downstream in the mammalian target of rapamycin (mTOR) pathway, and this may subsequently reduce measures of muscle protein catabolism (i.e., leucine oxidation, urinary 3-methylhistidine) [25,31]. Finally, creatine supplementation could reduce inflammation (i.e., cytokines) [42,43] and oxidative stress [44–46], and again, this may help reduce the loss of lean tissue mass with aging [4].

Incorporating a creatine-loading phase during the initial stages of a resistance training program was determined to be important for improving upper- and lower-body strength. It is well established that creatine-loading results in significant elevations in intramuscular creatine levels [47]. However, the magnitude of the effect on strength outcome measures may also depend on the maintenance dosage of creatine used for the remainder of the training program.

Regarding upper-body strength, older adults who loaded with creatine and then proceeded to ingest lower-dose creatine daily experienced greater upper body strength gains compared to those on placebo. However, independent of a creatine-loading phase, lower-dose creatine supplementation was no more effective than placebo. When all studies were included in the analysis, higher-dose creatine supplementation daily, with and without a creatine-loading phase, had no greater effect on upper-body strength compared to the placebo. However, sensitivity analysis showed that when the Candow et al. [27] study was removed, results became significant in favor of creatine. In this study, older males supplemented with higher-dose creatine daily during supervised, whole-body resistance training for 52 weeks. Results showed that changes in upper-body strength were similar between creatine and placebo over time. Both creatine and placebo groups experienced large increases in strength over time (creatine: ~69 kg; placebo: ~76 kg) which likely masked any effect from creatine supplementation.

Regarding lower-body strength, creatine-loading followed by higher-dose creatine daily had a favorable effect on strength whereas creatine-loading followed by lower-dose creatine daily had no greater effect compared to the placebo. The magnitude of responsiveness to creatine supplementation in older adults may depend on initial intramuscular creatine levels [10,48]. There is some evidence to suggest that phosphocreatine stores decrease with aging [10], especially in muscles of the lower limbs, possibly due to type-II muscle fiber atrophy, reduced participation in high-intensity activities and reduced meat consumption [32]. Furthermore, lower-body muscle groups are more negatively affected (i.e., greater strength deficit) by the aging process than upper-body muscle groups [49]. Therefore, to overcome possible age-related changes in muscle creatine content and lower-body muscle morphology, higher creatine dosages (as opposed to lower-creatine dosages) may be needed on a daily basis after a creatine-loading phase to improve lower-body strength in older adults.

Most importantly, all the studies identified as using a high dose (i.e., >5 g/day) were based on a relative dosing strategy (based on body mass; g/kg/day), while all the low dose studies used an absolute dosing strategy (g/day). As such, future research is required to directly compare an absolute and relative strategy to determine which method is superior.

Older adults who ingested creatine only on resistance training days experienced greater gains in measures of lean tissue mass and strength compared to the placebo. One study implemented a creatine-loading phase prior to lower-dose creatine daily [30] whereas the other studies implemented a higher-dose daily strategy [25,26]. A common theme across all studies was that creatine was consumed within 60 min' post-exercise. While the mechanistic actions of creatine were not determined in these studies, previous research has shown that prior muscle contractions (i.e., resistance training sessions) stimulate greater creatine uptake into muscle [50] possibly through increased activation of creatine transport kinetics [51,52]. These results may be important, as compliance to a creatine supplementation program may be higher when creatine is only consumed on training days. However, it is unknown whether older adults experience the same muscle benefits when consuming creatine supplementation daily vs. only on training days during a resistance training program. In addition, a provision of creatine from a regular diet should be accounted for a total exposure to creatine in this population since creatine consumption varies in the elderly [53].

Although the focus of this review was on combining creatine with resistance exercise, there appears to be some benefits of creatine without concomitant exercise in older adults [54,55]. Future research may be warranted to examine the dose of creatine to enhance muscle performance without exercise.

5. Conclusions

Increasing whole-body lean tissue mass and strength is fundamental for mitigating sarcopenia and associated conditions of osteoporosis and physical frailty [3]. Similar to previous meta-analyses [9,32], our results showed that creatine supplementation and resistance training increases measures of lean tissue mass and strength in older adults vs. placebo. However, unique and important results from our sub-analyses indicate that a creatine-loading phase is important for older adults wanting to improve muscle strength. In addition to a creatine-loading phase, a lower daily dosage of creatine ($\leq$5 g) appears sufficient to improve upper-body strength. However, a higher daily dosage of creatine (>5 g) after the loading phase is needed to increase lower-body strength. Regarding the effects of creatine ingestion frequency, creatine supplementation only on resistance training days significantly increased measures of lean tissue mass and strength compared to placebo.

Supplementary Materials: The following are available online at https://www.mdpi.com/article/10.3390/nu13061912/s1, Figures S1–S8.

Author Contributions: Conceptualization, S.C.F. and D.G.C.; methodology, S.C.F. and D.G.C.; formal analysis, D.G.C. and S.C.F.; writing—original draft preparation, S.C.F. and D.G.C.; writing—review and editing, S.C.F., D.G.C., S.M.O., M.D.R., P.D.C. All authors have read and agreed to the published version of the manuscript.

Funding: This research received no external funding.

Institutional Review Board Statement: Not applicable.

Informed Consent Statement: Not applicable.

Conflicts of Interest: S.C.F. has previously served as a scientific advisor for a company that sold creatine. D.G.C. has conducted industry-sponsored research involving creatine supplementation, received creatine donations for scientific studies and travel support for presentations involving creatine supplementation at scientific conferences. In addition, D.G.C. serves on the Scientific Advisory Board for AlzChem (a company which manufactures creatine) and has previously served as the Chief Scientific Officer for a company that sells creatine products. S.M.O. serves as a member of the Scientific Advisory Board on creatine in health and medicine (AlzChem LLC). S.M.O. owns patent "Sports Supplements Based on Liquid Creatine" at European Patent Office (WO2019150323 A1), and active patent application "Synergistic Creatine" at UK Intellectual Property Office (GB2012773.4). S.M.O. has served as a speaker at Abbott Nutrition, a consultant of Allied Beverages Adriatic and IMLEK, and an advisory board member for the University of Novi Sad School of Medicine, and has received research funding related to creatine from the Serbian Ministry of Education, Science, and Technological Development, Provincial Secretariat for Higher Education and Scientific Research, AlzChem GmbH, KW Pfannenschmidt GmbH, ThermoLife International LLC, and Monster Energy Company. S.M.O. is an employee of the University of Novi Sad and does not own stocks and shares in any organization. The funders had no role in the design of the study; in the collection, analyses, or interpretation of data; in the writing of the manuscript, or in the decision to publish the results.

References

1. Cruz-Jentoft, A.J.; Bahat, G.; Bauer, J.; Boirie, Y.; Bruyere, O.; Cederholm, T.; Cooper, C.; Landi, F.; Rolland, Y.; Sayer, A.A.; et al. Writing Group for the European Working Group on Sarcopenia in Older People 2 (EWGSOP2), and the Extended Group for EWGSOP2 Sarcopenia: Revised European consensus on definition and diagnosis. *Age Ageing* **2019**, *48*, 16–31. [CrossRef]
2. Shafiee, G.; Keshtkar, A.; Soltani, A.; Ahadi, Z.; Larijani, B.; Heshmat, R. Prevalence of sarcopenia in the world: A systematic review and meta- analysis of general population studies. *J. Diabetes Metab. Disord.* **2017**, *16*, 21. [CrossRef]
3. Candow, D.G.; Forbes, S.C.; Kirk, B.; Duque, G. Current Evidence and Possible Future Applications of Creatine Supplementation for Older Adults. *Nutrients* **2021**, *13*, 745. [CrossRef]
4. Candow, D.G.; Forbes, S.C.; Chilibeck, P.D.; Cornish, S.M.; Antonio, J.; Kreider, R.B. Effectiveness of Creatine Supplementation on Aging Muscle and Bone: Focus on Falls Prevention and Inflammation. *J. Clin. Med.* **2019**, *8*, 488. [CrossRef]
5. Mitchell, W.K.; Williams, J.; Atherton, P.; Larvin, M.; Lund, J.; Narici, M. Sarcopenia, dynapenia, and the impact of advancing age on human skeletal muscle size and strength; a quantitative review. *Front. Physiol.* **2012**, *3*, 260. [CrossRef] [PubMed]
6. Tournadre, A.; Vial, G.; Capel, F.; Soubrier, M.; Boirie, Y. Sarcopenia. *Jt. Bone Spine* **2019**, *86*, 309–314. [CrossRef] [PubMed]
7. Kirk, B.; Prokopidis, K.; Duque, G. Nutrients to mitigate osteosarcopenia: The role of protein, vitamin D and calcium. *Curr. Opin. Clin. Nutr. Metab. Care* **2021**, *24*, 25–32. [CrossRef]
8. Brosnan, J.T.; Brosnan, M.E. Creatine: Endogenous metabolite, dietary, and therapeutic supplement. *Annu. Rev. Nutr.* **2007**, *27*, 241–261. [CrossRef] [PubMed]
9. Candow, D.G.; Chilibeck, P.D.; Forbes, S.C. Creatine supplementation and aging musculoskeletal health. *Endocrine* **2014**, *45*, 354–361. [CrossRef] [PubMed]
10. Candow, D.G.; Forbes, S.C.; Chilibeck, P.D.; Cornish, S.M.; Antonio, J.; Kreider, R.B. Variables Influencing the Effectiveness of Creatine Supplementation as a Therapeutic Intervention for Sarcopenia. *Front. Nutr.* **2019**, *6*, 124. [CrossRef]
11. Alves, C.R.; Merege Filho, C.A.; Benatti, F.B.; Brucki, S.; Pereira, R.M.; de Sa Pinto, A.L.; Lima, F.R.; Roschel, H.; Gualano, B. Creatine supplementation associated or not with strength training upon emotional and cognitive measures in older women: A randomized double-blind study. *PLoS ONE* **2013**, *8*, e76301. [CrossRef]
12. Aguiar, A.F.; Januario, R.S.; Junior, R.P.; Gerage, A.M.; Pina, F.L.; do Nascimento, M.A.; Padovani, C.R.; Cyrino, E.S. Long-term creatine supplementation improves muscular performance during resistance training in older women. *Eur. J. Appl. Physiol.* **2013**, *113*, 987–996. [CrossRef] [PubMed]
13. Bemben, M.G.; Witten, M.S.; Carter, J.M.; Eliot, K.A.; Knehans, A.W.; Bemben, D.A. The effects of supplementation with creatine and protein on muscle strength following a traditional resistance training program in middle-aged and older men. *J. Nutr. Health Aging* **2010**, *14*, 155–159. [CrossRef] [PubMed]
14. Eliot, K.A.; Knehans, A.W.; Bemben, D.A.; Witten, M.S.; Carter, J.; Bemben, M.G. The effects of creatine and whey protein supplementation on body composition in men aged 48 to 72 years during resistance training. *J. Nutr. Health Aging* **2008**, *12*, 208–212. [CrossRef]
15. Bermon, S.; Venembre, P.; Sachet, C.; Valour, S.; Dolisi, C. Effects of creatine monohydrate ingestion in sedentary and weight-trained older adults. *Acta Physiol. Scand.* **1998**, *164*, 147–155. [CrossRef] [PubMed]
16. Brose, A.; Parise, G.; Tarnopolsky, M.A. Creatine supplementation enhances isometric strength and body composition improvements following strength exercise training in older adults. *J. Gerontol. A Biol. Sci. Med. Sci.* **2003**, *58*, 11–19. [CrossRef] [PubMed]
17. Deacon, S.J.; Vincent, E.E.; Greenhaff, P.L.; Fox, J.; Steiner, M.C.; Singh, S.J.; Morgan, M.D. Randomized controlled trial of dietary creatine as an adjunct therapy to physical training in chronic obstructive pulmonary disease. *Am. J. Respir. Crit. Care Med.* **2008**, *178*, 233–239. [CrossRef] [PubMed]
18. Eijnde, B.O.; Van Leemputte, M.; Goris, M.; Labarque, V.; Taes, Y.; Verbessem, P.; Vanhees, L.; Ramaekers, M.; Vanden Eynde, B.; Van Schuylenbergh, R.; et al. Effects of creatine supplementation and exercise training on fitness in men 55–75 yr old. *J. Appl. Physiol.* **2003**, *95*, 818–828. [CrossRef] [PubMed]
19. Gualano, B.; DE Salles Painneli, V.; Roschel, H.; Artioli, G.G.; Neves, M.; De Sa Pinto, A.L.; Da Silva, M.E.; Cunha, M.R.; Otaduy, M.C.; Leite Cda, C.; et al. Creatine in type 2 diabetes: A randomized, double-blind, placebo-controlled trial. *Med. Sci. Sports Exerc.* **2011**, *43*, 770–778. [CrossRef] [PubMed]
20. Gualano, B.; Macedo, A.R.; Alves, C.R.; Roschel, H.; Benatti, F.B.; Takayama, L.; de Sa Pinto, A.L.; Lima, F.R.; Pereira, R.M. Creatine supplementation and resistance training in vulnerable older women: A randomized double-blind placebo-controlled clinical trial. *Exp. Gerontol.* **2014**, *53*, 7–15. [CrossRef]
21. Hass, C.J.; Collins, M.A.; Juncos, J.L. Resistance training with creatine monohydrate improves upper-body strength in patients with Parkinson disease: A randomized trial. *Neurorehabil. Neural Repair* **2007**, *21*, 107–115. [CrossRef]
22. Neves, M.; Gualano, B.; Roschel, H.; Fuller, R.; Benatti, F.B.; Pinto, A.L.; Lima, F.R.; Pereira, R.M.; Lancha, A.H.; Bonfa, E. Beneficial effect of creatine supplementation in knee osteoarthritis. *Med. Sci. Sports Exerc.* **2011**, *43*, 1538–1543. [CrossRef]
23. Pinto, C.L.; Botelho, P.B.; Carneiro, J.A.; Mota, J.F. Impact of creatine supplementation in combination with resistance training on lean mass in the elderly. *J. Cachexia Sarcopenia Muscle* **2016**, *7*, 413–421. [CrossRef]
24. Bernat, P.; Candow, D.G.; Gryzb, K.; Butchart, S.; Schoenfeld, B.J.; Bruno, P. Effects of high-velocity resistance training and creatine supplementation in untrained healthy aging males. *Appl. Physiol. Nutr. Metab.* **2019**, *44*, 1246–1253. [CrossRef]

25. Candow, D.G.; Little, J.P.; Chilibeck, P.D.; Abeysekara, S.; Zello, G.A.; Kazachkov, M.; Cornish, S.M.; Yu, P.H. Low-dose creatine combined with protein during resistance training in older men. *Med. Sci. Sports Exerc.* **2008**, *40*, 1645–1652. [CrossRef]
26. Candow, D.G.; Vogt, E.; Johannsmeyer, S.; Forbes, S.C.; Farthing, J.P. Strategic creatine supplementation and resistance training in healthy older adults. *Appl. Physiol. Nutr. Metab.* **2015**, *40*, 689–694. [CrossRef] [PubMed]
27. Candow, D.G.; Chilibeck, P.D.; Gordon, J.; Vogt, E.; Landeryou, T.; Kaviani, M.; Paus-Jensen, L. Effect of 12 months of creatine supplementation and whole-body resistance training on measures of bone, muscle and strength in older males. *Nutr. Health* **2020**. [CrossRef]
28. Chilibeck, P.D.; Candow, D.G.; Landeryou, T.; Kaviani, M.; Paus-Jenssen, L. Effects of Creatine and Resistance Training on Bone Health in Postmenopausal Women. *Med. Sci. Sports Exerc.* **2015**, *47*, 1587–1595. [CrossRef] [PubMed]
29. Chrusch, M.J.; Chilibeck, P.D.; Chad, K.E.; Davison, K.S.; Burke, D.G. Creatine supplementation combined with resistance training in older men. *Med. Sci. Sports Exerc.* **2001**, *33*, 2111–2117. [CrossRef] [PubMed]
30. Cooke, M.B.; Brabham, B.; Buford, T.W.; Shelmadine, B.D.; McPheeters, M.; Hudson, G.M.; Stathis, C.; Greenwood, M.; Kreider, R.; Willoughby, D.S. Creatine supplementation post-exercise does not enhance training-induced adaptations in middle to older aged males. *Eur. J. Appl. Physiol.* **2014**, *114*, 1321–1332. [CrossRef] [PubMed]
31. Johannsmeyer, S.; Candow, D.G.; Brahms, C.M.; Michel, D.; Zello, G.A. Effect of creatine supplementation and drop-set resistance training in untrained aging adults. *Exp. Gerontol.* **2016**, *83*, 112–119. [CrossRef] [PubMed]
32. Chilibeck, P.D.; Kaviani, M.; Candow, D.G.; Zello, G.A. Effect of creatine supplementation during resistance training on lean tissue mass and muscular strength in older adults: A meta-analysis. *Open Access J. Sports Med.* **2017**, *8*, 213–226. [CrossRef] [PubMed]
33. Devries, M.C.; Phillips, S.M. Creatine supplementation during resistance training in older adults-a meta-analysis. *Med. Sci. Sports Exerc.* **2014**, *46*, 1194–1203. [CrossRef] [PubMed]
34. Yu, P.H.; Deng, Y. Potential cytotoxic effect of chronic administration of creatine, a nutrition supplement to augment athletic performance. *Med. Hypotheses* **2000**, *54*, 726–728. [CrossRef] [PubMed]
35. Antonio, J.; Candow, D.G.; Forbes, S.C.; Gualano, B.; Jagim, A.R.; Kreider, R.B.; Rawson, E.S.; Smith-Ryan, A.E.; VanDusseldorp, T.A.; Willoughby, D.S.; et al. Common questions and misconceptions about creatine supplementation: What does the scientific evidence really show? *J. Int. Soc. Sports Nutr.* **2021**, *18*, 13. [CrossRef]
36. Dalbo, V.J.; Roberts, M.D.; Stout, J.R.; Kerksick, C.M. Putting to rest the myth of creatine supplementation leading to muscle cramps and dehydration. *Br. J. Sports Med.* **2008**, *42*, 567–573. [CrossRef] [PubMed]
37. Kreider, R.B.; Kalman, D.S.; Antonio, J.; Ziegenfuss, T.N.; Wildman, R.; Collins, R.; Candow, D.G.; Kleiner, S.M.; Almada, A.L.; Lopez, H.L. International Society of Sports Nutrition position stand: Safety and efficacy of creatine supplementation in exercise, sport, and medicine. *J. Int. Soc. Sports Nutr.* **2017**, *14*, 18. [CrossRef]
38. Ju, J.S.; Smith, J.L.; Oppelt, P.J.; Fisher, J.S. Creatine feeding increases GLUT4 expression in rat skeletal muscle. *Am. J. Physiol. Endocrinol. Metab.* **2005**, *288*, 347. [CrossRef]
39. Roberts, P.A.; Fox, J.; Peirce, N.; Jones, S.W.; Casey, A.; Greenhaff, P.L. Creatine ingestion augments dietary carbohydrate mediated muscle glycogen supercompensation during the initial 24 h of recovery following prolonged exhaustive exercise in humans. *Amino Acids* **2016**, *48*, 1831–1842. [CrossRef]
40. Burke, D.G.; Candow, D.G.; Chilibeck, P.D.; MacNeil, L.G.; Roy, B.D.; Tarnopolsky, M.A.; Ziegenfuss, T. Effect of creatine supplementation and resistance-exercise training on muscle insulin-like growth factor in young adults. *Int. J. Sport Nutr. Exerc. Metab.* **2008**, *18*, 389–398. [CrossRef]
41. Safdar, A.; Yardley, N.J.; Snow, R.; Melov, S.; Tarnopolsky, M.A. Global and targeted gene expression and protein content in skeletal muscle of young men following short-term creatine monohydrate supplementation. *Physiol. Genom.* **2008**, *32*, 219–228. [CrossRef] [PubMed]
42. Bassit, R.A.; Curi, R.; Costa Rosa, L.F. Creatine supplementation reduces plasma levels of pro-inflammatory cytokines and PGE2 after a half-ironman competition. *Amino Acids* **2008**, *35*, 425–431. [CrossRef]
43. Santos, R.V.; Bassit, R.A.; Caperuto, E.C.; Costa Rosa, L.F. The effect of creatine supplementation upon inflammatory and muscle soreness markers after a 30km race. *Life Sci.* **2004**, *75*, 1917–1924. [CrossRef] [PubMed]
44. Saraiva, A.L.; Ferreira, A.P.; Silva, L.F.; Hoffmann, M.S.; Dutra, F.D.; Furian, A.F.; Oliveira, M.S.; Fighera, M.R.; Royes, L.F. Creatine reduces oxidative stress markers but does not protect against seizure susceptibility after severe traumatic brain injury. *Brain Res. Bull.* **2012**, *87*, 180–186. [CrossRef] [PubMed]
45. Rahimi, R. Creatine supplementation decreases oxidative DNA damage and lipid peroxidation induced by a single bout of resistance exercise. *J. Strength Cond. Res.* **2011**, *25*, 3448–3455. [CrossRef] [PubMed]
46. Deminice, R.; Rosa, F.T.; Franco, G.S.; Jordao, A.A.; de Freitas, E.C. Effects of creatine supplementation on oxidative stress and inflammatory markers after repeated-sprint exercise in humans. *Nutrition* **2013**, *29*, 1127–1132. [CrossRef]
47. Harris, R.C.; Soderlund, K.; Hultman, E. Elevation of creatine in resting and exercised muscle of normal subjects by creatine supplementation. *Clin. Sci.* **1992**, *83*, 367–374. [CrossRef] [PubMed]
48. Syrotuik, D.G.; Bell, G.J. Acute creatine monohydrate supplementation. A descriptive physiological profile of responders vs. nonresponders. *J. Strength Cond. Res.* **2004**, *18*, 610–617. [PubMed]
49. Candow, D.G.; Chilibeck, P.D. Differences in size, strength, and power of upper and lower body muscle groups in young and older men. *J. Gerontol. A Biol. Sci. Med. Sci.* **2005**, *60*, 148–156. [CrossRef]

50. Robinson, T.M.; Sewell, D.A.; Hultman, E.; Greenhaff, P.L. Role of submaximal exercise in promoting creatine and glycogen accumulation in human skeletal muscle. *J. Appl. Physiol.* **1999**, *87*, 598–604. [CrossRef]
51. Persky, A.M.; Brazeau, G.A.; Hochhaus, G. Pharmacokinetics of the dietary supplement creatine. *Clin. Pharmacokinet.* **2003**, *42*, 557–574. [CrossRef] [PubMed]
52. Forbes, S.C.; Candow, D.G. Timing of creatine supplementation and resistance training: A brief review. *J. Exerc. Nutr.* **2018**, *1*, 1.
53. Ostojic, S.M.; Korovljev, D.; Stajer, V. Dietary creatine and cognitive function in U.S. adults aged 60 years and over. *Aging Clin. Exp. Res.* **2021**. [CrossRef] [PubMed]
54. Forbes, S.C.; Candow, D.G.; Ferreira, L.H.B.; Souza-Junior, T.P. Effects of Creatine Supplementation on Properties of Muscle, Bone, and Brain Function in Older Adults: A Narrative Review. *J. Diet. Suppl.* **2021**, 1–18. [CrossRef]
55. Moon, A.; Heywood, L.; Rutherford, S.; Cobbold, C. Creatine supplementation: Can it improve quality of life in the elderly without associated resistance training? *Curr. Aging Sci.* **2013**, *6*, 251–257. [CrossRef]

Article

Creatine Monohydrate Supplementation Increases White Adipose Tissue Mitochondrial Markers in Male and Female Rats in a Depot Specific Manner

Chantal R. Ryan [1], Michael S. Finch [1], Tyler C. Dunham [2], Jensen E. Murphy [2], Brian D. Roy [2] and Rebecca E. K. MacPherson [1,*]

[1] Department of Health Sciences, Brock University, St. Catharines, ON L2S 3A1, Canada; cr13ai@brocku.ca (C.R.R.); mf15kj@brocku.ca (M.S.F.)
[2] Department of Kinesiology, Brock University, St. Catharines, ON L2S 3A1, Canada; td13nb@brocku.ca (T.C.D.); jm15ln@brocku.ca (J.E.M.); broy@brocku.ca (B.D.R.)
* Correspondence: rmacpherson@brocku.ca; Tel.: +1-905-688-5550

Citation: Ryan, C.R.; Finch, M.S.; Dunham, T.C.; Murphy, J.E.; Roy, B.D.; MacPherson, R.E.K. Creatine Monohydrate Supplementation Increases White Adipose Tissue Mitochondrial Markers in Male and Female Rats in a Depot Specific Manner. *Nutrients* 2021, 13, 2406. https://doi.org/10.3390/nu13072406

Academic Editors: Richard B. Kreider and Jeffrey R. Stout

Received: 26 May 2021
Accepted: 25 June 2021
Published: 14 July 2021

Abstract: White adipose tissue (WAT) is a dynamic endocrine organ that can play a significant role in thermoregulation. WAT has the capacity to adopt structural and functional characteristics of the more metabolically active brown adipose tissue (BAT) and contribute to non-shivering thermogenesis under specific stimuli. Non-shivering thermogenesis was previously thought to be uncoupling protein 1 (UCP1)-dependent however, recent evidence suggests that UCP1-independent mechanisms of thermogenesis exist. Namely, futile creatine cycling has been identified as a contributor to WAT thermogenesis. The purpose of this study was to examine the efficacy of creatine supplementation to alter mitochondrial markers as well as adipocyte size and multilocularity in inguinal (iWAT), gonadal (gWAT), and BAT. Thirty-two male and female Sprague-Dawley rats were treated with varying doses (0 g/L, 2.5 g/L, 5 g/L, and 10 g/L) of creatine monohydrate for 8 weeks. We demonstrate that mitochondrial markers respond in a sex and depot specific manner. In iWAT, female rats displayed significant increases in COXIV, PDH-E1alpha, and cytochrome C protein content. Male rats exhibited gWAT specific increases in COXIV and PDH E1alpha protein content. This study supports creatine supplementation as a potential method of UCP1-independant thermogenesis and highlights the importance of taking a sex-specific approach when examining the efficacy of browning therapeutics in future research.

Keywords: adipose tissue; creatine; mitochondria; thermogenesis

1. Introduction

The activation of white adipose tissue (WAT) into a more metabolically active tissue has become a burning topic in obesity prevention and treatment. Today, WAT is no longer regarded as an inert storage depot for triacylglycerides, but is considered to be a highly plastic endocrine organ that can undergo dramatic phenotypic changes in response to different stresses (i.e., exercise, cold) [1–4]. Recently, there has been a large focus on the mechanisms by which WAT can be recruited to resemble the phenotypical and functional characteristics of the more metabolically active brown adipose tissue (BAT). This process is known as adipose tissue "browning" and results in the activation of beige adipocytes that reside in WAT depots [5]. Lineage tracing and molecular analysis indicates these beige adipocytes are distinct thermogenic fat cells [6], however beige adipocytes can convert to a WAT phenotype when the stimulus, such as cold exposure, is terminated [7], indicating that further work is necessary to fully distinguish white and beige adipocyte lineages. Considering the vast amount of WAT that individuals possess, the activation of WAT into the more metabolically active form of beige adipose tissue has the potential to significantly increase daily energy expenditure [4,8]. Thus, it could be regarded as a potential therapeutic treatment to combat the obesity epidemic and its related comorbidities.

493

Morphologically, mature differentiated white adipocytes are described as unilocular—containing a single large lipid droplet, few mitochondria, and a nucleus that has been pushed to the border of the cell membrane [9]. In contrast, brown adipose tissue (BAT) is described as multilocular with several smaller lipid droplets, a more central nucleus, and a dense population of mitochondria, ultimately giving the tissue its characteristic brown appearance [9]. Beige adipocytes lie along a continuum between WAT and BAT. The thermogenic capacity of brown and beige fat relies predominantly on a fatty acid/H+ symport mechanism mediated by uncoupling protein 1 (UCP1) [10–13]. This inner mitochondrial membrane protein stimulates thermogenesis by uncoupling the electron transport chain. Specifically, UCP1 dissipates the proton motive force and increases the rate of substrate flux through the mitochondrial electron transport chain [14]. This UCP1 dependent process is the most studied mechanism underlying the thermogenic capabilities of BAT, however, recently, UCP1-independent modes of thermogenesis have been uncovered.

In comparing the response of wildtype and UCP1 knockout mice to adrenergic stimulation, it was found that the thermogenic response (pharmacological and cold-induced) was similar [15–18], implying the presence of UCP1-independent thermogenic mechanisms. Previous work has found that reductions in high energy-phosphate compounds, such as creatine, result in a dysregulation of thermogenesis [19,20] and a role for creatine cycling in adipose tissue thermogenesis was suggested almost four decades ago [1]. In 2015, Kazak et al. [14] performed quantitative mitochondrial proteomics and identified creatine metabolism as a signature of beige fat from cold-exposed mice. Interestingly, both genetic and pharmacological depletion of adipose creatine potentiates diet-induced obesity [21,22], and inactivation of creatine transport results in fat accumulation [21]. In adipose tissue, the creatine pool is regulated by intracellular synthesis and by influx from circulation. The forward and reverse phospho-transfer reactions of phosphocreatine (PCr)/creatine in most cells occur in a 1:1 stoichiometry with the ATP/ADP coupled. However, in thermogenic adipocytes, it has been estimated that there is an excessive release of ADP with respect to creatine [14,23]. Therefore, it is thought that creatine facilitates the regeneration of ADP through futile hydrolysis of PCr [14]. Importantly, it was recently demonstrated that creatine kinase B (CKB) traffics to the mitochondria where it plays an important role in the futile creatine cycle [24] and that tissue-nonspecific alkaline phosphatase (TNAP) localizes to the mitochondria where it acts as a robust PCr phosphatase in thermogenic fat [25]. Together, these studies demonstrate that adipocyte creatine energetics can be a key regulator of thermogenesis [14,20–23].

Creatine monohydrate (CM) is a stable form of creatine with an attached molecule of water and is one of the most widely used and researched oral supplements (for review [26,27]. CM supplementation is well known for its effects on enhancing body composition, muscle mass and health, and exercise performance [28,29]. Supplementation is commonly in the range of 5 to 20 g/day which all lie within a range that is safe and tolerable for consumption [27,30]. While most of the work in the area has focused on CM supplementation and skeletal muscle outcomes, limited investigations have been performed to demonstrate that CM supplementation can increase creatine concentrations in other tissues. CM supplementation increases creatine concentrations in cardiac muscle, brain, kidney, liver, and lung tissue female rodents [31]—however, adipose tissue was not analyzed.

The purpose of the current investigation was to determine the effect of creatine supplementation on WAT browning in male and female Sprague-Dawley rats. To examine this question, male and female rats were separated into four different groups and supplemented with different doses of creatine (0, 2.5, 5 and 10 g/L) for 8 weeks. The results of this study provide novel information about the potential of creatine supplementation to induce WAT browning and further provide insight into the sex-specific responses of creatine supplementation.

2. Materials and Methods

2.1. Animals and Study Design

Experimental protocols were approved by the Brock University Animal Care Committee (file #19-02-01) and are in compliance with the Canadian Council on Animal Care. Thirty-two Sprague-Dawley rats (16 male, average body weight 372.4 g $\pm$ 12.1 g; 16 female, average body weight 311.8 g $\pm$ 21.0 g) were ordered from Charles River Laboratories (Wilmington, MA, USA) at 11 weeks of age. Rats acclimatized for 7 days in the Brock University Comparative Biosciences Facility. All rats were kept on a 12-h light:12-h dark cycle and had ad libitum access to food (AIN-93G pellets) and water throughout the duration of the study. Rats were separated by sex, housed in pairs, and randomly assigned into one of four experimental groups: (1) control (1% sucrose via drinking water), (2) 2.5 g/L, (3) 5 g/L, and (4) 10 g/L of creatine monohydrate (CM) (Sigma-Aldrich; CAT# C3630) and 1% sucrose via drinking water. Incremental doses of creatine monohydrate were selected to determine if there would be a dose response in adipose tissue adaptations. Rat weight, food, and water intake were measured three times a week for 8 weeks [32].

2.2. Tissue Collection and Homogenization

After the 8-week feeding period, rats underwent non-survival surgeries using isoflurane gas anesthesia. WAT samples were collected from inguinal subcutaneous fat depots (iWAT) and gonadal fat depots (gWAT; epididymal and ovarian), while BAT samples were collected from the interscapular fat pads. Samples were divided and placed in either formalin for histological analysis or snap frozen in liquid nitrogen and stored at $-80\ ^\circ$C for analysis via Western blotting. Blood samples were collected from the heart and rats were euthanized via exsanguination.

2.3. Western Blotting

Samples were homogenized via FAST prep (FastPrep®, MP Biomedicals, Santa Ana, CA, USA) in 3 and 10 volumes (WAT and BAT samples respectively) of NP40 Cell Lysis Buffer (Life Technologies; CAT# FNN0021) supplemented with 34 µL phenylmethylsulfonyl fluoride and 50 µL protease inhibitor cocktail (Sigma; CAT# 7626-5G, CAT# P274-1BIL). Homogenates were then centrifuged at 4 $^\circ$C for 5 min at 1500$\times$ g, after which the infranatant was collected and protein concentration was determined using a Bicinchoninic acid assay (Sigma-Aldrich—B9643, VWR—BDH9312). Homogenates were prepared in 2$\times$ Laemmli sample buffer (1 µg/µL) and were denatured at 100 $^\circ$C for 5 min. Equal amounts of sample (10–20 µg) were loaded to undergo protein separation via SDS-PAGE (4% stacking, 10–15% resolving gel) for 90 min at 120 V). Protein was wet transferred to 0.45 µm nitrocellulose membranes (CAT# 10600001, Milipore Sigma Burlington, MA, USA) at 100 V for 60 min. Membranes were blocked in tris buffered saline/0.1% tween 20 (TBST) prepared with 5% non-fat dry milk for 1 h followed by overnight incubation at 4 $^\circ$C with the appropriate primary antibody. Following primary incubation, membranes were rinsed with TBST and incubated with the appropriate Horseradish peroxidase-conjugated secondary antibodies for 1 h at room temperature. Ponceau staining was used to confirm equal protein loading (<10% variability across the membrane). Signals were detected using enhanced chemiluminescence (Western Lightening ® Plus ECL, Perkin Elmer, MA, USA) and were subsequently quantified by densitometry using a FluorChem HD imaging system (Alpha Innotech, Santa Clara, CA, USA). The primary antibodies included: cytochrome C (ABCAM, CAT# ab76237), citrate synthase (ABCAM, CAT# ab96600), COXIV (Molecular Probes, CAT# A-21348), PDH (Millipore, CAT# ABS2082), UCP-1 (ABCAM, CAT# ab10983), PGC-1α (Millipore, CAT# AB3242), GAMT (ABCAM, CAT# ab126736), and CKB (Abclonal; cat. no. ab12631).

2.4. Real-Time qPCR

Adipose mRNA was extracted and reverse transcribed into cDNA and changes in mRNA expression were determined using real-time quantitative PCR as described previously [33]. RNA was isolated from adipose tissue following homogenization in Trizol reagent using an RNeasy kit according to the manufacturer's instructions (RNeasy Kit 74106; Qiagen). RNA yield and purity were determined using a Nano-drop system (NanoVue plus; GE healthcare). RNA samples were prepared at 1 μg/μL using RNase free water. cDNA was synthesized using random primers and dNTP (Invitrogen) at a 1:1 ratio as well as a master mix (5 × FSB, DTT, RNase out and SuperScript II Reverse Transcriptase). 7500 Fast Real-Time PCR system (Applied Biosystems) was used to perform the RT qPCR. Samples were loaded in duplicate and contained 10 μL of PCR master mix, 4 μL of RNase free water, 1 μL of gene expression assay, and 5 μL of cDNA. Gene expression assays were purchased for *Slc6a8* (Rn00506029_mL). *Gapdh* was used as a housekeeping gene and was not different between groups. Relative differences in *Slc6a8* were determined using the $2^{-\Delta\Delta CT}$ method and normalized to the respective control group [34].

2.5. Histology

Samples (gWAT, iWAT, and BAT) underwent fixation in 10% neutral buffered formalin (Millipore Sigma, CAT#HT501128) for 62 h. Following fixation, samples were transferred into 70% ethanol for future processing. Samples underwent dehydration via ethanol (1 × 90% 30 min, 3 × 100% 40 min) and xylene (Fischer Scientific) (3 × 45 min). Samples were embedded in paraffin and 10 μm sections were mounted on 1.2 mm Superfrost™ slides. Slides were stained with Harris hematoxylin and eosin (H&E), imaged using a Nikon Eclipse 80i microscope (CAT#PL-D655CU-CYL), and images were captured with Pixelink software. Three images from each animal (~150 cells/image) were sampled to determine cross-sectional area and percent multilocular (ImageJ software, National Institute of Mental Health, Bethesda, MD, USA).

2.6. Statistical Analysis

Control male and female comparisons were made with a one-way analysis of variance (ANOVA). Comparisons within sex and across doses were made with a one-way ANOVA with all measurements being made relative to the control groups. Post-hoc analysis was completed with a Tukey's multiple comparisons test. Statistical significance was assumed at $p \leq 0.05$, and GraphPad Prism 8 software (GraphPad Software, La Jolla, CA, USA) was used to perform all statistical analyses. Results are stated and presented as mean ± SEM for all groups.

3. Results

3.1. Animal and Adipose Tissue Depot Characteristics

No differences were observed for male or female rat body mass at the end of the intervention between creatine doses. Female rats had a lower body mass compared to males in each creatine supplementation group ($p < 0.05$). Food and water intake were not different for either male or female rats in each creatine supplementation group. Normalizing for body mass, female rats ate more when compared to males in each treatment group ($p < 0.05$). Total creatine consumption relative to body mass was different between doses for both male and female rats. Relative creatine consumption was higher in the 2.5 g/L ($p = 0.001$), 5 g/L ($p < 0.001$), and 10 g/L ($p < 0.001$) groups compared to the control group. Compared to the 2.5 g/L group, the 5 g/L ($p = 0.005$) and 10 g/L ($p < 0.001$) groups had higher relative creatine consumption. The 10 g/L group consumed more creatine relative to bodyweight than the 5 g/L group ($p < 0.001$). No differences were observed across treatment groups for male or female adipose tissue protein content (ug protein/mg tissue; Table 1). Using rat weights and water consumption data, it is estimated that rats consumed 0.22 g/kg/day (2.5 g/L group), 0.41 g/kg/day (5 g/L group), and 0.79 g/kg/day (10 g/L group) with no difference in relative creatine intake between sexes. The Human Equivalent Dose based on body surface area adjustments [35] suggests that these doses would be equivalent to 2.5, 4.7, and 9.0 g/day in a 7-kg human—normal ranges of CM supplementation.

Table 1. Animal and adipose depot characteristics.

Animal Characteristics			
Creatine (g·L^{-1})	0	2.5 $^\$$	5 $^\&$	10
End point body mass (g)			
Males	587.8 ± 9.1	602.8 ± 14.1	610.3 ± 25.3	584.5 ±23.0
Females	376.8 ± 13.5 *	416.3 ±13.5 *	397.0 ± 7.1 *	404.8 ± 28.3 *
Food Intake (g·day^{-1})			
Males	24.4 ± 0.3	24.3 ± 0.9	23.7 ± 1.7	23.8 ± 1.3
Females	17.7 ± 1.2 *	19.0 ± 0.7 *	18.3 ± 0.1 *	18.75 ± 1.2 *
Water Intake (mL·day^{-1})			
Males	57.9 ± 14.0	44.9 ± 8.7	36.1 ± 2.2 $^\#$	43.5 ± 1.0
Females	34.7 ± 4.6 *	35.8 ± 5.4 *	35.4 ± 1.9 *	27.9 ± 1.3 *
Gonadal Adipose Tissue Protein Content (ug protein/mg tissue)			
Males	19.4 ± 1.3	19.4 ± 1.2	18.2 ± 0.3	19.1 ± 1.3
Females	19.5 ± 1.0	18.0 ± 1.1	18.1 ± 0.3	18.1 ± 0.7
Inguinal Adipose Tissue Protein Content (ug protein/mg tissue)			
Males	28.2 ± 1.2	27.5 ± 3.9	33.9 ± 6.0	29.0 ± 0.9
Females	29.0 ± 4.6	25.1 ± 1.0	29.8 ± 1.7	29.9 ± 0.9
Interscapular Brown Adipose Tissue Protein Content (ug protein/mg tissue)			
Males	93.4 ± 5.4	111.6 ± 12.2	99.6 ± 9.4	107.2 ± 10.8
Females	112.0 ± 9.4	106.3 ± 12.2	117.2 ± 7.6	120.9 ± 3.1

End point body mass, total food intake, water intake, and creatine intake. Data are presented as means ± SD. * $p < 0.05$ compared to males in the same treatment group. $^\#$ $p < 0.05$ compared to control of same sex. $^\$$ $p < 0.05$ compared to 2.5 g/L group of same sex. $^\&$ $p < 0.05$ compared to 5 g/L group of same sex.

3.2. Creatine Supplementation Increases Mitochondrial Markers in Female Rat Inguinal White Adipose Tissue

Female rats had lower PDH-E1 alpha and citrate synthase protein content in inguinal subcutaneous adipose tissue when compared to male rats ($p < 0.05$; Figure 1A). Creatine supplementation did not alter mitochondrial protein content in male rat inguinal subcutaneous adipose tissue (Figure 1B). In female rats, creatine supplementation resulted in a higher COVIX (5 g/L), PDH-E1alpha (2.5 g/L, 5 g/L, and 10 g/L), and cytochrome C (10 g/L) content compared to control females ($p < 0.05$; Figure 1B). No differences were observed in UCP1 content in male or female rats (Figure 1C). No differences were observed for male or female adipocyte area or % multilocular in the 10 g/L creatine supplemented group compared to control (Figure 1D).

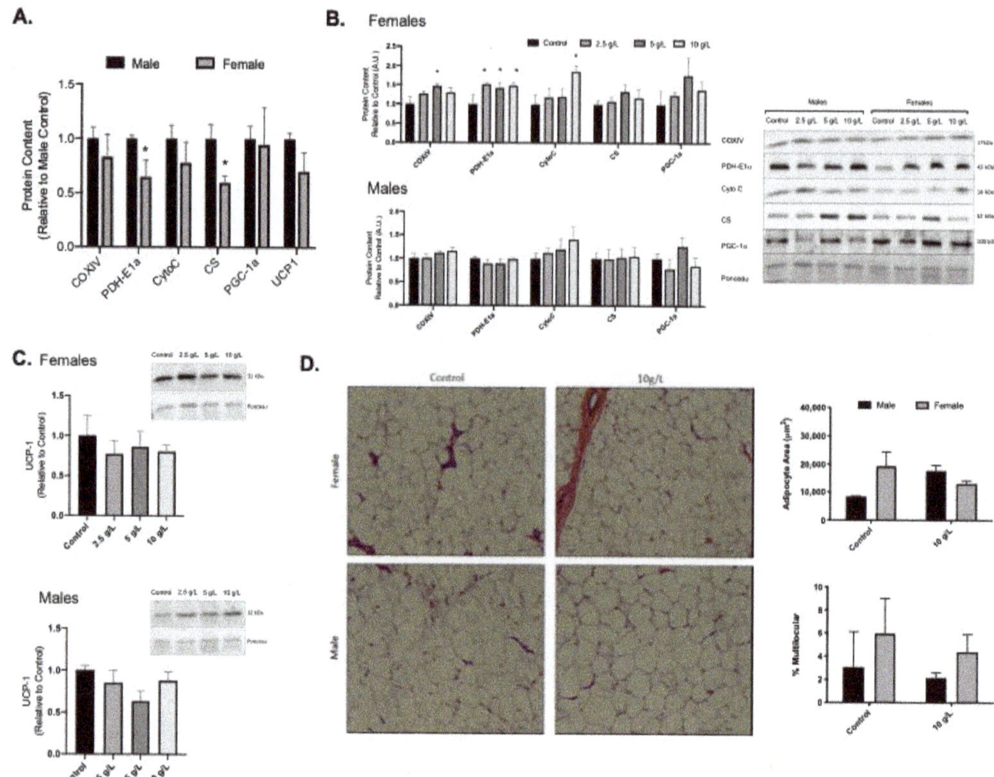

Figure 1. Subcutaneous inguinal white adipose tissue (iWAT) mitochondrial protein content and morphology. (**A**) Quantified Western blot data for iWAT protein content of COXIV, PDH-E1α, Cytochrome C (Cyto C), citrate synthase (CS), PGC-1α, and uncoupling protein-1 (UCP1) for control male and female rats. (**B**) Quantified Western blot data for mitochondrial proteins across creatine supplemented groups and representative Western blot images. (**C**) UCP-1 content in male and female rats across creatine supplemented groups. (**D**). Representative images of H&E stained slides. Scale bar on H&E representative image represents 100 μm. Data were analyzed by ANOVA and are presented as mean ± SEM, * denotes significantly different from control group, $p < 0.05$.

3.3. Creatine Supplementation increases Mitochondrial Markers in Male Rat Gonadal White Adipose Tissue

Female rats had lower PGC-1alpha content in gonadal visceral adipose tissue compared to male rats ($p < 0.05$; Figure 2A). Creatine supplementation did not alter mitochondrial protein content in female rat gonadal visceral adipose tissue (Figure 2B). In male rats, creatine supplementation resulted in a higher COXIV content (10 g/L vs. all other groups, $p < 0.05$) and PDH-E1alpha content (10 g/L vs. control and 2.5 g/L, and 5 g/L vs. 2.5 g/L, $p < 0.05$) (Figure 2B). No differences were observed for male or female adipocyte area (Figure 2C). Control male rats had higher % multilocular adipocytes compared to control females (Figure 2C). No differences in % multilocular adipocytes were observed in the 10 g/L creatine supplemented group compared to control or between sexes (Figure 2C).

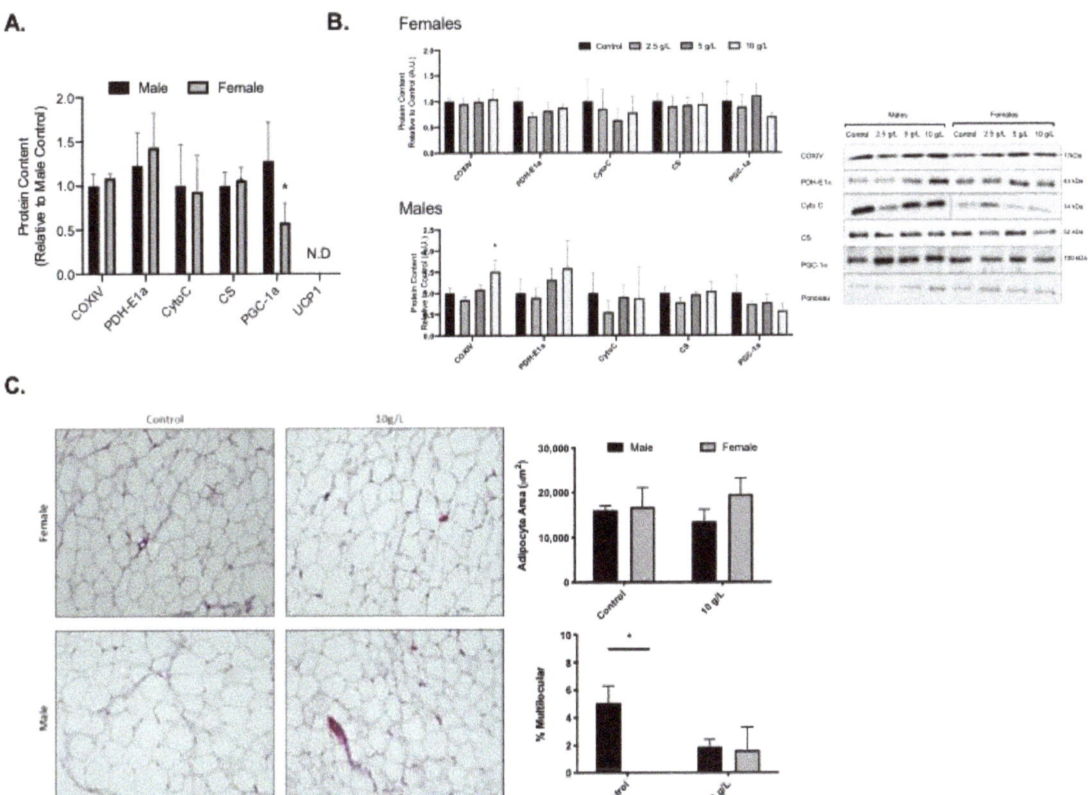

Figure 2. Visceral gonadal white adipose tissue (gWAT) mitochondrial protein content and morphology. (**A**) Quantified Western blot data for gWAT protein content of COXIV, PDH-E1α, Cytochrome C (Cyto C), citrate synthase (CS), PGC-1α, and uncoupling protein-1 (UCP1) for control male and female rats. (**B**) Quantified Western blot data for mitochondrial proteins across creatine supplemented groups and representative Western blot images. (**C**) Representative images of H&E stained slides. Scale bar on H&E representative image represents 100 μm. Data were analyzed by ANOVA and are presented as mean ± SEM, * denotes significantly different from control group, $p < 0.05$.

3.4. Creatine Supplementation does Not increase Mitochondrial Markers Brown Adipose Tissue in either Male or Female Rats

Female rats displayed higher PGC-1alpha content compared to male rats (Figure 3A). Creatine supplementation did not alter mitochondrial protein content in either male or female BAT at any of the supplemented doses (Figure 3B, $p < 0.05$). No differences were observed for UCP1 content in male or female rats across all creatine supplemented groups (Figure 3C). No differences were observed for male or female adipocyte area or % multilocular in the 10 g/L creatine supplemented group compared to control (Figure 3D).

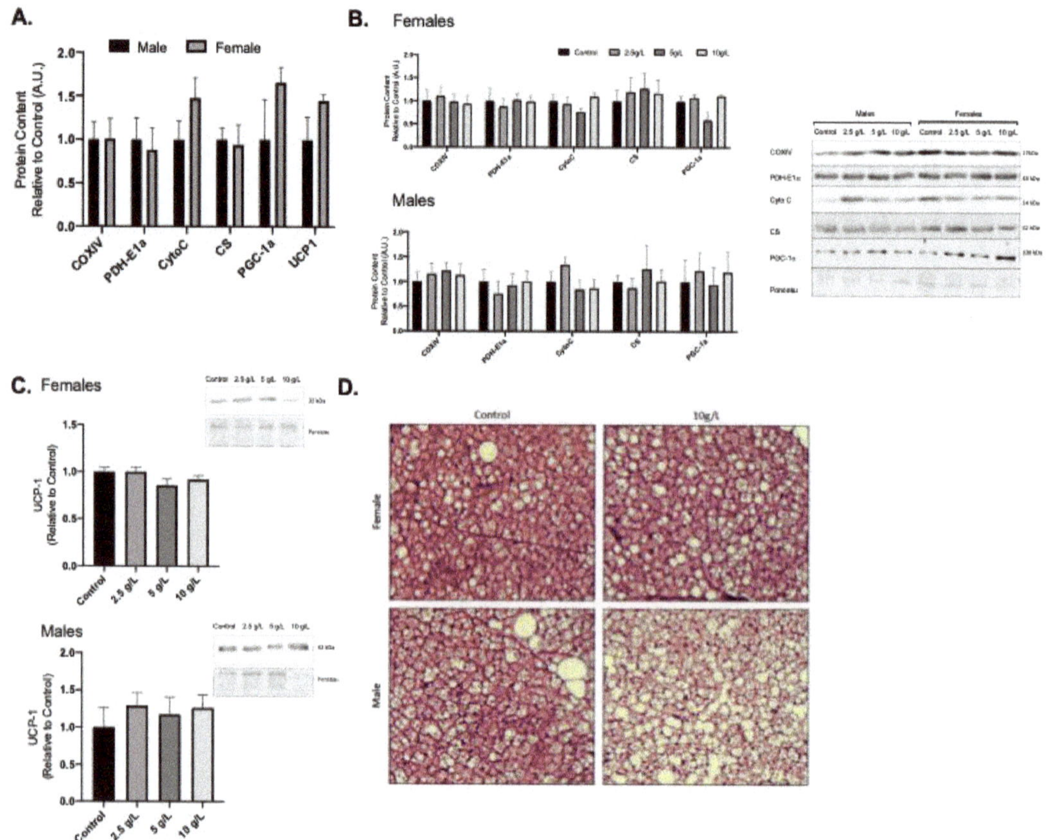

Figure 3. Brown adipose tissue (BAT) mitochondrial protein content and morphology. (**A**) Quantified Western blot data for BAT protein content of COXIV, PDH-E1α, Cytochrome C (Cyto C), citrate synthase (CS), PGC-1α, and uncoupling protein-1 (UCP1) for control male and female rats. (**B**) Quantified Western blot data for mitochondrial proteins across creatine supplemented groups and representative Western blot images. (**C**) UCP-1 content in male and female rats across creatine supplemented groups. (**D**) Representative images of H&E stained slides. Scale bar on H&E representative image represents 100 μm. Data were analyzed by ANOVA and are presented as mean ± SEM, denotes significantly different from control group, $p < 0.05$.

3.5. Creatine Supplementation Alters Markers of Creatine Uptake and Metabolism

Examination of GAMT protein content, a catalyst enzyme responsible for mediating intrinsic creatine synthesis, revealed no differences for GAMT content across any of the creatine dosages or sexes (Figure 4A). CKB was analyzed as a marker of creatine cycling. CKB content was higher in female iWAT at 5 g/L compared to control, this did not reach significance in the 10 g/L group. There were no differences observed in the other depots and no differences observed in the male samples (Figure 4B). As a marker of alterations in creatine transport, *Slc6a8* expression was examined. *Slc6a8* expression was higher in gWAT of female rats treated with 10 g/L creatine compared to controls with no differences in the other adipose depots (Figure 4C). In male rats, 5 and 10 g/L creatine supplementation resulted in higher *Slc6a8* expression in iWAT (Figure 4C). Finally, creatine content was examined in the adipose depots of the 10 g/L groups and compared to the control groups. Creatine content was measured with a commercially available assay kit and was performed as outlined in the assay kit instructions (Biovision, Catalog #K635-100). Creatine content was higher in female and male iWAT within the 10 g/L dose with no differences in the other adipose depots (Figure 4D).

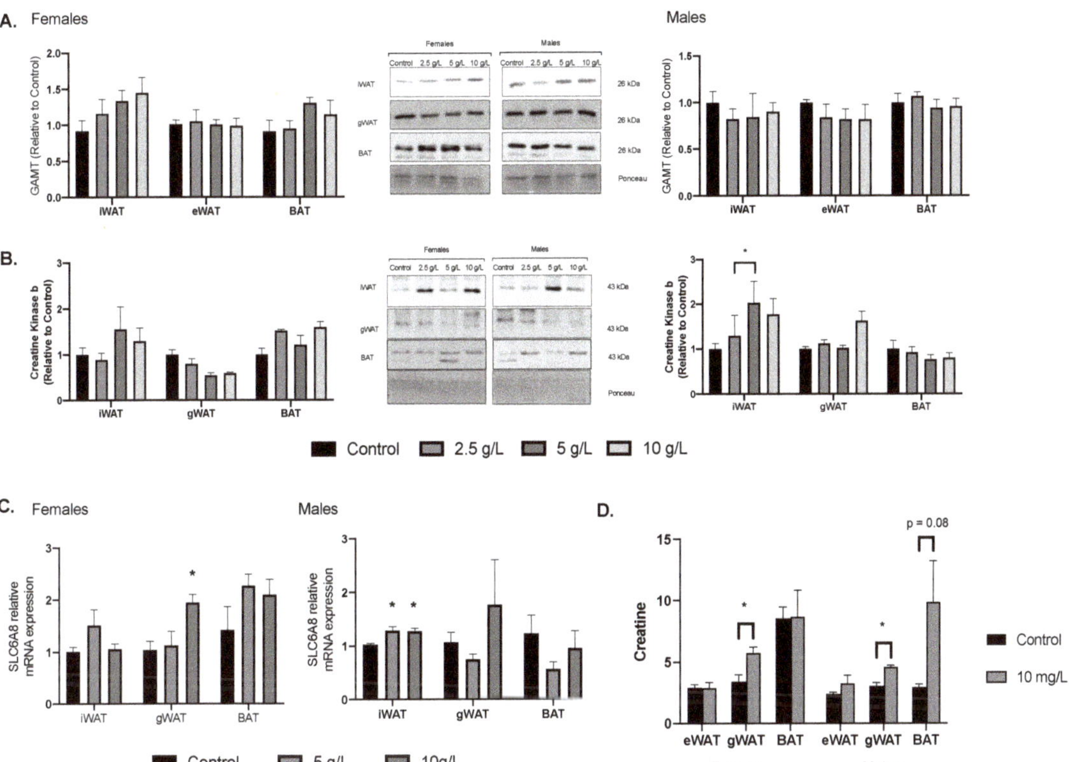

Figure 4. Adipose tissue markers of creatine uptake and metabolism. (**A**) Adipose tissue depot GAMT protein content in female (left) and male (right) rats. (**B**) Creatine kinase B content. (**C**) Slc6A8 mRNA expression. (**D**) Creatine content. Data were analyzed by ANOVA and are presented as mean ± SEM, * denotes significantly different from control group, $p < 0.05$.

4. Discussion

Futile creatine cycling has been identified as a mechanism involved in adipose tissue browning and thermogenesis [14,20–24]. This study examined the potential for creatine supplementation to alter mitochondrial markers in WAT in both male and female Sprague-Dawley rats. Our novel results demonstrate that male and female WAT have a depot specific and sex dependent response to creatine supplementation. Creatine supplementation resulted in an increased creatine content in iWAT in the female and male rats. This was accompanied by increased mitochondrial markers (COXIV, PDH-E1alpha, and cytochrome C) in female rats. No effects on mitochondrial protein markers were observed in male iWAT depots, despite an increase in creatine content. Alternatively, there were no changes in response to creatine supplementation in female rat gWAT, however in male rats, creatine supplementation increased mitochondrial markers (COXIV and PDH-E1alpha) in gWAT. Together, these results highlight creatine supplementation as a potential means to increase WAT mitochondrial content and further highlight the importance of examining sex differences when studying adipose tissue.

Historically, adipose mediated thermogenesis has focused on the role of UCP1, however, it has become apparent that there are UCP1 independent mechanisms that contribute to adipose tissue browning and thermogenesis. Creatine supports energy expenditure in adipocytes and recent work has highlighted a role for futile creatine cycling in adipose tissue thermogenesis [14,19–24,36,37]. The potential underlying mechanisms that link creatine to thermogenic respiration have been recently reviewed [8,37]. However, it was not until recently that work has shed some light into the exact underlying mechanisms. Briefly, it has been determined that creatine elicits a substrate cycle of mitochondrial ATP turnover in a sub-stoichiometric fashion [37] and that the recruitment of mitochondrial CKB plays a crucial role in the transfer of phosphate between ATP and creatine [24]. Recently, TNAP was also identified to play an important role [25]. In thermogenic adipocytes, TNAP localizes to mitochondria where it initiates the futile cycling of creatine dephosphorylation and phosphorylation [25]. The potential of dietary creatine supplementation to stimulate adipose tissue browning has yet to be fully investigated. Kazak et al. examined creatine supplementation in Adipo-Gatm KO mice and found that supplementation rescued impaired adrenergic thermogenesis in these mice [22]. Here, we demonstrated that 8 weeks of creatine supplementation results in increased mitochondrial markers in WAT depots with no change in UCP1 content. This is an important and novel finding as it suggests that dietary creatine may be a means to improve adipose tissue health and possibly increase thermogenesis, independent of UCP1. One limitation of this work is the lack of functional outcomes at the tissue and whole-body level. Future work should determine if these increases in mitochondrial protein content result in enhanced mitochondria respiration and in turn if this enhances whole-body energy expenditure. While there were no differences in body mass amongst groups at the end of the intervention, it is possible that longer term supplementation or supplementation in conjunction with exercise training may result in significant reductions in body mass.

Interestingly, the observed increases in mitochondrial markers were sex- and depot-specific. Our findings show that in female rats, 8 weeks of creatine supplementation resulted in mitochondrial adaptations in inguinal subcutaneous WAT with no changes in the visceral WAT, while the opposite was true for male rats. Much information about adipose tissue depots specific differences has accumulated over the past few decades. It is known that the type of adipose tissue and the location in which it accumulates is important with regard to disease risk. For example, the accumulation of visceral WAT is associated with an increased risk of insulin resistance, type 2 diabetes, dyslipidemia, and atherosclerosis [38,39], while subcutaneous WAT is associated with higher insulin sensitivity and a reduced risk of type 2 diabetes [40–42]. The underlying mechanisms for the varying responses and metabolic effects of subcutaneous and visceral fat are most likely due to unique properties within the depots. Subcutaneous and visceral adipocytes develop from different progenitor cell lines, differentiate at varying rates, and can develop unique gene expression profiles [43,44]. For example, the expression of PRDM16, a transcription coregulatory protein responsible for adipose tissue browning, is much higher in subcutaneous WAT compared to visceral WAT [45]. It is known that beige cells are found interspersed in the WAT of humans and rodents [46–48], and the browning occurs predominantly in subcutaneous WAT. These differences between adipose tissue depots could account for the diverse responses to dietary creatine supplementation in our study. However, further investigation is needed to fully determine the molecular mechanisms underlying the creatine-induced mitochondrial changes and how this regulation is specific to each adipose tissue depot in both sexes. Future studies should explore differences in creatine transporter content and uptake across depots and sexes.

Animal models have demonstrated that sex and sex hormones can influence adipose tissue development, adipogenesis, gene expression profiles regulating insulin resistance and lipolysis, as well as the inflammatory tone and remodeling responses to obesity [49]. It is possible that the observed differences in response to creatine supplementation are due to circulating sex hormones. Previous work has shown that subcutaneous WAT has a higher concentration of estrogen receptors and progesterone receptors compared to androgen receptors in females. In contrast, visceral WAT has a higher concentration of androgen receptors [49,50]. This differential expression of sex hormone receptors could have influenced the response to creatine supplementation observed here in female and male rats. Interestingly, in differentiated 3T3L1 adipocytes, estradiol stimulated the specific activity of creatine kinase [51,52]. This highlights the differential response in adipose depots across sexes and together with our results sets the groundwork for future work in the area.

In the current study, mRNA analysis was conducted to examine if creatine supplementation had an effect on the expression of the creatine transporter (*Slc6a8*). Differences were demonstrated in a sex- and depot-specific manner. Females exhibited increases of the *Slc6a8* gene within the gWAT adipose depot, whereas males exhibited this increase in the iWAT depot. This finding is compelling as it contrasts our other findings, which demonstrated that females experienced mitochondrial protein increases in iWAT and males in the gWAT. As explained previously, subcutaneous iWAT has higher concentrations of estrogen receptors in females, whereas visceral gWAT has higher concentrations of androgen receptors. Therefore, it is possible this finding may be explained as an adaptive physiological response; the *Slc6a8* gene may be upregulated in female gWAT depots and male iWAT depots in an attempt to produce equal physiological responses among different adipose depots within the same rat, however, this is purely speculative.

The current study provides new information demonstrating the potential of dietary creatine supplementation on improving WAT health and increasing mitochondrial markers. Of note are the novel sex- and depot- specific responses to creatine supplementation. These highlight the importance of examining sex-differences in adipose tissue. Future work should explore the depot- and sex- specific responses to creatine. Importantly, the results presented here highlight the efficacy of creatine supplementation to increase mitochondrial proteins and highlight the potential as a preventative or therapeutic treatment for obesity and related metabolic diseases.

Author Contributions: Conceptualization, B.D.R. and R.E.K.M.; methodology, C.R.R., M.S.F., T.C.D., J.E.M.; formal analysis, C.R.R., M.S.F., T.C.D., J.E.M.; resources, B.D.R. and R.E.K.M.; data curation, C.R.R., M.S.F., T.C.D., J.E.M.; writing—original draft preparation, C.R.R.; writing—review and editing, M.S.F., T.C.D., J.E.M., B.D.R., R.E.K.M.; supervision, B.D.R. and R.E.K.M.; funding acquisition, B.D.R. and R.E.K.M. All authors have read and agreed to the published version of the manuscript.

Funding: This work was supported by Brock University. C. Ryan and M. Finch are supported by an Ontario Graduate Scholarship (OGS).

Institutional Review Board Statement: The study was conducted according to the guidelines of the Declaration of Helsinki, and approved by the Institutional Review Board of Brock University.

Informed Consent Statement: Not applicable.

Data Availability Statement: Data available upon request.

Conflicts of Interest: The authors declare no conflict of interest.

References

1. Berlet, H.H.; Bonsmann, I.; Birringer, H. Occurrence of free creatine, phosphocreatine and creatine phosphokinase in adipose tissue. *Biochim. Biophys. Acta* **1976**, *437*, 166–174. [CrossRef]
2. Bertholet, A.M.; Kazak, L.; Chouchani, E.T.; Bogaczynska, M.G.; Paranjpe, I.; Wainwright, G.L.; Betourne, A.; Kajimura, S.; Spiegelman, B.M.; Kirichok, Y. Mitochondrial Patch Clamp of Beige Adipocytes Reveals UCP1-Positive and UCP1-Negative Cells Both Exhibiting Futile Creatine Cycling. *Cell Metab.* **2017**, *25*, 811–822.e4. [CrossRef]
3. Buford, T.W.; Kreider, R.B.; Stout, J.R.; Greenwood, M.; Campbell, B.; Spano, M.; Ziegenfuss, T.; Lopez, H.; Landis, J.; Antonio, J. International Society of Sports Nutrition position stand: Creatine supplementation and exercise. *J. Int. Soc. Sports Nutr.* **2007**, *4*, 6. [CrossRef]
4. Cannon, B.; Nedergaard, J. Brown adipose tissue: Function and physiological significance. *Physiol. Rev.* **2004**, *84*, 277–359. [CrossRef] [PubMed]
5. Chang, E.; Varghese, M.; Singer, K. Gender and Sex Differences in Adipose Tissue. *Curr. Diabetes Rep.* **2018**, *18*, 69. [CrossRef]
6. Choi, C.H.J.; Cohen, P. Adipose crosstalk with other cell types in health and disease. *Exp. Cell Res.* **2017**, *360*, 6–11. [CrossRef] [PubMed]

7.   Chouchani, E.T.; Kazak, L.; Spiegelman, B.M. New Advances in Adaptive Thermogenesis: UCP1 and Beyond. *Cell Metab.* **2019**, *29*, 27–37. [CrossRef]
8.   Cinti, S. Transdifferentiation properties of adipocytes in the adipose organ. *Am. J. Physiol. Endocrinol. Metab.* **2009**, *297*, E977–E986. [CrossRef] [PubMed]
9.   Cohen, P.; Levy, J.D.; Zhang, Y.; Frontini, A.; Kolodin, D.P.; Svensson, K.J.; Lo, J.C.; Zeng, X.; Ye, L.; Khandekar, M.J.; et al. Ablation of PRDM16 and beige adipose causes metabolic dysfunction and a subcutaneous to visceral fat switch. *Cell* **2014**, *156*, 304–316. [CrossRef] [PubMed]
10.   Cooper, R.; Naclerio, F.; Allgrove, J.; Jimenez, A. Creatine supplementation with specific view to exercise/sports performance: An update. *J. Int. Soc. Sports Nutr.* **2012**, *9*, 33. [CrossRef]
11.   Enerback, S. The origins of brown adipose tissue. *N. Engl. J. Med.* **2009**, *360*, 2021–2023. [CrossRef] [PubMed]
12.   Fedorenko, A.; Lishko, P.V.; Kirichok, Y. Mechanism of fatty-acid-dependent UCP1 uncoupling in brown fat mitochondria. *Cell* **2012**, *151*, 400–413. [CrossRef]
13.   Gesta, S.; Bluher, M.; Yamamoto, Y.; Norris, A.W.; Berndt, J.; Kralisch, S.; Boucher, J.; Lewis, C.; Kahn, C.R. Evidence for a role of developmental genes in the origin of obesity and body fat distribution. *Proc. Natl. Acad. Sci. USA* **2006**, *103*, 6676–6681. [CrossRef] [PubMed]
14.   Granneman, J.G.; Burnazi, M.; Zhu, Z.; Schwamb, L.A. White adipose tissue contributes to UCP1-independent thermogenesis. *Am. J. Physiol. Endocrinol. Metab.* **2003**, *285*, E1230–E1236. [CrossRef]
15.   Grimpo, K.; Volker, M.N.; Heppe, E.N.; Braun, S.; Heverhagen, J.T.; Heldmaier, G. Brown adipose tissue dynamics in wild-type and UCP1-knockout mice: In vivo insights with magnetic resonance. *J. Lipid Res.* **2014**, *55*, 398–409. [CrossRef] [PubMed]
16.   Hirshman, M.F.; Wardzala, L.J.; Goodyear, L.J.; Fuller, S.P.; Horton, E.D.; Horton, E.S. Exercise training increases the number of glucose transporters in rat adipose cells. *Am. J. Physiol.* **1989**, *257*, E520–E530. [CrossRef] [PubMed]
17.   Ikeda, K.; Kang, Q.; Yoneshiro, T.; Camporez, J.P.; Maki, H.; Homma, M.; Shinoda, K.; Chen, Y.; Lu, X.; Maretich, P.; et al. UCP1-independent signaling involving SERCA2b-mediated calcium cycling regulates beige fat thermogenesis and systemic glucose homeostasis. *Nat. Med.* **2017**, *23*, 1454–1465. [CrossRef]
18.   Ipsiroglu, O.S.; Stromberger, C.; Ilas, J.; Hoger, H.; Muhl, A.; Stockler-Ipsiroglu, S. Changes of tissue creatine concentrations upon oral supplementation of creatine-monohydrate in various animal species. *Life Sci.* **2001**, *69*, 1805–1815. [CrossRef]
19.   Kazak, L.; Chouchani, E.T.; Jedrychowski, M.P.; Erickson, B.K.; Shinoda, K.; Cohen, P.; Vetrivelan, R.; Lu, G.Z.; Laznik-Bogoslavski, D.; Hasenfuss, S.C.; et al. A creatine-driven substrate cycle enhances energy expenditure and thermogenesis in beige fat. *Cell* **2015**, *163*, 643–655. [CrossRef]
20.   Kazak, L.; Chouchani, E.T.; Lu, G.Z.; Jedrychowski, M.P.; Bare, C.J.; Mina, A.I.; Kumari, M.; Zhang, S.; Vuckovic, I.; Laznik-Bogoslavski, D.; et al. Genetic Depletion of Adipocyte Creatine Metabolism Inhibits Diet-Induced Thermogenesis and Drives Obesity. *Cell Metab.* **2017**, *26*, 693. [CrossRef]
21.   Kazak, L.; Rahbani, J.F.; Samborska, B.; Lu, G.Z.; Jedrychowski, M.P.; Lajoie, M.; Zhang, S.; Ramsay, L.C.; Dou, F.Y.; Tenen, D.; et al. Ablation of adipocyte creatine transport impairs thermogenesis and causes diet-induced obesity. *Nat. Metab.* **2019**, *1*, 360–370. [CrossRef]
22.   Kopecky, J.; Clarke, G.; Enerback, S.; Spiegelman, B.; Kozak, L.P. Expression of the mitochondrial uncoupling protein gene from the aP2 gene promoter prevents genetic obesity. *J. Clin. Investig.* **1995**, *96*, 2914–2923. [CrossRef]
23.   Kreider, R.B. Effects of creatine supplementation on performance and training adaptations. *Mol. Cell. Biochem.* **2003**, *244*, 89–94. [CrossRef]
24.   Kreider, R.B.; Melton, C.; Rasmussen, C.J.; Greenwood, M.; Lancaster, S.; Cantler, E.C.; Milnor, P.; Almada, A.L. Long-term creatine supplementation does not significantly affect clinical markers of health in athletes. *Mol. Cell. Biochem.* **2003**, *244*, 95–104. [CrossRef]
25.   Livak, K.J.; Schmittgen, T.D. Analysis of relative gene expression data using real-time quantitative PCR and the $2^{-\Delta\Delta C_T}$ Method. *Methods* **2001**, *25*, 402–408. [CrossRef] [PubMed]
26.   Lowell, B.B.; S-Susulic, V.; Hamann, A.; Lawitts, J.A.; Himms-Hagen, J.; Boyer, B.B.; Kozak, L.P.; Flier, J.S. Development of obesity in transgenic mice after genetic ablation of brown adipose tissue. *Nature* **1993**, *366*, 740–742. [CrossRef]
27.   Macotela, Y.; Boucher, J.; Tran, T.T.; Kahn, C.R. Sex and depot differences in adipocyte insulin sensitivity and glucose metabolism. *Diabetes* **2009**, *58*, 803–812. [CrossRef] [PubMed]
28.   MacPherson, R.E.; Baumeister, P.; Peppler, W.T.; Wright, D.C.; Little, J.P. Reduced cortical BACE1 content with one bout of exercise is accompanied by declines in AMPK, Akt, and MAPK signaling in obese, glucose-intolerant mice. *J. Appl. Physiol.* **2015**, *119*, 1097–1104. [CrossRef] [PubMed]
29.   McKie, G.L.; Wright, D.C. Biochemical adaptations in white adipose tissue following aerobic exercise: From mitochondrial biogenesis to browning. *Biochem. J.* **2020**, *477*, 1061–1081. [CrossRef]
30.   Misra, A.; Garg, A.; Abate, N.; Peshock, R.M.; Stray-Gundersen, J.; Grundy, S.M. Relationship of anterior and posterior subcutaneous abdominal fat to insulin sensitivity in nondiabetic men. *Obes. Res.* **1997**, *5*, 93–99. [CrossRef]
31.   Muller, S.; Balaz, M.; Stefanicka, P.; Varga, L.; Amri, E.Z.; Ukropec, J.; Wollscheid, B.; Wolfrum, C. Proteomic Analysis of Human Brown Adipose Tissue Reveals Utilization of Coupled and Uncoupled Energy Expenditure Pathways. *Sci. Rep.* **2016**, *6*, 30030. [CrossRef]

32. Dunham, T.C.; Murphy, J.E.; MacPherson, R.E.; Fajardo, V.A.; Ward, W.E.; Roy, B.D. Sex-and tissue-dependent creatine uptake in response to different creatine monohydrate doses in male and female Sprague-Dawley rats. *Appl. Physiol. Nutr. Metab.* **2021**. [CrossRef]
Nair, A.B.; Jacob, S. A simple practice guide for dose conversion between animals and human. *J. Basic Clin. Pharm.* **2016**, *7*, 27–31. [CrossRef]
33. Petrovic, N.; Walden, T.B.; Shabalina, I.G.; Timmons, J.A.; Cannon, B.; Nedergaard, J. Chronic peroxisome proliferator-activated receptor γ (PPARγ) activation of epididymally derived white adipocyte cultures reveals a population of thermogenically competent, UCP1-containing adipocytes molecularly distinct from classic brown adipocytes. *J. Biol. Chem.* **2010**, *285*, 7153–7164. [CrossRef] [PubMed]
34. Rahbani, J.F.; Roesler, A.; Hussain, M.F.; Samborska, B.; Dykstra, C.B.; Tsai, L.; Jedrychowski, M.P.; Vergnes, L.; Reue, K.; Spiegelman, B.M.; et al. Creatine kinase B controls futile creatine cycling in thermogenic fat. *Nature* **2021**, *590*, 480–485. [CrossRef] [PubMed]
35. Roesler, A.; Kazak, L. UCP1-independent thermogenesis. *Biochem. J.* **2020**, *477*, 709–725. [CrossRef] [PubMed]
36. Rosenwald, M.; Perdikari, A.; Rulicke, T.; Wolfrum, C. Bi-directional interconversion of brite and white adipocytes. *Nat. Cell Biol.* **2013**, *15*, 659–667. [CrossRef] [PubMed]
37. Sebo, Z.L.; Rodeheffer, M.S. Assembling the adipose organ: Adipocyte lineage segregation and adipogenesis in vivo. *Development* **2019**, *146*, dev172098. [CrossRef] [PubMed]
38. Snijder, M.B.; Dekker, J.M.; Visser, M.; Bouter, L.M.; Stehouwer, C.D.; Kostense, P.J.; Yudkin, J.S.; Heine, R.J.; Nijpels, G.; Seidell, J.C. Associations of hip and thigh circumferences independent of waist circumference with the incidence of type 2 diabetes: The Hoorn Study. *Am. J. Clin. Nutr.* **2003**, *77*, 1192–1197. [CrossRef] [PubMed]
39. Somjen, D.; Tordjman, K.; Waisman, A.; Mor, G.; Amir-Zaltsman, Y.; Kohen, F.; Kaye, A.M. Estrogen stimulation of creatine kinase B specific activity in 3T3L1 adipocytes after their differentiation in culture: Dependence on estrogen receptor. *J. Steroid Biochem. Mol. Biol.* **1997**, *62*, 401–408. [CrossRef]
40. Stallknecht, B.; Vinten, J.; Ploug, T.; Galbo, H. Increased activities of mitochondrial enzymes in white adipose tissue in trained rats. *Am. J. Physiol.* **1991**, *261*, E410–E414. [CrossRef] [PubMed]
41. Sun, Y.; Rahbani, J.F.; Jedrychowski, M.P.; Riley, C.L.; Vidoni, S.; Bogoslavski, D.; Hu, B.; Dumesic, P.A.; Zeng, X.; Wang, A.B.; et al. Mitochondrial TNAP controls thermogenesis by hydrolysis of phosphocreatine. *Nature* **2021**, *593*, 580–585. [CrossRef]
42. Tchkonia, T.; Lenburg, M.; Thomou, T.; Giorgadze, N.; Frampton, G.; Pirtskhalava, T.; Cartwright, A.; Cartwright, M.; Flanagan, J.; Karagiannides, I.; et al. Identification of depot-specific human fat cell progenitors through distinct expression profiles and developmental gene patterns. *Am. J. Physiol. Endocrinol. Metab.* **2007**, *292*, E298–E307. [CrossRef]
43. Townsend, L.K.; Wright, D.C. Looking on the "brite" side exercise-induced browning of white adipose tissue. *Pflug. Arch. Eur. J. Physiol.* **2019**, *471*, 455–465. [CrossRef]
44. Tran, T.T.; Yamamoto, Y.; Gesta, S.; Kahn, C.R. Beneficial effects of subcutaneous fat transplantation on metabolism. *Cell Metab.* **2008**, *7*, 410–420. [CrossRef] [PubMed]
45. Trayhurn, P. Endocrine and signalling role of adipose tissue: New perspectives on fat. *Acta Physiol. Scand.* **2005**, *184*, 285–293. [CrossRef] [PubMed]
46. Ukropec, J.; Anunciado, R.P.; Ravussin, Y.; Hulver, M.W.; Kozak, L.P. UCP1-independent thermogenesis in white adipose tissue of cold-acclimated Ucp1$^{-/-}$ mice. *J. Biol. Chem.* **2006**, *281*, 31894–31908. [CrossRef] [PubMed]
47. Van Loon, L.J.; Oosterlaar, A.M.; Hartgens, F.; Hesselink, M.K.; Snow, R.J.; Wagenmakers, A.J. Effects of creatine loading and prolonged creatine supplementation on body composition, fuel selection, sprint and endurance performance in humans. *Clin. Sci.* **2003**, *104*, 153–162. [CrossRef]
48. Veniant, M.M.; Sivits, G.; Helmering, J.; Komorowski, R.; Lee, J.; Fan, W.; Moyer, C.; Lloyd, D.J. Pharmacologic Effects of FGF21 Are Independent of the "Browning" of White Adipose Tissue. *Cell Metab.* **2015**, *21*, 731–738. [CrossRef]
49. Wakatsuki, T.; Hirata, F.; Ohno, H.; Yamamoto, M.; Sato, Y.; Ohira, Y. Thermogenic responses to high-energy phosphate contents and/or hindlimb suspension in rats. *Jpn. J. Physiol.* **1996**, *46*, 171–175. [CrossRef] [PubMed]
50. Wang, Y.; Rimm, E.B.; Stampfer, M.J.; Willett, W.C.; Hu, F.B. Comparison of abdominal adiposity and overall obesity in predicting risk of type 2 diabetes among men. *Am. J. Clin. Nutr.* **2005**, *81*, 555–563. [CrossRef]
51. Yamashita, H.; Ohira, Y.; Wakatsuki, T.; Yamamoto, M.; Kizaki, T.; Oh-ishi, S.; Ohno, H. Increased growth of brown adipose tissue but its reduced thermogenic activity in creatine-depleted rats fed β-guanidinopropionic acid. *Biochim. Biophys. Acta* **1995**, *1230*, 69–73. [CrossRef]
52. Zhang, C.; Rexrode, K.M.; van Dam, R.M.; Li, T.Y.; Hu, F.B. Abdominal obesity and the risk of all-cause, cardiovascular, and cancer mortality: Sixteen years of follow-up in US women. *Circulation* **2008**, *117*, 1658–1667. [CrossRef] [PubMed]

Article

Creatine Enhances the Effects of Cluster-Set Resistance Training on Lower-Limb Body Composition and Strength in Resistance-Trained Men: A Pilot Study

Diego A. Bonilla [1,2,3,4,*], Richard B. Kreider [5], Jorge L. Petro [1,3], Ramón Romance [6], Manuel García-Sillero [7], Javier Benítez-Porres [8] and Salvador Vargas-Molina [7,8]

[1] Research Division, Dynamical Business & Science Society–DBSS International SAS, Bogotá 110861, Colombia; jlpetros@hotmail.com
[2] Research Group in Biochemistry and Molecular Biology, Universidad Distrital Francisco José de Caldas, Bogotá 110311, Colombia
[3] Research Group in Physical Activity, Sports and Health Sciences (GICAFS), Universidad de Córdoba, Montería 230002, Colombia
[4] kDNA Genomics®, Joxe Mari Korta Research Center, University of the Basque Country UPV/EHU, 20018 Donostia-San Sebastián, Spain
[5] Exercise & Sport Nutrition Lab, Human Clinical Research Facility, Texas A&M University, College Station, TX 77843, USA; rbkreider@tamu.edu
[6] Body Composition and Biodynamic Laboratory, Faculty of Education Sciences, University of Málaga, 29071 Málaga, Spain; arromance@uma.es
[7] Faculty of Sport Sciences, EADE-University of Wales Trinity Saint David, 29018 Málaga, Spain; manugarciasillero@gmail.com (M.G.-S.); salvadorvargasmolina@gmail.com (S.V.-M.)
[8] Physical Education and Sports, Faculty of Medicine, University of Málaga, 29071 Málaga, Spain; benitez@uma.es
* Correspondence: dabonilla@dbss.pro; Tel.: +57-320-335-2050

Citation: Bonilla, D.A.; Kreider, R.B.; Petro, J.L.; Romance, R.; García-Sillero, M.; Benítez-Porres, J.; Vargas-Molina, S. Creatine Enhances the Effects of Cluster-Set Resistance Training on Lower-Limb Body Composition and Strength in Resistance-Trained Men: A Pilot Study. *Nutrients* **2021**, *13*, 2303. https://doi.org/10.3390/nu13072303

Academic Editor: Julien Louis

Received: 12 June 2021
Accepted: 1 July 2021
Published: 4 July 2021

Publisher's Note: MDPI stays neutral with regard to jurisdictional claims in published maps and institutional affiliations.

Abstract: Creatine monohydrate (CrM) supplementation has been shown to improve body composition and muscle strength when combined with resistance training (RT); however, no study has evaluated the combination of this nutritional strategy with cluster-set resistance training (CS-RT). The purpose of this pilot study was to evaluate the effects of CrM supplementation during a high-protein diet and a CS-RT program on lower-limb fat-free mass (LL-FFM) and muscular strength. Twenty-three resistance-trained men (>2 years of training experience, 26.6 ± 8.1 years, 176.3 ± 6.8 cm, 75.6 ± 8.9 kg) participated in this study. Subjects were randomly allocated to a CS-RT+CrM (n = 8), a CS-RT (n = 8), or a control group (n = 7). The CS-RT+CrM group followed a CrM supplementation protocol with 0.1 g·kg^{-1}·day^{-1} over eight weeks. Two sessions per week of lower-limb CS-RT were performed. LL-FFM corrected for fat-free adipose tissue (dual-energy X-ray absorptiometry) and muscle strength (back squat 1 repetition maximum (SQ-1RM) and countermovement jump (CMJ)) were measured pre- and post-intervention. Significant improvements were found in whole-body fat mass, fat percentage, LL-fat mass, LL-FFM, and SQ-1RM in the CS-RT+CrM and CS-RT groups; however, larger effect sizes were obtained in the CS-RT+CrM group regarding whole body FFM (0.64 versus 0.16), lower-limb FFM (0.62 versus 0.18), and SQ-1RM (1.23 versus 0.75) when compared to the CS-RT group. CMJ showed a significant improvement in the CS-RT+CrM group with no significant changes in CS-RT or control groups. No significant differences were found between groups. Eight weeks of CrM supplementation plus a high-protein diet during a CS-RT program has a higher clinical meaningfulness on lower-limb body composition and strength-related variables in trained males than CS-RT alone. Further research might study the potential health and therapeutic effects of this nutrition and exercise strategy.

Keywords: phosphocreatine; muscle fatigue; adipose tissue; muscle strength; dietary supplements; physiological adaptation

1. Introduction

Several changes with regard to the synthesis and hydrolysis of adenosine triphosphate (ATP) occur inside muscle cells during all-out short-term physical exercise [1]. This relationship between energy production and consumption (myocellular ATP/ADP ratio) is crucial for the onset of muscle fatigue, which is characterized by an acute reduction in force and power in response to contractile activity [2]. In parallel to the decrease of the ATP/ADP ratio, muscle concentrations of inorganic phosphate (Pi) and hydrogen ions (H^+) significantly rise over the course of high-intensity physical exercise. This metabolic stress is also currently considered one of the main mechanisms evoking muscle fatigue [3]. Affecting mainly the muscle fibers expressing myosin heavy chain isoform II (fast twitch or type II), the intracellular accumulations of Pi and H^+ impair the function of the contractile machinery via several mechanisms, including inhibition of actomyosin and sarco/endoplasmic reticulum Ca^{2+}-ATPases (SERCA), reduction of the Gibbs free energy of ATP hydrolysis, alteration of the state of actin–myosin cross-bridges, and uncoupling of dihydropyridine and ryanodine receptors [3–5]. In fact, it has been proposed that initial phase of force reduction is accompanied by an increase in Pi concentration and dysregulation of Ca^{2+} handling (i.e., disruptions in the Ca^{2+} release–reuptake process in the sarcoplasmic reticulum), suggesting a possible precipitation of calcium phosphate in the sarcoplasmic reticulum [6].

It is well known that the metabolism of phosphagen compounds (high-energy phosphates), such as phosphocreatine (PCr), provides an immediate and predominant energy source for the rapid and localized regeneration of ATP during explosive short-duration physical exercise [7]. Interestingly, during periods of muscle inactivity (rest between sets), PCr resynthesis requires energy from ATP hydrolysis to allow the transphosphorylation of free creatine (Cr) by different creatine kinase (CK) isoforms [8]. In comparison to ATP and ADP, PCr and Cr are smaller and less negatively charged molecules that can be found in much higher concentrations in the myocyte, allowing greater intracellular flux through the mitochondrial reticulum [9] and CK/PCr system [10,11]. Hence, the CK/PCr system might be considered as a crucial spatio-temporal energy and metabolic buffer during high-intensity short-duration physical exercise [8]. Optimization of the CK/PCr system can be attained by supplementation with creatine monohydrate (CrM), which is not only the most studied, safe, and effective nutritional ergogenic aid [12–15] but also has several potential health/therapeutic benefits [16–21].

As a type of explosive strength training, progressive heavy resistance training (RT) is recognized by promoting positive musculoskeletal and functional adaptations in a wide range of populations [22–25]. Traditional set configurations of RT usually involve a number of repetitions performed in a continuous fashion without any pause in between. However, several configurations of cluster-set RT (CS-RT), which includes intra-set rest periods, have been studied with interest in the recent years [26–29]. This strategy has been reported as a time-efficient tool to attenuate the loss in mean propulsive velocity, power, and peak force [30,31], and to improve the exercise adaptations in strength-trained individuals [32,33].

The intra-set rest periods of CS-RT might hypothetically lead to a greater PCr resynthesis, which in turn could optimize the energy and metabolic buffering role of the CK/PCr system. Indeed, it has been proposed that this would control intramuscular pH, avoid accumulation of metabolic products, restore membrane potential to resting values, and increase blood flow reperfusion into the muscle and consequently increase oxygen transport to the tissues [34]. Previous work has shown that during rest PCr resynthesis has a biphasic time course behavior (fast phase during the first 21–22 s and slow phase from 170 s and beyond) [35]. Although different studies have assessed the effects of CS-RT with different configurations on both progressive overload [36,37] and plyometric training [38,39], to our knowledge there is no study that has evaluated the combination of this strategy with CrM supplementation on body composition and strength. Thus, the aim of this pilot study was to evaluate the effects of chronic CrM supplementation ($0.1 \text{ g·kg}^{-1} \cdot \text{day}^{-1}$) plus a high-protein diet during an eight-week CS-RT program on lower-limb body composition and strength-related variables in resistance-trained men. We hypothesize that the combination of CrM, a high-protein diet, and CS-RT will have a great impact on exercise adaptations in advanced exercisers.

2. Materials and Methods

2.1. Trial Design

This was a triple-arm single-blinded and repeated-measures randomized controlled pilot study in resistance-trained men. The study was designed following the Consolidated Standards of Reporting Trials (CONSORT) extension to pilot and feasibility trials with suitable adaptations [40]. We evaluated the effects of CrM supplementation on a high-protein diet during eight weeks of a CS-RT program on lower-limb body composition and strength-related parameters. All variables were measured at baseline (Week 0) and after the CS-RT program (Week 8) (Figure 1).

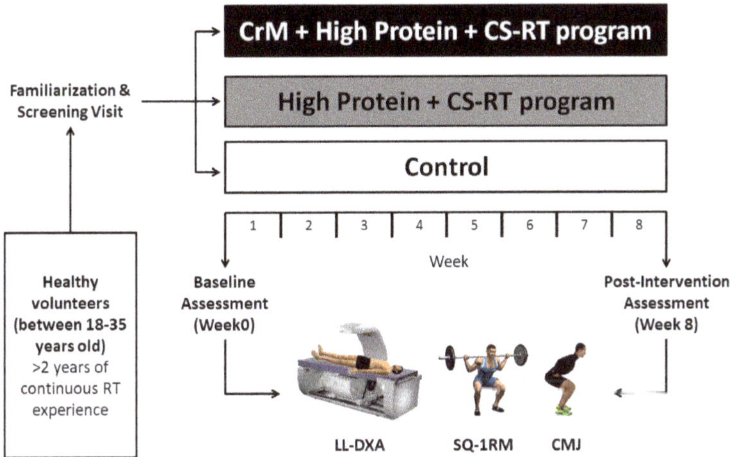

Figure 1. Experimental design of the study. CMJ: countermovement jump; CrM: creatine monohydrate; CS-RT: cluster-set resistance training; LL-DXA: lower-limb assessment with dual-energy X-ray absorptiometry; SQ-1RM: 1 repetition maximum for the back squat.

2.2. Participants

A total of 27 men (26.1 $\pm$ 7.6 years; 177.2 $\pm$ 7.1 cm; 75.8 $\pm$ 8.4 kg; 24.1 $\pm$ 2.1 kg·m^{-2}) volunteered to participate in this pilot study. Subjects were suitable for eligibility if: (i) >18 years old; (ii) >2 years of RT experience; (iii) were committed to adhere to their respective prescribed training protocol with no other physical activity performed nor extra dietary supplements consumed during the 8-week study period. Taking any performance and image-enhancing drugs in the 2-year period prior to the study were considered as exclusion criteria. The participants were informed about the experimental protocol and the potential associated risks. The research protocol was approved by the Ethics Committee of the EADE-University of Wales Trinity Saint David (code: EADECAFYD-2019) and developed in accordance to the ethical guidelines of the Declaration of Helsinki [41].

2.3. Intervention Procedures

Experimental procedures were conducted as previous studies carried out by our research group [26,42]. Both experimental groups (CS-RT+CrM and CS-RT) trained to volitional failure twice a week with 72 h of rest between sessions. The exercise intervention program was performed within the infrastructure of the fitness and strength conditioning center 'Physical Training' (Málaga, Spain) while measurements were taken in the Human Kinetics and Body Composition Laboratory at the University of Málaga.

2.3.1. Anthropometry

All anthropometric data were collected during the first visit to the laboratory during the familiarization weeks. Body mass was measured with a digital scale to the nearest 50 g (Tanita RD-545, Tokyo, Japan). A fixed stadiometer was used to measure the stature (SECA 220, Hamburg, Germany).

2.3.2. Exercise Protocol

All participants were familiarized with advanced resistance training strategies. The CS-RT+CrM and CS-RT groups performed the same training protocol during the eight weeks of the study while subjects of the control group were asked to maintain their habitual training regimes throughout the study (Table 1). Three exercises were performed in the following order: squat (3 sets), deadlift (3 sets), and leg press (2 sets). An unsupervised standardized training program for upper limbs was prescribed in the experimental groups. Training sessions were monitored by strength and conditioning specialists, adjusting the loads whenever necessary.

Table 1. Eight-week cluster-set resistance training protocol.

Group	Exercises	Sets per Exercise	Reps per Set	Clusters per Set	Intra-Set Rest (s)	Inter-Set Rest (s)
CS-RT+CrM	Squat, deadlift, and leg press *	3	12	$4 \times 3RM$	20	180
CS-RT		3	12	$4 \times 3RM$	20	180
Control	Followed their habitual diet and training programs (no recorded)					

Participants were encouraged to train on Monday and Thursday to favor recovery given all reported having lower energy levels and mechanical expenditure during the weekend. * Two sets were performed for the leg press exercise.

2.3.3. Dietary Intervention

All participants were prescribed to consume $\sim$39 kcal·kg^{-1} FFM per day in order to optimize changes in body composition [43]. In the CS-RT+CrM and CS-RT groups, participants were instructed to consume the following macronutrient distribution: 5.0 g·kg^{-1} FFM·day^{-1} of carbohydrates, 2.5 g·kg^{-1} FFM·day^{-1} of protein, and 1.0 g·kg^{-1} FFM·day^{-1} of fat [44]. Similar foods were recommended for the diets of participants in the CS-RT+CrM and CS-RT groups while subjects in the control group maintained their habitual diet. The CS-RT+CrM group followed an eight-week CrM supplementation protocol of 0.1 g·kg^{-1}·day^{-1} (Creatine-Red Gold Series, MTX Corporation, Irún, Spain). Several studies have shown that this chronic supplementation protocol (with or without loading phase) is safe and effective for improving exercise performance capacity and training adaptations in trained men [12,13]. The individualized amount of CrM was fully dissolved in $\approx$500 mL of a shake with 0.5 g·kg^{-1} beef protein (Carnicode, MTX Corporation, Irún, Spain) and given to the participants immediately after each training session (in the morning on non-training days). Both CS-RT+CrM and CS-RT groups consumed 0.5 g·kg^{-1} protein post-exercise and, therefore, we equated protein consumption between experimental groups. A sport nutritionist designed and supervised all individualized protocols to optimize dietary adherence and confirm the dietary compliance (total daily energy intake and macronutrient distribution).

2.4. Outcomes

Primary and secondary outcomes were measured following laboratory procedures reported in previous articles published by our research group [26,42].

2.4.1. Primary Outcome Measures

Whole and regional body composition were estimated using dual-energy X-ray absorptiometry (DXA). Each subject was scanned by a certified technician, and the distinguished bone and soft tissue, edge detection, and regional demarcations were performed by computer algorithms (software version APEX 3.0, Hologic QDR 4500, Bedford, MA, USA). The lower-limb region included the foot, lower leg, and upper leg and was defined by an inclined line passing just below the pelvis crossing the neck of the femur. For each scan, subjects wore sports clothes and were asked to remove all materials that could attenuate the X-ray beam. This included jewelry items and underwear containing a wire. Calibration of the densitometer was checked daily against the standard calibration block supplied by the manufacturer. To determine intertester reliability, two different observers manually selected the area for each subject (coefficient of variation ranged from 1.0 to 2.0%).

To eliminate the influence of the fat-free adipose tissue (FFAT), we adjusted the DXA measurements based on the model proposed by Heymsfield et al. (2002) [45]. This model describes that 85% of adipose tissue is fat while the remaining 15% is the estimated fat-free component. Eliminating the influence of FFAT on DXA-derived fat-free mass (FFM) has been shown to provide more accurate values to detect changes in body composition [46]. Therefore, to calculate adipose tissue mass we based our method on the one employed by Abe et al. (2018) [47] and adjusted the DXA-derived vales as follows: first, we estimated the adipose tissue as DXA fat mass ÷ 0.85; FFAT was then calculated as adipose tissue × 0.15; finally, DXA FFM was adjusted with the elimination of FFAT.

2.4.2. Secondary Outcome Measures

Muscle Strength (Repetition Maximum Test)

The 1 repetition maximum (1RM) for the back squat (SQ) was assessed in a Smith machine (Gervasport, Madrid, Spain) following procedures reported previously [26,42]. Participants reported to the laboratory having abstained from any exercise other than activities of daily living for at least 72 h before the reference test and at least 72 h before post-study testing. All men performed a general warm-up prior to testing, which consisted of 7 to 10 min of light cardiovascular exercise. A specific warm-up set of the given exercise was then provided for 12 to 15 repetitions with approximately 40% of the 1RM perceived by the participants, with a load progression for each exercise of 3 to 6 load increments. The increases in each load were approximately 10% 1RM until reaching a mean propulsive velocity of 0.5 m·s^{-1} [8], followed by increments of 5 to 10 kg until attainment of 1RM. A rest interval of three to five minutes was afforded between each successive attempt. Participants had to reach parallel in the 1RM SQ for the trial to be considered a successful attempt. The protocol followed the recommendations described by McGuigan [48] and the technical execution of the squat according to Caulfield & Berninger [49].

Muscle Power (Countermovement Jump Test)

Participants were instructed to avoid vigorous exercise for 72 h before the tests in both the pre- and post-test periods. Prior to testing, all men performed a general warm-up consisting of light stretching and stationary cycling for 10–12 min. The countermovement jump (CMJ) test was performed on a jump mat (Smart Jump; Fusion Sport, Coopers Plains, Australia) as we have reported previously [26]. Participants were instructed to initiate each jump by squatting to 90° of knee flexion while keeping their hands at the waist and trunk erect, emphasizing that the movement should be performed without interruption from beginning to end. A total of 3–5 attempts were permitted for familiarization. Thereafter, two jumps were recorded with a rest interval of 1 min between each trial; the highest value was used for analysis (coefficient of variation of the technician was 4.65%).

2.5. Sample Size

Non-probability sampling (convenience sampling) was implemented as it is often a strategy used in pilot studies. This leads to gain insight before a full-fledged research activity takes place [50]. After the call to participate in this study, 24 subjects were suitable for eligibility from the available population (i.e., resistance trained men attending the fitness and strength conditioning center 'Physical Training' located in Malaga, Spain).

2.6. Randomization

Subjects were randomly assigned (www.randomizer.org accessed on 3 July 2021) to three groups using a 1:1 allocation ratio design: the CS-RT+CrM group ($n = 9$), the CS-RT group ($n = 9$), and the control group ($n = 9$). Subsequently, familiarization and baseline measurements were performed.

2.7. Statistical Analysis

The descriptive statistics are expressed as mean and standard deviation (SD). To determine statistical significance, we examined the 95% CIs for the difference between the mean change scores (Δ = Week 8–Week 0). If the 95% CI excludes zero, the difference will attain significance at the $p < 0.05$ level. Effect size was calculated as unbiased Cohen's *d* (d_{unb}), considering a result of ≤ 0.2 as a small, 0.5 as a moderate, ≥ 0.8 as a large effect, and ≥ 1.30 as a very large effect [51]. Estimation plots were generated to display the repeated measures data across two time points (at baseline and after eight weeks). A difference-in-differences (Diff-in-Diff) analysis was performed to compare changes in the outcome variables between the groups, as we have implemented previously [52]. Complementarily, the Kruskall–Wallis test was used for the pairwise comparisons of the Δ between groups. Statistical analyses were performed using IBM SPSS, version 26 (IBM Corp., Armonk, NY, USA), and the Exploratory Software for Confidence Intervals [53].

3. Results

Twenty-three men (26.6 ± 8.1 years; 176.3 ± 6.8 cm; 75.6 ± 8.9 kg; 24.3 ± 2.0 kg·m^{-2}) completed the study and were included in the analysis. One participant from the control group discontinued intervention due to injury. Three participants (one per group) were excluded from the analysis because of personal adverse events (Figure 2).

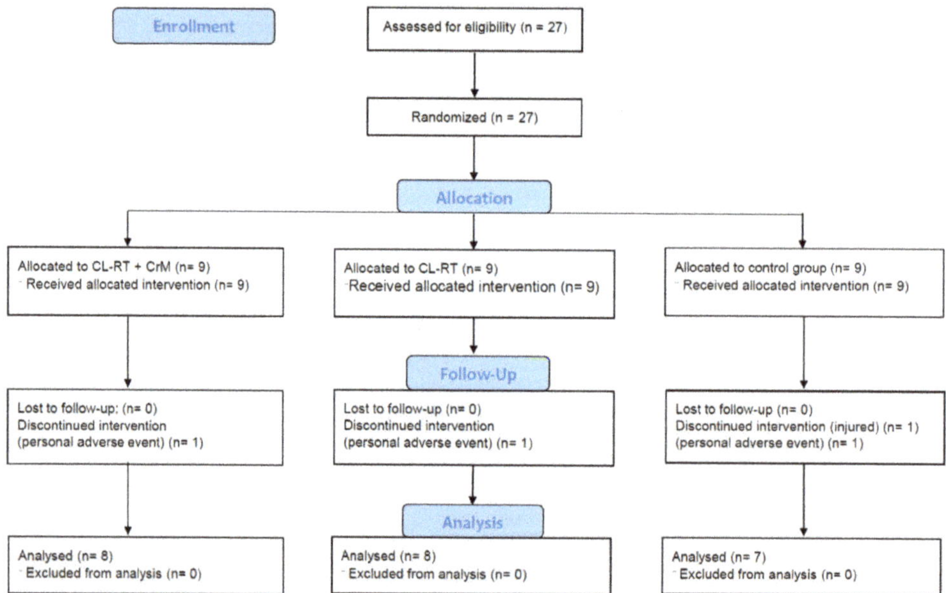

Figure 2. CONSORT flow diagram.

3.1. Baseline Data

Analysis of baseline characteristics showed that there were no statistical differences (Kruskal–Wallis test, $p > 0.05$) between the groups for the studied variables (Table 2).

Table 2. Descriptive information of participants at baseline.

Variable	CS-RT+CrM ($n = 8$) $\bar{x}$ (SD)	95% CI (min, max)	CS-RT ($n = 8$) $\bar{x}$ (SD)	95% CI (min, max)	Control ($n = 7$) $\bar{x}$ (SD)	95% CI (min, max)
Body mass (kg)	76.41 (6.72)	70.79, 82.03	75.09 (11.69)	65.31, 84.86	75.22 (8.83)	67.04, 83.39
Stature (cm)	179.41 (5.07)	175.16, 183.65	173.31 (7.72)	166.85, 179.77	176.45 (7.47)	169.54, 183.36
BMI (kg·m^{-2})	23.89 (1.71)	22.45, 25.32	25.02 (2.73)	22.73, 27.31	24.25 (2.51)	22.82, 25.68
Body fat$_{DXA}$ (%)	18.00 (4.45)	14.27, 21.73	16.63 (3.27)	13.89, 19.36	15.72 (3.99)	12.02, 19.41
FM$_{DXA}$ (kg)	13.89 (4.15)	10.42, 17.37	12.62 (3.65)	9.56, 15.67	12.04 (4.32)	8.04, 16.05
FFAT$_{total}$ (kg)	2.45 (0.73)	1.84, 3.06	2.22 (0.64)	1.68, 2.76	2.12 (0.76)	1.42, 2.83
FFM$_{DXA}$ (kg)	62.51 (4.69)	58.59, 66.43	62.47 (8.98)	54.95, 69.98	63.17 (5.69)	57.90, 68.44
FFM$_{DXA}$ minus FFAT$_{total}$ (kg)	60.06 (4.63)	56.18, 63.93	60.24 (8.58)	53.06, 67.42	61.04 (5.32)	56.12, 65.96
LL-mass (kg)	38.70 (3.83)	35.49, 41.90	37.10 (6.55)	31.6, 42.58	38.10 (4.97)	33.50, 42.71
LL-FM$_{DXA}$ (kg)	8.03 (2.83)	5.67, 10.40	7.11 (2.17)	5.30, 8.93	6.66 (2.58)	4.27, 9.05
LL-FFAT (kg)	1.41 (0.49)	1.00, 1.83	1.25 (0.38)	0.93, 1.57	1.17 (0.45)	0.75, 1.59
LL-FFM$_{DXA}$ (kg)	30.66 (2.42)	28.63, 32.69	29.98 (5.01)	25.78, 34.18	31.44 (3.01)	28.65, 34.23
LL-FFM minus LL-FFAT (kg)	29.24 (2.45)	27.19, 31.29	28.72 (4.79)	24.71, 32.74	30.27 (2.77)	27.70, 32.84
SQ-1RM (kg)	110.62 (13.36)	99.45, 121.79	113.00 (16.82)	98.93, 127.06	113.00 (16.02)	98.18, 127.81
CMJ (cm)	38.92 (5.64)	34.20, 43.64	39.14 (4.69)	35.21, 43.06	38.47 (5.91)	33.00, 43.95

Data are expressed as mean (standard deviation) and 95% confidence interval. FFM and LL-FFM were corrected for fat-free adipose tissue. BMI, body mass; CI, confidence interval; CMJ, countermovement jump; CrM, creatine monohydrate; CS-RT, cluster-set resistance training; FFAT, fat-free adipose tissue; FFM, fat-free mass; FM, fat mass; LL-FFAT; lower-limb fat-free adipose tissue; LL-FFM; lower-limb fat-free mass; LL-FM; lower-limb fat mass; LL-mass, lower-limb mass; SQ-1RM, 1 repetition maximum for the back squat.

3.2. Outcomes

The results of all variables are expressed as $\Delta \pm$ SD [95% CI]; d_{unb} [95% CI] and presented in Table 3. After eight weeks, there were no significant differences in body mass in any group. Whole-body fat mass and fat percentage had a moderately significant decrease in the groups CS-RT+CrM (-2.18 ± 0.82 (-2.88, -1.49); -0.44 (-0.77, -0.20) and -1.75 ± 1.41 (-2.94, -0.56); -0.47 (-0.93, -0.11), respectively) and CS-RT (-1.75 ± 1.41 (-2.94, -0.56); -0.47 (-0.93, -0.11) and -1.32 ± 1.12 (-2.26, -0.37); -0.36 (-0.64, -0.07), respectively) while moderate and small effects were detected for these variables in the control group. FFM (corrected for FFAT) increased significantly in all groups although a higher effect size was found in the participants of the CS-RT+CrM group (2.95 ± 1.68 (1.54, 4.35); 0.64 (0.24, 1.19)) in comparison to the CS-RT (1.57 ± 1.09 (0.65, 2.48); 0.16 (0.05, 0.30)) and control (0.87 ± 1.91 (-0.89, 2.64); 0.13 (-0.11, 0.41)) groups. A statistically significant and moderate reduction in lower-limb fat mass was observed in the CS-RT+CrM (-0.87 ± 0.44 (-1.24, -0.51); -0.28 (-0.51, -0.11)) and CS-RT (-0.83 ± 0.72 (-1.44, -0.22); -0.35 (-0.71, -0.07)) groups with a small effect for the control group (-0.45 ± 0.50 (-0.92, 0.006); -0.15 (-0.34, 0.002)). Finally, both experimental groups presented significant changes in lower-limb FFM (corrected for lower-limb FFAT), although the participants supplemented with CrM had a higher effect size (0.62 versus 0.18 for the CS-RT+CrM and CS-RT groups, respectively). No significant change was detected in the control group for lower-limb FFM.

Table 3. Pre- and post-intervention data on the main study variables.

Variable	Group	Week 0 $\bar{x}$ (SD)	Week 8 $\bar{x}$(SD)	Δ $\bar{x}$ (SD) [95% CI]	d_{unb} δ [95% CI]
Body mass (kg)	CS-RT+CrM	76.41 (6.72)	77.59 (4.89)	1.18 (2.33) [−0.76, 3.13]	0.17 [−0.09, 0.48]
	CS-RT	75.09 (11.69)	75.11 (11.34)	0.01 (1.17) [−0.96, 1.00]	0.00 [−0.06, 0.07]
	Control	75.22 (8.83)	74.96 (9.52)	−0.25 (2.16) [−2.25, 1.74]	−0.02 [−0.20, 0.14]
Body fat$_{DXA}$ (%)	CS-RT+CrM	18.00 (4.45)	15.81 (4.37)	−2.18 (0.82) [−2.88, −1.49] *	−0.44 [−0.77, −0.20]
	CS-RT	16.63 (3.27)	14.87 (3.28)	−1.75 (1.41) [−2.94, −0.56] *	−0.47 [−0.93, −0.11]
	Control	15.72 (3.99)	14.49 (3.91)	−1.22 (1.08) [−2.23, −0.22] *	−0.27 [−0.57, −0.03]
FM$_{DXA}$ (kg)	CS-RT+CrM	13.89 (4.15)	12.39 (3.92)	−1.50 (0.88) [−2.24, −0.76] *	−0.33 [−0.61, −0.12]
	CS-RT	12.62 (3.65)	11.30 (3.56)	−1.32 (1.12) [−2.26, −0.37] *	−0.36 [−0.64, −0.07]
	Control	12.04 (4.32)	11.08 (4.35)	−0.96 (0.97) [−1.86, −0.06] *	−0.19 [−0.42, −0.01]
FFM$_{DXA}$ minus FFAT$_{total}$ (kg)	CS-RT+CrM	60.06 (4.63)	63.01 (3.35)	2.95 (1.68) [1.54, 4.35] *	0.64 [0.24, 1.19]
	CS-RT	60.24 (8.58)	61.81 (8.62)	1.57 (1.09) [0.65, 2.48] *	0.16 [0.05, 0.30]
	Control	61.04 (5.32)	61.92 (6.04)	0.87 (1.91) [−0.89, 2.64]	0.13 [−0.11, 0.41]
LL-mass (kg)	CS-RT+CrM	38.70 (3.83)	39.26 (3.44)	0.55 (0.99) [−0.27, 1.38]	0.13 [−0.05, 0.35]
	CS-RT	37.10 (6.55)	37.17 (6.53)	0.07 (1.00) [−0.76, 0.91]	0.01 [−0.09, 0.11]
	Control	38.10 (4.97)	38.09 (5.34)	−0.01 (1.08) [−1.01, 0.99]	−0.002 [−0.15, 0.15]
LL-FM$_{DXA}$ (kg)	CS-RT+CrM	8.03 (2.83)	7.16 (2.65)	−0.87 (0.44) [−1.24, −0.51] *	−0.28 [−0.51, −0.11]
	CS-RT	7.11 (2.17)	6.28 (1.96)	−0.83 (0.72) [−1.44, −0.22] *	−0.35 [−0.71, −0.07]
	Control	6.66 (2.58)	6.20 (2.64)	−0.45 (0.50) [−0.92, 0.006] *	−0.15 [−0.34, 0.002]
LL-FFM minus LL-FFAT (kg)	CS-RT+CrM	29.24 (2.45)	30.83 (2.09)	1.59 (0.70) [1.00, 2.18] *	0.62 [0.27, 1.11]
	CS-RT	28.72 (4.79)	29.78 (5.10)	1.05 (0.43) [0.69, 1.41] *	0.18 [0.08, 0.33]
	Control	30.27 (2.77)	30.79 (3.15)	0.52 (1.12) [−0.51, 1.56]	0.15 [−0.12, 0.46]
SQ-1RM (kg)	CS-RT+CrM	110.62 (13.36)	132.16 (17.27)	21.53 (11.19) [12.17, 30.89] *	1.23 [0.50, 2.25]
	CS-RT	113.00 (16.82)	127.50 (17.48)	14.50 (12.27) [4.24, 24.75] *	0.75 [0.17, 1.48]
	Control	113.00 (16.02)	121.31 (19.87)	8.31 (9.02) [−0.03, 16.66]	0.40 [−0.001, 0.89]
CMJ (cm)	CS-RT+CrM	38.92 (5.64)	41.65 (5.23)	2.72 (1.99) [1.06, 4.39] *	0.44 [0.13, 0.85]
	CS-RT	39.14 (4.69)	40.50 (6.01)	1.36 (4.42) [−2.34, 5.06]	0.22 [−0.33, 0.82]
	Control	38.47 (5.91)	37.63 (5.71)	−0.84 (3.23) [−3.82, 2.14]	−0.12 [−0.55, 0.28]

Data is presented as mean ($\bar{x}$) and standard deviation (SD). FFM and LL-FFM were corrected for fat-free adipose tissue. BMI, body mass index; CI, confidence interval; CMJ, countermovement jump; CrM, creatine monohydrate; CS-RT, cluster-set resistance training; FFAT, fat-free adipose tissue; FFM, fat-free mass; FM, fat mass; LL-FFAT; lower-limb fat-free adipose tissue; LL-FFM; lower-limb fat-free mass; LL-FM; lower-limb fat mass; LL-mass, lower-limb mass; SQ-1RM, 1 repetition maximum for the back squat. * Statistically significant change ($p < 0.05$).

In regard to lower-limb strength, a statistically significant increase in SQ-1RM was found in the CS-RT+CrM and CS-RT groups with no significant changes in the control participants (8.31 ± 9.02 (−0.03, 16.66); 0.40 (−0.001, 0.89)). However, men supplemented with CrM showed a large effect size (21.53 ± 11.19 (12.17, 30.89); 1.23 (0.50, 2.25)) in comparison to the moderate effect seen in the CS-RT group (14.50 ± 12.27 (4.24, 24.75); 0.75 (0.17, 1.48)). Only participants of the CS-RT+CrM group showed a moderate statistically significant improvement in lower-limb muscle power measured as CMJ (2.72 ± 1.99 (1.06, 4.39); 0.44 (0.13, 0.85)) while no significant changes were observed in the CS-RT (1.36 ± 4.42 (−2.34, 5.06); 0.22 (−0.33, 0.82)) and control (−0.84 ± 3.23 (−3.82, 2.14); −0.12 (−0.55, 0.28)) groups. Figure 3 shows paired results between initial and final measurements.

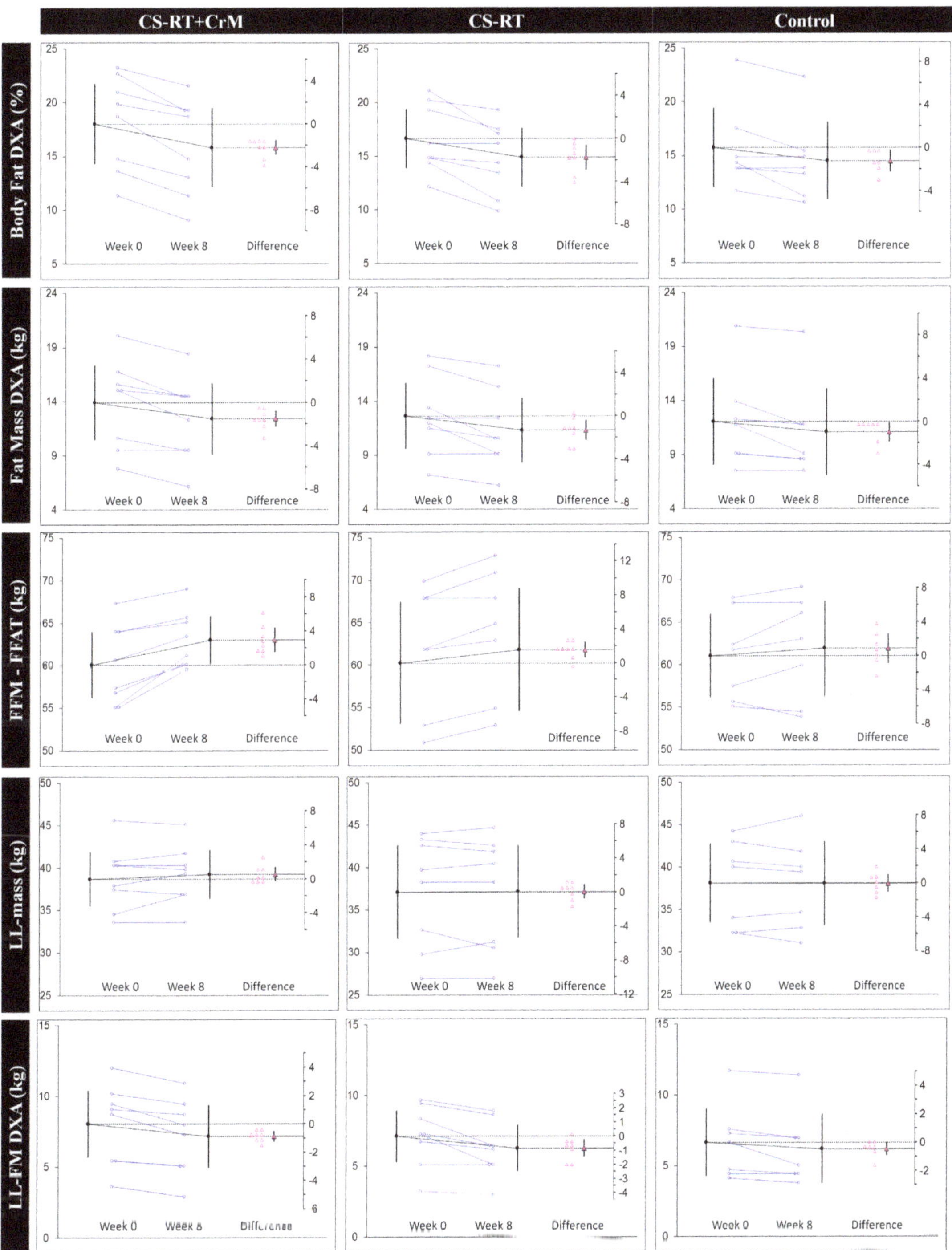

Figure 3. *Cont.*

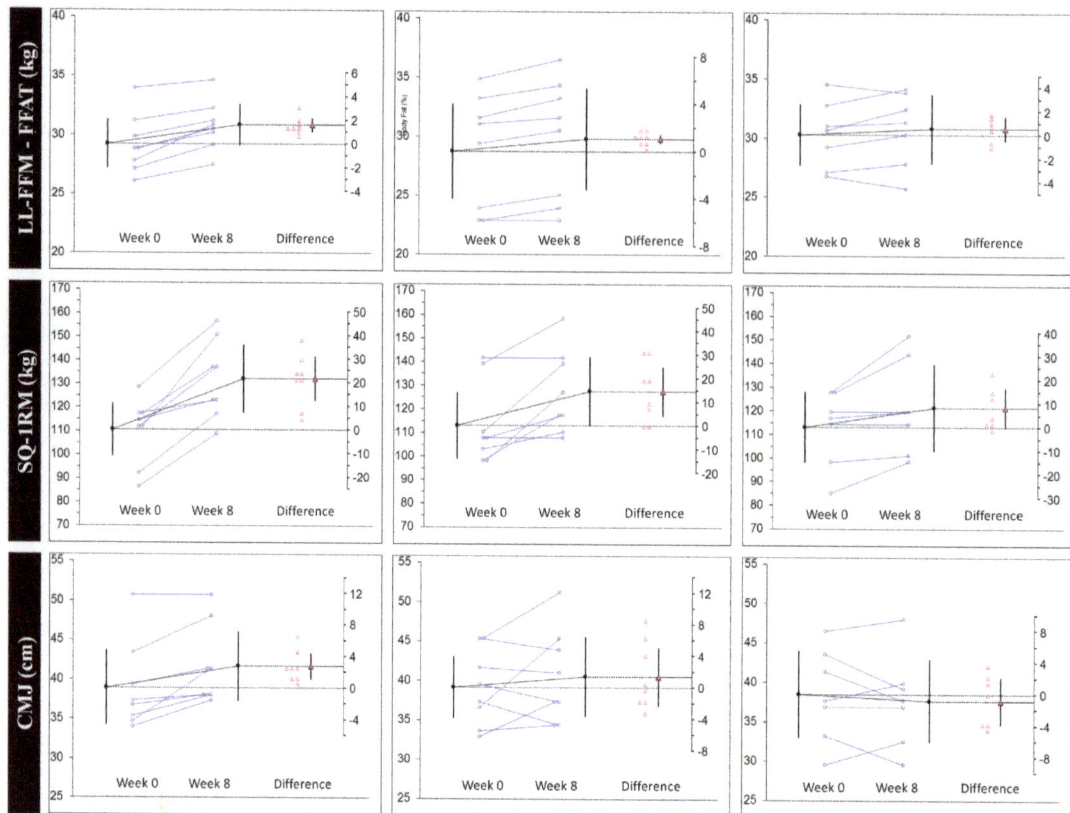

Figure 3. Estimation plots showing pre- and post-intervention values on analyzed variables. Paired data from CS-RT+CrM (left), CS-RT (middle), and control (right) groups are shown as small circles joined by blue lines. The differences between the initial (Week 0) and final (Week 8) means are plotted on a floating difference axis whose zero is aligned with the pre-test mean. The filled pink triangle marks the difference on that axis and the 95% CI on that difference is displayed. The differences are shown as open triangles on the difference axis. FFM and LL-FFM were corrected for fat-free adipose tissue. BMI, body mass index; CI, confidence interval; CMJ, countermovement jump; CrM, creatine monohydrate; CS-RT, cluster-set resistance training; FFAT, fat-free adipose tissue; FFM, fat-free mass; FM, fat mass; LL-FFAT; lower-limb fat-free adipose tissue; LL-FFM; lower-limb fat-free mass; LL-FM; lower-limb fat mass; LL-mass, lower-limb mass; SQ-1RM, 1 repetition maximum for the back squat.

The independent between-group Diff-in-Diff analysis (Table 4) showed no statistical differences. This was confirmed by performing a Kruskall–Wallis test for the pairwise comparisons of the Δ between groups (all $p > 0.05$). Figure 4 shows the change in the outcome variables in the experimental groups compared to the change in the outcome in the control group.

Table 4. Difference of differences between the studied groups.

Variable	DID for CS-RT+CrM (Δ_1) and CS-RT (Δ_2)				DID for CS-RT+CrM (Δ_1) and Control (Δ_3)				DID for CS-RT (Δ_2) and Control (Δ_3)			
	Mean ($\Delta_2-\Delta_1$)	DID	95% CI	p	Mean ($\Delta_3-\Delta_1$)	DID	95% CI	p	Mean ($\Delta_3-\Delta_2$)	DID	95% CI	p
Body mass (kg)	0.01–1.18	−1.16	−14.41, 12.08	0.858	−0.25–1.18	−1.44	−12.86, 9.97	0.797	−0.25–0.01	−0.27	−16.09, 15.53	0.972
Body fat$_{DXA}$ (%)	−1.75–−2.18	0.43	−5.20, 6.07	0.876	−1.22–−2.18	0.96	−5.37, 7.29	0.758	−1.22–−1.75	0.52	−4.90, 5.95	0.843
FM$_{DXA}$ (kg)	−1.32–−1.50	0.18	−5.36, 5.73	0.947	−0.96–−1.50	0.53	−5.75, 6.83	0.862	−0.96–−1.32	0.35	−5.60, 6.31	0.903
FFM$_{DXA}$ minus FFAT total (kg)	1.57–2.95	−1.38	−11.12, 8.34	0.774	0.87–2.95	−2.07	−9.41, 5.25	0.566	0.87–1.57	−0.69	−11.84, 10.44	0.899
LL-mass (kg)	0.07–0.55	−0.48	−8.15, 7.18	0.898	−0.01–0.55	−0.57	−7.21, 6.06	0.861	−0.01–0.07	−0.087	−9.03, 8.86	0.984
LL-FM$_{DXA}$ (kg)	−0.83–−0.87	0.04	−3.47 3.56	0.980	−0.45–−0.87	0.41	−3.62, 4.46	0.833	−0.45–−0.83	0.37	−3.14, 3.89	0.828
LL-FFM minus LL-FFAT (kg)	1.05–1.59	−0.53	−6.12, 5.04	0.845	0.52––1.59	−1.06	−5.01, 2.88	0.583	0.52–1.05	−0.53	−6.78, 5.72	0.863
SQ-1RM (kg)	14.50–21.53	−7.03	−30.68, 16.60	0.547	8.31–21.53	−13.22	−38.34, 11.89	0.289	8.31–14.50	−6.18	−32.63, 20.25	0.635
CMJ (cm)	1.36–2.72	−1.36	−9.21, 6.48	0.724	−0.84–2.72	−3.56	−12.02, 4.88	0.394	−0.84–1.36	−2.203	−10.62, 6.21	0.595

DID: Difference of differences. The p value is two-tailed with statistical significance when <0.05.

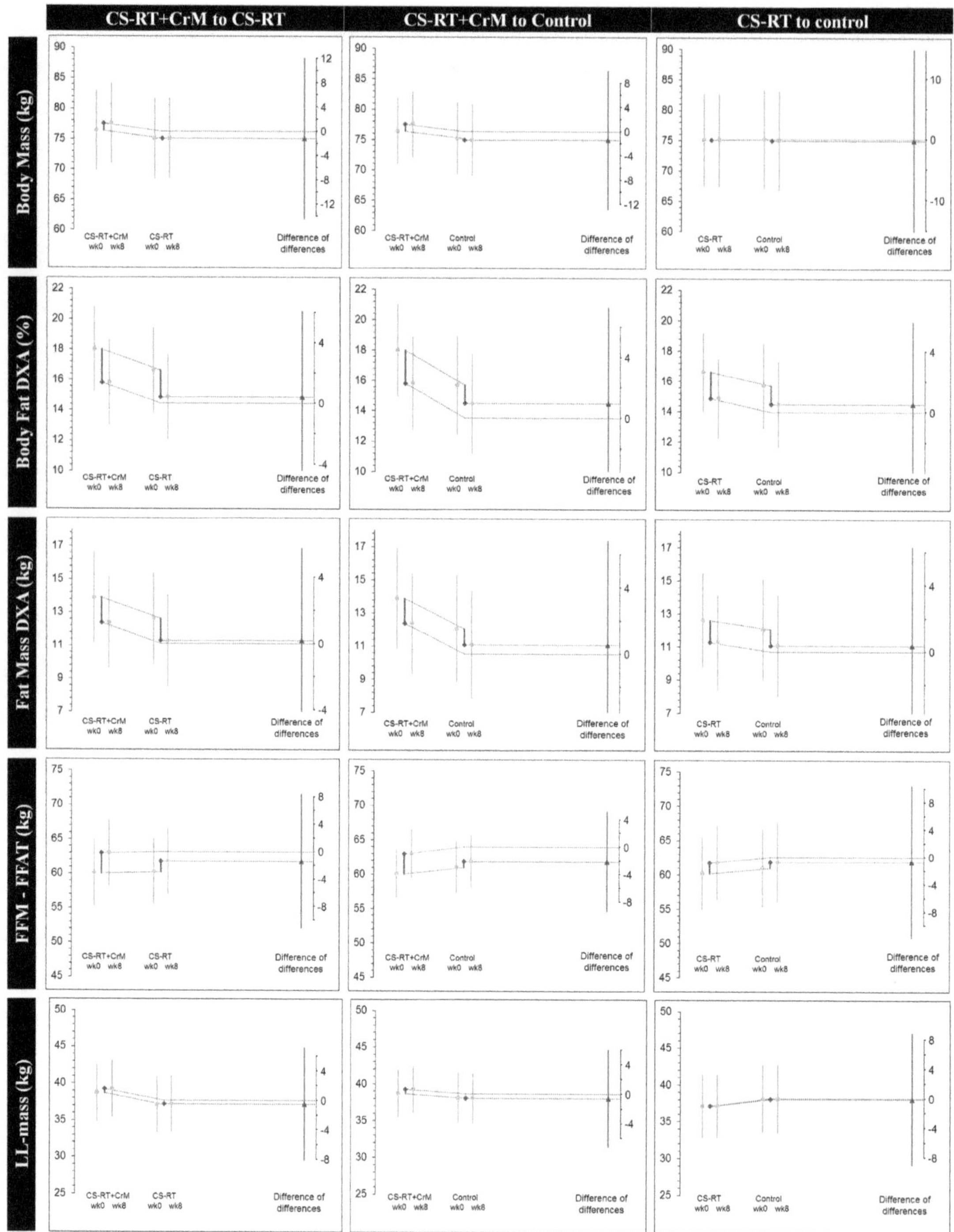

Figure 4. *Cont.*

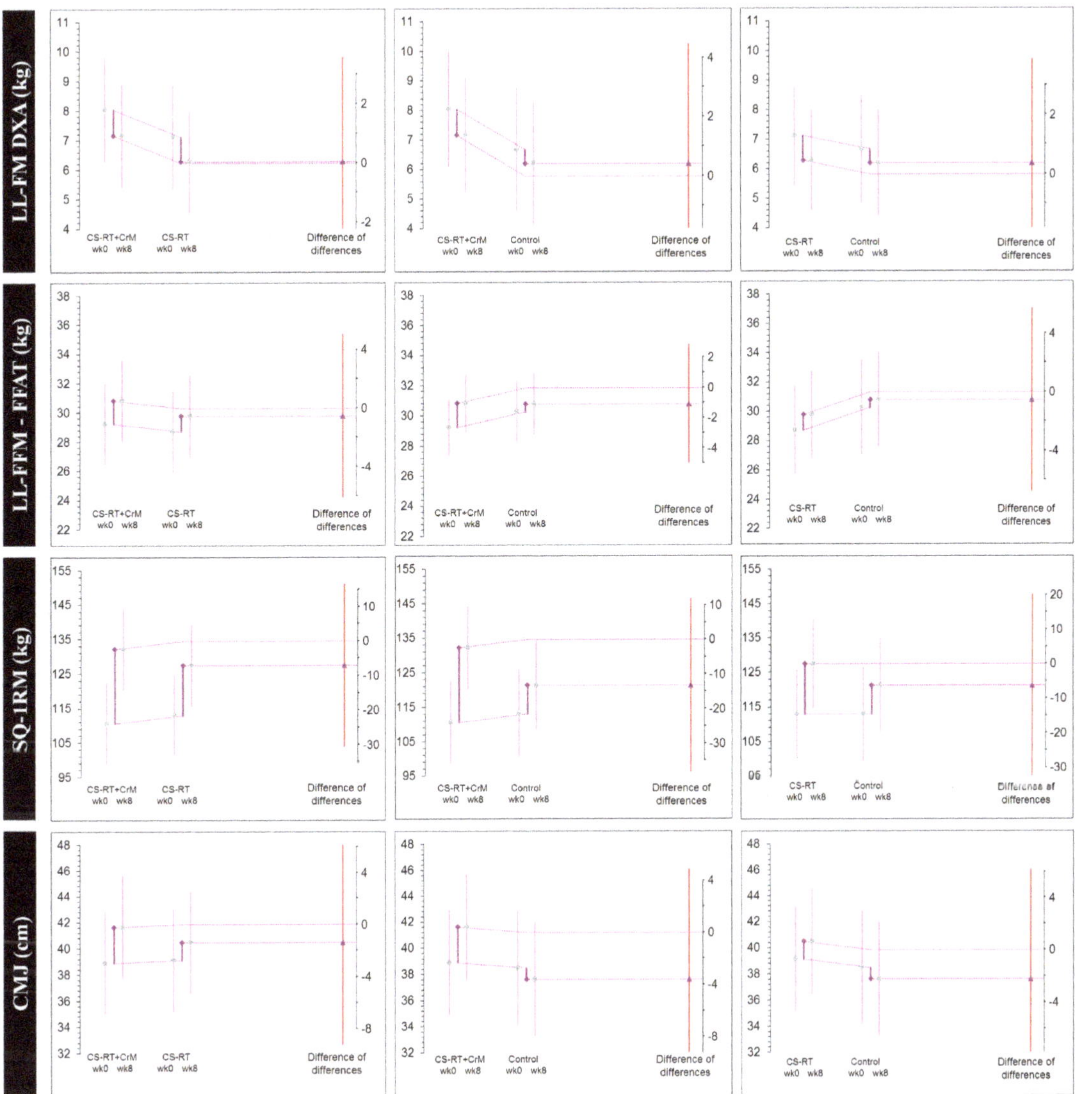

Figure 4. Difference-in-difference estimation plots for all variables. This graphic shows the difference (Δ = Week 8–Week 0) of the differences, which is the calculation of the group means: CS-RT+CrM (Δ_1), CS-RT (Δ_2), and control (Δ_3) groups on lower-limb body composition and strength-related variables. The effect chosen for examination is displayed as the triangle, with its 95% CI, against a floating different axis.

4. Discussion

The effects of CrM supplementation on body composition and RT performance have been ratified over 30 years of clinical research in different populations [12,14–16]. However, this study is expected to be the first literature contribution on the combination of CrM supplementation and a CS-RT program. An 8-week lower-limb CS-RT program (four clusters of 3RM per set with 20 s of intra-set rest period and 180 s of inter-set rest) was carried out twice per week with 72 h of recovery between sessions. We measured the effects of a CrM supplementation protocol ($0.1 \text{ g·kg}^{-1}\text{·day}^{-1}$) during this CS-RT program on lower-limb body composition and strength/power-related variables in resistance-trained men. Since CS-RT takes advantage of the intra-set rest periods to allow PCr resynthesis, which can evoke potentially greater benefits than traditional RT [32,33], we hypothesized that CrM would have a superior impact on the exercise adaptations to this training strategy in resistance-trained men that aim to optimize body composition and muscle strength.

Significant improvements were observed after the CS-RT program in the participants supplemented with or without CrM on the main studied variables (i.e., whole-body fat percentage, fat mass, whole-body FFM, lower-limb fat mass, and lower-limb FFM). However, higher clinical meaningfulness (larger effect sizes) was obtained in the CS-RT+CrM group regarding whole-body FFM (0.64 versus 0.16), lower-limb FFM (0.62 versus 0.18), and SQ-1RM (1.23 versus 0.75) when compared to the CS-RT group. Interestingly, lower-limb muscle power only improved significantly after CrM supplementation (d_{unb} = 0.44) with no significant changes in the CS-RT nor the control group. Thus, although no significant difference between groups was detected, we partially confirmed our initial hypothesis. In fact, a similar but not identical RT methodology known as drop-set RT has also been shown to be benefited by CrM supplementation [54]; nevertheless, the referred study included untrained aging males, which remarks the novel findings of our research.

It is highly possible that optimization of the CK/PCr system after CrM supplementation allow energy and metabolic buffering (regulation of Pi, H^+ and Ca^{2+} concentrations), which might result in a higher training volume with the same mechanical output. Some researchers have suggested recently that the CK/PCr system might act as a dynamic biosensor of the cellular chemo-mechanical energy transduction (cellular allostasis) [8]. This is important at the whole-body level if we consider that the altered phenotype of an individual is a result of an allostatic load that is sustained for an appropriate interval of time; hence, the faster the recovery, the sooner the desired alteration in the phenotype [55]. More studies in female and untrained exercisers are needed to confirm the potential optimization of CS-RT adaptations by CrM, especially if the similar effects of traditional RT and CS-RT in postmenopausal and elderly women are taken into account [56].

We evaluated effects on lower-limb body composition and strength due to the marked response in this group of muscles after CrM administration [57] and the potential sports- and health-related benefits that can be derived from CS-RT [58]. Hence, the clinical importance of our findings (moderate and large effect sizes) open an interesting line of research in other several areas. Thus, further research is warranted on a wide range of phenotypes that can be benefited from the combination of CS-RT and CrM supplementation, including age-related loss of lower-limb mass (sarcopenia) [59], age-related loss of lower-limb muscle strength/power (dynapenia) [60], cancer-related impairments of lower-limb neurological function and skeletal muscle mass [61], osteoarthritis-related low skeletal muscle mass in the lower limbs [62], and the potential of CrM to reduce hemodialysis-related sarcopenia and improve quality of life in hemodialysis patients [63,64].

It is noteworthy that we not only used an evidence-based nutritional strategy to optimize increases in lean body mass and strength (high-protein hyperenergetic diet) [43,65,66] but also analyzed DXA body composition data after adjusting for FFAT to accurately estimate the changes in lean tissue [46]. We are aware that as a result of the skewed selection of participants this study is susceptible to bias and other forms of selection errors [67]. However, besides serving as a time-efficient strategy with lower financial expenditures, this small-scale feasibility study allows us to evaluate the practicability of carrying out an intervention in a larger future study [50], in this case on the effects of CrM supplementation during a CS-RT program.

5. Conclusions

Resistance-trained men following a high-protein diet and a CS-RT program with or without CrM supplementation improved body composition and strength-related variables in lower limbs after eight weeks, while no changes were detected in participants following an unsupervised nutrition and RT exercise regime. Notwithstanding, the supplementation with CrM promoted greater exercise adaptations considering the higher clinical meaningfulness (larger effect sizes) on whole-body FFM, lower-limb FFM, and SQ-1RM in comparison to the CS-RT and control groups. Although double-blinded clinical trials with a larger sample are needed to confirm these findings, the combination of CrM supplementation and CS-RT seems a practical strategy to optimize training adaptation in advanced exercisers that aim to increase lower-limb FFM and strength. Further research might study the potential health and therapeutic effects of this nutrition and exercise strategy.

Author Contributions: Conceptualization, D.A.B. M.G.-S. and S.V.-M.; methodology, D.A.B. and S.V.-M.; data collection, S.V.-M., R.R. and M.G.-S.; formal analysis, D.A.B. and J.L.P.; data curation, D.A.B. and J.L.P.; writing—original draft preparation, D.A.B. and S.V.-M.; writing—review and editing, R.B.K., M.G.-S., J.L.P. and J.B.-P.; visualization, D.A.B.; project administration, S.V.-M. and D.A.B. All authors have read and agreed to the published version of the manuscript.

Funding: This study was performed within a framework of a "Cluster Training" project developed by several research members of the Dynamical Business & Science Society—DBSS International SAS. The dietary supplements were provided by the I+D+i+d department of MTX Corporation (Irún, Spain). The APC was partially funded by the research division of DBSS International SAS.

Institutional Review Board Statement: The study was conducted according to the guidelines of the Declaration of Helsinki, and approved by the Ethics Committee of the EADE-University of Wales Trinity Saint David (code: EADECAFYD-2019).

Informed Consent Statement: Informed consent was obtained from all subjects involved in the study.

Data Availability Statement: The data that support the findings of this study are available upon request from the corresponding author.

Acknowledgments: The authors would like to thank the study participants for their active involvement and sharing to make this project possible.

Conflicts of Interest: D.A.B. serves as science product manager for MTX Corporation®, a company that produces, distributes, sells, and does research on dietary supplements (including creatine) in Europe, has acted as a scientific consultant for MET-Rx and Healthy Sports in Colombia, and has received honoraria for speaking about creatine at international conferences. He also is a current member of the "Creatine in Health" scientific advisory board for AlzChem Tostberg GmbH, who sponsored this special issue on "Creatine Supplementation for Health and Clinical Diseases". R.B.K. has conducted industry-sponsored research on creatine, received financial support for presenting on creatine at industry-sponsored scientific conferences, and has served as an expert witness on cases related to creatine. Additionally, he serves as Chair of the "Creatine in Health" Scientific Advisory Board for AlzChem Tostberg GmbH. The other authors declare no conflicts of interest.

References

1. Sundberg, C.W.; Fitts, R.H. Bioenergetic basis of skeletal muscle fatigue. *Curr. Opin. Physiol.* **2019**, *10*, 118–127. [CrossRef]
2. Hunter, S.K. Performance fatigability: Mechanisms and task specificity. *Cold Spring Harb. Perspect. Med.* **2018**, *8*. [CrossRef]
3. Sundberg, C.W.; Hunter, S.K.; Trappe, S.W.; Smith, C.S.; Fitts, R.H. Effects of elevated H(+) and Pi on the contractile mechanics of skeletal muscle fibres from young and old men: Implications for muscle fatigue in humans. *J. Physiol.* **2018**, *596*, 3993–4015. [CrossRef]
4. Davies, R.C.; Eston, R.G.; Fulford, J.; Rowlands, A.V.; Jones, A.M. Muscle damage alters the metabolic response to dynamic exercise in humans: A 31P-MRS study. *J. Appl. Physiol.* **2011**, *111*, 782–790. [CrossRef]
5. Jarvis, K.; Woodward, M.; Debold, E.P.; Walcott, S. Acidosis affects muscle contraction by slowing the rates myosin attaches to and detaches from actin. *J. Muscle Res. Cell Motil.* **2018**, *39*, 135–147. [CrossRef]
6. Allen, D.G.; Trajanovska, S. The multiple roles of phosphate in muscle fatigue. *Front. Physiol.* **2012**, *3*, 463. [CrossRef]
7. Bonilla, D.A.; Moreno, Y. Molecular and metabolic insights of creatine supplementation on resistance training. *Rev. Colomb. Química* **2015**, *44*, 11–18. [CrossRef]
8. Bonilla, D.A.; Kreider, R.B.; Stout, J.R.; Forero, D.A.; Kerksick, C.M.; Roberts, M.D.; Rawson, E.S. Metabolic basis of creatine in health and disease: A Bioinformatics-assisted review. *Nutrients* **2021**, *13*, 1238. [CrossRef] [PubMed]
9. Wallimann, T. The extended, dynamic mitochondrial reticulum in skeletal muscle and the creatine kinase (CK)/phosphocreatine (PCr) shuttle are working hand in hand for optimal energy provision. *J. Muscle Res. Cell Motil.* **2015**, *36*, 297–300. [CrossRef] [PubMed]

10. Wallimann, T.; Tokarska-Schlattner, M.; Schlattner, U. The creatine kinase system and pleiotropic effects of creatine. *Amino Acids* **2011**, *40*, 1271–1296. [CrossRef]

11. Schlattner, U.; Kay, L.; Tokarska-Schlattner, M. Mitochondrial proteolipid complexes of creatine kinase. In *Membrane Protein Complexes: Structure and Function*; Springer: Berlin, Germany, 2018; pp. 365–408.

12. Kreider, R.B.; Kalman, D.S.; Antonio, J.; Ziegenfuss, T.N.; Wildman, R.; Collins, R.; Candow, D.G.; Kleiner, S.M.; Almada, A.L.; Lopez, H.L. International Society of Sports Nutrition position stand: Safety and efficacy of creatine supplementation in exercise, sport, and medicine. *J. Int. Soc. Sports Nutr.* **2017**, *14*. [CrossRef]

13. Antonio, J.; Candow, D.G.; Forbes, S.C.; Gualano, B.; Jagim, A.R.; Kreider, R.B.; Rawson, E.S.; Smith-Ryan, A.E.; Van Dusseldorp, T.A.; Willoughby, D.S.; et al. Common questions and misconceptions about creatine supplementation: What does the scientific evidence really show? *J. Int. Soc. Sports Nutr.* **2021**, *18*, 13. [CrossRef] [PubMed]

14. Wax, B.; Kerksick, C.M.; Jagim, A.R.; Mayo, J.J.; Lyons, B.C.; Kreider, R.B. Creatine for exercise and sports performance, with recovery considerations for healthy populations. *Nutrients* **2021**, *13*, 1915. [CrossRef] [PubMed]

15. Forbes, S.C.; Candow, D.G.; Ostojic, S.M.; Roberts, M.D.; Chilibeck, P.D. Meta-analysis examining the importance of creatine ingestion strategies on lean tissue mass and strength in older adults. *Nutrients* **2021**, *13*, 1912. [CrossRef]

16. Kreider, R.B.; Stout, J.R. Creatine in health and disease. *Nutrients* **2021**, *13*, 447. [CrossRef] [PubMed]

17. Harmon, K.K.; Stout, J.R.; Fukuda, D.H.; Pabian, P.S.; Rawson, E.S.; Stock, M.S. The Application of creatine supplementation in medical rehabilitation. *Nutrients* **2021**, *13*, 1825. [CrossRef]

18. Li, B.; Yang, L. Creatine in T Cell antitumor immunity and cancer immunotherapy. *Nutrients* **2021**, *13*, 1633. [CrossRef] [PubMed]

19. Wallimann, T.; Hall, C.H.T.; Colgan, S.P.; Glover, L.E. Creatine supplementation for patients with inflammatory bowl diseases: A scientific rationale for a clinical trial. *Nutrients* **2021**, *13*, 1429. [CrossRef]

20. Balestrino, M. Role of creatine in the heart: Health and disease. *Nutrients* **2021**, *13*, 1215. [CrossRef] [PubMed]

21. Muccini, A.M.; Tran, N.T.; de Guingand, D.L.; Philip, M.; Della Gatta, P.A.; Galinsky, R.; Sherman, L.S.; Kelleher, M.A.; Palmer, K.R.; Berry, M.J.; et al. Creatine metabolism in female reproduction, pregnancy and newborn health. *Nutrients* **2021**, *13*, 490. [CrossRef]

22. Stricker, P.R.; Faigenbaum, A.D.; McCambridge, T.M. Resistance training for children and adolescents. *Pediatrics* **2020**, *145*. [CrossRef]

23. Talar, K.; Hernández-Belmonte, A.; Vetrovsky, T.; Steffl, M.; Kałamacka, E.; Courel-Ibáñez, J. Benefits of Resistance training in early and late stages of frailty and sarcopenia: A systematic review and meta-analysis of randomized controlled studies. *J. Clin. Med.* **2021**, *10*, 1630. [CrossRef] [PubMed]

24. Evans, J.W. Periodized resistance training for enhancing skeletal muscle hypertrophy and strength: A mini-review. *Front. Physiol.* **2019**, *10*, 13. [CrossRef] [PubMed]

25. Bonilla, D.A.; Benítez-Porres, J.; Romance, R.; Medina, I.; Petro, J.L.; Schoenfeld, B.J.; García-Sillero, M.; Kreider, R.B.; Vargas-Molina, S. Effects of a non-linear resistance training program on biochemical and physiological health parameters in elderly. *Med. Sci. Sports Exerc.* **2020**, *52*, 749. [CrossRef]

26. Vargas-Molina, S.; Romance, R.; Schoenfeld, B.J.; García, M.; Petro, J.L.; Bonilla, D.A.; Kreider, R.B.; Martín-Rivera, F.; Benítez-Porres, J. Effects of cluster training on body composition and strength in resistance-trained men. *Isokinet. Exerc. Sci.* **2020**, *28*, 391–399. [CrossRef]

27. Garcia-Ramos, A.; Gonzalez-Hernandez, J.M.; Banos-Pelegrin, E.; Castano-Zambudio, A.; Capelo-Ramirez, F.; Boullosa, D.; Haff, G.G.; Jimenez-Reyes, P. Mechanical and metabolic responses to traditional and cluster set configurations in the bench press exercise. *J. Strength Cond. Res.* **2020**, *34*, 663–670. [CrossRef] [PubMed]

28. Morales-Artacho, A.J.; Padial, P.; Garcia-Ramos, A.; Perez-Castilla, A.; Feriche, B. Influence of a cluster set configuration on the adaptations to short-term power training. *J. Strength Cond. Res.* **2018**, *32*, 930–937. [CrossRef]

29. Tufano, J.J.; Halaj, M.; Kampmiller, T.; Novosad, A.; Buzgo, G. Cluster sets vs. traditional sets: Levelling out the playing field using a power-based threshold. *PLoS ONE* **2018**, *13*, e0208035. [CrossRef]

30. Latella, C.; Teo, W.-P.; Drinkwater, E.J.; Kendall, K.; Haff, G.G. The acute neuromuscular responses to cluster set resistance training: A systematic review and meta-analysis. *Sports Med.* **2019**, *49*, 1861–1877. [CrossRef]

31. Marshall, J.; Bishop, C.; Turner, A.; Haff, G.G. Optimal training sequences to develop lower body force, velocity, power, and jump height: A systematic review with meta-analysis. *Sports Med.* **2021**, *51*, 1245–1271. [CrossRef]

32. Arazi, H.; Khoshnoud, A.; Asadi, A.; Tufano, J.J. The effect of resistance training set configuration on strength and muscular performance adaptations in male powerlifters. *Sci. Rep.* **2021**, *11*, 7844. [CrossRef]

33. Jukic, I.; Van Hooren, B.; Ramos, A.G.; Helms, E.R.; McGuigan, M.R.; Tufano, J.J. The effects of set structure manipulation on chronic adaptations to resistance training: A systematic review and meta-analysis. *Sports Med.* **2021**, *51*, 1061–1086. [CrossRef]

34. Willardson, J.M. A Brief Review: Factors affecting the length of the rest interval between resistance exercise sets. *J. Strength Cond. Res.* **2006**, *20*. [CrossRef]

35. Harris, R.C.; Edwards, R.H.; Hultman, E.; Nordesjo, L.O.; Nylind, B.; Sahlin, K. The time course of phosphorylcreatine resynthesis during recovery of the quadriceps muscle in man. *Pflugers Arch.* **1976**, *367*, 137–142. [CrossRef] [PubMed]

36. Hansen, K.T.; Cronin, J.B.; Newton, M.J. The effect of cluster loading on force, velocity, and power during ballistic jump squat training. *Int. J. Sports Physiol. Perform.* **2011**, *6*, 455–468. [CrossRef] [PubMed]

37. Iglesias-Soler, E.; Carballeira, E.; Sanchez-Otero, T.; Mayo, X.; Fernandez-del-Olmo, M. Performance of maximum number of repetitions with cluster-set configuration. *Int. J. Sports Physiol. Perform.* **2014**, *9*, 637–642. [CrossRef]

38. Moreno, S.D.; Brown, L.E.; Coburn, J.W.; Judelson, D.A. Effect of cluster sets on plyometric jump power. *J. Strength Cond. Res.* **2014**, *28*, 2424–2428. [CrossRef] [PubMed]

39. Asadi, A.; Ramirez-Campillo, R. Effects of cluster vs. traditional plyometric training sets on maximal-intensity exercise performance. *Medicina* **2016**, *52*, 41–45. [CrossRef] [PubMed]

40. Eldridge, S.M.; Chan, C.L.; Campbell, M.J.; Bond, C.M.; Hopewell, S.; Thabane, L.; Lancaster, G.A. CONSORT 2010 statement: Extension to randomised pilot and feasibility trials. *Pilot Feasibility Stud.* **2016**, *2*. [CrossRef]

41. World Medical Association. World Medical Association Declaration of Helsinki: Ethical principles for medical research involving human subjects. *J. Postgrad. Med.* **2002**, *48*, 206–208.

42. Vargas-Molina, S.; Petro, J.L.; Romance, R.; Kreider, R.B.; Schoenfeld, B.J.; Bonilla, D.A.; Benitez-Porres, J. Effects of a ketogenic diet on body composition and strength in trained women. *J. Int. Soc. Sports Nutr.* **2020**, *17*, 19. [CrossRef]

43. Rozenek, R.; Ward, P.; Long, S.; Garhammer, J. Effects of high-calorie supplements on body composition and muscular strength following resistance training. *J. Sports Med. Phys. Fit.* **2002**, *42*, 340–347.

44. Campbell, B.; Kreider, R.B.; Ziegenfuss, T.; La Bounty, P.; Roberts, M.; Burke, D.; Landis, J.; Lopez, H.; Antonio, J. International Society of Sports Nutrition position stand: Protein and exercise. *J. Int. Soc. Sports Nutr.* **2007**, *4*. [CrossRef]

45. Heymsfield, S.B.; Gallagher, D.; Kotler, D.P.; Wang, Z.; Allison, D.B.; Heshka, S. Body-size dependence of resting energy expenditure can be attributed to nonenergetic homogeneity of fat-free mass. *Am. J. Physiol. Endocrinol. Metab.* **2002**, *282*, E132–E138. [CrossRef] [PubMed]

46. Abe, T.; Dankel, S.J.; Loenneke, J.P. Body fat loss automatically reduces lean mass by changing the fat-free component of adipose tissue. *Obesity* **2019**, *27*, 357–358. [CrossRef]

47. Abe, T.; Loenneke, J.P.; Thiebaud, R.S.; Fujita, E.; Akamine, T. The impact of DXA-derived fat-free adipose tissue on the prevalence of low muscle mass in older adults. *Eur. J. Clin. Nutr.* **2018**, *73*, 757–762. [CrossRef] [PubMed]

48. McGuigan, M. Administration, scoring, and interpretation of selected tests. In *Essentials of Strength Training and Conditioning*; Haff, G., Triplett, N., Eds.; Human Kinetics: Champaign, IL, USA, 2016; pp. 265–266.

49. Caulfield, S.; Berninger, D. Exercise technique for free weight and machine training. In *Essentials of Strength Training and Conditioning—National Strength and Conditioning Association*; Haff, G., Triplett, T., Eds.; Human Kinetics: Champaign, IL, USA, 2016; Volume 4, p. 735.

50. Morris, N.S.; Rosenbloom, D.A. Defining and understanding pilot and other feasibility studies. *Am. J. Nurs.* **2017**, *117*, 38–45. [CrossRef] [PubMed]

51. Rosenthal, J.A. Qualitative Descriptors of strength of association and effect size. *J. Soc. Serv. Res.* **1996**, *21*, 37–59. [CrossRef]

52. Cannataro, R.; Cione, E.; Gallelli, L.; Marzullo, N.; Bonilla, D.A. Acute effects of supervised making weight on health markers, hormones and body composition in muay thai fighters. *Sports* **2020**, *8*, 137. [CrossRef]

53. Cumming, G. *Understanding the New Statistics; Effect Sizes, Confidence Intervals, and Meta-Analysis*, 1st ed.; Routledge: New York, NY, USA, 2013.

54. Johannsmeyer, S.; Candow, D.G.; Brahms, C.M.; Michel, D.; Zello, G.A. Effect of creatine supplementation and drop-set resistance training in untrained aging adults. *Exp. Gerontol.* **2016**, *83*, 112–119. [CrossRef] [PubMed]

55. Bonilla, D.A.; Pérez-Idárraga, A.; Odriozola-Martínez, A.; Kreider, R.B. The 4R's framework of nutritional strategies for post-exercise recovery: A review with emphasis on new generation of carbohydrates. *Int. J. Environ. Res. Public Health* **2020**, *18*, 103. [CrossRef]

56. Dias, R.K.N.; Penna, E.M.; Noronha, A.S.N.; de Azevedo, A.B.C.; Barbalho, M.; Gentil, P.V.; Coswig, V.S. Cluster-sets resistance training induce similar functional and strength improvements than the traditional method in postmenopausal and elderly women. *Exp. Gerontol.* **2020**, *138*. [CrossRef]

57. Lanhers, C.; Pereira, B.; Naughton, G.; Trousselard, M.; Lesage, F.-X.; Dutheil, F. Creatine supplementation and lower limb strength performance: A systematic review and meta-analyses. *Sports Med.* **2015**, *45*, 1285–1294. [CrossRef]

58. Latella, C.; Peddle-McIntyre, C.; Marcotte, L.; Steele, J.; Kendall, K.; Fairman, C.M. Strengthening the case for cluster set resistance training in aged and clinical settings: Emerging evidence, proposed benefits and suggestions. *Sports Med.* **2021**. [CrossRef]

59. Larsson, L.; Degens, H.; Li, M.; Salviati, L.; Lee, Y.i.; Thompson, W.; Kirkland, J.L.; Sandri, M. Sarcopenia: Aging-related loss of muscle mass and function. *Physiol. Rev.* **2019**, *99*, 427–511. [CrossRef]

60. Newman, A.; Suetta, C.; Frandsen, U.; Alegre, L.M.; Prescott, E.; Hansen, S.K.; Kamper, R.S.; Haddock, B.; Aagaard, P.; Alcazar, J. Age-and sex-specific changes in lower-limb muscle power throughout the lifespan. *J. Gerontol. Ser. A* **2020**, *75*, 1369–1378. [CrossRef]

61. Morris, R.; Lewis, A. Falls and Cancer. *Clin. Oncol.* **2020**, *32*, 569–578. [CrossRef]

62. Alway, S.E.; Lee, S.Y.; Ro, H.J.; Chung, S.G.; Kang, S.H.; Seo, K.M.; Kim, D.-K. Low skeletal muscle mass in the lower limbs is independently associated to knee osteoarthritis. *PLoS ONE* **2016**, *11*. [CrossRef]

63. Marini, A.C.B.; Pimentel, G.D. Creatine supplementation plus neuromuscular electrical stimulation improves lower-limb muscle strength and quality of life in hemodialysis men. *Einstein Sao Paolo* **2020**, *18*. [CrossRef]

64. Post, A.; Schutten, J.C.; Kremer, D.; van der Veen, Y.; Groothof, D.; Sotomayor, C.G.; Koops, C.A.; de Blaauw, P.; Kema, I.P.; Westerhuis, R.; et al. Creatine homeostasis and protein energy wasting in hemodialysis patients. *J. Transl. Med.* **2021**, *19*. [CrossRef] [PubMed]
65. Jäger, R.; Kerksick, C.M.; Campbell, B.I.; Cribb, P.J.; Wells, S.D.; Skwiat, T.M.; Purpura, M.; Ziegenfuss, T.N.; Ferrando, A.A.; Arent, S.M.; et al. International Society of Sports Nutrition Position Stand: Protein and exercise. *J. Int. Soc. Sports Nutr.* **2017**, *14*. [CrossRef] [PubMed]
66. Helms, E.R.; Aragon, A.A.; Fitschen, P.J. Evidence-based recommendations for natural bodybuilding contest preparation: Nutrition and supplementation. *J. Int. Soc. Sports Nutr.* **2014**, *11*. [CrossRef] [PubMed]
67. Etikan, I.; Babtope, O. A basic approach in sampling methodology and sample size calculation. *MedLife Clin.* **2019**, *1*, 1006.

Article

Supplementing Soy-Based Diet with Creatine in Rats: Implications for Cardiac Cell Signaling and Response to Doxorubicin

Laurence Kay [1], Lucia Potenza [2], Isabelle Hininger-Favier [1], Hubert Roth [1], Stéphane Attia [1], Cindy Tellier [1], Christian Zuppinger [3], Cinzia Calcabrini [2], Piero Sestili [2], Theo Wallimann [4], Uwe Schlattner [1,*] and Malgorzata Tokarska-Schlattner [1,*]

[1] University Grenoble Alpes and Inserm U1055, Laboratory of Fundamental and Applied Bioenergetics (LBFA) and SFR Environmental and Systems Biology (BEeSy), 38059 Grenoble, France; Laurence.Kay@univ-grenoble-alpes.fr (L.K.); isabelle.hininger@univ-grenoble-alpes.fr (I.H.-F.); hubert.roth@laposte.net (H.R.); stephane.attia@univ-grenoble-alpes.fr (S.A.); cindy.tellier@univ-grenoble-alpes.fr (C.T.)
[2] Dipartimento di Scienze Biomolecolari, Università Degli Studi di Urbino "Carlo Bo", 61029 Urbino, Italy; lucia.potenza@uniurb.it (L.P.); cinzia.calcabrini@uniurb.it (C.C.); piero.sestili@uniurb.it (P.S.)
[3] Inselspital, Cardiology Department, Bern University Hospital, 3010 Bern, Switzerland; christian.zuppinger@dbmr.unibe.ch
[4] Department of Biology, ETH-Zurich, Emeritus, 8962 Bergdietikon, Switzerland; theo.wallimann@cell.biol.ethz.ch
* Correspondence: uwe.schlattner@univ-grenoble-alpes.fr (U.S.); malgorzata.tokarska-schlattner@univ-grenoble-alpes.fr (M.T.-S.); Tel.: +33-476-51-46-71 (U.S.); +33-476-51-46-70 (M.T.-S.)

Citation: Kay, L.; Potenza, L.; Hininger-Favier, I.; Roth, H.; Attia, S.; Tellier, C.; Zuppinger, C.; Calcabrini, C.; Sestili, P.; Wallimann, T.; et al. Supplementing Soy-Based Diet with Creatine in Rats: Implications for Cardiac Cell Signaling and Response to Doxorubicin. *Nutrients* **2022**, *14*, 583. https://doi.org/10.3390/nu14030583

Academic Editors: Richard B. Kreider and Jeffrey R. Stout

Received: 19 December 2021
Accepted: 21 January 2022
Published: 28 January 2022

Publisher's Note: MDPI stays neutral with regard to jurisdictional claims in published maps and institutional affiliations.

Abstract: Nutritional habits can have a significant impact on cardiovascular health and disease. This may also apply to cardiotoxicity caused as a frequent side effect of chemotherapeutic drugs, such as doxorubicin (DXR). The aim of this work was to analyze if diet, in particular creatine (Cr) supplementation, can modulate cardiac biochemical (energy status, oxidative damage and antioxidant capacity, DNA integrity, cell signaling) and functional parameters at baseline and upon DXR treatment. Here, male Wistar rats were fed for 4 weeks with either standard rodent diet (NORMAL), soy-based diet (SOY), or Cr-supplemented soy-based diet (SOY + Cr). Hearts were either freeze-clamped in situ or following ex vivo Langendorff perfusion without or with 25 μM DXR and after recording cardiac function. The diets had distinct cardiac effects. Soy-based diet (SOY vs. NORMAL) did not alter cardiac performance but increased phosphorylation of acetyl-CoA carboxylase (ACC), indicating activation of rather pro-catabolic AMP-activated protein kinase (AMPK) signaling, consistent with increased ADP/ATP ratios and lower lipid peroxidation. Creatine addition to the soy-based diet (SOY + Cr vs. SOY) slightly increased left ventricular developed pressure (LVDP) and contractility dp/dt, as measured at baseline in perfused heart, and resulted in activation of the rather pro-anabolic protein kinases Akt and ERK. Challenging perfused heart with DXR, as analyzed across all nutritional regimens, deteriorated most cardiac functional parameters and also altered activation of the AMPK, ERK, and Akt signaling pathways. Despite partial reprogramming of cell signaling and metabolism in the rat heart, diet did not modify the functional response to supraclinical DXR concentrations in the used acute cardiotoxicity model. However, the long-term effect of these diets on cardiac sensitivity to chronic and clinically relevant DXR doses remains to be established.

Keywords: adenosine 5′-monopnophosphate-activated protein kinase; anthracyclines; creatine supplementation; cardiac signaling; cardiotoxicity; doxorubicin; soy; vegetarian/vegan diet

1. Introduction

Nutritional habits are increasingly recognized for their impact on cardiovascular health and disease, including prevention of cancer relapse or different comorbidities [1]. This is

525

of particular interest for acute or chronic cardiotoxicity caused by anticancer chemother-apeutics, in particular anthracyclines, such as doxorubicin (DXR; reviewed in [2–4]), for which efficient preventive or therapeutic strategies are still lacking. Currently, cardiotoxic side effects are minimized by limiting total drug dose, slow administration by infusion rather than bolus injection, liposomal drug encapsulation, or co-administration of the iron chelator dexrazoxane, a protective adjuvant only approved in the USA [5–7]. More recent recommendations to patients have started to emphasize lifestyle changes, including physical activity (exercise) and nutritional habits [1,8–12].

Creatine (Cr) is one of the most popular dietary supplements [13]. Humans synthesize only about 50% of their daily Cr requirement in the kidney, pancreas, and liver [14] and certain brain cells [15]. The remainder has to be taken up from Cr-containing non-vegan nutrition, especially fish and meat [16]. Cr then enters cells, such as cardiomyocytes, via the plasma membrane Cr transporter. Cr constitutes a cellular energy precursor, transformed within the cell into the "energy-rich" phosphocreatine (PCr). Cr and PCr, together with isoforms of creatine kinase (CK), confer bioenergetic advantages by providing an efficient energy buffer and transfer system for cells and tissues with high and fluctuating energy requirements, such as the heart [17]. Supplementation with chemically pure Cr shows protective effects in different pathologies, such as cardiac ischemia and reperfusion injury ([18]; for a review, see [19]) or neurodegenerative and muscular disorders (reviewed in [16]). Some protective effects were also observed with DXR-induced injury, including reduced cardiac damage in DXR-treated animals [20], improved viability of DXR-treated cultured cardiomyocytes and H9c2 cells [21,22], and protection against DXR-induced RNA damage in non-cardiac cells [23]. CK overexpression in a murine model of DXR cardiotoxicity improved myocardial energetics, contractile dysfunction, and survival [24]. Beyond the bioenergetic functions of Cr and PCr, pleiotropic anti-oxidative and anti-apoptotic effects were reported, including reduced mitochondrial ROS, increased oxidative stress defense, and inhibition of mitochondrial permeability transition (reviewed by [16,25]). Cr and PCr also interact with anionic membrane phospholipids, increasing phospholipid packing and thus stabilizing and protecting biomembranes [26].

Cr supplementation is of particular interest in combination with vegetarian/vegan diets that naturally do not contain Cr. Using soy-derived products (soy meal or soy protein isolate) to replace meat- and fish-derived compounds as the main protein source yields a Cr-free diet, which can serve as an experimental control for Cr supplementation. However, soy-based products have been reported to also mediate protective metabolic effects on their own, including on the cardiovascular system [27–30]. Some of the bioactive compounds in soy-derived products are supposed to be isoflavones (genistein, daidzein, and equol) that may act as phytoestrogens [28].

Here, we investigated whether a soy-based vegan chow (SOY) as compared to a standard rodent chow (NORMAL), and a Cr-supplemented soy chow (SOY + CR) as compared to SOY, can affect cardiac function, biochemistry, and cell signaling in general, and the cardiac response to DXR in particular. In our case, NORMAL contained soy meal and fish hydrolysate as protein sources, with the latter containing a variable amount of naturally occurring Cr. SOY was entirely vegan, with soy products as the exclusive protein source and entirely lacking Cr, while SOY + CR was supplemented with 2% (*w*/*w*) chemically pure synthetic Cr monohydrate. Our data show that one month of being fed the differential diet has modest but significant functional and biochemical effects on the rat heart, in particular on specific cell signaling pathways. Some of them are potentially relevant for cardiac health and the response to DXR, but they did not confer functional improvement in the perfused heart model of acute DXR cardiotoxicity.

2. Materials and Methods

2.1. Materials

Doxorubicin hydrochloride (DXR) was purchased from Sigma (Saint Louis, MO, USA) or Selleck Chemicals (Houston, TX, USA). A stock solution (10 mM) was prepared in water and kept frozen until use. Further dilutions were prepared in Krebs-Henseleit buffer [31] just before heart perfusion. Protease inhibitor cocktail tablets were obtained from Roche (Mannheim, Germany) and phosphatase inhibitor cocktail was obtained from Pierce (Rockford, IL, USA). Creatine (creatine monohydrate, Creapure®) was a gift from AlzChem Trostberg GmbH (Trostberg, Germany).

2.2. Animals

All procedures involving animals were approved by the Grenoble Ethics Committee for Animal Experimentation (15_LBFA-U884-HD-01). Male Wistar rats initially fed a standard chow for young rats (A03 reference U8200, Safe, Augy, France; 3237 kcal/kg) were then differentially fed for 4 weeks starting from 2 months of age. One group of animals continued to receive the standard chow for adult rats (NORMAL; A04 reference U8220, Safe, Augy, France; 2791 kcal/kg) containing 4% (*w/w*) fish hydrolysate and 8% (*w/w*) soy meal. The second group was fed a Cr-free soy-based chow (SOY; modified A04 reference U8220 version 149, Safe, Augy, France; 2711 kcal/kg) where fish hydrolysate was replaced by the same percentage of soy isolate. The third group was fed the latter chow supplemented with 2% (*w/w*) creatine (SOY + Cr, modified A04 reference U8220 version 150, Safe, Augy, France; 2711 kcal/kg). Diets were purchased from Safe, Augy, France). After 4 weeks of the differential diets, animals were anaesthetized with sodium pentobarbital (50 mg/kg i.p.), and hearts were freeze-clamped in situ (immediately after thoracotomy of respirator-ventilated animals) or following ex vivo Langendorff perfusion with or without 25 µM DXR (see the experimental scheme in Figure 1). Frozen hearts were stored at −80 °C.

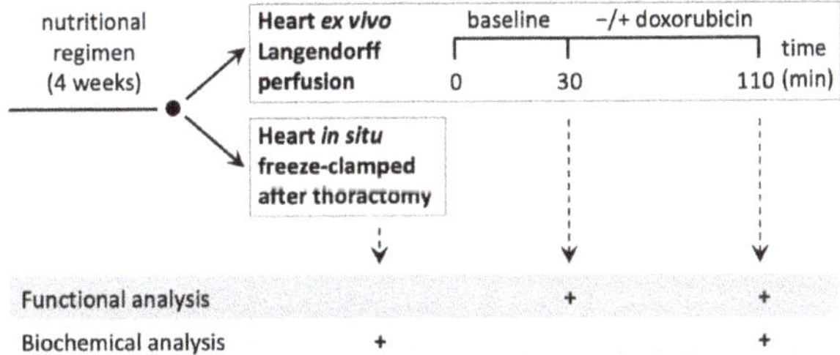

Figure 1. Scheme of the experimental procedure (for details, see Material and Methods).

2.3. Rat Heart Perfusion

Perfusion experiments were essentially performed according to the protocol described earlier [31–34]. Briefly, rats were anaesthetized with sodium pentobarbital (50 mg/kg i.p.) and heparinized (1500 IU/kg i.v.). Hearts were quickly removed and perfused at constant pressure in a non-circulating Langendorff apparatus with Krebs-Henseleit buffer, first for 30 min for stabilization (baseline) with Krebs-Henseleit buffer alone, and then for 80 min either with Krebs-Henseleit buffer without (control) or with 25 µM DXR. The DXR concentration was chosen on the basis of our previous studies [31–34]. During perfusion, systolic pressure, end diastolic pressure, dp/dt and –dp/dt, and heart rate were recorded every 10 min.

2.4. Metabolites, Oxidative Damage/Antioxidant Status, and DNA Integrity

Protein-free extracts were obtained by perchloric acid precipitation, and metabolites were quantified using HPLC (AMP, ADP, ATP) or a spectrophotometric assay (Cr and PCr) as described earlier [32,34]. Markers of oxidative damage and antioxidant status were quantified in heart extracts prepared as described earlier [35]. Reduced thiol (SH) groups were assayed according to [36]. N-acetyl cysteine (NAC) in the range of 0.125 to 1 mM (prepared from a 100 mM stock solution) was used for calibration. Standards and heart extracts were diluted in 50 mM phosphate buffer, 1 mM EDTA, pH 8, and 2.5 mM 5,5'-

dithio-bis-(2-nitrobenzoic acid) (DTNB), and subsequently the absorbance was measured at 412 nm. Antioxidant status was evaluated using the ferric reducing ability power (FRAP) assay [37]. Plasma thiobarbituric acid reactive substance (TBARS) concentrations were assessed as described [38]. Total genomic DNA was isolated with the QIAamp DNA mini kit (Qiagen) according to the manufacturer's instructions. The final concentration and quality of DNA were estimated both spectrophotometrically (DU-640; Beckman Instruments, Milan, Italy) at 260 nm, and by agarose gel electrophoresis. Nuclear and mitochondrial DNA damage were evaluated using a two-step strategy based on a long PCR and real-time PCR as described in detail elsewhere [34].

2.5. Immunoblotting

SDS-PAGE separation of heart homogenates (40–50 µg) and immunoblotting were performed according to standard procedures [34]. The transfer quality and equality of loading were checked by Ponceau staining. The blots were developed with chemiluminescence reagent (ECL Prime, GE Healthcare) using a CCD camera (ImageQuant LAS 4000, GE Healthcare). The quantification of signals was conducted using ImageQuantTL software (GE Healthcare). Tubulin or total protein were used for normalization for the phosphorylated proteins (probed on different membranes). The following primary antibodies obtained from Cell Signaling (Beverly, MA) were used: anti-AMPKα (Cat# 2532, RRID:AB_330331), anti-P(Thr172)AMPKα Cat# 2535, RRID:AB_331250), anti-ACC (Cat# 3662, RRID:AB_2219400), anti-P(Ser79)ACC (Cat# 3661, RRID:AB_330337), anti-Akt (Cat# 4691, RRID:AB_915783), anti-P(Ser473)Akt (Cat# 4060, RRID:AB_2315049), anti-P(Thr308)Akt (Cat# 13038, RRID:AB_2629447), anti-ERK-1/2 (Cat# 4695, RRID:AB_390779), anti-P(Thr202/Tyr204)ERK-1/2 (Cat# 4370, RRID:AB_2315112), anti-tubulin (Cat# 2128, RRID:AB_823664), anti-P(Ser/Thr)Akt Substrate Motif RXXS/T (Cat# 9614, RRID:AB_331810).

2.6. Data Analysis

Results are expressed as means $\pm$ SEM, if not stated otherwise. Depending on the experimental design, statistical analysis was performed using one- or two-way ANOVA (Sigma Plot; Systat Software, San Jose, CA, USA) or linear regression with dummy variables and robust standard errors (Stata 13; Stata Corp., College Station, TX, USA) to deal with the heterogeneity of variance. When appropriate, these were followed by the Student–Newman–Keuls or Bonferroni test, respectively, for pairwise comparisons. The one-way ANOVA P-values are not reported; the results of pairwise comparisons are reported in the case of significant one-way ANOVA P-values. For two factorial analysis, we report significance values for the effects of diet (P_{Diet}, independent of DXR), DXR (P_{DXR}, independent of diet), and the interaction of both diet and DXR treatment ($P_{Diet*DXR}$; indicating if the effect of one factor depends on the level of the second factor), and the results of pairwise comparisons (p). The P or p values are given in the graphs with 3 decimal places (and in bold characters) if significant, and with 2 decimal places if not. A value of P or $p < 0.05$ (for interaction $P < 0.1$) was considered statistically significant.

3. Results

The effects of the three nutritional regimens, standard rodent chow (NORMAL), soy-based diet (SOY), and Cr-supplemented soy-based diet (SOY + Cr), on the heart under control and DXR-challenged conditions were studied in a rat model (Figure 1). After one month of the differential diet, there was no significant difference in animal body weight (NORMAL 366 $\pm$ 6 g ($n = 36$), SOY 387 $\pm$ 12 g ($n = 23$), SOY + Cr 370 $\pm$ 11 g ($n = 21$)). Biochemical parameters were determined both in hearts freeze-clamped in situ immediately after thoracotomy, and after ex vivo Langendorff perfusion, consisting of a 30 min stabilization period followed by 80 min of perfusion without (control group) or with 25 µM DXR (DXR group). Functional parameters were measured ex vivo during Langendorff perfusion.

3.1. Heart Function

Cardiac function was first determined in perfused hearts at baseline (after 30 min of stabilization, Figure 2A), and then after 80 min of subsequent perfusion without or with DXR (Figure 2B). Data at baseline are considered to reflect the in vivo situation. The Cr-supplemented diet (SOY + Cr vs. SOY, Figure 2A) affected cardiac function, with a slight increase in the cardiac developed pressure LVDP ($p = 0.043$) and contractility dp/dt ($p = 0.039$). There was no significant effect of soy diet (SOY vs. NORMAL). DXR perfusion impaired almost all cardiac functional parameters ($P_{DXR} < 0.001$, Figure 2B) except heart rate, with a time-course (Figure S1) consistent with previous studies [31–34]. A statistically significant interaction between diet and DXR was only seen for diastolic pressure at the end of perfusion (EDP; $P_{Diet*DXR} = 0.047$, Figure 2B), increasing more in the group fed the Cr-free soy chow (SOY vs. NORMAL, $p = 0.002$; SOY vs. SOY + Cr, $p = 0.005$).

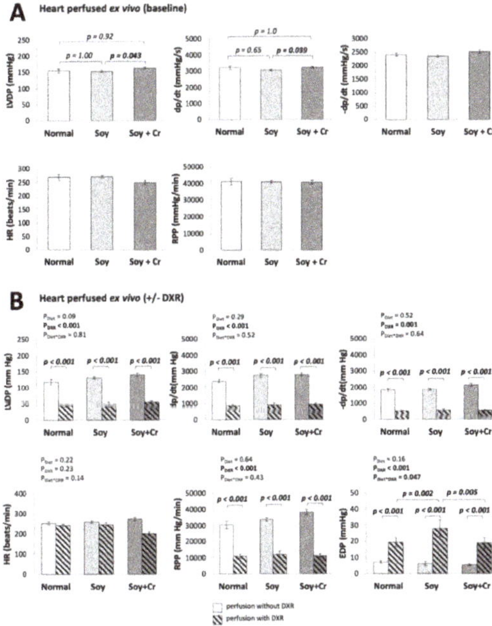

Figure 2. Heart function: effect of diet and DXR. Hemodynamic parameters: left ventricular developed pressure (LVDP), end-diastolic pressure (EDP), dp/dt, −dp/dt, heart rate (HR), and rate pressure product (RPP) measured in Langendorff perfused hearts after 30 min of stabilization (**A**) or after 30 min of stabilization followed by an additional 80 min of perfusion (**B**) without or with 25 µM DXR (empty or hatched bars, respectively). EDP values are given only in (**B**), as during the stabilization period shown in (**A**), EDP was adjusted to 5 mm Hg and thereafter the volume of the balloon rest unchanged. Statistical analysis with linear regression followed by the Bonferroni test for pairwise comparisons. For −dp/dt, HR, RPP in (**A**), and HR in (**B**), there are no statistically significant differences between groups. Mean ± SEM, $n = 11$–28 (**A**), $n = 4$–14 (**B**).

3.2. Creatine and Adenylate Levels

Cellular Cr availability and energy state were studied by determination of the Cr and adenylates in heart in situ and after ex vivo perfusion. A deteriorated energy state is often observed in DXR cardiotoxicity [39]. The diets affected the free and total Cr content in situ (Figure 3A) and also after ex vivo perfusion ($P_{Diet} = 0.019$ and 0.058, respectively, Figure 3B). As expected, 4 weeks of oral Cr supplementation (SOY + Cr vs. SOY) increased free and total Cr ($p = 0.036$ and $p = 0.028$, respectively, in ex vivo perfused hearts).

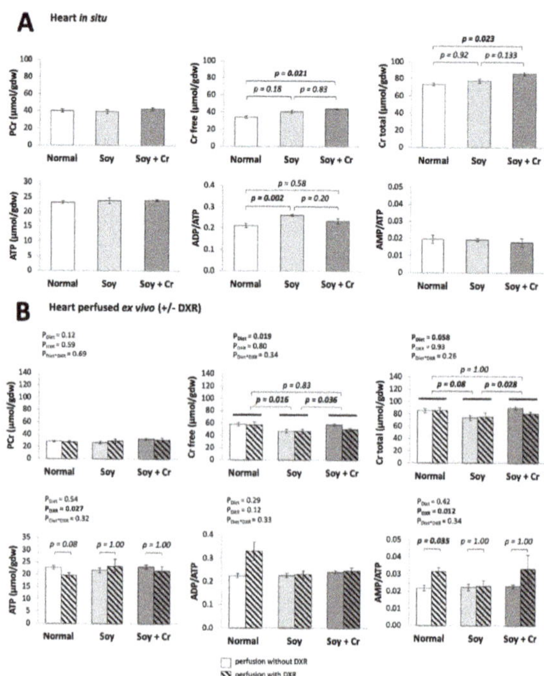

Figure 3. Energy metabolite levels: effect of diet and DXR. PCr, free Cr, total Cr, ATP, ADP/ATP, and AMP/ATP ratios in hearts freeze-clamped in situ immediately after thoracotomy (**A**) or following ex vivo Langendorff perfusion (**B**) without or with 25 μM DXR (empty or hatched bars, respectively). Statistical analysis with linear regression followed by the Bonferroni test for pairwise comparisons. For PCr, ATP, and AMP/ATP in (**A**) and PCr and ADP/ATP in (**B**), there are no statistically significant differences between groups. Mean ± SEM, n = 3–6 (**A**), n = 4–16 (**B**).

Interestingly, standard chow-fed animals also showed increased Cr in comparison to SOY (NORMAL vs. SOY), but the effect was seen only in the group used for ex vivo perfusion (p = 0.016 for free Cr and strong tendency p = 0.08 for total Cr), possibly due to a higher Cr content in the batch of NORMAL chow used here. Cardiac adenylate levels and ADP/ATP and AMP/ATP ratios remained largely unchanged between the diets (Figure 3A,B, lower rows), except for the soy chow, where the ADP/ATP ratio increased in hearts clamped in situ (SOY vs. NORMAL, p = 0.002). DXR perfusion across all nutritional regimens affected the ATP content and increased AMP/ATP ratios (P_{DXR} = 0.027 and 0.012, respectively, Figure 3B). No statistically significant interference was found between diet and DXR (Figure 3B).

3.3. Cell Signaling Pathways

Nutrition can lead to sustained alterations in cell signaling, and this can also occur with DXR treatment as we have shown earlier [31–34]. We therefore analyzed the activation of specific key signaling pathways involved in stress and pro-survival responses: AMP-activated protein kinase (AMPK; determined by phosphorylation of AMPK itself and its substrate acetyl-CoA carboxylase, ACC), extracellular signal-regulated kinase (ERK, determined by ERK phosphorylation), and Akt (determined by either Akt phosphorylation or global phosphorylation of Akt substrates) in heart in situ (Figure 4A) and after ex vivo perfusion (Figure 4B). Our experiments revealed a differential cardiac activation pattern of these signaling pathways, dependent on the diet (Figure 4A,B). The soy-based diet (SOY vs. NORMAL) almost doubled ACC phosphorylation (p = 0.004 in situ; p = 0.002

after ex vivo perfusion), consistent with the above-described increase in the ADP/ATP ratio. Changes in P-AMPK were similar but weaker and did not reach significance. The addition of Cr to the soy-based diet (SOY + Cr vs. SOY) led to no further change in P-ACC but activated ERK ($p = 0.014$ in situ) and Akt ($p = 0.015$ in situ; $p < 0.001$ after ex vivo perfusion at Ser473, and a tendency with $p = 0.08$ at Thr308). Perfusion with DXR changed the phosphorylation of AMPK, ACC, ERK, and Akt at Ser473 ($P_{DXR} = 0.022$, $P_{DXR} = 0.003$, $P_{DXR} < 0.001$, strong tendency with $P_{DXR} = 0.06$, respectively, Figure 4B). This confirmed our previous observations in animals fed a NORMAL diet [31,34], namely a DXR-induced inactivation of AMPK signaling with a decrease of P-AMPK and P-ACC (tendencies of $p = 0.11$ and 0.10, respectively), together with an activation of Akt ($p = 0.005$ at Ser473) as one factor potentially involved in AMPK inactivation [34]. For phosphorylation of Akt at Ser473, the interaction between diet and DXR was significant ($P_{Diet*DXR} = 0.09$, Figure 4B), with an increase only observed in the NORMAL group.

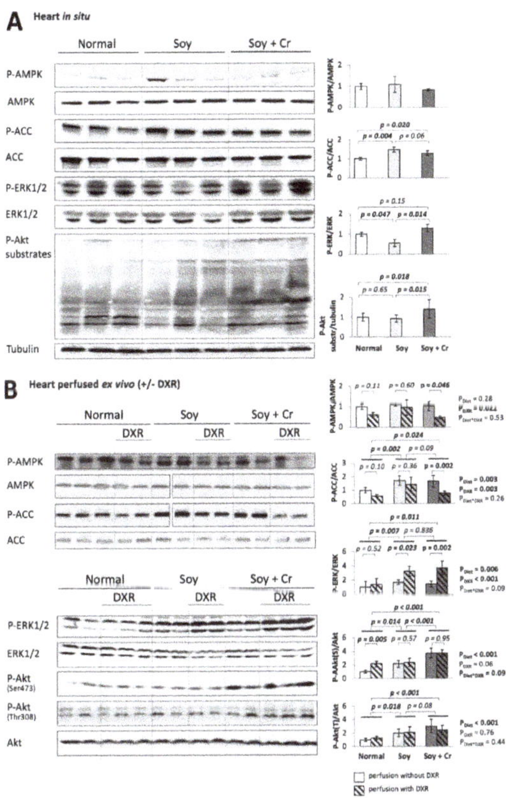

Figure 4. Activation of signaling pathways: effect of diet and DXR. Activation of cardiac pro-survival signaling pathways (AMPK, Akt, ERK) probed by immunoblot in total homogenates of hearts freeze-clamped in situ immediately after thoracotomy (**A**) or following ex vivo Langendorff perfusion (**B**) without or with 25 µM DXR. The P-ACC/ACC ratio and phosphorylated Akt substrates are a readout for the activation of the AMPK and Akt pathways, respectively. The quantification of the bands is given in the right panel; in (**B**), empty or hatched bars correspond to perfusion without or with 25 µM DXR, respectively. For each protein, all signals originate from the same blot. Total protein or tubulin signals were used for normalization. Statistical analysis with one-way (**A**) or two-way (**B**) ANOVA followed by the Student–Newman–Keuls test for pairwise comparisons. For P-AMPK/AMPK in (**A**), there is no statistically significant difference between groups. Mean ± SD, $n = 3$–5.

3.4. Oxidative Damage, Antioxidant Status, and DNA Integrity

Diet and DXR can affect the cellular oxidative/antioxidant balance. As a readout, we determined peroxidized lipids (TBA reactive substances, TBARS) and antioxidant status (reduced thiols; ferric reducing antioxidant power, FRAP), along with the integrity of mtDNA and nDNA in hearts in situ (Figures 5A and 6A) and after ex vivo perfusion (Figures 5B and 6B). In Situ, the soy-based diet diminished FRAP (SOY vs. NORMAL; $p = 0.032$, Figure 5A). In the ex vivo perfused heart, the diets affected both TBARS and FRAP ($P_{Diet} < 0.001$ and 0.001, respectively, Figure 5B). Again, the soy-based diet reduced both parameters (SOY vs. NORMAL; $p < 0.001$ for both; SOY vs. SOY + Cr; $p = 0.023$ also for both, Figure 5B). Despite these differences, the integrity of nuclear and mitochondrial DNAs was not affected by diet, neither in situ (Figure 6A) nor after perfusion (Figure 6B). Oxidative and genotoxic stress are molecular hallmarks of DXR toxicity [34,40,41].

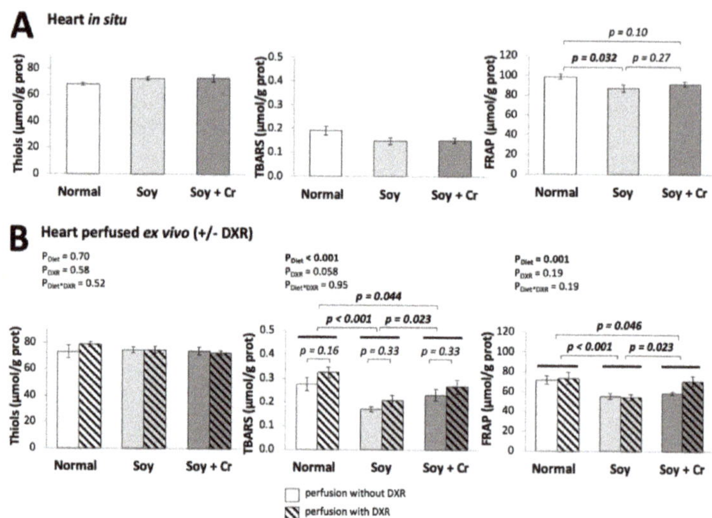

Figure 5. Oxidative/antioxidant status: effect of diet and DXR. Reduced thiols, peroxidized lipids (TBARS), and total antioxidant power (FRAP) measured in hearts freeze-clamped in situ immediately after thoracotomy (**A**) or following ex vivo Langendorff perfusion (**B**) without or with 25 µM DXR (empty or hatched bars, respectively). Statistical analysis with one-way (**A**) or two-way (**B**) ANOVA followed by the Student–Newman–Keuls test for pairwise comparisons. For thiols, TBARS in (**A**) and thiols in (**B**), there are no statistically significant differences between groups. Mean ± SEM, $n = 5$–6 (**A**), $n = 4$–7 (**B**).

After DXR perfusion, TBARS tended to increase as compared to the control ($P_{DXR} = 0.058$, Figure 5B), but the lower TBARS and FRAP values in the SOY group as compared to NORMAL were preserved (Figure 5B). Consistent with our previous study in ex vivo Langendorff perfused heart [34], DXR caused extensive mitochondrial and nuclear DNA damage, but again, the extent of damage was unaffected by the three diet regimens (Figure 6B).

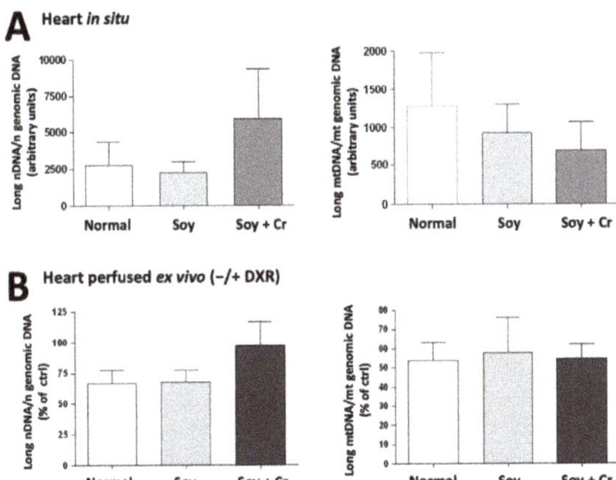

Figure 6. DNA integrity: effect of diet and DXR. Integrity of nuclear and mitochondrial DNA measured in hearts freeze-clamped in situ immediately after thoracotomy (**A**) or following ex vivo Langendorff perfusion (**B**). The bars in (**B**) represent the ratio of amplification values of DXR-perfused hearts to controls (perfusion without DXR) and are expressed as a percent. Statistical analysis with one-way ANOVA. For any parameter, there is no statistically significant difference between groups. Mean ± SD, $n = 4$–5.

4. Discussion

This study reveals nutrition-induced alterations in cardiac function, cell signaling, and some biochemical markers after only 4 weeks of differential feeding of young male rats. The Cr-free soy-based diet (SOY) as compared to standard rodent diet (NORMAL) activated AMPK signaling as revealed by increased ACC phosphorylation, slightly increased the ADP/ATP ratio, and lowered both lipid peroxidation and the total antioxidant capacity. Supplementation of SOY with 2% Cr (SOY + Cr) as compared to SOY moderately increased cellular Cr, predominantly affected signaling pathways by activating Akt and ERK, and slightly increased cardiac developed pressure and contractility (LVDP and dp/dt) at baseline. These alterations are, in principle, relevant for cardiac health and its response to DXR, but they did not alleviate cardiac dysfunction induced by acute DXR challenge in the perfused heart model applied here.

Three key signaling pathways with fundamental importance for cardiovascular health were altered by diet: AMPK, ERK, and Akt. This may be critical to many functional and biochemical changes detected in our study. AMPK is a central energy sensor and regulator of the cell. During energy stress, it is activated allosterically by AMP and ADP, favors catabolism, and maintains cellular energy homeostasis [42]. Soy-based diet (SOY) as compared to standard chow (NORMAL) led to strong phosphorylation of ACC, an AMPK substrate, reporting activation of this pathway in the heart, consistent with a slightly reduced energy state in situ. Perfusion with DXR is known to induce a drop in the cardiac energy state [39], but paradoxically, this often occurs without activation of AMPK signaling, as reported by us [31,34] and others [43]. In the present study, we also observed signs of bioenergetic impairment by DXR in perfused heart, together with decreased AMPK activation. Only the SOY group maintained AMPK energy signaling as seen at the level of P-AMPK and P-ACC. Indeed, some treatments known to activate AMPK were shown to mitigate DXR cardiotoxic effects, including diet restriction [43,44]. Different soy components were implicated in AMPK activation in tissues other than heart, such as phytoestrogens in rat [45], genistein in cultured cancer cells [46], and different types of polyphenols [47,48]. The addition of Cr (SOY + Cr) did not (further) activate cardiac AMPK, consistent with a study on skeletal muscle [49], and not supporting earlier data on muscle cells [50].

Phosphorylation and activation of the rather pro-anabolic Akt and ERK by diet in heart in situ occurred rather inversely relative to AMPK. While the soy diet (SOY vs. NORMAL) increased AMPK activity and left Akt activity unchanged or tended to diminish ERK activity, the addition of Cr (SOY + Cr vs. SOY) led to activation of Akt and ERK, with a trend of lower AMPK activity. This supports a negative cross-talk of these kinases as described by us [34] and others [34,51–53], and by which AMPK is inhibited via Akt-dependent phosphorylation in its catalytic α-subunit. This cross-talk can modulate AMPK activity in the heart both under basal conditions in situ and during DXR perfusion [34]. The inhibitory effects of the soy-based diet on Akt and ERK could be mediated by genistein, known to inhibit Tyr kinases (for a review, see [54]) and to have an anti-proliferative effect, consistent with indirect AMPK activation [55–59]. The activation of Akt and ERK seen with Cr supplementation was also reported for skeletal muscle [60,61] and suggested by recent database meta-analysis [62]. Such rather pro-anabolic effects could mediate many cytoprotective aspects of Cr supplementation, such as in cardioprotection [18,19], muscular dystrophies, neuromuscular and neurodegenerative disorders [16,63], brain health [64], or wound healing [65]. Notably, upregulation of Akt was shown to confer significant cardioprotection in DXR-treated animals [66].

Slightly altered performance of the perfused heart was observed only after Cr supplementation (SOY + Cr vs. SOY) under baseline conditions. Contractility (dp/dt) and developed pressure (LVDP) were modestly increased, together with a trend of a decreased heart rate, with the latter resulting in an unchanged rate pressure product (RPP). Beyond bioenergetics, Cr enhances the expression of muscle myogenic regulatory factors as reported for skeletal muscle [60,61,67,68] and affects signaling pathways, such as the Akt activation mentioned above. An earlier study did not detect Cr-induced changes in cardiac function [69], possibly because the reference diet plays an important role. Soy is not only Cr-free but may itself have additional cardio-vascular effects not examined here, such as blood pressure-lowering effects [70–74]. Moreover, the higher basal activity of the AMPK pathway may potentiate Cr effects, since both are directed to improve cell energetics. Perfusion with DXR deteriorated cardiac function as expected, but diet did not modulate the functional response, except for a lower increase in diastolic pressure in the SOY + Cr vs. SOY group. Cr was also not effective in a perfused heart model for acute oxidative stress [75]. Possibly, the acute insult at a supraclinical DXR dose is too strong, and long-term exposure of animals to low clinical DXR doses would be more suitable for an analysis of dietary effects.

A striking feature of the soy-based diet (SOY vs. NORMAL) was the low cardiac level of both lipid peroxidation and total antioxidant capacity. This was most pronounced in the ex vivo perfused heart, likely because of perfusion-associated oxidative stress. The antioxidant properties of soy include the main soy isoflavone genistein [76] and other phytoestrogens or polyphenolic compounds. They share a high reactivity as hydrogen or electron donors, can stabilize unpaired electrons as polyphenol radicals, chelate transition metal ions, modulate the expression of antioxidant defense genes, and activate signaling pathways [77,78]. However, a general lipid-lowering capacity of soy could also reduce detectable lipid peroxides [27,54,79], consistent with AMPK-induced reduction of lipid anabolism and an increase of their catabolism. The combined reduction of both lipid peroxidation and total antioxidant capacity may seem surprising, but the latter is likely an adaptation to the lower oxidative stress levels in the SOY group, as also indicated by literature data [80]. Cr supplementation had no such dramatic effects. Further, diet-related differences in oxidative stress were not reflected in oxidative DNA lesions, because these were either below the detection threshold, or they were rapidly removed by repair systems. Thus, mt and nDNA are unlikely targets and/or mediators of the diet-related effects described herein.

Regarding Cr, it should be emphasized that supplementation can only modestly increase cardiac Cr levels, as observed here and in earlier studies [69]. With increasing cellular Cr, a feedback mechanism downregulates the creatine transporter in charge of cellular Cr uptake [69]. Nevertheless, even this moderate increase in intracellular Cr was sufficient to trigger some significant cardiac effects.

Finally, our study calls for a note of caution with respect to diets used in animal studies. Already basic formulations likely contain ingredients with considerable biological activity. In particular, the soy-derived products commonly used in rodent chow (soy meal or protein isolate) must be considered as bioactive agents or even nutraceuticals [28]. The present study used a soy-based diet as a genuine Cr-free control chow for Cr supplementation studies. However, replacing 4% fish hydrolysate with 4% soy protein isolate already generates a bio-active diet. Soy-derived products contain isoflavones (genistein, daidzein, and equol) that are qualified as phytoestrogens due to their ability to act in the body as estrogens or selective estrogen receptor modulators [72]. Thus, caution is advised when translating results obtained with animals fed a high-soy diet directly to humans, especially those consuming a traditional Western diet. The quantity of circulating phytoestrogens in rats ingesting a soy-based diet may be comparable to Asian people who eat a soy-based diet [28]. Even if soy dietary supplements became popular in vegetarian/vegan cuisine, the overall benefit and/or safety of this diet is still a matter of debate (for a review, see [27–30]). Obviously, phytoestrogens may negatively affect the reproduction system, mainly in males and children [28]. In view of these controversies, the controlled use of soy or other bioactive compounds in animal diets and the explicit analysis of their effects is highly desirable. In this context, the choice of animal gender should also be considered. For example, practically all animal studies on DXR cardiotoxicity were conducted with male animals and these are then compared to human studies performed with both genders, although gender may affect the cardiac response to DXR. Earlier literature suggested that female sex is a risk factor for DXR cardiotoxicity [81–83], but most recent reports show that female sex hormones may protect against DXR cardiotoxicity by reducing oxidative stress and proinflammatory responses [84,85]. One may even ask whether phytoestrogens could successfully mimic this effect.

5. Conclusions

In conclusion, a soy-based diet alone or supplemented with Cr, fed for four weeks to rats, is sufficient to alter cardiac function, cell signaling, and biochemical markers of the energy state and oxidative stress. These effects are relevant for cardiovascular health but were not sufficient to alleviate cardiac dysfunction induced by a supraclinical DXR concentration in the perfused rat heart model. However, whether these diets could affect the long-term response to chronic and clinically relevant DXR doses in the rat model described here, or in human patients treated with DXR, remains to be established.

Supplementary Materials: The following are available online at https://www.mdpi.com/article/10.3390/nu14030583/s1, Figure S1: Time course of DXR effect on heart function.

Author Contributions: Conceptualization and design, M.T.-S., U.S., T.W., L.K., P.S.; Investigation and data acquisition, L.K., M.T.-S., L.P., I.H.-F., S.A., C.T., C.C., C.Z.; Data Curation, M.T.-S., L.K., I.H.-F., L.P.; Statistical analysis, M.T.-S., H.R.; Writing of Original Draft, M.T.-S., U.S., L.K.; Manuscript Review & Editing, M.T.-S., U.S., L.K., P.S., T.W.; Project Administration, M.T.-S., U.S.; Funding Acquisition, M.T.-S., U.S., C.Z. All authors have read and agreed to the published version of the manuscript.

Funding: This work was supported by the EU FP6 and 7 Marie Curie Intraeuropean Fellowship and Reintegration Grant (ANTHRAWES, 041870 and ANTHRAPLUS, 249202 to MTS.), the French Agence Nationale de Recherche ("chaire d'excellence" to U.S.) and the Germaine de Stael Program for Franco-Swiss collaboration (32771QE-to US and CZ).

Institutional Review Board Statement: The study was approved by the Grenoble Ethics Committee for Animal Experimentation (15_LBFA–U884-HD-01).

Data Availability Statement: Data supporting the findings of this manuscript are available from the corresponding authors upon reasonable request.

Acknowledgments: The authors thank Mireille Osman for excellent technical assistance, and Séverine Gratia and Kusumita Chuttor for experimental help.

Conflicts of Interest: The authors declare no conflict of interest.

References

1. Jones, L.W.; Demark-Wahnefried, W. Diet, exercise, and complementary therapies after primary treatment for cancer. *Lancet Oncol.* **2006**, *7*, 1017–1026. [CrossRef]
2. Minotti, G.; Menna, P.; Salvatorelli, E.; Cairo, G.; Gianni, L. Anthracyclines: Molecular advances and pharmacologic developments in antitumor activity and cardiotoxicity. *Pharmacol. Rev.* **2004**, *56*, 185–229. [CrossRef] [PubMed]
3. Eschenhagen, T.; Force, T.; Ewer, M.S.; de Keulenaer, G.W.; Suter, T.M.; Anker, S.D.; Avkiran, M.; de Azambuja, E.; Balligand, J.L.; Brutsaert, D.L.; et al. Cardiovascular side effects of cancer therapies: A position statement from the Heart Failure Association of the European Society of Cardiology. *Eur. J. Heart Fail.* **2011**, *13*, 1–10. [CrossRef] [PubMed]
4. Gianni, L.; Herman, E.H.; Lipshultz, S.E.; Minotti, G.; Sarvazyan, N.; Sawyer, D.B. Anthracycline cardiotoxicity: From bench to bedside. *J. Clin. Oncol.* **2008**, *26*, 3777–3784. [CrossRef]
5. Sterba, M.; Popelova, O.; Vavrova, A.; Jirkovsky, E.; Kovarikova, P.; Gersl, V.; Simunek, T. Oxidative stress, redox signaling, and metal chelation in anthracycline cardiotoxicity and pharmacological cardioprotection. *Antioxid. Redox Signal.* **2013**, *18*, 899–929. [CrossRef]
6. Lipshultz, S.E.; Cochran, T.R.; Franco, V.I.; Miller, T.L. Treatment-related cardiotoxicity in survivors of childhood cancer. *Nat. Rev. Clin. Oncol.* **2013**, *10*, 697–710. [CrossRef]
7. Lipshultz, S.E.; Karnik, R.; Sambatakos, P.; Franco, V.I.; Ross, S.W.; Miller, T.L. Anthracycline-related cardiotoxicity in childhood cancer survivors. *Curr. Opin. Cardiol.* **2014**, *29*, 103–112. [CrossRef]
8. Jones, L.W.; Alfano, C.M. Exercise-oncology research: Past, present, and future. *Acta Oncol.* **2013**, *52*, 195–215. [CrossRef]
9. Jones, L.W.; Dewhirst, M.W. Therapeutic properties of aerobic training after a cancer diagnosis: More than a one-trick pony? *J. Natl. Cancer Inst.* **2014**, *106*, dju042. [CrossRef]
10. Scott, J.M.; Koelwyn, G.J.; Hornsby, W.E.; Khouri, M.; Peppercorn, J.; Douglas, P.S.; Jones, L.W. Exercise therapy as treatment for cardiovascular and oncologic disease after a diagnosis of early-stage cancer. *Semin. Oncol.* **2013**, *40*, 218–228. [CrossRef]
11. Scott, J.M.; Lakoski, S.; Mackey, J.R.; Douglas, P.S.; Haykowsky, M.J.; Jones, L.W. The potential role of aerobic exercise to modulate cardiotoxicity of molecularly targeted cancer therapeutics. *Oncologist* **2013**, *18*, 221–231. [CrossRef] [PubMed]
12. Lipshultz, S.E.; Adams, M.J.; Colan, S.D.; Constine, L.S.; Herman, E.H.; Hsu, D.T.; Hudson, M.M.; Kremer, L.C.; Landy, D.C.; Miller, T.L.; et al. Long-term cardiovascular toxicity in children, adolescents, and young adults who receive cancer therapy: Pathophysiology, course, monitoring, management, prevention, and research directions: A scientific statement from the American Heart Association. *Circulation* **2013**, *128*, 1927–1995. [CrossRef] [PubMed]
13. Kreider, R.B.; Stout, J.R. Creatine in Health and Disease. *Nutrients* **2021**, *13*, 447. [CrossRef] [PubMed]
14. Wyss, M.; Kaddurah-Daouk, R. Creatine and creatinine metabolism. *Physiol. Rev.* **2000**, *80*, 1107–1213. [CrossRef] [PubMed]
15. Hanna-El-Daher, L.; Braissant, O. Creatine synthesis and exchanges between brain cells: What can be learned from human creatine deficiencies and various experimental models? *Amino Acids* **2016**, *48*, 1877–1895. [CrossRef] [PubMed]
16. Wallimann, T.; Tokarska-Schlattner, M.; Schlattner, U. The creatine kinase system and pleiotropic effects of creatine. *Amino Acids* **2011**, *40*, 1271–1296. [CrossRef] [PubMed]
17. Schlattner, U.; Tokarska-Schlattner, M.; Wallimann, T. Mitochondrial creatine kinase in human health and disease. *Biochim. Biophys. Acta* **2006**, *1762*, 164–180. [CrossRef]
18. Lygate, C.A.; Bohl, S.; ten Hove, M.; Faller, K.M.; Ostrowski, P.J.; Zervou, S.; Medway, D.J.; Aksentijevic, D.; Sebag-Montefiore, L.; Wallis, J.; et al. Moderate elevation of intracellular creatine by targeting the creatine transporter protects mice from acute myocardial infarction. *Cardiovasc. Res.* **2012**, *96*, 466–475. [CrossRef]
19. Balestrino, M. Role of Creatine in the Heart: Health and Disease. *Nutrients* **2021**, *13*, 1215. [CrossRef]
20. Santos, R.V.; Batista, M.L., Jr.; Caperuto, E.C.; Costa Rosa, L.F. Chronic supplementation of creatine and vitamins C and E increases survival and improves biochemical parameters after Doxorubicin treatment in rats. *Clin. Exp. Pharm. Physiol* **2007**, *34*, 1294–1299. [CrossRef]
21. Caretti, A.; Bianciardi, P.; Sala, G.; Terruzzi, C.; Lucchina, F.; Samaja, M. Supplementation of creatine and ribose prevents apoptosis in ischemic cardiomyocytes. *Cell. Physiol. Biochem.* **2010**, *26*, 831–838. [CrossRef] [PubMed]
22. Santacruz, L.; Darrabie, M.D.; Mantilla, J.G.; Mishra, R.; Feger, B.J.; Jacobs, D.O. Creatine supplementation reduces doxorubicin-induced cardiomyocellular injury. *Cardiovasc. Toxicol.* **2015**, *15*, 180–188. [CrossRef] [PubMed]
23. Fimognari, C.; Sestili, P.; Lenzi, M.; Bucchini, A.; Cantelli-Forti, G.; Hrelia, P. RNA as a new target for toxic and protective agents. *Mutat. Res.* **2008**, *648*, 15–22. [CrossRef] [PubMed]
24. Gupta, A.; Rohlfsen, C.; Leppo, M.K.; Chacko, V.P.; Wang, Y.; Steenbergen, C.; Weiss, R.G. Creatine kinase-overexpression improves myocardial energetics, contractile dysfunction and survival in murine doxorubicin cardiotoxicity. *PLoS ONE* **2013**, *8*, e74675. [CrossRef]
25. Sestili, P.; Martinelli, C.; Colombo, E.; Barbieri, E.; Potenza, L.; Sartini, S.; Fimognari, C. Creatine as an antioxidant. *Amino Acids* **2011**, *40*, 1385–1396. [CrossRef]

26. Tokarska-Schlattner, M.; Epand, R.F.; Meiler, F.; Zandomeneghi, G.; Neumann, D.; Widmer, H.R.; Meier, B.H.; Epand, R.M.; Saks, V.; Wallimann, T.; et al. Phosphocreatine interacts with phospholipids, affects membrane properties and exerts membrane-protective effects. *PLoS ONE* **2012**, *7*, e43178. [CrossRef]

27. Messina, M.; Messina, V. The role of soy in vegetarian diets. *Nutrients* **2010**, *2*, 855–888. [CrossRef]

28. Konhilas, J.P.; Leinwand, L.A. The effects of biological sex and diet on the development of heart failure. *Circulation* **2007**, *116*, 2747–2759. [CrossRef]

29. Sacks, F.M.; Lichtenstein, A.; Van Horn, L.; Harris, W.; Kris-Etherton, P.; Winston, M.; for the American Heart Association Nutrition Committee. Soy protein, isoflavones, and cardiovascular health: An American Heart Association Science Advisory for professionals from the Nutrition Committee. *Circulation* **2006**, *113*, 1034–1044. [CrossRef]

30. Erdman, J.W., Jr. AHA Science Advisory: Soy protein and cardiovascular disease: A statement for healthcare professionals from the Nutrition Committee of the AHA. *Circulation* **2000**, *102*, 2555–2559. [CrossRef]

31. Tokarska-Schlattner, M.; Zaugg, M.; da Silva, R.; Lucchinetti, E.; Schaub, M.C.; Wallimann, T.; Schlattner, U. Acute toxicity of doxorubicin on isolated perfused heart: Response of kinases regulating energy supply. *Am. J. Physiol. Heart Circ. Physiol.* **2005**, *289*, H37–H47. [CrossRef] [PubMed]

32. Tokarska-Schlattner, M.; Lucchinetti, E.; Zaugg, M.; Kay, L.; Gratia, S.; Guzun, R.; Saks, V.; Schlattner, U. Early effects of doxorubicin in perfused heart: Transcriptional profiling reveals inhibition of cellular stress response genes. *Am. J. Physiol. Regul. Integr. Comp. Physiol.* **2010**, *298*, R1075–R1088. [CrossRef] [PubMed]

33. Gratia, S.; Kay, L.; Michelland, S.; Seve, M.; Schlattner, U.; Tokarska-Schlattner, M. Cardiac phosphoproteome reveals cell signaling events involved in doxorubicin cardiotoxicity. *J. Proteom.* **2012**, *75*, 4705–4716. [CrossRef] [PubMed]

34. Gratia, S.; Kay, L.; Potenza, L.; Seffouh, A.; Novel-Chate, V.; Schnebelen, C.; Sestili, P.; Schlattner, U.; Tokarska-Schlattner, M. Inhibition of AMPK signalling by doxorubicin: At the crossroads of the cardiac responses to energetic, oxidative, and genotoxic stress. *Cardiovasc. Res.* **2012**, *95*, 290–299. [CrossRef]

35. Hininger-Favier, I.; Benaraba, R.; Coves, S.; Anderson, R.A.; Roussel, A.M. Green tea extract decreases oxidative stress and improves insulin sensitivity in an animal model of insulin resistance, the fructose-fed rat. *J. Am. Coll. Nutr.* **2009**, *28*, 355–361. [CrossRef]

36. Faure, P.; Lafond, J.L. Measurement of plasma sulfhydryl and carbonyl groups as a possible indicator of protein oxydation. In *Analysis of Free Radicals in Biological Systems*; Favier, A.E., Cadet, J., Kalnyanaraman, M., Fontecave, M., Pierre, J.L., Eds.; Birkäuser Basel: Boston, MA, USA; Berlin, Germany, 1995; pp. 237–248.

37. Benzie, I.F.; Strain, J.J. The ferric reducing ability of plasma (FRAP) as a measure of "antioxidant power": The FRAP assay. *Anal. Biochem.* **1996**, *239*, 70–76. [CrossRef]

38. Richard, M.J.; Portal, B.; Meo, J.; Coudray, C.; Hadjian, A.; Favier, A. Malondialdehyde kit evaluated for determining plasma and lipoprotein fractions that react with thiobarbituric acid. *Clin. Chem.* **1992**, *38*, 704–709. [CrossRef]

39. Tokarska-Schlattner, M.; Zaugg, M.; Zuppinger, C.; Wallimann, T.; Schlattner, U. New insights into doxorubicin-induced cardiotoxicity: The critical role of cellular energetics. *J. Mol. Cell. Cardiol.* **2006**, *41*, 389–405. [CrossRef]

40. Tokarska-Schlattner, M.; Wallimann, T.; Schlattner, U. Multiple interference of anthracyclines with mitochondrial creatine kinases: Preferential damage of the cardiac isoenzyme and its implications for drug cardiotoxicity. *Mol. Pharmacol.* **2002**, *61*, 516–523. [CrossRef]

41. Mousseau, M.; Faure, H.; Hininger, I.; Bayet-Robert, M.; Favier, A. Leukocyte 8-oxo-7,8-dihydro-2'-deoxyguanosine and comet assay in epirubicin-treated patients. *Free Radic. Res.* **2005**, *39*, 837–843. [CrossRef]

42. Hardie, D.G. Sensing of energy and nutrients by AMP-activated protein kinase. *Am. J. Clin. Nutr.* **2011**, *93*, 891S–896S. [CrossRef] [PubMed]

43. Kawaguchi, T.; Takemura, G.; Kanamori, H.; Takeyama, T.; Watanabe, T.; Morishita, K.; Ogino, A.; Tsujimoto, A.; Goto, K.; Maruyama, R.; et al. Prior starvation mitigates acute doxorubicin cardiotoxicity through restoration of autophagy in affected cardiomyocytes. *Cardiovasc. Res.* **2012**, *96*, 456–465. [CrossRef] [PubMed]

44. Mitra, M.S.; Donthamsetty, S.; White, B.; Latendresse, J.R.; Mehendale, H.M. Mechanism of protection of moderately diet restricted rats against doxorubicin-induced acute cardiotoxicity. *Toxicol. Appl. Pharm.* **2007**, *225*, 90–101. [CrossRef] [PubMed]

45. Cederroth, C.R.; Vinciguerra, M.; Gjinovci, A.; Kuhne, F.; Klein, M.; Cederroth, M.; Caille, D.; Suter, M.; Neumann, D.; James, R.W.; et al. Dietary phytoestrogens activate AMP-activated protein kinase with improvement in lipid and glucose metabolism. *Diabetes* **2008**, *57*, 1176–1185. [CrossRef] [PubMed]

46. Park, C.E.; Yun, H.; Lee, E.B.; Min, B.I.; Bae, H.; Choe, W.; Kang, I.; Kim, S.S.; Ha, J. The antioxidant effects of genistein are associated with AMP-activated protein kinase activation and PTEN induction in prostate cancer cells. *J. Med. Food* **2010**, *13*, 815–820. [CrossRef]

47. Hawley, S.A.; Ross, F.A.; Chevtzoff, C.; Green, K.A.; Evans, A.; Fogarty, S.; Towler, M.C.; Brown, L.J.; Ogunbayo, O.A.; Evans, A.M.; et al. Use of cells expressing gamma subunit variants to identify diverse mechanisms of AMPK activation. *Cell Metab.* **2010**, *11*, 554–565. [CrossRef]

48. Hardie, D.G. AMP-activated protein kinase: An energy sensor that regulates all aspects of cell function. *Genes Dev.* **2011**, *25*, 1895–1908. [CrossRef]

49. Eijnde, B.O.; Derave, W.; Wojtaszewski, J.F.; Richter, E.A.; Hespel, P. AMP kinase expression and activity in human skeletal muscle: Effects of immobilization, retraining, and creatine supplementation. *J. Appl. Physiol.* **2005**, *98*, 1228–1233. [CrossRef]

50. Ceddia, R.B.; Sweeney, G. Creatine supplementation increases glucose oxidation and AMPK phosphorylation and reduces lactate production in L6 rat skeletal muscle cells. *J. Physiol.* **2004**, *555*, 409–421. [CrossRef]

51. Petti, C.; Vegetti, C.; Molla, A.; Bersani, I.; Cleris, L.; Mustard, K.J.; Formelli, F.; Hardie, G.D.; Sensi, M.; Anichini, A. AMPK activators inhibit the proliferation of human melanomas bearing the activated MAPK pathway. *Melanoma Res.* **2012**, *22*, 341–350. [CrossRef]

52. Hawley, S.A.; Ross, F.A.; Gowans, G.J.; Tibarewal, P.; Leslie, N.R.; Hardie, D.G. Phosphorylation by Akt within the ST loop of AMPK-alpha1 down-regulates its activation in tumour cells. *Biochem. J.* **2014**, *459*, 275–287. [CrossRef] [PubMed]

53. Esteve-Puig, R.; Canals, F.; Colome, N.; Merlino, G.; Recio, J.A. Uncoupling of the LKB1-AMPKalpha energy sensor pathway by growth factors and oncogenic BRAF. *PLoS ONE* **2009**, *4*, e4771. [CrossRef]

54. Messina, M. A brief historical overview of the past two decades of soy and isoflavone research. *J. Nutr.* **2010**, *140*, 1350S–1354S. [CrossRef] [PubMed]

55. Wei, H.; Bowen, R.; Cai, Q.; Barnes, S.; Wang, Y. Antioxidant and antipromotional effects of the soybean isoflavone genistein. *Proc. Soc. Exp. Biol. Med.* **1995**, *208*, 124–130. [CrossRef] [PubMed]

56. Polkowski, K.; Mazurek, A.P. Biological properties of genistein. A review of in vitro and in vivo data. *Acta Pol. Pharm.* **2000**, *57*, 135–155.

57. Nakamura, Y.; Yogosawa, S.; Izutani, Y.; Watanabe, H.; Otsuji, E.; Sakai, T. A combination of indol-3-carbinol and genistein synergistically induces apoptosis in human colon cancer HT-29 cells by inhibiting Akt phosphorylation and progression of autophagy. *Mol. Cancer* **2009**, *8*, 100. [CrossRef]

58. Gong, L.; Li, Y.; Nedeljkovic-Kurepa, A.; Sarkar, F.H. Inactivation of NF-kappaB by genistein is mediated via Akt signaling pathway in breast cancer cells. *Oncogene* **2003**, *22*, 4702–4709. [CrossRef]

59. El Touny, L.H.; Banerjee, P.P. Akt GSK-3 pathway as a target in genistein-induced inhibition of TRAMP prostate cancer progression toward a poorly differentiated phenotype. *Carcinogenesis* **2007**, *28*, 1710–1717. [CrossRef]

60. Deldicque, L.; Theisen, D.; Bertrand, L.; Hespel, P.; Hue, L.; Francaux, M. Creatine enhances differentiation of myogenic C2C12 cells by activating both p38 and Akt/PKB pathways. *Am. J. Physiol. Cell Physiol.* **2007**, *293*, C1263–C1271. [CrossRef]

61. Hespel, P.; Derave, W. Ergogenic effects of creatine in sports and rehabilitation. *Subcell. Biochem.* **2007**, *46*, 245–259.

62. Bonilla, D.A.; Moreno, Y.; Rawson, E.S.; Forero, D.A.; Stout, J.R.; Kerksick, C.M.; Roberts, M.D.; Kreider, R.B. A Convergent Functional Genomics Analysis to Identify Biological Regulators Mediating Effects of Creatine Supplementation. *Nutrients* **2021**, *13*, 2521. [CrossRef] [PubMed]

63. Kreider, R.B.; Kalman, D.S.; Antonio, J.; Ziegenfuss, T.N.; Wildman, R.; Collins, R.; Candow, D.G.; Kleiner, S.M.; Almada, A.L.; Lopez, H.L. International Society of Sports Nutrition position stand: Safety and efficacy of creatine supplementation in exercise, sport, and medicine. *J. Int. Soc. Sports Nutr.* **2017**, *14*, 18. [CrossRef] [PubMed]

64. Roschel, H.; Gualano, B.; Ostojic, S.M.; Rawson, E.S. Creatine Supplementation and Brain Health. *Nutrients* **2021**, *13*, 586. [CrossRef] [PubMed]

65. Schlattner, U.; Mockli, N.; Speer, O.; Werner, S.; Wallimann, T. Creatine kinase and creatine transporter in normal, wounded, and diseased skin. *J. Investig. Dermatol.* **2002**, *118*, 416–423. [CrossRef] [PubMed]

66. Taniyama, Y.; Walsh, K. Elevated myocardial Akt signaling ameliorates doxorubicin-induced congestive heart failure and promotes heart growth. *J. Mol. Cell. Cardiol.* **2002**, *34*, 1241–1247. [CrossRef]

67. Eijnde, B.O.; Lebacq, J.; Ramaekers, M.; Hespel, P. Effect of muscle creatine content manipulation on contractile properties in mouse muscles. *Muscle Nerve* **2004**, *29*, 428–435. [CrossRef]

68. Sestili, P.; Barbieri, E.; Martinelli, C.; Battistelli, M.; Guescini, M.; Vallorani, L.; Casadei, L.; D'Emilio, A.; Falcieri, E.; Piccoli, G.; et al. Creatine supplementation prevents the inhibition of myogenic differentiation in oxidatively injured C2C12 murine myoblasts. *Mol. Nutr. Food Res.* **2009**, *53*, 1187–1204. [CrossRef]

69. Boehm, E.; Chan, S.; Monfared, M.; Wallimann, T.; Clarke, K.; Neubauer, S. Creatine transporter activity and content in the rat heart supplemented by and depleted of creatine. *Am. J. Physiol. Endocrinol. Metab.* **2003**, *284*, E399–E406. [CrossRef]

70. Mahn, K.; Borras, C.; Knock, G.A.; Taylor, P.; Khan, I.Y.; Sugden, D.; Poston, L.; Ward, J.P.; Sharpe, R.M.; Vina, J.; et al. Dietary soy isoflavone induced increases in antioxidant and eNOS gene expression lead to improved endothelial function and reduced blood pressure in vivo. *FASEB J.* **2005**, *19*, 1755–1757. [CrossRef]

71. Douglas, G.; Armitage, J.A.; Taylor, P.D.; Lawson, J.R.; Mann, G.E.; Poston, L. Cardiovascular consequences of life-long exposure to dietary isoflavones in the rat. *J. Physiol.* **2006**, *571*, 477–487. [CrossRef]

72. Siow, R.C.; Mann, G.E. Dietary isoflavones and vascular protection: Activation of cellular antioxidant defenses by SERMs or hormesis? *Mol. Asp. Med.* **2010**, *31*, 468–477. [CrossRef] [PubMed]

73. Si, H.; Liu, D. Genistein, a soy phytoestrogen, upregulates the expression of human endothelial nitric oxide synthase and lowers blood pressure in spontaneously hypertensive rats. *J. Nutr.* **2008**, *138*, 297–304. [CrossRef] [PubMed]

74. Mozaffarian, D.; Appel, L.J.; Van Horn, L. Components of a cardioprotective diet: New insights. *Circulation* **2011**, *123*, 2870–2891. [CrossRef]

75. Aksentijevic, D.; Zervou, S.; Faller, K.M.; McAndrew, D.J.; Schneider, J.E.; Neubauer, S.; Lygate, C.A. Myocardial creatine levels do not influence response to acute oxidative stress in isolated perfused heart. *PLoS ONE* **2014**, *9*, e109021. [CrossRef]

76. Record, I.R.; Dreosti, I.E.; McInerney, J.K. The antioxidant activity of genistein. *J. Nutr. Biochem.* **1995**, *6*, 481–485. [CrossRef]

77. Blokhina, O.; Virolainen, E.; Fagerstedt, K.V. Antioxidants, oxidative damage and oxygen deprivation stress: A review. *Ann. Bot.* **2003**, *91*, 179–194. [CrossRef]
78. Borras, C.; Gambini, J.; Gomez-Cabrera, M.C.; Sastre, J.; Pallardo, F.V.; Mann, G.E.; Vina, J. Genistein, a soy isoflavone, up-regulates expression of antioxidant genes: Involvement of estrogen receptors, ERK1/2, and NFkappaB. *FASEB J.* **2006**, *20*, 2136–2138. [CrossRef]
79. Zang, M.; Xu, S.; Maitland-Toolan, K.A.; Zuccollo, A.; Hou, X.; Jiang, B.; Wierzbicki, M.; Verbeuren, T.J.; Cohen, R.A. Polyphenols stimulate AMP-activated protein kinase, lower lipids, and inhibit accelerated atherosclerosis in diabetic LDL receptor-deficient mice. *Diabetes* **2006**, *55*, 2180–2191. [CrossRef]
80. Davies, K.J. Oxidative stress, antioxidant defenses, and damage removal, repair, and replacement systems. *IUBMB Life* **2000**, *50*, 279–289. [CrossRef]
81. Altieri, P.; Barisione, C.; Lazzarini, E.; Garuti, A.; Bezante, G.P.; Canepa, M.; Spallarossa, P.; Tocchetti, C.G.; Bollini, S.; Brunelli, C.; et al. Testosterone Antagonizes Doxorubicin-Induced Senescence of Cardiomyocytes. *J. Am. Heart Assoc.* **2016**, *5*, e002383. [CrossRef]
82. Green, D.M.; Grigoriev, Y.A.; Nan, B.; Takashima, J.R.; Norkool, P.A.; D'Angio, G.J.; Breslow, N.E. Congestive heart failure after treatment for Wilms' tumor: A report from the National Wilms' Tumor Study group. *J. Clin. Oncol.* **2001**, *19*, 1926–1934. [CrossRef]
83. Lipshultz, S.E.; Lipsitz, S.R.; Mone, S.M.; Goorin, A.M.; Sallan, S.E.; Sanders, S.P.; Orav, E.J.; Gelber, R.D.; Colan, S.D. Female sex and higher drug dose as risk factors for late cardiotoxic effects of doxorubicin therapy for childhood cancer. *N. Engl. J. Med.* **1995**, *332*, 1738–1743. [CrossRef]
84. Moulin, M.; Piquereau, J.; Mateo, P.; Fortin, D.; Rucker-Martin, C.; Gressette, M.; Lefebvre, F.; Gresikova, M.; Solgadi, A.; Veksler, V.; et al. Sexual dimorphism of doxorubicin-mediated cardiotoxicity: Potential role of energy metabolism remodeling. *Circ. Heart Fail.* **2015**, *8*, 98–108. [CrossRef]
85. Zhang, J.; Knapton, A.; Lipshultz, S.E.; Cochran, T.R.; Hiraragi, H.; Herman, E.H. Sex-related differences in mast cell activity and doxorubicin toxicity: A study in spontaneously hypertensive rats. *Toxicol. Pathol.* **2014**, *42*, 361–375. [CrossRef]

nutrients

Article

The Dietary Supplement Creatyl-L-Leucine Does Not Bioaccumulate in Muscle, Brain or Plasma and Is Not a Significant Bioavailable Source of Creatine

Robin P. da Silva

Department of Physiology and Pathophysiology, Max Rady College of Medicine, University of Manitoba, Basic Medical Sciences Building, 745 Bannatyne Avenue, Winnipeg, MB R3E 0J9, Canada; robin.dasilva@umanitoba.ca

Abstract: Creatine is an important energy metabolite that is concentrated in tissues such as the muscles and brain. Creatine is reversibly converted to creatine phosphate through a reaction with ATP or ADP, which is catalyzed by the enzyme creatine kinase. Dietary supplementation with relatively large amounts of creatine monohydrate has been proven as an effective sports supplement that can enhances athletic performance during acute high-energy demand physical activity. Some side effects have been reported with creatine monohydrate supplementation, which have stimulated research into new potential molecules that could be used as supplements to potentially provide bioavailable creatine. Recently, a popular supplement, creatyl-L-leucine, has been proposed as a potential dietary ingredient that may potentially provide bioavailable creatine. This study tests whether creatyl-L-leucine is a bioavailable compound and determines whether it can furnish creatine as a dietary supplement. Rats were deprived of dietary creatine for a period of two weeks and then given one of three treatments: a control AIN-93G creatine-free diet, AIN-93G supplemented with creatine monohydrate or AIN-93G with an equimolar amount of creatyl-L-leucine supplement in the diet for one week. When compared to the control and the creatine monohydrate-supplemented diet, creatyl-L-leucine supplementation resulted in no bioaccumulation of either creatyl-L-leucine or creatine in tissue.

Keywords: creatine; amino acids; dietary supplement

Citation: da Silva, R.P. The Dietary Supplement Creatyl-L-Leucine Does Not Bioaccumulate in Muscle, Brain or Plasma and Is Not a Significant Bioavailable Source of Creatine. *Nutrients* **2022**, *14*, 701. https://doi.org/10.3390/nu14030701

Academic Editors: Richard B. Kreider and Jeffrey R. Stout

Received: 18 January 2022
Accepted: 4 February 2022
Published: 8 February 2022

Publisher's Note: MDPI stays neutral with regard to jurisdictional claims in published maps and institutional affiliations.

1. Introduction

Creatine is a critical energy metabolite that is synthesized from the amino acids arginine, methionine, and glycine via a two-enzyme, interorgan, metabolic pathway [1–3]. Creatine serves particularly important function in the muscle and brain and is concentrated in these tissues by a sodium-dependent transporter to levels approximately 100-fold greater than plasma. Via a reaction catalyzed by the enzyme creatine kinase, creatine is in equilibrium with creatine phosphate, which effectively serves to buffer cellular ATP concentrations during periods of acute demand for large amounts of energy such as depolarization or contractile events. Approximately 2 g of creatine is lost from the body per day per 70 kg body weight and this creatine is replaced either endogenous synthesis or by absorption from the diet; creatine is obtained in the diet mainly from meat and dairy products [4].

Creatine monohydrate (CrM) has been used as a sports supplement for decades and there is clear evidence that dietary creatine supplements improve muscle performance during burst-type exercises such as sprinting, track and field events or powerlifting [5]. Numerous animal studies have documented that dietary creatine supplementation can increase total muscle and brain creatine content [1]. Albeit brain creatine content does not increase to the same extent as does muscle in response to dietary CrM supplementation since creatine does not easily cross the blood–brain barrier. Indeed it has been shown, using P-31 NMR, that two weeks of dietary supplementation with creatine increased muscle concentrations of creatine phosphate by approximately 20% in young male subjects [6]. Given the significant literature demonstrating the ergogenic benefit of creatine supplementation it is not surprising that creatine is a major component of the more than 5.4 billion dollar (in 2015) sports supplement industry [7].

The most common form of dietary creatine supplement is CrM [1,8]. CrM is usually taken as a loading dose of 20 g per day for 5–7 days followed by a maintenance dose of 5 g per day. However, some studies have suggested that the relatively large amount of creatine taken to achieve benefits can cause an osmotic induced gastrointestinal system [1]. This has stimulated some to investigate whether alternate forms of creatine could be better tolerated, stable and bioavailable. Jäger et al. published an extensive analysis of potential sources of creatine and concluded that CrM was the most favorable of the compounds tested [9]. Recently it has been suggested that the molecule creatyl-L-leucine (CLL) (Figure 1) may potentially serve as a bioavailable form of creatine that could be used as a dietary ingredient [10]. In the same study, it was shown that CLL had no toxicity when provided for 90 days at 5 g/kg body weight in rats. However, it was also noted that the biological fate of ingested CLL was unknown. The purpose of the present study is to test the bioavailability of CLL and to determine whether creatine is a digestive or metabolic product of dietary CLL.

(A)

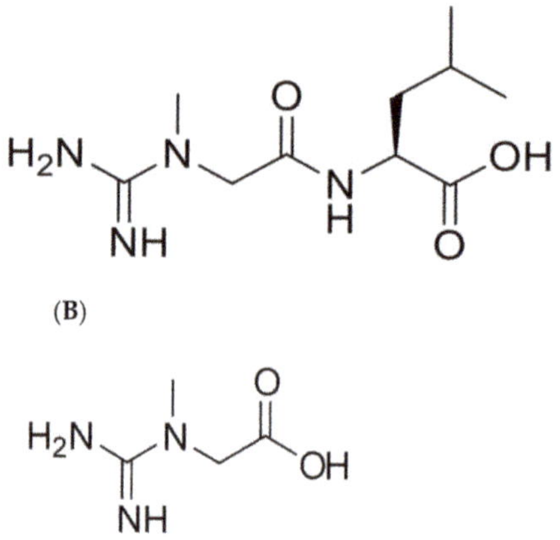

(B)

Figure 1. Chemical structures of (**A**) creatyl-L-leucine and (**B**) creatine.

2. Materials and Methods

All experimental procedures involving animals were conducted by CARE Research LLC in an AAALAC-accredited, USDA-certified, and OLAW-accredited facility. The study design and animal usage were reviewed and approved by the CARE Research Institutional Animal Care and Use Committee (IACUC) for compliance with regulations prior to study initiation (IACUC number 2042). Animal welfare for this study was in compliance with the U.S. Department of Agriculture's (USDA) Animal Welfare Act (9 CFR Parts 1, 2, and 3), the Guide for the Care and Use of Laboratory Animals, and CARE Research SOPs.

All chemicals were purchased from Sigma-Aldrich, except for HPLC grade Methanol and Formic acid which were purchased from Fisher Scientific. Purified AIN-93G diets [11] that were free of creatine or had supplemented CrM or an equimolar equivalent of CLL were custom ordered by CARE Research, LLC from Harlan Teklad (Envigo). Diets composition can be found in Table 1. The presence of CrM and CLL in the respective experimental diets was verified by aqueous acidic extraction and HPLC analysis as described below.

This study was designed to compare the bioavailability of CLL to that of the common creatine supplement CrM against a creatine-free control diet. To give the most favorable condition for absorption and accretion of creatine or CLL, 24 rats were provided with a control creatine-free AIN-93G diet (CON) for 14 days to deplete body creatine to levels supported by an endogenous synthetic capacity. At day 15, n = 8 rats were randomly chosen and switched to an AIN-93G diet supplemented with 0.4% *w/w* CrM (CrMD) and n = 8 different rats were randomly chosen and switched to an AIN-93G diet supplemented with 0.656% *w/w* CLL (CLLD). The remaining n = 8 rats were maintained on the creatine-free AIN-93G diet. Experimental diets were fed until tissues were collected on day 22. An equal number of male and female rats were used in each group.

Table 1. Composition of experimental diets.

	CON	CrMD *	CLLD *
Ingredient	**g/Kg Diet**	**g/Kg Diet**	**g/Kg Diet**
Casein	200.0	200.0	200.0
L-Cystine	3.0	3.0	3.0
Corn starch	397.49	393.49	390.93
Maltodextrin	132.0	132.0	132.0
Sucrose	100.0	100.0	100.0
Soybean oil	70.0	70.0	70.0
Cellulose	50.0	50.0	50.0
Mineral Mix (AIN-93G-MX, 94046)	35.0	35.0	35.0
Vitamin Mix (AIN-93-VX, 94047)	10.0	10.0	10.0
Choline bitartrate	2.5	2.5	2.5
Tert-butylhydroquinone	0.014	0.014	0.014
Creatine monohydrate	0	4.0	0
Creatyl-L-leucine	0	0	6.56

* 0.4% Creatine monohydrate (CrMD); 0.656% creatyl-L-leucine (CLLD). American Institute of Nutrition (AIN).

Rats were anesthetized using inhaled isoflurane and blood was collected from the portal vein and abdominal aorta. A portion of the quadricep muscle and the whole brain were removed quickly and snap-frozen in liquid nitrogen. Body weights were recorded every 7 days and food consumption was recorded daily.

Guanidine compounds were analyzed using the method of Buchnerger and Ferdig [12]. Briefly, plasma or tissues were deproteinized using ice-cold perchloric acid and neutralized with potassium carbonate and potassium hydroxide. After deproteinization, the guanidine compounds creatine, guanidinoacetate and CLL were reacted with ninhydrin to produce fluorescent derivatives. These derivatives were separated with an aqueous formic acid and methanol gradient using a Thermo Ultimate UHPLC system equipped with a reverse-phase (C18) HPLC column (Sigma). Pure reference standards were obtained from Sigma-Aldrich, except for CLL which was obtained from Hueston Hennigan LLP.

Validation of CLL measurement was accomplished by derivatizing pure CLL dissolved in water with the method of Buchberger and Ferdig. A distinct peak was observed for CLL that eluted later than creatine. No peak corresponding to CLL was present when a deproteinized water blank or mouse plasma were treated in the same manner. Mouse plasma was spiked with standard CLL and greater than 90% recovery was obtained. A standard curve for CLL was generated to determine the concentration of CLL in samples. The limit of detection for CLL was calculated to be 126 μM and the limit of quantification for CLL was calculated to be 383 μM.

Data were analyzed using a One-way Analysis of Variance (One-way ANOVA) with Tukey's post hoc test for multiple comparisons. A statistical p value less than 0.05 was used as a cut-off to determine significant statistical differences in means. Linear regression analysis was performed on body weight and food intake data. Data are presented as means +/− one standard deviation from the mean. n = 8 per group.

3. Results

Rats in all groups gained weight and there was no statistical difference in body weight between groups (Figure 2). Food intake was not different between any of the groups (Figure 3) and the average food intake over the course of the experiment was 21.7 ± 3.0,

21.0 ± 2.7 and 22.6 ± 2.4 g per day for the control, CrM and CLL diets, respectively. This amounts to an average daily intake of 0.56 mmoles of creatine by the CrMD rats and 0.61 mmoles of CLL by the CLLD group. For the last two days of the feeding period the rats consumed an average of 0.55 moles of creatine per day in the CrMD group and 0.59 mmoles of CLL per day in the CLLD group. Thus, rats consumed approximately equimolar quantity of creatine and CLL during the feeding period.

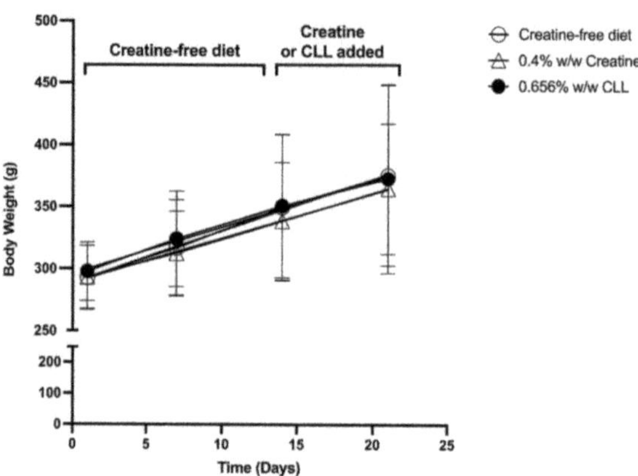

Figure 2. Body weight over the course of the feeding trial. Open circles represent weights for rats fed creatine-free AIN-93G diet for the entire 21 days; open triangles represent weights for rats fed creatine-free diet for 14 days and then switched to AIN-93G diet supplemented with 0.4% (w/w) creatine monohydrate for the remaining 7 days; black circles represent weights for rats fed creatine-free diet for 14 days and then switched to AIN-93G diet supplemented with 0.656% (w/w) creatyl-L-leucine (CLL) for the remaining 7 days. Values are presented as the mean ± one standard deviation, n = 8 per group.

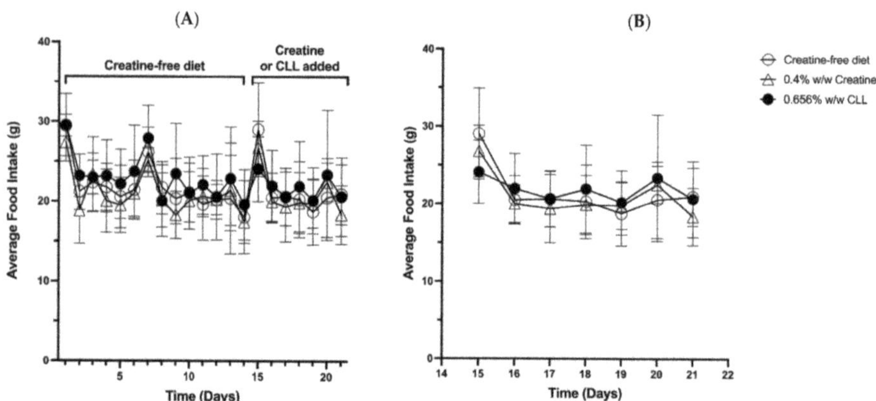

Figure 3. Food intake for all groups. Panel (**A**) Data for all days of experiment and panel (**B**) expanded data for days 15–21. Open circles represent the average daily food intake for rats fed creatine-free AIN-93G diet for the entire 21 days; open triangles represent average daily food intake for rats fed creatine-free diet for 14 days and then switched to AIN-93G diet supplemented with 0.4% (w/w) creatine monohydrate for days 15–21; black circles represent average daily food intake for rats fed creatine-free diet for 14 days and then switched to AIN-93G diet supplemented with 0.656% (w/w) creatyl-L-leucine (CLL) for days 15–21. Values are presented as the mean ± one standard deviation, n = 8 per group.

The creatine concentration in plasma from the abdominal aorta (AA) was approximately 7-fold higher in CrMD-fed rats when compared to either CON or CLLD groups (Figure 4A). Creatine content of portal venous (PV) plasma was 10-fold higher in CMD rats compared to CON and CLLD rats (Figure 4B). The content of creatine in AA and PV plasma from CON and CLLD groups did not differ.

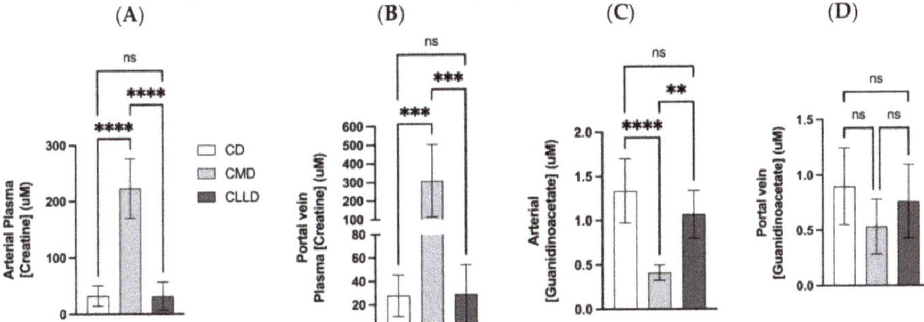

Figure 4. Plasma concentrations of creatine and guanidinoacetate. Panel (**A**) arterial plasma concentration of creatine; panel (**B**) portal venous plasma concentration of creatine; panel (**C**) arterial plasma concentration of guanidinoacetate; panel (**D**) portal venous concentration of guanidinoacetate. The open bar represents rats fed creatine-free diet (CON), light gray bar represents rats fed 0.4% *w/w* creatine monohydrate supplemented diet, dark gray bar represents rats fed 0.656% CLL supplemented diet. Values are presented as the mean ± one standard deviation, n = 8 per group. Asterisks indicate significance by *p*-value; $p < 0.0001$ = ****, $p < 0.001$ = ***, $p < 0.01$ = **, ns = not significant.

Guanidinoacetate (GAA) is the immediate precursor to creatine that is synthesized from arginine and glycine, primarily in the kidney. GAA can then undergo methylation to form creatine, primarily in the liver [3]. Renal synthesis of GAA is downregulated by dietary creatine supplementation, and we observe that GAA concentration in the AA plasma of CrMD rats was approximately one-third that of CON and CLLD (Figure 4C). GAA in the PV plasma was not significantly different between groups (Figure 4D).

The creatine content of the quadricep muscle from CMD rats was 1.63-fold higher than in CON rats and 1.53-fold higher than in CLLD rats (Figure 5A).

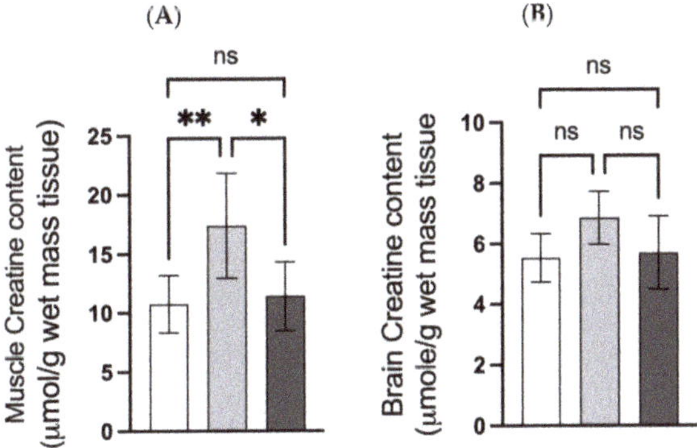

Figure 5. Muscle and brain creatine content. Panel (**A**) Creatine content of quadricep muscle; panel (**B**); creatine content of whole brain. The open bar represents rats fed creatine-free diet (CON), light gray bar represents rats fed 0.4% *w/w* creatine monohydrate-supplemented diet, dark gray bar represents rats fed 0.656% CLL supplemented diet. Values are presented as the mean ± one standard deviation, n = 8 per group. Asterisks indicate significance by *p*-value; $p < 0.01$ = **, $p < 0.05$ = *, ns = not significant.

The creatine content of the brain was not significantly different between groups (Figure 5B). However, multiple comparison analysis gave an adjusted P-value of 0.052 (1.26-fold difference in mean) between the CON and CrMD rats, indicating a trend toward higher creatine content in the brain of rats supplemented with CrM.

CLL was not detectable in any of the samples assayed.

4. Discussion

CrM remains one of the most widely used dietary supplements that provides significant ergogenic benefit during sport [5]. CrM supplements are also used as a therapy to treat several muscular and neurodegenerative disorders [13–15]. CrM is a stable powder that is soluble in water but is somewhat unstable in solution, depending on temperature and pH [1]. Scientific consensus is that CrM as a supplement is safe and the only adverse effect reported was gastrointestinal discomfort [1,16]. Efforts have been made to identify an alternative supplement that has more favorable characteristics and could provide bioavailable creatine. Although several creatine salts have been assessed for favorable properties and bioavailability, CrM was still found to be the most favorable form of creatine supplement [9]. Covalent modification of creatine has also been used to produce new compounds that could potentially be converted to creatine in vivo. The ethyl ester of creatine has been synthesized and tested for bioavailability and muscle performance in humans [17]. However, it was concluded that the ethyl ester of creatine did not increase muscle creatine content or enhance physical performance. CLL has been tested for toxicity and is used as an additive in some sports supplements and beverages under the trademark Super Creatine [18]. The present study is the first controlled study to independently test the bioavailability of CLL and to what extent CLL could be a bioavailable form of creatine.

The intent of this study was to provide the most favorable condition for the absorption and tissue uptake of the test compounds. Provision of a creatine-free diet reduces body creatine stores to a basal level that is sustained by endogenous synthesis of creatine from the amino acids, arginine, glycine, and methionine. Thereafter, relatively high doses of creatine or an equimolar amount of CLL were provided in the diet; a controlled design that compares both compounds equally. It was found that CLL supplementation in the diet did not increase plasma, muscle or brain creatine levels after 7 days of supplementation. In comparison, CrM supplementation significantly increased creatine levels in plasma and muscle while brain creatine content displayed a trend toward an increase. Since the portal vein drains blood from the intestine, it is expected that compounds that are well absorbed from the intestine would be more concentrated in plasma from this vessel. The observation that portal venous plasma creatine concentration from CLLD rats was not different from CON rats indicates that CLL is not converted to creatine in the intestine during digestion. Unpublished data suggested creatyl-L-glutamine was only 27% hydrolyzed to creatine in an in vitro model of digestion [10]. However, we conclude CLL is not a source of creatine during digestion as there was no bioaccumulation of creatine in tissues over the 7 days of CLL supplementation. Moreover, CLL was not detected in the portal venous plasma indicating that CLL is not well absorbed through the intestine. The limit of detection of CLL using the HPLC assay was 126 uM, therefore it cannot be concluded that CLL is not absorbed. However, there was no bioaccumulation of CLL detected in our study indicating that CLL is poorly absorbed even under high levels of dietary supplementation. For comparison, the level of supplement provided in our study would equate to a 70 kg human consuming approximately 17.6 g of creatine or 28.9 g of CLL per day.

Peptide bonds are a similar bond to that found in CLL. Peptide bonds are normally stable under physiological conditions, only hydrolyzed by protease or peptidase enzymes. However, the chemical structure of CLL is not likely to be recognized by the active sites of a protease or peptidase enzymes given the very different structure of creatine when compared to amino acid residues in a peptide. Thus, it is highly unlikely that a peptidase enzyme or other enzyme would bind to CLL to catalyze hydrolysis to creatine and leucine. If an appreciable amount of CLL is either hydrolyzed to creatine and then absorbed, or absorbed and then hydrolyzed in tissues, it would be reasonable to expect to see a significant increase in plasma or tissue creatine concentrations upon dietary supplementation with high doses of CLL, but this was not observed.

In summary, CLL has been proposed as a potential source of creatine when supplemented in the diet and this study tested the bioavailability of CLL in this capacity. The previous literature examining CLL did not test the bioavailability or offer insight into the biological fate of CLL [10,18]. It is now found that rats provided with large doses of CLL in the diet did not yield increases in creatine concentrations in plasma, muscle or brain tissue. CLL was also not detectable in any of the biological samples assayed from rats supplemented with CLL. Thus, it is concluded that CLL did not bioaccumulate, is poorly absorbed by the intestine and is not a bioavailable source of creatine. These data would also suggest that analogous creatine, amino acid-amides, would also have similar properties as dietary supplements.

Funding: This study was funded by Monster Energy Company.

Data Availability Statement: Data are available upon request.

Conflicts of Interest: The author received compensation for the time taken to complete this study. The author has no financial interests tied to the outcomes or findings of this study. The study design contained all appropriate controls, and the conclusions of this study were made based solely on unbiased scientific analysis of the data.

References

1. Wyss, M.; Kaddurah-Daouk, R. Creatine and creatinine metabolism. *Physiol. Rev.* **2000**, *80*, 1107–1213. [CrossRef] [PubMed]
2. Da Silva, R.P.; Clow, K.; Brosnan, J.T.; Brosnan, M.E. Synthesis of guanidinoacetate and creatine from amino acids by rat pancreas. *Br. J. Nutr.* **2014**, *111*, 571–577. [CrossRef] [PubMed]
3. da Silva, R.P.; Nissim, I.; Brosnan, M.E.; Brosnan, J.T. Creatine synthesis: Hepatic metabolism of guanidinoacetate and creatine in the rat in vitro and in vivo. *Am. J. Physiol.* **2009**, *296*, E256–E261. [CrossRef] [PubMed]
4. Brosnan, J.T.; da Silva, R.P.; Brosnan, M.E. The metabolic burden of creatine synthesis. *Amino Acids* **2011**, *40*, 1325–1331. [CrossRef] [PubMed]
5. Wax, B.; Kerksick, C.M.; Jagim, A.R.; Mayo, J.J.; Lyons, B.C.; Kreider, R.B. Creatine for Exercise and Sports Performance, with Recovery Considerations for Healthy Populations. *Nutrients* **2021**, *13*, 1915. [CrossRef] [PubMed]
6. Francaux, M.; Demeure, R.; Goudemant, J.F.; Poortmans, J.R. Effect of exogenous creatine supplementation on muscle PCr metabolism. *Int. J. Sports Med.* **2000**, *21*, 139–145. [CrossRef] [PubMed]
7. Butts, J.; Jacobs, B.; Silvis, M. Creatine Use in Sports. *Sport Health* **2018**, *10*, 31–34. [CrossRef] [PubMed]
8. Benzi, G.; Ceci, A. Creatine as nutritional supplementation and medicinal product. *J. Sports Med. Phys. Fit.* **2001**, *41*, 1–10.
9. Jager, R.; Purpura, M.; Shao, A.; Inoue, T.; Kreider, R.B. Analysis of the efficacy, safety, and regulatory status of novel forms of creatine. *Amino Acids* **2011**, *40*, 1369–1383. [CrossRef] [PubMed]
10. Reddeman, R.A.; Glavits, R.; Endres, J.R.; Murbach, T.S.; Hirka, G.; Vertesi, A.; Beres, E.; Szakonyine, I.P. A Toxicological Assessment of Creatyl-L-leucine. *Int. J. Toxicol.* **2018**, *37*, 171–187. [CrossRef] [PubMed]
11. Reeves, P.G.; Nielsen, F.H.; Fahey, G.C., Jr. AIN-93 purified diets for laboratory rodents: Final report of the American Institute of Nutrition ad hoc writing committee on the reformulation of the AIN-76A rodent diet. *J. Nutr.* **1993**, *123*, 1939–1951. [CrossRef] [PubMed]
12. Buchberger, W.; Ferdig, M. Improved high-performance liquid chromatographic determination of guanidino compounds by precolumn dervatization with ninhydrin and fluorescence detection. *J. Sep. Sci.* **2004**, *27*, 1309–1312. [CrossRef]
13. Brosnan, J.T.; Brosnan, M.E. Creatine: Endogenous metabolite, dietary, and therapeutic supplement. *Annu. Rev. Nutr.* **2007**, *27*, 241–261. [CrossRef] [PubMed]
14. Louis, M.; Lebacq, J.; Poortmans, J.R.; Belpaire-Dethiou, M.C.; Devogelaer, J.P.; Van Hecke, P.; Goubel, F.; Francaux, M. Beneficial effects of creatine supplementation in dystrophic patients. *Muscle Nerve* **2003**, *27*, 604–610. [CrossRef]
15. Wyss, M.; Schulze, A. Health implications of creatine: Can oral creatine supplementation protect against neurological and atherosclerotic disease? *Neuroscience* **2002**, *112*, 243–260. [CrossRef]
16. Poortmans, J.R.; Francaux, M. Adverse effects of creatine supplementation: Fact or fiction? *Sports Med.* **2000**, *30*, 155–170. [CrossRef] [PubMed]

17. Spillane, M.; Schoch, R.; Cooke, M.; Harvey, T.; Greenwood, M.; Kreider, R.; Willoughby, D.S. The effects of creatine ethyl ester supplementation combined with heavy resistance training on body composition, muscle performance, and serum and muscle creatine levels. *J. Int. Soc. Sports Nutr.* **2009**, *6*, 6. [CrossRef] [PubMed]
18. Schwarz, N.A.; McKinley-Barnard, S.K.; Blahnik, Z.J. Effect of Bang(R) Pre-Workout Master Blaster(R) combined with four weeks of resistance training on lean body mass, maximal strength, mircoRNA expression, and serum IGF-1 in men: A randomized, double-blind, placebo-controlled trial. *J. Int. Soc. Sports Nutr.* **2019**, *16*, 54. [CrossRef] [PubMed]

MDPI

St. Alban-Anlage 66

4052 Basel

Switzerland

Tel. +41 61 683 77 34

Fax +41 61 302 89 18

www.mdpi.com

Nutrients Editorial Office

E-mail: nutrients@mdpi.com

www.mdpi.com/journal/nutrients